Atomic Masses of the Elements

Name	Symbol	Atomic Number	Atomic Mass[a]	Name	Symbol	Atomic Number	Atomic Mass[a]
Actinium	Ac	89	(227)[b]	Mendelevium	Md	101	(258)
Aluminum	Al	13	26.98	Mercury	Hg	80	200.6
Americium	Am	95	(243)	Molybdenum	Mo	42	95.94
Antimony	Sb	51	121.8	Neodymium	Nd	60	144.2
Argon	Ar	18	39.95	Neon	Ne	10	20.18
Arsenic	As	33	74.92	Neptunium	Np	93	(237)
Astatine	At	85	(210)	Nickel	Ni	28	58.69
Barium	Ba	56	137.3	Niobium	Nb	41	92.91
Berkelium	Bk	97	(247)	Nitrogen	N	7	14.01
Beryllium	Be	4	9.012	Nobelium	No	102	(259)
Bismuth	Bi	83	209.0	Osmium	Os	76	190.2
Bohrium	Bh	107	(264)	Oxygen	O	8	16.00
Boron	B	5	10.81	Palladium	Pd	46	106.4
Bromine	Br	35	79.90	Phosphorus	P	15	30.97
Cadmium	Cd	48	112.4	Platinum	Pt	78	195.1
Calcium	Ca	20	40.08	Plutonium	Pu	94	(244)
Californium	Cf	98	(251)	Polonium	Po	84	(209)
Carbon	C	6	12.01	Potassium	K	19	39.10
Cerium	Ce	58	140.1	Praseodymium	Pr	59	140.9
Cesium	Cs	55	132.9	Promethium	Pm	61	(145)
Chlorine	Cl	17	35.45	Protactinium	Pa	91	231.0
Chromium	Cr	24	52.00	Radium	Ra	88	(226)
Cobalt	Co	27	58.93	Radon	Rn	86	(222)
Copernicium	Cn	112	(285)	Rhenium	Re	75	186.2
Copper	Cu	29	63.55	Rhodium	Rh	45	102.9
Curium	Cm	96	(247)	Roentgenium	Rg	111	(272)
Darmstadtium	Ds	110	(271)	Rubidium	Rb	37	85.47
Dubnium	Db	105	(262)	Ruthenium	Ru	44	101.1
Dysprosium	Dy	66	162.5	Rutherfordium	Rf	104	(261)
Einsteinium	Es	99	(252)	Samarium	Sm	62	150.4
Erbium	Er	68	167.3	Scandium	Sc	21	44.96
Europium	Eu	63	152.0	Seaborgium	Sg	106	(266)
Fermium	Fm	100	(257)	Selenium	Se	34	78.96
Flerovium	Fl	114	(289)	Silicon	Si	14	28.09
Fluorine	F	9	19.00	Silver	Ag	47	107.9
Francium	Fr	87	(223)	Sodium	Na	11	22.99
Gadolinium	Gd	64	157.3	Strontium	Sr	38	87.62
Gallium	Ga	31	69.72	Sulfur	S	16	32.07
Germanium	Ge	32	72.64	Tantalum	Ta	73	180.9
Gold	Au	79	197.0	Technetium	Tc	43	(99)
Hafnium	Hf	72	178.5	Tellurium	Te	52	127.6
Hassium	Hs	108	(265)	Terbium	Tb	65	158.9
Helium	He	2	4.003	Thallium	Tl	81	204.4
Holmium	Ho	67	164.9	Thorium	Th	90	232.0
Hydrogen	H	1	1.008	Thulium	Tm	69	168.9
Indium	In	49	114.8	Tin	Sn	50	118.7
Iodine	I	53	126.9	Titanium	Ti	22	47.87
Iridium	Ir	77	192.2	Tungsten	W	74	183.8
Iron	Fe	26	55.85	Uranium	U	92	238.0
Krypton	Kr	36	83.80	Vanadium	V	23	50.94
Lanthanum	La	57	138.9	Xenon	Xe	54	131.3
Lawrencium	Lr	103	(262)	Ytterbium	Yb	70	173.0
Lead	Pb	82	207.2	Yttrium	Y	39	88.91
Lithium	Li	3	6.941	Zinc	Zn	30	65.41
Livermorium	Lv	116	(293)	Zirconium	Zr	40	91.22
Lutetium	Lu	71	175.0	—	—	113	(284)
Magnesium	Mg	12	24.31	—	—	115	(288)
Manganese	Mn	25	54.94	—	—	117	(293)
Meitnerium	Mt	109	(268)	—	—	118	(294)

[a]Values for atomic masses are given to four significant figures.
[b]Values in parentheses are the mass number of an important radioactive isotope.

BASIC CHEMISTRY

BASIC CHEMISTRY

Fifth Edition

KAREN TIMBERLAKE
WILLIAM TIMBERLAKE

PEARSON

Editor in Chief: Jeanne Zalesky
Executive Editor: Terry Haugen
Senior Acquisitions Editor: Scott Dustan
Executive Field Marketing Manager: Chris Barker
Project Manager: Laura Perry
Program Manager: Lisa Pierce
Editorial Assistant: Lindsey Pruett
Marketing Assistant: Megan Riley
Executive Content Producer: Kristin Mayo
Media Content Producer: Jenny Moryan
Project Management Team Lead: David Zielonka
Program Management Team Lead: Kristen Flathman
Senior Project Manager, FSV: Lumina Datamatics
Compositor: Lumina Datamatics
Illustrator: Imagineering
Rights & Permissions Project Manager: Maya Gomez
Photo Researcher: Cordes Hoffman
Text Researcher: Erica Gordon
Design Manager: Marilyn Perry
Interior Designer: Gary Hespenheide
Cover Designer: Gary Hespenheide
Manufacturing Buyer: Maura Zaldivar-Garcia
Cover Photo Credit: Sergio Stakhnyk/Shutterstock (front), Kip Peticolas/Fundamental Photographs (back)

Credits and acknowledgments borrowed from other sources and reproduced, with permission, in this textbook appear on pp. C-1 to C-4.

Library of Congress Cataloging-in-Publication Data

Timberlake, Karen C.
  Basic Chemistry.—Fifth edition/Karen Timberlake, Los Angeles Valley College,
William Timberlake, Los Angeles Harbor College.
    pages cm.
  ISBN-13: 978-0-134-13804-6
  ISBN-10: 0-13-413804-X
  1. Chemistry—Textbooks. I. Timberlake, William E. II. Title.
  QD31.3.T54 2014
  540—dc23
                                      2012031269

2 16

www.pearsonhighered.com

ISBN-10: 0-13-413804-X
ISBN-13: 978-0-134-13804-6

Brief Contents

Table of Contents

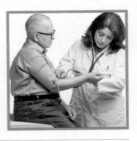
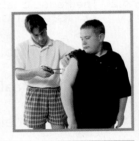

4
Atoms and Elements   102

5
Electronic Structure of Atoms and Periodic Trends   131

6
Ionic and Molecular Compounds   165

7
Chemical Quantities   194

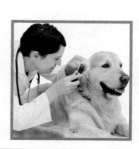

8

Chemical Reactions   224

9

Chemical Quantities in Reactions   249

10

Bonding and Properties of Solids and Liquids   279

11
Gases   323

12
Solutions   360

13
Reaction Rates and Chemical Equilibrium   409

14

Acids and Bases   443

15

Oxidation and Reduction   491

16

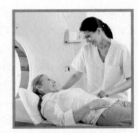

Nuclear Chemistry   523

Applications and Activities

Key Math Skills

Core Chemistry Skills

Guide to Problem Solving

About the Authors

KAREN TIMBERLAKE is Professor Emerita of Chemistry at Los Angeles Valley College, where she taught chemistry for allied health and preparatory chemistry for 36 years. She received her bachelor's degree in chemistry from the University of Washington and her master's degree in biochemistry from the University of California at Los Angeles.

Professor Timberlake has been writing chemistry textbooks for 40 years. During that time, her name **has become associated with the strategic use of pedagogical tools that promote student success in chemistry and the application of chemistry to real-life situations.** More than one million students have learned chemistry using texts, laboratory manuals, and study guides written by Karen Timberlake. In addition to *Basic Chemistry*, fifth edition, she is also the author of *General, Organic, and Biological Chemistry: Structures of Life*, fifth edition, with the accompanying *Study Guide*, and *Chemistry: An Introduction to General, Organic, and Biological Chemistry*, twelfth edition, with the accompanying *Study Guide*, and *Selected Solutions Manual*, *Laboratory Manual*, and *Essential Laboratory Manual*.

Professor Timberlake belongs to numerous scientific and educational organizations including the American Chemical Society (ACS) and the National Science Teachers Association (NSTA). She has been the Western Regional Winner of Excellence in College Chemistry Teaching Award given by the Chemical Manufacturers Association. She received the McGuffey Award in Physical Sciences from the Textbook Authors Association for her textbook *Chemistry: An Introduction to General, Organic, and Biological Chemistry*, eighth edition. She received the "Texty" Textbook Excellence Award from the Textbook Authors Association for the first edition of *Basic Chemistry*. She has participated in education grants for science teaching including the Los Angeles Collaborative for Teaching Excellence (LACTE) and a Title III grant at her college. She speaks at conferences and educational meetings on the use of student-centered teaching methods in chemistry to promote the learning success of students.

Her husband, William Timberlake, who is the coauthor of this text, is Professor Emeritus of Chemistry at Los Angeles Harbor College, where he taught preparatory and organic chemistry for 36 years. He received his bachelor's degree in chemistry from Carnegie Mellon University and his master's degree in organic chemistry from the University of California at Los Angeles.

When the Professors Timberlake are not writing textbooks, they relax by playing tennis, ballroom dancing, hiking, traveling, trying new restaurants, cooking, and taking care of their grandchildren, Daniel and Emily.

DEDICATION

- *Our son, John, daughter-in-law, Cindy, grandson, Daniel, and granddaughter, Emily, for the precious things in life*

- *The wonderful students over many years whose hard work and commitment always motivated us and put purpose in our writing*

Preface

Welcome to the fifth edition of *Basic Chemistry*. This chemistry text was written and designed to prepare you for science-related professions, such as engineering, nursing, medicine, environmental or agricultural science, or for careers such as laboratory technology. This text assumes no prior knowledge of chemistry. Our main objective in writing this text is to make the study of chemistry an engaging and a positive experience for you by relating the structure and behavior of matter to real life. This new edition introduces more problem-solving strategies, more problem-solving guides, new Analyze the Problem with Connect features, new Try It First and Engage features, conceptual and challenge problems, and new sets of combined problems.

It is our goal to help you become a critical thinker by understanding scientific concepts that will form a basis for making important decisions about issues concerning health and the environment. Thus, we have utilized materials that

- help you to learn and enjoy chemistry
- relate chemistry to careers that interest you
- develop problem-solving skills that lead to your success in chemistry
- promote learning and success in chemistry

New for the Fifth Edition

New and updated features have been added throughout this fifth edition, including the following:

- **NEW AND UPDATED! Chapter Openers** provide timely examples and engaging, topical issues of the chemistry that is part of contemporary professions.
- **NEW! A Follow Up** story continues with material and questions related to the chapter opener.
- **NEW! Engage** feature asks students to think about the paragraph they are reading and to test their understanding by answering the Engage question in the margin, which is related to the topic.
- **NEW! Try It First** now precedes the Solution section of each Sample Problem to encourage the student to work on the problem before reading the given Solution.
- **NEW! Connect** feature added to **Analyze** the **Problem** boxes indicates the relationships between *Given* and *Need*.
- **NEW! Applications** are added to Questions and Problems sets that show the relevance between the chemistry content and the chapter opener story.
- **NEW!** A new topic with questions and problems on Hess's Law, was added to Chapter 9.

- **NEW! Interactive Videos** give students the experience of step-by-step problem solving for problems from the text.
- **UPDATED! Chapter Readiness** sections at the beginning of each chapter list the Key Math Skills and Core Chemistry Skills from the previous chapters, which provide the foundation for learning new chemistry principles in the current chapter.
- **UPDATED! Key Math Skills** review basic math relevant to the chemistry the students are learning throughout the text. A **Key Math Skill Review** at the end of each chapter summarizes and gives additional examples.
- **UPDATED! Core Chemistry Skills** identify the key chemical principles in each chapter that are required for successfully learning chemistry. A **Core Chemistry Skill Review** at the end of each chapter helps reinforce the material and gives additional examples.
- **UPDATED! Analyze the Problem** features included in the Solutions of the Sample Problems strengthen critical-thinking skills and illustrate the breakdown of a word problem into the components required to solve it.
- **UPDATED! Questions and Problems, Sample Problems, and art** demonstrate the connection between the chemistry being discussed and how these skills will be needed in professional experience.
- **UPDATED! Combining Ideas** features offer sets of integrated problems that test students' understanding by integrating topics from two or more previous chapters.

Chapter Organization of the Fifth Edition

In each textbook we write, we consider it essential to relate every chemical concept to real-life issues. Because a chemistry course may be taught in different time frames, it may be difficult to cover all the chapters in this text. However, each chapter is a complete package, which allows some chapters to be skipped or the order of presentation to be changed.

Chapter 1, Chemistry in Our Lives, discusses the Scientific Method in everyday terms, guides students in developing a study plan for learning chemistry, with a section of Key Math Skills that reviews the basic math, including scientific notation needed in chemistry calculations.

- The Chapter Opener and Follow Up feature the work and career of a forensic scientist.
- "Scientific Method: Thinking Like a Scientist" discusses the scientific method in everyday terms.

- A new Sample Problem requires the interpretation of a graph to determine the decrease in a child's temperature when given Tylenol.
- Key Math Skills are: Identifying Place Values, Using Positive and Negative Numbers in Calculations including a new feature Calculator Operations, Calculating Percentages, Solving Equations, Interpreting Graphs, and Converting between Standard Numbers and Scientific Notation.

Chapter 2, Chemistry and Measurements, looks at measurement and emphasizes the need to understand numerical relationships of the metric system. Significant figures are discussed in the determination of final answers. Prefixes from the metric system are used to write equalities and conversion factors for problem-solving strategies. Density is discussed and used as a conversion factor.

- The Chapter Opener and Follow Up feature the work and career of a registered nurse.
- New photos, including an endoscope, a urine dipstick, a pint of blood, Keflex capsules, and salmon for omega-3 fatty acids, are added to improve visual introduction to clinical applications of chemistry.
- Updated Sample Problems relate questions and problem solving to health-related topics such as the measurements of blood volume, omega-3 fatty acids, radiological imaging, and medication orders.
- New Applications feature questions about measurements of daily values for minerals and vitamins, equalities and conversion factors for medications.
- A new Key Math Skill, Rounding Off, has been added.
- Core Chemistry Skills are: Counting Significant Figures, Using Significant Figures in Calculations, Using Prefixes, Writing Conversion Factors from Equalities, Using Conversion Factors, and Using Density as a Conversion Factor.

Chapter 3, Matter and Energy, classifies matter and states of matter, describes temperature measurement, and discusses energy, specific heat, and energy in nutrition. Physical and chemical changes and physical and chemical properties are discussed.

- The Chapter Opener and Follow Up feature the work and career of a dietitian.
- New Questions and Problems and Sample Problems include high temperatures used in cancer treatment, the energy produced by a high-energy shock output of a defibrillator, body temperature lowering using a cooling cap, and ice bag therapy for muscle injury.
- Core Chemistry Skills are: Classifying Matter, Identifying Physical and Chemical Changes, Converting between

Temperature Scales, Using Energy Units, Calculating Specific Heat, and Using the Heat Equation.

- The interchapter problem set, Combining Ideas from Chapters 1 to 3, completes the chapter.

Chapter 4, Atoms and Elements, introduces elements and atoms and the periodic table element. The names and symbols of element 114, Flerovium, Fl, and element 116, Livermorium, Lv, are part of the periodic table. Atomic numbers and mass number are determined for isotopes. Atomic mass is calculated using the masses of the naturally occurring isotopes and their abundances.

- The Chapter Opener and Follow Up feature the work and career of a farmer.
- Atomic number and mass number are used to calculate the number of protons and neutrons in an atom.
- The number of protons and neutrons are used to calculate the mass number and to write the atomic symbol for an isotope.
- A weighted average analogy uses 8-lb and 14-lb bowling balls and the percent abundance of each to calculate weighted average of a bowling ball.
- Core Chemistry Skills are: Counting Protons and Neutrons, and Writing Atomic Symbols for Isotopes.

Chapter 5, Electronic Structure of Atoms and Periodic Trends, uses the electromagnetic spectrum to explain atomic spectra and develop the concept of energy levels and sublevels. Electrons in sublevels and orbitals are represented using orbital diagrams and electron configurations. Periodic properties of elements, including atomic radius and ionization energy, are related to their valence electrons. Small periodic tables illustrate the trends of periodic properties.

- The Chapter Opener and Follow Up feature the work and career of a materials engineer.
- The diagram for the electromagnetic spectrum has been updated.
- The three-dimensional representations of the s, p, and d orbitals are drawn.
- The trends in periodic properties are described for valence electrons, atomic size, ionization energy, and metallic character.
- Core Chemistry Skills are: Writing Electron Configurations, Using the Periodic Table to Write Electron Configurations, Identifying Trends in Periodic Properties, and Drawing Lewis Symbols.

Chapter 6, Ionic and Molecular Compounds, describes the formation of ionic and covalent bonds. Chemical formulas are written, and ionic compounds—including those with polyatomic ions—and molecular compounds are named.

- The Chapter Opener and Follow Up feature the work and career of a pharmacist.
- Core Chemistry Skills are: Writing Positive and Negative Ions, Writing Ionic Formulas, Naming Ionic Compounds, and Writing the Names and Formulas for Molecular Compounds.

Chapter 7, Chemical Quantities, discusses Avogadro's number, the mole, and molar masses of compounds, which are used in calculations to determine the mass or number of particles in a quantity of a substance. The mass percent composition of a compound is calculated and used to determine its empirical and molecular formula.

- The Chapter Opener and Follow Up feature the work and career of a veterinarian.
- New and updated Guides to Problem Solving are: Converting the Moles (or Particles) of a Substance to Particles (or Moles), Calculating Moles of a Compound or Element, Calculating the Grams of an Element (or Compound) from the Grams of a Compound (or Element), and Calculating Mass Percent Composition from Molar Mass.
- Core Chemistry Skills are: Converting Particles to Moles, Calculating Molar Mass, Using Molar Mass as a Conversion Factor, Calculating Mass Percent Composition, Calculating an Empirical Formula, and Calculating a Molecular Formula.
- The interchapter problem set, Combining Ideas from Chapters 4 to 7, completes the chapter.

Chapter 8, Chemical Reactions introduces the method of balancing chemical equations, and discusses how to classify chemical reactions into types: combination, decomposition, single replacement, double replacement, and combustion reactions. A new section, Oxidation–Reduction Reactions, has been added.

- The Chapter Opener and Follow Up feature the work and career of an exercise physiologist.
- Core Chemistry Skills are: Balancing a Chemical Equation, Classifying Types of Chemical Reactions, and Identifying Oxidized and Reduced Substances.

Chapter 9, Chemical Quantities in Reactions, describes the mole and mass relationships among the reactants and products and provides calculations of limiting reactants and percent yields. A first section was divided into two new sections with an emphasis on the Law of Conservation of Mass.

- The Chapter Opener and Follow Up feature the work and career of an environmental scientist.
- Mole and mass relationships among the reactants and products are examined along with calculations of percent yield and limiting reactants.

- A new subsection, with questions and problems on Hess's Law, was added.
- Core Chemistry Skills are: Using Mole–Mole Factors, Converting Grams to Grams, Calculating Quantity of Product from a Limiting Reactant, Calculating Percent Yield, and Using the Heat of Reaction.

Chapter 10, Properties of Solids and Liquids, introduces Lewis structures for molecules and ions with single and multiple bonds as well as resonance structures. Electronegativity leads to a discussion of the polarity of bonds and molecules. Lewis structures and VSEPR theory illustrate covalent bonding and the three-dimensional shapes of molecules and ions. The intermolecular forces between particles and their impact on states of matter and changes of state are described. The energy involved with changes of state is calculated.

- The Chapter Opener and Follow Up feature the work and career of a histologist.
- Lewis structures are drawn for molecules and ions with single, double, and triple bonds. Resonance structures are drawn if two or more Lewis structures are possible.
- Shapes and polarity of bonds and molecules are predicted using VSEPR theory.
- Intermolecular forces in compounds are discussed including ionic bonds, hydrogen bonds, dipole–dipole attractions, and dispersion forces.
- Core Chemistry Skills are Drawing Lewis Structures, Drawing Resonance Structures, Predicting Shape, Using Electronegativity, Identifying Polarity of Molecules, Identifying Intermolecular Forces, and Calculating Heat for Change of State.
- The interchapter problem set, Combining Ideas from Chapters 8 to 10, completes the chapter.

Chapter 11, Gases, discusses the properties of gases and calculates changes in gases using the gas laws: Boyle's, Charles's, Gay-Lussac's, Avogadro's, Dalton's, and the Ideal Gas Law. Problem-solving strategies enhance the discussion and calculations with gas laws including chemical reactions using the ideal gas law.

- The Chapter Opener and Follow Up feature the work and career of a respiratory therapist.
- Applications includes calculations of mass or pressure of oxygen in uses of hyperbaric chambers.
- Core Chemistry Skills are: Using the Gas Laws, Using the Ideal Gas Law, Calculating Mass or Volume of a Gas in a Chemical Reaction, and Calculating Partial Pressure.

Chapter 12, Solutions, describes solutions, electrolytes, saturation and solubility, insoluble ionic compounds, concentrations, and osmosis. New problem-solving strategies clarify the use of concentrations to determine volume or mass of solute. The volumes and concentrations of solutions are used in calculations of dilutions, reactions, and titrations. Properties of solutions, osmosis in the body, dialysis and changes in the freezing and boiling points of a solvent are discussed.

- The Chapter Opener and Follow Up feature the work and career of a dialysis nurse.
- Core Chemistry Skills are: Using Solubility Rules, Calculating Concentration, Using Concentration as a Conversion Factor, Calculating the Quantity of a Reactant or Product for a Chemical Reaction in Solution, and Calculating the Freezing Point/Boiling Point of a Solution.

Chapter 13, Reaction Rates and Chemical Equilibrium,

looks at the rates of reactions and the equilibrium condition when forward and reverse rates for a reaction become equal. Equilibrium expressions for reactions are written and equilibrium constants are calculated. The equilibrium constant is used to calculate the concentration of a reactant or product at equilibrium. Le Châtelier's principle is used to evaluate the impact on concentrations when stress is placed on a system at equilibrium. The concentrations of solutes in a solution is used to calculate the solubility product constant (K_{sp}).

- The Chapter Opener and Follow Up feature the work and career of a chemical oceanographer.
- New problems that visually represent equilibrium situations are added.
- Core Chemistry Skills are: Writing the Equilibrium Expression, Calculating an Equilibrium Constant, Calculating Equilibrium Concentrations, Using Le Châtelier's Principle, Writing the Solubility Product Expression, Calculating a Solubility Product Constant, and Calculating the Molar Solubility.

Chapter 14, Acids and Bases,

discusses acids and bases and their strengths, and conjugate acid–base pairs. The dissociation of strong and weak acids and bases is related to their strengths as acids or bases. The dissociation of water leads to the water dissociation expression, K_w, the pH scale, and the calculation of pH. Chemical equations for acids in reactions are balanced and titration of an acid is illustrated. Buffers are discussed along with their role in the blood. The pH of a buffer is calculated.

- The Chapter Opener and Follow Up feature work and career of a clinical laboratory technician.
- A new Guide to Writing the Acid Dissociation Expression has been added.
- Key Math Skills are: Calculating pH from [H_3O^+], and Calculating [H_3O^+] from pH.
- Core Chemistry Skills are: Identifying Conjugate Acid–Base Pairs, Calculating [H_3O^+] and [OH^-] in Solutions, Writing Equations for Reactions of Acids and Bases,

Calculating Molarity or Volume of an Acid or Base in a Titration, and Calculating the pH of a Buffer.
- The interchapter problem set, Combining Ideas from Chapters 11 to 14, completes the chapter.

Chapter 15, Oxidation and Reduction,

looks at the characteristics of oxidation and reduction reactions. Oxidation numbers are assigned to the atoms in elements, molecules, and ions to determine the components that lose electrons during oxidation and gain electrons during reduction. The half-reaction method is utilized to balance oxidation–reduction reactions. The production of electrical energy in voltaic cells and the requirement of electrical energy in electrolytic cells are diagrammed using half-cells. The activity series is used to determine the spontaneous direction of an oxidation–reduction reaction.

- The Chapter Opener and Follow Up feature the work and career of a dentist.
- A new Guide to Identifying Oxidizing and Reducing Agents has been added.
- Core Chemistry Skills are: Assigning Oxidation Numbers, Using Oxidation Numbers, Identifying Oxidizing and Reducing Agents, Using Half-Reactions to Balance Redox Equations, and Identifying Spontaneous Reactions.

Chapter 16, Nuclear Chemistry,

looks at the types of radiation emitted from the nuclei of radioactive atoms. Nuclear equations are written and balanced for both naturally occurring radioactivity and artificially produced radioactivity. The half-lives of radioisotopes are discussed, and the amount of time for a sample to decay is calculated. Radioisotopes important in the field of nuclear medicine are described. Fission and fusion and their role in energy production are discussed.

- The Chapter Opener and Follow Up feature the work and career of a radiologist.
- Core Chemistry Skills are: Writing Nuclear Equations and Using Half-Lives.
- The interchapter problem set, Combining Ideas from Chapters 15 and 16, completes the chapter.

Chapter 17, Organic Chemistry,

compares inorganic and organic compounds, and describes the condensed and line-angle structural formulas of alkanes, alkenes, alcohols, ethers, aldehydes, ketones, carboxylic acids, esters, amines, and amides.

- The Chapter Opener and Follow Up feature the work and career of a firefighter/emergency medical technician.
- The properties of organic and inorganic compounds are now compared in Table 17.1.

- Line-angle structural formulas were added to Table 17.2 IUPAC Names, Molecular Formulas, Condensed and Line-Angle Structural Formulas of the First Ten Alkanes.
- More line-angle structures are included in text examples, sample problems, questions and problems.
- The two-dimensional and three-dimensional representations of methane and ethane are illustrated using condensed structural formulas, expanded structural formulas, ball-and-stick models, space-filling models, and wedge–dash models.
- The topic of structural isomers was added using condensed and line-angle structural formulas.
- Common substituents butyl, isobutyl, *sec*-butyl and *tert*-butyl were added to Table 17.3.
- Properties of solubility and density of alkanes were added.
- The chemical reaction of hydrogenation of alkenes and unsaturated fats was added.
- Updated recycling symbols for polymers were added.
- Core Chemistry Skills are: Naming and Drawing Alkanes, Writing Equations for Hydrogenation and Polymerization, Naming Aldehydes and Ketones, Naming Carboxylic Acids, Forming Esters, and Forming Amides.

Chapter 18, Biochemistry, looks at the chemical structures and reactions of chemicals that occur in living systems. We focus on four types of biomolecules—carbohydrates, lipids, proteins, and nucleic acids—as well as their biochemical reactions.

- The Chapter Opener and Follow Up feature the work and career of a clinical lipid specialist.
- Fischer projections with and D and L notations are described.
- Monosaccharides are classified as aldo or keto pentoses or hexoses.
- Haworth structures are drawn for monosaccharides, disaccharides, and polysaccharides.
- The Guide to Drawing Haworth Structures has been rewritten.
- Lipids distinguishes between the structures of fatty acids, waxes, triacylglycerols, and steroids.
- The shapes of proteins are related to the activity and regulation of enzyme activity.
- The genetic code is described and utilized in the process of protein synthesis.
- Core Chemistry Skills are: Drawing Haworth Structures, Identifying Fatty Acids, Drawing Structures for Triacylglycerols, Drawing the Products for the Hydrogenation and Saponification of a Triacylglycerol, Drawing the Ionized Form for an Amino Acid, Identifying the Primary, Secondary, Tertiary, and Quaternary Structures of Proteins, Writing the Complementary DNA Strand, and Writing the mRNA Segment for a DNA Template.
- The interchapter problem set, Combining Ideas from Chapters 17 and 18, completes the chapter.

Acknowledgments

The preparation of a new text is a continuous effort of many people. As in our work on other textbooks, we are thankful for the support, encouragement, and dedication of many people who put in hours of tireless effort to produce a high-quality book that provides an outstanding learning package. The editorial team at Pearson Publishing has done an exceptional job. We want to thank, Jeanne Zalesky, editor in chief, and Editors Terry Haugen and Scott Dustan, who supported our vision of this fifth edition and the development of new problem-solving strategies.

We much appreciate all the wonderful work of project manager Laura Perry, who was like an angel encouraging us at each step, while skillfully coordinating reviews, art, web site materials, and all the things it takes to make a book come together. We appreciate the work of Lisa Pierce, program manager, and Lindsay Bethoney of Lumina Datamatics, who brilliantly coordinated all phases of the manuscript to the final pages of a beautiful book. Thanks to Mark Quirie, manuscript and accuracy reviewer, and copy editors of Lumina Datamatics, Inc., who precisely analyzed and edited the initial and final manuscripts and pages to make sure the words and problems were correct to help students learn chemistry. Their keen eyes and thoughtful comments were extremely helpful in the development of this text.

We are especially proud of the art program in this text, which lends beauty and understanding to chemistry. We would like to thank Marilyn Perry and Gary Hespenheide, interior and cover design managers and book designer, whose creative ideas provided the outstanding design for the cover and pages of the book. Erica Gordon, photo researcher, was invaluable in researching and selecting vivid photos for the text so that students can see the beauty of chemistry. Thanks also to *Bio-Rad Laboratories* for their courtesy and use of *Know-ItAll ChemWindows* drawing software that helped us produce chemical structures for the manuscript. The macro-to-micro illustrations designed by Production Solutions and Precision Graphics give students visual impressions of the atomic and molecular organization of everyday things and are a fantastic learning tool. We also appreciate all the hard work put in by the marketing team in the field and Executive Marketing Manager, Chris Barker.

We are extremely grateful to an incredible group of peers for their careful assessment of all the new ideas for the text; for their suggested additions, corrections, changes, and deletions; and for providing an incredible amount of feedback about improvements for the book.

If you would like to share your experience with chemistry, or have questions and comments about this text, we would appreciate hearing from you.

Karen and Bill Timberlake
Email: khemist@aol.com

FAVORITE QUOTES

The whole art of teaching is only the art of awakening the natural curiosity of young minds.

—Anatole France

One must learn by doing the thing; though you think you know it, you have no certainty until you try.

—Sophocles

Discovery consists of seeing what everybody has seen and thinking what nobody has thought.

—Albert Szent-Györgyi

I never teach my pupils; I only attempt to provide the conditions in which they can learn.

—Albert Einstein

Reviewers

Fifth Edition Reviewers

David Atwood
University of Kentucky

Nathan Barrows
Grand Valley State University

Derek Behmke
Georgia Gwinnett College

Nancy Christensen
Waubonsee Community College

David Dollar
Tarrant Community College - SE

Maegan Harris
Waubonsee Community College

Yohani Kayinamura
Daytona State College

Andrew Knight
Florida Institute of Technology

Danica Nowosielski
Hudson Valley Community College

Mark Quirie
Algonquin College

Erin Rennells
Hudson Valley Community College

Kathy Wall
Waubonsee Community College

Mingming Xu
West Virginia University

Accuracy Reviewer

Mark Quirie
Algonquin College

Previous Edition Reviewers

Edward Alexander
San Diego Mesa College

Kristen Casey
Anne Arundel Community College

James Falender
Central Michigan University

Tamara Hanna
Texas Tech University

Shawn Korman
Rio Salado Community College

Robin Lasey
Arkansas Tech University

Lynda Nelson
University of Arkansas Fort Smith

Mary Repaske
Cincinnati State Technical and Community College

Mitchell Robertson
Southwestern Illinois College

Alan Sherman
Middlesex County College

Trent Vorlicek
Minnesota State University-Mankato

Joy Walker
Truman College

Marie Wolff
Joliet Junior College

Regina Zibuck
Wayne State University

Feature	Description	Benefit	Page
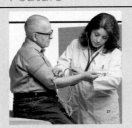	**Chapter Openers** discuss careers that involve chemistry.	Connects the chemistry in the chapter with the professionals who use chemistry every day.	27
CHEMISTRY LINK TO HEALTH — Toxicology and Risk–Benefit Assessment	**Chemistry Links to Health** apply chemical concepts to health and medicine such as weight loss and weight gain, alcohol abuse, blood buffers, and kidney dialysis.	Provide you with connections that illustrate the importance of understanding chemistry in real-life situations.	52
Follow Up — GREG'S VISIT WITH HIS DOCTOR	**Follow Ups** give a follow up to the discussion in the chapter opener and include Application questions.	Continue a theme through the entire chapter utilizing the new chemistry content.	60
	**Macro-to-Micro Art** utilizes photographs and drawings to illustrate the atomic structure of chemical phenomenas.	Helps you connect the world of atoms and molecules to the macroscopic world.	41
CHEMISTRY LINK TO THE ENVIRONMENT — Carbon Dioxide and Climate Change	**Chemistry Links to the Environment** relate chemistry to environmental topics such as climate change, radon in our homes, and fuel cells.	Help you extend your understanding of the impact of chemistry on the environment.	84
ENGAGE — Why is the 0.5 ppm mercury in tuna written as the equality 0.5 mg of mercury = 1 kg of tuna?	**Engage** feature is a question in the margin next to a paragraph of new material.	Reminds you to think about the paragraph you are reading and to test your understanding by answering the Engage question.	47

Engage students in the world of chemistry

Feature	Description	Benefit	Page
LEARNING GOAL Describe the intermolecular forces between ions, polar covalent molecules, and nonpolar covalent molecules.	**Learning Goals** at the beginning and end of each section identify the key concepts for that section and provide a roadmap for your study.	Help you focus your studying by emphasizing what is most important in each section.	311
Volume (*V*) The volume of gas equals the size of the container in which the gas is placed. When you inflate a tire or a basketball, you are adding more gas particles. The increase in the number of particles hitting the walls of the tire or basketball increases the volume. Sometimes, on a cold morning, a tire looks flat. The volume of the tire has decreased because a lower temperature decreases the speed of the molecules, which in turn reduces the force of their impacts on the walls of the tire. The most common units for volume measurement are liters (L) and milliliters (mL).	Timberlake's accessible **Writing Style** is based on careful development of chemical concepts suited to the skills and backgrounds of students in chemistry.	Helps you understand new terms and chemical concepts.	326
KEY MATH SKILL Calculating pH from $[H_3O^+]$	**Key Math Skills** review the basic math needed for chemistry. Instructors can also assign these through MasteringChemistry.	Help you master the basic quantitative skills to succeed in chemistry.	462
CORE CHEMISTRY SKILL Using Significant Figures in Calculations	**Core Chemistry Skills** identify content crucial to problem-solving strategies. Instructors can also assign these through MasteringChemistry.	Help you master the basic problem-solving skills needed to succeed in chemistry.	36
TRY IT FIRST	The **TRY IT FIRST** feature encourages you to try to solve the problem before you compare your work with the Solution.	Helps you identify what you know about the solution and what you need to learn.	30
CHAPTER REVIEW **17.1 Alkanes** **LEARNING GOAL** Write the IUPAC names and draw the condensed or line-angle structural formulas for alkanes.	The **Chapter Reviews** include Learning Goals and visual thumbnails to summarize the key points in each section.	Help you determine your mastery of the chapter concepts and study for your tests.	600
KEY TERMS **Avogadro's number** The number of items in a mole, equal to 6.022×10^{23}. **empirical formula** The simplest or smallest whole-number ratio of the atoms in a formula. **formula unit** The group of ions represented by the formula of an ionic compound.	**Key Terms** with definitions are listed at the end of each chapter as well as in the **Glossary/Index** at the end of the text.	Help you recall the important new terms in each chapter.	217
CONCEPT MAP CHEMISTRY AND MEASUREMENTS	**Concept Maps** at the end of each chapter show how all the key concepts fit together.	Encourage learning by giving you a visual guide to the interrelationship among all the concepts new to each chapter.	61

Tools to engage students in chemistry and show them how to solve problems

Feature	Description	Benefit	Page

Applications
 	
2.23 Identify the number of significant figures in each of the following:
  a. The mass of a neonate is 1.607 kg.
  b. The Daily Value (DV) for iodine for an infant is 130 mcg.
  c. There are 4.02×10^6 red blood cells in a blood sample.
	Applications in Questions and Problems show the relevance to the chemistry concepts in the chapter.	Show you how the chemistry you are learning is related to real life.	35

Guide to Determining the Polarity of a Molecule **STEP 1** Determine if the bonds are polar covalent or nonpolar covalent.	**Guides to Problem Solving** (GPS) illustrate the steps needed to solve problems.	Visually guide you step-by-step through each problem-solving strategy.	298
ANALYZE THE PROBLEM — Given: 260. g of ice at 0 °C · Need: joules to melt ice at 0 °C · Connect: heat of fusion	**Analyze the Problems** convert a word problem into components for problem solving. New **Connect** features specify information that relates the *Given* and *Need* sections.	Help you identify and connect the components within a word problem to set up a solution strategy.	303
QUESTIONS AND PROBLEMS 10.4 Electronegativity and Bond Polarity **LEARNING GOAL** Use electronegativity to determine the polarity of a bond. 10.23 Describe the trend in electronegativity as *increases* or *decreases* for each of the following: **a.** from B to F **b.** from Mg to Ba **c.** from F to I	**Questions and Problems** placed at the end of each section are paired. The **Answers** to the odd-numbered problems are given at the end of each chapter.	Encourage you to become involved immediately in the process of problem solving.	296
SAMPLE PROBLEM 2.4 Rounding Off ...	**Sample Problems** illustrate worked-out solutions with explanations and required calculations. **Study Checks** associated with each Sample Problem allow you to check your problem-solving strategies with the **Answer**.	Provide the intermediate steps to guide you successfully through each type of problem.	36
UNDERSTANDING THE CONCEPTS *The chapter sections to review are shown in parentheses at the end of each question.* 8.35 Balance each of the following by adding coefficients, and identify the type of reaction for each: (8.1, 8.2, 8.3) **a.** ___ + ___ → ___	**Understanding the Concepts** are questions with visual representations placed at the end of each chapter.	Build an understanding of newly learned chemical concepts.	244

ADDITIONAL QUESTIONS AND PROBLEMS
8.43 Identify the type of reaction for each of the following as combination, decomposition, single replacement, double replacement, or combustion: (8.3)
  a. A metal and a nonmetal form an ionic compound.
  b. A compound of hydrogen and carbon reacts with oxygen to produce carbon dioxide and water.
	Additional Questions and Problems at the end of each chapter provide further study and application of the topics from the entire chapter. Problems are paired and the **Answers** to the odd-numbered problems are given at the end of each chapter.	Promote critical thinking.	245

CHALLENGE QUESTIONS
The following groups of questions are related to the topics in ter. However, they do not all follow the chapter order, and t you to combine concepts and skills from several sections. T tions will help you increase your critical thinking skills and your next exam.
8.53 Balance each of the following chemical equations, a the type of reaction: (8.1, 8.2, 8.3)
  a. $K_2O(s) + H_2O(g) \longrightarrow KOH(aq)$
  b. $C_8H_{18}(l) + O_2(g) \xrightarrow{\Delta} CO_2(g) + H_2O(g)$
	Challenge Questions at the end of each chapter provide complex questions. **Answers** to the odd-numbered questions are given at the end of each chapter.	Promote critical thinking, group work, and cooperative learning environments.	246

COMBINING IDEAS from CHAPTERS 4 to 7 ...	**Combining Ideas** are sets of integrated problems placed after every two to four chapters that are useful as practice exams. Answers to the odd-numbered problems are given at the end of each Combining Ideas.	Test your understanding of the concepts from previous chapters by integrating topics.	222

Resources

Basic Chemistry, fifth edition, provides an integrated teaching and learning package of support material for both students and professors.

Name of Supplement	Available in Print	Available Online	Instructor or Student Supplement	Description
Study Guide and Selected Solutions Manual (ISBN 0134167260)	✓		Resource for Students	The Study Guide and Selected Solutions Manual, by Karen Timberlake and Mark Quirie, promotes active learning through a variety of exercises with answers as well as practice tests that are connected directly to the learning goals of the textbook. Complete solutions to odd-numbered problems are included.
MasteringChemistry® (www.masteringchemistry.com) (ISBN 0134177150)		✓	Resource for Students and Instructors	MasteringChemistry® from Pearson is the leading online teaching and learning system designed to improve results by engaging students before, during, and after class with powerful content. Ensure that students arrive ready to learn by assigning educationally effective content before class, and encourage critical thinking and retention with in-class resources such as Learning Catalytics. Students can further master concepts after class through traditional homework assignments that provide hints and answer-specific feedback. The Mastering gradebook records scores for all automatically graded assignments while diagnostic tools give instructors access to rich data to assess student understanding and misconceptions.
MasteringChemistry with Pearson eText (ISBN 0133899306)		✓	Resource for Students	The fifth edition of *Basic Chemistry* features a Pearson eText enhanced with media within Mastering. In conjunction with Mastering assessment capabilities, Interactive Videos, and 3D animations will improve student engagement and knowledge retention. Each chapter will contain a balance of interactive animations, videos, sample calculations, and self-assessments/quizzes embedded directly in the eText. Additionally, the Pearson eText offers students the power to create notes, highlight text in different colors, create bookmarks, zoom, and view single or multiple pages.
Instructor's Solutions Manual–Download Only (ISBN 0134167279)		✓	Resource for Instructors	Prepared by Mark Quirie, the solutions manual highlights chapter topics, and includes answers and solutions for all questions and problems in the text.
Instructor Resource Materials–Download Only (ISBN 0134167252)		✓	Resource for Instructors	Includes all the art, photos, and tables from the book in JPEG format for use in classroom projection or when creating study materials and tests. In addition, the instructors can access modifiable PowerPoint™ lecture outlines. Also available are downloadable files of the Instructor's Solutions Manual and a set of "clicker questions" designed for use with classroom-response systems. Also visit the Pearson Education catalog page for Timberlake's *Basic Chemistry*, fifth edition, at www.pearsonhighered.com to download available instructor supplements.
TestGen Test Bank–Download Only (ISBN 0133891895)		✓	Resource for Instructors	Prepared by William Timberlake, this resource includes more than 2000 questions in multiple-choice, matching, true/false, and short-answer format.
Laboratory Manual by Karen Timberlake (ISBN 0321811852)	✓		Resource for Students	This best-selling lab manual coordinates 35 experiments with the topics in *Basic Chemistry*, fifth edition, uses laboratory investigations to explore chemical concepts, develop skills of manipulating equipment, reporting data, solving problems, making calculations, and drawing conclusions.
Online Instructor Manual for Laboratory Manual (ISBN 0321812859)		✓	Resource for Students	This manual contains answers to report sheet pages for the Laboratory Manual and a list of the materials needed for each experiment with amounts given for 20 students working in pairs, available for download at www.pearsonhighered.com.

Highlighting Relevancy and Applications

Designed to prepare students for science-related careers, *Basic Chemistry* organizes chemical concepts and problem solving into clear, manageable pieces, ensuring students follow along and stay motivated throughout their first chemistry course. Timberlake's friendly writing style, student focus, challenging problems, and engaging applications continue to help students make connections between chemistry and their future careers as they develop problem-solving skills they'll need beyond the classroom.

Follow Ups and Applications

Chapter Openers throughout the text connect chemistry to real life. Each chapter begins with an image and details of a profession such as engineering, medicine, environmental science or agriculture science, or exercise physiology. **Follow Ups** at the end of chapter discuss the chemistry in the Chapter Opener and include Applications. These questions show students how the chemistry they are learning applies specifically to their professional careers.

Follow Up
FORENSIC EVIDENCE SOLVES THE MURDER

Using a variety of laboratory tests, Sarah finds ethylene glycol in the victim's blood. The quantitative tests indicate that the victim had ingested 125 g of ethylene glycol. Sarah determines that the liquid in a glass found at the crime scene was ethylene glycol that had been added to an alcoholic beverage. Ethylene glycol is a clear, sweet-tasting, thick liquid that is odorless and mixes with water. It is easy to obtain since it is used as antifreeze in automobiles and in brake fluid. Because the initial symptoms of ethylene glycol poisoning are similar to being intoxicated, the victim is often unaware of its presence.

If ingestion of ethylene glycol occurs, it can cause depression of the central nervous system, cardiovascular damage, and kidney failure. If discovered quickly, hemodialysis may be used to remove ethylene glycol from the blood. A toxic amount of ethylene glycol is 1.5 g of ethylene glycol/kg of body mass. Thus, 75 g could be fatal for a 50-kg (110 lb) person.

Mark determines that fingerprints on the glass containing the ethylene glycol were those of the victim's husband. This evidence along with the container of antifreeze found in the home led to the arrest and conviction of the husband for poisoning his wife.

Applications

1.33 A container was found in the home of the victim that contained 120 g of ethylene glycol in 450 g of liquid. What was the percentage of ethylene glycol? Express your answer to the ones place.

1.34 If the toxic quantity is 1.5 g of ethylene glycol per 1000 g of body mass, what percentage of ethylene glycol is fatal?

Chemistry in Our Lives

1

A CALL CAME IN TO 911 from a man who found his wife lying on the floor of their home. When the police arrived, they determined that the woman was dead. The husband said he had worked late, and just arrived home. The victim's body was lying on the floor of the living room. There was no blood at the scene, but the police did find a glass on the side table that contained a small amount of liquid. In an adjacent laundry room/garage, the police found a half-empty bottle of antifreeze. The bottle, glass, and liquid were bagged and sent to the forensic laboratory.

In another 911 call, a man was found lying on the grass outside his home. Blood was present on his body, and some bullet casings were found on the grass. Inside the victim's home, a weapon was recovered. The bullet casings and the weapon were bagged and sent to the forensic laboratory.

Sarah and Mark, forensic scientists, use scientific procedures and chemical tests to examine the evidence from law enforcement agencies. Sarah proceeds to analyze blood, stomach contents, and the unknown liquid from the first victim's home. She will look for the presence of drugs, poisons, and alcohol. Her

lab partner Mark will analyze the fingerprints on the glass. He will also match the characteristics of the bullet casings to the weapon that was found at the second crime scene.

Evidence from a crime scene is sent to the forensic laboratory.

CAREER
Forensic Scientist

Most forensic scientists work in crime laboratories that are part of city or county legal systems where they analyze bodily fluids and tissue samples collected by crime scene investigators. In analyzing these samples, forensic scientists identify the presence or absence of specific chemicals within the body to help solve the criminal case. Some of the chemicals they look for include alcohol, illegal or prescription drugs, poisons, arson debris, metals, and various gases such as carbon monoxide. In order to identify these substances, a variety of chemical instruments and highly specific methodologies are used. Forensic scientists also analyze samples from criminal suspects, athletes, and potential employees. They also work on cases involving environmental contamination and animal samples for wildlife crimes. Forensic scientists usually have a bachelor's degree that includes courses in math, chemistry, and biology.

Focusing on New Problem-Solving Strategies

This new edition introduces more problem-solving strategies, more problem-solving guides, new Analyze the Problem with Connect features, new Try It First and Engage features, conceptual and challenge problems, and new sets of combined problems.

- **NEW! Connect** feature has been added to the **Analyze the Problem** boxes, which specifies the information that relates the *Given* and *Need* sections.
- **NEW! Try It First** now precedes the Solution section of each Sample Problem to encourage the student to work on the problem before reading the given Solution.
- **NEW! Engage** feature asks students to think about the paragraph they are reading and to test their understanding by answering the Engage question in the margin, which is related to the topic.

ANALYZE THE PROBLEM	Given	Need	Connect
	standard number	scientific notation	coefficient is at least 1 but less than 10

SAMPLE PROBLEM 1.7 **Scientific Notation**

Write each of the following in scientific notation:

a. 3500   **b.** 0.000 016

TRY IT FIRST

ENGAGE

What is different and what is the same for an atom of Sn-105 and an atom of Sn-132?

Interactive Videos

Interactive videos and demonstrations help students through some of the more challenging topics by showing how chemistry works in real life and introducing a bit of humor into chemical problem solving and demonstrations. Topics include Using Conversion Factors, Balancing Nuclear Equations, and Chemical v. Physical Change.

Sample Calculations walk students through the most challenging chemistry problems and provide a fresh perspective on how to approach individual problems and plan solutions. Topics include Using Conversion Factors, Mass Calculations for Reactions, and Concentration of Solutions.

 Green play button icons appear in the margins throughout the text. In the eText, the icons link to new interactive videos that the student can use to clarify and reinforce important concepts. All Interactive Videos are available in web and mobile-friendly formats through the eText, and are assignable activities in MasteringChemistry.

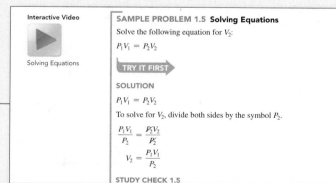

Interactive Video

Solving Equations

SAMPLE PROBLEM 1.5 Solving Equations

Solve the following equation for V_2:

$$P_1V_1 = P_2V_2$$

TRY IT FIRST

SOLUTION

$$P_1V_1 = P_2V_2$$

To solve for V_2, divide both sides by the symbol P_2.

$$\frac{P_1V_1}{P_2} = \frac{P_2V_2}{P_2}$$

$$V_2 = \frac{P_1V_1}{P_2}$$

STUDY CHECK 1.5

MasteringChemistry®

MasteringChemistry® from Pearson is the leading online teaching and learning system designed to improve results by engaging students before, during, and after class with powerful content. Instructors may ensure that students arrive ready to learn by assigning educationally effective content before class, and encourage critical thinking and retention with in-class resources such as Learning Catalytics. Students can further master concepts after class through traditional homework assignments that provide hints and answer-specific feedback. The Mastering gradebook records scores for all automatically graded assignments while diagnostic tools give instructors access to rich data to assess student understanding and misconceptions.

Mastering brings learning full circle by continuously adapting to each student and making learning more personal than ever—before, during, and after class.

Before Class

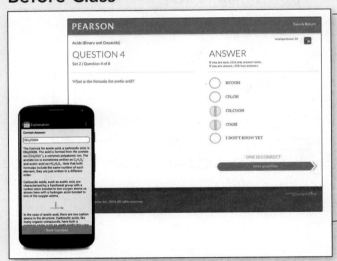

Dynamic Study Modules

Help students quickly learn chemistry!
Now assignable, Dynamic Study Modules (DSMs) enable your students to study on their own and be better prepared with the basic math and chemistry skills needed to succeed in the course. The mobile app is available for iOS and Android devices for study on the go and results can be tracked in the MasteringChemistry gradebook.

Reading Quizzes
Reading Quizzes give instructors the opportunity to assign reading and test students on their comprehension of chapter content.

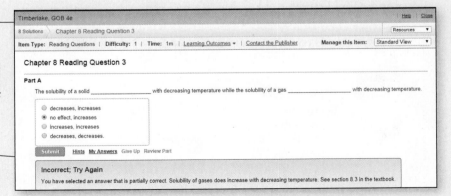

During Class

▶ Learning Catalytics

Learning Catalytics is a "bring your own device" student engagement, assessment, and classroom intelligence system. With Learning Catalytics you can:

- Assess students in real time, using open-ended tasks to probe student understanding.

- Understand immediately where students are and adjust your lecture accordingly.

- Manage student interactions with intelligent grouping and timing.

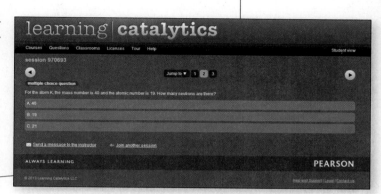

After Class

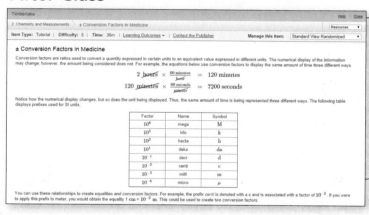

Tutorials and Coaching

Students learn chemistry by practicing chemistry.

Tutorials, featuring specific wrong-answer feedback, hints, and a wide variety of educationally effective content, guide your students through the toughest topics in Basic Chemistry.

Chemistry in Our Lives

A CALL CAME IN TO 911 from a man who found his wife lying on the floor of their home. When the police arrived, they determined that the woman was dead. The husband said he had worked late, and just arrived home. The victim's body was lying on the floor of the living room. There was no blood at the scene, but the police did find a glass on the side table that contained a small amount of liquid. In an adjacent laundry room/garage, the police found a half-empty bottle of antifreeze. The bottle, glass, and liquid were bagged and sent to the forensic laboratory.

In another 911 call, a man was found lying on the grass outside his home. Blood was present on his body, and some bullet casings were found on the grass. Inside the victim's home, a weapon was recovered. The bullet casings and the weapon were bagged and sent to the forensic laboratory.

Sarah and Mark, forensic scientists, use scientific procedures and chemical tests to examine the evidence from law enforcement agencies. Sarah proceeds to analyze blood, stomach contents, and the unknown liquid from the first victim's home. She will look for the presence of drugs, poisons, and alcohol. Her

lab partner Mark will analyze the fingerprints on the glass. He will also match the characteristics of the bullet casings to the weapon that was found at the second crime scene.

Evidence from a crime scene is sent to the forensic laboratory.

CAREER

Forensic Scientist

Most forensic scientists work in crime laboratories that are part of city or county legal systems where they analyze bodily fluids and tissue samples collected by crime scene investigators. In analyzing these samples, forensic scientists identify the presence or absence of specific chemicals within the body to help solve the criminal case. Some of the chemicals they look for include alcohol, illegal or prescription drugs, poisons, arson debris, metals, and various gases such as carbon monoxide. In order to identify these substances, a variety of chemical instruments and highly specific methodologies are used. Forensic scientists also analyze samples from criminal suspects, athletes, and potential employees. They also work on cases involving environmental contamination and animal samples for wildlife crimes. Forensic scientists usually have a bachelor's degree that includes courses in math, chemistry, and biology.

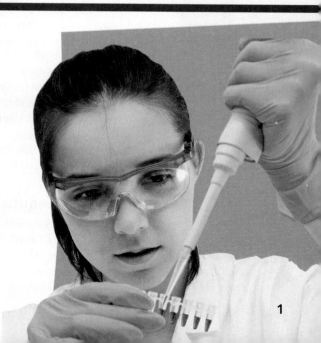

1.1 Chemistry and Chemicals

LEARNING GOAL Define the term chemistry and identify substances as chemicals.

Now that you are in a chemistry class, you may be wondering what you will be learning. What questions in science have you been curious about? Perhaps you are interested in how smog is formed or how aspirin relieves a headache. Just like you, chemists are curious about the world we live in.

How does car exhaust produce the smog that hangs over our cities? One component of car exhaust is nitrogen oxide (NO), which forms in car engines where high temperatures convert nitrogen gas (N_2) and oxygen gas (O_2) to NO. In the atmosphere, the $NO(g)$ reacts with $O_2(g)$ to form $NO_2(g)$, which has a reddish brown color of smog. In chemistry, reactions are written in the form of equations:

$$N_2(g) + O_2(g) \longrightarrow 2NO(g)$$
$$2NO(g) + O_2(g) \longrightarrow 2NO_2(g)$$
$$\text{Smog}$$

Why does aspirin relieve a headache? When a part of the body is injured, substances called prostaglandins are produced, which cause inflammation and pain. Aspirin acts to block the production of prostaglandins, thereby reducing inflammation, pain, and fever. Chemists in the medical field develop new treatments for diabetes, genetic defects, cancer, AIDS, and other diseases. Chemists in the environmental field study the ways in which human development impacts the environment and develop processes that help reduce environmental degradation. For the chemist in the forensic laboratory, the nurse in the dialysis unit, the dietitian, the chemical engineer, or the agricultural scientist, chemistry plays a central role in understanding problems, assessing possible solutions, and making important decisions.

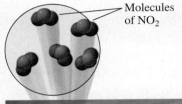

Molecules of NO_2

The chemical reaction of NO with oxygen in the air forms NO_2, which produces the reddish brown color of smog.

Chemistry

Chemistry is the study of the composition, structure, properties, and reactions of matter. *Matter* is another word for all the substances that make up our world. Perhaps you imagine that chemistry takes place only in a laboratory where a chemist is working in a white coat and goggles. Actually, chemistry happens all around you every day and has an impact on everything you use and do. You are doing chemistry when you cook food, add bleach to your laundry, or start your car. A chemical reaction has taken place when silver tarnishes or an antacid tablet fizzes when dropped into water. Plants grow because chemical reactions convert carbon dioxide, water, and energy to carbohydrates. Chemical reactions take place when you digest food and break it down into substances that you need for energy and health.

Antacid tablets undergo a chemical reaction when dropped into water.

Chemists working in research laboratories test new products and develop new pharmaceuticals.

Chemicals

A **chemical** is a substance that always has the same composition and properties wherever it is found. All the things you see around you are composed of one or more chemicals. Chemical processes take place in chemistry laboratories, manufacturing plants, and pharmaceutical labs as well as every day in nature and in our bodies. Often the terms *chemical* and *substance* are used interchangeably to describe a specific type of matter.

Every day, you use products containing substances that were developed and prepared by chemists. Soaps and shampoos contain chemicals that remove oils on your skin and scalp. When you brush your teeth, the substances in toothpaste clean your teeth, prevent plaque formation, and stop tooth decay. Some of the chemicals used to make toothpaste are listed in **TABLE 1.1**.

In cosmetics and lotions, chemicals are used to moisturize, prevent deterioration of the product, fight bacteria, and thicken the product. Your clothes may be made of natural materials, such as cotton, or synthetic substances, such as nylon or polyester. Perhaps you wear a ring or watch made of gold, silver, or platinum. Your breakfast cereal is probably fortified with iron, calcium, and phosphorus, whereas the milk you drink is enriched with vitamins A and D. Antioxidants are chemicals added to food to prevent it from spoiling. Some of the chemicals you may encounter when you cook in the kitchen are shown in **FIGURE 1.1**.

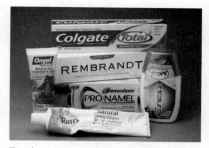

Toothpaste is a combination of many chemicals.

TABLE 1.1 Chemicals Commonly Used in Toothpaste

Chemical	Function
Calcium carbonate	Used as an abrasive to remove plaque
Sorbitol	Prevents loss of water and hardening of toothpaste
Sodium lauryl sulfate	Used to loosen plaque
Titanium dioxide	Makes toothpaste white and opaque
Triclosan	Inhibits bacteria that cause plaque and gum disease
Sodium fluorophosphate	Prevents formation of cavities by strengthening tooth enamel with fluoride
Methyl salicylate	Gives toothpaste a pleasant wintergreen flavor

Silicon dioxide (glass)

Chemically treated water

Metal alloy

Natural polymers

Natural gas

Fruits grown with fertilizers and pesticides

FIGURE 1.1 ▶ Many of the items found in a kitchen are chemicals or products of chemical reactions.
Q What are some other chemicals found in a kitchen?

Branches of Chemistry

The field of chemistry is divided into several branches. *General chemistry* is the study of the composition, properties, and reactions of matter. *Organic chemistry* is the study of substances that contain the element carbon. *Biological chemistry* is the study of the chemical reactions that take place in biological systems. Today chemistry is often combined with other sciences, such as geology and physics, to form cross-disciplines such as geochemistry and physical chemistry. *Geochemistry* is the study of the chemical composition of ores, soils, and minerals of the surface of the Earth and other planets. *Physical chemistry* is the study of the physical nature of chemical systems, including energy changes.

A geochemist collects newly erupted lava samples from Kilauea Volcano, Hawaii.

QUESTIONS AND PROBLEMS

1.1 Chemistry and Chemicals

LEARNING GOAL Define the term chemistry and identify substances as chemicals.

In every chapter, odd-numbered exercises in the *Questions and Problems* are paired with even-numbered exercises. The answers for the magenta, odd-numbered *Questions and Problems* are given at the end of each chapter. The complete solutions to the odd-numbered *Questions and Problems* are in the *Study Guide*.

1.1 Write a one-sentence definition for each of the following:
   a. chemistry
   b. chemical

1.2 Ask two of your friends (not in this class) to define the terms in problem 1.1. Do their answers agree with the definitions you provided?

Applications

1.3 Obtain a bottle of multivitamins and read the list of ingredients. What are four chemicals from the list?

1.4 Obtain a box of breakfast cereal and read the list of ingredients. What are four chemicals from the list?

1.5 Read the labels on some items found in your medicine cabinet. What are the names of some chemicals contained in those items?

1.6 Read the labels on products used to wash your dishes. What are the names of some chemicals contained in those products?

1.2 Scientific Method: Thinking Like a Scientist

LEARNING GOAL Describe the activities that are part of the scientific method.

When you were very young, you explored the things around you by touching and tasting. As you grew, you asked questions about the world in which you live. What is lightning? Where does a rainbow come from? Why is water blue? As an adult, you may have wondered how antibiotics work or why vitamins are important to your health. Every day, you ask questions and seek answers to organize and make sense of the world around you.

When the late Nobel Laureate Linus Pauling described his student life in Oregon, he recalled that he read many books on chemistry, mineralogy, and physics. "I mulled over the properties of materials: why are some substances colored and others not, why are some minerals or inorganic compounds hard and others soft?" He said, "I was building up this tremendous background of empirical knowledge and at the same time asking a great number of questions." Linus Pauling won two Nobel Prizes: the first, in 1954, was in chemistry for his work on the nature of chemical bonds and the determination of the structures of complex substances; the second, in 1962, was the Peace Prize.

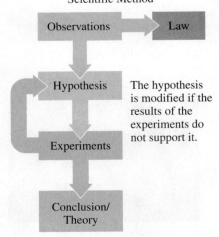

Linus Pauling won the Nobel Prize in Chemistry in 1954.

The Scientific Method

The process of trying to understand nature is unique to each scientist. However, the **scientific method** is a process that scientists use to make observations in nature, gather data, and explain natural phenomena.

1. **Observations** The first step in the scientific method is to make observations about nature and ask questions about what you observe. When an observation always seems to be true, it may be stated as a *law* that predicts that behavior and is often measurable. However, a law does not explain that observation. For example, we can use the *Law of Gravity* to predict that if we drop our chemistry book it would fall on the table or the floor but this law does not explain why our book falls.

2. **Hypothesis** A scientist forms a hypothesis, which gives a possible explanation of an observation or a law. The hypothesis must be stated in such a way that it can be tested by experiments.

3. **Experiments** To determine if a hypothesis is true or false, experiments are done to find a relationship between the hypothesis and the observations. The results of the experiments may confirm the hypothesis. However, if the experiments do not confirm the hypothesis, it is modified or discarded. Then new experiments will be designed to test the hypothesis.

4. **Conclusion/Theory** When the results of the experiments are analyzed, a conclusion is made as to whether the hypothesis is *true* or *false*. When experiments give consistent results, the hypothesis may be stated to be true. Even then, the hypothesis

Scientific Method

Observations → Law

Hypothesis — The hypothesis is modified if the results of the experiments do not support it.

Experiments

Conclusion/ Theory

The scientific method develops a conclusion or theory about nature using observations, hypotheses, and experiments.

continues to be tested and, based on new experimental results, may need to be modified or replaced. If many additional experiments by a group of scientists continue to support the hypothesis, it may become a *scientific theory*, which gives an explanation for the initial observations.

CHEMISTRY LINK TO HEALTH
Early Chemist: Paracelsus

For many centuries, chemistry has been the study of changes in matter. From the time of the ancient Greeks to about the sixteenth century, alchemists described matter in terms of four components of nature: earth, air, fire, and water, with the qualities of hot, cold, damp, or dry. By the eighth century, alchemists believed that they could rearrange these qualities in such a way as to change metals such as copper and lead into gold and silver. Although these efforts failed, the alchemists provided information on the chemical reactions involved in the extraction of metals from ores. The alchemists also designed some of the first laboratory equipment and developed early laboratory procedures. These early efforts were some of the first observations and experiments using the scientific method.

Paracelsus (1493–1541) was a physician and an alchemist who thought that alchemy should be about preparing new medicines. Using observation and experimentation, he proposed that a healthy body was regulated by a series of chemical processes that could be unbalanced by certain chemical compounds and rebalanced by using minerals and medicines. For example, he determined that inhaled dust, not underground spirits, caused lung disease in miners. He also thought that goiter was a problem caused by contaminated water, and he treated syphilis with compounds of mercury. His opinion of medicines was that the right dose makes the difference between a poison and a cure. Paracelsus changed alchemy in ways that helped establish modern medicine and chemistry.

Swiss physician and alchemist Paracelsus (1493–1541) believed that chemicals and minerals could be used as medicines.

Using the Scientific Method in Everyday Life

You may be surprised to realize that you use the scientific method in your everyday life. Suppose you visit a friend in her home. Soon after you arrive, your eyes start to itch and you begin to sneeze. Then you observe that your friend has a new cat. Perhaps you ask yourself why you are sneezing and you form the hypothesis that you are allergic to cats. To test your hypothesis, you leave your friend's home. If the sneezing stops, perhaps your hypothesis is correct. You test your hypothesis further by visiting another friend who also has a cat. If you start to sneeze again, your experimental results support your hypothesis and you come to the conclusion that you are allergic to cats. However, if you continue sneezing after you leave your friend's home, your hypothesis is not supported. Now you need to form a new hypothesis, which could be that you have a cold.

Students make observations in the chemistry laboratory.

Through observation you may conclude that you are allergic to cats.

SAMPLE PROBLEM 1.1 Scientific Method

Identify each of the following statements as an observation (O), a hypothesis (H), or an experiment (E):

a. A silver tray turns a dull gray color when left uncovered.
b. When a silver tray is covered with plastic wrap, it does not tarnish.
c. Oxygen reacts with silver when the tray is exposed to air.

> **TRY IT FIRST**
>
> **SOLUTION**
>
> **a.** observation (O)   **b.** experiment (E)   **c.** hypothesis (H)
>
> **STUDY CHECK 1.1**
>
> The following statements are found in a student's notebook. Identify each of the following as an observation (O), a hypothesis (H), or an experiment (E):
>
> **a.** "Today I placed two tomato seedlings in the garden, and two more in a closet. I will give all the plants the same amount of water and fertilizer."
> **b.** "After 50 days, the tomato plants in the garden are 3 ft high with green leaves. The plants in the closet are 8 in. tall and yellow."
> **c.** "Tomato plants need sunlight to grow."
>
> **ANSWER**
>
> **a.** experiment (E)   **b.** observation (O)   **c.** hypothesis (H)

Tomato plants grow faster when placed in the sun.

QUESTIONS AND PROBLEMS

1.2 Scientific Method: Thinking Like a Scientist

LEARNING GOAL Describe the activities that are part of the scientific method.

1.7 Define each of the following terms of the scientific method:
  a. hypothesis   **b.** experiment
  c. theory   **d.** observation

1.8 Identify each of the following activities in the scientific method as an observation (O), a hypothesis (H), an experiment (E), or a conclusion (C):
  a. Formulate a possible explanation for your experimental results.
  b. Make notes about nature.
  c. Design an experimental plan that will give new information about a problem.
  d. State a generalized summary of your experimental results.

Applications

1.9 Identify each activity, **a** to **f**, as an observation (O), a hypothesis (H), an experiment (E), or a conclusion (C). At a popular restaurant, where Chang is the head chef, the following occurred:
  a. Chang determined that sales of the house salad had dropped.
  b. Chang decided that the house salad needed a new dressing.
  c. In a taste test, Chang prepared four bowls of lettuce, each with a new dressing: sesame seed, olive oil and balsamic vinegar, creamy Italian, and blue cheese.
  d. The tasters rated the sesame seed salad dressing as the favorite.
  e. After two weeks, Chang noted that the orders for the house salad with the new sesame seed dressing had doubled.
  f. Chang decided that the sesame seed dressing improved the sales of the house salad because the sesame seed dressing enhanced the taste.

Customers rated the sesame seed dressing as the best.

1.10 Identify each activity, **a** to **f**, as an observation (O), a hypothesis (H), an experiment (E), or a conclusion (C).
  Lucia wants to develop a process for dyeing shirts so that the color will not fade when the shirt is washed. She proceeds with the following activities:
  a. Lucia notices that the dye in a design fades when the shirt is washed.
  b. Lucia decides that the dye needs something to help it combine with the fabric.
  c. She places a spot of dye on each of four shirts and then places each one separately in water, salt water, vinegar, and baking soda and water.
  d. After one hour, all the shirts are removed and washed with a detergent.
  e. Lucia notices that the dye has faded on the shirts in water, salt water, and baking soda, whereas the dye did not fade on the shirt soaked in vinegar.
  f. Lucia thinks that the vinegar binds with the dye so it does not fade when the shirt is washed.

1.11 Identify each of the following as an observation (O), a hypothesis (H), an experiment (E), or a conclusion (C):
  a. One hour after drinking a glass of regular milk, Jim experienced stomach cramps.
  b. Jim thinks he may be lactose intolerant.
  c. Jim drinks a glass of lactose-free milk and does not have any stomach cramps.
  d. Jim drinks a glass of regular milk to which he has added lactase, an enzyme that breaks down lactose, and has no stomach cramps.

1.12 Identify each of the following as an observation (O), a hypothesis (H), an experiment (E), or a conclusion (C):
  a. Sally thinks she may be allergic to shrimp.
  b. Yesterday, one hour after Sally ate a shrimp salad, she broke out in hives.
  c. Today, Sally had some soup that contained shrimp, but she did not break out in hives.
  d. Sally realizes that she does not have an allergy to shrimp.

1.3 Learning Chemistry: A Study Plan

LEARNING GOAL Develop a study plan for learning chemistry.

Here you are taking chemistry, perhaps for the first time. Whatever your reasons for choosing to study chemistry, you can look forward to learning many new and exciting ideas.

Features in This Text Help You Study Chemistry

This text has been designed with study features to complement your individual learning style. On the inside of the front cover is a periodic table of the elements. On the inside of the back cover are tables that summarize useful information needed throughout your study of chemistry. Each chapter begins with *Looking Ahead*, which outlines the topics in the chapter. *Key Terms* are bolded when they first appear in the text, and are summarized at the end of each chapter. They are also listed and defined in the comprehensive *Glossary and Index*, which appears at the end of the text. *Key Math Skills* and *Core Chemistry Skills* that are critical to learning chemistry are indicated by icons in the margin, and summarized at the end of each chapter. In the *Chapter Readiness* list at the beginning of every chapter, the *Key Math Skills* and *Core Chemistry Skills* from previous chapters related to the current chapter concepts are highlighted for your review.

Before you begin reading, obtain an overview of a chapter by reviewing the topics in *Looking Ahead*. As you prepare to read a section of the chapter, look at the section title and turn it into a question. For example, for section 1.1, "Chemistry and Chemicals," you could ask, "What is chemistry?" or "What are chemicals?" When you come to a *Sample Problem*, take the time to work it through and compare your solution to the one provided. As you read the text, you will see *Engage* features in the margin, which remind you to pause your reading and interact with a question related to the material.

The *Try It First* feature above the Solution of each Sample Problem is a reminder for you to work out the problem before you look at the Solution. Many *Sample Problems* are accompanied by a *Guide to Problem Solving*, which gives the steps needed to work the problem. The *Analyze the Problem* feature in some Sample Problems includes *Given*, the information we have; *Need*, what we are going to accomplish; and *Connect*, how we proceed from Given to Need. When you compare your answer with the Solution provided, you know what you need to correct or change. This process of trying the problem first will help you develop successful problem solving techniques. Then work the associated *Study Check*. The answers to all the Study Checks are included and you can compare your answer to the one provided.

At the end of each chapter section, you will find a set of *Questions and Problems* that allows you to apply problem solving immediately to the new concepts. The problems are paired, which means that each of the odd-numbered problems is matched to the following even-numbered problem. At the end of each chapter, the answers to all the odd-numbered problems are provided. If the answers match yours, you most likely understand the topic; if not, you need to study the section again.

Throughout each chapter, boxes titled "Chemistry Link to Health" and "Chemistry Link to the Environment" help you connect the chemical concepts you are learning to real-life situations. Many of the figures and diagrams use macro-to-micro illustrations to depict the atomic level of organization of ordinary objects, such as the atoms in aluminum foil. These visual models illustrate the concepts described in the text and allow you to "see" the world in a microscopic way.

At the end of each chapter, you will find several study aids that complete the chapter. *Chapter Reviews* provide a summary in easy-to-read bullet points and *Concept Maps* visually show the connections between important topics. The *Key Terms*, which are in boldface type within the chapter, are listed with their definitions. *Understanding the Concepts*, a set of questions that use art and models, helps you visualize concepts. *Additional Questions and Problems* and *Challenge Problems* provide additional exercises to test your understanding of the topics in the chapter. *Applications* are groups of problems that apply section content to current topics. *Answers* to all of the odd-numbered problems complete the chapter and you can compare your answers to the ones provided.

After some chapters, problem sets called *Combining Ideas* test your ability to solve problems containing material from more than one chapter.

KEY MATH SKILL

CORE CHEMISTRY SKILL

ENGAGE

What is different and what is the same for an atom of Sn-105 and an atom of Sn-132?

TRY IT FIRST

ANALYZE THE PROBLEM	Given	Need	Connect
	165 lb	kilograms	conversion factor

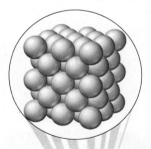

Using Active Learning

A student who is an active learner continually interacts with the chemical ideas while reading the text, working problems, and attending lectures. Let's see how this is done.

As you read and practice problem solving, you remain actively involved in studying, which enhances the learning process. In this way, you learn a small amount of information and establish the necessary foundation for understanding the next section. You may also note questions you have about the reading, which you can discuss with your professor or laboratory instructor. **TABLE 1.2** summarizes these steps for active learning. The time you spend in a lecture is a useful learning time. By keeping track of the class schedule and reading the assigned material before a lecture, you become aware of the new terms and concepts you need to learn. Some questions that occur during your reading may be answered during the lecture. If not, you can ask your professor for further clarification.

TABLE 1.2 Steps in Active Learning

1. Read each *Learning Goal* for an overview of the material.
2. Form a question from the title of the section you are going to read.
3. Read the section, looking for answers to your question.
4. Self-test by working *Sample Problems* and *Study Checks*.
5. Complete the *Questions and Problems* that follow that section, and check the answers for the magenta odd-numbered problems at the end of the chapter.
6. Proceed to the next section and repeat the steps.

Many students find that studying with a group can be beneficial to learning. In a group, students motivate each other to study, fill in gaps, and correct misunderstandings by teaching and learning together. Studying alone does not allow the process of peer correction. In a group, you can cover the ideas more thoroughly as you discuss the reading and problem solve with other students. You may find that it is easier to retain new material and new ideas if you study in short sessions throughout the week rather than all at once. Waiting to study until the night before an exam does not give you time to understand concepts and practice problem solving.

Studying in a group can be beneficial to learning.

Making a Study Plan

As you embark on your journey into the world of chemistry, think about your approach to studying and learning chemistry. You might consider some of the ideas in the following list. Check those ideas that will help you successfully learn chemistry. Commit to them now. *Your* success depends on *you*.

My study plan for learning chemistry will include the following:

_____ reading the chapter before lecture

_____ going to lecture

_____ reviewing the *Learning Goals*

_____ keeping a problem notebook

_____ reading the text as an active learner

_____ answering the *Engage* questions

_____ trying to work the *Sample Problem* before looking at the *Solution*

_____ working the *Questions and Problems* at the end of each section and checking answers

Students discuss a chemistry problem with their professor during office hours.

_____ being an active learner in lecture

_____ organizing a study group

_____ seeing the professor during office hours

_____ reviewing *Key Math Skills* and *Core Chemistry Skills*

_____ attending review sessions

_____ organizing my own review sessions

_____ studying as often as I can

SAMPLE PROBLEM 1.2 A Study Plan for Learning Chemistry

Which of the following activities would you include in your study plan for learning chemistry successfully?

a. skipping lecture
b. going to the professor's office hours
c. keeping a problem notebook
d. waiting to study until the night before the exam
e. trying to work the Sample Problem before looking at the Solution

> **TRY IT FIRST**

SOLUTION

Your success in chemistry can be improved by:

b. going to the professor's office hours
c. keeping a problem notebook
e. trying to work the Sample Problem before looking at the Solution

STUDY CHECK 1.2

Which of the following will help you learn chemistry?

a. skipping review sessions
b. working assigned problems
c. staying up all night before an exam
d. reading the assignment before a lecture

ANSWER

b and **d**

QUESTIONS AND PROBLEMS

1.3 Learning Chemistry: A Study Plan

LEARNING GOAL Develop a study plan for learning chemistry.

1.13 What are four things you can do to help yourself to succeed in chemistry?

1.14 What are four things that would make it difficult for you to learn chemistry?

1.15 A student in your class asks you for advice on learning chemistry. Which of the following might you suggest?
a. forming a study group
b. skipping a lecture
c. visiting the professor during office hours
d. waiting until the night before an exam to study
e. answering the Engage question

1.16 A student in your class asks you for advice on learning chemistry. Which of the following might you suggest?
a. doing the assigned problems
b. not reading the text; it's never on the test
c. attending review sessions
d. reading the assignment before a lecture
e. keeping a problem notebook

1.4 Key Math Skills for Chemistry

LEARNING GOAL Review math concepts used in chemistry: place values, positive and negative numbers, percentages, solving equations, and interpreting graphs.

During your study of chemistry, you will work many problems that involve numbers. You will need various math skills and operations. We will review some of the key math skills that are particularly important for chemistry. As we move through the chapters, we will also reference the key math skills as they apply.

KEY MATH SKILL
Identifying Place Values

Identifying Place Values

For any number, we can identify the *place value* for each of the digits in that number. These place values have names such as the ones place (first place to the left of the decimal point) or the tens place (second place to the left of the decimal point). A premature baby has a mass of 2518 g. We can indicate the place values for the number 2518 as follows:

Digit	Place Value
2	thousands
5	hundreds
1	tens
8	ones

ENGAGE

In the number 8.034, how do you know the 0 is in the tenths place?

We also identify place values such as the tenths place (first place to the right of the decimal point) and the hundredths place (second place to the right of the decimal place). A silver coin has a mass of 6.407 g. We can indicate the place values for the number 6.407 as follows:

Digit	Place Value
6	ones
4	ten**ths**
0	hundred**ths**
7	thousand**ths**

Note that place values ending with the suffix *ths* refer to the decimal places to the right of the decimal point.

> ### SAMPLE PROBLEM 1.3 Identifying Place Values
>
> A bullet found at a crime scene has a mass of 15.24 g. What are the place values for the digits in the mass of the bullet?
>
> **TRY IT FIRST**
>
> **SOLUTION**
>
> | Digit | Place Value |
> |-------|-------------|
> | 1 | tens |
> | 5 | ones |
> | 2 | tenths |
> | 4 | hundredths |
>
> **STUDY CHECK 1.3**
>
> A bullet found at a crime scene contains 0.925 g of lead. What are the place values for the digits in the mass of the lead?

ANSWER

Digit	Place Value
9	tenths
2	hundredths
5	thousandths

Using Positive and Negative Numbers in Calculations

KEY MATH SKILL
Using Positive and Negative Numbers in Calculations

A *positive number* is any number that is greater than zero and has a positive sign (+). Often the positive sign is understood and not written in front of the number. For example, the number +8 can also be written as 8. A *negative number* is any number that is less than zero and is written with a negative sign (−). For example, a negative eight is written as −8.

Multiplication and Division of Positive and Negative Numbers

When two positive numbers or two negative numbers are multiplied, the answer is positive (+).

$$2 \times 3 = +6$$
$$(-2) \times (-3) = +6$$

When a positive number and a negative number are multiplied, the answer is negative (−).

$$2 \times (-3) = -6$$
$$(-2) \times 3 = -6$$

The rules for the division of positive and negative numbers are the same as the rules for multiplication. When two positive numbers or two negative numbers are divided, the answer is positive (+).

$$\frac{6}{3} = 2 \qquad \frac{-6}{-3} = 2$$

When a positive number and a negative number are divided, the answer is negative (−).

$$\frac{-6}{3} = -2 \qquad \frac{6}{-3} = -2$$

Addition of Positive and Negative Numbers

When positive numbers are added, the sign of the answer is positive.

$$3 + 4 = 7 \quad \text{The + sign (+7) is understood.}$$

When negative numbers are added, the sign of the answer is negative.

$$(-3) + (-4) = -7$$

When a positive number and a negative number are added, the smaller number is subtracted from the larger number, and the result has the same sign as the larger number.

$$12 + (-15) = -3$$

Subtraction of Positive and Negative Numbers

When two numbers are subtracted, change the sign of the number to be subtracted and follow the rules for addition shown above.

$$12 - (+5) = 12 - 5 = 7$$
$$12 - (-5) = 12 + 5 = 17$$
$$-12 - (-5) = -12 + 5 = -7$$
$$-12 - (+5) = -12 - 5 = -17$$

Calculator Operations

On your calculator, there are four keys that are used for basic mathematical operations. The change sign $\boxed{+/-}$ key is used to change the sign of a number.

To practice these basic calculations on the calculator, work through the problem going from the left to the right doing the operations in the order they occur. If your calculator has a change sign $\boxed{+/-}$ key, a negative number is entered by pressing the number and then pressing the change sign $\boxed{+/-}$ key. At the end, press the equals $\boxed{=}$ key or ANS or ENTER.

Multiplication
Division
Subtraction
Equals
Change sign
Addition

Addition and Subtraction

Example 1: $15 - 8 + 2 =$

Solution: $15\;\boxed{-}\;8\;\boxed{+}\;2\;\boxed{=}\;9$

Example 2: $4 + (-10) - 5 =$

Solution: $4\;\boxed{+}\;10\;\boxed{+/-}\;\boxed{-}\;5\;\boxed{=}\;-11$

Multiplication and Division

Example 3: $2 \times (-3) =$

Solution: $2\;\boxed{\times}\;3\;\boxed{+/-}\;\boxed{=}\;-6$

Example 4: $\dfrac{8 \times 3}{4} =$

Solution: $8\;\boxed{\times}\;3\;\boxed{\div}\;4\;\boxed{=}\;6$

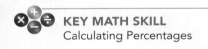

KEY MATH SKILL
Calculating Percentages

A bullet casing at a crime scene is marked as evidence.

Calculating Percentages

To determine a percentage, divide the parts by the total (whole) and multiply by 100%. For example, if an aspirin tablet contains 325 mg of aspirin (active ingredient) and the tablet has a mass of 545 mg, what is the percentage of aspirin in the tablet?

$$\frac{325 \text{ mg aspirin}}{545 \text{ mg tablet}} \times 100\% = 59.6\% \text{ aspirin}$$

When a value is described as a percentage (%), it represents the number of parts of an item in 100 of those items. If the percentage of red balls is 5, it means there are 5 red balls in every 100 balls. If the percentage of green balls is 50, there are 50 green balls in every 100 balls.

$$5\% \text{ red balls} = \frac{5 \text{ red balls}}{100 \text{ balls}} \qquad 50\% \text{ green balls} = \frac{50 \text{ green balls}}{100 \text{ balls}}$$

SAMPLE PROBLEM 1.4 Calculating a Percentage

A bullet found at a crime scene may be used as evidence in a trial if the percentage of metals is a match to the composition of metals in a bullet from the suspect's ammunition. If a bullet found at a crime scene contains 13.9 g of lead, 0.3 g of tin, and 0.9 g of antimony, what is the percentage of each metal in the bullet? Express your answers to the ones place.

TRY IT FIRST

SOLUTION

Total mass $= 13.9 \text{ g} + 0.3 \text{ g} + 0.9 \text{ g} = 15.1 \text{ g}$

Percentage of lead

$$\frac{13.9 \text{ g}}{15.1 \text{ g}} \times 100\% = 92\% \text{ lead}$$

Percentage of tin

$$\frac{0.3 \text{ g}}{15.1 \text{ g}} \times 100\% = 2\% \text{ tin}$$

Percentage of antimony

$$\frac{0.9 \text{ g}}{15.1 \text{ g}} \times 100\% = 6\% \text{ antimony}$$

STUDY CHECK 1.4

A bullet seized from the suspect's ammunition has a composition of lead 11.6 g, tin 0.5 g, and antimony 0.4 g.

a. What is the percentage of each metal in the bullet? Express your answers to the ones place.
b. Could the bullet removed from the suspect's ammunition be considered as evidence that the suspect was at the crime scene mentioned in Sample Problem 1.4?

ANSWER

a. The bullet from the suspect's ammunition is lead 93%, tin 4%, and antimony 3%.
b. The composition of this bullet does not match the bullet from the crime scene and cannot be used as evidence.

Solving Equations

In chemistry, we use equations that express the relationship between certain variables. Let's look at how we would solve for x in the following equation:

$$2x + 8 = 14$$

Our overall goal is to rearrange the items in the equation to obtain x on one side.

1. *Place all like terms on one side.* The numbers 8 and 14 are like terms. To remove the 8 from the left side of the equation, we subtract 8. To keep a balance, we need to subtract 8 from the 14 on the other side.

$$2x + 8 - 8 = 14 - 8$$
$$2x \qquad\quad = 6$$

2. *Isolate the variable you need to solve for.* In this problem, we obtain x by dividing both sides of the equation by 2. The value of x is the result when 6 is divided by 2.

$$\frac{2x}{2} = \frac{6}{2}$$
$$x = 3$$

3. *Check your answer.* Check your answer by substituting your value for x back into the original equation.

$$2(3) + 8 = 14$$
$$6 + 8 = 14$$
$$14 = 14 \quad \text{Your answer } x = 3 \text{ is correct.}$$

Summary: To solve an equation for a particular variable, be sure you perform the same mathematical operations on *both* sides of the equation.

If you eliminate a symbol or number by subtracting, you need to subtract that same symbol or number on the opposite side.

If you eliminate a symbol or number by adding, you need to add that same symbol or number on the opposite side.

KEY MATH SKILL
Solving Equations

ENGAGE

Why is the number 8 subtracted from both sides of this equation?

A plastic strip thermometer changes color to indicate body temperature.

Interactive Video

Solving Equations

If you cancel a symbol or number by dividing, you need to divide both sides by that same symbol or number.

If you cancel a symbol or number by multiplying, you need to multiply both sides by that same symbol or number.

When we work with temperature, we may need to convert between degrees Celsius and degrees Fahrenheit using the following equation:

$$T_F = 1.8(T_C) + 32$$

To obtain the equation for converting degrees Fahrenheit to degrees Celsius, we subtract 32 from both sides.

$$T_F = 1.8(T_C) + 32$$
$$T_F - 32 = 1.8(T_C) + \cancel{32} - \cancel{32}$$
$$T_F - 32 = 1.8(T_C)$$

To obtain T_C by itself, we divide both sides by 1.8.

$$\frac{T_F - 32}{1.8} = \frac{\cancel{1.8}(T_C)}{\cancel{1.8}} = T_C$$

SAMPLE PROBLEM 1.5 Solving Equations

Solve the following equation for V_2:

$$P_1 V_1 = P_2 V_2$$

TRY IT FIRST

SOLUTION

$$P_1 V_1 = P_2 V_2$$

To solve for V_2, divide both sides by the symbol P_2.

$$\frac{P_1 V_1}{P_2} = \frac{\cancel{P_2} V_2}{\cancel{P_2}}$$

$$V_2 = \frac{P_1 V_1}{P_2}$$

STUDY CHECK 1.5

Solve the following equation for m:

$$\text{heat} = m \times \Delta T \times SH$$

ANSWER

$$m = \frac{\text{heat}}{\Delta T \times SH}$$

KEY MATH SKILL
Interpreting Graphs

Interpreting Graphs

A graph represents the relationship between two variables. These quantities are plotted along two perpendicular axes, which are the *x* axis (horizontal) and *y* axis (vertical).

Example

In the graph Volume of a Balloon Versus Temperature, the volume of a gas in a balloon is plotted against its temperature.

Title

Look at the title. What does it tell us about the graph? The title indicates that the volume of a balloon was measured at different temperatures.

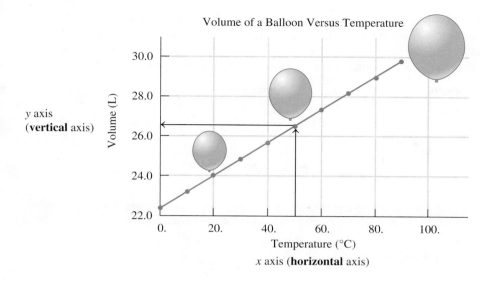

Volume of a Balloon Versus Temperature

y axis
(**vertical** axis)

x axis (**horizontal** axis)

Vertical Axis

Look at the label and the numbers on the vertical (*y*) axis. The label indicates that the volume of the balloon was measured in liters (L). The numbers, which are chosen to include the low and high measurements of the volume of the gas, are evenly spaced from 22.0 L to 30.0 L.

Horizontal Axis

The label on the horizontal (*x*) axis indicates that the temperature of the balloon was measured in degrees Celsius (°C). The numbers are measurements of the Celsius temperature, which are evenly spaced from 0. °C to 100. °C.

Points on the Graph

Each point on the graph represents a volume in liters that was measured at a specific temperature. When these points are connected, a line is obtained.

Interpreting the Graph

From the graph, we see that the volume of the gas increases as the temperature of the gas increases. This is called a *direct relationship*. Now we use the graph to determine the volume at various temperatures. For example, suppose we want to know the volume of the gas at 50. °C. We would start by finding 50. °C on the *x* axis and then drawing a line up to the plotted line. From there, we would draw a horizontal line that intersects the *y* axis and read the volume value where the line crosses the *y* axis as shown on the graph above.

SAMPLE PROBLEM 1.6 Interpreting a Graph

A nurse administers Tylenol to lower a child's fever. The graph shows the body temperature of the child plotted against time.

a. What is measured on the vertical axis?
b. What is the range of values on the vertical axis?
c. What is measured on the horizontal axis?
d. What is the range of values on the horizontal axis?

TRY IT FIRST

SOLUTION

a. body temperature, in degrees Celsius
b. 37.0 °C to 39.4 °C
c. time, in minutes, after Tylenol was given
d. 0 min to 30 min

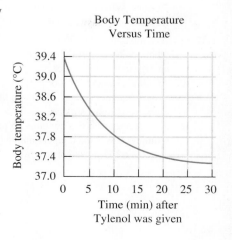

Body Temperature
Versus Time

STUDY CHECK 1.6

a. Using the graph in Sample Problem 1.6, what was the child's temperature 15 min after Tylenol was given?

b. How many minutes elapsed before the temperature decreased to 38.0 °C?

ANSWER

a. 37.6 °C   **b.** 8 min

QUESTIONS AND PROBLEMS

1.4 Key Math Skills for Chemistry

LEARNING GOAL Review math concepts used in chemistry: place values, positive and negative numbers, percentages, solving equations, and interpreting graphs.

1.17 What is the place value for the bold digit?
  a. 7.3**2**88
  b. 16.12**3**4
  c. 46**7**5.99

1.18 What is the place value for the bold digit?
  a. 97.5**6**89
  b. **3**75.88
  c. 46.1**0**00

1.19 Evaluate each of the following:
  a. $15 - (-8) = $ _____
  b. $-8 + (-22) = $ _____
  c. $4 \times (-2) + 6 = $ _____

1.20 Evaluate each of the following:
  a. $-11 - (-9) = $ _____
  b. $34 + (-55) = $ _____
  c. $\dfrac{-56}{8} = $ _____

Applications

1.21 a. A clinic had 25 patients on Friday morning. If 21 patients were given flu shots, what percentage of the patients received flu shots? Express your answer to the ones place.
  b. An alloy contains 56 g of pure silver and 22 g of pure copper. What is the percentage of silver in the alloy? Express your answer to the ones place.
  c. A collection of coins contains 11 nickels, 5 quarters, and 7 dimes. What is the percentage of dimes in the collection? Express your answer to the ones place.

1.22 a. At a local hospital, 35 babies were born. If 22 were boys, what percentage of the newborns were boys? Express your answer to the ones place.
  b. An alloy contains 67 g of pure gold and 35 g of pure zinc. What is the percentage of zinc in the alloy? Express your answer to the ones place.
  c. A collection of coins contains 15 pennies, 14 dimes, and 6 quarters. What is the percentage of pennies in the collection? Express your answer to the ones place.

1.23 Solve each of the following for a:
  a. $4a + 4 = 40$ **b.** $\dfrac{a}{6} = 7$

1.24 Solve each of the following for b:
  a. $2b + 7 = b + 10$ **b.** $3b - 4 = 24 - b$

Use the following graph for questions 1.25 and 1.26:

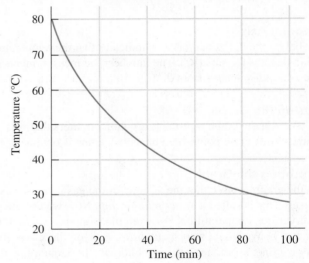

Time for Cooling of Tea Versus Temperature

1.25 a. What does the title indicate about the graph?
  b. What is measured on the vertical axis?
  c. What is the range of values on the vertical axis?
  d. Does the temperature increase or decrease with an increase in time?

1.26 a. What is measured on the horizontal axis?
  b. What is the range of values on the horizontal axis?
  c. What is the temperature of the tea after 20 min?
  d. How many minutes were needed to reach a temperature of 45 °C?

1.5 Writing Numbers in Scientific Notation

LEARNING GOAL Write a standard number in scientific notation and vice versa.

In chemistry, we use numbers that are very large and very small. We might measure something as tiny as the width of a human hair, which is about 0.000 008 m. Or perhaps we want to count the number of hairs on the average human scalp, which is about 100 000 hairs. In this text, we add spaces between sets of three digits when it helps make the places easier to count. However, we will see that it is more convenient to write large and small numbers in *scientific notation*.

Standard Number	Scientific Notation
0.000 008 m	8×10^{-6} m
100 000 hairs	1×10^5 hairs

Humans have an average of 1×10^5 hairs on their scalps. Each hair is about 8×10^{-6} m wide.

A number written in **scientific notation** has two parts: a coefficient and a power of 10. For example, the number 2400 is written in scientific notation as 2.4×10^3. The coefficient, 2.4, is obtained by moving the decimal point to the left to give a number that is at least 1 but less than 10. Because we moved the decimal point three places to the left, the power of 10 is a positive 3, which is written as 10^3. When a number greater than 1 is converted to scientific notation, the power of 10 is positive.

KEY MATH SKILL
Converting between Standard Numbers and Scientific Notation

Standard Number   **Scientific Notation**

$$2400. \quad = \quad 2.4 \quad \times \quad 10^3$$

← 3 places   Coefficient   Power of 10

In another example, 0.000 86 is written in scientific notation as 8.6×10^{-4}. The coefficient, 8.6, is obtained by moving the decimal point to the right. Because the decimal point is moved four places to the right, the power of 10 is a negative 4, written as 10^{-4}. When a number less than 1 is written in scientific notation, the power of 10 is negative.

Standard Number   **Scientific Notation**

$$0.00086 \quad = \quad 8.6 \quad \times \quad 10^{-4}$$

4 places →   Coefficient   Power of 10

TABLE 1.3 gives some examples of numbers written as positive and negative powers of 10. The powers of 10 are a way of keeping track of the decimal point in the number. **TABLE 1.4** gives several examples of writing measurements in scientific notation.

TABLE 1.3 Some Powers of 10

Standard Number	Multiples of 10	Scientific Notation	
10 000	$10 \times 10 \times 10 \times 10$	1×10^4	Some positive powers of 10
1 000	$10 \times 10 \times 10$	1×10^3	
100	10×10	1×10^2	
10	10	1×10^1	
1	0	1×10^0	
0.1	$\dfrac{1}{10}$	1×10^{-1}	Some negative powers of 10
0.01	$\dfrac{1}{10} \times \dfrac{1}{10} = \dfrac{1}{100}$	1×10^{-2}	
0.001	$\dfrac{1}{10} \times \dfrac{1}{10} \times \dfrac{1}{10} = \dfrac{1}{1\,000}$	1×10^{-3}	
0.0001	$\dfrac{1}{10} \times \dfrac{1}{10} \times \dfrac{1}{10} \times \dfrac{1}{10} = \dfrac{1}{10\,000}$	1×10^{-4}	

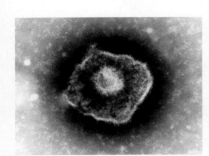

A chickenpox virus has a diameter of 3×10^{-7} m.

TABLE 1.4 Some Measurements Written as Standard Numbers and in Scientific Notation

Measured Quantity	Standard Number	Scientific Notation
Volume of gasoline used in the United States each year	550 000 000 000 L	5.5×10^{11} L
Diameter of the Earth	12 800 000 m	1.28×10^7 m
Average volume of blood pumped in 1 day	8500 L	8.5×10^3 L
Time for light to travel from the Sun to the Earth	500 s	5×10^2 s
Mass of a typical human	68 kg	6.8×10^1 kg
Mass of stirrup bone in ear	0.003 g	3×10^{-3} g
Diameter of a chickenpox (*Varicella zoster*) virus	0.000 000 3 m	3×10^{-7} m
Mass of bacterium (mycoplasma)	0.000 000 000 000 000 000 1 kg	1×10^{-19} kg

Scientific Notation and Calculators

You can enter a number in scientific notation on many calculators using the [EE or EXP] key. After you enter the coefficient, press the [EE or EXP] key and enter only the power of 10, because the [EE or EXP] key already includes the $\times$ 10 value. To enter a negative power of 10, press the [+/−] key or the [−] key, depending on your calculator.

Number to Enter	Procedure	Calculator Display
4×10^6	4 [EE or EXP] 6	$4\,06$ or 4^{06} or $4E06$
2.5×10^{-4}	2.5 [EE or EXP] [+/−] 4	$2.5-04$ or 2.5^{-04} or $2.5E-04$

When a calculator display appears in scientific notation, it is shown as a number that is at least 1 but less than 10, followed by a space and the power of 10. To express this display in scientific notation, write the coefficient value, write $\times$ 10, and use the power of 10 as an exponent.

Calculator Display	Expressed in Scientific Notation
7.52 04 or 7.52⁰⁴ or 7.52E04	7.52×10^4
5.8−02 or 5.8⁻⁰² or 5.8E−02	5.8×10^{-2}

On many scientific calculators, a number is converted into scientific notation using the appropriate keys. For example, the number 0.000 52 is entered, followed by pressing the 2nd or 3rd function key and the SCI key. The scientific notation appears in the calculator display as a coefficient and the power of 10.

0.000 52  [SCI] = 5.2−04 or 5.2⁻⁰⁴ or 5.2E−04 = 5.2×10^{-4}

Calculator display

SAMPLE PROBLEM 1.7 Scientific Notation

Write each of the following in scientific notation:

a. 3500   **b.** 0.000 016

TRY IT FIRST

SOLUTION

ANALYZE THE PROBLEM	Given	Need	Connect
	standard number	scientific notation	coefficient is at least 1 but less than 10

a. 3500

STEP 1 Move the decimal point to obtain a coefficient that is at least 1 but less than 10. For a number greater than 1, the decimal point is moved to the left three places to give a coefficient of 3.5.

STEP 2 Express the number of places moved as a power of 10. Moving the decimal point three places to the left gives a power of 3, written as 10^3.

STEP 3 Write the product of the coefficient multiplied by the power of 10.
3.5×10^3

b. 0.000 016

STEP 1 Move the decimal point to obtain a coefficient that is at least 1 but less than 10. For a number less than 1, the decimal point is moved to the right five places to give a coefficient of 1.6.

STEP 2 Express the number of places moved as a power of 10. Moving the decimal point five places to the right gives a power of negative 5, written as 10^{-5}.

STEP 3 Write the product of the coefficient multiplied by the power of 10.
1.6×10^{-5}

Guide to Writing a Number in Scientific Notation

STEP 1
Move the decimal point to obtain a coefficient that is at least 1 but less than 10.

STEP 2
Express the number of places moved as a power of 10.

STEP 3
Write the product of the coefficient multiplied by the power of 10.

STUDY CHECK 1.7

Write each of the following in scientific notation:

a. 425 000   **b.** 0.000 000 86

ANSWER

a. 4.25×10^5 **b.** 8.6×10^{-7}

Converting Scientific Notation to a Standard Number

When a number written in scientific notation has a positive power of 10, the standard number is obtained by moving the decimal point to the right for the same number of places as the power of 10. Placeholder zeros are used, as needed, to give additional places.

Scientific Notation		Standard Number
4.3×10^2	$=$ 4.30 $=$	430

For a number with a negative power of 10, the standard number is obtained by moving the decimal point to the left for the same number of places as the power of 10. Placeholder zeros are added in front of the coefficient as needed.

Scientific Notation		Standard Number
2.5×10^{-5}	$=$ 000002.5 $=$	0.000 025

SAMPLE PROBLEM 1.8 Writing Scientific Notation as a Standard Number

Write each of the following as a standard number:

a. 7.2×10^{-3} **b.** 2.4×10^5

TRY IT FIRST

SOLUTION

a. To write the standard number for an exponential number with a negative power of 10, move the decimal point to the left the same number of places (three) as the power of 10. Add placeholder zeros before the coefficient as needed.

7.2×10^{-3} $=$ 0007.2 m $=$ 0.0072

b. To write the standard number for an exponential number with a positive power of 10, move the decimal point to the right the same number of places (five) as the power of 10. Add placeholder zeros following the coefficient as needed.

2.4×10^5 $=$ 2.40000 $=$ 240 000

STUDY CHECK 1.8

Write 7.25×10^{-4} as a standard number.

ANSWER

0.000 725

QUESTIONS AND PROBLEMS

1.5 Writing Numbers in Scientific Notation

LEARNING GOAL Write a standard number in scientific notation and vice versa.

1.27 Write each of the following in scientific notation:
  a. 55 000   **b.** 480
  c. 0.000 005   **d.** 0.000 14
  e. 0.0072   **f.** 670 000

1.28 Write each of the following in scientific notation:
  a. 180 000 000   **b.** 0.000 06
  c. 750   **d.** 0.15
  e. 0.024   **f.** 1500

1.29 Write each of the following as a standard number:
  a. 1.2×10^4 **b.** 8.25×10^{-2}
  c. 4×10^6 **d.** 5.8×10^{-3}

1.30 Write each of the following as a standard number:
  a. 3.6×10^{-5} **b.** 8.75×10^4
  c. 3×10^{-2} **d.** 2.12×10^5

1.31 Which number in each of the following pairs is larger?
  a. 7.2×10^3 or 8.2×10^2 **b.** 4.5×10^{-4} or 3.2×10^{-2}
  c. 1×10^4 or 1×10^{-4} **d.** 0.000 52 or 6.8×10^{-2}

1.32 Which number in each of the following pairs is smaller?
  a. 4.9×10^{-3} or 5.5×10^{-9} **b.** 1250 or 3.4×10^2
  c. 0.000 000 4 or 5.0×10^2 **d.** 2.50×10^2 or 4×10^5

Follow Up

FORENSIC EVIDENCE SOLVES THE MURDER

Using a variety of laboratory tests, Sarah finds ethylene glycol in the victim's blood. The quantitative tests indicate that the victim had ingested 125 g of ethylene glycol. Sarah determines that the liquid in a glass found at the crime scene was ethylene glycol that had been added to an alcoholic beverage. Ethylene glycol is a clear, sweet-tasting, thick liquid that is odorless and mixes with water. It is easy to obtain since it is used as antifreeze in automobiles and in brake fluid. Because the initial symptoms of ethylene glycol poisoning are similar to being intoxicated, the victim is often unaware of its presence.

If ingestion of ethylene glycol occurs, it can cause depression of the central nervous system, cardiovascular damage, and kidney failure. If discovered quickly, hemodialysis may be used to remove ethylene glycol from the blood. A toxic amount of ethylene glycol is 1.5 g of ethylene glycol/kg of body mass. Thus, 75 g could be fatal for a 50-kg (110 lb) person.

Mark determines that fingerprints on the glass containing the ethylene glycol were those of the victim's husband. This evidence along with the container of antifreeze found in the home led to the arrest and conviction of the husband for poisoning his wife.

Applications

1.33 A container was found in the home of the victim that contained 120 g of ethylene glycol in 450 g of liquid. What was the percentage of ethylene glycol? Express your answer to the ones place.

1.34 If the toxic quantity is 1.5 g of ethylene glycol per 1000 g of body mass, what percentage of ethylene glycol is fatal?

CONCEPT MAP

CHEMISTRY IN OUR LIVES

deals with
- Substances
  - **called**
  - Chemicals

uses the
- Scientific Method
  - **starting with**
  - Observations
    - **that lead to**
    - Hypothesis
    - Experiments
    - Conclusion/Theory

is learned by
- Reading the Text
- Practicing Problem Solving
- Self-Testing
- Working with a Group
- Engaging
- Trying It First

uses key math skills
- Identifying Place Values
- Using Positive and Negative Numbers
- Calculating Percentages
- Solving Equations
- Interpreting Graphs
- Converting between Standard Numbers and Scientific Notation

CHAPTER REVIEW

1.1 Chemistry and Chemicals

LEARNING GOAL Define the term chemistry and identify substances as chemicals.

- Chemistry is the study of the composition, structure, properties, and reactions of matter.
- A chemical is any substance that always has the same composition and properties wherever it is found.

1.2 Scientific Method: Thinking Like a Scientist

LEARNING GOAL Describe the activities that are part of the scientific method.

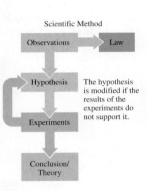

Scientific Method

Observations → Law

Hypothesis — The hypothesis is modified if the results of the experiments do not support it.

Experiments

Conclusion/ Theory

- The scientific method is a process of explaining natural phenomena beginning with making observations, forming a hypothesis, and performing experiments.
- After repeated successful experiments, a hypothesis may become a theory.

1.3 Learning Chemistry: A Study Plan

LEARNING GOAL Develop a study plan for learning chemistry.

- A study plan for learning chemistry utilizes the features in the text and develops an active learning approach to study.

- By using the *Learning Goals* in the chapter and working the *Sample Problems*, *Study Checks*, and the *Questions and Problems* at the end of each section, you can successfully learn the concepts of chemistry.

1.4 Key Math Skills for Chemistry

LEARNING GOAL Review math concepts used in chemistry: place values, positive and negative numbers, percentages, solving equations, and interpreting graphs.

- Solving chemistry problems involves a number of math skills: identifying place values, using positive and negative numbers, calculating percentages, solving equations, and interpreting graphs.

1.5 Writing Numbers in Scientific Notation

LEARNING GOAL Write a standard number in scientific notation and vice versa.

1×10^5 hairs 8×10^{-6} m

- A number written in scientific notation has two parts, a coefficient and a power of 10.
- When a number greater than 1 is converted to scientific notation, the power of 10 is positive.
- When a number less than 1 is written in scientific notation, the power of 10 is negative.

KEY TERMS

chemical A substance that has the same composition and properties wherever it is found.

chemistry The study of the composition, structure, properties, and reactions of matter.

conclusion An explanation of an observation that has been validated by repeated experiments that support a hypothesis.

experiment A procedure that tests the validity of a hypothesis.

hypothesis An unverified explanation of a natural phenomenon.

observation Information determined by noting and recording a natural phenomenon.

scientific method The process of making observations, proposing a hypothesis, and testing the hypothesis; after repeated experiments validate the hypothesis, it may become a theory.

scientific notation A form of writing large and small numbers using a coefficient that is at least 1 but less than 10, followed by a power of 10.

theory An explanation for an observation supported by additional experiments that confirm the hypothesis.

KEY MATH SKILLS

The chapter section containing each Key Math Skill is shown in parentheses at the end of each heading.

Identifying Place Values (1.4)

- The place value identifies the numerical value of each digit in a number.

Example: Identify the place value for each of the digits in the number 456.78.

Answer:

Digit	Place Value
4	hundreds
5	tens
6	ones
7	tenths
8	hundredths

Using Positive and Negative Numbers in Calculations (1.4)

- A *positive number* is any number that is greater than zero and has a positive sign (+). A *negative number* is any number that is less than zero and is written with a negative sign (−).
- When two positive numbers are added, multiplied, or divided, the answer is positive.
- When two negative numbers are multiplied or divided, the answer is positive. When two negative numbers are added, the answer is negative.
- When a positive and a negative number are multiplied or divided, the answer is negative.
- When a positive and a negative number are added, the smaller number is subtracted from the larger number and the result has the same sign as the larger number.
- When two numbers are subtracted, change the sign of the number to be subtracted then follow the rules for addition.

Example: Evaluate each of the following:

    a. $-8 - 14 =$ _____

    b. $6 \times (-3) =$ _____

Answer: **a.** $-8 - 14 = -22$

    b. $6 \times (-3) = -18$

Calculating Percentages (1.4)

- A percentage is the part divided by the total (whole) multiplied by 100%.

Example: A drawer contains 6 white socks and 18 black socks. What is the percentage of white socks?

Answer: $\dfrac{6 \text{ white socks}}{24 \text{ total socks}} \times 100\% = 25\%$ white socks

Solving Equations (1.4)

An equation in chemistry often contains an unknown. To rearrange an equation to obtain the unknown factor by itself, you keep it balanced by performing matching mathematical operations on both sides of the equation.

- If you eliminate a number or symbol by subtracting, subtract that same number or symbol on the opposite side.
- If you eliminate a number or symbol by adding, add that same number or symbol on the opposite side.
- If you cancel a number or symbol by dividing, divide both sides by that same number or symbol.
- If you cancel a number or symbol by multiplying, multiply both sides by that same number or symbol.

Example: Solve the equation for a: $3a - 8 = 28$

Answer: *Add 8 to both sides* $3a - 8 + 8 = 28 + 8$

$$3a = 36$$

Divide both sides by 3 $\dfrac{3a}{3} = \dfrac{36}{3}$

$$a = 12$$

Check: $3(12) - 8 = 28$

$$36 - 8 = 28$$

$$28 = 28$$

Your answer $a = 12$ is correct.

Interpreting Graphs (1.4)

- A graph represents the relationship between two variables.
- The quantities are plotted along two perpendicular axes, which are the x axis (horizontal) and y axis (vertical).
- The title indicates the components of the x and y axes.
- Numbers on the x and y axes show the range of values of the variables.
- The graph shows the relationship between the component on the y axis and that on the x axis.

Example: Solubility of Sugar in Water Versus Temperature

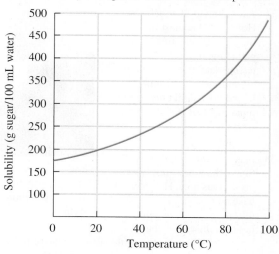

  a. Does the amount of sugar that dissolves in 100 mL of water increase or decrease when the temperature increases?

  b. How many grams of sugar dissolve in 100 mL of water at 70 °C?

  c. At what temperature (°C) will 275 g of sugar dissolve in 100 mL of water?

Answer: **a.** increase   **b.** 320 g   **c.** 55 °C

Converting between Standard Numbers and Scientific Notation (1.5)

- A number written in scientific notation consists of a coefficient and a power of 10.

A number is written in scientific notation by:

- Moving the decimal point to obtain a coefficient that is at least 1 but less than 10.
- Expressing the number of places moved as a power of 10. The power of 10 is positive if the decimal point is moved to the left, negative if the decimal point is moved to the right.
- The standard number is obtained by moving the decimal point for the same number of places as the power of 10.

Example: Write the number 28 000 in scientific notation.

Answer: Moving the decimal point four places to the left gives a coefficient of 2.8 and a positive power of 10, 10^4. The number 28 000 written in scientific notation is 2.8×10^4.

Example: Write 5.6×10^{-5} as a standard number.

Answer: Moving the decimal point five places to the left and adding placeholder zeros needed gives 0.000 056.

UNDERSTANDING THE CONCEPTS

The chapter sections to review are shown in parentheses at the end of each question.

1.35 A "chemical-free" shampoo includes the following ingredients: water, cocamide, glycerin, and citric acid. Is the shampoo truly "chemical-free"? (1.1)

1.36 A "chemical-free" sunscreen includes the following ingredients: titanium dioxide, vitamin E, and vitamin C. Is the sunscreen truly "chemical-free"? (1.1)

1.37 According to Sherlock Holmes, "One must follow the rules of scientific inquiry, gathering, observing, and testing data, then formulating, modifying, and rejecting hypotheses, until only one remains." Did Holmes use the scientific method? Why or why not? (1.2)

1.38 In *A Scandal in Bohemia*, Sherlock Holmes receives a mysterious note. He states, "I have no data yet. It is a capital mistake to theorize before one has data. Insensibly one begins to twist facts to suit theories, instead of theories to suit facts." What do you think Holmes meant? (1.2)

Sherlock Holmes is a fictional detective in novels written by Arthur Conan Doyle.

1.39 Classify each of the following statements as an observation (O) or a hypothesis (H): (1.2)
  a. A patient breaks out in hives after receiving penicillin.
  b. Dinosaurs became extinct when a large meteorite struck the Earth and caused a huge dust cloud that severely decreased the amount of light reaching the Earth.
  c. The 100-yd dash was run in 9.8 s.

1.40 Classify each of the following statements as an observation (O) or a hypothesis (H): (1.2)
  a. Analysis of 10 ceramic dishes showed that four dishes contained lead levels that exceeded federal safety standards.
  b. Marble statues undergo corrosion in acid rain.
  c. A child with a high fever and a rash may have chickenpox.

1.41 For each of the following, indicate if the answer has a positive or negative sign: (1.4)
  a. Two negative numbers are added.
  b. A positive and negative number are multiplied.

1.42 For each of the following, indicate if the answer has a positive or negative sign: (1.4)
  a. A negative number is subtracted from a positive number.
  b. Two negative numbers are divided.

ADDITIONAL QUESTIONS AND PROBLEMS

1.43 Identify each of the following as an observation (O), a hypothesis (H), an experiment (E), or a conclusion (C): (1.2)
  a. During an assessment in the emergency room, a nurse writes that the patient has a resting pulse of 30 beats/min.
  b. A nurse thinks that an incision from a recent surgery that is red and swollen is infected.
  c. Repeated studies show that lowering sodium in the diet leads to a decrease in blood pressure.

Nurses make observations in the hospital.

1.44 Identify each of the following as an observation (O), a hypothesis (H), an experiment (E), or a conclusion (C): (1.2)
  a. Drinking coffee at night keeps me awake.
  b. If I stop drinking coffee in the afternoon, I will be able to sleep at night.
  c. I will try drinking coffee only in the morning.

1.45 Select the correct phrase(s) to complete the following statement: If experimental results do not support your hypothesis, you should: (1.2)
  a. pretend that the experimental results support your hypothesis
  b. modify your hypothesis
  c. do more experiments

1.46 Select the correct phrase(s) to complete the following statement: A hypothesis is confirmed when: (1.2)
  a. one experiment proves the hypothesis
  b. many experiments validate the hypothesis
  c. you think your hypothesis is correct

1.47 Which of the following will help you develop a successful study plan? (1.3)
  a. skipping lecture and just reading the text
  b. working the *Sample Problems* as you go through a chapter
  c. going to your professor's office hours
  d. reading through the chapter, but working the problems later

1.48 Which of the following will help you develop a successful study plan? (1.3)
  a. studying all night before the exam
  b. forming a study group and discussing the problems together
  c. working problems in a notebook for easy reference
  d. copying the homework answers from a friend

1.49 Evaluate each of the following: (1.4)
  a. $4 \times (-8) =$ _____
  b. $-12 - 48 =$ _____
  c. $\dfrac{-168}{-4} =$ _____

1.50 Evaluate each of the following: (1.4)
  a. $-95 - (-11) =$ _____
  b. $\dfrac{152}{-19} =$ _____
  c. $4 - 56 =$ _____

1.51 A bag of gumdrops contains 16 orange gumdrops, 8 yellow gumdrops, and 16 black gumdrops. (1.4)
   a. What is the percentage of yellow gumdrops? Express your answer to the ones place.
   b. What is the percentage of black gumdrops? Express your answer to the ones place.

1.52 On the first chemistry test, 12 students got As, 18 students got Bs, and 20 students got Cs. (1.4)
   a. What is the percentage of students who received Bs? Express your answer to the ones place.
   b. What is the percentage of students who received Cs? Express your answer to the ones place.

1.53 Write each of the following in scientific notation: (1.5)
   a. 120 000   **b.** 0.000 000 34
   c. 0.066   **d.** 2700

1.54 Write each of the following in scientific notation: (1.5)
   a. 0.0042   **b.** 310
   c. 890 000 000   **d.** 0.000 000 056

1.55 Write each of the following as a standard number: (1.5)
   a. 2.6×10^{-5} **b.** 6.5×10^{2}
   c. 3.7×10^{-1} **d.** 5.3×10^{5}

1.56 Write each of the following as a standard number: (1.5)
   a. 7.2×10^{-2} **b.** 1.44×10^{3}
   c. 4.8×10^{-4} **d.** 9.1×10^{6}

CHALLENGE QUESTIONS

The following groups of questions are related to the topics in this chapter. However, they do not all follow the chapter order, and they require you to combine concepts and skills from several sections. These questions will help you increase your critical thinking skills and prepare for your next exam.

1.57 Classify each of the following as an observation (O), a hypothesis (H), or an experiment (E): (1.2)
   a. The bicycle tire is flat.
   b. If I add air to the bicycle tire, it will expand to the proper size.
   c. When I added air to the bicycle tire, it was still flat.
   d. The bicycle tire must have a leak in it.

1.58 Classify each of the following as an observation (O), a hypothesis (H), or an experiment (E): (1.2)
   a. A big log in the fire does not burn well.
   b. If I chop the log into smaller wood pieces, it will burn better.
   c. The small wood pieces burn brighter and make a hotter fire.
   d. The small wood pieces are used up faster than burning the big log.

1.59 Solve each of the following for x: (1.4)
   a. $2x + 5 = 41$ **b.** $\dfrac{5x}{3} = 40$

1.60 Solve each of the following for z: (1.4)
   a. $3z - (-6) = 12$ **b.** $\dfrac{4z}{-12} = -8$

Use the following graph for problems 1.61 and 1.62:

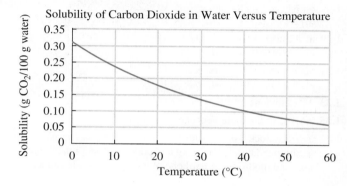

Solubility of Carbon Dioxide in Water Versus Temperature

1.61 **a.** What does the title indicate about the graph? (1.4)
   b. What is measured on the vertical axis?
   c. What is the range of values on the vertical axis?
   d. Does the solubility of carbon dioxide increase or decrease with an increase in temperature?

1.62 **a.** What is measured on the horizontal axis? (1.4)
   b. What is the range of values on the horizontal axis?
   c. What is the solubility of carbon dioxide in water at 25 °C?
   d. At what temperature does carbon dioxide have a solubility of 0.20 g/100 g water?

ANSWERS

Answers to Selected Questions and Problems

1.1 a. Chemistry is the study of the composition, structure, properties, and reactions of matter.
   b. A chemical is a substance that has the same composition and properties wherever it is found.

1.3 Many chemicals are listed on a vitamin bottle such as vitamin A, vitamin B_3, vitamin B_{12}, vitamin C, and folic acid.

1.5 Typical items found in a medicine cabinet and some of the chemicals they contain are as follows:

Antacid tablets: calcium carbonate, cellulose, starch, stearic acid, silicon dioxide

Mouthwash: water, alcohol, thymol, glycerol, sodium benzoate, benzoic acid

Cough suppressant: menthol, beta-carotene, sucrose, glucose

1.7 a. A hypothesis proposes a possible explanation for a natural phenomenon.
   b. An experiment is a procedure that tests the validity of a hypothesis.
   c. A theory is a hypothesis that has been validated many times by many scientists.
   d. An observation is a description or measurement of a natural phenomenon.

1.9 a. observation (O)   **b.** hypothesis (H)
   c. experiment (E)   **d.** observation (O)
   e. observation (O)   **f.** conclusion (C)

1.11 a. observation (O)   **b.** hypothesis (H)
   c. experiment (E)   **d.** experiment (E)

1.13 There are several things you can do that will help you success-fully learn chemistry: forming a study group, going to lecture, working *Sample Problems* and *Study Checks*, working *Questions and Problems* and checking answers, reading the assignment ahead of class, and keeping a problem notebook.

1.15 **a**, **c**, and **e**

1.17 **a.** thousandths   **b.** ones   **c.** hundreds

1.19 **a.** 23   **b.** −30   **c.** −2

1.21 **a.** 84%   **b.** 72%   **c.** 30%

1.23 **a.** 9   **b.** 42

1.25 **a.** The graph shows the relationship between the temperature of a cup of tea and time.
b. temperature, in °C
c. 20 °C to 80 °C
d. decrease

1.27 **a.** 5.5×10^4 **b.** 4.8×10^2 **c.** 5×10^{-6}
d. 1.4×10^{-4} **e.** 7.2×10^{-3} **f.** 6.7×10^5

1.29 **a.** 12 000   **b.** 0.0825
c. 4 000 000   **d.** 0.0058

1.31 **a.** 7.2×10^3 **b.** 3.2×10^{-2}
c. 1×10^4 **d.** 6.8×10^{-2}

1.33 27% ethylene glycol

1.35 No. All of the ingredients are chemicals.

1.37 Yes. Sherlock's investigation includes making observations (gathering data), formulating a hypothesis, testing the hypothesis, and modifying it until one of the hypotheses is validated.

1.39 **a.** observation (O)   **b.** hypothesis (H)
c. observation (O)

1.41 **a.** negative   **b.** negative

1.43 **a.** observation (O)   **b.** hypothesis (H)
c. conclusion (C)

1.45 **b**

1.47 **b** and **c**

1.49 **a.** −32   **b.** −60   **c.** 42

1.51 **a.** 20%   **b.** 40%

1.53 **a.** 1.2×10^5 **b.** 3.4×10^{-7}
c. 6.6×10^{-2} **d.** 2.7×10^3

1.55 **a.** 0.000 026   **b.** 650
c. 0.37   **d.** 530 000

1.57 **a.** observation (O)   **b.** hypothesis (H)
c. experiment (E)   **d.** hypothesis (H)

1.59 **a.** 18   **b.** 24

1.61 **a.** The graph shows the relationship between the solubility of carbon dioxide in water and temperature.
b. solubility of carbon dioxide (g CO_2/100 g water)
c. 0 to 0.35 g of CO_2/100 g of water
d. decrease

Chemistry and Measurements

DURING THE PAST FEW months, Greg has been experiencing an increased number of headaches, and frequently feels dizzy and nauseous. He goes to his doctor's office where, Sandra, the registered nurse completes the initial part of the exam by recording several measurements: weight 74.8 kg, height 171 cm, temperature 37.2 °C, and blood pressure 155/95. Normal blood pressure is 120/80 or below.

When Greg sees his doctor, he is diagnosed as having high blood pressure (hypertension). The doctor prescribes 80. mg of Inderal (propranolol). Inderal is a beta blocker, which relaxes the muscles of the heart. It is used to treat hypertension, angina (chest pain), arrhythmia, and migraine headaches.

Two weeks later, Greg visits his doctor again, who determines that Greg's blood pressure is now 152/90. The doctor increases the dosage of Inderal to 160 mg. The registered nurse, Sandra, informs Greg that he needs to increase his daily dosage from two tablets to four tablets.

CAREER

Registered Nurse

In addition to assisting physicians, registered nurses work to promote patient health, and prevent and treat disease. They provide patient care and help patients cope with illness. They take measurements such as a patient's weight, height, temperature, and blood pressure; make conversions; and calculate drug dosage rates. Registered nurses also maintain detailed medical records of patient symptoms and prescribed medications.

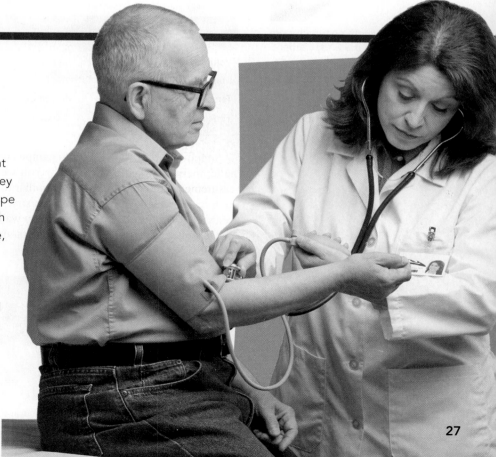

⊗⊕⊝ KEY MATH SKILLS

- Identifying Place Values (1.4)
- Using Positive and Negative Numbers in Calculations (1.4)
- Calculating Percentages (1.4)
- Converting between Standard Numbers and Scientific Notation (1.5)

*These Key Math Skills from the previous chapter are listed here for your review as you proceed to the new material in this chapter.

2.1 Units of Measurement

LEARNING GOAL Write the names and abbreviations for the metric or SI units used in measurements of length, volume, mass, temperature, and time.

Think about your day. You probably took some measurements. Perhaps you checked your weight by stepping on a bathroom scale. If you made some rice for dinner, you added two cups of water to one cup of rice. If you did not feel well, you may have taken your temperature. Whenever you take a measurement, you use a measuring device such as a scale, a measuring cup, or a thermometer.

Scientists and health professionals throughout the world use the **metric system** of measurement. It is also the common measuring system in all but a few countries in the world. The **International System of Units (SI)**, or Système International, is the official system of measurement throughout the world except for the United States. In chemistry, we use metric units and SI units for length, volume, mass, temperature, and time, as listed in **TABLE 2.1**.

Your weight on a bathroom scale is a measurement.

TABLE 2.1 Units of Measurement and Their Abbreviations

Measurement	Metric	SI
Length	meter (m)	meter (m)
Volume	liter (L)	cubic meter (m^3)
Mass	gram (g)	kilogram (kg)
Temperature	degree Celsius (°C)	kelvin (K)
Time	second (s)	second (s)

Suppose you walked 1.3 mi to campus today, carrying a backpack that weighs 26 lb. The temperature was 72 °F. Perhaps you weigh 128 lb and your height is 65 in. These measurements and units may seem familiar to you because they are stated in the U.S. system of measurement. However, in chemistry, we use the *metric system* in making our measurements. Using the metric system, you walked 2.1 km to campus, carrying a backpack that has a mass of 12 kg, when the temperature was 22 °C. You have a mass of 58.0 kg and a height of 1.7 m.

1.7 m (65 in.)

22 °C (72 °F)

58.0 kg (128 lb)

12 kg (26 lb)

2.1 km (1.3 mi)

There are many measurements in everyday life.

Length

The metric and SI unit of length is the **meter (m)**. A meter is 39.37 inches (in.), which makes it slightly longer than a yard (yd). The **centimeter (cm)**, a smaller unit of length, is commonly used in chemistry and is about equal to the width of your little finger. For comparison, there are 2.54 cm in 1 in. (see **FIGURE 2.1**). Some relationships between units for length are

$$1 \text{ m } = 100 \text{ cm}$$
$$1 \text{ m } = 39.37 \text{ in.}$$
$$1 \text{ m } = 1.094 \text{ yd}$$
$$2.54 \text{ cm } = 1 \text{ in.}$$

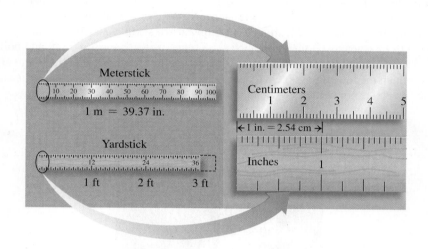

FIGURE 2.1 ▶ Length in the metric (SI) system is based on the meter, which is slightly longer than a yard.

 How many centimeters are in a length of 1 inch?

Volume

Volume (V) is the amount of space a substance occupies. A **liter (L)** is slightly larger than a quart (qt), (1 L = 1.057 qt). The SI unit of volume, the **cubic meter (m^3)** is the volume of a cube that has sides that measure 1 m in length. In a laboratory or a hospital, chemists work with metric units of volume that are smaller and more convenient, such as the **milliliter (mL)**. The volume of 1 mL is the same as 1 cm^3. A liter contains 1000 mL, as shown in **FIGURE 2.2**. A cubic meter is the same volume as 1000 L. Some relationships between units for volume are

$$1 \text{ m}^3 = 1000 \text{ L}$$
$$1 \text{ L } = 1000 \text{ mL}$$
$$1 \text{ mL } = 1 \text{ cm}^3$$
$$1 \text{ L } = 1.057 \text{ qt}$$
$$946.4 \text{ mL } = 1 \text{ qt}$$

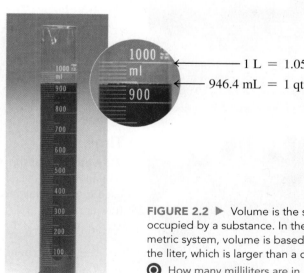

FIGURE 2.2 ▶ Volume is the space occupied by a substance. In the metric system, volume is based on the liter, which is larger than a quart.

 How many milliliters are in 1 quart?

Mass

The **mass** of an object is a measure of the quantity of material it contains. The SI unit of mass, the **kilogram (kg)**, is used for larger masses such as body mass. In the metric system, the unit for mass is the **gram (g)**, which is used for smaller masses. There are 1000 g in 1 kg. It takes 2.205 lb to make 1 kg, and 453.6 g is equal to 1 lb. Some relationships between units for mass are

$$1 \text{ kg } = 1000 \text{ g}$$
$$1 \text{ kg } = 2.205 \text{ lb}$$
$$453.6 \text{ g } = 1 \text{ lb}$$

The standard kilogram for the United States is stored at the National Institute of Standards and Technology (NIST).

You may be more familiar with the term *weight* than with mass. *Weight* is a measure of the gravitational pull on an object. On the Earth, an astronaut with a mass of 75.0 kg has a weight of 165 lb. On the Moon where the gravitational pull is one-sixth that of the Earth, the astronaut has a weight of 27.5 lb. However, the mass of the astronaut is the same as on Earth, 75.0 kg. Scientists measure mass rather than weight because mass does not depend on gravity.

In a chemistry laboratory, an electronic balance is used to measure the mass in grams of a substance (see **FIGURE 2.3**).

Temperature

Temperature tells us how hot something is, how cold it is outside, or helps us determine if we have a fever (see **FIGURE 2.4**). In the metric system, temperature is measured using Celsius temperature. On the **Celsius (°C) temperature scale**, water freezes at 0 °C and boils at 100 °C, whereas on the Fahrenheit (°F) scale, water freezes at 32 °F and boils at 212 °F. In the SI system, temperature is measured using the **Kelvin (K) temperature scale** on which the lowest possible temperature is 0 K. A unit on the Kelvin scale is called a kelvin (K) and is not written with a degree sign.

FIGURE 2.3 ▶ On an electronic balance, the digital readout gives the mass of a nickel, which is 5.01 g.

Ⓠ What is the mass of 10 nickels?

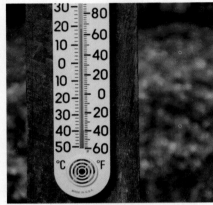

FIGURE 2.4 ▶ A thermometer is used to determine temperature.

Ⓠ What kinds of temperature readings have you made today?

Time

We typically measure time in units such as years (yr), days, hours (h), minutes (min), or seconds (s). Of these, the SI and metric unit of time is the **second (s)**. The standard now used to determine a second is an atomic clock. Some relationships between units for time are

$$1 \text{ day} = 24 \text{ h}$$
$$1 \text{ h} = 60 \text{ min}$$
$$1 \text{ min} = 60 \text{ s}$$

A stopwatch is used to measure the time of a race.

SAMPLE PROBLEM 2.1 Units of Measurement

On a typical day, a nurse encounters several situations involving measurement. State the name and type of measurement indicated by the units in each of the following:

a. A patient has a temperature of 38.5 °C.
b. A physician orders 1.5 g of cefuroxime for injection.
c. A physician orders 1 L of a sodium chloride solution to be given intravenously.
d. A medication is to be given to a patient every 4 h.

▶ **TRY IT FIRST** ▶

SOLUTION

a. A degree Celsius is a unit of temperature.
b. A gram is a unit of mass.
c. A liter is a unit of volume.
d. An hour is a unit of time.

STUDY CHECK 2.1

State the name and type of measurement indicated by an infant that is 32.5 cm long.

ANSWER

A centimeter is a unit of length.

QUESTIONS AND PROBLEMS

2.1 Units of Measurement

LEARNING GOAL Write the names and abbreviations for the metric or SI units used in measurements of length, volume, mass, temperature, and time.

2.1 Give the abbreviation for each of the following:
   a. gram   b. liter
   c. degree Celsius   d. pound
   e. second

2.2 Give the abbreviation for each of the following:
   a. kilogram   b. kelvin
   c. quart   d. meter
   e. cubic meter

2.3 State the type of measurement in each of the following statements:
   a. I put 12 L of gasoline in my gas tank.
   b. My friend is 170 cm tall.
   c. We are 385 000 km away from the Moon.
   d. The horse won the race by 1.2 s.

2.4 State the type of measurement in each of the following statements:
   a. I rode my bicycle 15 km today.
   b. My dog weighs 12 kg.
   c. It is hot today. It is 30 °C.
   d. I added 2 L of water to my fish tank.

2.5 State the name of the unit and the type of measurement indicated for each of the following quantities:
   a. 4.8 m   b. 325 g   c. 1.5 mL
   d. 4.8×10^2 s   e. 28 °C

2.6 State the name of the unit and the type of measurement indicated for each of the following quantities:
   a. 0.8 L   b. 3.6 cm   c. 4 kg
   d. 3.5 h   e. 373 K

Applications

2.7 On a typical day, medical personnel may encounter several situations involving measurement. State the name and type of measurement indicated by the units in each of the following:
   a. The clotting time for a blood sample is 12 s.
   b. A premature baby weighs 2.0 kg.
   c. An antacid tablet contains 1.0 g of calcium carbonate.
   d. An infant has a temperature 39.2 °C.

2.8 On a typical day, medical personnel may encounter several situations involving measurement. State the name and type of measurement indicated by the units in each of the following:
   a. During open-heart surgery, the temperature of a patient is lowered to 29 °C.
   b. The circulation time of a red blood cell through the body is 20 s.
   c. A patient with a persistent cough is given 10. mL of cough syrup.
   d. The amount of iron in the red blood cells of the body is 2.5 g.

A volume of 10. mL of cough syrup is measured for a patient.

2.2 Measured Numbers and Significant Figures

LEARNING GOAL Identify a number as measured or exact; determine the number of significant figures in a measured number.

When you make a measurement, you use some type of measuring device. For example, you may use a meterstick to measure your height, a scale to check your weight, or a thermometer to take your temperature.

FIGURE 2.5 ▶ The lengths of the rectangular objects are measured as **(a)** 4.5 cm and **(b)** 4.55 cm.

🔘 What is the length of the object in **(c)**?

CORE CHEMISTRY SKILL
Counting Significant Figures

Measured Numbers

Measured numbers are the numbers you obtain when you measure a quantity such as your height, weight, or temperature. Suppose you are going to measure the lengths of the objects in **FIGURE 2.5**. The metric ruler that you use may have lines marked in 1-cm divisions or perhaps in divisions of 0.1 cm. To report the length of the object, you observe the numerical values of the marked lines at the end of the object. Then you can *estimate* by visually dividing the space between the marked lines. This estimated value is the final digit in a measured number.

For example, in Figure 2.5a, the end of the object is between the marks of 4 cm and 5 cm, which means that the length is more than 4 cm but less than 5 cm. This is written as 4 cm plus an estimated digit. If you estimate that the end of the object is halfway between 4 cm and 5 cm, you would report its length as 4.5 cm. The last digit in a measured number may differ because people do not always estimate in the same way. Thus, someone else might report the length of the same object as 4.4 cm.

The metric ruler shown in Figure 2.5b is marked at every 0.1 cm. With this ruler, you can estimate to the hundredths place (0.01 cm). Now you can determine that the end of the object is between 4.5 cm and 4.6 cm. Perhaps you report its length as 4.55 cm, whereas another student reports its length as 4.56 cm. Both results are acceptable.

In Figure 2.5c, the end of the object appears to line up with the 3-cm mark. Because the divisions are marked in units of 1 cm, the length of the object is between 3 cm and 4 cm. Because the end of the object is on the 3-cm mark, the estimated digit is 0, which means the measurement is reported as 3.0 cm.

Significant Figures

In a measured number, the **significant figures (SFs)** are all the digits including the estimated digit. Nonzero numbers are always counted as significant figures. However, a zero may or may not be a significant figure depending on its position in a number. **TABLE 2.2** gives the rules and examples of counting significant figures.

TABLE 2.2 Significant Figures in Measured Numbers

Rule	Measured Number	Number of Significant Figures
1. A number is a *significant figure* if it is		
a. not a zero	4.5 g 122.35 m	2 5
b. a zero between nonzero digits	205 m 5.008 kg	3 4
c. a zero at the end of a decimal number	50. L 25.0 °C 16.00 g	2 3 4
d. in the coefficient of a number written in scientific notation	4.8×10^5 m 5.70×10^{-3} g	2 3
2. A zero is *not significant* if it is		
a. at the beginning of a decimal number	0.0004 s 0.075 cm	1 2
b. used as a placeholder in a large number without a decimal point	850 000 m 1 250 000 g	2 3

Significant Zeros and Scientific Notation

When one or more zeros in a large number are significant, they are shown clearly by writing the number in scientific notation. For example, if the first zero in the measurement 300 m is significant, but the second zero is not, the measurement is written as 3.0×10^2 m. *In this text, we will place a decimal point after a significant zero at the end of a number.* For example, if a measurement is written as 500. g, the decimal point after the second zero

ENGAGE

Why is the zero in the measurement 3.20×10^4 cm a significant figure?

indicates that *both zeros* are significant. To show this more clearly, we can write it as 5.00×10^2 g. We will assume that zeros at the end of large standard numbers without a decimal point are not significant. Therefore, we write 400 000 g as 4×10^5 g, which has only one significant figure.

SAMPLE PROBLEM 2.2 Significant Zeros

Identify the significant zeros in each of the following measured numbers:

a. 0.000 250 m **b.** 70.040 g **c.** 1 020 000 L

SOLUTION

ANALYZE THE PROBLEM	Given	Need	Connect
	measured number	identify significant zeros	rules for significant figures

a. The zero in the last decimal place following the 5 is significant. The zeros preceding the 2 are not significant.

b. Zeros between nonzero digits or at the end of decimal numbers are significant. All the zeros in 70.040 g are significant.

c. Zeros between nonzero digits are significant. Zeros at the end of a large number with no decimal point are placeholders but not significant. The zero between 1 and 2 is significant, but the four zeros following the 2 are not significant.

STUDY CHECK 2.2

Identify the significant zeros in each of the following measured numbers:

a. 0.040 08 m **b.** 6.00×10^3 g

ANSWER

a. The zeros between 4 and 8 are significant. The zeros preceding the first nonzero digit 4 are not significant.

b. All zeros in the coefficient of a number written in scientific notation are significant.

Exact Numbers

Exact numbers *are those numbers obtained by counting items or using a definition that compares two units in the same measuring system.* Suppose a friend asks you how many classes you are taking. You would answer by counting the number of classes in your schedule. It would not use any measuring tool. Suppose you are asked to state the number of seconds in one minute. Without using any measuring device, you would give the definition: There are 60 s in 1 min. *Exact numbers are not measured, do not have a limited number of significant figures, and do not affect the number of significant figures in a calculated answer.* For more examples of exact numbers, see **TABLE 2.3**.

TABLE 2.3 Examples of Some Exact Numbers

Counted Numbers	Defined Equalities	
Items	Metric System	U.S. System
8 doughnuts	1 L = 1000 mL	1 ft = 12 in.
2 baseballs	1 m = 100 cm	1 qt = 4 cups
5 capsules	1 kg = 1000 g	1 lb = 16 oz

For example, a mass of 42.2 g and a length of 5.0×10^{-3} cm are measured numbers because they are obtained using measuring tools. There are three SFs in 42.2 g because all nonzero digits are always significant. There are two SFs in 5.0×10^{-3} cm because all the

The number of baseballs is counted, which means 2 is an exact number.

digits in the coefficient of a number written in scientific notation are significant. However, a quantity of three eggs is an exact number that is obtained by counting the eggs. In the equality 1 kg = 1000 g, the masses of 1 kg and 1000 g are both exact numbers because this equality is a definition in the metric system of measurement.

SAMPLE PROBLEM 2.3 Measured and Exact Numbers

Identify each of the following numbers as measured or exact and give the number of significant figures (SFs) in each of the measured numbers:

a. 0.170 L

b. 4 knives

c. 6.3×10^{-6} s

d. 1 m = 100 cm

TRY IT FIRST

SOLUTION

ANALYZE THE PROBLEM	Given	Need	Connect
	numbers	identify exact or measured, number of SFs	rules for SFs in measured numbers

a. The volume of 0.170 L is a measured number because it is obtained with a measuring tool. There are three SFs in 0.170 L because nonzero digits are always significant and a zero after nonzero digits in a decimal number is significant.

b. The value of 4 knives is an exact number because it is obtained by counting rather than using a measuring tool.

c. The time of 6.3×10^{-6} s is a measured number because it is obtained with a measuring tool. There are two SFs in 6.3×10^{-6} s because all the numbers in the coefficient of a number written in scientific notation are significant.

d. The lengths of 1 m and 100 cm are both exact numbers because the relationship 1 m = 100 cm is a definition in the metric system.

STUDY CHECK 2.3

Identify each of the following numbers as measured or exact and give the number of significant figures (SFs) in each of the measured numbers:

a. 0.020 80 kg

b. 5.06×10^4 h

c. 4 chemistry books

ANSWER

a. measured; four SFs

b. measured; three SFs

c. exact

QUESTIONS AND PROBLEMS

2.2 Measured Numbers and Significant Figures

LEARNING GOAL Identify a number as measured or exact; determine the number of significant figures in a measured number.

2.9 What is the estimated digit in each of the following measured numbers?
a. 8.6 m   **b.** 45.25 g   **c.** 29 °C

2.10 What is the estimated digit in each of the following measured numbers?
a. 125.04 g   **b.** 5.057 m   **c.** 525.8 °C

2.11 Identify the numbers in each of the following statements as measured or exact:
a. A patient has a mass of 67.5 kg.
b. A patient is given 2 tablets of medication.
c. In the metric system, 1 L is equal to 1000 mL.
d. The distance from Denver, Colorado, to Houston, Texas, is 1720 km.

2.12 Identify the numbers in each of the following statements as measured or exact:
a. There are 31 students in the laboratory.
b. The oldest known flower lived 1.20×10^8 yr ago.
c. The largest gem ever found, an aquamarine, has a mass of 104 kg.
d. A laboratory test shows a blood cholesterol level of 184 mg/dL.

2.13 Identify the measured number(s), if any, in each of the following pairs of numbers:
a. 3 hamburgers and 6 oz of hamburger
b. 1 table and 4 chairs
c. 0.75 lb of grapes and 350 g of butter
d. 60 s = 1 min

2.14 Identify the exact number(s), if any, in each of the following pairs of numbers:
a. 5 pizzas and 50.0 g of cheese
b. 6 nickels and 16 g of nickel
c. 3 onions and 3 lb of onions
d. 5 miles and 5 cars

2.15 Indicate if the zeros are significant in each of the following measurements:
a. 0.0038 m
b. 5.04 cm
c. 800. L
d. 3.0×10^{-3} kg
e. 85 000 g

2.16 Indicate if the zeros are significant in each of the following measurements:
a. 20.05 °C
b. 5.00 m
c. 0.000 02 g
d. 120 000 yr
e. 8.05×10^{2} L

2.17 How many significant figures are in each of the following?
a. 11.005 g
b. 0.000 32 m
c. 36 000 000 km
d. 1.80×10^{4} kg
e. 0.8250 L
f. 30.0 °C

2.18 How many significant figures are in each of the following?
a. 20.60 mL
b. 1036.48 kg
c. 4.00 m
d. 20.8 °C
e. 60 800 000 g
f. 5.0×10^{-3} L

2.19 In which of the following pairs do both numbers contain the same number of significant figures?
a. 11.0 m and 11.00 m
b. 0.0250 m and 0.205 m
c. 0.000 12 s and 12 000 s
d. 250.0 L and 2.5×10^{-2} L

2.20 In which of the following pairs do both numbers contain the same number of significant figures?
a. 0.005 75 g and 5.75×10^{-3} g
b. 405 K and 405.0 K
c. 150 000 s and 1.50×10^{4} s
d. 3.8×10^{-2} L and 3.0×10^{5} L

2.21 Write each of the following in scientific notation with two significant figures:
a. 5000 L
b. 30 000 g
c. 100 000 m
d. 0.000 25 cm

2.22 Write each of the following in scientific notation with two significant figures:
a. 5 100 000 g
b. 26 000 s
c. 40 000 m
d. 0.000 820 kg

Applications

2.23 Identify the number of significant figures in each of the following:
a. The mass of a neonate is 1.607 kg.
b. The Daily Value (DV) for iodine for an infant is 130 mcg.
c. There are 4.02×10^{6} red blood cells in a blood sample.

2.24 Identify the number of significant figures in each of the following:
a. An adult with the flu has a temperature of 103.5 °F.
b. A brain contains 1.20×10^{10} neurons.
c. The time for a nerve impulse to travel from the feet to the brain is 0.46 s.

2.3 Significant Figures in Calculations

LEARNING GOAL Adjust calculated answers to give the correct number of significant figures.

In the sciences, we measure many things: the length of a bacterium, the volume of a gas sample, the temperature of a reaction mixture, or the mass of iron in a sample. The numbers obtained from these types of measurements are often used in calculations. The number of significant figures in the measured numbers determines the number of significant figures in the calculated answer.

Using a calculator will help you perform calculations faster. However, calculators cannot think for you. It is up to you to enter the numbers correctly, press the correct function keys, and give an answer with the correct number of significant figures.

Rounding Off

Suppose you decide to buy carpeting for a room that measures 5.52 m by 3.58 m. Each of these measurements has three significant figures because the measuring tape limits your estimated place to 0.01 m. To determine how much carpeting you need, you would calculate the area of the room by multiplying 5.52 times 3.58 on your calculator. The calculator shows the number 19.7616 in its display. However, this is not the correct final answer because there are too many digits, which is the result of the multiplication process. Because each of the original measurements has only three significant figures, the displayed number (19.7616) must be *rounded off* to three significant figures, 19.8. Therefore, you can order carpeting that will cover an area of 19.8 m^{2}.

KEY MATH SKILL
Rounding Off

A technician uses a calculator in the laboratory.

Each time you use a calculator, it is important to look at the original measurements and determine the number of significant figures that can be used for the answer. You can use the following rules to round off the numbers shown in a calculator display.

Rules for Rounding Off

1. If the first digit to be dropped is *4 or less*, then it and all following digits are simply dropped from the number.
2. If the first digit to be dropped is *5 or greater*, then the last retained digit of the number is increased by 1.

Number to Round Off	Three Significant Figures	Two Significant Figures
8.4234	8.42 (drop 34)	8.4 (drop 234)
14.780	14.8 (drop 80, increase the last retained digit by 1)	15 (drop 780, increase the last retained digit by 1)
3256	3260* (drop 6, increase the last retained digit by 1, add 0) (3.26×10^3)	3300* (drop 56, increase the last retained digit by 1, add 00) (3.3×10^3)

*The value of a large number is retained by using placeholder zeros to replace dropped digits.

SAMPLE PROBLEM 2.4 Rounding Off

Round off each of the following numbers to three significant figures:

a. 35.7823 m   **b.** 0.002 621 7 L   **c.** 3.8268×10^3 g

TRY IT FIRST

SOLUTION

a. To round off 35.7823 m to three significant figures, drop 823 and increase the last retained digit by 1 to give 35.8 m.
b. To round off 0.002 621 7 L to three significant figures, drop 17 to give 0.002 62 L.
c. To round off 3.8268×10^3 g to three significant figures, drop 68 and increase the last retained digit by 1 to give 3.83×10^3 g.

STUDY CHECK 2.4

Round off each of the numbers in Sample Problem 2.4 to two significant figures.

ANSWER

a. 36 m   **b.** 0.0026 L   **c.** 3.8×10^3 g

Multiplication and Division with Measured Numbers

In multiplication or division, the final answer is written so that it has the same number of significant figures (SFs) as the measurement with the fewest SFs. An example of rounding off a calculator display follows:

Perform the following operations with measured numbers:

$$\frac{2.8 \times 67.40}{34.8} =$$

When the problem has multiple steps, the numbers in the numerator are multiplied and then divided by each of the numbers in the denominator.

2.8   ×   67.40   ÷   34.8   =   5.422988506   =   5.4

Two SFs   Four SFs   Three SFs   Calculator display   Answer, rounded off to two SFs

Because the calculator display has more digits than the significant figures in the measured numbers allow, we need to round off. Using the measured number that has the fewest number (two) of significant figures, 2.8, we round off the calculator display to the answer with two SFs.

A calculator is helpful in working problems and doing calculations faster.

Adding Significant Zeros

Sometimes, a calculator display gives a small whole number. Then we add one or more significant zeros to the calculator display to obtain the correct number of significant figures. For example, suppose the calculator display is 4, but you used measurements that have three significant numbers. Then two significant zeros are *added* to give 4.00 as the correct answer.

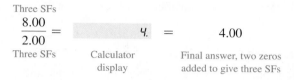

Three SFs
$$\frac{8.00}{2.00} = \qquad 4. \qquad = \qquad 4.00$$
Three SFs   Calculator   Final answer, two zeros
                   display   added to give three SFs

SAMPLE PROBLEM 2.5 Significant Figures in Multiplication and Division

Perform the following calculations with measured numbers. Round off the calculator display or add zeros to give each answer with the correct number of significant figures.

a. 56.8×0.37 **b.** $\dfrac{(2.075)(0.585)}{(8.42)(0.0245)}$ **c.** $\dfrac{25.0}{5.00}$

TRY IT FIRST

SOLUTION

	Given	Need	Connect
ANALYZE THE PROBLEM	multiplication and division	answer with SFs	rules for rounding off, adding zeros

STEP 1 Determine the number of significant figures in each measured number.

Three SFs Two SFs
a. 56.8×0.37

Four SFs Three SFs
b. $\dfrac{(2.075)(0.585)}{(8.42)(0.0245)}$
Three SFs Three SFs

Three SFs
c. $\dfrac{25.0}{5.00}$
Three SFs

STEP 2 Perform the indicated calculation.

a. 21.016
Calculator display

b. 5.884313345
Calculator display

c. 5.
Calculator display

STEP 3 Round off (or add zeros) to give the same number of significant figures as the measurement having the fewest significant figures.

a. 21 (rounded off to two SFs)   **b.** 5.88 (rounded off to three SFs)
c. 5.00 (two zeros added)

STUDY CHECK 2.5

Perform the following calculations with measured numbers and give the answers with the correct number of significant figures:

a. $45.26 \times 0.010\ 88$ **b.** $2.6 \div 324$ **c.** $\dfrac{4.0 \times 8.00}{16}$

ANSWER

a. 0.4924   **b.** 0.0080 or 8.0×10^{-3} **c.** 2.0

Guide to Calculating Answers with the Correct Number of Significant Figures

STEP 1
Determine the number of significant figures in each measured number.

STEP 2
Perform the indicated calculation.

STEP 3
Round off (or add zeros) to give the same number of significant figures as the measurement having the fewest significant figures.

Addition and Subtraction with Measured Numbers

In addition or subtraction, the final answer is written so that it has the same number of decimal places as the measurement having the fewest decimal places.

2.045	Thousandths place
$+$ 34.1	Tenths place
36.145	Calculator display
36.1	Answer, rounded off to the tenths place

When numbers are added or subtracted to give an answer ending in zero, the zero does not appear after the decimal point in the calculator display. For example, 14.5 g − 2.5 g = 12.0 g. However, if you do the subtraction on your calculator, the display shows 12. To write the correct answer, a significant zero is written after the decimal point.

SAMPLE PROBLEM 2.6 Significant Figures in Addition and Subtraction

Perform the following calculations and give each answer with the correct number of decimal places:

a. 104.45 mL + 0.838 mL + 46 mL **b.** 153.247 g − 14.82 g

TRY IT FIRST

SOLUTION

ANALYZE THE PROBLEM	Given	Need	Connect
	addition and subtraction	correct number of decimal places	rules for rounding off

STEP 1 Determine the number of decimal places in each measured number.

a. 104.45 mL   Hundredths place
0.838 mL   Thousandths place
$+$ 46 mL   Ones place

b. 153.247 g   Thousandths place
$-$ 14.82 g   Hundredths place

STEP 2 Perform the indicated calculation.

a. *151.288* Calculator display **b.** *138.427* Calculator display

STEP 3 Round off the answer to give the same number of decimal places as the measurement having the fewest decimal places.

a. 151 mL (rounded off to the ones place) **b.** 138.43 g (rounded off to the hundredths place)

STUDY CHECK 2.6

Perform the following calculations and give each answer with the correct number of decimal places:

a. 82.45 mg + 1.245 mg + 0.000 56 mg **b.** 4.259 L − 3.8 L

ANSWER

a. 83.70 mg **b.** 0.5 L

Guide to Calculating Answers with the Correct Number of Decimal Places

STEP 1
Determine the number of decimal places in each measured number.

STEP 2
Perform the indicated calculation.

STEP 3
Round off the answer to give the same number of decimal places as the measurement having the fewest decimal places.

QUESTIONS AND PROBLEMS

2.3 Significant Figures in Calculations

LEARNING GOAL Adjust calculated answers to give the correct number of significant figures.

2.25 Why do we usually need to round off calculations that use measured numbers?

2.26 Why do we sometimes add a zero to a number in a calculator display?

2.27 Round off each of the following measurements to three significant figures:
    a. 1.854 kg   **b.** 88.2038 L
    c. 0.004 738 265 cm   **d.** 8807 m
    e. 1.832×10^5 s

2.28 Round off each of the measurements in problem 2.27 to two significant figures.

2.29 Round off or add zeros to each of the following to three significant figures:
    a. 56.855 m   **b.** 0.002 282 g
    c. 11 527 s   **d.** 8.1 L

2.30 Round off or add zeros to each of the following to two significant figures:
    a. 3.2805 m   **b.** 1.855×10^2 g
    c. 0.002 341 mL   **d.** 2 L

2.31 Perform each of the following calculations, and give an answer with the correct number of significant figures:
    a. 45.7×0.034 **b.** $0.002\ 78 \times 5$
    c. $\dfrac{34.56}{1.25}$ **d.** $\dfrac{(0.2465)(25)}{1.78}$

e. $(2.8 \times 10^4)(5.05 \times 10^{-6})$

f. $\dfrac{(3.45 \times 10^{-2})(1.8 \times 10^5)}{(8 \times 10^3)}$

2.32 Perform each of the following calculations, and give an answer with the correct number of significant figures:
    a. 400×185 **b.** $\dfrac{2.40}{(4)(125)}$
    c. $0.825 \times 3.6 \times 5.1$ **d.** $\dfrac{(3.5)(0.261)}{(8.24)(20.0)}$
    e. $\dfrac{(5 \times 10^{-5})(1.05 \times 10^4)}{(8.24 \times 10^{-8})}$
    f. $\dfrac{(4.25 \times 10^2)(2.56 \times 10^{-3})}{(2.245 \times 10^{-3})(56.5)}$

2.33 Perform each of the following calculations, and give an answer with the correct number of decimal places:
    a. 45.48 cm + 8.057 cm
    b. 23.45 g + 104.1 g + 0.025 g
    c. 145.675 mL − 24.2 mL
    d. 1.08 L − 0.585 L

2.34 Perform each of the following calculations, and give an answer with the correct number of decimal places:
    a. 5.08 g + 25.1 g
    b. 85.66 cm + 104.10 cm + 0.025 cm
    c. 24.568 mL − 14.25 mL
    d. 0.2654 L − 0.2585 L

2.4 Prefixes and Equalities

LEARNING GOAL Use the numerical values of prefixes to write a metric equality.

The special feature of the SI as well as the metric system is that a **prefix** can be placed in front of any unit to increase or decrease its size by some factor of 10. For example, the prefixes *milli* and *micro* are used to make the smaller units, milligram (mg) and microgram (μg).

The U.S. Food and Drug Administration has determined the Daily Values (DV) for nutrients for adults and children aged 4 or older. Examples of these recommended Daily Values, some of which use prefixes, are listed in **TABLE 2.4**.

The prefix *centi* is like cents in a dollar. One cent would be a "centidollar" or 0.01 of a dollar. That also means that one dollar is the same as 100 cents. The prefix *deci* is like dimes in a dollar. One dime would be a "decidollar" or 0.1 of a dollar. That also means that one dollar is the same as 10 dimes. **TABLE 2.5** lists some of the metric prefixes, their symbols, and their numerical values.

The relationship of a prefix to a unit can be expressed by replacing the prefix with its numerical value. For example, when the prefix *kilo* in kilometer is replaced with its value of 1000, we find that a kilometer is equal to 1000 m. Other examples follow:

1 **kilo**meter (1 km) = **1000** meters (1000 m = 10^3 m)

1 **kilo**liter (1 kL) = **1000** liters (1000 L = 10^3 L)

1 **kilo**gram (1 kg) = **1000** grams (1000 g = 10^3 g)

TABLE 2.4 Daily Values for Selected Nutrients

Nutrient	Amount Recommended
Protein	50 g
Vitamin C	60 mg
Vitamin B_{12}	6 μg (6 mcg)
Calcium	1000 mg
Copper	2 mg
Iodine	150 μg (150 mcg)
Iron	18 mg
Magnesium	400 mg
Niacin	20 mg
Potassium	4700 mg
Selenium	70 μg (70 mcg)
Sodium	2400 mg
Zinc	15 mg

CORE CHEMISTRY SKILL
Using Prefixes

TABLE 2.5 Metric and SI Prefixes

Prefix	Symbol	Numerical Value	Scientific Notation	Equality
Prefixes That Increase the Size of the Unit				
peta	P	1 000 000 000 000 000	10^{15}	$1\ Pg = 1 \times 10^{15}\ g$ $1\ g = 1 \times 10^{-15}\ Pg$
tera	T	1 000 000 000 000	10^{12}	$1\ Ts = 1 \times 10^{12}\ s$ $1\ s = 1 \times 10^{-12}\ Ts$
giga	G	1 000 000 000	10^{9}	$1\ Gm = 1 \times 10^{9}\ m$ $1\ m = 1 \times 10^{-9}\ Gm$
mega	M	1 000 000	10^{6}	$1\ Mg = 1 \times 10^{6}\ g$ $1\ g = 1 \times 10^{-6}\ Mg$
kilo	k	1 000	10^{3}	$1\ km = 1 \times 10^{3}\ m$ $1\ m = 1 \times 10^{-3}\ km$
Prefixes That Decrease the Size of the Unit				
deci	d	0.1	10^{-1}	$1\ dL = 1 \times 10^{-1}\ L$ $1\ L = 10\ dL$
centi	c	0.01	10^{-2}	$1\ cm = 1 \times 10^{-2}\ m$ $1\ m = 100\ cm$
milli	m	0.001	10^{-3}	$1\ ms = 1 \times 10^{-3}\ s$ $1\ s = 1 \times 10^{3}\ ms$
micro	μ^*	0.000 001	10^{-6}	$1\ \mu g = 1 \times 10^{-6}\ g$ $1\ g = 1 \times 10^{6}\ \mu g$
nano	n	0.000 000 001	10^{-9}	$1\ nm = 1 \times 10^{-9}\ m$ $1\ m = 1 \times 10^{9}\ nm$
pico	p	0.000 000 000 001	10^{-12}	$1\ ps = 1 \times 10^{-12}\ s$ $1\ s = 1 \times 10^{12}\ ps$
femto	f	0.000 000 000 000 001	10^{-15}	$1\ fs = 1 \times 10^{-15}\ s$ $1\ s = 1 \times 10^{15}\ fs$

*In medicine, the abbreviation mc for the prefix micro is used because the symbol μ may be misread, which could result in a medication error. Thus, 1 μg would be written as 1 mcg.

ENGAGE

Why is 60. mg of vitamin C the same as 0.060 g of vitamin C?

SAMPLE PROBLEM 2.7 Prefixes and Equalities

An endoscopic camera has a width of 1 mm. Complete each of the following equalities involving millimeters:

a. 1 m = _____ mm

b. 1 cm = _____ mm

TRY IT FIRST

SOLUTION

a. 1 m = 1000 mm

b. 1 cm = 10 mm

STUDY CHECK 2.7

What is the relationship between millimeters and micrometers?

ANSWER

1 mm = 1000 μm (mcm)

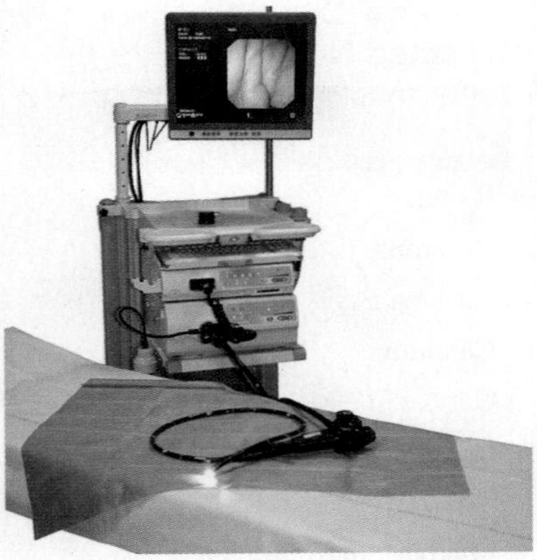

An endoscope, used to see inside the body, has a tiny video camera with a width of 1 mm attached to the end of a long, thin cable.

Measuring Length

An ophthalmologist may measure the diameter of the retina of an eye in centimeters (cm), whereas a surgeon may need to know the length of a nerve in millimeters (mm). When the prefix *centi* is used with the unit meter, it becomes *centimeter*, a length that is one-hundredth of a meter (0.01 m). When the prefix *milli* is used with the unit meter, it becomes *millimeter*, a length that is one-thousandth of a meter (0.001 m). There are 100 cm and 1000 mm in a meter.

If we compare the lengths of a millimeter and a centimeter, we find that 1 mm is 0.1 cm; there are 10 mm in 1 cm. These comparisons are examples of **equalities**, which show the relationship between two units that measure the same quantity. In every equality, each quantity has both a number and a unit. Examples of equalities between different metric units of length follow:

$$1\ m = 100\ cm = 1 \times 10^2\ cm$$
$$1\ m = 1000\ mm = 1 \times 10^3\ mm$$
$$1\ cm = 10\ mm = 1 \times 10^1\ mm$$

Some metric units for length are compared in **FIGURE 2.6**.

Using a retinal camera, an ophthalmologist photographs the retina of an eye.

First quantity — Second quantity: 1 m = 100 cm. Number + unit. This example of an equality shows the relationship between meters and centimeters.

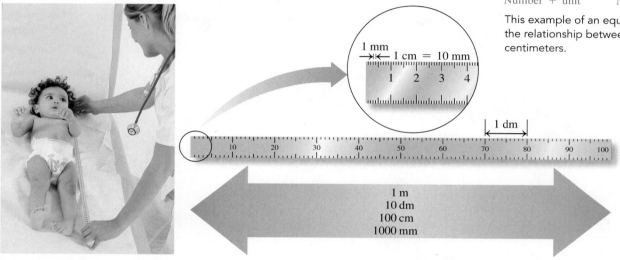

FIGURE 2.6 ▶ The metric length of 1 m is the same length as 10 dm, 100 cm, or 1000 mm.
How many millimeters (mm) are in 1 centimeter (cm)?

Measuring Volume

Volumes of 1 L or smaller are common in the health sciences. When a liter is divided into 10 equal portions, each portion is a deciliter (dL). There are 10 dL in 1 L. Laboratory results for blood work are often reported in mass per deciliter. **TABLE 2.6** lists typical laboratory test values for some substances in the blood.

TABLE 2.6 Some Typical Laboratory Test Values

Substance in Blood	Typical Range
Albumin	3.5–5.0 g/dL
Ammonia	20–70 μg/dL (mcg/dL)
Calcium	8.5–10.5 mg/dL
Cholesterol	105–250 mg/dL
Iron (male)	80–160 μg/dL (mcg/dL)
Protein (total)	6.0–8.0 g/dL

FIGURE 2.7 ▶ A laboratory technician transfers small volumes of solutions using a micropipette.

🔘 How many milliliters are equal to 200 mcL?

When a liter is divided into a thousand parts, each of the smaller volumes is called a milliliter (mL). In a 1-L container of physiological saline, there are 1000 mL of solution. In 1 mL of a laboratory sample, there are 1000 mcL (see **FIGURE 2.7**). Examples of equalities between different metric units of volume follow:

$$1 \text{ L} = 10 \text{ dL} \qquad = 1 \times 10^1 \text{ dL}$$
$$1 \text{ L} = 1000 \text{ mL} \qquad = 1 \times 10^3 \text{ mL}$$
$$1 \text{ dL} = 100 \text{ mL} \qquad = 1 \times 10^2 \text{ mL}$$
$$1 \text{ mL} = 1000 \, \mu\text{L (mcL)} = 1 \times 10^3 \, \mu\text{L (mcL)}$$

The **cubic centimeter** (abbreviated as **cm³** or **cc**) is the volume of a cube whose dimensions are 1 cm on each side. A cubic centimeter has the same volume as a milliliter, and the units are often used interchangeably.

$$1 \text{ cm}^3 = 1 \text{ cc} = 1 \text{ mL}$$

When you see *1 cm*, you are reading about length; when you see *1 cm³* or *1 cc* or *1 mL*, you are reading about volume. A comparison of units of volume is illustrated in **FIGURE 2.8**.

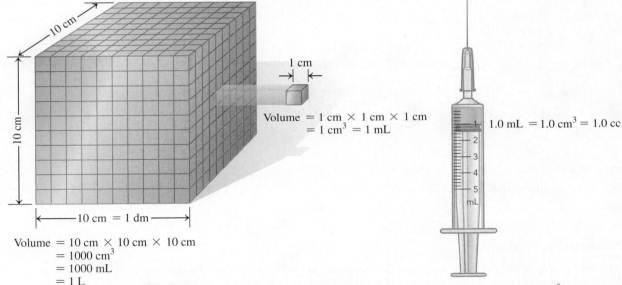

$$\text{Volume} = 10 \text{ cm} \times 10 \text{ cm} \times 10 \text{ cm}$$
$$= 1000 \text{ cm}^3$$
$$= 1000 \text{ mL}$$
$$= 1 \text{ L}$$

FIGURE 2.8 ▶ A cube measuring 10 cm on each side has a volume of 1000 cm³, or 1 L; a cube measuring 1 cm on each side has a volume of 1 cm³ (cc) or 1 mL.

🔘 What is the relationship between a milliliter (mL) and a cubic centimeter (cm³)?

Measuring Mass

When you go to the doctor for a physical examination, your mass is recorded in kilograms, whereas the results of your laboratory tests are reported in grams, milligrams (mg), or micrograms (μg or mcg). A kilogram is equal to 1000 g. One gram represents the same mass as 1000 mg, and one mg equals 1000 μg (or 1000 mcg). Examples of equalities between different metric units of mass follow:

$$1 \text{ kg} = 1000 \text{ g} \qquad = 1 \times 10^3 \text{ g}$$
$$1 \text{ g} = 1000 \text{ mg} \qquad = 1 \times 10^3 \text{ mg}$$
$$1 \text{ mg} = 1000 \, \mu\text{g (mcg)} = 1 \times 10^3 \, \mu\text{g (mcg)}$$

QUESTIONS AND PROBLEMS

2.4 Prefixes and Equalities

LEARNING GOAL Use the numerical values of prefixes to write a metric equality.

2.35 The speedometer is marked in both km/h and mi/h, or mph. What is the meaning of each abbreviation?

2.36 In a French car, the odometer reads 22269. What units would this be? What units would it be if this were an odometer in a car made for the United States?

2.37 Write the abbreviation for each of the following units:
  a. milligram
  b. deciliter
  c. kilometer
  d. femtogram

2.38 Write the abbreviation for each of the following units:
  a. gigagram
  b. megameter
  c. microliter
  d. nanosecond

2.39 Write the complete name for each of the following units:
  a. cL
  b. kg
  c. ms
  d. Gm

2.40 Write the complete name for each of the following units:
  a. dL
  b. Ts
  c. mcg
  d. pm

2.41 Write the numerical value for each of the following prefixes:
  a. centi
  b. tera
  c. milli
  d. deci

2.42 Write the numerical value for each of the following prefixes:
  a. giga
  b. micro
  c. mega
  d. nano

2.43 Use a prefix to write the name for each of the following:
  a. 0.1 g
  b. 10^{-6} g
  c. 1000 g
  d. 0.01 g

2.44 Use a prefix to write the name for each of the following:
  a. 10^{9} m
  b. 10^{6} m
  c. 0.001 m
  d. 10^{-15} m

2.45 Complete each of the following metric relationships:
  a. 1 m = _____ cm
  b. 1 m = _____ nm
  c. 1 mm = _____ m
  d. 1 L = _____ mL

2.46 Complete each of the following metric relationships:
  a. 1 Mg = _____ g
  b. 1 mL = _____ μL
  c. 1 g = _____ kg
  d. 1 g = _____ mg

2.47 For each of the following pairs, which is the larger unit?
  a. milligram or kilogram
  b. milliliter or microliter
  c. m or km
  d. kL or dL
  e. nanometer or picometer

2.48 For each of the following pairs, which is the smaller unit?
  a. mg or g
  b. centimeter or nanometer
  c. millimeter or micrometer
  d. mL or dL
  e. centigram or megagram

2.5 Writing Conversion Factors

LEARNING GOAL Write a conversion factor for two units that describe the same quantity.

Many problems in chemistry and the health sciences require you to change from one unit to another unit. You make changes in units every day. For example, suppose you worked 2.0 h on your homework, and someone asked you how many minutes that was. You would answer 120 min. You must have multiplied 2.0 h $\times$ 60 min/h because you knew the equality (1 h = 60 min) that related the two units. When you expressed 2.0 h as 120 min, you did not change the amount of time you spent studying. You changed only the unit of measurement used to express the time. *Any equality can be written as fractions called* **conversion factors** *with one of the quantities in the numerator and the other quantity in the denominator.* Be sure to include the units when you write the conversion factors. Two conversion factors are always possible from any equality.

Two Conversion Factors for the Equality: 1 h = 60 min

$$\frac{\text{Numerator}}{\text{Denominator}} \longrightarrow \frac{60 \text{ min}}{1 \text{ h}} \quad \text{and} \quad \frac{1 \text{ h}}{60 \text{ min}}$$

CORE CHEMISTRY SKILL
Writing Conversion Factors from Equalities

These factors are read as "60 minutes per 1 hour" and "1 hour per 60 minutes." The term *per* means "divide." Some common relationships are given in **TABLE 2.7**. It is important that the equality you select to form a conversion factor is a *true* relationship.

The numbers in any equality between two metric units or between two U.S. system units are obtained by definition. Because numbers in a definition are exact, they are not used to determine significant figures. For example, the equality of 1 g = 1000 mg is

TABLE 2.7 Some Common Equalities

Quantity	Metric (SI)	U.S.	Metric–U.S.
Length	1 km = 1000 m	1 ft = 12 in.	2.54 cm = 1 in. (exact)
	1 m = 1000 mm	1 yd = 3 ft	1 m = 39.37 in.
	1 cm = 10 mm	1 mi = 5280 ft	1 km = 0.6214 mi
Volume	1 L = 1000 mL	1 qt = 4 cups	946.4 mL = 1 qt
	1 dL = 100 mL	1 qt = 2 pt	1 L = 1.057 qt
	1 mL = 1 cm^3	1 gal = 4 qt	473.2 mL = 1 pt
	1 mL = 1 cc*		3.785 L = 1 gal
			5 mL = 1 tsp*
			15 mL = 1 T (tbsp)*
Mass	1 kg = 1000 g	1 lb = 16 oz	1 kg = 2.205 lb
	1 g = 1000 mg		453.6 g = 1 lb
	1 mg = 1000 mcg*		28.35 g = 1 oz
Time	1 h = 60 min	1 h = 60 min	
	1 min = 60 s	1 min = 60 s	

*Used in medicine.

defined, which means that both of the numbers 1 and 1000 are exact. *However, when an equality consists of a metric unit and a U.S. unit, one of the numbers in the equality is obtained by measurement and counts toward the significant figures in the answer.* For example, the equality of 1 lb = 453.6 g is obtained by measuring the grams in exactly 1 lb. In this equality, the measured quantity 453.6 g has four significant figures, whereas the 1 is exact. An exception is the relationship of 1 in. = 2.54 cm, which has been defined as exact.

Metric Conversion Factors

We can write metric conversion factors for any of the metric relationships. For example, from the equality for meters and centimeters, we can write the following factors:

ENGAGE

Why does the equality 1 day = 24 h have two conversion factors?

Metric Equality	Conversion Factors	
1 m = 100 cm	$\dfrac{100 \text{ cm}}{1 \text{ m}}$ and	$\dfrac{1 \text{ m}}{100 \text{ cm}}$

Both are proper conversion factors for the relationship; one is just the inverse of the other. *The usefulness of conversion factors is enhanced by the fact that we can turn a conversion factor over and use its inverse.* The numbers 100 and 1 in this equality and its conversion factors are both *exact* numbers.

Metric–U.S. System Conversion Factors

Suppose you need to convert from pounds, a unit in the U.S. system, to kilograms in the metric (or SI) system. A relationship you could use is

1 kg = 2.205 lb

The corresponding conversion factors would be

$$\frac{2.205 \text{ lb}}{1 \text{ kg}} \quad \text{and} \quad \frac{1 \text{ kg}}{2.205 \text{ lb}}$$

FIGURE 2.9 ▶ In the United States, the contents of many packaged foods are listed in both U.S. and metric units.

 What are some advantages of using the metric system?

FIGURE 2.9 illustrates the contents of some packaged foods in both U.S. and metric units.

SAMPLE PROBLEM 2.8 Writing Conversion Factors

Identify the correct conversion factor(s) for the equality: 1 pt of blood contains 473.2 mL of blood.

a. $\dfrac{473.2 \text{ pt}}{1 \text{ mL}}$ **b.** $\dfrac{1 \text{ pt}}{473.2 \text{ mL}}$ **c.** $\dfrac{473.2 \text{ mL}}{1 \text{ pt}}$ **d.** $\dfrac{1 \text{ mL}}{473.2 \text{ pt}}$

TRY IT FIRST

SOLUTION

The equality for pints and milliliters is 1 pt = 473.2 mL. Answers **b** and **c** are correctly written conversion factors for the equality.

STUDY CHECK 2.8

What are the two correctly written conversion factors for the equality: 1000 mm = 1 m?

ANSWER

$\dfrac{1000 \text{ mm}}{1 \text{ m}}$ and $\dfrac{1 \text{ m}}{1000 \text{ mm}}$

1 pt of blood contains 473.2 mL.

Conversion Factors with Powers

Sometimes we use a conversion factor that is squared or cubed. This is the case when we need to calculate an area or a volume.

Distance = length

Area　　= length $\times$ length = length^2

Volume = length $\times$ length $\times$ length = length^3

Suppose you want to write the equality and the conversion factors for the relationship between an area in square centimeters and in square meters. To square the equality 1 m = 100 cm, we square both the number and the unit on each side.

Equality: 1 m = 100 cm

Area: $(1 \text{ m})^2 = (100 \text{ cm})^2$　or　$1 \text{ m}^2 = (100)^2 \text{ cm}^2$

From the new equality, we can write two conversion factors as follows:

Conversion factors: $\dfrac{(100 \text{ cm})^2}{(1 \text{ m})^2}$ and $\dfrac{(1 \text{ m})^2}{(100 \text{ cm})^2}$

In the following example, we show that the equality 1 in. = 2.54 cm can be squared to give area or can be cubed to give volume. *Both the number and the unit must be squared or cubed.*

Measurement	Equality	Conversion Factors		
Length	1 in. = 2.54 cm	$\dfrac{2.54 \text{ cm}}{1 \text{ in.}}$	and	$\dfrac{1 \text{ in.}}{2.54 \text{ cm}}$
Area	$(1 \text{ in.})^2 = (2.54 \text{ cm})^2$ $(1 \text{ in.})^2 = (2.54)^2 \text{ cm}^2 = 6.45 \text{ cm}^2$	$\dfrac{(2.54 \text{ cm})^2}{(1 \text{ in.})^2}$	and	$\dfrac{(1 \text{ in.})^2}{(2.54 \text{ cm})^2}$
Volume	$(1 \text{ in.})^3 = (2.54 \text{ cm})^3$ $(1 \text{ in.})^3 = (2.54)^3 \text{ cm}^3 = 16.4 \text{ cm}^3$	$\dfrac{(2.54 \text{ cm})^3}{(1 \text{ in.})^3}$	and	$\dfrac{(1 \text{ in.})^3}{(2.54 \text{ cm})^3}$

Equalities and Conversion Factors Stated Within a Problem

An equality may also be stated within a problem that applies only to that problem. For example, the speed of a car in kilometers per hour or the milligrams of vitamin C in a tablet would be specific relationships for that problem only. However, it is possible to identify these relationships within a problem and to write corresponding conversion factors.

From each of the following statements, we can write an equality, and two conversion factors, and identify each number as exact or give the number of significant figures.

The car was traveling at a speed of 85 km/h.

Equality	Conversion Factors	Significant Figures or Exact
85 km = 1 h	$\dfrac{85 \text{ km}}{1 \text{ h}}$ and $\dfrac{1 \text{ h}}{85 \text{ km}}$	The 85 km is measured: It has two significant figures. The 1 h is exact.

One tablet contains 500 mg of vitamin C.

Equality	Conversion Factors	Significant Figures or Exact
1 tablet = 500 mg of vitamin C	$\dfrac{500 \text{ mg vitamin C}}{1 \text{ tablet}}$ and $\dfrac{1 \text{ tablet}}{500 \text{ mg vitamin C}}$	The 500 mg is measured: It has one significant figure. The 1 tablet is exact.

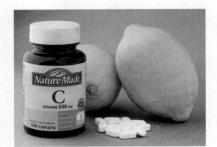

Vitamin C, an antioxidant needed by the body, is found in fruits such as lemons.

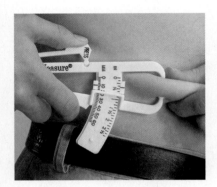

The thickness of the skin fold at the abdomen is used to determine the percentage of body fat.

Conversion Factors from a Percentage, ppm, and ppb

A percentage (%) is written as a conversion factor by choosing a unit and expressing the numerical relationship of the parts of this unit to 100 parts of the whole. For example, a person might have 18% body fat by mass. The percentage quantity can be written as 18 mass units of body fat in every 100 mass units of body mass. Different mass units such as grams (g), kilograms (kg), or pounds (lb) can be used, but both units in the factor must be the same.

Equality	Conversion Factors	Significant Figures or Exact
18 kg of body fat = 100 kg of body mass	$\dfrac{18 \text{ kg body fat}}{100 \text{ kg body mass}}$ and $\dfrac{100 \text{ kg body mass}}{18 \text{ kg body fat}}$	The 18 kg is measured: It has two significant figures. The 100 kg is exact.

When scientists want to indicate very small ratios, they use numerical relationships called *parts per million* (ppm) or *parts per billion* (ppb). The ratio of parts per million is the same as the milligrams of a substance per kilogram (mg/kg). The ratio of parts per billion equals the micrograms per kilogram (μg/kg, mcg/kg).

Ratio	Units
parts per million (ppm)	milligrams per kilogram (mg/kg)
parts per billion (ppb)	micrograms per kilogram (μg/kg, mcg/kg)

For example, the maximum amount of lead that is allowed by the Food and Drug Administration (FDA) in glazed pottery bowls is 2 ppm, which is 2 mg/kg.

Equality	Conversion Factors	Significant Figures or Exact
2 mg of lead = 1 kg of glaze	$\dfrac{2 \text{ mg lead}}{1 \text{ kg glaze}}$ and $\dfrac{1 \text{ kg glaze}}{2 \text{ mg lead}}$	The 2 mg is measured: It has one significant figure. The 1 kg is exact.

SAMPLE PROBLEM 2.9 Equalities and Conversion Factors Stated in a Problem

Write the equality and two conversion factors, and identify each number as exact or give the number of significant figures for each of the following:

a. The medication that Greg takes for his high blood pressure contains 40. mg of propranolol in 1 tablet.

b. Cold water fish such as salmon contains 1.9% omega-3 fatty acids by mass.

c. The U.S. Environmental Protection Agency (EPA) has set the maximum level for mercury in tuna at 0.5 ppm.

TRY IT FIRST

SOLUTION

a. There are 40. mg of propranolol in 1 tablet.

Equality	Conversion Factors	Significant Figures or Exact
1 tablet = 40. mg of propranolol	$\dfrac{40.\ \text{mg propranolol}}{1\ \text{tablet}}$ and $\dfrac{1\ \text{tablet}}{40.\ \text{mg propranolol}}$	The 40. mg is measured: It has two significant figures. The 1 tablet is exact.

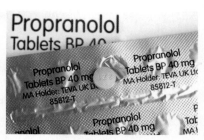

Propranolol is used to lower high blood pressure.

b. Cold water fish such as salmon contains 1.9% omega-3 fatty acids by mass.

Equality	Conversion Factors	Significant Figures or Exact
1.9 g of omega-3 fatty acids = 100 g of salmon	$\dfrac{1.9\ \text{g omega-3 fatty acids}}{100\ \text{g salmon}}$ and $\dfrac{100\ \text{g salmon}}{1.9\ \text{g omega-3 fatty acids}}$	The 1.9 g is measured: It has two significant figures. The 100 g is exact.

Salmon contains high levels of omega-3 fatty acids.

c. The EPA has set the maximum level for mercury in tuna at 0.5 ppm.

Equality	Conversion Factors	Significant Figures or Exact
0.5 mg of mercury = 1 kg of tuna	$\dfrac{0.5\ \text{mg mercury}}{1\ \text{kg tuna}}$ and $\dfrac{1\ \text{kg tuna}}{0.5\ \text{mg mercury}}$	The 0.5 mg is measured: It has one significant figure. The 1 kg is exact.

The maximum amount of mercury allowed in tuna is 0.5 ppm.

STUDY CHECK 2.9

Levsin (hyoscyamine), used to treat stomach and bladder problems, is available as drops with 0.125 mg Levsin per 1 mL of solution. Write the equality and two conversion factors, and identify each number as exact or give the number of significant figures.

ANSWER

0.125 mg of Levsin = 1 mL of solution

$$\dfrac{0.125\ \text{mg Levsin}}{1\ \text{mL solution}} \quad \text{and} \quad \dfrac{1\ \text{mL solution}}{0.125\ \text{mg Levsin}}$$

The 0.125 mg is measured: It has three SFs. The 1 mL is exact.

ENGAGE

Why is the 0.5 ppm mercury in tuna written as the equality 0.5 mg of mercury = 1 kg of tuna?

QUESTIONS AND PROBLEMS

2.5 Writing Conversion Factors

LEARNING GOAL Write a conversion factor for two units that describe the same quantity.

2.49 Why can two conversion factors be written for an equality such as 1 m = 100 cm?

2.50 How can you check that you have written the correct conversion factors for an equality?

2.51 Write the equality and two conversion factors for each of the following pairs of units:
  a. centimeters and meters
  b. nanograms and grams
  c. liters and kiloliters
  d. seconds and milliseconds
  e. cubic meters and cubic centimeters

2.52 Write the equality and two conversion factors for each of the following pairs of units:
a. centimeters and inches
b. kilometers and miles
c. pounds and grams
d. gallons and liters
e. square centimeters and square inches

2.53 Write the equality and two conversion factors, and identify the numbers as exact or give the number of significant figures for each of the following:
a. One yard is 3 ft.
b. One kilogram is 2.205 lb.
c. One minute is 60 s.
d. A car goes 27 mi on 1 gal of gas.
e. Sterling silver is 93% silver by mass.

2.54 Write the equality and two conversion factors, and identify the numbers as exact or give the number of significant figures for each of the following:
a. One liter is 1.057 qt.
b. At the store, oranges are $1.29 per lb.
c. There are 7 days in 1 week.
d. One deciliter contains 100 mL.
e. An 18-carat gold ring contains 75% gold by mass.

2.55 Write the equality and two conversion factors, and identify the numbers as exact or give the number of significant figures for each of the following:
a. A bee flies at an average speed of 3.5 m per second.
b. The Daily Value (DV) for potassium is 4700 mg.
c. An automobile traveled 46.0 km on 1 gal of gasoline.
d. The pesticide level in plums was 29 ppb.
e. Silicon makes up 28.2% by mass of the Earth's crust.

2.56 Write the equality and two conversion factors, and identify the numbers as exact or give the number of significant figures for each of the following:
a. The Daily Value (DV) for iodine is 150 mcg.
b. The nitrate level in well water was 32 ppm.

c. Gold jewelry contains 58% gold by mass.
d. The price of a liter of milk is $1.65.
e. A metric ton is 1000 kg.

Applications

2.57 Write the equality and two conversion factors, and identify the numbers as exact or give the number of significant figures for each of the following:
a. A calcium supplement contains 630 mg of calcium per tablet.
b. The Daily Value (DV) for vitamin C is 60 mg.
c. The label on a bottle reads 50 mg of atenolol per tablet.
d. A low-dose aspirin contains 81 mg of aspirin per tablet.

2.58 Write the equality and two conversion factors, and identify the numbers as exact or give the number of significant figures for each of the following:
a. The label on a bottle reads 10 mg of furosemide per 1 mL.
b. The Daily Value (DV) for selenium is 70. mcg.
c. An IV of normal saline solution has a flow rate of 85 mL per hour.
d. One capsule of fish oil contains 360 mg of omega-3 fatty acids.

2.59 Write an equality and two conversion factors for each of the following medications in stock:
a. 10 mg of Atarax per 5 mL of Atarax syrup
b. 0.25 g of Lanoxin per 1 tablet of Lanoxin
c. 300 mg of Motrin per 1 tablet of Motrin

2.60 Write an equality and two conversion factors for each of the following medications in stock:
a. 2.5 mg of Coumadin per 1 tablet of Coumadin
b. 100 mg of Clozapine per 1 tablet of Clozapine
c. 1.5 g of Cefuroxime per 1 mL of Cefuroxime

2.6 Problem Solving Using Unit Conversion

LEARNING GOAL Use conversion factors to change from one unit to another.

The process of problem solving in chemistry often requires one or more conversion factors to change a given unit to the needed unit. For the problem, the unit of the given quantity and the unit of the needed quantity are identified. From there, the problem is set up with one or more conversion factors used to convert the given unit to the needed unit as seen in the following Sample Problem.

Given unit × one or more conversion factors = needed unit

SAMPLE PROBLEM 2.10 Problem Solving Using Conversion Factors

Greg's doctor has ordered a PET scan of his heart. In radiological imaging such as PET or CT scans, dosages of pharmaceuticals are based on body mass. If Greg weighs 165 lb, what is his body mass in kilograms?

TRY IT FIRST

SOLUTION

STEP 1 State the given and needed quantities.

ANALYZE THE PROBLEM	Given	Need	Connect
	165 lb	kilograms	conversion factor (kg/lb)

STEP 2 Write a plan to convert the given unit to the needed unit. The conversion factor relates the given unit in the U.S. system of measurement and the needed unit in the metric system.

pounds → U.S.–Metric factor → kilograms

STEP 3 State the equalities and conversion factors.

$$1 \text{ kg} = 2.205 \text{ lb}$$
$$\frac{2.205 \text{ lb}}{1 \text{ kg}} \quad \text{and} \quad \frac{1 \text{ kg}}{2.205 \text{ lb}}$$

STEP 4 Set up the problem to cancel units and calculate the answer. Write the given, 165 lb, and multiply by the conversion factor that has the unit lb in the denominator (bottom number) to cancel out the given unit (lb) in the numerator.

Unit for answer goes here

$$165 \text{ lb} \times \frac{1 \text{ kg}}{2.205 \text{ lb}} = 74.8 \text{ kg}$$

Given   Conversion factor   Answer

Look at how the units cancel. The given unit lb cancels out and the needed unit kg is in the numerator. *The unit you want in the final answer is the one that remains after all the other units have canceled out.* This is a helpful way to check that you set up a problem properly.

$$\text{lb} \times \frac{\text{kg}}{\text{lb}} = \text{kg} \quad \text{Unit needed for answer}$$

The calculator display gives the numerical answer, which is adjusted to give a final answer with the proper number of significant figures (SFs).

$$165 \times \frac{1}{2.205} = 165 \div 2.205 = 74.82993197 = 74.8$$

Three SFs   Four SFs   Calculator display   Three SFs (rounded off)

The value of 74.8 combined with the unit, kg, gives the final answer of 74.8 kg. With few exceptions, answers to numerical problems contain a number and a unit.

STUDY CHECK 2.10

A total of 2500 mL of a boric acid antiseptic solution is prepared from boric acid concentrate. How many quarts of boric acid have been prepared?

ANSWER

2.6 qt

Guide to Problem Solving Using Conversion Factors

STEP 1 State the given and needed quantities.
STEP 2 Write a plan to convert the given unit to the needed unit.
STEP 3 State the equalities and conversion factors.
STEP 4 Set up the problem to cancel units and calculate the answer.

ENGAGE

If you need an answer of kilograms, how do you know the following set up is incorrect?

$$\text{lb} \times \frac{\text{lb}}{\text{kg}} =$$

Interactive Video

Conversion Factors

Using Two or More Conversion Factors

In problem solving, two or more conversion factors are often needed to complete the change of units. In setting up these problems, one factor follows the other. Each factor is arranged to cancel the preceding unit until the needed unit is obtained. Once the problem is set up to cancel units properly, the calculations can be done without writing intermediate results. The process is worth practicing until you understand unit cancellation, the steps on the calculator, and rounding off to give a final answer. In this text, when two or more conversion factors are required, the final answer will be based on obtaining a final calculator display and rounding off (or adding zeros) to give the correct number of significant figures.

SAMPLE PROBLEM 2.11 Using Two or More Conversion Factors

A doctor's order for 0.50 g of Keflex is available as 250-mg tablets. How many tablets of Keflex are needed? In the following setup, fill in the missing parts of the conversion factors, show the canceled units, and give the correct answer:

$$0.50 \text{ g Keflex} \times \frac{\_\_\_ \text{ mg Keflex}}{1 \text{ g Keflex}} \times \frac{1 \text{ tablet}}{\_\_\_ \text{ mg Keflex}} =$$

TRY IT FIRST

SOLUTION

$$0.50 \text{ g } \cancel{\text{Keflex}} \times \frac{1000 \text{ mg } \cancel{\text{Keflex}}}{1 \text{ g } \cancel{\text{Keflex}}} \times \frac{1 \text{ tablet}}{250 \text{ mg } \cancel{\text{Keflex}}} = 2 \text{ tablets}$$

STUDY CHECK 2.11

A newborn has a length of 450 mm. In the following setup that calculates the baby's length in inches, fill in the missing parts of the conversion factors, show the canceled units, and give the correct answer:

$$450 \text{ mm} \times \frac{1 \text{ cm}}{\_\_\_ \text{ mm}} \times \frac{1 \text{ in.}}{\_\_\_ \text{ cm}} =$$

ANSWER

$$450 \cancel{\text{ mm}} \times \frac{1 \cancel{\text{ cm}}}{10 \cancel{\text{ mm}}} \times \frac{1 \text{ in.}}{2.54 \cancel{\text{ cm}}} = 18 \text{ in.}$$

SAMPLE PROBLEM 2.12 Problem Solving Using Two Conversion Factors

During a volcanic eruption on Mauna Loa, the lava flowed at a rate of 33 m/min. At this rate, what distance, in kilometers, will the lava travel in 45 min?

TRY IT FIRST

SOLUTION

STEP 1 State the given and needed quantities.

ANALYZE THE PROBLEM	Given	Need	Connect
	45 min	kilometers	conversion factors (m/min, km/m)

STEP 2 Write a plan to convert the given unit to the needed unit.

minutes → Rate factor → meters → Metric factor → kilometers

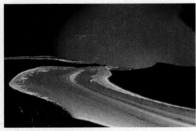
Lava flows from an eruption of Mauna Loa volcano, Hawaii.

STEP 3 State the equalities and conversion factors. The rate of lava flow (33 m/min) is used as an equality as well as the metric equality for meters and kilometers. Then conversion factors can be written for both.

$$1 \text{ min} = 33 \text{ m}$$

$$\frac{33 \text{ m}}{1 \text{ min}} \text{ and } \frac{1 \text{ min}}{33 \text{ m}}$$

$$1 \text{ km} = 1000 \text{ m}$$

$$\frac{1000 \text{ m}}{1 \text{ km}} \text{ and } \frac{1 \text{ km}}{1000 \text{ m}}$$

STEP 4 Set up the problem to cancel units and calculate the answer.

$$45 \text{ min} \times \frac{33 \text{ m}}{1 \text{ min}} \times \frac{1 \text{ km}}{1000 \text{ m}}$$

The calculation is done as follows:

$$45 \quad \boxed{\times} \quad 33 \quad \boxed{\div} \quad 1000 \quad \boxed{=} \qquad 1.485$$

Calculator display

Because there are two significant figures in the measured quantities, we write the needed answer with two significant figures, 1.5, and the unit km to give the final answer, 1.5 km.

$$45 \underset{\text{Two SFs}}{\text{min}} \times \frac{\overset{\text{Two SFs}}{33 \text{ m}}}{\underset{\text{Exact}}{1 \text{ min}}} \times \frac{\overset{\text{Exact}}{1 \text{ km}}}{\underset{\text{Exact}}{1000 \text{ m}}} = \underset{\text{Two SFs}}{1.5 \text{ km}}$$

STUDY CHECK 2.12

How many hours are required for the lava in Sample Problem 2.12 to flow a distance of 5.0 km?

ANSWER

2.5 h

SAMPLE PROBLEM 2.13 Using a Percentage as a Conversion Factor

A person who exercises regularly has 16% body fat by mass. If this person weighs 155 lb, what is the mass, in kilograms, of body fat?

TRY IT FIRST

SOLUTION

STEP 1 State the given and needed quantities.

ANALYZE THE PROBLEM	Given	Need	Connect
	155 lb body weight	kilograms of body fat	conversion factors (kg/lb, percent body fat)

Exercising regularly helps reduce body fat.

STEP 2 Write a plan to convert the given unit to the needed unit.

pounds of body weight → U.S.–Metric factor → kilograms of body mass → Percentage factor → kilograms of body fat

| STEP **3** State the equalities and conversion factors.

$$1 \text{ kg of body mass} = 2.205 \text{ lb of body weight}$$

$$\frac{2.205 \text{ lb body weight}}{1 \text{ kg body mass}} \text{ and } \frac{1 \text{ kg body mass}}{2.205 \text{ lb body weight}}$$

$$16 \text{ kg of body fat} = 100 \text{ kg of body mass}$$

$$\frac{16 \text{ kg body fat}}{100 \text{ kg body mass}} \text{ and } \frac{100 \text{ kg body mass}}{16 \text{ kg body fat}}$$

| STEP **4** Set up the problem to cancel units and calculate the answer. We set up the problem using conversion factors to cancel each unit, starting with lb of body weight, until we obtain the final unit, kg of body fat, in the numerator.

$$155 \text{ lb body weight} \times \frac{1 \text{ kg body mass}}{2.205 \text{ lb body weight}} \times \frac{16 \text{ kg body fat}}{100 \text{ kg body mass}} = 11 \text{ kg of body fat}$$

Exact (above 1 kg body mass); Two SFs (above 16 kg body fat); Three SFs (below 155); Four SFs (below 2.205); Exact (below 100); Two SFs (below 11)

| **STUDY CHECK 2.13**

Uncooked lean ground beef can contain up to 22% fat by mass. How many grams of fat would be contained in 0.25 lb of the ground beef?

ANSWER

25 g of fat

CHEMISTRY LINK TO **HEALTH**
Toxicology and Risk–Benefit Assessment

Each day, we make choices about what we do or what we eat, often without thinking about the risks associated with these choices. We are aware of the risks of cancer from smoking or the risks of lead poisoning, and we know there is a greater risk of having an accident if we cross a street where there is no light or crosswalk.

A basic concept of toxicology is the statement of Paracelsus that the dose is the difference between a poison and a cure. To evaluate the level of danger from various substances, natural or synthetic, a risk assessment is made by exposing laboratory animals to the substances and monitoring the health effects. Often, doses very much greater than humans might ordinarily encounter are given to the test animals.

Many hazardous chemicals or substances have been identified by these tests. One measure of toxicity is the LD_{50}, or lethal dose, which is the concentration of the substance that causes death in 50% of the test animals. A dosage is typically measured in milligrams per kilogram (mg/kg) of body mass or micrograms per kilogram (mcg/kg) of body mass.

Other evaluations need to be made, but it is easy to compare LD_{50} values. Parathion, a pesticide, with an LD_{50} of 3 mg/kg, would be highly toxic. This means that 3 mg of parathion per kg of body mass would be fatal to half the test animals. Table salt (sodium chloride) with an LD_{50} of 3300 mg/kg would have a much lower toxicity. You would need to ingest a huge amount of salt before any toxic

The LD_{50} of caffeine is 192 mg/kg.

effect would be observed. Although the risk to animals can be evaluated in the laboratory, it is more difficult to determine the impact in the environment since there is also a difference between continued exposure and a single, large dose of the substance.

TABLE 2.8 lists some LD_{50} values and compares substances in order of increasing toxicity.

TABLE 2.8 Some LD_{50} Values for Substances Tested in Rats

Substance	LD_{50} (mg/kg)
Table sugar	29 700
Boric acid	5140
Baking soda	4220
Table salt	3300
Ethanol	2080
Aspirin	1100
Codeine	800
Oxycodone	480
Caffeine	192
DDT	113
Cocaine (injected)	95
Dichlorvos (pesticide strips)	56
Ricin	30
Sodium cyanide	6
Parathion	3

QUESTIONS AND PROBLEMS

2.6 Problem Solving Using Unit Conversion

LEARNING GOAL Use conversion factors to change from one unit to another.

2.61 When you convert one unit to another, how do you know which unit of the conversion factor to place in the denominator?

2.62 When you convert one unit to another, how do you know which unit of the conversion factor to place in the numerator?

2.63 Perform each of the following conversions using metric conversion factors:
a. 44.2 mL to liters
b. 8.65 m to nanometers
c. 5.2×10^8 g to megagrams
d. 0.72 ks to milliseconds

2.64 Perform each of the following conversions using metric conversion factors:
a. 4.82×10^{-5} L to picoliters
b. 575.2 dm to kilometers
c. 5×10^{-4} kg to micrograms
d. 6.4×10^{10} ps to seconds

2.65 Perform each of the following conversions using metric and U.S. conversion factors:
a. 3.428 lb to kilograms
b. 1.6 m to inches
c. 4.2 L to quarts
d. 0.672 ft to millimeters

2.66 Perform each of the following conversions using metric and U.S. conversion factors:
a. 0.21 lb to grams
b. 11.6 in. to centimeters
c. 0.15 qt to milliliters
d. 35.41 kg to pounds

2.67 Use metric conversion factors to solve each of the following problems:
a. The height of a student is 175 cm. How tall is the student in meters?
b. A cooler has a volume of 5000 mL. What is the capacity of the cooler in liters?
c. A hummingbird has a mass of 0.0055 kg. What is the mass, in grams, of the hummingbird?
d. A balloon has a volume of 3500 cm³. What is the volume, in cubic meters?

2.68 Use metric conversion factors to solve each of the following problems:
a. The Daily Value (DV) for phosphorus is 800 mg. How many grams of phosphorus are recommended?
b. A glass of orange juice contains 3.2 dL of juice. How many milliliters of orange juice are in the glass?
c. A package of chocolate instant pudding contains 2840 mg of sodium. How many grams of sodium are in the pudding?
d. A park has an area of 150 000 m². What is the area, in square kilometers?

2.69 Solve each of the following problems using one or more conversion factors:
a. A container holds 0.500 qt of liquid. How many milliliters of lemonade will it hold?
b. What is the mass, in kilograms, of a person who weighs 175 lb?
c. An athlete has 15% body fat by mass. What is the weight of fat, in pounds, of a 74-kg athlete?
d. A plant fertilizer contains 15% nitrogen (N) by mass. In a container of soluble plant food, there are 10.0 oz of fertilizer. How many grams of nitrogen are in the container?

Agricultural fertilizers applied to a field provide nitrogen for plant growth.

2.70 Solve each of the following problems using one or more conversion factors:
a. Wine is 12% alcohol by volume. How many milliliters of alcohol are in a 0.750-L bottle of wine?
b. Blueberry high-fiber muffins contain 51% dietary fiber by mass. If a package with a net weight of 12 oz contains six muffins, how many grams of fiber are in each muffin?
c. A jar of crunchy peanut butter contains 1.43 kg of peanut butter. If you use 8.0% of the peanut butter for a sandwich, how many ounces of peanut butter did you take out of the container?
d. In a candy factory, the nutty chocolate bars contain 22.0% pecans by mass. If 5.0 kg of pecans were used for candy last Tuesday, how many pounds of nutty chocolate bars were made?

Applications

2.71 Using conversion factors, solve each of the following clinical problems:
a. You have used 250 L of distilled water for a dialysis patient. How many gallons of water is that?
b. A patient needs 0.024 g of a sulfa drug. There are 8-mg tablets in stock. How many tablets should be given?
c. The daily dose of ampicillin for the treatment of an ear infection is 115 mg/kg of body weight. What is the daily dose for a 34-lb child?
d. You need 4.0 oz of a steroid ointment. How many grams of ointment does the pharmacist need to prepare?

2.72 Using conversion factors, solve each of the following clinical problems:
a. The physician has ordered 1.0 g of tetracycline to be given every six hours to a patient. If your stock on hand is 500-mg tablets, how many will you need for one day's treatment?
b. An intramuscular medication is given at 5.00 mg/kg of body weight. What is the dose for a 180-lb patient?
c. A physician has ordered 0.50 mg of atropine, intramuscularly. If atropine were available as 0.10 mg/mL of solution, how many milliliters would you need to give?
d. During surgery, a patient receives 5.0 pt of plasma. How many milliliters of plasma were given?

2.73 Using conversion factors, solve each of the following clinical problems:
a. A nurse practitioner prepares 500. mL of an IV of normal saline solution to be delivered at a rate of 80. mL/h. What is the infusion time, in hours, to deliver 500. mL?
b. A nurse practitioner orders Medrol to be given 1.5 mg/kg of body weight. Medrol is an anti-inflammatory administered as an intramuscular injection. If a child weighs 72.6 lb and the available stock of Medrol is 20. mg/mL, how many milliliters does the nurse administer to the child?

2.74 Using conversion factors, solve each of the following clinical problems:

a. A nurse practitioner prepares an injection of promethazine, an antihistamine used to treat allergic rhinitis. If the stock bottle is labeled 25 mg/mL and the order is a dose of 12.5 mg, how many milliliters will the nurse draw up in the syringe?

b. You are to give ampicillin 25 mg/kg to a child with a mass of 62 lb. If stock on hand is 250 mg/capsule, how many capsules should be given?

2.75 Using Table 2.8, calculate the number of grams that would provide the LD_{50} for a 175-lb person for each of the following:
a. table salt
b. sodium cyanide
c. aspirin

2.76 Using Table 2.8, calculate the number of grams that would provide the LD_{50} for a 148-lb person for each of the following:
a. ethanol
b. ricin
c. baking soda

2.7 Density

LEARNING GOAL Calculate the density of a substance; use the density to calculate the mass or volume of a substance.

The mass and volume of any object can be measured. If we compare the mass of the object to its volume, we obtain a relationship called **density**.

$$\text{Density} = \frac{\text{mass of substance}}{\text{volume of substance}}$$

Every substance has a unique density, which distinguishes it from other substances. For example, lead has a density of 11.3 g/mL, whereas cork has a density of 0.26 g/mL. From these densities, we can predict if these substances will sink or float in water. *If an object is less dense than a liquid, the object floats when placed in the liquid.* If a substance, such as cork, is less dense than water, it will float. However, a lead object sinks because its density is greater than that of water (see **FIGURE 2.10**).

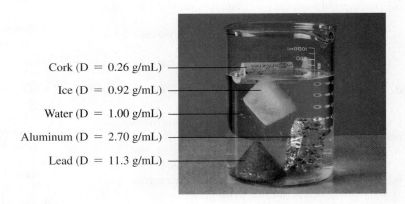

Cork (D = 0.26 g/mL)
Ice (D = 0.92 g/mL)
Water (D = 1.00 g/mL)
Aluminum (D = 2.70 g/mL)
Lead (D = 11.3 g/mL)

FIGURE 2.10 ▶ Objects that sink in water are more dense than water; objects that float are less dense.

Q Why does an ice cube float and a piece of aluminum sink?

Density is used in chemistry in many ways. For example, density can be used to identify a specific substance. If we calculate a density of a pure metal as 10.5 g/mL, then we could identify it as silver, but not gold or aluminum. Metals such as gold and silver have higher densities, whereas gases have low densities. In the metric system, the densities of solids and liquids are usually expressed as grams per cubic centimeter (g/cm^3) or grams per milliliter (g/mL). The densities of gases are usually stated as grams per liter (g/L). **TABLE 2.9** gives the densities of some common substances.

Calculating Density

We can calculate the density of a substance from its mass and volume as shown in Sample Problem 2.14.

TABLE 2.9 Densities of Some Common Substances

Solids (at 25 °C)	Density (g/mL)	Liquids (at 25 °C)	Density (g/mL)	Gases (at 0 °C)	Density (g/L)
Cork	0.26	Gasoline	0.74	Hydrogen	0.090
Body fat	0.909	Ethanol	0.79	Helium	0.179
Ice (at 0 °C)	0.92	Olive oil	0.92	Methane	0.714
Muscle	1.06	Water (at 4 °C)	1.00	Neon	0.902
Sugar	1.59	Urine	1.003–1.030	Nitrogen	1.25
Bone	1.80	Plasma (blood)	1.03	Air (dry)	1.29
Salt (NaCl)	2.16	Milk	1.04	Oxygen	1.43
Aluminum	2.70	Blood	1.06	Carbon dioxide	1.96
Iron	7.86	Mercury	13.6		
Copper	8.92				
Silver	10.5				
Lead	11.3				
Gold	19.3				

SAMPLE PROBLEM 2.14 Calculating Density

High-density lipoprotein (HDL) is a type of cholesterol, sometimes called "good cholesterol," that is measured in a routine blood test. If a 0.258-g sample of HDL has a volume of 0.215 cm^3, what is the density of the HDL sample?

TRY IT FIRST

SOLUTION

STEP 1 State the given and needed quantities.

ANALYZE THE PROBLEM	Given	Need	Connect
	0.258 g of HDL, 0.215 cm^3	density (g/cm^3) of HDL	density expression

STEP 2 Write the density expression.

$$\text{Density} = \frac{\text{mass of substance}}{\text{volume of substance}}$$

STEP 3 Express mass in grams and volume in cm^3.

Mass of HDL sample $= 0.258$ g

Volume of HDL sample $= 0.215$ cm^3

STEP 4 Substitute mass and volume into the density expression and calculate the density.

Three SFs

$$\text{Density} = \frac{0.258 \text{ g}}{0.215 \text{ cm}^3} = \frac{1.20 \text{ g}}{1 \text{ cm}^3} = 1.20 \text{ g/cm}^3$$

Three SFs   Three SFs

Guide to Calculating Density

STEP 1
State the given and needed quantities.

STEP 2
Write the density expression.

STEP 3
Express mass in grams and volume in milliliters (mL) or cm^3.

STEP 4
Substitute mass and volume into the density expression and calculate the density.

STUDY CHECK 2.14

Low-density lipoprotein (LDL), sometimes called "bad cholesterol," is also measured in a routine blood test. If a 0.380-g sample of LDL has a volume of 0.362 cm^3, what is the density of the LDL sample?

ANSWER

1.05 g/cm^3

Density of Solids Using Volume Displacement

The volume of a solid can be determined by volume displacement. When a solid is completely submerged in water, it displaces a volume that is equal to the volume of the solid. In **FIGURE 2.11**, the water level rises from 35.5 mL to 45.0 mL after the zinc object is added. This means that 9.5 mL of water is displaced and that the volume of the object is 9.5 mL. The density of the zinc is calculated using volume displacement as follows:

$$\text{Density} = \frac{\overset{\text{Four SFs}}{68.60 \text{ g Zn}}}{\underset{\text{Two SFs}}{9.5 \text{ mL}}} = \underset{\text{Two SFs}}{7.2 \text{ g/mL}}$$

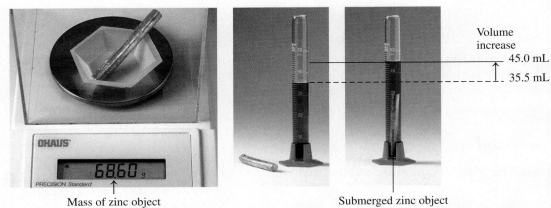

Mass of zinc object   Submerged zinc object

FIGURE 2.11 ▶ The density of a solid can be determined by volume displacement because a submerged object displaces a volume of water equal to its own volume.

◉ How is the volume of the zinc object determined?

SAMPLE PROBLEM 2.15 Using Volume Displacement to Calculate Density

A lead weight used in the belt of a scuba diver has a mass of 226 g. When the weight is carefully placed in a large graduated cylinder containing 200.0 cm^3 of water, the water level rises to 220.0 cm^3. What is the density (g/cm^3) of the lead weight?

TRY IT FIRST

SOLUTION

STEP 1 State the given and needed quantities.

	Given	Need	Connect
ANALYZE THE PROBLEM	226 g of lead, water level + lead = 220.0 cm^3 water level (initial) = 200.0 cm^3	density (g/cm^3) of lead	density expression

STEP 2 Write the density expression.

$$\text{Density} = \frac{\text{mass of substance}}{\text{volume of substance}}$$

Lead weights in a belt counteract the buoyancy of a scuba diver.

STEP **3** Express mass in grams and volume in cm^3.

Mass of lead weight = 226 g

The volume of the lead weight is equal to the volume of water it displaced, which is calculated as follows:

Water level after object submerged	=	220.0 cm^3
Water level before object submerged	=	−200.0 cm^3
Water displaced (volume of lead)	=	20.0 cm^3

STEP **4** Substitute mass and volume into the density expression and calculate the density. Be sure to use the volume of water displaced and *not* the initial volume of water.

$$\text{Density} = \frac{\overset{\text{Three SFs}}{226\ \text{g}}}{\underset{\text{Three SFs}}{20.0\ \text{cm}^3}} = \underset{\text{Three SFs}}{11.3\ \text{g/cm}^3}$$

STUDY CHECK 2.15

A total of 0.500 lb of glass marbles is added to 425 mL of water. The water level rises to a volume of 528 mL. What is the density (g/cm^3) of the glass marbles?

ANSWER

2.20 g/cm^3

CHEMISTRY LINK TO **HEALTH**
Bone Density

The density of our bones determines their health and strength. Our bones are constantly gaining and losing minerals such as calcium, magnesium, and phosphate. In childhood, bones form at a faster rate than they break down. As we age, the breakdown of bone occurs more rapidly than new bone forms. As the loss of bone minerals increases, bones begin to thin, causing a decrease in mass and density. Thinner bones lack strength, which increases the risk of fracture. Hormonal changes, disease, and certain medications can also contribute to the thinning of bone. Eventually, a condition of severe thinning of bone known as *osteoporosis* may occur. *Scanning electron micrographs* (SEMs) show (**a**) normal bone and (**b**) bone with osteoporosis due to loss of bone minerals.

Bone density is often determined by passing low-dose X-rays through the narrow part at the top of the femur (hip) and the spine (**c**). These locations are where fractures are more likely to occur, especially as we age. Bones with high density will block more of the X-rays compared to bones that are less dense. The results of a bone density test are compared to a healthy young adult as well as to other people of the same age.

Recommendations to improve bone strength include supplements of calcium and vitamin D. Weight-bearing exercise such as walking and lifting weights can also improve muscle strength, which in turn increases bone strength.

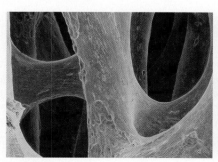

(**a**) Normal bone

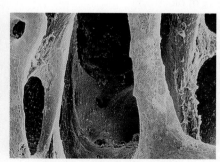

(**b**) Bone with osteoporosis

(**c**) Viewing a low-dose X-ray of the spine

Guide to Using Density

STEP 1
State the given and needed
quantities.

STEP 2
Write a plan to calculate the
needed quantity.

STEP 3
Write the equalities and their
conversion factors including
density.

STEP 4
Set up the problem to
calculate the needed quantity.

Problem Solving Using Density

Density can be used as a conversion factor. For example, if the volume and the density of a sample are known, the mass in grams of the sample can be calculated as shown in Sample Problem 2.16.

SAMPLE PROBLEM 2.16 Problem Solving with Density

Greg has a blood volume of 5.9 qt. If the density of blood is 1.06 g/mL, what is the mass, in grams, of Greg's blood?

▶ **TRY IT FIRST**

SOLUTION

STEP 1 State the given and needed quantities.

	Given	**Need**	**Connect**
ANALYZE THE PROBLEM	5.9 qt of blood	grams of blood	conversion factors (qt/mL, density of blood)

STEP 2 Write a plan to calculate the needed quantity.

quarts → U.S.–Metric factor → milliliters → Density factor → grams

STEP 3 Write the equalities and their conversion factors including density.

1 qt = 946.4 mL	1 mL of blood = 1.06 g of blood
$\dfrac{946.4 \text{ mL}}{1 \text{ qt}}$ and $\dfrac{1 \text{ qt}}{946.4 \text{ mL}}$	$\dfrac{1.06 \text{ g blood}}{1 \text{ mL blood}}$ and $\dfrac{1 \text{ mL blood}}{1.06 \text{ g blood}}$

STEP 4 Set up the problem to calculate the needed quantity.

$$7.5 \text{ qt blood} \times \underset{\text{Exact}}{\frac{946.4 \text{ mL}}{1 \text{ qt}}} \times \underset{\text{Exact}}{\frac{1.06 \text{ g blood}}{1 \text{ mL blood}}} = 7500 \text{ g of blood}$$

Two SFs   Four SFs   Three SFs   Two SFs

STUDY CHECK 2.16

During surgery, a patient receives 3.0 pt of blood. How many kilograms of blood (density = 1.06 g/mL) were needed for the transfusion?

ANSWER

1.5 kg of blood

Specific Gravity

Specific gravity (sp gr) is a relationship between the density of a substance and the density of water. Specific gravity is calculated by dividing the density of a sample by the density of water, which is 1.00 g/mL at 4 °C. A substance with a specific gravity of 1.00 has the same density as water (1.00 g/mL).

$$\text{Specific gravity} = \frac{\text{density of sample}}{\text{density of water}}$$

Specific gravity is one of the few unitless values you will encounter in chemistry. An instrument called a hydrometer is often used to measure the specific gravity of fluids such as wine or a sample of urine. The specific gravity of the urine helps evaluate the water balance

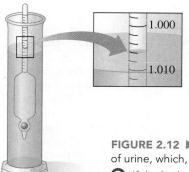

FIGURE 2.12 ▶ A hydrometer is used to measure the specific gravity of urine, which, for adults, is 1.003 to 1.030.

🎯 If the hydrometer reading is 1.006, what is the density of the urine?

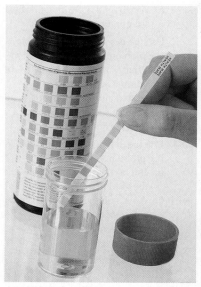

A dipstick is used to measure the specific gravity of a urine sample.

in the body and the substances in the urine. In **FIGURE 2.12**, a hydrometer is used to measure the specific gravity of urine. The normal range of specific gravity for urine is 1.003 to 1.030. The specific gravity can decrease with *type 2 diabetes* and kidney disease. Increased specific gravity may occur with dehydration, kidney infection, and liver disease. In a clinic or hospital, a dipstick containing chemical pads is used to evaluate specific gravity.

QUESTIONS AND PROBLEMS

2.7 Density

LEARNING GOAL Calculate the density of a substance; use the density to calculate the mass or volume of a substance.

2.77 Determine the density (g/mL) for each of the following:
   a. A 20.0-mL sample of a salt solution that has a mass of 24.0 g.
   b. A cube of butter weighs 0.250 lb and has a volume of 130.3 mL.
   c. A gem has a mass of 4.50 g. When the gem is placed in a graduated cylinder containing 12.00 mL of water, the water level rises to 13.45 mL.
   d. A medication, if 3.00 mL has a mass of 3.85 g.

2.78 Determine the density (g/mL) for each of the following:
   a. The fluid in a car battery has a volume of 125 mL and a mass of 155 g.
   b. A plastic material weighs 2.68 lb and has a volume of 3.5 L.
   c. A 5.00-mL urine sample from a person suffering from diabetes mellitus has a mass of 5.025 g.
   d. A solid object with a mass of 1.65 lb and a volume of 170 mL.

2.79 What is the density (g/mL) of each of the following samples?
   a. A lightweight head on a golf club is made of titanium. The volume of a sample of titanium is 114 cm³ and the mass is 514.1 g.

Lightweight heads on golf clubs are made of titanium.

   b. A syrup is added to an empty container with a mass of 115.25 g. When 0.100 pt of syrup is added, the total mass of the container and syrup is 182.48 g.

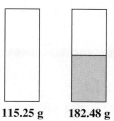

115.25 g   **182.48 g**

   c. A block of aluminum metal has a volume of 3.15 L and a mass of 8.51 kg.

2.80 What is the density (g/mL) of each of the following samples?
   a. An ebony carving has a mass of 275 g and a volume of 207 cm³.
   b. A 14.3-cm³ sample of tin has a mass of 0.104 kg.
   c. A bottle of acetone (fingernail polish remover) contains 55.0 mL of acetone with a mass of 43.5 g.

2.81 Use the density values in Table 2.9 to solve each of the following problems:
   a. How many liters of ethanol contain 1.50 kg of ethanol?
   b. How many grams of mercury are present in a barometer that holds 6.5 mL of mercury?
   c. A sculptor has prepared a mold for casting a silver figure. The figure has a volume of 225 cm³. How many ounces of silver are needed in the preparation of the silver figure?

2.82 Use the density values in Table 2.9 to solve each of the following problems:
   a. A graduated cylinder contains 18.0 mL of water. What is the new water level, in milliliters, after 35.6 g of silver metal is submerged in the water?
   b. A thermometer containing 8.3 g of mercury has broken. What volume, in cubic centimeters, of mercury spilled?

c. A fish tank holds 35 gal of water. How many kilograms of water are in the fish tank?

2.83 Use the density values in Table 2.9 to solve each of the following problems:
   a. What is the mass, in grams, of a cube of copper that has a volume of 74.1 cm^3?
   b. How many kilograms of gasoline fill a 12.0-gal gas tank?
   c. What is the volume, in cubic centimeters, of an ice cube that has a mass of 27 g?

2.84 Use the density values in Table 2.9 to solve each of the following problems:
   a. If a bottle of olive oil contains 1.2 kg of olive oil, what is the volume, in milliliters, of the olive oil?
   b. A cannon ball made of iron has a volume of 115 cm^3. What is the mass, in kilograms, of the cannon ball?
   c. A balloon filled with helium has a volume of 7.3 L. What is the mass, in grams, of helium in the balloon?

2.85 In an old trunk, you find a piece of metal that you think may be aluminum, silver, or lead. You take it to a lab, where you find it has a mass of 217 g and a volume of 19.2 cm^3. Using Table 2.9, what is the metal you found?

2.86 Suppose you have two 100-mL graduated cylinders. In each cylinder, there is 40.0 mL of water. You also have two cubes: one is lead, and the other is aluminum. Each cube measures 2.0 cm on each side. After you carefully lower each cube into the water of its own cylinder, what will the new water level be in each of the cylinders?

Applications

2.87 Solve each of the following problems:
   a. A urine sample has a density of 1.030 g/mL. What is the specific gravity of the sample?
   b. A 20.0-mL sample of a glucose IV solution that has a mass of 20.6 g. What is the density of the glucose solution?
   c. The specific gravity of a vegetable oil is 0.92. What is the mass, in grams, of 750 mL of vegetable oil?
   d. A bottle containing 325 g of cleaning solution is used to clean hospital equipment. If the cleaning solution has a specific gravity of 0.850, what volume, in milliliters, of solution was used?

2.88 Solve each of the following problems:
   a. A glucose solution has a density of 1.02 g/mL. What is its specific gravity?
   b. A 0.200-mL sample of high-density lipoprotein (HDL) has a mass of 0.230 g. What is the density of the HDL?
   c. Butter has a specific gravity of 0.86. What is the mass, in grams, of 2.15 L of butter?
   d. A 5.000-mL urine sample has a mass of 5.025 g. If the normal range for the specific gravity of urine is 1.003 to 1.030, would the specific gravity of this urine sample indicate that the patient could have type 2 diabetes?

Follow Up
GREG'S VISIT WITH HIS DOCTOR

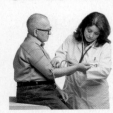

On Greg's last visit to his doctor, he complained of feeling tired. His doctor orders a blood test for iron. Sandra, the registered nurse, does a venipuncture and withdraws 8.0 mL of blood. About 70% of the iron in the body is used to form hemoglobin, which is a protein in the red blood cells that carries oxygen to the cells of the body. About 30% is stored in ferritin, bone marrow, and the liver. When the iron level is low, a person may have fatigue and decreased immunity.

The normal range for serum iron in men is 80 to 160 mcg/dL. Greg's iron test showed a blood serum iron level of 42 mcg/dL, which indicates that Greg has *iron deficiency anemia*. His doctor orders an iron supplement to be taken twice daily. One tablet of the iron supplement contains 65 mg of iron.

Applications

2.89 a. Write an equality and two conversion factors for Greg's serum iron level.
   b. How many micrograms of iron were in the 8.0-mL sample of Greg's blood?

2.90 a. Write an equality and two conversion factors for one tablet of the iron supplement.
   b. How many grams of iron will Greg consume in one week?

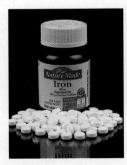

Each tablet contains 65 mg of iron, which is given for iron supplementation.

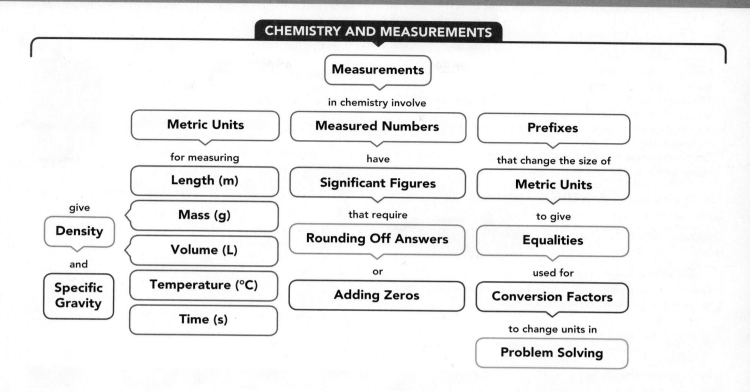

CHEMISTRY AND MEASUREMENTS

Measurements

in chemistry involve

Metric Units — Measured Numbers — Prefixes

for measuring / have / that change the size of

Length (m) — Significant Figures — Metric Units

Mass (g) / that require / to give

Volume (L) — Rounding Off Answers — Equalities

give → Density

Temperature (°C) / or / used for

and → Specific Gravity

Adding Zeros — Conversion Factors

Time (s)

to change units in

Problem Solving

CHAPTER REVIEW

2.1 Units of Measurement

LEARNING GOAL Write the names and abbreviations for the metric or SI units used in measurements of length, volume, mass, temperature, and time.

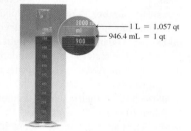

1 L = 1.057 qt
946.4 mL = 1 qt

- In science, physical quantities are described in units of the metric or International System of Units (SI).
- Some important units are meter (m) for length, liter (L) for volume, gram (g) and kilogram (kg) for mass, degree Celsius (°C) and kelvin (K) for temperature, and second (s) for time.

2.2 Measured Numbers and Significant Figures

LEARNING GOAL Identify a number as measured or exact; determine the number of significant figures in a measured number.

(a)

(b)

- A measured number is any number obtained by using a measuring device.
- An exact number is obtained by counting items or from a definition; no measuring device is needed.
- Significant figures are the numbers reported in a measurement including the estimated digit.
- Zeros in front of a decimal number or at the end of a nondecimal number are not significant.

2.3 Significant Figures in Calculations

LEARNING GOAL Adjust calculated answers to give the correct number of significant figures.

- In multiplication and division, the final answer is written so that it has the same number of significant figures as the measurement with the fewest significant figures.
- In addition and subtraction, the final answer is written so that it has the same number of decimal places as the measurement with the fewest decimal places.

2.4 Prefixes and Equalities

LEARNING GOAL Use the numerical values of prefixes to write a metric equality.

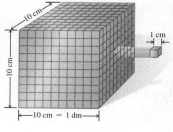

10 cm
10 cm
1 cm
10 cm = 1 dm

- A prefix placed in front of a metric or SI unit changes the size of the unit by factors of 10.
- Prefixes such as *centi*, *milli*, and *micro* provide smaller units; prefixes such as *kilo*, *mega*, and *tera* provide larger units.
- An equality shows the relationship between two units that measure the same quantity of length, volume, mass, or time.
- Examples of metric equalities are 1 m = 100 cm, 1 L = 1000 mL, 1 kg = 1000 g, and 1 min = 60 s.

2.5 Writing Conversion Factors

LEARNING GOAL Write a conversion factor for two units that describe the same quantity.

- Conversion factors are used to express a relationship in the form of a fraction.
- Two conversion factors can be written for any relationship in the metric or U.S. system.
- A percentage is written as a conversion factor by expressing matching units as the parts in 100 parts of the whole.
- Small ratios are expressed as parts per million (ppm) and parts per billion (ppb).

2.6 Problem Solving Using Unit Conversion

LEARNING GOAL Use conversion factors to change from one unit to another.

- Conversion factors are useful when changing a quantity expressed in one unit to a quantity expressed in another unit.

- In the problem-solving process, a given unit is multiplied by one or more conversion factors that cancel units until the needed answer is obtained.

2.7 Density

LEARNING GOAL Calculate the density of a substance; use the density to calculate the mass or volume of a substance.

- The density of a substance is a ratio of its mass to its volume, usually g/mL or g/cm³.
- The units of density can be used to write conversion factors that convert between the mass and volume of a substance.
- Specific gravity (sp gr) compares the density of a substance to the density of water, 1.00 g/mL.

KEY TERMS

Celsius (°C) temperature scale A temperature scale on which water has a freezing point of 0 °C and a boiling point of 100 °C.

centimeter (cm) A unit of length in the metric system; there are 2.54 cm in 1 in.

conversion factor A ratio in which the numerator and denominator are quantities from an equality or given relationship. For example, the two conversion factors for the equality 1 kg = 2.205 lb are written as

$$\frac{2.205 \text{ lb}}{1 \text{ kg}} \quad \text{and} \quad \frac{1 \text{ kg}}{2.205 \text{ lb}}$$

cubic centimeter (cm³, cc) The volume of a cube that has 1-cm sides; 1 cm³ is equal to 1 mL.

cubic meter (m³) The volume of a cube that has sides that measure 1 m in length.

density The relationship of the mass of an object to its volume expressed as grams per cubic centimeter (g/cm^3), grams per milliliter (g/mL), or grams per liter (g/L).

equality A relationship between two units that measure the same quantity.

exact number A number obtained by counting or by definition.

gram (g) The metric unit used in measurements of mass.

International System of Units (SI) The official system of measurement throughout the world except for the United States that modifies the metric system.

Kelvin (K) temperature scale A temperature scale on which the lowest possible temperature is 0 K.

kilogram (kg) A metric mass of 1000 g, equal to 2.205 lb. The kilogram is the SI standard unit of mass.

liter (L) The metric unit for volume that is slightly larger than a quart.

mass A measure of the quantity of material in an object.

measured number A number obtained when a quantity is determined by using a measuring device.

meter (m) The metric unit for length that is slightly longer than a yard. The meter is the SI standard unit of length.

metric system A system of measurement used by scientists and in most countries of the world.

milliliter (mL) A metric unit of volume equal to one-thousandth of a liter (0.001 L).

prefix The part of the name of a metric unit that precedes the base unit and specifies the size of the measurement. All prefixes are related on a decimal scale.

second (s) A unit of time used in both the SI and metric systems.

SI See International System of Units (SI).

significant figures (SFs) The numbers recorded in a measurement.

specific gravity (sp gr) A relationship between the density of a substance and the density of water:

$$\text{sp gr} = \frac{\text{density of sample}}{\text{density of water}}$$

temperature An indicator of the hotness or coldness of an object.

volume (V) The amount of space occupied by a substance.

KEY MATH SKILL

The chapter section containing each Key Math Skill is shown in parentheses at the end of each heading.

Rounding Off (2.3)

Calculator displays are rounded off to give the correct number of significant figures.

- If the first digit to be dropped is *4 or less*, then it and all following digits are simply dropped from the number.
- If the first digit to be dropped is *5 or greater*, then the last retained digit of the number is increased by 1.

One or more significant zeros are added when the calculator display has fewer digits than the needed number of significant figures.

Example: Round off each of the following to three significant figures:

a. 3.608 92 L	**Answer:**	**a.** 3.61 L
b. 0.003 870 298 m		**b.** 0.003 87 m
c. 6 g		**c.** 6.00 g

 CORE CHEMISTRY SKILLS

The chapter section containing each Core Chemistry Skill is shown in parentheses at the end of each heading.

Counting Significant Figures (2.2)

The significant figures (SFs) are all the *measured* numbers including the last, estimated digit:

- All nonzero digits
- Zeros between nonzero digits
- Zeros within a decimal number
- All digits in a coefficient of a number written in scientific notation

An *exact* number is obtained from counting or a definition and has no effect on the number of significant figures in the final answer.

Example: State the number of significant figures in each of the following:

	Answer:	
a. 0.003 045 mm		**a.** four SFs
b. 15 000 m		**b.** two SFs
c. 45.067 kg		**c.** five SFs
d. 5.30×10^3 g		**d.** three SFs
e. 2 cans of soda		**e.** exact

Using Significant Figures in Calculations (2.3)

- In multiplication or division, the final answer is written so that it has the same number of significant figures as the measurement with the fewest SFs.
- In addition or subtraction, the final answer is written so that it has the same number of decimal places as the measurement having the fewest decimal places.

Example: Perform the following calculations using measured numbers and give answers with the correct number of SFs:

a. 4.05 m × 0.6078 m	**b.** $\dfrac{4.50 \text{ g}}{3.27 \text{ mL}}$
c. 0.758 g + 3.10 g	**d.** 13.538 km − 8.6 km

Answer:
a. 2.46 m² **b.** 1.38 g/mL
c. 3.86 g **d.** 4.9 km

Using Prefixes (2.4)

- In the metric and SI systems of units, a prefix attached to any unit increases or decreases its size by some factor of 10.
- When the prefix *centi* is used with the unit meter, it becomes centimeter, a length that is one-hundredth of a meter (0.01 m).
- When the prefix *milli* is used with the unit meter, it becomes millimeter, a length that is one-thousandth of a meter (0.001 m).

Example: Complete the following statements with the correct prefix symbol:

a. 1000 m = 1 ____ m	**b.** 0.01 g = 1 ____ g

Answer: **a.** 1000 m = 1 km **b.** 0.01 g = 1 cg

Writing Conversion Factors from Equalities (2.5)

- A conversion factor allows you to change from one unit to another.
- Two conversion factors can be written for any equality in the metric, U.S., or metric–U.S. systems of measurement.
- Two conversion factors can be written for a relationship stated within a problem.

Example: Write two conversion factors for the equality:
1 L = 1000 mL.

Answer: $\dfrac{1000 \text{ mL}}{1 \text{ L}}$ and $\dfrac{1 \text{ L}}{1000 \text{ mL}}$

Using Conversion Factors (2.6)

In problem solving, conversion factors are used to cancel the given unit and to provide the needed unit for the answer.

- State the given and needed quantities.
- Write a plan to convert the given unit to the needed unit.
- State the equalities and conversion factors.
- Set up the problem to cancel units and calculate the answer.

Example: A computer chip has a width of 0.75 in. What is that distance in millimeters?

Answer: $0.75 \text{ in.} \times \dfrac{2.54 \text{ cm}}{1 \text{ in.}} \times \dfrac{10 \text{ mm}}{1 \text{ cm}} = 19 \text{ mm}$

Using Density as a Conversion Factor (2.7)

Density is an equality of mass and volume for a substance, which is written as the *density expression*.

$$\text{Density} = \frac{\text{mass of substance}}{\text{volume of substance}}$$

Density is useful as a conversion factor to convert between mass and volume.

Example: The element tungsten used in light bulb filaments has a density of 19.3 g/cm³. What is the volume, in cubic centimeters, of 250 g of tungsten?

Answer: $250 \text{ g} \times \dfrac{1 \text{ cm}^3}{19.3 \text{ g}} = 13 \text{ cm}^3$

UNDERSTANDING THE CONCEPTS

The chapter sections to review are shown in parentheses at the end of each question.

2.91 In which of the following pairs do both numbers contain the same number of significant figures? (2.2)
  a. 2.0500 m and 0.0205 m
  b. 600.0 K and 60 K
  c. 0.000 75 s and 75 000 s
  d. 6.240 L and 6.240×10^{-2} L

2.92 In which of the following pairs do both numbers contain the same number of significant figures? (2.2)
  a. 3.44×10^{-3} g and 0.0344 g
  b. 0.0098 s and 9.8×10^{4} s
  c. 6.8×10^{3} m and 68 000 m
  d. 258.000 g and 2.58×10^{-2} g

2.93 Indicate if each of the following is answered with an exact number or a measured number: (2.2)

  a. number of legs
  b. height of table
  c. number of chairs at the table
  d. area of tabletop

2.94 Measure the length of each of the objects in diagrams **(a)**, **(b)**, and **(c)** using the metric ruler in the figure. Indicate the number of significant figures for each and the estimated digit for each. (2.2)

(a)

(b)

(c)

2.95 State the temperature on the Celsius thermometer **A** to the correct number of significant figures: (2.3)

A

2.96 State the temperature on the Celsius thermometer **B** to the correct number of significant figures: (2.3)

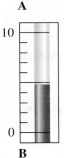

B

2.97 The length of this rug is 38.4 in. and the width is 24.2 in. (2.3)

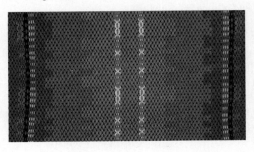

  a. What is the length of this rug, in centimeters?
  b. What is the width of this rug, in centimeters?
  c. How many significant figures are in the length measurement?
  d. Calculate the area of the rug, in square centimeters, to the correct number of significant figures.

2.98 A shipping box has a length of 7.00 in., a width of 6.00 in., and a height of 4.00 in. (2.3)

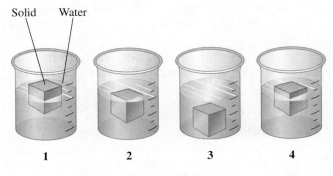

  a. What is the length of the box, in centimeters?
  b. What is the width of the box, in centimeters?
  c. How many significant figures are in the width measurement?
  d. Calculate the volume of the box, in cubic centimeters, to the correct number of significant figures.

2.99 Each of the following diagrams represents a container of water and a cube. Some cubes float while others sink. Match diagrams **1**, **2**, **3**, or **4** with one of the following descriptions and explain your choices: (2.7)

Solid   Water

1   **2**   **3**   **4**

  a. The cube has a greater density than water.
  b. The cube has a density that is 0.80 g/mL.
  c. The cube has a density that is one-half the density of water.
  d. The cube has the same density as water.

2.100 What is the density of the solid object that is weighed and submerged in water? (2.7)

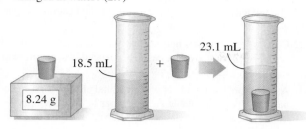

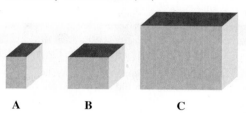

2.101 Consider the following solids. The solids **A**, **B**, and **C** represent aluminum (D = 2.70 g/mL), gold (D = 19.3 g/mL), and silver (D = 10.5 g/mL). If each has a mass of 10.0 g, what is the identity of each solid? (2.7)

A   B   C

2.102 A graduated cylinder contains three liquids **A**, **B**, and **C**, which have different densities and do not mix: mercury (D = 13.6 g/mL), vegetable oil (D = 0.92 g/mL), and water (D = 1.00 g/mL). Identify the liquids **A**, **B**, and **C** in the cylinder. (2.7)

2.103 The gray cube has a density of 4.5 g/cm³. Is the density of the green cube the same, lower than, or higher than that of the gray cube? (2.7)

2.104 The gray cube has a density of 4.5 g/cm³. Is the density of the green cube the same, lower than, or higher than that of the gray cube? (2.7)

ADDITIONAL QUESTIONS AND PROBLEMS

2.105 Round off or add zeros to the following calculated answers to give a final answer with three significant figures: (2.2)
a. 0.000 012 58 L
b. 3.528×10^2 kg
c. 125 111 m
d. 34.9673 s

2.106 Round off or add zeros to the following calculated answers to give a final answer with three significant figures: (2.2)
a. 58.703 g
b. 3×10^{-3} s
c. 0.010 826 g
d. 1.7484×10^3 ms

2.107 A dessert contains 137.25 g of vanilla ice cream, 84 g of fudge sauce, and 43.7 g of nuts. (2.3, 2.6)
a. What is the total mass, in grams, of the dessert?
b. What is the total weight, in pounds, of the dessert?

2.108 A fish company delivers 22 kg of salmon, 5.5 kg of crab, and 3.48 kg of oysters to your seafood restaurant. (2.3, 2.6)
a. What is the total mass, in kilograms, of the seafood?
b. What is the total number of pounds?

2.109 In France, grapes are 1.95 euros per kilogram. What is the cost of grapes, in dollars per pound, if the exchange rate is 1.14 dollars/euro? (2.6)

2.110 In Mexico, avocados are 48 pesos per kilogram. What is the cost, in cents, of an avocado that weighs 0.45 lb if the exchange rate is 15 pesos to the dollar? (2.6)

2.111 Bill's recipe for onion soup calls for 4.0 lb of thinly sliced onions. If an onion has an average mass of 115 g, how many onions does Bill need? (2.6)

2.112 The price of 1 lb of potatoes is $1.75. If all the potatoes sold today at the store bring in $1420, how many kilograms of potatoes did grocery shoppers buy? (2.6)

2.113 During a workout at the gym, you set the treadmill at a pace of 55.0 m/min. How many minutes will you walk if you cover a distance of 7500 ft? (2.6)

2.114 The distance between two cities is 1700 km. How long will it take, in hours, to drive from one city to the other if your average speed is 63 mi/h? (2.6)

2.115 The water level in a graduated cylinder initially at 215 mL rises to 285 mL after a piece of lead is submerged. What is the mass, in grams, of the lead (see Table 2.9)? (2.7)

2.116 A graduated cylinder contains 155 mL of water. A 15.0-g piece of iron and a 20.0-g piece of lead are added. What is the new water level, in milliliters, in the cylinder (see Table 2.9)? (2.7)

2.117 How many cubic centimeters (cm³) of gasoline have a mass 1.2 kg (see Table 2.9)? (2.7)

2.118 What is the volume, in quarts, of 3.40 kg of ethanol (see Table 2.9)? (2.7)

Applications

2.119 The following nutrition information is listed on a box of crackers: (2.6)

Serving size 0.50 oz (6 crackers)

Fat 4 g per serving; Sodium 140 mg per serving

a. If the box has a net weight (contents only) of 8.0 oz, about how many crackers are in the box?

b. If you ate 10 crackers, how many ounces of fat did you consume?

c. How many servings of crackers in part **a** would it take to obtain the Daily Value (DV) for sodium, which is 2400 mg?

2.120 A dialysis unit requires 75 000 mL of distilled water. How many gallons of water are needed? (2.6)

2.121 To prevent bacterial infection, a doctor orders 4 tablets of amoxicillin per day for 10 days. If each tablet contains 250 mg of amoxicillin, how many ounces of the medication are given in 10 days? (2.6)

2.122 Celeste's diet restricts her intake of protein to 24 g per day. If she eats 1.2 oz of protein, has she exceeded her protein limit for the day? (2.6)

2.123 A doctor orders 5.0 mL of phenobarbital elixir. If the phenobarbital elixir is available as 30. mg per 7.5 mL, how many milligrams is given to the patient? (2.6)

2.124 A doctor orders 2.0 mg of morphine. The vial of morphine on hand is 10. mg/mL. How many milliliters of morphine should you administer to the patient? (2.6)

CHALLENGE QUESTIONS

The following groups of questions are related to the topics in this chapter. However, they do not all follow the chapter order, and they require you to combine concepts and skills from several sections. These questions will help you increase your critical thinking skills and prepare for your next exam.

2.125 A balance measures mass to 0.001 g. If you determine the mass of an object that weighs about 31 g would you record the mass as 31 g, 31.1 g, 31.08 g, 31.075 g, or 31.0750? Explain your choice by writing two to three complete sentences that describe your thinking. (2.3)

2.126 When three students use the same meterstick to measure the length of a paper clip, they obtain results of 5.8 cm, 5.75 cm, and 5.76 cm. If the meterstick has millimeter markings, what are some reasons for the different values? (2.3)

2.127 A car travels at 55 mi/h and gets 11 km/L of gasoline. How many gallons of gasoline are needed for a 3.0-h trip? (2.6)

2.128 A sunscreen preparation contains 2.50% benzyl salicylate by mass. If a tube contains 4.0 oz of sunscreen, how many kilograms of benzyl salicylate are needed to manufacture 325 tubes of sunscreen? (2.6)

2.129 How many cubic centimeters of olive oil have the same mass as 1.50 L of gasoline (see Table 2.9)? (2.7)

2.130 A 50.0-g silver object and a 50.0-g gold object are both added to 75.5 mL of water contained in a graduated cylinder. What is the new water level, in milliliters, in the cylinder (see Table 2.9)? (2.7)

2.131 In the manufacturing of computer chips, cylinders of silicon are cut into thin wafers that are 3.00 in. in diameter and have a mass of 1.50 g of silicon. How thick, in millimeters, is each wafer if silicon has a density of 2.33 g/cm^3? (The volume of a cylinder is $V = \pi r^2 h$.) (2.6, 2.7)

2.132 A package of aluminum foil is 66.7 yd long, 12 in. wide, and 0.000 30 in. thick. If aluminum has a density of 2.70 g/cm^3, what is the mass, in grams, of the foil? (2.6, 2.7)

Applications

2.133 For a 180-lb person, calculate the quantity of each of the following that must be ingested to provide the LD$_{50}$ for caffeine given in Table 2.8: (2.6)

a. cups of coffee if one cup is 12 fl oz and there is 100. mg of caffeine per 6 fl oz of drip-brewed coffee

b. cans of cola if one can contains 50. mg of caffeine

c. tablets of NoDoz if one tablet contains 200. mg of caffeine

2.134 A dietary supplement contains the following components. If a person uses the dietary supplement once a day, how many milligrams of each component would that person consume in one week? (2.6)

a. calcium 0.20 g

b. iron 0.50 mg

c. iodine 53 mcg

d. chromium 45 mcg

e. threonine 0.285 g

2.135 a. Some athletes have as little as 3.0% body fat. If such a person has a body mass of 65 kg, how many pounds of body fat does that person have? (2.6)

b. In liposuction, a doctor removes fat deposits from a person's body. If body fat has a density of 0.94 g/mL and 3.0 L of fat is removed, how many pounds of fat were removed from the patient?

2.136 A mouthwash is 21.6% ethanol by mass. If each bottle contains 0.358 pt of mouthwash with a density of 0.876 g/mL, how many kilograms of ethanol are in 180 bottles of the mouthwash? (2.6, 2.7)

A mouthwash may contain over 20% ethanol.

ANSWERS

Answers to Selected Questions and Problems

2.1 a. g   **b.** L   **c.** °C   **d.** lb   **e.** s

2.3 a. volume   **b.** length   **c.** length   **d.** time

2.5 a. meter, length   **b.** gram, mass
   c. milliliter, volume   **d.** second, time
   e. degree Celsius, temperature

2.7 a. second, time   **b.** kilogram, mass
   c. gram, mass   **d.** degree Celsius, temperature

2.9 a. the 6 in the tenths place (0.6)
   b. the 5 in the hundredths place (0.05)
   c. the 9 in the ones place

2.11 a. measured   **b.** exact
   c. exact   **d.** measured

2.13 a. 6 oz   **b.** none
   c. 0.75 lb, 350 g   **d.** none (definitions are exact)

2.15 a. not significant   **b.** significant
   c. significant   **d.** significant
   e. not significant

2.17 a. five SFs   **b.** two SFs   **c.** two SFs
   d. three SFs   **e.** four SFs   **f.** three SFs

2.19 b and **c**

2.21 a. 5.0×10^3 L   **b.** 3.0×10^4 g
   c. 1.0×10^5 m   **d.** 2.5×10^{-4} cm

2.23 a. 4   **b.** 2   **c.** 3

2.25 A calculator often gives more digits than the number of significant figures allowed in the answer.

2.27 a. 1.85 kg   **b.** 88.2 L
   c. 0.004 74 cm   **d.** 8810 m
   e. 1.83×10^5 s

2.29 a. 56.9 m   **b.** 0.002 28 g
   c. 11 500 s (1.15×10^4 s)   **d.** 8.10 L

2.31 a. 1.6   **b.** 0.01
   c. 27.6   **d.** 3.5
   e. 0.14 (1.4×10^{-1})   **f.** 0.8 (8×10^{-1})

2.33 a. 53.54 cm   **b.** 127.6 g
   c. 121.5 mL   **d.** 0.50 L

2.35 km/h is kilometers per hour; mi/h is miles per hour.

2.37 a. mg   **b.** dL   **c.** km   **d.** fg

2.39 a. centiliter   **b.** kilogram
   c. millisecond   **d.** gigameter

2.41 a. 0.01   **b.** 10^{12} **c.** 0.001   **d.** 0.1

2.43 a. decigram   **b.** microgram
   c. kilogram   **d.** centigram

2.45 a. 100 cm   **b.** 1×10^9 nm
   c. 0.001 m   **d.** 1000 mL

2.47 a. kilogram   **b.** milliliter   **c.** km
   d. kL   **e.** nanometer

2.49 A conversion factor can be inverted to give a second conversion factor.

2.51 a. 1 m = 100 cm; $\dfrac{100 \text{ cm}}{1 \text{ m}}$ and $\dfrac{1 \text{ m}}{100 \text{ cm}}$

   b. 1 g = 1×10^9 ng; $\dfrac{1 \times 10^9 \text{ ng}}{1 \text{ g}}$ and $\dfrac{1 \text{ g}}{1 \times 10^9 \text{ ng}}$

   c. 1 kL = 1000 L; $\dfrac{1000 \text{ L}}{1 \text{ kL}}$ and $\dfrac{1 \text{ kL}}{1000 \text{ L}}$

   d. 1 s = 1000 ms; $\dfrac{1000 \text{ ms}}{1 \text{ s}}$ and $\dfrac{1 \text{ s}}{1000 \text{ ms}}$

   e. $(1 \text{ m})^3 = (100 \text{ cm})^3$; $\dfrac{(100 \text{ cm})^3}{(1 \text{ m})^3}$ and $\dfrac{(1 \text{ m})^3}{(100 \text{ cm})^3}$

2.53 a. 1 yd = 3 ft; $\dfrac{3 \text{ ft}}{1 \text{ yd}}$ and $\dfrac{1 \text{ yd}}{3 \text{ ft}}$

   The 1 yd and 3 ft are both exact.

   b. 1 kg = 2.205 lb; $\dfrac{2.205 \text{ lb}}{1 \text{ kg}}$ and $\dfrac{1 \text{ kg}}{2.205 \text{ lb}}$

   The 2.205 lb is measured: It has four SFs. The 1 kg is exact.

   c. 1 min = 60 s; $\dfrac{60 \text{ s}}{1 \text{ min}}$ and $\dfrac{1 \text{ min}}{60 \text{ s}}$

   The 1 min and 60 s are both exact.

   d. 1 gal = 27 mi; $\dfrac{27 \text{ mi}}{1 \text{ gal}}$ and $\dfrac{1 \text{ gal}}{27 \text{ mi}}$

   The 27 mi is measured: It has two SFs. The 1 gal is exact.

   e. 93 g of silver = 100 g of sterling;

   $\dfrac{93 \text{ g silver}}{100 \text{ g sterling}}$ and $\dfrac{100 \text{ g sterling}}{93 \text{ g silver}}$

   The 93 g is measured: It has two SFs. The 100 g is exact.

2.55 a. 3.5 m = 1 s; $\dfrac{3.5 \text{ m}}{1 \text{ s}}$ and $\dfrac{1 \text{ s}}{3.5 \text{ m}}$

   The 3.5 m is measured: It has two SFs. The 1 s is exact.

   b. 4700 mg of potassium = 1 day;

   $\dfrac{4700 \text{ mg potassium}}{1 \text{ day}}$ and $\dfrac{1 \text{ day}}{4700 \text{ mg potassium}}$

   The 4700 mg is measured: It has two SFs. The 1 day is exact.

   c. 46.0 km = 1 gal; $\dfrac{46.0 \text{ km}}{1 \text{ gal}}$ and $\dfrac{1 \text{ gal}}{46.0 \text{ km}}$

   The 46.0 km is measured: It has three SFs. The 1 gal is exact.

   d. 29 mcg of pesticide = 1 kg of plums;

   $\dfrac{29 \text{ mcg pesticide}}{1 \text{ kg plums}}$ and $\dfrac{1 \text{ kg plums}}{29 \text{ mcg pesticide}}$

   The 29 mcg is measured: It has two SFs. The 1 kg is exact.

   e. 28.2 g of silicon = 100 g of crust; $\dfrac{28.2 \text{ g silicon}}{100 \text{ g crust}}$ and $\dfrac{100 \text{ g crust}}{28.2 \text{ g silicon}}$

   The 28.2 g is measured: It has three SFs. The 100 g is exact.

2.57 a. 1 tablet = 630 mg of calcium;

   $\dfrac{630 \text{ mg calcium}}{1 \text{ tablet}}$ and $\dfrac{1 \text{ tablet}}{630 \text{ mg calcium}}$

   The 630 mg is measured: It has two SFs. The 1 tablet is exact.

   b. 60 mg of vitamin C = 1 day;

   $\dfrac{60 \text{ mg vitamin C}}{1 \text{ day}}$ and $\dfrac{1 \text{ day}}{60 \text{ mg vitamin C}}$

   The 60 mg is measured: It has one SF. The 1 day is exact.

   c. 1 tablet = 50 mg of atenolol;

   $\dfrac{50 \text{ mg atenolol}}{1 \text{ tablet}}$ and $\dfrac{1 \text{ tablet}}{50 \text{ mg atenolol}}$

   The 50 mg is measured: It has one SF. The 1 tablet is exact.

d. 1 tablet = 81 mg of aspirin;

$$\frac{81 \text{ mg aspirin}}{1 \text{ tablet}} \quad \text{and} \quad \frac{1 \text{ tablet}}{81 \text{ mg aspirin}}$$

The 81 mg is measured: It has two SFs. The 1 tablet is exact.

2.59 a. 5 mL of syrup = 10 mg of Atarax;

$$\frac{10 \text{ mg Atarax}}{5 \text{ mL syrup}} \quad \text{and} \quad \frac{5 \text{ mL syrup}}{10 \text{ mg Atarax}}$$

b. 1 tablet = 0.25 g of Lanoxin;

$$\frac{0.25 \text{ g Lanoxin}}{1 \text{ tablet}} \quad \text{and} \quad \frac{1 \text{ tablet}}{0.25 \text{ g Lanoxin}}$$

c. 1 tablet = 300 mg of Motrin;

$$\frac{300 \text{ mg Motrin}}{1 \text{ tablet}} \quad \text{and} \quad \frac{1 \text{ tablet}}{300 \text{ mg Motrin}}$$

2.61 The unit in the denominator must cancel with the preceding unit in the numerator.

2.63 a. 0.0442 L   **b.** 8.65×10^9 nm
    c. 5.2×10^2 Mg   **d.** 7.2×10^5 ms

2.65 a. 1.555 kg   **b.** 63 in.
    c. 4.4 qt   **d.** 205 mm

2.67 a. 1.75 m   **b.** 5 L
    c. 5.5 g   **d.** 3.5×10^{-3} m^3

2.69 a. 473 mL   **b.** 79.4 kg
    c. 24 lb   **d.** 43 g

2.71 a. 66 gal   **b.** 3 tablets
    c. 1800 mg   **d.** 110 g

2.73 a. 6.3 h   **b.** 2.5 mL

2.75 a. 260 g   **b.** 0.5 g   **c.** 88 g

2.77 a. 1.20 g/mL   **b.** 0.870 g/mL
    c. 3.10 g/mL   **d.** 1.28 g/mL

2.79 a. 4.51 g/mL   **b.** 1.42 g/mL   **c.** 2.70 g/mL

2.81 a. 1.91 L of ethanol   **b.** 88 g of mercury
    c. 83.3 oz of silver

2.83 a. 661 g   **b.** 34 kg   **c.** 29 cm^3

2.85 Since we calculate the density to be 11.3 g/cm^3, we identify the metal as lead.

2.87 a. 1.03   **b.** 1.03 g/mL
    c. 690 g   **d.** 382 mL

2.89 a. 42 mcg of iron = 1 dL of blood;

$$\frac{42 \text{ mcg iron}}{1 \text{ dL blood}} \quad \text{and} \quad \frac{1 \text{ dL blood}}{42 \text{ mcg iron}}$$

b. 3.4 mcg of iron

2.91 c and **d**

2.93 a. exact   **b.** measured
    c. exact   **d.** measured

2.95 61.5 °C

2.97 a. 97.5 cm   **b.** 61.5 cm
    c. three SFs   **d.** 6.00×10^3 cm^2

2.99 a. Diagram 3; a cube that has a greater density than the water will sink to the bottom.

    b. Diagram 4; a cube with a density of 0.80 g/mL will be about two-thirds submerged in the water.

    c. Diagram 1; a cube with a density that is one-half the density of water will be one-half submerged in the water.

    d. Diagram 2; a cube with the same density as water will float just at the surface of the water.

2.101 A would be gold; it has the highest density (19.3 g/mL) and the smallest volume.

    B would be silver; its density is intermediate (10.5 g/mL) and the volume is intermediate.

    C would be aluminum; it has the lowest density (2.70 g/mL) and the largest volume.

2.103 The green cube has the same volume as the gray cube. However, the green cube has a larger mass on the scale, which means that its mass/volume ratio is larger. Thus, the density of the green cube is higher than the density of the gray cube.

2.105 a. 0.000 012 6 L (1.26×10^{-5} L)
    b. 353 kg (3.53×10^2 kg)
    c. 125 000 m (1.25×10^5 m)
    d. 35.0 s

2.107 a. 265 g   **b.** 0.584 lb

2.109 $1.01 per lb

2.111 16 onions

2.113 42 min

2.115 790 g

2.117 1600 cm^3 (1.6×10^3 cm^3)

2.119 a. 96 crackers   **b.** 0.2 oz of fat
    c. 17 servings

2.121 0.35 oz

2.123 20. mg

2.125 You would record the mass as 31.075 g. Since the balance will weigh to the nearest 0.001 g, the mass value would be reported to 0.001 g.

2.127 6.4 gal

2.129 1200 cm^3

2.131 0.141 mm

2.133 a. 78 cups   **b.** 310 cans   **c.** 80 tablets

2.135 a. 4.3 lb of body fat   **b.** 6.2 lb

Matter and Energy

CHARLES IS 13 YEARS OLD and overweight. His doctor is worried that Charles is at risk for type 2 diabetes and advises his mother to make an appointment with a dietitian. Daniel, a dietitian, explains to them that choosing the appropriate foods is important to living a healthy lifestyle, losing weight, and preventing or managing diabetes.

Daniel also explains that food contains potential or stored energy and different foods contain different amounts of potential energy. For instance, carbohydrates contain 4 kcal/g (17 kJ/g) whereas fats contain 9 kcal/g (38 kJ/g). He then explains that diets high in fat require more exercise to burn the fats, as they contain more potential energy. When Daniel looks at Charles' typical daily diet, he calculates that Charles is obtaining 3800 kcal for one day. The American Heart Association recommends 1800 kcal for boys 9 to 13 years of age. Daniel encourages Charles and his mother to include whole grains, fruits, and vegetables in their diet instead of foods high in fat. They also discuss food labels and the fact that smaller serving sizes of healthy foods are necessary to lose weight. Daniel also recommends that Charles participate in at least 60 min of exercise every day. Before leaving, Charles and his mother are given a menu for the following two weeks, as well as a diary to keep track of what, and how much, they actually consume.

CAREER

Dietitian

Dietitians specialize in helping individuals learn about good nutrition and the need for a balanced diet. This requires them to understand biochemical processes, the importance of vitamins, and food labels, as well as the differences between carbohydrates, fats, and proteins in terms of their energy value and how they are metabolized. Dietitians work in a variety of environments, including hospitals, nursing homes, school cafeterias, and public health clinics. In these environments, they create specialized diets for individuals diagnosed with a specific disease or create meal plans for those in a nursing home.

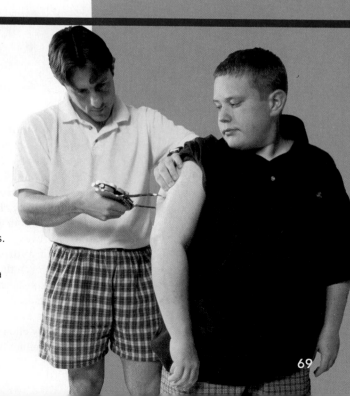

 **KEY MATH SKILLS**

- Using Positive and Negative Numbers in Calculations (1.4)
- Solving Equations (1.4)
- Interpreting Graphs (1.4)
- Converting between Standard Numbers and Scientific Notation (1.5)
- Rounding Off (2.3)

 **CORE CHEMISTRY SKILLS**

- Counting Significant Figures (2.2)
- Using Significant Figures in Calculations (2.3)
- Writing Conversion Factors from Equalities (2.5)
- Using Conversion Factors (2.6)

*These Key Math Skills and Core Chemistry Skills from previous chapters are listed here for your review as you proceed to the new material in this chapter.

LOOKING AHEAD

 CORE CHEMISTRY SKILL
Classifying Matter

A molecule of water, H_2O, consists of two atoms of hydrogen (white) for one atom of oxygen (red).

A molecule of hydrogen peroxide, H_2O_2, consists of two atoms of hydrogen (white) for every two atoms of oxygen (red).

ENGAGE

Why are elements and compounds both pure substances?

3.1 Classification of Matter

LEARNING GOAL Classify examples of matter as pure substances or mixtures.

Matter is anything that has mass and occupies space. Matter is everywhere around us: the orange juice we had for breakfast, the water we put in the coffee maker, the plastic bag we put our sandwich in, our toothbrush and toothpaste, the oxygen we inhale, and the carbon dioxide we exhale. To a scientist, all of this material is matter. The different types of matter are classified by their composition.

Pure Substances: Elements and Compounds

A **pure substance** is matter that has a fixed or definite composition. There are two kinds of pure substances: *elements* and *compounds*. An **element**, the simplest type of a pure substance, is composed of only one type of material such as silver, iron, or aluminum. Every element is composed of *atoms*, which are extremely tiny particles that make up each type of matter. Silver is composed of silver atoms, iron of iron atoms, and aluminum of aluminum atoms. A full list of the elements is found on the inside front cover of this text.

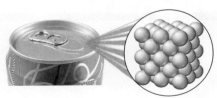

An aluminum can consists of many atoms of aluminum.

A **compound** is also a pure substance, but it consists of atoms of two or more elements always chemically combined in the same proportion. In compounds, the atoms are held together by attractions called *bonds*, which form small groups of atoms called *molecules*. For example, a molecule of the compound water has two hydrogen atoms for every one oxygen atom and is represented by the formula H_2O. This means that water found anywhere always has the same composition of H_2O. Another compound that consists of a chemical combination of hydrogen and oxygen is hydrogen peroxide. It has two hydrogen atoms for every two oxygen atoms and is represented by the formula H_2O_2. Thus, water (H_2O) and hydrogen peroxide (H_2O_2) are different compounds that have different properties even though they contain the same elements, hydrogen and oxygen.

Pure substances that are compounds can be broken down by chemical processes into their elements. They cannot be broken down through physical methods such as boiling or sifting. For example, ordinary table salt consists of the compound NaCl, which can be separated by chemical processes into sodium metal and chlorine gas, as seen in **FIGURE 3.1**. Elements cannot be broken down further.

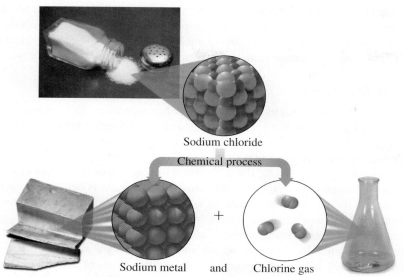

FIGURE 3.1 ▶ The decomposition of salt, NaCl, produces the elements sodium and chlorine.

Q How do elements and compounds differ?

Mixtures

In a **mixture**, two or more different substances are physically mixed, but not chemically combined. Much of the matter in our everyday lives consists of mixtures. The air we breathe is a mixture of mostly oxygen and nitrogen gases. The steel in buildings and railroad tracks is a mixture of iron, nickel, carbon, and chromium. The brass in doorknobs and fixtures is a mixture of zinc and copper (see **FIGURE 3.2**). Tea, coffee, and ocean water are mixtures too. Unlike compounds, the proportions of substances in a mixture are not consistent but can vary. For example, two sugar–water mixtures may look the same, but the one with the higher ratio of sugar to water would taste sweeter.

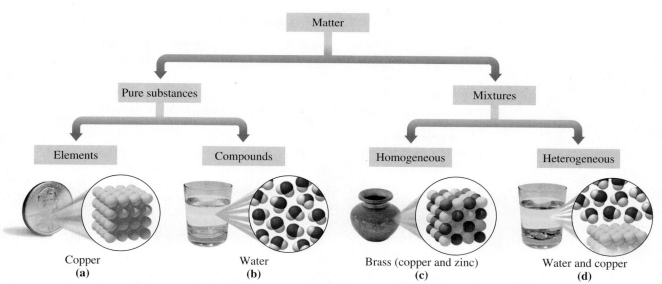

FIGURE 3.2 ▶ Matter is organized by its components: elements, compounds, and mixtures. **(a)** The element copper consists of copper atoms. **(b)** The compound water consists of H_2O molecules. **(c)** Brass is a homogeneous mixture of copper and zinc atoms. **(d)** Copper metal in water is a heterogeneous mixture of copper atoms and H_2O molecules.

Q Why are copper and water pure substances, but brass is a mixture?

FIGURE 3.3 ▶ A mixture of spaghetti and water is separated using a strainer, a physical method of separation.

ⓠ Why can physical methods be used to separate mixtures but not compounds?

Physical processes can be used to separate mixtures because there are no chemical interactions between the components. For example, different coins, such as nickels, dimes, and quarters, can be separated by size; iron particles mixed with sand can be picked up with a magnet; and water is separated from cooked spaghetti by using a strainer (see **FIGURE 3.3**).

Types of Mixtures

Mixtures are classified further as homogeneous or heterogeneous. In a *homogeneous mixture*, also called a *solution*, the composition is uniform throughout the sample. Familiar examples of homogeneous mixtures are air, which contains oxygen and nitrogen gases, and seawater, a solution of salt and water.

In a *heterogeneous mixture*, the components do not have a uniform composition throughout the sample. For example, a mixture of oil and water is heterogeneous because the oil floats on the surface of the water. Other examples of heterogeneous mixtures are a cookie with raisins and orange juice with pulp.

In the chemistry laboratory, mixtures are separated by various methods. Solids are separated from liquids by *filtration*, which involves pouring a mixture through a filter paper set in a funnel. In *chromatography*, different components of a liquid mixture separate as they move at different rates up the surface of a piece of chromatography paper.

(a) A mixture of a liquid and a solid is separated by filtration. **(b)** Different substances are separated as they travel at different rates up the surface of chromatography paper.

(a)   **(b)**

SAMPLE PROBLEM 3.1 Classifying Mixtures

Classify each of the following as a pure substance (element or compound) or a mixture (homogeneous or heterogeneous):

a. copper in copper wire
b. a chocolate-chip cookie
c. nitrox, a combination of oxygen and nitrogen used to fill scuba tanks

 **TRY IT FIRST**

SOLUTION

a. Copper is an element, which is a pure substance.
b. A chocolate-chip cookie does not have a uniform composition, which makes it a heterogeneous mixture.
c. The gases oxygen and nitrogen have a uniform composition in nitrox, which makes it a homogeneous mixture.

STUDY CHECK 3.1

A salad dressing is prepared with oil, vinegar, and chunks of blue cheese. Is this a homogeneous or heterogeneous mixture?

ANSWER

heterogeneous mixture

CHEMISTRY LINK TO **HEALTH**
Breathing Mixtures

The air we breathe is composed mostly of the gases oxygen (21%) and nitrogen (79%). The homogeneous breathing mixtures used by scuba divers differ from the air we breathe depending on the depth of the dive. Nitrox is a mixture of oxygen and nitrogen, but with more oxygen gas (up to 32%) and less nitrogen gas (68%) than air. A breathing mixture with less nitrogen gas decreases the risk of *nitrogen narcosis* associated with breathing regular air while diving. Heliox contains oxygen and helium, which is typically used for diving to more than 200 ft. By replacing nitrogen with helium, nitrogen narcosis does not occur. However, at dive depths over 300 ft, helium is associated with severe shaking and body temperature drop.

A breathing mixture used for dives over 400 ft is trimix, which contains oxygen, helium, and some nitrogen. The addition of some nitrogen lessens the problem of shaking that comes with breathing high levels of helium. Heliox and tri-mix are used only by professional, military, or other highly trained divers.

In hospitals, heliox may be used as a treatment for respiratory disorders and lung constriction in adults and premature infants. Heliox is less dense than air, which reduces the effort of breathing and helps distribute the oxygen gas to the tissues.

A nitrox mixture is used to fill scuba tanks.

QUESTIONS AND PROBLEMS

3.1 Classification of Matter

LEARNING GOAL Classify examples of matter as pure substances or mixtures.

3.1 Classify each of the following as a pure substance or a mixture:
  a. baking soda ($NaHCO_3$)   **b.** a blueberry muffin
  c. ice (H_2O)   **d.** zinc (Zn)
  e. trimix (oxygen, nitrogen, and helium) in a scuba tank

3.2 Classify each of the following as a pure substance or a mixture:
  a. a soft drink   **b.** propane (C_3H_8)
  c. a cheese sandwich   **d.** an iron (Fe) nail
  e. salt substitute (KCl)

3.3 Classify each of the following pure substances as an element or a compound:
  a. a silicon (Si) chip   **b.** hydrogen peroxide (H_2O_2)
  c. oxygen (O_2)   **d.** rust (Fe_2O_3)
  e. methane (CH_4) in natural gas

3.4 Classify each of the following pure substances as an element or a compound:
  a. helium gas (He)   **b.** sulfur (S)
  c. sugar ($C_{12}H_{22}O_{11}$)   **d.** mercury (Hg) in a
  e. lye (NaOH)   thermometer

3.5 Classify each of the following mixtures as homogeneous or heterogeneous:
  a. vegetable soup   **b.** seawater
  c. tea   **d.** tea with ice and lemon slices
  e. fruit salad

3.6 Classify each of the following mixtures as homogeneous or heterogeneous:
  a. nonfat milk   **b.** chocolate-chip ice cream
  c. gasoline   **d.** peanut butter sandwich
  e. cranberry juice

3.2 States and Properties of Matter

LEARNING GOAL Identify the states and the physical and chemical properties of matter.

On Earth, matter exists in one of three *physical forms* called the **states of matter**: *solids*, *liquids*, and *gases*. Water is a familiar example that we routinely observe in all three states. In the solid state, water can be an ice cube or a snowflake. It is a liquid when it comes out of a faucet or fills a pool. Water forms a gas, or vapor, when it evaporates from wet clothes or boils in a pan. A **solid**, such as a pebble or a baseball, has a definite shape and volume. You can probably recognize several solids within your reach right now such as books, pencils, or a computer mouse. In a solid, strong attractive forces hold the particles such as atoms or molecules close together. The particles in a solid are arranged in such a rigid pattern, their only movement is to vibrate slowly in fixed positions. For many solids, this rigid structure produces a crystal such as that seen in amethyst.

Amethyst, a solid, is a purple form of quartz (SiO_2).

A **liquid** has a definite volume, but not a definite shape. In a liquid, the particles move in random directions but are sufficiently attracted to each other to maintain a definite volume, although not a rigid structure. Thus, when water, oil, or vinegar is poured from one container to another, the liquid maintains its own volume but takes the shape of the new container.

A **gas** does not have a definite shape or volume. In a gas, the particles are far apart, have little attraction to each other, and move at high speeds, taking the shape and volume of their container. When you inflate a bicycle tire, the air, which is a gas, fills the entire volume of the tire. The propane gas in a tank fills the entire volume of the tank. **TABLE 3.1** compares the three states of matter.

Water as a liquid takes the shape of its container.

A gas takes the shape and volume of its container.

TABLE 3.1 A Comparison of Solids, Liquids, and Gases

Characteristic	Solid	Liquid	Gas
Shape	Has a definite shape	Takes the shape of the container	Takes the shape of the container
Volume	Has a definite volume	Has a definite volume	Fills the volume of the container
Arrangement of Particles	Fixed, very close	Random, close	Random, far apart
Interaction between Particles	Very strong	Strong	Essentially none
Movement of Particles	Very slow	Moderate	Very fast
Examples	Ice, salt, iron	Water, oil, vinegar	Water vapor, helium, air

TABLE 3.2 Some Physical Properties of Copper

State at 25 °C	Solid
Color	Orange-red
Odor	Odorless
Melting Point	1083 °C
Boiling Point	2567 °C
Luster	Shiny
Conduction of Electricity	Excellent
Conduction of Heat	Excellent

Physical Properties and Physical Changes

One way to describe matter is to observe its properties. For example, if you were asked to describe yourself, you might list your characteristics such as the color of your eyes and skin or the length, color, and texture of your hair.

Physical properties are those characteristics that can be observed or measured without affecting the identity of a substance. In chemistry, typical physical properties include the shape, color, melting point, boiling point, and physical state of a substance. For example, a penny has the physical properties of a round shape, an orange-red color, solid state, and a shiny luster. **TABLE 3.2** gives more examples of physical properties of copper found in pennies, electrical wiring, and copper pans.

Water is a substance that is commonly found in all three states: solid, liquid, and gas. When matter undergoes a **physical change**, its state, size, or its appearance will change, but its composition remains the same. The solid state of water, snow or ice, has a different appearance than its liquid or gaseous state, but all three states are water.

Copper, used in cookware, is a good conductor of heat.

The evaporation of water from seawater gives white, solid crystals of salt (sodium chloride).

The physical appearance of a substance can change in other ways too. Suppose that you dissolve some salt in water. The appearance of the salt changes, but you could re-form the salt crystals by evaporating the water. Thus, in a physical change of state, no new substances are produced.

Chemical Properties and Chemical Changes

Chemical properties are those that describe the ability of a substance to change into a new substance. When a **chemical change** takes place, the original substance is converted into one or more new substances, which have different physical and chemical properties. For example, the rusting or corrosion of a metal, such as iron, is a chemical property. In the rain, an iron (Fe) nail undergoes a chemical change when it reacts with oxygen (O_2) to form rust (Fe_2O_3). A chemical change has taken place: Rust is a new substance with new physical and chemical properties. **TABLE 3.3** gives examples of some physical and chemical changes. **TABLE 3.4** summarizes physical and chemical properties and changes.

ENGAGE

Why is bending an iron nail a physical change?

CORE CHEMISTRY SKILL
Identifying Physical and Chemical Changes

TABLE 3.3 Examples of Some Physical and Chemical Changes

Physical Changes	Chemical Changes
Water boils to form water vapor.	Shiny, silver metal reacts in air to give a black, grainy coating.
Copper is drawn into thin copper wires.	A piece of wood burns with a bright flame and produces heat, ashes, carbon dioxide, and water vapor.
Sugar dissolves in water to form a solution.	Heating white, granular sugar forms a smooth, caramel-colored substance.
Paper is cut into tiny pieces of confetti.	Iron, which is gray and shiny, combines with oxygen to form orange-red rust.

Flan has a topping of caramelized sugar.

TABLE 3.4 Summary of Physical and Chemical Properties and Changes

	Physical	Chemical
Property	A characteristic of a substance: color, shape, odor, luster, size, melting point, or density.	A characteristic that indicates the ability of a substance to form another substance: paper can burn, iron can rust, silver can tarnish.
Change	A change in a physical property that retains the identity of the substance: a change of state, a change in size, or a change in shape.	A change in which the original substance is converted to one or more new substances: paper burns, iron rusts, silver tarnishes.

A gold ingot is hammered to form gold leaf.

Interactive Video

Chemical vs. Physical Changes

SAMPLE PROBLEM 3.2 Physical and Chemical Changes

Classify each of the following as a physical or chemical change:

a. A gold ingot is hammered to form gold leaf.
b. Gasoline burns in air.
c. Garlic is chopped into small pieces.

TRY IT FIRST

SOLUTION

a. A physical change occurs when the gold ingot changes shape.
b. A chemical change occurs when gasoline burns and forms different substances with new properties.
c. A physical change occurs when the size of the garlic pieces changes.

STUDY CHECK 3.2

Classify each of the following as a physical or chemical change:

a. Water freezes on a pond.
b. Gas bubbles form when baking powder is placed in vinegar.
c. A log is cut for firewood.

ANSWER

a. physical change   **b.** chemical change   **c.** physical change

QUESTIONS AND PROBLEMS

3.2 States and Properties of Matter

LEARNING GOAL Identify the states and the physical and chemical properties of matter.

3.7 Indicate whether each of the following describes a gas, a liquid, or a solid:
  a. The breathing mixture in a scuba tank has no definite volume or shape.
  b. The neon atoms in a lighting display do not interact with each other.
  c. The particles in an ice cube are held in a rigid structure.

3.8 Indicate whether each of the following describes a gas, a liquid, or a solid:
  a. Lemonade has a definite volume but takes the shape of its container.
  b. The particles in a tank of oxygen are very far apart.
  c. Helium occupies the entire volume of a balloon.

3.9 Describe each of the following as a physical or chemical property:
  a. Chromium is a steel-gray solid.
  b. Hydrogen reacts readily with oxygen.
  c. A patient has a temperature of 40.2 °C.
  d. Milk will sour when left in a warm room.
  e. Butane gas in an igniter burns in oxygen.

3.10 Describe each of the following as a physical or chemical property:
  a. Neon is a colorless gas at room temperature.
  b. Apple slices turn brown when they are exposed to air.
  c. Phosphorus will ignite when exposed to air.
  d. At room temperature, mercury is a liquid.
  e. Propane gas is compressed to a liquid for placement in a small cylinder.

3.11 What type of change, physical or chemical, takes place in each of the following?
  a. Water vapor condenses to form rain.
  b. Cesium metal reacts explosively with water.
  c. Gold melts at 1064 °C.
  d. A puzzle is cut into 1000 pieces.
  e. Cheese is grated.

3.12 What type of change, physical or chemical, takes place in each of the following?
  a. Pie dough is rolled into thin pieces for a crust.
  b. A silver pin tarnishes in the air.
  c. A tree is cut into boards at a saw mill.
  d. Food is digested.
  e. A chocolate bar melts.

3.13 Describe each property of the element fluorine as physical or chemical.
  a. is highly reactive
  b. is a gas at room temperature
  c. has a pale, yellow color
  d. will explode in the presence of hydrogen
  e. has a melting point of −220 °C

3.14 Describe each property of the element zirconium as physical or chemical.
  a. melts at 1852 °C
  b. is resistant to corrosion
  c. has a grayish white color
  d. ignites spontaneously in air when finely divided
  e. is a shiny metal

3.3 Temperature

LEARNING GOAL Given a temperature, calculate a corresponding temperature on another scale.

The small particles, atoms and molecules, in matter are constantly moving due to heat or thermal energy. At higher temperatures, they move faster; at lower temperatures, they move slower.

Temperatures in science are measured and reported in *Celsius* (°C) units. On the Celsius scale, the reference points are the freezing point of water, defined as 0 °C, and the boiling point, 100 °C. In the United States, everyday temperatures are commonly reported in *Fahrenheit* (°F) units. On the Fahrenheit scale, water freezes at 32 °F and boils at 212 °F. A typical room temperature of 22 °C would be the same as 72 °F. Normal human body temperature is 37.0 °C, which is the same temperature as 98.6 ° F.

On the Celsius and Fahrenheit temperature scales, the temperature difference between freezing and boiling is divided into smaller units called *degrees*. On the Celsius scale, there are 100 degrees Celsius between the freezing and boiling points of water, whereas the Fahrenheit scale has 180 degrees Fahrenheit between the freezing and boiling points of water. That makes a degree Celsius almost twice the size of a degree Fahrenheit: $1\,°C = 1.8\,°F$ (see **FIGURE 3.4**).

$$180 \text{ degrees Fahrenheit} = 100 \text{ degrees Celsius}$$

$$\frac{180 \text{ degrees Fahrenheit}}{100 \text{ degrees Celsius}} = \frac{1.8\,°F}{1\,°C}$$

We can write a temperature equation that relates a Fahrenheit temperature and its corresponding Celsius temperature.

$$T_F = 1.8(T_C) + 32$$

Changes   Adjusts
°C to °F   freezing
        point

Temperature equation to obtain degrees Fahrenheit

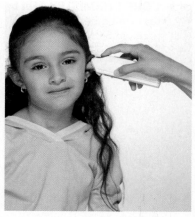

A digital ear thermometer is used to measure body temperature.

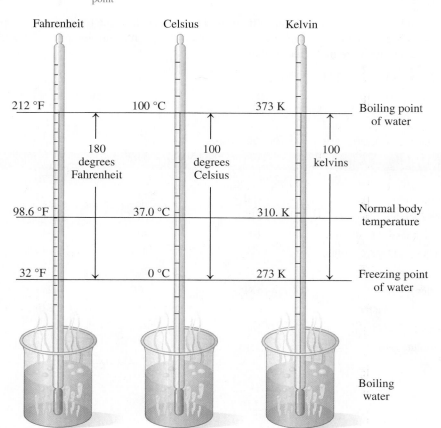

Fahrenheit	Celsius	Kelvin	
212 °F	100 °C	373 K	Boiling point of water
180 degrees Fahrenheit	100 degrees Celsius	100 kelvins	
98.6 °F	37.0 °C	310. K	Normal body temperature
32 °F	0 °C	273 K	Freezing point of water

Boiling water

FIGURE 3.4 ◄ A comparison of the Fahrenheit, Celsius, and Kelvin temperature scales between the freezing and boiling points of water.

Q What is the difference in the freezing points of water on the Celsius and Fahrenheit temperature scales?

ENGAGE

Why is a degree Celsius a larger unit of temperature than a degree Fahrenheit?

CORE CHEMISTRY SKILL
Converting between Temperature Scales

In this equation, the Celsius temperature is multiplied by 1.8 to change °C to °F; then 32 is added to adjust the freezing point from 0 °C to the Fahrenheit freezing point, 32 °F. The values, 1.8 and 32, used in the temperature equation are exact numbers and are not used to determine significant figures in the answer.

To convert from degrees Fahrenheit to degrees Celsius, the temperature equation is rearranged to solve for T_C. First, we subtract 32 from both sides since we must apply the same operation to both sides of the equation.

$$T_F - 32 = 1.8(T_C) + \cancel{32} - \cancel{32}$$
$$T_F - 32 = 1.8(T_C)$$

Second, we solve the equation for T_C by dividing both sides by 1.8.

$$\frac{T_F - 32}{1.8} = \frac{\cancel{1.8}(T_C)}{\cancel{1.8}}$$

$$\frac{T_F - 32}{1.8} = T_C \qquad \text{Temperature equation to obtain degrees Celsius}$$

Scientists have learned that the coldest temperature possible is −273 °C (more precisely, −273.15 °C). On the *Kelvin* scale, this temperature, called *absolute zero*, has the value of 0 K. Units on the Kelvin scale are called kelvins (K); *no degree symbol is used*. Because there are no lower temperatures, the Kelvin scale has no negative temperature values. Between the freezing point of water, 273 K, and the boiling point, 373 K, there are 100 kelvins, which makes a kelvin equal in size to a degree Celsius.

$$1 \text{ K} = 1 \text{ °C}$$

We can write an equation that relates a Celsius temperature to its corresponding Kelvin temperature by adding 273 to the Celsius temperature. **TABLE 3.5** gives a comparison of some temperatures on the three scales.

$$T_K = T_C + 273 \qquad \text{Temperature equation to obtain kelvins}$$

An antifreeze mixture in a car radiator will not freeze until the temperature drops to −37 °C. We can calculate the temperature of the antifreeze mixture in kelvins by adding 273 to the temperature in degrees Celsius.

$$T_K = -37 \text{ °C} + 273 = 236 \text{ K}$$

TABLE 3.5 A Comparison of Temperatures

Example	Fahrenheit (°F)	Celsius (°C)	Kelvin (K)
Sun	9937	5503	5776
A hot oven	450	232	505
Water boils	212	100	373
A high fever	104	40	313
Normal body temperature	98.6	37.0	310.
Room temperature	70	21	294
Water freezes	32	0	273
A northern winter	−66	−54	219
Nitrogen liquefies	−346	−210	63
Absolute zero	−459	−273	0

SAMPLE PROBLEM 3.3 Converting from Degrees Celsius to Degrees Fahrenheit

The typical temperature in a room is set at 21 °C. What is that temperature in degrees Fahrenheit?

TRY IT FIRST

SOLUTION

STEP 1 State the given and needed quantities.

ANALYZE THE PROBLEM	Given	Need	Connect
	21 °C	T in degrees Fahrenheit	temperature equation

STEP 2 Write a temperature equation.

$$T_F = 1.8(T_C) + 32$$

STEP 3 Substitute in the known values and calculate the new temperature.

$T_F = 1.8(21) + 32$ 1.8 is exact; 32 is exact

Two SFs

$= 38 + 32$

$= 70. °F$ Answer to the ones place

In the equation, *the values of 1.8 and 32 are exact numbers*, which do not affect the number of SFs used in the answer.

STUDY CHECK 3.3

In the process of making ice cream, rock salt is added to crushed ice to chill the ice cream mixture. If the temperature drops to −11 °C, what is it in degrees Fahrenheit?

ANSWER

12 °F

Guide to Calculating Temperature

STEP 1
State the given and needed quantities.

STEP 2
Write a temperature equation.

STEP 3
Substitute in the known values and calculate the new temperature.

ENGAGE

Show that −40. °C is also −40. °F.

SAMPLE PROBLEM 3.4 Converting from Degrees Fahrenheit to Degrees Celsius

In a type of cancer treatment called *thermotherapy*, temperatures as high as 113 °F are used to destroy cancer cells or make them more sensitive to radiation. What is that temperature in degrees Celsius?

TRY IT FIRST

SOLUTION

STEP 1 State the given and needed quantities.

ANALYZE THE PROBLEM	Given	Need	Connect
	113 °F	T in degrees Celsius	temperature equation

STEP 2 Write a temperature equation.

$$T_C = \frac{T_F - 32}{1.8}$$

STEP **3** Substitute in the known values and calculate the new temperature.

$$T_C = \frac{(113 - 32)}{1.8}$$ 32 is exact; 1.8 is exact

$$= \frac{81}{1.8} = 45 \,°C$$
Two SFs

STUDY CHECK 3.4

A child has a temperature of 103.6 °F. What is this temperature on a Celsius thermometer?

ANSWER

39.8 °C

SAMPLE PROBLEM 3.5 Converting from Degrees Celsius to Kelvins

A dermatologist uses cryogenic liquid nitrogen at −196 °C to remove skin lesions and some skin cancers. What is the temperature, in kelvins, of the liquid nitrogen?

TRY IT FIRST

SOLUTION

STEP **1** State the given and needed quantities.

ANALYZE THE PROBLEM	Given	Need	Connect
	−196 °C	T in kelvins	temperature equation

STEP **2** Write a temperature equation.

$$T_K = T_C + 273$$

STEP **3** Substitute in the known values and calculate the new temperature.

$$T_K = -196 + 273$$
$$= 77 \,K$$ Answer to the ones place

STUDY CHECK 3.5

On the planet Mercury, the average night temperature is 13 K and the average day temperature is 683 K. What are these temperatures in degrees Celsius?

ANSWER

night −260. °C; day 410. °C

CHEMISTRY LINK TO **HEALTH**
Variation in Body Temperature

Normal body temperature is considered to be 37.0 °C, although it varies throughout the day and from person to person. Oral temperatures of 36.1 °C are common in the morning and climb to a high of 37.2 °C between 6 P.M. and 10 P.M. Temperatures above 37.2 °C for a person at rest are usually an indication of illness. Individuals who are involved in prolonged exercise may also experience elevated temperatures. Body temperatures of marathon runners can range from 39 °C to 41 °C as heat production during exercise exceeds the body's ability to lose heat.

Changes of more than 3.5 °C from the normal body temperature begin to interfere with bodily functions. Body temperatures above 41 °C, *hyperthermia*, can lead to convulsions, particularly in children, which may cause permanent brain damage. Heatstroke occurs above 41.1 °C. Sweat production stops, and the skin becomes hot and dry. The pulse rate is elevated, and respiration becomes weak and rapid. The person can become lethargic and lapse into a coma. Damage to internal organs is a major concern, and treatment, which must be immediate, may include immersing the person in an ice-water bath.

At the low temperature extreme of *hypothermia*, body temperature can drop as low as 28.5 °C. The person may appear cold and pale and have an irregular heartbeat. Unconsciousness can occur if the body temperature drops below 26.7 °C. Respiration becomes slow and shallow, and oxygenation of the tissues decreases. Treatment involves providing oxygen and increasing blood volume with glucose and saline fluids. Injecting warm fluids (37.0 °C) into the peritoneal cavity may restore the internal temperature.

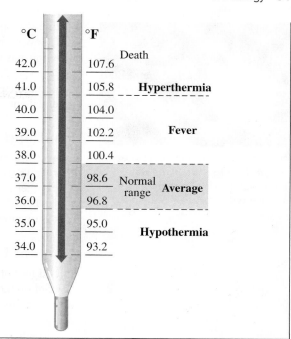

QUESTIONS AND PROBLEMS

3.3 Temperature

LEARNING GOAL Given a temperature, calculate a corresponding temperature on another scale.

3.15 Your friend who is visiting from Canada just took her temperature. When she reads 99.8 °F, she becomes concerned that she is quite ill. How would you explain this temperature to your friend?

3.16 You have a friend who is using a recipe for flan from a Mexican cookbook. You notice that he set your oven temperature at 175 °F. What would you advise him to do?

3.17 Calculate the unknown temperature in each of the following:
  a. 37.0 °C = _____ °F   **b.** 65.3 °F = _____ °C
  c. −27 °C = _____ K   **d.** 62 °C = _____ K
  e. 114 °F = _____ °C

3.18 Calculate the unknown temperature in each of the following:
  a. 25 °C = _____ °F   **b.** 155 °C = _____ °F
  c. −25 °F = _____ °C   **d.** 224 K = _____ °C
  e. 145 °C = _____ K

Applications

3.19 **a.** A person with hyperthermia has a temperature of 106 °F. What does this read on a Celsius thermometer?
  b. Because high fevers can cause convulsions in children, the doctor needs to be called if the child's temperature goes over 40.0 °C. Should the doctor be called if a child has a temperature of 103 °F?

3.20 **a.** Water is heated to 145 °F. What is the temperature of the hot water in degrees Celsius?
  b. During extreme hypothermia, a boy's temperature dropped to 20.6 °C. What was his temperature in degrees Fahrenheit?

3.4 Energy

LEARNING GOAL Identify energy as potential or kinetic; convert between units of energy.

Almost everything you do involves energy. When you are running, walking, dancing, or thinking, you are using energy to do *work*, any activity that requires energy. In fact, **energy** is defined as the ability to do work. Suppose you are climbing a steep hill and you become too tired to go on. At that moment, you do not have the energy to do any more work. Now suppose you sit down and have lunch. In a while, you will have obtained some energy from the food, and you will be able to do more work and complete the climb.

Kinetic and Potential Energy

Energy can be classified as kinetic energy or potential energy. **Kinetic energy** is the energy of motion. Any object that is moving has kinetic energy. **Potential energy** is determined by the position of an object or by the chemical composition of a substance. A boulder

ENGAGE

Why does a book have more potential energy when it is on the top of a high table than when it is on the floor?

Water at the top of the dam stores potential energy. When the water flows over the dam, potential energy is converted to kinetic energy.

resting on top of a mountain has potential energy because of its location. If the boulder rolls down the mountain, the potential energy becomes kinetic energy. Water stored in a reservoir has potential energy. When the water goes over the dam and falls to the stream below, its potential energy is converted to kinetic energy. Foods and fossil fuels have potential energy in their molecules. When you digest food or burn gasoline in your car, potential energy is converted to kinetic energy to do work.

Heat and Energy

Heat is the energy associated with the motion of particles. An ice cube feels cold because heat flows from your hand into the ice cube. The faster the particles move, the greater the heat or thermal energy of the substance. In the ice cube, the particles are moving very slowly. As heat is added, the motion of the particles in the ice cube increases. Eventually, the particles have enough energy to make the ice cube melt as it changes from a solid to a liquid.

Units of Energy

The SI unit of energy and work is the **joule (J)** (pronounced "jewel"). The joule is a small amount of energy, so scientists often use the kilojoule (kJ), 1000 joules. To heat water for one cup of tea, you need about 75 000 J or 75 kJ of heat. **TABLE 3.6** shows a comparison of energy in joules for several energy sources or uses.

You may be more familiar with the unit **calorie (cal)**, from the Latin *caloric*, meaning "heat." The calorie was originally defined as the amount of energy (heat) needed to raise the temperature of 1 g of water by 1 °C. Now, one calorie is defined as exactly 4.184 J. This equality can be written as two conversion factors:

$$1 \text{ cal} = 4.184 \text{ J (exact)} \qquad \frac{4.184 \text{ J}}{1 \text{ cal}} \quad \text{and} \quad \frac{1 \text{ cal}}{4.184 \text{ J}}$$

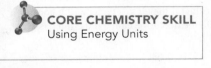

CORE CHEMISTRY SKILL
Using Energy Units

One *kilocalorie* (kcal) is equal to 1000 calories, and one *kilojoule* (kJ) is equal to 1000 joules. The equalities and conversion factors follow:

$$1 \text{ kcal} = 1000 \text{ cal} \qquad \frac{1000 \text{ cal}}{1 \text{ kcal}} \quad \text{and} \quad \frac{1 \text{ kcal}}{1000 \text{ cal}}$$

$$1 \text{ kJ} = 1000 \text{ J} \qquad \frac{1000 \text{ J}}{1 \text{ kJ}} \quad \text{and} \quad \frac{1 \text{ kJ}}{1000 \text{ J}}$$

TABLE 3.6   A Comparison of Energy for Various Resources and Uses

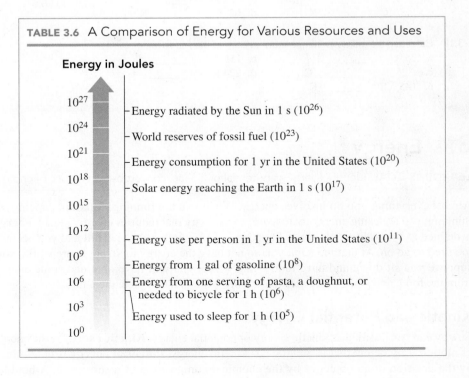

Energy in Joules

10^{27}

10^{24} — Energy radiated by the Sun in 1 s (10^{26})

10^{21} — World reserves of fossil fuel (10^{23})

10^{18} — Energy consumption for 1 yr in the United States (10^{20})

10^{15} — Solar energy reaching the Earth in 1 s (10^{17})

10^{12}

10^{9} — Energy use per person in 1 yr in the United States (10^{11})

10^{6} — Energy from 1 gal of gasoline (10^{8})

10^{3} — Energy from one serving of pasta, a doughnut, or needed to bicycle for 1 h (10^{6})

10^{0} — Energy used to sleep for 1 h (10^{5})

SAMPLE PROBLEM 3.6 Energy Units

A defibrillator gives a high-energy-shock output of 360 J. What is this quantity of energy in calories?

TRY IT FIRST

SOLUTION

STEP 1 State the given and needed quantities.

ANALYZE THE PROBLEM	Given	Need	Connect
	360 J	calories	energy factor

STEP 2 Write a plan to convert the given unit to the needed unit.

joules → Energy factor → calories

STEP 3 State the equalities and conversion factors.

$$1 \text{ cal} = 4.184 \text{ J}$$

$$\frac{4.184 \text{ J}}{1 \text{ cal}} \quad \text{and} \quad \frac{1 \text{ cal}}{4.184 \text{ J}}$$

STEP 4 Set up the problem to calculate the needed quantity.

$$360 \text{ J} \times \underset{\text{Exact}}{\frac{1 \text{ cal}}{4.184 \text{ J}}} = 86 \text{ cal}$$

Two SFs   Exact   Two SFs

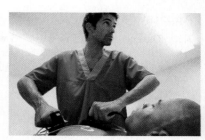

A defibrillator provides electrical energy to heart muscle to re-establish normal rhythm.

STUDY CHECK 3.6

When 1.0 g of glucose is metabolized in the body, it produces 3.9 kcal. How many kilojoules are produced?

ANSWER

16 kJ

QUESTIONS AND PROBLEMS

3.4 Energy

LEARNING GOAL Identify energy as potential or kinetic; convert between units of energy.

3.21 Discuss the changes in the potential and kinetic energy of a roller-coaster ride as the roller-coaster car climbs to the top and goes down the other side.

3.22 Discuss the changes in the potential and kinetic energy of a ski jumper taking the elevator to the top of the jump and going down the ramp.

3.23 Indicate whether each of the following statements describes potential or kinetic energy:
 a. water at the top of a waterfall
 b. kicking a ball
 c. the energy in a lump of coal
 d. a skier at the top of a hill

3.24 Indicate whether each of the following statements describes potential or kinetic energy:
 a. the energy in your food
 b. a tightly wound spring
 c. an earthquake
 d. a car speeding down the freeway

3.25 Convert each of the following energy units:
 a. 3500 cal to kcal
 b. 415 J to cal
 c. 28 cal to J
 d. 4.5 kJ to cal

3.26 Convert each of the following energy units:
 a. 8.1 kcal to cal
 b. 325 J to kJ
 c. 2550 cal to kJ
 d. 2.50 kcal to J

Applications

3.27 The energy needed to keep a 75-watt light bulb burning for 1.0 h is 270 kJ. Calculate the energy required to keep the light bulb burning for 3.0 h in each of the following energy units:
a. joules
b. kilocalories

3.28 A person uses 750 kcal on a long walk. Calculate the energy used for the walk in each of the following energy units:
a. joules
b. kilojoules

CHEMISTRY LINK TO THE **ENVIRONMENT**
Carbon Dioxide and Climate Change

The Earth's climate is a product of interactions between sunlight, the atmosphere, and the oceans. The Sun provides us with energy in the form of solar radiation. Some of this radiation is reflected back into space. The rest is absorbed by the clouds, atmospheric gases including carbon dioxide (CO_2), and the Earth's surface. For millions of years, concentrations of carbon dioxide have fluctuated. However in the last 100 years, the amount of CO_2 gas in our atmosphere has increased significantly. From the years 1000 to 1800, the atmospheric carbon dioxide level averaged 280 ppm. The concentration of ppm indicates the parts per million by volume, which for gases is the same as mL of CO_2 per kL of air. But since the beginning of the Industrial Revolution in 1800 up until 2005, the level of atmospheric carbon dioxide has risen from about 280 ppm to about 394 ppm in 2013, a 40% increase.

As the atmospheric CO_2 level increases, more solar radiation is trapped by the atmospheric gases, which raises the temperature at the surface of the Earth. Some scientists have estimated that if the carbon dioxide level doubles from its level before the Industrial Revolution, the average global temperature could increase by 2.0 to 4.4 °C. Although this seems to be a small temperature change, it could have dramatic impact worldwide. Even now, glaciers and snow cover in much of the world have diminished. Ice sheets in

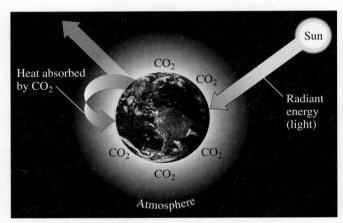

Heat from the Sun is trapped by the CO_2 layer in the atmosphere.

Antarctica and Greenland are melting faster and breaking apart. Although no one knows for sure how rapidly the ice in the polar regions is melting, this accelerating change will contribute to a rise in sea level. In the twentieth century, the sea level rose by 15 to 23 cm. Some scientists predict the sea level will rise by 1 m in this century. Such an increase will have a major impact on coastal areas.

Until recently, the carbon dioxide level was maintained as algae in the oceans and the trees in the forests utilized the carbon dioxide. However, the ability of these and other forms of plant life to absorb carbon dioxide is not keeping up with the increase in carbon dioxide. Most scientists agree that the primary source of the increase of carbon dioxide is the burning of fossil fuels such as gasoline, coal, and natural gas. The cutting and burning of trees in the rainforests (deforestation) also reduces the amount of carbon dioxide removed from the atmosphere.

Worldwide efforts are being made to reduce the carbon dioxide produced by burning fossil fuels that heat our homes, run our cars, and provide energy for industries. Scientists are exploring ways to provide alternative energy sources and to reduce the effects of deforestation. Meanwhile, we can reduce energy use in our homes by using appliances that are more energy efficient such as replacing incandescent light bulbs with fluorescent lights. Such an effort worldwide will reduce the possible impact of climate change and at the same time save our fuel resources.

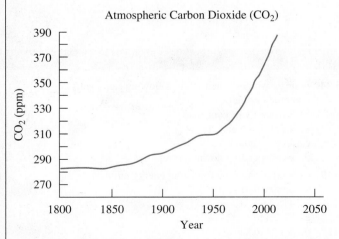

Atmospheric carbon dioxide levels are shown for the years from 1800 C.E. to 2010 C.E.

3.5 Specific Heat

LEARNING GOAL Calculate the specific heat for a substance. Use specific heat to calculate heat loss or gain.

Every substance has its own characteristic ability to absorb heat. When you bake a potato, you place it in a hot oven. If you are cooking pasta, you add the pasta to boiling water. You already know that adding heat to water increases its temperature until it boils. Certain substances must absorb more heat than others to reach a certain temperature.

The energy requirements for different substances are described in terms of a physical property called *specific heat*. The **specific heat (SH)** for a substance is defined as the amount of heat *(q)* in calories (or joules) needed to change the temperature of exactly 1 g of a substance by exactly 1 °C. To calculate the specific heat for a substance, we measure the heat in calories or joules, the mass in grams, and the temperature change written as ΔT.

$$\text{Specific heat } (SH) = \frac{q}{m \times \Delta T} = \frac{\text{cal (or J)}}{\text{g °C}}$$

The specific heat for water is written using our definition of the calorie and joule.

$$SH \text{ for } H_2O(l) = \frac{1.00 \text{ cal}}{\text{g °C}} = \frac{4.184 \text{ J}}{\text{g °C}}$$

If we look at **TABLE 3.7**, we see that 1 g of water requires 1.00 cal or 4.184 J to increase its temperature by 1 °C. Water has a large specific heat that is about five times the specific heat of aluminum. Aluminum has a specific heat that is about twice that of copper. However, adding the same amount of heat (1.00 cal or 4.184 J) will raise the temperature of 1 g of aluminum by about 5 °C and 1 g of copper by about 10 °C. The low specific heats of aluminum and copper mean they transfer heat efficiently, which makes them useful in cookware.

The high specific heat of water has a major impact on the temperatures in a coastal city compared to an inland city. A large mass of water near a coastal city can absorb or release five times the energy absorbed or released by the same mass of rock near an inland city. This means that in the summer a body of water absorbs large quantities of heat, which cools a coastal city, and then in the winter that same body of water releases large quantities of heat, which provides warmer temperatures. A similar effect happens with our bodies, which contain 70% water by mass. Water in the body absorbs or releases large quantities of heat to maintain an almost constant body temperature.

Symbols Used in Specific Heat Equation		
Symbol	**Meaning**	**Unit**
SH	specific heat	$\frac{\text{cal (or J)}}{\text{g °C}}$
q	heat	calories (cal), joules (J)
m	mass	grams (g)
ΔT	temperature change	degrees Celsius (°C)

TABLE 3.7 Specific Heats for Some Substances

Substance	cal/g °C	J/g °C
Elements		
Aluminum, Al(*s*)	0.214	0.897
Copper, Cu(*s*)	0.0920	0.385
Gold, Au(*s*)	0.0308	0.129
Iron, Fe(*s*)	0.108	0.452
Silver, Ag(*s*)	0.0562	0.235
Titanium, Ti(*s*)	0.125	0.523
Compounds		
Ammonia, NH₃(*g*)	0.488	2.04
Ethanol, C₂H₆O(*l*)	0.588	2.46
Sodium chloride, NaCl(*s*)	0.207	0.864
Water, H₂O(*l*)	1.00	4.184
Water, H₂O(*s*)	0.485	2.03

The high specific heat of water keeps temperatures more moderate in summer and winter.

SAMPLE PROBLEM 3.7 Calculating Specific Heat

What is the specific heat, in J/g °C, of lead if 57.0 J raises the temperature of 35.6 g of lead by 12.5 °C?

SOLUTION

STEP 1 State the given and needed quantities.

	Given	Need	Connect
ANALYZE THE PROBLEM	q = 57.0 J, m = 35.6 g of lead, ΔT = 12.5 °C	specific heat of lead (J/g °C)	specific heat equation

ENGAGE

Why will 1 g of silver have a greater increase of temperature than 1 g of iron when each absorbs 400 J of heat?

CORE CHEMISTRY SKILL
Calculating Specific Heat

CORE CHEMISTRY SKILL
Using the Heat Equation

STEP 2 Write the relationship for specific heat. The specific heat (*SH*) is calculated by dividing the heat (*q*) by the mass (*m*) and by the temperature change (ΔT).

$$SH = \frac{\text{heat}}{\text{mass} \quad \Delta T} = \frac{q}{m \quad \Delta T}$$

STEP 3 Set up the problem to calculate the specific heat.

$$SH = \frac{\overset{\text{Three SFs}}{57.0 \text{ J}}}{\underset{\text{Three SFs} \quad \text{Three SFs}}{35.6 \text{ g} \quad 12.5 \text{ °C}}} = \underset{\text{Three SFs}}{\frac{0.128 \text{ J}}{\text{g °C}}}$$

STUDY CHECK 3.7

What is the specific heat, in J/g °C, of sodium if 92.3 J is needed to raise the temperature of 3.00 g of sodium by 25.0 °C?

ANSWER

1.23 J/g °C

Heat Equation

When we know the specific heat of a substance, we can calculate the heat lost or gained by measuring the mass of the substance and the initial and final temperatures. We can substitute these measurements into the specific heat equation that is rearranged to solve for heat, which we call the **heat equation**.

$$SH = \frac{q}{m \times \Delta T}$$

$$m \times \Delta T \times SH = \frac{q}{\cancel{m} \times \cancel{\Delta T}} \times \cancel{m} \times \cancel{\Delta T}$$

Now we can write the heat equation as

$$q = m \times \Delta T \times SH$$

The heat lost or gained, in calories or joules, is obtained when the units of grams and °C in the numerator cancel grams and °C in the denominator of specific heat in the heat equation.

$$\text{cal} = \cancel{g} \times \cancel{°C} \times \frac{\text{cal}}{\cancel{g} \, \cancel{°C}}$$

$$\text{J} = \cancel{g} \times \cancel{°C} \times \frac{\text{J}}{\cancel{g} \, \cancel{°C}}$$

A cooling cap lowers the body temperature to reduce the oxygen required by the tissues.

SAMPLE PROBLEM 3.8 Calculating Heat Loss

During surgery or when a patient has suffered a cardiac arrest or stroke, lowering the body temperature will reduce the amount of oxygen needed by the body. Some methods used to lower body temperature include cooled saline solution, cool water blankets, or cooling caps worn on the head. How many kilojoules are lost when the body temperature of a surgery patient with a blood volume of 5500 mL is cooled from 38.5 °C to 33.2 °C? (Assume that the specific heat and density of blood is the same as for water.)

TRY IT FIRST

SOLUTION

STEP 1 State the given and needed quantities.

ANALYZE THE PROBLEM	Given	Need	Connect
	5500 mL of blood = 5500 g of blood, cooled from 38.5 °C to 33.2 °C	kilojoules removed	specific heat of water

STEP 2 Calculate the temperature change (ΔT).

$$\Delta T = 38.5\,°C - 33.2\,°C = 5.3\,°C$$

STEP 3 Write the heat equation and needed conversion factors.

$$q = m \times \Delta T \times SH$$

$$SH_{water} = \frac{4.184\ J}{g\ °C}$$

$$\frac{4.184\ J}{g\ °C} \text{ and } \frac{g\ °C}{4.184\ J}$$

$$1\ kJ = 1000\ J$$

$$\frac{1000\ J}{1\ kJ} \text{ and } \frac{1\ kJ}{1000\ J}$$

STEP 4 Substitute in the given values and calculate the heat, making sure units cancel.

$$q = 5500\ g \times 5.3\,°C \times \frac{4.184\ J}{g\ °C} \times \frac{1\ kJ}{1000\ J} = 120\ kJ$$

Two SFs   Two SFs   Exact   Exact   Two SFs

STUDY CHECK 3.8

Some cooking pans have a layer of copper on the bottom. How many kilojoules are needed to raise the temperature of 125 g of copper from 22 °C to 325 °C (see Table 3.7)?

ANSWER

14.6 kJ

Guide to Using Specific Heat

STEP 1 State the given and needed quantities.

STEP 2 Calculate the temperature change (ΔT).

STEP 3 Write the heat equation and needed conversion factors.

STEP 4 Substitute in the given values and calculate the heat, making sure units cancel.

The copper on a pan conducts heat rapidly to the food in the pan.

ENGAGE

Rearrange the heat equation to solve for mass (m).

Another use of the heat equation is to calculate the mass, in grams, of a substance by rearranging the heat equation to solve for mass (m) as shown in Sample Problem 3.9.

SAMPLE PROBLEM 3.9 Using the Heat Equation

When 655 J is added to a sample of ethanol, its temperature rises from 18.2 °C to 32.8 °C. What is the mass, in grams, of the ethanol sample (see Table 3.7)?

TRY IT FIRST

SOLUTION

STEP 1 State the given and needed quantities.

ANALYZE THE PROBLEM	Given	Need	Connect
	q = 655 J, heated from 18.2 °C to 32.8 °C	grams of ethanol	heat equation

STEP 2 Calculate the temperature change (ΔT).

$$\Delta T = 32.8\,°C - 18.2\,°C = 14.6\,°C$$

STEP **3** Write the heat equation.

$$q = m \times \Delta T \times SH$$

When the heat equation is rearranged for mass (m), the heat is divided by the temperature change and the specific heat.

$$m = \frac{q}{\Delta T \ SH}$$

STEP **4** Substitute in the given values and solve, making sure units cancel.

$$m = \frac{655 \ \cancel{J}}{14.6 \ \cancel{°C} \ \dfrac{2.46 \ \cancel{J}}{g \ \cancel{°C}}} = 18.2 \ g$$

Three SFs

Three SFs   Three SFs

Three SFs

STUDY CHECK 3.9

When 8.81 kJ is absorbed by a piece of iron, its temperature rises from 15 °C to 122 °C. What is the mass, in grams, of the piece of iron (see Table 3.7)?

ANSWER

182 g

QUESTIONS AND PROBLEMS

3.5 Specific Heat

LEARNING GOAL Calculate the specific heat for a substance. Use specific heat to calculate heat loss or gain, temperature change, or mass of a sample.

3.29 If the same amount of heat is supplied to samples of 10.0 g each of aluminum, iron, and copper all at 15.0 °C, which sample would reach the highest temperature (see Table 3.7)?

3.30 Substances A and B are the same mass and at the same initial temperature. When the same amount of heat is added to each, the final temperature of A is 75 °C and B is 35 °C. What does this tell you about the specific heats of A and B?

3.31 Calculate the specific heat (J/g °C) for each of the following:
a. a 13.5-g sample of zinc (Zn) heated from 24.2 °C to 83.6 °C that absorbs 312 J of heat
b. a 48.2-g sample of a metal that absorbs 345 J with a temperature increase from 35.0 °C to 57.9 °C

3.32 Calculate the specific heat (J/g °C) for each of the following:
a. an 18.5-g sample of tin (Sn) that absorbs 183 J of heat when its temperature increases from 35.0 °C to 78.6 °C
b. a 22.5-g sample of a metal that absorbs 645 J when its temperature increases from 36.2 °C to 92.0 °C

3.33 Use the heat equation to calculate the energy, in joules, for each of the following (see Table 3.7):
a. required to heat 25.0 g of water, H_2O, from 12.5 °C to 25.7 °C
b. required to heat 38.0 g of copper (Cu) from 122 °C to 246 °C

3.34 Use the heat equation to calculate the energy, in joules, for each of the following (see Table 3.7):
a. required to heat 5.25 g of water, H_2O, from 5.5 °C to 64.8 °C
b. required to heat 10.0 g of silver (Ag) from 112 °C to 275 °C

3.35 Calculate the mass, in grams, for each of the following using Table 3.7:
a. a sample of gold (Au) that absorbs 225 J to increase its temperature from 15.0 °C to 47.0 °C
b. a sample of iron (Fe) that loses 8.40 kJ when its temperature decreases from 168.0 °C to 82.0 °C

3.36 Calculate the mass, in grams, for each of the following using Table 3.7:
a. a sample of water, H_2O, that absorbs 8250 J when its temperature increases from 18.4 °C to 92.6 °C
b. a sample of silver (Ag) that loses 3.22 kJ when its temperature decreases from 145 °C to 24 °C

3.37 Calculate the change in temperature (ΔT) for each of the following using Table 3.7:
a. 20.0 g of iron (Fe) that absorbs 1580 J
b. 150.0 g of water, H_2O, that absorbs 7.10 kJ

3.38 Calculate the change in temperature (ΔT) for each of the following using Table 3.7:
a. 115 g of copper (Cu) that loses 2.45 kJ
b. 22.0 g of silver (Ag) that loses 625 J

3.6 Energy and Nutrition

LEARNING GOAL Use the energy values to calculate the kilocalories (kcal) or kilojoules (kJ) for a food.

The food we eat provides energy to do work in the body, which includes the growth and repair of cells. Carbohydrates are the primary fuel for the body, but if the carbohydrate reserves are exhausted, fats and then proteins are used for energy.

For many years in the field of nutrition, the energy from food was measured as Calories or kilocalories. The nutritional unit *Calorie, Cal* (with an uppercase C), is the same as 1000 cal, or 1 kcal. The international unit, kilojoule (kJ), is becoming more prevalent. For example, a baked potato has an energy value of 100 Calories, which is 100 kcal or 440 kJ. A typical diet that provides 2100 Cal (kcal) is the same as an 8800 kJ diet.

$$1 \text{ Cal} = 1 \text{ kcal} = 1000 \text{ cal}$$
$$1 \text{ Cal} = 4.184 \text{ kJ} = 4184 \text{ J}$$

In the nutrition laboratory, foods are burned in a *calorimeter* to determine their *energy value* (kcal/g or kJ/g)(see **FIGURE 3.6**). A sample of food is placed in a steel container called a calorimeter filled with oxygen. A measured amount of water is added to fill the area surrounding the combustion chamber. The food sample is ignited, releasing heat that increases the temperature of the water. From the known mass of the food and water as well as the measured temperature increase, the energy value for the food is calculated. We will assume that the energy absorbed by the calorimeter is negligible.

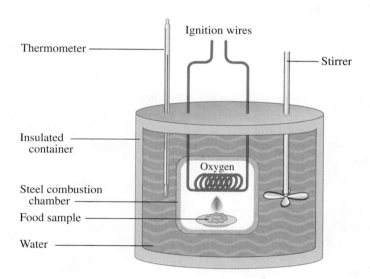

FIGURE 3.6 ▶ Heat released from burning a food sample in a calorimeter is used to determine the energy value for the food.

Ⓠ What happens to the temperature of water in a calorimeter during the combustion of a food sample?

Energy Values for Foods

The **energy values** for food are the kilocalories or kilojoules obtained from burning 1 g of carbohydrate, fat, or protein, which are listed in **TABLE 3.8**. Using these energy values, we can calculate the total energy for a food if the mass of each food type is known.

$$\text{kilocalories} = g \times \frac{\text{kcal}}{g} \qquad \text{kilojoules} = g \times \frac{\text{kJ}}{g}$$

On packaged food, the energy content is listed on the Nutrition Facts label, usually in terms of the number of Calories or kilojoules for one serving. The general composition and energy content for some foods are given in **TABLE 3.9**. The total energy, in kilocalories, for each food type was calculated using energy values in kilocalories. Total energy, in kilojoules, was calculated using energy values in kilojoules. The energy for each food type was rounded off to the tens place.

TABLE 3.8 Typical Energy Values for the Three Food Types

Food Type	kcal/g	kJ/g
Carbohydrate	4	17
Fat	9	38
Protein	4	17

ENGAGE

What type of food provides the most energy per gram for the human body?

Snack Crackers

Nutrition Facts

Serving Size 14 crackers (31g)
Servings Per Container About 7

Amount Per Serving

Calories 130   Calories from Fat 40
Kilojoules 500   kJ from Fat   150

	% Daily Value*
Total Fat 4g	**6%**
Saturated Fat 0.5g	**3%**
Trans Fat 0g	
Polyunsaturated Fat 0.5%	
Monounsaturated Fat 1.5g	
Cholesterol 0mg	**0%**
Sodium 310mg	**13%**
Total Carbohydrate 19g	**6%**
Dietary Fiber Less than 1g	**4%**
Sugars 2g	
Proteins 2g	

Vitamin A 0%	•	Vitamin C 0%
Calcium 4%	•	Iron 6%

*Percent Daily Values are based on a 2,000 calorie diet. Your daily values may be higher or lower depending on your calorie needs.

	Calories:	2,000	2,500
Total Fat	Less than	65g	80g
Sat Fat	Less than	20g	25g
Cholesterol	Less than	300mg	300mg
Sodium	Less than	2,400mg	2,400mg
Total Carbohydrate		300g	375g
Dietary Fiber		25g	30g

Calories per gram:
Carbohydrate 4 • Fat 9 • Protein 4

The Nutrition Facts include the total Calories and kilojoules, and the grams of carbohydrate, fat, and protein per serving.

Guide to Calculating the Energy from a Food

STEP 1
State the given and needed quantities.

STEP 2
Use the energy value for each food type and calculate the kcal or kJ rounded off to the tens place.

STEP 3
Add the energy for each food type to give the total energy from the food.

TABLE 3.9 General Composition and Energy Content for Some Foods

Food	Carbohydrate (g)	Fat (g)	Protein (g)	Energy
Apple, 1 medium	15	0	0	60 kcal (260 kJ)
Banana, 1 medium	26	0	1	110 kcal (460 kJ)
Beef, ground, 3 oz	0	14	22	220 kcal (900 kJ)
Broccoli, 3 oz	4	0	3	30 kcal (120 kJ)
Carrots, 1 cup	11	0	2	50 kcal (220 kJ)
Chicken, no skin, 3 oz	0	3	20	110 kcal (450 kJ)
Egg, 1 large	0	6	6	70 kcal (330 kJ)
Milk, nonfat, 1 cup	12	0	9	90 kcal (350 kJ)
Potato, baked	23	0	3	100 kcal (440 kJ)
Salmon, 3 oz	0	5	16	110 kcal (460 kJ)
Steak, 3 oz	0	27	19	320 kcal (1350 kJ)

SAMPLE PROBLEM 3.10 Energy Content for a Food Using Nutrition Facts

The Nutrition Facts label for crackers states that 1 serving contains 19 g of carbohydrate, 4 g of fat, and 2 g of protein. What is the energy, in kilocalories, from each food type and the total kilocalories for one serving of crackers? Round off the kilocalories for each food type to the tens place.

TRY IT FIRST

SOLUTION

STEP 1 State the given and needed quantities.

	Given	Need	Connect
ANALYZE THE PROBLEM	19 g of carbohydrate, 4 g of fat, 2 g of protein	total number of kilocalories	energy values

STEP 2 Use the energy value for each food type and calculate the kcal rounded off to the tens place. Using the energy values for carbohydrate, fat, and protein (see Table 3.8), we can calculate the energy for each type of food.

Food Type	Mass		Energy Value		Energy
Carbohydrate	19 g	×	$\dfrac{4 \text{ kcal}}{1 \text{ g}}$	=	80 kcal
Fat	4 g	×	$\dfrac{9 \text{ kcal}}{1 \text{ g}}$	=	40 kcal
Protein	2 g	×	$\dfrac{4 \text{ kcal}}{1 \text{ g}}$	=	10 kcal

STEP 3 Add the energy for each food type to give the total energy from the food.

Total energy = 80 kcal + 40 kcal + 10 kcal = 130 kcal

STUDY CHECK 3.10

a. Using the nutrition values for one serving of crackers in Sample Problem 3.10, calculate the energy, in kilojoules, for each food type. Round off the kilojoules for each food type to the tens place.

b. What is the total energy, in kilojoules, for one serving of crackers?

ANSWER

a. carbohydrate, 630 kJ; fat, 300 kJ; protein, 50 kJ

b. 980 kJ

CHEMISTRY LINK TO HEALTH
Losing and Gaining Weight

The number of kilocalories or kilojoules needed in the daily diet of an adult depends on gender, age, and level of physical activity. Some typical levels of energy needs are given in **TABLE 3.10**.

TABLE 3.10 Typical Energy Requirements for Adults

Gender	Age	Moderately Active kcal (kJ)	Highly Active kcal (kJ)
Female	19–30	2100 (8800)	2400 (10 000)
	31–50	2000 (8400)	2200 (9200)
Male	19–30	2700 (11 300)	3000 (12 600)
	31–50	2500 (10 500)	2900 (12 100)

One hour of swimming uses 2100 kJ of energy.

A person gains weight when food intake exceeds energy output. The amount of food a person eats is regulated by the hunger center in the hypothalamus, which is located in the brain. Food intake is normally proportional to the nutrient stores in the body. If these nutrient stores are low, you feel hungry; if they are high, you do not feel like eating.

A person loses weight when food intake is less than energy output. Many diet products contain cellulose, which has no nutritive value but provides bulk and makes you feel full. Some diet drugs depress the hunger center and must be used with caution, because they excite the nervous system and can elevate blood pressure. Because muscular exercise is an important way to expend energy, an increase in daily exercise aids weight loss. **TABLE 3.11** lists some activities and the amount of energy they require.

TABLE 3.11 Energy Expended by a 70.0-kg (154-lb) Adult

Activity	Energy (kcal/h)	Energy (kJ/h)
Sleeping	60	250
Sitting	100	420
Walking	200	840
Swimming	500	2100
Running	750	3100

QUESTIONS AND PROBLEMS
3.6 Energy and Nutrition

LEARNING GOAL Use the energy values to calculate the kilocalories (kcal) or kilojoules (kJ) for a food.

3.39 Calculate the kilocalories for each of the following:
   a. one stalk of celery that produces 125 kJ when burned in a calorimeter
   b. a waffle that produces 870. kJ when burned in a calorimeter

3.40 Calculate the kilocalories for each of the following:
   a. one cup of popcorn that produces 131 kJ when burned in a calorimeter
   b. a sample of butter that produces 23.4 kJ when burned in a calorimeter

3.41 Using the energy values for foods (see Table 3.8), determine each of the following (round off the answer for each food type to the tens place):
   a. the total kilojoules for one cup of orange juice that contains 26 g of carbohydrate, no fat, and 2 g of protein
   b. the grams of carbohydrate in one apple if the apple has no fat and no protein and provides 72 kcal of energy
   c. the kilocalories in one tablespoon of vegetable oil, which contains 14 g of fat and no carbohydrate or protein
   d. the grams of fat in one avocado that has 405 kcal, 13 g of carbohydrate, and 5 g of protein

3.42 Using the energy values for foods (see Table 3.8), determine each of the following (round off the answer for each food type to the tens place):
   a. the total kilojoules in two tablespoons of crunchy peanut butter that contains 6 g of carbohydrate, 16 g of fat, and 7 g of protein
   b. the grams of protein in one cup of soup that has 110 kcal with 9 g of carbohydrate and 6 g of fat
   c. the kilocalories in one can of cola if it has 40. g of carbohydrate and no fat or protein
   d. the total kilocalories for a diet that consists of 68 g of carbohydrate, 9 g of fat, and 150 g of protein

3.43 One cup of clam chowder contains 16 g of carbohydrate, 12 g of fat, and 9 g of protein. How much energy, in kilocalories and kilojoules, is in the clam chowder? (Round off the answer for each food type to the tens place.)

3.44 A high-protein diet contains 70.0 g of carbohydrate, 5.0 g of fat, and 150 g of protein. How much energy, in kilocalories and kilojoules, does this diet provide? (Round off the answer for each food type to the tens place.)

Applications

3.45 A patient receives 3.2 L of glucose solution intravenously (IV). If 100. mL of the solution contains 5.0 g of glucose (carbohydrate), how many kilocalories did the patient obtain from the glucose solution?

3.46 For lunch, a patient consumed 3 oz of skinless chicken, 3 oz of broccoli, 1 medium apple, and 1 cup of nonfat milk (see Table 3.9). How many kilocalories did the patient obtain from the lunch?

Follow Up

A DIET AND EXERCISE PROGRAM FOR CHARLES

It has been two weeks since Charles met with Daniel, a dietitian, who provided Charles with a menu for weight loss. Charles and his mother are going back to see Daniel again with a chart of the food Charles has eaten. The following is what Charles ate in one day:

Breakfast
1 banana, 1 cup of nonfat milk, 1 egg

Lunch
1 cup of carrots, 3 oz of ground beef, 1 apple, 1 cup of nonfat milk

Dinner
6 oz of skinless chicken, 1 baked potato, 3 oz of broccoli, 1 cup of nonfat milk

Applications

3.47 Using energy values from Table 3.9, determine each of the following:
   a. the total kilocalories for each meal
   b. the total kilocalories for one day
   c. If Charles consumes 1800 kcal per day, he will maintain his weight. Would he lose weight on his new diet?
   d. If expending 3500 kcal is equal to a loss of 1.0 lb, how many days will it take Charles to lose 5.0 lb?

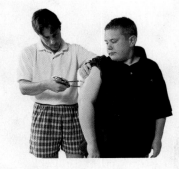

3.48 a. During one week, Charles swam for a total of 2.5 h and walked for a total of 8.0 h. If Charles expends 340 kcal/h swimming and 160 kcal/h walking, how many total kilocalories did he expend for one week?
   b. For the amount of exercise that Charles did for one week in part **a**, if expending 3500 kcal is equal to a loss of 1.0 lb, how many pounds did he lose?
   c. How many hours would Charles have to walk to lose 1.0 lb?
   d. How many hours would Charles have to swim to lose 1.0 lb?

CONCEPT MAP

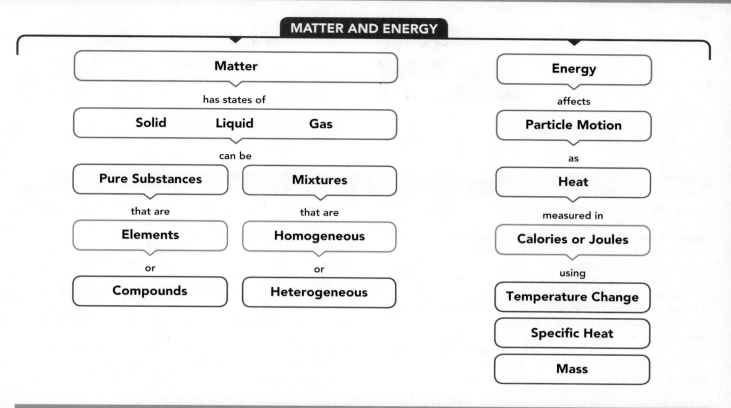

MATTER AND ENERGY

Matter

has states of

Solid   Liquid   Gas

can be

Pure Substances — **Mixtures**

that are — **Elements** | that are — **Homogeneous**

or | or

Compounds | **Heterogeneous**

Energy

affects

Particle Motion

as

Heat

measured in

Calories or Joules

using

Temperature Change

Specific Heat

Mass

CHAPTER REVIEW

3.1 Classification of Matter

LEARNING GOAL Classify examples of matter as pure substances or mixtures.

- Matter is anything that has mass and occupies space.
- Matter is classified as pure substances or mixtures.
- Pure substances, which are elements or compounds, have fixed compositions, and mixtures have variable compositions.
- The substances in mixtures can be separated using physical methods.

3.2 States and Properties of Matter

LEARNING GOAL Identify the states and the physical and chemical properties of matter.

- The three states of matter are solid, liquid, and gas.
- A physical property is a characteristic of a substance that can be observed or measured without affecting the identity of the substance.
- A physical change occurs when physical properties change, but not the composition of the substance.
- A chemical property indicates the ability of a substance to change into another substance.
- A chemical change occurs when one or more substances react to form a substance with new physical and chemical properties.

3.3 Temperature

LEARNING GOAL Given a temperature, calculate a corresponding temperature on another scale.

- In science, temperature is measured in degrees Celsius (°C) or kelvins (K).
- On the Celsius scale, there are 100 units between the freezing point (0 °C) and the boiling point (100 °C) of water.
- On the Fahrenheit scale, there are 180 units between the freezing point (32 °F) and the boiling point (212 °F) of water. A Fahrenheit temperature is related to its Celsius temperature by the equation $T_F = 1.8(T_C) + 32$.
- The SI unit, kelvin, is related to the Celsius temperature by the equation $T_K = T_C + 273$.

3.4 Energy

LEARNING GOAL Identify energy as potential or kinetic; convert between units of energy.

- Energy is the ability to do work.
- Potential energy is stored energy; kinetic energy is the energy of motion.
- Common units of energy are the calorie (cal), kilocalorie (kcal), joule (J), and kilojoule (kJ).
- One calorie is equal to 4.184 J.

3.5 Specific Heat

LEARNING GOAL Calculate the specific heat for a substance. Use specific heat to calculate heat loss or gain.

- Specific heat is the amount of energy required to raise the temperature of exactly 1 g of a substance by exactly 1 °C.
- The heat lost or gained by a substance is determined by multiplying its mass, the temperature change, and its specific heat.

3.6 Energy and Nutrition

LEARNING GOAL Use the energy values to calculate the kilocalories (kcal) or kilojoules (kJ) for a food.

- The nutritional Calorie is the same amount of energy as 1 kcal or 1000 calories.
- The energy of a food is the sum of kilocalories or kilojoules from carbohydrate, fat, and protein.

KEY TERMS

calorie (cal) The amount of heat energy that raises the temperature of exactly 1 g of water by exactly 1 °C.

change of state The transformation of one state of matter to another; for example, solid to liquid, liquid to solid, liquid to gas.

chemical change A change during which the original substance is converted into a new substance that has a different composition and new physical and chemical properties.

chemical properties The properties that indicate the ability of a substance to change into a new substance.

compound A pure substance consisting of two or more elements, with a definite composition, that can be broken down into simpler substances only by chemical methods.

element A pure substance containing only one type of matter, which cannot be broken down by chemical methods.

energy The ability to do work.

energy value The kilocalories (or kilojoules) obtained per gram of the food types: carbohydrate, fat, and protein.

gas A state of matter that does not have a definite shape or volume.

heat The energy associated with the motion of particles in a substance.

heat equation A relationship that calculates heat (q) given the mass, specific heat, and temperature change for a substance.

joule (J) The SI unit of heat energy; $4.184 \text{ J} = 1 \text{ cal}$.

kinetic energy The energy of moving particles.

liquid A state of matter that takes the shape of its container but has a definite volume.

matter The material that makes up a substance and has mass and occupies space.

mixture The physical combination of two or more substances that does not change the identities of the mixed substances.

physical change A change in which the physical properties of a substance change but its identity stays the same.

physical properties The properties that can be observed or measured without affecting the identity of a substance.

potential energy A type of energy related to position or composition of a substance.

pure substance A type of matter that has a definite composition.

solid A state of matter that has its own shape and volume.

specific heat (SH) A quantity of heat that changes the temperature of exactly 1 g of a substance by exactly 1 °C.

states of matter Three forms of matter: solid, liquid, and gas.

CORE CHEMISTRY SKILLS

The chapter section containing each Core Chemistry Skill is shown in parentheses at the end of each heading.

Classifying Matter (3.1)

- A pure substance is matter that has a fixed or constant composition.
- An element, the simplest type of a pure substance, is composed of only one type of matter, such as silver, iron, or aluminum.
- A compound is also a pure substance, but it consists of two or more elements chemically combined in the same proportion.
- In a *homogeneous mixture*, also called a *solution*, the composition of the substances in the mixture is uniform.
- In a *heterogeneous mixture*, the components are visible and do not have a uniform composition throughout the sample.

Example: Classify each of the following as a pure substance (element or compound) or a mixture (homogeneous or heterogeneous):

   a. iron in a nail
   b. black coffee
   c. carbon dioxide, a greenhouse gas

Answer: **a.** Iron is a pure substance, which is an element.

   b. Black coffee contains different substances with uniform composition, which makes it a homogeneous mixture.

   c. The gas carbon dioxide, which is a pure substance that contains two elements chemically combined, is a compound.

Identifying Physical and Chemical Changes (3.2)

- When matter undergoes a physical change, its state or its appearance changes, but its composition remains the same.
- When a chemical change takes place, the original substance is converted into a new substance, which has different physical and chemical properties.

Example: Classify each of the following as a physical or chemical change:

   a. Silver metal is melted and poured into a mold to make a ring.
   b. Methane, in natural gas, burns in air.
   c. Potassium reacts spontaneously with water to form hydrogen gas.

Answer: **a.** A change from a solid to a liquid (melting) is a physical change.

   b. When methane burns, it changes to different substances with new properties, which is a chemical change.

   c. The reaction of potassium with water produces new substances (including hydrogen gas), which makes this a chemical change.

Converting between Temperature Scales (3.3)

- The temperature equation $T_F = 1.8(T_C) + 32$ is used to convert from Celsius to Fahrenheit and can be rearranged to convert from Fahrenheit to Celsius.
- The temperature equation $T_K = T_C + 273$ is used to convert from Celsius to Kelvin and can be rearranged to convert from Kelvin to Celsius.

Example: Convert 75.0 °C to degrees Fahrenheit.

Answer: $T_F = 1.8(T_C) + 32$

$T_F = 1.8(75.0) + 32 = 135 + 32$

$= 167 °F$

Example: Convert 355 K to degrees Celsius.

Answer: $T_K = T_C + 273$

To solve the equation for T_C, subtract 273 from both sides.

$T_K - 273 = T_C + \cancel{273} - \cancel{273}$

$T_C = T_K - 273$

$T_C = 355 - 273$

$= 82 °C$

Using Energy Units (3.4)

- Equalities for energy units include 1 cal = 4.184 J, 1 kcal = 1000 cal, and 1 kJ = 1000 J.
- Each equality for energy units can be written as two conversion factors:

$$\frac{4.184 \text{ J}}{1 \text{ cal}} \text{ and } \frac{1 \text{ cal}}{4.184 \text{ J}} \qquad \frac{1000 \text{ cal}}{1 \text{ kcal}} \text{ and } \frac{1 \text{ kcal}}{1000 \text{ cal}} \qquad \frac{1000 \text{ J}}{1 \text{ kJ}} \text{ and } \frac{1 \text{ kJ}}{1000 \text{ J}}$$

- The energy unit conversion factors are used to cancel given units of energy and to obtain the needed unit of energy.

Example: Convert 45 000 J to kilocalories.

Answer: Using the conversion factors above, we start with the given 45 000 J and convert it to kilocalories.

$$45\,000 \text{ J} \times \frac{1 \text{ cal}}{4.184 \text{ J}} \times \frac{1 \text{ kcal}}{1000 \text{ cal}} = 11 \text{ kcal}$$

Calculating Specific Heat (3.5)

- Specific heat (SH) is the amount of heat (q) that raises the temperature of 1 g of a substance by 1 °C.

$$SH = \frac{q}{m \times \Delta T} = \frac{\text{cal (or J)}}{\text{g °C}}$$

- To calculate specific heat, the heat lost or gained is divided by the mass of the substance and the change in temperature (ΔT).

Example: Calculate the specific heat, in J/g °C, for a 4.0-g sample of tin that absorbs 66 J when heated from 125 °C to 197 °C.

Answer: $q = 66 \text{ J} \quad m = 4.0 \text{ g} \quad \Delta T = 197 °C - 125 °C = 72 °C$

$$SH = \frac{q}{m \times \Delta T} = \frac{66 \text{ J}}{4.0 \text{ g} \times 72 °C} = \frac{0.23 \text{ J}}{\text{g °C}}$$

Using the Heat Equation (3.5)

- The quantity of heat (q) absorbed or lost by a substance is calculated using the heat equation.
$q = m \times \Delta T \times SH$
- Heat, in calories or joules, is obtained when the specific heat of a substance is used.
- To cancel, the unit grams is used for mass, and the unit °C is used for temperature change.

Example: How many joules are required to heat 5.25 g of titanium from 85.5 °C to 132.5 °C?

Answer: $m = 5.25 \text{ g} \quad \Delta T = 132.5 °C - 85.5 °C = 47.0 °C$

SH for titanium $= 0.523$ J/g °C

The known values are substituted into the heat equation making sure units cancel.

$q = m \times \Delta T \times SH$

$= 5.25 \cancel{\text{ g}} \times 47.0 \cancel{°C} \times \dfrac{0.523 \text{ J}}{\cancel{\text{g}} \cancel{°C}}$

$= 129 \text{ J}$

UNDERSTANDING THE CONCEPTS

The chapter sections to review are shown in parentheses at the end of each question.

3.49 Identify each of the following as an element, a compound, or a mixture. Explain your choice. (3.1)

a. 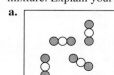 **b.**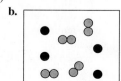

c.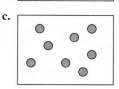

3.50 Identify each of the following as a homogeneous or heterogeneous mixture. Explain your choice. (3.1)

a. 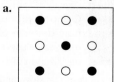 **b.**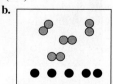

c.

3.51 Classify each of the following as a homogeneous or heterogeneous mixture: (3.1)
 a. lemon-flavored water
 b. stuffed mushrooms
 c. eye drops

3.52 Classify each of the following as a homogeneous or heterogeneous mixture: (3.1)
 a. ketchup **b.** hard-boiled egg **c.** tortilla soup

3.53 State the temperature on the Celsius thermometer and convert to Fahrenheit. (3.3)

3.54 State the temperature on the Celsius thermometer and convert to Fahrenheit. (3.3)

3.55 Compost can be made at home from grass clippings, some kitchen scraps, and dry leaves. As microbes break down organic matter, heat is generated and the compost can reach a temperature of 155 °F, which kills most pathogens. What is this temperature in degrees Celsius? In kelvins? (3.3)

Compost produced from decayed plant material is used to enrich the soil.

3.56 After a week, biochemical reactions in compost slow, and the temperature drops to 45 °C. The dark brown organic-rich mixture is ready for use in the garden. What is this temperature in degrees Fahrenheit? In kelvins? (3.3)

3.57 Calculate the energy to heat two cubes (gold and aluminum) each with a volume of 10.0 cm³ from 15 °C to 25 °C. Refer to Tables 2.9 and 3.7. (3.5)

3.58 Calculate the energy to heat two cubes (silver and copper), each with a volume of 10.0 cm³ from 15 °C to 25 °C. Refer to Tables 2.9 and 3.7. (3.5)

 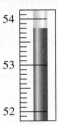

Applications

3.59 A 70.0-kg person had a quarter-pound cheeseburger, french fries, and a chocolate shake. (3.6)

Item	Carbohydrate (g)	Fat (g)	Protein (g)
Cheeseburger	46	40.	47
French fries	47	16	4
Chocolate shake	76	10.	10.

 a. Using Table 3.8, calculate the total kilocalories for each food type in this meal (round off the kilocalories to the tens place).
 b. Determine the total kilocalories for the meal (round off to the tens place).
 c. Using Table 3.11, determine the number of hours of sleep needed to burn off the kilocalories in this meal.
 d. Using Table 3.11, determine the number of hours of running needed to burn off the kilocalories in this meal.

3.60 Your friend, who has a mass of 70.0 kg, has a slice of pizza, a cola soft drink, and ice cream. (3.6)

Item	Carbohydrate (g)	Fat (g)	Protein (g)
Pizza	29	10.	13
Cola	51	0	0
Ice cream	44	28	8

 a. Using Table 3.8, calculate the total kilocalories for each food type in this meal (round off the kilocalories to the tens place).
 b. Determine the total kilocalories for the meal (round off to the tens place).
 c. Using Table 3.11, determine the number of hours of sitting needed to burn off the kilocalories in this meal.
 d. Using Table 3.11, determine the number of hours of swimming needed to burn off the kilocalories in this meal.

ADDITIONAL QUESTIONS AND PROBLEMS

3.61 Classify each of the following as an element, a compound, or a mixture: (3.1)
   a. carbon in pencils
   b. carbon monoxide (CO) in automobile exhaust
   c. orange juice

3.62 Classify each of the following as an element, a compound, or a mixture: (3.1)
   a. neon gas in lights
   b. a salad dressing of oil and vinegar
   c. sodium hypochlorite (NaOCl) in bleach

3.63 Classify each of the following mixtures as homogeneous or heterogeneous: (3.1)
   a. hot fudge sundae   **b.** herbal tea   **c.** vegetable oil

3.64 Classify each of the following mixtures as homogeneous or heterogeneous: (3.1)
   a. water and sand   **b.** mustard   **c.** blue ink

3.65 Identify each of the following as solid, liquid, or gas: (3.2)
   a. vitamin tablets in a bottle   **b.** helium in a balloon
   c. milk in a bottle   **d.** the air you breathe
   e. charcoal briquettes on a barbecue

3.66 Identify each of the following as solid, liquid, or gas: (3.2)
   a. popcorn in a bag   **b.** water in a garden hose
   c. a computer mouse   **d.** air in a tire
   e. hot tea in a teacup

3.67 Identify each of the following as a physical or chemical property: (3.2)
   a. Gold is shiny.
   b. Gold melts at 1064 °C.
   c. Gold is a good conductor of electricity.
   d. When gold reacts with yellow sulfur, a black sulfide compound forms.

3.68 Identify each of the following as a physical or chemical property: (3.2)
   a. A candle is 10 in. high and 2 in. in diameter.
   b. A candle burns.
   c. The wax of a candle softens on a hot day.
   d. A candle is blue.

3.69 Identify each of the following as a physical or chemical change: (3.2)
   a. A plant grows a new leaf.
   b. Chocolate is melted for a dessert.
   c. Wood is chopped for the fireplace.
   d. Wood burns in a woodstove.

3.70 Identify each of the following as a physical or chemical change: (3.2)
   a. Aspirin tablets are broken in half.
   b. Carrots are grated for use in a salad.
   c. Malt undergoes fermentation to make beer.
   d. A copper pipe reacts with air and turns green.

3.71 Calculate each of the following temperatures in degrees Celsius and kelvins: (3.3)
   a. The highest recorded temperature in the continental United States was 134 °F in Death Valley, California, on July 10, 1913.
   b. The lowest recorded temperature in the continental United States was −69.7 °F in Rodgers Pass, Montana, on January 20, 1954.

3.72 Calculate each of the following temperatures in kelvins and degrees Fahrenheit: (3.3)
   a. The highest recorded temperature in the world was 58.0 °C in El Azizia, Libya, on September 13, 1922.
   b. The lowest recorded temperature in the world was −89.2 °C in Vostok, Antarctica, on July 21, 1983.

3.73 What is −15 °F in degrees Celsius and in kelvins? (3.3)

3.74 On a hot sunny day, you get out of the swimming pool and sit in a metal chair, which is very hot. Would you predict that the specific heat of the metal is higher or lower than that of water? Explain. (3.5)

3.75 On a hot day, the beach sand gets hot but the water stays cool. Would you predict that the specific heat of sand is higher or lower than that of water? Explain. (3.5)

On a sunny day, the sand gets hot but the water stays cool.

3.76 A large bottle containing 883 g of water at 4 °C is removed from the refrigerator. How many kilojoules are absorbed to warm the water to room temperature of 22 °C? (3.5)

3.77 A 0.50-g sample of vegetable oil is placed in a calorimeter. When the sample is burned, 18.9 kJ is given off. What is the energy value, in kcal/g, for the oil? (3.6)

3.78 A 1.3-g sample of rice is placed in a calorimeter. When the sample is burned, 22.00 kJ is given off. What is the energy value, in kcal/g, for the rice? (3.6)

3.79 When a substance with a mass of 45.6 g absorbs 575 J, its temperature increases from 21 °C to 35 °C. Calculate the specific heat, in J/g °C, of the substance. Identify the substance from Table 3.7. (3.5)

3.80 When a substance with a mass of 8.36 g absorbs 378 J, its temperature increases from 17.2 °C to 35.6 °C. Calculate the specific heat, in J/g °C, of the substance. Identify the substance from Table 3.7. (3.5)

3.81 Use the heat equation to calculate the energy, in joules, for each of the following (see Table 3.7): (3.5)
   a. lost when 15.0 g of ethanol, C_2H_6O, cools from 60.5 °C to −42.0 °C
   b. lost when 125 g of iron (Fe) cools from 118 °C to 55 °C

3.82 Use the heat equation to calculate the energy, in joules, for each of the following (see Table 3.7): (3.5)
   a. lost when 75.0 g of water, H_2O, cools from 86.4 °C to 2.1 °C
   b. lost when 18.0 g of gold (Au) cools from 224 °C to 118 °C

3.83 Calculate the mass, in grams, for each of the following using Table 3.7: (3.5)
   a. a sample of aluminum (Al) that absorbs 8.80 kJ when heated from 12.5 °C to 26.8 °C
   b. a sample of titanium (Ti) that loses 14 200 J when it cools from 185 °C to 42 °C

3.84 Calculate the mass, in grams, for each of the following using Table 3.7: (3.5)
  a. a sample of silver (Ag) that absorbs 1650 J when its temperature increases from 65 °C to 187 °C
  b. an iron (Fe) bar that loses 2.52 kJ when its temperature decreases from 252 °C to 75 °C

3.85 Calculate the change in temperature (ΔT) for each of the following using Table 3.7: (3.5)
  a. 85.0 g of gold (Au) that absorbs 7680 J
  b. 50.0 g of copper (Cu) that absorbs 6.75 kJ

3.86 Calculate the change in temperature (ΔT) for each of the following using Table 3.7: (3.5)
  a. 0.650 kg of water, H_2O, that loses 5.48 kJ
  b. 35.0 g of silver (Ag) that loses 472 J

Applications

3.87 If you want to lose 1 lb of "body fat," which is 15% water, how many kilocalories do you need to expend? (3.6)

3.88 The highest recorded body temperature that a person has survived is 46.5 °C. Calculate that temperature in degrees Fahrenheit and in kelvins. (3.3)

3.89 A hot-water bottle for a patient contains 725 g of water at 65 °C. If the water cools to body temperature (37 °C), how many kilojoules of heat could be transferred to sore muscles? (3.5)

3.90 A young patient drinks whole milk as part of her diet. Calculate the total kilocalories if the glass of milk contains 12 g of carbohydrate, 9 g of fat, and 9 g of protein. (Round off answers for each food type to the tens place.) (3.6)

CHALLENGE QUESTIONS

The following groups of questions are related to the topics in this chapter. However, they do not all follow the chapter order, and they require you to combine concepts and skills from several sections. These questions will help you increase your critical thinking skills and prepare for your next exam.

3.91 When a 0.66-g sample of olive oil is burned in a calorimeter, the heat released increases the temperature of 370 g of water from 22.7 °C to 38.8 °C. What is the energy value, in kcal/g, for the olive oil? (3.5, 3.6)

3.92 When a 0.47-g sample of brown sugar is burned in a calorimeter, the heat released increases the temperature of 260 g of water from 21.6 °C to 28.9 °C. What is the energy value, in kcal/g, for the brown sugar? (3.5, 3.6)

3.93 In a large building, oil is used in a steam boiler heating system. The combustion of 1.0 lb of oil provides 2.4×10^7 J. How many kilograms of oil are needed to heat 150 kg of water from 22 °C to 100 °C? (3.4, 3.5)

3.94 When 1.0 g of gasoline burns, it releases 11 kcal of heat. The density of gasoline is 0.74 g/mL. (3.4, 3.5)
  a. How many megajoules are released when 1.0 gal of gasoline burns?
  b. If a television requires 150 kJ/h to run, how many hours can the television run on the energy provided by 1.0 gal of gasoline?

3.95 A 70.0-g piece of copper metal at 54.0 °C is placed in 50.0 g of water at 26.0 °C. If the final temperature of the water and metal is 29.2 °C, what is the specific heat (J/g °C) of copper? (3.5)

3.96 A 125-g piece of metal is heated to 288 °C and dropped into 85.0 g of water at 12.0 °C. The metal and water come to the same temperature of 24.0 °C. What is the specific heat, in J/g °C, of the metal? (3.5)

3.97 A metal is thought to be titanium or aluminum. When 4.7 g of the metal absorbs 11 J, its temperature rises by 4.5 °C. (3.5)
  a. What is the specific heat, in J/g °C, of the metal?
  b. Would you identify the metal as titanium or aluminum (see Table 3.7)?

3.98 A metal is thought to be copper or gold. When 18 g of the metal absorbs 58 cal, its temperature rises by 35 °C. (3.5)
  a. What is the specific heat, in cal/g °C, of the metal?
  b. Would you identify the metal as copper or gold (see Table 3.7)?

ANSWERS

Answers to Selected Questions and Problems

3.1 a. pure substance   **b.** mixture
  c. pure substance   **d.** pure substance
  e. mixture

3.3 a. element   **b.** compound
  c. element   **d.** compound
  e. compound

3.5 a. heterogeneous   **b.** homogeneous
  c. homogeneous   **d.** heterogeneous
  e. heterogeneous

3.7 a. gas   **b.** gas   **c.** solid

3.9 a. physical   **b.** chemical   **c.** physical
  d. chemical   **e.** chemical

3.11 a. physical   **b.** chemical   **c.** physical
  d. physical   **e.** physical

3.13 a. chemical   **b.** physical   **c.** physical
  d. chemical   **e.** physical

3.15 In the United States, we still use the Fahrenheit temperature scale. In °F, normal body temperature is 98.6. On the Celsius scale, her temperature would be 37.7 °C, a mild fever.

3.17 a. 98.6 °F   **b.** 18.5 °C   **c.** 246 K
  d. 335 K   **e.** 46 °C

3.19 a. 41 °C   **b.** No. The temperature is equivalent to 39 °C.

3.21 When the roller-coaster car is at the top of the ramp, it has its maximum potential energy. As it descends, potential energy changes to kinetic energy. At the bottom, all the energy is kinetic.

3.23 a. potential   **b.** kinetic
     c. potential   **d.** potential

3.25 a. 3.5 kcal   **b.** 99.2 cal
     c. 120 J   **d.** 1100 cal

3.27 a. 8.1×10^5 J   **b.** 190 kcal

3.29 Copper, which has the lowest specific heat, would reach the highest temperature.

3.31 a. 0.389 J/g °C   **b.** 0.313 J/g °C

3.33 a. 1380 J   **b.** 1810 J

3.35 a. 54.5 g   **b.** 216 g

3.37 a. 175 °C   **b.** 11.3 °C

3.39 a. 29.9 kcal   **b.** 208 kcal

3.41 a. 470 kJ   **b.** 18 g
     c. 130 kcal   **d.** 37 g

3.43 210 kcal, 880 kJ

3.45 640 kcal

3.47 a. Breakfast 270 kcal; lunch 420 kcal; dinner 440 kcal
     b. 1130 kcal total
     c. Yes. Charles should be losing weight.
     d. 26 days

3.49 a. compound, the molecules have a definite 2:1 ratio of atoms
     b. mixture, has two different kinds of atoms and molecules
     c. element, has a single kind of atom

3.51 a. homogeneous   **b.** heterogeneous   **c.** homogeneous

3.53 61.4 °C, 143 °F

3.55 68.3 °C, 341 K

3.57 gold, 250 J or 60. cal; aluminum, 240 J or 58 cal

3.59 a. carbohydrate, 680 kcal; fat, 590 kcal; protein, 240 kcal
     b. 1510 kcal
     c. 25 h
     d. 2.0 h

3.61 a. element   **b.** compound   **c.** mixture

3.63 a. heterogeneous   **b.** homogeneous
     c. homogeneous

3.65 a. solid   **b.** gas
     c. liquid   **d.** gas
     e. solid

3.67 a. physical   **b.** physical
     c. physical   **d.** chemical

3.69 a. chemical   **b.** physical
     c. physical   **d.** chemical

3.71 a. 56.7 °C, 330. K   **b.** −56.5 °C, 217 K

3.73 −26 °C, 247 K

3.75 The same amount of heat causes a greater temperature change in the sand than in the water; thus the sand must have a lower specific heat than that of water.

3.77 9.0 kcal/g

3.79 0.90 J/g °C

3.81 a. 3780 J   **b.** 3600 J

3.83 a. 686 g   **b.** 190. g

3.85 a. 700. °C   **b.** 351 °C

3.87 3500 kcal

3.89 85 kJ

3.91 9.1 kcal/g

3.93 0.93 kg

3.95 0.385 J/g °C

3.97 a. 0.52 J/g °C   **b.** titanium

CI.1 Gold, one of the most sought-after metals in the world, has a density of 19.3 g/cm³, a melting point of 1064 °C, and a specific heat of 0.129 J/g °C. A gold nugget found in Alaska in 1998 weighs 20.17 lb. (2.4, 2.6, 2.7, 3.3, 3.5)

Gold nuggets, also called native gold, can be found in streams and mines.

  a. How many significant figures are in the measurement of the weight of the nugget?
  b. Which is the mass of the nugget in kilograms?
  c. If the nugget were pure gold, what would its volume be in cubic centimeters?
  d. What is the melting point of gold in degrees Fahrenheit and kelvins?
  e. How many kilocalories are required to raise the temperature of the nugget from 500. °C to 1064 °C?
  f. If the price of gold is $45.98 per gram, what is the nugget worth in dollars?

CI.2 The mileage for a motorcycle with a fuel-tank capacity of 22 L is 35 mi/gal. (2.5, 2.6, 2.7, 3.4)

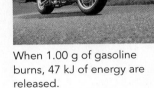

  a. How long a trip, in kilometers, can be made on one full tank of gasoline?
  b. If the price of gasoline is $3.82 per gallon, what would be the cost of fuel for the trip?
  c. If the average speed during the trip is 44 mi/h, how many hours will it take to reach the destination?

When 1.00 g of gasoline burns, 47 kJ of energy are released.

  d. If the density of gasoline is 0.74 g/mL, what is the mass, in grams, of the fuel in the tank?
  e. When 1.00 g of gasoline burns, 47 kJ of energy is released. How many kilojoules are produced when the fuel in one full tank is burned?

CI.3 Answer the following questions for the water samples **A** and **B** shown in the diagrams: (3.1, 3.2, 3.5)

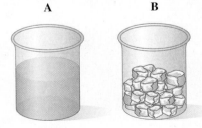

a. In which sample (**A** or **B**) does the water have its own shape?
b. Which diagram (**1** or **2** or **3**) represents the arrangement of particles in water sample **A**?
c. Which diagram (**1** or **2** or **3**) represents the arrangement of particles in water sample **B**?

Answer the following for diagrams **1**, **2**, and **3**: (3.2, 3.3)

d. The state of matter indicated in diagram **1** is a _____; in diagram **2**, it is a _____; and in diagram **3**, it is a _____.
e. The motion of the particles is slowest in diagram _____.
f. The arrangement of particles is farthest apart in diagram _____.
g. The particles fill the volume of the container in diagram _____.
h. If the water in diagram **2** has a mass of 19 g and a temperature of 45 °C, how much heat, in kilojoules, is removed to cool the liquid to 0 °C?

CI.4 The label of a black cherry almond energy bar with a mass of 68 g lists the nutrition facts as 39 g of carbohydrate, 5 g of fat, and 10. g of protein. (2.5, 2.6, 3.4, 3.6)

An energy bar contains carbohydrate, fat, and protein.

  a. Using the energy values for carbohydrates, fats, and proteins (see Table 3.8), what are the total kilocalories listed for a black cherry almond bar? (Round off answers for each food type to the tens place.)
  b. What are the kilojoules for the black cherry almond bar? (Round off answers for each food type to the tens place.)
  c. If you obtain 160 kJ, how many grams of the black cherry almond bar did you eat?
  d. If you are walking and using energy at a rate of 840 kJ/h, how many minutes will you need to walk to expend the energy from two black cherry almond bars?

CI.5 In one box of nails, there are 75 iron nails weighing 0.250 lb. The density of iron is 7.86 g/cm³. The specific heat of iron is 0.452 J/g °C. The melting point of iron is 1535 °C. (2.5, 2.6, 2.7, 3.4, 3.5)

Nails made of iron have a density of 7.86 g/cm³.

a. What is the volume, in cubic centimeters, of the iron nails in the box?

b. If 30 nails are added to a graduated cylinder containing 17.6 mL of water, what is the new level of water, in milliliters, in the cylinder?

c. How much heat, in joules, must be added to the nails in the box to raise their temperature from 16 °C to 125 °C?

d. How much heat, in joules, is required to heat one nail from 25 °C to its melting point?

CI.6 A hot tub is filled with 450 gal of water. (2.5, 2.6, 2.7, 3.3, 3.4, 3.5)

A hot tub filled with water is heated to 105 °F.

a. What is the volume of water, in liters, in the tub?

b. What is the mass, in kilograms, of water in the tub?

c. How many kilocalories are needed to heat the water from 62 °F to 105 °F?

d. If the hot-tub heater provides 5900 kJ/min, how long, in minutes, will it take to heat the water in the hot tub from 62 °F to 105 °F?

ANSWERS

CI.1 **a.** 4 significant figures
b. 9.147 kg
c. 474 cm³
d. 1947 °F; 1337 K
e. 159 kcal
f. $421 000

CI.3 **a.** B
b. A is represented by diagram **2**.
c. B is represented by diagram **1**.
d. solid; liquid; gas

e. diagram **1**
f. diagram **3**
g. diagram **3**
h. 3.6 kJ

CI.5 **a.** 14.4 cm³
b. 23.4 mL
c. 5590 J
d. 1030 J

4

Atoms and Elements

JOHN IS PREPARING FOR the next growing season as he decides how much of each crop should be planted and their location on his farm. Part of this decision is determined by the quality of the soil, including the pH, the amount of moisture, and the nutrient content in the soil. He begins by sampling the soil and performing a few chemical tests on the soil. John determines that several of his fields need additional fertilizer before the crops can be planted. John considers several different types of fertilizers as each supplies different nutrients to the soil to help increase crop production. Plants need three basic elements for growth. These elements are potassium, nitrogen, and phosphorus. Potassium (K on the periodic table) is a metal, whereas nitrogen (N) and phosphorus (P) are nonmetals. Fertilizers may also contain several other elements including calcium (Ca), magnesium (Mg), and sulfur (S). John applies a fertilizer containing a mixture of all of these elements to his soil and plans to re-check the soil nutrient content in a few days.

CAREER

Farmer

Farming involves much more than growing crops and raising animals. Farmers must understand how to perform chemical tests and how to apply fertilizer to soil and pesticides or herbicides to crops. Pesticides are chemicals used to kill insects that could destroy the crop, whereas herbicides are chemicals used to kill weeds that would compete for the crop's water and nutrient supply. This requires knowledge of how these chemicals work, their safety, effectiveness, and their storage. In using this information, farmers are able to grow crops that produce a higher yield, greater nutritional value, and better taste.

KEY MATH SKILLS
- Using Positive and Negative Numbers in Calculations (1.4)
- Calculating Percentages (1.4)
- Rounding Off (2.3)

CORE CHEMISTRY SKILLS
- Counting Significant Figures (2.2)
- Using Significant Figures in Calculations (2.3)

*These Key Math Skills and Core Chemistry Skills from previous chapters are listed here for your review as you proceed to the new material in this chapter.

LOOKING AHEAD

4.1 Elements and Symbols
4.2 The Periodic Table
4.3 The Atom
4.4 Atomic Number and Mass Number
4.5 Isotopes and Atomic Mass

4.1 Elements and Symbols

LEARNING GOAL Given the name of an element, write its correct symbol; from the symbol, write the correct name.

All matter is composed of *elements*, of which there are 118 different kinds. Of these, 88 elements occur naturally and make up all the substances in our world. Many elements are already familiar to you. Perhaps you use aluminum in the form of foil or drink soft drinks from aluminum cans. You may have a ring or necklace made of gold, silver, or perhaps platinum. If you play tennis or golf, then you may have noticed that your racket or clubs may be made from the elements titanium or carbon. In our bodies, calcium and phosphorus form the structure of bones and teeth, iron and copper are needed in the formation of red blood cells, and iodine is required for the proper functioning of the thyroid.

Elements are pure substances from which all other things are built. Elements cannot be broken down into simpler substances. Over the centuries, elements have been named for planets, mythological figures, colors, minerals, geographic locations, and famous people. Some sources of names of elements are listed in **TABLE 4.1**. A complete list of all the elements and their symbols are found on the inside front cover of this text.

TABLE 4.1 Some Elements, Symbols, and Source of Names

Element	Symbol	Source of Name
Uranium	U	The planet Uranus
Titanium	Ti	Titans (mythology)
Chlorine	Cl	*Chloros*: "greenish yellow" (Greek)
Iodine	I	*Ioeides*: "violet" (Greek)
Magnesium	Mg	Magnesia, a mineral
Californium	Cf	California
Curium	Cm	Marie and Pierre Curie
Copernicium	Cn	Nicolaus Copernicus

Chemical Symbols

Chemical symbols are one- or two-letter abbreviations for the names of the elements. Only the first letter of an element's symbol is capitalized. If the symbol has a second letter, it is lowercase so that we know when a different element is indicated. If two letters are capitalized, they represent the symbols of two different elements. For example, the element cobalt has the symbol Co. However, the two capital letters CO specify two elements, carbon (C) and oxygen (O).

Although most of the symbols use letters from the current names, some are derived from their ancient names. For example, Na, the symbol for sodium, comes from the Latin

One-Letter Symbols		Two-Letter Symbols	
C	Carbon	Co	Cobalt
S	Sulfur	Si	Silicon
N	Nitrogen	Ne	Neon
H	Hydrogen	He	Helium

What are the names and symbols of some elements that you encounter every day?

Aluminum

Carbon

Gold

Silver

Sulfur

TABLE 4.2 Names and Symbols of Some Common Elements

Name*	Symbol	Name*	Symbol	Name*	Symbol
Aluminum	Al	Gallium	Ga	Oxygen	O
Argon	Ar	Gold (*aurum*)	Au	Phosphorus	P
Arsenic	As	Helium	He	Platinum	Pt
Barium	Ba	Hydrogen	H	Potassium (*kalium*)	K
Boron	B	Iodine	I	Radium	Ra
Bromine	Br	Iron (*ferrum*)	Fe	Silicon	Si
Cadmium	Cd	Lead (*plumbum*)	Pb	Silver (*argentum*)	Ag
Calcium	Ca	Lithium	Li	Sodium (*natrium*)	Na
Carbon	C	Magnesium	Mg	Strontium	Sr
Chlorine	Cl	Manganese	Mn	Sulfur	S
Chromium	Cr	Mercury (*hydrargyrum*)	Hg	Tin (*stannum*)	Sn
Cobalt	Co	Neon	Ne	Titanium	Ti
Copper (*cuprum*)	Cu	Nickel	Ni	Uranium	U
Fluorine	F	Nitrogen	N	Zinc	Zn

*Names given in parentheses are ancient Latin or Greek words from which the symbols are derived.

word *natrium*. The symbol for iron, Fe, is derived from the Latin name *ferrum*. **TABLE 4.2** lists the names and symbols of some common elements. Learning their names and symbols will greatly help your learning of chemistry.

SAMPLE PROBLEM 4.1 **Names and Symbols of Chemical Elements**

Complete the following table with the correct name or symbol of each element:

Name	Symbol
nickel	_____
nitrogen	_____
_____	Zn
_____	K
iron	_____

TRY IT FIRST

SOLUTION

Name	Symbol
nickel	Ni
nitrogen	N
zinc	Zn
potassium	K
iron	Fe

STUDY CHECK 4.1

Write the chemical symbols for the elements silicon, sulfur, and silver.

ANSWER

Si, S, and Ag

CHEMISTRY LINK TO THE ENVIRONMENT
Many Forms of Carbon

Carbon has the symbol C. However, its atoms can be arranged in different ways to give several different substances. Two forms of carbon—diamond and graphite—have been known since prehistoric times. A diamond is transparent and harder than any other substance, whereas graphite is black and soft. In diamond, carbon atoms are arranged in a rigid structure. In graphite, carbon atoms are arranged in sheets that slide over each other. Graphite is used as pencil lead, as lubricants, and as carbon fibers for the manufacture of lightweight golf clubs and tennis rackets.

Two other forms of carbon have been discovered more recently. In the form called *Buckminsterfullerene* or *buckyball* (named after

R. Buckminster Fuller, who popularized the geodesic dome), 60 carbon atoms are arranged as rings of five and six atoms to give a spherical, cage-like structure. When a fullerene structure is stretched out, it produces a cylinder with a diameter of only a few nanometers called a *nanotube*. Practical uses for buckyballs and nanotubes are not yet developed, but they are expected to find use in lightweight structural materials, heat conductors, computer parts, and medicine. Recent research has shown that carbon nanotubes (CNT) can carry many drug molecules that can be released once the CNT enter the targeted cells.

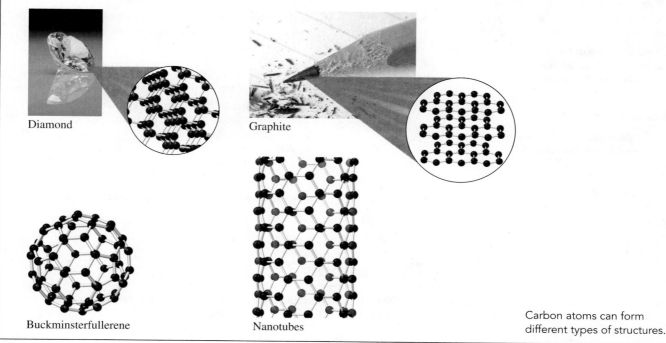

Diamond

Graphite

Buckminsterfullerene

Nanotubes

Carbon atoms can form different types of structures.

CHEMISTRY LINK TO HEALTH
Toxicity of Mercury

Mercury (Hg) is a silvery, shiny element that is a liquid at room temperature. Mercury can enter the body through inhaled mercury vapor, contact with the skin, or ingestion of foods or water contaminated with mercury. In the body, mercury destroys proteins and disrupts cell function. Long-term exposure to mercury can damage the brain and kidneys, cause mental retardation, and decrease physical development. Blood, urine, and hair samples are used to test for mercury.

In both freshwater and seawater, bacteria convert mercury into toxic methylmercury, which attacks the central nervous system. Because fish absorb methylmercury, we are exposed to mercury by consuming mercury-contaminated fish. As levels of mercury ingested from fish became a concern, the Food and Drug Administration set a maximum level of one part mercury per million parts seafood (1 ppm), which is the same as 1 mg of mercury in every

The element mercury is a silvery, shiny liquid at room temperature.

kilogram of seafood. Fish higher in the food chain, such as swordfish and shark, can have such high levels of mercury that the U.S. Environmental Protection Agency (EPA) recommends they be consumed no more than once a week.

One of the worst incidents of mercury poisoning occurred in Minamata and Niigata, Japan, in 1965. At that time, the ocean was polluted with high levels of mercury from industrial wastes. Because

fish were a major food in the diet, more than 2000 people were affected with mercury poisoning and died or developed neural damage. In the United States between 1988 and 1997, the use of mercury decreased by 75% when the use of mercury was banned in paint and pesticides, and regulated in batteries and other products. Certain batteries and compact fluorescent light bulbs (CFL) contain mercury, and instructions for their safe disposal should be followed.

QUESTIONS AND PROBLEMS

4.1 Elements and Symbols

LEARNING GOAL Given the name of an element, write its correct symbol; from the symbol, write the correct name.

4.1 Write the symbols for the following elements:
   a. copper   **b.** platinum   **c.** calcium   **d.** manganese
   e. iron   **f.** barium   **g.** lead   **h.** strontium

4.2 Write the symbols for the following elements:
   a. oxygen   **b.** lithium   **c.** uranium   **d.** titanium
   e. hydrogen   **f.** chromium   **g.** tin   **h.** gold

Applications

4.3 Write the name for the symbol of each of the following elements essential in the body:
   a. C   **b.** Cl   **c.** I   **d.** Se
   e. N   **f.** S   **g.** Zn   **h.** Co

4.4 Write the name for the symbol of each of the following elements essential in the body:
   a. V   **b.** P   **c.** Na   **d.** As
   e. Ca   **f.** Mo   **g.** Mg   **h.** Si

4.5 Write the names for the elements in each of the following formulas of compounds used in medicine:
   a. table salt, $NaCl$
   b. plaster casts, $CaSO_4$
   c. Demerol, $C_{15}H_{22}ClNO_2$
   d. treatment of bipolar disorder, Li_2CO_3

4.6 Write the names for the elements in each of the following formulas of compounds used in medicine:
   a. salt substitute, KCl
   b. dental cement, $Zn_3(PO_4)_2$
   c. antacid, $Mg(OH)_2$
   d. contrast agent for X-ray, $BaSO_4$

4.2 The Periodic Table

LEARNING GOAL Use the periodic table to identify the group and the period of an element; identify the element as a metal, a nonmetal, or a metalloid.

As more elements were discovered, it became necessary to organize them into some type of classification system. By the late 1800s, scientists recognized that certain elements looked alike and behaved much the same way. In 1869, a Russian chemist, Dmitri Mendeleev, arranged the 60 elements known at that time into groups with similar properties and placed them in order of increasing atomic masses. Today, this arrangement of 118 elements is known as the **periodic table** (see **FIGURE 4.1**).

Periods and Groups

Each horizontal row in the periodic table is a **period** (see **FIGURE 4.2**). The periods are counted from the top of the table as Periods 1 to 7. The first period contains two elements: hydrogen (H) and helium (He). The second period contains eight elements: lithium (Li), beryllium (Be), boron (B), carbon (C), nitrogen (N), oxygen (O), fluorine (F), and neon (Ne). The third period also contains eight elements beginning with sodium (Na) and ending with argon (Ar). The fourth period, which begins with potassium (K), and the fifth period, which begins with rubidium (Rb), have 18 elements each. The sixth period, which begins with cesium (Cs), has 32 elements. The seventh period, as of today, contains 32 elements, for a total of 118 elements.

Each vertical column on the periodic table contains a **group** (or family) of elements that have similar properties. A **group number** is written at the top of each vertical column (group) in the periodic table. For many years, the **representative elements** have had group

ENGAGE

If element 119 were discovered, in what group and period would it be placed?

Periodic Table of Elements

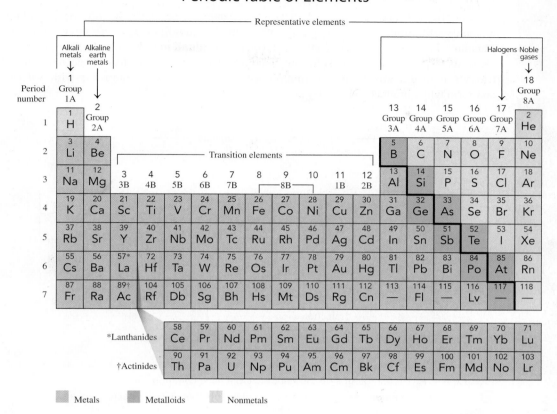

FIGURE 4.1 ▶ On the periodic table, groups are the elements arranged as vertical columns, and periods are the elements in each horizontal row.

◉ What is the symbol of the alkali metal in Period 3?

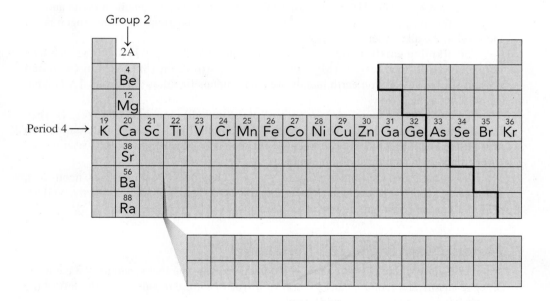

FIGURE 4.2 ▶ On the periodic table, each vertical column represents a group of elements and each horizontal row of elements represents a period.

◉ Are the elements Si, P, and S part of a group or a period?

numbers 1A to 8A. In the center of the periodic table is a block of elements known as the **transition elements**, which have had group numbers followed by the letter "B." A newer system assigns numbers of 1 to 18 to all of the groups going left to right across the periodic table. Because both systems are currently in use, they are both shown on the periodic table in this text and are included in our discussions of elements and group numbers. The two rows of 14 elements called the *lanthanides* and *actinides* (or the inner transition elements), which are part of Periods 6 and 7, are placed at the bottom of the periodic table to allow it to fit on a page.

Group
1A (1)

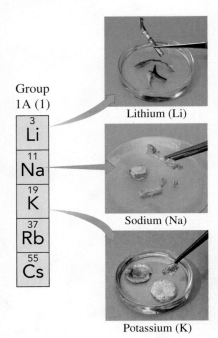

FIGURE 4.4 ▶ Lithium (Li), sodium (Na), and potassium (K) are some alkali metals from Group 1A (1).

◉ What physical properties do these alkali metals have in common?

Group
7A (17)

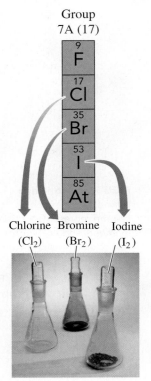

FIGURE 4.5 ▶ Chlorine (Cl_2), bromine (Br_2), and iodine (I_2) are examples of halogens from Group 7A (17).

◉ What elements are in the halogen group?

Names of Groups

Several groups in the periodic table have special names (see **FIGURE 4.3**). Group 1A (1) elements—lithium (Li), sodium (Na), potassium (K), rubidium (Rb), cesium (Cs), and francium (Fr)—are a family of elements known as the **alkali metals** (see **FIGURE 4.4**). The elements within this group are soft, shiny metals that are good conductors of heat and electricity and have relatively low melting points. Alkali metals react vigorously with water and form white products when they combine with oxygen.

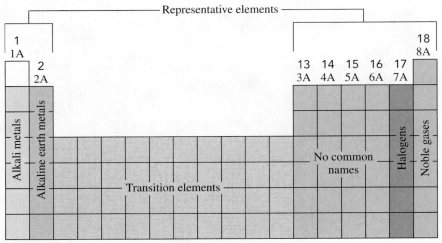

FIGURE 4.3 ▶ Certain groups on the periodic table have common names.

◉ What is the common name for the group of elements that includes helium and argon?

Although hydrogen (H) is at the top of Group 1A (1), it is not an alkali metal and has very different properties than the rest of the elements in this group. Thus, hydrogen is not included in the alkali metals.

The **alkaline earth metals** are found in Group 2A (2). They include the elements beryllium (Be), magnesium (Mg), calcium (Ca), strontium (Sr), barium (Ba), and radium (Ra). The alkaline earth metals are shiny metals like those in Group 1A (1), but they are not as reactive.

The **halogens** are found on the right side of the periodic table in Group 7A (17). They include the elements fluorine (F), chlorine (Cl), bromine (Br), iodine (I), and astatine (At) (see **FIGURE 4.5**). The halogens, especially fluorine and chlorine, are highly reactive and form compounds with most of the elements.

The **noble gases** are found in Group 8A (18). They include helium (He), neon (Ne), argon (Ar), krypton (Kr), xenon (Xe), and radon (Rn). They are quite unreactive and are seldom found in combination with other elements.

Metals, Nonmetals, and Metalloids

Another feature of the periodic table is the heavy zigzag line that separates the elements into the *metals* and the *nonmetals*. *Except for hydrogen*, the metals are to the left of the line with the nonmetals to the right (see **FIGURE 4.6**).

In general, most **metals** are shiny solids, such as copper (Cu), gold (Au), and silver (Ag). Metals can be shaped into wires (ductile) or hammered into a flat sheet (malleable). Metals are good conductors of heat and electricity. They usually melt at higher temperatures than nonmetals. All the metals are solids at room temperature, except for mercury (Hg), which is a liquid.

Nonmetals are not especially shiny, ductile, or malleable, and they are often poor conductors of heat and electricity. They typically have low melting points and low densities. Some examples of nonmetals are hydrogen (H), carbon (C), nitrogen (N), oxygen (O), chlorine (Cl), and sulfur (S).

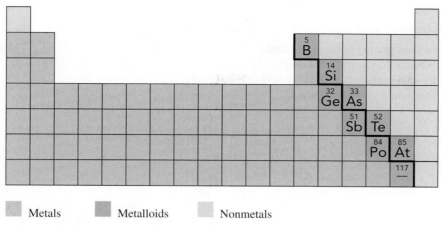

Metals   Metalloids   Nonmetals

FIGURE 4.6 ▶ Along the heavy zigzag line on the periodic table that separates the metals and nonmetals are metalloids, which exhibit characteristics of both metals and nonmetals.

Q On which side of the heavy zigzag line are the nonmetals located?

Except for aluminum, the elements located along the heavy line are **metalloids**: B, Si, Ge, As, Sb, Te, Po, and At. Metalloids are elements that exhibit some properties that are typical of the metals and other properties that are characteristic of the nonmetals. For example, they are better conductors of heat and electricity than the nonmetals, but not as good as the metals. The metalloids are *semiconductors* because they can be modified to function as conductors or insulators. **TABLE 4.3** compares some characteristics of silver, a metal, with those of antimony, a metalloid, and sulfur, a nonmetal.

TABLE 4.3 Some Characteristics of a Metal, a Metalloid, and a Nonmetal

Silver (Ag)	Antimony (Sb)	Sulfur (S)
Metal	Metalloid	Nonmetal
Shiny	Blue-gray, shiny	Dull, yellow
Extremely ductile	Brittle	Brittle
Can be hammered into sheets (malleable)	Shatters when hammered	Shatters when hammered
Good conductor of heat and electricity	Poor conductor of heat and electricity	Poor conductor of heat and electricity, good insulator
Used in coins, jewelry, tableware	Used to harden lead, color glass and plastics	Used in gunpowder, rubber, fungicides
Density 10.5 g/mL	Density 6.7 g/mL	Density 2.1 g/mL
Melting point 962 °C	Melting point 630 °C	Melting point 113 °C

A silver cup is shiny, antimony is blue-gray, and sulfur is a dull, yellow color.

SAMPLE PROBLEM 4.2 Metals, Nonmetals, and Metalloids

Use the periodic table to classify each of the following elements by its group and period, group name (if any), and as a metal, a nonmetal, or a metalloid:

a. Na, important in nerve impulses, regulates blood pressure
b. I, needed to produce thyroid hormones
c. Si, needed for tendons and ligaments

TRY IT FIRST

SOLUTION

a. Na (sodium), Group 1A (1), Period 3, is an alkali metal.
b. I (iodine), Group 7A (17), Period 5, halogen, is a nonmetal.
c. Si (silicon), Group 4A (14), Period 3, is a metalloid.

Strontium provides the red color in fireworks.

STUDY CHECK 4.2

Strontium is an element that gives a brilliant red color to fireworks.

a. In what group is strontium found?
b. What is the name of this chemical family?
c. In what period is strontium found?
d. Is strontium a metal, a nonmetal, or a metalloid?

ANSWER

a. Group 2A (2)
c. Period 5

b. alkaline earth metals
d. metal

CHEMISTRY LINK TO HEALTH
Elements Essential to Health

Of all the elements, only about 20 are essential for the well-being and survival of the human body. Of those, four elements—oxygen, carbon, hydrogen, and nitrogen—which are representative elements in Period 1 and Period 2 on the periodic table, make up 96% of our body mass. Most of the food in our daily diet provides these elements to maintain a healthy body. These elements are found in carbohydrates, fats, and proteins. Most of the hydrogen and oxygen is found in water, which makes up 55% to 60% of our body mass.

The *macrominerals*—Ca, P, K, Cl, S, Na, and Mg—are representative elements located in Period 3 and Period 4 of the periodic table. They are involved in the formation of bones and teeth, maintenance of heart and blood vessels, muscle contraction, nerve impulses, acid–base balance of body fluids, and regulation of cellular metabolism. The macrominerals are present in lower amounts than the major elements, so that smaller amounts are required in our daily diets.

The other essential elements, called *microminerals* or *trace elements*, are mostly transition elements in Period 4 along with Si in Period 3 and Mo and I in Period 5. They are present in the human body in very small amounts, some less than 100 mg. In recent years, the detection of such small amounts has improved so that researchers can more easily identify the roles of trace elements. Some trace elements such as arsenic, chromium, and selenium are toxic at high levels in the body but are still required by the body. Other elements, such as tin and nickel, are thought to be essential, but their metabolic role has not yet been determined. Some examples and the amounts present in a 60.-kg person are listed in **TABLE 4.4**.

☐ Major elements in the human body   ☐ Macrominerals   ☐ Microminerals (trace elements)

TABLE 4.4 Typical Amounts of Essential Elements in a 60.-kg Adult

Element	Quantity	Function
Major Elements		
Oxygen (O)	39 kg	Building block of biomolecules and water (H_2O)
Carbon (C)	11 kg	Building block of organic and biomolecules
Hydrogen (H)	6 kg	Component of biomolecules, water (H_2O), regulates pH of body fluids, stomach acid (HCl)
Nitrogen (N)	2 kg	Component of proteins and nucleic acids
Macrominerals		
Calcium (Ca)	1000 g	Needed for bones and teeth, muscle contraction, nerve impulses
Phosphorus (P)	600 g	Needed for bones and teeth, nucleic acids
Potassium (K)	120 g	Most prevalent positive ion (K^+) in cells, muscle contraction, nerve impulses
Chlorine (Cl)	100 g	Most prevalent negative ion (Cl^-) in fluids outside cells, stomach acid (HCl)
Sulfur (S)	86 g	Component of proteins, liver, vitamin B_1, insulin
Sodium (Na)	60 g	Most prevalent positive ion (Na^+) in fluids outside cells, water balance, muscle contraction, nerve impulses
Magnesium (Mg)	36 g	Component of bones, required for metabolic reactions
Microminerals (Trace Elements)		
Iron (Fe)	3600 mg	Component of oxygen carrier hemoglobin
Silicon (Si)	3000 mg	Needed for growth and maintenance of bones and teeth, tendons and ligaments, hair and skin
Zinc (Zn)	2000 mg	Needed for metabolic reactions in cells, DNA synthesis, growth of bones, teeth, connective tissue, immune system
Copper (Cu)	240 mg	Needed for blood vessels, blood pressure, immune system
Manganese (Mn)	60 mg	Needed for growth of bones, blood clotting, metabolic reactions
Iodine (I)	20 mg	Needed for proper thyroid function
Molybdenum (Mo)	12 mg	Needed to process Fe and N from food
Arsenic (As)	3 mg	Needed for growth and reproduction
Chromium (Cr)	3 mg	Needed for maintenance of blood sugar levels, synthesis of biomolecules
Cobalt (Co)	3 mg	Component of vitamin B_{12}, red blood cells
Selenium (Se)	2 mg	Used in the immune system, health of heart and pancreas
Vanadium (V)	2 mg	Needed in the formation of bones and teeth, energy from food

QUESTIONS AND PROBLEMS

4.2 The Periodic Table

LEARNING GOAL Use the periodic table to identify the group and the period of an element; identify the element as a metal, a nonmetal, or a metalloid.

4.7 Identify the group or period number described by each of the following:
  a. contains C, N, and O
  b. begins with helium
  c. contains the alkali metals
  d. ends with neon

4.8 Identify the group or period number described by each of the following:
  a. contains Na, K, and Rb
  b. begins with Be
  c. contains the noble gases
  d. contains B, N, and F

4.9 Give the symbol of the element described by each of the following:
  a. Group 4A (14), Period 2
  b. the noble gas in Period 1
  c. the alkali metal in Period 3
  d. Group 2A (2), Period 4
  e. Group 3A (13), Period 3

4.10 Give the symbol of the element described by each of the following:
  a. the alkaline earth metal in Period 2
  b. Group 5A (15), Period 3
  c. the noble gas in Period 4
  d. the halogen in Period 5
  e. Group 4A (14), Period 4

4.11 Identify each of the following elements as a metal, a nonmetal, or a metalloid:
  a. calcium
  b. sulfur
  c. a shiny element
  d. an element that is a gas at room temperature
  e. located in Group 8A (18)
  f. bromine
  g. boron
  h. silver

4.12 Identify each of the following elements as a metal, a nonmetal, or a metalloid:
  a. located in Group 2A (2)
  b. a good conductor of electricity
  c. chlorine
  d. arsenic
  e. an element that is not shiny
  f. oxygen
  g. nitrogen
  h. tin

Applications

4.13 Using Table 4.4, identify the function of each of the following in the body and classify each as an alkali metal, an alkaline earth metal, a transition element, or a halogen:
  a. Ca   b. Fe   c. K   d. Cl

4.14 Using Table 4.4, identify the function of each of the following in the body and classify each as an alkali metal, an alkaline earth metal, a transition element, or a halogen:
  a. Mg   b. Cu   c. I   d. Na

4.15 Using the Chemistry Link to Health: Elements Essential to Health, answer each of the following:
  a. What is a macromineral?
  b. What is the role of sulfur in the human body?
  c. How many grams of sulfur would be a typical amount in a 60.-kg adult?

4.16 Using the Chemistry Link to Health: Elements Essential to Health, answer each of the following:
  a. What is a micromineral?
  b. What is the role of iodine in the human body?
  c. How many milligrams of iodine would be a typical amount in a 60.-kg adult?

4.3 The Atom

LEARNING GOAL Describe the electrical charge and location in an atom for a proton, a neutron, and an electron.

All the elements listed on the periodic table are made up of atoms. An **atom** is the smallest particle of an element that retains the characteristics of that element. Imagine that you are tearing a piece of aluminum foil into smaller and smaller pieces. Now imagine that you have a microscopic piece so small that it cannot be divided any further. Then you would have a single atom of aluminum.

The concept of the atom is relatively recent. Although the Greek philosophers in 500 B.C.E. reasoned that everything must contain minute particles they called *atomos*, the idea of atoms did not become a scientific theory until 1808. Then, John Dalton (1766–1844) developed an atomic theory that proposed that atoms were responsible for the combinations of elements found in compounds.

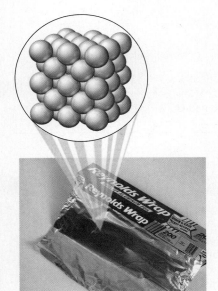

Aluminum foil consists of atoms of aluminum.

Dalton's Atomic Theory

1. All matter is made up of tiny particles called atoms.
2. All atoms of a given element are similar to one another and different from atoms of other elements.
3. Atoms of two or more different elements combine to form compounds. A particular compound is always made up of the same kinds of atoms and always has the same number of each kind of atom.
4. A chemical reaction involves the rearrangement, separation, or combination of atoms. Atoms are neither created nor destroyed during a chemical reaction.

Dalton's atomic theory formed the basis of current atomic theory, although we have modified some of Dalton's statements. We now know that atoms of the same element are not completely identical to each other and consist of even smaller particles. However, an atom is still the smallest particle that retains the properties of an element.

Although atoms are the building blocks of everything we see around us, we cannot see an atom or even a billion atoms with the naked eye. However, when billions and billions of atoms are packed together, the characteristics of each atom are added to those of the next until we can see the characteristics we associate with the element. For example, a small piece of the element gold consists of many, many gold atoms. A special kind of microscope called a *scanning tunneling microscope* (STM) produces images of individual atoms (see **FIGURE 4.7**).

Electrical Charges in an Atom

By the end of the 1800s, experiments with electricity showed that atoms were not solid spheres but were composed of even smaller bits of matter called **subatomic particles**, three of which are the *proton, neutron,* and *electron.* Two of these subatomic particles were discovered because they have electrical charges.

An electrical charge can be positive or negative. Experiments show that like charges repel or push away from each other. When you brush your hair on a dry day, electrical charges that are alike build up on the brush and in your hair. As a result, your hair flies away from the brush. However, opposite or unlike charges attract. The crackle of clothes taken from the clothes dryer indicates the presence of electrical charges. The clinginess of the clothing results from the attraction of opposite, unlike charges (see **FIGURE 4.8**).

Structure of the Atom

In 1897, J. J. Thomson, an English physicist, applied electricity to a glass tube, which produced streams of small particles called *cathode rays.* Because these rays were attracted to a positively charged electrode, Thomson realized that the particles in the rays must be negatively charged. In further experiments, these particles called **electrons** were found to be much smaller than the atom and to have extremely small masses. Because atoms are neutral, scientists soon discovered that atoms contained positively charged particles called **protons** that were much heavier than the electrons.

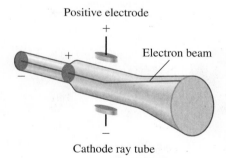

Negatively charged cathode rays (electrons) are attracted to the positive electrode.

Thomson proposed a "plum-pudding" model for the atom in which the electrons and protons were randomly distributed in a positively charged cloud like "plums in a pudding" or "chocolate chips in a cookie." In 1911, Ernest Rutherford worked with Thomson to test this model. In Rutherford's experiment, positively charged particles were aimed at a thin sheet of gold foil (see **FIGURE 4.9**). If the Thomson model were correct, the particles would travel in straight paths through the gold foil. Rutherford was greatly surprised to find that some of the particles were deflected as they passed through the gold foil, and a few particles were deflected so much that they went back in the opposite direction. According to Rutherford, it was as though he had shot a cannonball at a piece of tissue paper, and it bounced back at him.

From his gold-foil experiments, Rutherford realized that the protons must be contained in a small, positively charged region at the center of the atom, which he called the **nucleus**. He proposed that the electrons in the atom occupy the space surrounding the nucleus through which most of the particles traveled undisturbed. Only the particles that

FIGURE 4.7 ▶ Images of gold atoms are produced when magnified 16 million times by a scanning microscope.

Q Why is a microscope with extremely high magnification needed to see these atoms?

Positive charges repel

Negative charges repel

Unlike charges attract

FIGURE 4.8 ▶ Like charges repel and unlike charges attract.

Q Why are the electrons attracted to the protons in the nucleus of an atom?

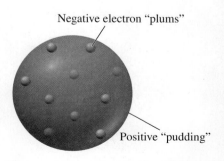

Thomson's "plum-pudding" model had protons and electrons scattered throughout the atom.

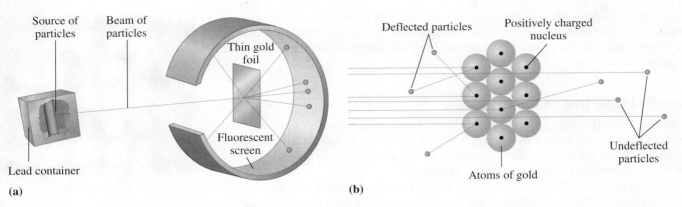

Source of particles

Beam of particles

Thin gold foil

Fluorescent screen

Lead container

(a)

Deflected particles

Positively charged nucleus

Undeflected particles

Atoms of gold

(b)

FIGURE 4.9 ▶ **(a)** Positive particles are aimed at a piece of gold foil. **(b)** Particles that come close to the atomic nuclei are deflected from their straight path.

Ⓠ Why are some particles deflected whereas most pass through the gold foil undeflected?

Interactive Video

Rutherford's Gold-Foil Experiment

came near this dense, positive center were deflected. If an atom were the size of a football stadium, the nucleus would be about the size of a golf ball placed in the center of the field.

Scientists knew that the nucleus was heavier than the mass of the protons, so they looked for another subatomic particle. Eventually, they discovered that the nucleus also contained a particle that is neutral, which they called a **neutron**. Thus, the masses of the protons and neutrons in the nucleus determine the mass of an atom (see **FIGURE 4.10**).

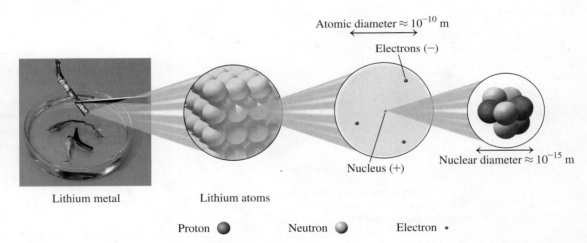

Atomic diameter ≈ 10^{-10} m

Electrons (−)

Nucleus (+)

Nuclear diameter ≈ 10^{-15} m

Lithium metal

Lithium atoms

Proton ⬤   Neutron ◯   Electron •

FIGURE 4.10 ▶ In an atom, the protons and neutrons that make up almost all the mass are packed into the tiny volume of the nucleus. The rapidly moving electrons (negative charge) surround the nucleus and account for the large volume of the atom.

Ⓠ Why can we say that the atom is mostly empty space?

Mass of the Atom

All the subatomic particles are extremely small compared with the things you see around you. One proton has a mass of 1.67×10^{-24} g, and the neutron is about the same. However, the electron has a mass 9.11×10^{-28} g, which is much less than the mass of either a proton or neutron. Because the masses of subatomic particles are so small, chemists use a very small unit of mass called an **atomic mass unit (amu)**. An amu is defined as one-twelfth of the mass of a carbon atom, which has a nucleus containing six protons and six neutrons. In biology, the atomic mass unit is called a *Dalton* (Da) in honor of John Dalton. On the amu scale, the proton and neutron each have a mass of about 1 amu. Because the electron mass is so small, it is usually ignored in atomic mass calculations. TABLE 4.5 summarizes some information about the subatomic particles in an atom.

TABLE 4.5 Subatomic Particles in the Atom

Particle	Symbol	Charge	Mass (amu)	Location in Atom
Proton	p or p^+	1+	1.007	Nucleus
Neutron	n or n^0	0	1.008	Nucleus
Electron	e^-	1−	0.000 55	Outside nucleus

SAMPLE PROBLEM 4.3 Subatomic Particles

Indicate whether each of the following is *true* or *false*:

a. A proton is heavier than an electron.
b. An electron is attracted to a neutron.
c. The nucleus contains all the protons and neutrons of an atom.

 TRY IT FIRST

SOLUTION

a. True
b. False; an electron is attracted to a proton.
c. True

STUDY CHECK 4.3

Is the following statement *true* or *false*?
The nucleus occupies a large volume in an atom.

ANSWER

False, most of the volume of the atom is outside the nucleus.

QUESTIONS AND PROBLEMS

4.3 The Atom

LEARNING GOAL Describe the electrical charge and location in an atom for a proton, a neutron, and an electron.

4.17 Identify each of the following as describing either a proton, a neutron, or an electron:
  a. has the smallest mass
  b. has a 1+ charge
  c. is found outside the nucleus
  d. is electrically neutral

4.18 Identify each of the following as describing either a proton, a neutron, or an electron:
  a. has a mass about the same as a proton
  b. is found in the nucleus
  c. is attracted to the protons
  d. has a 1− charge

4.19 What did Rutherford determine about the structure of the atom from his gold-foil experiment?

4.20 How did Thomson determine that the electrons have a negative charge?

4.21 Is each of the following statements *true* or *false*?
  a. A proton and an electron have opposite charges.
  b. The nucleus contains most of the mass of an atom.
  c. Electrons repel each other.
  d. A proton is attracted to a neutron.

4.22 Is each of the following statements *true* or *false*?
  a. A proton is attracted to an electron.
  b. A neutron has twice the mass of a proton.
  c. Neutrons repel each other.
  d. Electrons and neutrons have opposite charges.

4.23 On a dry day, your hair flies apart when you brush it. How would you explain this?

4.24 Sometimes clothes cling together when removed from a dryer. What kinds of charges are on the clothes?

4.4 Atomic Number and Mass Number

LEARNING GOAL Given the atomic number and the mass number of an atom, state the number of protons, neutrons, and electrons.

All the atoms of the same element always have the same number of protons. This feature distinguishes atoms of one element from atoms of all the other elements.

CORE CHEMISTRY SKILL
Counting Protons and Neutrons

Atomic Number

The **atomic number** of an element is equal to the number of protons in every atom of that element. The atomic number is the whole number that appears above the symbol of each element on the periodic table.

Atomic number = number of protons in an atom

The periodic table on the inside front cover of this text shows the elements in order of atomic number from 1 to 118. We can use an atomic number to identify the number of protons in an atom of any element. For example, a lithium atom, with atomic number 3, has 3 protons. Every lithium atom has 3 and only 3 protons. Any atom with 3 protons is always a lithium atom. In the same way, we determine that a carbon atom, with atomic number 6, has 6 protons. Every carbon atom has 6 protons, and any atom with 6 protons is carbon; every copper atom, with atomic number 29, has 29 protons and any atom with 29 protons is copper.

An atom is electrically neutral. That means that the number of protons in an atom is equal to the number of electrons, which gives every atom an overall charge of zero. Thus, for every atom, the atomic number also gives the number of electrons.

ENGAGE

Why does every atom of barium have 56 protons and 56 electrons?

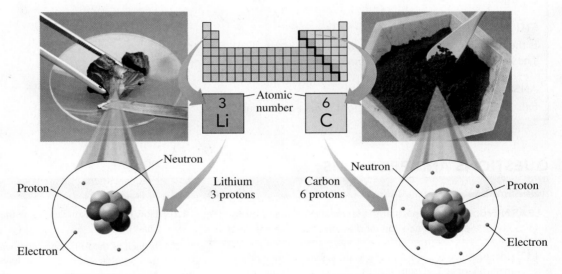

All atoms of lithium (left) contain three protons, and all atoms of carbon (right) contain six protons.

Mass Number

We now know that the protons and neutrons determine the mass of the nucleus. Thus, for a single atom, we assign a **mass number**, which is the total number of protons and neutrons in its nucleus. However, the mass number does not appear on the periodic table because it applies to single atoms only.

Mass number = number of protons + number of neutrons

For example, the nucleus of a single oxygen atom that contains 8 protons and 8 neutrons has a mass number of 16. An atom of iron that contains 26 protons and 32 neutrons has a mass number of 58.

If we are given the mass number of an atom and its atomic number, we can calculate the number of neutrons in its nucleus.

Number of neutrons in a nucleus = mass number − number of protons

For example, if we are given a mass number of 37 for an atom of chlorine (atomic number 17), we can calculate the number of neutrons in its nucleus.

Number of neutrons = 37 (mass number) − 17 (protons) = 20 neutrons

TABLE 4.6 illustrates these relationships between atomic number, mass number, and the number of protons, neutrons, and electrons in examples of single atoms for different elements.

ENGAGE

How many neutrons are in an atom of tin that has a mass number of 102?

TABLE 4.6 Composition of Some Atoms of Different Elements

Element	Symbol	Atomic Number	Mass Number	Number of Protons	Number of Neutrons	Number of Electrons
Hydrogen	H	1	1	1	0	1
Nitrogen	N	7	14	7	7	7
Oxygen	O	8	16	8	8	8
Chlorine	Cl	17	37	17	20	17
Potassium	K	19	39	19	20	19
Iron	Fe	26	58	26	32	26
Gold	Au	79	197	79	118	79

SAMPLE PROBLEM 4.4 Calculating Numbers of Protons, Neutrons, and Electrons

Zinc, a micromineral, is needed for metabolic reactions in cells, DNA synthesis, the growth of bones, teeth, and connective tissue, and the proper functioning of the immune system. For an atom of zinc that has a mass number of 68, determine the following:

a. the number of protons   **b.** the number of neutrons   **c.** the number of electrons

SOLUTION

	Given	Need	Connect
ANALYZE THE PROBLEM	zinc (Zn), mass number 68	number of protons, number of neutrons, number of electrons	periodic table, atomic number

a. Zinc (Zn), with an atomic number of 30, has 30 protons.
b. The number of neutrons in this atom is found by subtracting number of protons (atomic number) from the mass number.

Mass number − atomic number = number of neutrons
     68   −   30   =   38

c. Because the zinc atom is neutral, the number of electrons is equal to the number of protons. A zinc atom has 30 electrons.

STUDY CHECK 4.4

How many neutrons are in the nucleus of a bromine atom that has a mass number of 80?

ANSWER

45

QUESTIONS AND PROBLEMS

4.4 Atomic Number and Mass Number

LEARNING GOAL Given the atomic number and the mass number of an atom, state the number of protons, neutrons, and electrons.

4.25 Would you use the atomic number, mass number, or both to determine each of the following?
a. number of protons in an atom
b. number of neutrons in an atom
c. number of particles in the nucleus
d. number of electrons in a neutral atom

4.26 Identify the type of subatomic particles described by each of the following:
a. atomic number
b. mass number
c. mass number − atomic number
d. mass number + atomic number

4.27 Write the names and symbols for the elements with the following atomic numbers:
a. 3 b. 9 c. 20 d. 30
e. 10 f. 14 g. 53 h. 8

4.28 Write the names and symbols for the elements with the following atomic numbers:
a. 1 b. 11 c. 19 d. 82
e. 35 f. 47 g. 15 h. 2

4.29 How many protons and electrons are there in a neutral atom of each of the following elements?
a. argon
b. manganese
c. iodine
d. cadmium

4.30 How many protons and electrons are there in a neutral atom of each of the following elements?
a. carbon
b. fluorine
c. tin
d. nickel

Applications

4.31 Complete the following table for atoms of essential elements in the body:

Name of the Element	Symbol	Atomic Number	Mass Number	Number of Protons	Number of Neutrons	Number of Electrons
	Zn		66			
		12			12	
Potassium					20	
				16	15	
			56			26

4.32 Complete the following table for atoms of essential elements in the body:

Name of the Element	Symbol	Atomic Number	Mass Number	Number of Protons	Number of Neutrons	Number of Electrons
	N		15			
Calcium			42			
				53	72	
		14			16	
		29	65			

4.5 Isotopes and Atomic Mass

LEARNING GOAL Determine the number of protons, neutrons, and electrons in one or more of the isotopes of an element; calculate the atomic mass of an element using the percent abundance and mass of its naturally occurring isotopes.

We have seen that all atoms of the same element have the same number of protons and electrons. However, the atoms of any one element are not entirely identical because the atoms of most elements have different numbers of neutrons. When a sample of an element consists of two or more atoms with differing numbers of neutrons, those atoms are called *isotopes*.

Atoms and Isotopes

Isotopes are atoms of the same element that have the same atomic number but different numbers of neutrons. For example, all atoms of the element magnesium (Mg) have an atomic number of 12. Thus, every magnesium atom always has 12 protons. However, some naturally occurring magnesium atoms have 12 neutrons, others have 13 neutrons, and still others have 14 neutrons. The different numbers of neutrons give the magnesium atoms different mass numbers but do not change their chemical behavior. The three isotopes of magnesium have the same atomic number but different mass numbers.

To distinguish between the different isotopes of an element, we write an **atomic symbol** for a particular isotope that indicates the mass number in the upper left corner and the atomic number in the lower left corner.

An isotope may be referred to by its name or symbol, followed by its mass number, such as magnesium-24 or Mg-24. Magnesium has three naturally occurring isotopes, as

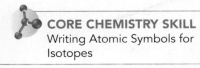

CORE CHEMISTRY SKILL
Writing Atomic Symbols for Isotopes

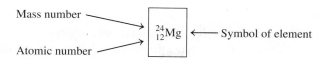

Atomic symbol for an isotope of magnesium, Mg-24.

shown in **TABLE 4.7**. In a large sample of naturally occurring magnesium atoms, each type of isotope can be present as a low percentage or a high percentage. For example, the Mg-24 isotope makes up almost 80% of the total sample, whereas Mg-25 and Mg-26 each make up only about 10% of the total number of magnesium atoms.

TABLE 4.7 Isotopes of Magnesium

Atomic Symbol	$^{24}_{12}Mg$	$^{25}_{12}Mg$	$^{26}_{12}Mg$
Name	Mg-24	Mg-25	Mg-26
Number of Protons	12	12	12
Number of Electrons	12	12	12
Mass Number	24	25	26
Number of Neutrons	12	13	14
Mass of Isotope (amu)	23.99	24.99	25.98
Percent Abundance	78.70	10.13	11.17

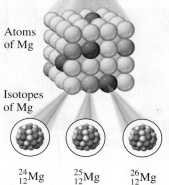

The nuclei of three naturally occurring magnesium isotopes have the same number of protons but different numbers of neutrons.

SAMPLE PROBLEM 4.5 **Identifying Protons and Neutrons in Isotopes**

Chromium, a micromineral needed for maintenance of blood sugar levels, has four naturally occurring isotopes. Calculate the number of protons and number of neutrons in each of the following isotopes:

a. $^{50}_{24}Cr$ **b.** $^{52}_{24}Cr$ **c.** $^{53}_{24}Cr$ **d.** $^{54}_{24}Cr$

 TRY IT FIRST

SOLUTION

ANALYZE THE PROBLEM	Given	Need	Connect
	atomic symbols for Cr isotopes	number of protons, number of neutrons	periodic table, atomic number

In the atomic symbol, the mass number is shown in the upper left corner of the symbol, and the atomic number is shown in the lower left corner of the symbol. Thus, each isotope of Cr, atomic number 24, has 24 protons. The number of neutrons is found by subtracting the number of protons (24) from the mass number of each isotope.

Atomic Symbol	Atomic Number	Mass Number	Number of Protons	Number of Neutrons
a. $^{50}_{24}Cr$	24	50	24	26 (50 − 24)
b. $^{52}_{24}Cr$	24	52	24	28 (52 − 24)
c. $^{53}_{24}Cr$	24	53	24	29 (53 − 24)
d. $^{54}_{24}Cr$	24	54	24	30 (54 − 24)

STUDY CHECK 4.5

Vanadium is a micromineral needed in the formation of bones and teeth. Write the atomic symbol for the single naturally occurring isotope of vanadium, which has 27 neutrons.

ANSWER

$^{50}_{23}V$

ENGAGE

What is different and what is the same for an atom of Sn-105 and an atom of Sn-132?

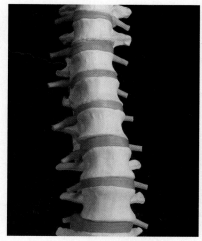

Vanadium along with calcium is important in forming strong bones and preventing osteoporosis.

Atomic Mass

In laboratory work, a chemist generally uses samples with many atoms that contain all the different atoms or isotopes of an element. Because each isotope has a different mass, chemists have calculated an **atomic mass** for an "average atom," which is a *weighted average* of the masses of all the naturally occurring isotopes of that element. On the periodic table, the atomic mass is the number including decimal places that is given below the symbol of each element. Most elements consist of two or more isotopes, which is one reason that the atomic masses on the periodic table are seldom whole numbers.

Weighted Average Analogy

To understand how the atomic mass as a weighted average for a group of isotopes is calculated, we will use an analogy of bowling balls with different weights. Suppose that a bowling alley has ordered five bowling balls that weigh 8 lb each and 20 bowling balls that weigh 14 lb each. In this group of bowling balls, there are more 14-lb balls than 8-lb balls. The abundance of the 14-lb balls is 80.% (20/25), and the abundance of the 8-lb balls is 20.% (5/25). Now we can calculate a weighted average for the "average" bowling ball using the weight and percent abundance of the two types of bowling balls.

The weighted average of 8-lb and 14-lb bowling balls is calculated using their percent abundance.

Item	Weight		Percent Abundance		Weight from Each Type
14-lb Bowling balls	14 lb	×	$\frac{80.}{100}$	=	11.2 lb
8-lb Bowling balls	8 lb	×	$\frac{20.}{100}$	=	1.6 lb
	Weighted average of a bowling ball			=	12.8 lb
	"Atomic mass" of a bowling ball			=	12.8 lb

Calculating Atomic Mass

To calculate the atomic mass of an element, we need to know the percentage abundance and the mass of each isotope, both of which must be determined experimentally. For example, a large sample of naturally occurring chlorine atoms consists of 75.76% of $^{35}_{17}\text{Cl}$ atoms and 24.24% of $^{37}_{17}\text{Cl}$ atoms. The $^{35}_{17}\text{Cl}$ isotope has a mass of 34.97 amu and the $^{37}_{17}\text{Cl}$ isotope has a mass of 36.97 amu.

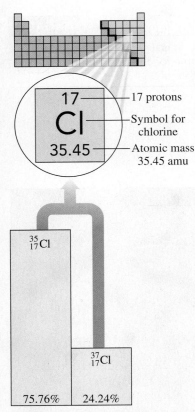

$$\text{Atomic mass of Cl} = \underbrace{\text{mass of } ^{35}_{17}\text{Cl} \times \frac{^{35}_{17}\text{Cl}\%}{100\%}}_{\text{amu from } ^{35}_{17}\text{Cl}} + \underbrace{\text{mass of } ^{37}_{17}\text{Cl} \times \frac{^{37}_{17}\text{Cl}\%}{100\%}}_{\text{amu from } ^{37}_{17}\text{Cl}}$$

Atomic Symbol	Mass (amu)	×	Abundance (%)	=	Contribution to Average Cl Atom
$^{35}_{17}\text{Cl}$	34.97	×	$\frac{75.76}{100}$	=	26.49 amu
$^{37}_{17}\text{Cl}$	36.97	×	$\frac{24.24}{100}$	=	8.962 amu
			Atomic mass of Cl	=	35.45 amu (weighted average mass)

Chlorine, with two naturally occurring isotopes, has an atomic mass of 35.45 amu.

The atomic mass of 35.45 amu is the weighted average mass of a sample of Cl atoms, although no individual Cl atom actually has this mass. An atomic mass of 35.45, which is closer to the mass number of Cl-35, also indicates there is a higher percentage of $^{35}_{17}\text{Cl}$ atoms in the chlorine sample. In fact, there are about three atoms of $^{35}_{17}\text{Cl}$ for every one atom of $^{37}_{17}\text{Cl}$ in a sample of chlorine atoms.

TABLE 4.8 lists the naturally occurring isotopes of some selected elements and their atomic masses along with their most prevalent isotopes.

TABLE 4.8 The Atomic Mass of Some Elements

Element	Atomic Symbols	Atomic Mass (Weighted Average)	Most Prevalent Isotope
Lithium	$^{6}_{3}\text{Li}, {}^{7}_{3}\text{Li}$	6.941 amu	$^{7}_{3}\text{Li}$
Carbon	$^{12}_{6}\text{C}, {}^{13}_{6}\text{C}, {}^{14}_{6}\text{C}$	12.01 amu	$^{12}_{6}\text{C}$
Oxygen	$^{16}_{8}\text{O}, {}^{17}_{8}\text{O}, {}^{18}_{8}\text{O}$	16.00 amu	$^{16}_{8}\text{O}$
Fluorine	$^{19}_{9}\text{F}$	19.00 amu	$^{19}_{9}\text{F}$
Sulfur	$^{32}_{16}\text{S}, {}^{33}_{16}\text{S}, {}^{34}_{16}\text{S}, {}^{36}_{16}\text{S}$	32.07 amu	$^{32}_{16}\text{S}$
Potassium	$^{39}_{19}\text{K}, {}^{40}_{19}\text{K}, {}^{41}_{19}\text{K}$	39.10 amu	$^{39}_{19}\text{K}$
Copper	$^{63}_{29}\text{Cu}, {}^{65}_{29}\text{Cu}$	63.55 amu	$^{63}_{29}\text{Cu}$

SAMPLE PROBLEM 4.6 **Calculating Atomic Mass**

Magnesium is a macromineral, which is a component of bone and needed for metabolic reactions. Using Table 4.7, calculate the atomic mass for magnesium using the weighted average mass method.

TRY IT FIRST

SOLUTION

STEP 1 Multiply the mass of each isotope by its percent abundance divided by 100.

Atomic Symbol	Mass (amu)		Abundance (%)		Contribution to the Atomic Mass
$^{24}_{12}\text{Mg}$	23.99	×	$\dfrac{78.70}{100}$	=	18.88 amu
$^{25}_{12}\text{Mg}$	24.99	×	$\dfrac{10.13}{100}$	=	2.531 amu
$^{26}_{12}\text{Mg}$	25.98	×	$\dfrac{11.17}{100}$	=	2.902 amu

STEP 2 Add the contribution of each isotope to obtain the atomic mass.

Atomic mass of Mg = 18.88 amu + 2.531 amu + 2.902 amu

= 24.31 amu (weighted average mass)

STUDY CHECK 4.6

There are two naturally occurring isotopes of boron. The isotope $^{10}_{5}\text{B}$ has a mass of 10.01 amu with an abundance of 19.80%, and the isotope $^{11}_{5}\text{B}$ has a mass of 11.01 amu with an abundance of 80.20%. Calculate the atomic mass for boron using the weighted average mass method.

ANSWER

10.81 amu

Guide to Calculating Atomic Mass

STEP 1
Multiply the mass of each isotope by its percent abundance divided by 100.

STEP 2
Add the contribution of each isotope to obtain the atomic mass.

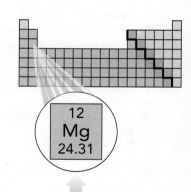

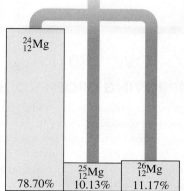

Magnesium, with three naturally occurring isotopes, has an atomic mass of 24.31 amu.

QUESTIONS AND PROBLEMS

4.5 Isotopes and Atomic Mass

LEARNING GOAL Determine the number of protons, neutrons, and electrons in one or more of the isotopes of an element; calculate the atomic mass of an element using the percent abundance and mass of its naturally occurring isotopes.

4.33 What are the number of protons, neutrons, and electrons in the following isotopes?

a. $^{89}_{38}Sr$ b. $^{52}_{24}Cr$
c. $^{34}_{16}S$ d. $^{81}_{35}Br$

4.34 What are the number of protons, neutrons, and electrons in the following isotopes?

a. $^{2}_{1}H$ b. $^{14}_{7}N$
c. $^{26}_{14}Si$ d. $^{70}_{30}Zn$

4.35 Write the atomic symbol for the isotope with each of the following characteristics:

a. 15 protons and 16 neutrons
b. 35 protons and 45 neutrons
c. 50 electrons and 72 neutrons
d. a chlorine atom with 18 neutrons
e. a mercury atom with 122 neutrons

4.36 Write the atomic symbol for the isotope with each of the following characteristics:

a. an oxygen atom with 10 neutrons
b. 4 protons and 5 neutrons
c. 25 electrons and 28 neutrons
d. a mass number of 24 and 13 neutrons
e. a nickel atom with 32 neutrons

4.37 Argon has three naturally occurring isotopes, with mass numbers 36, 38, and 40.

a. Write the atomic symbol for each of these atoms.
b. How are these isotopes alike?
c. How are they different?
d. Why is the atomic mass of argon listed on the periodic table not a whole number?
e. Which isotope is the most prevalent in a sample of argon?

4.38 Strontium has four naturally occurring isotopes, with mass numbers 84, 86, 87, and 88.

a. Write the atomic symbol for each of these atoms.
b. How are these isotopes alike?
c. How are they different?
d. Why is the atomic mass of strontium listed on the periodic table not a whole number?
e. Which isotope is the most prevalent in a sample of strontium?

4.39 What is the difference between the mass of an isotope and the atomic mass of an element?

4.40 What is the difference between the mass number and the atomic mass of an element?

4.41 Two isotopes of gallium are naturally occurring, with $^{69}_{31}Ga$ at 60.11% (68.93 amu) and $^{71}_{31}Ga$ at 39.89% (70.92 amu). Calculate the atomic mass for gallium using the weighted average mass method.

4.42 Two isotopes of rubidium occur naturally, with $^{85}_{37}Rb$ at 72.17% (84.91 amu) and $^{87}_{37}Rb$ at 27.83% (86.91 amu). Calculate the atomic mass for rubidium using the weighted average mass method.

4.43 Copper consists of two isotopes, $^{63}_{29}Cu$ and $^{65}_{29}Cu$. If the atomic mass for copper on the periodic table is 63.55, are there more atoms of $^{63}_{29}Cu$ or $^{65}_{29}Cu$ in a sample of copper?

4.44 A fluorine sample consists of only one type of atom, $^{19}_{9}F$, which has a mass of 19.00 amu. How would the mass of a $^{19}_{9}F$ atom compare to the atomic mass listed on the periodic table?

4.45 There are two naturally occurring isotopes of thallium: $^{203}_{81}Tl$ and $^{205}_{81}Tl$. Use the atomic mass of thallium listed on the periodic table to identify the more prevalent isotope.

4.46 Zinc consists of five naturally occurring isotopes: $^{64}_{30}Zn$, $^{66}_{30}Zn$, $^{67}_{30}Zn$, $^{68}_{30}Zn$, and $^{70}_{30}Zn$. None of these isotopes has the atomic mass of 65.41 listed for zinc on the periodic table. Explain.

Follow Up

IMPROVING CROP PRODUCTION

In plants, potassium (K) is needed for metabolic processes including the regulation of plant growth. Potassium is required by plants for protein synthesis, photosynthesis, enzymes, and ionic balance. Potassium-deficient potato plants may show purple or brown spots and plant, root, and seed growth is reduced. John has noticed that the leaves of his recent crop of potatoes had brown spots, the potatoes were undersized, and the crop yield was low.

Tests on soil samples showed that the potassium levels were below 100 ppm, which indicated that supplemental potassium was needed. John applied a fertilizer containing potassium chloride (KCl). To apply the correct amount of potassium, John needed to apply 170 kg of fertilizer per hectare.

Applications

4.47 a. What is the group number and name of the group that contains potassium?
b. Is potassium a metal, a nonmetal, or a metalloid?
c. How many protons are in an atom of potassium?
d. What is the most prevalent isotope of potassium (see Table 4.8)?

e. Potassium has three naturally occurring isotopes. They are K-39 (93.26%, 38.964 amu), K-40 (0.0117%, 39.964 amu), and K-41 (6.73%, 40.962 amu). Calculate the atomic mass of potassium from the naturally occurring isotopes and their abundance using the weighted average mass method.

4.48 a. How many neutrons are in K-41?
b. Write the electron configuration for potassium.
c. Which is the larger atom, K or Cs?
d. Which is the smallest atom, K, As, or Br?
e. If John's potato farm has an area of 34.5 hectares, how many pounds of fertilizer does John need to use?

A mineral deficiency of potassium causes brown spots on potato leaves.

CONCEPT MAP

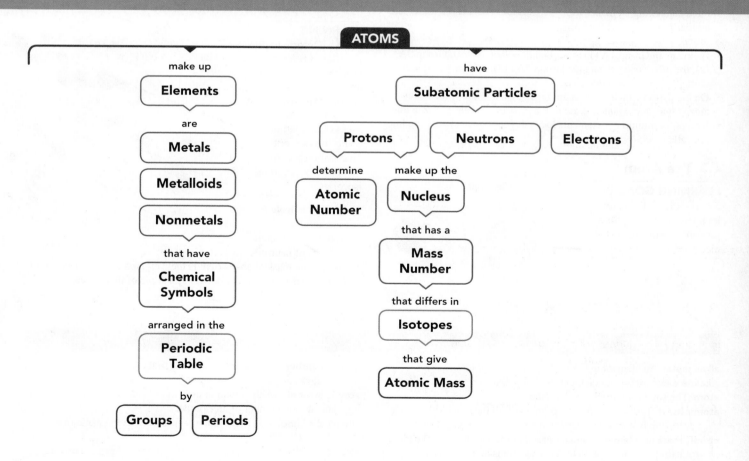

CHAPTER REVIEW

4.1 Elements and Symbols

LEARNING GOAL Given the name of an element, write its correct symbol; from the symbol, write the correct name.

• Elements are the primary substances of matter.
• Chemical symbols are one- or two-letter abbreviations of the names of the elements.

4.2 The Periodic Table

LEARNING GOAL Use the periodic table to identify the group and the period of an element; identify the element as a metal, a nonmetal, or a metalloid.

• The periodic table is an arrangement of the elements by increasing atomic number.
• A horizontal row is called a *period*.
• A vertical column on the periodic table containing elements with similar properties is called a *group*.
• Elements in Group 1A (1) are called the *alkali metals*; Group 2A (2), the *alkaline earth metals*; Group 7A (17), the *halogens*; and Group 8A (18), the *noble gases*.
• On the periodic table, *metals* are located on the left of the heavy zigzag line, and *nonmetals* are to the right of the heavy zigzag line.
• Except for aluminum, elements located along the heavy zigzag line are called *metalloids*.

4.3 The Atom

LEARNING GOAL Describe the electrical charge and location in an atom for a proton, a neutron, and an electron.

Electrons (−)

Nucleus (+)

• An atom is the smallest particle that retains the characteristics of an element.
• Atoms are composed of three types of subatomic particles.
• Protons have a positive charge (+), electrons carry a negative charge (−), and neutrons are electrically neutral.
• The protons and neutrons are found in the tiny, dense nucleus; electrons are located outside the nucleus.

4.4 Atomic Number and Mass Number

LEARNING GOAL Given the atomic number and the mass number of an atom, state the number of protons, neutrons, and electrons.

^{3}Li

Proton
Neutron
Lithium
3 protons
Electron

• The atomic number gives the number of protons in all the atoms of the same element.
• In a neutral atom, the number of protons and electrons is equal.
• The mass number is the total number of protons and neutrons in an atom.

4.5 Isotopes and Atomic Mass

LEARNING GOAL Determine the number of protons, neutrons, and electrons in one or more of the isotopes of an element; calculate the atomic mass of an element using the percent abundance and mass of its naturally occurring isotopes.

Atomic structure of Mg

Isotopes of Mg

$^{24}_{12}$Mg $^{25}_{12}$Mg $^{26}_{12}$Mg

• Atoms that have the same number of protons but different numbers of neutrons are called *isotopes*.
• The atomic mass of an element is the weighted average mass of all the isotopes in a naturally occurring sample of that element.

KEY TERMS

alkali metal An element in Group 1A (1), except hydrogen.
alkaline earth metal An element in Group 2A (2).
atom The smallest particle of an element.
atomic mass The weighted average mass of all the naturally occurring isotopes of an element.
atomic mass unit (amu) A small mass unit used to describe the mass of extremely small particles such as atoms and subatomic particles; 1 amu is equal to one-twelfth the mass of a $^{12}_{6}$C atom.

atomic number A number that is equal to the number of protons in an atom.
atomic symbol An abbreviation used to indicate the mass number and atomic number of an isotope.
chemical symbol An abbreviation that represents the name of an element.
electron A negatively charged subatomic particle having a minute mass that is usually ignored in mass calculations; its symbol is e^{-}.

group A vertical column in the periodic table that contains elements having similar physical and chemical properties.

group number A number that appears at the top of each vertical column (group) in the periodic table.

halogen An element in Group 7A (17).

isotope An atom that differs only in mass number from another atom of the same element. Isotopes have the same atomic number (number of protons), but different numbers of neutrons.

mass number The total number of protons and neutrons in the nucleus of an atom.

metal An element that is shiny, malleable, ductile, and a good conductor of heat and electricity. The metals are located to the left of the heavy zigzag line on the periodic table.

metalloid Elements with properties of both metals and nonmetals located along the heavy zigzag line on the periodic table.

neutron A neutral subatomic particle having a mass of about 1 amu and found in the nucleus of an atom; its symbol is n or n^0.

noble gas An element in Group 8A (18) of the periodic table.

nonmetal An element with little or no luster that is a poor conductor of heat and electricity. The nonmetals are located to the right of the heavy zigzag line on the periodic table.

nucleus The compact, extremely dense center of an atom, containing the protons and neutrons of the atom.

period A horizontal row of elements in the periodic table.

periodic table An arrangement of elements by increasing atomic number such that elements having similar chemical behavior are grouped in vertical columns.

proton A positively charged subatomic particle having a mass of about 1 amu and found in the nucleus of an atom; its symbol is p or p^+.

representative element An element in the first two columns on the left of the periodic table and the last six columns on the right that has a group number of 1A through 8A or 1, 2, and 13 through 18.

subatomic particle A particle within an atom; protons, neutrons, and electrons are subatomic particles.

transition element An element in the center of the periodic table that is designated with the letter "B" or the group number of 3 through 12.

CORE CHEMISTRY SKILLS

The chapter section containing each Core Chemistry Skill is shown in parentheses at the end of each heading.

Counting Protons and Neutrons (4.4)

- The atomic number of an element is equal to the number of protons in every atom of that element. The atomic number is the whole number that appears above the symbol of each element on the periodic table.

    Atomic number = number of protons in an atom

- Because atoms are neutral, the number of electrons is equal to the number of protons. Thus, the atomic number gives the number of electrons.

- The mass number is the total number of protons and neutrons in the nucleus of an atom.

    Mass number = number of protons + number of neutrons

- The number of neutrons is calculated from the mass number and atomic number.

    Number of neutrons = mass number − number of protons

Example: Calculate the number of protons, neutrons, and electrons in a krypton atom with a mass number of 80.

Answer:

Element	Atomic Number	Mass Number	Number of Protons	Number of Neutrons	Number of Electrons
Kr	36	80	equal to atomic number 36	equal to mass number − number of protons 80 − 36 = 44	equal to number of protons 36

Writing Atomic Symbols for Isotopes (4.5)

- Isotopes are atoms of the same element that have the same atomic number but different numbers of neutrons.
- An atomic symbol is written for a particular isotope, with its mass number (protons and neutrons) shown in the upper left corner and its atomic number (protons) shown in the lower left corner.

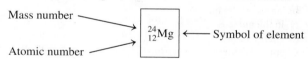

Example: Calculate the number of protons and neutrons in the cadmium isotope $^{112}_{48}\text{Cd}$.

Answer:

Atomic Symbol	Atomic Number	Mass Number	Number of Protons	Number of Neutrons
$^{112}_{48}\text{Cd}$	number in lower left corner 48	number in upper left corner 112	equal to atomic number 48	equal to mass number − number of protons 112 − 48 = 64

UNDERSTANDING THE CONCEPTS

The chapter sections to review are shown in parentheses at the end of each question.

4.49 According to Dalton's atomic theory, which of the following are *true* or *false*? If false, correct the statement to make it true. (4.3)
   a. Atoms of an element are identical to atoms of other elements.
   b. Every element is made of atoms.
   c. Atoms of different elements combine to form compounds.
   d. In a chemical reaction, some atoms disappear and new atoms appear.

4.50 Use Rutherford's gold-foil experiment to answer each of the following: (4.3)
   a. What did Rutherford expect to happen when he aimed particles at the gold foil?
   b. How did the results differ from what he expected?
   c. How did he use the results to propose a model of the atom?

4.51 Match the subatomic particles (**1** to **3**) to each of the descriptions below: (4.4)
   1. protons   **2.** neutrons   **3.** electrons

   a. atomic mass
   b. atomic number
   c. positive charge
   d. negative charge
   e. mass number – atomic number

4.52 Match the subatomic particles (**1** to **3**) to each of the descriptions below: (4.4)
   1. protons   **2.** neutrons   **3.** electrons

   a. mass number
   b. surround the nucleus
   c. in the nucleus
   d. charge of 0
   e. equal to number of electrons

4.53 Consider the following atoms in which X represents the chemical symbol of the element: (4.4, 4.5)

   $^{16}_{8}X$ $^{16}_{9}X$ $^{18}_{10}X$ $^{17}_{8}X$ $^{18}_{8}X$

   a. What atoms have the same number of protons?
   b. Which atoms are isotopes? Of what element?
   c. Which atoms have the same mass number?

4.54 Consider the following atoms in which X represents the chemical symbol of the element: (4.4, 4.5)

   $^{124}_{47}X$ $^{116}_{49}X$ $^{116}_{50}X$ $^{124}_{50}X$ $^{116}_{48}X$

   a. What atoms have the same number of protons?
   b. Which atoms are isotopes? Of what element?
   c. Which atoms have the same mass number?

4.55 Complete the following table for three of the naturally occurring isotopes of germanium, which is a metalloid used in semiconductors: (4.4, 4.5)

	Atomic Symbol		
	$^{70}_{32}Ge$	$^{73}_{32}Ge$	$^{76}_{32}Ge$
Atomic Number			
Mass Number			
Number of Protons			
Number of Neutrons			
Number of Electrons			

4.56 Complete the following table for the three naturally occurring isotopes of silicon, the major component in computer chips: (4.4, 4.5)

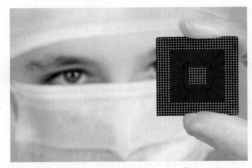

Computer chips consist primarily of the element silicon.

	Atomic Symbol		
	$^{28}_{14}Si$	$^{29}_{14}Si$	$^{30}_{14}Si$
Atomic Number			
Mass Number			
Number of Protons			
Number of Neutrons			
Number of Electrons			

4.57 For each representation of a nucleus **A** through **E**, write the atomic symbol, and identify which are isotopes. (4.4, 4.5)

   Proton ●
   Neutron ●

   A   **B**   **C**   **D**   **E**

4.58 Identify the element represented by each nucleus **A** through **E** in problem 4.57 as a metal, a nonmetal, or a metalloid. (4.2)

ADDITIONAL QUESTIONS AND PROBLEMS

4.59 Why is Co the symbol for cobalt, not CO? (4.1)

4.60 Which of the following is correct? Write the correct symbol if needed. (4.1)

   a. copper, Cp   **b.** silicon, SI
   c. iron, Fe   **d.** fluorine, Fl
   e. potassium, P   **f.** sodium, Na
   g. gold, Au   **h.** lead, PB

4.61 Give the group and period number for each of the following elements: (4.2)

   a. bromine   **b.** argon
   c. lithium   **d.** radium

4.62 Give the group and period number for each of the following elements: (4.2)

   a. radon   **b.** tin
   c. carbon   **d.** magnesium

4.63 The following trace elements have been found to be crucial to the functions of the body. Indicate each as a metal, a nonmetal, or a metalloid. (4.2)

   a. zinc   **b.** cobalt
   c. manganese   **d.** iodine

4.64 The following trace elements have been found to be crucial to the functions of the body. Indicate each as a metal, a nonmetal, or a metalloid. (4.2)

   a. copper   **b.** selenium
   c. arsenic   **d.** chromium

4.65 Indicate if each of the following statements is *true* or *false*: (4.3)

   a. The proton is a negatively charged particle.
   b. The neutron is 2000 times as heavy as a proton.
   c. The atomic mass unit is based on a carbon atom with 6 protons and 6 neutrons.
   d. The nucleus is the largest part of the atom.
   e. The electrons are located outside the nucleus.

4.66 Indicate if each of the following statements is *true* or *false*: (4.3)

   a. The neutron is electrically neutral.

   b. Most of the mass of an atom is because of the protons and neutrons.
   c. The charge of an electron is equal, but opposite, to the charge of a neutron.
   d. The proton and the electron have about the same mass.
   e. The mass number is the number of protons.

4.67 For the following atoms, give the number of protons, neutrons, and electrons: (4.3)

   a. $^{114}_{48}Cd$ **b.** $^{98}_{43}Tc$ **c.** $^{199}_{79}Au$
   d. $^{222}_{86}Rn$ **e.** $^{136}_{54}Xe$

4.68 For the following atoms, give the number of protons, neutrons, and electrons: (4.3)

   a. $^{202}_{80}Hg$ **b.** $^{127}_{53}I$ **c.** $^{75}_{35}Br$
   d. $^{133}_{55}Cs$ **e.** $^{195}_{78}Pt$

4.69 Complete the following table: (4.3)

Name of the Element	Atomic Symbol	Number of Protons	Number of Neutrons	Number of Electrons
	$^{80}_{34}Se$			
		28	34	
Magnesium			14	
	$^{228}_{88}Ra$			

4.70 Complete the following table: (4.3)

Name of the Element	Atomic Symbol	Number of Protons	Number of Neutrons	Number of Electrons
Potassium			22	
	$^{51}_{23}V$			
		48	64	
Barium			82	

CHALLENGE QUESTIONS

The following groups of questions are related to the topics in this chapter. However, they do not all follow the chapter order, and they require you to combine concepts and skills from several sections. These questions will help you increase your critical thinking skills and prepare for your next exam.

4.71 Complete the following statements: (4.2, 4.4)

   a. The atomic number gives the number of _____ in the nucleus.
   b. In an atom, the number of electrons is equal to the number of _____.
   c. Sodium and potassium are examples of elements called _____.

4.72 Complete the following statements: (4.2, 4.4)

   a. The number of protons and neutrons in an atom is the _____ number.
   b. The elements in Group 7A (17) are called the _____.
   c. Elements that are shiny and conduct heat are called _____.

4.73 Provide the following information: (4.2, 4.4)

   a. the atomic number and symbol of the lightest alkali metal
   b. the atomic number and symbol of the heaviest noble gas
   c. the atomic mass and symbol of the alkaline earth metal in Period 3
   d. the atomic mass and symbol of the halogen with the fewest electrons

4.74 Provide the following information: (4.2, 4.4)

   a. the atomic number and symbol of the heaviest metalloid in Group 4A (14)
   b. the atomic number and symbol of the element in Group 5A (15), Period 6
   c. the atomic mass and symbol of the alkali metal in Period 4
   d. the metalloid in Group 3A (13)

4.75 Write the name and symbol of the element with the following atomic number: (4.4)

   a. 28   **b.** 56   **c.** 88
   d. 33   **e.** 50   **f.** 55

4.76 Write the name and symbol of the element with the following atomic number: (4.4)
a. 22 b. 48 c. 26
d. 54 e. 78 f. 83

4.77 Give the number of protons and electrons in neutral atoms of each of the following: (4.4)
a. Mn b. phosphorus c. Sr
d. Co e. uranium

4.78 Give the number of protons and electrons in neutral atoms of each of the following: (4.4)
a. chromium b. Cs c. copper
d. chlorine e. Cd

4.79 The most abundant isotope of lead is $^{208}_{82}$Pb. (4.4, 4.5)
a. How many protons, neutrons, and electrons are in $^{208}_{82}$Pb?
b. What is the atomic symbol of another isotope of lead with 132 neutrons?
c. What is the name and symbol of an atom with the same mass number as in part **b** and 131 neutrons?

4.80 The most abundant isotope of silver is $^{107}_{47}$Ag. (4.4, 4.5)
a. How many protons, neutrons, and electrons are in $^{107}_{47}$Ag?
b. What is the symbol of another isotope of silver with 62 neutrons?
c. What is the name and symbol of an atom with the same mass number as in part **b** and 61 neutrons?

4.81 Write the atomic symbol for each of the following: (4.5)
a. an atom with 4 protons and 5 neutrons
b. an atom with 12 protons and 14 neutrons
c. a calcium atom with a mass number of 46
d. an atom with 30 electrons and 40 neutrons

4.82 Write the atomic symbol for each of the following: (4.5)
a. an aluminum atom with 14 neutrons
b. an atom with atomic number 26 and 32 neutrons
c. a strontium atom with 50 neutrons
d. an atom with a mass number of 72 and atomic number 33

4.83 Silicon has three naturally occurring isotopes: Si-28 that has a percent abundance of 92.23% and a mass of 27.977 amu, Si-29 that has a 4.68% abundance and a mass of 28.976 amu, and Si-30 that has a 3.09% abundance and a mass of 29.974 amu. Calculate the atomic mass for silicon using the weighted average mass method. (4.5)

4.84 Antimony (Sb) has two naturally occurring isotopes: Sb-121 that has a percent abundance of 57.21% and a mass of 120.90 amu, and Sb-123 that has a 42.79% abundance and a mass of 122.90 amu. Calculate the atomic mass for antimony using the weighted average mass method. (4.5)

4.85 The most prevalent isotope of gold is Au-197. (4.5)
a. How many protons, neutrons, and electrons are in this isotope?
b. What is the atomic symbol of another isotope of gold with 116 neutrons?
c. What is the atomic symbol of an atom with an atomic number of 78 and 116 neutrons?

4.86 Cadmium, atomic number 48, consists of eight naturally occurring isotopes. Do you expect any of the isotopes to have the atomic mass listed on the periodic table for cadmium? Explain. (4.5)

4.87 Lead consists of four naturally occurring isotopes. Calculate the atomic mass for lead using the weighted average mass method. (4.5)

Isotope	Mass (amu)	Abundance (%)
$^{204}_{82}$Pb	204.0	1.40
$^{206}_{82}$Pb	206.0	24.10
$^{207}_{82}$Pb	207.0	22.10
$^{208}_{82}$Pb	208.0	52.40

4.88 Indium (In) has two naturally occurring isotopes: In-113 and In-115. In-113 has a 4.30% abundance and a mass of 112.9 amu, and In-115 has a 95.70% abundance and a mass of 114.9 amu. Calculate the atomic mass for indium using the weighted average mass method. (4.5)

4.89 If the diameter of a sodium atom is 3.14×10^{-8} cm, how many sodium atoms would fit along a line exactly 1 in. long? (4.3)

4.90 A lead atom has a mass of 3.4×10^{-22} g. How many lead atoms are in a cube of lead that has a volume of 2.00 cm^3 if the density of lead is 11.3 g/cm^3? (4.3)

ANSWERS

Answers to Selected Questions and Problems

4.1 a. Cu b. Pt c. Ca d. Mn
e. Fe f. Ba g. Pb h. Sr

4.3 a. carbon b. chlorine c. iodine
d. selenium e. nitrogen f. sulfur
g. zinc h. cobalt

4.5 a. sodium, chlorine
b. calcium, sulfur, oxygen
c. carbon, hydrogen, chlorine, nitrogen, oxygen
d. lithium, carbon, oxygen

4.7 a. Period 2 b. Group 8A (18)
c. Group 1A (1) d. Period 2

4.9 a. C b. He c. Na
d. Ca e. Al

4.11 a. metal b. nonmetal c. metal
d. nonmetal e. nonmetal f. nonmetal
g. metalloid h. metal

4.13 a. needed for bones and teeth, muscle contraction, nerve impulses; alkaline earth metal
b. component of hemoglobin; transition element
c. muscle contraction, nerve impulses; alkali metal
d. found in fluids outside cells; halogen

4.15 a. A macromineral is an element essential to health, which is present in the body in amounts from 5 to 1000 g.
b. Sulfur is a component of proteins, liver, vitamin B$_1$, and insulin.
c. 86 g

4.17 a. electron b. proton
c. electron d. neutron

4.19 Rutherford determined that an atom contains a small, compact nucleus that is positively charged.

4.21 a. True
    c. True
b. True
d. False; a proton is attracted to an electron

4.23 In the process of brushing hair, strands of hair become charged with like charges that repel each other.

4.25 a. atomic number
    c. mass number
b. both
d. atomic number

4.27 a. lithium, Li
    c. calcium, Ca
    e. neon, Ne
    g. iodine, I
b. fluorine, F
d. zinc, Zn
f. silicon, Si
h. oxygen, O

4.29 a. 18 protons and 18 electrons
    b. 25 protons and 25 electrons
    c. 53 protons and 53 electrons
    d. 48 protons and 48 electrons

4.31

Name of the Element	Symbol	Atomic Number	Mass Number	Number of Protons	Number of Neutrons	Number of Electrons
Zinc	Zn	30	66	30	36	30
Magnesium	Mg	12	24	12	12	12
Potassium	K	19	39	19	20	19
Sulfur	S	16	31	16	15	16
Iron	Fe	26	56	26	30	26

4.33 a. 38 protons, 51 neutrons, 38 electrons
    b. 24 protons, 28 neutrons, 24 electrons
    c. 16 protons, 18 neutrons, 16 electrons
    d. 35 protons, 46 neutrons, 35 electrons

4.35 a. $^{31}_{15}P$
    d. $^{35}_{17}Cl$
b. $^{80}_{35}Br$
e. $^{202}_{80}Hg$
c. $^{122}_{50}Sn$

4.37 a. $^{36}_{18}Ar$ $^{38}_{18}Ar$ $^{40}_{18}Ar$
    b. They all have the same number of protons and electrons.
    c. They have different numbers of neutrons, which gives them different mass numbers.
    d. The atomic mass of Ar listed on the periodic table is the weighted average atomic mass of all the naturally occurring isotopes.
    e. The isotope Ar-40 is most prevalent because its mass is closest to the atomic mass of Ar on the periodic table.

4.39 The mass of an isotope is the mass of an individual atom. The atomic mass is the weighted average of all the naturally occurring isotopes of that element.

4.41 69.72 amu

4.43 Since the atomic mass of copper is closer to 63 amu, there are more atoms of $^{63}_{29}Cu$.

4.45 Since the atomic mass of thallium is 204.4 amu, the most prevalent isotope is $^{205}_{81}Tl$.

4.47 a. Group 1A (1), alkali metals
    c. 19 protons
    e. 39.10 amu
b. metal
d. K-39

4.49 a. False; All atoms of a given element are different from atoms of other elements.
    b. True
    c. True
    d. False; In a chemical reaction, atoms are neither created nor destroyed.

4.51 a. 1 + 2
    d. 3
b. 1
e. 2
c. 1

4.53 a. $^{16}_{8}X$, $^{17}_{8}X$, $^{18}_{8}X$ all have eight protons.
    b. $^{16}_{8}X$, $^{17}_{8}X$, $^{18}_{8}X$ are all isotopes of oxygen.
    c. $^{16}_{8}X$ and $^{16}_{9}X$ have mass numbers of 16, and $^{18}_{10}X$ and $^{18}_{8}X$ have mass numbers of 18.

4.55

	Atomic Symbol		
	$^{70}_{32}Ge$	$^{73}_{32}Ge$	$^{76}_{32}Ge$
Atomic Number	32	32	32
Mass Number	70	73	76
Number of Protons	32	32	32
Number of Neutrons	38	41	44
Number of Electrons	32	32	32

4.57 a. $^{9}_{4}Be$
    d. $^{10}_{5}B$
b. $^{11}_{5}B$
e. $^{12}_{6}C$
c. $^{13}_{6}C$

Representations **B** and **D** are isotopes of boron; **C** and **E** are isotopes of carbon.

4.59 The first letter of a symbol is a capital, but a second letter is lowercase. The symbol Co is for cobalt, but the symbols in CO are for carbon and oxygen.

4.61 a. Group 7A (17), Period 4
    b. Group 8A (18), Period 3
    c. Group 1A (1), Period 2
    d. Group 2A (2), Period 7

4.63 a. metal
    c. metal
b. metal
d. nonmetal

4.65 a. false
    d. false
b. false
e. true
c. true

4.67 a. 48 protons, 66 neutrons, 48 electrons
    b. 43 protons, 55 neutrons, 43 electrons
    c. 79 protons, 120 neutrons, 79 electrons
    d. 86 protons, 136 neutrons, 86 electrons
    e. 54 protons, 82 neutrons, 54 electrons

4.69

Name of the Element	Atomic Symbol	Number of Protons	Number of Neutrons	Number of Electrons
Selenium	$^{80}_{34}Se$	34	46	34
Nickel	$^{62}_{28}Ni$	28	34	28
Magnesium	$^{26}_{12}Mg$	12	14	12
Radium	$^{228}_{88}Ra$	88	140	88

4.71 a. protons
    c. alkali metals
b. protons

4.73 a. 3, Li
c. 24.31 amu, Mg

b. 86, Rn
d. 19.00 amu, F

4.75 a. nickel, Ni
d. arsenic, As

b. barium, Ba
e. tin, Sn

c. radium, Ra
f. cesium, Cs

4.77 a. 25 protons, 25 electrons
c. 38 protons, 38 electrons
e. 92 protons, 92 electrons

b. 15 protons, 15 electrons
d. 27 protons, 27 electrons

4.79 a. 82 protons, 126 neutrons, 82 electrons

b. $^{214}_{82}Pb$

c. $^{214}_{83}Bi$

4.81 a. $^{9}_{4}Be$

c. $^{46}_{20}Ca$

b. $^{26}_{12}Mg$

d. $^{70}_{30}Zn$

4.83 28.09 amu

4.85 a. 79 protons, 118 neutrons, 79 electrons

b. $^{195}_{79}Au$

c. $^{194}_{78}Pt$

4.87 207.3 amu

4.89 8.09×10^7 atoms of Na

Electronic Structure of Atoms and Periodic Trends

5

ROBERT AND JENNIFER

work in the materials science department of a research laboratory. As materials engineers, they deal with developing and testing various materials that are used in the manufacturing of consumer goods like computer chips, television sets, golf clubs, and snow skis. These engineers create new materials out of metals, ceramics, plastics, semiconductors, and combinations of materials called composites, which are needed for mechanical, chemical, and electrical industries. Often materials engineers study materials at an atomic level to learn how to improve the characteristics of materials.

In their research, Robert and Jennifer are working with some of the elements in Groups 3A (13), 4A (14), and 5A (15) of the periodic table. These elements, such as silicon, have properties that make them good semiconductors. A microchip requires growing a single crystal of a semiconductor such as pure silicon. When small amounts of impurities are added to the crystalline structure, holes form through which electrons can travel with little obstruction. Microchips are manufactured for use in computers, cell phones, satellites, televisions, calculators, GPS, and many other devices. Robert and Jennifer are working on materials that may be used to develop more complex microchips that lead to new applications.

CAREER

Materials Engineer

Materials engineers work with metals, ceramics, plastics, semiconductors, and composites to develop new or improved products such as computer chips, aircraft material, and tennis racquets.

Engineers use mathematics and the principles of chemistry and other sciences to solve technical problems. They may also work in the development and testing of new materials. Materials engineers typically have bachelor's degrees in mechanical, electrical, or chemical engineering, or related fields.

✦ KEY MATH SKILL

- Converting between Standard Numbers and Scientific Notation (1.5)

⚛ CORE CHEMISTRY SKILL

- Using Prefixes (2.4)

*This Key Math Skill and this Core Chemistry Skill from previous chapters are listed here for your review as you proceed to the new material in this chapter.

LOOKING AHEAD

5.1 Electromagnetic Radiation

LEARNING GOAL Compare the wavelength, frequency, and energy of electromagnetic radiation.

When we listen to a radio, use a microwave oven, turn on a light, see the colors of a rainbow, or have an X-ray taken, we are experiencing various forms of **electromagnetic radiation**. All of these types of electromagnetic radiation, including light, consist of particles that move as waves of energy.

Wavelength and Frequency

You are probably familiar with the action of waves in the ocean. If you were at a beach, you might notice that the water in each wave rises and falls as the wave comes in to shore. The highest point on the wave is called a *crest*, whereas the lowest point is a *trough*. On a calm day, there might be long distances between crests or troughs. However, if there is a storm with a lot of energy, the crests or troughs are much closer together.

The waves of electromagnetic radiation also have crests and troughs. The **wavelength** (symbol λ, lambda) is the distance from a crest or trough in the wave to the next crest or trough in the wave (see **FIGURE 5.1**). In some types of radiation, the crests or troughs are far apart, while in others, they are close together.

The wavelength is the distance between adjacent crests or troughs in a wave.

(a)

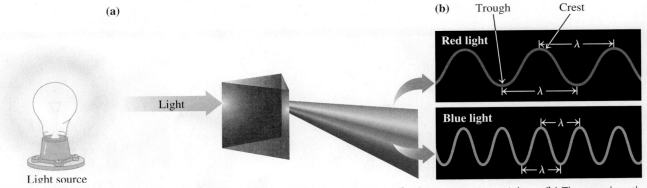

FIGURE 5.1 ▶ (a) Light passing through a prism is separated into a spectrum of colors we see in a rainbow. **(b)** The wavelength (λ) is the distance from a crest or trough on a wave to the next crest or trough.

❓ How does the wavelength of red light compare to that of blue light?

The **frequency** (symbol ν, nu) is the number of times the crests of a wave pass a point in 1 s. All electromagnetic radiation travels at the speed of light (c), which is a constant value equal to 3.00×10^8 m/s. Mathematically, the *wave equation* expresses the relationship of the speed of light (m/s) to wavelength (m) and frequency (s^{-1}).

$c = \lambda\nu$ Wave equation

Speed of light (c) = 3.00×10^8 m/s = wavelength (λ) × frequency (ν)

The speed of light is about a million times faster than the speed of sound, which is the reason we see lightning before we hear thunder during a storm.

ENGAGE

How many hours will it take light to travel from Rome to Los Angeles, a distance of 10 200 km?

Electromagnetic Spectrum

The **electromagnetic spectrum** is an arrangement of different types of electromagnetic radiation from longest wavelength to shortest wavelength. According to the wave equation, as the wavelength decreases, the frequency increases. Or as the wavelength increases, the frequency decreases. This type of relationship is called an *inverse relationship*.

Scientists have shown that the energy of electromagnetic radiation is directly related to frequency, which means that energy is also inversely related to the wavelength. Thus, as the wavelength of radiation increases, the frequency and, therefore, the energy decrease.

At one end of the electromagnetic spectrum are radiations with long wavelengths such as *radio waves* that are used for AM and FM radio bands, cellular phones, and TV signals. The wavelength of a typical AM radio wave can be as long as a football field. *Microwaves* have shorter wavelengths and higher frequencies than radio waves. *Infrared radiation* (IR) is responsible for the heat we feel from sunlight and the infrared lamps used to warm food in restaurants. When we change the volume or the station on a TV set, we use a remote control to send infrared impulses to the receiver in the TV. Wireless technology uses radiation with higher frequencies than infrared to connect many electronic devices, including mobile and cell phones and laptops (see **FIGURE 5.2**).

Visible light with wavelengths from 700 to 400 nm is the only radiation our eyes can detect. Red light has the longest wavelength at 700 nm; orange is about 600 nm; green is about 500 nm; and violet at 400 nm has the shortest wavelength of visible light. We see objects as different colors because the objects reflect only certain wavelengths, which are absorbed by our eyes.

Ultraviolet (UV) light has shorter wavelengths and higher frequencies than violet light of the visible range. The UV radiation in sunlight can cause serious sunburn, which may lead to skin cancer. While some UV light from the Sun is blocked by the ozone layer, the cosmetic industry has developed sunscreens to prevent the absorption of UV light by the skin. *X-rays* have shorter wavelengths than ultraviolet light, which means they have some of the highest frequencies. X-rays can pass through soft substances but not metals or bone, which allows us to see images of the bones and teeth in the body.

ENGAGE

Which type of electromagnetic radiation has lower frequency, ultraviolet or infrared?

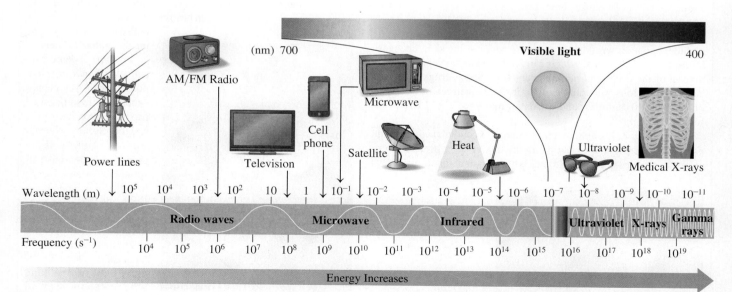

FIGURE 5.2 ▶ The electromagnetic spectrum shows the arrangement of wavelengths of electromagnetic radiation. The visible portion consists of wavelengths from 700 nm to 400 nm.

Q How does the wavelength of ultraviolet light compare to that of a microwave?

Microwaves with wavelengths of about 1 cm heat water molecules, which heats food.

SAMPLE PROBLEM 5.1 The Electromagnetic Spectrum

Arrange the following in order of decreasing wavelengths: X-rays, ultraviolet light, FM radio waves, and microwaves.

TRY IT FIRST

SOLUTION

The electromagnetic radiation with the longest wavelength is FM radio waves, then microwaves, followed by ultraviolet light, and then X-rays, which have the shortest wavelengths.

STUDY CHECK 5.1

Visible light contains colors from red to violet.

a. What color of light has the shortest wavelength?
b. What color of light has the lowest frequency?

ANSWER

a. violet light
b. red light

CHEMISTRY LINK TO **HEALTH**
Biological Reactions to UV Light

Our everyday life depends on sunlight, but exposure to sunlight can have damaging effects on living cells, and too much exposure can even cause their death. Light energy, especially ultraviolet (UV), excites electrons and may lead to unwanted chemical reactions. The list of damaging effects of sunlight includes sunburn; wrinkling; premature aging of the skin; changes in the DNA of the cells, which can lead to skin cancers; inflammation of the eyes; and perhaps cataracts. Some drugs, like the acne medications Accutane and Retin-A, as well as antibiotics, diuretics, sulfonamides, and estrogen, make the skin extremely sensitive to light.

Phototherapy uses light to treat certain skin conditions, including psoriasis, eczema, and dermatitis. In the treatment of psoriasis, for example, oral drugs are given to make the skin more photosensitive; then exposure to UV radiation follows. Low-energy radiation (blue light) with wavelengths from 390 to 470 nm is used to treat babies with neonatal jaundice, which converts high levels of bilirubin to water-soluble compounds that can be excreted from the body. Sunlight is also a factor in stimulating the immune system.

In a type of depression called *seasonal affective disorder* or SAD, people experience mood swings and depression during the winter. Some research suggests that SAD is the result of a decrease in serotonin, or an increase in melatonin, when there are fewer hours of sunlight. One treatment for SAD is therapy using bright light provided by a lamp called a light box. A daily exposure to blue light (460 nm) for 30 to 60 min seems to reduce symptoms of SAD.

Babies with neonatal jaundice are treated with UV light.

A light box is used to provide light, which reduces symptoms of SAD.

QUESTIONS AND PROBLEMS

5.1 Electromagnetic Radiation

LEARNING GOAL Compare the wavelength, frequency, and energy of electromagnetic radiation.

5.1 What is meant by the wavelength of UV light?

5.2 How are the wavelength and frequency of light related?

5.3 What is the difference between "white" light and blue or red light?

5.4 Why can we use X-rays, but not radio waves or microwaves, to give an image of bones and teeth?

5.5 Ultraviolet radiation (UVB) used to treat psoriasis has a wavelength of 3×10^{-7} m, whereas infrared radiation used in a thermal imaging camera has a wavelength of 1×10^{-5} m. Which has a higher frequency? Which has a higher energy?

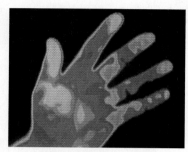

A thermal image of the hand of a patient with multiple sclerosis shows the temperature differences and poor blood flow (blue) to the ends of the fingers.

5.6 AM radio waves have a frequency of $8 \times 10^5 \text{ s}^{-1}$, whereas TV Channel 2 has a frequency of $6 \times 10^7 \text{ s}^{-1}$. Which has a shorter wavelength? Which has a lower energy?

5.7 If orange light has a wavelength of 6.3×10^{-5} cm, what is its wavelength in meters and nanometers?

5.8 A wavelength of 850 nm is used for fiber-optic transmission. What is its wavelength in meters?

5.9 Which type of electromagnetic radiation, cell phones, AM radio, or infrared light has the longest wavelengths?

5.10 Of radio waves, infrared light, and UV light, which has the shortest wavelengths?

5.11 Place the following types of electromagnetic radiation in order of increasing wavelengths: the blue color in a rainbow, X-rays, and microwaves from an oven.

5.12 Place the following types of electromagnetic radiation in order of decreasing wavelengths: FM radio station, red color in neon lights, and cell phone.

5.13 Place the following types of electromagnetic radiation in order of increasing frequencies: TV signal, X-rays, and microwaves.

5.14 Place the following types of electromagnetic radiation in order of decreasing frequencies: AM music station, UV radiation from the Sun, and infrared radiation from a heat lamp.

5.2 Atomic Spectra and Energy Levels

LEARNING GOAL Explain how atomic spectra correlate with the energy levels in atoms.

When the white light from the Sun or a light bulb is passed through a prism or raindrops, it produces a *continuous spectrum*, like a rainbow. When atoms of elements are heated, they also produce light. At night, you may have seen the yellow color of sodium streetlamps or the red color of neon lights.

A rainbow forms when light passes through water droplets.

Photons

The light emitted from a streetlamp or by atoms that are heated is a stream of particles called **photons**. Every photon is a packet of energy with both particle and wave characteristics that travels at the speed of light. High-frequency photons have high energy and short wavelengths, whereas low-frequency photons have low energy and long wavelengths.

Photons play an important role in our modern world, particularly in the use of lasers, which use a narrow range of wavelengths. For example, lasers use photons of a single frequency to read pits on compact discs (CDs) and digital versatile discs (DVDs) or to scan bar codes on labels when we buy groceries. A CD is read by a laser with a wavelength of 780 nm. The newer DVDs are read by a blue laser with a wavelength of 405 nm, hence the name Blu-ray. The shorter wavelength allows a smaller pit size on the disc, which means that the disc has a greater storage capacity. In hospitals, high-energy photons are used in treatments to reach tumors within the tissues without damaging the surrounding tissue.

A CD or DVD is read when light from a laser is reflected from the pits on the surface.

Atomic Spectra

When the light emitted from heated elements is passed through a prism, it does not produce a continuous spectrum. Instead, an **atomic spectrum** is produced that consists of lines of different colors separated by dark areas (see **FIGURE 5.3**). This separation of colors indicates that only certain wavelengths of light are produced when an element is heated, which gives each element a unique atomic spectrum.

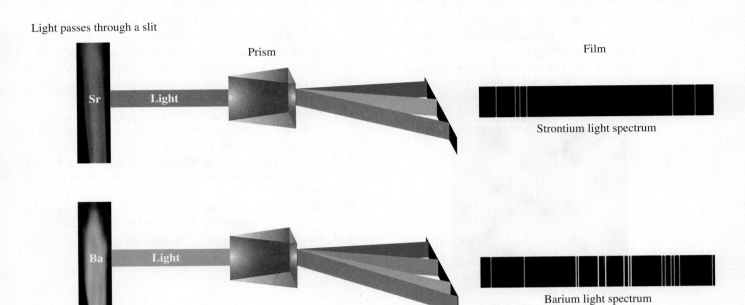

Light passes through a slit

FIGURE 5.3 ▶ In an atomic spectrum, light from a heated element separates into distinct lines.
🔍 Why don't the elements form a continuous spectrum as seen with white light?

Electron Energy Levels

Scientists have now determined that the lines in atomic spectra are associated with changes in the energies of the electrons. In an atom, each electron has a specific **energy level**, which is assigned a value called the **principal quantum number (*n*)**, ($n = 1, n = 2, \ldots$). Generally, electrons in the lower energy levels are closer to the nucleus, while electrons in the higher energy levels are farther away. The energy of an electron is *quantized*, which means that the energy of an electron can only have specific energy values, but cannot have values between these values.

Order of Increasing Energy of Levels

$$1 < 2 < 3 < 4 < 5 < 6 < 7$$

Lowest   ⟶   Highest
energy   energy

As an analogy, we can think of the energy levels of an atom as similar to the shelves in a bookcase. The first shelf is the lowest energy level; the second shelf is the second energy level; and so on. If we are arranging books on the shelves, it would take less energy to fill the bottom shelf first, and then the second shelf, and so on. However, we could never get any book to stay in the space between any of the shelves. Similarly, the energy of an electron must be at a specific energy level, and not between.

Unlike standard bookcases, however, there is a large difference between the energy of the first and second energy levels, but then the higher energy levels are closer together. Another difference is that the lower electron energy levels hold fewer electrons than the higher energy levels.

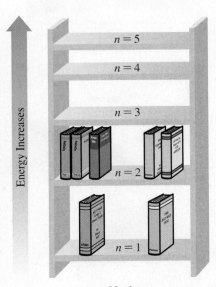

An electron can have only the energy of one of the energy levels in an atom.

Changes in Energy Levels

An electron can change from one energy level to a higher energy level only if it absorbs the energy equal to the difference in energy levels. When an electron changes to a lower energy level, it emits energy equal to the difference between the two energy levels. If the energy emitted is in the visible range, we see one of the colors of visible light (see FIGURE 5.4). The yellow color of sodium streetlights and the red color of neon lights are examples of electrons emitting energy in the visible color range.

Colors are produced when electricity excites electrons in noble gases.

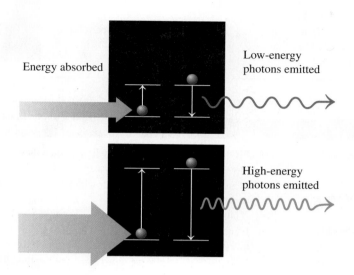

FIGURE 5.4 ▶ Electrons absorb a specific amount of energy to move to a higher energy level. When electrons lose energy, photons with specific energies are emitted.

Q What causes electrons to move to higher energy levels?

SAMPLE PROBLEM 5.2 Change in Energy Levels

a. How does an electron move to a higher energy level?
b. When an electron drops to a lower energy level, how is energy lost?

TRY IT FIRST

SOLUTION

a. An electron moves to a higher energy level when it absorbs an amount of energy equal to the difference in energy levels.
b. Energy, equal to the difference in energy levels, is emitted as a photon when an electron drops to a lower energy level.

STUDY CHECK 5.2

Why did scientists propose that electrons occupy specific energy levels in an atom?

ANSWER

Because the spectra of elements consisted of only discrete, separated lines, scientists concluded that electrons occupied only certain energy levels in the atom.

CHEMISTRY LINK TO THE ENVIRONMENT
Energy-Saving Fluorescent Bulbs

Compact fluorescent lights (CFL) are replacing the standard light bulb we use in our homes and workplaces. Compared to a standard light bulb, the CFL has a longer life and uses less electricity. Within about 20 days of use, the fluorescent bulb saves enough money in electricity to pay for its higher initial cost.

A standard incandescent light bulb has a thin tungsten filament inside a sealed glass bulb. When the light is switched on, electricity flows through this filament, and electrical energy is converted to heat energy. When the filament reaches a temperature around 2300 °C, we see white light.

A fluorescent bulb produces light in a different way. When the switch is turned on, electrons move between two electrodes and collide with mercury atoms in a mixture of mercury and argon gas inside the bulb. When the electrons in the mercury atoms absorb energy from the collisions, they are raised to higher energy levels. As electrons fall to lower energy levels emitting ultraviolet radiation, they strike the phosphor coating inside the tube, and fluorescence occurs as visible light is emitted.

The production of light is more efficient in a fluorescent bulb than in an incandescent light bulb. A 75-watt incandescent bulb can be replaced by a 20-watt CFL that gives the same amount of light, providing a 70% reduction in electricity costs. A typical incandescent light bulb lasts one to two months, whereas a compact fluorescent light bulb lasts from 1 to 2 yr. One drawback of the CFL is that each contains about 4 mg of mercury. As long as the bulb stays intact, no mercury is released. However, used CFL bulbs should not be disposed of in household trash but rather should be taken to a recycling center.

A compact fluorescent light (CFL) uses up to 70% less energy.

QUESTIONS AND PROBLEMS

5.2 Atomic Spectra and Energy Levels

LEARNING GOAL Explain how atomic spectra correlate with the energy levels in atoms.

5.15 What feature of an atomic spectrum indicates that the energy emitted by heating an element is not continuous?

5.16 How can we explain the distinct lines that appear in an atomic spectrum?

5.17 Electrons can jump to higher energy levels when they _____ (absorb/emit) a photon.

5.18 Electrons drop to lower energy levels when they _____ (absorb/emit) a photon.

5.19 Identify the photon in each pair with the greater energy.
a. green light or yellow light
b. red light or blue light

5.20 Identify the photon in each pair with the greater energy.
a. orange light or violet light
b. infrared light or ultraviolet light

5.3 Sublevels and Orbitals

LEARNING GOAL Describe the sublevels and orbitals for the electrons in an atom.

We have seen that the protons and neutrons are contained in the small, dense nucleus of an atom. However, it is the electrons within the atoms that determine the physical and chemical properties of the elements. Therefore, we will look at the arrangement of electrons within the large volume of space surrounding the nucleus.

There is a limit to the number of electrons allowed in each energy level. Only a few electrons can occupy the lower energy levels, while more electrons can be accommodated in higher energy levels. The maximum number of electrons allowed in any energy level is calculated using the formula $2n^2$ (two times the square of the principal quantum number). **TABLE 5.1** shows the maximum number of electrons allowed in the first four energy levels.

TABLE 5.1 Maximum Number of Electrons Allowed in Energy Levels 1 to 4

Energy Level (n)	1	2	3	4
$2n^2$	$2(1)^2$	$2(2)^2$	$2(3)^2$	$2(4)^2$
Maximum Number of Electrons	2	8	18	32

Sublevels

Each of the energy levels consists of one or more **sublevels**, in which electrons with identical energy are found. The sublevels are identified by the letters s, p, d, and f. The number of sublevels within an energy level is equal to the principal quantum number, n. For example, the first energy level ($n = 1$) has only one sublevel, $1s$. The second energy level ($n = 2$) has two sublevels, $2s$ and $2p$. The third energy level ($n = 3$) has three sublevels, $3s$, $3p$, and $3d$. The fourth energy level ($n = 4$) has four sublevels, $4s$, $4p$, $4d$, and $4f$. Energy levels $n = 5$, $n = 6$, and $n = 7$ also have as many sublevels as the value of n, but only s, p, d, and f sublevels are needed to hold the electrons in atoms of the 118 known elements (see **FIGURE 5.5**).

ENGAGE

How many sublevels are in energy level $n = 5$?

Energy Level	Number of Sublevels	Types of Sublevels			
		s	p	d	f
$n = 4$	4	▫	▫▫▫	▫▫▫▫▫	▫▫▫▫▫▫▫
$n = 3$	3	▫	▫▫▫	▫▫▫▫▫	
$n = 2$	2	▫	▫▫▫		
$n = 1$	1	▫			

FIGURE 5.5 ▶ The number of sublevels in an energy level is the same as the principal quantum number, n.

Q How many sublevels are in the n = 3 energy level?

Within each energy level, the *s* sublevel has the lowest energy. If there are additional sublevels, the *p* sublevel has the next lowest energy, then the *d* sublevel, and finally the *f* sublevel.

Order of Increasing Energy of Sublevels in an Energy Level

$s < p < d < f$

Lowest ⟶ Highest
Energy   Energy

Orbitals

There is no way to know the exact location of an electron in an atom. Instead, scientists describe the location of an electron in terms of probability. The **orbital** is the three-dimensional volume in which electrons have the highest probability of being found.

As an analogy, imagine that you could draw a circle with a 100-m radius around your chemistry classroom. There is a high probability of finding you within that circle when your chemistry class is in session. But once in a while, you may be outside that circle because you were sick or your car did not start.

Shapes of Orbitals

Each type of orbital has a unique three-dimensional shape. Electrons in an *s* orbital are most likely found in a region with a spherical shape. Imagine that you take a picture of the location of an electron in an *s* orbital every second for an hour. When all these pictures are overlaid, the result, called a *probability density*, would look like the electron cloud shown in **FIGURE 5.6a**. For convenience, we draw this electron cloud as a sphere called an *s* orbital. There is one *s* orbital for every energy level starting with $n = 1$. For example, in the first, second, and third energy levels there are *s* orbitals designated as 1*s*, 2*s*, and 3*s*. As the principal quantum number increases, there is an increase in the size of the *s* orbitals, although the shape is the same (see **FIGURE 5.6b**).

The orbitals occupied by *p*, *d*, and *f* electrons have three-dimensional shapes different from those of the *s* electrons. There are three *p* orbitals, starting with $n = 2$. Each *p* orbital has two lobes like a balloon tied in the middle. The three *p* orbitals are arranged in three perpendicular directions, along the *x*, *y*, and *z* axes around the nucleus (see **FIGURE 5.7**). As with *s* orbitals, the shape of *p* orbitals is the same, but the volume increases at higher energy levels.

In summary, the $n = 2$ energy level, which has 2*s* and 2*p* sublevels, consists of one *s* orbital and three *p* orbitals.

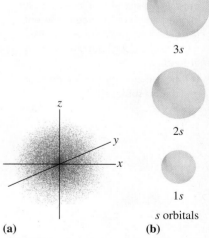

(a) **(b)**

FIGURE 5.6 ▶ **(a)** The electron cloud of an *s* orbital represents the highest probability of finding an *s* electron. **(b)** The *s* orbitals are shown as spheres. The sizes of the *s* orbitals increase because they contain electrons at higher energy levels.

Q Is the probability high or low of finding an *s* electron outside an *s* orbital?

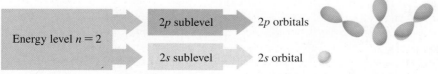

Energy level $n = 2$ consists of one 2*s* orbital and three 2*p* orbitals.

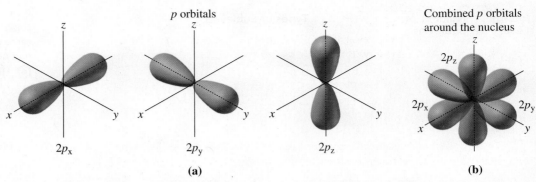

(a)   **(b)**

FIGURE 5.7 ▶ A *p* orbital has two regions of high probability, which gives a "dumb-bell" shape. **(a)** Each *p* orbital is aligned along a different axis from the other *p* orbitals. **(b)** All three *p* orbitals are shown around the nucleus.

Q What are some similarities and differences of the *p* orbitals in the $n = 3$ energy level?

Energy level $n = 3$ consists of three sublevels *s*, *p*, and *d*. The *d* sublevels contain five *d* orbitals (see **FIGURE 5.8**).

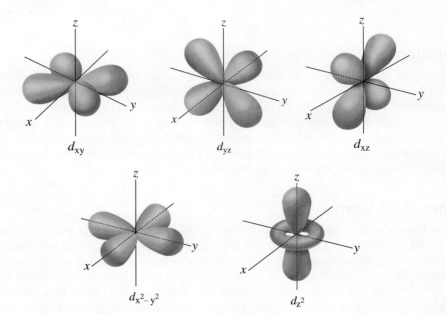

FIGURE 5.8 ▶ Four of the five *d* orbitals consist of four lobes that are aligned along or between different axes. One *d* orbital consists of two lobes and a doughnut-shaped ring around its center.

Q How many orbitals are there in the 5*d* sublevel?

Energy level $n = 4$ consists of four sublevels *s*, *p*, *d*, and *f*. In the *f* sublevel, there are seven *f* orbitals. The shapes of *f* orbitals are complex and so we have not included them in this text.

SAMPLE PROBLEM 5.3 Energy Levels, Sublevels, and Orbitals

Indicate the type and number of orbitals in each of the following energy levels or sublevels:

a. 3*p* sublevel   **b.** $n = 2$ **c.** $n = 3$ **d.** 4*d* sublevel

TRY IT FIRST ▶

SOLUTION

	Given	Need	Connect
ANALYZE THE PROBLEM	energy level, sublevel	orbitals	*n*, orbitals in *s*, *p*, *d*, *f*

a. The 3p sublevel contains three 3p orbitals.
b. The $n = 2$ energy level consists of one 2s and three 2p orbitals.
c. The $n = 3$ energy level consists of one 3s, three 3p, and five 3d orbitals.
d. The 4d sublevel contains five 4d orbitals.

STUDY CHECK 5.3

What is similar and what is different for 1s, 2s, and 3s orbitals?

ANSWER

The 1s, 2s, and 3s orbitals are all spherical, but they increase in volume because the electron is most likely to be found farther from the nucleus for higher energy levels.

Orbital Capacity and Electron Spin

The *Pauli exclusion principle* states that each orbital can hold a maximum of two electrons. According to a model for electron behavior, an electron is seen as spinning on its axis, which generates a magnetic field. When two electrons are in the same orbital, they will repel each other unless their magnetic fields cancel. This happens only when the two electrons spin in opposite directions. We can represent the spins of the electrons in the same orbital with one arrow pointing up and the other pointing down.

Electron spinning counterclockwise   Electron spinning clockwise

Opposite spins of electrons in an orbital

An orbital can hold up to two electrons with opposite spins.

Number of Electrons in Sublevels

There is a maximum number of electrons that can occupy each sublevel. An *s* sublevel holds one or two electrons. Because each *p* orbital can hold up to two electrons, the three *p* orbitals in a *p* sublevel can accommodate six electrons. A *d* sublevel with five *d* orbitals can hold a maximum of 10 electrons. With seven *f* orbitals, an *f* sublevel can hold up to 14 electrons.

As mentioned earlier, higher energy levels such as $n = 5, 6$, and 7 would have 5, 6, and 7 sublevels, but those beyond sublevel *f* are not utilized by the atoms of the elements known today. The total number of electrons in all the sublevels adds up to give the electrons allowed in an energy level. The number of sublevels, the number of orbitals, and the maximum number of electrons for energy levels 1 to 4 are shown in **TABLE 5.2**.

TABLE 5.2 Electron Capacity in Sublevels for Energy Levels 1 to 4

Energy Level (n)	Number of Sublevels	Type of Sublevel	Number of Orbitals	Maximum Number of Electrons	Total Electrons $(2n^2)$
4	4	4f	7	14	
		4d	5	10	
		4p	3	6	32
		4s	1	2	
3	3	3d	5	10	
		3p	3	6	18
		3s	1	2	
2	2	2p	3	6	
		2s	1	2	8
1	1	1s	1	2	2

QUESTIONS AND PROBLEMS

5.3 Sublevels and Orbitals

LEARNING GOAL Describe the sublevels and orbitals for the electrons in an atom.

5.21 Describe the shape of each of the following orbitals:
    a. 1s   **b.** 2p   **c.** 5s

5.22 Describe the shape of each of the following orbitals:
    a. 3p   **b.** 6s   **c.** 4p

5.23 Match statements **1** to **3** with **a** to **d**:
    1. They have the same shape.
    2. The maximum number of electrons is the same.
    3. They are in the same energy level.

    a. 1s and 2s orbitals   **b.** 3s and 3p sublevels
    c. 3p and 4p sublevels   **d.** three 3p orbitals

5.24 Match statements **1** to **3** with **a** to **d**:
    1. They have the same shape.
    2. The maximum number of electrons is the same.
    3. They are in the same energy level.

    a. 5s and 6s orbitals   **b.** 3p and 4p orbitals
    c. 3s and 4s sublevels   **d.** 2s and 2p orbitals

5.25 Indicate the number of each in the following:
    a. orbitals in the 3d sublevel
    b. sublevels in the $n = 1$ energy level
    c. orbitals in the 6s sublevel
    d. orbitals in the $n = 3$ energy level

5.26 Indicate the number of each in the following:
    a. orbitals in the $n = 2$ energy level
    b. sublevels in the $n = 4$ energy level
    c. orbitals in the 5f sublevel
    d. orbitals in the 6p sublevel

5.27 Indicate the maximum number of electrons in the following:
    a. 2p orbital
    b. 3p sublevel
    c. $n = 4$ energy level
    d. 5d sublevel

5.28 Indicate the maximum number of electrons in the following:
    a. 3s sublevel
    b. 4p orbital
    c. $n = 3$ energy level
    d. 4f sublevel

5.4 Orbital Diagrams and Electron Configurations

LEARNING GOAL Draw the orbital diagram and write the electron configuration for an element.

We can now look at how electrons are arranged within an atom. Electrons are added first to orbitals with the lowest energy levels, building progressively by adding electrons to levels with higher energies. This process of building the electrons in an atom is known as the *Aufbau principle*.

In an **orbital diagram**, electrons are shown as arrows that are placed in boxes that represent the orbitals in order of increasing energy (see **FIGURE 5.9**).

To draw an orbital diagram, the lowest energy orbitals are filled first. For example, we can draw the orbital diagram for carbon. The atomic number of carbon is 6, which means that a carbon atom has six electrons. The first two electrons go into the 1s orbital; the next two electrons go into the 2s orbital. In the orbital diagram, the two electrons in the 1s and 2s orbitals are shown with opposite spins; the first arrow is up and the second is down. The last two electrons in carbon begin to fill the 2p sublevel, which has the next lowest energy. However, there are three 2p orbitals of equal energy. Because the negatively charged electrons repel each other, they are placed in separate 2p orbitals. *Hund's rule* states that there is less repulsion when electrons are placed in separate orbitals of the same sublevel. With few exceptions (which will be noted later in this chapter) lower energy sublevels are filled first, and then the "building" of electrons continues to the next lowest energy sublevel that is available until all the electrons are placed.

ENGAGE

When do we need to use Hund's rule in drawing orbital diagrams?

Orbital diagram for carbon

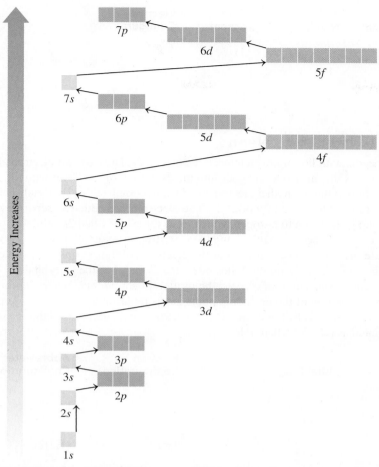

FIGURE 5.9 ▶ The orbitals in an atom fill in order of increasing energy beginning with the lowest energy level, 1s.

Q Why do 3d orbitals fill after the 4s orbital?

Electron Configurations

Chemists use a notation called the **electron configuration** to indicate the placement of the electrons of an atom in order of increasing energy. As in the orbital diagrams, an electron configuration is written with the lowest energy sublevel first, followed by the next lowest energy sublevel. The number of electrons in each sublevel is shown as a superscript.

> **CORE CHEMISTRY SKILL**
> Writing Electron Configurations

Electron Configuration for Carbon

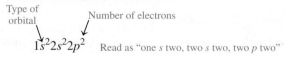

Type of orbital   Number of electrons

$1s^2 2s^2 2p^2$ Read as "one *s* two, two *s* two, two *p* two"

Period 1: Hydrogen and Helium

We will draw orbital diagrams and write the corresponding electron configurations for the elements H and He in Period 1. The 1s orbital (which is the 1s sublevel) is written first because it has the lowest energy. Hydrogen has one electron in the 1s sublevel; helium has two. In the orbital diagram, the electrons for helium are shown as arrows pointing in opposite directions.

Element	Atomic Number	Orbital Diagram	Electron Configuration
		$1s$	
H	1	[↑]	$1s^1$
He	2	[↑↓]	$1s^2$

Period 2: Lithium to Neon

Period 2 begins with lithium, which has three electrons. The first two electrons fill the $1s$ orbital, whereas the third electron goes into the $2s$ orbital, the sublevel with the next lowest energy. In beryllium, another electron is added to complete the $2s$ orbital. The next six electrons are used to fill the $2p$ orbitals. The electrons are added to separate p orbitals (Hund's rule) from boron to nitrogen, which gives three half-filled $2p$ orbitals.

From oxygen to neon, the remaining three electrons are paired up using opposite spins to complete the $2p$ sublevel. In writing the electron configurations for the elements in Period 2, begin with the $1s$ orbital followed by the $2s$ and then the $2p$ orbitals.

An electron configuration can also be written in an *abbreviated configuration*. The electron configuration of the preceding noble gas is replaced by writing its element symbol inside square brackets. For example, the electron configuration for lithium, $1s^2 2s^1$, can be abbreviated as $[He]2s^1$ where $[He]$ replaces $1s^2$.

Element	Atomic Number	Orbital Diagram	Electron Configuration	Abbreviated Electron Configuration
		$1s$ $2s$		
Li	3	[↑↓] [↑]	$1s^2 2s^1$	$[He]2s^1$
Be	4	[↑↓] [↑↓]	$1s^2 2s^2$	$[He]2s^2$
		$2p$		
B	5	[↑↓] [↑↓] [↑][ ][ ]	$1s^2 2s^2 2p^1$	$[He]2s^2 2p^1$
C	6	[↑↓] [↑↓] [↑][↑][ ]	$1s^2 2s^2 2p^2$	$[He]2s^2 2p^2$
		Unpaired electrons		
N	7	[↑↓] [↑↓] [↑][↑][↑]	$1s^2 2s^2 2p^3$	$[He]2s^2 2p^3$
O	8	[↑↓] [↑↓] [↑↓][↑][↑]	$1s^2 2s^2 2p^4$	$[He]2s^2 2p^4$
F	9	[↑↓] [↑↓] [↑↓][↑↓][↑]	$1s^2 2s^2 2p^5$	$[He]2s^2 2p^5$
Ne	10	[↑↓] [↑↓] [↑↓][↑↓][↑↓]	$1s^2 2s^2 2p^6$	$[He]2s^2 2p^6$

Guide to Drawing Orbital Diagrams

STEP 1
Draw boxes to represent the occupied orbitals.

STEP 2
Place a pair of electrons with opposite spins in each filled orbital.

STEP 3
Place the remaining electrons in the last occupied sublevel in separate orbitals.

SAMPLE PROBLEM 5.4 Drawing Orbital Diagrams

Nitrogen is an element that is used in the formation of amino acids, proteins, and nucleic acids. Draw the orbital diagram for nitrogen.

TRY IT FIRST

SOLUTION

STEP 1 Draw boxes to represent the occupied orbitals. Nitrogen has atomic number 7, which means it has seven electrons. For the orbital diagram, we draw boxes to represent the $1s$, $2s$, and $2p$ orbitals.

$1s$ $2s$ $2p$
[ ] [ ] [ ][ ][ ]

STEP 2 Place a pair of electrons with opposite spins in each filled orbital. First, we place a pair of electrons with opposite spins in both the 1s and 2s orbitals.

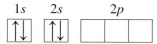

STEP 3 Place the remaining electrons in the last occupied sublevel in separate orbitals. Then we place the three remaining electrons in three separate 2p orbitals with arrows drawn in the same direction.

Orbital diagram for nitrogen (N)

STUDY CHECK 5.4

Draw the orbital diagram for fluorine, which is used to make nonstick coatings for cookware.

ANSWER

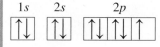

Fluorine is used to make nonstick coatings for cookware.

Period 3: Sodium to Argon

In Period 3, electrons enter the orbitals of the 3s and 3p sublevels, but not the 3d sublevel. We notice that the elements sodium to argon, which are directly below the elements lithium to neon in Period 2, have a similar pattern of filling their s and p orbitals. In sodium and magnesium, one and two electrons go into the 3s orbital. The electrons for aluminum, silicon, and phosphorus go into separate 3p orbitals. The remaining electrons in sulfur, chlorine, and argon are paired up (opposite spins) with the electrons already in the 3p orbitals. For the abbreviated electron configurations of Period 3, the symbol [Ne] replaces $1s^2 2s^2 2p^6$.

Element	Atomic Number	Orbital Diagram (3s and 3p orbitals only)	Electron Configuration	Abbreviated Electron Configuration
Na	11	[Ne]	$1s^2 2s^2 2p^6 3s^1$	$[Ne]3s^1$
Mg	12	[Ne]	$1s^2 2s^2 2p^6 3s^2$	$[Ne]3s^2$
Al	13	[Ne]	$1s^2 2s^2 2p^6 3s^2 3p^1$	$[Ne]3s^2 3p^1$
Si	14	[Ne]	$1s^2 2s^2 2p^6 3s^2 3p^2$	$[Ne]3s^2 3p^2$
P	15	[Ne]	$1s^2 2s^2 2p^6 3s^2 3p^3$	$[Ne]3s^2 3p^3$
S	16	[Ne]	$1s^2 2s^2 2p^6 3s^2 3p^4$	$[Ne]3s^2 3p^4$
Cl	17	[Ne]	$1s^2 2s^2 2p^6 3s^2 3p^5$	$[Ne]3s^2 3p^5$
Ar	18	[Ne]	$1s^2 2s^2 2p^6 3s^2 3p^6$	$[Ne]3s^2 3p^6$

Guide to Writing Electron Configurations

STEP 1
State the number of electrons from the atomic number on the periodic table.

STEP 2
Write the number of electrons for each orbital in order of increasing energy until filling is complete.

STEP 3
Write an abbreviated electron configuration by replacing the configuration of the preceding noble gas with its symbol.

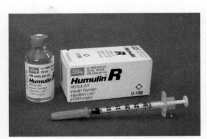

Sulfur atoms are part of the structure of human insulin.

SAMPLE PROBLEM 5.5 Electron Configurations

Silicon is the basis of semiconductors. Write the complete and abbreviated electron configurations for silicon.

TRY IT FIRST

SOLUTION

	Given	Need	Connect
ANALYZE THE PROBLEM	silicon (Si)	electron configuration, abbreviated electron configuration	periodic table, atomic number, order of filling orbitals

STEP 1 State the number of electrons from the atomic number on the periodic table. Silicon has an atomic number of 14, which means it has 14 electrons.

STEP 2 Write the number of electrons for each orbital in order of increasing energy until filling is complete.

$1s^2 2s^2 2p^6 3s^2 3p^2$ Electron configuration for Si

STEP 3 Write an abbreviated electron configuration by replacing the configuration of the preceding noble gas with its symbol.

$[Ne]3s^2 3p^2$ Abbreviated electron configuration for Si

STUDY CHECK 5.5

Write the complete and abbreviated electron configurations for sulfur, which is a macro-mineral in proteins, vitamin B_1, and insulin.

ANSWER

$1s^2 2s^2 2p^6 3s^2 3p^4$, $[Ne]3s^2 3p^4$

QUESTIONS AND PROBLEMS

5.4 Orbital Diagrams and Electron Configurations

LEARNING GOAL Draw the orbital diagram and write the electron configuration for an element.

5.29 Compare the terms *electron configuration* and *abbreviated electron configuration*.

5.30 Compare the terms *orbital diagram* and *electron configuration*.

5.31 Draw the orbital diagram for each of the following:
a. boron
b. aluminum
c. phosphorus
d. argon

5.32 Draw the orbital diagram for each of the following:
a. carbon
b. sulfur
c. magnesium
d. beryllium

5.33 Write the complete electron configuration for each of the following:
a. nickel
b. sodium
c. lithium
d. titanium

5.34 Write the complete electron configuration for each of the following:
a. nitrogen
b. chlorine
c. strontium
d. neon

5.35 Write the abbreviated electron configuration for each of the following:
a. tin
b. cadmium
c. selenium
d. fluorine

5.36 Write the abbreviated electron configuration for each of the following:
a. barium
b. oxygen
c. manganese
d. arsenic

5.37 Give the symbol of the element with each of the following electron or abbreviated electron configurations:
a. $1s^2 2s^1$
b. $1s^2 2s^2 2p^6 3s^2 3p^6 4s^2 3d^2$
c. $[Ar]4s^2 3d^{10} 4p^2$
d. $[He]2s^2 2p^5$

5.38 Give the symbol of the element with each of the following electron or abbreviated electron configurations:
a. $1s^2 2s^2 2p^4$
b. $[Ne]3s^2$
c. $1s^2 2s^2 2p^6 3s^2 3p^6$
d. $[Ne]3s^2 3p^1$

5.39 Give the symbol of the element that meets the following conditions:
a. has three electrons in the $n = 3$ energy level
b. has two $2p$ electrons
c. completes the $3p$ sublevel
d. completes the $2s$ sublevel

5.40 Give the symbol of the element that meets the following conditions:
a. has five electrons in the $3p$ sublevel
b. has four $2p$ electrons
c. completes the $3s$ sublevel
d. has one electron in the $3s$ sublevel

5.5 Electron Configurations and the Periodic Table

LEARNING GOAL Write the electron configuration for an atom using the sublevel blocks on the periodic table.

Up to now, we have written electron configurations using their energy level diagrams. As configurations involve more energy levels, this can become tedious. However, the electron configurations of the elements are related to their position on the periodic table. Different sections or blocks within the periodic table correspond to the s, p, d, and f sublevels (see **FIGURE 5.10**). Therefore, we can "build" the electron configurations of atoms by reading the periodic table in order of increasing atomic number.

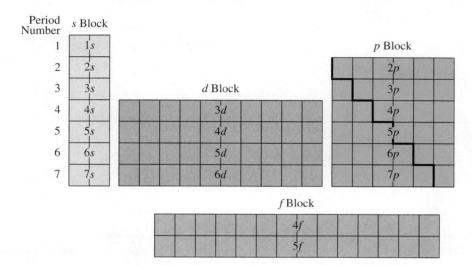

FIGURE 5.10 ▶ Electron configurations follow the order of occupied sublevels on the periodic table.

◉ If neon is in Group 8A (18), Period 2, how many electrons are in the 1s, 2s, and 2p sublevels of neon?

Blocks on the Periodic Table

1. The *s* **block** includes hydrogen and helium as well as the elements in Group 1A (1) and Group 2A (2). This means that the final one or two electrons in the elements of the s block are located in an s orbital. The period number indicates the particular s orbital that is filling: $1s$, $2s$, and so on.

2. The *p* **block** consists of the elements in Group 3A (13) to Group 8A (18). There are six p block elements in each period because three p orbitals can hold up to six electrons. The period number indicates the particular p sublevel that is filling: $2p$, $3p$, and so on.

3. The *d* **block**, containing the transition elements, first appears after calcium (atomic number 20). There are 10 elements in each period of the d block because five d orbitals can hold up to 10 electrons. The particular d sublevel is one less $(n - 1)$ than the period number. For example, in Period 4, the d block is the $3d$ sublevel. In Period 5, the d block is the $4d$ sublevel.

4. The *f* **block**, the inner transition elements, are the two rows at the bottom of the periodic table. There are 14 elements in each f block because seven f orbitals can hold up to 14 electrons. Elements that have atomic numbers higher than 57 (La) have electrons in the $4f$ block. The particular f sublevel is two less $(n - 2)$ than the period number. For example, in Period 6, the f block is the $4f$ sublevel. In Period 7, the f block is the $5f$ sublevel.

Writing Electron Configurations Using Sublevel Blocks

Now we can write electron configurations using the sublevel blocks on the periodic table. As before, each configuration begins at H. But now we move across the table from left to right, writing down each sublevel block we come to until we reach the element for which

ENGAGE

On the periodic table, what are the group numbers that make up the *p* block? the *s* block?

CORE CHEMISTRY SKILL
Using the Periodic Table to Write Electron Configurations

we are writing an electron configuration. We will show how to write the electron configuration for chlorine (atomic number 17) from the sublevel blocks on the periodic table in Sample Problem 5.6.

SAMPLE PROBLEM 5.6 Using the Sublevel Blocks to Write Electron Configurations

Use the sublevel blocks on the periodic table to write the electron configuration for chlorine.

TRY IT FIRST

SOLUTION

Guide to Writing Electron Configurations Using Sublevel Blocks

STEP 1
Locate the element on the periodic table.

STEP 2
Write the filled sublevels in order, going across each period.

STEP 3
Complete the configuration by counting the electrons in the last occupied sublevel block.

STEP 1 Locate the element on the periodic table. Chlorine (atomic number 17) is in Group 7A (17) and Period 3.

ANALYZE THE PROBLEM	Given	Need	Connect
	element chlorine	electron configuration	sublevel blocks

STEP 2 Write the filled sublevels in order, going across each period.

Period	Sublevel Block Filling	Sublevel Block Notation (filled)
1	$1s$ (H $\longrightarrow$ He)	$1s^2$
2	$2s$ (Li $\longrightarrow$ Be) and $2p$ (B $\longrightarrow$ Ne)	$2s^2 2p^6$
3	$3s$ (Na $\longrightarrow$ Mg)	$3s^2$

STEP 3 Complete the configuration by counting the electrons in the last occupied sublevel block. Because chlorine is the fifth element in the $3p$ block, there are five electrons in the $3p$ sublevel.

Period	Sublevel Block Filling	Sublevel Block Notation
3	$3p$ (Al $\longrightarrow$ Cl)	$3p^5$

The electron configuration for chlorine (Cl) is: $1s^2 2s^2 2p^6 3s^2 3p^5$.

STUDY CHECK 5.6

Use the sublevel blocks on the periodic table to write the electron configuration for argon.

ANSWER

$1s^2 2s^2 2p^6 3s^2 3p^6$

ENGAGE

What sublevel blocks are used to add electrons to the elements from K to Zn?

Electron Configurations for Period 4 and Above

Up to Period 4, the filling of the sublevels has progressed in order. However, if we look at the sublevel blocks in Period 4, we see that the $4s$ sublevel fills before the $3d$ sublevel. This occurs because the electrons in the $4s$ sublevel have slightly lower energy than the electrons in the $3d$ sublevel. This order occurs again in Period 5 when the $5s$ sublevel fills before the $4d$ sublevel and again in Period 6 when the $6s$ fills before the $5d$.

At the beginning of Period 4, the electrons in potassium (19) and calcium (20) go into the $4s$ sublevel. In scandium, the next electron added after the $4s$ sublevel is filled goes into the $3d$ block. The $3d$ block continues to fill until it is complete with 10 electrons at zinc (30). Once the $3d$ block is complete, the next six electrons, gallium to krypton, go into the $4p$ block.

Element	Atomic Number	Electron Configuration	Abbreviated Electron Configuration	
4s Block				
K	19	$1s^2 2s^2 2p^6 3s^2 3p^6 4s^1$	$[Ar]4s^1$	
Ca	20	$1s^2 2s^2 2p^6 3s^2 3p^6 4s^2$	$[Ar]4s^2$	
3d Block				
Sc	21	$1s^2 2s^2 2p^6 3s^2 3p^6 4s^2 3d^1$	$[Ar]4s^2 3d^1$	
Ti	22	$1s^2 2s^2 2p^6 3s^2 3p^6 4s^2 3d^2$	$[Ar]4s^2 3d^2$	
V	23	$1s^2 2s^2 2p^6 3s^2 3p^6 4s^2 3d^3$	$[Ar]4s^2 3d^3$	
Cr*	24	$1s^2 2s^2 2p^6 3s^2 3p^6 4s^1 3d^5$	$[Ar]4s^1 3d^5$	(half-filled d sublevel is stable)
Mn	25	$1s^2 2s^2 2p^6 3s^2 3p^6 4s^2 3d^5$	$[Ar]4s^2 3d^5$	
Fe	26	$1s^2 2s^2 2p^6 3s^2 3p^6 4s^2 3d^6$	$[Ar]4s^2 3d^6$	
Co	27	$1s^2 2s^2 2p^6 3s^2 3p^6 4s^2 3d^7$	$[Ar]4s^2 3d^7$	
Ni	28	$1s^2 2s^2 2p^6 3s^2 3p^6 4s^2 3d^8$	$[Ar]4s^2 3d^8$	
Cu*	29	$1s^2 2s^2 2p^6 3s^2 3p^6 4s^1 3d^{10}$	$[Ar]4s^1 3d^{10}$	(filled d sublevel is stable)
Zn	30	$1s^2 2s^2 2p^6 3s^2 3p^6 4s^2 3d^{10}$	$[Ar]4s^2 3d^{10}$	
4p Block				
Ga	31	$1s^2 2s^2 2p^6 3s^2 3p^6 4s^2 3d^{10} 4p^1$	$[Ar]4s^2 3d^{10} 4p^1$	
Ge	32	$1s^2 2s^2 2p^6 3s^2 3p^6 4s^2 3d^{10} 4p^2$	$[Ar]4s^2 3d^{10} 4p^2$	
As	33	$1s^2 2s^2 2p^6 3s^2 3p^6 4s^2 3d^{10} 4p^3$	$[Ar]4s^2 3d^{10} 4p^3$	
Se	34	$1s^2 2s^2 2p^6 3s^2 3p^6 4s^2 3d^{10} 4p^4$	$[Ar]4s^2 3d^{10} 4p^4$	
Br	35	$1s^2 2s^2 2p^6 3s^2 3p^6 4s^2 3d^{10} 4p^5$	$[Ar]4s^2 3d^{10} 4p^5$	
Kr	36	$1s^2 2s^2 2p^6 3s^2 3p^6 4s^2 3d^{10} 4p^6$	$[Ar]4s^2 3d^{10} 4p^6$	

*Exceptions to the order of filling.

SAMPLE PROBLEM 5.7 Using Sublevel Blocks to Write Electron Configurations

Selenium is used in making glass and in pigments. Use the sublevel blocks on the periodic table to write the electron configuration for selenium.

TRY IT FIRST

SOLUTION

ANALYZE THE PROBLEM	Given	Need	Connect
	selenium	electron configuration	sublevel blocks

STEP 1 Locate the element on the periodic table. Selenium is in Group 6A (16) and Period 4.

STEP 2 Write the filled sublevels in order, going across each period.

Period	Sublevel Block Filling	Sublevel Block Notation (filled)
1	$1s$ (H $\longrightarrow$ He)	$1s^2$
2	$2s$ (Li $\longrightarrow$ Be) and $2p$ (B $\longrightarrow$ Ne)	$2s^22p^6$
3	$3s$ (Na $\longrightarrow$ Mg) and $3p$ (Al $\longrightarrow$ Ar)	$3s^23p^6$
4	$4s$ (K $\longrightarrow$ Ca) and $3d$ (Sc $\longrightarrow$ Zn)	$4s^23d^{10}$

STEP 3 Complete the configuration by counting the electrons in the last occu-pied sublevel block. Because selenium is the fourth element in the $4p$ block, there are four electrons in the $4p$ sublevel.

Period	Sublevel Block Filling	Sublevel Block Notation
4	$4p$ (Ga $\longrightarrow$ Se)	$4p^4$

The electron configuration for selenium (Se) is: $1s^22s^22p^63s^23p^64s^23d^{10}4p^4$.

STUDY CHECK 5.7

Iodine is a micromineral needed for thyroid function. Use the sublevel blocks on the periodic table to write the electron configuration for iodine.

ANSWER
$1s^22s^22p^63s^23p^64s^23d^{10}4p^65s^24d^{10}5p^5$

Some dietary sources of iodine include seafood, eggs, and milk.

Some Exceptions in Sublevel Block Order

Within the filling of the $3d$ sublevel, exceptions occur for chromium and copper. In Cr and Cu, the $3d$ sublevel is close to being a half-filled or filled sublevel, which is particularly stable. Thus, the electron configuration for chromium has only one electron in the $4s$ and five electrons in the $3d$ sublevel to give the added stability of a half-filled d sublevel. This is shown in the abbreviated orbital diagram for chromium:

Abbreviated orbital diagram for chromium

A similar exception occurs when copper achieves a stable, filled $3d$ sublevel with 10 elec-trons and only one electron in the $4s$ orbital. This is shown in the abbreviated orbital dia-gram for copper:

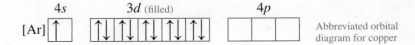

Abbreviated orbital diagram for copper

After the $4s$ and $3d$ sublevels are completed, the $4p$ sublevel fills as expected from gallium to krypton, the noble gas that completes Period 4. There are also exceptions in filling for the higher d and f electron sublevels, some caused by the added stability of half-filled shells and others where the cause is not known.

QUESTIONS AND PROBLEMS

5.5 Electron Configurations and the Periodic Table

LEARNING GOAL Write the electron configuration for an atom using the sublevel blocks on the periodic table.

5.41 Use the sublevel blocks on the periodic table to write a complete electron configuration for an atom of each of the following:
a. arsenic
b. iron
c. cobalt
d. krypton

5.42 Use the sublevel blocks on the periodic table to write a complete electron configuration for an atom of each of the following:
a. calcium
b. manganese
c. gallium
d. cadmium

5.43 Use the sublevel blocks on the periodic table to write an abbreviated electron configuration for an atom of each of the following:
a. titanium
b. bromine
c. barium
d. lead

5.44 Use the sublevel blocks on the periodic table to write an abbreviated electron configuration for an atom of each of the following:
a. vanadium
b. palladium
c. zinc
d. cesium

5.45 Use the periodic table to give the symbol of the element with each of the following electron configurations:
a. $1s^2 2s^2 2p^6 3s^2 3p^3$
b. $1s^2 2s^2 2p^6 3s^2 3p^6 4s^2 3d^7$
c. $[Ar]4s^2 3d^{10}$
d. $[Xe]6s^2 4f^{14} 5d^{10} 6p^3$

5.46 Use the periodic table to give the symbol of the element with each of the following electron configurations:
a. $1s^2 2s^2 2p^6 3s^2 3p^6 4s^2 3d^8$
b. $[Kr]5s^2 4d^{10} 5p^4$
c. $1s^2 2s^2 2p^6 3s^2 3p^6 4s^2 3d^{10} 4p^2$
d. $[Ar]4s^2 3d^{10} 4p^5$

5.47 Use the periodic table to give the symbol of the element that meets the following conditions:
a. has three electrons in the $n = 4$ energy level
b. has three $2p$ electrons
c. completes the $5p$ sublevel
d. has two electrons in the $4d$ sublevel

5.48 Use the periodic table to give the symbol of the element that meets the following conditions:
a. has five electrons in the $n = 3$ energy level
b. has one electron in the $6p$ sublevel
c. completes the $7s$ sublevel
d. has four $5p$ electrons

5.49 Use the periodic table to give the number of electrons in the indicated sublevels for the following:
a. $3d$ in zinc
b. $2p$ in sodium
c. $4p$ in arsenic
d. $5s$ in rubidium

5.50 Use the periodic table to give the number of electrons in the indicated sublevels for the following:
a. $3d$ in manganese
b. $5p$ in antimony
c. $6p$ in lead
d. $3s$ in magnesium

5.6 Trends in Periodic Properties

LEARNING GOAL Use the electron configurations of elements to explain the trends in periodic properties.

CORE CHEMISTRY SKILL
Identifying Trends in Periodic Properties

The electron configurations of atoms are an important factor in the physical and chemical properties of the elements and in the properties of the compounds that they form. In this section, we will look at the *valence electrons* in atoms, the trends in *atomic size*, *ionization energy*, and *metallic character*. Going across a period, there is a pattern of regular change in these properties from one group to the next. Known as *periodic properties*, each property increases or decreases across a period, and then the trend is repeated in each successive period. We can use the seasonal changes in temperatures as an analogy for periodic properties. In the winter, temperatures are cold and become warmer in the spring. By summer, the outdoor temperatures are hot but begin to cool in the fall. By winter, we expect cold temperatures again as the pattern of decreasing and increasing temperatures repeats for another year.

SPRING SUMMER AUTUMN WINTER

The change in temperature with the seasons is a periodic property.

Group Number and Valence Electrons

The chemical properties of representative elements are mostly due to the **valence electrons**, which are the electrons in the outermost energy level. These valence electrons occupy the *s* and *p* sublevels with the highest principal quantum number *n*. The group numbers indicate the number of valence (outer) electrons for the elements in each vertical column. For example, the elements in Group 1A (1), such as lithium, sodium, and potassium, all have one electron in an s orbital. Looking at the sublevel block, we can represent the valence electron in the alkali metals of Group 1A (1) as ns^1. All the elements in Group 2A (2), the alkaline earth metals, have two valence electrons, ns^2. The halogens in Group 7A (17) have seven valence electrons, ns^2np^5.

We can see the repetition of the outermost s and p electrons for the representative elements for Periods 1 to 4 in TABLE 5.3. Helium is included in Group 8A (18) because it is a noble gas, but it has only two electrons in its complete energy level.

TABLE 5.3 Valence Electron Configuration for Representative Elements in Periods 1 to 4

1A (1)	2A (2)	3A (13)	4A (14)	5A (15)	6A (16)	7A (17)	8A (18)
1 H $1s^1$							2 He $1s^2$
3 Li $2s^1$	4 Be $2s^2$	5 B $2s^22p^1$	6 C $2s^22p^2$	7 N $2s^22p^3$	8 O $2s^22p^4$	9 F $2s^22p^5$	10 Ne $2s^22p^6$
11 Na $3s^1$	12 Mg $3s^2$	13 Al $3s^23p^1$	14 Si $3s^23p^2$	15 P $3s^23p^3$	16 S $3s^23p^4$	17 Cl $3s^23p^5$	18 Ar $3s^23p^6$
19 K $4s^1$	20 Ca $4s^2$	31 Ga $4s^24p^1$	32 Ge $4s^24p^2$	33 As $4s^24p^3$	34 Se $4s^24p^4$	35 Br $4s^24p^5$	36 Kr $4s^24p^6$

SAMPLE PROBLEM 5.8 Using Group Numbers

Using the periodic table, write the group number, the period, the number of valence electrons, and the valence electron configuration for each of the following:

a. calcium   **b.** iodine   **c.** lead

> **TRY IT FIRST**

SOLUTION

The valence electrons are the outermost s and p electrons. Although there may be electrons in the d or f sublevels, they are not valence electrons.

a. Calcium is in Group 2A (2), Period 4, has two valence electrons, and has a valence electron configuration of $4s^2$.

b. Iodine is in Group 7A (17), Period 5, has seven valence electrons, and has a valence electron configuration of $5s^25p^5$.

c. Lead is in Group 4A (14), Period 6, has four valence electrons, and has a valence electron configuration of $6s^26p^2$.

STUDY CHECK 5.8

What are the group numbers, the periods, the number of valence electrons, and the valence electron configurations for sulfur and strontium?

ANSWER

Sulfur is in Group 6A (16), Period 3, has six valence electrons, and a $3s^23p^4$ valence electron configuration. Strontium is in Group 2A (2), Period 5, has two valence electrons, and a $5s^2$ valence electron configuration.

CORE CHEMISTRY SKILL
Drawing Lewis Symbols

Lewis Symbols

A **Lewis symbol** is a convenient way to represent the valence electrons, which are shown as dots placed on the sides, top, or bottom of the symbol for the element. One to four valence electrons are arranged as single dots. When there are five to eight electrons, one or more electrons are paired. Any of the following would be an acceptable Lewis symbol for magnesium, which has two valence electrons:

Lewis Symbols for Magnesium

Mg· Mg ·Mg ·Mg· Mg· ·Mg

Lewis symbols for selected elements are given in TABLE 5.4.

TABLE 5.4 Lewis Symbols for Selected Elements in Periods 1 to 4

				Group Number				
	IA (1)	2A (2)	3A (13)	4A (14)	5A (15)	6A (16)	7A (17)	8A (18)
Number of Valence Electrons	1	2	3	4	5	6	7	8*
				Number of Valence Electrons Increases ⟶				
Lewis Symbol	H·							He:
	Li·	Be·	·B·	·C·	·N·	·O·	·F:	:Ne:
	Na·	Mg·	·Al·	·Si·	·P·	·S:	·Cl:	:Ar:
	K·	Ca·	·Ga·	·Ge·	·As·	·Se:	·Br:	:Kr:

*Helium (He) is stable with two valence electrons.

SAMPLE PROBLEM 5.9 **Drawing Lewis Symbols**

Draw the Lewis symbol for each of the following:

a. bromine **b.** aluminum

TRY IT FIRST

SOLUTION

a. The Lewis symbol for bromine, which is in Group 7A (17), has seven valence electrons. Thus, three pairs of dots and one single dot are drawn on the sides of the Br symbol.

·Br:

b. The Lewis symbol for aluminum, which is in Group 3A (13), has three valence electrons drawn as single dots on the sides of the Al symbol.

·Al·

STUDY CHECK 5.9

Draw the Lewis symbol for phosphorus, a macromineral needed for bones and teeth.

ANSWER

·P·

Atomic Size

The **atomic size** of an atom is determined by the distance of the valence electrons from the nucleus. For each group of representative elements, the atomic size *increases* going from the top to the bottom because the outermost electrons in each energy level are farther from the nucleus. For example, in Group 1A (1), Li has a valence electron in energy level 2; Na has a valence electron in energy level 3; and K has a valence electron in energy level 4. This means that a K atom is larger than a Na atom and a Na atom is larger than a Li atom (see **FIGURE 5.11**).

ENGAGE

Why is a phosphorus atom larger than a nitrogen atom but smaller than a silicon atom?

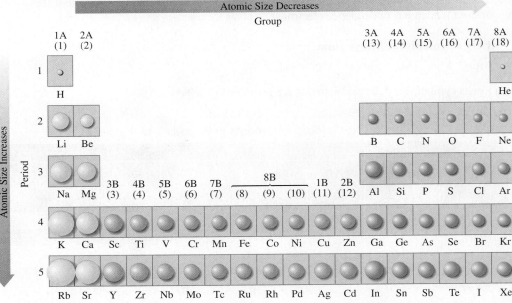

FIGURE 5.11 ▶ The atomic size increases going down a group but decreases going from left to right across a period.

🎯 Why does the atomic size increase going down a group?

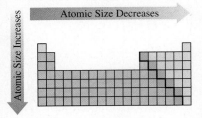

Atomic size increases going down a group and decreases going from left to right across a period.

The atomic size of representative elements is affected by the attractive forces of the protons in the nucleus on the electrons in the outermost level. For the elements going across a period, the increase in the number of protons in the nucleus increases the positive charge of the nucleus. As a result, the electrons are pulled closer to the nucleus, which means that the atomic size of representative elements decreases going from left to right across a period.

The size of atoms of transition elements within the same period changes only slightly because electrons are filling *d* orbitals rather than the outermost energy level. Because the increase in nuclear charge is canceled by an increase in *d* electrons, the attraction of the valence electrons by the nucleus remains about the same. Because there is little change in the nuclear attraction for the valence electrons, the atomic size remains relatively constant for the transition elements.

SAMPLE PROBLEM 5.10 Sizes of Atoms

Identify the smaller atom in each of the following pairs:

a. N or F   **b.** K or Kr   **c.** Ca and Sr

▶ **TRY IT FIRST**

SOLUTION

a. The F atom has a greater positive charge on the nucleus, which pulls electrons closer, and makes the F atom smaller than the N atom. Atomic size decreases going from left to right across a period.

b. The Kr atom has a greater positive charge on the nucleus, which pulls electrons closer, and makes the Kr atom smaller than the K atom. Atomic size decreases going from left to right across a period.

c. The outer electrons in the Ca atom are closer to the nucleus than in the Sr atom, which makes the Ca atom smaller than the Sr atom. Atomic size increases going down a group.

STUDY CHECK 5.10

Which atom has the largest atomic size, P, As, or Se?

ANSWER

As

Ionization Energy

In an atom, negatively charged electrons are attracted to the positive charge of the protons in the nucleus. Therefore, energy is required to remove an electron from an atom. The **ionization energy** is the energy needed to remove one electron from an atom in the gaseous (*g*) state. When an electron is removed from a neutral atom, a cation with a 1+ charge is formed.

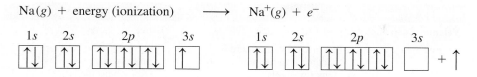

The attraction of a nucleus for the outermost electrons decreases as those electrons are farther from the nucleus. Thus the ionization energy decreases going down a group (see **FIGURE 5.12**). However, going across a period from left to right, the positive charge of the nucleus increases because there is an increase in the number of protons. Thus the ionization energy increases going from left to right across the periodic table.

In summary, the ionization energy is low for the metals and high for the nonmetals. The high ionization energies of the noble gases indicate that their electron configurations are especially stable.

SAMPLE PROBLEM 5.11 Ionization Energy

Indicate the element in each set that has the higher ionization energy and explain your choice.

a. K or Na **b.** Mg or Cl **c.** F, N, or C

> TRY IT FIRST

SOLUTION

a. Na. In Na, an electron is removed from a sublevel closer to the nucleus, which requires a higher ionization energy for Na compared with K.
b. Cl. The increased nuclear charge of Cl increases the attraction for the valence electrons, which requires a higher ionization energy for Cl compared to Mg.
c. F. The increased nuclear charge of F increases the attraction for the valence electrons, which requires a higher ionization energy for F compared to C or N.

STUDY CHECK 5.11

Arrange Sr, I, and Sn in order of increasing ionization energy.

ANSWER

Ionization energy increases going from left to right across a period: Sr, Sn, I.

Metallic Character

An element that has **metallic character** is an element that loses valence electrons easily. Metallic character is more prevalent in the elements on the left side of the periodic table (metals) and decreases going from left to right across a period. The elements on the right side of the periodic table (nonmetals) do not easily lose electrons, which means they are less metallic. Most of the metalloids between the metals and nonmetals tend to lose electrons, but not as easily as the metals. Thus, in Period 3, sodium, which loses electrons most easily, would be the most metallic. Going across from left to right in Period 3, metallic character decreases to argon, which has the least metallic character.

For elements in the same group of representative elements, metallic character increases going from top to bottom. Atoms at the bottom of any group have more electron levels, which makes it easier to lose electrons. Thus, the elements at the bottom of a group

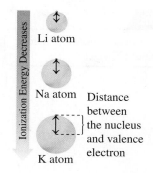

FIGURE 5.12 ▶ As the distance from the nucleus to the valence electron in Li, Na, and K atoms increases in Group 1A (1), the ionization energy decreases and less energy is required to remove the valence electron.

Q Why would Cs have a lower ionization energy than K?

> ENGAGE

When a magnesium atom ionizes to form a magnesium ion Mg^{2+}, which electrons are lost?

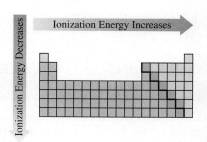

Ionization energy decreases going down a group and increases going from left to right across a period.

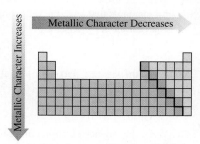

Metallic character increases going down a group and decreases going from left to right across a period.

ENGAGE

Why is magnesium more metallic than aluminum?

on the periodic table have lower ionization energy and are more metallic compared to the elements at the top.

A summary of the trends in periodic properties we have discussed is given in **TABLE 5.5**.

TABLE 5.5 Summary of Trends in Periodic Properties of Representative Elements

Periodic Property	Top to Bottom within a Group	Left to Right across a Period
Valence Electrons	Remains the same	Increases
Atomic Size	Increases because there is an increase in the number of energy levels	Decreases as the number of protons increases, which strengthens the attraction of the nucleus for the valence electrons, and pulls them closer to the nucleus
Ionization Energy	Decreases because the valence electrons are easier to remove when they are farther from the nucleus	Increases as the number of protons increases, which strengthens the attraction of the nucleus for the valence electrons, and more energy is needed to remove a valence electron
Metallic Character	Increases because the valence electrons are easier to remove when they are farther from the nucleus	Decreases as the number of protons increases, which strengthens the attraction of the nucleus for the valence electrons, and makes it more difficult to remove a valence electron

QUESTIONS AND PROBLEMS

5.6 Trends in Periodic Properties

LEARNING GOAL Use the electron configurations of elements to explain the trends in periodic properties.

5.51 What do the group numbers from 1A (1) to 8A (18) for the elements indicate about electron configurations of those elements?

5.52 What is similar and what is different about the valence electrons of the elements in a group?

5.53 Write the group number using both A/B and 1 to 18 notations for elements that have the following outer electron configuration:
a. $2s^2$
b. $3s^2 3p^3$
c. $4s^2 3d^5$
d. $5s^2 4d^{10} 5p^4$

5.54 Write the group number using both A/B and 1 to 18 notations for elements that have the following outer electron configuration:
a. $4s^2 3d^{10} 4p^5$
b. $4s^1$
c. $4s^2 3d^8$
d. $5s^2 4d^{10} 5p^2$

5.55 Write the valence electron configuration for each of the following:
a. alkali metals
b. Group 4A (14)
c. Group 7A (17)
d. Group 5A (15)

5.56 Write the valence electron configuration for each of the following:
a. halogens
b. Group 6A (16)
c. Group 13
d. alkaline earth metals

5.57 Indicate the number of valence electrons in each of the following:
a. aluminum
b. Group 5A (15)
c. barium
d. F, Cl, Br, and I

5.58 Indicate the number of valence electrons in each of the following:
a. Li, Na, K, Rb, and Cs
b. Se
c. C, Si, Ge, Sn, and Pb
d. Group 8A (18)

5.59 Write the group number and draw the Lewis symbol for each of the following elements:
a. sulfur
b. nitrogen
c. calcium
d. sodium
e. gallium

5.60 Write the group number and draw the Lewis symbol for each of the following elements:
a. carbon
b. oxygen
c. argon
d. lithium
e. chlorine

5.61 Select the larger atom in each pair.
a. Na or Cl
b. Na or Rb
c. Na or Mg
d. Rb or I

5.62 Select the larger atom in each pair.
a. S or Ar
b. S or O
c. S or K
d. S or Mg

5.63 Place the elements in each set in order of decreasing atomic size.
a. Al, Si, Mg
b. Cl, I, Br
c. Sb, Sr, I
d. P, Si, Na

5.64 Place the elements in each set in order of decreasing atomic size.
  a. Cl, S, P
  b. Ge, Si, C
  c. Ba, Ca, Sr
  d. S, O, Se

5.65 Select the element in each pair with the higher ionization energy.
  a. Br or I
  b. Mg or Sr
  c. Si or P
  d. I or Xe

5.66 Select the element in each pair with the higher ionization energy.
  a. O or Ne
  b. K or Br
  c. Ca or Ba
  d. N or Ne

5.67 Arrange each set of elements in order of increasing ionization energy.
  a. F, Cl, Br
  b. Na, Cl, Al
  c. Na, K, Cs
  d. As, Ca, Br

5.68 Arrange each set of elements in order of increasing ionization energy.
  a. O, N, C
  b. S, P, Cl
  c. P, As, N
  d. Al, Si, P

5.69 Place the following in order of decreasing metallic character:
  Br, Ge, Ca, Ga

5.70 Place the following in order of increasing metallic character:
  Na, P, Al, Ar

5.71 Fill in each of the following blanks using *higher* or *lower*, *more* or *less*: Sr has a _____ ionization energy and is _____ metallic than Sb.

5.72 Fill in each of the following blanks using *higher* or *lower*, *more* or *less*: N has a _____ ionization energy and is _____ metallic than As.

5.73 Complete each of the following statements **a** to **d** using **1**, **2**, or **3**:
  1. decreases
  2. increases
  3. remains the same

Going down Group 6A (16),
  a. the ionization energy _____
  b. the atomic size _____
  c. the metallic character _____
  d. the number of valence electrons _____

5.74 Complete each of the following statements **a** to **d** using **1**, **2**, or **3**:
  1. decreases
  2. increases
  3. remains the same

Going from left to right across Period 3,
  a. the ionization energy _____
  b. the atomic size _____
  c. the metallic character _____
  d. the number of valence electrons _____

5.75 Which statements completed with **a** to **e** will be *true* and which will be *false*?

In Period 2, an atom of N compared to an atom of Li has a larger (greater)
  a. atomic size
  b. ionization energy
  c. number of protons
  d. metallic character
  e. number of valence electrons

5.76 Which statements completed with **a** to **e** will be *true* and which will be *false*?

In Group 4A (14), an atom of C compared to an atom of Sn has a larger (greater)
  a. atomic size
  b. ionization energy
  c. number of protons
  d. metallic character
  e. number of valence electrons

Follow Up

DEVELOPING NEW MATERIALS FOR COMPUTER CHIPS

As part of a new research project, Jennifer and Robert were assigned to investigate the properties of indium (In) and tellurium (Te). Both these elements are used to make computer chips. As a result of their research, Jennifer and Robert hope to develop new computer chips that will be faster and less expensive to manufacture.

5.77 **a.** What is the atomic number of In?
  b. How many electrons are in an atom of In?
  c. Use the sublevel blocks on the periodic table to write the electron configuration and abbreviated electron configuration for an atom of In.
  d. Write the group number and draw the Lewis symbol for In.
  e. Which is larger, an atom of indium or an atom of iodine?
  f. Which has a higher ionization energy, an atom of indium or an atom of iodine?

5.78 **a.** What is the atomic number of Te?
  b. How many electrons are in an atom of Te?
  c. Use the sublevel blocks on the periodic table to write the electron configuration and abbreviated electron configuration for an atom of Te.
  d. Write the group number and draw the Lewis symbol for Te.
  e. Which is smaller, an atom of selenium or an atom of tellurium?
  f. Which has a lower ionization energy, an atom of selenium or an atom of tellurium?

CONCEPT MAP

ELECTRONIC STRUCTURE OF ATOMS AND PERIODIC TRENDS

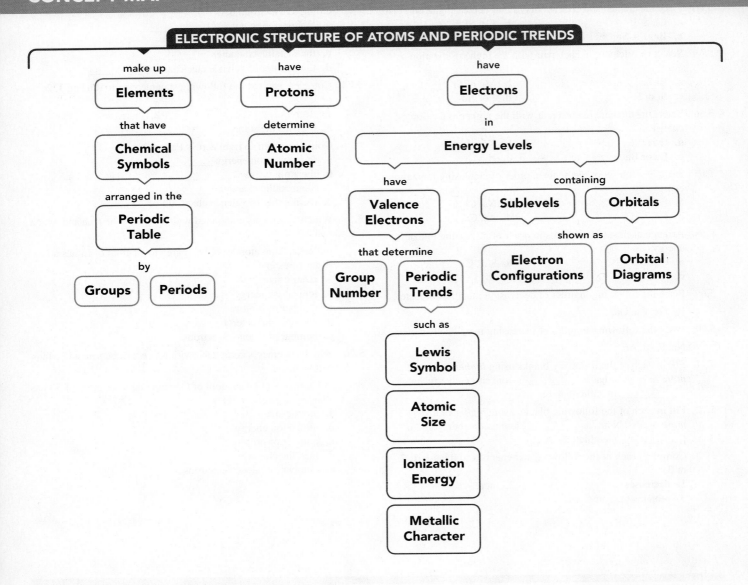

make up → **Elements** → that have → **Chemical Symbols** → arranged in the → **Periodic Table** → by → **Groups** **Periods**

have → **Protons** → determine → **Atomic Number**

have → **Electrons** → in → **Energy Levels**

Energy Levels → have → **Valence Electrons** → that determine → **Group Number** **Periodic Trends**

Periodic Trends → such as → **Lewis Symbol**, **Atomic Size**, **Ionization Energy**, **Metallic Character**

Energy Levels → containing → **Sublevels** **Orbitals** → shown as → **Electron Configurations** **Orbital Diagrams**

CHAPTER REVIEW

5.1 Electromagnetic Radiation

LEARNING GOAL Compare the wavelength, frequency, and energy of electromagnetic radiation.

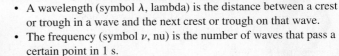

- Electromagnetic radiation such as radio waves and visible light is energy that travels at the speed of light.
- Each particular type of radiation has a specific wavelength and frequency.

- A wavelength (symbol λ, lambda) is the distance between a crest or trough in a wave and the next crest or trough on that wave.
- The frequency (symbol ν, nu) is the number of waves that pass a certain point in 1 s.
- All electromagnetic radiation travels at the speed of light (c), which is 3.00×10^8 m/s.
- Mathematically, the relationship of the speed of light, wavelength, and frequency is expressed as $c = \lambda\nu$.
- Long-wavelength radiation has low frequencies, while short-wavelength radiation has high frequencies.
- Radiation with a high frequency has high energy.

5.2 Atomic Spectra and Energy Levels

LEARNING GOAL Explain how atomic spectra correlate with the energy levels in atoms.

- The atomic spectra of elements are related to the specific energy levels occupied by electrons.
- Light consists of photons, which are particles of a specific energy.
- When an electron absorbs a photon of a particular energy, it attains a higher energy level. When an electron drops to a lower energy level, a photon of a particular energy is emitted.
- Each element has its own unique atomic spectrum.

5.3 Sublevels and Orbitals

LEARNING GOAL Describe the sublevels and orbitals for the electrons in an atom.

- An orbital is a region around the nucleus where an electron with a specific energy is most likely to be found.
- Each orbital holds a maximum of two electrons, which must have opposite spins.
- In each energy level (n), electrons occupy orbitals of identical energy within sublevels.
- An s sublevel contains one s orbital, a p sublevel contains three p orbitals, a d sublevel contains five d orbitals, and an f sublevel contains seven f orbitals.
- Each type of orbital has a unique shape.

5.4 Orbital Diagrams and Electron Configurations

LEARNING GOAL Draw the orbital diagram and write the electron configuration for an element.

- Within a sublevel, electrons enter orbitals in the same energy level one at a time until all the orbitals are half-filled.
- Additional electrons enter with opposite spins until the orbitals in that sublevel are filled with two electrons each.

$1s$
↑ $1s^1$
↑↓ $1s^2$

- The orbital diagram for an element such as silicon shows the orbitals that are occupied by paired and unpaired electrons:

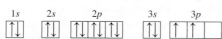

- The electron configuration for an element such as silicon shows the number of electrons in each sublevel: $1s^2 2s^2 2p^6 3s^2 3p^2$.

- An abbreviated electron configuration for an element such as silicon places the symbol of a noble gas in brackets to represent the filled sublevels: $[Ne]3s^2 3p^2$.

5.5 Electron Configurations and the Periodic Table

LEARNING GOAL Write the electron configuration for an atom using the sublevel blocks on the periodic table.

- The periodic table consists of s, p, d, and f sublevel blocks.
- An electron configuration can be written following the order of the sublevel blocks on the periodic table.
- Beginning with $1s$, an electron configuration is obtained by writing the sublevel blocks in order going across each period on the periodic table until the element is reached.

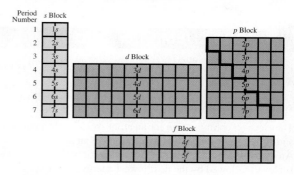

5.6 Trends in Periodic Properties

LEARNING GOAL Use the electron configurations of elements to explain the trends in periodic properties.

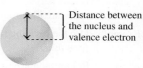

- The properties of elements are related to the valence electrons of the atoms.
- With only a few exceptions, each group of elements has the same arrangement of valence electrons differing only in the energy level.
- Valence electrons are represented as dots around the symbol of the element in the Lewis symbol.
- The size of an atom increases going down a group and decreases going from left to right across a period.
- The energy required to remove a valence electron is the ionization energy, which decreases going down a group, and increases going from left to right across a period.
- The metallic character of an element increases going down a group and decreases going from left to right across a period.

KEY TERMS

atomic size The distance between the outermost electrons and the nucleus.

atomic spectrum A series of lines specific for each element produced by photons emitted by electrons dropping to lower energy levels.

d block The 10 elements in Groups 3B (3) to 2B (12) in which electrons fill the five d orbitals.

electromagnetic radiation Forms of energy such as visible light, microwaves, radio waves, infrared, ultraviolet light, and X-rays that travel as waves at the speed of light.

electromagnetic spectrum The arrangement of types of radiation from long wavelengths to short wavelengths.

electron configuration A list of the number of electrons in each sublevel within an atom, arranged by increasing energy.

energy level A group of electrons with similar energy.

f block The 14 elements in the rows at the bottom of the periodic table in which electrons fill the seven 4*f* and 5*f* orbitals.

frequency The number of times the crests of a wave pass a point in 1 s.

ionization energy The energy needed to remove the least tightly bound electron from the outermost energy level of an atom.

Lewis symbol The representation of an atom that shows valence electrons as dots around the symbol of the element.

metallic character A measure of how easily an element loses a valence electron.

orbital The region around the nucleus of an atom where electrons of certain energy are most likely to be found: *s* orbitals are spherical; *p* orbitals have two lobes.

orbital diagram A diagram that shows the distribution of electrons in the orbitals of the energy levels.

p block The elements in Groups 3A (13) to 8A (18) in which electrons fill the *p* orbitals.

photon A packet of energy that has both particle and wave characteristics and travels at the speed of light.

principal quantum number (n) The number ($n = 1, n = 2, \ldots$) assigned to an energy level.

s block The elements in Groups 1A (1) and 2A (2) in which electrons fill the *s* orbitals.

sublevel A group of orbitals of equal energy within energy levels. The number of sublevels in each energy level is the same as the principal quantum number (n).

valence electrons The electrons in the highest energy level of an atom.

wavelength The distance between adjacent crests or troughs in a wave.

CORE CHEMISTRY SKILLS

The chapter section containing each Core Chemistry Skill is shown in parentheses at the end of each heading.

Writing Electron Configurations (5.4)

- The electron configuration for an atom specifies the energy levels and sublevels occupied by the electrons of an atom.
- An electron configuration is written starting with the lowest energy sublevel, followed by the next lowest energy sublevel.
- The number of electrons in each sublevel is shown as a superscript.

Example: Write the electron configuration for palladium.

Answer: Palladium has atomic number 46, which means it has 46 protons and 46 electrons.

$$1s^2 2s^2 2p^6 3s^2 3p^6 4s^2 3d^{10} 4p^6 5s^2 4d^8$$

Using the Periodic Table to Write Electron Configurations (5.5)

- An electron configuration corresponds to the location of an element on the periodic table, where different blocks within the periodic table are identified as the *s*, *p*, *d*, and *f* sublevels.

Example: Use the periodic table to write the electron configuration for sulfur.

Answer: Sulfur (atomic number 16) is in Group 6A (16) and Period 3.

Period	Sublevel Block Filling	Sublevel Block Notation
1	1*s* (H $\longrightarrow$ He)	$1s^2$
2	2*s* (Li $\longrightarrow$ Be) and 2*p* (B $\longrightarrow$ Ne)	$2s^2 2p^6$
3	3*s* (Na $\longrightarrow$ Mg)	$3s^2$
	3*p* (Al $\longrightarrow$ S)	$3p^4$

The electron configuration for sulfur (S) is: $1s^2 2s^2 2p^6 3s^2 3p^4$.

Identifying Trends in Periodic Properties (5.6)

- The size of an atom increases going down a group and decreases going from left to right across a period.
- The ionization energy decreases going down a group and increases going from left to right across a period.
- The metallic character of an element increases going down a group and decreases going from left to right across a period.

Example: For Mg, P, and Cl, identify which has the
  a. largest atomic size
  b. highest ionization energy
  c. greatest metallic character

Answer: **a.** Mg   **b.** Cl   **c.** Mg

Drawing Lewis Symbols (5.6)

- The valence electrons are the electrons in the *s* and *p* sublevels with the highest principal quantum number *n*.
- The number of valence electrons is the same as the group number for the representative elements.
- A Lewis symbol represents the number of valence electrons as dots placed around the symbol for the element.

Example: Give the group number and number of valence electrons, and draw the Lewis symbol for each of the following:
  a. Rb   **b.** Se   **c.** Xe

Answer: **a.** Group 1A (1), one valence electron, Rb·
  b. Group 6A (16), six valence electrons, ·S̈e:
  c. Group 8A (18), eight valence electrons, :Ẍe:

UNDERSTANDING THE CONCEPTS

The chapter sections to review are shown in parentheses at the end of each question.

Use the following diagram for problems 5.79 and 5.80:

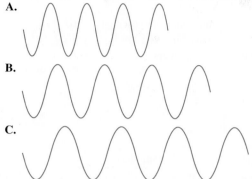

A.

B.

C.

5.79 Select diagram **A**, **B**, or **C** that (5.1)
  a. has the longest wavelength
  b. has the shortest wavelength
  c. has the highest frequency
  d. has the lowest frequency

5.80 Select diagram **A**, **B**, or **C** that (5.1)
  a. has the highest energy
  b. has the lowest energy
  c. would represent blue light
  d. would represent red light

5.81 Match the following with an *s* or *p* orbital: (5.3)
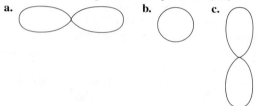
  a.   **b.** **c.**

5.82 Match the following with *s* or *p* orbitals: (5.3)
  a. two lobes
  b. spherical shape
  c. found in $n = 1$
  d. found in $n = 3$

5.83 Indicate whether or not the following orbital diagrams are possible and explain. When possible, indicate the element it represents. (5.4)

a.
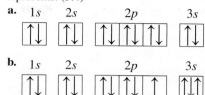

b.
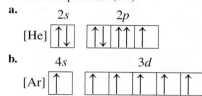

5.84 Indicate whether or not the following abbreviated orbital diagrams are possible and explain. When possible, indicate the element it represents. (5.4)

a.
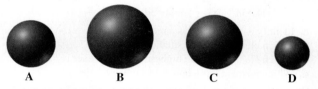

b.
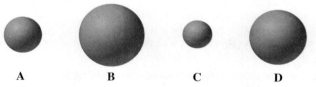

5.85 Match the spheres **A** through **D** with atoms of Li, Na, K, and Rb. (5.6)

5.86 Match the spheres **A** through **D** with atoms of K, Ge, Ca, and Kr. (5.6)

ADDITIONAL QUESTIONS AND PROBLEMS

5.87 What is the difference between a continuous spectrum and an atomic spectrum? (5.1)

5.88 Why does a neon sign give off red light? (5.1)

5.89 What is the Pauli exclusion principle? (5.3)

5.90 Why would there be five unpaired electrons in a *d* sublevel but no paired electrons? (5.3)

5.91 Which of the following orbitals are possible in an atom: 4*p*, 2*d*, 3*f*, and 5*f*? (5.3)

5.92 Which of the following orbitals are possible in an atom: 1*p*, 4*f*, 6*s*, and 4*d*? (5.3)

5.93 **a.** What electron sublevel starts to fill after completion of the 3*s* sublevel? (5.4)
  b. What electron sublevel starts to fill after completion of the 4*p* sublevel?
  c. What electron sublevel starts to fill after completion of the 3*d* sublevel?
  d. What electron sublevel starts to fill after completion of the 3*p* sublevel?

5.94 **a.** What electron sublevel starts to fill after completion of the 5*s* sublevel? (5.4)
  b. What electron sublevel starts to fill after completion of the 4*d* sublevel?
  c. What electron sublevel starts to fill after completion of the 4*f* sublevel?
  d. What electron sublevel starts to fill after completion of the 5*p* sublevel?

5.95 a. How many 3*d* electrons are in Fe? (5.4)
  b. How many 5*p* electrons are in Ba?
  c. How many 4*d* electrons are in I?
  d. How many 7*s* electrons are in Ra?

5.96 a. How many 4*d* electrons are in Cd? (5.4)
  b. How many 4*p* electrons are in Br?
  c. How many 6*p* electrons are in Bi?
  d. How many 4*s* electrons are in Zn?

5.97 What do the elements Ca, Sr, and Ba have in common in terms of their electron configuration? Where are they located on the periodic table? (5.4, 5.5)

5.98 What do the elements O, S, and Se have in common in terms of their electron configuration? Where are they located on the periodic table? (5.4, 5.5)

5.99 Consider three elements with the following abbreviated electron configurations: (5.4, 5.5, 5.6)

  X = [Ar]4s^2 Y = [Ne]3$s^2$3p^4 Z = [Ar]4$s^2$3d^{10}4p^4

  a. Identify each element as a metal, nonmetal, or metalloid.
  b. Which element has the largest atomic size?
  c. Which element has the highest ionization energy?
  d. Which element has the smallest atomic size?

5.100 Consider three elements with the following abbreviated electron configurations: (5.4, 5.5, 5.6)

  X = [Ar]4$s^2$3d^5 Y = [Ar]4$s^2$3d^{10}4p^1 Z = [Ar]4$s^2$3d^{10}4p^6

  a. Identify each element as a metal, nonmetal, or metalloid.
  b. Which element has the smallest atomic size?
  c. Which element has the highest ionization energy?
  d. Which element has a half-filled sublevel?

5.101 Name the element that corresponds to each of the following: (5.4, 5.5, 5.6)
  a. 1$s^2$2$s^2$2$p^6$3$s^2$3p^3
  b. alkali metal with the smallest atomic size
  c. [Kr]5$s^2$4d^{10}
  d. Group 5A (15) element with the highest ionization energy
  e. Period 3 element with the largest atomic size

5.102 Name the element that corresponds to each of the following: (5.4, 5.5, 5.6)
  a. 1$s^2$2$s^2$2$p^6$3$s^2$3$p^6$4$s^1$3d^5
  b. [Xe]6$s^2$4f^{14}5d^{10}6p^5
  c. halogen with the highest ionization energy

  d. Group 2A (2) element with the lowest ionization energy
  e. Period 4 element with the smallest atomic size

5.103 Why is the ionization energy of Ca higher than that of K but lower than that of Mg? (5.6)

5.104 Why is the ionization energy of Br lower than that of Cl but higher than that of Se? (5.6)

5.105 Select the more metallic element in each pair. (5.6)
  a. As or Sb   **b.** Sn or Sb   **c.** Cl or P   **d.** O or P

5.106 Select the more metallic element in each pair. (5.6)
  a. Sn or As   **b.** Cl or I   **c.** Ca or Ba   **d.** Ba or Hg

5.107 Of the elements Na, P, Cl, and F, which (5.6)
  a. is a metal?
  b. is in Group 5A (15)?
  c. has the highest ionization energy?
  d. loses an electron most easily?
  e. is found in Group 7A (17), Period 3?

5.108 Of the elements K, Ca, Br, Kr, which (5.6)
  a. is a halogen?
  b. has the smallest atomic size?
  c. has the lowest ionization energy?
  d. requires the most energy to remove an electron?
  e. is found in Group 2A (2), Period 4?

5.109 Write the abbreviated electron configuration and group number for each of the following elements: (5.4)
  a. Si   **b.** Se   **c.** Mn   **d.** Sb

5.110 Write the abbreviated electron configuration and group number for each of the following elements: (5.4)
  a. Br   **b.** Rh   **c.** Tc   **d.** Ra

5.111 Write the group number and draw the Lewis symbol for each of the following elements: (5.6)
  a. barium   **b.** fluorine   **c.** krypton   **d.** arsenic

5.112 Write the group number and draw the Lewis symbol for each of the following elements: (5.6)
  a. neon   **b.** iodine   **c.** bismuth   **d.** tin

CHALLENGE QUESTIONS

The following groups of questions are related to the topics in this chapter. However, they do not all follow the chapter order, and they require you to combine concepts and skills from several sections. These questions will help you increase your critical thinking skills and prepare for your next exam.

5.113 Give the symbol of the element that has the (5.6)
  a. smallest atomic size in Group 6A (16)
  b. smallest atomic size in Period 3
  c. highest ionization energy in Group 3A (13)
  d. lowest ionization energy in Period 3
  e. abbreviated electron configuration [Kr]5$s^2$4d^6

5.114 Give the symbol of the element that has the (5.6)
  a. largest atomic size in Period 5
  b. largest atomic size in Group 2A (2)
  c. highest ionization energy in Group 8A (18)
  d. lowest ionization energy in Period 2
  e. abbreviated electron configuration [Kr]5$s^2$4d^{10}5p^2

5.115 How do scientists explain the colored lines observed in the spectra of heated atoms? (5.2)

5.116 Even though H has only one electron, there are many lines in the atomic spectrum of H. Explain. (5.2)

5.117 What is meant by an energy level, a sublevel, and an orbital? (5.3)

5.118 In some periodic tables, H is placed in Group 1A (1). In other periodic tables, H is also placed in Group 7A (17). Why? (5.4, 5.5)

5.119 Compare F, S, and Cl in terms of atomic size and ionization energy. (5.6)

5.120 Compare K, Cs, and Li in terms of atomic size and ionization energy. (5.6)

ANSWERS

Answers to Selected Questions and Problems

5.1 The wavelength of UV light is the distance between crests of the wave.

5.3 White light has all the colors of the spectrum, including red and blue light.

5.5 Ultraviolet radiation has a higher frequency and higher energy.

5.7 6.3×10^{-7} m; 630 nm

5.9 AM radio has a longer wavelength than cell phones or infrared.

5.11 Order of increasing wavelengths: X-rays, blue light, microwaves.

5.13 Order of increasing frequencies: TV, microwaves, X-rays

5.15 Atomic spectra consist of a series of lines separated by dark sections, indicating that the energy emitted by the elements is not continuous.

5.17 absorb

5.19 The photon with greater energy is
a. green light   **b.** blue light

5.21 a. spherical   **b.** two lobes   **c.** spherical

5.23 a. 1 and **2**   **b. 3**
c. 1 and **2**   **d. 1, 2,** and **3**

5.25 a. There are five orbitals in the $3d$ sublevel.
b. There is one sublevel in the $n = 1$ energy level.
c. There is one orbital in the $6s$ sublevel.
d. There are nine orbitals in the $n = 3$ energy level.

5.27 a. There is a maximum of two electrons in a $2p$ orbital.
b. There is a maximum of six electrons in the $3p$ sublevel.
c. There is a maximum of 32 electrons in the $n = 4$ energy level.
d. There is a maximum of 10 electrons in the $5d$ sublevel.

5.29 The electron configuration shows the number of electrons in each sublevel of an atom. The abbreviated electron configuration uses the symbol of the preceding noble gas to show completed sublevels.

5.31 a.

b.

c.

d.

5.33 a. $1s^22s^22p^63p^64s^23d^8$
b. $1s^22s^22p^63s^1$
c. $1s^22s^1$
d. $1s^22s^22p^63s^23p^64s^23d^2$

5.35 a. $[Kr]5s^24d^{10}5p^2$ **b.** $[Kr]5s^24d^{10}$
c. $[Ar]4s^23d^{10}4p^4$ **d.** $[He]2s^22p^5$

5.37 a. Li   **b.** Ti   **c.** Ge   **d.** F

5.39 a. Al   **b.** C   **c.** Ar   **d.** Be

5.41 a. $1s^22s^22p^63s^23p^64s^23d^{10}4p^3$
b. $1s^22s^22p^63s^23p^64s^23d^6$
c. $1s^22s^22p^63s^23p^64s^23d^7$
d. $1s^22s^22p^63s^23p^64s^23d^{10}4p^6$

5.43 a. $[Ar]4s^23d^2$ **b.** $[Ar]4s^23d^{10}4p^5$
c. $[Xe]6s^2$ **d.** $[Xe]6s^24f^{14}5d^{10}6p^2$

5.45 a. P   **b.** Co   **c.** Zn   **d.** Bi

5.47 a. Ga   **b.** N   **c.** Xe   **d.** Zr

5.49 a. 10   **b.** 6   **c.** 3   **d.** 1

5.51 The group numbers 1A to 8A indicate the number of valence electrons from 1 to 8.

5.53 a. 2A (2)   **b.** 5A (15)   **c.** 7B (7)   **d.** 6A (16)

5.55 a. ns^1 **b.** ns^2np^2 **c.** ns^2np^5 **d.** ns^2np^3

5.57 a. 3   **b.** 5   **c.** 2   **d.** 7

5.59 a. Sulfur is in Group 6A (16); $\cdot\overset{\displaystyle\cdot\cdot}{\underset{\displaystyle\cdot\cdot}{S}}\cdot$
b. Nitrogen is in Group 5A (15); $\cdot\overset{\displaystyle\cdot}{N}\cdot$
c. Calcium is in Group 2A (2); $\overset{\displaystyle\cdot}{Ca}\cdot$
d. Sodium is in Group 1A (1); Na$\cdot$
e. Gallium is in Group 3A (13); $\cdot\overset{\displaystyle\cdot}{Ga}\cdot$

5.61 a. Na   **b.** Rb   **c.** Na   **d.** Rb

5.63 a. Mg, Al, Si   **b.** I, Br, Cl
c. Sr, Sb, I   **d.** Na, Si, P

5.65 a. Br   **b.** Mg   **c.** P   **d.** Xe

5.67 a. Br, Cl, F   **b.** Na, Al, Cl
c. Cs, K, Na   **d.** Ca, As, Br

5.69 Ca, Ga, Ge, Br

5.71 lower, more

5.73 a. 1. decreases
b. 2. increases
c. 2. increases
d. 3. remains the same

5.75 a. false   **b.** true   **c.** true   **d.** false   **e.** true

5.77 a. 49   **b.** 49
c. $1s^22s^22p^63s^23p^64s^23d^{10}4p^65s^24d^{10}5p^1$; $[Kr]5s^24d^{10}5p^1$
d. 3A (13), $\cdot\overset{\displaystyle\cdot}{In}\cdot$ **e.** indium   **f.** iodine

5.79 a. C has the longest wavelength.
b. A has the shortest wavelength.
c. A has the highest frequency.
d. C has the lowest frequency.

5.81 a. p **b.** s **c.** p

5.83 a. This is possible. This element is magnesium.
b. Not possible. The $2p$ sublevel would fill before the $3s$, and only two electrons are allowed in an s orbital.

5.85 Li is **D**, Na is **A**, K is **C**, and Rb is **B**.

5.87 A continuous spectrum from white light contains wavelengths of all energies. Atomic spectra are line spectra in which a series of lines corresponds to energy emitted when electrons drop from a higher energy level to a lower level.

5.89 The Pauli exclusion principle states that two electrons in the same orbital must have opposite spins.

5.91 A 4p orbital is possible because the $n = 4$ energy level has four sublevels, including a p sublevel. A 2d orbital is not possible because the $n = 2$ energy level has only s and p sublevels. There are no 3f orbitals because only s, p, and d sublevels are allowed for the $n = 3$ energy level. A 5f sublevel is possible in the $n = 5$ energy level because five sublevels are allowed, including an f sublevel.

5.93 a. 3p   **b.** 5s   **c.** 4p   **d.** 4s

5.95 a. 6   **b.** 6   **c.** 10   **d.** 2

5.97 Ca, Sr, and Ba all have two valence electrons, ns^2, which place them in Group 2A (2).

5.99 a. X is a metal; Y and Z are nonmetals.
b. X has the largest atomic size.
c. Y has the highest ionization energy.
d. Y has the smallest atomic size.

5.101 a. phosphorus   **b.** lithium (H is a nonmetal)
c. cadmium   **d.** nitrogen
e. sodium

5.103 Calcium has a greater number of protons than K. The least tightly bound electron in Ca is farther from the nucleus than in Mg and needs less energy to remove.

5.105 a. Sb   **b.** Sn   **c.** P   **d.** P

5.107 a. Na   **b.** P   **c.** F
d. Na   **e.** Cl

5.109 a. [Ne]$3s^2 3p^2$; Group 4A (14)
b. [Ar]$4s^2 3d^{10} 4p^4$; Group 6A (16)
c. [Ar]$4s^2 3d^5$; Group 7B (7)
d. [Kr]$5s^2 4d^{10} 5p^3$; Group 5A (15)

5.111 a. Barium in Group 2A (2); Ba·

b. Fluorine is in Group 7A (17); ·F̈:

c. Krypton is in Group 8A (18); :K̈r:

d. Arsenic is in Group 5A (15); ·Äs·

5.113 a. O   **b.** Ar   **c.** B
d. Na   **e.** Ru

5.115 The series of lines separated by dark sections in atomic spectra indicate that the energy emitted by the elements is not continuous and that electrons are moving between discrete energy levels.

5.117 The energy level contains all the electrons with similar energy. A sublevel contains electrons with the same energy, while an orbital is the region around the nucleus where electrons of a certain energy are most likely to be found.

5.119 S has a larger atomic size than Cl; Cl is larger than F: S > Cl > F. F has a higher ionization energy than Cl; Cl has a higher ionization energy than S: F > Cl > S.

Ionic and Molecular Compounds

RICHARD'S DOCTOR HAS recommended that he take a low-dose aspirin (81 mg) every day to prevent a heart attack or stroke. Richard is concerned about taking aspirin and asks Sarah, a pharmacist working at a local pharmacy, about the effects of aspirin. Sarah explains to Richard that aspirin is acetylsalicylic acid and has the chemical formula $C_9H_8O_4$. Aspirin is a molecular compound, often referred to as an organic molecule because it contains the nonmetals carbon (C), hydrogen (H), and oxygen (O). Sarah explains to Richard that aspirin is used to relieve minor pains, to reduce inflammation and fever, and to slow blood clotting. Aspirin is one of several nonsteroidal anti-inflammatory drugs (NSAIDs) that reduce pain and fever by blocking the formation of prostaglandins, which are chemical messengers that transmit pain signals to the brain and cause fever. Some potential side effects of aspirin may include heartburn, upset stomach, nausea, and an increased risk of a stomach ulcer.

CAREER
Pharmacist

Pharmacists work in hospitals, pharmacies, clinics, and long-term care facilities where they are responsible for the preparation and distribution of pharmaceutical medications based on a doctor's orders. They obtain the proper medication, and also calculate, measure, and label the patient's medication. Pharmacists advise clients and health care practitioners on the selection of both prescription and over-the-counter drugs, proper dosages, and geriatric considerations, as well as possible side effects and interactions. They may also administer vaccinations; prepare sterile intravenous solutions; and advise clients about health, diet, and home medical equipment. Pharmacists also prepare insurance claims, and create and maintain patient profiles.

KEY MATH SKILLS

- Using Positive and Negative Numbers in Calculations (1.4)
- Solving Equations (1.4)

CORE CHEMISTRY SKILLS

- Writing Electron Configurations (5.4)
- Drawing Lewis Symbols (5.6)

*These Key Math Skills and Core Chemistry Skills from previous chapters are listed here for your review as you proceed to the new material in this chapter.

6.1 Ions: Transfer of Electrons

LEARNING GOAL Write the symbols for the simple ions of the representative elements.

Most of the elements, except the noble gases, are found in nature combined as compounds. The noble gases are so stable that they form compounds only under extreme conditions. One explanation for the stability of noble gases is that they have a filled valence electron energy level.

Compounds form when electrons are transferred or shared to give stable electron configuration to the atoms. In the formation of either an *ionic bond* or a *covalent bond*, atoms lose, gain, or share valence electrons to acquire an **octet** of eight valence electrons. This tendency of atoms to attain a stable electron configuration is known as the **octet rule** and provides a key to our understanding of the ways in which atoms bond and form compounds. A few elements achieve the stability of helium with two valence electrons. However, we do not use the octet rule with transition elements.

Ionic bonds occur when the valence electrons of atoms of a metal are transferred to atoms of nonmetals. For example, sodium atoms lose electrons and chlorine atoms gain electrons to form the ionic compound NaCl. *Covalent bonds* form when atoms of nonmetals share valence electrons. In the molecular compounds H_2O and C_3H_8, atoms share electrons.

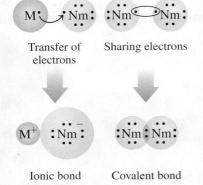

Transfer of electrons

Sharing electrons

Ionic bond

Covalent bond

M is a metal
Nm is a nonmetal

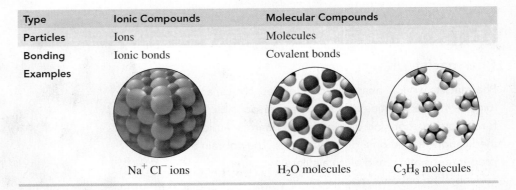

Type	Ionic Compounds	Molecular Compounds
Particles	Ions	Molecules
Bonding	Ionic bonds	Covalent bonds
Examples		

Na^+ Cl^- ions

H_2O molecules

C_3H_8 molecules

Positive Ions: Loss of Electrons

In ionic bonding, **ions**, which have electrical charges, form when atoms lose or gain electrons to form a stable electron configuration. Because the ionization energies of metals of Groups 1A (1), 2A (2), and 3A (13) are low, metal atoms readily lose their valence electrons. In doing so, they form ions with positive charges. A metal atom obtains the same electron configuration as its nearest noble gas (usually eight valence electrons). For example, when a sodium atom loses its single valence electron, the remaining electrons have a stable electron configuration. By losing an electron, sodium has 10 negatively charged electrons instead of 11. Because there are still 11 positively charged protons in its nucleus, the atom is no longer neutral. It is now a sodium ion with a positive electrical charge, called an **ionic charge**, of 1+. In the Lewis symbol for the sodium ion, the ionic charge of

1+ is written in the upper right-hand corner, Na⁺, where the 1 is understood. The sodium ion is smaller than the sodium atom because the ion has lost its outermost electron from the third energy level. A positively charged ion of a metal is called a **cation** (pronounced *cat-eye-un*) and uses the name of the element.

Ionic charge = Charge of protons + Charge of electrons

1+ = (11+)   + (10−)

ENGAGE

A cesium atom has 55 protons and 55 electrons. A cesium ion has 55 protons and 54 electrons. Why does the cesium ion have a 1+ charge?

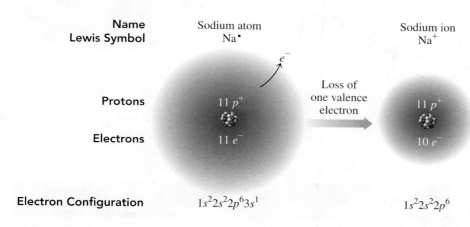

Name Lewis Symbol	Sodium atom Na•		Sodium ion Na⁺
Protons	11 p^+	Loss of one valence electron	11 p^+
Electrons	11 e^-		10 e^-
Electron Configuration	$1s^2 2s^2 2p^6 3s^1$		$1s^2 2s^2 2p^6$

Magnesium, a metal in Group 2A (2), obtains a stable electron configuration by losing two valence electrons to form a magnesium ion with a 2+ ionic charge, Mg^{2+}. The magnesium ion is smaller than the magnesium atom because the outermost electrons in the third energy level were removed. The octet in the magnesium ion is made up of electrons that fill its second energy level.

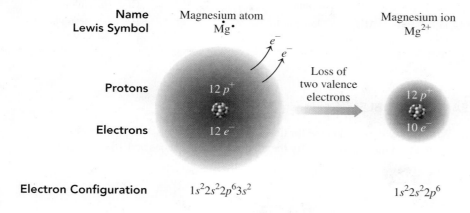

Name Lewis Symbol	Magnesium atom Mg•		Magnesium ion Mg^{2+}
Protons	12 p^+	Loss of two valence electrons	12 p^+
Electrons	12 e^-		10 e^-
Electron Configuration	$1s^2 2s^2 2p^6 3s^2$		$1s^2 2s^2 2p^6$

Negative Ions: Gain of Electrons

The ionization energy of a nonmetal atom in Groups 5A (15), 6A (16), or 7A (17) is high. In an ionic compound, a nonmetal atom gains one or more valence electrons to obtain a stable electron configuration. By gaining electrons, a nonmetal atom forms a negatively charged ion. For example, an atom of chlorine with seven valence electrons gains one electron to form an octet. Because it now has 18 electrons and 17 protons in its nucleus, the chlorine atom is no longer neutral. It is a *chloride ion* with an ionic charge of 1−, which is written as Cl⁻, with the 1 understood. A negatively charged ion, called an **anion** (pronounced *an-eye-un*), is named by using the first syllable of its element name followed by *ide*. The chloride ion is larger than the chlorine atom because the ion has an additional electron, which completes its outermost energy level.

Ionic charge = Charge of protons + Charge of electrons

1− = (17+)   + (18−)

CORE CHEMISTRY SKILL
Writing Positive and Negative Ions

ENGAGE

Why does Li form a positive ion, Li⁺, whereas Br forms a negative ion, Br⁻?

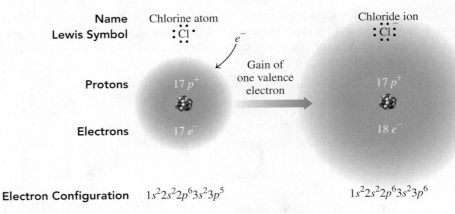

Name	Chlorine atom		Chloride ion
Lewis Symbol	$:\overset{..}{\underset{..}{Cl}}\cdot$		$:\overset{..}{\underset{..}{Cl}}:$
Protons	$17\,p^+$	Gain of one valence electron	$17\,p^+$
Electrons	$17\,e^-$		$18\,e^-$
Electron Configuration	$1s^2 2s^2 2p^6 3s^2 3p^5$		$1s^2 2s^2 2p^6 3s^2 3p^6$

TABLE 6.1 lists the names of some important metal and nonmetal ions.

TABLE 6.1 Formulas and Names of Some Common Ions

Metals			Nonmetals		
Group Number	Cation	Name of Cation	Group Number	Anion	Name of Anion
1A (1)	Li^+	Lithium	5A (15)	N^{3-}	Nitride
	Na^+	Sodium		P^{3-}	Phosphide
	K^+	Potassium	6A (16)	O^{2-}	Oxide
2A (2)	Mg^{2+}	Magnesium		S^{2-}	Sulfide
	Ca^{2+}	Calcium	7A (17)	F^-	Fluoride
	Ba^{2+}	Barium		Cl^-	Chloride
3A (13)	Al^{3+}	Aluminum		Br^-	Bromide
				I^-	Iodide

SAMPLE PROBLEM 6.1 Ions

a. Write the symbol and name for the ion that has 7 protons and 10 electrons.
b. Write the symbol and name for the ion that has 20 protons and 18 electrons.

TRY IT FIRST

SOLUTION

a. The element with 7 protons is nitrogen. In an ion of nitrogen with 10 electrons, the ionic charge would be $3-$, $[(7+) + (10-) = 3-]$. The ion, written as N^{3-}, is the *nitride* ion.
b. The element with 20 protons is calcium. In an ion of calcium with 18 electrons, the ionic charge would be $2+$, $[(20+) + (18-) = 2+]$. The ion, written as Ca^{2+}, is the *calcium* ion.

STUDY CHECK 6.1

How many protons and electrons are in each of the following ions?

a. Sr^{2+} **b.** Cl^-

ANSWER

a. 38 protons, 36 electrons   **b.** 17 protons, 18 electrons

Ionic Charges from Group Numbers

In ionic compounds, representative elements usually lose or gain electrons to give eight valence electrons like their nearest noble gas (or two for helium). We can use the group numbers in the periodic table to determine the charges for the ions of the representative elements. The elements in Group 1A (1) lose one electron to form ions with a 1+ charge. The elements in Group 2A (2) lose two electrons to form ions with a 2+ charge. The elements in Group 3A (13) lose three electrons to form ions with a 3+ charge. In this text, we do not use the group numbers of the transition elements to determine their ionic charges.

In ionic compounds, the elements in Group 7A (17) gain one electron to form ions with a 1− charge. The elements in Group 6A (16) gain two electrons to form ions with a 2− charge. The elements in Group 5A (15) gain three electrons to form ions with a 3− charge.

The nonmetals of Group 4A (14) do not typically form ions. However, the metals Sn and Pb in Group 4A (14) lose electrons to form positive ions. **TABLE 6.2** lists the ionic charges for some common monatomic ions of representative elements.

> **ENGAGE**
>
> Why do all the atoms in Group 2A (2) form ions with 2+ charges?

TABLE 6.2 Examples of Monatomic Ions and Their Nearest Noble Gases

Noble Gases		Metals Lose Valence Electrons			Nonmetals Gain Valence Electrons				Noble Gases
		1A (1)	2A (2)	3A (13)	5A (15)	6A (16)	7A (17)		
He	⇐	Li^+			N^{3-}	O^{2-}	F^-	⇒	Ne
Ne	⇐	Na^+	Mg^{2+}	Al^{3+}	P^{3-}	S^{2-}	Cl^-	⇒	Ar
Ar	⇐	K^+	Ca^{2+}			Se^{2-}	Br^-	⇒	Kr
Kr	⇐	Rb^+	Sr^{2+}				I^-	⇒	Xe
Xe	⇐	Cs^+	Ba^{2+}				At^-	⇒	Rn

SAMPLE PROBLEM 6.2 Writing Symbols for Ions

Consider the elements aluminum and oxygen.

a. Identify each as a metal or a nonmetal.
b. State the number of valence electrons for each.
c. State the number of electrons that must be lost or gained for each to achieve an octet.
d. Write the symbol, including its ionic charge, and the name for each resulting ion.

> **TRY IT FIRST**

SOLUTION

Aluminum	Oxygen
a. metal	nonmetal
b. three valence electrons	six valence electrons
c. loses 3 e^-	gains 2 e^-
d. Al^{3+}, [(13+) + (10−) = 3+], aluminum ion	O^{2-}, [(8+) + (10−) = 2−], oxide ion

CHEMISTRY LINK TO **HEALTH**
Some Important Ions in the Body

Several ions in body fluids have important physiological and metabolic functions. Some are listed in **TABLE 6.3**.

Foods such as bananas, milk, cheese, and potatoes provide the body with ions that are important in regulating body functions.

Milk, cheese, bananas, cereal, and potatoes provide ions for the body.

TABLE 6.3 Ions in the Body

Ion	Occurrence	Function	Source	Result of Too Little	Result of Too Much
Na^+	Principal cation outside the cell	Regulation and control of body fluids	Salt, cheese, pickles	Hyponatremia, anxiety, diarrhea, circulatory failure, decrease in body fluid	Hypernatremia, little urine, thirst, edema
K^+	Principal cation inside the cell	Regulation of body fluids and cellular functions	Bananas, orange juice, milk, prunes, potatoes	Hypokalemia (hypopotassemia), lethargy, muscle weakness, failure of neurological impulses	Hyperkalemia (hyperpotassemia), irritability, nausea, little urine, cardiac arrest
Ca^{2+}	Cation outside the cell; 90% of calcium in the body in bones	Major cation of bones; needed for muscle contraction	Milk, yogurt, cheese, greens, spinach	Hypocalcemia, tingling fingertips, muscle cramps, osteoporosis	Hypercalcemia, relaxed muscles, kidney stones, deep bone pain
Mg^{2+}	Cation outside the cell; 50% of magnesium in the body in bone structure	Essential for certain enzymes, muscles, nerve control	Widely distributed (part of chlorophyll of all green plants), nuts, whole grains	Disorientation, hypertension, tremors, slow pulse	Drowsiness
Cl^-	Principal anion outside the cell	Gastric juice, regulation of body fluids	Salt	Same as for Na^+	Same as for Na^+

QUESTIONS AND PROBLEMS

6.1 Ions: Transfer of Electrons

LEARNING GOAL Write the symbols for the simple ions of the representative elements.

6.1 State the number of electrons that must be lost by atoms of each of the following to achieve a stable electron configuration:
    a. Li   **b.** Ca   **c.** Ga   **d.** Cs   **e.** Ba

6.2 State the number of electrons that must be gained by atoms of each of the following to achieve a stable electron configuration:
    a. Cl   **b.** Se   **c.** N   **d.** I   **e.** S

6.3 State the number of electrons lost or gained when the following elements form ions:
    a. Sr   **b.** P   **c.** Group 7A (17)
    d. Na   **e.** Br

6.4 State the number of electrons lost or gained when the following elements form ions:
    a. O   **b.** Group 2A (2)   **c.** F
    d. K   **e.** Rb

6.5 Write the symbols for the ions with the following number of protons and electrons:
    a. 3 protons, 2 electrons   **b.** 9 protons, 10 electrons
    c. 12 protons, 10 electrons   **d.** 26 protons, 23 electrons

6.6 Write the symbols for the ions with the following number of protons and electrons:
    a. 8 protons, 10 electrons   **b.** 19 protons, 18 electrons
    c. 35 protons, 36 electrons   **d.** 50 protons, 46 electrons

6.7 State the number of protons and electrons in each of the following:
    a. Cu^{2+} **b.** Se^{2-}
    c. Br^- **d.** Fe^{3+}

6.8 State the number of protons and electrons in each of the following:
    a. P^{3-} **b.** Ni^{2+}
    c. Au^{3+} **d.** Ag^+

6.9 Write the symbol for the ion of each of the following:
    a. chlorine   **b.** cesium
    c. nitrogen   **d.** radium

6.10 Write the symbol for the ion of each of the following:
    a. fluorine   **b.** calcium
    c. sodium   **d.** iodine

6.11 Write the names for each of the following ions:
    a. Li^+ **b.** Ca^{2+}
    c. Ga^{3+} **d.** P^{3-}

6.12 Write the names for each of the following ions:
    a. Rb^+ **b.** Sr^{2+}
    c. S^{2-} **d.** F^-

Applications

6.13 State the number of protons and electrons in each of the following ions:
    a. O^{2-}, used to build biomolecules and water
    b. K^+, most prevalent positive ion in cells, needed for muscle contraction, nerve impulses
    c. I^-, needed for thyroid function
    d. Na^+, most prevalent positive ion in extracellular fluid

6.14 State the number of protons and electrons in each of the following ions:
    a. P^{3-}, needed for bones and teeth
    b. F^-, used to strengthen tooth enamel
    c. Mg^{2+}, needed for bones and teeth
    d. Ca^{2+}, needed for bones and teeth

6.2 Ionic Compounds

LEARNING GOAL Using charge balance, write the correct formula for an ionic compound.

We utilize ionic compounds such as salt, NaCl, and baking soda, $NaHCO_3$, every day. Milk of magnesia, $Mg(OH)_2$, or calcium carbonate, $CaCO_3$, may be taken to settle an upset stomach. In a mineral supplement, iron may be present as iron(II) sulfate, $FeSO_4$, iodine as potassium iodide, KI, and manganese as manganese(II) sulfate, $MnSO_4$. Some sunscreens contain zinc oxide, ZnO, while tin(II) fluoride, SnF_2, in toothpaste provides fluoride to help prevent tooth decay. Gemstones are ionic compounds that are cut and polished to make jewelry. For example, sapphires and rubies are aluminum oxide, Al_2O_3. Impurities of chromium ions make rubies red, and iron and titanium ions make sapphires blue.

Rubies and sapphires are the ionic compound aluminum oxide, with chromium ions in rubies, and titanium and iron ions in sapphires.

Properties of Ionic Compounds

In an **ionic compound**, one or more electrons are transferred from metals to nonmetals, which form positive and negative ions. The attraction between these ions is called an *ionic bond*.

The physical and chemical properties of an ionic compound such as NaCl are very different from those of the original elements. For example, the original elements of NaCl were sodium, which is a soft, shiny metal, and chlorine, which is a yellow-green poisonous gas. However, when they react and form positive and negative ions, they produce NaCl, which is ordinary table salt, a hard, white, crystalline substance that is important in our diet.

In a crystal of NaCl, the larger Cl^- ions are arranged in a three-dimensional structure in which the smaller Na^+ ions occupy the spaces between the Cl^- ions (see **FIGURE 6.1**). In this crystal, every Na^+ ion is surrounded by six Cl^- ions, and every Cl^- ion is surrounded by six Na^+ ions. Thus, there are many strong attractions between the positive and negative ions, which account for the high melting points of ionic compounds. For example, the melting point of NaCl is 801 °C. At room temperature, ionic compounds are solids.

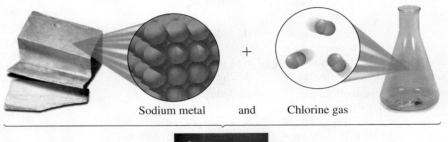

Sodium metal   and   Chlorine gas

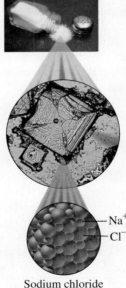

Na^+
Cl^-

Sodium chloride

FIGURE 6.1 ▶ The elements sodium and chlorine react to form the ionic compound sodium chloride, the compound that makes up table salt. The magnification of NaCl crystals shows the arrangement of Na^+ and Cl^- ions.

Ⓠ What is the type of bonding between Na^+ and Cl^- ions in NaCl?

Chemical Formulas of Ionic Compounds

The **chemical formula** of a compound represents the symbols and subscripts in the lowest whole-number ratio of the atoms or ions. In the formula of an ionic compound, the sum of the ionic charges in the formula is always zero. *Thus, the total amount of positive charge is equal to the total amount of negative charge.* For example, to achieve a stable electron configuration, one Na atom (metal) loses its one valence electron to form Na^+, and one Cl atom (nonmetal) gains one electron to form a Cl^- ion. The formula NaCl indicates that the compound has charge balance because there is one sodium ion, Na^+, for every chloride ion, Cl^-. Although the ions are positively or negatively charged, they are not shown in the formula of the compound.

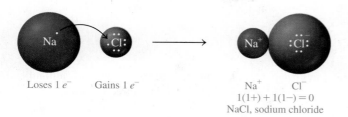

Loses 1 e^- Gains 1 e^-

Na^+ Cl^-
$1(1+) + 1(1-) = 0$
NaCl, sodium chloride

Subscripts in Formulas

Consider a compound of magnesium and chlorine. To achieve a stable electron configuration, one Mg atom (metal) loses its two valence electrons to form Mg^{2+}. Two Cl atoms (nonmetals) each gain one electron to form two Cl^- ions. The two Cl^- ions are needed to balance the positive charge of Mg^{2+}. This gives the formula $MgCl_2$, magnesium chloride, in which the subscript 2 shows that two Cl^- ions are needed for charge balance.

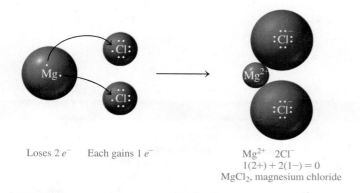

Loses 2 e^- Each gains 1 e^-

Mg^{2+} $2Cl^-$
$1(2+) + 2(1-) = 0$
$MgCl_2$, magnesium chloride

Writing Ionic Formulas from Ionic Charges

The subscripts in the formula of an ionic compound represent the number of positive and negative ions that give an overall charge of zero. Thus, we can now write a formula directly from the ionic charges of the positive and negative ions. Suppose we wish to write the formula for the ionic compound containing Na^+ and S^{2-} ions. To balance the ionic charge of the S^{2-} ion, we will need to place two Na^+ ions in the formula. This gives the formula Na_2S, which has an overall charge of zero. In the formula of an ionic compound, the cation is written first followed by the anion. Appropriate subscripts are used to show the number of each of the ions. This formula is the lowest ratio of ions in the ionic compound. Since ionic compounds do not exist as molecules, this lowest ratio of ions is called a *formula unit*.

CORE CHEMISTRY SKILL
Writing Ionic Formulas

ENGAGE

How are the charges of ions used to write an ionic formula?

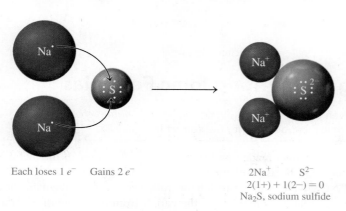

Each loses 1 e^- Gains 2 e^-

$2Na^+$ S^{2-}
$2(1+) + 1(2-) = 0$
Na_2S, sodium sulfide

SAMPLE PROBLEM 6.3 Writing Formulas from Ionic Charges

Write the symbols for the ions, and the correct formula for the ionic compound formed when lithium and nitrogen react.

TRY IT FIRST

SOLUTION

Lithium, which is a metal in Group 1A (1), forms Li^+; nitrogen, which is a nonmetal in Group 5A (15), forms N^{3-}. The charge of $3-$ is balanced by three Li^+ ions.

$$3(1+) + 1(3-) = 0$$

Writing the cation (positive ion) first and the anion (negative ion) second gives the formula Li_3N.

STUDY CHECK 6.3

Write the symbols for the ions, and the correct formula for the ionic compound that would form when calcium and oxygen react.

ANSWER

Ca^{2+}, O^{2-}, CaO

QUESTIONS AND PROBLEMS

6.2 Ionic Compounds

LEARNING GOAL Using charge balance, write the correct formula for an ionic compound.

6.15 Which of the following pairs of elements are likely to form an ionic compound?
  a. lithium and chlorine
  b. oxygen and bromine
  c. potassium and oxygen
  d. sodium and neon
  e. cesium and magnesium
  f. nitrogen and fluorine

6.16 Which of the following pairs of elements are likely to form an ionic compound?
  a. helium and oxygen
  b. magnesium and chlorine
  c. chlorine and bromine
  d. potassium and sulfur
  e. sodium and potassium
  f. nitrogen and iodine

6.17 Write the correct ionic formula for the compound formed between each of the following pairs of ions:
  a. Na^+ and O^{2-}
  b. Al^{3+} and Br^-
  c. Ba^{2+} and N^{3-}
  d. Mg^{2+} and F^-
  e. Al^{3+} and S^{2-}

6.18 Write the correct ionic formula for the compound formed between each of the following pairs of ions:
  a. Al^{3+} and Cl^-
  b. Ca^{2+} and S^{2-}
  c. Li^+ and S^{2-}
  d. Rb^+ and P^{3-}
  e. Cs^+ and I^-

6.19 Write the symbols for the ions, and the correct formula for the ionic compound formed by each of the following:
  a. potassium and sulfur
  b. sodium and nitrogen
  c. aluminum and iodine
  d. gallium and oxygen

6.20 Write the symbols for the ions, and the correct formula for the ionic compound formed by each of the following:
  a. calcium and chlorine
  b. rubidium and bromine
  c. sodium and phosphorus
  d. magnesium and oxygen

6.3 Naming and Writing Ionic Formulas

LEARNING GOAL Given the formula of an ionic compound, write the correct name; given the name of an ionic compound, write the correct formula.

CORE CHEMISTRY SKILL
Naming Ionic Compounds

In the name of an ionic compound made up of two elements, the name of the metal ion, which is written first, is the same as its element name. The name of the nonmetal ion is obtained by using the first syllable of its element name followed by *ide*. In the name of any ionic compound, a space separates the name of the cation from the name of the anion. Subscripts are not used; they are understood because of the charge balance of the ions in the compound (see **TABLE 6.4**).

TABLE 6.4 Names of Some Ionic Compounds

Compound	Metal Ion	Nonmetal Ion	Name
KI	K^+ Potassium	I^- Iodide	Potassium iodide
$MgBr_2$	Mg^{2+} Magnesium	Br^- Bromide	Magnesium bromide
Al_2O_3	Al^{3+} Aluminum	O^{2-} Oxide	Aluminum oxide

Iodized salt contains KI to prevent iodine deficiency.

SAMPLE PROBLEM 6.4 Naming Ionic Compounds

Write the name for the ionic compound Mg_3N_2.

TRY IT FIRST

SOLUTION

ANALYZE THE PROBLEM	Given	Need	Connect
	Mg_3N_2	name	cation, anion

STEP 1 Identify the cation and anion. The cation is Mg^{2+} and the anion is N^{3-}.

STEP 2 Name the cation by its element name. The cation Mg^{2+} is magnesium.

STEP 3 Name the anion by using the first syllable of its element name followed by *ide*. The anion N^{3-} is nitride.

STEP 4 Write the name for the cation first and the name for the anion second.
magnesium nitride

STUDY CHECK 6.4

Name the compound Ga_2S_3.

ANSWER
gallium sulfide

Guide to Naming Ionic Compounds with Metals That Form a Single Ion

STEP 1
Identify the cation and anion.

STEP 2
Name the cation by its element name.

STEP 3
Name the anion by using the first syllable of its element name followed by *ide*.

STEP 4
Write the name for the cation first and the name for the anion second.

Metals with Variable Charges

We have seen that the charge of an ion of a representative element can be obtained from its group number. However, we cannot determine the charge of a transition element because it typically forms two or more positive ions. The transition elements lose electrons, but they are lost from the highest energy level and sometimes from a lower energy level as well. This is also true for metals of representative elements in Groups 4A (14) and 5A (15), such as Pb, Sn, and Bi.

In some ionic compounds, iron is in the Fe^{2+} form, but in other compounds, it has the Fe^{3+} form. Copper also forms two different ions, Cu^+ and Cu^{2+}. When a metal can form two or more types of ions, it has *variable charge*. Then we cannot predict the ionic charge from the group number.

For metals that form two or more ions, a naming system is used to identify the particular cation. To do this, a Roman numeral that is equal to the ionic charge is placed in parentheses immediately after the name of the metal. For example, Fe^{2+} is iron(II), and Fe^{3+} is iron(III). **TABLE 6.5** lists the ions of some metals that produce more than one ion.

ENGAGE

Why is a Roman numeral placed after the name of the cations of most transition elements?

<table>
<tr><td colspan="3">TABLE 6.5 Some Metals That Form More Than One Positive Ion</td></tr>
</table>

Element	Possible Ions	Name of Ion
Bismuth	Bi^{3+} Bi^{5+}	Bismuth(III) Bismuth(V)
Chromium	Cr^{2+} Cr^{3+}	Chromium(II) Chromium(III)
Cobalt	Co^{2+} Co^{3+}	Cobalt(II) Cobalt(III)
Copper	Cu^{+} Cu^{2+}	Copper(I) Copper(II)
Gold	Au^{+} Au^{3+}	Gold(I) Gold(III)
Iron	Fe^{2+} Fe^{3+}	Iron(II) Iron(III)
Lead	Pb^{2+} Pb^{4+}	Lead(II) Lead(IV)
Manganese	Mn^{2+} Mn^{3+}	Manganese(II) Manganese(III)
Mercury	Hg_2^{2+} Hg^{2+}	Mercury(I)* Mercury(II)
Nickel	Ni^{2+} Ni^{3+}	Nickel(II) Nickel(III)
Tin	Sn^{2+} Sn^{4+}	Tin(II) Tin(IV)

*Mercury(I) ions form an ion pair with a 2+ charge.

The transition elements form more than one positive ion except for zinc (Zn^{2+}), cadmium (Cd^{2+}), and silver (Ag^+), which form only one ion. Thus, no Roman numerals are used with zinc, cadmium, and silver when naming their cations in ionic compounds. Metals in Groups 4A (14) and 5A (15) also form more than one type of positive ion. For example, lead and tin in Group 4A (14) form cations with charges of 2+ and 4+, and bismuth in Group 5A (15) forms cations with charges of 3+ and 5+.

Determination of Variable Charge

When you name an ionic compound, you need to determine if the metal is a representative element or a transition element. If it is a transition element, except for zinc, cadmium, or silver, you will need to use its ionic charge as a Roman numeral as part of its name. The calculation of ionic charge depends on the negative charge of the anions in the formula. For example, we use charge balance to determine the charge of a copper cation in the ionic compound $CuCl_2$. Because there are two chloride ions, each with a 1− charge, the total negative charge is 2−. To balance this 2− charge, the copper ion must have a charge of 2+, or Cu^{2+}:

$CuCl_2$

$$Cu \text{ charge } + 2\,Cl^- \text{ charge } = 0$$
$$? \qquad\qquad + 2(1-) \qquad = 0$$
$$2+ \qquad\qquad + 2- \qquad\quad = 0$$

To indicate the 2+ charge for the copper ion Cu^{2+}, we place the Roman numeral (II) immediately after copper when naming this compound: copper(II) chloride. Some ions and their location on the periodic table are seen in **FIGURE 6.2**.

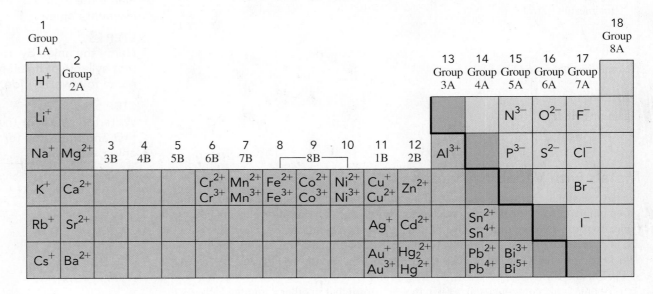

FIGURE 6.2 ▶ In ionic compounds, metals form positive ions, and nonmetals form negative ions.

What are the typical ions of calcium, copper, and oxygen in ionic compounds?

SAMPLE PROBLEM 6.5 Naming Ionic Compounds with Variable Charge Metal Ions

Antifouling paint contains Cu_2O, which prevents the growth of barnacles and algae on the bottoms of boats. What is the name of Cu_2O?

 TRY IT FIRST

SOLUTION

ANALYZE THE PROBLEM	Given	Need	Connect
	Cu_2O	name	cation, anion

The growth of barnacles is prevented by using a paint with Cu_2O on the bottom of a boat.

STEP 1 Determine the charge of the cation from the anion.

	Metal	**Nonmetal**
Element	copper (Cu)	oxygen (O)
Group	transition element	6A (16)
Ion	Cu?	O^{2-}
Charge Balance	2Cu? +	2− = 0
	$\dfrac{2Cu?}{2} = \dfrac{2+}{2} = 1+$	
Ion	Cu^{+}	O^{2-}

STEP 2 Name the cation by its element name and use a Roman numeral in parentheses for the charge. copper(I)

STEP 3 Name the anion by using the first syllable of its element name followed by *ide*. oxide

STEP 4 Write the name for the cation first and the name for the anion second. copper(I) oxide

STUDY CHECK 6.5

Write the name for the compound with the formula Mn_2S_3.

ANSWER

manganese(III) sulfide

> ### Guide to Naming Ionic Compounds with Variable Charge Metals
>
> **STEP 1**
> Determine the charge of the cation from the anion.
>
> **STEP 2**
> Name the cation by its element name and use a Roman numeral in parentheses for the charge.
>
> **STEP 3**
> Name the anion by using the first syllable of its element name followed by *ide*.
>
> **STEP 4**
> Write the name for the cation first and the name for the anion second.

TABLE 6.6 lists the names of some ionic compounds in which the transition elements and metals from Groups 4A (14) and 5A (15) have more than one positive ion.

Writing Formulas from the Name of an Ionic Compound

The formula for an ionic compound is written from the first part of the name that describes the metal ion, including its charge, and the second part of the name that specifies the nonmetal ion. Subscripts are added, as needed, to balance the charge. The steps for writing a formula from the name of an ionic compound are shown in Sample Problem 6.6.

TABLE 6.6 Some Ionic Compounds of Metals That Form Two Kinds of Positive Ions

Compound	Systematic Name
$FeCl_2$	Iron(II) chloride
Fe_2O_3	Iron(III) oxide
Cu_3P	Copper(I) phosphide
$CrBr_2$	Chromium(II) bromide
$SnCl_2$	Tin(II) chloride
PbS_2	Lead(IV) sulfide
BiF_3	Bismuth(III) fluoride

SAMPLE PROBLEM 6.6 Writing Formulas for Ionic Compounds

Write the correct formula for iron(III) chloride.

SOLUTION

ANALYZE THE PROBLEM	Given	Need	Connect
	iron(III) chloride	formula	cation, anion, charge balance

STEP 1 Identify the cation and anion.

Type of Ion	Cation	Anion
Name	iron(III)	chloride
Group	transition element	7A (17)
Symbol of Ion	Fe^{3+}	Cl^-

STEP 2 Balance the charges. The charge of 3+ is balanced by three Cl^- ions.

$$1(3+) + 3(1-) = 0$$

STEP 3 Write the formula, cation first, using subscripts from the charge balance.
$FeCl_3$

STUDY CHECK 6.6

Write the correct formula for chromium(III) oxide.

ANSWER

Cr_2O_3

Guide to Writing Formulas from the Name of an Ionic Compound

STEP 1
Identify the cation and anion.

STEP 2
Balance the charges.

STEP 3
Write the formula, cation first, using subscripts from the charge balance.

The pigment chrome oxide green contains chromium(III) oxide.

QUESTIONS AND PROBLEMS

6.3 Naming and Writing Ionic Formulas

LEARNING GOAL Given the formula of an ionic compound, write the correct name; given the name of an ionic compound, write the correct formula.

6.21 Write the name for each of the following ionic compounds:
a. Al_2O_3
b. $CaCl_2$
c. Na_2O
d. Mg_3P_2
e. KI
f. BaF_2

6.22 Write the name for each of the following ionic compounds:
a. $MgCl_2$
b. K_3P
c. Li_2S
d. CsF
e. MgO
f. $SrBr_2$

6.23 Write the name for each of the following ions (include the Roman numeral when necessary):
a. Fe^{2+}
b. Cu^{2+}
c. Zn^{2+}
d. Pb^{4+}
e. Cr^{3+}
f. Mn^{2+}

6.24 Write the name for each of the following ions (include the Roman numeral when necessary):
a. Ag^+
b. Cu^+
c. Bi^{3+}
d. Sn^{2+}
e. Au^{3+}
f. Ni^{2+}

6.25 Write the name for each of the following ionic compounds:
a. $SnCl_2$
b. FeO
c. Cu_2S
d. CuS
e. $CdBr_2$
f. $HgCl_2$

6.26 Write the name for each of the following ionic compounds:
a. Ag_3P
b. PbS
c. SnO_2
d. $MnCl_3$
e. Bi_2O_3
f. $CoCl_2$

6.27 Write the symbol for the cation in each of the following ionic compounds:
a. $AuCl_3$
b. Fe_2O_3
c. PbI_4
d. $SnCl_2$

6.28 Write the symbol for the cation in each of the following ionic compounds:
a. $FeCl_2$
b. CrO
c. Ni_2S_3
d. AlP

6.29 Write the formula for each of the following ionic compounds:
a. magnesium chloride
b. sodium sulfide
c. copper(I) oxide
d. zinc phosphide
e. gold(III) nitride
f. cobalt(III) fluoride

6.30 Write the formula for each of the following ionic compounds:
a. nickel(III) oxide
b. barium fluoride
c. tin(IV) chloride
d. silver sulfide
e. bismuth(V) chloride
f. potassium nitride

6.31 Write the formula for each of the following ionic compounds:
a. cobalt(III) chloride
b. lead(IV) oxide
c. silver iodide
d. calcium nitride
e. copper(I) phosphide
f. chromium(II) chloride

6.32 Write the formula for each of the following ionic compounds:
a. zinc bromide
b. iron(III) sulfide
c. manganese(IV) oxide
d. chromium(III) iodide
e. lithium nitride
f. gold(I) oxide

Applications

6.33 The following compounds contain ions that are required in small amounts by the body. Write the formula for each:
a. potassium phosphide
b. copper(II) chloride
c. iron(III) bromide
d. magnesium oxide

6.34 The following compounds contain ions that are required in small amounts by the body. Write the formula for each:
a. calcium chloride
b. nickel(II) iodide
c. manganese(II) oxide
d. zinc nitride

Vitamins and minerals are substances required in small amounts by the body for growth and activity.

6.4 Polyatomic Ions

LEARNING GOAL Write the name and formula for an ionic compound containing a polyatomic ion.

An ionic compound may also contain a *polyatomic ion* as one of its cations or anions. A **polyatomic ion** is a group of covalently bonded atoms that has an overall ionic charge. Most polyatomic ions consist of a nonmetal such as phosphorus, sulfur, carbon, or nitrogen covalently bonded to oxygen atoms.

Almost all the polyatomic ions are anions with charges $1-$, $2-$, or $3-$. Only one common polyatomic ion, NH_4^+, has a positive charge. Some models of common polyatomic ions are shown in **FIGURE 6.3**.

Plaster cast
$CaSO_4$

Fertilizer
NH_4NO_3

Ca^{2+} SO_4^{2-}
Sulfate ion

NH_4^+ NO_3^-
Ammonium ion   Nitrate ion

FIGURE 6.3 ▶ Many products contain polyatomic ions, which are groups of atoms that have an ionic charge.

Q What is the charge of a sulfate ion?

Names of Polyatomic Ions

The names of the most common polyatomic ions end in *ate*, such as nitrate and sulfate. When a related ion has one less oxygen atom, the *ite* ending is used for its name such as nitrite and sulfite. Recognizing these endings will help you identify polyatomic ions in the name of a compound. The hydroxide ion (OH^-) and cyanide ion (CN^-) are exceptions to this naming pattern.

By learning the formulas, charges, and names of the polyatomic ions shown in bold type in **TABLE 6.7**, you can derive the related ions. Note that both the *ate* ion and *ite* ion of a particular nonmetal have the same ionic charge. For example, the sulfate ion is SO_4^{2-}, and the sulfite ion, which has one less oxygen atom, is SO_3^{2-}. Phosphate and phosphite ions each have a 3− charge; nitrate and nitrite each have a 1− charge; and perchlorate, chlorate, chlorite, and hypochlorite all have a 1− charge. The halogens form four different polyatomic ions with oxygen.

TABLE 6.7 Names and Formulas of Some Common Polyatomic Ions

Nonmetal	Formula of Ion*	Name of Ion
Hydrogen	OH^-	Hydroxide
Nitrogen	NH_4^+	Ammonium
	NO_3^-	**Nitrate**
	NO_2^-	Nitrite
Chlorine	ClO_4^-	Perchlorate
	ClO_3^-	**Chlorate**
	ClO_2^-	Chlorite
	ClO^-	Hypochlorite
Carbon	**CO_3^{2-}**	**Carbonate**
	HCO_3^-	Hydrogen carbonate (or bicarbonate)
	CN^-	Cyanide
	$C_2H_3O_2^-$	Acetate
Sulfur	**SO_4^{2-}**	**Sulfate**
	HSO_4^-	Hydrogen sulfate (or bisulfate)
	SO_3^{2-}	Sulfite
	HSO_3^-	Hydrogen sulfite (or bisulfite)
Phosphorus	**PO_4^{3-}**	**Phosphate**
	HPO_4^{2-}	Hydrogen phosphate
	$H_2PO_4^-$	Dihydrogen phosphate
	PO_3^{3-}	Phosphite

*Formulas and names in bold type indicate the most common polyatomic ion for that element.

The formula of hydrogen carbonate, or *bicarbonate*, is written by placing a hydrogen in front of the polyatomic ion formula for carbonate (CO_3^{2-}), and the charge is decreased from 2− to 1− to give HCO_3^-.

$$H^+ + CO_3^{2-} = HCO_3^-$$

Writing Formulas for Compounds Containing Polyatomic Ions

No polyatomic ion exists by itself. Like any ion, a polyatomic ion must be associated with ions of opposite charge. The bonding between polyatomic ions and other ions is one of electrical attraction. For example, the compound sodium chlorite consists of sodium ions (Na^+) and chlorite ions (ClO_2^-) held together by ionic bonds.

To write correct formulas for compounds containing polyatomic ions, we follow the same rules of charge balance that we used for writing the formulas for simple ionic compounds. The total negative and positive charges must equal zero. For example, consider the formula for a compound containing sodium ions and chlorite ions. The ions are written as

$$Na^+ \qquad ClO_2^-$$

Sodium ion   Chlorite ion

$$(1+) + (1-) = 0$$

Because one ion of each balances the charge, the formula is written as

$$NaClO_2$$

Sodium chlorite

Sodium chlorite is used in the processing and bleaching of pulp from wood fibers and recycled cardboard.

When more than one polyatomic ion is needed for charge balance, parentheses are used to enclose the formula of the ion. A subscript is written outside the right parenthesis of the polyatomic ion to indicate the number needed for charge balance. Consider the formula for magnesium nitrate. The ions in this compound are the magnesium ion and the nitrate ion, a polyatomic ion.

$$Mg^{2+} \qquad NO_3^-$$

Magnesium ion   Nitrate ion

To balance the positive charge of 2+ on the magnesium ion, two nitrate ions are needed. In the formula of the compound, parentheses are placed around the nitrate ion, and the subscript 2 is written outside the right parenthesis.

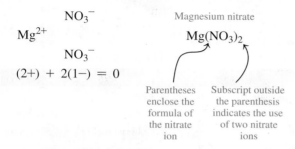

$$NO_3^-$$
$$Mg^{2+}$$
$$NO_3^-$$
$$(2+) + 2(1-) = 0$$

Magnesium nitrate

$$Mg(NO_3)_2$$

Parentheses enclose the formula of the nitrate ion

Subscript outside the parenthesis indicates the use of two nitrate ions

ENGAGE

Why does the formula for magnesium nitrate contain two NO_3^- ions?

SAMPLE PROBLEM 6.7 Writing Formulas Containing Polyatomic Ions

An antacid called Amphojel contains aluminum hydroxide, which treats acid indigestion and heartburn. Write the formula for aluminum hydroxide.

TRY IT FIRST

SOLUTION

ANALYZE THE PROBLEM	Given	Need	Connect
	aluminium hydroxide	formula	cation, polyatomic anion

STEP 1 Identify the cation and polyatomic ion (anion).

Cation	**Polyatomic Ion (anion)**
aluminum	hydroxide
Al^{3+}	OH^-

STEP 2 Balance the charges. The charge of 3+ is balanced by three OH^- ions.

$$1(3+) + 3(1-) = 0$$

STEP 3 Write the formula, cation first, using the subscripts from charge balance. The formula for the compound is written by enclosing the formula of the hydroxide ion, OH^-, in parentheses and writing the subscript 3 outside the right parenthesis.

$$Al(OH)_3$$

STUDY CHECK 6.7

Write the formula for a compound containing ammonium ions and phosphate ions.

ANSWER

$(NH_4)_3PO_4$

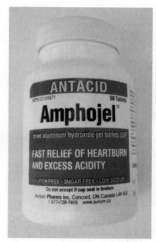

Aluminum hydroxide is an antacid used to treat acid indigestion.

Guide to Writing Formulas with Polyatomic Ions

STEP 1
Identify the cation and polyatomic ion (anion).

STEP 2
Balance the charges.

STEP 3
Write the formula, cation first, using the subscripts from charge balance.

Naming Ionic Compounds Containing Polyatomic Ions

When naming ionic compounds containing polyatomic ions, we first write the positive ion, usually a metal, and then we write the name for the polyatomic ion. It is important that you learn to recognize the polyatomic ion in the formula and name it correctly. As with other ionic compounds, no prefixes are used.

Na_2SO_4 $FePO_4$ $Al_2(CO_3)_3$

$Na_2\boxed{SO_4}$ $Fe\boxed{PO_4}$ $Al_2(\boxed{CO_3})_3$

Sodium sulfate   Iron(III) phosphate   Aluminum carbonate

TABLE 6.8 lists the formulas and names of some ionic compounds that include polyatomic ions and also gives their uses in medicine and industry.

TABLE 6.8 Some Ionic Compounds That Contain Polyatomic Ions

Formula	Name	Medical Use
$AlPO_4$	Aluminum phosphate	Antacid
$Al_2(SO_4)_3$	Aluminum sulfate	Antiperspirant, anti-infective
$BaSO_4$	Barium sulfate	Contrast medium for X-rays
$CaCO_3$	Calcium carbonate	Antacid, calcium supplement
$Ca_3(PO_4)_2$	Calcium phosphate	Calcium dietary supplement
$CaSO_4$	Calcium sulfate	Plaster casts
$MgSO_4$	Magnesium sulfate	Cathartic, Epsom salts
K_2CO_3	Potassium carbonate	Alkalizer, diuretic
$AgNO_3$	Silver nitrate	Topical anti-infective
$NaHCO_3$	Sodium bicarbonate or sodium hydrogen carbonate	Antacid
$Zn_3(PO_4)_2$	Zinc phosphate	Dental cement

A solution of Epsom salts, magnesium sulfate, $MgSO_4$, may be used to soothe sore muscles.

SAMPLE PROBLEM 6.8 Naming Compounds Containing Polyatomic Ions

Name the following ionic compounds:

a. $Cu(NO_2)_2$ **b.** $KClO_3$

▶ **TRY IT FIRST**

SOLUTION

ANALYZE THE PROBLEM	Given	Need	Connect
	formula	name	cation, polyatomic ion

	STEP 1		**STEP 2**	**STEP 3**	**STEP 4**
Formula	Cation	Anion	Name of Cation	Name of Anion	Name of Compound
a. $Cu(NO_2)_2$	Cu^{2+}	NO_2^-	Copper(II) ion	Nitrite ion	Copper(II) nitrite
b. $KClO_3$	K^+	ClO_3^-	Potassium ion	Chlorate ion	Potassium chlorate

STUDY CHECK 6.8

What is the name of $Co_3(PO_4)_2$?

ANSWER

cobalt(II) phosphate

Guide to Naming Ionic Compounds with Polyatomic Ions

STEP 1
Identify the cation and polyatomic ion (anion).

STEP 2
Name the cation using a Roman numeral, if needed.

STEP 3
Name the polyatomic ion.

STEP 4
Write the name for the compound, cation first and the polyatomic ion second.

QUESTIONS AND PROBLEMS

6.4 Polyatomic Ions

LEARNING GOAL Write the name and formula for an ionic compound containing a polyatomic ion.

6.35 Write the formula including the charge for each of the following polyatomic ions:
a. hydrogen carbonate (bicarbonate)
b. ammonium
c. phosphite
d. chlorate

6.36 Write the formula including the charge for each of the following polyatomic ions:
a. nitrite
b. sulfite
c. hydroxide
d. acetate

6.37 Name the following polyatomic ions:
a. SO_4^{2-}
b. CO_3^{2-}
c. HSO_3^{-}
d. NO_3^{-}

6.38 Name the following polyatomic ions:
a. OH^{-}
b. PO_4^{3-}
c. CN^{-}
d. NO_2^{-}

6.39 Complete the following table with the formula and name of the compound that forms between each pair of ions:

	NO_2^{-}	CO_3^{2-}	HSO_4^{-}	PO_4^{3-}
Li^+				
Cu^{2+}				
Ba^{2+}				

6.40 Complete the following table with the formula and name of the compound that forms between each pair of ions:

	NO_3^{-}	HCO_3^{-}	SO_3^{2-}	HPO_4^{2-}
NH_4^+				
Al^{3+}				
Pb^{4+}				

6.41 Write the correct formula for the following ionic compounds:
a. barium hydroxide
b. sodium hydrogen sulfate
c. iron(II) nitrite
d. zinc phosphate
e. iron(III) carbonate

6.42 Write the correct formula for the following ionic compounds:
a. aluminum chlorate
b. ammonium oxide
c. magnesium bicarbonate
d. sodium nitrite
e. copper(I) sulfate

6.43 Write the formula for the polyatomic ion and name each of the following compounds:
a. Na_2CO_3
b. $(NH_4)_2S$
c. $Ba_3(PO_4)_2$
d. $Sn(NO_2)_2$

6.44 Write the formula for the polyatomic ion and name each of the following compounds:
a. $MnCO_3$
b. Au_2SO_4
c. $Mg_3(PO_4)_2$
d. $Fe(HCO_3)_3$

Applications

6.45 Name each of the following ionic compounds:
a. $Zn(C_2H_3O_2)_2$, cold remedy
b. $Mg_3(PO_4)_2$, antacid
c. NH_4Cl, expectorant
d. $NaHCO_3$, corrects pH imbalance

6.46 Name each of the following ionic compounds:
a. Li_2CO_3, antidepressant
b. $MgSO_4$, Epsom salts
c. $NaClO$, disinfectant
d. Na_3PO_4, laxative

6.5 Molecular Compounds: Sharing Electrons

LEARNING GOAL Given the formula of a molecular compound, write its correct name; given the name of a molecular compound, write its formula.

A **molecular compound** consists of atoms of two or more nonmetals that share one or more valence electrons. The shared atoms are held together by **covalent bonds** that form a **molecule**. There are many more molecular compounds than there are ionic ones. For example, water (H_2O) and carbon dioxide (CO_2) are both molecular compounds. Molecular compounds consist of molecules, which are discrete groups of atoms in a definite proportion. A molecule of water (H_2O) consists of two atoms of hydrogen and one atom of oxygen. When you have iced tea, perhaps you add molecules of sugar ($C_{12}H_{22}O_{11}$), which is a molecular compound. Other familiar molecular compounds include propane (C_3H_8), alcohol (C_2H_6O), the antibiotic amoxicillin ($C_{16}H_{19}N_3O_5S$), and the antidepressant Prozac ($C_{17}H_{18}F_3NO$).

Names and Formulas of Molecular Compounds

When naming a molecular compound, the first nonmetal in the formula is named by its element name; the second nonmetal is named using the first syllable of its element name,

CORE CHEMISTRY SKILL
Writing the Names and Formulas for Molecular Compounds

TABLE 6.9 Prefixes Used in Naming Molecular Compounds

1 mono	6 hexa
2 di	7 hepta
3 tri	8 octa
4 tetra	9 nona
5 penta	10 deca

ENGAGE

Why are prefixes used to name molecular compounds?

followed by *ide*. When a subscript indicates two or more atoms of an element, a prefix is shown in front of its name. **TABLE 6.9** lists prefixes used in naming molecular compounds.

The names of molecular compounds need prefixes because several different compounds can be formed from the same two nonmetals. For example, carbon and oxygen can form two different compounds, carbon monoxide, CO, and carbon dioxide, CO_2, in which the number of atoms of oxygen in each compound is indicated by the prefixes *mono* or *di* in their names.

When the vowels *o* and *o* or *a* and *o* appear together, the first vowel is omitted, as in carbon monoxide. In the name of a molecular compound, the prefix *mono* is usually omitted, as in NO, nitrogen oxide. Traditionally, however, CO is named carbon monoxide. **TABLE 6.10** lists the formulas, names, and commercial uses of some molecular compounds.

TABLE 6.10 Some Common Molecular Compounds

Formula	Name	Commercial Uses
CO_2	Carbon dioxide	Fire extinguishers, dry ice, propellant in aerosols, carbonation of beverages
CS_2	Carbon disulfide	Manufacture of rayon
N_2O	Dinitrogen oxide	Inhalation anesthetic, "laughing gas"
NO	Nitrogen oxide	Stabilizer, biochemical messenger in cells
SO_2	Sulfur dioxide	Preserving fruits, vegetables; disinfectant in breweries; bleaching textiles
SF_6	Sulfur hexafluoride	Electrical circuits
SO_3	Sulfur trioxide	Manufacture of explosives

SAMPLE PROBLEM 6.9 Naming Molecular Compounds

Name the molecular compound NCl_3.

TRY IT FIRST

SOLUTION

ANALYZE THE PROBLEM	Given	Need	Connect
	NCl_3	name	prefixes

Guide to Naming Molecular Compounds

STEP 1
Name the first nonmetal by its element name.

STEP 2
Name the second nonmetal by using the first syllable of its element name followed by *ide*.

STEP 3
Add prefixes to indicate the number of atoms (subscripts).

STEP 1 Name the first nonmetal by its element name. In NCl_3, the first nonmetal (N) is nitrogen.

STEP 2 Name the second nonmetal by using the first syllable of its element name followed by *ide*. The second nonmetal (Cl) is named chloride.

STEP 3 Add prefixes to indicate the number of atoms (subscripts). Because there is one nitrogen atom, no prefix is needed. The subscript 3 for the Cl atoms is shown as the prefix *tri*.

Symbol of Element	N	Cl
Name	nitrogen	chloride
Subscript	1	3
Prefix	none	tri

The name of NCl_3 is nitrogen trichloride.

STUDY CHECK 6.9

Write the name for each of the following molecular compounds:

a. $SiBr_4$ **b.** Br_2O

ANSWER

a. silicon tetrabromide **b.** dibromine oxide

Writing Formulas from the Names of Molecular Compounds

In the name of a molecular compound, the names of two nonmetals are given along with prefixes for the number of atoms of each. To write the formula from the name, we use the symbol for each element and a subscript if a prefix indicates two or more atoms.

SAMPLE PROBLEM 6.10 Writing Formulas for Molecular Compounds

Write the formula for the molecular compound diboron trioxide.

> **TRY IT FIRST**

SOLUTION

ANALYZE THE PROBLEM	Given	Need	Connect
	diboron trioxide	formula	subscripts from prefixes

STEP 1 Write the symbols in the order of the elements in the name.

Name of Element	boron	oxygen
Symbol of Element	B	O
Subscript	2 (di)	3 (tri)

STEP 2 Write any prefixes as subscripts. The prefix *di* in *di*boron indicates that there are two atoms of boron, shown as a subscript 2 in the formula. The prefix *tri* in *tri*oxide indicates that there are three atoms of oxygen, shown as a subscript 3 in the formula.

B_2O_3

STUDY CHECK 6.10

Write the formula for the molecular compound iodine pentafluoride.

ANSWER

IF_5

> **Guide to Writing Formulas for Molecular Compounds**
>
> **STEP 1**
> Write the symbols in the order of the elements in the name.
>
> **STEP 2**
> Write any prefixes as subscripts.

Summary of Naming Ionic and Molecular Compounds

We have now examined strategies for naming ionic and molecular compounds. In general, compounds having two elements are named by stating the first element name followed by the name of the second element with an *ide* ending. If the first element is a metal, the compound is usually ionic; if the first element is a nonmetal, the compound is usually molecular. For ionic compounds, it is necessary to determine whether the metal can form more than one type of positive ion; if so, a Roman numeral following the name of the metal indicates the particular ionic charge. One exception is the ammonium ion, NH_4^+, which is also written first as a positively charged polyatomic ion. Ionic compounds having three or more elements include some type of polyatomic ion. They are named by ionic rules but have an *ate* or *ite* ending when the polyatomic ion has a negative charge.

In naming molecular compounds having two elements, prefixes are necessary to indicate two or more atoms of each nonmetal as shown in that particular formula (see **FIGURE 6.4**).

SAMPLE PROBLEM 6.11 Naming Ionic and Molecular Compounds

Identify each of the following compounds as ionic or molecular and give its name:

a. K_3P **b.** $NiSO_4$ **c.** SO_3

> **TRY IT FIRST**

SOLUTION

a. K_3P, consisting of a metal and a nonmetal, is an ionic compound. As a representative element in Group 1A (1), K forms the potassium ion, K^+. Phosphorus, as a representative element in Group 5A (15), forms a phosphide ion, P^{3-}. Writing the name of the cation followed by the name of the anion gives the name potassium phosphide.

ENGAGE

Why is sodium phosphate an ionic compound and diphosphorus pentoxide a molecular compound?

b. $NiSO_4$, consisting of a cation of a transition element and a polyatomic ion SO_4^{2-}, is an ionic compound. As a transition element, Ni forms more than one type of ion. In this formula, the $2-$ charge of SO_4^{2-} is balanced by one nickel ion, Ni^{2+}. In the name, a Roman numeral written after the metal name, nickel(II), specifies the $2+$ charge. The anion SO_4^{2-} is a polyatomic ion named sulfate. The compound is named nickel(II) sulfate.

c. SO_3 consists of two nonmetals, which indicates that it is a molecular compound. The first element S is sulfur (no prefix is needed). The second element O, oxide, has subscript 3, which requires a prefix *tri* in the name. The compound is named sulfur trioxide.

STUDY CHECK 6.11

What is the name of $Fe(NO_3)_3$?

ANSWER

iron(III) nitrate

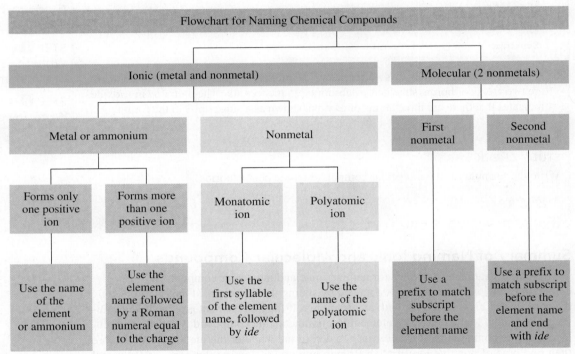

FIGURE 6.4 ▶ A flowchart illustrates naming for ionic and molecular compounds.

Why are the names of some metal ions followed by a Roman numeral in the name of a compound?

QUESTIONS AND PROBLEMS

6.5 Molecular Compounds: Sharing Electrons

LEARNING GOAL Given the formula of a molecular compound, write its correct name; given the name of a molecular compound, write its formula.

6.47 Name each of the following molecular compounds:
  a. PBr_3 **b.** Cl_2O **c.** CBr_4
  d. HF   **e.** NF_3

6.48 Name each of the following molecular compounds:
  a. CS_2 **b.** P_2O_5 **c.** SiO_2
  d. PCl_3 **e.** CO

6.49 Name each of the following molecular compounds:
  a. N_2O_3 **b.** Si_2Br_6 **c.** P_4S_3
  d. PCl_5 **e.** SeF_6

6.50 Name each of the following molecular compounds:
  a. SiF_4 **b.** IBr_3 **c.** CO_2
  d. N_2F_2 **e.** N_2S_3

6.51 Write the formula for each of the following molecular compounds:
  a. carbon tetrachloride   **b.** carbon monoxide
  c. phosphorus trichloride   **d.** dinitrogen tetroxide

6.52 Write the formula for each of the following molecular compounds:
    a. sulfur dioxide   **b.** silicon tetrachloride
    c. iodine trifluoride   **d.** dinitrogen oxide

6.53 Write the formula for each of the following molecular compounds:
    a. oxygen difluoride   **b.** boron trichloride
    c. dinitrogen trioxide   **d.** sulfur hexafluoride

6.54 Write the formula for each of the following molecular compounds:
    a. sulfur dibromide   **b.** carbon disulfide
    c. tetraphosphorus hexoxide   **d.** dinitrogen pentoxide

Applications

6.55 Name each of the following ionic or molecular compounds:
    a. $Al_2(SO_4)_3$, antiperspirant
    b. $CaCO_3$, antacid
    c. N_2O, "laughing gas," inhaled anesthetic
    d. $Mg(OH)_2$, laxative

6.56 Name each of the following ionic or molecular compounds:
    a. $Al(OH)_3$, antacid
    b. $FeSO_4$, iron supplement in vitamins
    c. NO, vasodilator
    d. $Cu(OH)_2$, fungicide

Follow Up

COMPOUNDS AT THE PHARMACY

A few days ago, Richard went back to the pharmacy to pick up aspirin, $C_9H_8O_4$, and acetaminophen, $C_8H_9NO_2$. He also wanted to talk to Sarah about a way to treat his sore toe. Sarah recommended soaking his foot in a solution of Epsom salts, which is magnesium sulfate. Richard also asked Sarah to recommend an antacid for his upset stomach and an iron supplement. Sarah suggested an antacid that contains calcium carbonate and aluminum hydroxide, and iron(II) sulfate as an iron supplement. Richard also picked up toothpaste containing tin(II) fluoride, and carbonated water, which contains carbon dioxide.

Applications

6.57 Write the chemical formula for each of the following:
    a. magnesium sulfate   **b.** tin(II) fluoride
    c. aluminum hydroxide

6.58 Write the chemical formula for each of the following:
    a. calcium carbonate   **b.** carbon dioxide
    c. iron(II) sulfate

6.59 Identify each of the compounds in problem 6.57 as ionic or molecular.

6.60 Identify each of the compounds in problem 6.58 as ionic or molecular.

CONCEPT MAP

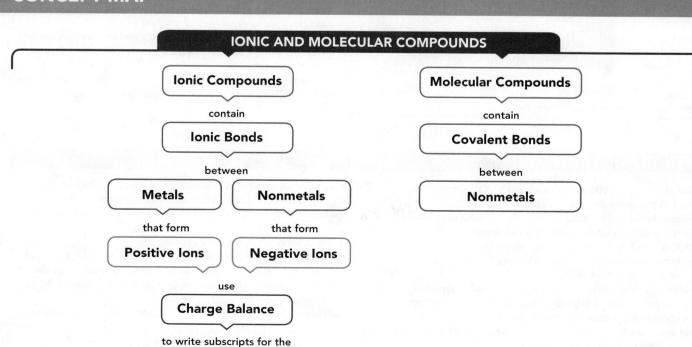

IONIC AND MOLECULAR COMPOUNDS

Ionic Compounds — contain — Ionic Bonds — between — Metals / Nonmetals — that form — Positive Ions / Negative Ions — use — Charge Balance — to write subscripts for the — Chemical Formula

Molecular Compounds — contain — Covalent Bonds — between — Nonmetals

CHAPTER REVIEW

6.1 Ions: Transfer of Electrons

LEARNING GOAL Write the symbols for the simple ions of the representative elements.

Transfer of electrons

- The stability of the noble gases is associated with a stable electron configuration in the outermost energy level.
- With the exception of helium, which has two electrons, noble gases have eight valence electrons, which is an octet.
- Atoms of elements in Groups 1A to 7A (1, 2, 13 to 17) achieve stability by losing, gaining, or sharing their valence electrons in the formation of compounds.
- Metals of the representative elements lose valence electrons to form positively charged ions (cations): Group 1A (1), 1+, Group 2A (2), 2+, and Group 3A (13), 3+.
- When reacting with metals, nonmetals gain electrons to form octets and form negatively charged ions (anions): Groups 5A (15), 3−, 6A (16), 2−, and 7A (17), 1−.

6.2 Ionic Compounds

LEARNING GOAL Using charge balance, write the correct formula for an ionic compound.

- The total positive and negative ionic charge is balanced in the formula of an ionic compound.
- Charge balance in a formula is achieved by using subscripts after each symbol so that the overall charge is zero.

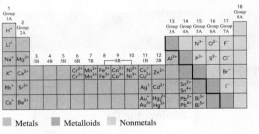

Sodium chloride

6.3 Naming and Writing Ionic Formulas

LEARNING GOAL Given the formula of an ionic compound, write the correct name; given the name of an ionic compound, write the correct formula.

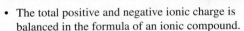
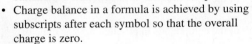

- In naming ionic compounds, the positive ion is given first followed by the name of the negative ion.
- The names of ionic compounds containing two elements end with *ide*.
- Except for Ag, Cd, and Zn, transition elements form cations with two or more ionic charges.
- The charge of the cation is determined from the total negative charge in the formula and included as a Roman numeral immediately following the name of the metal that has a variable charge.

6.4 Polyatomic Ions

LEARNING GOAL Write the name and formula for an ionic compound containing a polyatomic ion.

- A polyatomic ion is a covalently bonded group of atoms with an electrical charge; for example, the carbonate ion has the formula CO_3^{2-}.
- Most polyatomic ions have names that end with *ate* or *ite*.
- Most polyatomic ions contain a nonmetal and one or more oxygen atoms.
- The ammonium ion, NH_4^+, is a positive polyatomic ion.
- When more than one polyatomic ion is used for charge balance, parentheses enclose the formula of the polyatomic ion.

Ca^{2+} SO_4^{2-}
Sulfate ion

6.5 Molecular Compounds: Sharing Electrons

LEARNING GOAL Given the formula of a molecular compound, write its correct name; given the name of a molecular compound, write its formula.

1	mono
2	di
3	tri
4	tetra
5	penta

- In a covalent bond, atoms of nonmetals share valence electrons such that each atom has a stable electron configuration.
- The first nonmetal in a molecular compound uses its element name; the second nonmetal uses the first syllable of its element name followed by *ide*.
- The name of a molecular compound with two different atoms uses prefixes to indicate the subscripts in the formula.

KEY TERMS

anion A negatively charged ion such as Cl^-, O^{2-}, or SO_4^{2-}.

cation A positively charged ion such as Na^+, Mg^{2+}, Al^{3+}, or NH_4^+.

chemical formula The group of symbols and subscripts that represents the atoms or ions in a compound.

covalent bond A sharing of valence electrons by atoms.

ion An atom or group of atoms having an electrical charge because of a loss or gain of electrons.

ionic charge The difference between the number of protons (positive) and the number of electrons (negative) written in the upper right corner of the symbol for the element or polyatomic ion.

ionic compound A compound of positive and negative ions held together by ionic bonds.

molecular compound A combination of atoms in which stable electron configurations are attained by sharing electrons.

molecule The smallest unit of two or more atoms held together by covalent bonds.

octet A set of eight valence electrons.

octet rule Elements in Groups 1A to 7A (1, 2, 13 to 17) react with other elements by forming ionic or covalent bonds to produce a stable electron configuration, usually eight electrons in the outer shell.

polyatomic ion A group of covalently bonded nonmetal atoms that has an overall electrical charge.

CORE CHEMISTRY SKILLS

The chapter section containing each Core Chemistry Skill is shown in parentheses at the end of each heading.

Writing Positive and Negative Ions (6.1)

- In the formation of an ionic bond, atoms of a metal lose and atoms of a nonmetal gain valence electrons to acquire a stable electron configuration, usually eight valence electrons.
- This tendency of atoms to attain a stable electron configuration is known as the octet rule.

Example: State the number of electrons lost or gained by atoms and the ion formed for each of the following to obtain a stable electron configuration:

    a. Br   **b.** Ca   **c.** S

Answer: **a.** Br atoms gain one electron to achieve a stable electron configuration, Br^-.

        b. Ca atoms lose two electrons to achieve a stable electron configuration, Ca^{2+}.

        c. S atoms gain two electrons to achieve a stable electron configuration, S^{2-}.

Writing Ionic Formulas (6.2)

- The chemical formula of a compound represents the lowest whole-number ratio of the atoms or ions.
- In the chemical formula of an ionic compound, the sum of the positive and negative charges is always zero.
- Thus, in a chemical formula of an ionic compound, the total positive charge is equal to the total negative charge.

Example: Write the formula for magnesium phosphide.

Answer: Magnesium phosphide is an ionic compound that contains the ions Mg^{2+} and P^{3-}.

      Using charge balance, we determine the number(s) of each type of ion.

$$3(2+) + 2(3-) = 0$$

      $3Mg^{2+}$ and $2P^{3-}$ give the formula Mg_3P_2.

Naming Ionic Compounds (6.3)

- In the name of an ionic compound made up of two elements, the name of the metal ion, which is written first, is the same as its element name.
- For metals that form two or more ions, a Roman numeral that is equal to the ionic charge is placed in parentheses immediately after the name of the metal.
- The name of a nonmetal ion is obtained by using the first syllable of its element name followed by *ide*.

Example: What is the name of PbS?

Answer: This compound contains the S^{2-} ion which has a 2− charge.

      For charge balance, the positive ion must have a charge of 2+.

$$Pb? + (2-) = 0;\quad Pb = 2+$$

      Because lead can form two different positive ions, a Roman numeral (II) is used in the name of the compound: lead(II) sulfide.

Writing the Names and Formulas for Molecular Compounds (6.5)

- When naming a molecular compound, the first nonmetal in the formula is named by its element name; the second nonmetal is named using the first syllable of its element name followed by *ide*.
- When a subscript indicates two or more atoms of an element, a prefix is shown in front of its name.

Example: Name the molecular compound BrF_5.

Answer: Two nonmetals share electrons and form a molecular compound. Br (first nonmetal) is bromine; F (second nonmetal) is fluoride. In the name for a molecular compound, prefixes indicate the subscripts in the formulas. The subscript 1 is understood for Br. The subscript 5 for fluoride is written with the prefix *penta*. The name is bromine pentafluoride.

UNDERSTANDING THE CONCEPTS

The chapter sections to review are shown in parentheses at the end of each question.

6.61 **a.** How does the octet rule explain the formation of a magnesium ion? (6.1)

      b. What noble gas has the same electron configuration as the magnesium ion?

      c. Why are Group 1A (1) and Group 2A (2) elements found in many compounds, but not Group 8A (18) elements?

6.62 **a.** How does the octet rule explain the formation of a chloride ion? (6.1)

      b. What noble gas has the same electron configuration as the chloride ion?

      c. Why are Group 7A (17) elements found in many compounds, but not Group 8A (18) elements?

6.63 Identify each of the following atoms or ions: (6.1)

$18\,e^-$	$8\,e^-$	$28\,e^-$	$23\,e^-$
$15\,p^+$	$8\,p^+$	$30\,p^+$	$26\,p^+$
$16\,n$	$8\,n$	$35\,n$	$28\,n$
A	**B**	**C**	**D**

6.64 Identify each of the following atoms or ions: (6.1)

$2\,e^-$	$0\,e^-$	$3\,e^-$	$10\,e^-$
$3\,p^+$	$1\,p^+$	$3\,p^+$	$7\,p^+$
$4\,n$		$4\,n$	$8\,n$
A	**B**	**C**	**D**

6.65 Consider the following Lewis symbols for elements X and Y: (6.1, 6.2, 6.5)

X·   ·Ÿ·

a. What are the group numbers of X and Y?
b. Will a compound of X and Y be ionic or molecular?
c. What ions would be formed by X and Y?
d. What would be the formula of a compound of X and Y?
e. What would be the formula of a compound of X and sulfur?
f. What would be the formula of a compound of Y and chlorine?
g. Is the compound in part **f** ionic or molecular?

6.66 Consider the following Lewis symbols for elements X and Y: (6.1, 6.2, 6.5)

Ẋ·   ·Ẏ·

a. What are the group numbers of X and Y?
b. Will a compound of X and Y be ionic or molecular?
c. What ions would be formed by X and Y?
d. What would be the formula of a compound of X and Y?
e. What would be the formula of a compound of X and sulfur?
f. What would be the formula of a compound of Y and chlorine?
g. Is the compound in part **f** ionic or molecular?

6.67 Using each of the following electron configurations, give the formulas for the cation and anion that form, the formula for the compound they form, and its name. (6.2, 6.3)

Electron Configurations		Cation	Anion	Formula of Compound	Name of Compound
$1s^22s^22p^63s^2$	$1s^22s^22p^3$				
$1s^22s^22p^63s^23p^64s^1$	$1s^22s^22p^4$				
$1s^22s^22p^63s^23p^1$	$1s^22s^22p^63s^23p^5$				

6.68 Using each of the following electron configurations, give the formulas for the cation and anion that form, the formula for the compound they form, and its name. (6.2, 6.3)

Electron Configurations		Cation	Anion	Formula of Compound	Name of Compound
$1s^22s^22p^63s^1$	$1s^22s^22p^5$				
$1s^22s^22p^63s^23p^64s^2$	$1s^22s^22p^63s^23p^4$				
$1s^22s^22p^63s^23p^1$	$1s^22s^22p^63s^23p^3$				

ADDITIONAL QUESTIONS AND PROBLEMS

6.69 Write the name for the following: (6.1)
a. N^{3-} b. Mg^{2+} c. O^{2-} d. Al^{3+}

6.70 Write the name for the following: (6.1)
a. K^+ b. Na^+ c. Ba^{2+} d. Cl^-

6.71 Consider an ion with the symbol X^{2+} formed from a representative element. (6.1, 6.2, 6.3)
a. What is the group number of the element?
b. What is the Lewis symbol of the element?
c. If X is in Period 3, what is the element?
d. What is the formula of the compound formed from X and the nitride ion?

6.72 Consider an ion with the symbol Y^{3-} formed from a representative element. (6.1, 6.2, 6.3)
a. What is the group number of the element?
b. What is the Lewis symbol of the element?
c. If Y is in Period 3, what is the element?
d. What is the formula of the compound formed from the barium ion and Y?

6.73 One of the ions of tin is tin(IV). (6.1, 6.2, 6.3, 6.4)
a. What is the symbol for this ion?
b. How many protons and electrons are in the ion?
c. What is the formula of tin(IV) oxide?
d. What is the formula of tin(IV) phosphate?

6.74 One of the ions of gold is gold(III). (6.1, 6.2, 6.3, 6.4)
a. What is the symbol for this ion?
b. How many protons and electrons are in the ion?
c. What is the formula of gold(III) sulfate?
d. What is the formula of gold(III) nitrate?

6.75 Write the formula for each of the following ionic compounds: (6.2, 6.3)
a. tin(II) sulfide   b. lead(IV) oxide
c. silver chloride   d. calcium nitride
e. copper(I) phosphide   f. chromium(II) bromide

6.76 Write the formula for each of the following ionic compounds: (6.2, 6.3)
a. nickel(III) oxide   b. iron(III) sulfide
c. lead(II) sulfate   d. chromium(III) iodide
e. lithium nitride   f. gold(I) oxide

6.77 Name each of the following molecular compounds: (6.5)
a. NCl_3 b. N_2S_3 c. N_2O
d. IF e. BF_3 f. P_2O_5

6.78 Name each of the following molecular compounds: (6.5)
a. CBr_4 b. SF_6 c. $BrCl$
d. N_2O_4 e. SO_2 f. CS_2

6.79 Write the formula for each of the following molecular compounds: (6.5)

    a. carbon sulfide   **b.** diphosphorus pentoxide

    c. dihydrogen sulfide   **d.** sulfur dichloride

6.80 Write the formula for each of the following molecular compounds: (6.5)

    a. silicon dioxide   **b.** carbon tetrabromide

    c. diphosphorus tetraiodide   **d.** dinitrogen trioxide

6.81 Classify each of the following as ionic or molecular, and give its name: (6.3, 6.5)

    a. $FeCl_3$ **b.** Na_2SO_4 **c.** NO_2

    d. Rb_2S **e.** PF_5 **f.** CF_4

6.82 Classify each of the following as ionic or molecular, and give its name: (6.3, 6.5)

    a. $Al_2(CO_3)_3$ **b.** ClF_5 **c.** BCl_3

    d. Mg_3N_2 **e.** ClO_2 **f.** $CrPO_4$

6.83 Write the formula for each of the following: (6.3, 6.4, 6.5)

    a. tin(II) carbonate   **b.** lithium phosphide

    c. silicon tetrachloride   **d.** manganese(III) oxide

    e. tetraphosphorus   **f.** calcium bromide

        triselenide

6.84 Write the formula for each of the following: (6.3, 6.4, 6.5)

    a. sodium carbonate   **b.** nitrogen dioxide

    c. aluminum nitrate   **d.** copper(I) nitride

    e. potassium phosphate   **f.** cobalt(III) sulfate

CHALLENGE QUESTIONS

The following groups of questions are related to the topics in this chapter. However, they do not all follow the chapter order, and they require you to combine concepts and skills from several sections. These questions will help you increase your critical thinking skills and prepare for your next exam.

6.85 Complete the following table for atoms or ions: (6.1)

Atom or Ion	Number of Protons	Number of Electrons	Electrons Lost/ Gained
K^+			
	$12\,p^+$	$10\,e^-$	
	$8\,p^+$		$2\,e^-$ gained
		$10\,e^-$	$3\,e^-$ lost

6.86 Complete the following table for atoms or ions: (6.1)

Atom or Ion	Number of Protons	Number of Electrons	Electrons Lost/ Gained
	$30\,p^+$		$2\,e^-$ lost
	$36\,p^+$	$36\,e^-$	
	$16\,p^+$		$2\,e^-$ gained
		$46\,e^-$	$4\,e^-$ lost

6.87 Identify the group number in the periodic table of X, a representative element, in each of the following ionic compounds: (6.2)

    a. XCl_3 **b.** Al_2X_3 **c.** XCO_3

6.88 Identify the group number in the periodic table of X, a representative element, in each of the following ionic compounds: (6.2)

    a. X_2O_3 **b.** X_2SO_3 **c.** Na_3X

6.89 Classify each of the following as ionic or molecular, and name each: (6.2, 6.3, 6.4, 6.5)

    a. Li_2HPO_4 **b.** ClF_3 **c.** $Mg(ClO_2)_2$

    d. NF_3 **e.** $Ca(HSO_4)_2$ **f.** $KClO_4$

    g. $Au_2(SO_3)_3$

6.90 Classify each of the following as ionic or molecular, and name each: (6.2, 6.3, 6.4, 6.5)

    a. $FePO_3$ **b.** Cl_2O_7 **c.** $Ca_3(PO_4)_2$

    d. PCl_3 **e.** $Al(ClO_2)_3$ **f.** $Pb(C_2H_3O_2)_2$

    g. $MgCO_3$

ANSWERS

Answers to Selected Questions and Problems

6.1 a. 1   **b.** 2   **c.** 3   **d.** 1   **e.** 2

6.3 a. $2\,e^-$ lost   **b.** $3\,e^-$ gained   **c.** $1\,e^-$ gained

    d. $1\,e^-$ lost   **e.** $1\,e^-$ gained

6.5 a. Li^+ **b.** F^- **c.** Mg^{2+} **d.** Fe^{3+}

6.7 a. 29 protons, 27 electrons   **b.** 34 protons, 36 electrons

    c. 35 protons, 36 electrons   **d.** 26 protons, 23 electrons

6.9 a. Cl^- **b.** Cs^+ **c.** N^{3-} **d.** Ra^{2+}

6.11 a. lithium   **b.** calcium

    c. gallium   **d.** phosphide

6.13 a. 8 protons, 10 electrons   **b.** 19 protons, 18 electrons

    c. 53 protons, 54 electrons   **d.** 11 protons, 10 electrons

6.15 a and **c**

6.17 a. Na_2O **b.** $AlBr_3$ **c.** Ba_3N_2

    d. MgF_2 **e.** Al_2S_3

6.19 a. K^+ and S^{2-}, K_2S **b.** Na^+ and N^{3-}, Na_3N
    c. Al^{3+} and I^-, AlI_3 **d.** Ga^{3+} and O^{2-}, Ga_2O_3

6.21 a. aluminum oxide   **b.** calcium chloride
    c. sodium oxide   **d.** magnesium phosphide
    e. potassium iodide   **f.** barium fluoride

6.23 a. iron(II)   **b.** copper(II)
    c. zinc   **d.** lead(IV)
    e. chromium(III)   **f.** manganese(II)

6.25 a. tin(II) chloride   **b.** iron(II) oxide
    c. copper(I) sulfide   **d.** copper(II) sulfide
    e. cadmium bromide   **f.** mercury(II) chloride

6.27 a. Au^{3+} **b.** Fe^{3+} **c.** Pb^{4+} **d.** Sn^{2+}

6.29 a. $MgCl_2$ **b.** Na_2S **c.** Cu_2O **d.** Zn_3P_2
    e. AuN **f.** CoF_3

6.31 a. $CoCl_3$ **b.** PbO_2 **c.** AgI **d.** Ca_3N_2
    e. Cu_3P **f.** $CrCl_2$

6.33 a. K_3P **b.** $CuCl_2$ **c.** $FeBr_3$ **d.** MgO

6.35 a. HCO_3^- **b.** NH_4^+ **c.** PO_3^{3-} **d.** ClO_3^-

6.37 a. sulfate   **b.** carbonate
    c. hydrogen sulfite (bisulfite)   **d.** nitrate

6.39

	NO_2^-	CO_3^{2-}	HSO_4^-	PO_4^{3-}
Li^+	$LiNO_2$ Lithium nitrite	Li_2CO_3 Lithium carbonate	$LiHSO_4$ Lithium hydrogen sulfate	Li_3PO_4 Lithium phosphate
Cu^{2+}	$Cu(NO_2)_2$ Copper(II) nitrite	$CuCO_3$ Copper(II) carbonate	$Cu(HSO_4)_2$ Copper(II) hydrogen sulfate	$Cu_3(PO_4)_2$ Copper(II) phosphate
Ba^{2+}	$Ba(NO_2)_2$ Barium nitrite	$BaCO_3$ Barium carbonate	$Ba(HSO_4)_2$ Barium hydrogen sulfate	$Ba_3(PO_4)_2$ Barium phosphate

6.41 a. $Ba(OH)_2$ **b.** $NaHSO_4$ **c.** $Fe(NO_2)_2$
    d. $Zn_3(PO_4)_2$ **e.** $Fe_2(CO_3)_3$

6.43 a. CO_3^{2-}, sodium carbonate   **b.** NH_4^+, ammonium sulfide
    c. PO_4^{3-}, barium phosphate   **d.** NO_2^-, tin(II) nitrite

6.45 a. zinc acetate   **b.** magnesium phosphate
    c. ammonium chloride
    d. sodium bicarbonate or sodium hydrogen bicarbonate

6.47 a. phosphorus tribromide   **b.** dichlorine oxide
    c. carbon tetrabromide   **d.** hydrogen fluoride
    e. nitrogen trifluoride

6.49 a. dinitrogen trioxide   **b.** disilicon hexabromide
    c. tetraphosphorus trisulfide   **d.** phosphorus pentachloride
    e. selenium hexafluoride

6.51 a. CCl_4 **b.** CO **c.** PCl_3 **d.** N_2O_4

6.53 a. OF_2 **b.** BCl_3 **c.** N_2O_3 **d.** SF_6

6.55 a. aluminum sulfate   **b.** calcium carbonate
    c. dinitrogen oxide   **d.** magnesium hydroxide

6.57 a. $MgSO_4$ **b.** SnF_2 **c.** $Al(OH)_3$

6.59 a. ionic   **b.** ionic   **c.** ionic

6.61 a. By losing two valence electrons from the third energy level, magnesium achieves an octet in the second energy level.
    b. The magnesium ion Mg^{2+} has the same electron configuration as Ne ($1s^2 2s^2 2p^6$).
    c. Group 1A (1) and 2A (2) elements achieve octets by losing electrons to form compounds. Group 8A (18) elements are stable with octets (or two electrons for helium).

6.63 a. P^{3-} ion   **b.** O atom
    c. Zn^{2+} ion   **d.** Fe^{3+} ion

6.65 a. X = Group 1A (1), Y = Group 6A (16)
    b. ionic   **c.** X^+ and Y^{2-} **d.** X_2Y
    e. X_2S **f.** YCl_2 **g.** molecular

6.67

Electron Configurations		Cation	Anion	Formula of Compound	Name of Compound
$1s^2 2s^2 2p^6 3s^2$	$1s^2 2s^2 2p^3$	Mg^{2+}	N^{3-}	Mg_3N_2	Magnesium nitride
$1s^2 2s^2 2p^6 3s^2 3p^6 4s^1$	$1s^2 2s^2 2p^4$	K^+	O^{2-}	K_2O	Potassium oxide
$1s^2 2s^2 2p^6 3s^2 3p^1$	$1s^2 2s^2 2p^6 3s^2 3p^5$	Al^{3+}	Cl^-	$AlCl_3$	Aluminum chloride

6.69 a. nitride   **b.** magnesium
    c. oxide   **d.** aluminum

6.71 a. 2A (2)   **b.** $\dot{X}\cdot$ **c.** Mg   **d.** X_3N_2

6.73 a. Sn^{4+} **b.** 50 protons and 46 electrons
    c. SnO_2 **d.** $Sn_3(PO_4)_4$

6.75 a. SnS **b.** PbO_2 **c.** $AgCl$
    d. Ca_3N_2 **e.** Cu_3P **f.** $CrBr_2$

6.77 a. nitrogen trichloride   **b.** dinitrogen trisulfide
    c. dinitrogen oxide   **d.** iodine fluoride
    e. boron trifluoride   **f.** diphosphorus pentoxide

6.79 a. CS **b.** P_2O_5 **c.** H_2S **d.** SCl_2

6.81 a. ionic, iron(III) chloride   **b.** ionic, sodium sulfate
    c. molecular, nitrogen dioxide   **d.** ionic, rubidium sulfide
    e. molecular, phosphorus pentafluoride   **f.** molecular, carbon tetrafluoride

6.83 a. $SnCO_3$ **b.** Li_3P **c.** $SiCl_4$
    d. Mn_2O_3 **e.** P_4Se_3 **f.** $CaBr_2$

6.85

Atom or Ion	Number of Protons	Number of Electrons	Electrons Lost/ Gained
K^+	$19\,p^+$	$18\,e^-$	$1\,e^-$ lost
Mg^{2+}	$12\,p^+$	$10\,e^-$	$2\,e^-$ lost
O^{2-}	$8\,p^+$	$10\,e^-$	$2\,e^-$ gained
Al^{3+}	$13\,p^+$	$10\,e^-$	$3\,e^-$ lost

6.87 a. Group 3A (13)   **b.** Group 6A (16)
    c. Group 2A (2)

6.89 a. ionic, lithium hydrogen phosphate
    b. molecular, chlorine trifluoride
    c. ionic, magnesium chlorite
    d. molecular, nitrogen trifluoride
    e. ionic, calcium bisulfate or calcium hydrogen sulfate
    f. ionic, potassium perchlorate
    g. ionic, gold(III) sulfite

7 Chemical Quantities

MAX, A SIX-YEAR-OLD
dog, is listless, drinking large amounts of water, and not eating his food. His owner takes Max to his veterinarian, Chris, for an examination. Chris weighs Max and obtains a blood sample for a blood chemistry profile, which determines the overall health, detects any metabolic disorders, and measures the concentration of electrolytes.

The results of the lab tests for Max indicate that the white blood count is elevated, which may indicate an infection or inflammation. His electrolytes, which are also indicators of good health, were all in the normal ranges. The electrolyte chloride (Cl^-) was in the normal range of 0.106 to 0.118 mol/L, and the electrolyte sodium (Na^+) has a normal range of 0.144 to 0.160 mol/L.

Potassium (K^+), another important electrolyte, has a normal range of 0.0035 to 0.0058 mol/L. To treat the possible infection, Chris ordered 375-mg tablets of Clavamox, which is a broad-spectrum antibiotic approved for veterinary use in dogs.

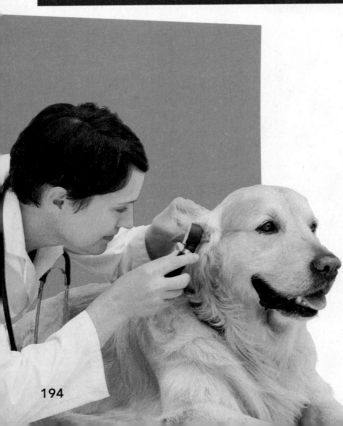

CAREER

Veterinarian

Veterinarians care for domesticated pets such as rats, birds, dogs, and cats. Some veterinarians specialize in the treatment of large animals, such as horses and cattle. Veterinarians interact with pet owners as they give advice about feeding, behavior, and breeding. In the assessment of an ill animal, they record the animal's symptoms and medical history including dietary intake, medications, eating habits, weight, and any signs of disease.

To diagnose health problems, veterinarians perform laboratory tests on animals including a complete blood count and urinalysis. They also obtain tissue and blood samples, as well as vaccinate against distemper, rabies, and other diseases. Veterinarians prescribe medication if an animal has an infection or illness. If an animal is injured, a veterinarian treats wounds, and sets fractures. A veterinarian also performs surgery such as neutering and spaying, provides dental cleanings, removes tumors, and euthanizes animals.

- Calculating Percentages (1.4)
- Solving Equations (1.4)
- Converting between Standard Numbers and Scientific Notation (1.5)
- Rounding Off (2.3)

 CORE CHEMISTRY SKILLS

- Counting Significant Figures (2.2)
- Using Significant Figures in Calculations (2.3)
- Writing Conversion Factors from Equalities (2.5)
- Using Conversion Factors (2.6)

*These Key Math Skills and Core Chemistry Skills from previous chapters are listed here for your review as you proceed to the new material in this chapter.

LOOKING AHEAD

7.1 The Mole

LEARNING GOAL Use Avogadro's number to calculate the number of particles in a given number of moles. Calculate the number of moles of an element in a given number of moles of a compound.

At the grocery store, you buy eggs by the dozen or soda by the case. In an office-supply store, pencils are ordered by the gross and paper by the ream. Common terms such as *dozen, case, gross,* and *ream* are used to count the number of items present. For example, when you buy a dozen eggs, you know you will get 12 eggs in the carton.

Avogadro's Number

In chemistry, particles such as atoms, molecules, and ions are counted by the **mole** (abbreviated *mol* in calculations), which contains 6.022×10^{23} items. This value, known as **Avogadro's number**, is a very big number because atoms are so small that it takes an extremely large number of atoms to provide a sufficient amount to weigh and use in chemical reactions. Avogadro's number is named for Amedeo Avogadro (1776–1856), an Italian physicist.

 CORE CHEMISTRY SKILL
Converting Particles to Moles

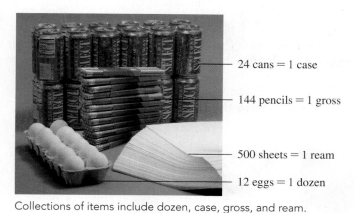

- 24 cans = 1 case
- 144 pencils = 1 gross
- 500 sheets = 1 ream
- 12 eggs = 1 dozen

Collections of items include dozen, case, gross, and ream.

$6.022 \times 10^{23} = 602\ 200\ 000\ 000\ 000\ 000\ 000\ 000$

One mole of any element always contains Avogadro's number of atoms. For example, 1 mol of carbon contains 6.022×10^{23} carbon atoms; 1 mol of aluminum contains 6.022×10^{23} aluminum atoms; 1 mol of sulfur contains 6.022×10^{23} sulfur atoms.

$$1 \text{ mol of an element} = 6.022 \times 10^{23} \text{ atoms of that element}$$

Avogadro's number tells us that 1 mol of a compound contains 6.022×10^{23} of the particular type of particles that make up that compound. One mole of a molecular

One mole of sulfur contains 6.022×10^{23} sulfur atoms.

compound contains Avogadro's number of molecules. For example, 1 mol of CO_2 contains 6.022×10^{23} molecules of CO_2. For an ionic compound, 1 mol contains Avogadro's number of **formula units**, which are the groups of ions represented by its formula. For the ionic formula, NaCl, 1 mol contains 6.022×10^{23} formula units of NaCl (Na^+, Cl^-). **TABLE 7.1** gives examples of the number of particles in some 1-mol quantities.

TABLE 7.1 Number of Particles in 1-Mol Quantities

Substance	Number and Type of Particles
1 mol of Al	6.022×10^{23} atoms of Al
1 mol of Fe	6.022×10^{23} atoms of Fe
1 mol of water (H_2O)	6.022×10^{23} molecules of H_2O
1 mol of vitamin C ($C_6H_8O_6$)	6.022×10^{23} molecules of vitamin C
1 mol of NaCl	6.022×10^{23} formula units of NaCl

Using Avogadro's Number as a Conversion Factor

We can use Avogadro's number as a conversion factor to convert between the moles of a substance and the number of particles it contains.

$$\frac{6.022 \times 10^{23} \text{ particles}}{1 \text{ mol}} \quad \text{and} \quad \frac{1 \text{ mol}}{6.022 \times 10^{23} \text{ particles}}$$

For example, we use Avogadro's number to convert 4.00 mol of iron to atoms of iron.

$$4.00 \text{ mol Fe} \times \frac{6.022 \times 10^{23} \text{ atoms Fe}}{1 \text{ mol Fe}} = 2.41 \times 10^{24} \text{ atoms of Fe}$$

Avogadro's number as a conversion factor

We can also use Avogadro's number to convert 3.01×10^{24} molecules of CO_2 to moles of CO_2.

$$3.01 \times 10^{24} \text{ molecules } CO_2 \times \frac{1 \text{ mol } CO_2}{6.022 \times 10^{23} \text{ molecules } CO_2} = 5.00 \text{ mol of } CO_2$$

Avogadro's number as a conversion factor

In calculations that convert between moles and particles, the number of moles will be small compared to the number of atoms or molecules, which will be large as shown in Sample Problem 7.1.

ENGAGE

Why is 0.20 mol of aluminum a small number, but the number of atoms in 0.20 mol is a large number: 1.2×10^{23} atoms of aluminum?

SAMPLE PROBLEM 7.1 **Calculating the Number of Molecules**

How many molecules are present in 1.75 mol of carbon dioxide?

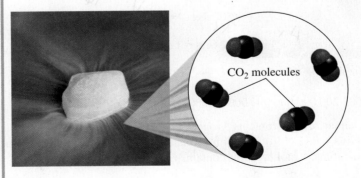

CO$_2$ molecules

The solid form of carbon dioxide is known as "dry ice."

TRY IT FIRST

SOLUTION

STEP 1 State the given and needed quantities.

ANALYZE THE PROBLEM	Given	Need	Connect
	1.75 mol of CO_2	molecules of CO_2	Avogadro's number

STEP 2 Write a plan to convert moles to particles.

moles of CO_2 **Avogadro's number**   molecules of CO_2

STEP 3 Use Avogadro's number to write conversion factors.

$$1 \text{ mol of } CO_2 = 6.022 \times 10^{23} \text{ molecules of } CO_2$$

$$\frac{6.022 \times 10^{23} \text{ molecules } CO_2}{1 \text{ mol } CO_2} \quad \text{and} \quad \frac{1 \text{ mol } CO_2}{6.022 \times 10^{23} \text{ molecules } CO_2}$$

STEP 4 Set up the problem to calculate the number of particles.

$$1.75 \text{ mol } CO_2 \times \frac{6.022 \times 10^{23} \text{ molecules } CO_2}{1 \text{ mol } CO_2} = 1.05 \times 10^{24} \text{ molecules of } CO_2$$

STUDY CHECK 7.1

How many moles of water, H_2O, contain 2.60×10^{23} molecules of water?

ANSWER

0.432 mol of H_2O

Guide to Converting the Moles (or Particles) of a Substance to Particles (or Moles)

STEP 1
State the given and needed quantities.

STEP 2
Write a plan to convert moles (or particles) to particles (or moles).

STEP 3
Use Avogadro's number to write conversion factors.

STEP 4
Set up the problem to calculate the number of particles (or moles).

Moles of Elements in a Chemical Compound

We have seen that the subscripts in a chemical formula indicate the number of atoms of each type of element in a compound. For example, aspirin, $C_9H_8O_4$, is a drug used to reduce pain and inflammation in the body. Using the subscripts in the chemical formula of aspirin shows that there are 9 carbon atoms, 8 hydrogen atoms, and 4 oxygen atoms. The subscripts of the formula of aspirin, $C_9H_8O_4$, also tell us the number of moles of each element in 1 mol of aspirin: 9 mol of C atoms, 8 mol of H atoms, and 4 mol of O atoms.

ENGAGE

Why does 1 mol of $Zn(C_2H_3O_2)_2$, a dietary supplement, contain 1 mol of Zn, 4 mol of C, 6 mol of H, and 4 mol of O?

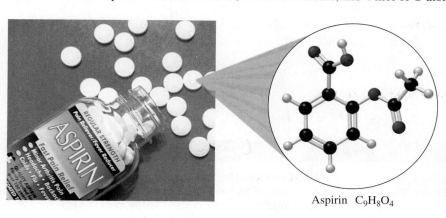

Number of atoms in 1 molecule
Carbon (C)   Hydrogen (H)   Oxygen (O)

Aspirin $C_9H_8O_4$

The chemical formula subscripts specify the $C_9H_8O_4$

	Carbon	**Hydrogen**	**Oxygen**
Atoms in 1 molecule	9 atoms of C	8 atoms of H	4 atoms of O
Moles of each element in 1 mol	9 mol of C	8 mol of H	4 mol of O

Using a Chemical Formula to Derive Conversion Factors

Using the subscripts from the formula, $C_9H_8O_4$, we can write the conversion factors for each of the elements in 1 mol of aspirin:

$$\frac{9 \text{ mol C}}{1 \text{ mol C}_9\text{H}_8\text{O}_4} \qquad \frac{8 \text{ mol H}}{1 \text{ mol C}_9\text{H}_8\text{O}_4} \qquad \frac{4 \text{ mol O}}{1 \text{ mol C}_9\text{H}_8\text{O}_4}$$

$$\frac{1 \text{ mol C}_9\text{H}_8\text{O}_4}{9 \text{ mol C}} \qquad \frac{1 \text{ mol C}_9\text{H}_8\text{O}_4}{8 \text{ mol H}} \qquad \frac{1 \text{ mol C}_9\text{H}_8\text{O}_4}{4 \text{ mol O}}$$

SAMPLE PROBLEM 7.2 Calculating the Moles of an Element in a Compound

How many moles of carbon are present in 1.50 mol of aspirin, $C_9H_8O_4$?

TRY IT FIRST

SOLUTION

STEP 1 State the given and needed quantities.

ANALYZE THE PROBLEM	Given	Need	Connect
	1.50 mol of aspirin, $C_9H_8O_4$	moles of C	subscripts in formula

STEP 2 Write a plan to convert moles of a compound to moles of an element.

moles of $C_9H_8O_4$ | Subscript | moles of C

STEP 3 Write the equalities and conversion factors using subscripts.

1 mol of $C_9H_8O_4$ = 9 mol of C

$$\frac{9 \text{ mol C}}{1 \text{ mol C}_9\text{H}_8\text{O}_4} \quad \text{and} \quad \frac{1 \text{ mol C}_9\text{H}_8\text{O}_4}{9 \text{ mol C}}$$

STEP 4 Set up the problem to calculate the moles of an element.

$$1.50 \text{ mol } \cancel{C_9H_8O_4} \times \frac{9 \text{ mol C}}{1 \text{ mol } \cancel{C_9H_8O_4}} = 13.5 \text{ mol of C}$$

STUDY CHECK 7.2

How many moles of aspirin, $C_9H_8O_4$, contain 0.480 mol of O?

ANSWER

0.120 mol of aspirin

Guide to Calculating Moles of a Compound or Element

STEP 1
State the given and needed quantities.

STEP 2
Write a plan to convert moles of a compound to moles of an element (or moles of an element to moles of a compound).

STEP 3
Write the equalities and conversion factors using subscripts.

STEP 4
Set up the problem to calculate the moles of an element (or moles of a compound).

QUESTIONS AND PROBLEMS

7.1 The Mole

LEARNING GOAL Use Avogadro's number to calculate the number of particles in a given number of moles. Calculate the number of moles of an element in a given number of moles of a compound.

7.1   What is a mole?

7.2   What is Avogadro's number?

7.3   Calculate each of the following:
   a. number of C atoms in 0.500 mol of C
   b. number of SO_2 molecules in 1.28 mol of SO_2
   c. moles of Fe in 5.22×10^{22} atoms of Fe
   d. moles of C_2H_6O in 8.50×10^{24} molecules of C_2H_6O

7.4   Calculate each of the following:
   a. number of Li atoms in 4.5 mol of Li
   b. number of CO_2 molecules in 0.0180 mol of CO_2
   c. moles of Cu in 7.8×10^{21} atoms of Cu
   d. moles of C_2H_6 in 3.75×10^{23} molecules of C_2H_6

7.5   Calculate each of the following quantities in 2.00 mol of H_3PO_4:
   a. moles of H   **b.** moles of O
   c. atoms of P   **d.** atoms of O

7.6   Calculate each of the following quantities in 0.185 mol of $C_6H_{14}O$:
   a. moles of C   **b.** moles of O
   c. atoms of H   **d.** atoms of C

Applications

7.7 Quinine, $C_{20}H_{24}N_2O_2$, is a component of tonic water and bitter lemon.
   a. How many moles of H are in 1.5 mol of quinine?
   b. How many moles of C are in 5.0 mol of quinine?
   c. How many moles of N are in 0.020 mol of quinine?

7.8 Aluminum sulfate, $Al_2(SO_4)_3$, is used in some antiperspirants.
   a. How many moles of O are present in 3.0 mol of $Al_2(SO_4)_3$?
   b. How many moles of aluminum ions (Al^{3+}) are present in 0.40 mol of $Al_2(SO_4)_3$?
   c. How many moles of sulfate ions ($SO_4{}^{2-}$) are present in 1.5 mol of $Al_2(SO_4)_3$?

7.9 Naproxen is used to treat pain and inflammation caused by arthritis. Naproxen has a formula of $C_{14}H_{14}O_3$.
   a. How many moles of C are present in 2.30 mol of naproxen?
   b. How many moles of H are present in 0.444 mol of naproxen?
   c. How many moles of O are present in 0.0765 mol of naproxen?

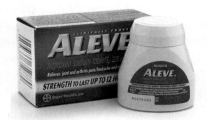

Naproxen is used to treat pain and inflammation caused by arthritis.

7.10 Benadryl is an over-the-counter drug used to treat allergy symptoms. The formula of Benadryl is $C_{17}H_{21}NO$.
   a. How many moles of C are present in 0.733 mol of Benadryl?
   b. How many moles of H are present in 2.20 mol of Benadryl?
   c. How many moles of N are present in 1.54 mol of Benadryl?

7.2 Molar Mass

LEARNING GOAL Given the chemical formula of a substance, calculate its molar mass.

A single atom or molecule is much too small to weigh, even on the most accurate balance. In fact, it takes a huge number of atoms or molecules to make enough of a substance for you to see. An amount of water that contains Avogadro's number of water molecules is only a few sips. However, in the laboratory, we can use a balance to weigh out Avogadro's number of particles for 1 mol of substance.

For any element, the quantity called **molar mass** is the quantity in grams that equals the atomic mass of that element. We are counting 6.022×10^{23} atoms of an element when we weigh out the number of grams equal to its molar mass. For example, carbon has an atomic mass of 12.01 on the periodic table. This means 1 mol of carbon atoms has a mass of 12.01 g. Then to obtain 1 mol of carbon atoms, we would need to weigh out 12.01 g of carbon. Thus, the molar mass of carbon is found by looking at its atomic mass on the periodic table.

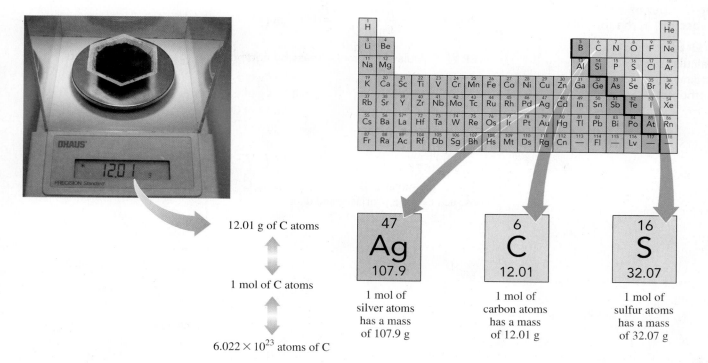

12.01 g of C atoms

1 mol of C atoms

6.022×10^{23} atoms of C

| 47 |
| Ag |
| 107.9 |

1 mol of silver atoms has a mass of 107.9 g

| 6 |
| C |
| 12.01 |

1 mol of carbon atoms has a mass of 12.01 g

| 16 |
| S |
| 32.07 |

1 mol of sulfur atoms has a mass of 32.07 g

CORE CHEMISTRY SKILL
Calculating Molar Mass

Molar Mass of a Compound

To determine the molar mass of a compound, multiply the molar mass of each element by its subscript in the formula and add the results as shown in Sample Problem 7.3. **In this text, we round the molar mass of an element to the hundredths place (0.01) or use at least four significant figures for calculations.**

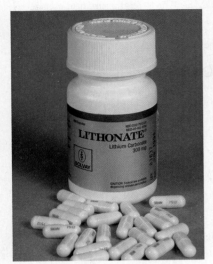

Lithium carbonate is used to treat bipolar disorder.

Guide to Calculating Molar Mass

STEP 1
Obtain the molar mass of each element.

STEP 2
Multiply each molar mass by the number of moles (subscript) in the formula.

STEP 3
Calculate the molar mass by adding the masses of the elements.

SAMPLE PROBLEM 7.3 Calculating the Molar Mass of a Compound

Calculate the molar mass for lithium carbonate, Li_2CO_3, used to treat bipolar disorder.

> **TRY IT FIRST**

SOLUTION

ANALYZE THE PROBLEM	Given	Need	Connect
	formula Li_2CO_3	molar mass of Li_2CO_3	periodic table

STEP 1 Obtain the molar mass of each element.

$$\frac{6.941 \text{ g Li}}{1 \text{ mol Li}} \qquad \frac{12.01 \text{ g C}}{1 \text{ mol C}} \qquad \frac{16.00 \text{ g O}}{1 \text{ mol O}}$$

STEP 2 Multiply each molar mass by the number of moles (subscript) in the formula.

Grams from 2 mol of Li:

$$2 \text{ mol Li} \times \frac{6.941 \text{ g Li}}{1 \text{ mol Li}} = 13.88 \text{ g of Li}$$

Grams from 1 mol of C:

$$1 \text{ mol C} \times \frac{12.01 \text{ g C}}{1 \text{ mol C}} = 12.01 \text{ g of C}$$

Grams from 3 mol of O:

$$3 \text{ mol O} \times \frac{16.00 \text{ g O}}{1 \text{ mol O}} = 48.00 \text{ g of O}$$

STEP 3 Calculate the molar mass by adding the masses of the elements.

2 mol of Li	= 13.88 g of Li
1 mol of C	= 12.01 g of C
3 mol of O	= 48.00 g of O
Molar mass of Li_2CO_3	= 73.89 g

STUDY CHECK 7.3

Calculate the molar mass for salicylic acid, $C_7H_6O_3$, which is used to treat skin conditions such as acne, psoriasis, and dandruff.

ANSWER

138.12 g

FIGURE 7.1 shows some 1-mol quantities of substances.

1-Mol Quantities

| S | Fe | NaCl | $K_2Cr_2O_7$ | $C_{12}H_{22}O_{11}$ |

FIGURE 7.1 ▶ 1-mol samples: sulfur, S (32.07 g); iron, Fe (55.85 g); salt, NaCl (58.44 g); potassium dichromate, $K_2Cr_2O_7$ (294.2 g); and sucrose, $C_{12}H_{22}O_{11}$ (342.3 g).

Q How is the molar mass for $K_2Cr_2O_7$ obtained?

QUESTIONS AND PROBLEMS

7.2 Molar Mass

LEARNING GOAL Given the chemical formula of a substance, calculate its molar mass.

7.11 Calculate the molar mass for each of the following:
  a. Cl_2 **b.** $C_3H_6O_3$
  c. $Mg_3(PO_4)_2$

7.12 Calculate the molar mass for each of the following:
  a. O_2 **b.** KH_2PO_4
  c. $Fe(ClO_4)_3$

7.13 Calculate the molar mass for each of the following:
  a. AlF_3 **b.** $C_2H_4Cl_2$
  c. SnF_2

7.14 Calculate the molar mass for each of the following:
  a. $C_4H_8O_4$ **b.** $Ga_2(CO_3)_3$
  c. $KBrO_4$

Applications

7.15 Calculate the molar mass for each of the following:
  a. NaCl, table salt
  b. Fe_2O_3, rust
  c. $C_{19}H_{20}FNO_3$, Paxil, an antidepressant

7.16 Calculate the molar mass for each of the following:
  a. $FeSO_4$, iron supplement
  b. Al_2O_3, absorbent and abrasive
  c. $C_7H_5NO_3S$, saccharin, an artificial sweetener

7.17 Calculate the molar mass for each of the following:
  a. $Al_2(SO_4)_3$, antiperspirant
  b. $KC_4H_5O_6$, cream of tartar
  c. $C_{16}H_{19}N_3O_5S$, amoxicillin, an antibiotic

7.18 Calculate the molar mass for each of the following:
  a. C_3H_8O, rubbing alcohol
  b. $(NH_4)_2CO_3$, baking powder
  c. $Zn(C_2H_3O_2)_2$, a zinc supplement

7.19 Calculate the molar mass for each of the following:
  a. $C_8H_9NO_2$, acetaminophen used in Tylenol
  b. $Ca_3(C_6H_5O_7)_2$, a calcium supplement
  c. $C_{17}H_{18}FN_3O_3$, Cipro, used to treat a range of bacterial infections

7.20 Calculate the molar mass for each of the following:
  a. $CaSO_4$, calcium sulfate, used to make casts to protect broken bones
  b. $C_6H_{12}N_2O_4Pt$, Carboplatin, used in chemotherapy
  c. $C_{12}H_{18}O$, Propofol, used to induce anesthesia during surgery

7.3 Calculations Using Molar Mass

LEARNING GOAL Given the number of moles (or grams) of a substance, calculate the grams (or moles). Given the grams of a compound, calculate the grams of one of the elements.

The molar mass of an element is one of the most useful conversion factors in chemistry because it converts moles of a substance to grams, or grams to moles. For example, 1 mol of silver has a mass of 107.9 g. To express molar mass of Ag as an equality, we write

  1 mol of Ag $=$ 107.9 g of Ag

From this equality for a molar mass, two conversion factors can be written as

$$\frac{107.9 \text{ g Ag}}{1 \text{ mol Ag}} \quad \text{and} \quad \frac{1 \text{ mol Ag}}{107.9 \text{ g Ag}}$$

Sample Problem 7.4 shows how the molar mass of silver is used as a conversion factor.

SAMPLE PROBLEM 7.4 Converting Moles of an Element to Grams

Silver metal is used in the manufacture of tableware, mirrors, jewelry, and dental alloys. If the design for a piece of jewelry requires 0.750 mol of silver, how many grams of silver are needed?

> **TRY IT FIRST**

SOLUTION

STEP 1 State the given and needed quantities.

ANALYZE THE PROBLEM	Given	Need	Connect
	0.750 mol of Ag	grams of Ag	molar mass

STEP 2 Write a plan to convert moles to grams.

moles of Ag | Molar mass | grams of Ag

STEP 3 Determine the molar mass and write conversion factors.

$$1 \text{ mol of Ag} = 107.9 \text{ g of Ag}$$
$$\frac{107.9 \text{ g Ag}}{1 \text{ mol Ag}} \quad \text{and} \quad \frac{1 \text{ mol Ag}}{107.9 \text{ g Ag}}$$

STEP 4 Set up the problem to convert moles to grams.

$$0.750 \text{ mol Ag} \times \frac{107.9 \text{ g Ag}}{1 \text{ mol Ag}} = 80.9 \text{ g of Ag}$$

STUDY CHECK 7.4

A dentist orders 24.4 g of gold (Au) to prepare dental crowns and fillings. Calculate the number of moles of gold in the order.

ANSWER

0.124 mol of Au

CORE CHEMISTRY SKILL
Using Molar Mass as a Conversion Factor

Silver metal is used to make jewelry.

Guide to Calculating the Moles (or Grams) of a Substance from Grams (or Moles)

STEP 1
State the given and needed quantities.

STEP 2
Write a plan to convert moles to grams (or grams to moles).

STEP 3
Determine the molar mass and write conversion factors.

STEP 4
Set up the problem to convert moles to grams (or grams to moles).

Writing Conversion Factors for the Molar Mass of a Compound

The conversion factors for a compound are also written from the molar mass. For example, the molar mass of the compound H_2O is written

$$1 \text{ mol of } H_2O = 18.02 \text{ g of } H_2O$$

From this equality, conversion factors for the molar mass of H_2O are written as

$$\frac{18.02 \text{ g } H_2O}{1 \text{ mol } H_2O} \quad \text{and} \quad \frac{1 \text{ mol } H_2O}{18.02 \text{ g } H_2O}$$

We can now change from moles to grams, or grams to moles, using the conversion factors derived from the molar mass of a compound shown in Sample Problem 7.5. (Remember, you must determine the molar mass first.)

SAMPLE PROBLEM 7.5 **Converting Mass of a Compound to Moles**

A box of salt contains 737 g of NaCl. How many moles of NaCl are present in the box?

 **TRY IT FIRST**

SOLUTION

STEP 1 State the given and needed quantities.

ANALYZE THE PROBLEM	Given	Need	Connect
	737 g of NaCl	moles of NaCl	molar mass

STEP 2 Write a plan to convert grams to moles.

grams of NaCl Molar mass moles of NaCl

STEP 3 Determine the molar mass and write conversion factors.

$(1 \times 22.99) + (1 \times 35.45) = 58.44$ g/mol

$$1 \text{ mol of NaCl} = 58.44 \text{ g of NaCl}$$

$$\frac{58.44 \text{ g NaCl}}{1 \text{ mol NaCl}} \quad \text{and} \quad \frac{1 \text{ mol NaCl}}{58.44 \text{ g NaCl}}$$

STEP 4 Set up the problem to convert grams to moles.

$$737 \text{ g NaCl} \times \frac{1 \text{ mol NaCl}}{58.44 \text{ g NaCl}} = 12.6 \text{ mol of NaCl}$$

STUDY CHECK 7.5

One tablet of an antacid contains 680. mg of $CaCO_3$. How many moles of $CaCO_3$ are present?

ANSWER

0.00679 or 6.79×10^{-3} mol of $CaCO_3$

Table salt is sodium chloride, NaCl.

FIGURE 7.2 gives a summary of the calculations to show the connections between the moles of a compound, its mass in grams, the number of molecules (or formula units if ionic), and the moles and atoms of each element in that compound in the following flowchart:

ENGAGE

Why are there more grams of chlorine than grams of fluorine in 1 mol of Freon-12, CCl_2F_2?

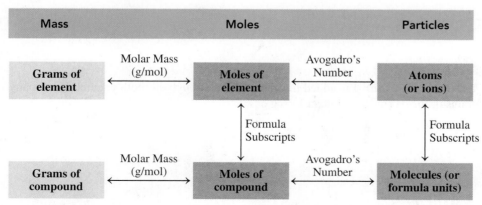

FIGURE 7.2 ▶ The moles of a compound are related to its mass in grams by molar mass, to the number of molecules (or formula units) by Avogadro's number, and to the moles of each element by the subscripts in the formula.

❶ What steps are needed to calculate the number of atoms of H in 5.00 g of CH_4?

We can now convert the mass in grams of a compound to the mass of one of the elements in a compound as shown in Sample Problem 7.6.

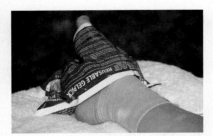

In some hot packs, heat is generated when CaCl$_2$ dissolves in water.

SAMPLE PROBLEM 7.6 Converting Grams of Compound to Grams of Element

Hot packs are used to reduce muscle aches, inflammation, and muscle spasms. A hot pack consists of a bag of water and an inner bag containing 10.2 g of CaCl$_2$. When the bag is smashed, the CaCl$_2$ dissolves in the water and heat is released. How many grams of Cl are in the CaCl$_2$ in the inner bag?

TRY IT FIRST

SOLUTION

STEP 1 State the given and needed quantities.

ANALYZE THE PROBLEM	Given	Need	Connect
	10.2 g of CaCl$_2$	grams of Cl	molar mass, subscript for Cl

STEP 2 Write a plan to convert grams of a compound to grams of an element.

grams of CaCl$_2$ | Molar mass | moles of CaCl$_2$ | Mole factor | moles of Cl

Molar mass | grams of Cl

STEP 3 Write the equalities and conversion factors for molar mass and mole factors.

$$1 \text{ mol of CaCl}_2 = 110.98 \text{ g of CaCl}_2$$
$$\frac{110.98 \text{ g CaCl}_2}{1 \text{ mol CaCl}_2} \quad \text{and} \quad \frac{1 \text{ mol CaCl}_2}{110.98 \text{ g CaCl}_2}$$

$$1 \text{ mol of CaCl}_2 = 2 \text{ mol of C}$$
$$\frac{2 \text{ mol Cl}}{1 \text{ mol CaCl}_2} \quad \text{and} \quad \frac{1 \text{ mol CaCl}_2}{2 \text{ mol Cl}}$$

$$1 \text{ mol of Cl} = 35.45 \text{ g of Cl}$$
$$\frac{35.45 \text{ g Cl}}{1 \text{ mol Cl}} \quad \text{and} \quad \frac{1 \text{ mol Cl}}{35.45 \text{ g Cl}}$$

STEP 4 Set up the problem to convert grams of a compound to grams of an element.

$$10.2 \text{ g CaCl}_2 \times \frac{1 \text{ mol CaCl}_2}{110.98 \text{ g CaCl}_2} \times \frac{2 \text{ mol Cl}}{1 \text{ mol CaCl}_2} \times \frac{35.45 \text{ g Cl}}{1 \text{ mol Cl}} = 6.52 \text{ g of Cl}$$

STUDY CHECK 7.6

Tin(II) fluoride (SnF$_2$) is added to toothpaste to strengthen tooth enamel. How many grams of tin(II) fluoride contain 1.46 g of F?

ANSWER

6.02 g of SnF$_2$

Guide to Calculating the Grams of an Element (or Compound) from the Grams of a Compound (or Element)

STEP 1
State the given and needed quantities.

STEP 2
Write a plan to convert grams of a compound (or element) to grams of an element (or compound).

STEP 3
Write the equalities and conversion factors for molar mass and mole factors.

STEP 4
Set up the problem to convert grams of a compound (or element) to grams of an element (or compound).

QUESTIONS AND PROBLEMS

7.3 Calculations Using Molar Mass

LEARNING GOAL Given the number of moles (or grams) of a substance, calculate the grams (or moles). Given the grams of a compound, calculate the grams of one of the elements.

7.21 Calculate the mass, in grams, for each of the following:
   a. 1.50 mol of Na
   b. 2.80 mol of Ca
   c. 0.125 mol of CO_2
   d. 0.0485 mol of Na_2CO_3
   e. 7.14×10^2 mol of PCl_3

7.22 Calculate the mass, in grams, for each of the following:
   a. 5.12 mol of Al
   b. 0.75 mol of Cu
   c. 3.52 mol of $MgBr_2$
   d. 0.145 mol of C_2H_6O
   e. 2.08 mol of $(NH_4)_2SO_4$

7.23 Calculate the mass, in grams, in 0.150 mol of each of the following:
   a. Ne
   b. I_2
   c. Na_2O
   d. $Ca(NO_3)_2$
   e. C_6H_{14}

7.24 Calculate the mass, in grams, in 2.28 mol of each of the following:
   a. N_2
   b. SO_3
   c. $C_3H_6O_3$
   d. $Mg(HCO_3)_2$
   e. SF_6

7.25 Calculate the number of moles in each of the following:
   a. 82.0 g of Ag
   b. 0.288 g of C
   c. 15.0 g of ammonia, NH_3
   d. 7.25 g of CH_4
   e. 245 g of Fe_2O_3

7.26 Calculate the number of moles in each of the following:
   a. 85.2 g of Ni
   b. 144 g of K
   c. 6.4 g of H_2O
   d. 308 g of $BaSO_4$
   e. 252.8 g of fructose, $C_6H_{12}O_6$

7.27 Calculate the number of moles in 25.0 g of each of the following:
   a. He
   b. O_2
   c. $Al(OH)_3$
   d. Ga_2S_3
   e. C_4H_{10}, butane

7.28 Calculate the number of moles in 4.00 g of each of the following:
   a. Au
   b. SnO_2
   c. CS_2
   d. Ca_3N_2
   e. $C_6H_8O_6$, vitamin C

7.29 Calculate the mass, in grams, of C in each of the following:
   a. 0.688 g of CO_2
   b. 275 g of C_3H_6
   c. 1.84 g of C_2H_6O
   d. 73.4 g of $C_8H_{16}O_2$

7.30 Calculate the mass, in grams, of N in each of the following:
   a. 0.82 g of $NaNO_3$
   b. 40.0 g of $(NH_4)_3P$
   c. 0.464 g of N_2H_4
   d. 1.46 g of N_4O_6

Applications

7.31 Propane gas, C_3H_8, is used as a fuel for many barbecues.
   a. How many moles of H are in 34.0 g of propane?
   b. How many grams of C are in 1.50 mol of propane?
   c. How many grams of C are in 34.0 g of propane?
   d. How many grams of H are in 0.254 g of propane?

7.32 Allyl sulfide, $(C_3H_5)_2S$, gives garlic, onions, and leeks their characteristic odor.

The characteristic odor of garlic is due to a sulfur-containing compound.

   a. How many moles of S are in 23.2 g of $(C_3H_5)_2S$?
   b. How many grams of H are in 0.75 mol of $(C_3H_5)_2S$?
   c. How many grams of C are in 4.66 g of $(C_3H_5)_2S$?
   d. How many grams of S are in 15.0 g of $(C_3H_5)_2S$?

7.33 **a.** The compound $MgSO_4$, Epsom salts, is used to soothe sore feet and muscles. How many grams will you need to prepare a bath containing 5.00 mol of Epsom salts?
   b. Potassium iodide, KI, is used as an expectorant. How many grams of KI are in 0.450 mol of KI?

7.34 **a.** Cyclopropane, C_3H_6, is an anesthetic given by inhalation. How many grams are in 0.25 mol of cyclopropane?
   b. The sedative Demerol hydrochloride has the formula $C_{15}H_{22}ClNO_2$. How many grams are in 0.025 mol of Demerol hydrochloride?

7.35 Dinitrogen oxide (or nitrous oxide), N_2O, also known as laughing gas, is widely used as an anesthetic in dentistry.
   a. How many grams of the compound are in 1.50 mol of N_2O?
   b. How many moles of the compound are in 34.0 g of N_2O?
   c. How many grams of N are in 34.0 g of N_2O?

7.36 Chloroform, $CHCl_3$, was formerly used as an anesthetic but its use was discontinued due to respiratory and cardiac failure.
   a. How many grams of the compound are in 0.122 mol of $CHCl_3$?
   b. How many moles of the compound are in 15.6 g of $CHCl_3$?
   c. How many grams of Cl are in 26.7 g of $CHCl_3$?

7.4 Mass Percent Composition

LEARNING GOAL Given the formula of a compound, calculate the mass percent composition.

Because the atoms of the elements in a compound are combined in a definite mole ratio, they are also combined in a definite proportion by mass. When we know the mass of an element in the mass of a sample of a compound, we can calculate its **mass percent composition** or **mass percent**, which is the mass of an element divided by the total mass of the compound and multiplied by 100%. For example, we can calculate the mass percent of N if we find from experiment that 7.64 g of N are present in 12.0 g of N_2O, "laughing gas," which is used as an anesthetic for surgery and in dentistry.

From the grams of N and the grams of N_2O, we calculate the mass percent of nitrogen as follows:

$$\text{Mass percent of an element} = \frac{\text{mass of an element}}{\text{total mass of the compound}} \times 100\%$$

$$\text{Mass percent of N} = \frac{7.64 \text{ g N}}{12.0 \text{ g } N_2O} \times 100\% = 63.7\% \text{ N}$$

Mass Percent Composition Using Molar Mass

CORE CHEMISTRY SKILL
Calculating Mass Percent Composition

We can also calculate mass percent composition of a compound by using its molar mass. Then the total mass of each element is divided by the molar mass of the compound and multiplied by 100%.

$$\text{Mass percent composition} = \frac{\text{mass of each element}}{\text{molar mass of the compound}} \times 100\%$$

The odor of pears is due to propyl acetate, $C_5H_{10}O_2$.

Guide to Calculating Mass Percent Composition from Molar Mass

STEP 1
Determine the total mass of each element in the molar mass of a formula.

STEP 2
Divide the total mass of each element by the molar mass and multiply by 100%.

SAMPLE PROBLEM 7.7 Calculating Mass Percent Composition from Molar Mass

The odor of pears is due to the compound propyl acetate, which has a formula of $C_5H_{10}O_2$. What is its mass percent composition?

TRY IT FIRST

SOLUTION

ANALYZE THE PROBLEM	Given	Need		Connect
	$C_5H_{10}O_2$	mass percent composition: %C, %H, %O		periodic table

STEP 1 Determine the total mass of each element in the molar mass of a formula.

$$5 \text{ mol C} \times \frac{12.01 \text{ g C}}{1 \text{ mol C}} = 60.05 \text{ g of C}$$

$$10 \text{ mol H} \times \frac{1.008 \text{ g H}}{1 \text{ mol H}} = 10.08 \text{ g of H}$$

$$2 \text{ mol O} \times \frac{16.00 \text{ g O}}{1 \text{ mol O}} = \underline{32.00 \text{ g of O}}$$

Molar mass of $C_5H_{10}O_2 = 102.13$ g of $C_5H_{10}O_2$

STEP 2 Divide the total mass of each element by the molar mass and multiply by 100%.

$$\text{Mass \% C} = \frac{60.05 \text{ g C}}{102.13 \text{ g } C_5H_{10}O_2} \times 100\% = 58.80\% \text{ C}$$

$$\text{Mass \% H} = \frac{10.08 \text{ g H}}{102.13 \text{ g } C_5H_{10}O_2} \times 100\% = 9.870\% \text{ H}$$

$$\text{Mass \% O} = \frac{32.00 \text{ g O}}{102.13 \text{ g } C_5H_{10}O_2} \times 100\% = 31.33\% \text{ O}$$

The total mass percent for all the elements in the compound should equal 100%. In some cases, because of rounding off, the sum of the mass percents may not total exactly 100%.

$$58.80\% \text{ C} + 9.870\% \text{ H} + 31.33\% \text{ O} = 100.00\%$$

STUDY CHECK 7.7

Ethylene glycol, $C_2H_6O_2$, used as automobile antifreeze, is a sweet-tasting liquid, which is toxic to humans and animals. What is the mass percent composition of ethylene glycol?

ANSWER
38.70% C; 9.744% H; 51.56% O

CHEMISTRY LINK TO THE **ENVIRONMENT**
Fertilizers

Every year in the spring, homeowners and farmers add fertilizers to the soil to produce greener lawns and larger crops. Plants require several nutrients, but the major ones are nitrogen, phosphorus, and potassium. Nitrogen promotes green growth, phosphorus promotes strong root development for strong plants and abundant flowers, and potassium helps plants defend against diseases and weather extremes. The numbers on a package of fertilizer give the percentages each of N, P, and K by mass. For example, the set of numbers 30−3−4 describes a fertilizer that contains 30% N, 3% P, and 4% K.

The major nutrient, nitrogen, is present in huge quantities as N_2 in the atmosphere, but plants cannot utilize nitrogen in this form. Bacteria in the soil convert atmospheric N_2 to usable forms by *nitrogen fixation*. To provide additional nitrogen to plants, several types of nitrogen-containing chemicals, including ammonia, nitrates, and ammonium compounds, are added to soil. The nitrates are absorbed directly, but ammonia and ammonium salts are first converted to nitrates by the soil bacteria.

The percent nitrogen depends on the type of nitrogen compound used in the fertilizer. The percent nitrogen by mass in each type is calculated using mass percent composition.

Nitrogen (N) 30%
Phosphorus (P) 3%
Potassium (K) 4%

The label on a bag of fertilizer states the percentages of N, P, and K.

The choice of a fertilizer depends on its use and convenience. A fertilizer can be prepared as crystals or a powder, in a liquid solution, or as a gas such as ammonia. The ammonia and ammonium fertilizers are water soluble and quick-acting. Other forms may be made to slow-release by enclosing water-soluble ammonium salts in a thin plastic coating. The most commonly used fertilizer is NH_4NO_3 because it is easy to apply and has a high percent of N by mass.

Type of Fertilizer	Percent Nitrogen by Mass	
NH_3	$\dfrac{14.01 \text{ g N}}{17.03 \text{ g } NH_3}$	$\times\ 100\% = 82.27\% \text{ N}$
NH_4NO_3	$\dfrac{28.02 \text{ g N}}{80.05 \text{ g } NH_4NO_3}$	$\times\ 100\% = 35.00\% \text{ N}$
$(NH_4)_2HPO_4$	$\dfrac{28.02 \text{ g N}}{132.06 \text{ g } (NH_4)_2HPO_4}$	$\times\ 100\% = 21.22\% \text{ N}$
$(NH_4)_2SO_4$	$\dfrac{28.02 \text{ g N}}{132.15 \text{ g } (NH_4)_2SO_4}$	$\times\ 100\% = 21.20\% \text{ N}$

QUESTIONS AND PROBLEMS

7.4 Mass Percent Composition

LEARNING GOAL Given the formula of a compound, calculate the mass percent composition.

7.37 Calculate the mass percent composition of each of the following:
    a. 4.68 g of Si and 5.32 g of O
    b. 5.72 g of C and 1.28 g of H
    c. 16.1 of Na, 22.5 g of S, and 11.3 g of O
    d. 6.22 g of C, 1.04 g of H, and 4.14 g of O

7.38 Calculate the mass percent composition of each of the following:
    a. 0.890 g of Ba and 1.04 g of Br
    b. 3.82 g of K and 1.18 g of I
    c. 3.29 of N, 0.946 g of H, and 3.76 g of S
    d. 4.14 g of C, 0.695 g of H, and 2.76 g of O

Applications

7.39 Calculate the mass percent composition of each of the following:
    a. MgF_2, magnesium fluoride
    b. $Ca(OH)_2$, calcium hydroxide
    c. $C_4H_8O_4$, erythrose, a carbohydrate
    d. $(NH_4)_3PO_4$, ammonium phosphate, fertilizer
    e. $C_{17}H_{19}NO_3$, morphine, a painkiller

7.40 Calculate the mass percent composition of each of the following:
    a. $CaCl_2$, calcium chloride
    b. $Na_2Cr_2O_7$, sodium dichromate
    c. $C_2H_3Cl_3$, trichloroethane, a cleaning solvent
    d. $Ca_3(PO_4)_2$, calcium phosphate, found in bone and teeth
    e. $C_{18}H_{36}O_2$, stearic acid, a fatty acid

7.41 Calculate the mass percent of N in each of the following:
    a. N_2O_5, dinitrogen pentoxide
    b. NH_4Cl, expectorant in cough medicine
    c. $C_2H_8N_2$, dimethylhydrazine, rocket fuel
    d. $C_9H_{15}N_5O$, Rogaine, stimulates hair growth
    e. $C_{14}H_{22}N_2O$, lidocaine, local anesthetic

7.42 Calculate the mass percent of S in each of the following:
    a. Na_2SO_4, sodium sulfate
    b. Al_2S_3, aluminum sulfide
    c. SO_3, sulfur trioxide
    d. C_2H_6SO, dimethylsulfoxide, topical anti-inflammatory
    e. $C_{10}H_{10}N_4O_2S$, sulfadiazine, antibacterial

7.5 Empirical Formulas

LEARNING GOAL From the mass percent composition, calculate the empirical formula for a compound.

Up to now, the formulas you have seen have been **molecular formulas**, which are the actual formulas of compounds. If we write a formula that represents the lowest whole-number ratio of the atoms in a compound, it is called the simplest or **empirical formula**. For example, the compound benzene, with molecular formula C_6H_6, has the empirical formula CH. Some molecular formulas and their empirical formulas are shown in **TABLE 7.2.**

TABLE 7.2 Examples of Molecular and Empirical Formulas

Name	Molecular (actual formula)	Empirical (simplest formula)
Acetylene	C_2H_2	CH
Benzene	C_6H_6	CH
Ammonia	NH_3	NH_3
Hydrazine	N_2H_4	NH_2
Ribose	$C_5H_{10}O_5$	CH_2O
Glucose	$C_6H_{12}O_6$	CH_2O

CORE CHEMISTRY SKILL
Calculating an Empirical Formula

The empirical formula of a compound is determined by converting the number of grams of each element to moles and finding the lowest whole-number ratio to use as subscripts, as shown in Sample Problem 7.8.

SAMPLE PROBLEM 7.8 Calculating an Empirical Formula

A compound of iron and chlorine is used to purify water in water-treatment plants. What is the empirical formula of the compound if experimental analysis shows that a sample of the compound contains 6.87 g of iron and 13.1 g of chlorine?

TRY IT FIRST

SOLUTION

ANALYZE THE PROBLEM	Given	Need	Connect
	6.87 g of Fe, 13.1 g of Cl	empirical formula	molar mass

Water-treatment plants use chemicals to purify sewage.

STEP 1 Calculate the moles of each element.

$$6.87 \text{ g Fe} \times \frac{1 \text{ mol Fe}}{55.85 \text{ g Fe}} = 0.123 \text{ mol of Fe}$$

$$13.1 \text{ g Cl} \times \frac{1 \text{ mol Cl}}{35.45 \text{ g Cl}} = 0.370 \text{ mol of Cl}$$

STEP 2 Divide by the smallest number of moles. For this problem, the smallest number of moles is 0.123.

$$\frac{0.123 \text{ mol Fe}}{0.123} = 1.00 \text{ mol of Fe}$$

$$\frac{0.370 \text{ mol Cl}}{0.123} = 3.01 \text{ mol of Cl}$$

Guide to Calculating an Empirical Formula

STEP 1
Calculate the moles of each element.

STEP 2
Divide by the smallest number of moles.

STEP 3
Use the lowest whole-number ratio of moles as subscripts.

STEP 3 Use the lowest whole-number ratio of moles as subscripts. The relationship of moles of Fe to moles of Cl is 1 to 3, which we obtain by rounding off 3.01 to 3.

$$Fe_{1.00}Cl_{3.01} \longrightarrow Fe_1Cl_3, \text{ written as } FeCl_3 \quad \text{Empirical formula}$$

STUDY CHECK 7.8

Phosphine is a highly toxic compound used for pest and rodent control. If a sample of phosphine contains 0.456 g of P and 0.0440 g of H, what is its empirical formula?

ANSWER

PH_3

Often, the mass percent of each element in a compound is given. The mass percent composition is true for any quantity of the compound. For example, methane, CH_4, always has a mass percent composition of 74.9% C and 25.1% H. Thus, if we assume that we have a sample of 100. g of the compound, we can determine the grams of each element in that 100.-g sample and use these to calculate the empirical formula as shown in Sample Problem 7.9.

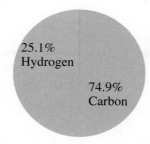

25.1%
Hydrogen

74.9%
Carbon

Mass percent composition of methane, CH_4

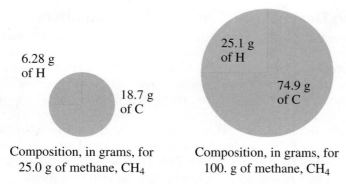

6.28 g
of H

18.7 g
of C

25.1 g
of H

74.9 g
of C

Composition, in grams, for
25.0 g of methane, CH_4

Composition, in grams, for
100. g of methane, CH_4

The mass percent composition for any amount of methane is always the same.

SAMPLE PROBLEM 7.9 Calculating an Empirical Formula from the Mass Percent Composition

Tetrachloroethene is a colorless liquid used for dry cleaning. What is the empirical formula of tetrachloroethene if its mass percent composition is 14.5% C and 85.5% Cl?

TRY IT FIRST

SOLUTION

	Given	Need	Connect
ANALYZE THE PROBLEM	14.5% C, 85.5% Cl	empirical formula	molar mass

STEP 1 Calculate the moles of each element. In a sample size of 100. g of this compound, there are 14.5 g of C and 85.5 g of Cl.

$$14.5 \text{ g C} \times \frac{1 \text{ mol C}}{12.01 \text{ g C}} = 1.21 \text{ mol of C}$$

$$85.5 \text{ g Cl} \times \frac{1 \text{ mol Cl}}{35.45 \text{ g Cl}} = 2.41 \text{ mol of Cl}$$

STEP 2 Divide by the smallest number of moles. For this problem, the smallest number of moles is 1.21.

$$\frac{1.21 \text{ mol C}}{1.21} = 1.00 \text{ mol of C}$$

$$\frac{2.41 \text{ mol Cl}}{1.21} = 1.99 \text{ mol of Cl}$$

STEP 3 Use the lowest whole-number ratio of moles as subscripts.

$$C_{1.00}Cl_{1.99} \longrightarrow C_1Cl_2 = CCl_2$$

STUDY CHECK 7.9

Sulfate of potash is the common name of a compound used in fertilizers to supply potassium and sulfur. What is the empirical formula of this compound if it has a mass percent composition of 44.9% K, 18.4% S, and 36.7% O?

ANSWER

K_2SO_4

Sulfate of potash supplies sulfur and potassium.

ENGAGE

If we obtain a ratio of 1.0 mol of P and 2.5 mol of O, why do we multiply by 2 to obtain an empirical formula of P_2O_5?

Converting Decimal Numbers to Whole Numbers

Sometimes the result of dividing by the smallest number of moles gives a decimal instead of a whole number. Decimal values that are very close to whole numbers can be rounded off. For example, 2.04 rounds off to 2 and 6.98 rounds off to 7. *However, a decimal that is greater than 0.1 or less than 0.9 should not be rounded off.* Instead, we multiply by a small integer to obtain a whole number. Some multipliers that are typically used are listed in **TABLE 7.3**.

Let us suppose the numbers of moles we obtain give subscripts in the ratio of $C_{1.00}H_{2.33}O_{0.99}$. While 0.99 rounds off to 1, we cannot round off 2.33. If we multiply 2.33 × 2, we obtain 4.66, which is still not a whole number. If we multiply 2.33 by 3, the answer is 6.99, which rounds off to 7. To complete the empirical formula, *all the other subscripts must be multiplied by 3.*

$$C_{(1.00 \times 3)}H_{(2.33 \times 3)}O_{(0.99 \times 3)} = C_{3.00}H_{6.99}O_{2.97} \longrightarrow C_3H_7O_3$$

TABLE 7.3 Some Multipliers That Convert Decimals to Whole-Number Subscripts

Decimal	Multiplier	Example		Whole Number
0.20	5	1.20 × 5	=	6
0.25	4	2.25 × 4	=	9
0.33	3	1.33 × 3	=	4
0.50	2	2.50 × 2	=	5
0.67	3	1.67 × 3	=	5

SAMPLE PROBLEM 7.10 Calculating an Empirical Formula Using Multipliers

Ascorbic acid (vitamin C), found in citrus fruits and vegetables, is important in metabolic reactions in the body, in the synthesis of collagen, and in the prevention of scurvy. If the mass percent composition of ascorbic acid is 40.9% C, 4.58% H, and 54.5% O, what is the empirical formula of ascorbic acid?

Citrus fruits are a good source of vitamin C.

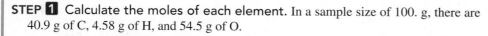

SOLUTION

	Given	Need	Connect
ANALYZE THE PROBLEM	40.9% C, 4.58% H, 54.5% O	empirical formula	molar mass

STEP 1 Calculate the moles of each element. In a sample size of 100. g, there are 40.9 g of C, 4.58 g of H, and 54.5 g of O.

$$40.9 \ \cancel{g \ C} \times \frac{1 \ \text{mol C}}{12.01 \ \cancel{g \ C}} = 3.41 \ \text{mol of C}$$

$$4.58 \ \cancel{g \ H} \times \frac{1 \ \text{mol H}}{1.008 \ \cancel{g \ H}} = 4.54 \ \text{mol of H}$$

$$54.5 \ \cancel{g \ O} \times \frac{1 \ \text{mol O}}{16.00 \ \cancel{g \ O}} = 3.41 \ \text{mol of O}$$

STEP 2 Divide by the smallest number of moles. For this problem, the smallest number of moles is 3.41.

$$\frac{3.41 \ \text{mol C}}{3.41} = 1.00 \ \text{mol of C}$$

$$\frac{4.54 \ \text{mol H}}{3.41} = 1.33 \ \text{mol of H}$$

$$\frac{3.41 \ \text{mol O}}{3.41} = 1.00 \ \text{mol of O}$$

STEP 3 Use the lowest whole-number ratio of moles as subscripts. As calculated, the ratio of moles gives the formula

$$C_{1.00}H_{1.33}O_{1.00}$$

Because the subscript for H has a decimal that is greater than 0.1 and less than 0.9, it should not be rounded off. Then, we multiply each of the subscripts by 3 to obtain a whole number for H, which is 4. Thus, the empirical formula of ascorbic acid is $C_3H_4O_3$.

$$C_{(1.00 \times 3)}H_{(1.33 \times 3)}O_{(1.00 \times 3)} = C_{3.00}H_{3.99}O_{3.00} \longrightarrow C_3H_4O_3$$

STUDY CHECK 7.10

Glyoxylic acid is used by plants and bacteria to convert fats into glucose. What is the empirical formula of glyoxylic acid if it has a mass percent composition of 32.5% C, 2.70% H, and 64.8% O?

ANSWER

$C_2H_2O_3$

QUESTIONS AND PROBLEMS

7.5 Empirical Formulas

LEARNING GOAL From the mass percent composition, calculate the empirical formula for a compound.

7.43 Calculate the empirical formula for each of the following:
a. 3.57 g of N and 2.04 g of O
b. 7.00 g of C and 1.75 g of H
c. 0.175 g of H, 2.44 g of N, and 8.38 g of O
d. 2.06 g of Ca, 2.66 g of Cr, and 3.28 g of O

7.44 Calculate the empirical formula for each of the following:
a. 2.90 g of Ag and 0.430 g of S
b. 2.22 g of Na and 0.774 g of O
c. 2.11 g of Na, 0.0900 g of H, 2.94 g of S, and 5.86 g of O
d. 5.52 g of K, 1.45 g of P, and 3.00 g of O

7.45 Calculate the empirical formula for each of the following:
a. 70.9% K and 29.1% S
b. 55.0% Ga and 45.0% F
c. 31.0% B and 69.0% O
d. 18.8% Li, 16.3% C, and 64.9% O
e. 51.7% C, 6.95% H, and 41.3% O

7.46 Calculate the empirical formula for each of the following:
a. 55.5% Ca and 44.5% S
b. 78.3% Ba and 21.7% F
c. 76.0% Zn and 24.0% P
d. 29.1% Na, 40.6% S, and 30.3% O
e. 19.8% C, 2.20% H, and 78.0% Cl

7.6 Molecular Formulas

LEARNING GOAL Determine the molecular formula of a substance from the empirical formula and molar mass.

An empirical formula represents the lowest whole-number ratio of atoms in a compound. However, empirical formulas do not necessarily represent the actual number of atoms in a molecule. A molecular formula is related to the empirical formula by a small integer such as 1, 2, or 3.

Molecular formula = small integer × empirical formula

For example, in **TABLE 7.4**, we see several different compounds that have the same empirical formula, CH_2O. The molecular formulas are related to the empirical formulas by small whole numbers (integers). The same relationship is true for the molar mass and empirical formula mass. The molar mass of each of the different compounds is related to the mass of the empirical formula (30.03 g) by the same small integer.

TABLE 7.4 Comparing the Molar Mass of Some Compounds with the Empirical Formula of CH_2O

Compound	Empirical Formula	Molecular Formula	Molar Mass (g)	Integer × Empirical Formula	Integer × Empirical Formula Mass
Acetaldehyde	CH_2O	CH_2O	30.03	$1(CH_2O)$	1 × 30.03
Acetic acid	CH_2O	$C_2H_4O_2$	60.06	$2(CH_2O)$	2 × 30.03
Lactic acid	CH_2O	$C_3H_6O_3$	90.09	$3(CH_2O)$	3 × 30.03
Erythrose	CH_2O	$C_4H_8O_4$	120.12	$4(CH_2O)$	4 × 30.03
Ribose	CH_2O	$C_5H_{10}O_5$	150.15	$5(CH_2O)$	5 × 30.03

Relating Empirical and Molecular Formulas

Once we determine the empirical formula, we can calculate the empirical formula mass in grams. If we are given the molar mass of the compound, we can calculate the value of the small integer.

$$\text{Small integer} = \frac{\text{molar mass of compound}}{\text{empirical formula mass}}$$

For example, when the molar mass of ribose is divided by the empirical formula mass, the integer is 5.

$$\text{Small integer} = \frac{\text{molar mass of ribose}}{\text{empirical formula mass of } CH_2O} = \frac{150.15 \text{ g}}{30.03 \text{ g}} = 5$$

Multiplying the subscripts in the empirical formula (CH_2O) by 5 gives the molecular formula of ribose, $C_5H_{10}O_5$.

$$5 \times \text{empirical formula } (CH_2O) = \text{molecular formula } (C_5H_{10}O_5)$$

ENGAGE

Why do the molecular formulas $C_2H_4O_2$ and $C_5H_{10}O_5$ both have an empirical formula of CH_2O?

Calculating a Molecular Formula

Earlier, in Sample Problem 7.10, we determined that the empirical formula of ascorbic acid (vitamin C) was $C_3H_4O_3$. If the molar mass for ascorbic acid is 176.12 g, we can calculate its molecular formula as follows:

The mass of the empirical formula $C_3H_4O_3$ is obtained in the same way as molar mass.

$$\text{Empirical formula} = 3 \text{ mol of C} + 4 \text{ mol of H} + 3 \text{ mol of O}$$
$$\text{Empirical formula mass} = (3 \times 12.01 \text{ g}) + (4 \times 1.008 \text{ g}) + (3 \times 16.00 \text{ g})$$
$$= 88.06 \text{ g}$$

$$\text{Small integer} = \frac{\text{molar mass of ascorbic acid}}{\text{empirical formula mass of } C_3H_4O_3} = \frac{176.12 \text{ g}}{88.06 \text{ g}} = 2$$

Multiplying all the subscripts in the empirical formula of ascorbic acid by 2 gives its molecular formula.

$$C_{(3\times2)}H_{(4\times2)}O_{(3\times2)} = C_6H_8O_6 \quad \text{Molecular formula}$$

CORE CHEMISTRY SKILL
Calculating a Molecular Formula

Brightly colored dishes are made of melamine.

SAMPLE PROBLEM 7.11 Determination of a Molecular Formula

Melamine, which is used to make plastic items such as dishes and toys, contains 28.57% C, 4.80% H, and 66.64% N. If the experimental molar mass is 125 g, what is the molecular formula of melamine?

TRY IT FIRST

SOLUTION

	Given	Need	Connect
ANALYZE THE PROBLEM	28.57% C, 4.80% H, 66.64% N, molar mass 125 g	molecular formula	empirical formula mass

STEP 1 Obtain the empirical formula and calculate the empirical formula mass. In a sample size of 100. g of this compound, there are 28.57 g of C, 4.80 g of H, and 66.64 g of N.

$$28.57 \text{ g C} \times \frac{1 \text{ mol C}}{12.01 \text{ g C}} = 2.38 \text{ mol of C}$$

Guide to Calculating a Molecular Formula from an Empirical Formula

STEP 1
Obtain the empirical formula and calculate the empirical formula mass.

STEP 2
Divide the molar mass by the empirical formula mass to obtain a small integer.

STEP 3
Multiply the empirical formula by the small integer to obtain the molecular formula.

$$4.80 \text{ g H} \times \frac{1 \text{ mol H}}{1.008 \text{ g H}} = 4.76 \text{ mol of H}$$

$$66.64 \text{ g N} \times \frac{1 \text{ mol N}}{14.01 \text{ g N}} = 4.76 \text{ mol of N}$$

Divide the moles of each element by the smallest number of moles, 2.38, to obtain the subscripts of each element in the formula.

$$\frac{2.38 \text{ mol C}}{2.38} = 1.00 \text{ mol of C}$$

$$\frac{4.76 \text{ mol H}}{2.38} = 2.00 \text{ mol of H}$$

$$\frac{4.76 \text{ mol N}}{2.38} = 2.00 \text{ mol of N}$$

Using these values as subscripts, $C_{1.00}H_{2.00}N_{2.00}$, we write the empirical formula for melamine as CH_2N_2.

$$C_{1.00}H_{2.00}N_{2.00} = CH_2N_2 \quad \text{Empirical formula}$$

Now we calculate the molar mass for this empirical formula as follows:

Empirical formula = 1 mol of C + 2 mol of H + 2 mol of N

Empirical formula mass = $(1 \times 12.01) + (2 \times 1.008) + (2 \times 14.01) = 42.05$ g

STEP 2 Divide the molar mass by the empirical formula mass to obtain a small integer.

$$\text{Small integer} = \frac{\text{molar mass of melamine}}{\text{empirical formula mass of } CH_2N_2} = \frac{125 \text{ g}}{42.05 \text{ g}} = 2.97$$

STEP 3 Multiply the empirical formula by the small integer to obtain the molecular formula. Because the experimental molar mass is close to 3 times the empirical formula mass, the subscripts in the empirical formula are multiplied by 3 to give the molecular formula.

$$C_{(1 \times 3)}H_{(2 \times 3)}N_{(2 \times 3)} = C_3H_6N_6 \quad \text{Molecular formula}$$

STUDY CHECK 7.11

The insecticide lindane has a mass percent composition of 24.78% C, 2.08% H, and 73.14% Cl. If its experimental molar mass is 290 g, what is the molecular formula?

ANSWER

$C_6H_6Cl_6$

QUESTIONS AND PROBLEMS

7.6 Molecular Formulas

LEARNING GOAL Determine the molecular formula of a substance from the empirical formula and molar mass.

Applications

7.47 Write the empirical formula for each of the following substances:
 a. H_2O_2, peroxide
 b. $C_{18}H_{12}$, chrysene, used in the manufacture of dyes
 c. $C_{10}H_{16}O_2$, chrysanthemic acid, in pyrethrum flowers

 d. $C_9H_{18}N_6$, altretamine, an anticancer medication
 e. $C_2H_4N_2O_2$, oxamide, a fertilizer

7.48 Write the empirical formula for each of the following substances:
 a. $C_6H_6O_3$, pyrogallol, a developer in photography
 b. $C_6H_{12}O_6$, galactose, a carbohydrate
 c. $C_8H_6O_4$, terephthalic acid, used in the manufacture of plastic bottles
 d. C_6Cl_6, hexachlorobenzene, a fungicide
 e. $C_{24}H_{16}O_{12}$, laccaic acid, a crimson dye

7.49 The carbohydrate fructose found in honey and fruits has an empirical formula of CH_2O. If the experimental molar mass of fructose is 180 g, what is its molecular formula?

7.50 Caffeine has an empirical formula of $C_4H_5N_2O$. If it has an experimental molar mass of 194 g, what is the molecular formula of caffeine?

Coffee beans are a source of caffeine.

7.51 Benzene and acetylene have the same empirical formula, CH. However, benzene has an experimental molar mass of 78 g, and acetylene has an experimental molar mass of 26 g. What are the molecular formulas of benzene and acetylene?

7.52 Glyoxal, used in textiles; maleic acid, used to retard oxidation of fats and oils; and aconitic acid, a plasticizer, all have the same empirical formula, CHO. However, the experimental molar masses are glyoxal 58 g, maleic acid 117 g, and aconitic acid 174 g. What are the molecular formulas of glyoxal, maleic acid, and aconitic acid?

7.53 Mevalonic acid is involved in the biosynthesis of cholesterol. Mevalonic acid contains 48.64% C, 8.16% H, and 43.20% O. If mevalonic acid has an experimental molar mass of 148 g, what is its molecular formula?

7.54 Chloral hydrate, a sedative, contains 14.52% C, 1.83% H, 64.30% Cl, and 19.35% O. If it has an experimental molar mass of 165 g, what is the molecular formula of chloral hydrate?

7.55 Vanillic acid contains 57.14% C, 4.80% H, and 38.06% O, and has an experimental molar mass of 168 g. What is the molecular formula of vanillic acid?

7.56 Lactic acid, the substance that builds up in muscles during exercise, has a mass percent composition of 40.0% C, 6.71% H, and 53.3% O, and an experimental molar mass of 90. g. What is the molecular formula of lactic acid?

7.57 A sample of nicotine, a poisonous compound found in tobacco leaves, contains 74.0% C, 8.70% H, and 17.3% N. If the experimental molar mass of nicotine is 162 g, what is its molecular formula?

7.58 Adenine, a nitrogen-containing compound found in DNA and RNA, contains 44.5% C, 3.70% H, and 51.8% N. If adenine has an experimental molar mass of 135 g, what is its molecular formula?

Follow Up

TWO PRESCRIPTIONS FOR MAX

After completing the examination and lab tests, Chris, the veterinarian, prescribed two drugs for Max: Clavamox and Proin.

Applications

7.59 Clavamox, a broad-spectrum antibiotic, contains clavulanic acid, which has a molecular formula of $C_8H_9NO_5$.
a. What is the molar mass of clavulanic acid?
b. What is the mass percent of C in clavulanic acid?
c. Max weighs 29 kg. If the dose of clavulanic acid is 2.5 mg/kg, how many moles of clavulanic acid were given?

7.60 Proin is used to treat urinary tract infections. The molecular formula of Proin is $C_9H_{13}NO$.
a. What is the molar mass of Proin?
b. What is the mass percent of N in Proin?
c. Max weighs 29 kg. If the dose of Proin is 2.0 mg/kg, how many moles of Proin were given?

CONCEPT MAP

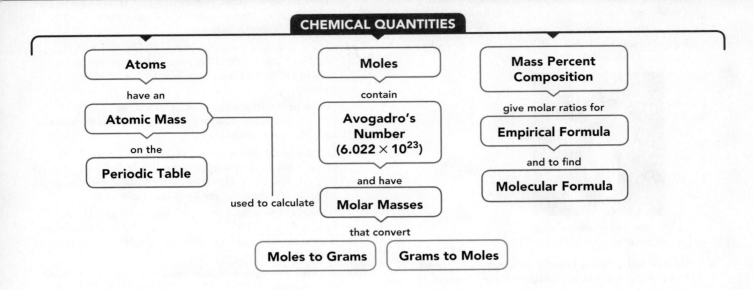

CHEMICAL QUANTITIES

- **Atoms**
  - have an
  - **Atomic Mass**
    - on the
    - **Periodic Table**

- **Moles**
  - contain
  - **Avogadro's Number (6.022×10^{23})**
    - and have
    - **Molar Masses**
      - that convert
      - **Moles to Grams**   **Grams to Moles**

- **Mass Percent Composition**
  - give molar ratios for
  - **Empirical Formula**
    - and to find
    - **Molecular Formula**

used to calculate

CHAPTER REVIEW

7.1 The Mole

LEARNING GOAL Use Avogadro's number to calculate the number of particles in a given number of moles. Calculate the number of moles of an element in a given number of moles of a compound.

- One mole of an element contains 6.022×10^{23} atoms.
- One mole of a compound contains 6.022×10^{23} molecules or formula units.

7.2 Molar Mass

LEARNING GOAL Given the chemical formula of a substance, calculate its molar mass.

- The molar mass (g/mol) of a substance is the mass in grams equal numerically to its atomic mass, or the sum of the atomic masses, which have been multiplied by their subscripts in a formula.

7.3 Calculations Using Molar Mass

LEARNING GOAL Given the number of moles (or grams) of a substance, calculate the grams (or moles). Given the grams of a compound, calculate the grams of one of the elements.

- The molar mass is used as a conversion factor to change a quantity from grams to moles, or from moles to grams.

7.4 Mass Percent Composition

LEARNING GOAL Given the formula of a compound, calculate the mass percent composition.

- The mass percent composition is obtained by dividing the mass in grams of each element in a compound by the mass of that compound.

7.5 Empirical Formulas

LEARNING GOAL From the mass percent composition, calculate the empirical formula for a compound.

- The empirical formula is calculated by determining the lowest whole-number mole ratio from the grams of the elements present in a sample.
- If mole ratios for an empirical formula are not all whole numbers, multiply all values by an integer to give whole numbers.

7.6 Molecular Formulas

LEARNING GOAL Determine the molecular formula of a substance from the empirical formula and molar mass.

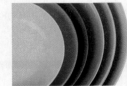

- A molecular formula is equal to, or a multiple of, the empirical formula.
- The experimental molar mass, which must be known, is divided by the mass of the empirical formula to obtain the small integer used to convert the empirical formula to the molecular formula.

KEY TERMS

Avogadro's number The number of items in a mole, equal to 6.022×10^{23}.

empirical formula The simplest or smallest whole-number ratio of the atoms in a formula.

formula unit The group of ions represented by the formula of an ionic compound.

mass percent composition The percent by mass of the elements in a formula.

molar mass The mass in grams of 1 mol of an element is equal numerically to its atomic mass. The molar mass of a compound is equal to the sum of the masses of the elements in the formula.

mole A group of atoms, molecules, or formula units that contains 6.022×10^{23} of these items.

molecular formula The actual formula that gives the number of atoms of each type of element in the compound.

CORE CHEMISTRY SKILLS

The chapter section containing each Core Chemistry Skill is shown in parentheses at the end of each heading.

Converting Particles to Moles (7.1)

- In chemistry, atoms, molecules, and ions are counted by the mole (abbreviated *mol* in calculations), a unit that contains 6.022×10^{23} items, which is Avogadro's number.
- For example, 1 mol of carbon contains 6.022×10^{23} atoms of carbon and 1 mol of H_2O contains 6.022×10^{23} molecules of H_2O.
- Avogadro's number is used to convert between particles and moles.

Example: How many moles of nickel contain 2.45×10^{24} Ni atoms?

Answer: $2.45 \times 10^{24} \text{ Ni atoms} \times \dfrac{1 \text{ mol Ni}}{6.022 \times 10^{23} \text{ Ni atoms}}$

$= 4.07$ mol of Ni

Calculating Molar Mass (7.2)

- The molar mass of a compound is the sum of the molar mass of each element in its chemical formula multiplied by its subscript in the formula.

Example: Calculate the molar mass for pinene, $C_{10}H_{16}$, a component of pine tree sap.

Pinene is a component of pine sap.

Answer: $10 \text{ mol C} \times \dfrac{12.01 \text{ g C}}{1 \text{ mol C}} = 120.1$ g of C

$16 \text{ mol H} \times \dfrac{1.008 \text{ g H}}{1 \text{ mol H}} = \underline{16.13 \text{ g of H}}$

Molar mass of $C_{10}H_{16}$ $= 136.2$ g

Using Molar Mass as a Conversion Factor (7.3)

- Molar mass is used as a conversion factor to convert between the moles and grams of a substance.

Example: The frame of a bicycle contains 6500 g of aluminum. How many moles of aluminum are in the bicycle frame?

A bicycle with an aluminum frame.

Answer: Equality: 1 mol of Al = 26.98 g of Al

Conversion Factors: $\dfrac{26.98 \text{ g Al}}{1 \text{ mol Al}}$ and $\dfrac{1 \text{ mol Al}}{26.98 \text{ g Al}}$

$6500 \text{ g Al} \times \dfrac{1 \text{ mol Al}}{26.98 \text{ g Al}} = 240$ mol of Al

Calculating Mass Percent Composition (7.4)

- The mass of a compound contains a definite proportion by mass of its elements.
- The mass percent of an element in a compound is calculated by dividing the mass of that element by the mass of the compound.

$$\text{Mass percent of an element} = \frac{\text{mass of an element}}{\text{mass of the compound}} \times 100\%$$

Example: Dinitrogen tetroxide, N_2O_4, is used in liquid fuels for rockets. If it has a molar mass of 92.02 g, what is the mass percent of nitrogen?

Answer: Mass % N $= \dfrac{28.02 \text{ g N}}{92.02 \text{ g } N_2O_4} \times 100\% = 30.45\%$ N

Calculating an Empirical Formula (7.5)

- The empirical formula or simplest formula represents the lowest whole-number ratio of the atoms and therefore moles of elements in a compound.
- For example, dinitrogen tetroxide, N_2O_4, has the empirical formula NO_2.
- To calculate the empirical formula, the grams of each element are converted to moles and divided by the smallest number of moles to obtain the lowest whole-number ratio.

Example: Calculate the empirical formula for a compound that contains 3.28 g of Cr and 6.72 g of Cl.

Answer: Convert the grams of each element to moles.

$$3.28 \text{ g Cr} \times \frac{1 \text{ mol Cr}}{52.00 \text{ g Cr}} = 0.0631 \text{ mol of Cr}$$

$$6.72 \text{ g Cl} \times \frac{1 \text{ mol Cl}}{35.45 \text{ g Cl}} = 0.190 \text{ mol of Cl}$$

Divide by the smallest number of moles (0.0631) to obtain the empirical formula.

$$\frac{0.0631 \text{ mol Cr}}{0.0631} = 1.00 \text{ mol of Cr}$$

$$\frac{0.190 \text{ mol Cl}}{0.0631} = 3.01 \text{ mol of Cl}$$

Write the empirical formula using the whole-number ratios of moles.

$$Cr_{1.00}Cl_{3.01} \longrightarrow CrCl_3$$

Calculating a Molecular Formula (7.6)

• A molecular formula is related to the empirical formula by a small whole number (integer) such as 1, 2, or 3.

Molecular formula = small integer × empirical formula

• If the empirical formula mass and the molar mass are known for a compound, an integer can be calculated by dividing the molar mass by the empirical formula mass.

$$\text{Small integer} = \frac{\text{molar mass of compound}}{\text{empirical formula mass}}$$

Example: Cymene, a component in oil of thyme, has an empirical formula C_5H_7 and an experimental molar mass of 135 g. What is the molecular formula of cymene?

Answer: Empirical formula = 5 mol of C + 7 mol of H

Empirical formula mass = $(5 \times 12.01 \text{ g}) + (7 \times 1.008 \text{ g})$

$$= 67.11 \text{ g}$$

$$\frac{\text{Molar mass of cymene}}{\text{Empirical formula mass of } C_5H_7} = \frac{135 \text{ g}}{67.11 \text{ g}}$$

$$= 2.01 \text{ (round off to 2)}$$

The molecular formula of cymene is calculated by multiplying each of the subscripts in the empirical formula by 2.

$$C_{(5 \times 2)}H_{(7 \times 2)} = C_{10}H_{14}$$

UNDERSTANDING THE CONCEPTS

The chapter sections to review are shown in parentheses at the end of each question.

7.61 A dandruff shampoo contains dipyrithione, $C_{10}H_8N_2O_2S_2$, an antibacterial and antifungal agent. (7.1, 7.2, 7.3, 7.4, 7.5)

Dandruff shampoo contains dipyrithione.

a. What is the empirical formula of dipyrithione?
b. What is the molar mass of dipyrithione?
c. What is the mass percent of O in dipyrithione?
d. How many grams of C are in 25.0 g of dipyrithione?
e. How many moles of dipyrithione are in 25.0 g of dipyrithione?

7.62 Ibuprofen, the anti-inflammatory drug in Advil, has the formula $C_{13}H_{18}O_2$. (7.1, 7.2, 7.3, 7.4)

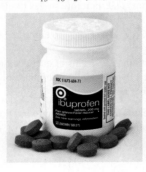

Ibuprofen is an anti-inflammatory drug.

a. What is the empirical formula of ibuprofen?
b. What is the molar mass of ibuprofen?
c. What is the mass percent of O in ibuprofen?
d. How many grams of C are in 0.425 g of ibuprofen?
e. How many moles of ibuprofen are in 2.45 g of ibuprofen?

7.63 Using the models of the molecules (black = C, white = H, yellow = S, green = Cl), determine each of the following for models of compounds **1** and **2**: (7.2, 7.4, 7.5, 7.6)

1.    **2.**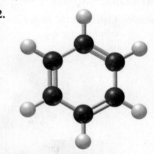

a. molecular formula   **b.** empirical formula
c. molar mass   **d.** mass percent composition

7.64 Using the models of the molecules (black = C, white = H, yellow = S, red = O), determine each of the following for models of compounds **1** and **2**: (7.2, 7.4, 7.5, 7.6)

1.   **2.**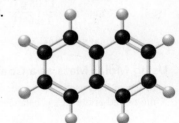

a. molecular formula   **b.** empirical formula
c. molar mass   **d.** mass percent composition

ADDITIONAL QUESTIONS AND PROBLEMS

7.65 Calculate the molar mass for each of the following: (7.2)
  a. $ZnSO_4$, zinc sulfate, zinc supplement
  b. $Ca(IO_3)_2$, calcium iodate, iodine source in table salt
  c. $C_5H_8NNaO_4$, monosodium glutamate, flavor enhancer
  d. $C_6H_{12}O_2$, isoamyl formate, used to make artificial fruit syrups

7.66 Calculate the molar mass for each of the following: (7.2)
  a. $MgCO_3$, magnesium carbonate, an antacid
  b. $Au(OH)_3$, gold(III) hydroxide, used in gold plating
  c. $C_{18}H_{34}O_2$, oleic acid, from olive oil
  d. $C_{21}H_{26}O_5$, prednisone, anti-inflammatory

7.67 Calculate the mass, in grams, of O in each of the following: (7.3)
  a. 50.0 g of H_2O
  b. 17.5 g of CO_2
  c. 48 g of $C_7H_6O_2$

7.68 Calculate the mass, in grams, of Cu in each of the following: (7.3)
  a. 4.92 g of $CuCO_3$
  b. 0.654 g of Cu_2S
  c. 89 g of $Cu_3(PO_4)_2$

7.69 Calculate the mass percent composition for each of the following compounds: (7.4)
  a. 3.85 g of Ca and 3.65 g of F
  b. 0.389 g of Na and 0.271 g of O
  c. 12.4 of K, 17.4 g of Mn, and 20.3 g of O

7.70 Calculate the mass percent composition for each of the following compounds: (7.4)
  a. 0.457 g of C and 0.043 g of H
  b. 3.65 g of Na, 2.54 g of S, and 3.81 g of O
  c. 0.907 g of Na, 1.40 g of Cl, and 1.89 g of O

7.71 Calculate the mass percent composition for each of the following compounds: (7.4)
  a. K_2CrO_4
  b. $Al(HCO_3)_3$
  c. $C_6H_{12}O_6$

7.72 Calculate the mass percent composition for each of the following compounds: (7.4)
  a. $CaCO_3$
  b. $NaC_2H_3O_2$
  c. $Ba(NO_3)_2$

7.73 Aspirin, $C_9H_8O_4$, is used to reduce inflammation and reduce fever. (7.1, 7.2, 7.3, 7.4)
  a. What is the mass percent composition of aspirin?
  b. How many moles of aspirin contain 5.0×10^{24} atoms of C?
  c. How many grams of O are in 7.50 g of aspirin?
  d. How many molecules of aspirin contain 2.50 g of H?

7.74 Ammonium sulfate, $(NH_4)_2SO_4$, is used in fertilizers. (7.1, 7.2, 7.3, 7.4)
  a. What is the mass percent composition of $(NH_4)_2SO_4$?
  b. How many atoms of H are in 0.75 mol of $(NH_4)_2SO_4$?
  c. How many grams of O are in 4.50×10^{23} formula units of $(NH_4)_2SO_4$?
  d. What mass of $(NH_4)_2SO_4$ contains 2.50 g of S?

7.75 A mixture contains 0.250 mol of Mn_2O_3 and 20.0 g of MnO_2. (7.1, 7.2, 7.3)
  a. How many moles of O are present in the mixture?
  b. How many grams of Mn are in the mixture?

7.76 A mixture contains 4.00×10^{23} molecules of PCl_3 and 0.250 mol of PCl_5. (7.1, 7.2, 7.3)
  a. How many grams of Cl are present in the mixture?
  b. How many moles of P are in the mixture?

7.77 Write the empirical formula for each of the following: (7.5)
  a. $C_5H_5N_5$, adenine, a nitrogen compound in RNA and DNA
  b. FeC_2O_4, iron(II) oxalate, a photographic developer
  c. $C_{16}H_{16}N_4$, stilbamidine, an antibiotic for animals
  d. $C_6H_{14}N_2O_2$, lysine, an amino acid needed for growth

7.78 Write the empirical formula for each of the following: (7.5)
  a. $C_{12}H_{24}N_2O_4$, carisoprodol, a skeletal muscle relaxant
  b. $C_{10}H_{10}O_5$, opianic acid, used to synthesize a drug to treat tuberculosis
  c. $CrCl_3$, chromium(III) chloride, used in chrome plating
  d. $C_{16}H_{16}N_2O_2$, lysergic acid, a controlled substance from ergot

7.79 Calculate the empirical formula for each of the following compounds: (7.5)
  a. 2.20 g of S and 7.81 g of F
  b. 6.35 g of Ag, 0.825 g of N, and 2.83 g of O
  c. 43.6% P and 56.4% O
  d. 22.1% Al, 25.4% P, and 52.5% O

7.80 Calculate the empirical formula for each of the following compounds: (7.5)
  a. 5.13 g of Cr and 2.37 g of O
  b. 2.82 g of K, 0.870 g of C, and 2.31 g of O
  c. 61.0% Sn and 39.0% F
  d. 25.9% N and 74.1% O

7.81 Oleic acid, a component of olive oil, is 76.54% C, 12.13% H, and 11.33% O. The experimental value of the molar mass is 282 g. (7.1, 7.2, 7.3, 7.4, 7.5, 7.6)
  a. What is the molecular formula of oleic acid?
  b. If oleic acid has a density of 0.895 g/mL, how many molecules of oleic acid are in 3.00 mL of oleic acid?

7.82 Iron pyrite, commonly known as "fool's gold," is 46.5% Fe and 53.5% S. (7.1, 7.2, 7.3, 7.4, 7.5, 7.6)

Iron pyrite is commonly known as "fool's gold."

  a. If the empirical formula and the molecular formula are the same, what is the molecular formula of the compound?
  b. If the crystal contains 4.85 g of iron, how many grams of S are in the crystal?

7.83 Succinic acid is 40.7% C, 5.12% H, and 54.2% O. If it has an experimental molar mass of 118 g, what are the empirical and molecular formulas? (7.4, 7.5, 7.6)

7.84 A compound is 70.6% Hg, 12.5% Cl, and 16.9% O. If it has an experimental molar mass of 568 g, what are the empirical and molecular formulas? (7.4, 7.5, 7.6)

7.85 A sample of a compound contains 1.65×10^{23} atoms of C, 0.552 g of H, and 4.39 g of O. If 1 mol of the compound contains 4 mol of O, what is the molecular formula and molar mass of the compound? (7.4, 7.5, 7.6)

7.86 What is the molecular formula of a compound if 0.500 mol of the compound contains 0.500 mol of Sr, 1.81×10^{24} atoms of O, and 35.5 g of Cl? (7.4, 7.5, 7.6)

CHALLENGE QUESTIONS

The following groups of questions are related to the topics in this chapter. However, they do not all follow the chapter order, and they require you to combine concepts and skills from several sections. These questions will help you increase your critical thinking skills and prepare for your next exam.

7.87 A toothpaste contains 0.240% by mass sodium fluoride used to prevent tooth decay and 0.30% by mass triclosan, $C_{12}H_7Cl_3O_2$, a preservative and antigingivitis agent. One tube contains 119 g of toothpaste. (7.1, 7.2, 7.3, 7.4)

Components in toothpaste include triclosan and NaF.

a. How many moles of NaF are in the tube of toothpaste?
b. How many fluoride ions, F^-, are in the tube of toothpaste?
c. How many grams of sodium ion, Na^+, are in 1.50 g of toothpaste?
d. How many molecules of triclosan are in the tube of toothpaste?
e. What is the mass percent composition of triclosan?

7.88 Sorbic acid, an inhibitor of mold in cheese, has a mass percent composition of 64.27% C, 7.19% H, and 28.54% O. If sorbic acid has an experimental molar mass of 112 g, what is its molecular formula? (7.4, 7.5, 7.6)

Cheese contains sorbic acid, which prevents the growth of mold.

7.89 Iron(III) chromate, a yellow powder used as a pigment in paints, contains 24.3% Fe, 33.9% Cr, and 41.8% O. If it has an experimental molar mass of 460 g, what are its empirical and molecular formulas? (7.4, 7.5, 7.6)

Iron(III) chromate is a yellow pigment used in paints.

7.90 A gold bar is 5.50 cm long, 3.10 cm wide, and 0.300 cm thick. (7.1, 7.2, 7.3, 7.4, 7.5, 7.6)

A gold bar consists of gold atoms.

a. If gold has a density of 19.3 g/cm^3, what is the mass of the gold bar?
b. How many atoms of gold are in the bar?
c. When the same mass of gold combines with oxygen, the oxide product has a mass of 111 g. How many moles of O are combined with the gold?
d. What is the molecular formula of the oxide product if it is the same as the empirical formula?

ANSWERS

Answers to Selected Questions and Problems

7.1 One mole contains 6.022×10^{23} atoms of an element, molecules of a molecular substance, or formula units of an ionic substance.

7.3 a. 3.01×10^{23} atoms of C
b. 7.71×10^{23} molecules of SO_2
c. 0.0867 mol of Fe
d. 14.1 mol of C_2H_6O

7.5 a. 6.00 mol of H
b. 8.00 mol of O
c. 1.20×10^{24} atoms of P
d. 4.82×10^{24} atoms of O

7.7 a. 36 mol of H
b. 1.0×10^2 mol of C
c. 0.040 mol of N

7.9 a. 32.2 mol of C
b. 6.22 mol of H
c. 0.230 mol of O

7.11 a. 70.90 g
b. 90.08 g
c. 262.9 g

7.13 a. 83.98 g
b. 98.95 g
c. 156.7 g

7.15 a. 58.44 g
b. 159.7 g
c. 329.4 g

7.17 a. 342.2 g
b. 188.18 g
c. 365.5 g

7.19 a. 151.16 g
b. 498.4 g
c. 331.3 g

7.21 a. 34.5 g
b. 112 g
c. 5.50 g
d. 5.14 g
e. 9.80×10^4 g

7.23 a. 3.03 g
b. 38.1 g
c. 9.30 g
d. 24.6 g
e. 12.9 g

7.25 a. 0.760 mol of Ag
b. 0.0240 mol of C
c. 0.881 mol of NH_3
d. 0.452 mol of CH_4
e. 1.53 mol of Fe_2O_3

7.27 a. 6.25 mol of He
b. 0.781 mol of O_2
c. 0.321 mol of $Al(OH)_3$
d. 0.106 mol of Ga_2S_3
e. 0.430 mol of C_4H_{10}

7.29 a. 0.188 g of C
b. 235 g of C
c. 0.959 g of C
d. 48.9 g of C

7.31 a. 6.17 mol of H
b. 54.0 g of C
c. 27.8 g of C
d. 0.0465 g or 4.65×10^{-2} g of H

7.33 a. 602 g   **b.** 74.7 g

7.35 a. 66.0 g of N_2O **b.** 0.772 mol of N_2O
     c. 21.6 g of N

7.37 a. 46.8% Si; 53.2% O
     b. 81.7% C; 18.3% H
     c. 32.3% Na; 45.1% S; 22.6% O
     d. 54.6% C; 9.12% H; 36.3% O

7.39 a. 39.01% Mg; 60.99% F
     b. 54.09% Ca; 43.18% O; 2.721% H
     c. 40.00% C; 6.714% H; 53.29% O
     d. 28.19% N; 8.115% H; 20.77% P; 42.92% O
     e. 71.55% C; 6.710% H; 4.909% N; 16.82% O

7.41 a. 25.94% N   **b.** 26.19% N
     c. 46.62% N   **d.** 33.47% N
     e. 11.96% N

7.43 a. N_2O **b.** CH_3
     c. HNO_3 **d.** $CaCrO_4$

7.45 a. K_2S **b.** GaF_3
     c. B_2O_3 **d.** Li_2CO_3
     e. $C_5H_8O_3$

7.47 a. HO   **b.** C_3H_2
     c. C_5H_8O **d.** $C_3H_6N_2$
     e. CH_2NO

7.49 $C_6H_{12}O_6$

7.51 benzene C_6H_6; acetylene C_2H_2

7.53 $C_6H_{12}O_4$

7.55 $C_8H_8O_4$

7.57 $C_{10}H_{14}N_2$

7.59 a. 199.16 g   **b.** 48.24% C
     c. 3.6×10^{-4} mol

7.61 a. C_5H_4NOS
     b. 252.3 g
     c. 12.68% O
     d. 11.9 g of C
     e. 0.0991 mol of dipyrithione

7.63 1. a. S_2Cl_2 **b.** SCl
       c. 135.04 g   **d.** 47.50% S; 52.50% Cl
   2. a. C_6H_6 **b.** CH
       c. 78.11 g   **d.** 92.25% C; 7.743% H

7.65 a. 161.48 g   **b.** 389.9 g
     c. 169.11 g   **d.** 116.16 g

7.67 a. 44.4 g   **b.** 12.7 g   **c.** 13 g

7.69 a. 51.3% Ca; 48.7% F
     b. 58.9% Na; 41.1% O
     c. 24.8% K; 34.7% Mn; 40.5% O

7.71 a. 40.27% K; 26.78% Cr; 32.96% O
     b. 12.85% Al; 1.440% H; 17.16% C; 68.57% O
     c. 40.00% C; 6.716% H; 53.29% O

7.73 a. 59.99% C; 4.475% H; 35.52% O
     b. 0.92 mol of aspirin
     c. 2.66 g of O
     d. 1.87×10^{23} molecules of aspirin

7.75 a. 1.210 mol of O
     b. 40.1 g of Mn

7.77 a. CHN   **b.** FeC_2O_4
     c. C_4H_4N **d.** C_3H_7NO

7.79 a. SF_6 **b.** $AgNO_3$
     c. P_2O_5 **d.** $AlPO_4$

7.81 a. $C_{18}H_{34}O_2$
     b. 5.72×10^{21} molecules of oleic acid

7.83 The empirical formula is $C_2H_3O_2$; the molecular formula is $C_4H_6O_4$.

7.85 The molecular formula is $C_4H_8O_4$; the molar mass is 120.10 g.

7.87 a. 0.00680 mol of NaF
     b. 4.10×10^{21} F^- ions
     c. 0.00197 g of Na^+ ions
     d. 7.4×10^{20} molecules of triclosan
     e. 49.76% C; 2.436% H; 36.74% Cl; 11.05% O

7.89 The empirical formula is $Fe_2Cr_3O_{12}$; the molecular formula is $Fe_2Cr_3O_{12}$.

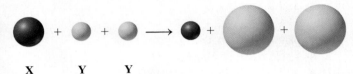

CI.7 For parts **a** to **f**, consider the loss of electrons by atoms of the element X, and a gain of electrons by atoms of the element Y. Element X is in Group 2A (2), Period 3, and Y is in Group 7A (17), Period 3. (4.2, 5.4, 5.5, 6.2, 6.3)

X   Y   Y

a. Which element is a metal, X or Y?
b. Which element is a nonmetal, X or Y?
c. What are the ionic charges of X and Y?
d. Write the electron configurations of the atoms X and Y.
e. Write the actual formula and name of the ionic compound indicated by the ions.

CI.8 A bracelet of sterling silver marked 925 contains 92.5% silver by mass and 7.5% other metals. It has a volume of 25.6 cm³ and a density of 10.2 g/cm³. (2.6, 2.8, 4.4, 7.1, 7.5, 7.6)

Silver is a shiny metal that tarnishes easily, is a good conductor of heat, and can be shaped into objects.

a. What is the mass, in kilograms, of the sterling silver bracelet?
b. How many atoms of silver are in the bracelet?
c. Determine the number of protons and neutrons in each of the two stable isotopes of silver:

$^{107}_{47}Ag$ and $^{109}_{47}Ag$

d. When silver combines with oxygen, a compound forms that contains 93.10% Ag by mass. What are the name and the molecular formula of the oxide product if the molecular formula is the same as the empirical formula?

CI.9 Oxalic acid, a compound found in plants and vegetables such as rhubarb, has a mass percent composition of 26.7% C, 2.24% H, and 71.1% O. Oxalic acid can interfere with respiration and cause kidney or bladder stones. If a large quantity of rhubarb leaves is ingested, the oxalic acid can be toxic. The lethal dose (LD$_{50}$) in rats for oxalic acid is 375 mg/kg. Rhubarb leaves contain about 0.5% by mass of oxalic acid. (2.7, 7.4, 7.5, 7.6)

Rhubarb leaves are a source of oxalic acid.

a. What is the empirical formula of oxalic acid?
b. If oxalic acid has an experimental molar mass of 90. g, what is its molecular formula?
c. Using the LD$_{50}$, how many grams of oxalic acid would be toxic for a 160-lb person?
d. How many kilograms of rhubarb leaves would the person in part **c** need to eat to reach the toxic level of oxalic acid?

CI.10 The active ingredient in Tums is calcium carbonate. One Tums tablet contains 500. mg of calcium carbonate. (2.7, 6.3, 7. 2, 7.3 7.4)

The active ingredient in Tums neutralizes excess stomach acid.

a. What is the chemical formula of calcium carbonate?
b. What is the molar mass of calcium carbonate?
c. How many moles of calcium carbonate are in one roll of Tums that contains 12 tablets?
d. If a person takes two Tums tablets, how many grams of calcium are obtained?
e. If the Daily Value (DV) for Ca^{2+} to maintain bone strength in older women is 1500 mg, how many Tums tablets are needed each day?
f. What is the mass percent composition of calcium carbonate?

CI.11 Tamiflu (oseltamivir), C$_{16}$H$_{28}$N$_2$O$_4$, is a drug that is used to treat influenza. The preparation of Tamiflu begins with the extraction of shikimic acid from the seedpods of star anise. From 2.6 g of star anise, 0.13 g of shikimic acid can be obtained and used to produce one capsule containing 75 mg of Tamiflu. The usual adult dosage for treatment of influenza is two capsules of Tamiflu daily for 5 days. (2.7, 6.5, 7.2, 7.3, 7.4, 7.5, 7.6)

The spice called star anise is a plant source of shikimic acid.

Shikimic acid is the basis for the antiviral drug in Tamiflu.

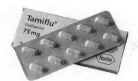

Each capsule contains
75 mg of Tamiflu.

a. What is the empirical formula of Tamiflu?
b. What is the mass percent composition of Tamiflu?
c. What is the molecular formula of shikimic acid? (Black
  spheres are carbon, white spheres are hydrogen, and red
  spheres are oxygen.)
d. How many moles of shikimic acid are contained in 1.3 g of
  shikimic acid?
e. How many capsules containing 75 mg of Tamiflu could be
  produced from 154 g of star anise?
f. How many grams of C are in one dose (75 mg)
  of Tamiflu?
g. How many kilograms of Tamiflu would be needed to
  treat all the people in a city with a population of 500 000
  people?

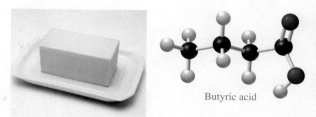

Butyric acid

Butyric acid gives the unpleasant odor to rancid butter.

CI.12 The compound butyric acid gives rancid butter its characteristic
  odor. (2.7, 2.8, 6.5, 7.1, 7.2, 7.3, 7.4, 7.5)
a. What is the molecular formula of butyric acid? (Black
  spheres are carbon, white spheres are hydrogen, and red
  spheres are oxygen.)
b. What is the empirical formula of butyric acid?
c. What is the mass percent composition of butyric acid?
d. How many grams of C are in 0.850 g of butyric acid?
e. How many grams of butyric acid contain 3.28×10^{23}
  O atoms?
f. Butyric acid has a density of 0.959 g/mL at 20 °C. How
  many moles of butyric acid are contained in 0.565 mL of
  butyric acid?

ANSWERS

CI.7 a. X is a metal; elements in Group 2A (2) are metals.
  b. Y is a nonmetal; elements in Group 7A (17) are nonmetals.
  c. X^{2+}, Y^-
  d. X = $1s^2 2s^2 2p^6 3s^2$ Y = $1s^2 2s^2 2p^6 3s^2 3p^5$
  e. $MgCl_2$, magnesium chloride

CI.9 a. CHO_2
  b. $C_2H_2O_4$
  c. 27 g of oxalic acid
  d. 5 kg of rhubarb

CI.11 a. $C_8H_{14}NO_2$
  b. 61.52% C; 9.033% H; 8.969% N; 20.49% O
  c. $C_7H_{10}O_5$
  d. 7.5×10^{-3} mol of shikimic acid
  e. 59 capsules
  f. 0.046 g of C
  g. 4×10^2 kg

8

Chemical Reactions

NATALIE WAS RECENTLY diagnosed with mild pulmonary emphysema due to secondhand cigarette smoke. She has been referred to Angela, an exercise physiologist, who begins to assess Natalie's condition by connecting her to an electrocardiogram (ECG or EKG), a pulse oximeter, and a blood pressure cuff. The ECG records the electrical activity of Natalie's heart, which is used to measure the rate and rhythm of her heartbeat, and possible presence of heart damage. The pulse oximeter measures her pulse and the saturation level of oxygen in her arterial blood (the percentage of hemoglobin that is saturated with O_2). The blood pressure cuff determines the pressure exerted by the heart in pumping her blood.

To determine possible heart disease, Natalie has an exercise stress test on a treadmill to measure how her heart rate and blood pressure respond to exertion by walking faster as the slope of the treadmill is increased. Electrical leads are attached to measure the heart rate and blood pressure first at rest and then on the treadmill. Additional equipment using a face mask collects expired air and measures Natalie's maximal volume of oxygen uptake, or $V_{O_2\,max}$.

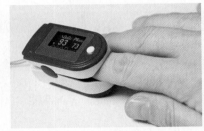

A pulse oximeter measures the pulse and the O_2 saturation in the blood.

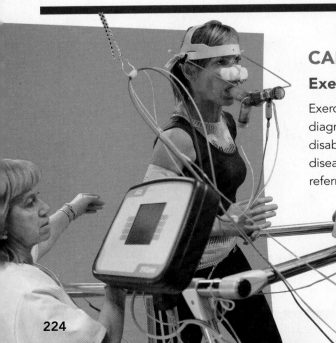

CAREER

Exercise Physiologist

Exercise physiologists work with athletes as well as patients who have been diagnosed with diabetes, heart disease, pulmonary disease, or other chronic disabilities or diseases. Patients who have been diagnosed with one of these diseases are often prescribed exercise as a form of treatment, and they are referred to an exercise physiologist. The exercise physiologist evaluates the patient's overall health and then creates a customized exercise program for that individual. The program for an athlete might focus on reducing the number of injuries, whereas a program for a cardiac patient would focus on strengthening the heart muscles. The exercise physiologist also monitors the patient for improvement and determines if the exercise is helping to reduce or reverse the progression of the disease.

CORE CHEMISTRY SKILLS

- Writing Ionic Formulas (6.2)
- Naming Ionic Compounds (6.3)

- Writing the Names and Formulas for Molecular Compounds (6.5)

*These Core Chemistry Skills from previous chapters are listed here for your review as you proceed to the new material in this chapter.

LOOKING AHEAD

8.1 Equations for Chemical Reactions

8.2 Balancing a Chemical Equation

8.3 Types of Chemical Reactions

8.4 Oxidation–Reduction Reactions

8.1 Equations for Chemical Reactions

LEARNING GOAL Identify a balanced chemical equation; determine the number of atoms in the reactants and products.

Chemical reactions occur everywhere. The fuel in our cars burns with oxygen to make the car move and run the air conditioner. When we cook our food or bleach our hair, chemical reactions take place. In our bodies, chemical reactions convert food into molecules that build muscles and move them. In the leaves of trees and plants, carbon dioxide and water are converted into carbohydrates. Some chemical reactions are simple, whereas others are quite complex. However, they can all be written with chemical equations that chemists use to describe chemical reactions. In every chemical reaction, the atoms in the reacting substances, called *reactants*, are rearranged to give new substances called *products*.

A *chemical change* occurs when a substance is converted into one or more new substances. For example, when silver tarnishes, the shiny silver metal (Ag) reacts with sulfur (S) to become the dull, black substance we call tarnish (Ag_2S) (see **FIGURE 8.1**).

A *chemical reaction* always involves chemical change because atoms of the reacting substances form new combinations with new properties. For example, a chemical reaction takes place when a piece of iron (Fe) combines with oxygen (O_2) in the air to produce a new substance, rust (Fe_2O_3), which has a reddish-brown color. During a chemical change, new properties become visible, which are an indication that a chemical reaction has taken place (see **TABLE 8.1**).

FIGURE 8.1 ▶ A chemical change produces new substances with new properties.

Ⓠ Why is the formation of tarnish a chemical change?

TABLE 8.1 Types of Evidence of a Chemical Reaction

1. Change in color

Fe Fe_2O_3

Iron nails change color when they react with oxygen to form rust.

2. Formation of a gas (bubbles)

Bubbles (gas) form when $CaCO_3$ reacts with acid.

3. Formation of a solid (precipitate)

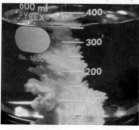

A yellow solid forms when potassium iodide is added to lead nitrate.

4. Heat (or a flame) produced or heat absorbed

Methane gas burns in the air with a hot flame.

Writing a Chemical Equation

When you build a model airplane, prepare a new recipe, or mix a medication, you follow a set of directions. These directions tell you what materials to use and the products you will obtain. In chemistry, a *chemical equation* tells us the materials we need and the products that will form.

Suppose you work in a bicycle shop, assembling wheels and frames into bicycles. You could represent this process by a simple equation:

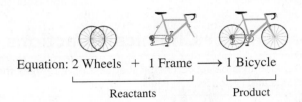

Equation: 2 Wheels + 1 Frame ⟶ 1 Bicycle

Reactants   Product

ENGAGE

What is the evidence for a chemical change in the reaction of carbon and oxygen to form carbon dioxide?

When you burn charcoal in a grill, the carbon in the charcoal combines with oxygen to form carbon dioxide. We can represent this reaction by a chemical equation.

Reactants   Product

Equation: $C(s) + O_2(g) \xrightarrow{\Delta} CO_2(g)$

TABLE 8.2 Some Symbols Used in Writing Equations

Symbol	Meaning
+	Separates two or more formulas
⟶	Reacts to form products
(s)	Solid
(l)	Liquid
(g)	Gas
(aq)	Aqueous
$\xrightarrow{\Delta}$	Reactants are heated

In a **chemical equation**, the formulas of the **reactants** are written on the left of the arrow and the formulas of the **products** on the right. The chemical equation for burning carbon is *balanced* because there is one carbon atom and two oxygen atoms in both the reactants and the products. When there are two or more formulas on the same side, they are separated by plus (+) signs.

Generally, each formula in an equation is followed by an abbreviation, in parentheses, that gives the physical state of the substance: solid (*s*), liquid (*l*), or gas (*g*). If a substance is dissolved in water, it is in an aqueous (*aq*) solution. The delta sign (Δ) indicates that heat was used to start the reaction. **TABLE 8.2** summarizes some of the symbols used in equations.

Identifying a Balanced Chemical Equation

When a chemical reaction takes place, the bonds between the atoms of the reactants are broken and new bonds are formed to give the products. All atoms are conserved, which means that atoms cannot be gained, lost, or changed into other types of atoms. Every chemical reaction must be written as a **balanced equation**, which shows the same number of atoms for each element in the reactants and in the products.

Now consider the balanced reaction in which hydrogen reacts with oxygen to form water written as follows:

$$2H_2(g) + O_2(g) \longrightarrow 2H_2O(g)$$

In the *balanced* equation, there are whole numbers called **coefficients** in front of the formulas. On the reactant side, the coefficient of 2 in front of the H_2 formula represents two molecules of hydrogen, which is 4 atoms of H. A coefficient of 1 is understood for O_2, which gives 2 atoms of O. On the product side, the coefficient of 2 in front of the H_2O formula represents 2 molecules of water. Because the coefficient of 2 multiplies all the atoms in H_2O, there are 4 hydrogen atoms and 2 oxygen atoms in the products. Because there are the same number of hydrogen atoms and oxygen atoms in the reactants as in the products, we know that the equation is *balanced*. This illustrates the *Law of Conservation of Matter*, which states that matter cannot be created or destroyed during a chemical reaction.

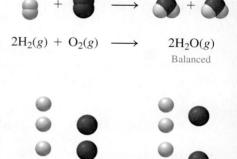

$$2H_2(g) + O_2(g) \longrightarrow 2H_2O(g)$$
Balanced

	H	O		H	O
Reactant atoms			=	Product atoms	

SAMPLE PROBLEM 8.1 **Number of Atoms in Balanced Chemical Equations**

Indicate the number of each type of atom in the following balanced chemical equation:

$$Fe_2S_3(s) + 6HCl(aq) \longrightarrow 2FeCl_3(aq) + 3H_2S(g)$$

	Reactants	Products
Fe		
S		
H		
Cl		

TRY IT FIRST

SOLUTION

The total number of atoms in each formula is obtained by multiplying the coefficient by each subscript in the chemical formula.

	Reactants	Products
Fe	2	2
S	3	3
H	6	6
Cl	6	6

STUDY CHECK 8.1

When ethane, C_2H_6, burns in oxygen, the products are carbon dioxide and water. The balanced equation is written as

$$2C_2H_6(g) + 7O_2(g) \xrightarrow{\Delta} 4CO_2(g) + 6H_2O(g)$$

Calculate the number of each type of atom in the reactants and in the products.

ANSWER

In both the reactants and products, there are 4 C atoms, 12 H atoms, and 14 O atoms.

QUESTIONS AND PROBLEMS

8.1 Equations for Chemical Reactions

LEARNING GOAL Identify a balanced chemical equation; determine the number of atoms in the reactants and products.

8.1 State the number of atoms of oxygen in the reactants and in the products for each of the following equations:
  a. $3NO_2(g) + H_2O(l) \longrightarrow NO(g) + 2HNO_3(aq)$
  b. $5C(s) + 2SO_2(g) \longrightarrow CS_2(g) + 4CO(g)$
  c. $2C_2H_2(g) + 5O_2(g) \xrightarrow{\Delta} 4CO_2(g) + 2H_2O(g)$
  d. $N_2H_4(g) + 2H_2O_2(g) \longrightarrow N_2(g) + 4H_2O(g)$

8.2 State the number of atoms of oxygen in the reactants and in the products for each of the following equations:
  a. $CH_4(g) + 2O_2(g) \xrightarrow{\Delta} CO_2(g) + 2H_2O(g)$
  b. $4P(s) + 5O_2(g) \longrightarrow P_4O_{10}(s)$
  c. $4NH_3(g) + 6NO(g) \longrightarrow 5N_2(g) + 6H_2O(g)$
  d. $6CO_2(g) + 6H_2O(l) \longrightarrow C_6H_{12}O_6(aq) + 6O_2(g)$
  Glucose

8.3 Determine whether each of the following equations is balanced or not balanced:
  a. $S(s) + O_2(g) \longrightarrow SO_3(g)$
  b. $Al(s) + Cl_2(g) \longrightarrow AlCl_3(s)$
  c. $2NaOH(aq) + H_2SO_4(aq) \longrightarrow 2H_2O(l) + Na_2SO_4(aq)$
  d. $C_3H_8(g) + 5O_2(g) \xrightarrow{\Delta} 3CO_2(g) + 4H_2O(g)$

8.4 Determine whether each of the following equations is balanced or not balanced:
  a. $PCl_3(l) + Cl_2(g) \longrightarrow PCl_5(s)$
  b. $CO(g) + 2H_2(g) \longrightarrow CH_3OH(l)$
  c. $2KClO_3(s) \longrightarrow 2KCl(s) + O_2(g)$
  d. $Mg(s) + N_2(g) \longrightarrow Mg_3N_2(s)$

8.5 All of the following are balanced equations. State the number of atoms of each element in the reactants and in the products.
  a. $2Na(s) + Cl_2(g) \longrightarrow 2NaCl(s)$
  b. $PCl_3(l) + 3H_2(g) \longrightarrow PH_3(g) + 3HCl(g)$
  c. $P_4O_{10}(s) + 6H_2O(l) \longrightarrow 4H_3PO_4(aq)$

8.6 All of the following are balanced equations. State the number of atoms of each element in the reactants and in the products.
  a. $2N_2(g) + 3O_2(g) \longrightarrow 2N_2O_3(g)$
  b. $Al_2O_3(s) + 6HCl(aq) \longrightarrow 3H_2O(l) + 2AlCl_3(aq)$
  c. $C_5H_{12}(l) + 8O_2(g) \xrightarrow{\Delta} 5CO_2(g) + 6H_2O(g)$

8.2 Balancing a Chemical Equation

CORE CHEMISTRY SKILL
Balancing a Chemical Equation

LEARNING GOAL Write a balanced chemical equation from the formulas of the reactants and products for a chemical reaction.

The chemical reaction that occurs in the flame of a gas burner you use in the laboratory or a gas cooktop is the reaction of methane gas, CH_4, and oxygen to produce carbon dioxide and water. We now show the process of writing and balancing the chemical equation for this reaction in Sample Problem 8.2.

Guide to Writing and Balancing a Chemical Equation

STEP 1
Write an equation using the correct formulas for the reactants and products.

STEP 2
Count the atoms of each element in the reactants and products.

STEP 3
Use coefficients to balance each element.

STEP 4
Check the final equation to confirm it is balanced.

SAMPLE PROBLEM 8.2 Writing and Balancing a Chemical Equation

The reaction of methane gas (CH_4) and oxygen gas (O_2) produces the gases carbon dioxide (CO_2) and water (H_2O). Write a balanced chemical equation for this reaction.

TRY IT FIRST

SOLUTION

	Given	Need	Connect
ANALYZE THE PROBLEM	reactants, products	balanced equation	equal numbers of atoms in reactants and products

STEP 1 Write an equation using the correct formulas for the reactants and products.

$$CH_4(g) + O_2(g) \xrightarrow{\Delta} CO_2(g) + H_2O(g)$$

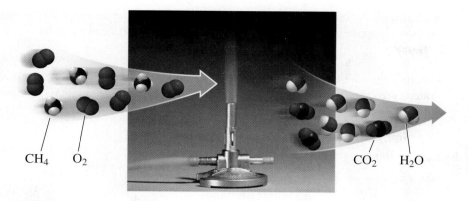

CH₄   O₂   CO₂   H₂O

STEP 2 Count the atoms of each element in the reactants and products. When we compare the atoms on the reactant side and the atoms on the product side, we see that there are more H atoms in the reactants and more O atoms in the products.

$$CH_4(g) + O_2(g) \xrightarrow{\Delta} CO_2(g) + H_2O(g)$$

Reactants	Products	
1 C atom	1 C atom	Balanced
4 H atoms	2 H atoms	Not balanced
2 O atoms	3 O atoms	Not balanced

STEP 3 Use coefficients to balance each element. We will start by balancing the H atoms in CH₄ because it has the most atoms. By placing a coefficient of 2 in front of the formula for H₂O, a total of 4 H atoms in the products is obtained. *Only use coefficients to balance an equation. Do not change any of the subscripts: This would alter the chemical formula of a reactant or product.*

$$CH_4(g) + O_2(g) \xrightarrow{\Delta} CO_2(g) + 2H_2O(g)$$

Reactants	Products	
1 C atom	1 C atom	Balanced
4 H atoms	4 H atoms	Balanced
2 O atoms	4 O atoms	Not balanced

We can balance the O atoms on the reactant side by placing a coefficient of 2 in front of the formula O₂. There are now 4 O atoms in both the reactants and products.

$$CH_4(g) + 2O_2(g) \xrightarrow{\Delta} CO_2(g) + 2H_2O(g) \quad \text{Balanced}$$

STEP 4 Check the final equation to confirm it is balanced. In the final equation, the numbers of atoms of C, H, and O are the same in both the reactants and the products. The equation is balanced.

$$CH_4(g) + 2O_2(g) \xrightarrow{\Delta} CO_2(g) + 2H_2O(g)$$

Reactants	Products	
1 C atom	1 C atom	Balanced
4 H atoms	4 H atoms	Balanced
4 O atoms	4 O atoms	Balanced

ENGAGE

How do you check that a chemical equation is balanced?

In a balanced chemical equation, the coefficients must be the *lowest possible whole numbers*. Suppose you had obtained the following for the balanced equation:

$$2CH_4(g) + 4O_2(g) \xrightarrow{\Delta} 2CO_2(g) + 4H_2O(g) \quad \text{Incorrect}$$

Although there are equal numbers of atoms on both sides of the equation, this is not written correctly. To obtain coefficients that are the lowest whole numbers, we divide all the coefficients by 2.

STUDY CHECK 8.2

Balance the following chemical equation:

$$Al(s) + Cl_2(g) \longrightarrow AlCl_3(s)$$

ANSWER

$$2Al(s) + 3Cl_2(g) \longrightarrow 2AlCl_3(s)$$

Whole-Number Coefficients

Sometimes the coefficients of the compounds in an equation need to be increased to give whole numbers for the coefficients. Then we need to adjust the coefficients and count the total atoms on both sides once again as shown in Sample Problem 8.3.

SAMPLE PROBLEM 8.3 Balancing a Chemical Equation with Whole-Number Coefficients

Acetylene, C_2H_2, is used to produce high temperatures for welding by reacting it with O_2 to produce CO_2 and H_2O. All of the compounds are gases. Write a balanced chemical equation with whole-number coefficients for this reaction.

TRY IT FIRST

SOLUTION

ANALYZE THE PROBLEM	Given	Need	Connect
	reactants, products	balanced equation	equal numbers of atoms in reactants and products

STEP 1 Write an equation using the correct formulas of the reactants and products.

$$C_2H_2(g) + O_2(g) \xrightarrow{\Delta} CO_2(g) + H_2O(g) \quad \text{Not balanced}$$
Acetylene

STEP 2 Count the atoms of each element in the reactants and products. When we compare the atoms on the reactant side and the product side, we see that there are more C atoms in the reactants and more O atoms in the products.

$$C_2H_2(g) + O_2(g) \xrightarrow{\Delta} CO_2(g) + H_2O(g)$$

Reactants	Products	
2 C atoms	1 C atom	Not balanced
2 H atoms	2 H atoms	Balanced
2 O atoms	3 O atoms	Not balanced

STEP 3 Use coefficients to balance each element. We start by balancing the C in C_2H_2 by placing a coefficient of 2 in front of the formula for CO_2. When we recheck all the atoms, there are 2 C atoms and 2 H atoms in both the reactants and products.

$$C_2H_2(g) + O_2(g) \xrightarrow{\Delta} 2CO_2(g) + H_2O(g)$$

Reactants	Products	
2 C atoms	2 C atoms	Balanced
2 H atoms	2 H atoms	Balanced
2 O atoms	5 O atoms	Not balanced

To balance the O atoms, we place a coefficient of 5/2 in front of the formula for O_2, which gives a total of 5 O atoms on each side of the equation.

$$C_2H_2(g) + \frac{5}{2}O_2(g) \xrightarrow{\Delta} 2CO_2(g) + H_2O(g)$$

Reactants	Products	
2 C atoms	2 C atoms	Balanced
2 H atoms	2 H atoms	Balanced
5 O atoms	5 O atoms	Balanced

Now the equation is balanced for atoms, but the coefficient for O_2 is a fraction. To obtain all whole-number coefficients, we multiply *all* the coefficients by 2.

$$2C_2H_2(g) + 5O_2(g) \xrightarrow{\Delta} 4CO_2(g) + 2H_2O(g)$$

STEP 4 Check the final equation to confirm it is balanced. A check of the total number of atoms indicates that the equation is now balanced with whole-number coefficients.

$$2C_2H_2(g) + 5O_2(g) \xrightarrow{\Delta} 4CO_2(g) + 2H_2O(g)$$

Reactants	Products	
4 C atoms	4 C atoms	Balanced
4 H atoms	4 H atoms	Balanced
10 O atoms	10 O atoms	Balanced

STUDY CHECK 8.3

Balance the following chemical equation:

$$NO_2(g) + H_2(g) \longrightarrow NH_3(g) + H_2O(g)$$

ANSWER

$$2NO_2(g) + 7H_2(g) \longrightarrow 2NH_3(g) + 4H_2O(g)$$

Equations with Polyatomic Ions

Sometimes an equation contains the same polyatomic ion in both the reactants and the products. Then we can balance the polyatomic ions as a group on both sides of the equation as shown in Sample Problem 8.4.

SAMPLE PROBLEM 8.4 Balancing Chemical Equations with Polyatomic Ions

Balance the following chemical equation:

$$Na_3PO_4(aq) + MgCl_2(aq) \longrightarrow Mg_3(PO_4)_2(s) + NaCl(aq)$$

TRY IT FIRST

SOLUTION

ANALYZE THE PROBLEM	Given	Need	Connect
	reactants, products	balanced equation	equal numbers of atoms in reactants and products

ENGAGE

What is the evidence for a chemical reaction when $Na_3PO_4(aq)$ and $MgCl_2(aq)$ are mixed?

STEP 1 Write an equation using the correct formulas for the reactants and products.

$$Na_3PO_4(aq) + MgCl_2(aq) \longrightarrow Mg_3(PO_4)_2(s) + NaCl(aq) \quad \text{Not balanced}$$

STEP 2 Count the atoms of each element in the reactants and products. When we compare the number of ions in the reactants and products, we find that the equation is not balanced. In this equation, we can balance the phosphate ion as a group of atoms because it appears on both sides of the equation.

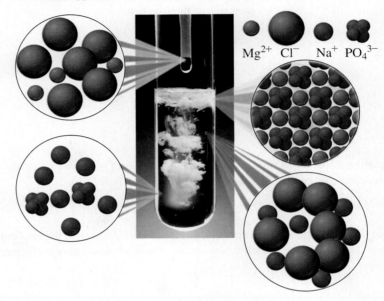

$$Mg^{2+} \quad Cl^- \quad Na^+ \quad PO_4^{3-}$$

$$Na_3PO_4(aq) + MgCl_2(aq) \longrightarrow Mg_3(PO_4)_2(s) + NaCl(aq)$$

Reactants	Products	
$3\,Na^+$	$1\,Na^+$	Not balanced
$1\,\mathbf{PO_4^{3-}}$	$2\,\mathbf{PO_4^{3-}}$	Not balanced
$1\,Mg^{2+}$	$3\,Mg^{2+}$	Not balanced
$2\,Cl^-$	$1\,Cl^-$	Not balanced

STEP 3 Use coefficients to balance each element. We begin with the formula that has the highest subscript values, which in this equation is $Mg_3(PO_4)_2$. The subscript 3 in $Mg_3(PO_4)_2$ is used as a coefficient for $MgCl_2$ to balance magnesium. The subscript 2 in $Mg_3(PO_4)_2$ is used as a coefficient for Na_3PO_4 to balance the phosphate ion.

$$2Na_3PO_4(aq) + 3MgCl_2(aq) \longrightarrow Mg_3(PO_4)_2(s) + NaCl(aq)$$

Reactants	Products	
$6\,Na^+$	$1\,Na^+$	Not balanced
$2\,\mathbf{PO_4^{3-}}$	$2\,\mathbf{PO_4^{3-}}$	Balanced
$3\,Mg^{2+}$	$3\,Mg^{2+}$	Balanced
$6\,Cl^-$	$1\,Cl^-$	Not balanced

In the reactants and products, we see that the sodium and chloride ions are not yet balanced. A coefficient of 6 is placed in front of the NaCl to balance the equation.

$$2Na_3PO_4(aq) + 3MgCl_2(aq) \longrightarrow Mg_3(PO_4)_2(s) + 6NaCl(aq)$$

STEP 4 Check the final equation to confirm it is balanced. A check of the total number of ions confirms the equation is balanced. A coefficient of 1 is understood and not usually written.

$$2Na_3PO_4(aq) + 3MgCl_2(aq) \longrightarrow Mg_3(PO_4)_2(s) + 6NaCl(aq) \quad \text{Balanced}$$

Reactants	Products	
$6\,Na^+$	$6\,Na^+$	Balanced
$2\,PO_4^{3-}$	$2\,PO_4^{3-}$	Balanced
$3\,Mg^{2+}$	$3\,Mg^{2+}$	Balanced
$6\,Cl^-$	$6\,Cl^-$	Balanced

STUDY CHECK 8.4

Balance the following chemical equation:

$$Pb(NO_3)_2(aq) + AlBr_3(aq) \longrightarrow PbBr_2(s) + Al(NO_3)_3(aq)$$

ANSWER

$$3Pb(NO_3)_2(aq) + 2AlBr_3(aq) \longrightarrow 3PbBr_2(s) + 2Al(NO_3)_3(aq)$$

QUESTIONS AND PROBLEMS

8.2 Balancing a Chemical Equation

LEARNING GOAL Write a balanced chemical equation from the formulas of the reactants and products for a chemical reaction.

8.7 Balance each of the following chemical equations:
   a. $N_2(g) + O_2(g) \longrightarrow NO(g)$
   b. $HgO(s) \xrightarrow{\Delta} Hg(l) + O_2(g)$
   c. $Fe(s) + O_2(g) \longrightarrow Fe_2O_3(s)$
   d. $Na(s) + Cl_2(g) \longrightarrow NaCl(s)$
   e. $Cu_2O(s) + O_2(g) \longrightarrow CuO(s)$

8.8 Balance each of the following chemical equations:
   a. $Ca(s) + Br_2(l) \longrightarrow CaBr_2(s)$
   b. $P_4(s) + O_2(g) \longrightarrow P_4O_{10}(s)$
   c. $C_4H_8(g) + O_2(g) \xrightarrow{\Delta} CO_2(g) + H_2O(g)$
   d. $Ca(OH)_2(aq) + HNO_3(aq) \longrightarrow H_2O(l) + Ca(NO_3)_2(aq)$
   e. $Fe_2O_3(s) + C(s) \longrightarrow Fe(s) + CO(g)$

8.9 Balance each of the following chemical equations:
   a. $Mg(s) + AgNO_3(aq) \longrightarrow Ag(s) + Mg(NO_3)_2(aq)$
   b. $CuCO_3(s) \longrightarrow CuO(s) + CO_2(g)$
   c. $Al(s) + CuSO_4(aq) \longrightarrow Cu(s) + Al_2(SO_4)_3(aq)$
   d. $Pb(NO_3)_2(aq) + NaCl(aq) \longrightarrow PbCl_2(s) + NaNO_3(aq)$
   e. $Al(s) + HCl(aq) \longrightarrow H_2(g) + AlCl_3(aq)$

8.10 Balance each of the following chemical equations:
   a. $Zn(s) + HNO_3(aq) \longrightarrow H_2(g) + Zn(NO_3)_2(aq)$
   b. $Al(s) + H_2SO_4(aq) \longrightarrow H_2(g) + Al_2(SO_4)_3(aq)$
   c. $K_2SO_4(aq) + BaCl_2(aq) \longrightarrow BaSO_4(s) + KCl(aq)$
   d. $CaCO_3(s) \longrightarrow CaO(s) + CO_2(g)$
   e. $AlCl_3(aq) + KOH(aq) \longrightarrow Al(OH)_3(s) + KCl(aq)$

8.11 Balance each of the following chemical equations:
   a. $Fe_2O_3(s) + CO(g) \longrightarrow Fe(s) + CO_2(g)$
   b. $Li_3N(s) \longrightarrow Li(s) + N_2(g)$
   c. $Al(s) + HBr(aq) \longrightarrow H_2(g) + AlBr_3(aq)$
   d. $Ba(OH)_2(aq) + Na_3PO_4(aq) \longrightarrow$
   $$Ba_3(PO_4)_2(s) + NaOH(aq)$$
   e. $As_4S_6(s) + O_2(g) \longrightarrow As_4O_6(s) + SO_2(g)$

8.12 Balance each of the following chemical equations:
   a. $K(s) + H_2O(l) \longrightarrow H_2(g) + KOH(aq)$
   b. $Cr(s) + S_8(s) \longrightarrow Cr_2S_3(s)$
   c. $BCl_3(s) + H_2O(l) \longrightarrow H_3BO_3(aq) + HCl(aq)$
   d. $Fe(OH)_3(s) + H_2SO_4(aq) \longrightarrow H_2O(l) + Fe_2(SO_4)_3(aq)$
   e. $BaCl_2(aq) + Na_3PO_4(aq) \longrightarrow Ba_3(PO_4)_2(s) + NaCl(aq)$

8.13 Write a balanced equation using the correct formulas and include conditions (*s*, *l*, *g*, or *aq*) for each of the following chemical reactions:
   a. Lithium metal reacts with liquid water to form hydrogen gas and aqueous lithium hydroxide.
   b. Solid phosphorus reacts with chlorine gas to form solid phosphorus pentachloride.
   c. Solid iron(II) oxide reacts with carbon monoxide gas to form solid iron and carbon dioxide gas.
   d. Liquid pentene (C_5H_{10}) burns in oxygen gas to form carbon dioxide gas and water vapor.
   e. Hydrogen sulfide gas and solid iron(III) chloride react to form solid iron(III) sulfide and hydrogen chloride gas.

8.14 Write a balanced equation using the correct formulas and include conditions (*s*, *l*, *g*, or *aq*) for each of the following chemical reactions:
   a. Solid sodium carbonate decomposes to produce solid sodium oxide and carbon dioxide gas.
   b. Nitrogen oxide gas reacts with carbon monoxide gas to produce nitrogen gas and carbon dioxide gas.
   c. Iron metal reacts with solid sulfur to produce solid iron(III) sulfide.
   d. Solid calcium reacts with nitrogen gas to produce solid calcium nitride.
   e. In the *Apollo* lunar module, hydrazine gas, N_2H_4, reacts with dinitrogen tetroxide gas to produce gaseous nitrogen and water vapor.

Applications

8.15 Dinitrogen oxide, also known as laughing gas, is a sweet-tasting gas used in dentistry. When solid ammonium nitrate is heated, the products are gaseous water and gaseous dinitrogen oxide. Write the balanced chemical equation for the reaction.

8.16 When ethanol $C_2H_6O(aq)$ is consumed, it reacts with oxygen gas in the body to produce gaseous carbon dioxide and liquid water. Write the balanced chemical equation for the reaction.

8.17 In the body, the amino acid alanine $C_3H_7NO_2(aq)$ reacts with oxygen gas to produce gaseous carbon dioxide, liquid water, and urea, $CH_4N_2O(aq)$. Write the balanced chemical equation for the reaction.

8.18 In the body, the amino acid asparagine $C_4H_8N_2O_3(aq)$ reacts with oxygen gas to produce gaseous carbon dioxide, liquid water, and urea, $CH_4N_2O(aq)$. Write the balanced chemical equation for the reaction.

8.3 Types of Chemical Reactions

LEARNING GOAL Identify a chemical reaction as a combination, decomposition, single replacement, double replacement, or combustion.

CORE CHEMISTRY SKILL
Classifying Types of Chemical Reactions

A great number of chemical reactions occur in nature, in biological systems, and in the laboratory. However, there are some general patterns among all reactions that help us classify reactions. Some reactions may fit into more than one reaction type.

Combination Reactions

In a **combination reaction**, two or more elements or compounds bond to form one product. For example, sulfur and oxygen combine to form the product sulfur dioxide.

In **FIGURE 8.2**, the elements magnesium and oxygen combine to form a single product, which is the ionic compound magnesium oxide formed from Mg^{2+} and O^{2-} ions.

$$2Mg(s) + O_2(g) \longrightarrow 2MgO(s)$$

Combination

Two or more reactants combine to yield a single product

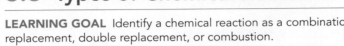

 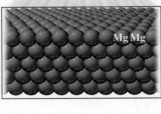 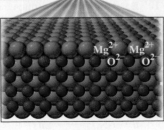

$$2Mg(s) \qquad + \qquad O_2(g) \qquad \overset{\Delta}{\longrightarrow} \qquad 2MgO(s)$$

Magnesium   Oxygen   Magnesium oxide

FIGURE 8.2 ▶ In a combination reaction, two or more substances combine to form one substance as product.

◉ What happens to the atoms in the reactants in a combination reaction?

In other examples of combination reactions, elements or compounds combine to form a single product.

$$N_2(g) + 3H_2(g) \longrightarrow 2NH_3(g)$$
Ammonia

$$Cu(s) + S(s) \longrightarrow CuS(s)$$

$$MgO(s) + CO_2(g) \longrightarrow MgCO_3(s)$$

Decomposition Reactions

In a **decomposition reaction**, a reactant splits into two or more simpler products. For example, when mercury(II) oxide is heated, the compound breaks apart into mercury atoms and oxygen (see **FIGURE 8.3**).

$$2HgO(s) \xrightarrow{\Delta} 2Hg(l) + O_2(g)$$

In another example of a decomposition reaction, when calcium carbonate is heated, it breaks apart into simpler compounds of calcium oxide and carbon dioxide.

$$CaCO_3(s) \xrightarrow{\Delta} CaO(s) + CO_2(g)$$

Decomposition

A reactant splits into two or more products

A B $\longrightarrow$ A + B

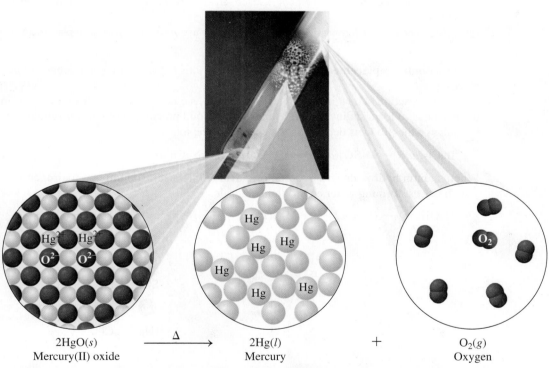

$$\begin{array}{ccc} 2HgO(s) & \xrightarrow{\Delta} & 2Hg(l) & + & O_2(g) \\ \text{Mercury(II) oxide} & & \text{Mercury} & & \text{Oxygen} \end{array}$$

FIGURE 8.3 ▶ In a decomposition reaction, one reactant breaks down into two or more products.
Q How do the differences in the reactant and products classify this as a decomposition reaction?

Replacement Reactions

In a replacement reaction, elements in a compound are replaced by other elements. In a **single replacement reaction**, a reacting element switches place with an element in the other reacting compound.

In the single replacement reaction shown in **FIGURE 8.4**, zinc replaces hydrogen in hydrochloric acid, HCl(*aq*).

$$Zn(s) + 2HCl(aq) \longrightarrow H_2(g) + ZnCl_2(aq)$$

In another single replacement reaction, chlorine replaces bromine in the compound potassium bromide.

$$Cl_2(g) + 2KBr(s) \longrightarrow 2KCl(s) + Br_2(l)$$

Single replacement

One element replaces another element

A + B C $\longrightarrow$ A C + B

How can you distinguish a single replacement reaction from a double replacement reaction?

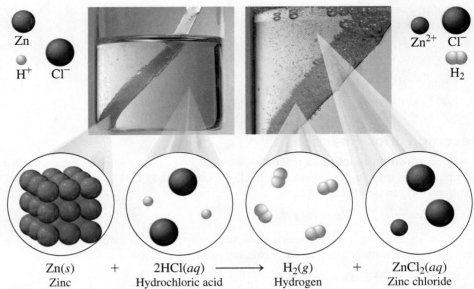

Zn(s)   +   2HCl(aq)   ⟶ $H_2(g)$ + $ZnCl_2(aq)$
Zinc   Hydrochloric acid   Hydrogen   Zinc chloride

FIGURE 8.4 ▶ In a single replacement reaction, an atom or ion replaces an atom or ion in a compound.

Q What changes in the formulas of the reactants identify this equation as a single replacement?

Double replacement

Two elements   replace   each other

A B + C D ⟶ A D + C B

In a **double replacement reaction**, the positive ions in the reacting compounds switch places.

In the reaction shown in **FIGURE 8.5**, barium ions change places with sodium ions in the reactants to form sodium chloride and a white solid precipitate of barium sulfate. The formulas of the products depend on the charges of the ions.

$$BaCl_2(aq) + Na_2SO_4(aq) \longrightarrow BaSO_4(s) + 2NaCl(aq)$$

When sodium hydroxide and hydrochloric acid (HCl) react, sodium and hydrogen ions switch places, forming water and sodium chloride.

$$NaOH(aq) + HCl(aq) \longrightarrow H_2O(l) + NaCl(aq)$$

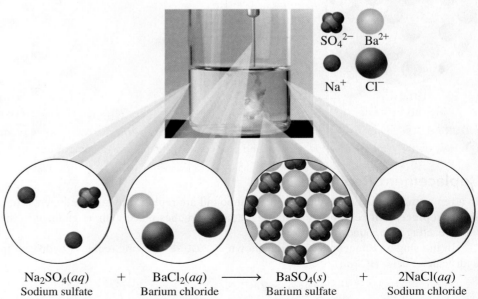

$Na_2SO_4(aq)$ + $BaCl_2(aq)$ ⟶ $BaSO_4(s)$ +   2NaCl(aq)
Sodium sulfate   Barium chloride   Barium sulfate   Sodium chloride

FIGURE 8.5 ▶ In a double replacement reaction, the positive ions in the reactants replace each other.

Q How do the changes in the formulas of the reactants identify this equation as a double replacement reaction?

Combustion Reactions

The burning of a candle and the burning of fuel in the engine of a car are examples of combustion reactions. In a **combustion reaction**, a carbon-containing compound, usually a fuel, burns in oxygen from the air to produce carbon dioxide (CO_2), water (H_2O), and energy in the form of heat or a flame. For example, methane gas (CH_4) undergoes combustion when used to cook our food on a gas cooktop and to heat our homes. In the equation for the combustion of methane, each element in the fuel (CH_4) forms a compound with oxygen.

$$CH_4(g) + 2O_2(g) \xrightarrow{\Delta} CO_2(g) + 2H_2O(g) + \text{energy}$$
Methane

The balanced equation for the combustion of propane (C_3H_8) is

$$C_3H_8(g) + 5O_2(g) \xrightarrow{\Delta} 3CO_2(g) + 4H_2O(g) + \text{energy}$$

Propane is the fuel used in portable heaters and gas barbecues. Gasoline, a mixture of liquid hydrocarbons, is the fuel that powers our cars, lawn mowers, and snow blowers.

TABLE 8.3 summarizes the reaction types and gives examples.

In a combustion reaction, a candle burns using the oxygen in the air.

TABLE 8.3 Summary of Reaction Types

Reaction Type	Example
Combination	
$A + B \longrightarrow AB$	$Ca(s) + Cl_2(g) \longrightarrow CaCl_2(s)$
Decomposition	
$AB \longrightarrow A + B$	$Fe_2S_3(s) \longrightarrow 2Fe(s) + 3S(s)$
Single Replacement	
$A + BC \longrightarrow AC + B$	$Cu(s) + 2AgNO_3(aq) \longrightarrow 2Ag(s) + Cu(NO_3)_2(aq)$
Double Replacement	
$AB + CD \longrightarrow AD + CB$	$BaCl_2(aq) + K_2SO_4(aq) \longrightarrow BaSO_4(s) + 2KCl(aq)$
Combustion	
$C_XH_Y + ZO_2(g) \xrightarrow{\Delta} XCO_2(g) + \frac{Y}{2}H_2O(g) + \text{energy}$	$CH_4(g) + 2O_2(g) \xrightarrow{\Delta} CO_2(g) + 2H_2O(g) + \text{energy}$

SAMPLE PROBLEM 8.5 Identifying Reactions

Classify each of the following as a combination, decomposition, single replacement, double replacement, or combustion reaction:

a. $2Fe_2O_3(s) + 3C(s) \longrightarrow 3CO_2(g) + 4Fe(s)$

b. $2KClO_3(s) \xrightarrow{\Delta} 2KCl(s) + 3O_2(g)$

c. $C_2H_4(g) + 3O_2(g) \xrightarrow{\Delta} 2CO_2(g) + 2H_2O(g) + \text{energy}$

TRY IT FIRST

SOLUTION

a. In this single replacement reaction, a C atom replaces Fe in Fe_2O_3 to form the compound CO_2 and Fe atoms.

b. When one reactant breaks down to produce two products, the reaction is decomposition.

c. The reaction of a carbon compound with oxygen to produce carbon dioxide, water, and energy makes this a combustion reaction.

STUDY CHECK 8.5

Nitrogen gas (N_2) and oxygen gas (O_2) react to form nitrogen dioxide gas. Write the balanced chemical equation using the correct chemical formulas of the reactants and product, and identify the reaction type.

ANSWER

$N_2(g) + 2O_2(g) \longrightarrow 2NO_2(g)$ Combination

CHEMISTRY LINK TO **HEALTH**
Incomplete Combustion: Toxicity of Carbon Monoxide

When a propane heater, fireplace, or woodstove is used in a closed room, there must be adequate ventilation. If the supply of oxygen is limited, incomplete combustion from burning gas, oil, or wood produces carbon monoxide. The incomplete combustion of methane in natural gas is written

$$2CH_4(g) + 3O_2(g) \xrightarrow{\Delta} 2CO(g) + 4H_2O(g) + \text{energy}$$

Limited   Carbon
oxygen   monoxide
supply

Carbon monoxide (CO) is a colorless, odorless, poisonous gas. When inhaled, CO passes into the bloodstream, where it attaches to

hemoglobin, which reduces the amount of oxygen (O_2) reaching the cells. As a result, a person can experience a reduction in exercise capability, visual perception, and manual dexterity.

Hemoglobin is the protein that transports O_2 in the blood. When the amount of hemoglobin bound to CO (COHb) is about 10%, a person may experience shortness of breath, mild headache, and drowsiness. Heavy smokers can have levels of COHb in their blood as high as 9%. When as much as 30% of the hemoglobin is bound to CO, a person may experience more severe symptoms, including dizziness, mental confusion, severe headache, and nausea. If 50% or more of the hemoglobin is bound to CO, a person could become unconscious and die if not treated immediately with oxygen.

When camping, a butane cartridge provides fuel for a portable burner.

SAMPLE PROBLEM 8.6 Writing an Equation for Combustion

A portable burner is fueled with butane, C_4H_{10}. Write the reactants and products for the complete combustion of butane, and balance the equation.

TRY IT FIRST

SOLUTION

ANALYZE THE PROBLEM	Given	Need	Connect
	C_4H_{10}	balanced equation for combustion	reactants: carbon compound + O_2, products: CO_2 + H_2O

In a combustion reaction, butane gas reacts with the gas O_2 to form the gases CO_2, H_2O, and energy. We write the unbalanced equation as

$$C_4H_{10}(g) + O_2(g) \xrightarrow{\Delta} CO_2(g) + H_2O(g) + \text{energy}$$

We can begin by using the subscripts in C_4H_{10} to balance the C atoms in CO_2 and the H atoms in H_2O. However, this gives a total of 13 O atoms in the products. This is balanced by placing a coefficient of 13/2 in front of the formula for O_2.

$$C_4H_{10}(g) + \frac{13}{2} O_2(g) \xrightarrow{\Delta} \underbrace{4CO_2(g) + 5H_2O(g)}_{\text{13 O atoms}} + \text{energy}$$

To obtain a whole number coefficient for O_2, we need to multiply all the coefficients by 2.

$$2C_4H_{10}(g) + 13O_2(g) \xrightarrow{\Delta} 8CO_2(g) + 10H_2O(g) + \text{energy}$$

STUDY CHECK 8.6

Ethene, used to ripen fruit, has the formula C_2H_4. Write a balanced chemical equation for the complete combustion of ethene.

ANSWER

$$C_2H_4(g) + 3O_2(g) \xrightarrow{\Delta} 2CO_2(g) + 2H_2O(g) + \text{energy}$$

QUESTIONS AND PROBLEMS

8.3 Types of Chemical Reactions

LEARNING GOAL Identify a chemical reaction as a combination, decomposition, single replacement, double replacement, or combustion.

8.19 Classify each of the following as a combination, decomposition, single replacement, double replacement, or combustion reaction:
    a. $2Al_2O_3(s) \xrightarrow{\Delta} 4Al(s) + 3O_2(g)$
    b. $Br_2(l) + BaI_2(s) \longrightarrow BaBr_2(s) + I_2(g)$
    c. $2C_2H_2(g) + 5O_2(g) \xrightarrow{\Delta} 4CO_2(g) + 2H_2O(g)$
    d. $BaCl_2(aq) + K_2CO_3(aq) \longrightarrow BaCO_3(s) + 2KCl(aq)$
    e. $Pb(s) + O_2(g) \longrightarrow PbO_2(s)$

8.20 Classify each of the following as a combination, decomposition, single replacement, double replacement, or combustion reaction:
    a. $H_2(g) + Br_2(l) \longrightarrow 2HBr(g)$
    b. $AgNO_3(aq) + NaCl(aq) \longrightarrow AgCl(s) + NaNO_3(aq)$
    c. $2H_2O_2(aq) \longrightarrow 2H_2O(l) + O_2(g)$
    d. $Zn(s) + CuCl_2(aq) \longrightarrow Cu(s) + ZnCl_2(aq)$
    e. $C_5H_8(g) + 7O_2(g) \xrightarrow{\Delta} 5CO_2(g) + 4H_2O(g)$

8.21 Classify each of the following as a combination, decomposition, single replacement, double replacement, or combustion reaction:
    a. $4Fe(s) + 3O_2(g) \longrightarrow 2Fe_2O_3(s)$
    b. $Mg(s) + 2AgNO_3(aq) \longrightarrow 2Ag(s) + Mg(NO_3)_2(aq)$
    c. $CuCO_3(s) \xrightarrow{\Delta} CuO(s) + CO_2(g)$

    d. $Al_2(SO_4)_3(aq) + 6KOH(aq) \longrightarrow$
                                    $2Al(OH)_3(s) + 3K_2SO_4(aq)$
    e. $C_4H_8(g) + 6O_2(g) \xrightarrow{\Delta} 4CO_2(g) + 4H_2O(g)$

8.22 Classify each of the following as a combination, decomposition, single replacement, double replacement, or combustion reaction:
    a. $CuO(s) + 2HCl(aq) \longrightarrow CuCl_2(aq) + H_2O(l)$
    b. $2Al(s) + 3Br_2(l) \longrightarrow 2AlBr_3(s)$
    c. $C_6H_{12}(l) + 9O_2(g) \xrightarrow{\Delta} 6CO_2(g) + 6H_2O(g)$
    d. $Fe_2O_3(s) + 3C(s) \longrightarrow 2Fe(s) + 3CO(g)$
    e. $C_6H_{12}O_6(aq) \longrightarrow 2C_2H_6O(aq) + 2CO_2(g)$

8.23 Using Table 8.3, predict the products that would result from each of the following reactions and balance:
    a. combination: $Mg(s) + Cl_2(g) \longrightarrow$
    b. decomposition: $HBr(g) \longrightarrow$
    c. single replacement: $Mg(s) + Zn(NO_3)_2(aq) \longrightarrow$
    d. double replacement: $K_2S(aq) + Pb(NO_3)_2(aq) \longrightarrow$
    e. combustion: $C_2H_6(g) + O_2(g) \xrightarrow{\Delta}$

8.24 Using Table 8.3, predict the products that would result from each of the following reactions and balance:
    a. combination: $Ca(s) + O_2(g) \longrightarrow$
    b. combustion: $C_6H_6(g) + O_2(g) \xrightarrow{\Delta}$
    c. decomposition: $PbO_2(s) \xrightarrow{\Delta}$
    d. single replacement: $KI(s) + Cl_2(g) \longrightarrow$
    e. double replacement: $CuCl_2(aq) + Na_2S(aq) \longrightarrow$

8.4 Oxidation–Reduction Reactions

LEARNING GOAL Define the terms oxidation and reduction; identify the reactants oxidized and reduced.

Perhaps you have never heard of an oxidation and reduction reaction. However, this type of reaction has many important applications in your everyday life. When you see a rusty nail, tarnish on a silver spoon, or corrosion on metal, you are observing oxidation.

$$4Fe(s) + 3O_2(g) \longrightarrow 2Fe_2O_3(s) \quad \text{Fe is oxidized}$$
                        Rust

When we turn the lights on in our automobiles, an oxidation–reduction reaction within the car battery provides the electricity. On a cold, wintry day, we might build a fire. As the wood burns, oxygen combines with carbon and hydrogen to produce carbon dioxide, water, and heat. In the previous section, we called this a combustion reaction, but it is also an *oxidation–reduction reaction*. When we eat foods with starches in them, the starches break down to give glucose, which is oxidized in our cells to give us energy along with carbon dioxide and water. Every breath we take provides oxygen to carry out oxidation in our cells.

$$C_6H_{12}O_6(aq) + 6O_2(g) \longrightarrow 6CO_2(g) + 6H_2O(l) + \text{energy}$$
Glucose

Oxidation–Reduction Reactions

In an **oxidation–reduction reaction** (*redox*), electrons are transferred from one substance to another. If one substance loses electrons, another substance must gain electrons. **Oxidation** is defined as the *loss* of electrons; **reduction** is the *gain* of electrons.

Rust forms when the oxygen in the air reacts with iron.

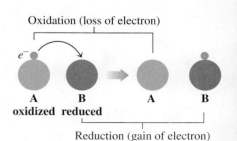
Oxidation (loss of electron)
e^-
A   **B**   **A**   **B**
oxidized reduced
Reduction (gain of electron)

One way to remember these definitions is to use the following acronym:

OIL RIG

Oxidation **I**s **L**oss of electrons

Reduction **I**s **G**ain of electrons

Oxidation–Reduction

In general, atoms of metals lose electrons to form positive ions, whereas nonmetals gain electrons to form negative ions. Now we can say that metals are oxidized and nonmetals are reduced.

The green color that appears on copper surfaces from weathering, known as *patina*, is a mixture of $CuCO_3$ and CuO. We can now look at the oxidation and reduction reactions that take place when copper metal reacts with oxygen in the air to produce copper(II) oxide.

$$2Cu(s) + O_2(g) \longrightarrow 2CuO(s)$$

The element Cu in the reactants has a charge of 0, but in the CuO product, it is present as Cu^{2+}, which has a 2+ charge. Because the Cu atom lost two electrons, the charge is more positive. This means that Cu was oxidized in the reaction.

$$Cu^0(s) \longrightarrow Cu^{2+}(s) + 2\,e^- \quad \text{Oxidation: loss of electrons by Cu}$$

At the same time, the element O in the reactants has a charge of 0, but in the CuO product, it is present as O^{2-}, which has a 2− charge. Because the O atom has gained two electrons, the charge is more negative. This means that O was reduced in the reaction.

$$O_2^0(g) + 4\,e^- \longrightarrow 2O^{2-}(s) \quad \text{Reduction: gain of electrons by O}$$

Thus, the overall equation for the formation of CuO involves an oxidation and a reduction that occur simultaneously. In every oxidation and reduction, the number of electrons lost must be equal to the number of electrons gained. Therefore, we multiply the oxidation reaction of Cu by 2. Canceling the $4\,e^-$ on each side, we obtain the overall oxidation–reduction equation for the formation of CuO.

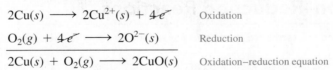

$2Cu(s) \longrightarrow 2Cu^{2+}(s) + \cancel{4e^-}$	Oxidation
$O_2(g) + \cancel{4e^-} \longrightarrow 2O^{2-}(s)$	Reduction
$2Cu(s) + O_2(g) \longrightarrow 2CuO(s)$	Oxidation–reduction equation

As we see in the next reaction between zinc and copper(II) sulfate, there is always an oxidation with every reduction (see **FIGURE 8.6**).

$$Zn(s) + CuSO_4(aq) \longrightarrow ZnSO_4(aq) + Cu(s)$$

We rewrite the equation to show the atoms and ions.

$$\mathbf{Zn}(s) + \mathbf{Cu^{2+}}(aq) + SO_4{}^{2-}(aq) \longrightarrow \mathbf{Zn^{2+}}(aq) + SO_4{}^{2-}(aq) + \mathbf{Cu}(s)$$

In this reaction, Zn atoms lose two electrons to form Zn^{2+}. The increase in positive charge indicates that Zn is oxidized. At the same time, Cu^{2+} gains two electrons. The decrease in charge indicates that Cu is reduced. The $SO_4{}^{2-}$ ions are *spectator ions*, which are present in both the reactants and products and do not change.

$$Zn(s) \longrightarrow Zn^{2+}(aq) + 2\,e^- \quad \text{Oxidation of Zn}$$

$$Cu^{2+}(aq) + 2\,e^- \longrightarrow Cu(s) \quad \text{Reduction of } Cu^{2+}$$

In this single replacement reaction, zinc was oxidized and copper(II) ion was reduced.

Oxidation and Reduction in Biological Systems

Oxidation may also involve the addition of oxygen or the loss of hydrogen, and reduction may involve the loss of oxygen or the gain of hydrogen. In the cells of the body, oxidation of organic (carbon) compounds involves the transfer of hydrogen atoms (H), which are composed of

The green patina on copper is due to oxidation.

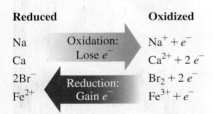

Reduced		Oxidized
Na	Oxidation: Lose e^-	$Na^+ + e^-$
Ca		$Ca^{2+} + 2\,e^-$
$2Br^-$	Reduction: Gain e^-	$Br_2 + 2\,e^-$
Fe^{2+}		$Fe^{3+} + e^-$

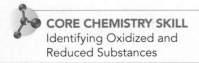

Oxidation is a loss of electrons; reduction is a gain of electrons.

CORE CHEMISTRY SKILL

Identifying Oxidized and Reduced Substances

ENGAGE

How can you determine that Cu^{2+} is reduced in this reaction?

FIGURE 8.6 ▶ In this single replacement reaction, Zn(s) is oxidized to $Zn^{2+}(aq)$ when it provides two electrons to reduce $Cu^{2+}(aq)$ to Cu(s): $Zn(s) + Cu^{2+}(aq) \longrightarrow Zn^{2+}(aq) + Cu(s)$

Q In the oxidation, does Zn(s) lose or gain electrons?

electrons and protons. For example, the oxidation of a typical biochemical molecule can involve the transfer of two hydrogen atoms (or $2H^+$ and $2 e^-$) to a hydrogen ion acceptor such as the coenzyme FAD (flavin adenine dinucleotide). The coenzyme is reduced to $FADH_2$.

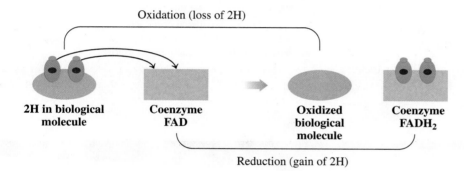

In many biochemical oxidation–reduction reactions, the transfer of hydrogen atoms is necessary for the production of energy in the cells. For example, methyl alcohol (CH_4O), a poisonous substance, is metabolized in the body by the following reactions:

$CH_4O \longrightarrow CH_2O + 2H$ Oxidation: loss of H atoms
Methyl   Formaldehyde
alcohol

The formaldehyde can be oxidized further, this time by the addition of oxygen, to produce formic acid.

$2CH_2O + O_2 \longrightarrow 2CH_2O_2$ Oxidation: addition of O atoms
Formaldehyde   Formic acid

Finally, formic acid is oxidized to carbon dioxide and water.

$2CH_2O_2 + O_2 \longrightarrow 2CO_2 + 2H_2O$ Oxidation: addition of O atoms
Formic acid

The intermediate products of the oxidation of methyl alcohol are quite toxic, causing blindness and possibly death as they interfere with key reactions in the cells of the body.

In summary, we find that the particular definition of oxidation and reduction we use depends on the process that occurs in the reaction. All these definitions are summarized in **TABLE 8.4**. Oxidation always involves a loss of electrons, but it may also be seen as an addition of oxygen or the loss of hydrogen atoms. A reduction always involves a gain of electrons and may also be seen as the loss of oxygen or the gain of hydrogen.

TABLE 8.4 Characteristics of Oxidation and Reduction

Always Involves	May Involve
Oxidation	
Loss of electrons	Addition of oxygen
	Loss of hydrogen
Reduction	
Gain of electrons	Loss of oxygen
	Gain of hydrogen

QUESTIONS AND PROBLEMS

8.4 Oxidation–Reduction Reactions

LEARNING GOAL Define the terms oxidation and reduction; identify the reactants oxidized and reduced.

8.25 Identify each of the following as an oxidation or a reduction:
a. $Na^+(aq) + e^- \longrightarrow Na(s)$
b. $Ni(s) \longrightarrow Ni^{2+}(aq) + 2\, e^-$
c. $Cr^{3+}(aq) + 3\, e^- \longrightarrow Cr(s)$
d. $2H^+(aq) + 2\, e^- \longrightarrow H_2(g)$

8.26 Identify each of the following as an oxidation or a reduction:
a. $O_2(g) + 4\, e^- \longrightarrow 2O^{2-}(aq)$
b. $Ag(s) \longrightarrow Ag^+(aq) + e^-$
c. $Fe^{3+}(aq) + e^- \longrightarrow Fe^{2+}(aq)$
d. $2Br^-(aq) \longrightarrow Br_2(l) + 2\, e^-$

8.27 In each of the following reactions, identify the reactant that is oxidized and the reactant that is reduced:
a. $Zn(s) + Cl_2(g) \longrightarrow ZnCl_2(s)$
b. $Cl_2(g) + 2NaBr(aq) \longrightarrow 2NaCl(aq) + Br_2(l)$
c. $2PbO(s) \longrightarrow 2Pb(s) + O_2(g)$
d. $2Fe^{3+}(aq) + Sn^{2+}(aq) \longrightarrow 2Fe^{2+}(aq) + Sn^{4+}(aq)$

8.28 In each of the following reactions, identify the reactant that is oxidized and the reactant that is reduced:
a. $2Li(s) + F_2(g) \longrightarrow 2LiF(s)$
b. $Cl_2(g) + 2KI(aq) \longrightarrow I_2(s) + 2KCl(aq)$
c. $2Al(s) + 3Sn^{2+}(aq) \longrightarrow 2Al^{3+}(aq) + 3Sn(s)$
d. $Fe(s) + CuSO_4(aq) \longrightarrow Cu(s) + FeSO_4(aq)$

Applications

8.29 In the mitochondria of human cells, energy is provided by the oxidation and reduction reactions of the iron ions in the cytochromes in electron transport. Identify each of the following reactions as an oxidation or a reduction:
a. $Fe^{3+} + e^- \longrightarrow Fe^{2+}$ b. $Fe^{2+} \longrightarrow Fe^{3+} + e^-$

8.30 Chlorine (Cl_2) is a strong germicide used to disinfect drinking water and to kill microbes in swimming pools. If the product is Cl^-, was the elemental chlorine oxidized or reduced?

8.31 When linoleic acid, an unsaturated fatty acid, reacts with hydrogen, it forms a saturated fatty acid. Is linoleic acid oxidized or reduced in the hydrogenation reaction?

$$C_{18}H_{32}O_2 + 2H_2 \longrightarrow C_{18}H_{36}O_2$$
Linoleic acid

8.32 In one of the reactions in the citric acid cycle, which provides energy, succinic acid is converted to fumaric acid.

$$C_4H_6O_4 \longrightarrow C_4H_4O_4 + 2H$$
Succinic acid   Fumaric acid

The reaction is accompanied by the reaction of a coenzyme, flavin adenine dinucleotide (FAD).

$$FAD + 2H \longrightarrow FADH_2$$

a. Is succinic acid oxidized or reduced?
b. Is FAD oxidized or reduced?
c. Why would the two reactions occur together?

Follow Up

IMPROVING NATALIE'S OVERALL FITNESS

Low-intensity exercises are used at the beginning of Natalie's exercise program.

Natalie's test results indicate that she has a blood oxygen level of 89%. The normal values for pulse oximeter readings are 95% to 100%, which means that Natalie's O_2 saturation is low. Thus, Natalie does not have an adequate amount of O_2 in her blood, and may be *hypoxic*. This may be the reason she has noticed a shortness of breath and a dry cough. Her doctor diagnosed her with *interstitial lung disease*, which is scarring of the tissue of the lungs.

Angela teaches Natalie to inhale and exhale slower and deeper to fill the lungs with more air and thus more oxygen. Angela also develops a workout program with the goal of increasing Natalie's overall fitness level. During the exercises, Angela continues to monitor Natalie's heart rate, blood O_2 level, and blood pressure to ensure that Natalie is exercising at a level that will enable her to become stronger without breaking down muscle due to a lack of oxygen.

Applications

8.33 a. During cellular respiration, $C_6H_{12}O_6$ (glucose) in the cells undergoes combustion. Write and balance the chemical equation for reaction of glucose in the human body.
b. In plants, carbon dioxide and oxygen gases are converted to glucose ($C_6H_{12}O_6$) and water. Write and balance the chemical equation for production of glucose in plants.

8.34 Fatty acids undergo reaction with O_2 and form CO_2 and H_2O when utilized for energy in the body.
a. Write and balance the equation for the combustion of the fatty acid capric acid, $C_{10}H_{20}O_2$.
b. Write and balance the equation for the combustion of the fatty acid myristic acid, $C_{14}H_{28}O_2$.

CONCEPT MAP

CHEMICAL REACTIONS

show a chemical change between

Reactants and Products

are balanced with

Coefficients

to give

Equal Numbers of Atoms on Each Side

are classified as

Combination, Decomposition, Single Replacement, Double Replacement, and Combustion

and those with

Loss or Gain of Electrons

are also

Oxidation–Reduction Reactions

CHAPTER REVIEW

8.1 Equations for Chemical Reactions

LEARNING GOAL Identify a balanced chemical equation; determine the number of atoms in the reactants and products.

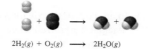

$2H_2(g) + O_2(g) \longrightarrow 2H_2O(g)$
Balanced

- A chemical reaction occurs when the atoms of the initial substances rearrange to form new substances.
- A chemical equation shows the formulas of the substances that react on the left side of a reaction arrow and the products that form on the right side of the reaction arrow.

8.2 Balancing a Chemical Equation

LEARNING GOAL Write a balanced chemical equation from the formulas of the reactants and products for a chemical reaction.

CH_4 O_2 CO_2 H_2O

- A chemical equation is balanced by writing coefficients, small whole numbers, in front of formulas to equalize the atoms of each of the elements in the reactants and the products.

8.3 Types of Chemical Reactions

LEARNING GOAL Identify a chemical reaction as a combination, decomposition, single replacement, double replacement, or combustion.

Single replacement

One element   replaces   another element

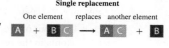

A + B C ⟶ A C + B

- Many chemical reactions can be organized by reaction type: combination, decomposition, single replacement, double replacement, or combustion.

8.4 Oxidation–Reduction Reactions

LEARNING GOAL Define the terms oxidation and reduction; identify the reactants oxidized and reduced.

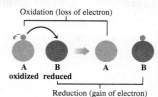

Oxidation (loss of electron)

e^-

A   B   A   B
oxidized reduced

Reduction (gain of electron)

- When electrons are transferred in a reaction, it is an oxidation–reduction reaction.
- One reactant loses electrons, and another reactant gains electrons.
- Overall, the number of electrons lost and gained is equal.

KEY TERMS

balanced equation The final form of a chemical equation that shows the same number of atoms of each element in the reactants and products.

chemical equation A shorthand way to represent a chemical reaction using chemical formulas to indicate the reactants and products and coefficients to show reacting ratios.

coefficients Whole numbers placed in front of the formulas to balance the number of atoms or moles of atoms of each element on both sides of an equation.

combination reaction A chemical reaction in which reactants combine to form a single product.

combustion reaction A chemical reaction in which a fuel containing carbon and hydrogen reacts with oxygen to produce CO_2, H_2O, and energy.

decomposition reaction A reaction in which a single reactant splits into two or more simpler substances.

double replacement reaction A reaction in which the positive ions in the reacting compounds exchange places.

oxidation The loss of electrons by a substance. Oxidation may involve the addition of oxygen or the loss of hydrogen.

oxidation–reduction reaction A reaction in which the oxidation of one reactant is always accompanied by the reduction of another reactant.

products The substances formed as a result of a chemical reaction.

reactants The initial substances that undergo change in a chemical reaction.

reduction The gain of electrons by a substance. Reduction may involve the loss of oxygen or the gain of hydrogen.

single replacement reaction A reaction in which an element replaces a different element in a compound.

 ## CORE CHEMISTRY SKILLS

The chapter section containing each Core Chemistry Skill is shown in parentheses at the end of each heading.

Balancing a Chemical Equation (8.2)

- In a balanced chemical equation, whole numbers called coefficients multiply each of the atoms in the chemical formulas so that the number of each type of atom in the reactants is equal to the number of the same type of atom in the products.

Example: Balance the following chemical equation:

$$SnCl_4(s) + H_2O(l) \longrightarrow Sn(OH)_4(s) + HCl(aq)$$

Answer: When we compare the atoms on the reactant side and the product side, we see that there are more Cl atoms in the reactants and more O and H atoms in the products.

To balance the equation, we need to use coefficients in front of the formulas containing the Cl atoms, H atoms, and O atoms.

- Place a 4 in front of the formula HCl to give a total of 8 H atoms and 4 Cl atoms in the products.

$$SnCl_4(s) + H_2O(l) \longrightarrow Sn(OH)_4(s) + 4HCl(aq)$$

- Place a 4 in front of the formula H_2O to give 8 H atoms and 4 O atoms in the reactants.

$$SnCl_4(s) + 4H_2O(l) \longrightarrow Sn(OH)_4(s) + 4HCl(aq)$$

- The total number of Sn (1), Cl (4), H (8), and O (4) atoms is now equal on both sides of the equation.

$$SnCl_4(s) + 4H_2O(l) \longrightarrow Sn(OH)_4(s) + 4HCl(aq)$$

Classifying Types of Chemical Reactions (8.3)

- Chemical reactions are classified by identifying general patterns in their equations.

- In a combination reaction, two or more elements or compounds bond to form one product.
- In a decomposition reaction, a single reactant splits into two or more products.
- In a single replacement reaction, an uncombined element takes the place of an element in a compound.
- In a double replacement reaction, the positive ions in the reacting compounds switch places.
- In a combustion reaction, a carbon-containing compound that is the fuel burns in oxygen from the air to produce carbon dioxide (CO_2), water (H_2O), and energy.

Example: Classify the type of the following reaction:

$$2Al(s) + Fe_2O_3(s) \xrightarrow{\Delta} Al_2O_3(s) + 2Fe(l)$$

Answer: The iron in iron(III) oxide is replaced by aluminum, which makes this a single replacement reaction.

Identifying Oxidized and Reduced Substances (8.4)

- In an oxidation–reduction reaction (abbreviated *redox*), one reactant is oxidized when it loses electrons, and another reactant is reduced when it gains electrons.
- Oxidation is the *loss* of electrons; reduction is the *gain* of electrons.

Example: For the following redox reaction, identify the reactant that is oxidized, and the reactant that is reduced:

$$Fe(s) + Cu^{2+}(aq) \longrightarrow Fe^{2+}(aq) + Cu(s)$$

Answer: $Fe^0(s) \longrightarrow Fe^{2+}(aq) + 2e^-$

Fe loses electrons; it is oxidized.

$$Cu^{2+}(aq) + 2e^- \longrightarrow Cu^0(s)$$

Cu^{2+} gains electrons; it is reduced.

UNDERSTANDING THE CONCEPTS

The chapter sections to review are shown in parentheses at the end of each question.

8.35 Balance each of the following by adding coefficients, and identify the type of reaction for each: (8.1, 8.2, 8.3)

a. __ ●●● + __ ●● ⟶ __ ●●

b. __ ●●● ⟶ __ ●● + __ ●●

8.36 Balance each of the following by adding coefficients, and identify the type of reaction for each: (8.1, 8.2, 8.3)

a. __ ●○ + __ ●● ⟶ __ ●● + __ ●○

b. __ ●● + __ ● ⟶ __ ●●●

8.37 If red spheres represent oxygen atoms, blue spheres represent nitrogen atoms, and all the molecules are gases, (8.1, 8.2, 8.3)

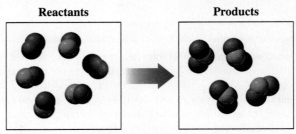

Reactants   Products

a. write the formula for each of the reactants and products.
b. write a balanced equation for the reaction.
c. indicate the type of reaction as combination, decomposition, single replacement, double replacement, or combustion.

8.38 If purple spheres represent iodine atoms, white spheres represent hydrogen atoms, and all the molecules are gases, (8.1, 8.2, 8.3)

Reactants   **Products**

   a. write the formula for each of the reactants and products.
   b. write a balanced equation for the reaction.
   c. indicate the type of reaction as combination, decomposition, single replacement, double replacement, or combustion.

8.39 If blue spheres represent nitrogen atoms, purple spheres represent iodine atoms, the reacting molecules are solid, and the products are gases, (8.1, 8.2, 8.3)

Reactants   **Products**

   a. write the formula for each of the reactants and products.
   b. write a balanced equation for the reaction.
   c. indicate the type of reaction as combination, decomposition, single replacement, double replacement, or combustion.

8.40 If green spheres represent chlorine atoms, yellow-green spheres represent fluorine atoms, white spheres represent hydrogen atoms, and all the molecules are gases, (8.1, 8.2, 8.3)

Reactants   **Products**

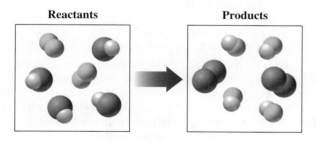

   a. write the formula for each of the reactants and products.
   b. write a balanced equation for the reaction.
   c. indicate the type of reaction as combination, decomposition, single replacement, double replacement, or combustion.

8.41 If green spheres represent chlorine atoms, red spheres represent oxygen atoms, and all the molecules are gases, (8.1, 8.2, 8.3)

Reactants   **Products**

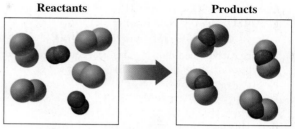

   a. write the formula for each of the reactants and products.
   b. write a balanced equation for the reaction.
   c. indicate the type of reaction as combination, decomposition, single replacement, double replacement, or combustion.

8.42 If blue spheres represent nitrogen atoms, purple spheres represent iodine atoms, the reacting molecules are gases, and the products are solid, (8.1, 8.2, 8.3)

Reactants   **Products**

   a. write the formula for each of the reactants and products.
   b. write a balanced equation for the reaction.

ADDITIONAL QUESTIONS AND PROBLEMS

8.43 Identify the type of reaction for each of the following as combination, decomposition, single replacement, double replacement, or combustion: (8.3)
   a. A metal and a nonmetal form an ionic compound.
   b. A compound of hydrogen and carbon reacts with oxygen to produce carbon dioxide and water.
   c. Heating calcium carbonate produces calcium oxide and carbon dioxide.
   d. Zinc replaces copper in $Cu(NO_3)_2$.

8.44 Identify the type of reaction for each of the following as combination, decomposition, single replacement, double replacement, or combustion: (8.3)
   a. A compound breaks apart into its elements.
   b. Copper and bromine form copper(II) bromide.
   c. Iron(II) sulfite breaks down to iron(II) oxide and sulfur dioxide.
   d. Silver ion from $AgNO_3(aq)$ forms a solid with bromide ion from $KBr(aq)$.

8.45 Balance each of the following chemical equations, and identify the type of reaction: (8.1, 8.2, 8.3)

a. $NH_3(g) + HCl(g) \longrightarrow NH_4Cl(s)$
b. $C_4H_8(g) + O_2(g) \xrightarrow{\Delta} CO_2(g) + H_2O(g)$
c. $Sb(s) + Cl_2(g) \longrightarrow SbCl_3(s)$
d. $NI_3(s) \longrightarrow N_2(g) + I_2(g)$
e. $KBr(aq) + Cl_2(aq) \longrightarrow KCl(aq) + Br_2(l)$
f. $Fe(s) + H_2SO_4(aq) \longrightarrow H_2(g) + Fe_2(SO_4)_3(aq)$
g. $Al_2(SO_4)_3(aq) + NaOH(aq) \longrightarrow$
$$Al(OH)_3(s) + Na_2SO_4(aq)$$

8.46 Balance each of the following chemical equations, and identify the type of reaction: (8.1, 8.2, 8.3)

a. $Si_3N_4(s) \longrightarrow Si(s) + N_2(g)$
b. $Mg(s) + N_2(g) \longrightarrow Mg_3N_2(s)$
c. $Al(s) + H_3PO_4(aq) \longrightarrow H_2(g) + AlPO_4(aq)$
d. $C_3H_4(g) + O_2(g) \xrightarrow{\Delta} CO_2(g) + H_2O(g)$
e. $Cr_2O_3(s) + H_2(g) \longrightarrow Cr(s) + H_2O(g)$
f. $Al(s) + Cl_2(g) \longrightarrow AlCl_3(s)$
g. $MgCl_2(aq) + AgNO_3(aq) \longrightarrow AgCl(s) + Mg(NO_3)_2(aq)$

8.47 Predict the products and write a balanced equation for each of the following: (8.1, 8.2, 8.3)

a. single replacement:
$Zn(s) + HCl(aq) \longrightarrow$ _____ + _____
b. decomposition: $BaCO_3(s) \xrightarrow{\Delta}$ _____ + _____
c. double replacement:
$NaOH(aq) + HCl(aq) \longrightarrow$ _____ + _____
d. combination: $Al(s) + F_2(g) \longrightarrow$ _____

8.48 Predict the products and write a balanced equation for each of the following: (8.1, 8.2, 8.3)

a. decomposition: $NaCl(s) \xrightarrow{\text{Electricity}}$ _____ + _____
b. combination: $Ca(s) + Br_2(l) \longrightarrow$ _____

c. combustion:
$C_2H_4(g) + O_2(g) \xrightarrow{\Delta}$ _____ + _____
d. double replacement:
$NiCl_2(aq) + NaOH(aq) \longrightarrow$ _____ + _____

8.49 Write a balanced equation for each of the following reactions and identify the type of reaction: (8.1, 8.2, 8.3)

a. Sodium metal reacts with oxygen gas to form solid sodium oxide.
b. Aqueous sodium chloride and aqueous silver nitrate react to form solid silver chloride and aqueous sodium nitrate.
c. Gasohol is a fuel that contains liquid ethanol, C_2H_6O, which burns in oxygen gas to form two gases, carbon dioxide and water.

8.50 Write a balanced equation for each of the following reactions and identify the type of reaction: (8.1, 8.2, 8.3)

a. Solid potassium chlorate is heated to form solid potassium chloride and oxygen gas.
b. Carbon monoxide gas and oxygen gas combine to form carbon dioxide gas.
c. Ethene gas, C_2H_4, reacts with chlorine gas, Cl_2, to form dichloroethane, $C_2H_4Cl_2$.

8.51 In each of the following reactions, identify the reactant that is oxidized and the reactant that is reduced: (8.4)

a. $N_2(g) + 2O_2(g) \longrightarrow 2NO_2(g)$
b. $CO(g) + 3H_2(g) \longrightarrow CH_4(g) + H_2O(g)$
c. $Mg(s) + Br_2(l) \longrightarrow MgBr_2(s)$

8.52 In each of the following reactions, identify the reactant that is oxidized and the reactant that is reduced: (8.4)

a. $2Al(s) + 3F_2(g) \longrightarrow 2AlF_3(s)$
b. $ZnO(s) + H_2(g) \longrightarrow Zn(s) + H_2O(g)$
c. $2CuS(s) + 3O_2(g) \longrightarrow 2CuO(s) + 2SO_2(g)$

CHALLENGE QUESTIONS

The following groups of questions are related to the topics in this chapter. However, they do not all follow the chapter order, and they require you to combine concepts and skills from several sections. These questions will help you increase your critical thinking skills and prepare for your next exam.

8.53 Balance each of the following chemical equations, and identify the type of reaction: (8.1, 8.2, 8.3)

a. $K_2O(s) + H_2O(g) \longrightarrow KOH(aq)$
b. $C_8H_{18}(l) + O_2(g) \xrightarrow{\Delta} CO_2(g) + H_2O(g)$
c. $Fe(OH)_3(s) \longrightarrow Fe_2O_3(s) + H_2O(g)$
d. $CuS(s) + HCl(aq) \longrightarrow CuCl_2(aq) + H_2S(g)$

8.54 Balance each of the following chemical equations, and identify the type of reaction: (8.1, 8.2, 8.3)

a. $TiCl_4(s) + Mg(s) \longrightarrow MgCl_2(s) + Ti(s)$
b. $P_4O_{10}(s) + H_2O(g) \xrightarrow{\Delta} H_3PO_4(aq)$
c. $KClO_3(s) \xrightarrow{\Delta} KCl(s) + O_2(g)$
d. $C_3H_6(g) + O_2(g) \xrightarrow{\Delta} CO_2(g) + H_2O(g)$

8.55 Complete and balance each of the following chemical equations: (8.1, 8.2, 8.3)

a. Single replacement: $Fe_3O_4(s) + H_2(g) \longrightarrow$
b. Combustion: $C_4H_{10}(g) + O_2(g) \xrightarrow{\Delta}$
c. Combination: $Al(s) + O_2(g) \longrightarrow$
d. Double replacement: $NaOH(aq) + ZnSO_4(aq) \longrightarrow$

8.56 Complete and balance each of the following chemical equations: (8.1, 8.2, 8.3)

a. Decomposition: $HgO(s) \xrightarrow{\Delta}$
b. Double replacement: $BaCl_2(aq) + AgNO_3(aq) \longrightarrow$
c. Single replacement: $Ca(s) + AlCl_3(s) \longrightarrow$
d. Combination: $Mg(s) + N_2(g) \longrightarrow$

8.57 Write the correct formulas for the reactants and products, the balanced equation for each of the following reaction descriptions, and identify each type of reaction: (8.1, 8.2, 8.3)

a. An aqueous solution of lead(II) nitrate is mixed with aqueous sodium phosphate to produce solid lead(II) phosphate and aqueous sodium nitrate.
b. Gallium metal heated in oxygen gas forms solid gallium(III) oxide.
c. When solid sodium nitrate is heated, solid sodium nitrite and oxygen gas are produced.

8.58 Write the correct formulas for the reactants and products, the balanced equation for each of the following reaction descriptions, and identify each type of reaction: (8.1, 8.2, 8.3)

a. Solid bismuth(III) oxide and solid carbon react to form bismuth metal and carbon monoxide gas.
b. Solid sodium bicarbonate is heated and forms solid sodium carbonate, gaseous carbon dioxide, and liquid water.
c. Liquid hexane, C_6H_{14}, reacts with oxygen gas to form two gaseous products: carbon dioxide and liquid water.

8.59 In the following diagram, if blue spheres are the element X and yellow spheres are the element Y: (8.1, 8.2, 8.3)

Reactants	Products

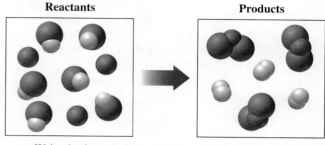

a. Write the formula for each of the reactants and products.
b. Write a balanced equation for the reaction.
c. Indicate the type of reaction as combination, decomposition, single replacement, double replacement, or combustion.

8.60 In the following diagram, if red spheres are the element A, white spheres are the element B, and green spheres are the element C: (8.1, 8.2, 8.3).

Reactants	Products

a. Write the formula for each of the reactants and products.
b. Write a balanced equation for the reaction.
c. Indicate the type of reaction as combination, decomposition, single replacement, double replacement, or combustion.

ANSWERS

Answers to Selected Questions and Problems

8.1 a. reactants/products: 7 O atoms
 b. reactants/products: 4 O atoms
 c. reactants/products: 10 O atoms
 d. reactants/products: 4 O atoms

8.3 a. not balanced   **b.** not balanced
 c. balanced   **d.** balanced

8.5 a. 2 Na atoms, 2 Cl atoms
 b. 1 P atom, 3 Cl atoms, 6 H atoms
 c. 4 P atoms, 16 O atoms, 12 H atoms

8.7 a. $N_2(g) + O_2(g) \longrightarrow 2NO(g)$
 b. $2HgO(s) \xrightarrow{\Delta} 2Hg(l) + O_2(g)$
 c. $4Fe(s) + 3O_2(g) \longrightarrow 2Fe_2O_3(s)$
 d. $2Na(s) + Cl_2(g) \longrightarrow 2NaCl(s)$
 e. $2Cu_2O(s) + O_2(g) \longrightarrow 4CuO(s)$

8.9 a. $Mg(s) + 2AgNO_3(aq) \longrightarrow 2Ag(s) + Mg(NO_3)_2(aq)$
 b. $CuCO_3(s) \longrightarrow CuO(s) + CO_2(g)$
 c. $2Al(s) + 3CuSO_4(aq) \longrightarrow 3Cu(s) + Al_2(SO_4)_3(aq)$
 d. $Pb(NO_3)_2(aq) + 2NaCl(aq) \longrightarrow$
   $PbCl_2(s) + 2NaNO_3(aq)$
 e. $2Al(s) + 6HCl(aq) \longrightarrow 3H_2(g) + 2AlCl_3(aq)$

8.11 a. $Fe_2O_3(s) + 3CO(g) \longrightarrow 2Fe(s) + 3CO_2(g)$
 b. $2Li_3N(s) \longrightarrow 6Li(s) + N_2(g)$
 c. $2Al(s) + 6HBr(aq) \longrightarrow 3H_2(g) + 2AlBr_3(aq)$
 d. $3Ba(OH)_2(aq) + 2Na_3PO_4(aq) \longrightarrow$
   $Ba_3(PO_4)_2(s) + 6NaOH(aq)$
 e. $As_4S_6(s) + 9O_2(g) \longrightarrow As_4O_6(s) + 6SO_2(g)$

8.13 a. $2Li(s) + 2H_2O(l) \longrightarrow H_2(g) + 2LiOH(aq)$
 b. $2P(s) + 5Cl_2(g) \longrightarrow 2PCl_5(s)$
 c. $FeO(s) + CO(g) \longrightarrow Fe(s) + CO_2(g)$
 d. $2C_5H_{10}(l) + 15O_2(g) \xrightarrow{\Delta} 10CO_2(g) + 10H_2O(g)$
 e. $3H_2S(g) + 2FeCl_3(s) \longrightarrow Fe_2S_3(s) + 6HCl(g)$

8.15 $NH_4NO_3(s) \xrightarrow{\Delta} 2H_2O(g) + N_2O(g)$

8.17 $2C_3H_7NO_2(aq) + 6O_2(g) \longrightarrow$
  $5CO_2(g) + 5H_2O(l) + CH_4N_2O(aq)$

8.19 a. decomposition   **b.** single replacement
 c. combustion   **d.** double replacement
 e. combination

8.21 a. combination   **b.** single replacement
 c. decomposition   **d.** double replacement
 e. combustion

8.23 a. $Mg(s) + Cl_2(g) \longrightarrow MgCl_2(s)$
 b. $2HBr(g) \longrightarrow H_2(g) + Br_2(l)$
 c. $Mg(s) + Zn(NO_3)_2(aq) \longrightarrow Zn(s) + Mg(NO_3)_2(aq)$
 d. $K_2S(aq) + Pb(NO_3)_2(aq) \longrightarrow PbS(s) + 2KNO_3(aq)$
 e. $2C_2H_6(g) + 7O_2(g) \xrightarrow{\Delta} 4CO_2(g) + 6H_2O(g)$

8.25 a. reduction   **b.** oxidation
 c. reduction   **d.** reduction

8.27 a. Zn is oxidized; Cl_2 is reduced.
 b. The Br^- in NaBr is oxidized; Cl_2 is reduced.
 c. The O^{2-} in PbO is oxidized; the Pb^{2+} in PbO is reduced.
 d. Sn^{2+} is oxidized; Fe^{3+} is reduced.

8.29 a. reduction   **b.** oxidation

8.31 Linoleic acid gains hydrogen atoms and is reduced.

8.33 a. $C_6H_{12}O_6(aq) + 6O_2(g) \longrightarrow 6CO_2(g) + 6H_2O(l)$
 b. $6CO_2(g) + 6H_2O(l) \longrightarrow C_6H_{12}O_6(aq) + 6O_2(g)$

8.35 a. 1,1,2 combination   **b.** 2,2,1 decomposition

8.37 a. reactants NO and O_2; product NO_2
 b. $2NO(g) + O_2(g) \longrightarrow 2NO_2(g)$
 c. combination

8.39 a. reactant NI_3; products N_2 and I_2
 b. $2NI_3(s) \longrightarrow N_2(g) + 3I_2(g)$
 c. decomposition

8.41 a. reactants Cl_2 and O_2; product OCl_2
 b. $2Cl_2(g) + O_2(g) \longrightarrow 2OCl_2(g)$
 c. combination

8.43 a. combination   **b.** combustion
 c. decomposition   **d.** single replacement

8.45 a. $NH_3(g) + HCl(g) \longrightarrow NH_4Cl(s)$ combination
 b. $C_4H_8(g) + 6O_2(g) \xrightarrow{\Delta} 4CO_2(g) + 4H_2O(g)$ combustion
 c. $2Sb(s) + 3Cl_2(g) \longrightarrow 2SbCl_3(s)$ combination
 d. $2NI_3(s) \longrightarrow N_2(g) + 3I_2(g)$ decomposition
 e. $2KBr(aq) + Cl_2(aq) \longrightarrow 2KCl(aq) + Br_2(l)$
   single replacement

   f. $2Fe(s) + 3H_2SO_4(aq) \longrightarrow 3H_2(g) + Fe_2(SO_4)_3(aq)$

single replacement

   g. $Al_2(SO_4)_3(aq) + 6NaOH(aq) \longrightarrow 2Al(OH)_3(s) +$

$3Na_2SO_4(aq)$ double replacement

8.47 a. $Zn(s) + 2HCl(aq) \longrightarrow H_2(g) + ZnCl_2(aq)$

   b. $BaCO_3(s) \xrightarrow{\Delta} BaO(s) + CO_2(g)$

   c. $NaOH(aq) + HCl(aq) \longrightarrow H_2O(l) + NaCl(aq)$

   d. $2Al(s) + 3F_2(g) \longrightarrow 2AlF_3(s)$

8.49 a. $4Na(s) + O_2(g) \longrightarrow 2Na_2(s)$ combination

   b. $NaCl(aq) + AgNO_3(aq) \longrightarrow AgCl(s) + NaNO_3(aq)$

double replacement

   c. $C_2H_6O(l) + 3O_2(g) \xrightarrow{\Delta} 2CO_2(g) + 3H_2O(g)$

combustion

8.51 a. N_2 is oxidized; O_2 is reduced.

   b. The C in CO is oxidized; H_2 is reduced.

   c. Mg is oxidized; Br_2 is reduced.

8.53 a. $K_2O(s) + H_2O(g) \longrightarrow 2KOH(s)$ combination

   b. $2C_8H_{18}(l) + 25O_2(g) \xrightarrow{\Delta} 16CO_2(g) + 18H_2O(g)$

combustion

   c. $2Fe(OH)_3(s) \longrightarrow Fe_2O_3(s) + 3H_2O(g)$ decomposition

   d. $CuS(s) + 2HCl(aq) \longrightarrow CuCl_2(aq) + H_2S(g)$

double replacement

8.55 a. $Fe_3O_4(s) + 4H_2(g) \longrightarrow 3Fe(s) + 4H_2O(g)$

   b. $2C_4H_{10}(g) + 13O_2(g) \xrightarrow{\Delta} 8CO_2(g) + 10H_2O(g)$

   c. $4Al(s) + 3O_2(g) \longrightarrow 2Al_2O_3(s)$

   d. $NaOH(aq) + ZnSO_4(aq) \longrightarrow Zn(OH)_2(s) + Na_2SO_4(aq)$

8.57 a. $3Pb(NO_3)_2(aq) + 2Na_3PO_4(aq) \longrightarrow Pb_3(PO_4)_2(s) +$

$6NaNO_3(aq)$ double replacement

   b. $4Ga(s) + 3O_2(g) \xrightarrow{\Delta} 2Ga_2O_3(s)$ combination

   c. $2NaNO_3(s) \xrightarrow{\Delta} 2NaNO_2(s) + O_2(g)$ decomposition

8.59 a. reactants: X and Y_2; product: XY_3

   b. $2X + 3Y_2 \longrightarrow 2XY_3$

   c. combination

Chemical Quantities in Reactions

9

LANCE, AN ENVIRONMENTAL scientist, is collecting soil and water samples at a nearby farm to test for the presence and concentration of any pesticides and pharmaceuticals. Farmers use pesticides to increase food production and pharmaceuticals to treat and prevent animal-related diseases. Due to the common use of these chemicals, they may pass into the soil and water supply, potentially contaminating the environment and causing health problems.

Recently, a farmer treated his cotton and bean fields with a pesticide called Sevin (carbaryl). A few days later, Lance collected samples of soil and water and detected small amounts of Sevin. In humans, Sevin, which is an acetylcholinesterase inhibitor, can cause headaches, nausea, and paralysis of the respiratory system. Because it is very soluble, it is important that the residue pesticide does not contaminate the water supply. Lance advised the farmer to decrease the amount of pesticide he uses on his crops to reduce the pesticide level. He indicates that he will return in a week to retest the soil and water for Sevin.

CAREER
Environmental Scientist

Environmental science is a multidisciplinary field that combines chemistry, biology, ecology, and geology to study environmental problems. Environmental scientists monitor environmental pollution to protect the health of the public. By using specialized equipment, environmental scientists measure pollution levels in soil, air, and water, as well as noise and radiation levels. They can specialize in a specific area, such as air quality or hazardous and solid waste. For instance, air-quality experts monitor indoor air for allergens, mold, and toxins; they measure outdoor air pollutants created by businesses, vehicles, and agriculture. Since environmental scientists obtain samples containing potentially hazardous materials, they must be knowledgeable about safety protocols and wear personal protective equipment. They may also recommend methods to diminish various pollutants, and may assist in cleanup and remediation efforts.

KEY MATH SKILL

- Calculating Percentages (1.4)

 CORE CHEMISTRY SKILLS

- Counting Significant Figures (2.2)
- Using Significant Figures in Calculations (2.3)
- Writing Conversion Factors from Equalities (2.5)
- Using Conversion Factors (2.6)
- Using Energy Units (3.4)

- Calculating Molar Mass (7.2)
- Using Molar Mass as a Conversion Factor (7.3)
- Balancing a Chemical Equation (8.2)
- Classifying Types of Chemical Reactions (8.3)

* These Key Math Skills and Core Chemistry Skills from previous chapters are listed here for your review as you proceed to the new material in this chapter.

LOOKING AHEAD

9.1 Conservation of Mass

LEARNING GOAL Calculate the total mass of reactants and the total mass of products in a balanced chemical equation.

In any chemical reaction, the total amount of matter in the reactants is equal to the total amount of matter in the products. Thus, the total mass of all the reactants must be equal to the total mass of all the products. This is known as the **law of conservation of mass**, which states that there is no change in the total mass of the substances reacting in a balanced chemical reaction. Thus, no material is lost or gained as original substances are changed to new substances.

$2Ag(s)$ $+$ $S(s)$ $\longrightarrow$ $Ag_2S(s)$

Mass of reactants $=$ Mass of product

In the chemical reaction of Ag and S, the mass of the reactants is the same as the mass of the product, Ag_2S.

For example, tarnish (Ag_2S) forms when silver reacts with sulfur to form silver sulfide.

$$2Ag(s) + S(s) \longrightarrow Ag_2S(s)$$

In this reaction, the number of silver atoms that reacts is twice the number of sulfur atoms. When 200 silver atoms react, 100 sulfur atoms are required. However, in the actual chemical reaction, many more atoms of both silver and sulfur would react. If we are dealing with molar amounts, then the coefficients in the equation can be interpreted in terms of moles. Thus, 2 mol of silver reacts with 1 mol of sulfur to produce 1 mol of Ag_2S. Because the molar mass of each can be determined, the moles of Ag, S, and Ag_2S can also be stated in terms of mass in grams of each. Thus, 215.8 g of Ag and 32.1 g of S react to form 247.9 g of Ag_2S. The total mass of the reactants (247.9 g) is equal to the mass of the product (247.9 g). The various ways in which a chemical equation can be interpreted are seen in **TABLE 9.1**.

ENGAGE

Why is the total mass of the reactants equal to the total mass of the products?

TABLE 9.1 Information Available from a Balanced Equation

	Reactants		Products
Equation	$2Ag(s)$	$+ S(s)$	$\longrightarrow Ag_2S(s)$
Atoms/Formula Units	2 Ag atoms	+ 1 S atom	$\longrightarrow$ 1 Ag_2S formula unit
	200 Ag atoms	+ 100 S atoms	$\longrightarrow$ 100 Ag_2S formula units
Avogadro's Number of Atoms	$2(6.022 \times 10^{23})$ Ag atoms	$+ 1(6.022 \times 10^{23})$ S atoms	$\longrightarrow 1(6.022 \times 10^{23})$ Ag_2S formula units
Moles	2 mol of Ag	+ 1 mol of S	$\longrightarrow$ 1 mol of Ag_2S
Mass (g)	2(107.9 g) of Ag	+ 1(32.07 g) of S	$\longrightarrow$ 1(247.9 g) of Ag_2S
Total Mass (g)	247.9 g		$\longrightarrow$ 247.9 g

SAMPLE PROBLEM 9.1 Conservation of Mass

The combustion of methane (CH_4) with oxygen produces carbon dioxide, water, and energy. Calculate the total mass of the reactants and the products for the following equation when 1 mol of CH_4 reacts:

$$CH_4(g) + 2O_2(g) \xrightarrow{\Delta} CO_2(g) + 2H_2O(g)$$

TRY IT FIRST

SOLUTION

Interpreting the coefficients in the equation as the number of moles of each substance and multiplying by its molar mass gives the total mass of reactants and products. The quantities of moles are exact because the coefficients in the balanced equation are exact.

	Reactants		Products
Equation	$CH_4(g) + 2O_2(g)$	$\xrightarrow{\Delta}$	$CO_2(g) + 2H_2O(g)$
Moles	1 mol of CH_4 + 2 mol of O_2	$\longrightarrow$	1 mol of CO_2 + 2 mol of H_2O
Mass	16.04 g of CH_4 + 64.00 g of O_2	$\longrightarrow$	44.01 g of CO_2 + 36.03 g of H_2O
Total Mass	80.04 g of reactants	$=$	80.04 g of products

STUDY CHECK 9.1

Calculate the total mass of the reactants and the products for the following equation:

$$4K(s) + O_2(g) \longrightarrow 2K_2O(s)$$

ANSWER

188.4 g of reactants and 188.4 g of products

QUESTIONS AND PROBLEMS

9.1 Conservation of Mass

LEARNING GOAL Calculate the total mass of reactants and the total mass of products in a balanced chemical equation.

9.1 Calculate the total mass of the reactants and the products for each of the following equations:

a. $2SO_2(g) + O_2(g) \longrightarrow 2SO_3(g)$

b. $4P(s) + 5O_2(g) \longrightarrow 2P_2O_5(s)$

9.2 Calculate the total mass of the reactants and the products for each of the following equations:

a. $2Al(s) + 3Cl_2(g) \longrightarrow 2AlCl_3(s)$

b. $4HCl(g) + O_2(g) \longrightarrow 2Cl_2(g) + 2H_2O(g)$

9.2 Calculating Moles Using Mole–Mole Factors

LEARNING GOAL Use a mole–mole factor from a balanced chemical equation to calculate the number of moles of another substance in the reaction.

When iron reacts with sulfur, the product is iron(III) sulfide.

$$2Fe(s) + 3S(s) \longrightarrow Fe_2S_3(s)$$

| Iron (Fe) | | Sulfur (S) | | Iron(III) sulfide (Fe_2S_3) |
| $2Fe(s)$ | + | $3S(s)$ | $\longrightarrow$ | $Fe_2S_3(s)$ |

From the balanced equation, we see that 2 mol of iron reacts with 3 mol of sulfur to form 1 mol of iron(III) sulfide. Actually, any amount of iron or sulfur may be used, but the *ratio* of iron reacting with sulfur will always be the same. From the coefficients, we can write **mole–mole factors** between reactants and between reactants and products. The coefficients used in the mole–mole factors are exact numbers; they do not limit the number of significant figures.

Fe and S: $\dfrac{2 \text{ mol Fe}}{3 \text{ mol S}}$ and $\dfrac{3 \text{ mol S}}{2 \text{ mol Fe}}$

Fe and Fe_2S_3: $\dfrac{2 \text{ mol Fe}}{1 \text{ mol Fe}_2S_3}$ and $\dfrac{1 \text{ mol Fe}_2S_3}{2 \text{ mol Fe}}$

S and Fe_2S_3: $\dfrac{3 \text{ mol S}}{1 \text{ mol Fe}_2S_3}$ and $\dfrac{1 \text{ mol Fe}_2S_3}{3 \text{ mol S}}$

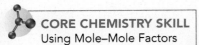

CORE CHEMISTRY SKILL
Using Mole–Mole Factors

Using Mole–Mole Factors in Calculations

Whenever you prepare a recipe, adjust an engine for the proper mixture of fuel and air, or prepare medicines in a pharmaceutical laboratory, you need to know the proper amounts of reactants to use and how much of the product will form. Now that we have written all the possible conversion factors for the balanced equation $2Fe(s) + 3S(s) \longrightarrow$ $Fe_2S_3(s)$, we will use those mole–mole factors in a chemical calculation in Sample Problem 9.2.

SAMPLE PROBLEM 9.2 Calculating Moles of a Reactant

In the chemical reaction of iron and sulfur, how many moles of sulfur are needed to react with 1.42 mol of iron?

$$2Fe(s) + 3S(s) \longrightarrow Fe_2S_3(s)$$

TRY IT FIRST

SOLUTION

STEP 1 State the given and needed quantities (moles).

ANALYZE THE PROBLEM	Given	Need	Connect
	1.42 mol of Fe	moles of S	mole–mole factor
	Equation		
	$2Fe(s) + 3S(s) \longrightarrow Fe_2S_3(s)$		

STEP 2 Write a plan to convert the given to the needed quantity (moles).

moles of Fe → Mole–mole factor → moles of S

STEP 3 Use coefficients to write mole–mole factors.

$$2 \text{ mol of Fe} = 3 \text{ mol of S}$$
$$\frac{2 \text{ mol Fe}}{3 \text{ mol S}} \quad \text{and} \quad \frac{3 \text{ mol S}}{2 \text{ mol Fe}}$$

STEP 4 Set up the problem to give the needed quantity (moles).

$$1.42 \text{ mol Fe} \times \underbrace{\frac{3 \text{ mol S}}{2 \text{ mol Fe}}}_{\text{Exact}} = 2.13 \text{ mol of S}$$

Three SFs   Exact   Three SFs

STUDY CHECK 9.2

Using the equation in Sample Problem 9.2, calculate the number of moles of iron needed to react with 2.75 mol of sulfur.

ANSWER

1.83 mol of iron

> ## Guide to Calculating the Quantities of Reactants and Products in a Chemical Reaction
>
> **STEP 1**
> State the given and needed quantities (moles or grams).
>
> **STEP 2**
> Write a plan to convert the given to the needed quantity (moles or grams).
>
> **STEP 3**
> Use coefficients to write mole–mole factors; write molar masses if needed.
>
> **STEP 4**
> Set up the problem to give the needed quantity (moles or grams).

SAMPLE PROBLEM 9.3 Calculating Moles of a Product

Propane gas (C_3H_8), a fuel used in camp stoves, soldering torches, and specially equipped automobiles, reacts with oxygen to produce carbon dioxide, water, and energy. How many moles of CO_2 can be produced when 2.25 mol of C_3H_8 reacts?

$$C_3H_8(g) + 5O_2(g) \xrightarrow{\Delta} 3CO_2(g) + 4H_2O(g)$$

TRY IT FIRST

Propane fuel reacts with O_2 in the air to produce CO_2, H_2O, and energy.

SOLUTION

STEP 1 State the given and needed quantities (moles).

ANALYZE THE PROBLEM	Given	Need	Connect
	2.25 mol of C_3H_8	moles of CO_2	mole–mole factor
	Equation		
	$C_3H_8(g) + 5O_2(g) \xrightarrow{\Delta} 3CO_2(g) + 4H_2O(g)$		

STEP 2 Write a plan to convert the given to the needed quantity (moles).

moles of C_3H_8 [Mole–mole factor] moles of CO_2

STEP 3 Use coefficients to write mole–mole factors.

$$1 \text{ mol of } C_3H_8 = 3 \text{ mol of } CO_2$$
$$\frac{3 \text{ mol } CO_2}{1 \text{ mol } C_3H_8} \quad \text{and} \quad \frac{1 \text{ mol } C_3H_8}{3 \text{ mol } CO_2}$$

STEP 4 Set up the problem to give the needed quantity (moles).

$$2.25 \text{ mol } C_3H_8 \times \underset{\text{Exact}}{\frac{3 \text{ mol } CO_2}{1 \text{ mol } C_3H_8}} = 6.75 \text{ mol of } CO_2$$

Three SFs   Exact   Three SFs

STUDY CHECK 9.3

Using the equation in Sample Problem 9.3, calculate the moles of oxygen that must react to produce 0.756 mol of water.

ANSWER

0.945 mol of O_2

QUESTIONS AND PROBLEMS

9.2 Calculating Moles Using Mole–Mole Factors

LEARNING GOAL Use a mole–mole factor from a balanced chemical equation to calculate the number of moles of another substance in the reaction.

9.3 Write all of the mole–mole factors for each of the following equations:
  a. $2SO_2(g) + O_2(g) \longrightarrow 2SO_3(g)$
  b. $4P(s) + 5O_2(g) \longrightarrow 2P_2O_5(s)$

9.4 Write all of the mole–mole factors for each of the following equations:
  a. $2Al(s) + 3Cl_2(g) \longrightarrow 2AlCl_3(s)$
  b. $4HCl(g) + O_2(g) \longrightarrow 2Cl_2(g) + 2H_2O(g)$

9.5 For the equations in problem 9.3, write the setup with the correct mole–mole factor:
  a. moles of SO_3 from the moles of SO_2
  b. moles of O_2 needed to react with moles of P

9.6 For the equations in problem 9.4, write the setup with the correct mole–mole factor:
  a. moles of $AlCl_3$ from the moles of Cl_2
  b. moles of O_2 needed to react with moles of HCl

9.7 The reaction of hydrogen with oxygen produces water.

$$2H_2(g) + O_2(g) \longrightarrow 2H_2O(g)$$

  a. How many moles of O_2 are required to react with 2.6 mol of H_2?
  b. How many moles of H_2 are needed to react with 5.0 mol of O_2?
  c. How many moles of H_2O form when 2.5 mol of O_2 reacts?

9.8 Ammonia is produced by the reaction of nitrogen and hydrogen.

$$N_2(g) + 3H_2(g) \longrightarrow 2NH_3(g)$$
$$\text{Ammonia}$$

a. How many moles of H_2 are needed to react with 1.8 mol of N_2?
b. How many moles of N_2 reacted if 0.60 mol of NH_3 is produced?
c. How many moles of NH_3 are produced when 1.4 mol of H_2 reacts?

9.9 Carbon disulfide and carbon monoxide are produced when carbon is heated with sulfur dioxide.

$$5C(s) + 2SO_2(g) \xrightarrow{\Delta} CS_2(l) + 4CO(g)$$

a. How many moles of C are needed to react with 0.500 mol of SO_2?
b. How many moles of CO are produced when 1.2 mol of C reacts?
c. How many moles of SO_2 are required to produce 0.50 mol of CS_2?
d. How many moles of CS_2 are produced when 2.5 mol of C reacts?

9.10 In the acetylene torch, acetylene gas (C_2H_2) burns in oxygen to produce carbon dioxide, water, and energy.

$$2C_2H_2(g) + 5O_2(g) \xrightarrow{\Delta} 4CO_2(g) + 2H_2O(g)$$

a. How many moles of O_2 are needed to react with 2.40 mol of C_2H_2?
b. How many moles of CO_2 are produced when 3.5 mol of C_2H_2 reacts?
c. How many moles of C_2H_2 are required to produce 0.50 mol of H_2O?
d. How many moles of CO_2 are produced when 0.100 mol of O_2 reacts?

9.3 Mass Calculations for Reactions

LEARNING GOAL Given the mass in grams of a substance in a reaction, calculate the mass in grams of another substance in the reaction.

When we have the balanced chemical equation for a reaction, we can use the mass of one of the substances (A) in the reaction to calculate the mass of another substance (B) in the reaction. However, the calculations require us to convert the mass of A to moles of A using the molar mass for A. Then we use the mole–mole factor that links substance A to substance B, which we obtain from the coefficients in the balanced equation. This mole–mole factor (B/A) will convert the moles of A to moles of B. Then the molar mass of B is used to calculate the grams of substance B.

Substance A			Substance B	
grams of A	**Molar mass A**	moles of A	**Mole–mole factor B/A**	moles of B	**Molar mass B**	grams of B

SAMPLE PROBLEM 9.4 Grams of Product from Grams of Reactant

In the engines of cars and trucks, nitrogen and oxygen from the air react at high temperature to produce nitrogen oxide, a component of smog. Complete the following to help answer the question: How many grams of NO can be produced when 12.5 g of O_2 reacts?

$$N_2(g) + O_2(g) \longrightarrow 2NO(g)$$

a. What molar mass is needed to convert grams of O_2 to moles of O_2?
b. What mole–mole factor is needed to convert moles of O_2 to moles of NO?
c. What molar mass is needed to convert moles of NO to grams of NO?

CORE CHEMISTRY SKILL
Converting Grams to Grams

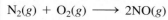
TRY IT FIRST

SOLUTION

a. The molar mass that gives the moles of O_2 is

$$\frac{1 \text{ mol } O_2}{32.00 \text{ g } O_2}$$

b. The mole–mole factor that gives the moles of NO is

$$\frac{2 \text{ mol NO}}{1 \text{ mol } O_2}$$

c. The molar mass that gives the grams of NO is

$$\frac{30.01 \text{ g NO}}{1 \text{ mol NO}}$$

STUDY CHECK 9.4

Using the equation in Sample Problem 9.4, what molar mass is needed to convert grams of N_2 to moles of N_2?

ANSWER

$$\frac{1 \text{ mol } N_2}{28.02 \text{ g } N_2}$$

SAMPLE PROBLEM 9.5 Grams of Product

When acetylene, C_2H_2, burns in oxygen, high temperatures are produced that are used for welding metals.

$$2C_2H_2(g) + 5O_2(g) \xrightarrow{\Delta} 4CO_2(g) + 2H_2O(g)$$

How many grams of CO_2 are produced when 54.6 g of C_2H_2 is burned?

▶ **TRY IT FIRST**

SOLUTION

STEP 1 State the given and needed quantities (grams).

A mixture of acetylene and oxygen undergoes combustion during the welding of metals.

	Given	Need	Connect
ANALYZE THE PROBLEM	54.6 g of C_2H_2	grams of CO_2	molar masses, mole–mole factor
	Equation		
	$2C_2H_2(g) + 5O_2(g) \xrightarrow{\Delta} 4CO_2(g) + 2H_2O(g)$		

STEP 2 Write a plan to convert the given to the needed quantity (grams).

grams of C_2H_2 → [Molar mass] → moles of C_2H_2 → [Mole–mole factor] → moles of CO_2 → [Molar mass] → grams of CO_2

STEP 3 Use coefficients to write mole–mole factors; write molar masses.

1 mol of C_2H_2 = 26.04 g of C_2H_2

$$\frac{26.04 \text{ g } C_2H_2}{1 \text{ mol } C_2H_2} \quad \text{and} \quad \frac{1 \text{ mol } C_2H_2}{26.04 \text{ g } C_2H_2}$$

2 mol of C_2H_2 = 4 mol of CO_2

$$\frac{2 \text{ mol } C_2H_2}{4 \text{ mol } CO_2} \quad \text{and} \quad \frac{4 \text{ mol } CO_2}{2 \text{ mol } C_2H_2}$$

1 mol of CO_2 = 44.01 g of CO_2

$$\frac{44.01 \text{ g } CO_2}{1 \text{ mol } CO_2} \quad \text{and} \quad \frac{1 \text{ mol } CO_2}{44.01 \text{ g } CO_2}$$

STEP 4 Set up the problem to give the needed quantity (grams).

Exact   Exact   Four SFs

$$54.6 \text{ g } C_2H_2 \times \frac{1 \text{ mol } C_2H_2}{26.04 \text{ g } C_2H_2} \times \frac{4 \text{ mol } CO_2}{2 \text{ mol } C_2H_2} \times \frac{44.01 \text{ g } CO_2}{1 \text{ mol } CO_2} = 185 \text{ g of } CO_2$$

Three SFs   Four SFs   Exact   Exact   Three SFs

ENGAGE

Why is a mole–mole factor needed in the problem setup that converts grams of a reactant to grams of a product?

STUDY CHECK 9.5

Using the equation in Sample Problem 9.5, calculate the grams of CO_2 that can be produced when 25.0 g of O_2 reacts.

ANSWER

27.5 g of CO_2

SAMPLE PROBLEM 9.6 Grams of Reactant

The fuel heptane (C_7H_{16}) is designated as the zero point in the octane rating of gasoline. Heptane is an undesirable compound in gasoline because it burns rapidly and causes engine knocking. How many grams of O_2 are required to react with 22.5 g of C_7H_{16}?

$$C_7H_{16}(l) + 11O_2(g) \xrightarrow{\Delta} 7CO_2(g) + 8H_2O(g)$$

TRY IT FIRST

SOLUTION

STEP 1 State the given and needed quantities (grams).

	Given	Need	Connect
ANALYZE THE PROBLEM	22.5 g of C_7H_{16}	grams of O_2	molar masses, mole–mole factor
	Equation		
	$C_7H_{16}(l) + 11O_2(g) \xrightarrow{\Delta} 7CO_2(g) + 8H_2O(g)$		

The octane rating is a measure of how well a fuel burns in an engine.

STEP 2 Write a plan to convert the given to the needed quantity (grams).

grams of C_7H_{16} → [Molar mass] → moles of C_7H_{16} → [Mole–mole factor] → moles of O_2 → [Molar mass] → grams of O_2

STEP 3 Use coefficients to write mole–mole factors; write molar masses.

$$1 \text{ mol of } C_7H_{16} = 100.2 \text{ g of } C_7H_{16}$$

$$\frac{100.2 \text{ g } C_7H_{16}}{1 \text{ mol } C_7H_{16}} \quad \text{and} \quad \frac{1 \text{ mol } C_7H_{16}}{100.2 \text{ g } C_7H_{16}}$$

$$1 \text{ mol of } C_7H_{16} = 11 \text{ mol of } O_2 \qquad\qquad 1 \text{ mol of } O_2 = 32.00 \text{ g of } O_2$$

$$\frac{11 \text{ mol } O_2}{1 \text{ mol } C_7H_{16}} \quad \text{and} \quad \frac{1 \text{ mol } C_7H_{16}}{11 \text{ mol } O_2} \qquad \frac{32.00 \text{ g } O_2}{1 \text{ mol } O_2} \quad \text{and} \quad \frac{1 \text{ mol } O_2}{32.00 \text{ g } O_2}$$

STEP 4 Set up the problem to give the needed quantity (grams).

Exact   Exact   Four SFs

$$22.5 \text{ g } C_7H_{16} \times \frac{1 \text{ mol } C_7H_{16}}{100.2 \text{ g } C_7H_{16}} \times \frac{11 \text{ mol } O_2}{1 \text{ mol } C_7H_{16}} \times \frac{32.00 \text{ g } O_2}{1 \text{ mol } O_2} = 79.1 \text{ g of } O_2$$

Three SFs   Four SFs   Exact   Exact   Three SFs

STUDY CHECK 9.6

Using the equation in Sample Problem 9.6, calculate the grams of C_7H_{16} that are needed to produce 15.0 g of H_2O.

ANSWER

10.4 g of C_7H_{16}

QUESTIONS AND PROBLEMS

9.3 Mass Calculations for Reactions

LEARNING GOAL Given the mass in grams of a substance in a reaction, calculate the mass in grams of another substance in the reaction.

9.11 Sodium reacts with oxygen to produce sodium oxide.

$$4Na(s) + O_2(g) \longrightarrow 2Na_2O(s)$$

a. How many grams of Na_2O are produced when 57.5 g of Na reacts?
b. If you have 18.0 g of Na, how many grams of O_2 are required for reaction?
c. How many grams of O_2 are needed in a reaction that produces 75.0 g of Na_2O?

9.12 Nitrogen gas reacts with hydrogen gas to produce ammonia.

$$N_2(g) + 3H_2(g) \longrightarrow 2NH_3(g)$$

a. If you have 3.64 g of H_2, how many grams of NH_3 can be produced?
b. How many grams of H_2 are needed to react with 2.80 g of N_2?
c. How many grams of NH_3 can be produced from 12.0 g of H_2?

9.13 Ammonia and oxygen react to form nitrogen and water.

$$4NH_3(g) + 3O_2(g) \longrightarrow 2N_2(g) + 6H_2O(g)$$

a. How many grams of O_2 are needed to react with 13.6 g of NH_3?
b. How many grams of N_2 can be produced when 6.50 g of O_2 reacts?
c. How many grams of H_2O are formed from the reaction of 34.0 g of NH_3?

9.14 Iron(III) oxide reacts with carbon to give iron and carbon monoxide.

$$Fe_2O_3(s) + 3C(s) \longrightarrow 2Fe(s) + 3CO(g)$$

a. How many grams of C are required to react with 16.5 g of Fe_2O_3?
b. How many grams of CO are produced when 36.0 g of C reacts?

c. How many grams of Fe can be produced when 6.00 g of Fe_2O_3 reacts?

9.15 Nitrogen dioxide and water react to produce nitric acid, HNO_3, and nitrogen oxide.

$$3NO_2(g) + H_2O(l) \longrightarrow 2HNO_3(aq) + NO(g)$$

a. How many grams of H_2O are required to react with 28.0 g of NO_2?
b. How many grams of NO are produced from 15.8 g of H_2O?
c. How many grams of HNO_3 are produced from 8.25 g of NO_2?

9.16 Calcium cyanamide, $CaCN_2$, reacts with water to form calcium carbonate and ammonia.

$$CaCN_2(s) + 3H_2O(l) \longrightarrow CaCO_3(s) + 2NH_3(g)$$

a. How many grams of H_2O are needed to react with 75.0 g of $CaCN_2$?
b. How many grams of NH_3 are produced from 5.24 g of $CaCN_2$?
c. How many grams of $CaCO_3$ form if 155 g of H_2O reacts?

9.17 When solid lead(II) sulfide reacts with oxygen gas, the products are solid lead(II) oxide and sulfur dioxide gas.
a. Write the balanced equation for the reaction.
b. How many grams of oxygen are required to react with 29.9 g of lead(II) sulfide?
c. How many grams of sulfur dioxide can be produced when 65.0 g of lead(II) sulfide reacts?
d. How many grams of lead(II) sulfide are used to produce 128 g of lead(II) oxide?

9.18 When the gases dihydrogen sulfide and oxygen react, they form the gases sulfur dioxide and water vapor.
a. Write the balanced equation for the reaction.
b. How many grams of oxygen are required to react with 2.50 g of dihydrogen sulfide?
c. How many grams of sulfur dioxide can be produced when 38.5 g of oxygen reacts?
d. How many grams of oxygen are required to produce 55.8 g of water vapor?

9.4 Limiting Reactants

LEARNING GOAL Identify a limiting reactant when given the quantities of two reactants; calculate the amount of product formed from the limiting reactant.

When we make peanut butter sandwiches for lunch, we need 2 slices of bread and 1 tablespoon of peanut butter for each sandwich. As an equation, we could write:

2 slices of bread + 1 tablespoon of peanut butter $\longrightarrow$ 1 peanut butter sandwich

If we have 8 slices of bread and a full jar of peanut butter, we will run out of bread after we make 4 peanut butter sandwiches. We cannot make any more sandwiches once the bread is used up, even though there is a lot of peanut butter left in the jar. The number of slices of bread has limited the number of sandwiches we can make.

On a different day, we might have 8 slices of bread but only a tablespoon of peanut butter left in the peanut butter jar. We will run out of peanut butter after we make just 1 peanut butter sandwich and have 6 slices of bread left over. The small amount of peanut butter available has limited the number of sandwiches we can make.

The reactant that is completely used up is the **limiting reactant**. The reactant that does not completely react and is left over is called the *excess reactant*.

Bread	Peanut Butter	Sandwiches	Limiting Reactant	Excess Reactant
1 loaf (20 slices)	1 tablespoon	1	peanut butter	bread
4 slices	1 full jar	2	bread	peanut butter
8 slices	1 full jar	4	bread	peanut butter

ENGAGE

For a picnic you have 10 spoons, 8 forks, and 6 knives. If each person requires 1 spoon, 1 fork, and 1 knife, why can you only serve 6 people at the picnic?

Calculating Moles of Product from a Limiting Reactant

In a similar way, the reactants in a chemical reaction do not always combine in quantities that allow each to be used up at exactly the same time. In many reactions, there is a limiting reactant that determines the amount of product that can be formed. When we know the quantities of the reactants of a chemical reaction, we calculate the amount of product that is possible from each reactant if it were completely consumed. We are looking for the *limiting reactant*, which is the one that runs out first, producing the smaller amount of product.

CORE CHEMISTRY SKILL
Calculating Quantity of Product from a Limiting Reactant

SAMPLE PROBLEM 9.7 Moles of Product from Limiting Reactant

Carbon monoxide and hydrogen are used to produce methanol (CH_4O). The balanced chemical reaction is

$$CO(g) + 2H_2(g) \longrightarrow CH_4O(g)$$

If 3.00 mol of CO and 5.00 mol of H_2 are the initial reactants, what is the limiting reactant and how many moles of methanol can be produced?

TRY IT FIRST

Guide to Calculating the Quantity (Moles or Grams) of Product from a Limiting Reactant

STEP 1
State the given and needed quantity (moles or grams).

STEP 2
Write a plan to convert the quantity (moles or grams) of each reactant to quantity (moles or grams) of product.

STEP 3
Write the mole–mole factors and molar masses (if needed).

STEP 4
Calculate the quantity (moles or grams) of product from each reactant and select the smaller quantity (moles or grams) as the limiting reactant.

SOLUTION

STEP 1 State the given and needed quantity (moles).

	Given	Need	Connect
ANALYZE THE PROBLEM	3.00 mol of CO, 5.00 mol of H_2	moles of CH_4O produced, limiting reactant	mole–mole factors
	Equation		
	$CO(g) + 2H_2(g) \longrightarrow CH_4O(g)$		

STEP 2 Write a plan to convert the quantity (moles) of each reactant to quantity (moles) of product.

moles of CO → Mole–mole factor → moles of CH_4O

moles of H_2 → Mole–mole factor → moles of CH_4O

STEP 3 Write the mole–mole factors.

$$\frac{1 \text{ mol of CO} = 1 \text{ mol of } CH_4O}{1 \text{ mol CO}} \quad \frac{1 \text{ mol } CH_4O}{1 \text{ mol CO}} \text{ and } \frac{1 \text{ mol } CH_4O}{1 \text{ mol CO}}$$

$$2 \text{ mol of } H_2 = 1 \text{ mol of } CH_4O \quad \frac{2 \text{ mol } H_2}{1 \text{ mol } CH_4O} \text{ and } \frac{1 \text{ mol } CH_4O}{2 \text{ mol } H_2}$$

STEP 4 Calculate the quantity (moles) of product from each reactant and select the smaller quantity (moles) as the limiting reactant.

Moles of CH_4O (product) from CO:

Exact

$$3.00 \text{ mol CO} \times \frac{1 \text{ mol } CH_4O}{1 \text{ mol CO}} = 3.00 \text{ mol of } CH_4O$$

Three SFs   Exact   Three SFs

Moles of CH_4O (product) from H_2:

Three SFs   Exact

$$5.00 \text{ mol } H_2 \times \frac{1 \text{ mol } CH_4O}{2 \text{ mol } H_2} = 2.50 \text{ mol of } CH_4O$$ Smaller amount of product

Limiting reactant   Exact   Three SFs

The smaller amount, 2.50 mol of CH_4O, is the maximum amount of methanol that can be produced from the limiting reactant, H_2, because it is completely consumed.

STUDY CHECK 9.7

If the initial mixture of reactants for Sample Problem 9.7 contains 4.00 mol of CO and 4.00 mol of H_2, what is the limiting reactant and how many moles of methanol can be produced?

ANSWER

H_2 is the limiting reactant; 2.00 mol of methanol can be produced.

Calculating Mass of Product from a Limiting Reactant

The quantities of the reactants can also be given in grams. The calculations to identify the limiting reactant are the same as before, but the grams of each reactant must first be converted to moles, then to moles of product, and finally to grams of product. Then select the smaller mass of product, which is from complete use of the limiting reactant. This calculation is shown in Sample Problem 9.8.

A ceramic brake disc in a sports car withstands temperatures of 1400 °C.

SAMPLE PROBLEM 9.8 Grams of Product from a Limiting Reactant

When silicon dioxide (sand) and carbon are heated, the products are silicon carbide, SiC, and carbon monoxide. Silicon carbide is a ceramic material that tolerates extreme temperatures and is used as an abrasive and in the brake discs of sports cars. How many grams of CO are formed from 70.0 g of SiO_2 and 50.0 g of C?

$$SiO_2(s) + 3C(s) \xrightarrow{\Delta} SiC(s) + 2CO(g)$$

 **TRY IT FIRST**

SOLUTION

STEP 1 State the given and needed quantity (grams).

ANALYZE THE PROBLEM	Given	Need	Connect
	70.0 g of SiO_2, 50.0 g of C	grams of CO from limiting reactant	molar masses, mole–mole factor
	Equation		
	$SiO_2(s) + 3C(s) \xrightarrow{Heat} SiC(s) + 2CO(g)$		

STEP 2 Write a plan to convert the quantity (grams) of each reactant to quantity (grams) of product.

grams of SiO_2 → Molar mass → moles of SiO_2 → Mole–mole factor → moles of CO → Molar mass → grams of CO

grams of C → Molar mass → moles of C → Mole–mole factor → moles of CO → Molar mass → grams of CO

STEP 3 Write the mole–mole factors and molar masses.

Molar masses:

1 mol of SiO_2 = 60.09 g of SiO_2

$\dfrac{60.09 \text{ g } SiO_2}{1 \text{ mol } SiO_2}$ and $\dfrac{1 \text{ mol } SiO_2}{60.09 \text{ g } SiO_2}$

1 mol of C = 12.01 g of C

$\dfrac{12.01 \text{ g C}}{1 \text{ mol C}}$ and $\dfrac{1 \text{ mol C}}{12.01 \text{ g C}}$

1 mol of CO = 28.01 g of CO

$\dfrac{28.01 \text{ g CO}}{1 \text{ mol CO}}$ and $\dfrac{1 \text{ mol CO}}{28.01 \text{ g CO}}$

Mole–mole factors:

2 mol of CO = 1 mol of SiO_2

$\dfrac{2 \text{ mol CO}}{1 \text{ mol } SiO_2}$ and $\dfrac{1 \text{ mol } SiO_2}{2 \text{ mol CO}}$

2 mol of CO = 3 mol of C

$\dfrac{2 \text{ mol CO}}{3 \text{ mol C}}$ and $\dfrac{3 \text{ mol C}}{2 \text{ mol CO}}$

STEP 4 Calculate the quantity (grams) of product from each reactant and select the smaller quantity (grams) as the limiting reactant.

Grams of CO (product) from SiO₂:

Three SFs   Exact   Exact   Four SFs   Three SFs

$$70.0 \text{ g SiO}_2 \times \frac{1 \text{ mol SiO}_2}{60.09 \text{ g SiO}_2} \times \frac{2 \text{ mol CO}}{1 \text{ mol SiO}_2} \times \frac{28.01 \text{ g CO}}{1 \text{ mol CO}} = 65.3 \text{ g of CO}$$

Limiting reactant   Four SFs   Exact   Exact   Smaller amount of product

Grams of CO (product) from C:

Three SFs   Exact   Exact   Four SFs   Three SFs

$$50.0 \text{ g C} \times \frac{1 \text{ mol C}}{12.01 \text{ g C}} \times \frac{2 \text{ mol CO}}{3 \text{ mol C}} \times \frac{28.01 \text{ g CO}}{1 \text{ mol CO}} = 77.7 \text{ g of CO}$$

Four SFs   Exact   Exact

The smaller amount, 65.3 g of CO, is the most CO that can be produced. This also means that the SiO₂ is the limiting reactant.

STUDY CHECK 9.8

Hydrogen sulfide burns with oxygen to give sulfur dioxide and water. How many grams of sulfur dioxide are formed from the reaction of 8.52 g of H_2S and 9.60 g of O_2?

$$2H_2S(g) + 3O_2(g) \xrightarrow{\Delta} 2SO_2(g) + 2H_2O(g)$$

ANSWER

12.8 g of SO_2

QUESTIONS AND PROBLEMS

9.4 Limiting Reactants

LEARNING GOAL Identify a limiting reactant when given the quantities of two reactants; calculate the amount of product formed from the limiting reactant.

9.19 A taxi company has 10 taxis.
  a. On a certain day, only eight taxi drivers show up for work. How many taxis can be used to pick up passengers?
  b. On another day, 10 taxi drivers show up for work but three taxis are in the repair shop. How many taxis can be driven?

9.20 A clock maker has 15 clock faces. Each clock requires one face and two hands.
  a. If the clock maker has 42 hands, how many clocks can be produced?
  b. If the clock maker has only eight hands, how many clocks can be produced?

9.21 Nitrogen and hydrogen react to form ammonia.
$$N_2(g) + 3H_2(g) \longrightarrow 2NH_3(g)$$
Determine the limiting reactant in each of the following mixtures of reactants:
  a. 3.0 mol of N_2 and 5.0 mol of H_2
  b. 8.0 mol of N_2 and 4.0 mol of H_2
  c. 3.0 mol of N_2 and 12.0 mol of H_2

9.22 Iron and oxygen react to form iron(III) oxide.
$$4Fe(s) + 3O_2(g) \longrightarrow 2Fe_2O_3(s)$$
Determine the limiting reactant in each of the following mixtures of reactants:
  a. 2.0 mol of Fe and 6.0 mol of O_2
  b. 5.0 mol of Fe and 4.0 mol of O_2
  c. 16.0 mol of Fe and 20.0 mol of O_2

9.23 For each of the following reactions, 20.0 g of each reactant is present initially. Determine the limiting reactant, and calculate the grams of product in parentheses that would be produced.
  a. $2Al(s) + 3Cl_2(g) \longrightarrow 2AlCl_3(s)$ (AlCl₃)
  b. $4NH_3(g) + 5O_2(g) \longrightarrow 4NO(g) + 6H_2O(g)$ (H₂O)
  c. $CS_2(g) + 3O_2(g) \xrightarrow{\Delta} CO_2(g) + 2SO_2(g)$ (SO₂)

9.24 For each of the following reactions, 20.0 g of each reactant is present initially. Determine the limiting reactant, and calculate the grams of product in parentheses that would be produced.
  a. $4Al(s) + 3O_2(g) \longrightarrow 2Al_2O_3(s)$ (Al₂O₃)
  b. $3NO_2(g) + H_2O(l) \longrightarrow 2HNO_3(aq) + NO(g)$ (HNO₃)
  c. $C_2H_6O(l) + 3O_2(g) \xrightarrow{\Delta} 2CO_2(g) + 3H_2O(g)$ (H₂O)

9.25 For each of the following reactions, calculate the grams of indicated product when 25.0 g of the first reactant and 40.0 g of the second reactant are used:

  a. $2SO_2(g) + O_2(g) \longrightarrow 2SO_3(g)$ (SO_3)
  b. $3Fe(s) + 4H_2O(l) \longrightarrow Fe_3O_4(s) + 4H_2(g)$ (Fe_3O_4)
  c. $C_7H_{16}(l) + 11O_2(g) \overset{\Delta}{\longrightarrow} 7CO_2(g) + 8H_2O(g)$ (CO_2)

9.26 For each of the following reactions, calculate the grams of indicated product when 15.0 g of the first reactant and 10.0 g of the second reactant are used:

  a. $4Li(s) + O_2(g) \longrightarrow 2Li_2O(s)$ (Li_2O)
  b. $Fe_2O_3(s) + 3H_2(g) \longrightarrow 2Fe(s) + 3H_2O(l)$ (Fe)
  c. $Al_2S_3(s) + 6H_2O(l) \longrightarrow$

$$2Al(OH)_3(aq) + 3H_2S(g) \quad (H_2S)$$

9.5 Percent Yield

LEARNING GOAL Given the actual quantity of product, determine the percent yield for a reaction.

In our problems up to now, we assumed that all of the reactants changed completely to product. Thus, we have calculated the amount of product as the maximum quantity possible, or 100%. While this would be an ideal situation, it does not usually happen. As we carry out a reaction and transfer products from one container to another, some product is usually lost. In the lab as well as commercially, the starting materials may not be completely pure, and side reactions may use some of the reactants to give unwanted products. Thus, 100% of the desired product is not actually obtained.

When we do a chemical reaction in the laboratory, we measure out specific quantities of the reactants. We calculate the **theoretical yield** for the reaction, which is the amount of product (100%) we would expect if all the reactants were converted to the desired product. When the reaction ends, we collect and measure the mass of the product, which is the **actual yield** for the product. Because some product is usually lost, the actual yield is less than the theoretical yield. Using the actual yield and the theoretical yield for a product, we can calculate the **percent yield**.

$$\text{Percent yield (\%)} = \frac{\text{actual yield}}{\text{theoretical yield}} \times 100\%$$

ENGAGE

For your chemistry class party, you have prepared cookie dough to make five dozen cookies. However, 12 of the cookies burned and you threw them away. Why is the percent yield of cookies for the chemistry party only 80%?

CORE CHEMISTRY SKILL
Calculating Percent Yield

SAMPLE PROBLEM 9.9 Calculating Percent Yield

On a space shuttle, LiOH is used to absorb exhaled CO_2 from breathing air to form $LiHCO_3$.

$$LiOH(s) + CO_2(g) \longrightarrow LiHCO_3(s)$$

What is the percent yield of $LiHCO_3$ for the reaction if 50.0 g of LiOH gives 72.8 g of $LiHCO_3$?

TRY IT FIRST

SOLUTION

STEP 1 State the given and needed quantities.

	Given	Need	Connect
ANALYZE THE PROBLEM	50.0 g of LiOH, 72.8 g of LiHCO$_3$	percent yield of LiHCO$_3$	molar masses, mole–mole factor, percent yield expression
	Equation		
	LiOH(s) + CO$_2$(g) $\longrightarrow$ LiHCO$_3$(s)		

On a space shuttle, the LiOH in the canisters removes CO_2 from the air.

STEP 2 Write a plan to calculate the theoretical yield and the percent yield.

Calculation of theoretical yield:

grams of LiOH $\xrightarrow{\text{Molar mass}}$ moles of LiOH $\xrightarrow{\text{Mole–mole factor}}$ moles of $LiHCO_3$ $\xrightarrow{\text{Molar mass}}$ grams of $LiHCO_3$
Theoretical yield

Calculation of percent yield:

$$\text{Percent yield (\%)} = \frac{\text{actual yield of } LiHCO_3}{\text{theoretical yield of } LiHCO_3} \times 100\%$$

STEP 3 Write the molar masses and the mole–mole factor from the balanced equation.

Molar masses:

1 mol of LiOH = 23.95 g of LiOH	1 mol of $LiHCO_3$ = 67.96 g of $LiHCO_3$
$\dfrac{23.95 \text{ g LiOH}}{1 \text{ mol LiOH}}$ and $\dfrac{1 \text{ mol LiOH}}{23.95 \text{ g LiOH}}$	$\dfrac{67.96 \text{ g } LiHCO_3}{1 \text{ mol } LiHCO_3}$ and $\dfrac{1 \text{ mol } LiHCO_3}{67.96 \text{ g } LiHCO_3}$

Mole–mole factor:

1 mol of $LiHCO_3$ = 1 mol of LiOH

$$\frac{1 \text{ mol } LiHCO_3}{1 \text{ mol LiOH}} \quad \text{and} \quad \frac{1 \text{ mol LiOH}}{1 \text{ mol } LiHCO_3}$$

STEP 4 Calculate the percent yield by dividing the actual yield (given) by the theoretical yield and multiplying the result by 100%.

Calculation of theoretical yield:

$$50.0 \text{ g LiOH} \times \underset{\text{Four SFs}}{\frac{1 \text{ mol LiOH}}{23.95 \text{ g LiOH}}} \times \underset{\text{Exact}}{\frac{1 \text{ mol } LiHCO_3}{1 \text{ mol LiOH}}} \times \underset{\text{Exact}}{\frac{67.96 \text{ g } LiHCO_3}{1 \text{ mol } LiHCO_3}}$$

(Three SFs) (Exact) (Four SFs) (Exact)

$$= 142 \text{ g of } LiHCO_3$$
Three SFs

Calculation of percent yield:

$$\frac{\text{actual yield (given)}}{\text{theoretical yield (calculated)}} \times 100\% = \frac{72.8 \text{ g } LiHCO_3}{142 \text{ g } LiHCO_3} \times 100\% = 51.3\%$$

A percent yield of 51.3% means that 72.8 g of the theoretical amount of 142 g of $LiHCO_3$ was actually produced by the reaction.

STUDY CHECK 9.9

For the reaction in Sample Problem 9.9, what is the percent yield of $LiHCO_3$ if 8.00 g of CO_2 produces 10.5 g of $LiHCO_3$?

ANSWER

84.7%

Guide to Calculating Percent Yield

STEP 1
State the given and needed quantities.

STEP 2
Write a plan to calculate the theoretical yield and the percent yield.

STEP 3
Write the molar masses and the mole–mole factor from the balanced equation.

STEP 4
Calculate the percent yield by dividing the actual yield (given) by the theoretical yield and multiplying the result by 100%.

QUESTIONS AND PROBLEMS

9.5 Percent Yield

LEARNING GOAL Given the actual quantity of product, determine the percent yield for a reaction.

9.27 Carbon disulfide is produced by the reaction of carbon and sulfur dioxide.

$$5C(s) + 2SO_2(g) \longrightarrow CS_2(g) + 4CO(g)$$

a. What is the percent yield of carbon disulfide if the reaction of 40.0 g of carbon produces 36.0 g of carbon disulfide?
b. What is the percent yield of carbon disulfide if the reaction of 32.0 g of sulfur dioxide produces 12.0 g of carbon disulfide?

9.28 Iron(III) oxide reacts with carbon monoxide to produce iron and carbon dioxide.

$$Fe_2O_3(s) + 3CO(g) \longrightarrow 2Fe(s) + 3CO_2(g)$$

a. What is the percent yield of iron if the reaction of 65.0 g of iron(III) oxide produces 15.0 g of iron?
b. What is the percent yield of carbon dioxide if the reaction of 75.0 g of carbon monoxide produces 85.0 g of carbon dioxide?

9.29 Aluminum reacts with oxygen to produce aluminum oxide.

$$4Al(s) + 3O_2(g) \longrightarrow 2Al_2O_3(s)$$

Calculate the mass of Al_2O_3 that can be produced if the reaction of 50.0 g of aluminum and sufficient oxygen has a 75.0% yield.

9.30 Propane (C_3H_8) burns in oxygen to produce carbon dioxide and water.

$$C_3H_8(g) + 5O_2(g) \xrightarrow{\Delta} 3CO_2(g) + 4H_2O(g)$$

Calculate the mass of CO_2 that can be produced if the reaction of 45.0 g of propane and sufficient oxygen has a 60.0% yield.

9.31 When 30.0 g of carbon is heated with silicon dioxide, 28.2 g of carbon monoxide is produced. What is the percent yield of carbon monoxide for this reaction?

$$SiO_2(s) + 3C(s) \xrightarrow{\Delta} SiC(s) + 2CO(g)$$

9.32 When 56.6 g of calcium is reacted with nitrogen gas, 32.4 g of calcium nitride is produced. What is the percent yield of calcium nitride for this reaction?

$$3Ca(s) + N_2(g) \longrightarrow Ca_3N_2(s)$$

9.6 Energy in Chemical Reactions

LEARNING GOAL Given the heat of reaction (enthalpy change), calculate the loss or gain of heat for an exothermic or endothermic reaction.

Almost every chemical reaction involves a loss or gain of energy. To discuss energy change (enthalpy change) for a reaction, we look at the energy that is absorbed or lost during a chemical reaction.

Energy Units for Chemical Reactions

The SI unit for energy is the *joule* (J). Often, the unit of *kilojoules* (kJ) is used to show the energy change in a reaction.

1 kilojoule (kJ) = 1000 joules (J)

Heat of Reaction (Enthalpy Change)

The **heat of reaction** is the amount of heat absorbed or released during a reaction that takes place at constant pressure. A change of energy occurs as reactants interact, bonds break apart, and products form. We determine a heat of reaction, symbol ΔH, as the difference in the energy of the products and the reactants.

$$\Delta H = H_{products} - H_{reactants}$$

Exothermic Reactions

In an **exothermic reaction** (*exo* means "out"), the energy of the products is lower than that of the reactants. This means that heat is released along with the products that form. Let us look at the equation for the exothermic reaction in which 185 kJ of heat is released when 1 mol of hydrogen and 1 mol of chlorine react to form 2 mol of hydrogen chloride. For an exothermic reaction, the heat of reaction can be written as one of the products. It can also be written as a ΔH value with a negative sign ($-$).

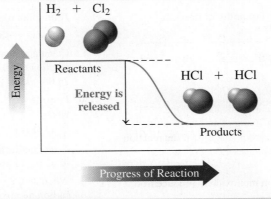

Exothermic, Heat Released	Heat Is a Product
$H_2(g) + Cl_2(g) \longrightarrow 2HCl(g) + 185 \text{ kJ}$	$\Delta H = -185 \text{ kJ}$ Negative sign

Endothermic Reactions

In an **endothermic reaction** (*endo* means "within"), the energy of the products is higher than that of the reactants. Heat is required to convert the reactants to products. Let us look at the equation for the endothermic reaction in which 180 kJ of heat is needed to convert 1 mol of nitrogen and 1 mol of oxygen to 2 mol of nitrogen oxide. For an endothermic reaction, the heat of reaction can be written as one of the reactants. It can also be written as a ΔH value with a positive sign ($+$).

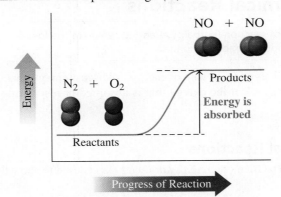

Endothermic, Heat Absorbed	Heat Is a Reactant
$N_2(g) + O_2(g) + 180 \text{ kJ} \longrightarrow 2NO(g)$	$\Delta H = +180 \text{ kJ}$ Positive sign

Reaction	Energy Change	Heat in the Equation	Sign of ΔH
Exothermic	Heat released	Product side	Negative sign ($-$)
Endothermic	Heat absorbed	Reactant side	Positive sign ($+$)

SAMPLE PROBLEM 9.10 Exothermic and Endothermic Reactions

In the reaction of 1 mol of solid carbon with oxygen gas, the energy of the carbon dioxide gas produced is 393 kJ less than that of the reactants.

a. Is the reaction exothermic or endothermic?
b. Write the balanced chemical equation including the heat of the reaction.
c. What is the value, in kilojoules, for the heat of reaction?

TRY IT FIRST

SOLUTION

a. When the energy of the products is lower than that of the reactants, the reaction gives off heat, which means that it is exothermic.

b. In an exothermic reaction, the heat is written as a product.

$$C(s) + O_2(g) \longrightarrow CO_2(g) + 393 \text{ kJ}$$

c. The heat of reaction for an exothermic reaction has a negative sign: $\Delta H = -393 \text{ kJ}$

STUDY CHECK 9.10

For the reaction $C(s) + 2H_2O(g) \longrightarrow CO_2(g) + 2H_2(g)$, the heat change is +90. kJ. Is this an exothermic or endothermic reaction?

ANSWER
endothermic

Calculations of Heat in Reactions

The value of ΔH refers to the heat change in kilojoules for the each substance in the balanced equation for the reaction. Consider the following decomposition reaction:

$$2H_2O(l) \longrightarrow 2H_2(g) + O_2(g) \quad \Delta H = +572 \text{ kJ}$$
$$2H_2O(l) + 572 \text{ kJ} \longrightarrow 2H_2(g) + O_2(g)$$

For this reaction, 572 kJ are absorbed by 2 mol of H_2O to produce 2 mol of H_2 and 1 mol of O_2. We can write heat conversion factors for each substance in this reaction as follows:

$$\frac{+572 \text{ kJ}}{2 \text{ mol } H_2O} \qquad \frac{+572 \text{ kJ}}{2 \text{ mol } H_2} \qquad \frac{+572 \text{ kJ}}{1 \text{ mol } O_2}$$

Suppose in this reaction that 9.00 g of H_2O undergoes reaction. We can calculate the quantity of heat absorbed as

$$9.00 \text{ g } H_2O \times \frac{1 \text{ mol } H_2O}{18.02 \text{ g } H_2O} \times \frac{+572 \text{ kJ}}{2 \text{ mol } H_2O} = +143 \text{ kJ}$$

> **ENGAGE**
>
> Why is there a single heat of reaction whereas there are two or more heat conversion factors possible for the reaction?

SAMPLE PROBLEM 9.11 Calculating Heat in a Reaction

How much heat, in kilojoules, is released when nitrogen and hydrogen react to form 50.0 g of ammonia?

$$N_2(g) + 3H_2(g) \longrightarrow 2NH_3(g) \qquad \Delta H = -92.2 \text{ kJ}$$

> **CORE CHEMISTRY SKILL**
> Using the Heat of Reaction

TRY IT FIRST

SOLUTION

STEP 1 State the given and needed quantities.

	Given	Need	Connect
ANALYZE THE PROBLEM	50.0 g of NH_3, $\Delta H = -92.2 \text{ kJ}$	heat released, in kilojoules	molar mass, heat conversion factor
	Equation		
	$N_2(g) + 3H_2(g) \longrightarrow 2NH_3(g)$		

Guide to Calculations Using the Heat of Reaction (ΔH)

STEP 1
State the given and needed quantities.

STEP 2
Write a plan using the heat of reaction and any molar mass needed.

STEP 3
Write the conversion factors including heat of reaction.

STEP 4
Set up the problem to calculate the heat.

STEP 2 Write a plan using the heat of reaction and any molar mass needed.

grams of NH_3 → | Molar mass | → moles of NH_3 → | Heat of reaction | → kilojoules

STEP 3 Write the conversion factors including heat of reaction.

$$1 \text{ mol of } NH_3 = 17.03 \text{ g of } NH_3$$

$$\frac{17.03 \text{ g } NH_3}{1 \text{ mol } NH_3} \quad \text{and} \quad \frac{1 \text{ mol } NH_3}{17.03 \text{ g } NH_3}$$

$$2 \text{ mol of } NH_3 = -92.2 \text{ kJ}$$

$$\frac{-92.2 \text{ kJ}}{2 \text{ mol } NH_3} \quad \text{and} \quad \frac{2 \text{ mol } NH_3}{-92.2 \text{ kJ}}$$

STEP 4 Set up the problem to calculate the heat.

$$\underset{\text{Three SFs}}{50.0 \text{ g } NH_3} \times \underset{\text{Four SFs}}{\frac{\overset{\text{Exact}}{1 \text{ mol } NH_3}}{17.03 \text{ g } NH_3}} \times \underset{\text{Exact}}{\frac{\overset{\text{Three SFs}}{-92.2 \text{ kJ}}}{2 \text{ mol } NH_3}} = \underset{\text{Three SFs}}{-135 \text{ kJ}}$$

STUDY CHECK 9.11

Mercury(II) oxide decomposes to mercury and oxygen.

$$2HgO(s) \longrightarrow 2Hg(l) + O_2(g) \quad \Delta H = +182 \text{ kJ}$$

a. Is the reaction exothermic or endothermic?
b. How many kilojoules are needed when 25.0 g of mercury(II) oxide reacts?

ANSWER
a. endothermic   **b.** 20.5 kJ

CHEMISTRY LINK TO **HEALTH**
Cold Packs and Hot Packs

In a hospital, at a first-aid station, or at an athletic event, an instant *cold pack* may be used to reduce swelling from an injury, remove heat from inflammation, or decrease capillary size to lessen the effect of hemorrhaging. Inside the plastic container of a cold pack, there is a compartment containing solid ammonium nitrate (NH_4NO_3) that is separated from a compartment containing water.

Cold packs use an endothermic reaction.

The pack is activated when it is hit or squeezed hard enough to break the walls between the compartments and cause the ammonium nitrate to mix with the water (shown as H_2O over the reaction arrow). In an endothermic process, 1 mol of NH_4NO_3 that dissolves absorbs 26 kJ of heat. The temperature drops to about 4 to 5 °C to give a cold pack that is ready to use.

Endothermic Reaction in a Cold Pack

$$NH_4NO_3(s) + 26 \text{ kJ} \xrightarrow{H_2O} NH_4NO_3(aq)$$

Hot packs are used to relax muscles, lessen aches and cramps, and increase circulation by expanding capillary size. Constructed in the same way as cold packs, a hot pack contains a salt such as $CaCl_2$. When 1 mol of $CaCl_2$ dissolves in water, 82 kJ are released as heat. The temperature increases as much as 66 °C to give a hot pack that is ready to use.

Exothermic Reaction in a Hot Pack

$$CaCl_2(s) \xrightarrow{H_2O} CaCl_2(aq) + 82 \text{ kJ}$$

Hess's Law

According to **Hess's law**, heat can be absorbed or released in a single chemical reaction or in several steps. When there are two or more steps in the reaction, the overall enthalpy change is the sum of the enthalpy changes of those steps, provided they all occur at the same temperature.

Steps in solving problems involving Hess's law:

1. If you reverse a chemical equation, you must also reverse the sign of ΔH.
2. If a chemical equation is multiplied by some factor, then ΔH must be multiplied by the same factor.

We can see how the enthalpy change for a specific reaction is the sum of two or more reactions in Sample Problem 9.12.

SAMPLE PROBLEM 9.12 Hess's Law and Calculating Heat of Reaction

Calculate the ΔH value for the following reaction using equations 1 and 2:

$$C(s) + 2H_2O(g) \longrightarrow CO_2(g) + 2H_2(g)$$

Equation 1 $C(s) + O_2(g) \longrightarrow CO_2(g)$ $\Delta H = -304 \text{ kJ}$

Equation 2 $H_2(g) + \dfrac{1}{2}O_2(g) \longrightarrow H_2O(g)$ $\Delta H = -242 \text{ kJ}$

TRY IT FIRST

SOLUTION

ANALYZE THE PROBLEM	**Given**	**Need**	**Connect**
	overall reaction, equations 1, 2	heat of reaction	combine equations, total heats of reactions

By combining, rearranging, and multiplying the substances including the ΔH in equations 1 and 2, we can obtain the overall equation for the reaction. Then the sum of the ΔH values involved in each will give the ΔH value for the reaction.

STEP 1 Arrange the given equations to place reactants on the left and products on the right.

$C(s) + O_2(g) \longrightarrow CO_2(g)$ $\Delta H = -304 \text{ kJ}$

$H_2O(g) \longrightarrow H_2(g) + \dfrac{1}{2}O_2(g)$ $\Delta H = +242 \text{ kJ}$ Changed sign

STEP 2 If an equation is multiplied to balance coefficients, multiply the ΔH by the same number. The second equation must be multiplied by 2 to give $2H_2O(g)$ on the reactant side.

$2 \times [H_2O(g) \longrightarrow H_2(g) + \dfrac{1}{2}O_2(g)$ $\Delta H = +242 \text{ kJ}]$

$2H_2O(g) \longrightarrow 2H_2(g) + O_2(g)$ $\Delta H = +484 \text{ kJ}$

STEP 3 Combine the equations and cancel any substances that are common to both sides. Add the ΔHs

$C(s) + \cancel{O_2(g)} \longrightarrow CO_2(g)$ $\Delta H = -304 \text{ kJ}$
$2H_2O(g) \longrightarrow 2H_2(g) + \cancel{O_2(g)}$ $\Delta H = +484 \text{ kJ}$
_____
$C(s) + 2H_2O(g) \longrightarrow CO_2(g) + 2H_2(g)$ $\Delta H = +180 \text{ kJ}$

STUDY CHECK 9.12

Calculate the ΔH value for the following reaction using equations 1 and 2:

$$2NO(g) + O_2(g) \longrightarrow N_2O_4(g)$$

Equation 1 $N_2O_4(g) \longrightarrow 2NO_2(g)$ $\Delta H = +57.2 \text{ kJ}$

Equation 2 $NO(g) + \dfrac{1}{2}O_2(g) \longrightarrow NO_2(g)$ $\Delta H = -57.0 \text{ kJ}$

ANSWER

$\Delta H = -171.2 \text{ kJ}$

Guide to Using Hess's Law

STEP 1
Arrange the given equations to place reactants on the left and products on the right.

STEP 2
If an equation is multiplied to balance coefficients, multiply the ΔH by the same number.

STEP 3
Combine the equations and cancel any substances that are common to both sides. Add the ΔHs.

QUESTIONS AND PROBLEMS

9.6 Energy in Chemical Reactions

LEARNING GOAL Given the heat of reaction (*enthalpy change*), calculate the loss or gain of heat for an exothermic or endothermic reaction.

9.33 In an exothermic reaction, is the energy of the products higher or lower than that of the reactants?

9.34 In an endothermic reaction, is the energy of the products higher or lower than that of the reactants?

9.35 Classify each of the following as exothermic or endothermic:
   a. A reaction releases 550 kJ.
   b. The energy level of the products is higher than that of the reactants.
   c. The metabolism of glucose in the body provides energy.

9.36 Classify each of the following as exothermic or endothermic:
   a. The energy level of the products is lower than that of the reactants.
   b. In the body, the synthesis of proteins requires energy.
   c. A reaction absorbs 125 kJ.

9.37 Classify each of the following as exothermic or endothermic and give the ΔH for each:
   a. $CH_4(g) + 2O_2(g) \longrightarrow CO_2(g) + 2H_2O(g) + 802\text{ kJ}$
   b. $Ca(OH)_2(s) + 65.3\text{ kJ} \longrightarrow CaO(s) + H_2O(l)$
   c. $2Al(s) + Fe_2O_3(s) \longrightarrow Al_2O_3(s) + 2Fe(l) + 850\text{ kJ}$

The thermite reaction of aluminum and iron(III) oxide produces very high temperatures used to cut or weld railroad tracks.

9.38 Classify each of the following as exothermic or endothermic and give the ΔH for each:
   a. $C_3H_8(g) + 5O_2(g) \longrightarrow 3CO_2(g) + 4H_2O(g) + 2220\text{ kJ}$
   b. $2Na(s) + Cl_2(g) \longrightarrow 2NaCl(s) + 819\text{ kJ}$
   c. $PCl_5(g) + 67\text{ kJ} \longrightarrow PCl_3(g) + Cl_2(g)$

9.39 **a.** How many kilojoules are released when 125 g of Cl_2 reacts with silicon?
   $$Si(s) + 2Cl_2(g) \longrightarrow SiCl_4(g) \quad \Delta H = -657\text{ kJ}$$
   b. How many kilojoules are absorbed when 278 g of PCl_5 reacts?
   $$PCl_5(g) \longrightarrow PCl_3(g) + Cl_2(g) \quad \Delta H = +67\text{ kJ}$$

9.40 **a.** How many kilojoules are released when 75.0 g of methanol reacts?
   $$2CH_4O(l) + 3O_2(g) \longrightarrow 2CO_2(g) + 4H_2O(l)$$
   $$\Delta H = -726\text{ kJ}$$
   b. How many kilojoules are absorbed when 315 g of $Ca(OH)_2$ reacts?
   $$Ca(OH)_2(s) \longrightarrow CaO(s) + H_2O(l) \quad \Delta H = +65.3\text{ kJ}$$

9.41 Calculate the enthalpy change for the reaction $N_2(g) + 2O_2(g) \longrightarrow 2NO_2(g)$ from the following:
   $$N_2(g) + O_2(g) \longrightarrow 2NO(g) \qquad \Delta H = +180\text{ kJ}$$
   $$2NO_2(g) \longrightarrow 2NO(g) + O_2(g) \qquad \Delta H = +112\text{ kJ}$$

9.42 Calculate the enthalpy change for the reaction $Fe_2O_3(s) + CO(g) \longrightarrow 2FeO(s) + CO_2(g)$ from the following:
   $$Fe_2O_3(s) + 3CO(g) \longrightarrow 2Fe(s) + 3CO_2(g) \quad \Delta H = -23.4\text{ kJ}$$
   $$FeO(s) + CO(g) \longrightarrow Fe(s) + CO_2(g) \qquad \Delta H = -10.9\text{ kJ}$$

9.43 Calculate the enthalpy change for the reaction $S(s) + O_2(g) \longrightarrow SO_2(g)$ from the following:
   $$S(s) + \frac{3}{2}O_2(g) \longrightarrow SO_3(g) \qquad \Delta H = -396\text{ kJ}$$
   $$SO_2(g) + \frac{1}{2}O_2(g) \longrightarrow SO_3(g) \quad \Delta H = -90\text{ kJ}$$

9.44 Calculate the enthalpy for the reaction $3C(s) + O_2(g) \longrightarrow C_3O_2(g)$ from the following:
   $$2CO(g) + C(s) \longrightarrow C_3O_2(g) \qquad \Delta H = +127\text{ kJ}$$
   $$C(s) + \frac{1}{2}O_2(g) \longrightarrow CO(g) \qquad \Delta H = -111\text{ kJ}$$

Follow Up

TESTING WATER SAMPLES FOR PESTICIDES

One of the problems that Lance monitored was water pollution by insecticides that exceed government regulations. These insecticides are made by organic synthesis, in which smaller molecules are combined to form larger molecules, in a stepwise fashion.

9.45 In one step in the synthesis of the insecticide Sevin, naphthol reacts with phosgene as shown.

$$C_{10}H_8O + COCl_2 \longrightarrow C_{11}H_7O_2Cl + HCl$$

Naphthol   Phosgene

a. How many kilograms of $C_{11}H_7O_2Cl$ form from 2.2×10^2 kg of naphthol?

b. If 100. g of naphthol and 100. g of phosgene react, what is the theoretical yield of $C_{11}H_7O_2Cl$?

c. If the actual yield of $C_{11}H_7O_2Cl$ in part **b** is 115 g, what is the percent yield?

9.46 Another widely used insecticide is carbofuran (Furadan), an extremely toxic insecticide. In one step in the synthesis of carbofuran, the reaction shown is used.

$$C_6H_6O_2 + C_4H_7Cl \longrightarrow C_{10}H_{12}O_2 + HCl$$

a. How many grams of $C_6H_6O_2$ are needed to produce 3.8×10^3 g of $C_{10}H_{12}O_2$?

b. If 67.0 g of $C_6H_6O_2$ and 51.0 g of C_4H_7Cl react, what is the theoretical yield of $C_{10}H_{12}O_2$?

c. If the actual yield of $C_{10}H_{12}O_2$ in part **b** is 85.7 g, what is the percent yield?

CONCEPT MAP

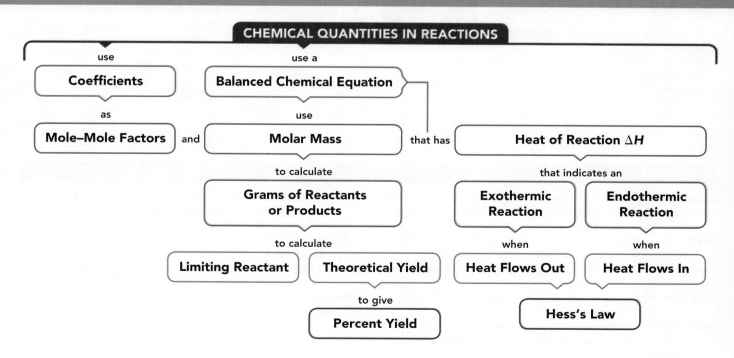

CHEMICAL QUANTITIES IN REACTIONS

use — Coefficients — as — Mole–Mole Factors

and

use a — Balanced Chemical Equation — use — Molar Mass — to calculate — Grams of Reactants or Products — to calculate — Limiting Reactant / Theoretical Yield — to give — Percent Yield

that has — Heat of Reaction ΔH — that indicates an — Exothermic Reaction / Endothermic Reaction — when — Heat Flows Out / Heat Flows In — Hess's Law

CHAPTER REVIEW

9.1 Conservation of Mass

LEARNING GOAL Calculate the total mass of reactants and the total mass of products in a balanced chemical equation.

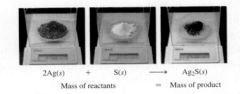

$$2Ag(s) + S(s) \longrightarrow Ag_2S(s)$$

Mass of reactants = Mass of product

- In a balanced equation, the total mass of the reactants is equal to the total mass of the products.

9.2 Calculating Moles Using Mole–Mole Factors

LEARNING GOAL Use a mole–mole factor from a balanced chemical equation to calculate the number of moles of another substance in the reaction.

- The coefficients in an equation describing the relationship between the moles of any two components are used to write mole–mole factors.
- When the number of moles for one substance is known, a mole–mole factor is used to find the moles of a different substance in the reaction.

9.3 Mass Calculations for Reactions

LEARNING GOAL Given the mass in grams of a substance in a reaction, calculate the mass in grams of another substance in the reaction.

- In calculations using equations, the molar masses of the substances and their mole–mole factors are used to change the number of grams of one substance to the corresponding grams of a different substance.

9.4 Limiting Reactants

LEARNING GOAL Identify a limiting reactant when given the quantities of two reactants; calculate the amount of product formed from the limiting reactant.

- A limiting reactant is the reactant that produces the smaller amount of product while the other reactant is left over.
- When the masses of two reactants are given, the mass of a product is calculated from the limiting reactant.

9.5 Percent Yield

LEARNING GOAL Given the actual quantity of product, determine the percent yield for a reaction.

- The percent yield for a reaction indicates the percent of product actually produced during a reaction.
- The percent yield is calculated by dividing the actual yield in grams of a product by the theoretical yield in grams and multiplying by 100%.

9.6 Energy in Chemical Reactions

LEARNING GOAL Given the heat of reaction (enthalpy change), calculate the loss or gain of heat for an exothermic or endothermic reaction.

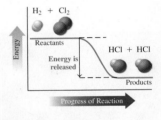

- In chemical reactions, the heat of reaction (ΔH) is the energy difference between the products and the reactants.
- In an exothermic reaction, the energy of the products is lower than that of the reactants. Heat is released, and ΔH is negative.
- In an endothermic reaction, the energy of the products is higher than that of the reactants; heat is absorbed, and ΔH is positive.
- Hess's law states that heat can be absorbed or released in a single step or in several steps.

KEY TERMS

actual yield The actual amount of product produced by a reaction.

endothermic reaction A reaction wherein the energy of the products is higher than that of the reactants.

exothermic reaction A reaction wherein the energy of the products is lower than that of the reactants.

heat of reaction The heat (symbol ΔH) absorbed or released when a reaction takes place at constant pressure.

Hess's law Heat can be absorbed or released in a single chemical reaction or in several steps.

law of conservation of mass In a chemical reaction, the total mass of the reactants is equal to the total mass of the products; matter is neither lost nor gained.

limiting reactant The reactant used up during a chemical reaction, which limits the amount of product that can form.

mole–mole factor A conversion factor that relates the number of moles of two compounds in an equation derived from their coefficients.

percent yield The ratio of the actual yield for a reaction to the theoretical yield possible for the reaction.

theoretical yield The maximum amount of product that a reaction can produce from a given amount of reactant.

CORE CHEMISTRY SKILLS

The chapter section containing each Core Chemistry Skill is shown in parentheses at the end of each heading.

Using Mole–Mole Factors (9.2)

Consider the balanced chemical equation

$$4Na(s) + O_2(g) \longrightarrow 2Na_2O(s)$$

- The coefficients in a balanced chemical equation represent the moles of reactants and the moles of products. Thus, 4 mol of Na react with 1 mol of O_2 to form 2 mol of Na_2O.
- From the coefficients, mole–mole factors can be written for any two substances as follows:

Na and O_2 $\dfrac{4 \text{ mol Na}}{1 \text{ mol } O_2}$ and $\dfrac{1 \text{ mol } O_2}{4 \text{ mol Na}}$

Na and Na_2O $\dfrac{4 \text{ mol Na}}{2 \text{ mol } Na_2O}$ and $\dfrac{2 \text{ mol } Na_2O}{4 \text{ mol Na}}$

O_2 and Na_2O $\dfrac{2 \text{ mol } Na_2O}{1 \text{ mol } O_2}$ and $\dfrac{1 \text{ mol } O_2}{2 \text{ mol } Na_2O}$

- A mole–mole factor is used to convert the number of moles of one substance in the reaction to the number of moles of another substance in the reaction.

Example: How many moles of sodium are needed to produce 3.5 mol of sodium oxide?

Answer:

$$3.5 \underset{\text{Two SFs}}{\text{mol Na}_2\text{O}} \times \underset{\text{Exact}}{\frac{4 \text{ mol Na}}{2 \text{ mol Na}_2\text{O}}} = 7.0 \underset{\text{Two SFs}}{\text{mol of Na}}$$

Converting Grams to Grams (9.3)

- When we have the balanced chemical equation for a reaction, we can use the mass of substance A and then calculate the mass of substance B. The process is as follows:
  - Use the molar mass of A to convert the mass, in grams, of A to moles of A.
  - Use the mole–mole factor that converts moles of A to moles of B.
  - Use the molar mass of B to calculate the mass, in grams, of B.

$$\text{grams of A} \xrightarrow{\substack{\text{Molar} \\ \text{mass A}}} \text{moles of A} \xrightarrow{\substack{\text{Mole–mole} \\ \text{factor}}} \text{moles of B} \xrightarrow{\substack{\text{Molar} \\ \text{mass B}}} \text{grams of B}$$

Example: How many grams of O_2 are needed to completely react with 14.6 g of Na?

$$4Na(s) + O_2(g) \longrightarrow 2Na_2O(s)$$

Answer:

$$14.6 \underset{\text{Three SFs}}{\text{g Na}} \times \underset{\text{Four SFs}}{\frac{1 \text{ mol Na}}{22.99 \text{ g Na}}} \times \underset{\text{Exact}}{\frac{1 \text{ mol } O_2}{4 \text{ mol Na}}} \times \underset{\text{Exact}}{\frac{32.00 \text{ g } O_2}{1 \text{ mol } O_2}} = 5.08 \underset{\text{Three SFs}}{\text{g of } O_2}$$

Calculating Quantity of Product from a Limiting Reactant (9.4)

Often in reactions, the reactants are not present in quantities that allow both reactants to be completely used up. Then one of the reactants, called the *limiting reactant*, determines the maximum amount of product that can form.

- To determine the limiting reactant, we calculate the amount of product that is possible from each reactant.
- The limiting reactant is the one that produces the smaller amount of product.

Example: If 12.5 g of S reacts with 17.2 g of O_2, what is the limiting reactant and the mass, in grams, of SO_3 produced?

$$2S(s) + 3O_2(g) \longrightarrow 2SO_3(g)$$

Answer:

Mass of SO_3 from S:

$$12.5 \underset{\text{Three SFs}}{\text{g S}} \times \underset{\text{Four SFs}}{\frac{1 \text{ mol S}}{32.07 \text{ g S}}} \times \underset{\text{Exact}}{\frac{2 \text{ mol } SO_3}{2 \text{ mol S}}} \times \underset{\text{Exact}}{\frac{80.07 \text{ g } SO_3}{1 \text{ mol } SO_3}} = 31.2 \underset{\text{Three SFs}}{\text{g of } SO_3}$$

Mass of SO_3 from O_2:

$$17.2 \underset{\text{Three SFs}}{\text{g } O_2} \times \underset{\text{Four SFs}}{\frac{1 \text{ mol } O_2}{32.00 \text{ g } O_2}} \times \underset{\text{Exact}}{\frac{2 \text{ mol } SO_3}{3 \text{ mol } O_2}} \times \underset{\text{Exact}}{\frac{80.07 \text{ g } SO_3}{1 \text{ mol } SO_3}}$$

Limiting reactant $= 28.7 \underset{\text{Three SFs}}{\text{g of } SO_3}$ Smaller amount of SO_3

Calculating Percent Yield (9.5)

- The *theoretical yield* for a reaction is the amount of product (100%) formed if all the reactants were converted to desired product.
- The *actual yield* for the reaction is the mass, in grams, of the product obtained at the end of the experiment. Because some product is usually lost, the actual yield is less than the theoretical yield.
- The *percent yield* is calculated from the actual yield divided by the theoretical yield and multiplied by 100%.

$$\text{Percent yield (\%)} = \frac{\text{actual yield}}{\text{theoretical yield}} \times 100\%$$

Example: If 22.6 g of Al reacts completely with O_2, and 37.8 g of Al_2O_3 is obtained, what is the percent yield of Al_2O_3 for the reaction?

$$4Al(s) + 3O_2(g) \longrightarrow 2Al_2O_3(s)$$

Answer:

Calculation of theoretical yield:

$$22.6 \underset{\text{Three SFs}}{\text{g Al}} \times \underset{\text{Four SFs}}{\frac{1 \text{ mol Al}}{26.98 \text{ g Al}}} \times \underset{\text{Exact}}{\frac{2 \text{ mol } Al_2O_3}{4 \text{ mol Al}}} \times \underset{\text{Exact}}{\frac{101.96 \text{ g } Al_2O_3}{1 \text{ mol } Al_2O_3}}$$

$$= 42.7 \underset{\text{Three SFs}}{\text{g of } Al_2O_3} \quad \text{Theoretical yield}$$

Calculation of percent yield:

$$\frac{\text{actual yield (given)}}{\text{theoretical yield (calculated)}} \times 100\%$$

$$= \frac{37.8 \underset{\text{Three SFs}}{\text{g } Al_2O_3}}{42.7 \underset{\text{Three SFs}}{\text{g } Al_2O_3}} \times 100\% = 88.5\% \underset{\text{Three SFs}}{}$$

Using the Heat of Reaction (9.6)

- The heat of reaction is the amount of heat, usually in kJ, that is absorbed or released during a reaction.
- The heat of reaction or *enthalpy change*, symbol ΔH, is the difference in the energy of the products and the reactants.

$$\Delta H = H_{\text{products}} - H_{\text{reactants}}$$

- In an exothermic reaction (*exo* means "out"), the energy of the products is lower than that of the reactants. This means that heat is released along with the products that form. Then the sign for the heat of reaction, ΔH, is negative.
- In an endothermic reaction (*endo* means "within"), the energy of the products is higher than that of the reactants. The heat is

required to convert the reactants to products. Then the sign for the heat of reaction, ΔH, is positive.

Example: How many kilojoules are released when 3.50 g of CH_4 undergoes combustion?

$$CH_4(g) + 2O_2(g) \longrightarrow CO_2(g) + 2H_2O(g) \quad \Delta H = -802 \text{ kJ}$$

$$\underset{\text{Exact}}{} \qquad \underset{\text{Three SFs}}{}$$

Answer: $3.50 \text{ g } CH_4 \times \dfrac{1 \text{ mol } CH_4}{16.04 \text{ g } CH_4} \times \dfrac{-802 \text{ kJ}}{1 \text{ mol } CH_4} = -175 \text{ kJ}$

$$\underset{\text{Three SFs}}{} \quad \underset{\text{Four SFs}}{} \quad \underset{\text{Exact}}{} \quad \underset{\text{Three SFs}}{}$$

UNDERSTANDING THE CONCEPTS

The chapter sections to review are shown in parentheses at the end of each question.

9.47 If red spheres represent oxygen atoms and blue spheres represent nitrogen atoms, and all the molecules are gases, (9.2, 9.4)

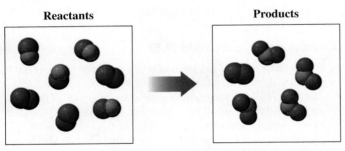

Reactants   **Products**

    a. write a balanced equation for the reaction
    b. identify the limiting reactant

9.48 If green spheres represent chlorine atoms, yellow-green spheres represent fluorine atoms, and white spheres represent hydrogen atoms, and all the molecules are gases, (9.2, 9.4)

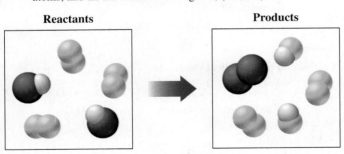

Reactants   **Products**

    a. write a balanced equation for the reaction
    b. identify the limiting reactant

9.49 If blue spheres represent nitrogen atoms and white spheres represent hydrogen atoms, and all the molecules are gases, (9.2, 9.4)

Reactants

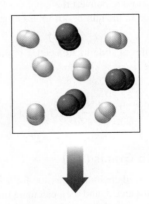

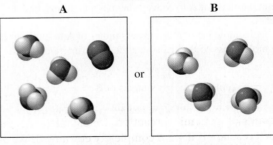

Products

A   or   B

    a. write a balanced equation for the reaction
    b. identify the diagram that shows the products

9.50 If purple spheres represent iodine atoms and white spheres represent hydrogen atoms, and all the molecules are gases, (9.2, 9.4)

    a. write a balanced equation for the reaction   **b.** identify the diagram that shows the products

Reactants

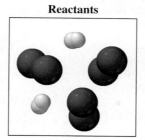

Products

    A   **B**   **C**

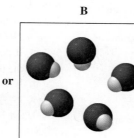 or 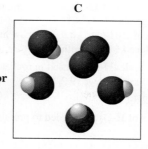 or

9.51 If blue spheres represent nitrogen atoms and purple spheres represent iodine atoms, and the reacting molecules are solid, and the products are gases, (9.2, 9.4, 9.5)

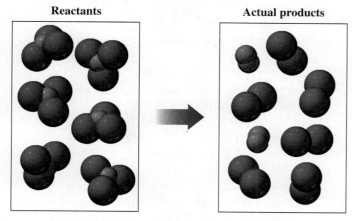

    a. write a balanced equation for the reaction
    b. from the diagram of the actual products that result, calculate the percent yield for the reaction

9.52 If green spheres represent chlorine atoms and red spheres represent oxygen atoms, and all the molecules are gases, (9.2, 9.4, 9.5)

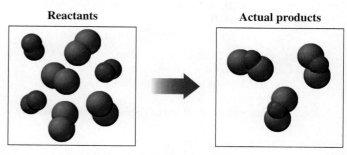

    a. write a balanced equation for the reaction
    b. identify the limiting reactant
    c. from the diagram of the actual products that result, calculate the percent yield for the reaction

9.53 Use the balanced chemical equation to complete the table: (9.1, 9.2)

$$2FeS(s) + 3O_2(g) \longrightarrow 2FeO(s) + 2SO_2(g)$$

FeS	O$_2$	FeO	SO$_2$
2.0 mol	___ mol	___ mol	___ mol
___ mol	___ mol	___ mol	4.6 mol

9.54 Use the balanced chemical equation to complete the table: (9.1, 9.2)

$$2C_2H_6(g) + 7O_2(g) \xrightarrow{\Delta} 4CO_2(g) + 6H_2O(g)$$

C$_2$H$_6$	O$_2$	CO$_2$	H$_2$O
___ mol	___ mol	3.2 mol	___ mol
___ mol	2.8 mol	___ mol	___ mol

ADDITIONAL QUESTIONS AND PROBLEMS

9.55 When ammonia (NH_3) reacts with fluorine (F_2), the products are dinitrogen tetrafluoride and hydrogen fluoride. (9.2, 9.3)

$$2NH_3(g) + 5F_2(g) \longrightarrow N_2F_4(g) + 6HF(g)$$

a. How many moles of each reactant are needed to produce 4.00 mol of HF?

b. How many grams of F_2 are required to react with 25.5 g of NH_3?

c. How many grams of N_2F_4 can be produced when 3.40 g of NH_3 reacts?

9.56 Gasohol is a fuel that contains ethanol (C_2H_6O) that burns in oxygen (O_2) to give carbon dioxide and water. (9.2, 9.3)

$$C_2H_6O(g) + 3O_2(g) \xrightarrow{\Delta} 2CO_2(g) + 3H_2O(g)$$

a. How many moles of O_2 are needed to completely react with 4.0 mol of C_2H_6O?

b. If a car produces 88 g of CO_2, how many grams of O_2 are used up in the reaction?

c. If you add 125 g of C_2H_6O to your fuel, how many grams of CO_2 can be produced from the ethanol?

9.57 When hydrogen peroxide (H_2O_2) is used in rocket fuels, it produces water and oxygen (O_2). (9.2, 9.3)

$$2H_2O_2(l) \longrightarrow 2H_2O(l) + O_2(g)$$

a. How many moles of H_2O_2 are needed to produce 3.00 mol of H_2O?

b. How many grams of H_2O_2 are required to produce 36.5 g of O_2?

c. How many grams of H_2O can be produced when 12.2 g of H_2O_2 reacts?

9.58 Propane gas, C_3H_8, reacts with oxygen to produce carbon dioxide and water. (9.2, 9.3)

$$C_3H_8(g) + 5O_2(g) \xrightarrow{\Delta} 3CO_2(g) + 4H_2O(g)$$

a. How many moles of H_2O form when 5.00 mol of C_3H_8 completely reacts?

b. How many grams of CO_2 are produced from 18.5 g of oxygen gas?

c. How many grams of H_2O can be produced when 46.3 g of C_3H_8 reacts?

9.59 When 12.8 g of Na and 10.2 g of Cl_2 react, what is the mass, in grams, of NaCl that is produced? (9.2, 9.3, 9.4)

$$2Na(s) + Cl_2(g) \longrightarrow 2NaCl(s)$$

9.60 If 35.8 g of CH_4 and 75.5 g of S react, how many grams of H_2S are produced? (9.2, 9.3, 9.4)

$$CH_4(g) + 4S(g) \longrightarrow CS_2(g) + 2H_2S(g)$$

9.61 Pentane gas, C_5H_{12}, reacts with oxygen to produce carbon dioxide and water. (9.2, 9.3, 9.4)

$$C_5H_{12}(g) + 8O_2(g) \xrightarrow{\Delta} 5CO_2(g) + 6H_2O(g)$$

a. How many moles of C_5H_{12} must react to produce 4.00 mol of water?

b. How many grams of CO_2 are produced from 32.0 g of O_2?

c. How many grams of CO_2 are formed if 44.5 g of C_5H_{12} is mixed with 108 g of O_2?

9.62 When nitrogen dioxide (NO_2) from car exhaust combines with water in the air, it forms nitrogen oxide and nitric acid (HNO_3), which causes acid rain. (9.2, 9.3, 9.4)

$$3NO_2(g) + H_2O(l) \longrightarrow NO(g) + 2HNO_3(aq)$$

a. How many moles of NO_2 are needed to react with 0.250 mol of H_2O?

b. How many grams of HNO_3 are produced when 60.0 g of NO_2 completely reacts?

c. How many grams of HNO_3 can be produced if 225 g of NO_2 is mixed with 55.2 g of H_2O?

9.63 The gaseous hydrocarbon acetylene, C_2H_2, used in welders' torches, burns according to the following equation: (9.2, 9.3, 9.4, 9.5)

$$2C_2H_2(g) + 5O_2(g) \xrightarrow{\Delta} 4CO_2(g) + 2H_2O(g)$$

a. What is the theoretical yield, in grams, of CO_2, if 22.0 g of C_2H_2 completely reacts?

b. If the actual yield in part **a** is 64.0 g of CO_2, what is the percent yield of CO_2 for the reaction?

9.64 The equation for the decomposition of potassium chlorate is written as (9.2, 9.3, 9.4, 9.5)

$$2KClO_3(s) \xrightarrow{\Delta} 2KCl(s) + 3O_2(g)$$

a. When 46.0 g of $KClO_3$ is completely decomposed, what is the theoretical yield, in grams, of O_2?

b. If the actual yield in part **a** is 12.1 g of O_2, what is the percent yield of O_2?

9.65 When 28.0 g of acetylene reacts with hydrogen, 24.5 g of ethane is produced. What is the percent yield of C_2H_6 for the reaction? (9.2, 9.3, 9.4, 9.5)

$$C_2H_2(g) + 2H_2(g) \xrightarrow{Pt} C_2H_6(g)$$

9.66 When 50.0 g of iron(III) oxide reacts with carbon monoxide, 32.8 g of iron is produced. What is the percent yield of Fe for the reaction? (9.2, 9.3, 9.4, 9.5)

$$Fe_2O_3(s) + 3CO(g) \longrightarrow 2Fe(s) + 3CO_2(g)$$

9.67 Nitrogen and hydrogen combine to form ammonia. (9.2, 9.3, 9.4, 9.5)

$$N_2(g) + 3H_2(g) \longrightarrow 2NH_3(g)$$

a. If 50.0 g of N_2 is mixed with 20.0 g of H_2, what is the theoretical yield, in grams, of NH_3?

b. If the reaction in part **a** has a percent yield of 62.0%, what is the actual yield, in grams, of ammonia?

9.68 Sodium and nitrogen combine to form sodium nitride. (9.2, 9.3, 9.4, 9.5)

$$6Na(s) + N_2(g) \longrightarrow 2Na_3N(s)$$

a. If 80.0 g of Na is mixed with 20.0 g of nitrogen gas, what is the theoretical yield, in grams, of Na_3N?

b. If the reaction in part **a** has a percent yield of 75.0%, what is the actual yield, in grams, of Na_3N?

9.69 The equation for the reaction of nitrogen and oxygen to form nitrogen oxide is written as (9.2, 9.6)

$$N_2(g) + O_2(g) \longrightarrow 2NO(g) \qquad \Delta H = +90.2 \text{ kJ}$$

a. How many kilojoules are required to form 3.00 g of NO?

b. What is the complete equation (including heat) for the decomposition of NO?

c. How many kilojoules are released when 5.00 g of NO decomposes to N_2 and O_2?

9.70 The equation for the reaction of iron and oxygen gas to form rust (Fe_2O_3) is written as (9.2, 9.6)

$$4Fe(s) + 3O_2(g) \longrightarrow 2Fe_2O_3(s) \quad \Delta H = -1.7 \times 10^3 \text{ kJ}$$

a. How many kilojoules are released when 2.00 g of Fe reacts?

b. How many grams of rust form when 150 kJ are released?

9.71 Each of the following is a reaction that occurs in the cells of the body. Identify which is exothermic and endothermic. (9.6)
  a. Succinyl CoA + $H_2O \longrightarrow$ succinate + CoA + 37 kJ
  b. GDP + P_i + 34 kJ $\longrightarrow$ GTP + H_2O

9.72 Each of the following is a reaction that occurs in the cells of the body. Identify which is exothermic and endothermic. (9.6)
  a. Phosphocreatine + $H_2O \longrightarrow$ creatine + P_i + 42.7 kJ
  b. Fructose-6-phosphate + P_i + 16 kJ $\longrightarrow$

fructose-1,6-bisphosphate

CHALLENGE QUESTIONS

The following groups of questions are related to the topics in this chapter. However, they do not all follow the chapter order, and they require you to combine concepts and skills from several sections. These questions will help you increase your critical thinking skills and prepare for your next exam.

9.73 Chromium and oxygen combine to form chromium(III) oxide. (9.2, 9.3, 9.4, 9.5)

$$4Cr(s) + 3O_2(g) \longrightarrow 2Cr_2O_3(s)$$

  a. How many moles of O_2 react with 4.50 mol of Cr?
  b. How many grams of Cr_2O_3 are produced when 24.8 g of Cr reacts?
  c. When 26.0 g of Cr reacts with 8.00 g of O_2, how many grams of Cr_2O_3 can form?
  d. If 74.0 g of Cr and 62.0 g of O_2 are mixed, and 87.3 g of Cr_2O_3 is actually obtained, what is the percent yield of Cr_2O_3 for the reaction?

9.74 Aluminum and chlorine combine to form aluminum chloride. (9.2, 9.3, 9.4, 9.5)

$$2Al(s) + 3Cl_2(g) \longrightarrow 2AlCl_3(s)$$

  a. How many moles of Cl_2 are needed to react with 4.50 mol of Al?
  b. How many grams of $AlCl_3$ are produced when 50.2 g of Al reacts?
  c. When 13.5 g of Al reacts with 8.00 g of Cl_2, how many grams of $AlCl_3$ can form?
  d. If 45.0 g of Al and 62.0 g of Cl_2 are mixed, and 66.5 g of $AlCl_3$ is actually obtained, what is the percent yield of $AlCl_3$ for the reaction?

9.75 The combustion of propyne, C_3H_4, releases heat when it burns according to the following equation: (9.2, 9.3, 9.4, 9.5)

$$C_3H_4(g) + 4O_2(g) \xrightarrow{\Delta} 3CO_2(g) + 2H_2O(g)$$

  a. How many moles of O_2 are needed to react completely with 0.225 mol of C_3H_4?
  b. How many grams of water are produced from the complete reaction of 64.0 g of O_2?
  c. How many grams of CO_2 are produced from the complete reaction of 78.0 g of C_3H_4?
  d. If the reaction in part **c** produces 186 g of CO_2, what is the percent yield of CO_2 for the reaction?

9.76 Butane gas, C_4H_{10}, burns according to the following equation: (9.2, 9.3, 9.4, 9.5)

$$2C_4H_{10}(g) + 13O_2(g) \xrightarrow{\Delta} 8CO_2(g) + 10H_2O(g)$$

  a. How many moles of H_2O are produced from the complete reaction of 2.50 mol of C_4H_{10}?
  b. How many grams of O_2 are needed to react completely with 22.5 g of C_4H_{10}?
  c. How many grams of CO_2 are produced from the complete reaction of 55.0 g of C_4H_{10}?

  d. If the reaction in part **c** produces 145 g of CO_2, what is the percent yield of CO_2 for the reaction?

9.77 Use the balanced chemical equation to complete the table: (9.1, 9.2, 9.3)

$$2FeS(s) + 3O_2(g) \longrightarrow 2FeO(s) + 2SO_2(g)$$

FeS	O_2	FeO	SO_2
26 g	___ g	___ g	___ g
___ g	___ g	___ g	7.94 g

9.78 Use the balanced chemical equation to complete the table: (9.1, 9.2, 9.3)

$$2C_2H_6(g) + 7O_2(g) \xrightarrow{\Delta} 4CO_2(g) + 6H_2O(g)$$

C_2H_6	O_2	CO_2	H_2O
___ g	___ g	39 g	___ g
___ g	73.3 g	___ g	___ g

9.79 Sulfur trioxide decomposes to sulfur and oxygen. (9.2, 9.6)

$$2SO_3(g) \longrightarrow 2S(s) + 3O_2(g) \quad \Delta H = +790\,kJ$$

  a. Is the reaction endothermic or exothermic?
  b. How many kilojoules are required when 1.5 mol of SO_3 reacts?
  c. How many kilojoules are required when 150 g of O_2 is formed?

9.80 When hydrogen peroxide (H_2O_2) is used in rocket fuels, it produces water, oxygen, and heat. (9.2, 9.6)

$$2H_2O_2(l) \longrightarrow 2H_2O(l) + O_2(g) \quad \Delta H = -196\,kJ$$

  a. Is the reaction endothermic or exothermic?
  b. How many kilojoules are released when 2.50 mol of H_2O_2 reacts?
  c. How many kilojoules are released when 275 g of O_2 is produced?

9.81 Calculate the enthalpy change for the reaction
$NH_4Cl(s) \longrightarrow NH_3(g) + HCl(g)$, from the following equations: (9.6)

$$\frac{1}{2}H_2(g) + \frac{1}{2}Cl_2(g) \longrightarrow HCl(g) \qquad \Delta H = -92\,kJ$$
$$N_2(g) + 4H_2(g) + Cl_2(g) \longrightarrow 2NH_4Cl(s) \quad \Delta H = -631\,kJ$$
$$N_2(g) + 3H_2(g) \longrightarrow 2NH_3(g) \qquad \Delta H = -296\,kJ$$

9.82 Calculate the enthalpy change for the reaction
$Mg(s) + N_2(g) + 3O_2(g) \longrightarrow Mg(NO_3)_2(s)$ from the following equations: (9.6)

$$8Mg(s) + Mg(NO_3)_2(s) \longrightarrow Mg_3N_2(s) + 6MgO(s)$$
$$\Delta H = -3281\,kJ$$
$$Mg_3N_2(s) \longrightarrow 3Mg(s) + N_2(g) \quad \Delta H = +461\,kJ$$
$$2MgO(s) \longrightarrow 2Mg(s) + O_2(g) \quad \Delta H = +1204\,kJ$$

ANSWERS

Answers to Selected Questions and Problems

9.1 a. 160.14 g of reactants = 160.14 g of products
b. 283.88 g of reactants = 283.88 g of products

9.3 a. $\dfrac{2\ \text{mol SO}_2}{1\ \text{mol O}_2}$ and $\dfrac{1\ \text{mol O}_2}{2\ \text{mol SO}_2}$

$\dfrac{2\ \text{mol SO}_2}{2\ \text{mol SO}_3}$ and $\dfrac{2\ \text{mol SO}_3}{2\ \text{mol SO}_2}$

$\dfrac{2\ \text{mol SO}_3}{1\ \text{mol O}_2}$ and $\dfrac{1\ \text{mol O}_2}{2\ \text{mol SO}_3}$

b. $\dfrac{4\ \text{mol P}}{5\ \text{mol O}_2}$ and $\dfrac{5\ \text{mol O}_2}{4\ \text{mol P}}$

$\dfrac{4\ \text{mol P}}{2\ \text{mol P}_2\text{O}_5}$ and $\dfrac{2\ \text{mol P}_2\text{O}_5}{4\ \text{mol P}}$

$\dfrac{5\ \text{mol O}_2}{2\ \text{mol P}_2\text{O}_5}$ and $\dfrac{2\ \text{mol P}_2\text{O}_5}{5\ \text{mol O}_2}$

9.5 a. $\text{mol SO}_2 \times \dfrac{2\ \text{mol SO}_3}{2\ \text{mol SO}_2} = \text{mol of SO}_3$

b. $\text{mol P} \times \dfrac{5\ \text{mol O}_2}{4\ \text{mol P}} = \text{mol of O}_2$

9.7 a. 1.3 mol of O_2 **b.** 10. mol of H_2
c. 5.0 mol of H_2O

9.9 a. 1.25 mol of C **b.** 0.96 mol of CO
c. 1.0 mol of SO_2 **d.** 0.50 mol of CS_2

9.11 a. 77.5 g of Na_2O **b.** 6.26 g of O_2
c. 19.4 g of O_2

9.13 a. 19.2 g of O_2 **b.** 3.79 g of N_2
c. 54.0 g of H_2O

9.15 a. 3.66 g of H_2O **b.** 26.3 g of NO
c. 7.53 g of HNO_3

9.17 a. $2PbS(s) + 3O_2(g) \longrightarrow 2PbO(s) + 2SO_2(g)$
b. 6.00 g of O_2 **c.** 17.4 g of SO_2
d. 137 g of PbS

9.19 a. Eight taxis can be used to pick up passengers.
b. Seven taxis can be driven.

9.21 a. 5.0 mol of H_2 **b.** 4.0 mol of H_2
c. 3.0 mol of N_2

9.23 a. Cl_2 is the limiting reactant, which would produce 25.1 g of $AlCl_3$.
b. O_2 is the limiting reactant, which would produce 13.5 g of H_2O.
c. O_2 is the limiting reactant, which would produce 26.7 g of SO_2.

9.25 a. 31.2 g of SO_3 **b.** 34.6 g of Fe_3O_4
c. 35.0 g of CO_2

9.27 a. 71.0% **b.** 63.2%

9.29 70.9 g of Al_2O_3

9.31 60.5%

9.33 In exothermic reactions, the energy of the products is lower than that of the reactants.

9.35 a. exothermic **b.** endothermic
c. exothermic

9.37 a. Heat is released, exothermic, $\Delta H = -802$ kJ
b. Heat is absorbed, endothermic, $\Delta H = +65.3$ kJ
c. Heat is released, exothermic, $\Delta H = -850$ kJ

9.39 a. 579 kJ **b.** 89 kJ

9.41 +68 kJ

9.43 −306 kJ

9.45 a. 3.2×10^2 kg **b.** 143 g **c.** 80.4%

9.47 a. $2NO(g) + O_2(g) \longrightarrow 2NO_2(g)$
b. NO is the limiting reactant.

9.49 a. $N_2(g) + 3H_2(g) \longrightarrow 2NH_3(g)$ **b.** A

9.51 a. $2NI_3(g) \longrightarrow N_2(g) + 3I_2(g)$ **b.** 67%

9.53

FeS	O_2	FeO	SO_2
2.0 mol	3.0 mol	2.0 mol	2.0 mol
4.6 mol	6.9 mol	4.6 mol	4.6 mol

9.55 a. 1.33 mol of NH_3 and 3.33 mol of F_2
b. 142 g of F_2
c. 10.4 g of N_2F_4

9.57 a. 3.00 mol of H_2O_2 **b.** 77.6 g of H_2O_2
c. 6.46 g of H_2O

9.59 16.8 g of NaCl

9.61 a. 0.667 mol of C_5H_{12} **b.** 27.5 g of CO_2
c. 92.8 g of CO_2

9.63 a. 74.4 g of CO_2 **b.** 86.0%

9.65 75.9%

9.67 a. 60.8 g of NH_3 **b.** 37.7 g of NH_3

9.69 a. 4.51 kJ
b. $2NO(g) \longrightarrow N_2(g) + O_2(g) + 90.2$ kJ
c. 7.51 kJ

9.71 a. exothermic **b.** endothermic

9.73 a. 3.38 mol of oxygen
b. 36.2 g of chromium(III) oxide
c. 25.3 g of chromium(III) oxide
d. 80.8%

9.75 a. 0.900 mol of oxygen **b.** 18.0 g of water
c. 257 g of carbon dioxide **d.** 72.4%

9.77

FeS	O_2	FeO	SO_2
26 g	14 g	21 g	19 g
10.9 g	5.95 g	8.90 g	7.94 g

9.79 a. endothermic **b.** 590 kJ **c.** 1200 kJ

9.81 76 kJ

Bonding and Properties of Solids and Liquids

10

BILL HAS BEEN DIAGNOSED with basal cell carcinoma, the most common form of skin cancer. He has an appointment to undergo Mohs surgery, a specialized procedure to remove the cancerous growth found on his shoulder. The surgeon begins the process by removing the abnormal growth, in addition to a thin layer of surrounding (margin) tissue, which he sends to Lisa, a histologist. Lisa prepares the tissue sample to be viewed by a pathologist. Tissue preparation requires Lisa to cut the tissue into a very thin section (normally about 0.001 cm), which is then mounted onto a microscope slide.

Lisa then treats the tissue with a dye to stain the cells, as this enables the pathologist to view any abnormal cells more easily. The pathologist examines the tissue sample and reports back to the surgeon that no abnormal cells were present in the margin tissue and that Bill's tumor has been completely removed. No further tissue removal is necessary.

CAREER

Histologist

Histologists study the microscopic make-up of tissues, cells, and bodily fluids in order to detect and identify the presence of a specific disease. They determine blood types and the concentrations of drugs and other substances in the blood. Histologists also help establish a rationale as to why a patient may not be responding to his or her treatment. Sample preparation is a critical component of a histologist's job, as they prepare tissue samples from humans, animals, and plants. The tissue samples are cut into extremely thin sections, which are then mounted and stained using various chemical dyes. The dyes provide contrast for the cells to be viewed and help highlight any abnormalities that may exist. Utilization of various dyes requires the histologist to be familiar with solution preparation and the handling of potentially hazardous chemicals.

 **KEY MATH SKILLS**

- Using Positive and Negative Numbers in Calculations (1.4)
- Solving Equations (1.4)

 CORE CHEMISTRY SKILLS

- Using the Heat Equation (3.5)
- Writing Electron Configurations (5.4)
- Drawing Lewis Symbols (5.6)

*These Key Math Skills and Core Chemistry Skills from previous chapters are listed here for your review as you proceed to the new material in this chapter.

LOOKING AHEAD

10.1 Lewis Structures for Molecules and Polyatomic Ions

LEARNING GOAL Draw the Lewis structures for molecular compounds or polyatomic ions with single and multiple bonds.

Now we can investigate more complex chemical bonds and how they contribute to the structure of a molecule or a polyatomic ion. Lewis structures use Lewis symbols to diagram the sharing of valence electrons in molecules and polyatomic ions. The presence of multiple bonds can be identified, and resonance structures can be drawn, if needed.

ENGAGE

Why do nitrogen and phosphorus have five dots in their Lewis symbols?

Lewis Symbols

A **Lewis symbol** is a convenient way to represent the valence electrons, which are shown as dots placed on the sides, top, or bottom of the symbol for the element.

Lewis symbols for selected elements are given in **TABLE 10.1**.

Number of Valence Electrons Increases →

TABLE 10.1 Lewis Symbols for Selected Elements in Periods 1 to 4

	Group Number							
	1A (1)	2A (2)	3A (13)	4A (14)	5A (15)	6A (16)	7A (17)	8A (18)
Number of Valence Electrons	1	2	3	4	5	6	7	8
Lewis Symbol	H·							He:*
	Li·	Be·	·B·	·C·	·N·	·O·	·F:	:Ne:
	Na·	Mg·	·Al·	·Si·	·P·	·S:	·Cl:	:Ar:
	K·	Ca·	·Ga·	·Ge·	·As·	·Se:	·Br:	:Kr:

*Helium (He) is stable with two valence electrons.

Lewis Structure for the Hydrogen Molecule

The simplest molecule is hydrogen, H_2. When two H atoms are far apart, there is no attraction between them. As the H atoms move closer, the positive charge of each nucleus attracts the electron of the other atom. This attraction, which is greater than the repulsion between the valence electrons, pulls the H atoms closer until they share a pair of valence electrons (see **FIGURE 10.1**). The result is called a *covalent bond*, in which the shared electrons give the stable electron configuration of He to *each* of the H atoms. When the H atoms form H_2, they are more stable than two individual H atoms.

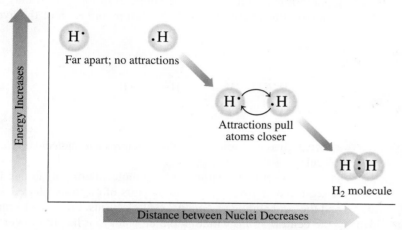

FIGURE 10.1 ▶ A covalent bond forms as H atoms move close together to share electrons.

◎ What determines the attraction between two H atoms?

Lewis Structures for Molecular Compounds

A molecule is represented by a **Lewis structure** in which the valence electrons of all the atoms are arranged to give octets except for hydrogen, which has two electrons. The shared electrons, or *bonding pairs*, are shown as two dots or a single line between atoms. The nonbonding pairs of electrons, or *lone pairs*, are placed on the outside. For example, a fluorine molecule, F_2, consists of two fluorine atoms, which are in Group 7A (17), each with seven valence electrons. In the Lewis structure for the F_2 molecule, each F atom achieves an octet by sharing its unpaired valence electron.

Hydrogen (H_2) and fluorine (F_2) are examples of nonmetal elements whose natural state is diatomic; that is, they contain two like atoms. The elements that exist as diatomic molecules are listed in **TABLE 10.2**.

Drawing Lewis Structures

The number of electrons that a nonmetal atom shares and the number of covalent bonds it forms are usually equal to the number of electrons it needs to achieve a stable electron configuration.

TABLE 10.2 Elements That Exist as Diatomic Molecules

Diatomic Molecule	Name
H_2	Hydrogen
N_2	Nitrogen
O_2	Oxygen
F_2	Fluorine
Cl_2	Chlorine
Br_2	Bromine
I_2	Iodine

CORE CHEMISTRY SKILL
Drawing Lewis Structures

To draw the Lewis structure for CH_4, we first draw the Lewis symbols for carbon and hydrogen.

$$\cdot\ddot{C}\cdot \qquad \cdot H$$

Then we determine the number of valence electrons needed for carbon and hydrogen. When a carbon atom shares its four electrons with four hydrogen atoms, carbon obtains an octet and each hydrogen atom is complete with two shared electrons. The Lewis structure is drawn with the carbon atom as the central atom, with the hydrogen atoms on each of the sides. The bonding pairs of electrons, which are single covalent bonds, may also be shown as single lines between the carbon atom and each of the hydrogen atoms.

TABLE 10.3 Lewis Structures for Some Molecular Compounds

CH_4	NH_3	H_2O

Lewis Structures

Molecular Models

Methane molecule	Ammonia molecule	Water molecule

TABLE 10.3 gives examples of Lewis structures and molecular models for some molecules.

When we draw a Lewis structure for a molecule or polyatomic ion, we show the sequence of atoms, the bonding pairs of electrons shared between atoms, and the nonbonding or *lone pairs* of electrons. From the formula, we identify the central atom, which is the element that has the fewer atoms. Then, the central atom is bonded to the other atoms, as shown in Sample Problem 10.1.

SAMPLE PROBLEM 10.1 Drawing Lewis Structures

Draw the Lewis structure for PCl_3, phosphorus trichloride, used commercially to prepare insecticides and flame retardants.

TRY IT FIRST

SOLUTION

Guide to Drawing Lewis Structures

STEP 1
Determine the arrangement of atoms.

STEP 2
Determine the total number of valence electrons.

STEP 3
Attach each bonded atom to the central atom with a pair of electrons.

STEP 4
Place the remaining electrons using single or multiple bonds to complete octets (two for H).

ANALYZE THE PROBLEM	Given	Need	Connect
	PCl_3	Lewis structure	total valence electrons

STEP 1 Determine the arrangement of atoms. In PCl_3, the central atom is P because there is only one P atom.

Cl P Cl

Cl

STEP 2 Determine the total number of valence electrons. We use the group number to determine the number of valence electrons for each of the atoms in the molecule.

Element	Group	Atoms		Valence Electrons	=	Total
P	5A (15)	1 P	×	$5\ e^-$	=	$5\ e^-$
Cl	7A (17)	3 Cl	×	$7\ e^-$	=	$21\ e^-$
				Total valence electrons for PCl_3	=	$26\ e^-$

STEP 3 Attach each bonded atom to the central atom with a pair of electrons. Each bonding pair can also be represented by a bond line.

$$\text{Cl}\!:\!\text{P}\!:\!\text{Cl} \quad \text{or} \quad \text{Cl}-\text{P}-\text{Cl}$$

Six electrons ($3 \times 2 \, e^-$) are used to bond the central P atom to three Cl atoms. Twenty valence electrons are left.

26 valence e^- − 6 bonding e^- = 20 e^- remaining

STEP 4 Place the remaining electrons using single bonds to complete octets. We use the remaining 20 electrons as lone pairs, which are placed around the outer Cl atoms and on the P atom, such that all the atoms have octets.

$$:\!\ddot{\text{Cl}}\!:\!\text{P}\!:\!\ddot{\text{Cl}}\!: \quad \text{or} \quad :\!\ddot{\text{Cl}}-\ddot{\text{P}}-\ddot{\text{Cl}}\!:$$

The ball-and-stick model of PCl₃ consists of P and Cl atoms connected by single bonds.

STUDY CHECK 10.1

Draw the Lewis structure for Cl₂O.

ANSWER

$$:\!\ddot{\text{Cl}}\!:\!\ddot{\text{O}}\!:\!\ddot{\text{Cl}}\!: \quad \text{or} \quad :\!\ddot{\text{Cl}}-\ddot{\text{O}}-\ddot{\text{Cl}}\!:$$

SAMPLE PROBLEM 10.2 Drawing Lewis Structures for Polyatomic Ions

Sodium chlorite, NaClO₂, is a component of mouthwashes, toothpastes, and contact lens cleaning solutions. Draw the Lewis structure for the chlorite ion, ClO₂⁻.

> **TRY IT FIRST**

SOLUTION

ANALYZE THE PROBLEM	Given	Need	Connect
	ClO₂⁻	Lewis structure	total valence electrons

STEP 1 Determine the arrangement of atoms. For the polyatomic ion ClO₂⁻, the central atom is Cl because there is only one Cl atom. For a polyatomic ion, the atoms and electrons are placed in brackets, and the charge is written outside to the upper right.

[O   Cl   O]⁻

In ClO₂⁻, the central Cl atom is bonded to two O atoms.

STEP 2 Determine the total number of valence electrons. We use the group numbers to determine the number of valence electrons for each of the atoms in the ion. Because the ion has a negative charge, one more electron is added to the valence electrons.

Element	Group	Atoms		Valence Electrons	=	Total
O	6A (16)	2 O	×	6 e^-	=	12 e^-
Cl	7A (17)	1 Cl	×	7 e^-	=	7 e^-
Ionic charge (negative) add				1 e^-	=	1 e^-
		Total valence electrons for ClO₂⁻			=	20 e^-

STEP 3 Attach each bonded atom to the central atom with a pair of electrons. Each bonding pair can also be represented by a line, which indicates a single bond.

[O $\!:\!$ Cl $\!:\!$ O]⁻   or   [O — Cl — O]⁻

Four electrons are used to bond the O atoms to the central Cl atom, which leaves 16 valence electrons.

STEP 4 Place the remaining electrons using single bonds to complete octets. We use the remaining 16 electrons as lone pairs; 12 electrons are drawn as lone pairs to complete the octets of the O atoms.

$$\left[:\ddot{\text{O}}:\text{Cl}:\ddot{\text{O}}:\right]^{-} \qquad \left[:\ddot{\text{O}}—\text{Cl}—\ddot{\text{O}}:\right]^{-}$$

The remaining four electrons are placed as two lone pairs on the central Cl atom.

$$\left[:\ddot{\text{O}}:\ddot{\text{Cl}}:\ddot{\text{O}}:\right]^{-} \quad \text{or} \quad \left[:\ddot{\text{O}}—\ddot{\text{Cl}}—\ddot{\text{O}}:\right]^{-}$$

STUDY CHECK 10.2

Draw the Lewis structure for the polyatomic ion NH_2^-.

ANSWER

$$\left[\text{H}:\ddot{\text{N}}:\text{H}\right]^{-} \quad \text{or} \quad \left[\text{H}—\ddot{\text{N}}—\text{H}\right]^{-}$$

ENGAGE

What is wrong with the following Lewis structure for ClO_2^-?

$$\left[:\ddot{\text{O}}=\text{Cl}=\ddot{\text{O}}:\right]^{-}$$

Double and Triple Bonds

Up to now, we have looked at bonding in molecules having only single bonds. In many molecular compounds, atoms share two or three pairs of electrons to complete their octets.

Double and triple bonds form when the number of valence electrons is not enough to complete the octets of all the atoms in the molecule. Then one or more lone pairs of electrons from the atoms attached to the central atom are shared with the central atom. A **double bond** occurs when two pairs of electrons are shared; in a **triple bond**, three pairs of electrons are shared. Atoms of carbon, oxygen, nitrogen, and sulfur are most likely to form multiple bonds.

Atoms of hydrogen and the halogens do not form double or triple bonds. The process of drawing a Lewis structure with multiple bonds is shown in Sample Problem 10.3.

SAMPLE PROBLEM 10.3 Drawing Lewis Structures with Multiple Bonds

Draw the Lewis structure for carbon dioxide, CO_2, in which the central atom is C.

SOLUTION

ANALYZE THE PROBLEM	Given	Need	Connect
	CO_2	Lewis structure	total valence electrons

STEP 1 Determine the arrangement of atoms. O C O

STEP 2 Determine the total number of valence electrons. We use the group number to determine the number of valence electrons for each of the atoms in the molecule.

Element	Group	Atoms		Valence Electrons	=	Total
C	4A (14)	1 C	×	$4\,e^-$	=	$4\,e^-$
O	6A (16)	2 O	×	$6\,e^-$	=	$12\,e^-$
			Total valence electrons for CO_2		=	$16\,e^-$

STEP 3 Attach each bonded atom to the central atom with a pair of electrons.

O:C:O   or   O—C—O

We use four valence electrons to attach the central C atom to two O atoms.

STEP 4 Place the remaining electrons using single or multiple bonds to complete octets.

The 12 remaining electrons are placed as six lone pairs of electrons on the outside O atoms. However, this does not complete the octet of the C atom.

:Ö:C:Ö: or   :Ö—C—Ö:

To obtain an octet, the C atom must share lone pairs of electrons from each of the O atoms. When two bonding pairs occur between atoms, it is known as a double bond.

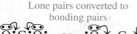

Lone pairs converted to bonding pairs

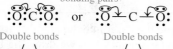

:Ö:C:Ö: or   :Ö—C—Ö:

Double bonds   Double bonds

:O::C::O: or   :O=C=O:

Molecule of carbon dioxide

STUDY CHECK 10.3

Draw the Lewis structure for HCN, which has a triple bond.

ANSWER

H:C⦂⦂N: or   H—C≡N:

Exceptions to the Octet Rule

Although the octet rule is useful for bonding in many compounds, there are exceptions. We have already seen that a hydrogen (H_2) molecule requires just two electrons or a single bond. Usually the nonmetals form octets. However, in BCl_3, the B atom has only three valence electrons to share. Boron compounds typically have six valence electrons on the central B atoms and form just three bonds. Although we will generally see compounds of P, S, Cl, Br, and I with octets, they can form molecules in which they share more of their valence electrons. This expands their valence electrons to 10, 12, or even 14 electrons. For example, we have seen that the P atom in PCl_3 has an octet, but in PCl_5, the P atom has five bonds with 10 valence electrons. In H_2S, the S atom has an octet, but in SF_6, there are six bonds to sulfur with 12 valence electrons.

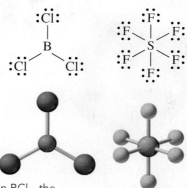

In BCl_3, the central B atom is bonded to three Cl atoms.

In SF_6, the central S atom is bonded to six F atoms.

QUESTIONS AND PROBLEMS

10.1 Lewis Structures for Molecules and Polyatomic Ions

LEARNING GOAL Draw the Lewis structures for molecular compounds or polyatomic ions with single and multiple bonds.

10.1 Determine the total number of valence electrons for each of the following:
  a. H_2S b. I_2 c. CCl_4 d. OH^-

10.2 Determine the total number of valence electrons for each of the following:
  a. SBr_2 b. NBr_3 c. CH_3OH d. NH_4^+

10.3 Draw the Lewis structure for each of the following molecules or polyatomic ions:
  a. HF   b. SF_2 c. NBr_3 d. BH_4^-

  e. CH_3OH (methyl alcohol)   H C O H

  f. N_2H_4 (hydrazine)   H N N H

10.4 Draw the Lewis structure for each of the following molecules or polyatomic ions:

  a. H_2O **b.** CCl_4 **c.** H_3O^+ **d.** SiF_4

  e. CF_2Cl_2

  f. C_2H_6

$$\begin{matrix} H & H \\ H\ C & C\ H \\ H & H \end{matrix}$$

10.5 When is it necessary to draw a multiple bond in a Lewis structure?

10.6 If the available number of valence electrons for a molecule or polyatomic ion does not complete all of the octets in a Lewis structure, what should you do?

10.7 Draw the Lewis structure for each of the following molecules or ions:

  a. CO   **b.** CN^- **c.** H_2CO (C is the central atom)

10.8 Draw the Lewis structure for each of the following molecules or ions:

  a. HCCH (acetylene)   **b.** CS_2 **c.** NO^+

10.2 Resonance Structures

LEARNING GOAL Draw Lewis structures for molecules or polyatomic ions that have two or more resonance structures.

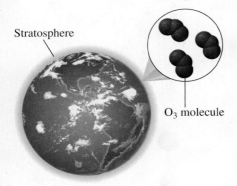

CORE CHEMISTRY SKILL
Drawing Resonance Structures

When a molecule or polyatomic ion contains multiple bonds, it may be possible to draw more than one Lewis structure. We can see how this happens when we draw the Lewis structure for ozone, O_3, a component in the stratosphere that protects us from the ultraviolet rays of the Sun.

To draw the Lewis structure for O_3, we determine the number of valence electrons for an O atom, and then the total number of valence electrons for O_3. Because O is in Group 6A (16), it has six valence electrons. Therefore, the compound O_3 would have a total of 18 valence electrons.

Element	Group	Atoms		Valence Electrons	=	Total
O	6A (16)	3 O	×	$6\ e^-$	=	$18\ e^-$

For the Lewis structure for O_3, we place three O atoms in a row. Using four of the valence electrons, we draw a bonding pair between each of the O atoms on the ends and the central O atom. Two bonding pairs require four valence electrons.

O—O—O

The remaining valence electrons (14) are placed as lone pairs of electrons around the O atoms on both ends of the Lewis structure, and one lone pair goes on the central O atom.

:Ö—Ö—Ö:

Stratosphere

O_3 molecule

Ozone, O_3, is a component in the stratosphere that protects us from the ultraviolet rays of the Sun.

To complete an octet for the central O atom, one lone pair of electrons from an end O atom needs to be shared. But which lone pair should be used? One possibility is to form a double bond between the central O atom and the O on the left, and the other possibility is to form a double bond between the central O atom and the O on the right.

:Ö—Ö—Ö: or   :Ö—Ö—Ö:

ENGAGE

Why are there two possible ways to form a double bond in ozone?

Thus it is possible to draw two or more Lewis structures for a molecule such as O_3 or for a polyatomic ion. When this happens, all the Lewis structures are called **resonance structures**, and their relationship is shown by drawing a double-headed arrow between them.

:Ö=Ö—Ö: ⟷ :Ö—Ö=Ö:

Resonance structures

Experiments show that the actual bond lengths in ozone are equivalent to a molecule with a "one-and-a-half" bond between the central O atom and each outside O atom. In an actual ozone molecule, the electrons are spread equally over all the O atoms. When we draw resonance structures for molecules or polyatomic ions, the true structure is really an average of those structures.

SAMPLE PROBLEM 10.4 Drawing Resonance Structures

Sulfur dioxide is produced by volcanic activity and the burning of sulfur-containing coal. Once in the atmosphere, the SO_2 is converted to SO_3, which combines with water, forming sulfuric acid, H_2SO_4, a component of acid rain. Draw two resonance structures for sulfur dioxide.

TRY IT FIRST

SOLUTION

ANALYZE THE PROBLEM	Given	Need	Connect
	SO_2	resonance structures	total valence electrons, double bonds

STEP 1 Determine the arrangement of atoms. In SO_2, the S atom is the central atom because there is only one S atom.

O   S   O

STEP 2 Determine the total number of valence electrons. We use the group number to determine the number of valence electrons for each of the atoms in the molecule.

Element	Group	Atoms		Valence Electrons	=	Total
S	6A (16)	1 S	×	$6\,e^-$	=	$6\,e^-$
O	6A (16)	2 O	×	$6\,e^-$	=	$12\,e^-$
			Total valence electrons for SO_2		=	$18\,e^-$

STEP 3 Attach each bonded atom to the central atom with a pair of electrons.

O—S—O

We use four electrons to form single bonds between the central S atom and the O atoms.

STEP 4 Place the remaining electrons using single or multiple bonds to complete octets. The remaining 14 electrons are drawn as lone pairs, which complete the octets for the O atoms but not the S atom.

$$:\!\ddot{O}\!—\!\ddot{S}\!—\!\ddot{O}\!:$$

To complete the octet for S, one lone pair of electrons from one of the O atoms is shared to form a double bond. One possibility is to form a double bond between the central S atom and the O on the left, and the other possibility is to form a double bond between the central S atom and the O on the right.

$$:\!O\!=\!\ddot{S}\!—\!\ddot{O}\!:\ \longleftrightarrow\ :\!\ddot{O}\!—\!\ddot{S}\!=\!O\!:$$

The ball-and-stick models of SO_2 with S (yellow) and O (red) atoms show the two resonance structures.

STUDY CHECK 10.4

Draw three resonance structures for SO_3.

ANSWER

$$:\!\ddot{O}\!—\!S\!=\!\ddot{O}\!:\ \longleftrightarrow\ :\!\ddot{O}\!—\!\ddot{S}\!—\!\ddot{O}\!:\ \longleftrightarrow\ :\!\ddot{O}\!=\!S\!—\!\ddot{O}\!:$$
$$\quad\ \ \ \ |\qquad\qquad\qquad\ \ \|\qquad\qquad\qquad\ \ \ |$$
$$\quad\ \ \ :\!\ddot{O}\!:\qquad\qquad\qquad\ :\!\ddot{O}\!:\qquad\qquad\qquad\ :\!\ddot{O}\!:$$

ENGAGE

Explain why SO_2 has resonance structures but SCl_2 does not.

TABLE 10.4 summarizes this method of drawing Lewis structures for several molecules and ions.

TABLE 10.4 Using Valence Electrons to Draw Lewis Structures

Molecule or Polyatomic Ion	Total Valence Electrons	Form Single Bonds to Attach Atoms (electrons used)	Electrons Remaining	Completed Octets (or H:)
Cl_2	$2(7) = 14$	$Cl—Cl\ (2\ e^-)$	$14 - 2 = 12$	$:\overset{..}{\underset{..}{Cl}}—\overset{..}{\underset{..}{Cl}}:$
HCl	$1 + 7 = 8$	$H—Cl\ (2\ e^-)$	$8 - 2 = 6$	$H—\overset{..}{\underset{..}{Cl}}:$
H_2O	$2(1) + 6 = 8$	$H—O—H\ (4\ e^-)$	$8 - 4 = 4$	$H—\overset{..}{\underset{..}{O}}—H$
PCl_3	$5 + 3(7) = 26$	$\begin{array}{c} Cl \\ | \\ Cl—P—Cl \end{array} (6\ e^-)$	$26 - 6 = 20$	$\begin{array}{c} :\overset{..}{\underset{..}{Cl}}: \\ | \\ :\overset{..}{\underset{..}{Cl}}—P—\overset{..}{\underset{..}{Cl}}: \end{array}$
ClO_3^-	$7 + 3(6) + 1 = 26$	$\left[\begin{array}{c} O \\ | \\ O—Cl—O \end{array} \right]^- (6\ e^-)$	$26 - 6 = 20$	$\left[\begin{array}{c} :\overset{..}{\underset{..}{O}}: \\ | \\ :\overset{..}{\underset{..}{O}}—Cl—\overset{..}{\underset{..}{O}}: \end{array} \right]^-$
NO_2^-	$5 + 2(6) + 1 = 18$	$[O—N—O]^-\ (4\ e^-)$	$18 - 4 = 14$	$[:\overset{..}{\underset{..}{O}}—\overset{..}{N}=\overset{..}{O}:]^-$ $\updownarrow$ $[:\overset{..}{O}=\overset{..}{N}—\overset{..}{\underset{..}{O}}:]^-$

QUESTIONS AND PROBLEMS

10.2 Resonance Structures

LEARNING GOAL Draw Lewis structures for molecules or polyatomic ions that have two or more resonance structures.

10.9 What is resonance?

10.10 When does a molecular compound have resonance?

10.11 Draw resonance structures for each of the following molecules or ions:
a. $ClNO_2$ (N is the central atom)
b. OCN^- (C is the central atom)

10.12 Draw resonance structures for each of the following molecules or ions:
a. $H_2CNO_2^-$
b. N_2O (N N O)

10.3 Shapes of Molecules and Polyatomic Ions (VSEPR Theory)

LEARNING GOAL Predict the three-dimensional structure of a molecule or a polyatomic ion.

CORE CHEMISTRY SKILL
Predicting Shape

Using the Lewis structures, we can predict the three-dimensional shapes of many molecules and polyatomic ions. The shape of a compound is important in our understanding of how molecules interact with enzymes or certain antibiotics or produce our sense of taste and smell.

The three-dimensional shape of a molecule is determined by drawing its Lewis structure and identifying the number of electron groups (one or more electron pairs) around the central atom. We count lone pairs of electrons, single, double, and triple bonds as *one* electron group. In the **valence shell electron-pair repulsion (VSEPR) theory**, the electron groups are arranged as far apart as possible around the central atom to minimize the repulsion between their negative charges. Once we have counted the number of electron groups surrounding the central atom, we can determine its specific shape from the number of atoms bonded to the central atom.

Central Atoms with Two Electron Groups

In the Lewis structure for CO_2, there are two electron groups (two double bonds) attached to the central atom. According to VSEPR theory, minimal repulsion occurs when two electron groups are on opposite sides of the central C atom. This gives the CO_2 molecule a *linear* electron-group geometry and a shape that is **linear** with a bond angle of 180°.

$$:\ddot{O}\!=\!C\!=\!\ddot{O}:$$

180°

Linear
electron-group
geometry

Linear shape

Central Atoms with Three Electron Groups

In the Lewis structure for formaldehyde, H_2CO, the central atom C is attached to two H atoms by single bonds and to the O atom by a double bond. Minimal repulsion occurs when three electron groups are as far apart as possible around the central C atom, which gives 120° bond angles. This type of electron-group geometry is *trigonal planar* and gives a shape for H_2CO called ***trigonal planar***.

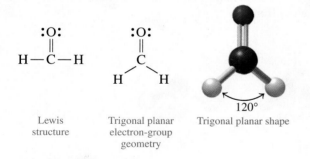

:O:
‖
H—C—H

Lewis
structure

:O:
‖
C
H   H

Trigonal planar
electron-group
geometry

120°

Trigonal planar shape

In the Lewis structure for SO_2, there are also three electron groups around the central S atom: a single bond to an O atom, a double bond to another O atom, and a lone pair of electrons. As in H_2CO, three electron groups have minimal repulsion when they form trigonal planar electron-group geometry. However, in SO_2 one of the electron groups is a lone pair of electrons. Therefore, the shape of the SO_2 molecule is determined by the two O atoms bonded to the central S atom, which gives the SO_2 molecule a shape that is **bent** with a bond angle of 120°. When there are one or more lone pairs on the central atom, the shape has a different name than that of the electron-group geometry.

$$:\ddot{O}\!-\!\ddot{S}\!=\!\ddot{O}:$$

S
:O: :O:

120°

Bent shape

Lewis
structure

Trigonal planar
electron-group
geometry

Central Atom with Four Electron Groups

In a molecule of methane, CH_4, the central C atom is bonded to four H atoms. From the Lewis structure, you may think that CH_4 is planar with 90° bond angles. However, the best geometry for minimal repulsion is *tetrahedral*, giving bond angles of 109°. When there are four atoms attached to four electron groups, the shape of the molecule is ***tetrahedral***.

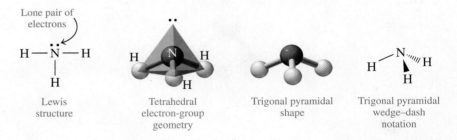

Lewis structure   Tetrahedral electron-group geometry   Tetrahedral shape   Tetrahedral wedge–dash notation

A way to represent the three-dimensional structure of methane is to use the *wedge–dash notation*. In this representation, the two bonds connecting carbon to hydrogen by solid lines are in the plane of the paper. The wedge represents a carbon-to-hydrogen bond coming out of the page toward us, whereas the dash represents a carbon-to-hydrogen bond going into the page away from us.

Now we can look at molecules that have four electron groups, of which one or more are lone pairs of electrons. Then the central atom is attached to only two or three atoms. For example, in the Lewis structure for ammonia, NH_3, four electron groups have a tetrahedral electron-group geometry. However, in NH_3 one of the electron groups is a lone pair of electrons. Therefore, the shape of NH_3 is determined by the three H atoms bonded to the central N atom. Therefore, the shape of the NH_3 molecule is ***trigonal pyramidal***, with a bond angle of 109°. The wedge–dash notation can also represent this three-dimensional structure of ammonia with one N—H bond in the plane, one N—H bond coming toward us, and one N—H bond going away from us.

ENGAGE

If the four electron groups in a PH_3 molecule have a tetrahedral geometry, why does a PH_3 molecule have a trigonal pyramidal shape and not a tetrahedral shape?

Lone pair of electrons

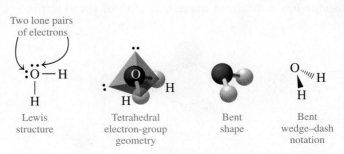

Lewis structure   Tetrahedral electron-group geometry   Trigonal pyramidal shape   Trigonal pyramidal wedge–dash notation

In the Lewis structure for water, H_2O, there are also four electron groups, which have minimal repulsion when the electron-group geometry is tetrahedral. However, in H_2O, two of the electron groups are lone pairs of electrons. Because the shape of H_2O is determined by the two H atoms bonded to the central O atom, the H_2O molecule has a *bent shape* with a bond angle of 109°. **TABLE 10.5** gives the shapes of molecules with two, three, or four bonded atoms.

Two lone pairs of electrons

Lewis structure   Tetrahedral electron-group geometry   Bent shape   Bent wedge–dash notation

TABLE 10.5 Molecular Shapes for a Central Atom with Two, Three, and Four Bonded Atoms

Electron Groups	Electron-Group Geometry	Bonded Atoms	Lone Pairs	Bond Angle*	Molecular Shape	Example	Three-Dimensional Model
2	Linear	2	0	180°	Linear	CO_2	
3	Trigonal planar	3	0	120°	Trigonal planar	H_2CO	
3	Trigonal planar	2	1	120°	Bent	SO_2	
4	Tetrahedral	4	0	109°	Tetrahedral	CH_4	
4	Tetrahedral	3	1	109°	Trigonal pyramidal	NH_3	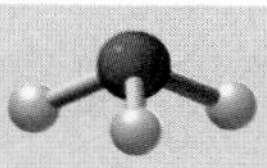
4	Tetrahedral	2	2	109°	Bent	H_2O	

*The bond angles in actual molecules may vary slightly.

SAMPLE PROBLEM 10.5 Shapes of Molecules

Use VSEPR theory to predict the shape of the molecule $SiCl_4$.

TRY IT FIRST

SOLUTION

ANALYZE THE PROBLEM	Given	Need	Connect
	$SiCl_4$	shape	Lewis structure, electron groups, bonded atoms

STEP 1 Draw the Lewis structure.

Name of Element	Silicon	Chlorine
Symbol of Element	Si	Cl
Atoms of Element	1	4
Valence Electrons	$4\,e^-$	$7\,e^-$
Total Electrons	$1(4\,e^-)$	$+ \quad 4(7\,e^-) = 32\,e^-$

Using $32\,e^-$, we draw the bonds and lone pairs for the Lewis structure of $SiCl_4$.

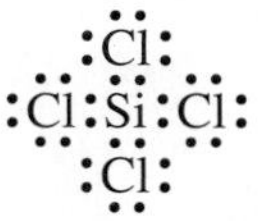

:Cl:
:Cl:Si:Cl:
:Cl:

STEP 2 Arrange the electron groups around the central atom to minimize repulsion. To minimize repulsion, the four electron groups would have a tetrahedral geometry.

Guide to Predicting Shape (VSEPR Theory)

STEP 1
Draw the Lewis structure.

STEP 2
Arrange the electron groups around the central atom to minimize repulsion.

STEP 3
Use the atoms bonded to the central atom to determine the shape.

STEP 3 Use the atoms bonded to the central atom to determine the shape.
Because the central Si atom is bonded to four atoms and no lone pairs of electrons, the $SiCl_4$ molecule has a tetrahedral shape.

STUDY CHECK 10.5

Use VSEPR theory to predict the shape of SCl_2.

ANSWER

The central atom S has four electron groups: two bonded atoms and two lone pairs of electrons. The shape of SCl_2 is bent, 109°.

SAMPLE PROBLEM 10.6 Predicting Shape of an Ion

Use VSEPR theory to predict the shape of the polyatomic ion NO_3^-.

TRY IT FIRST

SOLUTION

	Given	Need	Connect
ANALYZE THE PROBLEM	NO_3^-	shape	Lewis structure, electron groups, bonded atoms

STEP 1 Draw the Lewis structure.

Name of Element	Nitrogen	Oxygen
Symbol of Element	N	O
Atoms of Element	1	3
Valence Electrons	$5\,e^-$	$6\,e^-$
Total Electrons	$1(5\,e^-) + 3(6\,e^-) + 1\,e^-$ (charge) $= 24\,e^-$	

The polyatomic ion NO_3^- contains three electron groups (two single bonds between the central N atom and O atoms, and one double bond between N and O). Note that the double bond can be drawn to any of the O atoms, which results in three resonance structures. However, we need just one of the structures to predict its shape.

STEP 2 Arrange the electron groups around the central atom to minimize repulsion. To minimize repulsion, the three electron groups have a trigonal planar geometry.

STEP 3 Use the atoms bonded to the central atom to determine the shape.
Because NO_3^- has three bonded atoms and no lone pairs, it has a trigonal planar shape.

STUDY CHECK 10.6

Use VSEPR theory to predict the shape of ClO_2^- (Cl is the central atom).

ANSWER

The central atom Cl has four electron groups: two bonded atoms and two lone pairs of electrons. The shape of ClO_2^- is bent, 109°.

QUESTIONS AND PROBLEMS

10.3 Shapes of Molecules and Polyatomic Ions (VSEPR Theory)

LEARNING GOAL Predict the three-dimensional structure of a molecule or polyatomic ion.

10.13 Choose the shape (**1** to **6**) that matches each of the following descriptions (**a** to **c**):
  1. linear **2.** bent (109°) **3.** trigonal planar
  4. bent (120°) **5.** trigonal pyramidal **6.** tetrahedral
  a. a molecule with a central atom that has four electron groups and four bonded atoms
  b. a molecule with a central atom that has four electron groups and three bonded atoms
  c. a molecule with a central atom that has three electron groups and three bonded atoms

10.14 Choose the shape (**1** to **6**) that matches each of the following descriptions (**a** to **c**):
  1. linear **2.** bent (109°) **3.** trigonal planar
  4. bent (120°) **5.** trigonal pyramidal **6.** tetrahedral
  a. a molecule with a central atom that has four electron groups and two bonded atoms
  b. a molecule with a central atom that has two electron groups and two bonded atoms
  c. a molecule with a central atom that has three electron groups and two bonded atoms

10.15 Complete each of the following statements for a molecule of SeO_3:
  a. There are _____ electron groups around the central Se atom.
  b. The electron-group geometry is _____.
  c. The number of atoms attached to the central Se atom is _____.
  d. The shape of the molecule is _____.

10.16 Complete each of the following statements for a molecule of H_2S:
  a. There are _____ electron groups around the central S atom.
  b. The electron-group geometry is _____.
  c. The number of atoms attached to the central S atom is _____.
  d. The shape of the molecule is _____.

10.17 Compare the Lewis structures of CF_4 and NF_3. Why do these molecules have different shapes?

10.18 Compare the Lewis structures of CH_4 and H_2O. Why do these molecules have similar bond angles but different molecular shapes?

10.19 Use VSEPR theory to predict the shape of each of the following:
  a. GaH_3 **b.** OF_2 **c.** HCN **d.** CCl_4

10.20 Use VSEPR theory to predict the shape of each of the following:
  a. CF_4 **b.** NCl_3 **c.** $SeBr_2$ **d.** CS_2

10.21 Draw the Lewis structure and predict the shape for each of the following:
  a. AlH_4^- **b.** SO_4^{2-} **c.** NH_4^+ **d.** NO_2^+

10.22 Draw the Lewis structure and predict the shape for each of the following:
  a. NO_2^- **b.** PO_4^{3-} **c.** ClO_4^- **d.** SF_3^+

10.4 Electronegativity and Bond Polarity

LEARNING GOAL Use electronegativity to determine the polarity of a bond.

We can learn more about the chemistry of compounds by looking at how bonding electrons are shared between atoms. The bonding electrons are shared equally in a bond between identical nonmetal atoms. However, when a bond is between atoms of different elements, the electron pairs are usually shared unequally. Then the shared pairs of electrons are attracted to one atom in the bond more than the other.

The **electronegativity** of an atom is its ability to attract the shared electrons in a chemical bond. Nonmetals have higher electronegativities than do metals, because nonmetals have a greater attraction for electrons than metals. On the electronegativity scale, fluorine was assigned a value of 4.0, and the electronegativities for all other elements were determined relative to the attraction of fluorine for shared electrons. The nonmetal fluorine, which has the highest electronegativity (4.0), is located in the upper right corner of the periodic table. The metal cesium, which has the lowest electronegativity (0.7), is located in the lower left corner of the periodic table. The electronegativities for the representative elements are shown in **FIGURE 10.2**. Note that there are no electronegativity values for the noble gases because they do not typically form bonds. The electronegativity values for transition elements are also low, but we have not included them in our discussion.

CORE CHEMISTRY SKILL
Using Electronegativity

ENGAGE

Why is the electronegativity of a chlorine atom greater than that of an iodine atom?

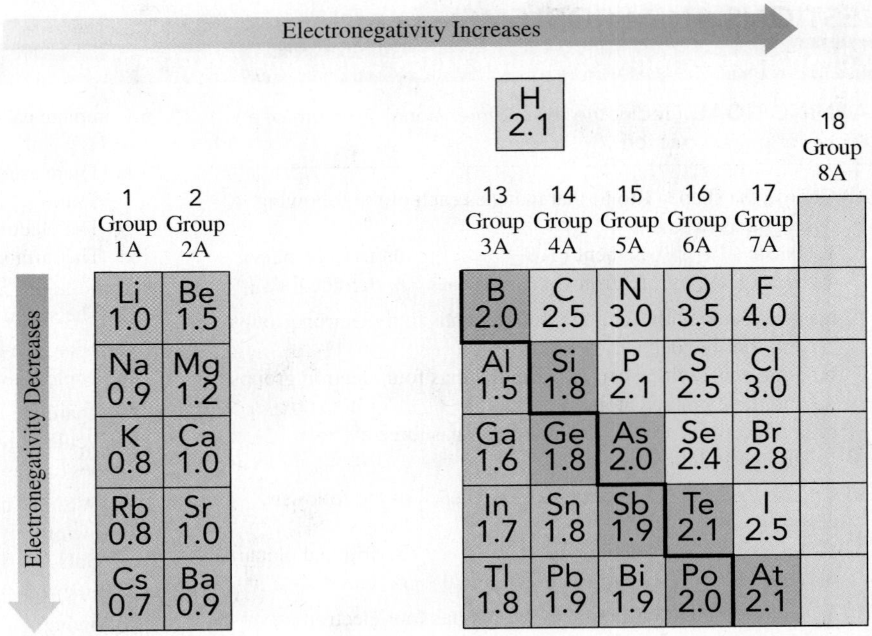

FIGURE 10.2 ▶ The electronegativity values of representative elements in Group 1A (1) to Group 7A (17), which indicate the ability of atoms to attract shared electrons, increase going across a period from left to right and decrease going down a group.

◎ What element on the periodic table has the strongest attraction for shared electrons?

Electronegativity Increases →

Electronegativity Decreases ↓

| H
2.1 | | 18
Group
8A |

1 Group 1A	2 Group 2A		13 Group 3A	14 Group 4A	15 Group 5A	16 Group 6A	17 Group 7A
Li 1.0	Be 1.5		B 2.0	C 2.5	N 3.0	O 3.5	F 4.0
Na 0.9	Mg 1.2		Al 1.5	Si 1.8	P 2.1	S 2.5	Cl 3.0
K 0.8	Ca 1.0		Ga 1.6	Ge 1.8	As 2.0	Se 2.4	Br 2.8
Rb 0.8	Sr 1.0		In 1.7	Sn 1.8	Sb 1.9	Te 2.1	I 2.5
Cs 0.7	Ba 0.9		Tl 1.8	Pb 1.9	Bi 1.9	Po 2.0	At 2.1

Polarity of Bonds

The difference in the electronegativity values of two atoms can be used to predict the type of chemical bond, ionic or covalent, that forms. For the H—H bond, the electronegativity difference is zero $(2.1 - 2.1 = 0)$, which means the bonding electrons are shared equally. We illustrate this by drawing a symmetrical electron cloud around the H atoms. A bond between atoms with identical or very similar electronegativity values is a **nonpolar covalent bond**. However, when covalent bonds are between atoms with different electronegativity values, the electrons are shared unequally; the bond is a **polar covalent bond**. The electron cloud for a polar covalent bond is unsymmetrical. For the H—Cl bond, there is an electronegativity difference of $3.0 (Cl) - 2.1 (H) = 0.9$, which means that the H—Cl bond is polar covalent (see **FIGURE 10.3**). When finding the electronegativity difference, the smaller electronegativity is always subtracted from the larger; thus, the difference is always a positive number.

FIGURE 10.3 ▶ In the nonpolar covalent bond of H_2, electrons are shared equally. In the polar covalent bond of HCl, electrons are shared unequally.

◎ H_2 has a nonpolar covalent bond, but HCl has a polar covalent bond. Explain.

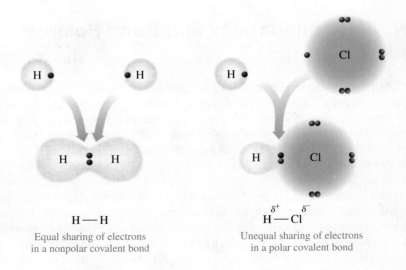

H—H
Equal sharing of electrons
in a nonpolar covalent bond

δ^+ δ^-
H—Cl
Unequal sharing of electrons
in a polar covalent bond

Dipoles and Bond Polarity

The **polarity** of a bond depends on the difference in the electronegativity values of its atoms. In a polar covalent bond, the shared electrons are attracted to the more electronegative atom, which makes it partially negative, because of the negatively charged electrons around that atom. At the other end of the bond, the atom with the lower electronegativity becomes partially positive because of the lack of electrons at that atom.

A bond becomes more *polar* as the electronegativity difference increases. A polar covalent bond that has a separation of charges is called a **dipole**. The positive and negative ends of the dipole are indicated by the lowercase Greek letter delta with a positive or negative sign, δ^+ and δ^-. Sometimes we use an arrow that points from the positive charge to the negative charge $\longmapsto$ to indicate the dipole.

Examples of Dipoles in Polar Covalent Bonds

$$\overset{\delta^+}{C}-\overset{\delta^-}{O} \qquad \overset{\delta^+}{N}-\overset{\delta^-}{O} \qquad \overset{\delta^+}{Cl}-\overset{\delta^-}{F}$$
$\longmapsto \qquad\qquad \longmapsto \qquad\qquad \longmapsto$

Variations in Bonding

The variations in bonding are continuous; there is no definite point at which one type of bond stops and the next starts. When the electronegativity difference is between 0.0 and 0.4, the electrons are considered to be shared equally in a *nonpolar covalent bond*. For example, the C—C bond (2.5 − 2.5 = 0.0) and the C—H bond (2.5 − 2.1 = 0.4) are classified as nonpolar covalent bonds.

As the electronegativity difference increases, the shared electrons are attracted more strongly to the more electronegative atom, which increases the polarity of the bond. When the electronegativity difference is from 0.5 to 1.8, the bond is a *polar covalent bond*. For example, the O—H bond (3.5 − 2.1 = 1.4) is classified as a polar covalent bond (see **TABLE 10.6**).

ENGAGE

Use electronegativity differences to explain why a Si—S bond is polar covalent and a Si—P bond is nonpolar covalent.

TABLE 10.6 Electronegativity Differences and Types of Bonds

Electronegativity Difference	0.0 to 0.4	0.5 to 1.8	1.9 to 3.3
Bond Type	Nonpolar covalent	Polar covalent	Ionic
Electron Bonding	Electrons shared equally	Electrons shared unequally	Electrons transferred

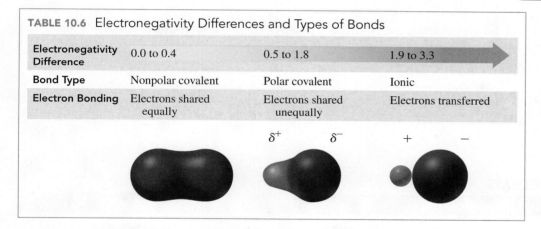

When the electronegativity difference is greater than 1.8, electrons are transferred from one atom to another, which results in an *ionic bond*. For example, the electronegativity difference for the ionic compound NaCl is 3.0 − 0.9 = 2.1. Thus, for large differences in electronegativity, we would predict an ionic bond (see **TABLE 10.7**).

TABLE 10.7 Predicting Bond Type from Electronegativity Differences

Molecule	Bond	Type of Electron Sharing	Electronegativity Difference*	Type of Bond	Reason
H_2	H—H	Shared equally	2.1 − 2.1 = 0.0	Nonpolar covalent	From 0.0 to 0.4
BrCl	Br—Cl	Shared about equally	3.0 − 2.8 = 0.2	Nonpolar covalent	From 0.0 to 0.4
HBr	H^{δ^+}—Br^{δ^-}	Shared unequally	2.8 − 2.1 = 0.7	Polar covalent	From 0.5 to 1.8
HCl	H^{δ^+}—Cl^{δ^-}	Shared unequally	3.0 − 2.1 = 0.9	Polar covalent	From 0.5 to 1.8
NaCl	Na^+Cl^-	Electron transfer	3.0 − 0.9 = 2.1	Ionic	From 1.9 to 3.3
MgO	$Mg^{2+}O^{2-}$	Electron transfer	3.5 − 1.2 = 2.3	Ionic	From 1.9 to 3.3

*Values are taken from Figure 10.2.

SAMPLE PROBLEM 10.7 Bond Polarity

Using electronegativity values, classify each of the following bonds as nonpolar covalent, polar covalent, or ionic:

O—K,   Cl—As,   N—N,   P—Br

TRY IT FIRST

SOLUTION

ANALYZE THE PROBLEM	Given	Need	Connect
	bonds	type of bonds	electronegativity values

For each bond, we obtain the electronegativity values and calculate the difference in electronegativity.

Bond	Electronegativity Difference	Type of Bond
O—K	$3.5 - 0.8 = 2.7$	Ionic
Cl—As	$3.0 - 2.0 = 1.0$	Polar covalent
N—N	$3.0 - 3.0 = 0.0$	Nonpolar covalent
P—Br	$2.8 - 2.1 = 0.7$	Polar covalent

STUDY CHECK 10.7

Using electronegativity values, classify each of the following bonds as nonpolar covalent, polar covalent, or ionic:

a. P—Cl   **b.** Br—Br   **c.** Na—O

ANSWER

a. polar covalent (0.9)   **b.** nonpolar covalent (0.0)   **c.** ionic (2.6)

QUESTIONS AND PROBLEMS

10.4 Electronegativity and Bond Polarity

LEARNING GOAL Use electronegativity to determine the polarity of a bond.

10.23 Describe the trend in electronegativity as *increases* or *decreases* for each of the following:
  a. from B to F   **b.** from Mg to Ba   **c.** from F to I

10.24 Describe the trend in electronegativity as *increases* or *decreases* for each of the following:
  a. from Al to Cl   **b.** from Br to K   **c.** from Li to Cs

10.25 What electronegativity difference (**a** to **c**) would you expect for a nonpolar covalent bond?
  a. from 0.0 to 0.4   **b.** from 0.5 to 1.8   **c.** from 1.9 to 3.3

10.26 What electronegativity difference (**a** to **c**) would you expect for a polar covalent bond?
  a. from 0.0 to 0.4   **b.** from 0.5 to 1.8   **c.** from 1.9 to 3.3

10.27 Using the periodic table, arrange the atoms in each of the following sets in order of increasing electronegativity:
  a. Li, Na, K   **b.** Na, Cl, P   **c.** Se, Ca, O

10.28 Using the periodic table, arrange the atoms in each of the following sets in order of increasing electronegativity:
  a. Cl, F, Br   **b.** B, O, N   **c.** Mg, F, S

10.29 Predict whether each of the following bonds is nonpolar covalent, polar covalent, or ionic:
  a. Si—Br   **b.** Li—F   **c.** Br—F
  d. I—I   **e.** N—P   **f.** C—P

10.30 Predict whether each of the following bonds is nonpolar covalent, polar covalent, or ionic:
  a. Si—O   **b.** K—Cl   **c.** S—F
  d. P—Br   **e.** Li—O   **f.** N—S

10.31 For each of the following bonds, indicate the positive end with δ^+ and the negative end with δ^-. Draw an arrow to show the dipole for each.
  a. N—F   **b.** Si—Br   **c.** C—O
  d. P—Br   **e.** N—P

10.32 For each of the following bonds, indicate the positive end with δ^+ and the negative end with δ^-. Draw an arrow to show the dipole for each.
  a. P—Cl   **b.** Se—F   **c.** Br—F
  d. N—H   **e.** B—Cl

10.5 Polarity of Molecules

LEARNING GOAL Use the three-dimensional structure of a molecule to classify it as polar or nonpolar.

We have seen that covalent bonds in molecules can be polar or nonpolar. Now we will look at how the bonds in a molecule and its shape determine whether that molecule is classified as polar or nonpolar.

CORE CHEMISTRY SKILL
Identifying Polarity of Molecules

Nonpolar Molecules

In a **nonpolar molecule**, all the bonds are nonpolar or the polar bonds cancel each other out. Molecules such as H_2, Cl_2, and CH_4 are nonpolar because they contain only nonpolar covalent bonds.

$$H-H \qquad Cl-Cl \qquad H-\underset{\underset{H}{|}}{\overset{\overset{H}{|}}{C}}-H$$

Nonpolar molecules

A *nonpolar molecule* also occurs when polar bonds (dipoles) cancel each other because they are in a symmetrical arrangement. For example, CO_2, a linear molecule, contains two equal polar covalent bonds whose dipoles point in opposite directions. As a result, the dipoles cancel out, making a CO_2 molecule nonpolar.

Another example of a nonpolar molecule is the CCl_4 molecule, which has four polar bonds symmetrically arranged around the central C atom. Each of the C—Cl bonds has the same polarity, but because they have a tetrahedral arrangement, their opposing dipoles cancel out. As a result, a molecule of CCl_4 is nonpolar.

$$O=C=O$$

Dipoles cancel

CO_2 is a nonpolar molecule.

The four C—Cl dipoles cancel out.

Polar Molecules

In a **polar molecule**, one end of the molecule is more negatively charged than the other end. Polarity in a molecule occurs when the dipoles from the individual polar bonds do not cancel each other. For example, HCl is a polar molecule because it has one covalent bond that is polar.

In molecules with two or more electron groups, the shape, such as bent or trigonal pyramidal, determines whether the dipoles cancel. For example, we have seen that H_2O has a bent shape. Thus, a water molecule is polar because the individual dipoles do not cancel.

$$H-Cl$$

A single dipole does not cancel.

More negative end of molecule

More positive end of molecule

H_2O is a polar molecule because its dipoles do not cancel.

The NH_3 molecule has a tetrahedral electron-group geometry with three bonded atoms, which gives it a trigonal pyramidal shape. Thus, the NH_3 molecule is polar because the individual N—H dipoles do not cancel.

More negative end of molecule

More positive end of molecule

NH_3 is a polar molecule because its dipoles do not cancel.

In the molecule CH_3F, the C—F bond is polar covalent, but the three C—H bonds are nonpolar covalent. Because there is only one dipole in CH_3F, CH_3F is a polar molecule.

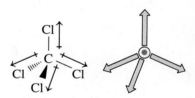

CH_3F is a polar molecule.

SAMPLE PROBLEM 10.8 Polarity of Molecules

Determine whether a molecule of OF_2 is polar or nonpolar.

 TRY IT FIRST

SOLUTION

	Given	Need	Connect
ANALYZE THE PROBLEM	OF_2	polarity	Lewis structure, bond polarity

STEP 1 Determine if the bonds are polar covalent or nonpolar covalent. From Figure 10.2, F and O have an electronegativity difference of 0.5 $(4.0 - 3.5 = 0.5)$, which makes each of the O—F bonds polar covalent.

STEP 2 If the bonds are polar covalent, draw the Lewis structure and determine if the dipoles cancel. The Lewis structure for OF_2 has four electron groups and two bonded atoms. The molecule has a bent shape in which the dipoles of the O—F bonds do not cancel. The OF_2 molecule would be polar.

OF_2 is a polar molecule.

STUDY CHECK 10.8

Would a PCl_3 molecule be polar or nonpolar?

ANSWER

polar

Guide to Determining the Polarity of a Molecule

STEP 1
Determine if the bonds are polar covalent or nonpolar covalent.

STEP 2
If the bonds are polar covalent, draw the Lewis structure and determine if the dipoles cancel.

QUESTIONS AND PROBLEMS

10.5 Polarity of Molecules

LEARNING GOAL Use the three-dimensional structure of a molecule to classify it as polar or nonpolar.

10.33 Why is F_2 a nonpolar molecule, but HF is a polar molecule?

10.34 Why is CCl_4 a nonpolar molecule, but PCl_3 is a polar molecule?

10.35 Identify each of the following molecules as polar or nonpolar:
a. CS_2 **b.** NF_3 **c.** CHF_3 **d.** SO_3

10.36 Identify each of the following molecules as polar or nonpolar:
a. SeF_2 **b.** PBr_3 **c.** SiF_4 **d.** SO_2

10.37 The molecule CO_2 is nonpolar, but CO is a polar molecule. Explain.

10.38 The molecules CH_4 and CH_3Cl both have tetrahedral shapes. Why is CH_4 nonpolar whereas CH_3Cl is polar?

10.6 Intermolecular Forces between Atoms or Molecules

LEARNING GOAL Describe the intermolecular forces between ions, polar covalent molecules, and nonpolar covalent molecules.

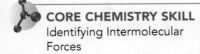 **CORE CHEMISTRY SKILL**
Identifying Intermolecular Forces

In gases, the interactions between particles are minimal, which allows gas molecules to move far apart from each other. In solids and liquids, there are sufficient interactions between the particles to hold them close together. Such differences in properties are explained by looking at the various kinds of *intermolecular* forces between particles including *dipole–dipole attractions*, *hydrogen bonding*, *dispersion forces*, as well as *ionic bonds*.

Ionic Bonds

Ionic bonds are the strongest of the attractive forces found in compounds. Thus, most ionic compounds are solids at room temperature. The ionic compound sodium chloride, $NaCl$, melts at 801 °C. Large amounts of energy are needed to overcome the strong attractive forces between positive and negative ions and change solid sodium chloride to a liquid.

Dipole–Dipole Attractions

All polar molecules are attracted to each other by **dipole–dipole attractions**. Because polar molecules have dipoles, the positively charged end of the dipole in one molecule is attracted to the negatively charged end of the dipole in another molecule.

Hydrogen Bonds

Polar molecules containing hydrogen atoms bonded to highly electronegative atoms of nitrogen, oxygen, or fluorine form especially strong dipole–dipole attractions. This type of attraction, called a **hydrogen bond**, occurs between the partially positive hydrogen atom in one molecule and the partially negative nitrogen, oxygen, or fluorine atom in another molecule. Hydrogen bonds are the strongest type of attractive forces between polar covalent molecules. They are a major factor in the formation and structure of biological molecules such as proteins and DNA.

Dispersion Forces

Very weak attractions called **dispersion forces** are the only intermolecular forces that occur between nonpolar molecules. Usually, the electrons in a nonpolar covalent molecule are distributed symmetrically. However, the movement of the electrons may place more of them in one part of the molecule than another, which forms a *temporary dipole*. These momentary dipoles align the molecules so that the positive end of one molecule is attracted to the negative end of another molecule. Although dispersion forces are very weak, they make it possible for nonpolar molecules to form liquids and solids.

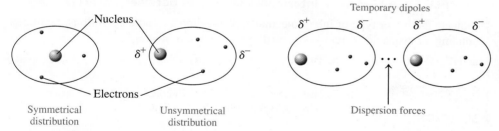

Nonpolar covalent molecules have weak attractions when they form temporary dipoles.

Attractive Forces and Melting Points

The melting point of a substance is related to the strength of the attractive forces between its particles. A compound with weak attractive forces, such as dispersion forces, has a low melting point because only a small amount of energy is needed to separate the molecules and form a liquid. A compound with dipole–dipole attractions requires more energy to break the attractive forces between the molecules. A compound that can form hydrogen bonds requires even more energy to overcome the attractive forces that exist between its molecules. Larger amounts of energy are needed to overcome the strong attractive forces between positive and negative ions and to melt an ionic solid. For example, the ionic solid MgF_2 melts at 1248 °C. **TABLE 10.8** compares the melting points of some substances with various kinds of attractive forces. The various types of attractions between particles in solids and liquids are summarized in **TABLE 10.9**.

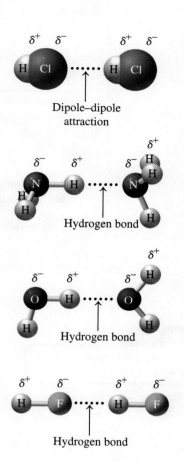

TABLE 10.8 Melting Points of Selected Substances

Substance	Melting Point (°C)
Ionic Bonds	
MgF_2	1248
$NaCl$	801
Hydrogen Bonds	
H_2O	0
NH_3	−78
Dipole–Dipole Attractions	
HI	−51
HBr	−89
HCl	−115
Dispersion Forces	
Br_2	−7
Cl_2	−101
F_2	−220

TABLE 10.9 Comparison of Bonding and Attractive Forces

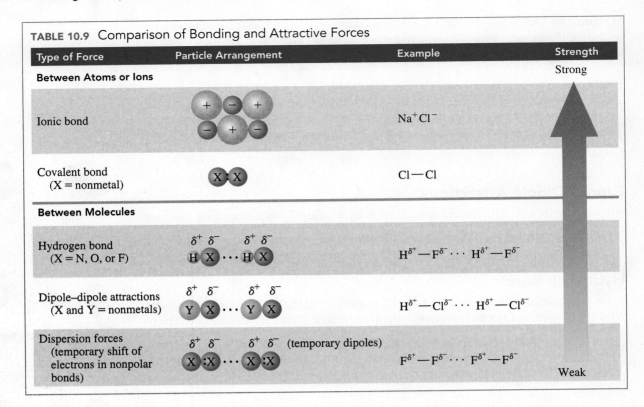

Type of Force	Particle Arrangement	Example	Strength
Between Atoms or Ions			Strong
Ionic bond		$Na^+ Cl^-$	
Covalent bond (X = nonmetal)		$Cl-Cl$	
Between Molecules			
Hydrogen bond (X = N, O, or F)		$H^{\delta+}-F^{\delta-} \cdots H^{\delta+}-F^{\delta-}$	
Dipole–dipole attractions (X and Y = nonmetals)		$H^{\delta+}-Cl^{\delta-} \cdots H^{\delta+}-Cl^{\delta-}$	
Dispersion forces (temporary shift of electrons in nonpolar bonds)	(temporary dipoles)	$F^{\delta+}-F^{\delta-} \cdots F^{\delta+}-F^{\delta-}$	Weak

Size, Mass, and Melting and Boiling Points

As the size and mass of similar types of molecular compounds increase, there are more electrons that produce stronger temporary dipoles. As the molar mass of similar compounds increases, the dispersion forces also increase due to the increase in the number of electrons. In general, larger nonpolar molecules with increased molar masses also have higher melting and boiling points.

> **ENGAGE**
>
> Why does GeH_4 have a higher boiling point than CH_4?

SAMPLE PROBLEM 10.9 Intermolecular Forces between Particles

Indicate the major type of intermolecular forces—dipole–dipole attractions, hydrogen bonding, or dispersion forces—expected of each of the following:

a. HF   **b.** Br_2 **c.** PCl_3

> **TRY IT FIRST**

SOLUTION

a. HF is a polar molecule that interacts with other HF molecules by hydrogen bonding.
b. Br_2 is nonpolar; only dispersion forces provide temporary intermolecular forces.
c. The polarity of PCl_3 molecules provides dipole–dipole attractions.

STUDY CHECK 10.9

Why is the melting point of H_2S lower than that of H_2O?

ANSWER

The intermolecular forces between H_2S molecules are dipole–dipole attractions, whereas the intermolecular forces between H_2O molecules are hydrogen bonds, which are stronger and require more energy to break.

QUESTIONS AND PROBLEMS

10.6 Intermolecular Forces between Atoms or Molecules

LEARNING GOAL Describe the intermolecular forces between ions, polar covalent molecules, and nonpolar covalent molecules.

10.39 Identify the major type of intermolecular forces between the particles of each of the following:
a. BrF   **b.** KCl   **c.** NF₃   **d.** Cl₂

10.40 Identify the major type of intermolecular forces between the particles of each of the following:
a. HCl   **b.** MgF₂   **c.** PBr₃   **d.** NH₃

10.41 Identify the strongest intermolecular forces between the particles of each of the following:
a. CH₃OH   **b.** CO
c. CF₄   **d.** CH₃—CH₃

10.42 Identify the strongest intermolecular forces between the particles of each of the following:
a. O₂   **b.** SiH₄
c. CH₃Cl   **d.** H₂O₂

10.43 Identify the substance in each of the following pairs that would have the higher boiling point and explain your choice:
a. HF or HBr   **b.** HF or NaF
c. MgBr₂ or PBr₃   **d.** CH₄ or CH₃OH

10.44 Identify the substance in each of the following pairs that would have the higher boiling point and explain your choice:
a. NaCl or HCl   **b.** H₂O or H₂Se
c. NH₃ or PH₃   **d.** F₂ or HF

10.7 Changes of State

LEARNING GOAL Describe the changes of state between solids, liquids, and gases; calculate the energy involved.

The states and properties of gases, liquids, and solids depend on the types of attractive forces between their particles. Matter undergoes a **change of state** when it is converted from one state to another state (see **FIGURE 10.4**).

Heat absorbed
Heat released

FIGURE 10.4 ▶ Changes of state include melting and freezing, vaporization and condensation, sublimation and deposition.

Q Is heat added or released when liquid water freezes?

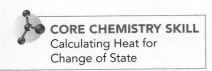

Solid   + Heat   Liquid

Melting

Freezing

− Heat

Melting and freezing are reversible processes.

Melting and Freezing

When heat is added to a solid, the particles move faster. At a temperature called the **melting point (mp)**, the particles of a solid gain sufficient energy to overcome the attractive forces that hold them together. The particles in the solid separate and move about in random patterns. The substance is **melting**, changing from a solid to a liquid.

If the temperature of a liquid is lowered, the reverse process takes place. Kinetic energy is lost, the particles slow down, and attractive forces pull the particles close together. The substance is **freezing**. A liquid changes to a solid at the **freezing point (fp)**, which is the same temperature as its melting point. Every substance has its own freezing (melting) point: Solid water (ice) melts at 0 °C when heat is added, and freezes at 0 °C when heat is removed. Gold melts at 1064 °C when heat is added, and freezes at 1064 °C when heat is removed.

During a change of state, the temperature of a substance remains constant. Suppose we have a glass containing ice and water. The ice melts when heat is added at 0 °C, forming liquid. When heat is removed at 0 °C, the liquid water freezes, forming solid. The process of melting requires heat; the process of freezing releases heat. Melting and freezing are reversible at 0 °C.

Heat of Fusion

During melting, the **heat of fusion** is the energy that must be added to convert exactly 1 g of solid to liquid at the melting point. For example, 334 J of heat is needed to melt exactly 1 g of ice at its melting point (0 °C).

$$H_2O(s) + 334 \text{ J/g} \longrightarrow H_2O(l) \quad \text{Endothermic}$$

Heat of Fusion for Water

$$\frac{334 \text{ J}}{1 \text{ g H}_2\text{O}} \quad \text{and} \quad \frac{1 \text{ g H}_2\text{O}}{334 \text{ J}}$$

The heat of fusion is also the quantity of heat that must be removed to freeze exactly 1 g of water at its freezing point (0 °C).

Water is sometimes sprayed in fruit orchards during subfreezing weather. If the air temperature drops to 0 °C, the water begins to freeze. Because heat is released as the water molecules form solid ice, the air warms above 0 °C and protects the fruit.

$$H_2O(l) \longrightarrow H_2O(s) + 334 \text{ J/g} \quad \text{Exothermic}$$

The heat of fusion can be used as a conversion factor in calculations. For example, to determine the heat needed to melt a sample of ice, the mass of ice, in grams, is multiplied by the heat of fusion. Because the temperature remains constant as long as the ice is melting, there is no temperature change given in the calculation, as shown in Sample Problem 10.10.

CORE CHEMISTRY SKILL

Calculating Heat for Change of State

Calculating Heat to Melt (or Freeze) Water

Heat = mass × heat of fusion

$$\text{J} = \cancel{\text{g}} \times \frac{334 \text{ J}}{\cancel{\text{g}}}$$

SAMPLE PROBLEM 10.10 Heat of Fusion

Ice bags are used by sports trainers to treat muscle injuries. If 260. g of ice are placed in an ice bag, how much heat, in joules, will be absorbed to melt all the ice at 0 °C?

TRY IT FIRST

An ice bag is used to treat a sports injury.

SOLUTION

STEP 1 State the given and needed quantities.

ANALYZE THE PROBLEM	Given	Need	Connect
	260. g of ice at 0 °C	joules to melt ice at 0 °C	heat of fusion

STEP 2 Write a plan to convert the given quantity to the needed quantity.

grams of ice → | Heat of fusion | → joules

STEP 3 Write the heat conversion factor.

$$1 \text{ g of } H_2O \ (s \longrightarrow l) = 334 \text{ J}$$

$$\frac{334 \text{ J}}{1 \text{ g } H_2O} \quad \text{and} \quad \frac{1 \text{ g } H_2O}{334 \text{ J}}$$

STEP 4 Set up the problem and calculate the needed quantity.

$$26.0 \text{ g } H_2O \ \times \ \underset{\text{Exact}}{\frac{\overset{\text{Three SFs}}{334 \text{ J}}}{1 \text{ g } H_2O}} \ = \ \underset{\text{Three SFs}}{8680 \text{ J } (8.68 \times 10^3 \text{ J})}$$

Three SFs

STUDY CHECK 10.10

In a freezer, 125 g of water at 0 °C is placed in an ice cube tray. How much heat, in kilojoules, must be removed to form ice cubes at 0 °C?

ANSWER

41.8 kJ

Guide to Using a Heat Conversion Factor

STEP 1
State the given and needed quantities.

STEP 2
Write a plan to convert the given quantity to the needed quantity.

STEP 3
Write the heat conversion factor and any metric factor.

STEP 4
Set up the problem and calculate the needed quantity.

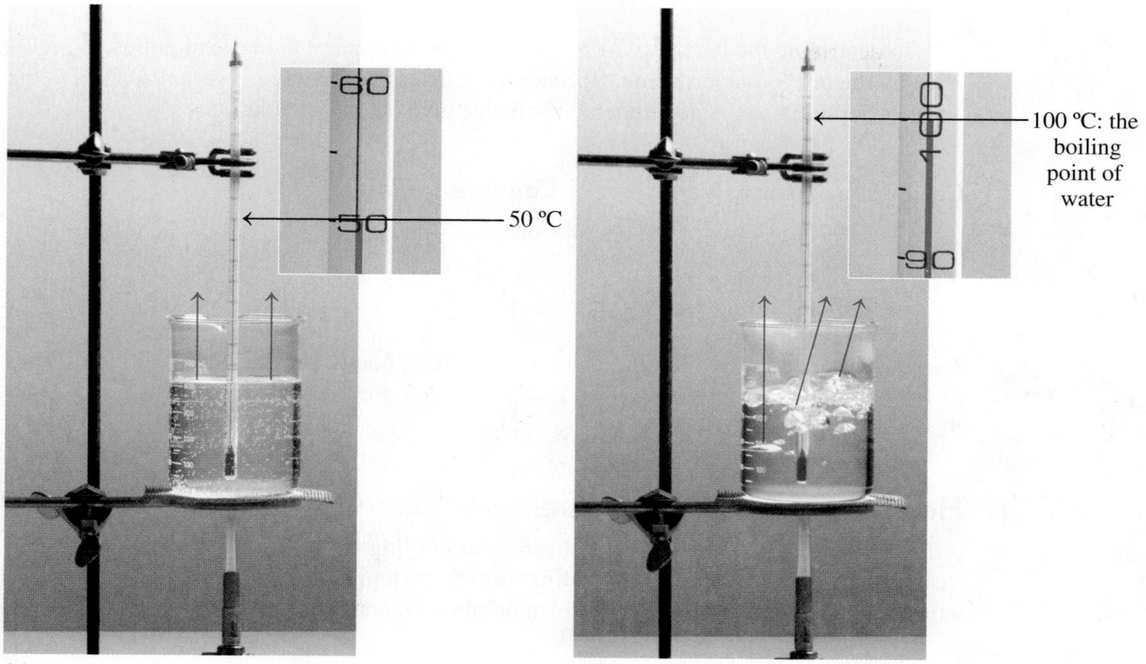

(a) (b)

FIGURE 10.5 ► (a) Evaporation occurs at the surface of a liquid. (b) Boiling occurs as bubbles of gas form throughout the liquid.

Q Why does water evaporate faster at 85 °C than at 15 °C?

100 °C: the boiling point of water

50 °C

Evaporation, Boiling, and Condensation

Water in a mud puddle disappears, unwrapped food dries out, and clothes hung on a line dry. **Evaporation** is taking place as water molecules with sufficient energy escape from the liquid surface and enter the gas phase (see **FIGURE 10.5a**). The loss of the "hot" water molecules removes heat, which cools the remaining liquid water. As heat is added, more and more water molecules gain sufficient energy to evaporate. At the **boiling point (bp)**, the molecules within a liquid have enough energy to overcome their attractive forces and become a gas. We observe the **boiling** of a liquid such as water as gas bubbles form throughout the liquid, rise to the surface, and escape (see **FIGURE 10.5b**).

When heat is removed, a reverse process takes place. In **condensation**, water vapor is converted to liquid as the water molecules lose kinetic energy and slow down. Condensation occurs at the same temperature as boiling but differs because heat is removed. You may have noticed that condensation occurs when you take a hot shower and the water vapor forms water droplets on a mirror. Because a substance loses heat as it condenses, its surroundings become warmer. That is why, when a rainstorm is approaching, you may notice a warming of the air as gaseous water molecules condense to rain.

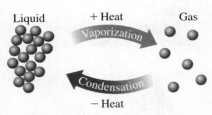

Vaporization and condensation are reversible processes.

Heat of Vaporization and Condensation

The **heat of vaporization** is the energy that must be added to convert exactly 1 g of liquid to gas at its boiling point. For water, 2260 J is needed to convert 1 g of water to vapor at 100 °C.

$$H_2O(l) + 2260 \text{ J/g} \longrightarrow H_2O(g) \quad \text{Endothermic}$$

This same amount of heat is released when 1 g of water vapor (gas) changes to liquid at 100 °C.

$$H_2O(g) \longrightarrow H_2O(l) + 2260 \text{ J/g} \quad \text{Exothermic}$$

Therefore, 2260 J/g is also the *heat of condensation* of water.

Heat of Vaporization for Water

$$\frac{2260 \text{ J}}{1 \text{ g H}_2\text{O}} \quad \text{and} \quad \frac{1 \text{ g H}_2\text{O}}{2260 \text{ J}}$$

To determine the heat needed to boil a sample of water, the mass, in grams, is multiplied by the heat of vaporization. Because the temperature remains constant as long as the water is boiling, there is no temperature change given in the calculation.

Calculating Heat to Vaporize (or Condense) Water

$$\text{Heat} = \text{mass} \times \text{heat of vaporization}$$

$$\text{J} = \cancel{g} \times \frac{2260 \text{ J}}{\cancel{g}}$$

Just as substances have different melting and boiling points, they also have different heats of fusion and vaporization. As seen in **FIGURE 10.6**, the heats of vaporization are larger than the heats of fusion.

Heating and Cooling Curves

All the changes of state during the heating or cooling of a substance can be illustrated visually. On a **heating curve** or **cooling curve**, the temperature is shown on the vertical axis and the loss or gain of heat is shown on the horizontal axis.

Heats of Fusion and Heats of Vaporization for Selected Substances

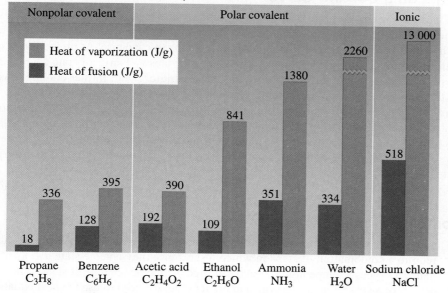

ENGAGE

A 5.0-g sample of ammonia has a boiling point of −33 °C and a heat of vaporization of 1380 J/g. The heat change when the sample is completely vaporized is 6900 J. Explain why the temperature −33 °C was not used in the calculation of the heat.

FIGURE 10.6 ▶ For any substance, the heat of vaporization is greater than the heat of fusion.

◉ Why does the formation of a gas require more energy than the formation of a liquid of the same compound?

Steps on a Heating Curve

The first diagonal line indicates a warming of a solid as heat is added. When the melting temperature is reached, a horizontal line, or plateau, indicates that the solid is melting. As melting takes place, the solid is changing to liquid without any change in temperature (see **FIGURE 10.7**).

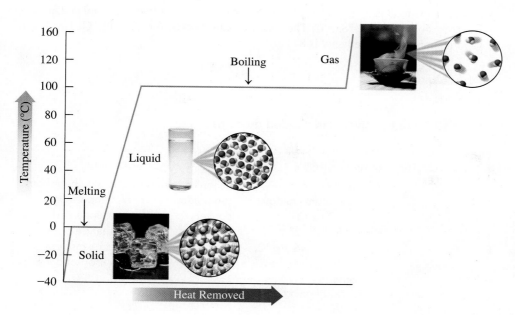

FIGURE 10.7 ▶ A heating curve diagrams the temperature increases and changes of state as heat is added.

◉ What does the plateau at 100 °C represent on the heating curve for water?

Once all the particles are in the liquid state, heat that is added will increase the temperature of the liquid. This increase is drawn as a diagonal line from the melting point temperature to the boiling point temperature. Once the liquid reaches its boiling point, a horizontal line indicates that the temperature is constant as liquid changes to gas. Because the heat of vaporization is greater than the heat of fusion, the horizontal line at the boiling point is longer than the line at the melting point. Once all the liquid becomes gas, adding more heat increases the temperature of the gas.

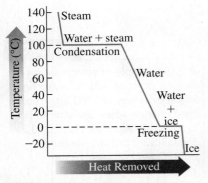

A cooling curve for water illustrates the temperature decreases and changes of state as heat is removed.

Steps on a Cooling Curve

A cooling curve is a diagram of the cooling process in which the temperature decreases as heat is removed. Initially, a diagonal line to the boiling (condensation) point is drawn to show that heat is removed from a gas until it begins to condense. At the boiling (condensation) point, a horizontal line is drawn that indicates a change of state as gas condenses to form a liquid. After all the gas has changed into liquid, further cooling lowers the temperature. The decrease in temperature is shown as a diagonal line from the condensation temperature to the freezing temperature. At the freezing point, another horizontal line indicates that liquid is changing to solid at the freezing point temperature. Once all the substance is frozen, the removal of more heat decreases the temperature of the solid below its freezing point, which is shown as a diagonal line below its freezing point.

Combining Energy Calculations

Up to now, we have calculated one step in a heating or cooling curve. However, a problem may require a combination of steps in which a substance may change state and then the new state undergoes a temperature change followed by another change of state. When the temperature of a substance is changing and not its state, we need to use the heat equation, which includes the ΔT.

$$\text{Heat (J)} = \text{mass (g)} \times \text{temperature change } (\Delta T) \times \text{specific heat (J/g °C)}$$

The heat is calculated for each step separately, and the results are added together to give the total energy, as seen in Sample Problem 10.11.

SAMPLE PROBLEM 10.11 Combining Heat Calculations

Calculate the total heat, in joules, needed to convert 15.0 g of liquid ethanol at 25.0 °C to gas at its boiling point, 78.0 °C. The specific heat of ethanol is 2.46 J/g °C. The heat of vaporization for liquid ethanol is 841 J/g.

TRY IT FIRST

SOLUTION

STEP 1 State the given and needed quantities.

ANALYZE THE PROBLEM	Given	Need	Connect
	15.0 g of ethanol, temperature change (25.0 °C to 78.0 °C), specific heat of ethanol = 2.46 J/g °C, heat of vaporization of ethanol = 841 J/g	total joules to raise temperature and convert to gas	combine heat from temperature change and change of state (heat of vaporization)

STEP 2 Write a plan to convert the given quantity to the needed quantity. When several changes occur, draw a diagram of heating and changes of state.

Total joules =

(1) joules to heat ethanol from 25.0 °C $\longrightarrow$ 78.0 °C

$$\text{Joules} = \boxed{\text{mass (g)}} \times \boxed{\text{temperature change } (\Delta T)} \times \boxed{\text{specific heat }\left(\frac{J}{g\,°C}\right)}$$

(2) joules to change liquid ethanol to gas at 78.0 °C

$$\text{Joules} = \boxed{\text{mass (g)}} \times \boxed{\text{heat of vaporization} \left(\dfrac{J}{g}\right)}$$

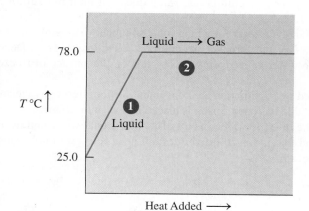

STEP 3 Write the heat conversion factors needed.

$$SH_{\text{Ethanol}} = \dfrac{2.46 \text{ J}}{g \text{ °C}} \qquad 1 \text{ g of ethanol } (l \longrightarrow g) = 841 \text{ J}$$

$$\dfrac{2.46 \text{ J}}{g \text{ °C}} \quad \text{and} \quad \dfrac{g \text{ °C}}{2.46 \text{ J}} \qquad \dfrac{841 \text{ J}}{1 \text{ g ethanol}} \quad \text{and} \quad \dfrac{1 \text{ g ethanol}}{841 \text{ J}}$$

STEP 4 Set up the problem and calculate the needed quantity.

$$\Delta T = 78.0 \text{ °C} - 25.0 \text{ °C} = \boxed{53.0 \text{ °C}}$$

Temperature change: Heat needed to warm ethanol (liquid) from 25.0 °C to 78.0 °C.

$$\boxed{15.0 \text{ g}} \times \boxed{53.0 \text{ °}\cancel{C}} \times \boxed{\dfrac{2.46 \text{ J}}{g \text{ °}\cancel{C}}} = \boxed{1960 \text{ J}}$$

Change of state at constant temperature: Heat needed to convert ethanol (liquid) to ethanol (gas) at 78.0 °C.

$$\boxed{15.0 \text{ g } \cancel{\text{ethanol}}} \times \boxed{\dfrac{841 \text{ J}}{1 \text{ g } \cancel{\text{ethanol}}}} = \boxed{12\,600 \text{ J}}$$

Calculate the total heat:

Heating ethanol (25.0 °C to 78.0 °C)	1 960 J
Changing liquid to gas (78.0 °C)	12 600 J
Total heat needed	14 600 J (rounded off)

STUDY CHECK 10.11

How many kilojoules are released when 75.0 g of ethanol at 78 °C condenses, cools to −114 °C, and freezes? The specific heat of ethanol is 2.46 J/g °C. For heats of fusion and vaporization, see Figure 10.6. (*Hint:* The solution will require three energy calculations.)

ANSWER

106.7 kJ

Sublimation and deposition are reversible processes.

Water vapor will change to solid on contact with a cold surface.

Sublimation and Deposition

In a process called **sublimation**, the particles on the surface of a solid change directly to a gas with no temperature change and without going through the liquid state. In the reverse process called **deposition**, gas particles change directly to a solid. For example, dry ice, which is solid carbon dioxide, sublimes at $-78\,°C$. It is called "dry" because it does not form a liquid as it warms. In extremely cold areas, snow does not melt but sublimes directly to water vapor.

When frozen foods are left in the freezer for a long time, so much water sublimes that foods, especially meats, become dry and shrunken, a condition called *freezer burn*. Deposition occurs in a freezer when water vapor forms ice crystals on the surface of freezer bags and frozen food.

Freeze-dried foods prepared by sublimation are convenient for long-term storage and for camping and hiking. A food that has been frozen is placed in a vacuum chamber where it dries as the ice sublimes. The dried food retains all of its nutritional value and needs only water to be edible. A food that is freeze-dried does not need refrigeration because bacteria cannot grow without moisture.

CHEMISTRY LINK TO **HEALTH**
Steam Burns

Hot water at $100\,°C$ will cause burns and damage to the skin. However, getting steam on the skin is even more dangerous. If 25 g of hot water at $100\,°C$ falls on a person's skin, the temperature of the water will drop to body temperature, $37\,°C$. The heat released during cooling can cause severe burns. The amount of heat can be calculated from the mass, the temperature change, $100\,°C - 37\,°C = 63\,°C$, and the specific heat of water, $4.184\ J/g\,°C$.

$$25\ \cancel{g} \times 63\ \cancel{°C} \times \frac{4.184\ J}{\cancel{g}\,\cancel{°C}} = 6600\ J \text{ of heat released when water cools}$$

However, getting steam on the skin is even more damaging. The condensation of the same quantity of steam to liquid at $100\,°C$ releases much more heat—almost ten times as much. This amount of heat can be calculated using the heat of vaporization, which is $2260\ J/g$ for water at $100\,°C$.

$$25\ \cancel{g} \times \frac{2260\ J}{1\ \cancel{g}} = 57\,000\ J \text{ released when water (gas) condenses to water (liquid) at } 100\,°C$$

The total heat released is calculated by combining the heat from the condensation at $100\,°C$ and the heat from cooling of the steam from $100\,°C$ to $37\,°C$ (body temperature). We can see that most of the heat is from the condensation of steam. This large amount of heat released on the skin is what causes damage from steam burns.

Condensation $(100\,°C)$	$= 57\,000\ J$
Cooling $(100\,°C \text{ to } 37\,°C)$	$= \underline{\ \ 6\,600\ J}$
Heat released	$= 64\,000\ J$ (rounded off)

The amount of heat released from steam is almost ten times greater than the heat from the same amount of hot water.

When 1 g of steam condenses, 2260 J is released.

QUESTIONS AND PROBLEMS

10.7 Changes of State

LEARNING GOAL Describe the changes of state between solids, liquids, and gases; calculate the energy involved.

10.45 Using Figure 10.6, calculate the heat change needed for each of the following at the melting/freezing point:
   a. joules to melt 65.0 g of ice
   b. joules to melt 17.0 g of benzene
   c. kilojoules to freeze 225 g of acetone
   d. kilojoules to freeze 0.0500 kg of water

10.46 Using Figure 10.6, calculate the heat change needed for each of the following at the melting/freezing point:
   a. joules to freeze 35.2 g of acetic acid
   b. joules to freeze 275 g of water
   c. kilojoules to melt 145 g of ammonia
   d. kilojoules to melt 5.00 kg of ice

10.47 Using Figure 10.6, calculate the heat change needed for each of the following at the boiling/condensation point:
   a. joules to vaporize 10.0 g of water
   b. kilojoules to vaporize 50.0 g of ethanol
   c. joules to condense 8.00 g of acetic acid
   d. kilojoules to condense 0.175 kg of ammonia

10.48 Using Figure 10.6, calculate the heat change needed for each of the following at the boiling/condensation point:
   a. joules to condense 10.0 g of steam
   b. kilojoules to condense 76.0 g of acetic acid
   c. joules to vaporize 44.0 g of ammonia
   d. kilojoules to vaporize 5.0 kg of water

10.49 Using Figure 10.6 and the specific heat of water, 4.184 J/g °C, calculate the total amount of heat for each of the following:
   a. joules needed to warm 20.0 g of water at 15 °C to 72 °C
   b. joules needed to melt 50.0 g of ice at 0.0 °C and to warm the liquid to 65.0 °C
   c. kilojoules released when 15.0 g of steam condenses at 100 °C and the liquid cools to 0 °C
   d. kilojoules needed to melt 24.0 g of ice at 0 °C, warm the liquid to 100 °C, and change it to steam at 100 °C

10.50 Using Figure 10.6 and the specific heat of water, 4.184 J/g °C, calculate the total amount of heat for each of the following:
   a. joules to condense 125 g of steam at 100 °C and to cool the liquid to 15.0 °C
   b. joules needed to melt a 525-g ice sculpture at 0 °C and to warm the liquid to 15.0 °C
   c. kilojoules released when 85.0 g of steam condenses at 100 °C, the liquid cools to 0 °C, and freezes
   d. joules to warm 55.0 mL of water (density = 1.00 g/mL) from 10.0 °C to 100 °C and vaporize it at 100 °C

Applications

10.51 An ice bag containing 275 g of ice at 0 °C was used to treat sore muscles. When the bag was removed, the ice had melted and the liquid water had a temperature of 24.0 °C. How many kilojoules of heat were absorbed?

10.52 A patient arrives in the emergency room with a burn caused by steam. Calculate the heat, in kilojoules, that is released when 18.0 g of steam at 100 °C hits the skin, condenses, and cools to body temperature of 37.0 °C.

Follow Up

HISTOLOGIST STAINS TISSUE WITH DYE

When Lisa received the sample of tissue from surgery, she places it in liquid nitrogen, which has a heat of vaporization of 198 J/g at the boiling point of nitrogen, which is −196 °C. When the tissue is frozen, Lisa places the tissue on a microtome, cuts thin slices, and mounts each on a glass slide. One of the stains that Lisa uses to allow the pathologist to see any abnormalities is Eosin, which has a formula, $C_{20}H_6Br_4Na_2O_5$. In preparing to stain the tissues, Lisa uses solutions of lithium carbonate and sodium bicarbonate.

The pathologist examines the tissue sample and reports back to the surgeon that no abnormal cells are present in the margin tissue. No further tissue removal is necessary.

Applications

10.53 At the boiling point of nitrogen, N_2, how many joules are needed to vaporize 15.8 g of N_2?

10.54 In the preparation of liquid nitrogen, how many kilojoules must be removed from 112 g of N_2 gas at its boiling point to form liquid N_2?

10.55 Using the electronegativity values in Figure 10.2, for each of the following bonds found in eosin, calculate the electronegativity difference and predict if the bond is nonpolar covalent, polar covalent, or ionic:
   a. C—C   **b.** C—H   **c.** C—O

10.56 Using the electronegativity values in Figure 10.2, for each of the following bonds found in eosin, calculate the electronegativity difference and predict if the bond is nonpolar covalent, polar covalent, or ionic:
   a. O—H   **b.** Na—O   **c.** C—Br

10.57 a. Draw the resonance structures for carbonate (CO_3^{2-}).
     b. Determine the shape of the carbonate ion.

10.58 a. Draw the resonance structures for bicarbonate (HCO_3^-) with a central C atom and an H atom attached to one of the O atoms.
     b. Determine the shape of the bicarbonate ion around the C atom.

CONCEPT MAP

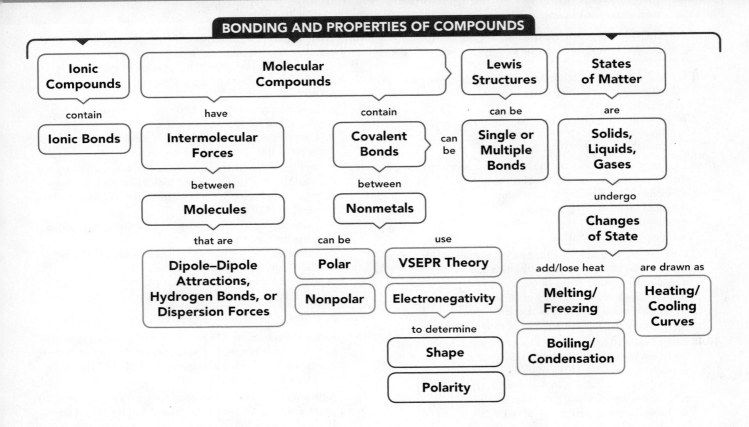

BONDING AND PROPERTIES OF COMPOUNDS

- **Ionic Compounds** — contain → **Ionic Bonds**
- **Molecular Compounds**
  - have → **Intermolecular Forces** — between → **Molecules** — that are → **Dipole–Dipole Attractions, Hydrogen Bonds, or Dispersion Forces**
  - contain → **Covalent Bonds** — between → **Nonmetals** — can be → **Polar** / **Nonpolar**
  - **Covalent Bonds** can be → **Single or Multiple Bonds**
- **Lewis Structures** — can be → **Single or Multiple Bonds**
  - use → **VSEPR Theory** / **Electronegativity** — to determine → **Shape** / **Polarity**
- **States of Matter** — are → **Solids, Liquids, Gases** — undergo → **Changes of State**
  - add/lose heat → **Melting/Freezing** / **Boiling/Condensation**
  - are drawn as → **Heating/Cooling Curves**

CHAPTER REVIEW

10.1 Lewis Structures for Molecules and Polyatomic Ions

LEARNING GOAL Draw the Lewis structures for molecular compounds or polyatomic ions with single and multiple bonds.

- The total number of valence electrons is determined for all the atoms in the molecule or polyatomic ion.
- Any negative charge is added to the total valence electrons, whereas any positive charge is subtracted.
- In the Lewis structure, a bonding pair of electrons is placed between the central atom and each of the attached atoms.
- Any remaining valence electrons are used as lone pairs to complete the octets of the surrounding atoms and then the central atom.
- If octets are not completed, one or more lone pairs of electrons are placed as bonding pairs forming double or triple bonds.

10.2 Resonance Structures

LEARNING GOAL Draw Lewis structures for molecules or polyatomic ions that have two or more resonance structures.

- Resonance structures can be drawn when there are two or more equivalent Lewis structures for a molecule or ion with multiple bonds.

10.3 Shapes of Molecules and Polyatomic Ions (VSEPR Theory)

LEARNING GOAL Predict the three-dimensional structure of a molecule or a polyatomic ion.

Tetrahedral shape

- The shape of a molecule is determined from the Lewis structure, the electron-group geometry, and the number of bonded atoms.
- The electron-group geometry around a central atom with two electron groups is linear; with three electron groups, the geometry is trigonal planar; and with four electron groups, the geometry is tetrahedral.
- When all the electron groups are bonded to atoms, the shape has the same name as the electron arrangement.
- A central atom with three electron groups and two bonded atoms has a bent shape, 120°.
- A central atom with four electron groups and three bonded atoms has a trigonal pyramidal shape.
- A central atom with four electron groups and two bonded atoms has a bent shape, 109°.

10.4 Electronegativity and Bond Polarity

LEARNING GOAL Use electronegativity to determine the polarity of a bond.

- Electronegativity is the ability of an atom to attract the electrons it shares with another atom. In general, the electronegativities of metals are low, whereas nonmetals have high electronegativities.

- In a nonpolar covalent bond, atoms share electrons equally.
- In a polar covalent bond, the electrons are unequally shared because they are attracted to the more electronegative atom.
- The atom in a polar bond with the lower electronegativity is partially positive (δ^+), and the atom with the higher electronegativity is partially negative (δ^-).
- Atoms that form ionic bonds have large differences in electronegativities.

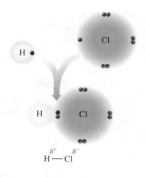

10.5 Polarity of Molecules

LEARNING GOAL Use the three-dimensional structure of a molecule to classify it as polar or nonpolar.

- Nonpolar molecules contain nonpolar covalent bonds or have an arrangement of bonded atoms that causes the dipoles to cancel out.
- In polar molecules, the dipoles do not cancel.

10.6 Intermolecular Forces Between Atoms or Molecules

LEARNING GOAL Describe the intermolecular forces between ions, polar covalent molecules, and nonpolar covalent molecules.

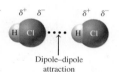

Dipole–dipole attraction

- In ionic solids, oppositely charged ions are held in a rigid structure by ionic bonds.

- Intermolecular forces called dipole–dipole attractions and hydrogen bonds hold the solid and liquid states of polar molecular compounds together.
- Nonpolar compounds form solids and liquids by weak attractions between temporary dipoles called dispersion forces.

10.7 Changes of State

LEARNING GOAL Describe the changes of state between solids, liquids, and gases; calculate the energy involved.

- Melting occurs when the particles in a solid absorb enough energy to break apart and form a liquid.
- The amount of energy required to convert exactly 1 g of solid to liquid is called the heat of fusion.
- For water, 334 J is needed to melt 1 g of ice or must be removed to freeze 1 g of water.
- Evaporation occurs when particles in a liquid state absorb enough energy to break apart and form gaseous particles.
- Boiling is the vaporization of liquid at its boiling point. The heat of vaporization is the amount of heat needed to convert exactly 1 g of liquid to vapor.
- For water, 2260 J is needed to vaporize 1 g of water or must be removed to condense 1 g of steam.
- A heating or cooling curve illustrates the changes in temperature and state as heat is added to or removed from a substance. Plateaus on the graph indicate changes of state.
- The total heat absorbed or removed from a substance undergoing temperature changes and changes of state is the sum of energy calculations for changes of state and changes in temperature.
- Sublimation is a process whereby a solid changes directly to a gas.

KEY TERMS

bent The shape of a molecule with two bonded atoms and one lone pair or two lone pairs.

boiling The formation of bubbles of gas throughout a liquid.

boiling point (bp) The temperature at which a liquid changes to gas (boils) and gas changes to liquid (condenses).

change of state The transformation of one state of matter to another, for example, solid to liquid, liquid to solid, liquid to gas.

condensation The change of state from a gas to a liquid.

cooling curve A diagram that illustrates temperature changes and changes of state for a substance as heat is removed.

deposition The change of a gas directly to a solid; the reverse of sublimation.

dipole The separation of positive and negative charge in a polar bond indicated by an arrow that is drawn from the more positive atom to the more negative atom.

dipole–dipole attractions Attractive forces between oppositely charged ends of polar molecules.

dispersion forces Weak dipole bonding that results from a momentary polarization of nonpolar molecules.

double bond A sharing of two pairs of electrons by two atoms.

electronegativity The relative ability of an element to attract electrons in a bond.

evaporation The formation of a gas (vapor) by the escape of high-energy molecules from the surface of a liquid.

freezing A change of state from liquid to solid.

freezing point (fp) The temperature at which a liquid changes to a solid (freezes) and a solid changes to a liquid (melts).

heat of fusion The energy required to melt exactly 1 g of a substance at its melting point. For water, 334 J is needed to melt 1 g of ice; 334 J is released when 1 g of water freezes.

heat of vaporization The energy required to vaporize 1 g of a substance at its boiling point. For water, 2260 J is needed to vaporize exactly 1 g of water; 1 g of steam gives off 2260 J when it condenses.

heating curve A diagram that illustrates the temperature changes and changes of state of a substance as it is heated.

hydrogen bond The attraction between a partially positive H atom and a strongly electronegative atom of N, O, or F.

Lewis structure A structure drawn in which the valence electrons of all the atoms are arranged to give octets except two electrons for hydrogen.

linear The shape of a molecule that has two bonded atoms and no lone pairs.

melting The change of state from a solid to a liquid.

melting point (mp) The temperature at which a solid becomes a liquid (melts). It is the same temperature as the freezing point.

nonpolar covalent bond A covalent bond in which the electrons are shared equally between atoms.

nonpolar molecule A molecule that has only nonpolar bonds or in which the bond dipoles cancel.

polar covalent bond A covalent bond in which the electrons are shared unequally between atoms.

polar molecule A molecule containing bond dipoles that do not cancel.

polarity A measure of the unequal sharing of electrons, indicated by the difference in electronegativities.

resonance structures Two or more Lewis structures that can be drawn for a molecule or polyatomic ion by placing a multiple bond between different atoms.

sublimation The change of state in which a solid is transformed directly to a gas without forming a liquid first.

tetrahedral The shape of a molecule with four bonded atoms.

trigonal planar The shape of a molecule with three bonded atoms and no lone pairs.

trigonal pyramidal The shape of a molecule that has three bonded atoms and one lone pair.

triple bond A sharing of three pairs of electrons by two atoms.

valence shell electron-pair repulsion (VSEPR) theory A theory that predicts the shape of a molecule by moving the electron groups on a central atom as far apart as possible to minimize the mutual repulsion of the electrons.

CORE CHEMISTRY SKILLS

The chapter section containing each Core Chemistry Skill is shown in parentheses at the end of each heading.

Drawing Lewis Structures (10.1)

- The Lewis structure for a molecule or polyatomic ion shows the sequence of atoms, the bonding pairs of electrons shared between atoms, and the nonbonding or *lone pairs* of electrons.
- Double or triple bonds result when a second or third electron pair is shared between the same atoms to complete octets.

Example: Draw the Lewis structure for CS_2.

Answer: The central atom in the atom arrangement is C.

   S   C   S

Determine the total number of valence electrons.

$$2\,S \times 6\,e^- = 12\,e^-$$
$$1\,C \times 4\,e^- = \underline{\;\;4\,e^-}$$
$$\text{Total} = 16\,e^-$$

Attach each bonded atom to the central atom using a pair of electrons. Two bonding pairs use four electrons.

   S:C:S

Place the 12 remaining electrons as lone pairs around the S atoms.

   :S:C:S:

To complete the octet for C, a lone pair of electrons from each of the S atoms is shared with C, which forms two double bonds.

   :S::C::S: or   :S=C=S:

Drawing Resonance Structures (10.2)

- When a molecule or polyatomic ion contains multiple bonds, it may be possible to draw more than one Lewis structure called resonance structures.

Example: Draw the resonance structures for NO_2^-.

Answer: The central atom in the atom arrangement is N.

   O   N   O

Determine the total number of valence electrons.

$$1\,N \times 5\,e^- = \;\;5\,e^-$$
$$2\,O \times 6\,e^- = 12\,e^-$$
$$\text{negative charge} = \underline{\;\;1\,e^-}$$
$$\text{Total} = 18\,e^-$$

Use electron pairs to attach each bonded atom to the central atom. Two bonding pairs use four electrons.

   O:N:O

Place the 14 remaining electrons as lone pairs around the O and N atoms.

$$\left[:\!\ddot{O}\!:\!N\!:\!\ddot{O}\!:\right]^-$$

To complete the octet for N, a lone pair from one O atom is shared with N, which forms one double bond. Because there are two O atoms, there are two resonance structures.

$$\left[:\!O\!=\!\ddot{N}\!-\!\ddot{O}\!:\right]^- \longleftrightarrow \left[:\!\ddot{O}\!-\!\ddot{N}\!=\!O\!:\right]^-$$

Predicting Shape (10.3)

- The three-dimensional shape of a molecule or polyatomic ion is determined by drawing a Lewis structure and identifying the number of electron groups (one or more electron pairs) around the central atom and the number of bonded atoms.
- In the valence shell electron-pair repulsion (VSEPR) theory, the electron groups are arranged as far apart as possible around a central atom to minimize the repulsion.
- A central atom with two electron groups bonded to two atoms is linear. A central atom with three electron groups bonded to three atoms is trigonal planar, and to two atoms is bent (120°). A central atom with four electron groups bonded to four atoms is tetrahedral, to three atoms is trigonal pyramidal, and to two atoms is bent (109°).

Example: Predict the shape for NO_2^-.

Answer: Using the resonance structures we drew for NO_2^-, we count three electron groups around the central N atom: one double bond, one single bond, and a lone pair of electrons.

$$\left[:\!O\!=\!\ddot{N}\!-\!\ddot{O}\!:\right]^- \quad \text{or} \quad \left[:\!\ddot{O}\!-\!\ddot{N}\!=\!O\!:\right]^-$$

The electron-group geometry is trigonal planar, but with the central N atom bonded to two O atoms, the shape is bent, 120°.

$$\left[\begin{array}{c} \ddot{N} \\ :\!\ddot{O}\!\diagup \quad \diagdown\!\ddot{O}\!: \end{array}\right]^-$$

Using Electronegativity (10.4)

- The electronegativity values indicate the ability of atoms to attract shared electrons.
- Electronegativity values increase going across a period from left to right, and decrease going down a group.
- A nonpolar covalent bond occurs between atoms with identical or very similar electronegativity values such that the electronegativity difference is 0.0 to 0.4.
- A polar covalent bond typically occurs when electrons are shared unequally between atoms with electronegativity differences from 0.5 to 1.8.
- An ionic bond typically occurs when the difference in electronegativity for two atoms is greater than 1.8.

Example: Use electronegativity values to classify each of the following bonds as nonpolar covalent, polar covalent, or ionic:

   a. Sr — Cl   **b.** C — S   **c.** O — Br

Answer: **a.** An electronegativity difference of 2.0 (Cl 3.0 − Sr 1.0) makes this an ionic bond.

   b. An electronegativity difference of 0.0 (C 2.5 − S 2.5) makes this a nonpolar covalent bond.

   c. An electronegativity difference of 0.7 (O 3.5 − Br 2.8) makes this a polar covalent bond.

Identifying Polarity of Molecules (10.5)

- A molecule is nonpolar if all of its bonds are nonpolar or it has polar bonds that cancel out. CCl_4 is a nonpolar molecule that consists of four polar bonds that cancel out.

- A molecule is polar if it contains polar bonds that do not cancel out. H_2O is a polar molecule that consists of polar bonds that do not cancel out.

Example: Predict whether $AsCl_3$ is polar or nonpolar.

Answer: From its Lewis structure, we see that $AsCl_3$ has four electron groups with three bonded atoms.

The shape of a molecule of $AsCl_3$ would be trigonal pyramidal with three polar bonds (As — Cl = 3.0 − 2.0 = 1.0) that do not cancel. Thus, it is a polar molecule.

Identifying Intermolecular Forces (10.6)

- Dipole–dipole attractions occur between the dipoles in polar compounds because the positively charged end of one molecule is attracted to the negatively charged end of another molecule.

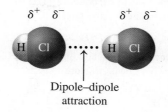

Dipole–dipole
attraction

- Strong dipole–dipole attractions called hydrogen bonds occur in compounds in which H is bonded to N, O, or F. The partially positive H atom in one molecule has a strong attraction to the partially negative N, O, or F in another molecule.
- Dispersion forces are very weak attractive forces between nonpolar molecules that occur when *temporary dipoles* form as electrons are unsymmetrically distributed.

Example: Identify the strongest type of intermolecular forces in each of the following:

   a. HF   **b.** F_2 **c.** NF_3

Answer: **a.** HF molecules, which are polar with H bonded to F, have hydrogen bonding.

   b. Nonpolar F_2 molecules have only dispersion forces.

   c. NF_3 molecules, which are polar, have dipole–dipole attractions.

- Substances with ionic bonds have the highest melting and boiling points. Substances with hydrogen bonds have higher melting and boiling points than compounds with only dipole–dipole attractions. Substances with only dispersion forces would typically have the lowest melting and boiling points, which increase as molar mass increases.

Example: Identify the compound with the highest boiling point in each of the following:

   a. HI, HBr, HF   **b.** F_2, Cl_2, I_2

Answer: **a.** HBr and HI have dipole–dipole attractions, but HF has hydrogen bonding, which gives HF the highest boiling point.

   b. Because F_2, Cl_2, and I_2 have only dispersion forces, I_2 with the greatest molar mass would have the highest boiling point.

Calculating Heat for Change of State (10.7)

- At the melting/freezing point, the heat of fusion is absorbed/released to convert 1 g of a solid to a liquid or 1 g of liquid to a solid.
- For example, 334 J of heat is needed to melt (freeze) exactly 1 g of ice at its melting (freezing) point (0 °C).
- At the boiling/condensation point, the heat of vaporization is absorbed/released to convert exactly 1 g of liquid to gas or 1 g of gas to liquid.
- For example, 2260 J of heat is needed to boil (condense) exactly 1 g of water/steam at its boiling (condensation) point, 100 °C.

Example: What is the quantity of heat, in kilojoules, released when 45.8 g of steam (water) condenses at its boiling (condensation) point?

Answer: $45.8 \text{ g steam} \times \dfrac{2260 \text{ J}}{1 \text{ g steam}} \times \dfrac{1 \text{ kJ}}{1000 \text{ J}} = 104 \text{ kJ}$

Three SFs — Exact — Exact — Three SFs

UNDERSTANDING THE CONCEPTS

The chapter sections to review are shown in parentheses at the end of each question.

10.59 State the number of valence electrons, bonding pairs, and lone pairs in each of the following Lewis structures: (10.1)

a. H:H   **b.** H:B̈r: **c.** :B̈r:B̈r:

10.60 State the number of valence electrons, bonding pairs, and lone pairs in each of the following Lewis structures: (10.1)

a. H:Ö: **b.** H:N̈:H   **c.** :B̈r:Ö:B̈r:
      H   H

10.61 Match each of the Lewis structures (**a** to **c**) with the correct diagram (**1** to **3**) of its shape, and name the shape; indicate if each molecule is polar or nonpolar. Assume X and Y are nonmetals and all bonds are polar covalent. (10.1, 10.3, 10.5)

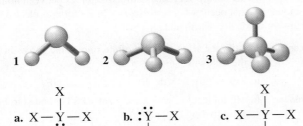

a. X—Ÿ—X   **b.** :Ÿ—X   **c.** X—Y—X
                     | |
                     X   X

10.62 Match each of the formulas (**a** to **c**) with the correct diagram (**1** to **3**) of its shape, and name the shape; indicate if each molecule is polar or nonpolar. (10.1, 10.3, 10.5)

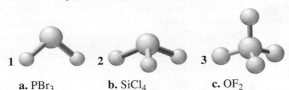

a. PBr_3 **b.** $SiCl_4$ **c.** OF_2

10.63 Consider the following bonds: Ca—O, C—O, K—O, O—O, and N—O. (10.4)
a. Which bonds are polar covalent?
b. Which bonds are nonpolar covalent?
c. Which bonds are ionic?
d. Arrange the covalent bonds in order of decreasing polarity.

10.64 Consider the following bonds: F—Cl, Cl—Cl, Cs—Cl, O—Cl, and Ca—Cl. (10.4)
a. Which bonds are polar covalent?
b. Which bonds are nonpolar covalent?
c. Which bonds are ionic?
d. Arrange the covalent bonds in order of decreasing polarity.

10.65 Identify the major intermolecular forces between each of the following atoms or molecules: (10.6)
a. PH_3 **b.** NO_2
c. CH_3—NH_2 **d.** Ar

10.66 Identify the major intermolecular forces between each of the following atoms or molecules: (10.6)
a. He   **b.** HBr
c. SnH_4 **d.** CH_3—CH_2—CH_2—OH

10.67 Use your knowledge of changes of state to explain the following: (10.7)

Perspiration forms on the skin during heavy exercise.

a. How does perspiration during heavy exercise cool the body?
b. Why do towels dry more quickly on a hot summer day than on a cold winter day?
c. Why do wet clothes stay wet in a plastic bag?

10.68 Use your knowledge of changes of state to explain the following: (10.7)

A spray is used to numb a sports injury.

a. Why is a spray that evaporates quickly, such as ethyl chloride, used to numb a sports injury during a game?
b. Why does water in a wide, flat, shallow dish evaporate more quickly than the same amount of water in a tall, narrow vase?
c. Why does a sandwich on a plate dry out faster than a sandwich in plastic wrap?

10.69 Draw a heating curve for a sample of ice that is heated from −20 °C to 150 °C. Indicate the segment of the graph that corresponds to each of the following: (10.7)
a. solid
b. melting
c. liquid
d. boiling
e. gas

10.70 Draw a cooling curve for a sample of steam that cools from 110 °C to −10 °C. Indicate the segment of the graph that corresponds to each of the following: (10.7)
a. solid
b. freezing
c. liquid
d. condensing
e. gas

10.71 The following is a heating curve for chloroform, a solvent for fats, oils, and waxes: (10.5)

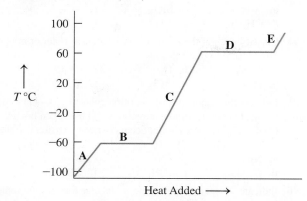

a. What is the approximate melting point of chloroform?
b. What is the approximate boiling point of chloroform?
c. On the heating curve, identify the segments **A**, **B**, **C**, **D**, and **E** as solid, liquid, gas, melting, or boiling.
d. At the following temperatures, is chloroform a solid, liquid, or gas? $-80\ °C$; $-40\ °C$; $25\ °C$; $80\ °C$

10.72 Associate the contents of the beakers (**1** to **5**) with segments (**A** to **E**) on the heating curve for water. (10.7)

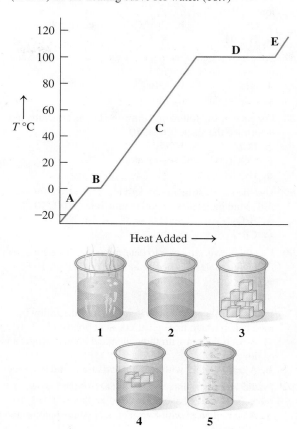

ADDITIONAL QUESTIONS AND PROBLEMS

10.73 Determine the total number of valence electrons in each of the following: (10.1)
a. HNO_2 **b.** CH_3CHO
c. PH_4^+ **d.** SO_3^{2-}

10.74 Determine the total number of valence electrons in each of the following: (10.1)
a. $COCl_2$ **b.** N_2O
c. BrO_2^- **d.** $SeCl_2$

10.75 Draw the Lewis structure for each of the following: (10.1)
a. BF_4^- **b.** Cl_2O
c. H_2NOH (N is the central atom)
d. H_2CCCl_2

10.76 Draw the Lewis structure for each of the following: (10.1)
a. H_3COCH_3 (the atoms are in the order C O C)
b. HNO_2 (the atoms are in the order HONO)
c. IO_3^- **d.** BrO^-

10.77 Draw resonance structures for each of the following: (10.2)
a. N_3^- **b.** NO_2^+ **c.** HCO_2^-

10.78 Draw resonance structures for each of the following: (10.2)
a. NO_3^- **b.** CO_3^{2-} **c.** SCN^-

10.79 Use the periodic table to arrange the following atoms in order of increasing electronegativity: (10.4)
a. I, F, Cl **b.** Li, K, S, Cl **c.** Mg, Sr, Ba, Be

10.80 Use the periodic table to arrange the following atoms in order of increasing electronegativity: (10.4)
a. Cl, Br, Se **b.** Na, Cs, O, S **c.** O, F, B, Li

10.81 Select the more polar bond in each of the following pairs: (10.4)
a. C—N or C—O **b.** N—F or N—Br
c. Br—Cl or S—Cl **d.** Br—Cl or Br—I
e. N—F or N—O

10.82 Select the more polar bond in each of the following pairs: (10.4)
a. C—C or C—O **b.** P—Cl or P—Br
c. Si—S or Si—Cl **d.** F—Cl or F—Br
e. P—O or P—S

10.83 Show the dipole arrow for each of the following bonds: (10.4)
a. Si—Cl **b.** C—N **c.** F—Cl
d. C—F **e.** N—O

10.84 Show the dipole arrow for each of the following bonds: (10.4)
a. P—O **b.** N—F **c.** O—Cl
d. S—Cl **e.** P—F

10.85 Calculate the electronegativity difference and classify each of the following bonds as nonpolar covalent, polar covalent, or ionic: (10.4)
a. Si—Cl **b.** C—C **c.** Na—Cl
d. C—H **e.** F—F

10.86 Calculate the electronegativity difference and classify each of the following bonds as nonpolar covalent, polar covalent, or ionic: (10.4)

a. C—N   **b.** Cl—Cl   **c.** K—Br
d. H—H   **e.** N—F

10.87 For each of the following, draw the Lewis structure and determine the shape: (10.1, 10.3)

a. NF_3 **b.** $SiBr_4$
c. CSe_2 **d.** SO_2

10.88 For each of the following, draw the Lewis structure and determine the shape: (10.1, 10.3)

a. PCl_4^+ **b.** O_2^{2-}
c. $COCl_2$ (C is the central atom)
d. HCCH

10.89 Use the Lewis structure to determine the shape for each of the following molecules or polyatomic ions: (10.1, 10.3)

a. BrO_2^- **b.** H_2O
c. CBr_4 **d.** PO_3^{3-}

10.90 Use the Lewis structure to determine the shape for each of the following molecules or polyatomic ions: (10.1, 10.3)

a. PH_3 **b.** NO_3^-
c. HCN   **d.** SO_3^{2-}

10.91 Predict the shape and polarity of each of the following molecules, which have polar covalent bonds: (10.3, 10.5)

a. A central atom with three identical bonded atoms and one lone pair.
b. A central atom with two bonded atoms and two lone pairs.

10.92 Predict the shape and polarity of each of the following molecules, which have polar covalent bonds: (10.3, 10.5)

a. A central atom with four identical bonded atoms and no lone pairs.
b. A central atom with four bonded atoms that are not identical and no lone pairs.

10.93 Classify each of the following molecules as polar or nonpolar: (10.3, 10.5)

a. HBr   **b.** SiO_2 **c.** NCl_3
d. CH_3Cl **e.** NI_3 **f.** H_2O

10.94 Classify each of the following molecules as polar or nonpolar: (10.3, 10.5)

a. GeH_4 **b.** I_2 **c.** CF_3Cl
d. PCl_3 **e.** BCl_3 **f.** SCl_2

10.95 Indicate the major type of intermolecular forces—(1) ionic bonds, (2) dipole–dipole attractions, (3) hydrogen bonds, (4) dispersion forces—that occurs between particles of the following: (10.6)

a. NF_3 **b.** ClF   **c.** Br_2
d. Cs_2O **e.** C_4H_{10} **f.** CH_3OH

10.96 Indicate the major type of intermolecular forces—(1) ionic bonds, (2) dipole–dipole attractions, (3) hydrogen bonds, (4) dispersion forces—that occurs between particles of the following: (10.6)

a. $CHCl_3$ **b.** H_2O **c.** LiCl
d. OBr_2 **e.** HBr   **f.** IBr

10.97 When it rains or snows, the air temperature seems warmer. Explain. (10.7)

10.98 Water is sprayed on the ground of an orchard when temperatures are near freezing to keep the fruit from freezing. Explain. (10.7)

10.99 Using Figure 10.6, calculate the grams of ice that will melt at 0 °C if 1540 J is absorbed. (10.7)

10.100 Using Figure 10.6, calculate the grams of ethanol that will vaporize at its boiling point if 4620 J is absorbed. (10.7)

10.101 Using Figure 10.6, calculate the grams of acetic acid that will freeze at its freezing point if 5.25 kJ is removed. (10.7)

10.102 Using Figure 10.6, calculate the grams of benzene that will condense at its boiling point if 8.46 kJ is removed. (10.7)

CHALLENGE QUESTIONS

The following groups of questions are related to the topics in this chapter. However, they do not all follow the chapter order, and they require you to combine concepts and skills from several sections. These questions will help you increase your critical thinking skills and prepare for your next exam.

10.103 Complete the Lewis structure for each of the following: (10.1)

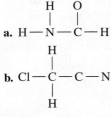

10.104 Identify the errors in each of the following Lewis structures and draw the correct formula: (10.1)

a. $:\ddot{Cl} = O = \ddot{Cl}:$

b. H—C—H (with :Ö: above and below C)

c. H—N̈=Ö—H (with H below N)

10.105 Predict the shape of each of the following molecules or ions: (10.3)

a. NH_2Cl (N is the central atom)
b. PH_4^+
c. SCN^-

10.106 Classify each of the following molecules as polar or nonpolar: (10.3, 10.5)

a. N_2
b. TeO_2
c. NH_2Cl (N is the central atom)

10.107 The melting point of dibromomethane is −53 °C and its boiling point is 97 °C. Sketch a heating curve for dibromomethane from −100 °C to 120 °C. (10.7)
  a. What is the state of dibromomethane at −75 °C?
  b. What happens on the curve at −53 °C?
  c. What is the state of dibromomethane at −18 °C?
  d. What is the state of dibromomethane at 110 °C?
  e. At what temperature will both solid and liquid be present?

10.108 The melting point of benzene is 5.5 °C and its boiling point is 80.1 °C. Draw a heating curve for benzene from 0 °C to 100 °C. (10.7)
  a. What is the state of benzene at 15 °C?
  b. What happens on the curve at 5.5 °C?
  c. What is the state of benzene at 63 °C?
  d. What is the state of benzene at 98 °C?
  e. At what temperature will both liquid and gas be present?

10.109 A 45.0-g piece of ice at 0.0 °C is added to a sample of water at 8.0 °C. All of the ice melts, and the temperature of the water decreases to 0.0 °C. How many grams of water were in the sample? (10.7)

10.110 An ice cube at 0 °C with a mass of 115 g is added to water in a beaker that has a temperature of 64.0 °C. If the final temperature of the mixture is 24.0 °C, what was the initial mass of the warm water? (10.7)

10.111 An ice cube tray holds 325 g of water. If the water initially has a temperature of 25 °C, how many kilojoules of heat must be removed to cool and freeze the water at 0 °C? (10.7)

10.112 A 3.0-kg block of lead is taken from a furnace at 300 °C and placed on a large block of ice at 0 °C. The specific heat of lead is 0.13 J/g °C. If all the heat given up by the lead is used to melt ice, how much ice is melted when the temperature of the lead drops to 0 °C? (10.7)

ANSWERS

Answers to Selected Questions and Problems

10.1 a. 8 valence electrons   **b.** 14 valence electrons
  c. 32 valence electrons   **d.** 8 valence electrons

10.3 a. HF $(8\ e^-)$ H:F̈:   or   H—F̈:

  b. SF$_2$ $(20\ e^-)$:F̈:S̈:F̈:   or   :F̈—S̈—F̈:

  c. NBr$_3$ $(26\ e^-)$:B̈r:N:B̈r: or   :B̈r—N—B̈r:
  (with :B̈r: above)

  d. BH$_4^-$ $(8\ e^-)$ [H:B̈:H with H above and H below]$^-$ or   [H—B—H with H above and H below]$^-$

  e. CH$_3$OH $(14\ e^-)$ H:C̈:Ö:H (with H above and below C)   or   H—C—Ö—H (with H above and below C)

  f. N$_2$H$_4$ $(14\ e^-)$ H:N̈:N̈:H (with H H above)   or   H—N—N—H (with H H above)

10.5 If complete octets cannot be formed by using all the valence electrons, it is necessary to draw multiple bonds.

10.7 a. CO $(10\ e^-)$:C::O: or   :C≡O:
  b. CN$^-$ $(10\ e^-)$ [:C::N:]$^-$ or   [:C≡N:]$^-$
  c. H$_2$CO $(12\ e^-)$ H:C:H (with :O: above)   or   H—C—H (with :O: double-bonded above)

10.9 Resonance occurs when we can draw two or more electron-dot formulas for the same molecule or ion.

10.11 a. ClNO$_2$ $(24\ e^-)$:C̈l—N—Ö: (with :O: double above) ⟷ :C̈l—N=Ö: (with :O: single above)

  b. OCN$^-$ $(16\ e^-)$ [:O≡C—N̈:]$^-$ ⟷ [:Ö=C=N̈:]$^-$ ⟷ [:Ö—C≡N:]$^-$

10.13 a. 6, tetrahedral
  b. 5, trigonal pyramidal
  c. 3, trigonal planar

10.15 a. three   **b.** trigonal planar
  c. three   **d.** trigonal planar

10.17 In CF$_4$, the central atom C has four bonded atoms and no lone pairs of electrons, which gives it a tetrahedral shape. In NF$_3$, the central atom N has three bonded atoms and one lone pair of electrons, which gives NF$_3$ a trigonal pyramidal shape.

10.19 a. trigonal planar   **b.** bent (109°)
  c. linear   **d.** tetrahedral

10.21 a. AlH$_4^-$ $(8\ e^-)$ [H—Al—H with H above and H below]$^-$ tetrahedral

  b. SO$_4^{2-}$ $(32\ e^-)$ [:Ö—S—Ö: with :O: above and :O: below]$^{2-}$ tetrahedral

  c. NH$_4^+$ $(8\ e^-)$ [H—N—H with H above and H below]$^+$ tetrahedral

  d. NO$_2^+$ $(16\ e^-)$ [:Ö=N=Ö:]$^+$ linear

10.23 a. increases   **b.** decreases   **c.** decreases

10.25 a. between 0.0 and 0.4

10.27 a. K, Na, Li   **b.** Na, P, Cl   **c.** Ca, Se, O

10.29 a. polar covalent   **b.** ionic
    c. polar covalent   **d.** nonpolar covalent
    e. polar covalent   **f.** nonpolar covalent

10.31 a. $\overset{\delta^+}{N}$—$\overset{\delta^-}{F}$ **b.** $\overset{\delta^+}{Si}$—$\overset{\delta^-}{Br}$

    c. $\overset{\delta^+}{C}$—$\overset{\delta^-}{O}$ **d.** $\overset{\delta^+}{P}$—$\overset{\delta^-}{Br}$

    e. $\overset{\delta^-}{N}$—$\overset{\delta^+}{P}$

10.33 Electrons are shared equally between two identical atoms and unequally between nonidentical atoms.

10.35 a. nonpolar   **b.** polar
    c. polar   **d.** nonpolar

10.37 In the molecule CO_2, the two C—O dipoles cancel; in CO, there is only one dipole.

10.39 a. dipole–dipole attractions   **b.** ionic bonds
    c. dipole–dipole attractions   **d.** dispersion forces

10.41 a. hydrogen bonds   **b.** dipole–dipole attractions
    c. dispersion forces   **d.** dispersion forces

10.43 a. HF; hydrogen bonds are stronger than the dipole–dipole attractions in HBr.
    b. NaF; ionic bonds are stronger than the hydrogen bonds in HF.
    c. $MgBr_2$; ionic bonds are stronger than the dipole–dipole attractions in PBr_3.
    d. CH_3OH; hydrogen bonds are stronger than the dispersion forces in CH_4.

10.45 a. 21 700 J   **b.** 2180 J
    c. 22.3 kJ   **d.** 16.7 kJ

10.47 a. 22 600 J   **b.** 42.1 kJ
    c. 3100 J   **d.** 242 kJ

10.49 a. 4800 J   **b.** 30 300 J
    c. 40.2 kJ   **d.** 72.2 kJ

10.51 119.5 kJ

10.53 3120 J

10.55 a. 0.0, nonpolar covalent
    b. 0.4, nonpolar covalent
    c. 1.0, polar covalent

10.57 a.

$$\left[\overset{:\ddot{O}:}{\underset{:\ddot{O}:}{\overset{\|}{C}}} \overset{\ddot{O}:}{} \right]^{2-} \longleftrightarrow \left[\overset{:\ddot{O}:}{\underset{:\ddot{O}:}{C}} \overset{\ddot{O}:}{} \right]^{2-} \longleftrightarrow \left[\overset{:\ddot{O}:}{\underset{:\ddot{O}:}{C}} \overset{\ddot{O}:}{} \right]^{2-}$$

    b. Carbonate ion is trigonal planar.

10.59 a. two valence electrons, one bonding pair, no lone pairs
    b. eight valence electrons, one bonding pair, three lone pairs
    c. 14 valence electrons, one bonding pair, six lone pairs

10.61 a. 2, trigonal pyramidal, polar
    b. 1, bent (109°), polar
    c. 3, tetrahedral, nonpolar

10.63 a. C—O and N—O
    b. O—O
    c. Ca—O and K—O
    d. C—O, N—O, O—O

10.65 a. dispersion forces   **b.** dipole–dipole attractions
    c. hydrogen bonds   **d.** dispersion forces

10.67 a. The heat from the skin is used to evaporate the water (perspiration). Therefore, the skin is cooled.
    b. On a hot day, there are more molecules with sufficient energy to become water vapor.
    c. In a closed bag, some molecules evaporate, but they cannot escape and will condense back to liquid; the clothes will not dry.

10.69

10.71 a. about −60 °C   **b.** about 60 °C
    c. The diagonal line **A** represents the solid state as temperature increases. The horizontal line **B** represents the change from solid to liquid or melting of the substance. The diagonal line **C** represents the liquid state as temperature increases. The horizontal line **D** represents the change from liquid to gas or boiling of the liquid. The diagonal line **E** represents the gas state as temperature increases.
    d. At −80 °C, solid; at −40 °C, liquid; at 25 °C, liquid; 80 °C, gas

10.73 a. 1 + 5 + 2(6) = 18 valence electrons
    b. 2(4) + 4(1) + 6 = 18 valence electrons
    c. 5 + 4(1) − 1 = 8 valence electrons
    d. 6 + 3(6) + 2 = 26 valence electrons

10.75 a. BF_4^- (32 e^-)

    b. Cl_2O (20 e^-)

    c. H_2NOH (14 e^-)

    d. H_2CCCl_2 (24 e^-)

10.77 a. (16 e^-)

    b. (16 e^-)

    c. (18 e^-)

10.79 a. I, Cl, F   **b.** K, Li, S, Cl   **c.** Ba, Sr, Mg, Be

10.81 a. C—O   **b.** N—F
     c. S—Cl   **d.** Br—I
     e. N—F

10.83 a. Si—Cl   **b.** C—N
       $\longleftrightarrow$ $\longleftrightarrow$
     c. F—Cl   **d.** C—F
       $\longleftarrow$ $\longleftrightarrow$
     e. N—O
       $\longleftrightarrow$

10.85 a. polar covalent   **b.** nonpolar covalent
     c. ionic   **d.** nonpolar covalent
     e. nonpolar covalent

10.87 a. NF_3 (26 e^-)   :F—N—F: trigonal pyramidal

b. $SiBr_4$ (32 e^-)   :Br—Si—Br: tetrahedral

c. CSe_2 (16 e^-)   :Se=C=Se: linear

d. SO_2 (18 e^-)   :O=S—O: $\longleftrightarrow$:O—S=O:
                                    bent (120°)

10.89 a. bent (109°)   **b.** bent (109°)
     c. tetrahedral   **d.** trigonal pyramidal

10.91 a. trigonal pyramidal, polar
     b. bent (109°), polar

10.93 a. polar   **b.** nonpolar
     c. nonpolar   **d.** polar
     e. polar   **f.** polar

10.95 a. (2) dipole–dipole attractions
     b. (2) dipole–dipole attractions
     c. (4) dispersion forces
     d. (1) ionic bonds
     e. (4) dispersion forces
     f. (3) hydrogen bonds

10.97 When water vapor condenses or liquid water freezes, heat is released, which warms the air.

10.99 4.61 g of water

10.101 27.3 g of acetic acid

10.103 a. (18 e^-)   [structure: H—N—C—H with H and :O: groups]

     b. (22 e^-)   :Cl—C—C≡N: [with H]

     c. (12 e^-)   H—N=N—H

     d. (30 e^-)   :Cl—C—O—C—H   [with :O:, H, H]

10.105 a. trigonal pyramidal   **b.** tetrahedral   **c.** linear

10.107

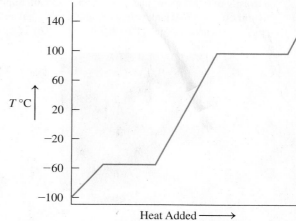

     a. solid   **b.** solid dibromomethane melts
     c. liquid   **d.** gas
     e. −53 °C

10.109 450 g of water

10.111 143 kJ of heat is removed.

CI.13 In an experiment, the mass of a piece of copper is determined to be 8.56 g. Then the copper is reacted with sufficient oxygen gas to produce solid copper(II) oxide. (7.1, 8.1, 8.2, 8.3, 9.1, 9.2, 9.3, 9.5)

— Cu

8.56 g

a. How many copper atoms are in the sample?
b. Write the balanced chemical equation for the reaction.
c. Classify the type of reaction.
d. How many grams of O_2 are required to completely react with the Cu?
e. How many grams of CuO will result from the reaction of 8.56 g of Cu and 3.72 g of oxygen?
f. How many grams of CuO will result in part **e**, if the yield for the reaction is 85.0%?

CI.14 One of the components in gasoline is octane, C_8H_{18}, which has a density of 0.803 g/cm^3 and $\Delta H = -510$ kJ/mol. Suppose a hybrid car has a fuel tank with a capacity of 11.9 gal and has a gas mileage of 45 mi/gal. (1.10, 6.7, 8.3, 8.5, 9.6)

Octane is one of the components of motor fuel.

a. Write a balanced chemical equation for the complete combustion of octane including the heat of reaction.
b. What is the energy, in kilojoules, produced from one tank of fuel assuming it is all octane?
c. How many molecules of C_8H_{18} are present in one tank of fuel assuming it is all octane?
d. If this hybrid car is driven 24 500 miles in one year, how many kilograms of carbon dioxide will be produced from the combustion of the fuel assuming it is all octane?

CI.15 When clothes have stains, bleach may be added to the wash to react with the soil and make the stains colorless. The bleach solution is prepared by bubbling chlorine gas into a solution of sodium hydroxide to produce a solution of sodium hypochlorite, sodium chloride, and water. One brand of bleach contains 5.25% sodium hypochlorite by mass (active ingredient) with a density of 1.08 g/mL. (6.4, 7.1, 7.2, 8.2, 9.3, 9.4, 9.5, 10.1)

a. What is the formula and molar mass of sodium hypochlorite?
b. Draw the Lewis structure for the hypochlorite ion.

Bleach is a solution of sodium hypochlorite.

c. How many hypochlorite ions are present in 1.00 gal of bleach solution?
d. Write the balanced chemical equation for the preparation of bleach.
e. How many grams of NaOH are required to produce the sodium hypochlorite for 1.00 gal of bleach?
f. If 165 g of Cl_2 is passed through a solution containing 275 g of NaOH and 162 g of sodium hypochlorite is produced, what is the percent yield of sodium hypochlorite for the reaction?

CI.16 Ethanol, C_2H_6O, is obtained from renewable crops such as corn, which use the Sun as their source of energy. In the United States, automobiles can now use a fuel known as E85 that contains 85.0% ethanol and 15.0% gasoline by volume. Ethanol has a melting point of $-115\,°C$, a boiling point of $78\,°C$, a heat of fusion of 109 J/g, and a heat of vaporization of 841 J/g. Liquid ethanol has a density of 0.796 g/mL and a specific heat of 2.46 J/g $°C$. (8.3, 8.5, 9.2, 9.3, 10.4, 10.5)

E85 fuel contains 85% ethanol.

a. Draw a heating curve for ethanol from $-150\,°C$ to $100\,°C$.

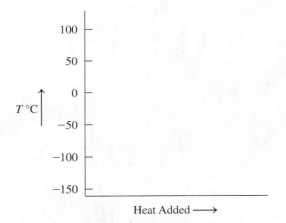

b. When 20.0 g of ethanol at $-62\,°C$ is heated and completely vaporized at $78\,°C$, how much energy, in kilojoules, is required?
c. If a 15.0-gal gas tank is filled with E85, how many liters of ethanol are in the gas tank?

d. Write the balanced chemical equation for the complete combustion of ethanol.

e. How many kilograms of CO_2 are produced by the complete combustion of the ethanol in a full 15.0-gal gas tank?

f. What would be the strongest intermolecular force between liquid ethanol molecules?

Applications

CI.17 Chloral hydrate, a sedative and hypnotic, was the first drug used to treat insomnia. Chloral hydrate has a melting point of 57 °C. At its boiling point of 98 °C, it breaks down to chloral and water. (7.4, 7.5, 8.4, 10.1)

Chloral hydrate   Chloral

a. Draw the Lewis structures for chloral hydrate and chloral.

b. What are the empirical formulas of chloral hydrate and chloral?

c. What is the mass percent of Cl in chloral hydrate?

CI.18 Ethylene glycol, $C_2H_6O_2$, used as a coolant and antifreeze, has a density of 1.11 g/mL. As a sweet-tasting liquid, it can be appealing to pets and small children, but it can be toxic, with an LD_{50} of 4700 mg/kg. Its accidental ingestion can cause kidney damage and difficulty with breathing. In the body, ethylene glycol is converted to another toxic substance, oxalic acid, $H_2C_2O_4$. (2.8, 7.4, 7.5, 8.2, 8.4, 10.1, 10.3, 10.4)

a. What are the empirical formulas of ethylene glycol and oxalic acid?

b. If ethylene glycol has a C—C single bond with two H atoms attached to each C atom, what is its Lewis structure?

c. Which bonds in ethylene glycol are polar covalent and which are nonpolar covalent?

d. How many milliliters of ethylene glycol could be toxic for an 11.0-lb cat?

e. What would be the strongest intermolecular force in ethylene glycol?

Antifreeze often contains ethylene glycol.

f. If oxalic acid has two carbon atoms attached by a C—C single bond with each carbon also attached to two oxygen atoms, what is its Lewis structure?

g. Write the balanced chemical equation for the reaction of ethylene glycol and oxygen (O_2) to give oxalic acid and water.

CI.19 Acetone (propanone), a clear liquid solvent with an acrid odor, is used to remove nail polish, paints, and resins. It has a low boiling point and is highly flammable. Acetone has a density of 0.786 g/mL and a heat of combustion of −1790 kJ/mol. (2.8, 7.2, 8.3, 8.4, 9.3, 9.6)

Acetone has carbon atoms (black), hydrogen atoms (white), and an oxygen atom (red).

a. Draw the Lewis structure for acetone.

b. What are the molecular formula and molar mass of acetone?

c. Write a balanced chemical equation for the complete combustion of acetone, including the heat of reaction.

d. Is the combustion of acetone an endothermic or exothermic reaction?

e. How much heat, in kilojoules, is released if 2.58 g of acetone reacts completely with oxygen?

f. How many grams of oxygen gas are needed to react with 15.0 mL of acetone?

CI.20 The compound dihydroxyacetone (DHA) is used in "sunless" tanning lotions, which darken the skin by reacting with the amino acids in the outer surface of the skin. A typical drugstore lotion contains 4.0% (mass/volume) DHA. (2.8, 7.2, 8.4)

a. Draw the Lewis structure for DHA.

b. DHA has C—C, C—H, C—O, and O—H bonds. Which of these bonds are polar covalent and which are nonpolar covalent?

c. What are the molecular formula and molar mass of DHA?

d. A bottle of sunless tanning lotion contains 177 mL of lotion. How many milligrams of DHA are in a bottle?

Dihydroxyacetone has carbon atoms (black), hydrogen atoms (white), and oxygen atoms (red).

ANSWERS

CI.13 **a.** 8.11×10^{22} atoms of copper

b. $2Cu(s) + O_2(g) \longrightarrow 2CuO(s)$

c. combination reaction

d. 2.16 g of O_2

e. 10.7 g of CuO

f. 9.10 g of CuO

CI.15 **a.** NaOCl, 74.44 g/mol

b. $(14\ e^-)$ $\left[:\ddot{\underset{..}{Cl}} - \ddot{\underset{..}{O}}: \right]^-$

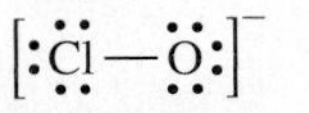

c. 1.74×10^{24} OCl^- ions

d. $2NaOH(aq) + Cl_2(g) \longrightarrow$

$$NaOCl(aq) + NaCl(aq) + H_2O(l)$$

e. 231 g of NaOH

f. 93.6%

CI.17 **a.** $(44\ e^-)$

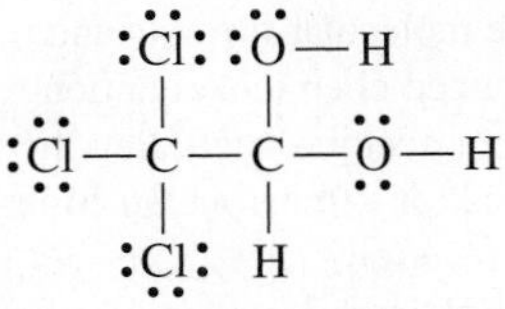

$$:\ddot{\underset{..}{Cl}} - \overset{\displaystyle :\ddot{\underset{..}{Cl}}: \quad :\ddot{\underset{..}{O}} - H}{\underset{\displaystyle :\ddot{\underset{..}{Cl}}: \quad H}{\underset{|}{C}} - \underset{|}{C} - \ddot{\underset{..}{O}} - H}$$

$(36\ e^-)$

$$:\ddot{\underset{..}{Cl}} - \overset{\displaystyle :\ddot{\underset{..}{Cl}}: \quad :\ddot{O}:}{\underset{\displaystyle :\ddot{\underset{..}{Cl}}:}{\underset{|}{C}} - \overset{\parallel}{C} - H}$$

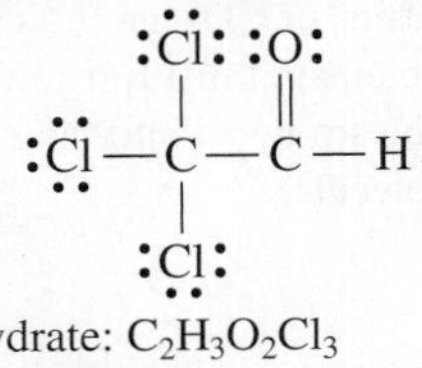

b. chloral hydrate: $C_2H_3O_2Cl_3$

chloral: C_2HOCl_3

c. 64.33% Cl (by mass)

CI.19 **a.** 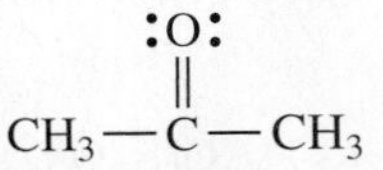 $CH_3 - \overset{\displaystyle :O:}{\overset{\parallel}{C}} - CH_3$

b. C_3H_6O; 58.08 g/mol

c. $C_3H_6O(g) + 4O_2(g) \overset{\Delta}{\longrightarrow}$

$$3CO_2(g) + 3H_2O(g) + 1790\ kJ$$

d. exothermic

e. 79.5 kJ

f. 26.0 g of O_2

Gases

AFTER SOCCER PRACTICE, Whitney complained that she was having difficulty breathing. Her father took her to the emergency room where she was seen by Sam, a respiratory therapist, who listened to Whitney's chest and then tested her breathing capacity using a spirometer. Based on her limited breathing capacity and the wheezing noise in her chest, Whitney was diagnosed as having asthma.

Sam gave Whitney a nebulizer containing a bronchodilator that opens the airways and allows more air to go into the lungs. During the breathing treatment, he measured the amount of oxygen (O_2) in her blood and explained to Whitney and her father that air is a mixture of gases containing 78% nitrogen (N_2) gas and 21% O_2 gas. Because Whitney had difficulty obtaining sufficient oxygen, Sam gave her supplemental oxygen through an oxygen mask. Within

a short period of time, Whitney's breathing returned to normal. The therapist then explained that the lungs work according to Boyle's law: The volume of the lungs increases upon inhalation, and the pressure decreases to allow air to flow in. However, during an asthma attack, the airways become restricted, and it becomes more difficult to expand the volume of the lungs.

CAREER

Respiratory Therapist

Respiratory therapists assess and treat a range of patients, including premature infants whose lungs have not developed and asthmatics or patients with emphysema or cystic fibrosis. In assessing patients, they perform a variety of diagnostic tests including breathing capacity, concentrations of oxygen and carbon dioxide in a patient's blood, as well as blood pH. In order to treat patients, therapists provide oxygen or aerosol medications to the patient, as well as chest physiotherapy to remove mucus from their lungs. Respiratory therapists also educate patients on how to correctly use their inhalers.

KEY MATH SKILL

- Solving Equations (1.4)

 CORE CHEMISTRY SKILLS

- Using Significant Figures in Calculations (2.3)
- Writing Conversion Factors from Equalities (2.5)
- Using Conversion Factors (2.6)

- Using Molar Mass as a Conversion Factor (7.3)
- Using Mole–Mole Factors (9.2)
- Converting Grams to Grams (9.3)

*These Key Math Skills and Core Chemistry Skills from previous chapters are listed here for your review as you proceed to the new material in this chapter.

LOOKING AHEAD

11.1 Properties of Gases

LEARNING GOAL Describe the kinetic molecular theory of gases and the units of measurement used for gases.

We all live at the bottom of a sea of gases called the atmosphere. The most important of these gases is oxygen, which constitutes about 21% of the atmosphere. Without oxygen, life on this planet would be impossible: Oxygen is vital to all life processes of plants and animals. Ozone (O_3), formed in the upper atmosphere by the interaction of oxygen with ultraviolet light, absorbs some of the harmful radiation before it can strike Earth's surface. The other gases in the atmosphere include nitrogen (78%), argon, carbon dioxide (CO_2), and water vapor. Carbon dioxide gas, a product of combustion and metabolism, is used by plants in photosynthesis, which produces the oxygen that is essential for humans and animals.

The behavior of gases is quite different from that of liquids and solids. Gas particles are far apart, whereas particles of both liquids and solids are held close together. A gas has no definite shape or volume and will completely fill any container. Because there are great distances between gas particles, a gas is less dense than a solid or liquid, and easy to compress. A model for the behavior of a gas, called the **kinetic molecular theory of gases**, helps us understand gas behavior.

Interactive Video

Kinetic Molecular Theory

Kinetic Molecular Theory of Gases

1. **A gas consists of small particles (atoms or molecules) that move randomly with high velocities.** Gas molecules moving in random directions at high speeds cause a gas to fill the entire volume of a container.
2. **The attractive forces between the particles of a gas are usually very small.** Gas particles are far apart and fill a container of any size and shape.
3. **The actual volume occupied by gas molecules is extremely small compared to the volume that the gas occupies.** The volume of the gas is considered equal to the volume of the container. Most of the volume of a gas is empty space, which allows gases to be easily compressed.
4. **Gas particles are in constant motion, moving rapidly in straight paths.** When gas particles collide, they rebound and travel in new directions. Every time they hit the walls of the container, they exert pressure. An increase in the number or force of collisions against the walls of the container causes an increase in the pressure of the gas.
5. **The average kinetic energy of gas molecules is proportional to the Kelvin temperature.** Gas particles move faster as the temperature increases. At higher temperatures, gas particles hit the walls of the container more often and with more force, producing higher pressures.

The kinetic molecular theory helps explain some of the characteristics of gases. For example, you can smell perfume when a bottle is opened on the other side of a room because its particles move rapidly in all directions. At room temperature, the molecules in the air are moving at about 450 m/s, which is 1000 mi/h. They move faster at higher temperatures and more slowly at lower temperatures. Sometimes tires and gas-filled containers explode when temperatures are too high. From the kinetic molecular theory, you know that gas particles move faster when heated, hit the walls of a container with more force, and cause a buildup of pressure inside a container.

When we talk about a gas, we describe it in terms of four properties: pressure, volume, temperature, and the amount of gas.

Pressure (*P*)

Gas particles are extremely small and move rapidly. When they hit the walls of a container, they exert a **pressure** (see **FIGURE 11.1**). If we heat the container, the molecules move faster and smash into the walls of the container more often and with increased force, thus increasing the pressure. The gas particles in the air, mostly oxygen and nitrogen, exert a pressure on us called **atmospheric pressure** (see **FIGURE 11.2**). As you go to higher altitudes, the atmospheric pressure is less because there are fewer particles in the air. The most common units used to measure gas pressure are the *atmosphere* (atm) and *millimeters of mercury* (mmHg). On the TV weather report, you may hear or see the atmospheric pressure given in inches of mercury, or in kilopascals in countries other than the United States. In a hospital, the unit torr or pounds per square inch (psi) may be used.

Use the kinetic molecular theory to explain why a gas completely fills a container of any size and shape.

FIGURE 11.1 ▶ Gas particles move in straight lines within a container. The gas particles exert pressure when they collide with the walls of the container.

◎ Why does heating the container increase the pressure of the gas within it?

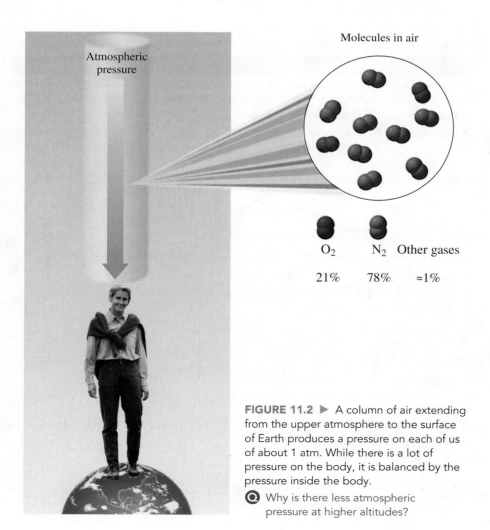

Molecules in air

O₂ N₂ Other gases

21% 78% ≈1%

FIGURE 11.2 ▶ A column of air extending from the upper atmosphere to the surface of Earth produces a pressure on each of us of about 1 atm. While there is a lot of pressure on the body, it is balanced by the pressure inside the body.

◎ Why is there less atmospheric pressure at higher altitudes?

Volume (*V*)

The volume of gas equals the size of the container in which the gas is placed. When you inflate a tire or a basketball, you are adding more gas particles. The increase in the number of particles hitting the walls of the tire or basketball increases the volume. Sometimes, on a cold morning, a tire looks flat. The volume of the tire has decreased because a lower temperature decreases the speed of the molecules, which in turn reduces the force of their impacts on the walls of the tire. The most common units for volume measurement are liters (L) and milliliters (mL).

Temperature (*T*)

The temperature of a gas is related to the kinetic energy of its particles. For example, if we have a gas at 200 K and heat it to a temperature of 400 K, the gas particles will have twice the kinetic energy that they did at 200 K. This also means that the gas at 400 K exerts twice the pressure of the gas at 200 K. Although you measure gas temperature using a Celsius thermometer, all comparisons of gas behavior and all calculations related to temperature must use the Kelvin temperature scale. No one has quite created the conditions for absolute zero (0 K), but we predict that the particles will have zero kinetic energy and exert zero pressure at absolute zero.

Amount of Gas (*n*)

When you add air to a bicycle tire, you increase the amount of gas, which results in a higher pressure in the tire. Usually, we measure the amount of gas by its mass, in grams. In gas law calculations, we need to change the grams of gas to moles.

A summary of the four properties of a gas is given in **TABLE 11.1**.

TABLE 11.1 Properties That Describe a Gas

Property	Description	Units of Measurement
Pressure (*P*)	The force exerted by a gas against the walls of the container	atmosphere (atm); millimeter of mercury (mmHg); torr; pascal (Pa)
Volume (*V*)	The space occupied by a gas	liter (L); milliliter (mL)
Temperature (*T*)	The determining factor of the kinetic energy and rate of motion of gas particles	degree Celsius (°C); kelvin (K) *is required in calculations*
Amount (*n*)	The quantity of gas present in a container	gram (g); mole (*n*) *is required in calculations*

SAMPLE PROBLEM 11.1 Properties of Gases

Identify the property of a gas that is described by each of the following:

a. increases the kinetic energy of gas particles
b. the force of the gas particles hitting the walls of the container
c. the space that is occupied by a gas

> **TRY IT FIRST**

SOLUTION

a. temperature   **b.** pressure   **c.** volume

STUDY CHECK 11.1

When helium is added to a balloon, the number of grams of helium increases. What property of a gas is described?

ANSWER

The mass, in grams, gives the amount of gas.

CHEMISTRY LINK TO **HEALTH**
Measuring Blood Pressure

Your blood pressure is one of the vital signs a doctor or nurse checks during a physical examination. It actually consists of two separate measurements. Acting like a pump, the heart contracts to create the pressure that pushes blood through the circulatory system. During contraction, the blood pressure is at its highest; this is your *systolic* pressure. When the heart muscles relax, the blood pressure falls; this is your *diastolic* pressure. The normal range for systolic pressure is 100 to 200 mmHg. For diastolic pressure, it is 60 to 80 mmHg. These two measurements are usually expressed as a ratio such as 100/80. These values are somewhat higher in older people. When blood pressures are elevated, such as 140/90, there is a greater risk of stroke, heart attack, or kidney damage. Low blood pressure prevents the brain from receiving adequate oxygen, causing dizziness and fainting.

The blood pressures are measured by a *sphygmomanometer*, an instrument consisting of a stethoscope and an inflatable cuff connected to a tube of mercury called a manometer. After the cuff is wrapped around the upper arm, it is pumped up with air until it cuts off the flow of blood through the arm. With the stethoscope over the artery, the air is slowly released from the cuff, decreasing the pressure on the artery. When the blood flow first starts again in the artery, a noise can be heard through the stethoscope signifying the systolic blood pressure as the pressure shown on the manometer. As air continues to be released, the cuff deflates until no sound is heard in the artery. A second pressure reading is taken at the moment of silence and denotes the diastolic pressure, the pressure when the heart is not contracting.

The use of digital blood pressure monitors is becoming more common. However, they have not been validated for use in all situations and can sometimes give inaccurate readings.

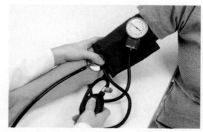

The measurement of blood pressure is part of a routine physical exam.

Measurement of Gas Pressure

When billions and billions of gas particles hit against the walls of a container, they exert pressure, which is a force acting on a certain area.

$$\text{Pressure } (P) = \frac{\text{force}}{\text{area}}$$

The atmospheric pressure can be measured using a barometer (see **FIGURE 11.3**). At a pressure of exactly 1 atmosphere (atm), a mercury column in an inverted tube would be *exactly* 760 mm high. One **atmosphere (atm)** is defined as *exactly* 760 mmHg (millimeters of mercury). One atmosphere is also 760 **Torr**, a pressure unit named to honor Evangelista Torricelli, the inventor of the barometer. Because the units of torr and mmHg are equal, they are used interchangeably. One atmosphere is also equivalent to 29.9 in. of mercury (inHg).

1 atm = 760 mmHg = 760 Torr (exact)

1 atm = 29.9 inHg

1 mmHg = 1 Torr (exact)

In SI units, pressure is measured in pascals (Pa); 1 atm is equal to 101 325 Pa. Because a pascal is a very small unit, pressures are usually reported in kilopascals.

1 atm = 101 325 Pa = 101.325 kPa

The U.S. equivalent of 1 atm is 14.7 lb/in.^{2} (psi). When you use a pressure gauge to check the air pressure in the tires of a car, it may read 30 to 35 psi. This measurement is actually 30 to 35 psi above the pressure that the atmosphere exerts on the outside of the tire.

1 atm = 14.7 lb/in.^{2}

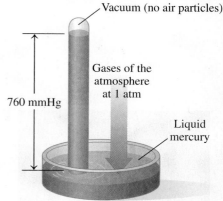

FIGURE 11.3 ▶ A barometer: the pressure exerted by the gases in the atmosphere is equal to the downward pressure of a mercury column in a closed glass tube. The height of the mercury column measured in mmHg is called atmospheric pressure.

Q Why does the height of the mercury column change from day to day?

TABLE 11.2 summarizes the various units used in the measurement of pressure.

TABLE 11.2 Units for Measuring Pressure

Unit	Abbreviation	Unit Equivalent to 1 atm
atmosphere	atm	1 atm (exact)
millimeters of Hg	mmHg	760 mmHg (exact)
torr	Torr	760 Torr (exact)
inches of Hg	inHg	29.9 inHg
pounds per square inch	lb/in.^{2} (psi)	14.7 lb/in.^{2}
pascal	Pa	101 325 Pa
kilopascal	kPa	101.325 kPa

Atmospheric pressure changes with variations in weather and altitude. On a hot, sunny day, the mercury column rises, indicating a higher atmospheric pressure. On a rainy day, the atmosphere exerts less pressure, which causes the mercury column to fall. In the weather report, this type of weather is called a *low-pressure system*. Above sea level, the density of the gases in the air decreases, which causes lower atmospheric pressures; the atmospheric pressure is greater than 760 mmHg at the Dead Sea because it is below sea level (see **TABLE 11.3**).

$P = 0.70$ atm
(530 mmHg)

$P = 1.0$ atm
(760 mmHg)

Sea level

The atmospheric pressure decreases as the altitude increases.

TABLE 11.3 Altitude and Atmospheric Pressure

Location	Altitude (km)	Atmospheric Pressure (mmHg)
Dead Sea	−0.40	800
Sea level	0.00	760
Los Angeles	0.09	752
Las Vegas	0.70	700
Denver	1.60	630
Mount Whitney	4.50	440
Mount Everest	8.90	253

Divers must be concerned about increasing pressures on their ears and lungs when they dive below the surface of the ocean. Because water is more dense than air, the pressure on a diver increases rapidly as the diver descends. At a depth of 33 ft below the surface of the ocean, an additional 1 atm of pressure is exerted by the water on a diver, which gives a total pressure of 2 atm. At 100 ft, there is a total pressure of 4 atm on a diver. The regulator that a diver uses continuously adjusts the pressure of the breathing mixture to match the increase in pressure.

SAMPLE PROBLEM 11.2 Units of Pressure

The oxygen in a tank in the hospital respiratory unit has a pressure of 4820 mmHg. Calculate the pressure, in atmospheres, of the oxygen gas.

TRY IT FIRST

SOLUTION

The equality 1 atm = 760 mmHg can be written as two conversion factors:

$$\frac{760 \text{ mmHg}}{1 \text{ atm}} \quad \text{and} \quad \frac{1 \text{ atm}}{760 \text{ mmHg}}$$

A patient with severe COPD obtains oxygen from an oxygen tank.

Using the conversion factor that cancels mmHg and gives atm, we can set up the problem as

$$4820 \; \underset{\text{Three SFs}}{\cancel{\text{mmHg}}} \times \underset{\substack{\text{Exact} \\ 760 \; \cancel{\text{mmHg}}}}{\frac{1 \; \text{atm}}{\text{Exact}}} = \underset{\text{Three SFs}}{6.34 \; \text{atm}}$$

STUDY CHECK 11.2

A tank of nitrous oxide (N_2O) used as an anesthetic has a pressure of 48 psi. What is that pressure in atmospheres?

ANSWER

3.3 atm

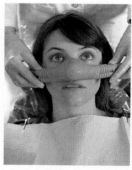

The anesthetic N_2O gas is used for pain relief.

QUESTIONS AND PROBLEMS

11.1 Properties of Gases

LEARNING GOAL Describe the kinetic molecular theory of gases and the units of measurement used for gases.

11.1 Use the kinetic molecular theory of gases to explain each of the following:
a. Gases move faster at higher temperatures.
b. Gases can be compressed much more easily than liquids or solids.
c. Gases have low densities.

11.2 Use the kinetic molecular theory of gases to explain each of the following:
a. A container of nonstick cooking spray explodes when thrown into a fire.
b. The air in a hot-air balloon is heated to make the balloon rise.
c. You can smell the odor of cooking onions from far away.

11.3 Identify the property of a gas that is measured in each of the following:
a. 350 K   **b.** 125 mL
c. 2.00 g of O_2 **d.** 755 mmHg

11.4 Identify the property of a gas that is measured in each of the following:
a. 425 K   **b.** 1.0 atm
c. 10.0 L   **d.** 0.50 mol of He

11.5 Which of the following statement(s) describes the pressure of a gas?
a. the force of the gas particles on the walls of the container
b. the number of gas particles in a container
c. 4.5 L of helium gas
d. 750 Torr
e. 28.8 lb/in.^{2}

11.6 Which of the following statement(s) describes the pressure of a gas?
a. 350 K
b. the volume of the container
c. 3.00 atm
d. 0.25 mol of O_2
e. 101 kPa

Applications

11.7 An tank contains oxygen (O_2) at a pressure of 2.00 atm. What is the pressure in the tank in terms of the following units?
a. torr   **b.** lb/in.^{2}
c. mmHg   **d.** kPa

11.8 On a climb up Mount Whitney, the atmospheric pressure drops to 467 mmHg. What is the pressure in terms of the following units?
a. atm   **b.** torr
c. inHg   **d.** Pa

11.2 Pressure and Volume (Boyle's Law)

LEARNING GOAL Use the pressure–volume relationship (Boyle's law) to calculate the unknown pressure or volume when the temperature and amount of gas are constant.

Imagine that you can see air particles hitting the walls inside a bicycle tire pump. What happens to the pressure inside the pump as you push down on the handle? As the volume decreases, there is a decrease in the surface area of the container. The air particles are crowded together, more collisions occur, and the pressure increases within the container.

When a change in one property (in this case, volume) causes a change in another property (in this case, pressure), the properties are related. If the change occurs in opposite directions, the properties have an **inverse relationship**. The inverse relationship between the pressure and volume of a gas is known as **Boyle's law**. The law states that

ENGAGE

Why does the pressure of a gas in a closed container increase when the volume of the container decreases?

Piston —

$V_1 = 4\text{ L}$
$P_1 = 1\text{ atm}$

$V_2 = 2\text{ L}$
$P_2 = 2\text{ atm}$

FIGURE 11.4 ▶ **Boyle's law:** As volume decreases, gas molecules become more crowded, which causes the pressure to increase. Pressure and volume are inversely related.

 If the volume of a gas increases, what will happen to its pressure?

CORE CHEMISTRY SKILL
Using the Gas Laws

Oxygen therapy increases the oxygen available to the tissues of the body.

Guide to Using the Gas Laws

STEP 1
State the given and needed quantities.

STEP 2
Rearrange the gas law equation to solve for the unknown quantity.

STEP 3
Substitute values into the gas law equation and calculate.

---

the volume (V) of a sample of gas changes inversely with the pressure (P) of the gas as long as there is no change in the temperature (T) or amount of gas (n), as illustrated in **FIGURE 11.4**.

If the volume or pressure of a gas changes without any change occurring in the temperature or in the amount of the gas, then the final pressure and volume will give the same PV product as the initial pressure and volume. Then we can set the initial and final PV products equal to each other. In the equation for Boyle's law, the initial pressure and volume are written as P_1 and V_1 and the final pressure and volume are written as P_2 and V_2.

Boyle's Law

$$P_1V_1 = P_2V_2 \quad \text{No change in temperature and number of moles}$$

SAMPLE PROBLEM 11.3 Calculating Volume When Pressure Changes

When Whitney had her asthma attack, she was given oxygen through a face mask. The gauge on a 12-L tank of compressed oxygen reads 3800 mmHg. How many liters would this same gas occupy at a final pressure of 570 mmHg when temperature and amount of gas do not change?

 TRY IT FIRST

SOLUTION

STEP 1 State the given and needed quantities. We place the gas data in a table by writing the initial pressure and volume as P_1 and V_1 and the final pressure and volume as P_2 and V_2. We see that the pressure decreases from 3800 mmHg to 570 mmHg. Using Boyle's law, we predict that the volume increases.

	Given		Need	Connect
ANALYZE THE PROBLEM	$P_1 = 3800$ mmHg $P_2 = 570$ mmHg $V_1 = 12$ L		V_2	Boyle's law, $P_1V_1 = P_2V_2$
	Factors that remain constant: T and n			**Predict:** P decreases, V increases

STEP 2 Rearrange the gas law equation to solve for the unknown quantity. For a PV relationship, we use Boyle's law and solve for V_2 by dividing both sides by P_2. According to Boyle's law, a decrease in the pressure will cause an increase in the volume when T and n remain constant.

$$P_1V_1 = P_2V_2$$

$$\frac{P_1V_1}{P_2} = \frac{P_2V_2}{P_2}$$

$$V_2 = V_1 \times \frac{P_1}{P_2}$$

STEP 3 Substitute values into the gas law equation and calculate. When we substitute in the values with pressures in units of mmHg, the ratio of pressures (pressure factor) is greater than 1, which increases the volume as predicted.

$$V_2 = 12\text{ L} \times \frac{3800\ \cancel{\text{mmHg}}}{570\ \cancel{\text{mmHg}}} = 80.\text{ L}$$

Pressure factor
increases volume

STUDY CHECK 11.3

In an underground gas reserve, a bubble of methane gas (CH_4) has a volume of 45.0 mL at 1.60 atm pressure. What volume, in milliliters, will the gas bubble occupy when it reaches the surface where the atmospheric pressure is 744 mmHg, if there is no change in the temperature and amount of gas?

ANSWER

73.5 mL

CHEMISTRY LINK TO **HEALTH**
Pressure–Volume Relationship in Breathing

The importance of Boyle's law becomes apparent when you consider the mechanics of breathing. Our lungs are elastic, balloon-like structures contained within an airtight chamber called the thoracic cavity. The diaphragm, a muscle, forms the flexible floor of the cavity.

Inspiration

The process of taking a breath of air begins when the diaphragm contracts and the rib cage expands, causing an increase in the volume of the thoracic cavity. The elasticity of the lungs allows them to expand when the thoracic cavity expands. According to Boyle's law, the pressure inside the lungs decreases when their volume increases, causing the pressure inside the lungs to fall below the pressure of the atmosphere. This difference in pressures produces a *pressure gradient* between the lungs and the atmosphere. In a pressure gradient, molecules flow from an area of higher pressure to an area of lower pressure. During the inhalation phase of breathing, air flows into the lungs (*inspiration*), until the pressure within the lungs becomes equal to the pressure of the atmosphere.

Expiration

Expiration, or the exhalation phase of breathing, occurs when the diaphragm relaxes and moves back up into the thoracic cavity to its

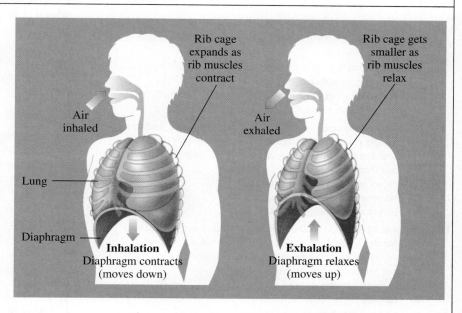

resting position. The volume of the thoracic cavity decreases, which squeezes the lungs and decreases their volume. Now the pressure in the lungs is higher than the pressure of the atmosphere, so air flows out of the lungs. Thus, breathing is a process in which pressure gradients are continuously created between the lungs and the environment because of the changes in the volume and pressure.

QUESTIONS AND PROBLEMS

11.2 Pressure and Volume (Boyle's Law)

LEARNING GOAL Use the pressure–volume relationship (Boyle's law) to calculate the unknown pressure or volume when the temperature and amount of gas are constant.

11.9 Why do scuba divers need to exhale air when they ascend to the surface of the water?

11.10 Why does a sealed bag of chips expand when you take it to a higher altitude?

11.11 The air in a cylinder with a piston has a volume of 220 mL and a pressure of 650 mmHg.
  a. To obtain a higher pressure inside the cylinder at constant temperature and amount of gas, would the cylinder change as shown in **A** or **B**? Explain your choice.

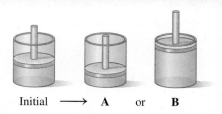

Initial ⟶ **A** or **B**

  b. If the pressure inside the cylinder increases to 1.2 atm, what is the final volume, in milliliters, of the cylinder?

11.12 A balloon is filled with helium gas. When each of the following changes are made at constant temperature, which of these diagrams (**A**, **B**, or **C**) shows the final volume of the balloon?

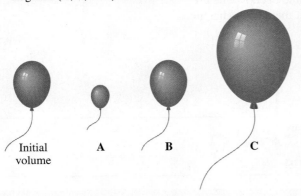

Initial volume   **A**   **B**   **C**

a. The balloon floats to a higher altitude where the outside pressure is lower.
b. The balloon is taken inside the house, but the atmospheric pressure does not change.
c. The balloon is put in a hyperbaric chamber in which the pressure is increased.

11.13 A gas with a volume of 4.0 L is in a closed container. Indicate the changes (*increases, decreases, does not change*) in its pressure when the volume undergoes the following changes at constant temperature and amount of gas:
a. The volume is compressed to 2.0 L.
b. The volume expands to 12 L.
c. The volume is compressed to 0.40 L.

11.14 A gas at a pressure of 2.0 atm is in a closed container. Indicate the changes (*increases, decreases, does not change*) in its volume when the pressure undergoes the following changes at constant temperature and amount of gas:
a. The pressure increases to 6.0 atm.
b. The pressure remains at 2.0 atm.
c. The pressure drops to 0.40 atm.

11.15 A 10.0-L balloon contains helium gas at a pressure of 655 mmHg. What is the final pressure, in millimeters of mercury, when the helium is placed in tanks that have the following volumes, if there is no change in temperature and amount of gas?
a. 20.0 L   **b.** 2.50 L
c. 13 800 mL   **d.** 1250 mL

11.16 The air in a 5.00-L tank has a pressure of 1.20 atm. What is the final pressure, in atmospheres, when the air is placed in tanks that have the following volumes, if there is no change in temperature and amount of gas?
a. 1.00 L   **b.** 2500. mL
c. 750. mL   **d.** 8.00 L

11.17 A sample of nitrogen (N_2) has a volume of 50.0 L at a pressure of 760. mmHg. What is the final volume, in liters, of the gas at each of the following pressures, if there is no change in temperature and amount of gas?
a. 725 mmHg   **b.** 2.0 atm
c. 0.500 atm   **d.** 850 Torr

11.18 A sample of methane (CH_4) has a volume of 25 mL at a pressure of 0.80 atm. What is the final volume, in milliliters, of the gas at each of the following pressures, if there is no change in temperature and amount of gas?
a. 0.40 atm   **b.** 2.00 atm
c. 2500 mmHg   **d.** 80.0 Torr

11.19 A sample of Ar gas has a volume of 5.40 L with an unknown pressure. The gas has a volume of 9.73 L when the pressure is 3.62 atm, with no change in temperature or amount of gas. What was the initial pressure, in atmospheres, of the gas?

11.20 A sample of Ne gas has a pressure of 654 mmHg with an unknown volume. The gas has a pressure of 345 mmHg when the volume is 495 mL, with no change in temperature or amount of gas. What was the initial volume, in milliliters, of the gas?

Applications

11.21 Cyclopropane, (C_3H_6), is a general anesthetic. A 5.0-L sample has a pressure of 5.0 atm. What is the final volume, in liters, of this gas given to a patient at a pressure of 1.0 atm with no change in temperature and amount of gas?

11.22 A tank holds 20.0 L of oxygen (O_2) at a pressure of 15.0 atm. What is the final volume, in liters, of this gas when it is released at a pressure of 1.00 atm with no change in temperature and amount of gas?

11.23 Use the words *inspiration* and *expiration* to describe the part of the breathing cycle that occurs as a result of each of the following:
a. The diaphragm contracts.
b. The volume of the lungs decreases.
c. The pressure within the lungs is less than that of the atmosphere.

11.24 Use the words *inspiration* and *expiration* to describe the part of the breathing cycle that occurs as a result of each of the following:
a. The diaphragm relaxes, moving up into the thoracic cavity.
b. The volume of the lungs expands.
c. The pressure within the lungs is higher than that of the atmosphere.

11.3 Temperature and Volume (Charles's Law)

LEARNING GOAL Use the temperature–volume relationship (Charles's law) to calculate the unknown temperature or volume when the pressure and amount of gas are constant.

Suppose that you are going to take a ride in a hot-air balloon. The captain turns on a propane burner to heat the air inside the balloon. As the air is heated, it expands and becomes less dense than the air outside, causing the balloon and its passengers to lift off. In 1787, Jacques Charles, a balloonist as well as a physicist, proposed that the volume of a gas is related to the temperature. This proposal became **Charles's law**, which states that the volume (V) of a gas is directly related to the temperature (T) when there is no change in the pressure (P) or amount (n) of gas. A **direct relationship** is one in which the related properties increase or decrease together. For two conditions, initial and final, we can write Charles's law as follows:

Charles's Law

$$\frac{V_1}{T_1} = \frac{V_2}{T_2}$$ No change in pressure and number of moles

All temperatures used in gas law calculations must be converted to their corresponding Kelvin (K) temperatures.

To determine the effect of changing temperature on the volume of a gas, the pressure and the amount of gas are kept constant. If we increase the temperature of a gas sample, we know from the kinetic molecular theory that the motion (kinetic energy) of the gas particles will also increase. To keep the pressure constant, the volume of the container must increase (see **FIGURE 11.5**). If the temperature of the gas is decreased, the volume of the container must also decrease to maintain the same pressure when the amount of gas is constant.

As the gas in a hot-air balloon is heated, it expands.

SAMPLE PROBLEM 11.4 Calculating Volume When Temperature Changes

Helium gas is used to inflate the abdomen during laparoscopic surgery. A sample of helium gas has a volume of 5.40 L and a temperature of 15 °C. What is the final volume, in liters, of the gas after the temperature has been increased to 42 °C at constant pressure and amount of gas?

TRY IT FIRST

SOLUTION

STEP 1 State the given and needed quantities. We place the gas data in a table by writing the initial temperature and volume as T_1 and V_1 and the final temperature and volume as T_2 and V_2. We see that the temperature increases from 15 °C to 42 °C. Using Charles's law, we predict that the volume increases.

$T_1 = 15\ °C + 273 = 288\ K$
$T_2 = 42\ °C + 273 = 315\ K$

ANALYZE THE PROBLEM	Given	Need	Connect
	$T_1 = 288\ K$ $T_2 = 315\ K$ $V_1 = 5.40\ L$	V_2	Charles's law, $\frac{V_1}{T_1} = \frac{V_2}{T_2}$
	Factors that remain constant: P and n		**Predict:** T increases, V increases

STEP 2 Rearrange the gas law equation to solve for the unknown quantity. In this problem, we want to know the final volume (V_2) when the temperature increases. Using Charles's law, we solve for V_2 by multiplying both sides by T_2.

$$\frac{V_1}{T_1} = \frac{V_2}{T_2}$$

$$\frac{V_1}{T_1} \times T_2 = \frac{V_2}{\cancel{T_2}} \times \cancel{T_2}$$

$$V_2 = V_1 \times \frac{T_2}{T_1}$$

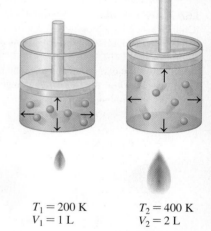

$T_1 = 200\ K$ $T_2 = 400\ K$
$V_1 = 1\ L$ $V_2 = 2\ L$

FIGURE 11.5 ▶ Charles's law: The Kelvin temperature of a gas is directly related to the volume of the gas when there is no change in the pressure and amount of gas. When the temperature increases, making the molecules move faster, the volume must increase to maintain constant pressure.

Q If the temperature of a gas decreases at a constant pressure, how will the volume change?

STEP 3 Substitute values into the gas law equation and calculate. From the table, we see that the temperature has increased. Because temperature is directly related to volume, the volume must increase. When we substitute in the values, we see that the ratio of the temperatures (temperature factor) is greater than 1, which increases the volume as predicted.

$$V_2 = 5.40\,\text{L} \times \frac{315\,\text{K}}{288\,\text{K}} = 5.91\,\text{L}$$

Temperature factor
increases volume

STUDY CHECK 11.4

A mountain climber inhales air that has a temperature of $-8\,°C$. If the final volume of air in the lungs is 569 mL at a body temperature of 37 °C, what was the initial volume of air, in milliliters, inhaled by the climber?

ANSWER

486 mL

QUESTIONS AND PROBLEMS

11.3 Temperature and Volume (Charles's Law)

LEARNING GOAL Use the temperature–volume relationship (Charles's law) to calculate the unknown temperature or volume when the pressure and amount of gas are constant.

11.25 Select the diagram that shows the final volume of a balloon when each of the following changes are made at constant pressure and amount of gas:

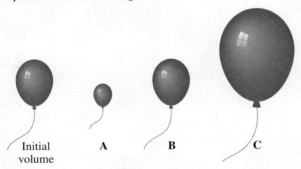

Initial volume   **A**   **B**   **C**

a. The temperature is changed from 100 K to 300 K.
b. The balloon is placed in a freezer.
c. The balloon is first warmed and then returned to its starting temperature.

11.26 Indicate whether the final volume of gas in each of the following is the *same*, *larger*, or *smaller* than the initial volume, if pressure and amount of gas do not change:
a. A volume of 505 mL of air on a cold winter day at $-15\,°C$ is breathed into the lungs, where body temperature is 37 °C.
b. The heater used to heat the air in a hot-air balloon is turned off.
c. A balloon filled with helium at the amusement park is left in a car on a hot day.

11.27 A sample of neon initially has a volume of 2.50 L at 15 °C. What final temperature, in degrees Celsius, is needed to change the volume of the gas to each of the following, if P and n do not change?
a. 5.00 L
b. 1250 mL
c. 7.50 L
d. 3550 mL

11.28 A gas has a volume of 4.00 L at 0 °C. What final temperature, in degrees Celsius, is needed to change the volume of the gas to each of the following, if P and n do not change?
a. 1.50 L
b. 1200 mL
c. 10.0 L
d. 50.0 mL

11.29 A balloon contains 2500 mL of helium gas at 75 °C. What is the final volume, in milliliters, of the gas when the temperature changes to each of the following, if P and n do not change?
a. 55 °C
b. 680. K
c. $-25\,°C$
d. 240. K

11.30 An air bubble has a volume of 0.500 L at 18 °C. What is the final volume, in liters, of the gas when the temperature changes to each of the following, if P and n do not change?
a. 0 °C
b. 425 K
c. $-12\,°C$
d. 575 K

11.31 A gas sample has a volume of 0.256 L with an unknown temperature. The same gas has a volume of 0.198 L when the temperature is 32 °C, with no change in the pressure or amount of gas. What was the initial temperature, in degrees Celsius, of the gas?

11.32 A gas sample has a temperature of 22 °C with an unknown volume. The same gas has a volume of 456 mL when the temperature is 86 °C, with no change in the pressure or amount of gas. What was the initial volume, in milliliters, of the gas?

11.4 Temperature and Pressure (Gay-Lussac's Law)

LEARNING GOAL Use the temperature–pressure relationship (Gay-Lussac's law) to calculate the unknown temperature or pressure when the volume and amount of gas are constant.

If we could observe the molecules of a gas as the temperature rises, we would notice that they move faster and hit the sides of the container more often and with greater force. If we maintain a constant volume and amount of gas, the pressure would increase. In the temperature–pressure relationship known as **Gay-Lussac's law**, the pressure of a gas is directly related to its Kelvin temperature. This means that an increase in temperature increases the pressure of a gas, and a decrease in temperature decreases the pressure of the gas as long as the volume and amount of gas do not change (see **FIGURE 11.6**).

Gay-Lussac's Law

$$\frac{P_1}{T_1} = \frac{P_2}{T_2}$$ No change in volume and number of moles

All temperatures used in gas law calculations must be converted to their corresponding Kelvin (K) temperatures.

$T_1 = 200$ K $T_2 = 400$ K
$P_1 = 1$ atm $P_2 = 2$ atm

FIGURE 11.6 ▶ Gay-Lussac's law: When the Kelvin temperature of a gas is doubled at constant volume and amount of gas, the pressure also doubles.

Q How does a decrease in the temperature of a gas affect its pressure at constant volume and amount of gas?

SAMPLE PROBLEM 11.5 Calculating Pressure When Temperature Changes

Home oxygen tanks, which provide an oxygen-rich environment, can be dangerous if they are heated, because they can explode. Suppose an oxygen tank has a pressure of 120 atm at a room temperature of 25 °C. If a fire in the room causes the temperature of the gas inside the oxygen tank to reach 402 °C, what will be its pressure in atmospheres if the volume and amount of gas do not change? The oxygen tank may rupture if the pressure inside exceeds 180 atm. Would you expect it to rupture?

▶ **TRY IT FIRST**

SOLUTION

STEP 1 **State the given and needed quantities.** We place the gas data in a table by writing the initial temperature and pressure as T_1 and P_1 and the final temperature and pressure as T_2 and P_2. We see that the temperature increases from 25 °C to 402 °C. Using Gay-Lussac's law, we predict that the pressure increases.

$T_1 = 25\,°C + 273 = 298$ K
$T_2 = 402\,°C + 273 = 675$ K

	Given	Need	Connect
ANALYZE THE PROBLEM	$P_1 = 4.0$ atm $T_1 = 298$ K $T_2 = 675$ K	P_2	Gay-Lussac's law, $\dfrac{P_1}{T_1} = \dfrac{P_2}{T_2}$
	Factors that remain constant: V and n		**Predict:** T increases, P increases

STEP 2 **Rearrange the gas law equation to solve for the unknown quantity.** Using Gay-Lussac's law, we solve for P_2 by multiplying both sides by T_2.

$$\frac{P_1}{T_1} = \frac{P_2}{T_2}$$

$$\frac{P_1}{T_1} \times T_2 = \frac{P_2}{\cancel{T_2}} \times \cancel{T_2}$$

$$P_2 = P_1 \times \frac{T_2}{T_1}$$

STEP 3 Substitute values into the gas law equation and calculate. When we substitute in the values, we see that the ratio of the temperatures (temperature factor) is greater than 1, which increases the pressure as predicted.

$$P_2 = 120 \text{ atm} \times \underbrace{\frac{675 \text{ K}}{298 \text{ K}}}_{\substack{\text{Temperature factor} \\ \text{increases pressure}}} = 270 \text{ atm}$$

Because the calculated pressure of 270 atm exceeds the limit of 180 atm, we would expect the oxygen tank to rupture.

STUDY CHECK 11.5

In a storage area of a hospital where the temperature has reached 55 °C, the pressure of oxygen gas in a 15.0-L steel cylinder is 965 Torr. To what temperature, in degrees Celsius, would the gas have to be cooled to reduce the pressure to 850 Torr, when the volume and the amount of the gas do not change?

ANSWER

16 °C

Vapor Pressure and Boiling Point

When liquid molecules with sufficient kinetic energy break away from the surface, they become gas particles or vapor. In an open container, all the liquid will eventually evaporate. In a closed container, the vapor accumulates and creates pressure called **vapor pressure**. Each liquid exerts its own vapor pressure at a given temperature. As temperature increases, more vapor forms, and vapor pressure increases.

A liquid reaches its boiling point when its vapor pressure becomes equal to the external pressure. As boiling occurs, bubbles of the gas form within the liquid and quickly rise to the surface. For example, at an atmospheric pressure of 760 mmHg, water will boil at 100 °C, the temperature at which its vapor pressure reaches 760 mmHg (see **TABLE 11.4**).

TABLE 11.4 Vapor Pressure of Water

Temperature (°C)	Vapor Pressure (mmHg)
0	5
10	9
20	18
30	32
37*	47
40	55
50	93
60	149
70	234
80	355
90	528
100	760

*At body temperature

100 °C

Atmospheric pressure 760 mmHg

760 mmHg

Vapor pressure in the bubble equals the atmospheric pressure

At high altitudes, where atmospheric pressures are lower than 760 mmHg, the boiling point of water is lower than 100 °C. For example, a typical atmospheric pressure in Denver is 630 mmHg. This means that water in Denver boils when the vapor pressure is 630 mmHg. TABLE 11.5 shows that an increase in the pressure for water increases the boiling point.

In a closed container such as a pressure cooker, a pressure greater than 1 atm can be obtained, which means that water boils at a temperature higher than 100 °C. Laboratories and hospitals use closed containers called *autoclaves* to sterilize laboratory and surgical equipment at temperature of 121 °C to 135 °C.

TABLE 11.5 Pressure and the Boiling Point of Water

Pressure (mmHg)	Boiling Point (°C)
270	70
467	87
630	95
752	99
760	100
800	100.4
1075	110
1520 (2 atm)	120
3800 (5 atm)	160
7600 (10 atm)	180

An autoclave used to sterilize equipment attains a temperature higher than 100 °C.

QUESTIONS AND PROBLEMS

11.4 Temperature and Pressure (Gay-Lussac's Law)

LEARNING GOAL Use the temperature–pressure relationship (Gay-Lussac's law) to calculate the unknown temperature or pressure when the volume and amount of gas are constant.

11.33 Calculate the final pressure, in millimeters of mercury, for each of the following, if V and n do not change:
  a. A gas with an initial pressure of 1200 Torr at 155 °C is cooled to 0 °C.
  b. A gas in an aerosol can at an initial pressure of 1.40 atm at 12 °C is heated to 35 °C.

11.34 Calculate the final pressure, in atmospheres, for each of the following, if V and n do not change:
  a. A gas with an initial pressure of 1.20 atm at 75 °C is cooled to −32 °C.
  b. A sample of N_2 with an initial pressure of 780. mmHg at −75 °C is heated to 28 °C.

11.35 Calculate the final temperature, in degrees Celsius, for each of the following, if V and n do not change:
  a. A sample of xenon gas at 25 °C and 740. mmHg is cooled to give a pressure of 620. mmHg.
  b. A tank of argon gas with a pressure of 0.950 atm at −18 °C is heated to give a pressure of 1250 Torr.

11.36 Calculate the final temperature, in degrees Celsius, for each of the following, if V and n do not change:
  a. A sample of helium gas with a pressure of 250 Torr at 0 °C is heated to give a pressure of 1500 Torr.
  b. A sample of air at 40 °C and 740. mmHg is cooled to give a pressure of 680. mmHg.

11.37 A gas sample has a pressure of 766 mmHg with an unknown temperature. The same gas has a pressure of 744 mmHg when the temperature is 22 °C, with no change in the volume or amount of gas. What was the initial temperature, in degrees Celsius, of the gas?

11.38 A gas sample has a unknown pressure with a temperature of 46 °C. The same gas has a pressure of 2.35 atm when the temperature is −15 °C, with no change in the volume or amount of gas. What was the initial pressure, in atmospheres, of the gas?

11.39 Explain each of the following observations:
  a. Water boils at 87 °C on the top of Mount Whitney.
  b. Food cooks more quickly in a pressure cooker than in an open pan.

11.40 Explain each of the following observations:
  a. Boiling water at sea level is hotter than boiling water in the mountains.
  b. Water used to sterilize surgical equipment is heated to 120 °C at 2.0 atm in an autoclave.

Applications

11.41 A tank contains isoflurane, an inhaled anesthetic, at a pressure of 1.8 atm and 5 °C. What is the pressure, in atmospheres, if the gas is warmed to a temperature of 22 °C, if V and n do not change?

11.42 Bacteria and viruses are inactivated by temperatures above 135 °C. An autoclave contains steam at 1.00 atm and 100 °C. At what pressure, in atmospheres, will the temperature of the steam in the autoclave reach 135 °C, if V and n do not change?

11.5 The Combined Gas Law

LEARNING GOAL Use the combined gas law to calculate the unknown pressure, volume, or temperature of a gas when changes in two of these properties are given and the amount of gas is constant.

All of the pressure–volume–temperature relationships for gases that we have studied may be combined into a single relationship called the **combined gas law**. This expression is useful for studying the effect of changes in two of these variables on the third as long as the amount of gas (number of moles) remains constant.

Combined Gas Law

$$\frac{P_1 V_1}{T_1} = \frac{P_2 V_2}{T_2} \quad \text{No change in number of moles of gas}$$

By using the combined gas law, we can derive any of the gas laws by omitting those properties that do not change, as seen in **TABLE 11.6**.

TABLE 11.6 Summary of Gas Laws

Combined Gas Law	Properties Held Constant	Relationship	Name of Gas Law
$\dfrac{P_1 V_1}{\cancel{T_1}} = \dfrac{P_2 V_2}{\cancel{T_2}}$	T, n	$P_1 V_1 = P_2 V_2$	Boyle's law
$\dfrac{\cancel{P_1} V_1}{T_1} = \dfrac{\cancel{P_2} V_2}{T_2}$	P, n	$\dfrac{V_1}{T_1} = \dfrac{V_2}{T_2}$	Charles's law
$\dfrac{P_1 \cancel{V_1}}{T_1} = \dfrac{P_2 \cancel{V_2}}{T_2}$	V, n	$\dfrac{P_1}{T_1} = \dfrac{P_2}{T_2}$	Gay-Lussac's law

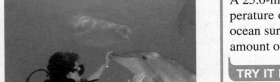

ENGAGE

Why does the pressure of a gas decrease to one-fourth of its initial pressure when the volume of the gas doubles and the Kelvin temperature decreases by half, if the amount of gas does not change?

Under water, the pressure on a diver is greater than the atmospheric pressure.

SAMPLE PROBLEM 11.6 Using the Combined Gas Law

A 25.0-mL bubble is released from a diver's air tank at a pressure of 4.00 atm and a temperature of 11 °C. What is the volume, in milliliters, of the bubble when it reaches the ocean surface where the pressure is 1.00 atm and the temperature is 18 °C? (Assume the amount of gas in the bubble does not change.)

▶ **TRY IT FIRST**

SOLUTION

STEP 1 State the given and needed quantities. We list the properties that change, which are the pressure, volume, and temperature. The property that remains constant, which is the amount of gas, is shown below the table. The temperatures in degrees Celsius must be changed to kelvins.

$T_1 = 11\,°\text{C} + 273 = 284\ \text{K}$
$T_2 = 18\,°\text{C} + 273 = 291\ \text{K}$

ANALYZE THE PROBLEM	Given		Need	Connect
	$P_1 = 4.00\ \text{atm}$	$P_2 = 1.00\ \text{atm}$	V_2	combined gas law, $\dfrac{P_1 V_1}{T_1} = \dfrac{P_2 V_2}{T_2}$
	$V_1 = 25.0\ \text{mL}$			
	$T_1 = 284\ \text{K}$	$T_2 = 291\ \text{K}$		
	Factor that remains constant: n			

STEP 2 Rearrange the gas law equation to solve for the unknown quantity. Using the combined gas law, we solve for V_2 by multiplying both sides by T_2 and dividing both sides by P_2.

$$\frac{P_1 V_1}{T_1} = \frac{P_2 V_2}{T_2}$$

$$\frac{P_1 V_1}{T_1} \times \frac{T_2}{P_2} = \frac{\cancel{P_2} V_2}{\cancel{T_2}} \times \frac{\cancel{T_2}}{\cancel{P_2}}$$

$$V_2 = V_1 \times \frac{P_1}{P_2} \times \frac{T_2}{T_1}$$

STEP 3 Substitute values into the gas law equation and calculate. From the data table, we determine that both the pressure decrease and the temperature increase will increase the volume.

$$V_2 = 25.0 \text{ mL} \times \frac{4.00 \text{ atm}}{1.00 \text{ atm}} \times \frac{291 \text{ K}}{284 \text{ K}} = 102 \text{ mL}$$

Pressure factor increases volume   Temperature factor increases volume

However, in situations where the unknown value is decreased by one change but increased by the second change, it is difficult to predict the overall change for the unknown.

STUDY CHECK 11.6

A weather balloon is filled with 15.0 L of helium at a temperature of 25 °C and a pressure of 685 mmHg. What is the pressure, in millimeters of mercury, of the helium in the balloon in the upper atmosphere when the final temperature is −35 °C and the final volume becomes 34.0 L, if the amount of He does not change?

ANSWER

241 mmHg

QUESTIONS AND PROBLEMS

11.5 The Combined Gas Law

LEARNING GOAL Use the combined gas law to calculate the unknown pressure, volume, or temperature of a gas when changes in two of these properties are given and the amount of gas is constant.

11.43 Rearrange the variables in the combined gas law to solve for T_2.

11.44 Rearrange the variables in the combined gas law to solve for P_2.

11.45 A sample of helium gas has a volume of 6.50 L at a pressure of 845 mmHg and a temperature of 25 °C. What is the final pressure of the gas, in atmospheres, when the volume and temperature of the gas sample are changed to the following, if the amount of gas does not change?
  a. 1850 mL and 325 K
  b. 2.25 L and 12 °C
  c. 12.8 L and 47 °C

11.46 A sample of argon gas has a volume of 735 mL at a pressure of 1.20 atm and a temperature of 112 °C. What is the final volume

of the gas, in milliliters, when the pressure and temperature of the gas sample are changed to the following, if the amount of gas does not change?
  a. 658 mmHg and 281 K
  b. 0.55 atm and 75 °C
  c. 15.4 atm and −15 °C

Applications

11.47 A 124-mL bubble of hot gas initially at 212 °C and 1.80 atm is emitted from an active volcano. What is the final temperature, in degrees Celsius, of the gas in the bubble outside the volcano if the final volume of the bubble is 138 mL and the pressure is 0.800 atm, if the amount of gas does not change?

11.48 A scuba diver 60 ft below the ocean surface inhales 50.0 mL of compressed air from a scuba tank at a pressure of 3.00 atm and a temperature of 8 °C. What is the final pressure of air, in atmospheres, in the lungs when the gas expands to 150.0 mL at a body temperature of 37 °C, if the amount of gas does not change?

11.6 Volume and Moles (Avogadro's Law)

LEARNING GOAL Use Avogadro's law to calculate the unknown amount or volume of a gas when the pressure and temperature are constant.

In our study of the gas laws, we have looked at changes in properties for a specified amount (n) of gas. Now we will consider how the properties of a gas change when there is a change in the number of moles or grams of the gas.

When you blow up a balloon, its volume increases because you add more air molecules. If the balloon has a small hole in it, air leaks out, causing its volume to decrease. In 1811, Amedeo Avogadro formulated **Avogadro's law**, which states that the volume of a gas is directly related to the number of moles of a gas when temperature and pressure do not change. For example, if the number of moles of a gas is doubled, then the volume will also double as long as we do not change the pressure or the temperature (see **FIGURE 11.7**). At constant pressure and temperature, we can write Avogadro's law as follows:

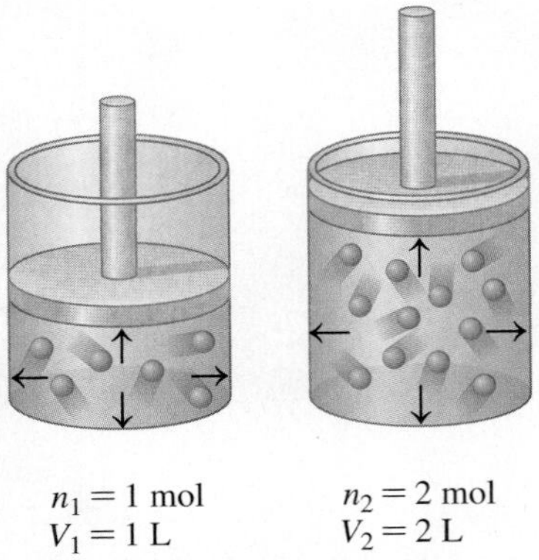

Avogadro's Law

$$\frac{V_1}{n_1} = \frac{V_2}{n_2} \quad \text{No change in pressure and temperature}$$

$n_1 = 1$ mol
$V_1 = 1$ L

$n_2 = 2$ mol
$V_2 = 2$ L

FIGURE 11.7 ▶ Avogadro's law: The volume of a gas is directly related to the number of moles of the gas. If the number of moles is doubled, the volume must double at constant temperature and pressure.

Q If a balloon has a leak, what happens to its volume?

SAMPLE PROBLEM 11.7 Calculating Volume for a Change in Moles

A weather balloon with a volume of 44 L is filled with 2.0 mol of helium. What is the final volume, in liters, if 3.0 mol of helium are added, to give a total of 5.0 mol of helium, if the pressure and temperature do not change?

TRY IT FIRST

SOLUTION

STEP 1 **State the given and needed quantities.** We list those properties that change, which are volume and amount (moles). The properties that do not change, which are pressure and temperature, are shown below the table. Because there is an increase in the number of moles of gas, we can predict that the volume increases.

	Given	Need	Connect
ANALYZE THE PROBLEM	$V_1 = 44$ L $n_1 = 2.0$ mol $n_2 = 5.0$ mol **Factors that remain constant:** P and T	V_2	Avogadro's law, $\dfrac{V_1}{n_1} = \dfrac{V_2}{n_2}$ **Predict:** n increases, V increases

STEP 2 **Rearrange the gas law equation to solve for the unknown quantity.** Using Avogadro's law, we can solve for V_2 by multiplying both sides of the equation by n_2.

$$\frac{V_1}{n_1} = \frac{V_2}{n_2}$$

$$\frac{V_1}{n_1} \times n_2 = \frac{V_2}{n_2} \times n_2$$

$$V_2 = V_1 \times \frac{n_2}{n_1}$$

STEP 3 Substitute values into the gas law equation and calculate. When we substitute in the values, we see that the ratio of the moles (mole factor) is greater than 1, which increases the volume as predicted.

$$V_2 = 44 \text{ L} \times \frac{5.0 \text{ mol}}{2.0 \text{ mol}} = 110 \text{ L}$$

Mole factor
increases volume

STUDY CHECK 11.7

A sample containing 8.00 g of oxygen gas has a volume of 5.00 L. What is the volume, in liters, after 4.00 g of oxygen gas is added to the 8.00 g of oxygen in the balloon, if the temperature and pressure do not change?

ANSWER

7.50 L

STP and Molar Volume

Using Avogadro's law, we can say that any two gases will have equal volumes if they contain the same number of moles of gas at the same temperature and pressure. To help us make comparisons between different gases, arbitrary conditions called *standard temperature* (273 K) and *standard pressure* (1 atm) together abbreviated **STP**, were selected by scientists:

STP Conditions

Standard temperature is *exactly* 0 °C (273 K).

Standard pressure is *exactly* 1 atm (760 mmHg).

At STP, one mole of any gas occupies a volume of 22.4 L, which is about the same as the volume of three basketballs. This volume, 22.4 L, of any gas is called the **molar volume** (see **FIGURE 11.8**).

The molar volume of a gas at STP is about the same as the volume of three basketballs.

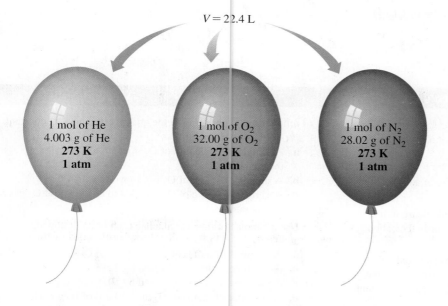

$V = 22.4$ L

1 mol of He
4.003 g of He
273 K
1 atm

1 mol of O_2
32.00 g of O_2
273 K
1 atm

1 mol of N_2
28.02 g of N_2
273 K
1 atm

ENGAGE

Why would the molar volume of a gas be greater than 22.4 L/mol if the pressure is 1 atm and the temperature is 100 °C?

FIGURE 11.8 ▶ Avogadro's law indicates that 1 mol of any gas at STP has a volume of 22.4 L.

Q What volume of gas is occupied by 16.0 g of methane gas, CH_4, at STP?

When a gas is at STP conditions (0 °C and 1 atm), its molar volume can be used to write conversion factors between the number of moles of gas and its volume, in liters.

Molar Volume Conversion Factors

1 mol of gas = 22.4 L (STP)

$$\frac{1 \text{ mol gas}}{22.4 \text{ L (STP)}} \quad \text{and} \quad \frac{22.4 \text{ L (STP)}}{1 \text{ mol gas}}$$

SAMPLE PROBLEM 11.8 Using Molar Volume to Find Volume at STP

What is the volume, in liters, of 64.0 g of O_2 gas at STP?

TRY IT FIRST

SOLUTION

STEP 1 State the given and needed quantities.

ANALYZE THE PROBLEM	Given	Need	Connect
	64.0 g of $O_2(g)$ at STP	liters of O_2 gas at STP	molar mass, molar volume (STP)

STEP 2 Write a plan to calculate the needed quantity.

grams of O_2 | Molar mass | moles of O_2 | Molar volume | liters of O_2

STEP 3 Write the equalities and conversion factors including 22.4 L/mol at STP.

$$1 \text{ mol of } O_2 = 32.00 \text{ g of } O_2$$
$$\frac{32.00 \text{ g } O_2}{1 \text{ mol } O_2} \quad \text{and} \quad \frac{1 \text{ mol } O_2}{32.00 \text{ g } O_2}$$

$$1 \text{ mol of } O_2 = 22.4 \text{ L of } O_2 \text{ (STP)}$$
$$\frac{22.4 \text{ L } O_2 \text{ (STP)}}{1 \text{ mol } O_2} \quad \text{and} \quad \frac{1 \text{ mol } O_2}{22.4 \text{ L } O_2 \text{ (STP)}}$$

STEP 4 Set up the problem with factors to cancel units.

$$64.0 \text{ g } O_2 \times \frac{1 \text{ mol } O_2}{32.00 \text{ g } O_2} \times \frac{22.4 \text{ L } O_2 \text{ (STP)}}{1 \text{ mol } O_2} = 44.8 \text{ L of } O_2 \text{ (STP)}$$

STUDY CHECK 11.8

How many grams of $Cl_2(g)$ are in 5.00 L of $Cl_2(g)$ at STP?

ANSWER

15.8 g of $Cl_2(g)$

Guide to Using Molar Volume

STEP 1
State the given and needed quantities.

STEP 2
Write a plan to calculate the needed quantity.

STEP 3
Write the equalities and conversion factors including 22.4 L/mol at STP.

STEP 4
Set up the problem with factors to cancel units.

QUESTIONS AND PROBLEMS

11.6 Volume and Moles (Avogadro's Law)

LEARNING GOAL Use Avogadro's law to calculate the unknown amount or volume of a gas when the pressure and temperature are constant.

11.49 What happens to the volume of a bicycle tire or a basketball when you use an air pump to add air?

11.50 Sometimes when you blow up a balloon and release it, it flies around the room. What is happening to the air in the balloon and its volume?

11.51 A sample containing 1.50 mol of Ne gas has an initial volume of 8.00 L. What is the final volume, in liters, when each of the following changes occur in the quantity of the gas at constant pressure and temperature?
   a. A leak allows one-half of Ne atoms to escape.
   b. A sample of 3.50 mol of Ne is added to the 1.50 mol of Ne gas in the container.
   c. A sample of 25.0 g of Ne is added to the 1.50 mol of Ne gas in the container.

11.52 A sample containing 4.80 g of O_2 gas has an initial volume of 15.0 L. What is the final volume, in liters, when each of the following changes occur in the quantity of the gas at constant pressure and temperature?
   a. A sample of 0.500 mol of O_2 is added to the 4.80 g of O_2 in the container.
   b. A sample of 2.00 g of O_2 is removed.
   c. A sample of 4.00 g of O_2 is added to the 4.80 g of O_2 gas in the container.

11.53 Use the molar volume to calculate each of the following at STP:
   a. the number of moles of O_2 in 44.8 L of O_2 gas
   b. the volume, in liters, occupied by 2.50 mol of N_2 gas
   c. the volume, in liters, occupied by 50.0 g of Ar gas
   d. the number of grams of H_2 in 1620 mL of H_2 gas

11.54 Use the molar volume to calculate each of the following at STP:
   a. the number of moles of CO_2 in 4.00 L of CO_2 gas
   b. the volume, in liters, occupied by 0.420 mol of He gas
   c. the volume, in liters, occupied by 6.40 g of O_2 gas
   d. the number of grams of Ne contained in 11.2 L of Ne gas

11.7 The Ideal Gas Law

LEARNING GOAL Use the ideal gas law equation to calculate the unknown *P, V, T,* or *n* of a gas when given three of the four values in the ideal gas law equation. Calculate the molar mass of a gas.

The **ideal gas law** is the relationship between the four properties used in the measurement of a gas—pressure (P), volume (V), temperature (T), and amount of a gas (n).

Ideal Gas Law

$$PV = nRT$$

Rearranging the ideal gas law equation shows that the four gas properties equal the gas law constant, R.

$$\frac{PV}{nT} = R$$

To calculate the value of R, we substitute the STP conditions for molar volume into the expression: 1.00 mol of any gas occupies 22.4 L at STP (273 K and 1.00 atm).

$$R = \frac{(1.00 \text{ atm})(22.4 \text{ L})}{(1.00 \text{ mol})(273 \text{ K})} = \frac{0.0821 \text{ L} \cdot \text{atm}}{\text{mol} \cdot \text{K}}$$

The value for the **ideal gas constant, R**, is 0.0821 L · atm per mol · K. If we use 760. mmHg for the pressure, we obtain another useful value for R of 62.4 L · mmHg per mol · K.

$$R = \frac{(760. \text{ mmHg})(22.4 \text{ L})}{(1.00 \text{ mol})(273 \text{ K})} = \frac{62.4 \text{ L} \cdot \text{mmHg}}{\text{mol} \cdot \text{K}}$$

The ideal gas law is a useful expression when you are given the quantities for any three of the four properties of a gas. Although real gases show some deviations in behavior, the ideal gas law closely approximates the behavior of real gases at typical conditions. In working problems using the ideal gas law, the units of each variable must match the units in the R you select.

Ideal Gas Component	Units	
Ideal Gas Constant (*R*)	$\frac{0.0821 \text{ L} \cdot \text{atm}}{\text{mol} \cdot \text{K}}$	$\frac{62.4 \text{ L} \cdot \text{mmHg}}{\text{mol} \cdot \text{K}}$
Pressure (*P*)	atm	mmHg
Volume (*V*)	L	L
Amount (*n*)	mol	mol
Temperature (*T*)	K	K

CORE CHEMISTRY SKILL
Using the Ideal Gas Law

SAMPLE PROBLEM 11.9 Using the Ideal Gas Law

Dinitrogen oxide, N_2O, which is an anesthetic also called laughing gas, is used in dentistry. What is the pressure, in atmospheres, of 0.350 mol of N_2O at 22 °C in a 5.00-L container?

TRY IT FIRST

SOLUTION

STEP 1 State the given and needed quantities. When three of the four quantities (*P, V, n,* and *T*) are known, we use the ideal gas law equation to solve for the unknown quantity. It is helpful to organize the data in a table. The temperature is converted from degrees Celsius to kelvins so that the units of *V, n,* and *T* match the unit of the gas constant *R*.

Dinitrogen oxide is used as an anesthetic in dentistry.

Guide to Using the Ideal Gas Law

STEP 1
State the given and needed quantities.

STEP 2
Rearrange the ideal gas law equation to solve for the needed quantity.

STEP 3
Substitute the gas data into the equation and calculate the needed quantity.

	Given	Need	Connect
ANALYZE THE PROBLEM	$V = 5.00$ L $n = 0.350$ mol $T = 22\,°C + 273 = 295$ K	P	ideal gas law, $PV = nRT$ $R = \dfrac{0.0821\ \text{L} \cdot \text{atm}}{\text{mol} \cdot \text{K}}$

STEP 2 **Rearrange the ideal gas law equation to solve for the needed quantity.** By dividing both sides of the ideal gas law by V, we solve for pressure, P.

$PV = nRT$ Ideal gas law equation

$$\frac{P\cancel{V}}{\cancel{V}} = \frac{nRT}{V}$$

$$P = \frac{nRT}{V}$$

STEP 3 **Substitute the gas data into the equation and calculate the needed quantity.**

$$P = \frac{0.350\ \cancel{\text{mol}} \times \dfrac{0.0821\ \cancel{\text{L}} \cdot \text{atm}}{\cancel{\text{mol}} \cdot \cancel{\text{K}}} \times 295\ \cancel{\text{K}}}{5.00\ \cancel{\text{L}}} = 1.70\ \text{atm}$$

STUDY CHECK 11.9

Chlorine gas, Cl_2, is used to purify water. How many moles of chlorine gas are in a 7.00-L tank if the gas has a pressure of 865 mmHg and a temperature of 24 °C?

ANSWER

0.327 mol of Cl_2

Many times we need to know the amount of gas, in grams. Then the ideal gas law equation can be rearranged to solve for the amount (n) of gas, which is converted to mass in grams using its molar mass as shown in Sample Problem 11.10.

When camping, butane is used as a fuel for a portable stove.

SAMPLE PROBLEM 11.10 Calculating Mass Using the Ideal Gas Law

Butane, C_4H_{10}, is used as a fuel for camping stoves. If you have 108 mL of butane gas at 715 mmHg and 25 °C, what is the mass, in grams, of butane?

TRY IT FIRST

SOLUTION

STEP 1 **State the given and needed quantities.** When three of the quantities (P, V, and T) are known, we use the ideal gas law equation to solve for the quantity moles (n). The volume given in milliliters (mL) is converted to a volume in liters (L).

	Given	Need	Connect
ANALYZE THE PROBLEM	$P = 715$ mmHg $V = 108$ mL (0.108 L) $T = 25\,°C + 273 = 298$ K	n	ideal gas law, $PV = nRT$ $R = \dfrac{62.4\ \text{L} \cdot \text{mmHg}}{\text{mol} \cdot \text{K}}$ molar mass

STEP 2 Rearrange the ideal gas law equation to solve for the needed quantity.
By dividing both sides of the ideal gas law equation by RT, we solve for moles, n.

$$PV = \boxed{n} \; RT \quad \text{Ideal gas law equation}$$

$$\frac{PV}{RT} = \frac{n \; \cancel{RT}}{\cancel{RT}}$$

$$\boxed{n} = \frac{PV}{RT}$$

STEP 3 Substitute the gas data into the equation and calculate the needed quantity.

$$\boxed{n} = \frac{715 \; \cancel{\text{mmHg}} \times 0.108 \; \cancel{L}}{\dfrac{62.4 \; \cancel{L} \cdot \cancel{\text{mmHg}}}{\text{mol} \cdot \cancel{K}} \times 298 \; \cancel{K}} = 0.00415 \text{ mol } (4.15 \times 10^{-3} \text{ mol})$$

Now we can convert the moles of butane to grams using its molar mass of 58.12 g/mol:

$$0.00415 \; \cancel{\text{mol } C_4H_{10}} \times \frac{58.12 \text{ g } C_4H_{10}}{1 \; \cancel{\text{mol } C_4H_{10}}} = 0.241 \text{ g of } C_4H_{10}$$

STUDY CHECK 11.10

What is the volume, in liters, of 1.20 g of carbon monoxide at 8 °C if it has a pressure of 724 mmHg?

ANSWER

1.04 L

Molar Mass of a Gas

Another use of the ideal gas law is to determine the molar mass of a gas. If the mass, in grams, of the gas is known, we calculate the number of moles of the gas using the ideal gas law equation. Then the molar mass (g/mol) can be determined.

SAMPLE PROBLEM 11.11 Molar Mass of a Gas Using the Ideal Gas Law

What is the molar mass, in grams per mole, of a gas if a 3.16-g sample of gas at 0.750 atm and 45 °C occupies a volume of 2.05 L?

TRY IT FIRST

SOLUTION

STEP 1 State the given and needed quantities.

	Given	Need	Connect
ANALYZE THE PROBLEM	$P = 0.750$ atm $V = 2.05$ L $T = 45\,°C + 273 = 318$ K mass = 3.16 g	n molar mass	ideal gas law, $PV = nRT$ $R = \dfrac{0.0821 \text{ L} \cdot \text{atm}}{\text{mol} \cdot \text{K}}$

Guide to Calculating the Molar Mass of a Gas

STEP 1
State the given and needed quantities.

STEP 2
Rearrange the ideal gas law equation to solve for the number of moles.

STEP 3
Obtain the molar mass by dividing the given number of grams by the number of moles.

STEP 2 Rearrange the ideal gas law equation to solve for the number of moles. To solve for moles, n, divide both sides of the ideal gas law equation by RT.

$$PV = \boxed{n} \, RT \quad \text{Ideal gas law equation}$$

$$\frac{PV}{RT} = \frac{n \, \cancel{RT}}{\cancel{RT}}$$

$$\boxed{n} = \frac{PV}{RT}$$

$$\boxed{n} = \frac{0.750 \, \cancel{atm} \times 2.05 \, \cancel{L}}{\dfrac{0.0821 \, \cancel{L} \cdot \cancel{atm}}{\text{mol} \cdot \cancel{K}} \times 318 \, \cancel{K}} = 0.0589 \, \text{mol}$$

STEP 3 Obtain the molar mass by dividing the given number of grams by the number of moles.

$$\text{Molar mass} = \frac{\text{mass}}{\text{moles}} = \frac{3.16 \, \text{g}}{0.0589 \, \text{mol}} = 53.7 \, \text{g/mol}$$

STUDY CHECK 11.11

What is the molar mass, in grams per mole, of an unknown gas in a 1.50-L container if 0.488 g of the gas has a pressure of 0.0750 atm at 19 °C?

ANSWER

104 g/mol

QUESTIONS AND PROBLEMS

11.7 The Ideal Gas Law

LEARNING GOAL Use the ideal gas law equation to calculate the unknown *P*, *V*, *T*, or *n* of a gas when given three of the four values in the ideal gas law equation. Calculate the molar mass of a gas.

11.55 Calculate the pressure, in atmospheres, of 2.00 mol of helium gas in a 10.0-L container at 27 °C.

11.56 What is the volume, in liters, of 4.00 mol of methane gas, CH_4, at 18 °C and 1.40 atm?

11.57 An oxygen gas container has a volume of 20.0 L. How many grams of oxygen are in the container if the gas has a pressure of 845 mmHg at 22 °C?

11.58 A 10.0-g sample of krypton has a temperature of 25 °C at 575 mmHg. What is the volume, in milliliters, of the krypton gas?

11.59 A 25.0-g sample of nitrogen, N_2, has a volume of 50.0 L and a pressure of 630. mmHg. What is the temperature, in kelvins and degrees Celsius, of the gas?

11.60 A 0.226-g sample of carbon dioxide, CO_2, has a volume of 525 mL and a pressure of 455 mmHg. What is the temperature, in kelvins and degrees Celsius, of the gas?

11.61 Determine the molar mass of each of the following gases:
  a. 0.84 g of a gas that occupies 450 mL at 0 °C and 1.00 atm (STP)

  b. 1.28 g of a gas that occupies 1.00 L at 0 °C and 760 mmHg (STP)
  c. 1.48 g of a gas that occupies 1.00 L at 685 mmHg and 22 °C
  d. 2.96 g of a gas that occupies 2.30 L at 0.95 atm and 24 °C

11.62 Determine the molar mass of each of the following gases:
  a. 2.90 g of a gas that occupies 0.500 L at 0 °C and 1.00 atm (STP)
  b. 1.43 g of a gas that occupies 2.00 L at 0 °C and 760 mmHg (STP)
  c. 0.726 g of a gas that occupies 855 mL at 1.20 atm and 18 °C
  d. 2.32 g of a gas that occupies 1.23 L at 685 mmHg and 25 °C

Applications

11.63 A single-patient hyperbaric chamber has a volume of 640 L. At a temperature of 24 °C, how many grams of oxygen are needed to give a pressure of 1.6 atm?

11.64 A multipatient hyperbaric chamber has a volume of 3400 L. At a temperature of 22 °C, how many grams of oxygen are needed to give a pressure of 2.4 atm?

11.8 Gas Laws and Chemical Reactions

LEARNING GOAL Calculate the mass or volume of a gas that reacts or forms in a chemical reaction.

Gases are involved as reactants and products in many chemical reactions. For example, we have seen that the combustion of organic fuels with oxygen gas produces carbon dioxide gas and water vapor. In combination reactions, we have seen that hydrogen gas and nitrogen gas react to form ammonia gas, and hydrogen gas and oxygen gas produce water. Typically, the information given for a gas in a reaction is its pressure (P), volume (V), and temperature (T). Then we can use the ideal gas law equation to determine the moles of a gas in a reaction. If we are given the number of moles for one of the gases in a reaction, we can use a mole–mole factor to determine the moles of any other substance.

CORE CHEMISTRY SKILL
Calculating Mass or Volume of a Gas in a Chemical Reaction

SAMPLE PROBLEM 11.12 Gases in Chemical Reactions

Limestone ($CaCO_3$) reacts with HCl to produce carbon dioxide gas, water, and aqueous calcium chloride.

$$CaCO_3(s) + 2HCl(aq) \longrightarrow CO_2(g) + H_2O(l) + CaCl_2(aq)$$

How many liters of CO_2 are produced at 752 mmHg and 24 °C from a 25.0-g sample of limestone?

TRY IT FIRST

SOLUTION

STEP 1 State the given and needed quantities.

ANALYZE THE PROBLEM	Given	Need	Connect
	25.0 g of $CaCO_3$ $P = 752$ mmHg $T = 24\,°C + 273 = 297$ K	V of $CO_2(g)$	ideal gas law, $PV = nRT$ $R = \dfrac{62.4\ L \cdot mmHg}{mol \cdot K}$ molar mass
	Equation		
	$CaCO_3(s) + 2HCl(aq) \longrightarrow CO_2(g) + H_2O(l) + CaCl_2(aq)$		

STEP 2 Write a plan to convert the given quantity to the needed moles.

grams of $CaCO_3$ → Molar mass → moles of $CaCO_3$ → Mole–mole factor → moles of CO_2

STEP 3 Write the equalities and conversion factors for molar mass and mole–mole factors.

1 mol of $CaCO_3$ = 100.09 g of $CaCO_3$
$$\dfrac{100.09\ g\ CaCO_3}{1\ mol\ CaCO_3} \quad \text{and} \quad \dfrac{1\ mol\ CaCO_3}{100.09\ g\ CaCO_3}$$

1 mol of $CaCO_3$ = 1 mol of CO_2
$$\dfrac{1\ mol\ CaCO_3}{1\ mol\ CO_2} \quad \text{and} \quad \dfrac{1\ mol\ CO_2}{1\ mol\ CaCO_3}$$

STEP 4 Set up the problem to calculate moles of needed quantity.

$$25.0\ \cancel{g\ CaCO_3} \times \dfrac{1\ \cancel{mol\ CaCO_3}}{100.09\ \cancel{g\ CaCO_3}} \times \dfrac{1\ mol\ CO_2}{1\ \cancel{mol\ CaCO_3}} = 0.250\ mol\ of\ CO_2$$

Guide to Using the Ideal Gas Law for Reactions

STEP 1
State the given and needed quantities.

STEP 2
Write a plan to convert the given quantity to the needed moles.

STEP 3
Write the equalities and conversion factors for molar mass and mole–mole factors.

STEP 4
Set up the problem to calculate moles of needed quantity.

STEP 5
Convert the moles of needed quantity to mass or volume using the molar mass or the ideal gas law equation.

When aluminum reacts with HCl, bubbles of H_2 gas form.

STEP 5 Convert the moles of needed quantity to volume using the ideal gas law equation.

$$V = \frac{nRT}{P}$$

$$V = \frac{0.250 \text{ mol} \times \dfrac{62.4 \text{ L} \cdot \text{mmHg}}{\text{mol} \cdot \text{K}} \times 297 \text{ K}}{752 \text{ mmHg}} = 6.16 \text{ L of } CO_2$$

STUDY CHECK 11.12

If 12.8 g of aluminum reacts with HCl, how many liters of H_2 would be formed at 715 mmHg and 19 °C?

$$2Al(s) + 6HCl(aq) \longrightarrow 3H_2(g) + 2AlCl_3(aq)$$

ANSWER

18.1 L of H_2

QUESTIONS AND PROBLEMS

11.8 Gas Laws and Chemical Reactions

LEARNING GOAL Calculate the mass or volume of a gas that reacts or forms in a chemical reaction.

11.65 Mg metal reacts with HCl to produce hydrogen gas.

$$Mg(s) + 2HCl(aq) \longrightarrow H_2(g) + MgCl_2(aq)$$

a. What volume, in liters, of hydrogen at 0 °C and 1.00 atm (STP) is released when 8.25 g of Mg reacts?
b. How many grams of magnesium are needed to prepare 5.00 L of H_2 at 735 mmHg and 18 °C?

11.66 When heated to 350 °C at 0.950 atm, ammonium nitrate decomposes to produce nitrogen, water, and oxygen gases.

$$2NH_4NO_3(s) \xrightarrow{\Delta} 2N_2(g) + 4H_2O(g) + O_2(g)$$

a. How many liters of water vapor are produced when 25.8 g of NH_4NO_3 decomposes?
b. How many grams of NH_4NO_3 are needed to produce 10.0 L of oxygen?

11.67 Butane undergoes combustion when it reacts with oxygen to produce carbon dioxide and water. What volume, in liters, of oxygen is needed to react with 55.2 g of butane at 0.850 atm and 25 °C?

$$2C_4H_{10}(g) + 13O_2(g) \xrightarrow{\Delta} 8CO_2(g) + 10H_2O(g)$$

11.68 Potassium nitrate decomposes to potassium nitrite and oxygen. What volume, in liters, of O_2 can be produced from the decomposition of 50.0 g of KNO_3 at 35 °C and 1.19 atm?

$$2KNO_3(s) \longrightarrow 2KNO_2(s) + O_2(g)$$

11.69 Aluminum and oxygen react to form aluminum oxide. How many liters of oxygen at 0 °C and 760 mmHg (STP) are required to completely react with 5.4 g of aluminum?

$$4Al(s) + 3O_2(g) \longrightarrow 2Al_2O_3(s)$$

11.70 Nitrogen dioxide reacts with water to produce oxygen and ammonia. How many grams of NH_3 can be produced when 4.00 L of NO_2 reacts at 415 °C and 725 mmHg?

$$4NO_2(g) + 6H_2O(g) \longrightarrow 7O_2(g) + 4NH_3(g)$$

11.9 Partial Pressures (Dalton's Law)

LEARNING GOAL Use Dalton's law of partial pressures to calculate the total pressure of a mixture of gases.

Many gas samples are a mixture of gases. For example, the air you breathe is a mixture of mostly oxygen and nitrogen gases. In ideal gas mixtures, scientists observed that all gas particles behave in the same way. Therefore, the total pressure of the gases in a mixture is a result of the collisions of the gas particles regardless of what type of gas they are.

In a gas mixture, each gas exerts its **partial pressure**, which is the pressure it would exert if it were the only gas in the container. **Dalton's law** states that the total pressure of a gas mixture is the sum of the partial pressures of the gases in the mixture.

Dalton's Law

$$P_{total} = P_1 + P_2 + P_3 + \cdots$$

Total pressure of = Sum of the partial pressures
a gas mixture   of the gases in the mixture

Suppose we have two separate tanks, one filled with helium at a pressure of 2.0 atm and the other filled with argon at a pressure of 4.0 atm. When the gases are combined in a single tank with the same volume and temperature, the number of gas molecules, not the type of gas, determines the pressure in the container. There the pressure of the gas mixture would be 6.0 atm, which is the sum of their individual or partial pressures.

$$
\begin{aligned}
P_{total} &= P_{He} + P_{Ar} \\
&= 2.0 \text{ atm} + 4.0 \text{ atm} \\
&= 6.0 \text{ atm}
\end{aligned}
$$

$P_{He} = 2.0$ atm $P_{Ar} = 4.0$ atm

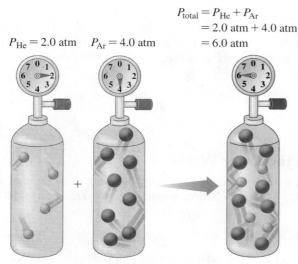

The total pressure of two gases is the sum of their partial pressures.

TABLE 11.7 Typical Composition of Air

Gas	Partial Pressure (mmHg)	Percentage (%)
Nitrogen, N_2	594	78.2
Oxygen, O_2	160.	21.0
Carbon dioxide, CO_2 Argon, Ar Water vapor, H_2O	6	0.8
Total air	760.	100

Air Is a Gas Mixture

The air you breathe is a mixture of gases. What we call the *atmospheric pressure* is actually the sum of the partial pressures of the gases in the air. **TABLE 11.7** lists partial pressures for the gases in air on a typical day.

SAMPLE PROBLEM 11.13 Partial Pressure of a Gas in a Mixture

A heliox breathing mixture of oxygen and helium is prepared for a patient with *chronic obstructive pulmonary disease* (COPD). The gas mixture has a total pressure of 7.00 atm. If the partial pressure of the oxygen in the tank is 1140 mmHg, what is the partial pressure, in atmospheres, of the helium in the breathing mixture?

TRY IT FIRST

SOLUTION

ANALYZE THE PROBLEM	Given	Need	Connect
	$P_{total} = 7.00$ atm $P_{O_2} = 1140$ mmHg	partial pressure of He	Dalton's law

STEP 1 Write the equation for the sum of the partial pressures.

$$P_{total} = P_{O_2} + P_{He} \quad \text{Dalton's law}$$

STEP 2 Rearrange the equation to solve for the unknown pressure. To solve for the partial pressure of helium (P_{He}), we rearrange the equation to give the following:

$$P_{He} = P_{total} - P_{O_2}$$

Convert units to match.

$$P_{O_2} = 1140 \ \cancel{\text{mmHg}} \times \frac{1 \text{ atm}}{760 \ \cancel{\text{mmHg}}} = 1.50 \text{ atm}$$

STEP 3 Substitute known pressures into the equation and calculate the unknown pressure.

$$P_{He} = P_{total} - P_{O_2}$$

$$P_{He} = 7.00 \text{ atm} - 1.50 \text{ atm} = 5.50 \text{ atm}$$

STUDY CHECK 11.13

An anesthetic consists of a mixture of cyclopropane gas, C_3H_6, and oxygen gas, O_2. If the mixture has a total pressure of 1.09 atm, and the partial pressure of the cyclopropane is 73 mmHg, what is the partial pressure, in millimeters of mercury, of the oxygen in the anesthetic?

ANSWER

755 mmHg

Gases Collected Over Water

In the laboratory, gases are often collected by bubbling them through water into a container (see **FIGURE 11.9**). In a reaction, magnesium (Mg) reacts with HCl to form H_2 gas and $MgCl_2$.

$$Mg(s) + 2HCl(aq) \longrightarrow H_2(g) + MgCl_2(aq)$$

As hydrogen is produced during the reaction, it displaces some of the water in the container. Because of the vapor pressure of water, the gas that is collected is a mixture of hydrogen and water vapor. For our calculation, we need the pressure of the dry hydrogen gas. We use the vapor pressure of water (see Table 11.4) at the experimental temperature, and subtract it from the total gas pressure. Then we can use the ideal gas law to determine the moles or grams of the hydrogen gas that were collected.

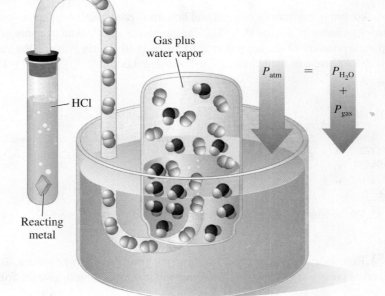

FIGURE 11.9 ▶ A gas from a reaction is collected by bubbling through water. Due to evaporation of water, the total pressure is equal to the partial pressure of the gas and the vapor pressure of water.

🅞 How is the pressure of the dry gas determined?

SAMPLE PROBLEM 11.14 Moles of Gas Collected Over Water

When magnesium reacts with HCl, a volume of 0.355 L of hydrogen gas is collected over water at 26 °C. The vapor pressure of water at 26 °C is 25 mmHg.

$$Mg(s) + 2HCl(aq) \longrightarrow H_2(g) + MgCl_2(aq)$$

If the total pressure is 752 mmHg, how many moles of $H_2(g)$ were collected?

TRY IT FIRST

SOLUTION

	Given	Need	Connect
ANALYZE THE PROBLEM	$V = 0.355$ L of H_2, $P = 752$ mmHg $T = 26\,°C + 273 = 299$ K $P_{H_2O} = 25$ mmHg	moles of H_2	ideal gas law, $PV = nRT$ $R = \dfrac{62.4 \text{ L} \cdot \text{mmHg}}{\text{mol} \cdot \text{K}}$

STEP 1 Obtain the vapor pressure of water. The vapor pressure of water at 26 °C is 25 mmHg.

STEP 2 Subtract the vapor pressure from the total gas pressure to give the partial pressure of the needed gas. Using Dalton's law of partial pressures, determine the partial pressure of H_2.

$$P_{total} = P_{H_2} + P_{H_2O}$$

Solving for the partial pressure of H_2 gives

$$P_{H_2} = P_{total} - P_{H_2O}$$

$$P_{H_2} = 752 \text{ mmHg} - 25 \text{ mmHg}$$

$$= 727 \text{ mmHg}$$

STEP 3 Use the ideal gas law to convert P_{gas} to moles of gas collected. By dividing both sides of the ideal gas law equation by RT, we solve for moles, n, of gas.

$$PV = n\,RT \quad \text{Ideal gas law equation}$$

$$\frac{PV}{RT} = \frac{n\,RT}{RT}$$

$$n = \frac{PV}{RT}$$

Calculate the moles of H_2 gas by placing the partial pressure of H_2 (727 mmHg), volume of gas container (0.355 L), temperature (26 °C + 273 = 299 K), and R, using mmHg, into the ideal gas law equation.

$$n = \frac{727 \text{ mmHg} \times 0.355 \text{ L}}{\dfrac{62.4 \text{ L} \cdot \text{mmHg}}{\text{mol} \cdot \text{K}} \times 299 \text{ K}} = 0.0138 \text{ mol of } H_2 \ (1.38 \times 10^{-2} \text{ mol of } H_2)$$

STUDY CHECK 11.14

A 456-mL sample of oxygen gas (O_2) is collected over water at a pressure of 744 mmHg and a temperature of 20. °C. How many grams of dry O_2 are collected?

ANSWER

0.579 g of O_2

Guide to Calculating Partial Pressure of Gases Collected Over Water

STEP 1
Obtain the vapor pressure of water.

STEP 2
Subtract the vapor pressure from the total gas pressure to give the partial pressure of the needed gas.

STEP 3
Use the ideal gas law to convert P_{gas} to moles or grams of gas collected.

CHEMISTRY LINK TO **HEALTH**
Hyperbaric Chambers

A burn patient may undergo treatment for burns and infections in a hyperbaric chamber, a device in which pressures can be obtained that are two to three times greater than atmospheric pressure. A greater oxygen pressure increases the level of dissolved oxygen in the blood and tissues, where it fights bacterial infections. High levels of oxygen are toxic to many strains of bacteria. The hyperbaric chamber may also be used during surgery, to help counteract carbon monoxide (CO) poisoning, and to treat some cancers.

The blood is normally capable of dissolving up to 95% of the oxygen. Thus, if the partial pressure of the oxygen in the hyperbaric chamber is 2280 mmHg (3 atm), about 2170 mmHg of oxygen can dissolve in the blood, saturating the tissues. In the treatment for carbon monoxide poisoning, oxygen at high pressure is used to displace the CO from the hemoglobin faster than breathing pure oxygen at 1 atm.

A patient undergoing treatment in a hyperbaric chamber must also undergo decompression (reduction of pressure) at a rate that slowly reduces the concentration of dissolved oxygen in the blood. If decompression is too rapid, the oxygen dissolved in the blood may form gas bubbles in the circulatory system.

Similarly, if a scuba diver does not decompress slowly, a condition called the "bends" may occur. While below the surface of the ocean, a diver uses a breathing mixture with higher pressures. If there is nitrogen in the mixture, higher quantities of nitrogen gas will dissolve in the blood. If the diver ascends to the surface too quickly, the dissolved nitrogen forms gas bubbles that can block a blood vessel and cut off the flow of blood in the joints and tissues of the body and be quite painful. A diver suffering from the bends is placed immediately into a hyperbaric chamber where pressure is first increased and then slowly decreased. The dissolved nitrogen can then diffuse through the lungs until atmospheric pressure is reached.

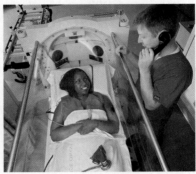

A hyperbaric chamber is used in the treatment of certain diseases.

QUESTIONS AND PROBLEMS

11.9 Partial Pressures (Dalton's Law)

LEARNING GOAL Use Dalton's law of partial pressures to calculate the total pressure of a mixture of gases.

11.71 In a gas mixture, the partial pressures are nitrogen 425 Torr, oxygen 115 Torr, and helium 225 Torr. What is the total pressure, in torr, exerted by the gas mixture?

11.72 In a gas mixture, the partial pressures are argon 415 mmHg, neon 75 mmHg, and nitrogen 125 mmHg. What is the total pressure, in millimeters of mercury, exerted by the gas mixture?

11.73 An air sample in the lungs contains oxygen at 93 mmHg, nitrogen at 565 mmHg, carbon dioxide at 38 mmHg, and water vapor at 47 mmHg. What is the total pressure, in atmospheres, exerted by the gas mixture?

11.74 A nitrox II gas mixture for scuba diving contains oxygen gas at 53 atm and nitrogen gas at 94 atm. What is the total pressure, in torr, of the scuba gas mixture?

11.75 A gas mixture containing oxygen, nitrogen, and helium exerts a total pressure of 925 Torr. If the partial pressures are oxygen 425 Torr and helium 75 Torr, what is the partial pressure, in torr, of the nitrogen in the mixture?

11.76 A gas mixture containing oxygen, nitrogen, and neon exerts a total pressure of 1.20 atm. If helium added to the mixture increases the pressure to 1.50 atm, what is the partial pressure, in atmospheres, of the helium?

11.77 When solid $KClO_3$ is heated, it decomposes to give solid KCl and O_2 gas. A volume of 256 mL of gas is collected over water at a total pressure of 765 mmHg and 24 °C. The vapor pressure of water at 24 °C is 22 mmHg.

$$2KClO_3(s) \xrightarrow{\Delta} 2KCl(s) + 3O_2(g)$$

a. What was the partial pressure of the O_2 gas?
b. How many moles of O_2 gas were produced in the reaction?

11.78 When Zn reacts with HCl solution, the products are H_2 gas and $ZnCl_2$. A volume of 425 mL of H_2 gas is collected over water at a total pressure of 758 mmHg and 16 °C. The vapor pressure of water at 16 °C is 14 mmHg.

$$Zn(s) + HCl(aq) \longrightarrow H_2(g) + ZnCl_2(aq)$$

a. What was the partial pressure of the H_2 gas?
b. How many moles of H_2 gas were produced in the reaction?

Follow Up

EXERCISE-INDUCED ASTHMA

Vigorous exercise with high levels of physical activity can induce asthma particularly in children. When Whitney had her asthma attack, her breathing became more rapid, the temperature within her airways increased, and the muscles around the bronchi contracted, causing a narrowing of the airways.  Whitney's symptoms, which may occur within 5 to 20 min after the start of vigorous exercise, include shortness of breath, wheezing, and coughing.

Whitney now does several things to prevent exercise-induced asthma. She uses a pre-exercise inhaled medication before she starts her activity. The medication relaxes the muscles that surround the airways and opens up the airways. Then she does a warm-up set of exercises. If pollen counts are high, she avoids exercising outdoors.

Applications

11.79 Whitney's lung capacity was measured as 3.2 L at a body temperature of 37 °C and a pressure of 745 mmHg. How many moles of oxygen are in her lungs if air contains 21% oxygen?

11.80 Whitney's tidal volume, which is the volume of air that she inhales and exhales, was 0.54 L. Her tidal volume was measured at a body temperature of 37 °C and a pressure of 745 mmHg. How many moles of nitrogen does she inhale in one breath if air contains 78% nitrogen?

CONCEPT MAP

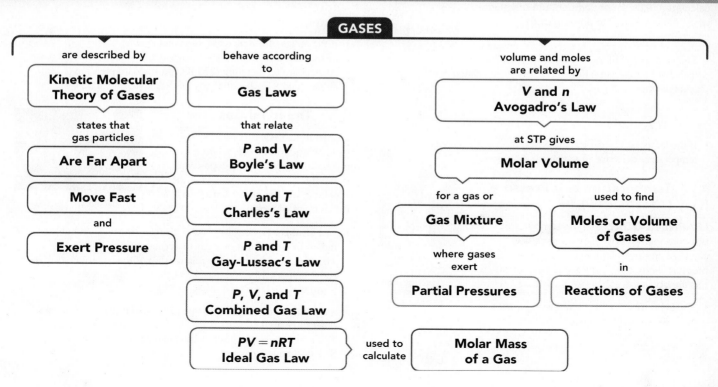

GASES

are described by
Kinetic Molecular Theory of Gases

states that gas particles
Are Far Apart
Move Fast
and
Exert Pressure

behave according to
Gas Laws

that relate
P and V Boyle's Law
V and T Charles's Law
P and T Gay-Lussac's Law
P, V, and T Combined Gas Law
PV = nRT Ideal Gas Law

used to calculate
Molar Mass of a Gas

volume and moles are related by
V and n Avogadro's Law

at STP gives
Molar Volume

for a gas or
Gas Mixture

where gases exert
Partial Pressures

used to find
Moles or Volume of Gases

in
Reactions of Gases

CHAPTER REVIEW

11.1 Properties of Gases

LEARNING GOAL Describe the kinetic molecular theory of gases and the units of measurement used for gases.

- In a gas, particles are so far apart and moving so fast that their attractions are negligible.
- A gas is described by the physical properties of pressure (P), volume (V), temperature (T), and amount in moles (n).
- A gas exerts pressure, the force of the gas particles striking the surface of a container.
- Gas pressure is measured in units such as torr, mmHg, atm, and Pa.

11.2 Pressure and Volume (Boyle's Law)

Piston

$V_1 = 4\,L$ $V_2 = 2\,L$
$P_1 = 1\,atm$ $P_2 = 2\,atm$

LEARNING GOAL Use the pressure–volume relationship (Boyle's law) to calculate the unknown pressure or volume when the temperature and amount of gas are constant.

- The volume (V) of a gas changes inversely with the pressure (P) of the gas if there is no change in the temperature and the amount of gas.

$$P_1V_1 = P_2V_2$$

- The pressure of a gas increases if its volume decreases; its pressure decreases if the volume increases.

11.3 Temperature and Volume (Charles's Law)

$T_1 = 200\,K$ $T_2 = 400\,K$
$V_1 = 1\,L$ $V_2 = 2\,L$

LEARNING GOAL Use the temperature–volume relationship (Charles's law) to calculate the unknown temperature or volume when the pressure and amount of gas are constant.

- The volume (V) of a gas is directly related to its Kelvin temperature (T) when there is no change in the pressure and the amount of gas.

$$\frac{V_1}{T_1} = \frac{V_2}{T_2}$$

- If the temperature of a gas increases, its volume increases; if its temperature decreases, the volume decreases.

11.4 Temperature and Pressure (Gay-Lussac's Law)

$T_1 = 200\,K$ $T_2 = 400\,K$
$P_1 = 1\,atm$ $P_2 = 2\,atm$

LEARNING GOAL Use the temperature–pressure relationship (Gay-Lussac's law) to calculate the unknown temperature or pressure when the volume and amount of gas are constant.

- The pressure (P) of a gas is directly related to its Kelvin temperature (T).

$$\frac{P_1}{T_1} = \frac{P_2}{T_2}$$

- An increase in temperature increases the pressure of a gas, and a decrease in temperature decreases the pressure, if there is no change in the volume and the amount of gas.
- Vapor pressure is the pressure of the gas that forms when a liquid evaporates.
- At the boiling point of a liquid, the vapor pressure equals the external pressure.

11.5 The Combined Gas Law

LEARNING GOAL Use the combined gas law to calculate the unknown pressure, volume, or temperature of a gas when changes in two of these properties are given and the amount of gas is constant.

- The combined gas law is the relationship of pressure (P), volume (V), and temperature (T) for a constant amount of gas.

$$\frac{P_1V_1}{T_1} = \frac{P_2V_2}{T_2}$$

- The combined gas law is used to determine the effect of changes in two of the variables on the third.

11.6 Volume and Moles (Avogadro's Law)

$V = 22.4\,L$

1 mol of O_2
32.00 g of O_2
273 K
1 atm

LEARNING GOAL Use Avogadro's law to calculate the unknown amount or volume of a gas when the pressure and temperature are constant.

- The volume (V) of a gas is directly related to the number of moles (n) of the gas when the pressure and temperature of the gas do not change.

$$\frac{V_1}{n_1} = \frac{V_2}{n_2}$$

- If the moles of gas increase, the volume must increase; if the moles of gas decrease, the volume must decrease.
- At standard temperature (273 K) and standard pressure (1 atm), abbreviated STP, 1 mol of any gas has a volume of 22.4 L.

11.7 The Ideal Gas Law

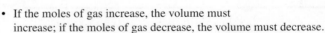

LEARNING GOAL Use the ideal gas law equation to calculate the unknown P, V, T, or n of a gas when given three of the four values in the ideal gas law equation. Calculate the molar mass of a gas.

- The ideal gas law can be used to calculate the mass, number of moles, or the volume of a gas in a reaction.
- The ideal gas law gives the relationship of the quantities P, V, n, and T that describe and measure a gas.

$$PV = nRT$$

- Any of the four variables can be calculated if the values of the other three are known.
- The molar mass of a gas can be calculated using molar volume at STP or the ideal gas law.

11.8 Gas Laws and Chemical Reactions

LEARNING GOAL Calculate the mass or volume of a gas that reacts or forms in a chemical reaction.

- The ideal gas law equation is used to convert the quantities (P, V, and T) of gases to moles in a chemical reaction.
- The moles of gases can be used to determine the number of moles or grams of other substances in the reaction.

11.9 Partial Pressures (Dalton's Law)

$P_{total} = P_{He} + P_{Ar}$
$= 2.0\ atm + 4.0\ atm$
$= 6.0\ atm$

LEARNING GOAL Use Dalton's law of partial pressures to calculate the total pressure of a mixture of gases.

- In a mixture of two or more gases, the total pressure is the sum of the partial pressures of the individual gases.

$$P_{total} = P_1 + P_2 + P_3 + \cdots$$

- The partial pressure of a gas in a mixture is the pressure it would exert if it were the only gas in the container.
- For gases collected over water, the vapor pressure of water is subtracted from the total pressure of the gas mixture to obtain the partial pressure of the dry gas.

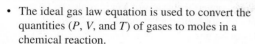

KEY TERMS

atmosphere (atm) A unit equal to the pressure exerted by a column of mercury 760 mm high.

atmospheric pressure The pressure exerted by the atmosphere.

Avogadro's law A gas law stating that the volume of a gas is directly related to the number of moles of gas when pressure and temperature do not change.

Boyle's law A gas law stating that the pressure of a gas is inversely related to the volume when temperature and moles of the gas do not change.

Charles's law A gas law stating that the volume of a gas is directly related to the Kelvin temperature when pressure and moles of the gas do not change.

combined gas law A relationship that combines several gas laws relating pressure, volume, and temperature when the amount of gas does not change.

$$\frac{P_1 V_1}{T_1} = \frac{P_2 V_2}{T_2}$$

Dalton's law A gas law stating that the total pressure exerted by a mixture of gases in a container is the sum of the partial pressures that each gas would exert alone.

direct relationship A relationship in which two properties increase or decrease together.

Gay-Lussac's law A gas law stating that the pressure of a gas is directly related to the Kelvin temperature when the number of moles of a gas and its volume do not change.

ideal gas constant, R A numerical value that relates the quantities P, V, n, and T in the ideal gas law, $PV = nRT$.

ideal gas law A law that combines the four measured properties of a gas: $PV = nRT$.

inverse relationship A relationship in which two properties change in opposite directions.

kinetic molecular theory of gases A model used to explain the behavior of gases.

molar volume A volume of 22.4 L occupied by 1 mol of a gas at STP conditions of 0 °C (273 K) and 1 atm.

partial pressure The pressure exerted by a single gas in a gas mixture.

pressure The force exerted by gas particles that hit the walls of a container.

STP Standard conditions of exactly 0 °C (273 K) temperature and 1 atm pressure used for the comparison of gases.

torr A unit of pressure equal to 1 mmHg; 760 Torr = 1 atm.

vapor pressure The pressure exerted by the particles of vapor above a liquid.

CORE CHEMISTRY SKILLS

The chapter section containing each Core Chemistry Skill is shown in parentheses at the end of each heading.

Using the Gas Laws (11.2)

- Boyle's, Charles's, Gay-Lussac's, and Avogadro's laws show the relationships between two properties of a gas.

$P_1 V_1 = P_2 V_2$ Boyle's law

$\dfrac{V_1}{T_1} = \dfrac{V_2}{T_2}$ Charles's law

$\dfrac{P_1}{T_1} = \dfrac{P_2}{T_2}$ Gay-Lussac's law

$\dfrac{V_1}{n_1} = \dfrac{V_2}{n_2}$ Avogadro's law

- The combined gas law shows the relationship between P, V, and T for a gas.

$$\frac{P_1 V_1}{T_1} = \frac{P_2 V_2}{T_2}$$

- When two properties of a gas vary and the other two are constant, we list the initial and final conditions of each property in a table.

Example: A sample of helium gas (He) has a volume of 6.8 L and a pressure of 2.5 atm. What is the final volume, in liters, if it has a final pressure of 1.2 atm with no change in temperature and amount of gas?

Answer:

	Given	Need	Connect
ANALYZE THE PROBLEM	$P_1 = 2.5$ atm $P_2 = 1.2$ atm $V_1 = 6.8$ L	V_2	Boyle's law, $P_1V_1 = P_2V_2$
	Factors that remain constant: *T* and *n*		**Predict:** *P* decreases, *V* increases

Using Boyle's law, we can write the relationship for V_2, which we predict will increase.

$$V_2 = V_1 \times \frac{P_1}{P_2}$$

$$V_2 = 6.8 \text{ L} \times \frac{2.5 \text{ atm}}{1.2 \text{ atm}} = 14 \text{ L}$$

Using the Ideal Gas Law (11.7)

- The ideal gas law equation combines the relationships of the four properties of a gas into one equation.

  $$PV = nRT$$

- When three of the four properties are given, we rearrange the ideal gas law equation for the needed quantity.

Example: What is the volume, in liters, of 0.750 mol of CO_2 at a pressure of 1340 mmHg and a temperature of 295 K?

Answer: $V = \dfrac{nRT}{P}$

$$= \frac{0.750 \text{ mol} \times \dfrac{62.4 \text{ L} \cdot \text{mmHg}}{\text{mol} \cdot \text{K}} \times 295 \text{ K}}{1340 \text{ mmHg}} = 10.3 \text{ L}$$

Calculating Mass or Volume of a Gas in a Chemical Reaction (11.8)

- The ideal gas law equation is used to calculate the volume or mass of a gas in a chemical reaction.

Example: What is the volume, in liters, of N_2 required to react with 18.5 g of magnesium at a pressure of 1.20 atm and a temperature of 303 K?

$$3Mg(s) + N_2(g) \longrightarrow Mg_3N_2(s)$$

Answer: Initially, we convert the grams of Mg to moles and use a mole–mole factor from the balanced equation to calculate the moles of N_2 gas.

$$18.5 \text{ g Mg} \times \frac{1 \text{ mol Mg}}{24.31 \text{ g Mg}} \times \frac{1 \text{ mol N}_2}{3 \text{ mol Mg}} = 0.254 \text{ mol of N}_2$$

Now, we use the ideal gas law equation and solve for liters, the needed quantity.

$$V = \frac{nRT}{P}$$

$$= \frac{0.254 \text{ mol N}_2 \times \dfrac{0.0821 \text{ L} \cdot \text{atm}}{\text{mol} \cdot \text{K}} \times 303 \text{ K}}{1.20 \text{ atm}} = 5.27 \text{ L}$$

Calculating Partial Pressure (11.9)

- In a gas mixture, each gas exerts its partial pressure, which is the pressure it would exert if it were the only gas in the container.
- Dalton's law states that the total pressure of a gas mixture is the sum of the partial pressures of the gases in the mixture.

  $$P_{total} = P_1 + P_2 + P_3 + \cdots$$

Example: A gas mixture with a total pressure of 1.18 atm contains helium gas at a partial pressure of 465 mmHg and nitrogen gas. What is the partial pressure, in atmospheres, of the nitrogen gas?

Answer: Initially, we convert the partial pressure of helium gas from mmHg to atm.

$$465 \text{ mmHg} \times \frac{1 \text{ atm}}{760 \text{ mmHg}} = 0.612 \text{ atm of He gas}$$

Using Dalton's law, we solve for the needed quantity, P_{N_2}.

$$P_{total} = P_{N_2} + P_{He}$$

$$P_{N_2} = P_{total} - P_{He}$$

$$P_{N_2} = 1.18 \text{ atm} - 0.612 \text{ atm} = 0.57 \text{ atm}$$

UNDERSTANDING THE CONCEPTS

The chapter sections to review are shown in parentheses at the end of each question.

11.81 Two flasks of equal volume and at the same temperature contain different gases. One flask contains 10.0 g of Ne, and the other flask contains 10.0 g of He. Is each of the following statements *true* or *false*? Explain. (11.1)
  a. The flask that contains He has a higher pressure than the flask that contains Ne.
  b. The densities of the gases are the same.

11.82 Two flasks of equal volume and at the same temperature contain different gases. One flask contains 5.0 g of O_2, and the other flask contains 5.0 g of H_2. Is each of the following statements *true* or *false*? Explain. (11.1)
  a. Both flasks contain the same number of molecules.
  b. The pressures in the flasks are the same.

11.83 At 100 °C, which of the following diagrams (**1**, **2**, or **3**) represents a gas sample that exerts the: (11.6)
  a. lowest pressure?   **b.** highest pressure?

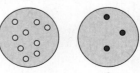

    1   2   3

11.84 Indicate which diagram (**1**, **2**, or **3**) represents the volume of the gas sample in a flexible container when each of the following changes (**a** to **e**) takes place: (11.2, 11.3)

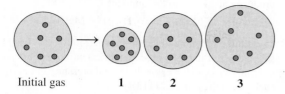

Initial gas   **1**   **2**   **3**

  a. Temperature increases at constant pressure.
  b. Temperature decreases at constant pressure.
  c. Atmospheric pressure decreases at constant temperature.
  d. Doubling the atmospheric pressure and doubling the Kelvin temperature.

11.85 A balloon is filled with helium gas with a partial pressure of 1.00 atm and neon gas with a partial pressure of 0.50 atm. For each of the following changes (**a** to **e**) of the initial balloon, select the diagram (**A**, **B**, or **C**) that shows the final volume of the balloon: (11.2, 11.3, 11.6)

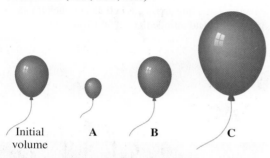

Initial   **A**   **B**   **C**
volume

  a. The balloon is put in a cold storage unit (*P* and *n* constant).
  b. The balloon floats to a higher altitude where the pressure is less (*T* and *n* constant).
  c. All of the helium gas is removed (*T* and *P* constant).
  d. The Kelvin temperature doubles, and half of the gas atoms leak out (*P* constant).
  e. 2.0 mol of O_2 gas is added (*T* and *P* constant).

11.86 Indicate if pressure *increases*, *decreases*, or *stays the same* in each of the following: (11.2, 11.4, 11.6)

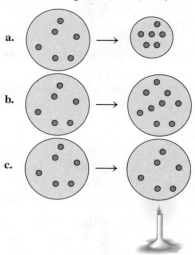

  a.
  b.
  c.

11.87 An airplane is pressurized to 650. mmHg. (11.9)
  a. If air is 21% oxygen, what is the partial pressure of oxygen on the plane?
  b. If the partial pressure of oxygen drops below 100. mmHg, passengers become drowsy. If this happens, oxygen masks are released. What is the total cabin pressure at which oxygen masks are dropped?

11.88 At a restaurant, a customer chokes on a piece of food. You put your arms around the person's waist and use your fists to push up on the person's abdomen, an action called the Heimlich maneuver. (11.2)
  a. How would this action change the volume of the chest and lungs?
  b. Why does it cause the person to expel the food item from the airway?

ADDITIONAL QUESTIONS AND PROBLEMS

11.89 In 1783, Jacques Charles launched his first balloon filled with hydrogen gas, which he chose because it was lighter than air. The balloon had a volume of 31 000 L when it was filled at a pressure of 755 mmHg and a temperature of 22 °C. When the balloon reached an altitude of 1000 m, the pressure was 658 mmHg and the temperature was −8 °C. What is the volume of the balloon at these conditions, if the amount of hydrogen remains the same? (11.5, 11.6)

Jacques Charles used hydrogen to launch his balloon in 1783.

11.90 Your spaceship has docked at a space station above Mars. The temperature inside the space station is a carefully controlled 24 °C at a pressure of 745 mmHg. A balloon with a volume of 425 mL drifts into the airlock where the temperature is −95 °C and the pressure is 0.115 atm. What is the final volume, in milliliters, of the balloon if *n* does not change and the balloon is very elastic? (11.5)

11.91 A gas sample has a volume of 4250 mL at 15 °C and 745 mmHg. What is the final temperature, in degrees Celsius, after the sample is transferred to a different container with a volume of 2.50 L and a pressure of 1.20 atm, when *n* does not change? (11.5)

11.92 A weather balloon has a volume of 750 L when filled with helium at 8 °C at a pressure of 380 Torr. What is the final volume, in liters, of the balloon when the pressure is 0.20 atm, the temperature is −45 °C, and *n* does not change? (11.5)

11.93 A 2.00-L container is filled with methane gas, CH_4, at a pressure of 2500. mmHg and a temperature of 18 °C. How many grams of methane are in the container? (11.7)

11.94 A steel cylinder with a volume of 15.0 L is filled with 50.0 g of nitrogen gas at 25 °C. What is the pressure, in atmospheres, of the N_2 gas in the cylinder? (11.7)

11.95 A sample of gas with a mass of 1.62 g occupies a volume of 941 mL at a pressure of 748 Torr and a temperature of 20.0 °C. What is the molar mass of the gas? (11.7)

11.96 What is the molar mass of a gas if 1.15 g of the gas has a volume of 8 mL at 0 °C and 1.00 atm (STP)? (11.7)

11.97 How many grams of CO_2 are in 35.0 L of $CO_2(g)$ at 1.20 atm and 5 °C? (11.7)

11.98 A container is filled with 0.644 g of O_2 at 5 °C and 845 mmHg. What is the volume, in milliliters, of the container? (11.7)

11.99 How many liters of H_2 gas can be produced at 0 °C and 1.00 atm (STP) from 25.0 g of Zn? (11.7, 11.8)

$$Zn(s) + 2HCl(aq) \longrightarrow H_2(g) + ZnCl_2(aq)$$

11.100 In the formation of smog, nitrogen and oxygen gas react to form nitrogen dioxide. How many grams of NO_2 will be produced when 2.0 L of nitrogen at 840 mmHg and 24 °C are completely reacted? (11.7, 11.8)

$$N_2(g) + 2O_2(g) \longrightarrow 2NO_2(g)$$

11.101 Nitrogen dioxide reacts with water to produce oxygen and ammonia. A 5.00-L sample of $H_2O(g)$ reacts at a temperature of 375 °C and a pressure of 725 mmHg. How many grams of NH_3 can be produced? (11.7, 11.8)

$$4NO_2(g) + 6H_2O(g) \longrightarrow 7O_2(g) + 4NH_3(g)$$

11.102 Hydrogen gas can be produced in the laboratory through the reaction of magnesium metal with hydrochloric acid. When 12.0 g of Mg reacts, what volume, in liters, of H_2 gas is produced at 24 °C and 835 mmHg? (11.7, 11.8)

$$Mg(s) + 2HCl(aq) \longrightarrow H_2(g) + MgCl_2(aq)$$

11.103 A gas mixture with a total pressure of 2400 Torr is used by a scuba diver. If the mixture contains 2.0 mol of helium and 6.0 mol of oxygen, what is the partial pressure, in torr, of each gas in the sample? (11.9)

11.104 A gas mixture with a total pressure of 4.6 atm is used in a hospital. If the mixture contains 5.4 mol of nitrogen and 1.4 mol of oxygen, what is the partial pressure, in atmospheres, of each gas in the sample? (11.9)

11.105 A gas mixture contains oxygen and argon at partial pressures of 0.60 atm and 425 mmHg. If nitrogen gas added to the sample increases the total pressure to 1250 Torr, what is the partial pressure, in torr, of the nitrogen added? (11.9)

11.106 What is the total pressure, in millimeters of mercury, of a gas mixture containing argon gas at 0.25 atm, helium gas at 350 mmHg, and nitrogen gas at 360 Torr? (11.9)

CHALLENGE QUESTIONS

The following groups of questions are related to the topics in this chapter. However, they do not all follow the chapter order, and they require you to combine concepts and skills from several sections. These questions will help you increase your critical thinking skills and prepare for your next exam.

11.107 Solid aluminum reacts with aqueous H_2SO_4 to form H_2 gas and aluminum sulfate. When a sample of Al is allowed to react, 415 mL of gas is collected over water at 23 °C, at a pressure of 755 mmHg. At 23 °C, the vapor pressure of water is 21 mmHg. (11.7, 11.8, 11.9)

$$2Al(s) + 3H_2SO_4(aq) \longrightarrow 3H_2(g) + Al_2(SO_4)_3(aq)$$

a. What is the pressure, in millimeters of mercury, of the dry H_2 gas?
b. How many moles of H_2 were produced?
c. How many grams of Al were reacted?

11.108 When heated, $KClO_3$ forms KCl and O_2. When a sample of $KClO_3$ is heated, 226 mL of gas with a pressure of 744 mmHg is collected over water at 26 °C. At 26 °C, the vapor pressure of water is 25 mmHg. (11.7, 11.8, 11.9)

$$2KClO_3(s) \xrightarrow{\Delta} 2KCl(s) + 3O_2(g)$$

a. What is the pressure, in millimeters of mercury, of the dry O_2 gas?
b. How many moles of O_2 were produced?
c. How many grams of $KClO_3$ were reacted?

11.109 A sample of gas with a mass of 1.020 g occupies a volume of 762 mL at 0 °C and 1.00 atm (STP). (7.4, 7.5, 11.6, 11.7)

a. What is the molar mass of the gas?
b. If the unknown gas is composed of 0.815 g of carbon and the rest is hydrogen, what is its molecular formula?

11.110 A sample of an unknown gas with a mass of 3.24 g occupies a volume of 1.88 L at a pressure of 748 mmHg and a temperature of 20. °C. (7.4, 7.5, 11.6, 11.7)

a. What is the molar mass of the gas?
b. If the unknown gas is composed of 2.78 g of carbon and the rest is hydrogen, what is its molecular formula?

11.111 The propane, C_3H_8, in a fuel cylinder, undergoes combustion with oxygen in the air. How many liters of CO_2 are produced at STP if the cylinder contains 881 g of propane? (11.6, 11.7)

$$C_3H_8(g) + 5O_2(g) \xrightarrow{\Delta} 3CO_2(g) + 4H_2O(g)$$

11.112 When sensors in a car detect a collision, they cause the reaction of sodium azide, NaN_3, which generates nitrogen gas to fill the airbags within 0.03 s. How many liters of N_2 are produced at STP if the airbag contains 132 g of NaN_3? (11.6, 11.7)

$$2NaN_3(s) \longrightarrow 2Na(s) + 3N_2(g)$$

11.113 Glucose, $C_6H_{12}O_6$, is metabolized in living systems. How many grams of water can be produced from the reaction of 18.0 g of glucose and 7.50 L of O_2 at 1.00 atm and 37 °C? (9.2, 9.3, 11.7, 11.8)

$$C_6H_{12}O_6(s) + 6O_2(g) \longrightarrow 6CO_2(g) + 6H_2O(l)$$

11.114 2.00 L of N_2, at 25 °C and 1.08 atm, is mixed with 4.00 L of O_2, at 25 °C and 0.118 atm, and the mixture is allowed to react. How much NO, in grams, is produced? (9.2, 9.3, 11.7, 11.8)

$$N_2(g) + O_2(g) \longrightarrow 2NO(g)$$

ANSWERS

Answers to Selected Questions and Problems

11.1 a. At a higher temperature, gas particles have greater kinetic energy, which makes them move faster.
  b. Because there are great distances between the particles of a gas, they can be pushed closer together and still remain a gas.
  c. Gas particles are very far apart, which means that the mass of a gas in a certain volume is very small, resulting in a low density.

11.3 a. temperature **b.** volume
  c. amount **d.** pressure

11.5 Statements **a**, **d**, and **e** describe the pressure of a gas.

11.7 a. 1520 Torr **b.** 29.4 lb/in.^{2}
  c. 1520 mmHg **d.** 203 kPa

11.9 As a diver ascends to the surface, external pressure decreases. If the air in the lungs were not exhaled, its volume would expand and severely damage the lungs. The pressure in the lungs must adjust to changes in the external pressure.

11.11 a. The pressure is greater in cylinder **A**. According to Boyle's law, a decrease in volume pushes the gas particles closer together, which will cause an increase in the pressure.
  b. 160 mL

11.13 a. increases **b.** decreases
  c. increases

11.15 a. 328 mmHg **b.** 2620 mmHg
  c. 475 mmHg **d.** 5240 mmHg

11.17 a. 52.4 L **b.** 25 L
  c. 100. L **d.** 45 L

11.19 6.52 atm

11.21 25 L of cyclopropane

11.23 a. inspiration **b.** expiration
  c. inspiration

11.25 a. C **b.** A
  c. B

11.27 a. 303 °C **b.** −129 °C
  c. 591 °C **d.** 136 °C

11.29 a. 2400 mL **b.** 4900 mL
  c. 1800 mL **d.** 1700 mL

11.31 121 °C

11.33 a. 770 mmHg **b.** 1150 mmHg

11.35 a. −23 °C **b.** 168 °C

11.37 31 °C

11.39 a. On top of a mountain, water boils below 100 °C because the atmospheric (external) pressure is less than 1 atm.
  b. Because the pressure inside a pressure cooker is greater than 1 atm, water boils above 100 °C. At a higher temperature, food cooks faster.

11.41 1.9 atm

11.43 $T_2 = T_1 \times \dfrac{P_2}{P_1} \times \dfrac{V_2}{V_1}$

11.45 a. 4.26 atm **b.** 3.07 atm
  c. 0.606 atm

11.47 −33 °C

11.49 The volume increases because the number of gas particles is increased.

11.51 a. 4.00 L **b.** 26.7 L
  c. 14.6 L

11.53 a. 2.00 mol of O_2 **b.** 56.0 L
  c. 28.0 L **d.** 0.146 g of H_2

11.55 4.93 atm

11.57 29.4 g of O_2

11.59 566 K (293 °C)

11.61 a. 42 g/mol **b.** 28.7 g/mol
  c. 39.8 g/mol **d.** 33 g/mol

11.63 1300 g of O_2

11.65 a. 7.60 L of H_2 **b.** 4.92 g of Mg

11.67 178 L of O_2

11.69 3.4 L of O_2

11.71 765 Torr

11.73 0.978 atm

11.75 425 Torr

11.77 a. 743 mmHg **b.** 0.0103 mol of O_2

11.79 0.025 mol of O_2

11.81 a. True. The flask containing helium has more moles of helium and thus more helium atoms.
  b. True. The mass and volume of each are the same, which means the mass/volume ratio or density is the same in both flasks.

11.83 a. 2 Fewest number of gas particles exerts the lowest pressure.
  b. 1 Greatest number of gas particles exerts the highest pressure.

11.85 a. A: Volume decreases when temperature decreases.
  b. C: Volume increases when pressure decreases.
  c. A: Volume decreases when the moles of gas decrease.
  d. B: Doubling the temperature, in kelvins, would double the volume, but when half of the gas escapes, the volume would decrease by half. These two opposing effects cancel each other, and there is no change in the volume.
  e. C: Increasing the moles increases the volume to keep T and P constant.

11.87 a. 140 mmHg **b.** 480 mmHg

11.89 32 000 L

11.91 −66 °C

11.93 4.41 g of CH_4

11.95 42.1 g/mol

11.97 81.0 g of CO_2

11.99 8.56 L of H_2

11.101 1.02 g of NH_3

11.103 He 600 Torr, O_2 1800 Torr

11.105 370 Torr

11.107 a. 734 mmHg **b.** 0.0165 mol of H_2
  c. 0.297 g of Al

11.109 a. 30.0 g/mol **b.** C_2H_6

11.111 1340 L of CO_2

11.113 5.31 g of water

Solutions

OUR KIDNEYS PRODUCE

urine, which carries waste products and excess fluid from the body. They also reabsorb electrolytes such as potassium and produce hormones that regulate blood pressure and the levels of calcium in the blood. Diseases such as diabetes and high blood pressure can cause a decrease in kidney function. Symptoms of kidney malfunction include protein in the urine, an abnormal level of urea nitrogen in the blood, frequent urination, and swollen feet. If kidney failure occurs, it may be treated with dialysis or transplantation.

Michelle has been suffering from kidney disease because of severe strep throat she contracted as a child. When her kidneys stopped functioning, Michelle was placed on dialysis three times a week. As she enters the dialysis unit, her dialysis nurse, Amanda, asks Michelle how she is feeling. Michelle indicates that she feels tired today and has considerable swelling around her ankles. Amanda informs her that these side effects occur because of her body's inability to regulate the amount of water in her cells. Amanda explains that the amount of water is regulated by the concentration of electrolytes in her body fluids and the rate at which waste products are removed from her body. Amanda explains that although water is essential for the many chemical reactions that occur in the body, the amount of water can become too high or too low because of various diseases and conditions. Because Michelle's kidneys no longer perform dialysis, she cannot regulate the amount of electrolytes or waste in her body fluids. As a result, she has an electrolyte imbalance and a buildup of waste products, so her body is retaining water. Amanda then explains that the dialysis machine does the work of her kidneys to reduce the high levels of electrolytes and waste products.

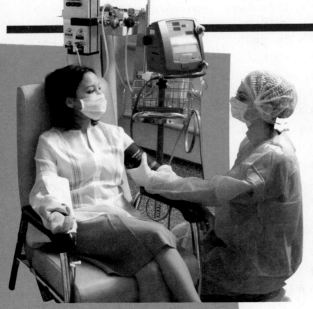

CAREER

Dialysis Nurse

A dialysis nurse specializes in assisting patients with kidney disease undergoing dialysis. This requires monitoring the patient before, during, and after dialysis for any complications such as a drop in blood pressure or cramping. The dialysis nurse connects the patient to the dialysis unit via a dialysis catheter that is inserted into the neck or chest, which must be kept clean to prevent infection. A dialysis nurse must have considerable knowledge about how the dialysis machine functions to ensure that it is operating correctly at all times.

KEY MATH SKILLS

- Calculating Percentages (1.4)
- Solving Equations (1.4)
- Interpreting Graphs (1.4)

CORE CHEMISTRY SKILLS

- Writing Conversion Factors from Equalities (2.5)
- Using Conversion Factors (2.6)
- Writing Positive and Negative Ions (6.1)
- Using Molar Mass as a Conversion Factor (7.3)

- Using Mole–Mole Factors (9.2)
- Converting Grams to Grams (9.3)
- Identifying Intermolecular Forces (10.6)

*These Key Math Skills and Core Chemistry Skills from previous chapters are listed here for your review as you proceed to the new material in this chapter.

LOOKING AHEAD

12.1 Solutions

LEARNING GOAL Identify the solute and solvent in a solution; describe the formation of a solution.

Solutions are everywhere around us. Most of the gases, liquids, and solids we see are mixtures of at least one substance dissolved in another. There are different types of solutions. The air we breathe is a solution that is primarily oxygen and nitrogen gases. Carbon dioxide gas dissolved in water makes carbonated drinks. When we make solutions of coffee or tea, we use hot water to dissolve substances from coffee beans or tea leaves. The ocean is also a solution, consisting of many ionic compounds such as sodium chloride dissolved in water. In your medicine cabinet, the antiseptic tincture of iodine is a solution of iodine dissolved in ethanol.

A **solution** is a homogeneous mixture in which one substance, called the **solute**, is uniformly dispersed in another substance called the **solvent**. Because the solute and the solvent do not react with each other, they can be mixed in varying proportions. A solution of a little salt dissolved in water tastes slightly salty. When a large amount of salt is dissolved in water, the solution tastes very salty. Usually, the solute (in this case, salt) is the substance present in the lesser amount, whereas the solvent (in this case, water) is present in the greater amount. For example, in a solution composed of 5.0 g of salt and 50. g of water, salt is the solute and water is the solvent. In a solution, the particles of the solute are evenly dispersed among the molecules within the solvent (see **FIGURE 12.1**).

Solute: The substance present in lesser amount

Salt

Water

Solvent: The substance present in greater amount

A solution has at least one solute dispersed in a solvent.

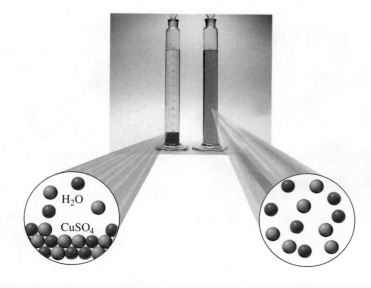

FIGURE 12.1 ▶ A solution of copper(II) sulfate ($CuSO_4$) forms as particles of solute dissolve, move away from the crystal, and become evenly dispersed among the solvent (water) molecules.

Q What does the uniform blue color in the graduated cylinder on the right indicate?

Types of Solutes and Solvents

Solutes and solvents may be solids, liquids, or gases. The solution that forms has the same physical state as the solvent. When sugar crystals are dissolved in water, the resulting sugar solution is liquid. Sugar is the solute, and water is the solvent. Soda water and soft drinks are prepared by dissolving carbon dioxide gas in water. The carbon dioxide gas is the solute, and water is the solvent. **TABLE 12.1** lists some solutes and solvents and their solutions.

TABLE 12.1   Some Examples of Solutions

Type	Example	Primary Solute	Solvent
Gas Solutions			
Gas in a gas	Air	$O_2(g)$	$N_2(g)$
Liquid Solutions			
Gas in a liquid	Soda water	$CO_2(g)$	$H_2O(l)$
	Household ammonia	$NH_3(g)$	$H_2O(l)$
Liquid in a liquid	Vinegar	$HC_2H_3O_2(l)$	$H_2O(l)$
Solid in a liquid	Seawater	$NaCl(s)$	$H_2O(l)$
	Tincture of iodine	$I_2(s)$	$C_2H_5OH(l)$
Solid Solutions			
Solid in a solid	Brass	$Zn(s)$	$Cu(s)$
	Steel	$C(s)$	$Fe(s)$

Water as a Solvent

Water is one of the most common solvents in nature. In the H_2O molecule, an oxygen atom shares electrons with two hydrogen atoms. Because oxygen is much more electronegative than hydrogen, the O—H bonds are polar. In each polar bond, the oxygen atom has a partial negative (δ^-) charge and the hydrogen atom has a partial positive (δ^+) charge. Because the shape of a water molecule is bent, not linear, its dipoles do not cancel out. Thus, water is polar, and is a *polar solvent*.

Attractive forces known as *hydrogen bonds* occur between molecules where partially positive hydrogen atoms are attracted to the partially negative atoms N, O, or F. As seen in the diagram, the hydrogen bonds are shown as a series of dots. Although hydrogen bonds are much weaker than covalent or ionic bonds, there are many of them linking water molecules together. Hydrogen bonds are important in the properties of biological compounds such as proteins, carbohydrates, and DNA.

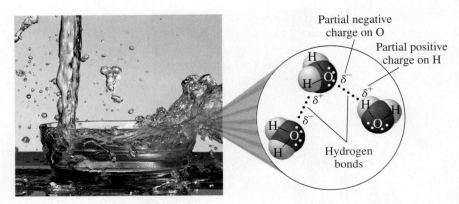

In water, hydrogen bonds form between an oxygen atom in one water molecule and the hydrogen atom in another.

CHEMISTRY LINK TO **HEALTH**
Water in the Body

The average adult is about 60% water by mass, and the average infant about 75%. About 60% of the body's water is contained within the cells as intracellular fluids; the other 40% makes up extracellular fluids, which include the interstitial fluid in tissue and the plasma in the blood. These external fluids carry nutrients and waste materials between the cells and the circulatory system.

Every day you lose between 1500 and 3000 mL of water from the kidneys as urine, from the skin as perspiration, from the lungs as you exhale, and from the gastrointestinal tract. Serious dehydration can occur in an adult if there is a 10% net loss in total body fluid; a 20% loss of fluid can be fatal. An infant suffers severe dehydration with only a 5 to 10% loss in body fluid.

Water loss is continually replaced by the liquids and foods in the diet and from metabolic processes that produce water in the cells of the body. **TABLE 12.2** lists the percentage by mass of water contained in some foods.

Typical water gain and loss during 24 hours

Water Gain		Water Loss	
Liquid	1000 mL	Urine	1500 mL
Food	1200 mL	Perspiration	300 mL
Metabolism	300 mL	Breath	600 mL
		Feces	100 mL
Total	2500 mL	Total	2500 mL

The water lost from the body is replaced by the intake of fluids.

TABLE 12.2 Percentage of Water in Some Foods

Food	Water (% by mass)	Food	Water (% by mass)
Vegetables		**Meats/Fish**	
Carrot	88	Chicken, cooked	71
Celery	94	Hamburger, broiled	60
Cucumber	96	Salmon	71
Tomato	94		
Fruits		**Milk Products**	
Apple	85	Cottage cheese	78
Cantaloupe	91	Milk, whole	87
Orange	86	Yogurt	88
Strawberry	90		
Watermelon	93		

Formation of Solutions

The interactions between solute and solvent will determine whether a solution will form. Initially, energy is needed to separate the particles in the solute and the solvent particles. Then energy is released as solute particles move between the solvent particles to form a solution. However, there must be attractions between the solute and the solvent particles to provide the energy for the initial separation. These attractions occur when the solute and the solvent have similar polarities. The expression "like dissolves like" is a way of saying that the polarities of a solute and a solvent must be similar in order for a solution to form (see **FIGURE 12.2**). In the absence of attractions between a solute and a solvent, there is insufficient energy to form a solution (see **TABLE 12.3**).

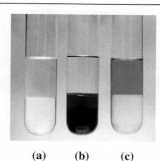

(a)   (b)   (c)

FIGURE 12.2 ▶ Like dissolves like. In each test tube, the lower layer is CH_2Cl_2 (more dense), and the upper layer is water (less dense). **(a)** CH_2Cl_2 is nonpolar and water is polar; the two layers do not mix. **(b)** The nonpolar solute I_2 (purple) is soluble in the nonpolar solvent CH_2Cl_2. **(c)** The ionic solute $Ni(NO_3)_2$ (green) is soluble in the polar solvent water.

Q In which layer would polar molecules of sucrose ($C_{12}H_{22}O_{11}$) be soluble?

TABLE 12.3 Possible Combinations of Solutes and Solvents

Solutions Will Form		Solutions Will Not Form	
Solute	Solvent	Solute	Solvent
Polar	Polar	Polar	Nonpolar
Nonpolar	Nonpolar	Nonpolar	Polar

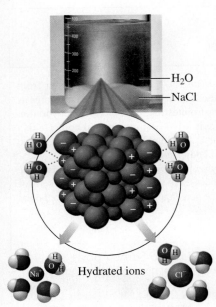

FIGURE 12.3 ▶ Ions on the surface of a crystal of NaCl dissolve in water as they are attracted to the polar water molecules that pull the ions into solution and surround them.

Q What helps keep Na^+ and Cl^- ions in solution?

ENGAGE

Why does KCl(s) form a solution with water, but nonpolar hexane (C_6H_{14}) does not form a solution with water?

Solutions with Ionic and Polar Solutes

In ionic solutes such as sodium chloride, NaCl, there are strong ionic bonds between positively charged Na^+ ions and negatively charged Cl^- ions. In water, a polar solvent, the hydrogen bonds provide strong solvent–solvent attractions. When NaCl crystals are placed in water, partially negative oxygen atoms in water molecules attract positive Na^+ ions, and the partially positive hydrogen atoms in other water molecules attract negative Cl^- ions (see **FIGURE 12.3**). As soon as the Na^+ ions and the Cl^- ions form a solution, they undergo **hydration** as water molecules surround each ion. Hydration of the ions diminishes their attraction to other ions and keeps them in solution.

In the equation for the formation of the NaCl solution, the solid and aqueous NaCl are shown with the formula H_2O over the arrow, which indicates that water is needed for the dissociation process but is not a reactant.

$$NaCl(s) \xrightarrow[\text{Dissociation}]{H_2O} Na^+(aq) + Cl^-(aq)$$

In another example, we find that a polar molecular compound such as methanol, $CH_3{-}OH$, is soluble in water because methanol has a polar $-OH$ group that forms hydrogen bonds with water (see **FIGURE 12.4**). Polar solutes require polar solvents for a solution to form.

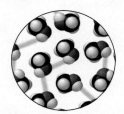 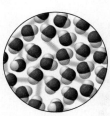 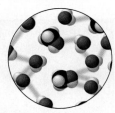

Methanol ($CH_3{-}OH$) solute   Water solvent   Methanol–water solution with hydrogen bonding

FIGURE 12.4 ▶ Polar molecules of methanol, $CH_3{-}OH$, form hydrogen bonds with polar water molecules to form a methanol–water solution.

Q Why are there attractions between the solute methanol and the solvent water?

Solutions with Nonpolar Solutes

Compounds containing nonpolar molecules, such as iodine (I_2), oil, or grease, do not dissolve in water because there are no attractions between the particles of a nonpolar solute and the polar solvent. Nonpolar solutes require nonpolar solvents for a solution to form.

QUESTIONS AND PROBLEMS

12.1 Solutions

LEARNING GOAL Identify the solute and solvent in a solution; describe the formation of a solution.

12.1 Identify the solute and the solvent in each solution composed of the following:
   a. 10.0 g of NaCl and 100.0 g of H_2O
   b. 50.0 mL of ethanol, C_2H_5OH, and 10.0 mL of H_2O
   c. 0.20 L of O_2 and 0.80 L of N_2

12.2 Identify the solute and the solvent in each solution composed of the following:
   a. 10.0 mL of acetic acid and 200. mL of water
   b. 100.0 mL of water and 5.0 g of sugar
   c. 1.0 g of Br_2 and 50.0 mL of methylene chloride(*l*)

12.3 Describe the formation of an aqueous KI solution, when solid KI dissolves in water.

12.4 Describe the formation of an aqueous LiBr solution, when solid LiBr dissolves in water.

Applications

12.5 Water is a polar solvent and carbon tetrachloride (CCl_4) is a nonpolar solvent. In which solvent is each of the following, which is found or used in the body, more likely to be soluble?
   a. $CaCO_3$ (calcium supplement), ionic
   b. retinol (vitamin A), nonpolar
   c. sucrose (table sugar), polar
   d. cholesterol (lipid), nonpolar

12.6 Water is a polar solvent and hexane (C_6H_{14}) is a nonpolar solvent. In which solvent is each of the following, which is found or used in the body, more likely to be soluble?
   a. vegetable oil, nonpolar
   b. oleic acid (lipid), nonpolar
   c. niacin (vitamin B_3), polar
   d. $FeSO_4$ (iron supplement), ionic

12.2 Electrolytes and Nonelectrolytes

LEARNING GOAL Identify solutes as electrolytes or nonelectrolytes.

Solutes can be classified by their ability to conduct an electrical current. When **electrolytes** dissolve in water, the process of *dissociation* separates them into ions forming solutions that conduct electricity. When **nonelectrolytes** dissolve in water, they do not separate into ions and their solutions do not conduct electricity.

To test solutions for the presence of ions, we can use an apparatus that consists of a battery and a pair of electrodes connected by wires to a light bulb. The light bulb glows when electricity can flow, which can only happen when electrolytes provide ions that move between the electrodes to complete the circuit.

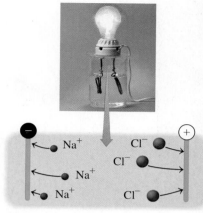

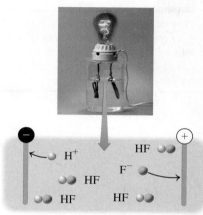

(a) Strong electrolyte

Types of Electrolytes

Electrolytes can be further classified as *strong electrolytes* or *weak electrolytes*. For a **strong electrolyte**, such as sodium chloride (NaCl), there is 100% dissociation of the solute into ions. When the electrodes from the light bulb apparatus are placed in the NaCl solution, the light bulb glows very brightly.

In an equation for dissociation of a compound in water, the charges must balance. For example, magnesium nitrate dissociates to give one magnesium ion for every two nitrate ions. However, only the ionic bonds between Mg^{2+} and NO_3^- are broken, not the covalent bonds within the polyatomic ion. The equation for the dissociation of $Mg(NO_3)_2$ is written as follows:

$$Mg(NO_3)_2(s) \xrightarrow[\text{Dissociation}]{H_2O} Mg^{2+}(aq) + 2NO_3^-(aq)$$

A **weak electrolyte** is a compound that dissolves in water mostly as molecules. Only a few of the dissolved solute molecules undergo *dissociation*, producing a small number of ions in solution. Thus, solutions of weak electrolytes do not conduct electrical current as well as solutions of strong electrolytes. When the electrodes are placed in a solution of a weak electrolyte, the glow of the light bulb is very dim. In an aqueous solution of the weak electrolyte HF, a few HF molecules dissociate to produce H^+ and F^- ions. As more H^+ and F^- ions form, some recombine to give HF molecules. These forward and reverse reactions of molecules to ions and back again are indicated by two arrows between reactant and products that point in opposite directions:

$$HF(aq) \underset{\text{Recombination}}{\overset{\text{Dissociation}}{\rightleftharpoons}} H^+(aq) + F^-(aq)$$

A nonelectrolyte such as methanol (CH_3OH) dissolves in water only as molecules, which do not ionize. When electrodes of the light bulb apparatus are placed in a solution of a nonelectrolyte, the light bulb does not glow, because the solution does not contain ions and cannot conduct electricity.

$$CH_3OH(l) \xrightarrow{H_2O} CH_3OH(aq)$$

TABLE 12.4 summarizes the classification of solutes in aqueous solutions.

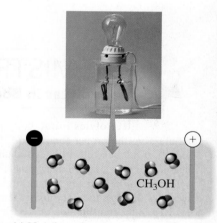

(b) Weak electrolyte

(c) Nonelectrolyte

TABLE 12.4 Classification of Solutes in Aqueous Solutions

Type of Solute	In Solution	Type(s) of Particles in Solution	Conducts Electricity?	Examples
Strong electrolyte	Dissociates completely	Only ions	Yes	Ionic compounds such as NaCl, KBr, $MgCl_2$, $NaNO_3$; bases such as NaOH, KOH; acids such as HCl, HBr, HI, HNO_3, $HClO_4$, H_2SO_4
Weak electrolyte	Ionizes partially	Mostly molecules and a few ions	Weakly	HF, H_2O, NH_3, $HC_2H_3O_2$ (acetic acid)
Nonelectrolyte	No ionization	Only molecules	No	Carbon compounds such as CH_3OH (methanol), C_2H_5OH (ethanol), $C_{12}H_{22}O_{11}$ (sucrose), CH_4N_2O (urea)

SAMPLE PROBLEM 12.1 Solutions of Electrolytes and Nonelectrolytes

Indicate whether solutions of each of the following contain only ions, only molecules, or mostly molecules and a few ions. Write the equation for the formation of a solution for each of the following:

a. $Na_2SO_4(s)$, a strong electrolyte
b. sucrose, $C_{12}H_{22}O_{11}(s)$, a nonelectrolyte
c. acetic acid, $HC_2H_3O_2(l)$, a weak electrolyte

TRY IT FIRST

SOLUTION

a. An aqueous solution of $Na_2SO_4(s)$ contains only the ions Na^+ and SO_4^{2-}.

$$Na_2SO_4(s) \xrightarrow{H_2O} 2Na^+(aq) + SO_4^{2-}(aq)$$

b. A nonelectrolyte such as sucrose, $C_{12}H_{22}O_{11}(s)$, produces only molecules when it dissolves in water.

$$C_{12}H_{22}O_{11}(s) \xrightarrow{H_2O} C_{12}H_{22}O_{11}(aq)$$

c. A weak electrolyte such as $HC_2H_3O_2(l)$ produces mostly molecules and a few ions when it dissolves in water.

$$HC_2H_3O_2(l) \xrightleftharpoons{H_2O} H^+(aq) + C_2H_3O_2^-(aq)$$

STUDY CHECK 12.1

Boric acid, $H_3BO_3(s)$, is a weak electrolyte. Would you expect a boric acid solution to contain only ions, only molecules, or mostly molecules and a few ions?

ANSWER

A solution of a weak electrolyte would contain mostly molecules and a few ions.

ENGAGE

Why does a solution of $LiNO_3$, a strong electrolyte, contain only ions whereas a solution of urea, CH_4N_2O, a nonelectrolyte, contains only molecules?

CHEMISTRY LINK TO **HEALTH**
Electrolytes in Body Fluids

Electrolytes in the body play an important role in maintaining the proper function of the cells and organs in the body. Typically, the electrolytes sodium, potassium, chloride, and bicarbonate are measured in a blood test. Sodium ions regulate the water content in the body and are important in carrying electrical impulses through the nervous system. Potassium ions are also involved in the transmission of electrical impulses and play a role in the maintenance of a regular heartbeat. Chloride ions control the balance of fluids in the body. Bicarbonate is important in maintaining the proper pH of the blood. There is a charge balance because the total number of positive charges is equal to the total number of negative charges. Sometimes when vomiting, diarrhea, or sweating is excessive, the concentrations of certain electrolytes may decrease. Then fluids such as Pedialyte may be given to return electrolyte levels to normal.

An intravenous solution is used to replace electrolytes in the body.

QUESTIONS AND PROBLEMS

12.2 Electrolytes and Nonelectrolytes

LEARNING GOAL Identify solutes as electrolytes or nonelectrolytes.

12.7 KF is a strong electrolyte, and HF is a weak electrolyte. How is the solution of KF different from that of HF?

12.8 NaOH is a strong electrolyte, and CH_3OH is a nonelectrolyte. How is the solution of NaOH different from that of CH_3OH?

12.9 Write a balanced equation for the dissociation of each of the following strong electrolytes in water:
  a. KCl   **b.** $CaCl_2$
  c. K_3PO_4 **d.** $Fe(NO_3)_3$

12.10 Write a balanced equation for the dissociation of each of the following strong electrolytes in water:
  a. LiBr   **b.** $NaNO_3$
  c. $CuCl_2$ **d.** K_2CO_3

12.11 Indicate whether aqueous solutions of each of the following solutes contain only ions, only molecules, or mostly molecules and a few ions:
  a. acetic acid, $HC_2H_3O_2$, a weak electrolyte
  b. NaBr, a strong electrolyte
  c. fructose, $C_6H_{12}O_6$, a nonelectrolyte

12.12 Indicate whether aqueous solutions of each of the following solutes contain only ions, only molecules, or mostly molecules and a few ions:
  a. NH_4Cl, a strong electrolyte
  b. ethanol, C_2H_5OH, a nonelectrolyte
  c. HCN, hydrocyanic acid, a weak electrolyte

12.13 Classify the solute represented in each of the following equations as a strong, weak, or nonelectrolyte:

  a. $K_2SO_4(s) \xrightarrow{H_2O} 2K^+(aq) + SO_4^{2-}(aq)$

  b. $NH_3(g) + H_2O(l) \rightleftarrows NH_4^+(aq) + OH^-(aq)$

  c. $C_6H_{12}O_6(s) \xrightarrow{H_2O} C_6H_{12}O_6(aq)$

12.14 Classify the solute represented in each of the following equations as a strong, weak, or nonelectrolyte:

  a. $CH_3OH(l) \xrightarrow{H_2O} CH_3OH(aq)$

  b. $MgCl_2(s) \xrightarrow{H_2O} Mg^{2+}(aq) + 2Cl^-(aq)$

  c. $HClO(aq) \rightleftarrows H^+(aq) + ClO^-(aq)$

12.3 Solubility

LEARNING GOAL Define solubility; distinguish between an unsaturated and a saturated solution. Identify an ionic compound as soluble or insoluble.

The term *solubility* is used to describe the amount of a solute that can dissolve in a given amount of solvent. Many factors, such as the type of solute, the type of solvent, and the temperature, affect the solubility of a solute. **Solubility**, usually expressed in grams of solute in 100. g of solvent, is the maximum amount of solute that can be dissolved at a certain temperature. If a solute readily dissolves when added to the solvent, the solution does not contain the maximum amount of solute. We call this solution an **unsaturated solution**.

A solution that contains all the solute that can dissolve is a **saturated solution**. When a solution is saturated, the rate at which the solute dissolves becomes equal to the rate at which solid forms, a process known as *recrystallization*. Then there is no further change in the amount of dissolved solute in solution.

$$\text{Solute} + \text{solvent} \underset{\text{Solute recrystallizes}}{\overset{\text{Solute dissolves}}{\rightleftarrows}} \text{saturated solution}$$

We can prepare a saturated solution by adding an amount of solute greater than that needed to reach solubility. Stirring the solution will dissolve the maximum amount of solute and leave the excess on the bottom of the container. Once we have a saturated solution, the addition of more solute will only increase the amount of undissolved solute.

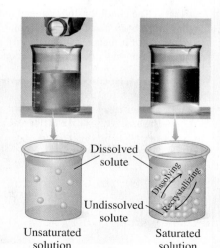

Additional solute can dissolve in an unsaturated solution, but not in a saturated solution.

SAMPLE PROBLEM 12.2 Saturated Solutions

At 20 °C, the solubility of KCl is 34 g/100. g of H_2O. In the laboratory, a student mixes 75 g of KCl with 200. g of H_2O at a temperature of 20 °C.

a. How much of the KCl will dissolve?
b. Is the solution saturated or unsaturated?
c. What is the mass, in grams, of any solid KCl left undissolved on the bottom of the container?

TRY IT FIRST

SOLUTION

a. At 20 °C, KCl has a solubility of 34 g of KCl in 100. g of water. Using the solubility as a conversion factor, we can calculate the maximum amount of KCl that can dissolve in 200. g of water as follows:

$$200. \text{ g } H_2O \times \frac{34 \text{ g KCl}}{100. \text{ g } H_2O} = 68 \text{ g of KCl}$$

b. Because 75 g of KCl exceeds the maximum amount (68 g) that can dissolve in 200. g of water, the KCl solution is saturated.

c. If we add 75 g of KCl to 200. g of water and only 68 g of KCl can dissolve, there is 7 g (75 g − 68 g) of solid (undissolved) KCl on the bottom of the container.

STUDY CHECK 12.2

At 40 °C, the solubility of KNO_3 is 65 g/100. g of H_2O. How many grams of KNO_3 will dissolve in 120 g of H_2O at 40 °C?

ANSWER

78 g of KNO_3

CHEMISTRY LINK TO **HEALTH**
Gout and Kidney Stones: A Problem of Saturation in Body Fluids

The conditions of gout and kidney stones involve compounds in the body that exceed their solubility levels and form solid products. Gout affects adults, primarily men, over the age of 40. Attacks of gout may occur when the concentration of uric acid in blood plasma exceeds its solubility, which is 7 mg/100 mL of plasma at 37 °C. Insoluble deposits of needle-like crystals of uric acid can form in the cartilage, tendons, and soft tissues where they cause painful gout attacks. They may also form in the tissues of the kidneys, where they can cause renal damage. High levels of uric acid in the body can be caused by an increase in uric acid production, failure of the kidneys to remove uric acid, or a diet with an overabundance of foods containing purines, which are metabolized to uric acid in the body. Foods in the diet that contribute to high levels of uric acid include certain meats, sardines, mushrooms, asparagus, and beans. Drinking alcoholic beverages may also significantly increase uric acid levels and bring about gout attacks.

Treatment for gout involves diet changes and drugs. Medications, such as probenecid, which helps the kidneys eliminate uric acid, or allopurinol, which blocks the production of uric acid by the body may be useful.

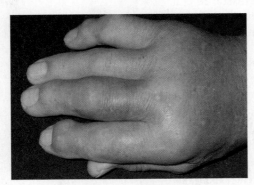

Gout occurs when uric acid exceeds its solubility.

Kidney stones are solid materials that form in the urinary tract. Most kidney stones are composed of calcium phosphate and calcium oxalate, although they can be solid uric acid. Insufficient water intake and high levels of calcium, oxalate, and phosphate in the urine can lead to the formation of kidney stones. When a kidney stone passes through the urinary tract, it causes considerable pain and discomfort, necessitating the use of painkillers and surgery. Sometimes ultrasound is used to break up kidney stones. Persons prone to kidney stones are advised to drink six to eight glasses of water every day to prevent saturation levels of minerals in the urine.

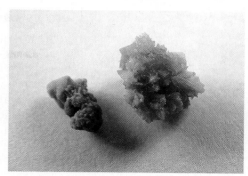

Kidney stones form when calcium phosphate exceeds its solubility.

Effect of Temperature on Solubility

The solubility of most solids is greater as temperature increases, which means that solutions usually contain more dissolved solute at higher temperature. A few substances show little change in solubility at higher temperatures, and a few are less soluble (see **FIGURE 12.5**). For example, when you add sugar to iced tea, some undissolved sugar may form on the bottom of the glass. But if you add sugar to hot tea, many teaspoons of sugar are needed before solid sugar appears. Hot tea dissolves more sugar than does cold tea because the solubility of sugar is much greater at a higher temperature.

When a saturated solution is carefully cooled, it becomes a *supersaturated solution* because it contains more solute than the solubility allows. Such a solution is unstable, and if the solution is agitated or if a solute crystal is added, the excess solute will recrystallize to give a saturated solution again.

Rock candy can be made from a saturated solution of sugar (sucrose).

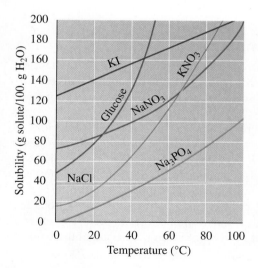

FIGURE 12.5 ▶ In water, most common solids are more soluble as the temperature increases.

Compare the solubility of NaNO$_3$ at 20 °C and 60 °C.

Conversely, the solubility of a gas in water decreases as the temperature increases. At higher temperatures, more gas molecules have the energy to escape from the solution. Perhaps you have observed the bubbles escaping from a cold carbonated soft drink as it warms. At high temperatures, bottles containing carbonated solutions may burst as more gas molecules leave the solution and increase the gas pressure inside the bottle. Biologists have found that increased temperatures in rivers and lakes cause the amount of dissolved oxygen to decrease until the warm water can no longer support a biological community. Electricity-generating plants are required to have their own ponds to use with their cooling towers to lessen the threat of thermal pollution in surrounding waterways.

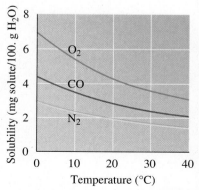

In water, gases are less soluble as the temperature increases.

Henry's Law

Henry's law states that the solubility of gas in a liquid is directly related to the pressure of that gas above the liquid. At higher pressures, there are more gas molecules available to enter and dissolve in the liquid. A can of soda is carbonated by using CO_2 gas at high pressure to increase the solubility of the CO_2 in the beverage. When you open the can at atmospheric pressure, the pressure on the CO_2 drops, which decreases the solubility of CO_2. As a result, bubbles of CO_2 rapidly escape from the solution. The burst of bubbles is even more noticeable when you open a warm can of soda.

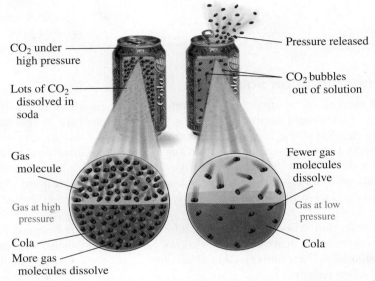

When the pressure of a gas above a solution decreases, the solubility of that gas in the solution also decreases.

Soluble and Insoluble Ionic Compounds

In our discussion up to now, we have considered ionic compounds that dissolve in water. However, some ionic compounds do not dissociate into ions and remain as solids even in contact with water. The **solubility rules** give some guidelines about the solubility of ionic compounds in water.

Ionic compounds that are soluble in water typically contain at least one of the ions in **TABLE 12.5**. *Only an ionic compound containing a soluble cation or anion will dissolve in water.* Most ionic compounds containing Cl^- are soluble, but $AgCl$, $PbCl_2$, and Hg_2Cl_2 are insoluble. Similarly, most ionic compounds containing SO_4^{2-} are soluble, but a few are insoluble. Most other ionic compounds are insoluble (see **FIGURE 12.6**). In an insoluble ionic compound, the ionic bonds between its positive and negative ions are too strong for the polar water molecules to break. We can use the solubility rules to predict whether a solid ionic compound would be soluble or not. **TABLE 12.6** illustrates the use of these rules.

TABLE 12.5 Solubility Rules for Ionic Compounds in Water

An ionic compound is soluble in water if it contains one of the following:	
Positive Ions:	Li^+, Na^+, K^+, Rb^+, Cs^+, NH_4^+
Negative Ions:	NO_3^-, $C_2H_3O_2^-$
	Cl^-, Br^-, I^- except when combined with Ag^+, Pb^{2+}, or Hg_2^{2+}
	SO_4^{2-} except when combined with Ba^{2+}, Pb^{2+}, Ca^{2+}, Sr^{2+}, or Hg_2^{2+}
Ionic compounds that do not contain at least one of these ions are usually insoluble.	

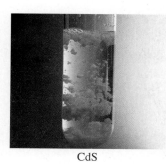
CdS

FeS

PbI_2

$Ni(OH)_2$

FIGURE 12.6 ▶ If an ionic compound contains a combination of a cation and an anion that are not soluble, that ionic compound is insoluble. For example, combinations of cadmium and sulfide, iron and sulfide, lead and iodide, and nickel and hydroxide do not contain any soluble ions. Thus, they form insoluble ionic compounds.

Q Why are each of these ionic compounds insoluble in water?

TABLE 12.6 Using Solubility Rules

Ionic Compound	Solubility in Water	Reasoning
K_2S	Soluble	Contains K^+
$Ca(NO_3)_2$	Soluble	Contains NO_3^-
$PbCl_2$	Insoluble	Is an insoluble chloride
NaOH	Soluble	Contains Na^+
$AlPO_4$	Insoluble	Contains no soluble ions

CORE CHEMISTRY SKILL
Using Solubility Rules

ENGAGE

Why is K_2S soluble in water whereas $PbCl_2$ is not soluble?

In medicine, insoluble $BaSO_4$ is used as an opaque substance to enhance X-rays of the gastrointestinal tract (see **FIGURE 12.7**). $BaSO_4$ is so insoluble that it does not dissolve in gastric fluids. Other ionic barium compounds cannot be used because they would dissolve in water, releasing Ba^{2+} which is poisonous.

SAMPLE PROBLEM 12.3 Soluble and Insoluble Ionic Compounds

Predict whether each of the following ionic compounds is soluble in water and explain why:

a. Na_3PO_4 **b.** $CaCO_3$

TRY IT FIRST ▶

SOLUTION

a. The ionic compound Na_3PO_4 is soluble in water because any compound that contains Na^+ is soluble.
b. The ionic compound $CaCO_3$ is not soluble. The compound does not contain a soluble positive ion, which means that ionic compound containing Ca^{2+} and CO_3^{2-} is not soluble.

STUDY CHECK 12.3

In some electrolyte drinks, $MgCl_2$ is added to provide magnesium. Why would you expect $MgCl_2$ to be soluble in water?

ANSWER

$MgCl_2$ is soluble in water because ionic compounds that contain chloride are soluble unless they contain Ag^+, Pb^{2+}, or Hg_2^{2+}.

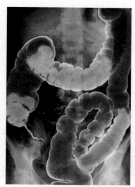

FIGURE 12.7 ▶ A barium sulfate-enhanced X-ray of the abdomen shows the lower gastrointestinal (GI) tract.

Q Is $BaSO_4$ a soluble or an insoluble substance?

Formation of a Solid

We can use solubility rules to predict whether a solid, called a *precipitate*, forms when two solutions containing soluble reactants are mixed as shown in Sample Problem 12.4.

SAMPLE PROBLEM 12.4 Writing Equations for the Formation of an Insoluble Ionic Compound

When solutions of NaCl and $AgNO_3$ are mixed, a white solid forms. Write the ionic and net ionic equations for the reaction.

TRY IT FIRST

SOLUTION

Guide to Writing an Equation for the Formation of an Insoluble Ionic Compound

STEP 1
Write the ions of the reactants.

STEP 2
Write the combinations of ions and determine if any are insoluble.

STEP 3
Write the ionic equation including any solid.

STEP 4
Write the net ionic equation.

STEP 1 Write the ions of the reactants.

**Reactants
(initial combinations)**

$$Ag^+(aq) + NO_3^-(aq)$$

$$Na^+(aq) + Cl^-(aq)$$

STEP 2 Write the combinations of ions and determine if any are insoluble. When we look at the ions of each solution, we see that the combination of Ag^+ and Cl^- forms an insoluble ionic compound.

Mixture (new combinations)	Product	Soluble
$Ag^+(aq) + Cl^-(aq)$	AgCl	No
$Na^+(aq) + NO_3^-(aq)$	$NaNO_3$	Yes

STEP 3 Write the ionic equation including any solid. In the *ionic equation*, we show all the ions of the reactants. The products include the solid AgCl that forms.

$$Ag^+(aq) + NO_3^-(aq) + Na^+(aq) + Cl^-(aq) \longrightarrow AgCl(s) + Na^+(aq) + NO_3^-(aq)$$

STEP 4 Write the net ionic equation. To write a *net ionic equation*, we remove the Na^+ and NO_3^- ions, known as *spectator ions*, which are unchanged.

$$Ag^+(aq) + \underbrace{NO_3^-(aq) + Na^+(aq)}_{\text{Spectator ions}} + Cl^-(aq) \longrightarrow AgCl(s) + \underbrace{Na^+(aq) + NO_3^-(aq)}_{\text{Spectator ions}}$$

$$Ag^+(aq) + Cl^-(aq) \longrightarrow AgCl(s) \quad \text{Net ionic equation}$$

ENGAGE

When mixing solutions of $Pb(NO_3)_2$ and NaBr, how do we know that $PbBr_2(s)$ is the solid that forms?

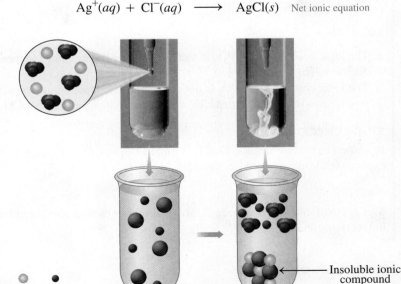

Cl^- NO_3^- Ag^+ Na^+

Insoluble ionic compound

STUDY CHECK 12.4

Predict whether a solid might form in each of the following mixtures of solutions. If so, write the net ionic equation for the reaction.

a. $NH_4Cl(aq) + Ca(NO_3)_2(aq)$ **b.** $Pb(NO_3)_2(aq) + KCl(aq)$

ANSWER

a. No solid forms because the products, $NH_4NO_3(aq)$ and $CaCl_2(aq)$, are soluble.
b. $Pb^{2+}(aq) + 2Cl^-(aq) \longrightarrow PbCl_2(s)$

QUESTIONS AND PROBLEMS

12.3 Solubility

LEARNING GOAL Define solubility; distinguish between an unsaturated and a saturated solution. Identify an ionic compound as soluble or insoluble.

12.15 State whether each of the following refers to a saturated or an unsaturated solution:
  a. A crystal added to a solution does not change in size.
  b. A sugar cube completely dissolves when added to a cup of coffee.

12.16 State whether each of the following refers to a saturated or an unsaturated solution:
  a. A spoonful of salt added to boiling water dissolves.
  b. A layer of sugar forms on the bottom of a glass of tea as ice is added.

Use the following table for problems 12.17 to 12.20:

	Solubility (g/100. g H_2O)	
Substance	**20 °C**	**50 °C**
KCl	34	43
$NaNO_3$	88	110
$C_{12}H_{22}O_{11}$ (sugar)	204	260

12.17 Determine whether each of the following solutions will be saturated or unsaturated at 20 °C:
  a. adding 25 g of KCl to 100. g of H_2O
  b. adding 11 g of $NaNO_3$ to 25 g of H_2O
  c. adding 400. g of sugar to 125 g of H_2O

12.18 Determine whether each of the following solutions will be saturated or unsaturated at 50 °C:
  a. adding 25 g of KCl to 50. g of H_2O
  b. adding 150. g of $NaNO_3$ to 75 g of H_2O
  c. adding 80. g of sugar to 25 g of H_2O

12.19 A solution containing 80. g of KCl in 200. g of H_2O at 50 °C is cooled to 20 °C.
  a. How many grams of KCl remain in solution at 20 °C?
  b. How many grams of solid KCl crystallized after cooling?

12.20 A solution containing 80. g of $NaNO_3$ in 75 g of H_2O at 50 °C is cooled to 20 °C.
  a. How many grams of $NaNO_3$ remain in solution at 20 °C?
  b. How many grams of solid $NaNO_3$ crystallized after cooling?

12.21 Explain the following observations:
  a. More sugar dissolves in hot tea than in iced tea.
  b. Champagne in a warm room goes flat.
  c. A warm can of soda has more spray when opened than a cold one.

12.22 Explain the following observations:
  a. An open can of soda loses its "fizz" faster at room temperature than in the refrigerator.
  b. Chlorine gas in tap water escapes as the sample warms to room temperature.
  c. Less sugar dissolves in iced coffee than in hot coffee.

12.23 Predict whether each of the following ionic compounds is soluble in water:
  a. LiCl   **b.** PbS   **c.** $BaCO_3$ **d.** K_2O **e.** $Fe(NO_3)_3$

12.24 Predict whether each of the following ionic compounds is soluble in water:
  a. AgCl   **b.** KI   **c.** Na_2S **d.** Ag_2O **e.** $CaSO_4$

12.25 Determine whether a solid forms when solutions containing the following ionic compounds are mixed. If so, write the ionic equation and the net ionic equation.
  a. $KCl(aq)$ and $Na_2S(aq)$
  b. $AgNO_3(aq)$ and $K_2S(aq)$
  c. $CaCl_2(aq)$ and $Na_2SO_4(aq)$
  d. $CuCl_2(aq)$ and $Li_3PO_4(aq)$

12.26 Determine whether a solid forms when solutions containing the following ionic compounds are mixed. If so, write the ionic equation and the net ionic equation.
  a. $Na_3PO_4(aq)$ and $AgNO_3(aq)$
  b. $K_2SO_4(aq)$ and $Na_2CO_3(aq)$
  c. $Pb(NO_3)_2(aq)$ and $Na_2CO_3(aq)$
  d. $BaCl_2(aq)$ and $KOH(aq)$

12.4 Solution Concentrations

LEARNING GOAL Calculate the concentration of a solute in a solution; use concentration as conversion factors to calculate the amount of solute or solution.

CORE CHEMISTRY SKILL
Calculating Concentration

Our body fluids contain water and dissolved substances including glucose, urea, and electrolytes such as K^+, Na^+, Cl^-, Mg^{2+}, HCO_3^-, and HPO_4^{2-}. Proper amounts of each of these dissolved substances and water must be maintained in the body fluids. Small changes in electrolyte levels can seriously disrupt cellular processes and endanger our health. Solutions can be described by their **concentration**, which is the amount of solute in a specific amount of that solution as shown in **TABLE 12.7**. The amount of solute dissolved in a certain amount of solution is called the *concentration* of the solution. We will look at ways to express a concentration as a ratio of a certain amount of solute in a given amount of solution. The amount of a solute may be expressed in units of grams, milliliters, or moles. The amount of a solution may be expressed in units of grams, milliliters, or liters.

$$\text{Concentration of a solution} = \frac{\text{amount of solute}}{\text{amount of solution}}$$

TABLE 12.7 Summary of Types of Concentration Expressions and Their Units

Concentration Units	Mass Percent (m/m)	Volume Percent (v/v)	Mass/Volume Percent (m/v)	Molarity (M)
Solute	g	mL	g	mole
Solution	g	mL	mL	L

Mass Percent (m/m) Concentration

Mass percent (m/m) describes the mass of the solute in grams for 100. g of solution. The mass percent is calculated by dividing the mass of a solute by the mass of the solution multiplied by 100% to give the percentage. In the calculation of mass percent (m/m), the units of mass of the solute and solution must be the same. If the mass of the solute is given as grams, then the mass of the solution must also be grams. The mass of the solution is the sum of the mass of the solute and the mass of the solvent.

Add 8.00 g of KCl

$$\text{Mass percent (m/m)} = \frac{\text{mass of solute (g)}}{\text{mass of solute (g)} + \text{mass of solvent (g)}} \times 100\%$$

$$= \frac{\text{mass of solute (g)}}{\text{mass of solution (g)}} \times 100\%$$

Suppose we prepared a solution by mixing 8.00 g of KCl (solute) with 42.00 g of water (solvent). Together, the mass of the solute and mass of solvent give the mass of the solution (8.00 g + 42.00 g = 50.00 g). Mass percent is calculated by substituting the mass of the solute and the mass of the solution into the mass percent expression.

Add water until the solution has a mass of 50.00 g

When water is added to 8.00 g of KCl to form 50.00 g of a KCl solution, the mass percent concentration is 16.0% (m/m).

$$\frac{8.00 \text{ g KCl}}{50.00 \text{ g solution}} \times 100\% = 16.0\% \text{ (m/m) KCl solution}$$

$$\overbrace{8.00 \text{ g KCl}}^{\text{Solute}} + \overbrace{42.00 \text{ g H}_2\text{O}}^{\text{Solvent}}$$

SAMPLE PROBLEM 12.5 Calculating Mass Percent (m/m) Concentration

What is the mass percent of NaOH in a solution prepared by dissolving 30.0 g of NaOH in 120.0 g of H_2O?

TRY IT FIRST

SOLUTION

STEP 1 State the given and needed quantities.

ANALYZE THE PROBLEM	Given	Need	Connect
	30.0 g of NaOH, 120.0 g of H_2O	mass percent (m/m)	$\dfrac{\text{mass of solute}}{\text{mass of solution}} \times 100\%$

STEP 2 Write the concentration expression.

$$\text{Mass percent (m/m)} = \frac{\text{grams of solute}}{\text{grams of solution}} \times 100\%$$

STEP 3 Substitute solute and solution quantities into the expression and calculate. The mass of the solution is obtained by adding the mass of the solute and the mass of the solution.

$$\text{mass of solution} = 30.0 \text{ g NaOH} + 120.0 \text{ g } H_2O = 150.0 \text{ g of NaOH solution}$$

$$\text{Mass percent (m/m)} = \frac{\overset{\text{Three SFs}}{30.0 \text{ g NaOH}}}{\underset{\text{Four SFs}}{150.0 \text{ g solution}}} \times 100\%$$

$$= \underset{\text{Three SFs}}{20.0\% \text{ (m/m) NaOH solution}}$$

STUDY CHECK 12.5

What is the mass percent (m/m) of NaCl in a solution made by dissolving 2.0 g of NaCl in 56.0 g of H_2O?

ANSWER

3.4% (m/m) NaCl solution

Using Mass Percent Concentration as a Conversion Factor

In the preparation of solutions, we often need to calculate the amount of solute or solution. Then the concentration of a solution is useful as a conversion factor as shown in Sample Problem 12.6.

CORE CHEMISTRY SKILL
Using Concentration as a Conversion Factor

SAMPLE PROBLEM 12.6 Using Mass Percent to Find Mass of Solute

The topical antibiotic ointment Neosporin is 3.5% (m/m) neomycin solution. How many grams of neomycin are in a tube containing 64 g of ointment?

TRY IT FIRST

SOLUTION

STEP 1 State the given and needed quantities.

ANALYZE THE PROBLEM	Given	Need	Connect
	64 g of 3.5% (m/m) neomycin solution	grams of neomycin	mass percent factor $\dfrac{\text{g of solute}}{100. \text{ g of solution}}$

Guide to Calculating Solution Concentration

STEP 1
State the given and needed quantities.

STEP 2
Write the concentration expression.

STEP 3
Substitute solute and solution quantities into the expression and calculate.

Guide to Using Concentration to Calculate Mass or Volume

STEP 1
State the given and needed quantities.

STEP 2
Write a plan to calculate the mass or volume.

STEP 3
Write equalities and conversion factors.

STEP 4
Set up the problem to calculate the mass or volume.

ENGAGE

How is the mass percent (m/m) of a solution used to convert the mass of the solution to the grams of solute?

The label indicates that vanilla extract contains 35% (v/v) alcohol.

Lemon extract is a solution of lemon flavor and alcohol.

STEP 2 Write a plan to calculate the mass.

grams of ointment → % (m/m) factor → grams of neomycin

STEP 3 Write equalities and conversion factors. The mass percent (m/m) indicates the grams of a solute in every 100. g of a solution. The mass percent (3.5% m/m) can be written as two conversion factors.

$$3.5 \text{ g of neomycin} = 100. \text{ g of ointment}$$

$$\frac{3.5 \text{ g neomycin}}{100. \text{ g ointment}} \quad \text{and} \quad \frac{100. \text{ g ointment}}{3.5 \text{ g neomycin}}$$

STEP 4 Set up the problem to calculate the mass.

$$64 \text{ g ointment} \times \underset{\text{Exact}}{\frac{3.5 \text{ g neomycin}}{100. \text{ g ointment}}} = 2.2 \text{ g of neomycin}$$

Two SFs · Two SFs · Two SFs

STUDY CHECK 12.6

Calculate the grams of KCl in 225 g of an 8.00% (m/m) KCl solution.

ANSWER

18.0 g of KCl

Volume Percent (v/v) Concentration

Because the volumes of liquids or gases are easily measured, the concentrations of their solutions are often expressed as **volume percent (v/v)**. The units of volume used in the ratio must be the same, for example, both in milliliters or both in liters.

$$\text{Volume percent (v/v)} = \frac{\text{volume of solute}}{\text{volume of solution}} \times 100\%$$

We interpret a volume percent as the volume of solute in 100. mL of solution. On a bottle of extract of vanilla, a label that reads alcohol 35% (v/v) means 35 mL of ethanol solute in 100. mL of vanilla solution.

SAMPLE PROBLEM 12.7 Calculating Volume Percent (v/v) Concentration

A bottle contains 59 mL of lemon extract solution. If the extract contains 49 mL of alcohol, what is the volume percent (v/v) of the alcohol in the solution?

TRY IT FIRST

SOLUTION

STEP 1 State the given and needed quantities.

ANALYZE THE PROBLEM	Given	Need	Connect
	49 mL of alcohol, 59 mL of solution	volume percent (v/v)	$\dfrac{\text{volume of solute}}{\text{volume of solution}} \times 100\%$

STEP 2 Write the concentration expression.

$$\text{Volume percent (v/v)} = \frac{\text{volume of solute}}{\text{volume of solution}} \times 100\%$$

STEP 3 Substitute solute and solution quantities into the expression and calculate.

$$\text{Volume percent (v/v)} = \frac{49 \text{ mL alcohol}}{59 \text{ mL solution}} \times 100\%$$

$$= 83\% \text{ (v/v) alcohol solution}$$

STUDY CHECK 12.7

What is the volume percent (v/v) of Br_2 in a solution prepared by dissolving 12 mL of liquid bromine (Br_2) in the solvent carbon tetrachloride (CCl_4) to make 250 mL of solution?

ANSWER

4.8% (v/v) Br_2 in CCl_4

Mass/Volume Percent (m/v) Concentration

Mass/volume percent (m/v) describes the mass of the solute in grams for exactly 100. mL of solution. In the calculation of mass/volume percent, the unit of mass of the solute is grams and the unit of the solution volume is milliliters.

$$\text{Mass/volume percent (m/v)} = \frac{\text{grams of solute}}{\text{milliliters of solution}} \times 100\%$$

The mass/volume percent is widely used in hospitals and pharmacies for the preparation of intravenous solutions and medicines. For example, a 5% (m/v) glucose solution contains 5 g of glucose in 100. mL of solution. The volume of solution represents the combined volumes of the glucose and H_2O.

SAMPLE PROBLEM 12.8 Calculating Mass/Volume Percent (m/v) Concentration

A potassium iodide solution may be used in a diet that is low in iodine. A KI solution is prepared by dissolving 5.0 g of KI in enough water to give a final volume of 250 mL. What is the mass/volume percent (m/v) of the KI solution?

TRY IT FIRST

SOLUTION

STEP 1 State the given and needed quantities.

	Given	Need	Connect
ANALYZE THE PROBLEM	5.0 g of KI solute, 250 mL of KI solution	mass/volume percent (m/v)	$\frac{\text{mass of solute}}{\text{volume of solution}} \times 100\%$

STEP 2 Write the concentration expression.

$$\text{Mass/volume percent (m/v)} = \frac{\text{mass of solute}}{\text{volume of solution}} \times 100\%$$

STEP 3 Substitute solute and solution quantities into the expression and calculate.

$$\text{Mass/volume percent (m/v)} = \frac{\overset{\text{Two SFs}}{5.0 \text{ g KI}}}{\underset{\text{Two SFs}}{250 \text{ mL solution}}} \times 100\% = \underset{\text{Two SFs}}{2.0\% \text{ (m/v) KI solution}}$$

Water added to make a solution   250 mL

5.0 g of KI   250 mL of KI solution

STUDY CHECK 12.8

What is the mass/volume percent (m/v) of NaOH in a solution prepared by dissolving 12 g of NaOH in enough water to make 220 mL of solution?

ANSWER

5.5% (m/v) NaOH solution

SAMPLE PROBLEM 12.9 Using Mass/Volume Percent to Find Mass of Solute

A topical antibiotic is 1.0% (m/v) clindamycin. How many grams of clindamycin are in 60. mL of the 1.0% (m/v) solution?

TRY IT FIRST

SOLUTION

STEP 1 State the given and needed quantities.

ANALYZE THE PROBLEM	Given	Need	Connect
	60. mL of 1.0% (m/v) clindamycin solution	grams of clindamycin	% (m/v) factor

STEP 2 Write a plan to calculate the mass.

milliliters of solution   % (m/v) factor   grams of clindamycin

STEP 3 Write equalities and conversion factors. The percent (m/v) indicates the grams of a solute in every 100. mL of a solution. The 1.0% (m/v) can be written as two conversion factors.

$$1.0 \text{ g of clindamycin} = 100. \text{ mL of solution}$$

$$\frac{1.0 \text{ g clindamycin}}{100. \text{ mL solution}} \quad \text{and} \quad \frac{100. \text{ mL solution}}{1.0 \text{ g clindamycin}}$$

STEP 4 Set up the problem to calculate the mass. The volume of the solution is converted to mass of solute using the conversion factor that cancels mL.

$$60. \text{ mL solution} \times \frac{1.0 \text{ g clindamycin}}{100. \text{ mL solution}} = 0.60 \text{ g of clindamycin}$$

STUDY CHECK 12.9

In 2010, the FDA approved a 2.0% (m/v) morphine oral solution to treat severe or chronic pain. How many grams of morphine does a patient receive if 0.60 mL of 2.0% (m/v) morphine solution was ordered?

ANSWER

0.012 g of morphine

Molarity (M) Concentration

When chemists work with solutions, they often use **molarity (M)**, a concentration that states the number of moles of solute in exactly 1 L of solution.

$$\text{Molarity (M)} = \frac{\text{moles of solute}}{\text{liters of solution}}$$

The molarity of a solution can be calculated when we know the moles of solute and the volume of solution in liters. For example, if 1.0 mol of NaCl were dissolved in enough water to prepare 1.0 L of solution, the resulting NaCl solution has a molarity of 1.0 M. The abbreviation M indicates the units of mole per liter (mol/L).

$$M = \frac{\text{moles of solute}}{\text{liters of solution}} = \frac{1.0 \text{ mol NaCl}}{1 \text{ L solution}} = 1.0 \text{ M NaCl solution}$$

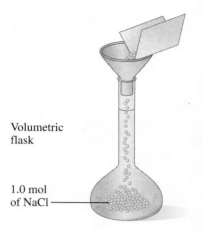

Volumetric flask

1.0 mol of NaCl

SAMPLE PROBLEM 12.10 Calculating Molarity

What is the molarity (M) of 60.0 g of NaOH in 0.250 L of NaOH solution?

 TRY IT FIRST

SOLUTION

STEP 1 State the given and needed quantities.

	Given	Need	Connect
ANALYZE THE PROBLEM	60.0 g of NaOH, 0.250 L of NaOH solution	molarity (mol/L)	molar mass of NaOH, moles of solute liters of solution

Add water until 1-liter mark is reached.

Mix

A 1.0 molar (M) NaCl solution

To calculate the moles of NaOH, we need to write the equality and conversion factors for the molar mass of NaOH. Then the moles in 60.0 g of NaOH can be determined.

$$1 \text{ mol of NaOH} = 40.00 \text{ g of NaOH}$$
$$\frac{40.00 \text{ g NaOH}}{1 \text{ mol NaOH}} \quad \text{and} \quad \frac{1 \text{ mol NaOH}}{40.00 \text{ g NaOH}}$$

$$\text{moles of NaOH} = 60.0 \text{ g NaOH} \times \frac{1 \text{ mol NaOH}}{40.00 \text{ g NaOH}}$$

$$= 1.50 \text{ mol of NaOH}$$

$$\text{volume of solution} = 0.250 \text{ L of NaOH solution}$$

STEP 2 Write the concentration expression.

$$\text{Molarity (M)} = \frac{\text{moles of solute}}{\text{liters of solution}}$$

STEP 3 Substitute solute and solution quantities into the expression and calculate.

$$M = \frac{1.50 \text{ mol NaOH}}{0.250 \text{ L solution}} = \frac{6.00 \text{ mol NaOH}}{1 \text{ L solution}} = 6.00 \text{ M NaOH solution}$$

STUDY CHECK 12.10

What is the molarity of a solution that contains 75.0 g of KNO_3 dissolved in 0.350 L of solution?

ANSWER

2.12 M KNO_3 solution

SAMPLE PROBLEM 12.11 Using Molarity to Calculate Volume of Solution

How many liters of a 2.00 M NaCl solution are needed to provide 67.3 g of NaCl?

TRY IT FIRST

SOLUTION

STEP 1 State the given and needed quantities.

ANALYZE THE PROBLEM	Given	Need	Connect
	67.3 g of NaCl, 2.00 M NaCl solution	liters of NaCl solution	molar mass of NaCl, molarity

STEP 2 Write a plan to calculate the volume.

grams of NaCl → Molar mass → moles of NaCl → Molarity → liters of NaCl solution

STEP 3 Write equalities and conversion factors.

$$1 \text{ mol of NaCl} = 58.44 \text{ g of NaCl}$$
$$\frac{58.44 \text{ g NaCl}}{1 \text{ mol NaCl}} \text{ and } \frac{1 \text{ mol NaCl}}{58.44 \text{ g NaCl}}$$

$$1 \text{ L of NaCl solution} = 2.00 \text{ mol of NaCl}$$
$$\frac{2.00 \text{ mol NaCl}}{1 \text{ L NaCl solution}} \text{ and } \frac{1 \text{ L NaCl solution}}{2.00 \text{ mol NaCl}}$$

STEP 4 Set up the problem to calculate the volume.

$$67.3 \text{ g NaCl} \times \frac{1 \text{ mol NaCl}}{58.44 \text{ g NaCl}} \times \frac{1 \text{ L NaCl solution}}{2.00 \text{ mol NaCl}} = 0.576 \text{ L of NaCl solution}$$

STUDY CHECK 12.11

How many milliliters of a 6.0 M HCl solution will provide 164 g of HCl?

ANSWER

750 mL of HCl solution

A summary of percent concentrations and molarity, their meanings, and conversion factors are given in TABLE 12.8.

TABLE 12.8 Conversion Factors from Concentrations

Percent Concentration	Meaning	Conversion Factors	
10% (m/m) KCl solution	10 g of KCl in 100. g of KCl solution	$\frac{10 \text{ g KCl}}{100. \text{ g solution}}$ and	$\frac{100. \text{ g solution}}{10 \text{ g KCl}}$
12% (v/v) ethanol solution	12 mL of ethanol in 100. mL of ethanol solution	$\frac{12 \text{ mL ethanol}}{100. \text{ mL solution}}$ and	$\frac{100. \text{ mL solution}}{12 \text{ mL ethanol}}$
5% (m/v) glucose solution	5 g of glucose in 100. mL of glucose solution	$\frac{5 \text{ g glucose}}{100. \text{ mL solution}}$ and	$\frac{100. \text{ mL solution}}{5 \text{ g glucose}}$
Molarity	**Meaning**	**Conversion Factors**	
6.0 M HCl solution	6.0 mol of HCl in 1 L of HCl solution	$\frac{6.0 \text{ mol HCl}}{1 \text{ L solution}}$ and	$\frac{1 \text{ L solution}}{6.0 \text{ mol HCl}}$

QUESTIONS AND PROBLEMS

12.4 Solution Concentrations

LEARNING GOAL Calculate the concentration of a solute in a solution; use concentration as conversion factors to calculate the amount of solute or solution.

12.27 What is the difference between a 5.00% (m/m) glucose solution and a 5.00% (m/v) glucose solution?

12.28 What is the difference between a 10.0% (v/v) methanol (CH_3OH) solution and a 10.0% (m/m) methanol solution?

12.29 Calculate the mass percent (m/m) for the solute in each of the following:
a. 25 g of KCl and 125 g of H_2O
b. 12 g of sucrose in 225 g of tea solution
c. 8.0 g of $CaCl_2$ in 80.0 g of $CaCl_2$ solution

12.30 Calculate the mass percent (m/m) for the solute in each of the following:
a. 75 g of NaOH in 325 g of NaOH solution
b. 2.0 g of KOH and 20.0 g of H_2O
c. 48.5 g of Na_2CO_3 in 250.0 g of Na_2CO_3 solution

12.31 Calculate the mass/volume percent (m/v) for the solute in each of the following:
a. 75 g of Na_2SO_4 in 250 mL of Na_2SO_4 solution
b. 39 g of sucrose in 355 mL of a carbonated drink

12.32 Calculate the mass/volume percent (m/v) for the solute in each of the following:
a. 2.50 g of LiCl in 40.0 mL of LiCl solution
b. 7.5 g of casein in 120 mL of low-fat milk

12.33 Calculate the grams or milliliters of solute needed to prepare the following:
a. 50. g of a 5.0% (m/m) KCl solution
b. 1250 mL of a 4.0% (m/v) NH_4Cl solution
c. 250. mL of a 10.0% (v/v) acetic acid solution

12.34 Calculate the grams or milliliters of solute needed to prepare the following:
a. 150. g of a 40.0% (m/m) $LiNO_3$ solution
b. 450 mL of a 2.0% (m/v) KOH solution
c. 225 mL of a 15% (v/v) isopropyl alcohol solution

12.35 A mouthwash contains 22.5% (v/v) alcohol. If the bottle of mouthwash contains 355 mL, what is the volume, in milliliters, of alcohol?

12.36 A bottle of champagne is 11% (v/v) alcohol. If there are 750 mL of champagne in the bottle, what is the volume, in milliliters, of alcohol?

12.37 For each of the following solutions, calculate the:
a. grams of 25% (m/m) $LiNO_3$ solution that contains 5.0 g of $LiNO_3$
b. milliliters of 10.0% (m/v) KOH solution that contains 40.0 g of KOH
c. milliliters of 10.0% (v/v) formic acid solution that contains 2.0 mL of formic acid

12.38 For each of the following solutions, calculate the:
a. grams of 2.0% (m/m) NaCl solution that contains 7.50 g of NaCl
b. milliliters of 25% (m/v) NaF solution that contains 4.0 g of NaF
c. milliliters of 8.0% (v/v) ethanol solution that contains 20.0 mL of ethanol

12.39 Calculate the molarity of each of the following:
a. 2.00 mol of glucose in 4.00 L of a glucose solution
b. 4.00 g of KOH in 2.00 L of a KOH solution
c. 5.85 g of NaCl in 400. mL of a NaCl solution

12.40 Calculate the molarity of each of the following:
a. 0.500 mol of glucose in 0.200 L of a glucose solution
b. 73.0 g of HCl in 2.00 L of a HCl solution
c. 30.0 g of NaOH in 350. mL of a NaOH solution

12.41 Calculate the grams of solute needed to prepare each of the following:
a. 2.00 L of a 1.50 M NaOH solution
b. 4.00 L of a 0.200 M KCl solution
c. 25.0 mL of a 6.00 M HCl solution

12.42 Calculate the grams of solute needed to prepare each of the following:
a. 2.00 L of a 6.00 M NaOH solution
b. 5.00 L of a 0.100 M $CaCl_2$ solution
c. 175 mL of a 3.00 M $NaNO_3$ solution

12.43 For each of the following solutions, calculate the:
a. liters of a 2.00 M KBr solution to obtain 3.00 mol of KBr
b. liters of a 1.50 M NaCl solution to obtain 15.0 mol of NaCl
c. milliliters of a 0.800 M $Ca(NO_3)_2$ solution to obtain 0.0500 mol of $Ca(NO_3)_2$

12.44 For each of the following solutions, calculate the:
a. liters of a 4.00 M KCl solution to obtain 0.100 mol of KCl
b. liters of a 6.00 M HCl solution to obtain 5.00 mol of HCl
c. milliliters of a 2.50 M K_2SO_4 solution to obtain 1.20 mol of K_2SO_4

12.45 Calculate the volume, in milliliters, for each of the following that provides the given amount of solute:
a. 12.5 g of Na_2CO_3 from a 0.120 M Na_2CO_3 solution
b. 0.850 mol of $NaNO_3$ from a 0.500 M $NaNO_3$ solution
c. 30.0 g of LiOH from a 2.70 M LiOH solution

12.46 Calculate the volume, in liters, for each of the following that provides the given amount of solute:
a. 5.00 mol of NaOH from a 12.0 M NaOH solution
b. 15.0 g of Na_2SO_4 from a 4.00 M Na_2SO_4 solution
c. 28.0 g of $NaHCO_3$ from a 1.50 M $NaHCO_3$ solution

Applications

12.47 A patient receives 100. mL of 20.% (m/v) mannitol solution every hour.
a. How many grams of mannitol are given in 1 h?
b. How many grams of mannitol does the patient receive in 12 h?

12.48 A patient receives 250 mL of a 4.0% (m/v) amino acid solution twice a day.
a. How many grams of amino acids are in 250 mL of solution?
b. How many grams of amino acids does the patient receive in 1 day?

12.49 A patient needs 100. g of glucose in the next 12 h. How many liters of a 5% (m/v) glucose solution must be given?

12.50 A patient received 2.0 g of NaCl in 8 h. How many milliliters of a 0.90% (m/v) NaCl (saline) solution were delivered?

12.5 Dilution of Solutions

LEARNING GOAL Describe the dilution of a solution; calculate the unknown concentration or volume when a solution is diluted.

In chemistry and biology, we often prepare diluted solutions from more concentrated solutions. In a process called **dilution**, a solvent, usually water, is added to a solution, which increases the volume. As a result, the concentration of the solution decreases. In an everyday example, you are making a dilution when you add three cans of water to a can of concentrated orange juice.

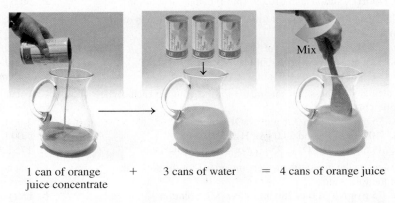

1 can of orange   +   3 cans of water   =   4 cans of orange juice
juice concentrate

Although the addition of solvent increases the volume, the amount of solute does not change; it is the same in the concentrated solution and the diluted solution (see **FIGURE 12.8**).

Grams or moles of solute = grams or moles of solute
Concentrated solution   Diluted solution

We can write this equality in terms of the concentration, C, and the volume, V. The concentration, C, may be percent concentration or molarity.

$$C_1V_1 = C_2V_2$$
Concentrated   Diluted
solution   solution

If we are given any three of the four variables (C_1, C_2, V_1, or V_2), we can rearrange the dilution expression to solve for the unknown quantity as seen in Sample Problem 12.12.

FIGURE 12.8 ▶ When water is added to a concentrated solution, there is no change in the number of particles. However, the solute particles spread out as the volume of the diluted solution increases.

Q What is the concentration of the diluted solution after an equal volume of water is added to a sample of a 6 M HCl solution?

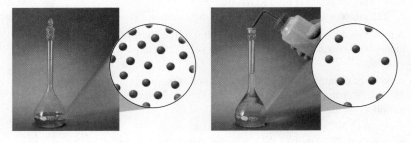

SAMPLE PROBLEM 12.12 Dilution of a Solution

A doctor orders 1000. mL of a 35.0% (m/v) dextrose solution. If you have a 50.0% (m/v) dextrose solution, how many milliliters would you use to prepare 1000. mL of 35.0% (m/v) dextrose solution?

▶ **TRY IT FIRST**

SOLUTION

STEP 1 Prepare a table of the concentrations and volumes of the solutions. For our problem analysis, we organize the solution data in a table, making sure that the units of concentration and volume are the same.

‎‎

	Given		Need	Connect
ANALYZE THE PROBLEM	$C_1 = 50.0\%$ (m/v)	$C_2 = 35.0\%$ (m/v)		$C_1V_1 = C_2V_2$
		$V_2 = 1000.\ \text{mL}$	V_1	C_1 increases, V_1 decreases

STEP 2 Rearrange the dilution expression to solve for the unknown quantity.

$$C_1V_1 = C_2V_2$$

$$\frac{\cancel{C_1}V_1}{\cancel{C_1}} = \frac{C_2V_2}{C_1} \qquad \text{Divide both sides by } C_1$$

$$V_1 = V_2 \times \frac{C_2}{C_1}$$

STEP 3 Substitute the known quantities into the dilution expression and calculate.

$$V_1 = 1000.\ \text{mL} \times \overset{\text{Three SFs}}{\frac{35.0\%}{50.0\%}} = 700.\ \text{mL of dextrose solution}$$

Four SFs   Three SFs   Three SFs

Concentration factor
decreases volume

When the final volume (V_2) is multiplied by a ratio of the percent concentrations (concentration factor) that is less than 1, the initial volume (V_1) is less than the final volume (V_2) as predicted in Step 1.

STUDY CHECK 12.12

What initial volume of a 15% (m/v) mannose solution is needed to prepare 125 mL of a 3.0% (m/v) mannose solution?

ANSWER

25 mL of a 15% (m/v) mannose solution

> **Guide to Calculating Dilution Quantities**
>
> **STEP 1**
> Prepare a table of the concentrations and volumes of the solutions.
>
> **STEP 2**
> Rearrange the dilution expression to solve for the unknown quantity.
>
> **STEP 3**
> Substitute the known quantities into the dilution expression and calculate.

SAMPLE PROBLEM 12.13 Molarity of a Diluted Solution

What is the molarity of a solution when 75.0 mL of a 4.00 M KCl solution is diluted to a volume of 500. mL?

TRY IT FIRST

SOLUTION

STEP 1 Prepare a table of the concentrations and volumes of the solutions.

	Given		Need	Connect
ANALYZE THE PROBLEM	$C_1 = 4.00\ \text{M}$		C_2	$C_1V_1 = C_2V_2$
	$V_1 = 75.0\ \text{mL}$	$V_2 = 500.\ \text{mL}$		V_2 increases, C_2 decreases
	$= 0.0750\ \text{L}$	$= 0.500\ \text{L}$		

STEP 2 Rearrange the dilution expression to solve for the unknown quantity.

$$C_1V_1 = C_2V_2$$

$$\frac{C_1V_1}{V_2} = \frac{C_2\cancel{V_2}}{\cancel{V_2}} \qquad \text{Divide both sides by } V_2$$

$$C_2 = C_1 \times \frac{V_1}{V_2}$$

STEP **3** Substitute the known quantities into the dilution expression and calculate.

$$C_2 = 4.00 \text{ M} \times \frac{75.0 \text{ mL}}{500. \text{ mL}} = 0.600 \text{ M (diluted KCl solution)}$$

Volume factor
decreases concentration

When the initial molarity (C_1) is multiplied by a ratio of the volumes (volume factor) that is less than 1, the molarity of the diluted solution decreases as predicted in Step 1.

STUDY CHECK 12.13

What is the molarity of a solution when 50.0 mL of a 4.00 M KOH solution is diluted to 200. mL?

ANSWER

1.00 M KOH solution

QUESTIONS AND PROBLEMS

12.5 Dilution of Solutions

LEARNING GOAL Describe the dilution of a solution; calculate the unknown concentration or volume when a solution is diluted.

12.51 To make tomato soup, you add one can of water to the condensed soup. Why is this a dilution?

12.52 A can of frozen lemonade calls for the addition of three cans of water to make a pitcher of the beverage. Why is this a dilution?

12.53 Calculate the final concentration of each of the following:
   a. 2.0 L of a 6.0 M HCl solution is added to water so that the final volume is 6.0 L.
   b. Water is added to 0.50 L of a 12 M NaOH solution to make 3.0 L of a diluted NaOH solution.
   c. A 10.0-mL sample of a 25% (m/v) KOH solution is diluted with water so that the final volume is 100.0 mL.
   d. A 50.0-mL sample of a 15% (m/v) H_2SO_4 solution is added to water to give a final volume of 250 mL.

12.54 Calculate the final concentration of each of the following:
   a. 1.0 L of a 4.0 M HNO_3 solution is added to water so that the final volume is 8.0 L.
   b. Water is added to 0.25 L of a 6.0 M NaF solution to make 2.0 L of a diluted NaF solution.
   c. A 50.0-mL sample of an 8.0% (m/v) KBr solution is diluted with water so that the final volume is 200.0 mL.
   d. A 5.0-mL sample of a 50.0% (m/v) acetic acid ($HC_2H_3O_2$) solution is added to water to give a final volume of 25 mL.

12.55 Determine the final volume, in milliliters, of each of the following:
   a. a 1.5 M HCl solution prepared from 20.0 mL of a 6.0 M HCl solution
   b. a 2.0% (m/v) LiCl solution prepared from 50.0 mL of a 10.0% (m/v) LiCl solution
   c. a 0.500 M H_3PO_4 solution prepared from 50.0 mL of a 6.00 M H_3PO_4 solution
   d. a 5.0% (m/v) glucose solution prepared from 75 mL of a 12% (m/v) glucose solution

12.56 Determine the final volume, in milliliters, of each of the following:
   a. a 1.00% (m/v) H_2SO_4 solution prepared from 10.0 mL of a 20.0% H_2SO_4 solution
   b. a 0.10 M HCl solution prepared from 25 mL of a 6.0 M HCl solution
   c. a 1.0 M NaOH solution prepared from 50.0 mL of a 12 M NaOH solution
   d. a 1.0% (m/v) $CaCl_2$ solution prepared from 18 mL of a 4.0% (m/v) $CaCl_2$ solution

12.57 Determine the initial volume, in milliliters, required to prepare each of the following:
   a. 255 mL of a 0.200 M HNO_3 solution using a 4.00 M HNO_3 solution
   b. 715 mL of a 0.100 M $MgCl_2$ solution using a 6.00 M $MgCl_2$ solution
   c. 0.100 L of a 0.150 M KCl solution using an 8.00 M KCl solution

12.58 Determine the initial volume, in milliliters, required to prepare each of the following:
   a. 20.0 mL of a 0.250 M KNO_3 solution using a 6.00 M KNO_3 solution
   b. 25.0 mL of a 2.50 M H_2SO_4 solution using a 12.0 M H_2SO_4 solution
   c. 0.500 L of a 1.50 M NH_4Cl solution using a 10.0 M NH_4Cl solution

Applications

12.59 You need 500. mL of a 5.0% (m/v) glucose solution. If you have a 25% (m/v) glucose solution on hand, how many milliliters do you need?

12.60 A doctor orders 100. mL of 2.0% (m/v) ibuprofen. If you have 8.0% (m/v) ibuprofen on hand, how many milliliters do you need?

12.6 Chemical Reactions in Solution

LEARNING GOAL Given the volume and concentration of a solution in a chemical reaction, calculate the amount of a reactant or product in the reaction.

When chemical reactions involve aqueous solutions, we use the balanced chemical equation, the molarity, and the volume to determine the moles or grams of a reactant or product. For example, we can determine the volume of a solution from the molarity and the grams of reactant as seen in Sample Problem 12.14.

SAMPLE PROBLEM 12.14 Volume of a Solution in a Reaction

Zinc reacts with HCl to produce hydrogen gas, H_2, and $ZnCl_2$.

$$Zn(s) + 2HCl(aq) \longrightarrow H_2(g) + ZnCl_2(aq)$$

How many liters of a 1.50 M HCl solution completely react with 5.32 g of zinc?

TRY IT FIRST

SOLUTION

STEP 1 State the given and needed quantities.

ANALYZE THE PROBLEM	Given	Need	Connect
	5.32 g of Zn, 1.50 M HCl solution	liters of HCl solution	molar mass of Zn, molarity, mole–mole factor
	Equation		
	$Zn(s) + 2HCl(aq) \longrightarrow H_2(g) + ZnCl_2(aq)$		

STEP 2 Write a plan to calculate the needed quantity.

grams of Zn → **Molar mass** → moles of Zn → **Mole–mole factor** → moles of HCl

Molarity → liters of HCl solution

STEP 3 Write equalities and conversion factors including mole–mole and concentration factors.

$$1 \text{ mol of Zn} = 65.41 \text{ g of Zn}$$
$$\frac{65.41 \text{ g Zn}}{1 \text{ mol Zn}} \quad \text{and} \quad \frac{1 \text{ mol Zn}}{65.41 \text{ g Zn}}$$

$$1 \text{ mol of Zn} = 2 \text{ mol of HCl}$$
$$\frac{2 \text{ mol HCl}}{1 \text{ mol Zn}} \quad \text{and} \quad \frac{1 \text{ mol Zn}}{2 \text{ mol HCl}}$$

$$1 \text{ L of solution} = 1.50 \text{ mol of HCl}$$
$$\frac{1.50 \text{ mol HCl}}{1 \text{ L solution}} \quad \text{and} \quad \frac{1 \text{ L solution}}{1.50 \text{ mol HCl}}$$

STEP 4 Set up the problem to calculate the needed quantity.

$$5.32 \text{ g Zn} \times \frac{1 \text{ mol Zn}}{65.41 \text{ g Zn}} \times \frac{2 \text{ mol HCl}}{1 \text{ mol Zn}} \times \frac{1 \text{ L solution}}{1.50 \text{ mol HCl}} = 0.108 \text{ L of HCl solution}$$

STUDY CHECK 12.14

Using the reaction in Sample Problem 12.14, how many grams of zinc can react with 225 mL of a 0.200 M HCl solution?

ANSWER

1.47 g of Zn

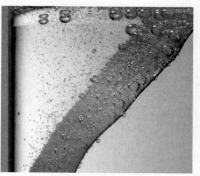

Zinc reacts when placed in a solution of HCl.

Guide to Calculations Involving Solutions in Chemical Reactions

STEP 1
State the given and needed quantities.

STEP 2
Write a plan to calculate the needed quantity or concentration.

STEP 3
Write equalities and conversion factors including mole–mole and concentration factors.

STEP 4
Set up the problem to calculate the needed quantity or concentration.

CORE CHEMISTRY SKILL
Calculating the Quantity of a Reactant or Product for a Chemical Reaction in Solution

When a $BaCl_2$ solution is added to a Na_2SO_4 solution, $BaSO_4$, a white solid, forms.

SAMPLE PROBLEM 12.15 Volume of a Reactant in a Solution

How many milliliters of a 0.250 M $BaCl_2$ solution are needed to react with 0.0325 L of a 0.160 M Na_2SO_4 solution?

$$Na_2SO_4(aq) + BaCl_2(aq) \longrightarrow BaSO_4(s) + 2NaCl(aq)$$

> TRY IT FIRST

SOLUTION

STEP 1 State the given and needed quantities.

ANALYZE THE PROBLEM	Given	Need	Connect
	0.0325 L of 0.160 M Na_2SO_4 solution, 0.250 M $BaCl_2$ solution	milliliters of $BaCl_2$ solution	mole–mole factor, metric factor
	Equation		
	$Na_2SO_4(aq) + BaCl_2(aq) \longrightarrow BaSO_4(s) + 2NaCl(aq)$		

STEP 2 Write a plan to calculate the needed quantity.

liters of Na_2SO_4 solution → Molarity → moles of Na_2SO_4 → Mole–mole factor → moles of $BaCl_2$

Molarity → liters of $BaCl_2$ solution → Metric factor → milliliters of $BaCl_2$ solution

STEP 3 Write equalities and conversion factors including mole–mole and concentration factors.

$$1 \text{ L of solution} = 0.160 \text{ mol of } Na_2SO_4$$
$$\frac{0.160 \text{ mol } Na_2SO_4}{1 \text{ L solution}} \quad \text{and} \quad \frac{1 \text{ L solution}}{0.160 \text{ mol } Na_2SO_4}$$

$$1 \text{ mol of } Na_2SO_4 = 1 \text{ mol of } BaCl_2$$
$$\frac{1 \text{ mol } BaCl_2}{1 \text{ mol } Na_2SO_4} \quad \text{and} \quad \frac{1 \text{ mol } Na_2SO_4}{1 \text{ mol } BaCl_2}$$

$$1 \text{ L of solution} = 0.250 \text{ mol of } BaCl_2$$
$$\frac{0.250 \text{ mol } BaCl_2}{1 \text{ L solution}} \quad \text{and} \quad \frac{1 \text{ L solution}}{0.250 \text{ mol } BaCl_2}$$

$$1 \text{ L} = 1000 \text{ mL}$$
$$\frac{1000 \text{ mL}}{1 \text{ L}} \quad \text{and} \quad \frac{1 \text{ L}}{1000 \text{ mL}}$$

STEP 4 Set up the problem to calculate the needed quantity.

$$0.0325 \text{ L solution} \times \frac{0.160 \text{ mol } Na_2SO_4}{1 \text{ L solution}} \times \frac{1 \text{ mol } BaCl_2}{1 \text{ mol } Na_2SO_4} \times \frac{1 \text{ L solution}}{0.250 \text{ mol } BaCl_2} \times \frac{1000 \text{ mL } BaCl_2 \text{ solution}}{1 \text{ L solution}}$$

$$= 20.8 \text{ mL of } BaCl_2 \text{ solution}$$

STUDY CHECK 12.15

For the reaction in Sample Problem 12.15, how many milliliters of a 0.330 M Na_2SO_4 solution are needed to react with 26.8 mL of a 0.216 M $BaCl_2$ solution?

ANSWER

17.5 mL of Na_2SO_4 solution

SAMPLE PROBLEM 12.16 Volume of a Gas from a Reaction in Solution

Acid rain results from the reaction of nitrogen dioxide with water in the air.

$$3NO_2(g) + H_2O(l) \longrightarrow 2HNO_3(aq) + NO(g)$$

At STP, how many liters of NO_2 gas are required to produce 0.275 L of a 0.400 M HNO_3 solution?

TRY IT FIRST

SOLUTION

STEP 1 State the given and needed quantities.

ANALYZE THE PROBLEM	Given	Need	Connect
	0.275 L of 0.400 M HNO_3 solution	liters of $NO_2(g)$ at STP	mole–mole factor, molar volume
	Equation		
	$3NO_2(g) + H_2O(l) \longrightarrow 2HNO_3(aq) + NO(g)$		

STEP 2 Write a plan to calculate the needed quantity. We start the problem with the volume and molarity of the HNO_3 solution to calculate moles. Then we can use the mole–mole factor and the molar volume to calculate the liters of NO_2 gas.

liters of solution → [Molarity] → moles of HNO_3 → [Mole–mole factor] → moles of NO_2 → [Molar volume] → liters of NO_2 (at STP)

STEP 3 Write equalities and conversion factors including mole–mole and concentration factors.

1 L of solution = 0.400 mol of HNO_3	3 mol of NO_2 = 2 mol of HNO_3
$\dfrac{0.400 \text{ mol } HNO_3}{1 \text{ L solution}}$ and $\dfrac{1 \text{ L solution}}{0.400 \text{ mol } HNO_3}$	$\dfrac{3 \text{ mol } NO_2}{2 \text{ mol } HNO_3}$ and $\dfrac{2 \text{ mol } HNO_3}{3 \text{ mol } NO_2}$

1 mol of NO_2 = 22.4 L of NO_2 (STP)

$\dfrac{22.4 \text{ L } NO_2 \text{ (STP)}}{1 \text{ mol } NO_2}$ and $\dfrac{1 \text{ mol } NO_2}{22.4 \text{ L } NO_2 \text{ (STP)}}$

STEP 4 Set up the problem to calculate the needed quantity.

$$0.275 \text{ L solution} \times \frac{0.400 \text{ mol } HNO_3}{1 \text{ L solution}} \times \frac{3 \text{ mol } NO_2}{2 \text{ mol } HNO_3} \times \frac{22.4 \text{ L } NO_2 \text{ (STP)}}{1 \text{ mol } NO_2} = 3.70 \text{ L of } NO_2 \text{ (STP)}$$

STUDY CHECK 12.16

Using the equation in Sample Problem 12.16, determine the volume of NO produced at 100 °C and 1.20 atm, when 2.20 L of a 1.50 M HNO_3 solution is produced.

ANSWER

42.1 L of NO

FIGURE 12.9 gives a summary of the pathways and conversion factors needed for substances including solutions involved in chemical reactions.

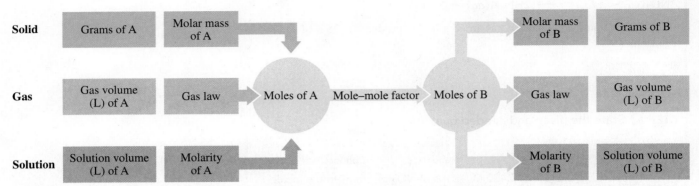

FIGURE 12.9 ▶ In calculations involving chemical reactions, substance A is converted to moles of A using molar mass (if solid), gas laws (if gas), or molarity (if solution). Then moles of A are converted to moles of substance B, which are converted to grams of solid, liters of gas, or liters of solution, as needed.

◉ What sequence of conversion factors would you use to calculate the number of grams of $CaCO_3$ needed to react with 1.50 L of a 2.00 M HCl solution in the reaction $CaCO_3(s) + 2HCl(aq)$?

QUESTIONS AND PROBLEMS

12.6 Chemical Reactions in Solution

LEARNING GOAL Given the volume and concentration of a solution in a chemical reaction, calculate the amount of a reactant or product in the reaction.

12.61 Answer the following for the reaction:

$$Pb(NO_3)_2(aq) + 2KCl(aq) \longrightarrow PbCl_2(s) + 2KNO_3(aq)$$

  a. How many grams of $PbCl_2$ will be formed from 50.0 mL of a 1.50 M KCl solution?

  b. How many milliliters of a 2.00 M $Pb(NO_3)_2$ solution will react with 50.0 mL of a 1.50 M KCl solution?

  c. What is the molarity of 20.0 mL of a KCl solution that reacts completely with 30.0 mL of a 0.400 M $Pb(NO_3)_2$ solution?

12.62 Answer the following for the reaction:

$$NiCl_2(aq) + 2NaOH(aq) \longrightarrow Ni(OH)_2(s) + 2NaCl(aq)$$

  a. How many milliliters of a 0.200 M NaOH solution are needed to react with 18.0 mL of a 0.500 M $NiCl_2$ solution?

  b. How many grams of $Ni(OH)_2$ are produced from the reaction of 35.0 mL of a 1.75 M NaOH solution and excess $NiCl_2$?

  c. What is the molarity of 30.0 mL of a $NiCl_2$ solution that reacts completely with 10.0 mL of a 0.250 M NaOH solution?

12.63 Answer the following for the reaction:

$$Mg(s) + 2HCl(aq) \longrightarrow H_2(g) + MgCl_2(aq)$$

  a. How many milliliters of a 6.00 M HCl solution are required to react with 15.0 g of magnesium?

  b. How many liters of hydrogen gas can form at STP when 0.500 L of a 2.00 M HCl solution reacts with excess magnesium?

  c. What is the molarity of a HCl solution if the reaction of 45.2 mL of the HCl solution with excess magnesium produces 5.20 L of H_2 gas at 735 mmHg and 25 °C?

12.64 Answer the following for the reaction:

$$CaCO_3(s) + 2HCl(aq) \longrightarrow CO_2(g) + H_2O(l) + CaCl_2(aq)$$

  a. How many milliliters of a 0.200 M HCl solution can react with 8.25 g of $CaCO_3$?

  b. How many liters of CO_2 gas can form at STP when 15.5 mL of a 3.00 M HCl solution reacts with excess $CaCO_3$?

  c. What is the molarity of a HCl solution if the reaction of 200. mL of the HCl solution with excess $CaCO_3$ produces 12.0 L of CO_2 gas at 725 mmHg and 18 °C?

12.65 Answer the following for the reaction:

$$Zn(s) + 2HBr(aq) \longrightarrow H_2(g) + ZnBr_2(aq)$$

  a. How many milliliters of a 3.50 M HBr solution are required to react with 8.56 g of zinc?

  b. How many liters of hydrogen gas can form at STP when 0.750 L of a 6.00 M HBr solution reacts with excess zinc?

  c. What is the molarity of a HBr solution if the reaction of 28.7 mL of the HBr solution with excess zinc produces 0.620 L of H_2 gas at 725 mmHg and 24 °C?

12.66 Answer the following for the reaction:

$$3AgNO_3(aq) + Na_3PO_4(aq) \longrightarrow Ag_3PO_4(s) + 3NaNO_3(aq)$$

  a. How many milliliters of a 0.265 M $AgNO_3$ solution are required to react with 31.2 mL of 0.154 M Na_3PO_4 solution?

  b. How many grams of silver phosphate are produced from the reaction of 25.0 mL of a 0.165 M $AgNO_3$ solution and excess Na_3PO_4?

  c. What is the molarity of 20.0 mL of a Na_3PO_4 solution that reacts completely with 15.0 mL of a 0.576 M $AgNO_3$ solution?

12.7 Molality and Freezing Point Lowering/Boiling Point Elevation

LEARNING GOAL Identify a mixture as a solution, a colloid, or a suspension. Using the molality, calculate the new freezing point and new boiling point for a solution.

The size and number of solute particles in different types of mixtures play an important role in determining the properties of that mixture.

Solutions

In the solutions discussed up to now, the solute was dissolved as small particles that are uniformly dispersed throughout the solvent to give a homogeneous solution. When you observe a solution, such as salt water, you cannot visually distinguish the solute from the solvent. The solution appears transparent, although it may have a color. The particles are so small that they go through filters and through *semipermeable membranes*. A semipermeable membrane allows solvent molecules such as water and very small solute particles to pass through, but does allow the passage of large solute molecules.

Colloids

The particles in a **colloid** are much larger than solute particles in a solution. Colloidal particles are large molecules, such as proteins, or groups of molecules or ions. Colloids, similar to solutions, are homogeneous mixtures that do not separate or settle out. Colloidal particles are small enough to pass through filters, but too large to pass through semipermeable membranes. **TABLE 12.9** lists several examples of colloids.

TABLE 12.9 Examples of Colloids

Colloid	Substance Dispersed	Dispersing Medium
Fog, clouds, hair sprays	Liquid	Gas
Dust, smoke	Solid	Gas
Shaving cream, whipped cream, soapsuds	Gas	Liquid
Styrofoam, marshmallows	Gas	Solid
Mayonnaise, homogenized milk	Liquid	Liquid
Cheese, butter	Liquid	Solid
Blood plasma, paints (latex), gelatin	Solid	Liquid

Suspensions

Suspensions are heterogeneous, nonuniform mixtures that are very different from solutions or colloids. The particles of a suspension are so large that they can often be seen with the naked eye. They are trapped by filters and semipermeable membranes.

The weight of the suspended solute particles causes them to settle out soon after mixing. If you stir muddy water, it mixes but then quickly separates as the suspended particles settle to the bottom and leave clear liquid at the top. You can find suspensions among the medications in a hospital or in your medicine cabinet. These include Kaopectate, calamine lotion, antacid mixtures, and liquid penicillin. It is important to follow the instructions on the label that states "shake well before using" so that the particles form a suspension.

Water-treatment plants make use of the properties of suspensions to purify water. When chemicals such as aluminum sulfate or iron(III) sulfate are added to untreated water, they react with impurities to form large suspended particles called *floc*. In the water-treatment plant, a system of filters traps the suspended particles, but clean water passes through.

TABLE 12.10 compares the different types of mixtures and **FIGURE 12.10** illustrates some properties of solutions, colloids, and suspensions.

TABLE 12.10 Comparison of Solutions, Colloids, and Suspensions

Type of Mixture	Type of Particle	Settling	Separation
Solution	Small particles such as atoms, ions, or small molecules	Particles do not settle	Particles cannot be separated by filters or semipermeable membranes
Colloid	Larger molecules or groups of molecules or ions	Particles do not settle	Particles can be separated by semipermeable membranes but not by filters
Suspension	Very large particles that may be visible	Particles settle rapidly	Particles can be separated by filters

FIGURE 12.10 ▶ Properties of different types of mixtures: **(a)** suspensions settle out; **(b)** suspensions are separated by a filter; **(c)** solution particles go through a semipermeable membrane, but colloids and suspensions do not.

Q A filter can be used to separate suspension particles from a solution, but a semipermeable membrane is needed to separate colloids from a solution. Explain.

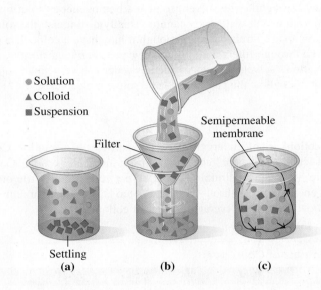

● Solution
▲ Colloid
■ Suspension

Semipermeable membrane

Filter

Settling
(a) **(b)** **(c)**

Vapor pressure   Lower vapor pressure

Pure solvent   Solute and solvent

The Alaska Upis beetle produces biological antifreeze to survive subfreezing temperatures.

Freezing Point Lowering and Boiling Point Elevation

When we add a solute to water, it changes the vapor pressure, freezing point, and boiling point of the solvent pure water. These types of changes in physical properties, known as *colligative properties*, depend only on the concentration of solute particles in the solution.

We can illustrate how these changes in physical properties occur by comparing the number of evaporating solvent molecules in pure solvent with those in a solution with a nonvolatile solute in the solvent. In the solution, there are fewer solvent molecules at the surface because the solute that has been added takes up some of the space at the surface. As a result, fewer solvent molecules can evaporate compared to the pure solvent. Vapor pressure is lowered for the solution. If we add more solute molecules, the vapor pressure will be lowered even more.

The freezing point of a solvent is lowered when a nonvolatile solute is added. In this case, the solute particles prevent the organization of solvent molecules needed to form the solid state. Thus, a lower temperature is required before the molecules of the solvent can become organized enough to freeze.

Insects and fish in climates with subfreezing temperatures control ice formation by producing biological antifreezes made of glycerol, proteins, and sugars, such as glucose, within their bodies. Some insects can survive temperatures below −60 °C. These forms of biological antifreezes may one day be applied to the long-term preservation of human organs.

The boiling point of a solvent is raised when a nonvolatile solute is added. The vapor pressure of a solvent must reach atmospheric pressure before it begins boiling. However, because the solute lowers the vapor pressure of the solvent, a temperature higher than the normal boiling point of the pure solvent is needed to cause the solution to boil. When you spread salt on an icy sidewalk when temperatures drop below freezing, the particles from the salt combine with water to lower the freezing point, which causes the ice to melt. Another example is the addition of antifreeze, such as ethylene glycol, $C_2H_6O_2$, to the water in a car radiator. If an ethylene glycol and water mixture is about 50–50% by mass, it does not freeze until the temperature drops to about −30 °F, and does not boil unless the temperature reaches about 225 °F. The solution in the radiator prevents the water in the radiator from forming ice in cold weather or boiling over on a hot desert highway.

Ethylene glycol is added to a radiator to form an aqueous solution that has a lower freezing point than water and a higher boiling point.

Particles in Solution

A solute that is a nonelectrolyte dissolves as molecules, whereas a solute that is a strong electrolyte dissolves entirely as ions. A common solute in antifreeze is the nonelectrolyte ethylene glycol, $C_2H_6O_2$.

   Nonelectrolyte: 1 mol of $C_2H_6O_2(l)$ = 1 mol of $C_2H_6O_2(aq)$

However, when 1 mol of a strong electrolyte, such as NaCl or $CaCl_2$, dissolves in water, the NaCl solution will contain 2 mol of particles and the $CaCl_2$ solution will contain 3 mol of particles.

   Strong electrolytes:

$$1 \text{ mol of NaCl}(s) = \underbrace{1 \text{ mol of Na}^+(aq) + 1 \text{ mol of Cl}^-(aq)}_{2 \text{ mol of particles } (aq)}$$

$$1 \text{ mol of CaCl}_2(s) = \underbrace{1 \text{ mol of Ca}^{2+}(aq) + 2 \text{ mol of Cl}^-(aq)}_{3 \text{ mol of particles } (aq)}$$

Molality (*m*)

The calculation for freezing point lowering or boiling point elevation uses a concentration unit of *molality*. The **molality**, abbreviation *m*, of a solution is the number of moles of solute particles per kilogram of solvent. This may seem similar to molarity, but the denominator for molality refers to the mass of the solvent, not the volume of the solution.

$$\text{Molality } (m) = \frac{\text{moles of solute particles}}{\text{kilograms of solvent}}$$

SAMPLE PROBLEM 12.17 Calculating Molality

Calculate the molality of a solution containing 35.5 g of the nonelectrolyte glucose ($C_6H_{12}O_6$) in 0.400 kg of water.

TRY IT FIRST

SOLUTION

STEP 1 State the given and needed quantities.

	Given	Need	Connect
ANALYZE THE PROBLEM	35.5 g of glucose ($C_6H_{12}O_6$), 0.400 kg of water	molality (*m*)	molar mass of glucose, $\dfrac{\text{moles}}{\text{kg}}$

STEP 2 Write the molality expression.

$$\text{Molality } (m) \text{ of glucose solution} = \frac{\text{moles of glucose}}{\text{kilograms of water}}$$

Guide to Calculating Molality

STEP 1
State the given and needed quantities.

STEP 2
Write the molality expression.

STEP 3
Substitute solute and solvent quantities into the expression and calculate.

STEP 3 Substitute solute and solvent quantities into the expression and calculate. To match the unit of moles in the molality expression, we convert the grams of glucose to moles of glucose, using its molar mass.

$$1 \text{ mol of glucose} = 180.16 \text{ g of glucose}$$

$$\frac{180.16 \text{ g glucose}}{1 \text{ mol glucose}} \quad \text{and} \quad \frac{1 \text{ mol glucose}}{180.16 \text{ g glucose}}$$

$$\text{moles of glucose} = 35.5 \text{ g glucose} \times \frac{1 \text{ mol glucose}}{180.16 \text{ g glucose}} = 0.197 \text{ mol of glucose}$$

$$\text{Molality } (m) = \frac{0.197 \text{ mol glucose}}{0.400 \text{ kg water}} = 0.493 \, m$$

STUDY CHECK 12.17

Calculate the molality of a solution containing 15.8 g of urea, CH_4N_2O, a nonelectrolyte, in 250. g of water.

ANSWER

1.05 m

Freezing Point Lowering

CORE CHEMISTRY SKILL
Calculating the Freezing Point/Boiling Point of a Solution

The change in the freezing point temperature of a solvent (ΔT_f) is determined from the molality (m) of the particles in the solution and the freezing point constant, K_f, for the solvent.

$$\Delta T_f = m \times K_f$$

The freezing point constant (K_f) for water is $\frac{1.86 \text{ °C}}{m}$.

For a 1 m solution. we can calculate the change in the freezing point temperature as

$$\Delta T_f = m \times K_f = 1 \, m \times \frac{1.86 \text{ °C}}{m} = 1.86 \text{ °C}$$

Then we can calculate the new, lower freezing point as

$$T_{solution} = T_{water} - \Delta T_f$$
$$= 0.00 \text{ °C} - 1.86 \text{ °C}$$
$$= -1.86 \text{ °C}$$

ENGAGE

Why does 1 m solution of KBr, a strong electrolyte, lower the freezing point more than 1 m solution of urea, a nonelectrolyte?

SAMPLE PROBLEM 12.18 Calculating Freezing Point of a Solution

In the northeastern United States during freezing weather, $CaCl_2$ is spread on icy highways to melt the ice. Calculate the freezing point lowering and freezing point of a solution containing 225 g of $CaCl_2$ in 500. g of water.

 TRY IT FIRST

SOLUTION

STEP 1 State the given and needed quantities.

	Given	Need	Connect
ANALYZE THE PROBLEM	225 g of $CaCl_2$, 500. g of water = 0.500 kg of water	ΔT_f, freezing point	molar mass, $\Delta T_f = m \times K_f$

A truck spreads calcium chloride on the road to melt ice and snow.

STEP 2 Determine the number of moles of solute particles and calculate the molality.

We use molar mass to calculate the moles of $CaCl_2$. Then we multiply by three to obtain the number of moles of ions (particles) produced by 1 mol of $CaCl_2$ in solution.

$$\text{moles of particles} = 225 \text{ g } \cancel{CaCl_2} \times \frac{1 \text{ mol } \cancel{CaCl_2}}{110.98 \text{ g } \cancel{CaCl_2}} \times \frac{3 \text{ mol particles}}{1 \text{ mol } \cancel{CaCl_2}}$$

$$= 6.08 \text{ mol of particles}$$

The molality (m) of the particles in solution is obtained by dividing the moles of particles by the number of kilograms of water in the solution.

$$m = \frac{\text{moles of particles}}{\text{kilograms of water}}$$

$$\text{Molality } (m) = \frac{6.08 \text{ mol particles}}{0.500 \text{ kg water}} = 12.2 \; m$$

STEP 3 Calculate the temperature change and subtract from the freezing point. The freezing point lowering is calculated using the molality and the freezing point constant. Finally, the freezing point lowering is subtracted from 0.00 °C to obtain the new freezing point of the $CaCl_2$ solution.

$$\Delta T_f = m \times K_f$$

$$\Delta T_f = 12.2 \; \cancel{m} \times \frac{1.86 \text{ °C}}{\cancel{m}} = 22.7 \text{ °C}$$

$$T_{\text{solution}} = T_{\text{water}} - \Delta T_f$$

$$= 0.0 \text{ °C} - 22.7 \text{ °C} = -22.7 \text{ °C}$$

STUDY CHECK 12.18

Ethylene glycol, $C_2H_6O_2$, a nonelectrolyte, is added to the water in a radiator to give a solution containing 515 g of ethylene glycol in 565 g of water (solvent). Calculate the freezing point lowering and freezing point of the solution.

ANSWER

$\Delta T_f = 27.2 \text{ °C}$;

freezing point is -27.2 °C

> **Guide to Calculating Freezing Point Lowering/Boiling Point Elevation**
>
> **STEP 1**
> State the given and needed quantities.
>
> **STEP 2**
> Determine the number of moles of solute particles and calculate the molality.
>
> **STEP 3**
> Calculate the temperature change and subtract from the freezing point or add to the boiling point.

Boiling Point Elevation

A similar change occurs with the boiling point of water. The boiling point elevation (ΔT_b) is determined from the molality (m) of the particles in the solution and the boiling point constant, K_b, which is $\frac{0.52 \text{ °C}}{m}$ for water.

$$\Delta T_b = m \times K_b$$

SAMPLE PROBLEM 12.19 Calculating the Boiling Point of a Solution

Propylene glycol, $C_3H_8O_2$, is a nonelectrolyte that is added to the water in a radiator to increase the boiling point. If 4.6 mol of propylene glycol is added to 1.55 kg of water (solvent) in a radiator, what is the boiling point, in degrees Celsius, of the solution?

TRY IT FIRST

SOLUTION

STEP 1 State the given and needed quantities.

	Given	Need	Connect
ANALYZE THE PROBLEM	4.6 mol of propylene glycol, 1.55 kg of water	boiling point	$\Delta T_b = m \times K_b$

STEP 2 Determine the number of moles of solute particles and calculate the molality. Since propylene glycol is a nonelectrolyte, the moles of propylene glycol is equal to the moles of particles.

$$m = \frac{\text{moles of particles}}{\text{kilograms of water}} = \frac{4.6 \text{ mol}}{1.55 \text{ kg}} = 3.0 \ m$$

STEP 3 Calculate the temperature change and add to the boiling point.

$$\Delta T_b = m \times K_b = 3.0 \ \cancel{m} \times \frac{0.52 \, ^\circ\text{C}}{\cancel{m}} = 1.6 \, ^\circ\text{C}$$

$$\begin{aligned} T_\text{solution} &= T_\text{water} + \Delta T_b \\ &= 100.0 \, ^\circ\text{C} + 1.6 \, ^\circ\text{C} \\ &= 101.6 \, ^\circ\text{C} \end{aligned}$$

STUDY CHECK 12.19

A 1.2-mol sample of potassium phosphate, K_3PO_4, a strong electrolyte, is added to 0.26 kg of water. What is the boiling point, in degrees Celsius, of this solution?

ANSWER

109.6 °C

The effect of some solutions on freezing and boiling point is summarized in **TABLE 12.11**.

TABLE 12.11 Effect of Solute Concentration on Freezing and Boiling Points of 1 kg of Water

Substance/kg water	Type of Solute	Molality of Particles	Freezing Point	Boiling Point
Pure water	None	0	0 °C	100 °C
1 mol of $C_2H_6O_2$	Nonelectrolyte	1 m	−1.86 °C	100.52 °C
1 mol of NaCl	Strong electrolyte	2 m	−3.72 °C	101.04 °C
1 mol of $CaCl_2$	Strong electrolyte	3 m	−5.58 °C	101.56 °C

QUESTIONS AND PROBLEMS

12.7 Molality and Freezing Point Lowering/Boiling Point Elevation

LEARNING GOAL Identify a mixture as a solution, a colloid, or a suspension. Using the molality, calculate the new freezing point and new boiling point for a solution.

12.67 Identify the following as characteristic of a solution, colloid, or suspension:
   a. a mixture that cannot be separated by a semipermeable membrane
   b. a mixture that settles out upon standing

12.68 Identify the following as characteristic of a solution, colloid, or suspension:
   a. Particles of this mixture remain inside a semipermeable membrane but pass through filters.
   b. The particles of solute in this mixture are very large and visible.

12.69 In each pair, identify the solution that will have a lower freezing point. Explain.
  a. 1.0 mol of glycerol (nonelectrolyte) and 2.0 mol of ethylene glycol (nonelectrolyte) each in 1.0 kg of water
  b. 0.50 mol of KCl (strong electrolyte) and 0.50 mol of $MgCl_2$ (strong electrolyte) each in 1.0 kg of water

12.70 In each pair, identify the solution that will have a higher boiling point. Explain.
  a. 1.50 mol of LiOH (strong electrolyte) and 3.00 mol of KOH (strong electrolyte) each in 1.0 kg of water
  b. 0.40 mol of $Al(NO_3)_3$ (strong electrolyte) and 0.40 mol of CsCl (strong electrolyte) each in 1.0 kg of water

12.71 Calculate the molality (m) of the following solutions:
  a. 325 g of methanol, CH_3OH, a nonelectrolyte, added to 455 g of water
  b. 640. g of the antifreeze propylene glycol, $C_3H_8O_2$, a nonelectrolyte, dissolved in 1.22 kg of water

12.72 Calculate the molality (m) of the following solutions:
  a. 65.0 g of glucose, $C_6H_{12}O_6$, a nonelectrolyte, dissolved in 112 g of water
  b. 110. g of sucrose, $C_{12}H_{22}O_{11}$, a nonelectrolyte, dissolved in 1.50 kg of water

12.73 Calculate the freezing point and boiling point of each solution in problem 12.71.

12.74 Calculate the freezing point and boiling point of each solution in problem 12.72.

12.8 Properties of Solutions: Osmosis

LEARNING GOAL Describe how the number of particles in solution affects osmosis.

The movement of water into and out of the cells of plants as well as the cells of our bodies is an important biological process that also depends on the solute concentration. In a process called **osmosis**, water molecules move through a semipermeable membrane from the solution with the lower concentration of solute into a solution with the higher solute concentration. In an osmosis apparatus, water is placed on one side of a semipermeable membrane and a sucrose (sugar) solution on the other side. The semipermeable membrane allows water molecules to flow back and forth but blocks the sucrose molecules because they cannot pass through the membrane. Because the sucrose solution has a higher solute concentration, more water molecules flow into the sucrose solution than out of the sucrose solution. The volume level of the sucrose solution rises as the volume level on the water side falls. The increase of water dilutes the sucrose solution to equalize (or attempt to equalize) the concentrations on both sides of the membrane.

Eventually the height of the sucrose solution creates sufficient pressure to equalize the flow of water between the two compartments. This pressure, called **osmotic pressure**, prevents the flow of additional water into the more concentrated solution. Then there is no

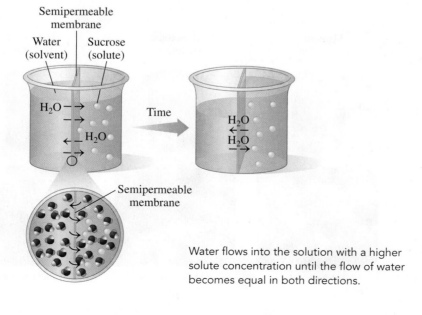

Water flows into the solution with a higher solute concentration until the flow of water becomes equal in both directions.

further change in the volumes of the two solutions. The osmotic pressure depends on the concentration of solute particles in the solution. The greater the number of particles dissolved, the higher its osmotic pressure. In this example, the sucrose solution has a higher osmotic pressure than pure water, which has an osmotic pressure of zero.

In a process called *reverse osmosis*, a pressure greater than the osmotic pressure is applied to a solution so that it is forced through a purification membrane. The flow of water is reversed because water flows from an area of lower water concentration to an area of higher water concentration. The molecules and ions in solution stay behind, trapped by the membrane, while water passes through the membrane. This process of reverse osmosis is used in a few desalination plants to obtain pure water from sea (salt) water. However, the pressure that must be applied requires so much energy that reverse osmosis is not yet an economical method for obtaining pure water in most parts of the world.

Isotonic Solutions

A 0.9% NaCl solution is isotonic with the solute concentration of the blood cells of the body.

Because the cell membranes in biological systems are semipermeable, osmosis is an ongoing process. The solutes in body solutions such as blood, tissue fluids, lymph, and plasma all exert osmotic pressure. Most intravenous (IV) solutions used in a hospital are *isotonic solutions*, which exert the same osmotic pressure as body fluids such as blood. The percent concentration typically used in IV solutions is mass/volume percent (m/v), which is a type of percent concentration we have already discussed. The most typical isotonic solutions are 0.9% (m/v) NaCl solution, or 0.9 g of NaCl/100. mL of solution, and 5% (m/v) glucose, or 5 g of glucose/100. mL of solution. Although they do not contain the same kinds of particles, a 0.9% (m/v) NaCl solution as well as a 5% (m/v) glucose solution both have the same osmotic pressure. A red blood cell placed in an isotonic solution retains its volume because there is an equal flow of water into and out of the cell (see **FIGURE 12.11a**).

Hypotonic and Hypertonic Solutions

If a red blood cell is placed in a solution that is not isotonic, the differences in osmotic pressure inside and outside the cell can drastically alter the volume of the cell. When a red blood cell is placed in a *hypotonic solution*, which has a lower solute concentration (*hypo* means "lower than"), water flows into the cell by osmosis. The increase in fluid causes the cell to swell, and possibly burst—a process called *hemolysis* (see **FIGURE 12.11b**). A similar process occurs when you place dehydrated food, such as raisins or dried fruit, in water. The water enters the cells, and the food becomes plump and smooth.

If a red blood cell is placed in a *hypertonic solution*, which has a higher solute concentration (*hyper* means "greater than"), water flows out of the cell into the hypertonic

FIGURE 12.11 ▶ **(a)** In an isotonic solution, a red blood cell retains its normal volume. **(b)** Hemolysis: In a hypotonic solution, water flows into a red blood cell, causing it to swell and burst. **(c)** Crenation: In a hypertonic solution, water leaves the red blood cell, causing it to shrink.

◉ What happens to a red blood cell placed in a 4% NaCl solution?

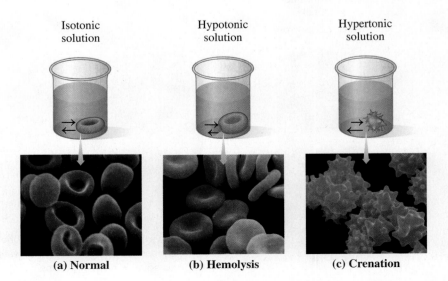

(a) **Normal**   (b) **Hemolysis**   (c) **Crenation**

solution by osmosis. Suppose a red blood cell is placed in a 10% (m/v) NaCl solution. Because the osmotic pressure in the red blood cell is the same as a 0.9% (m/v) NaCl solution, the 10% (m/v) NaCl solution has a much greater osmotic pressure. As water leaves the cell, it shrinks, a process called *crenation* (see **FIGURE 12.11c**). A similar process occurs when making pickles, which uses a hypertonic salt solution that causes the cucumbers to shrivel as they lose water.

SAMPLE PROBLEM 12.20 Isotonic, Hypotonic, and Hypertonic Solutions

Describe each of the following solutions as isotonic, hypotonic, or hypertonic. Indicate whether a red blood cell placed in each solution will undergo hemolysis, crenation, or no change.

a. a 5% (m/v) glucose solution
b. a 0.2% (m/v) NaCl solution

TRY IT FIRST

SOLUTION

a. A 5% (m/v) glucose solution is isotonic. A red blood cell will not undergo any change.
b. A 0.2% (m/v) NaCl solution is hypotonic. A red blood cell will undergo hemolysis.

STUDY CHECK 12.20

What will happen to a red blood cell placed in a 10% (m/v) glucose solution?

ANSWER

The red blood cell will shrink (crenate).

Dialysis

Dialysis is a process that is similar to osmosis. In dialysis, a semipermeable membrane, called a dialyzing membrane, permits small solute molecules and ions as well as solvent water molecules to pass through, but it retains large particles, such as colloids. Dialysis is a way to separate solution particles from colloids.

Suppose we fill a cellophane bag with a solution containing NaCl, glucose, starch, and protein and place it in pure water. Cellophane is a dialyzing membrane, and the sodium ions, chloride ions, and glucose molecules will pass through it into the surrounding water. However, large colloidal particles, like starch and protein, remain inside. Water molecules will flow into the cellophane bag. Eventually the concentrations of sodium ions, chloride ions, and glucose molecules inside and outside the dialysis bag become equal. To remove more NaCl or glucose, the cellophane bag must be placed in a fresh sample of pure water.

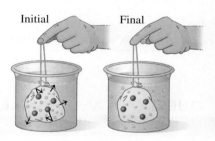

Initial   Final

● Solution particles such as Na⁺, Cl⁻, glucose

● Colloidal particles such as protein, starch

Solution particles pass through a dialyzing membrane, but colloidal particles are retained.

CHEMISTRY LINK TO HEALTH
Dialysis by the Kidneys and the Artificial Kidney

The fluids of the body undergo dialysis by the membranes of the kidneys, which remove waste materials, excess salts, and water. In an adult, each kidney contains about 2 million nephrons. At the top of each nephron, there is a network of arterial capillaries called the *glomerulus*.

As blood flows into the glomerulus, small particles, such as amino acids, glucose, urea, water, and certain ions, will move through the capillary membranes into the nephron. As this solution moves through the nephron, substances still of value to the body (such as amino acids, glucose, certain ions, and 99% of the water) are reabsorbed. The major waste product, urea, is excreted in the urine.

Hemodialysis

If the kidneys fail to dialyze waste products, increased levels of urea can become life-threatening in a relatively short time. A person with kidney failure must use an artificial kidney, which cleanses the blood by *hemodialysis*.

A typical artificial kidney machine contains a large tank filled with water containing selected electrolytes. In the center of this dialyzing bath (dialysate), there is a dialyzing coil or membrane made of cellulose tubing. As the patient's blood flows through the dialyzing coil, the highly concentrated waste products dialyze out of the blood. No blood is lost because the membrane is not permeable to large particles such as red blood cells.

Dialysis patients do not produce much urine. As a result, they retain large amounts of water between dialysis treatments, which produces a strain on the heart. The intake of fluids for a dialysis patient may be restricted to as little as a few teaspoons of water a day. In the dialysis procedure, the pressure of the blood is increased as it circulates through the dialyzing coil so water can be squeezed out of the blood. For some dialysis patients, 2 to 10 L of water may be removed during one treatment. Dialysis patients have from two to three treatments a week, each treatment requiring about 5 to 7 h. Some of the newer treatments require less time. For many patients, dialysis is done at home with a home dialysis unit.

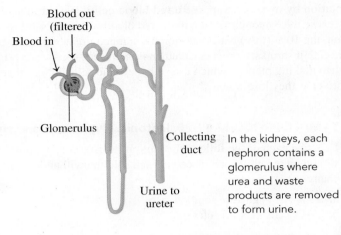

In the kidneys, each nephron contains a glomerulus where urea and waste products are removed to form urine.

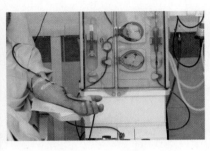

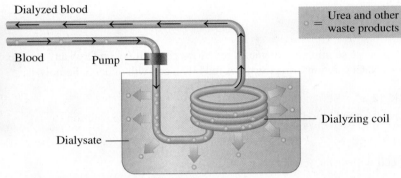

During dialysis, waste products and excess water are removed from the blood.

QUESTIONS AND PROBLEMS

12.8 Properties of Solutions: Osmosis

LEARNING GOAL Describe how the number of particles in solution affects osmosis.

12.75 A 10% (m/v) starch solution is separated from a 1% (m/v) starch solution by a semipermeable membrane. (Starch is a colloid.)
   a. Which compartment has the higher osmotic pressure?
   b. In which direction will water flow initially?
   c. In which compartment will the volume level rise?

12.76 A 0.1% (m/v) albumin solution is separated from a 2% (m/v) albumin solution by a semipermeable membrane. (Albumin is a colloid.)
   a. Which compartment has the higher osmotic pressure?
   b. In which direction will water flow initially?
   c. In which compartment will the volume level rise?

12.77 Indicate the compartment (**A** or **B**) that will increase in volume for each of the following pairs of solutions separated by a semipermeable membrane:

A	B
a. 5% (m/v) sucrose	10% (m/v) sucrose
b. 8% (m/v) albumin	4% (m/v) albumin
c. 0.1% (m/v) starch	10% (m/v) starch

12.78 Indicate the compartment (**A** or **B**) that will increase in volume for each of the following pairs of solutions separated by a semipermeable membrane:

A	B
a. 20% (m/v) starch	10% (m/v) starch
b. 10% (m/v) albumin	2% (m/v) albumin
c. 0.5% (m/v) sucrose	5% (m/v) sucrose

Applications

12.79 Are the following solutions isotonic, hypotonic, or hypertonic compared with a red blood cell?
   a. distilled H_2O
   b. 1% (m/v) glucose
   c. 0.9% (m/v) NaCl
   d. 15% (m/v) glucose

12.80 Will a red blood cell undergo crenation, hemolysis, or no change in each of the following solutions?
   a. 1% (m/v) glucose
   b. 2% (m/v) NaCl
   c. 5% (m/v) glucose
   d. 0.1% (m/v) NaCl

12.81 Each of the following mixtures is placed in a dialyzing bag and immersed in distilled water. Which substances will be found outside the bag in the distilled water?
a. NaCl solution
b. starch solution (colloid) and alanine (an amino acid) solution
c. NaCl solution and starch solution (colloid)
d. urea solution

12.82 Each of the following mixtures is placed in a dialyzing bag and immersed in distilled water. Which substances will be found outside the bag in the distilled water?
a. KCl solution and glucose solution
b. albumin solution (colloid)
c. an albumin solution (colloid), KCl solution, and glucose solution
d. urea solution and NaCl solution

Follow Up

USING DIALYSIS FOR RENAL FAILURE

As a dialysis patient, Michelle has a 4-h dialysis treatment three times a week. When she arrives at the dialysis clinic, her weight, temperature, and blood pressure are taken and blood tests are done to determine the level of electrolytes and urea in her blood. In the dialysis center, tubes to the dialyzer are connected to the catheter she has had implanted.

Blood is then pumped out of her body, through the dialyzer where it is filtered, and returned to her body. As Michelle's blood flows through the dialyzer, electrolytes from the dialysate move into her blood, and waste products in her blood move into the dialysate, which is continually renewed. To achieve normal serum electrolyte levels, dialysate fluid contains sodium, chloride, and magnesium levels that are equal to serum concentrations. These electrolytes are removed from the blood only if their concentrations are higher than normal. Typically, in dialysis patients, the potassium ion level is higher than normal. Therefore, initial dialysis may start

with a low concentration of potassium ion in the dialysate. During dialysis excess fluid is removed by osmosis. A 4-h dialysis session requires at least 120 L of dialysis fluid. During dialysis, the electrolytes in the dialysate are adjusted until the electrolytes have the same levels as normal serum. Initially the dialysate solution prepared for Michelle contains the following: HCO_3^-, K^+, Na^+, Ca^{2+}, Mg^{2+}, Cl^-, glucose.

Applications

12.83 A doctor orders 0.075 g of chlorpromazine, which is used to treat schizophrenia. If the stock solution is 2.5% (m/v), how many milliliters are administered to the patient?

12.84 A doctor orders 5.0 mg of compazine, which is used to treat nausea, vertigo, and migraine headaches. If the stock solution is 2.5% (m/v), how many milliliters are administered to the patient?

12.85 A $CaCl_2$ solution is given to increase blood levels of calcium. If a patient receives 5.0 mL of a 10.% (m/v) $CaCl_2$ solution, how many grams of $CaCl_2$ were given?

12.86 An intravenous solution of mannitol is used as a diuretic to increase the loss of sodium and chloride by a patient. If a patient receives 30.0 mL of a 25% (m/v) mannitol solution, how many grams of mannitol were given?

CONCEPT MAP

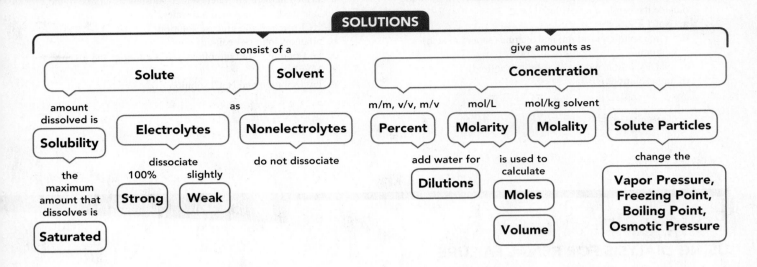

SOLUTIONS

consist of a

Solute **Solvent**

give amounts as

Concentration

amount dissolved is

Solubility

as

Electrolytes **Nonelectrolytes**

m/m, v/v, m/v

Percent

mol/L

Molarity

mol/kg solvent

Molality

Solute Particles

the maximum amount that dissolves is

Saturated

dissociate 100%

Strong

slightly

Weak

do not dissociate

add water for

Dilutions

is used to calculate

Moles

Volume

change the

Vapor Pressure, Freezing Point, Boiling Point, Osmotic Pressure

---

CHAPTER REVIEW

12.1 Solutions

LEARNING GOAL Identify the solute and solvent in a solution; describe the formation of a solution.

- A solution forms when a solute dissolves in a solvent.
- In a solution, the particles of solute are evenly dispersed in the solvent.
- The solute and solvent may be solid, liquid, or gas.
- The polar O—H bond leads to hydrogen bonding between water molecules.
- An ionic solute dissolves in water, a polar solvent, because the polar water molecules attract and pull the ions into solution, where they become hydrated.
- The expression *like dissolves like* means that a polar or an ionic solute dissolves in a polar solvent while a nonpolar solute dissolves in a nonpolar solvent.

11.2 Electrolytes and Nonelectrolytes

LEARNING GOAL Identify solutes as electrolytes or nonelectrolytes.

- Substances that produce ions in water are called electrolytes because their solutions will conduct an electrical current.
- Strong electrolytes are completely dissociated, whereas weak electrolytes are only partially ionized.
- Nonelectrolytes are substances that dissolve in water to produce only molecules and cannot conduct electrical currents.

Strong electrolyte

12.3 Solubility

LEARNING GOAL Define solubility; distinguish between an unsaturated and a saturated solution. Identify an ionic compound as soluble or insoluble.

Dissolved solute

Undissolved solute

Dissolving Recrystallizing

Saturated solution

- The solubility of a solute is the maximum amount of a solute that can dissolve in 100. g of solvent.
- A solution that contains the maximum amount of dissolved solute is a saturated solution.
- A solution containing less than the maximum amount of dissolved solute is unsaturated.
- An increase in temperature increases the solubility of most solids in water, but decreases the solubility of gases in water.
- Ionic compounds that are soluble in water usually contain Li^+, Na^+, K^+, NH_4^+, NO_3^-, or acetate, $C_2H_3O_2^-$.

12.4 Solution Concentrations

LEARNING GOAL Calculate the concentration of a solute in a solution; use concentration as conversion factors to calculate the amount of solute or solution.

Water added to make a solution   250 mL

5.0 g of KI   250 mL of KI solution

- Mass percent expresses the mass/mass (m/m) ratio of the mass of solute to the mass of solution multiplied by 100%.
- Percent concentration can also be expressed as volume/volume (v/v) and mass/volume (m/v) ratios.
- Molarity is the moles of solute per liter of solution.

- In calculations of grams or milliliters of solute or solution, the concentration is used as a conversion factor.
- Molarity (or mol/L) is written as conversion factors to solve for moles of solute or volume of solution.

12.5 Dilution of Solutions

LEARNING GOAL Describe the dilution of a solution; calculate the unknown concentration or volume when a solution is diluted.

- In dilution, a solvent such as water is added to a solution, which increases its volume and decreases its concentration.

12.6 Chemical Reactions in Solution

LEARNING GOAL Given the volume and concentration of a solution in a chemical reaction, calculate the amount of a reactant or product in the reaction.

- When solutions are involved in chemical reactions, the moles of a substance in solution can be determined from the volume and molarity of the solution.
- When mass, volume, and molarities of substances in a reaction are given, the balanced equation is used to determine the quantities or concentrations of other substances in the reaction.

12.7 Molality and Freezing Point Lowering/Boiling Point Elevation

LEARNING GOAL Identify a mixture as a solution, a colloid, or a suspension. Using the molality, calculate the new freezing point and new boiling point for a solution.

- Colloids contain particles that pass through most filters but do not settle out or pass through semipermeable membranes.
- Suspensions have very large particles that settle out.
- The particles in a solution lower the vapor pressure, lower the freezing point, and raise the boiling point.
- Molality is the moles of solute per kilogram of solvent, usually water.

12.8 Properties of Solutions: Osmosis

LEARNING GOAL Describe how the number of particles in solution affects osmosis.

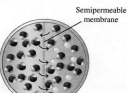
Semipermeable membrane

- The particles in a solution increase the osmotic pressure.
- In osmosis, solvent (water) passes through a semipermeable membrane from a solution with a lower osmotic pressure (lower solute concentration) to a solution with a higher osmotic pressure (higher solute concentration).
- Isotonic solutions have osmotic pressures equal to that of body fluids.
- A red blood cell maintains its volume in an isotonic solution but swells in a hypotonic solution, and shrinks in a hypertonic solution.
- In dialysis, water and small solute particles pass through a dialyzing membrane, while larger particles are retained.

KEY TERMS

colloid A mixture having particles that are moderately large. Colloids pass through filters but cannot pass through semipermeable membranes.

concentration A measure of the amount of solute that is dissolved in a specified amount of solution.

dialysis A process in which water and small solute particles pass through a semipermeable membrane.

dilution A process by which water (solvent) is added to a solution to increase the volume and decrease (dilute) the solute concentration.

electrolyte A substance that produces ions when dissolved in water; its solution conducts electricity.

Henry's law The solubility of a gas in a liquid is directly related to the pressure of that gas above the liquid.

hydration The process of surrounding dissolved ions by water molecules.

mass percent (m/m) The grams of solute in 100. g of solution.

mass/volume percent (m/v) The grams of solute in 100. mL of solution.

molality (*m*) The number of moles of solute particles in exactly 1 kg of solvent.

molarity (M) The number of moles of solute in exactly 1 L of solution.

nonelectrolyte A substance that dissolves in water as molecules; its solution does not conduct an electrical current.

osmosis The flow of a solvent, usually water, through a semipermeable membrane into a solution of higher solute concentration.

osmotic pressure The pressure that prevents the flow of water into the more concentrated solution.

saturated solution A solution containing the maximum amount of solute that can dissolve at a given temperature. Any additional solute will remain undissolved in the container.

solubility The maximum amount of solute that can dissolve in 100. g of solvent, usually water, at a given temperature.

solubility rules A set of guidelines that states whether an ionic compound is soluble or insoluble in water.

solute The component in a solution that is present in the lesser amount.

solution A homogeneous mixture in which the solute is made up of small particles (ions or molecules) that can pass through filters and semipermeable membranes.

solvent The substance in which the solute dissolves; usually the component present in greater amount.

strong electrolyte A compound that ionizes completely when it dissolves in water; its solution is a good conductor of electricity.

suspension A mixture in which the solute particles are large enough and heavy enough to settle out and be retained by both filters and semipermeable membranes.

unsaturated solution A solution that contains less solute than can be dissolved.

volume percent (v/v) The milliliters of solute in 100. mL of solution.

weak electrolyte A substance that produces only a few ions along with many molecules when it dissolves in water; its solution is a weak conductor of electricity.

CORE CHEMISTRY SKILLS

The chapter section containing each Core Chemistry Skill is shown in parentheses at the end of each heading.

Using Solubility Rules (12.3)

- Soluble ionic compounds contain Li^+, Na^+, K^+, NH_4^+, NO_3^-, or $C_2H_3O_2^-$ (acetate).
- Ionic compounds containing Cl^-, Br^-, or I^- are soluble, but if they are combined with Ag^+, Pb^{2+}, or Hg_2^{2+}, they are insoluble.
- Most ionic compounds containing SO_4^{2-} are soluble, but if they contain Ba^{2+}, Pb^{2+}, Ca^{2+}, Sr^{2+}, or Hg_2^{2+} they are insoluble.
- Most other ionic compounds including those containing the anions CO_3^{2-}, S^{2-}, PO_4^{3-}, or OH^- are insoluble.
- To write an equation or ionic equation for the formation of an insoluble ionic compound, we write the cations and anions to identify any combination that would form an insoluble ionic compound.

Example: Determine if an ionic compound forms when solutions of $CaCl_2$ and K_2CO_3 are mixed. If so, write the net ionic equation for the reaction.

Answer: In the ionic equation, we show all the ions of the reactants and show the ionic compound $CaCO_3$ as a solid.

$$Ca^{2+}(aq) + 2Cl^-(aq) + 2K^+(aq) + CO_3^{2-}(aq) \longrightarrow$$
$$CaCO_3(s) + 2K^+(aq) + 2Cl^-(aq)$$

For the net ionic equation, spectator ions that appear on both sides of the equation are removed.

$$Ca^{2+}(aq) + CO_3^{2-}(aq) \longrightarrow CaCO_3(s)$$
Net ionic equation

Calculating Concentration (12.4)

The amount of solute dissolved in a certain amount of solution is called the concentration of the solution.

- Mass percent (m/m) $= \dfrac{\text{mass of solute}}{\text{mass of solution}} \times 100\%$

- Volume percent (v/v) $= \dfrac{\text{volume of solute}}{\text{volume of solution}} \times 100\%$

- Mass/volume percent (m/v) $= \dfrac{\text{grams of solute}}{\text{milliliters of solution}} \times 100\%$

- Molarity (M) $= \dfrac{\text{moles of solute}}{\text{liters of solution}}$

Example: What is the mass/volume percent (m/v) and the molarity (M) of 225 mL (0.225 L) of a LiCl solution that contains 17.1 g of LiCl?

Answer: Mass/volume % (m/v) $= \dfrac{\text{grams of solute}}{\text{milliliters of solution}} \times 100\%$

$= \dfrac{17.1 \text{ g LiCl}}{225 \text{ mL solution}} \times 100\%$

$= 7.60\%$ (m/v) LiCl solution

moles of LiCl $= 17.1 \text{ g LiCl} \times \dfrac{1 \text{ mol LiCl}}{42.39 \text{ g LiCl}}$

$= 0.403$ mol of LiCl

Molarity (M) $= \dfrac{\text{moles of solute}}{\text{liters of solution}} = \dfrac{0.403 \text{ mol LiCl}}{0.225 \text{ L solution}}$

$= 1.79$ M LiCl solution

Using Concentration as a Conversion Factor (12.4)

- When we need to calculate the amount of solute or solution, we use the concentration as a conversion factor.
- For example, the concentration of a 4.50 M HCl solution means there are 4.50 mol of HCl in 1 L of HCl solution, which gives two conversion factors written as

$$\dfrac{4.50 \text{ mol HCl}}{1 \text{ L solution}} \quad \text{and} \quad \dfrac{1 \text{ L solution}}{4.50 \text{ mol HCl}}$$

Example: How many milliliters of a 4.50 M HCl solution will provide 41.2 g of HCl?

Answer:

$$41.2 \text{ g HCl} \times \dfrac{1 \text{ mol HCl}}{36.46 \text{ g HCl}} \times \dfrac{1 \text{ L solution}}{4.50 \text{ mol HCl}} \times \dfrac{1000 \text{ mL solution}}{1 \text{ L solution}}$$

$= 251$ mL of HCl solution

Calculating the Quantity of a Reactant or Product for a Chemical Reaction in Solution (12.6)

- When chemical reactions involve aqueous solutions of reactants or products, we use the balanced chemical equation, the molarity, and the volume to determine the moles or grams of the reactants or products.

Example: How many grams of zinc metal will react with 0.315 L of a 1.20 M HCl solution?

$$Zn(s) + 2HCl(aq) \longrightarrow H_2(g) + ZnCl_2(aq)$$

Answer:

$$0.315 \text{ L solution} \times \dfrac{1.20 \text{ mol HCl}}{1 \text{ L solution}} \times \dfrac{1 \text{ mol Zn}}{2 \text{ mol HCl}} \times \dfrac{65.41 \text{ g Zn}}{1 \text{ mol Zn}}$$

$= 12.4$ g of Zn

Calculating the Freezing Point/Boiling Point of a Solution (12.7)

- The particles in a solution lower the freezing point, raise the boiling point, and increase the osmotic pressure.
- The freezing point lowering (ΔT_f) is determined from the molality (m) of the particles in the solution and the freezing point constant, K_f.

$$\Delta T_f = m \times K_f$$

- The boiling point elevation (ΔT_b) is determined from the molality (m) of the particles in the solution and the boiling point constant, K_b.

$$\Delta T_b = m \times K_b$$

Example: What is the boiling point of a solution that contains 1.5 mol of the strong electrolyte KCl in 1 kg of water?

Answer: A solution of 1.5 mol of KCl in 1 kg of water, which contains 3.0 mol of particles (1.5 mol of K^+ and 1.5 mol of Cl^-), is a 3.0-m solution.

$$\Delta T_b = m \times K_b = 3.0 \, m \times \dfrac{0.52 \,^\circ C}{m} = 1.6 \,^\circ C$$

$$T_{\text{solution}} = T_{\text{water}} + \Delta T_b$$
$$= 100.0 \,^\circ C + 1.6 \,^\circ C$$
$$= 101.6 \,^\circ C$$

UNDERSTANDING THE CONCEPTS

The chapter sections to review are shown in parentheses at the end of each question.

12.87 Match the diagrams with the following: (12.1)
  a. a polar solute and a polar solvent
  b. a nonpolar solute and a polar solvent
  c. a nonpolar solute and a nonpolar solvent

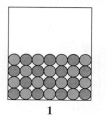
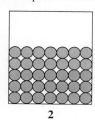

  1   2

12.88 If all the solute is dissolved in diagram **1**, how would heating or cooling the solution cause each of the following changes? (12.3)
  a. 2 to **3**   **b. 2** to **1**

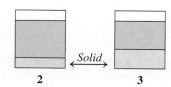

  1   2   ←*Solid*→   3

12.89 Select the diagram that represents the solution formed by a solute 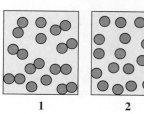 that is a (12.2)
  a. nonelectrolyte
  b. weak electrolyte
  c. strong electrolyte

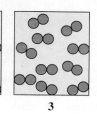

  1   2   3

12.90 Select the container that represents the dilution of a 4% (m/v) KCl solution to give each of the following: (12.5)
  a. a 2% (m/v) KCl solution
  b. a 1% (m/v) KCl solution

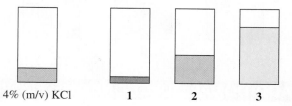

  4% (m/v) KCl   1   2   3

Use the following illustration of beakers and solutions for problems 12.91 and 12.92:

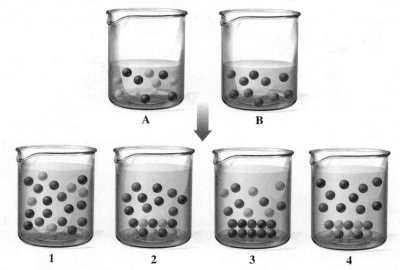

  A   B

  1   2   3   4

12.91 Use the following ions: (12.3)

  Na^+ ⚪ Cl^- ⚫ Ag^+ ⚫ NO_3^- ⚫

  a. Select the beaker (**1**, **2**, **3**, or **4**) that contains the products after the solutions in beakers **A** and **B** are mixed.
  b. If an insoluble ionic compound forms, write the ionic equation.
  c. If a reaction occurs, write the net ionic equation.

12.92 Use the following ions: (12.3)

  K^+ ⚪ NO_3^- ⚫ NH_4^+ ⚫ Br^- ⚫

  a. Select the beaker (**1**, **2**, **3**, or **4**) that contains the products after the solutions in beakers **A** and **B** are mixed.

  b. If an insoluble ionic compound forms, write the ionic equation.
  c. If a reaction occurs, write the net ionic equation.

12.93 A pickle is made by soaking a cucumber in brine, a salt-water solution. What makes the smooth cucumber become wrinkled like a prune? (12.8)

12.94 Why do lettuce leaves in a salad wilt after a vinaigrette dressing containing salt is added? (12.8)

12.95 A semipermeable membrane separates two compartments, **A** and **B**. If the levels in **A** and **B** are equal initially, select the diagram that illustrates the final levels in **a** to **d**: (12.8)

A B   A B   A B

1   2   3

Solution in A	Solution in B
a. 2% (m/v) starch	8% (m/v) starch
b. 1% (m/v) starch	1% (m/v) starch
c. 5% (m/v) sucrose	1% (m/v) sucrose
d. 0.1% (m/v) sucrose	1% (m/v) sucrose

12.96 Select the diagram that represents the shape of a red blood cell when placed in each of the following **a** to **e**: (12.8)

1   2   3

Normal red blood cell

a. 0.9% (m/v) NaCl solution
b. 10% (m/v) glucose solution
c. 0.01% (m/v) NaCl solution
d. 5% (m/v) glucose solution
e. 1% (m/v) glucose solution

ADDITIONAL QUESTIONS AND PROBLEMS

12.97 Why does iodine dissolve in hexane, but not in water? (12.1)

12.98 How do temperature and pressure affect the solubility of solids and gases in water? (12.3)

12.99 Potassium nitrate has a solubility of 32 g of KNO_3 in 100. g of H_2O at 20 °C. Determine if each of the following forms an unsaturated or saturated solution at 20 °C: (12.3)
  a. adding 32 g of KNO_3 to 200. g of H_2O
  b. adding 19 g of KNO_3 to 50. g of H_2O
  c. adding 68 g of KNO_3 to 150. g of H_2O

12.100 Potassium chloride has a solubility of 43 g of KCl in 100. g of H_2O at 50 °C. Determine if each of the following forms an unsaturated or saturated solution at 50 °C: (12.3)
  a. adding 25 g of KCl to 100. g of H_2O
  b. adding 15 g of KCl to 25 g of H_2O
  c. adding 86 g of KCl to 150. g of H_2O

12.101 Indicate whether each of the following ionic compounds is soluble or insoluble in water: (12.3)
  a. KCl   **b.** $MgSO_4$ **c.** CuS
  d. $AgNO_3$ **e.** $Ca(OH)_2$

12.102 Indicate whether each of the following ionic compounds is soluble or insoluble in water: (12.3)
  a. $CuCO_3$ **b.** FeO   **c.** $Mg_3(PO_4)_2$
  d. $(NH_4)_2SO_4$ **e.** $NaHCO_3$

12.103 Write the net ionic equation to show the formation of a solid (insoluble ionic compound) when the following solutions are mixed. Write *none* if no solid forms. (12.3)
  a. $AgNO_3(aq)$ and $LiCl(aq)$
  b. $NaCl(aq)$ and $KNO_3(aq)$
  c. $Na_2SO_4(aq)$ and $BaCl_2(aq)$

12.104 Write the net ionic equation to show the formation of a solid (insoluble ionic compound) when the following solutions are mixed. Write *none* if no solid forms. (12.3)
  a. $Ca(NO_3)_2(aq)$ and $Na_2S(aq)$
  b. $Na_3PO_4(aq)$ and $Pb(NO_3)_2(aq)$
  c. $FeCl_3(aq)$ and $NH_4NO_3(aq)$

12.105 If NaCl has a solubility of 36.0 g in 100. g of H_2O at 20 °C, how many grams of water are needed to prepare a saturated solution containing 80.0 g of NaCl? (12.3)

12.106 If the solid NaCl in a saturated solution of NaCl continues to dissolve, why is there no change in the concentration of the NaCl solution? (12.3)

12.107 Calculate the mass percent (m/m) of a solution containing 15.5 g of Na_2SO_4 and 75.5 g of H_2O. (12.4)

12.108 Calculate the mass percent (m/m) of a solution containing 26 g of K_2CO_3 and 724 g of H_2O. (12.4)

12.109 How many milliliters of a 12% (v/v) propyl alcohol solution would you need to obtain 4.5 mL of propyl alcohol? (12.4)

12.110 An 80-proof brandy is a 40.% (v/v) ethanol solution. The "proof" is twice the percent concentration of alcohol in the beverage. How many milliliters of alcohol are present in 750 mL of brandy? (12.4)

12.111 How many liters of a 12% (m/v) KOH solution would you need to obtain 86.0 g of KOH? (12.4)

12.112 How many liters of a 5.0% (m/v) glucose solution would you need to obtain 75 g of glucose? (12.4)

12.113 What is the molarity of a solution containing 8.0 g of NaOH in 400. mL of NaOH solution? (12.4)

12.114 What is the molarity of a solution containing 15.6 g of KCl in 274 mL of KCl solution? (12.4)

12.115 How many milliliters of a 1.75 M LiCl solution contain 15.2 g of LiCl? (12.4)

12.116 How many milliliters of a 1.50 M NaBr solution contain 75.0 g of NaBr? (12.4)

12.117 How many liters of a 2.50 M KNO_3 solution can be prepared from 60.0 g of KNO_3? (12.4)

12.118 How many liters of a 4.00 M NaCl solution will provide 25.0 g of NaCl? (12.4)

12.119 How many grams of solute are in each of the following solutions? (12.4)

a. 2.5 L of a 3.0 M $Al(NO_3)_3$ solution

b. 75 mL of a 0.50 M $C_6H_{12}O_6$ solution

c. 235 mL of a 1.80 M LiCl solution

12.120 How many grams of solute are in each of the following solutions? (12.4)

a. 0.428 L of a 0.450 M K_2CO_3 solution

b. 10.5 mL of a 2.50 M $AgNO_3$ solution

c. 28.4 mL of a 6.00 M H_3PO_4 solution

12.121 Calculate the final concentration of the solution when water is added to prepare each of the following: (12.5)

a. 25.0 mL of a 0.200 M NaBr solution is diluted to 50.0 mL

b. 15.0 mL of a 12.0% (m/v) K_2SO_4 solution is diluted to 40.0 mL

c. 75.0 mL of a 6.00 M NaOH solution is diluted to 255 mL

12.122 Calculate the final concentration of the solution when water is added to prepare each of the following: (12.5)

a. 25.0 mL of a 18.0 M HCl solution is diluted to 500. mL

b. 50.0 mL of a 15.0% (m/v) NH_4Cl solution is diluted to 125 mL

c. 4.50 mL of a 8.50 M KOH solution is diluted to 75.0 mL

12.123 What is the initial volume, in milliliters, needed to prepare each of the following diluted solutions? (12.5)

a. 250 mL of 3.0% (m/v) HCl from 10.0% (m/v) HCl

b. 500. mL of 0.90% (m/v) NaCl from 5.0% (m/v) NaCl

c. 350. mL of 2.00 M NaOH from 6.00 M NaOH

12.124 What is the initial volume, in milliliters, needed to prepare each of the following diluted solutions? (12.5)

a. 250 mL of 5.0% (m/v) glucose from 20.% (m/v) glucose

b. 45.0 mL of 1.0% (m/v) $CaCl_2$ from 5.0% (m/v) $CaCl_2$

c. 100. mL of 6.00 M H_2SO_4 from 18.0 M H_2SO_4

12.125 What is the final volume, in milliliters, when 25.0 mL of each of the following solutions is diluted to provide the given concentration? (12.5)

a. 10.0% (m/v) HCl solution to give a 2.50% (m/v) HCl solution

b. 5.00 M HCl solution to give a 1.00 M HCl solution

c. 6.00 M HCl solution to give a 0.500 M HCl solution

12.126 What is the final volume, in milliliters, when 5.00 mL of each of the following solutions is diluted to provide the given concentration? (12.5)

a. 20.0% (m/v) NaOH solution to give a 4.00% (m/v) NaOH solution

b. 0.600 M NaOH solution to give a 0.100 M NaOH solution

c. 16.0% (m/v) NaOH solution to give a 2.00% (m/v) NaOH solution

12.127 The antacid Amphojel contains aluminum hydroxide $Al(OH)_3$. How many milliliters of a 6.00 M HCl solution are required to react with 60.0 mL of a 2.00 M $Al(OH)_3$ solution? (12.6)

$$Al(OH)_3(s) + 3HCl(aq) \longrightarrow 3H_2O(l) + AlCl_3(aq)$$

12.128 Cadmium reacts with HCl to produce hydrogen gas and cadmium chloride. What is the molarity (M) of the HCl solution if 250. mL of the HCl solution reacts with excess cadmium to produce 4.20 L of H_2 gas measured at STP? (12.6)

$$Cd(s) + 2HCl(aq) \longrightarrow H_2(g) + CdCl_2(aq)$$

12.129 Calculate the freezing point of each of the following solutions: (12.7)

a. 0.580 mol of lactose, a nonelectrolyte, added to 1.00 kg of water

b. 45.0 g of KCl, a strong electrolyte, dissolved in 1.00 kg of water

c. 1.5 mol of K_3PO_4, a strong electrolyte, dissolved in 1.00 kg of water

12.130 Calculate the boiling point of each of the following solutions: (12.7)

a. 175 g of glucose, $C_6H_{12}O_6$, a nonelectrolyte, added to 1.00 kg of water

b. 1.8 mol of $CaCl_2$, a strong electrolyte, dissolved in 1.00 kg of water

c. 50.0 g of $LiNO_3$, a strong electrolyte, dissolved in 1.00 kg of water

Applications

12.131 An antacid tablet, such as Amphojel, may be taken to reduce excess stomach acid, which is 0.20 M HCl. If one dose of Amphojel contains 450 mg of $Al(OH)_3$, what volume, in milliliters, of stomach acid will be neutralized? (12.6)

$$Al(OH)_3(s) + 3HCl(aq) \longrightarrow 3H_2O(l) + AlCl_3(aq)$$

12.132 Calcium carbonate, $CaCO_3$, reacts with stomach acid, (HCl, hydrochloric acid) according to the following equation: (12.6)

$$CaCO_3(s) + 2HCl(aq) \longrightarrow CO_2(g) + H_2O(l) + CaCl_2(aq)$$

Tums, an antacid, contains $CaCO_3$. If Tums is added to 20.0 mL of a 0.400 M HCl solution, how many grams of CO_2 gas are produced?

CHALLENGE QUESTIONS

The following groups of questions are related to the topics in this chapter. However, they do not all follow the chapter order, and they require you to combine concepts and skills from several sections. These questions will help you increase your critical thinking skills and prepare for your next exam.

12.133 Write the net ionic equation to show the formation of a solid when the following solutions are mixed. Write *none* if no solid forms. (12.3)

a. $AgNO_3(aq) + Na_2SO_4(aq)$

b. $KCl(aq) + Pb(NO_3)_2(aq)$

c. $CaCl_2(aq) + (NH_4)_3PO_4(aq)$

d. $K_2SO_4(aq) + BaCl_2(aq)$

12.134 Write the net ionic equation to show the formation of a solid when the following solutions are mixed. Write *none* if no solid forms. (12.3)

a. $Pb(NO_3)_2(aq) + NaBr(aq)$

b. $AgNO_3(aq) + (NH_4)_2CO_3(aq)$

c. $Na_3PO_4(aq) + Al(NO_3)_3(aq)$

d. $NaOH(aq) + CuCl_2(aq)$

12.135 In a laboratory experiment, a 10.0-mL sample of NaCl solution is poured into an evaporating dish with a mass of 24.10 g. The combined mass of the evaporating dish and NaCl solution is 36.15 g. After heating, the evaporating dish and dry NaCl have a combined mass of 25.50 g. (12.4)

a. What is the mass percent (m/m) of the NaCl solution?
b. What is the molarity (M) of the NaCl solution?
c. If water is added to 10.0 mL of the initial NaCl solution to give a final volume of 60.0 mL, what is the molarity of the diluted NaCl solution?

12.136 In a laboratory experiment, a 15.0-mL sample of KCl solution is poured into an evaporating dish with a mass of 24.10 g. The combined mass of the evaporating dish and KCl solution is 41.50 g. After heating, the evaporating dish and dry KCl have a combined mass of 28.28 g. (12.4)
a. What is the mass percent (m/m) of the KCl solution?
b. What is the molarity (M) of the KCl solution?
c. If water is added to 10.0 mL of the initial KCl solution to give a final volume of 60.0 mL, what is the molarity of the diluted KCl solution?

12.137 Potassium fluoride has a solubility of 92 g of KF in 100. g of H_2O at 18 °C. Determine if each of the following mixtures forms an unsaturated or saturated solution at 18 °C: (12.3)
a. adding 35 g of KF to 25 g of H_2O
b. adding 42 g of KF to 50. g of H_2O
c. adding 145 g of KF to 150. g of H_2O

12.138 Lithium chloride has a solubility of 55 g of LiCl in 100. g of H_2O at 25 °C. Determine if each of the following mixtures forms an unsaturated or saturated solution at 25 °C: (12.3)
a. adding 10 g of LiCl to 15 g of H_2O
b. adding 25 g of LiCl to 50. g of H_2O
c. adding 75 g of LiCl to 150. g of H_2O

12.139 A solution is prepared with 70.0 g of HNO_3 and 130.0 g of H_2O. The HNO_3 solution has a density of 1.21 g/mL. (12.4)
a. What is the mass percent (m/m) of the HNO_3 solution?
b. What is the total volume, in milliliters, of the solution?
c. What is the mass/volume percent (m/v) of the solution?
d. What is the molarity (M) of the solution?

12.140 A solution is prepared by dissolving 22.0 g of NaOH in 118.0 g of water. The NaOH solution has a density of 1.15 g/mL. (12.4)
a. What is the mass percent (m/m) of the NaOH solution?
b. What is the total volume, in milliliters, of the solution?
c. What is the mass/volume percent (m/v) of the solution?
d. What is the molarity (M) of the solution?

12.141 A 355-mL sample of a HCl solution reacts with excess Mg to produce 4.20 L of H_2 gas measured at 745 mmHg and 35 °C. What is the molarity (M) of the HCl solution? (12.6)

$$Mg(s) + 2HCl(aq) \longrightarrow H_2(g) + MgCl_2(aq)$$

12.142 A 255-mL sample of a HCl solution reacts with excess Ca to produce 14.0 L of H_2 gas measured at 732 mmHg and 22 °C. What is the molarity (M) of the HCl solution? (12.6)

$$Ca(s) + 2HCl(aq) \longrightarrow H_2(g) + CaCl_2(aq)$$

12.143 How many moles of each of the following strong electrolytes are needed to give the same freezing point lowering as 1.2 mol of the nonelectrolyte ethylene glycol in 1 kg of water? (12.7)
a. NaCl
b. K_3PO_4

12.144 How many moles of each of the following are needed to give the same freezing point lowering as 3.0 mol of the nonelectrolyte ethylene glycol in 1 kg of water? (12.7)
a. CH_3OH, a nonelectrolyte
b. KNO_3, a strong electrolyte

12.145 The boiling point of a NaCl solution is 101.04 °C. (12.7)
a. What is the molality (m) of the solute particles in the NaCl solution?
b. What is the freezing point of the NaCl solution?

12.146 The freezing point of a $CaCl_2$ solution is −25 °C. (12.7)
a. What is the molality (m) of the solute particles in the $CaCl_2$ solution?
b. What is the boiling point of the $CaCl_2$ solution?

ANSWERS

Answers to Selected Questions and Problems

12.1 a. NaCl, solute; water, solvent
b. water, solute; ethanol, solvent
c. oxygen, solute; nitrogen, solvent

12.3 The polar water molecules pull the K^+ and I^- ions away from the solid and into solution, where they are hydrated.

12.5 a. water   b. CCl_4
c. water   d. CCl_4

12.7 In a solution of KF, only the ions of K^+ and F^- are present in the solvent. In a HF solution, there are a few ions of H^+ and F^- present but mostly dissolved HF molecules.

12.9 a. $KCl(s) \xrightarrow{H_2O} K^+(aq) + Cl^-(aq)$

b. $CaCl_2(s) \xrightarrow{H_2O} Ca^{2+}(aq) + 2Cl^-(aq)$

c. $K_3PO_4(s) \xrightarrow{H_2O} 3K^+(aq) + PO_4^{3-}(aq)$

d. $Fe(NO_3)_3(s) \xrightarrow{H_2O} Fe^{3+}(aq) + 3NO_3^-(aq)$

12.11 a. mostly molecules and a few ions
b. only ions
c. only molecules

12.13 a. strong electrolyte   b. weak electrolyte
c. nonelectrolyte

12.15 a. saturated   b. unsaturated

12.17 a. unsaturated   b. unsaturated
c. saturated

12.19 a. 68 g of KCl   b. 12 g of KCl

12.21 a. The solubility of solid solutes typically increases as temperature increases.
b. The solubility of a gas is less at a higher temperature.
c. Gas solubility is less at a higher temperature and the CO_2 pressure in the can is increased.

12.23 a. soluble   b. insoluble
c. insoluble   d. soluble
e. soluble

12.25 a. No solid forms.
b. $2Ag^+(aq) + 2NO_3^-(aq) + 2K^+(aq) + S^{2-}(aq) \longrightarrow$
$Ag_2S(s) + 2K^+(aq) + 2NO_3^-(aq)$
$2Ag^+(aq) + S^{2-}(aq) \longrightarrow Ag_2S(s)$
c. $Ca^{2+}(aq) + 2Cl^-(aq) + 2Na^+(aq) + SO_4^{2-}(aq) \longrightarrow$
$CaSO_4(s) + 2Na^+(aq) + 2Cl^-(aq)$
$Ca^{2+}(aq) + SO_4^{2-}(aq) \longrightarrow CaSO_4(s)$
d. $3Cu^{2+}(aq) + 6Cl^-(aq) + 6Li^+(aq) + 2PO_4^{3-}(aq) \longrightarrow$
$Cu_3(PO_4)_2(s) + 6Li^+(aq) + 6Cl^-(aq)$
$3Cu^{2+}(aq) + 2PO_4^{3-}(aq) \longrightarrow Cu_3(PO_4)_2(s)$

12.27 A 5.00% (m/m) glucose solution can be made by adding 5.00 g of glucose to 95.00 g of water, while a 5.00% (m/v) glucose solution can be made by adding 5.00 g of glucose to enough water to make 100.0 mL of solution.

12.29 a. 17% (m/m) KCl solution
  b. 5.3% (m/m) sucrose solution
  c. 10.% (m/m) $CaCl_2$ solution

12.31 a. 30.% (m/v) Na_2SO_4 solution
  b. 11% (m/v) sucrose solution

12.33 a. 2.5 g of KCl   **b.** 50. g of NH_4Cl
  c. 25.0 mL of acetic acid

12.35 79.9 mL of alcohol

12.37 a. 20. g of $LiNO_3$ solution
  b. 400. mL of KOH solution
  c. 20. mL of formic acid solution

12.39 a. 0.500 M glucose solution
  b. 0.0356 M KOH solution
  c. 0.250 M NaCl solution

12.41 a. 120. g of NaOH   **b.** 59.6 g of KCl
  c. 5.47 g of HCl

12.43 a. 1.50 L of KBr solution
  b. 10.0 L of NaCl solution
  c. 62.5 mL of $Ca(NO_3)_2$ solution

12.45 a. 983 mL   **b.** 1.70×10^3 mL
  c. 464 mL

12.47 a. 20. g of mannitol   **b.** 240 g of mannitol

12.49 2 L of glucose solution

12.51 Adding water (solvent) to the soup increases the volume and dilutes the tomato soup concentration.

12.53 a. 2.0 M HCl solution
  b. 2.0 M NaOH solution
  c. 2.5% (m/v) KOH solution
  d. 3.0% (m/v) H_2SO_4 solution

12.55 a. 80. mL of HCl solution
  b. 250 mL of LiCl solution
  c. 600. mL of H_3PO_4 solution
  d. 180 mL of glucose solution

12.57 a. 12.8 mL of 4.00 M HNO_3 solution
  b. 11.9 mL of 6.00 M $MgCl_2$ solution
  c. 1.88 mL of 8.00 M KCl solution

12.59 1.0×10^2 mL

12.61 a. 10.4 g of $PbCl_2$
  b. 18.8 mL of $Pb(NO_3)_2$ solution
  c. 1.20 M KCl solution

12.63 a. 206 mL of HCl solution
  b. 11.2 L of H_2 gas
  c. 9.09 M HCl solution

12.65 a. 74.8 mL of 3.50 M HBr solution
  b. 50.4 L of H_2
  c. 1.69 M HBr

12.67 a. solution   **b.** suspension

12.69 a. 2.0 mol of ethylene glycol in 1.0 kg of water will have a lower freezing point because it has more particles in solution.
  b. 0.50 mol of $MgCl_2$ in 1.0 kg of water has a lower freezing point because each formula unit of $MgCl_2$ dissociates in water to give three particles, whereas each formula unit of KCl dissociates to give only two particles.

12.71 a. 22.3 *m*   **b.** 6.89 *m*

12.73 a. freezing point: −41.5 °C; boiling point: 111.6 °C
  b. freezing point: −12.8 °C; boiling point: 103.6 °C

12.75 a. 10% (m/v) starch solution
  b. from the 1% (m/v) starch solution into the 10% (m/v) starch solution
  c. 10% (m/v) starch solution

12.77 a. B 10% (m/v) sucrose solution
  b. A 8% (m/v) albumin solution
  c. B 10% (m/v) starch solution

12.79 a. hypotonic   **b.** hypotonic
  c. isotonic   **d.** hypertonic

12.81 a. Na^+, Cl^- **b.** alanine
  c. Na^+, Cl^- **d.** urea

12.83 3.0 mL of chlorpromazine solution

12.85 0.50 g of $CaCl_2$

12.87 a. 1   **b.** 2
  c. 1

12.89 a. 3   **b.** 1
  c. 2

12.91 a. beaker 3
  b. $Na^+(aq) + Cl^-(aq) + Ag^+(aq) + NO_3^-(aq) \longrightarrow$
     $AgCl(s) + Na^+(aq) + NO_3^-(aq)$
  c. $Ag^+(aq) + Cl^-(aq) \longrightarrow AgCl(s)$

12.93 The skin of the cucumber acts like a semipermeable membrane, and the more dilute solution inside flows into the brine solution.

12.95 a. 2   **b.** 1
  c. 3   **d.** 2

12.97 Because iodine is a nonpolar molecule, it will dissolve in hexane, a nonpolar solvent. Iodine does not dissolve in water because water is a polar solvent.

12.99 a. unsaturated solution   **b.** saturated solution
  c. saturated solution

12.101 a. soluble   **b.** soluble
  c. insoluble   **d.** soluble
  e. insoluble

12.103 a. $Ag^+(aq) + Cl^-(aq) \longrightarrow AgCl(s)$
  b. none
  c. $Ba^{2+}(aq) + SO_4^{2-}(aq) \longrightarrow BaSO_4(s)$

12.105 222 g of water

12.107 17.0% (m/m) Na_2SO_4 solution

12.109 38 mL of propyl alcohol solution

12.111 0.72 L of KOH solution

12.113 0.500 M NaOH solution

12.115 205 mL of LiCl solution

12.117 0.237 L of KNO_3 solution

12.119 a. 1600 g of $Al(NO_3)_3$ **b.** 6.8 g of $C_6H_{12}O_6$
  c. 17.9 g of LiCl

12.121 a. 0.100 M NaBr solution
  b. 4.50% (m/v) K_2SO_4 solution
  c. 1.76 M NaOH solution

12.123 a. 75 mL of 10.0% (m/v) HCl solution
  b. 90. mL of 5.0% (m/v) NaCl solution
  c. 117 mL of 6.00 M NaOH solution

12.125 a. 100. mL   **b.** 125 mL
  c. 300. mL

12.127 60.0 mL of HCl solution

12.129 a. $-1.08\,°C$ **b.** $-2.25\,°C$
     c. $-11\,°C$

12.131 87 mL of HCl solution

12.133 a. $2Ag^+(aq) + SO_4^{2-}(aq) \longrightarrow Ag_2SO_4(s)$
     b. $Pb^{2+}(aq) + 2Cl^-(aq) \longrightarrow PbCl_2(s)$
     c. $3Ca^{2+}(aq) + 2PO_4^{3-}(aq) \longrightarrow Ca_3(PO_4)_2(s)$
     d. $Ba^{2+}(aq) + SO_4^{2-}(aq) \longrightarrow BaSO_4(s)$

12.135 a. 11.6% (m/m) NaCl solution
     b. 2.40 M NaCl solution
     c. 0.400 M NaCl solution

12.137 a. saturated   **b.** unsaturated
     c. saturated

12.139 a. 35.0% (m/m) HNO_3 solution
     b. 165 mL
     c. 42.4% (m/v) HNO_3 solution
     d. 6.73 M HNO_3 solution

12.141 0.917 M HCl solution

12.143 a. 0.60 mol of NaCl   **b.** 0.30 mol of K_3PO_4

12.145 a. 2.0 m **b.** $-3.7\,°C$

13

Reaction Rates and Chemical Equilibrium

PETER, A CHEMICAL oceanographer, is collecting data concerning the amount of dissolved gases, specifically carbon dioxide (CO_2) in the Atlantic Ocean. Studies indicate that CO_2 in the atmosphere has increased as much as 25% since the eighteenth century, which has resulted in a scientific debate regarding its effects. Peter's research involves measuring the amount of dissolved CO_2 in the oceans and trying to determine its impact. The oceans are a complex mixture of many different chemicals including gases, elements and minerals, and organic and particulate matter. As a result, the oceans have been called a "chemical soup," which can complicate a study like Peter's. Peter understands that CO_2 is absorbed in the ocean through a series of equilibrium reactions. An equilibrium reaction is a reversible reaction in which both the products and the reactants are present. If the equilibrium reactions shift according to Le Châtelier's principle, an increase in the CO_2 concentration could eventually increase the amount of dissolved calcium carbonate, $CaCO_3$, which makes up coral reefs and shells.

CAREER

Chemical Oceanographer

A chemical oceanographer, also called a marine chemist, studies the chemistry of the ocean. One area of study includes how chemicals or pollutants enter into and affect the ocean. These include sewage, oil or fuels, chemical fertilizers, and storm drain overflows. Oceanographers analyze how these chemicals interact with seawater, marine life, and sediments, as they can behave differently due to the ocean's varied environmental conditions. Chemical oceanographers also study how the various elements are cycled within the ocean. For instance, oceanographers quantify the amount and rate at which carbon dioxide is absorbed at the ocean's surface and eventually transferred to deep waters. Chemical oceanographers also aid ocean engineers in the development of instruments and vessels that enable researchers to collect data and discover previously unknown marine life.

⊗⊖⊕ KEY MATH SKILLS

- Solving Equations (1.4)
- Converting between Standard Numbers and Scientific Notation (1.5)

CORE CHEMISTRY SKILLS

- Using Significant Figures in Calculations (2.3)
- Balancing a Chemical Equation (8.2)
- Calculating Concentration (12.4)

*These Key Math Skills and Core Chemistry Skills from previous chapters are listed here for your review as you proceed to the new material in this chapter.

13.1 Rates of Reactions

LEARNING GOAL Describe how temperature, concentration, and catalysts affect the rate of a reaction.

Earlier we looked at chemical reactions and determined the amounts of substances that react and the products that form. Now we are interested in how fast a reaction goes. If we know how fast a medication acts on the body, we can adjust the time over which the medication is taken. In construction, substances are added to cement to make it dry faster so that work can continue. Some reactions such as explosions or the formation of precipitates in a solution are very fast. When we roast a turkey or bake a cake, the reaction is slower. Some reactions such as the tarnishing of silver and the aging of the body are much slower (see **FIGURE 13.1**). We will see that some reactions need energy while other reactions produce energy. In this chapter, we will also look at the effect of changing the concentrations of reactants or products on the rate of reaction.

Reaction Rate Increases

5 days

5 months

50 years

FIGURE 13.1 ▶ Reaction rates vary greatly for everyday processes. A banana ripens in a few days, silver tarnishes in a few months, while the aging process of humans takes many years.

Ⓠ How would you compare the rates of the reaction that forms sugars in plants by photosynthesis with the reactions that digest sugars in the body?

For a chemical reaction to take place, the molecules of the reactants must come in contact with each other. The **collision theory** indicates that a reaction takes place only when molecules collide with the proper orientation and sufficient energy. Many collisions can occur, but only a few actually lead to the formation of product. For example, consider the reaction of nitrogen and oxygen molecules (see **FIGURE 13.2**). To form nitrogen oxide (NO) product, the collisions between N_2 and O_2 molecules must place the atoms in the proper alignment. If the molecules are not aligned properly, no reaction takes place.

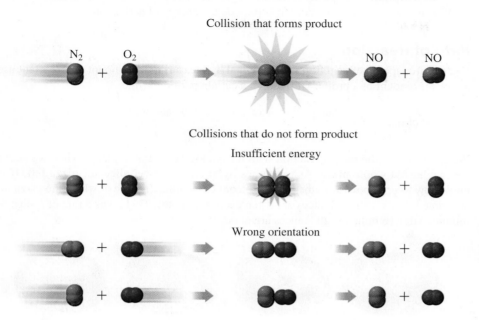

FIGURE 13.2 ▶ Reacting molecules must collide, have a minimum amount of energy, and have the proper orientation to form product.

❓ What happens when reacting molecules collide with the minimum energy but don't have the proper orientation?

Activation Energy

Even when a collision has the proper orientation, there still must be sufficient energy to break the bonds between the atoms of the reactants. The **activation energy** is the minimum amount of energy required to break the bonds between atoms of the reactants. In **FIGURE 13.3**, activation energy appears as an energy hill. The concept of activation energy is analogous to climbing a hill. To reach a destination on the other side, we must have the energy needed to climb to the top of the hill. Once we are at the top, we can run down the other side. The energy needed to get us from our starting point to the top of the hill would be our activation energy.

In the same way, a collision must provide enough energy to push the reactants to the top of the energy hill. Then the reactants may be converted to products. If the energy provided by the collision is less than the activation energy, the molecules simply bounce apart and no reaction occurs. The features that lead to a successful reaction are summarized next.

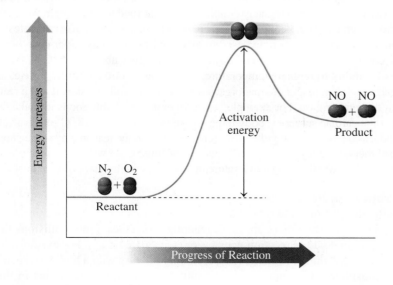

FIGURE 13.3 ▶ The activation energy is the minimum energy needed to convert the colliding molecules into product.

❓ What happens in a collision of reacting molecules that have the proper orientation, but not the energy of activation?

Three Conditions Required for a Reaction to Occur

1. **Collision**   The reactants must collide.
2. **Orientation** The reactants must align properly to break and form bonds.
3. **Energy**   The collision must provide the energy of activation.

Rate of Reaction

The **rate** (or speed) **of reaction** is determined by measuring the amount of a reactant used up, or the amount of a product formed, in a certain period of time.

$$\text{Rate of reaction} = \frac{\text{change in concentration of reactant or product}}{\text{change in time}}$$

We can describe the rate of reaction with the analogy of eating a pizza. When we start to eat, we have a whole pizza. As time goes by, there are fewer slices of pizza left. If we know how long it took to eat the pizza, we could determine the rate at which the pizza was consumed. Let's assume 4 slices are eaten every 8 minutes. That gives a rate of $\frac{1}{2}$ slice per minute. After 16 minutes, all 8 slices are gone.

Rate at Which Pizza Slices Are Eaten				
Slices Eaten	0	4 slices	6 slices	8 slices
Time (min)	0	8 min	12 min	16 min

Factors that Affect the Rate of a Reaction

Reactions with low activation energies go faster than reactions with high activation energies. Some reactions go very fast, while others are very slow. For any reaction, the rate is affected by changes in temperature, changes in the concentration of the reactants, and the addition of catalysts.

Temperature

At higher temperatures, the increase in kinetic energy of the reactants makes them move faster and collide more often, and it provides more collisions with the required energy of activation. Reactions almost always go faster at higher temperatures. For every 10 °C increase in temperature, most reaction rates approximately double. If we want food to cook faster, we increase the temperature. When body temperature rises, there is an increase in the pulse rate, rate of breathing, and metabolic rate. If we are exposed to extreme heat, we may experience *heat stroke*, which is a condition that occurs when body temperature goes above 40.5 °C (105 °F). If the body loses its ability to regulate temperature, body temperature continues to rise, and may cause damage to the brain and internal organs. On the other hand, we slow down a reaction by decreasing the temperature. For example, we refrigerate perishable foods to make them last longer. In some cardiac surgeries, body temperature is decreased to 28 °C so the heart can be stopped and less oxygen is required by the brain. This is also the reason why some people have survived submersion in icy lakes for long periods of time. Cool water or an ice blanket may also be used to decrease the body temperature of a person with hyperthermia or heat stroke.

Concentrations of Reactants

For virtually all reactions, the rate of a reaction increases when the concentration of the reactants increases. When there are more reacting molecules, more collisions that form products can occur, and the reaction goes faster (see **FIGURE 13.4**). For example, a patient having difficulty breathing may be given a breathing mixture with a higher oxygen content than the atmosphere. The increase in the number of oxygen molecules in the lungs

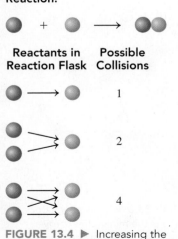

FIGURE 13.4 ▶ Increasing the concentration of a reactant increases the number of collisions that are possible.

Ⓠ Why does doubling the reactants increase the rate of reaction?

increases the rate at which oxygen combines with hemoglobin. The increased rate of oxygenation of the blood means that the patient can breathe more easily.

$$Hb(aq) + O_2(g) \longrightarrow HbO_2(aq)$$

Hemoglobin   Oxygen   Oxyhemoglobin

Catalysts

Another way to speed up a reaction is to lower the *energy of activation*. The energy of activation is the minimum energy needed to break apart the bonds of the reacting molecules. If a collision provides less than the activation energy, the bonds do not break and the reactant molecules bounce apart. A **catalyst** speeds up a reaction by providing an alternative pathway that has a lower energy of activation. When activation energy is lowered, more collisions provide sufficient energy for reactants to form product. During a reaction, a catalyst is not changed or consumed.

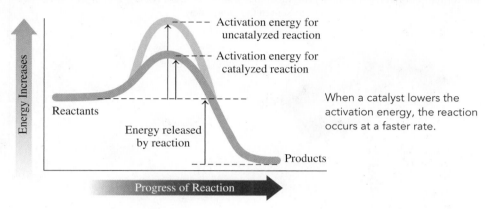

When a catalyst lowers the activation energy, the reaction occurs at a faster rate.

Catalysts have many uses in industry. In the manufacturing of margarine, hydrogen (H_2) is added to vegetable oils. Normally, the reaction is very slow because it has a high activation energy. However, when platinum (Pt) is used as a catalyst, the reaction occurs rapidly. In the body, bio-catalysts called *enzymes* make most metabolic reactions proceed at rates necessary for proper cellular activity. Enzymes are added to laundry detergents to break down proteins (proteases), starches (amylases), or greases (lipases) that have stained clothes. Such enzymes function at the low temperatures that are used in home washing machines, and they are biodegradable as well.

The factors affecting reaction rates are summarized in TABLE 13.1.

SAMPLE PROBLEM 13.1 Factors That Affect the Rate of Reaction

Indicate whether the following changes will increase, decrease, or have no effect on the rate of reaction:

a. increasing the temperature
b. increasing the number of reacting molecules
c. adding a catalyst

> TRY IT FIRST

SOLUTION

a. A higher temperature increases the kinetic energy of the particles, which increases the number of collisions and makes more collisions effective, causing an increase in the rate of reaction.
b. Increasing the number of reacting molecules increases the number of collisions and the rate of the reaction.
c. Adding a catalyst increases the rate of reaction by lowering the activation energy, which increases the number of collisions that form product.

STUDY CHECK 13.1

How does using an ice blanket on a patient affect the rate of metabolism in the body?

ANSWER

Lowering the temperature will decrease the rate of metabolism.

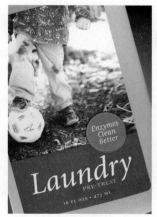

Enzymes in laundry detergent catalyze the removal of stains at low temperatures.

TABLE 13.1 Factors That Increase Reaction Rate

Factor	Reason
Increasing temperature	More collisions, more collisions with energy of activation
Increasing reactant concentration	More collisions
Adding a catalyst	Lowers energy of activation

CHEMISTRY LINK TO THE **ENVIRONMENT**
Catalytic Converters

For over 30 years, manufacturers have been required to include catalytic converters in the exhaust systems of gasoline automobile engines. When gasoline burns, the products found in the exhaust of a car contain high levels of pollutants. These include carbon monoxide (CO) from incomplete combustion, hydrocarbons such as C_8H_{18} (octane) from unburned fuel, and nitrogen oxide (NO) from the reaction of N_2 and O_2 at the high temperatures reached within the engine. Carbon monoxide is toxic, and unburned hydrocarbons and nitrogen oxide are involved in the formation of smog and acid rain.

A catalytic converter consists of solid-particle catalysts, such as platinum (Pt) and palladium (Pd), on a ceramic honeycomb that provides a large surface area and facilitates contact with pollutants. As the pollutants pass through the converter, they react with the catalysts. Today, we all use unleaded gasoline because lead interferes with the ability of the Pt and Pd catalysts in the converter to react with the pollutants. The purpose of a catalytic converter is to lower the activation energy for reactions that convert each of these pollutants into substances such as CO_2, N_2, O_2, and H_2O, which are already present in the atmosphere.

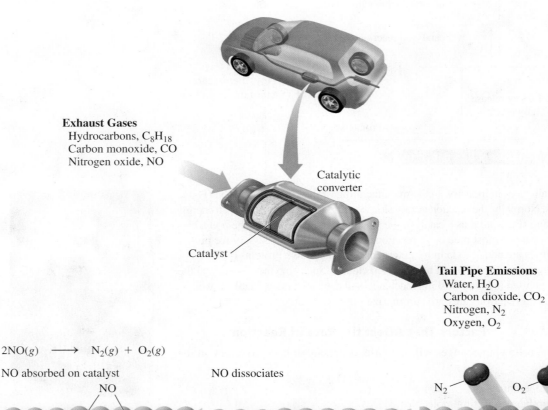

Exhaust Gases
Hydrocarbons, C_8H_{18}
Carbon monoxide, CO
Nitrogen oxide, NO

Catalytic converter

Catalyst

Tail Pipe Emissions
Water, H_2O
Carbon dioxide, CO_2
Nitrogen, N_2
Oxygen, O_2

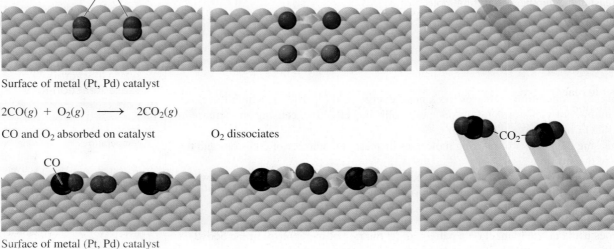

$$2NO(g) \longrightarrow N_2(g) + O_2(g)$$

NO absorbed on catalyst

NO

NO dissociates

N_2 O_2

Surface of metal (Pt, Pd) catalyst

$$2CO(g) + O_2(g) \longrightarrow 2CO_2(g)$$

CO and O_2 absorbed on catalyst

CO

O_2 dissociates

CO_2

Surface of metal (Pt, Pd) catalyst

QUESTIONS AND PROBLEMS

13.1 Rates of Reactions

LEARNING GOAL Describe how temperature, concentration, and catalysts affect the rate of a reaction.

13.1 **a.** What is meant by the rate of a reaction?
b. Why does bread grow mold more quickly at room temperature than in the refrigerator?

13.2 **a.** How does a catalyst affect the activation energy?
b. Why is pure oxygen used in respiratory distress?

13.3 In the following reaction, what happens to the number of collisions when more $Br_2(g)$ molecules are added?

$$H_2(g) + Br_2(g) \longrightarrow 2HBr(g)$$

13.4 In the following reaction, what happens to the number of collisions when the temperature of the reaction is decreased?

$$2H_2(g) + CO(g) \longrightarrow CH_4O(g)$$

13.5 How would each of the following change the rate of the reaction shown here?

$$2SO_2(g) + O_2(g) \longrightarrow 2SO_3(g)$$

a. adding some $SO_2(g)$
b. increasing the temperature
c. adding a catalyst
d. removing some $O_2(g)$

13.6 How would each of the following change the rate of the reaction shown here?

$$2NO(g) + 2H_2(g) \longrightarrow N_2(g) + 2H_2O(g)$$

a. adding some $NO(g)$
b. decreasing the temperature
c. removing some $H_2(g)$
d. adding a catalyst

13.2 Chemical Equilibrium

LEARNING GOAL Use the concept of reversible reactions to explain chemical equilibrium.

We consider the *forward reaction* in an equation and assumed that all of the reactants were converted to products. However, most of the time reactants are not completely converted to products because a *reverse reaction* takes place in which products collide to form the reactants. When a reaction proceeds in both a forward and reverse direction, it is said to be *reversible*. We have looked at other reversible processes. For example, the melting of solids to form liquids and the freezing of liquids to solids is a reversible physical change. Even in our daily life we have reversible events. We go from home to school and we return from school to home. We go up an escalator and we come back down. We put money in our bank account and we take money out.

An analogy for a forward and reverse reaction can be found in the phrase "We are going to the grocery store." Although we mention our trip in one direction, we know that we will also return home from the store. Because our trip has both a forward and reverse direction, we can say the trip is reversible. It is not very likely that we would stay at the store forever.

A trip to the grocery store can be used to illustrate another aspect of reversible reactions. Perhaps the grocery store is nearby and we usually walk. However, we can change our rate. Suppose that one day we drive to the store, which increases our rate and gets us to the store faster. Correspondingly, a car also increases the rate at which we return home.

Reversible Chemical Reactions

A **reversible reaction** proceeds in both the forward and reverse direction. That means there are two reaction rates: one is the rate of the forward reaction, and the other is the rate of the reverse reaction. When molecules begin to react, the rate of the forward reaction is faster than the rate of the reverse reaction. As reactants are consumed and products accumulate, the rate of the forward reaction decreases and the rate of the reverse reaction increases.

Equilibrium

Eventually, the rates of the forward and reverse reactions become equal; the reactants form products at the same rate that the products form reactants. A reaction reaches **chemical equilibrium** when no further change takes place in the concentrations of the reactants and products, even though the two reactions continue at equal but opposite rates.

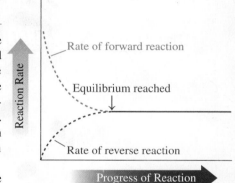

As a reaction progresses, the rate of the forward reaction decreases and that of the reverse reaction increases. At equilibrium, the rates of the forward and reverse reactions are equal.

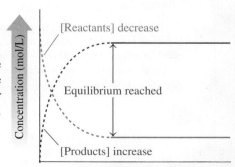

Equilibrium is reached when there are no further changes in the concentrations of reactants and products.

At Equilibrium:

The rate of the forward reaction is equal to the rate of the reverse reaction.

No further changes occur in the concentrations of reactants and products, even though the two reactions continue at equal but opposite rates.

ENGAGE

Why do the concentrations of the reactants decrease before equilibrium is reached?

Let us look at the process as the reaction of H_2 and I_2 proceeds to equilibrium. Initially, only the reactants H_2 and I_2 are present. Soon, a few molecules of HI are produced by the forward reaction. With more time, additional HI molecules are produced. As the concentration of HI increases, more HI molecules collide and react in the reverse direction. As HI product builds up, the rate of the reverse reaction increases, while the rate of the forward reaction decreases. Eventually, the rates become equal, which means the reaction has reached equilibrium. Even though the concentrations remain constant at equilibrium, the forward and reverse reactions continue to occur. The forward and reverse reactions are usually shown together in a single equation by using a double arrow. A reversible reaction is two opposing reactions that occur at the same time (see **FIGURE 13.5**).

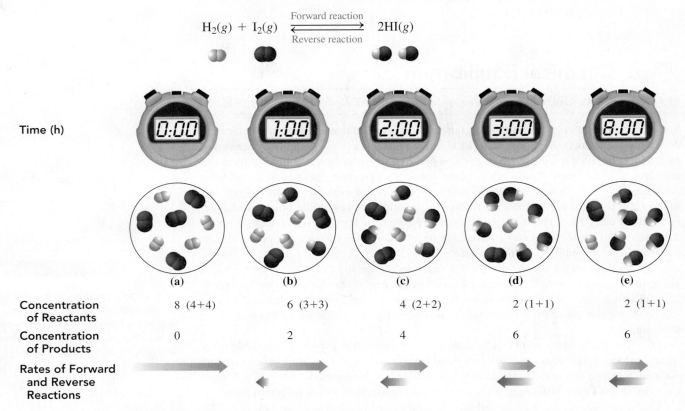

FIGURE 13.5 ▶ **(a)** Initially, the reaction flask contains only the reactants H_2 (white) and I_2 (purple). **(b)** The forward reaction between H_2 and I_2 begins to produce HI. **(c)** As the reaction proceeds, there are fewer molecules of H_2 and I_2 and more molecules of HI, which increases the rate of the reverse reaction. **(d)** At equilibrium, the concentrations of reactants H_2 and I_2 and product HI are constant. **(e)** The reaction continues with the rate of the forward reaction equal to the rate of the reverse reaction.

Q How do the rates of the forward and reverse reactions compare once a chemical reaction reaches equilibrium?

SAMPLE PROBLEM 13.2 Reaction Rates and Equilibrium

Complete each of the following with *equal* or *not equal*, *faster* or *slower*, *change* or *do not change*:

a. Before equilibrium is reached, the concentrations of the reactants and products _____.

b. Initially, reactants placed in a container have a _____ rate of reaction than the rate of reaction of the products.

c. At equilibrium, the rate of the forward reaction is _____ to the rate of the reverse reaction.

TRY IT FIRST

SOLUTION

a. Before equilibrium is reached, the concentrations of the reactants and products *change*.

b. Initially, reactants placed in a container have a *faster* rate of reaction than the rate of reaction of the products.

c. At equilibrium, the rate of the forward reaction is *equal* to the rate of the reverse reaction.

STUDY CHECK 13.2

Complete the following statement with *change* or *do not change*:
At equilibrium, the concentrations of the reactants and products _____.

ANSWER

At equilibrium, the concentrations of the reactants and products *do not change*.

We can also set up a reaction starting with only reactants or with only products. Let's look at the initial reactions in each, the forward and reverse reactions, and the equilibrium mixture that forms (see **FIGURE 13.6**).

$$2SO_2(g) + O_2(g) \rightleftharpoons 2SO_3(g)$$

If we start with only the reactants SO_2 and O_2 in the container, the reaction to form SO_3 takes place until equilibrium is reached. However, if we start with only the product SO_3 in the container, the reaction to form SO_2 and O_2 takes place until equilibrium is reached. In both containers, the equilibrium mixture contains the same concentrations of SO_2, O_2, and SO_3.

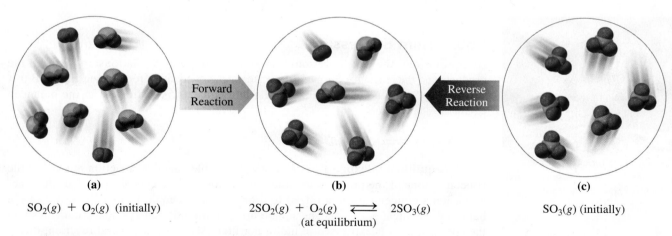

(a)	**(b)**	**(c)**
$SO_2(g) + O_2(g)$ (initially)	$2SO_2(g) + O_2(g) \rightleftharpoons 2SO_3(g)$ (at equilibrium)	$SO_3(g)$ (initially)

FIGURE 13.6 ▶ Sample **(a)** initially contains $SO_2(g)$ and $O_2(g)$. At equilibrium, sample **(b)** contains mostly $SO_3(g)$ and only small amounts of $SO_2(g)$ and $O_2(g)$, whereas sample **(c)** contains only $SO_3(g)$.

Q Why is the same equilibrium mixture obtained from $SO_2(g)$ and $O_2(g)$ as from $SO_3(g)$?

QUESTIONS AND PROBLEMS

13.2 Chemical Equilibrium

LEARNING GOAL Use the concept of reversible reactions to explain chemical equilibrium.

13.7 What is meant by the term reversible reaction?

13.8 When does a reversible reaction reach equilibrium?

13.9 Which of the following are at equilibrium?
  a. The rate of the forward reaction is twice as fast as the rate of the reverse reaction.
  b. The concentrations of the reactants and the products do not change.
  c. The rate of the reverse reaction does not change.

13.10 Which of the following are not at equilibrium?
   a. The rates of the forward and reverse reactions are equal.
   b. The rate of the forward reaction does not change.
   c. The concentrations of reactants and the products are not constant.

13.11 The following diagrams show the chemical reaction with time:

A ⇌ B

If A is blue and B is orange, state whether or not the reaction has reached equilibrium in this time period and explain why.

| 1 h | 2 h | 3 h | 4 h |

13.12 The following diagrams show the chemical reaction with time:

C ⇌ D

If C is blue and D is yellow, state whether or not the reaction has reached equilibrium in this time period and explain why.

| 1 h | 2 h | 3 h | 4 h |

13.3 Equilibrium Constants

LEARNING GOAL Calculate the equilibrium constant for a reversible reaction given the concentrations of reactants and products at equilibrium.

At equilibrium, the concentrations of the reactants and products are constant. We can use a ski lift as an analogy. Early in the morning, skiers at the bottom of the mountain begin to ride the ski lift up to the slopes. After the skiers reach the top of the mountain, they ski down. Eventually, the number of people riding up the ski lift becomes equal to the number of people skiing down the mountain. There is no further change in the number of skiers on the slopes; the system is at equilibrium.

At equilibrium, the number of people riding up the lift and the number of people skiing down the slope are constant.

Equilibrium Expression

At equilibrium, the concentrations can be used to set up a relationship between the products and the reactants. Suppose we write a general equation for reactants A and B that form products C and D. The small italic letters are the coefficients in the balanced equation.

$$a\text{A} + b\text{B} \rightleftharpoons c\text{C} + d\text{D}$$

An **equilibrium expression,** K_c, for a reversible chemical reaction multiplies the concentrations of the products together and divides by the concentrations of the reactants. Each concentration is raised to a power that is equal to its coefficient in the balanced chemical equation. The square bracket around each substance indicates that the concentration is expressed in moles per liter (M). For our general reaction, this is written as:

CORE CHEMISTRY SKILL
Writing the Equilibrium Expression

$$K_c = \frac{[\text{Products}]}{[\text{Reactants}]} = \frac{[\text{C}]^c\,[\text{D}]^d}{[\text{A}]^a\,[\text{B}]^b} \underset{}{\overset{}{\Big\rangle}} \text{Coefficients}$$

Equilibrium expression

We can now describe how to write the equilibrium expression for the reaction of H_2 and I_2 that forms HI. The balanced chemical equation is written with a double arrow between the reactants and the products.

$$H_2(g) + I_2(g) \rightleftharpoons 2HI(g)$$

We show the concentration of the products using brackets in the numerator and the concentrations of the reactants in brackets in the denominator and write any coefficient as an exponent of its concentration (a coefficient 1 is understood).

$$K_c = \frac{[HI]^2 \longleftarrow \text{Coefficient of HI}}{[H_2][I_2]}$$

SAMPLE PROBLEM 13.3 Writing Equilibrium Expressions

Write the equilibrium expression for the following reaction:

$$2SO_2(g) + O_2(g) \rightleftharpoons 2SO_3(g)$$

TRY IT FIRST

SOLUTION

ANALYZE THE PROBLEM	Given	Need	Connect
	equation	equilibrium expression	[products] [reactants]

STEP 1 Write the balanced chemical equation.

$$2SO_2(g) + O_2(g) \rightleftharpoons 2SO_3(g)$$

STEP 2 Write the concentrations of the products as the numerator and the reactants as the denominator.

$$\frac{[Products]}{[Reactants]} \longrightarrow \frac{[SO_3]}{[SO_2][O_2]}$$

STEP 3 Write any coefficient in the equation as an exponent.

$$K_c = \frac{[SO_3]^2}{[SO_2]^2[O_2]}$$

STUDY CHECK 13.3

Write the equilibrium expression for the following reaction:

$$2NO(g) + O_2(g) \rightleftharpoons 2NO_2(g)$$

ANSWER

$$K_c = \frac{[NO_2]^2}{[NO]^2[O_2]}$$

> **Guide to Writing the Equilibrium Expression**
>
> **STEP 1**
> Write the balanced chemical equation.
>
> **STEP 2**
> Write the concentrations of the products as the numerator and the reactants as the denominator.
>
> **STEP 3**
> Write any coefficient in the equation as an exponent.

Calculating Equilibrium Constants

The **equilibrium constant, K_c,** is the numerical value obtained by substituting experimentally measured molar concentrations at equilibrium into the equilibrium expression. For example, the equilibrium expression for the reaction of H_2 and I_2 is written

$$H_2(g) + I_2(g) \rightleftharpoons 2HI(g) \qquad K_c = \frac{[HI]^2}{[H_2][I_2]}$$

In the first experiment, the molar concentrations for the reactants and products at equilibrium are found to be $[H_2] = 0.10$ M, $[I_2] = 0.20$ M, and $[HI] = 1.04$ M. When we substitute these values into the equilibrium expression, we obtain the numerical value of K_c.

In additional experiments 2 and 3, the mixtures have different equilibrium concentrations for the system at equilibrium at the same temperature. However, when these concentrations are used to calculate the equilibrium constant, we obtain the same value of K_c for each (see **TABLE 13.2**). *Thus, a reaction at a specific temperature can have only one value for the equilibrium constant.*

CORE CHEMISTRY SKILL
Calculating an Equilibrium Constant

TABLE 13.2 Equilibrium Constant for $H_2(g) + I_2(g) \rightleftharpoons 2HI(g)$ at 427 °C

Experiment	Concentrations at Equilibrium			Equilibrium Constant
	$[H_2]$	$[I_2]$	$[HI]$	$K_c = \dfrac{[HI]^2}{[H_2][I_2]}$
1	0.10 M	0.20 M	1.04 M	$K_c = \dfrac{[1.04]^2}{[0.10][0.20]} = 54$
2	0.20 M	0.20 M	1.47 M	$K_c = \dfrac{[1.47]^2}{[0.20][0.20]} = 54$
3	0.30 M	0.17 M	1.66 M	$K_c = \dfrac{[1.66]^2}{[0.30][0.17]} = 54$

The units of K_c depend on the specific equation. In this example, the units of $[M]^2/[M]^2$ cancel out to give a value of 54. In other equations, the concentration units do not cancel. However, in this text, the numerical value will be given without any units as shown in Sample Problem 13.4.

SAMPLE PROBLEM 13.4 Calculating an Equilibrium Constant

The decomposition of dinitrogen tetroxide forms nitrogen dioxide.

$$N_2O_4(g) \rightleftharpoons 2NO_2(g)$$

What is the numerical value of K_c at 100 °C if a reaction mixture at equilibrium contains 0.45 M N_2O_4 and 0.31 M NO_2?

TRY IT FIRST

SOLUTION

STEP 1 State the given and needed quantities.

ANALYZE THE PROBLEM	Given	Need	Connect
	0.45 M N_2O_4, 0.31 M NO_2	K_c	equilibrium expression
	Equation		
	$N_2O_4(g) \rightleftharpoons 2NO_2(g)$		

STEP 2 Write the equilibrium expression, K_c.

$$K_c = \frac{[NO_2]^2}{[N_2O_4]}$$

STEP 3 Substitute equilibrium (molar) concentrations and calculate K_c.

$$K_c = \frac{[0.31]^2}{[0.45]} = 0.21$$

STUDY CHECK 13.4

Calculate the numerical value of K_c if an equilibrium mixture contains 0.040 M NH_3, 0.60 M H_2, and 0.20 M N_2.

$$2NH_3(g) \rightleftharpoons 3H_2(g) + N_2(g)$$

ANSWER

$K_c = 27$

Guide to Calculating the K_c Value

STEP 1
State the given and needed quantities.

STEP 2
Write the equilibrium expression, K_c.

STEP 3
Substitute equilibrium (molar) concentrations and calculate K_c.

Heterogeneous Equilibrium

Up to now, our examples have been reactions that involve only gases. A reaction in which all the reactants and products are in the same state reaches **homogeneous equilibrium**. When the reactants and products are in two or more states, the equilibrium is termed a **heterogeneous equilibrium**. In the following reaction, solid calcium carbonate reaches heterogeneous equilibrium with solid calcium oxide and carbon dioxide gas (see **FIGURE 13.7**).

$$CaCO_3(s) \rightleftharpoons CaO(s) + CO_2(g)$$

In contrast to gases, the concentrations of pure solids and pure liquids are constant; they do not change. Therefore, pure solids and liquids are not included in the equilibrium expression. For this heterogeneous equilibrium, the K_c expression does not include the concentration of $CaCO_3(s)$ or $CaO(s)$. It is written as $K_c = [CO_2]$.

$$CaCO_3(s) \rightleftharpoons CaO(s) + CO_2(g)$$

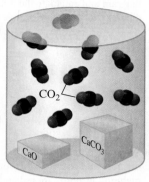

$T = 800\ °C$

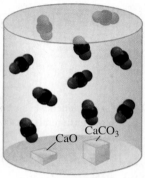

$T = 800\ °C$

FIGURE 13.7 ▶ At equilibrium at constant temperature, the concentration of CO_2 is the same regardless of the amounts of $CaCO_3(s)$ and $CaO(s)$ in the container.

🔵 Why are the concentrations of $CaO(s)$ and $CaCO_3(s)$ not included in K_c for the decomposition of $CaCO_3(s)$?

SAMPLE PROBLEM 13.5 Heterogeneous Equilibrium Expression

Write the equilibrium expression for the following reaction at equilibrium:

$$4HCl(g) + O_2(g) \rightleftharpoons 2H_2O(l) + 2Cl_2(g)$$

▶ **TRY IT FIRST**

SOLUTION

STEP 1 Write the balanced chemical equation.

$$4HCl(g) + O_2(g) \rightleftharpoons 2H_2O(l) + 2Cl_2(g)$$

STEP 2 Write the concentrations of the products as the numerator and the reactants as the denominator. In this heterogeneous reaction, the concentration of the H_2O, which is a pure liquid, is not included in the equilibrium expression.

$$\frac{[Products]}{[Reactants]} \longrightarrow \frac{[Cl_2]}{[HCl][O_2]}$$

STEP 3 Write any coefficient in the equation as an exponent.

$$K_c = \frac{[Cl_2]^2}{[HCl]^4[O_2]}$$

STUDY CHECK 13.5

Solid iron(II) oxide and carbon monoxide gas react to produce solid iron and carbon dioxide gas. Write the balanced chemical equation and the equilibrium expression for this reaction at equilibrium.

ANSWER

$$FeO(s) + CO(g) \rightleftharpoons Fe(s) + CO_2(g)$$

$$K_c = \frac{[CO_2]}{[CO]}$$

QUESTIONS AND PROBLEMS

13.3 Equilibrium Constants

LEARNING GOAL Calculate the equilibrium constant for a reversible reaction given the concentrations of reactants and products at equilibrium.

13.13 Write the equilibrium expression for each of the following reactions:
  a. $CH_4(g) + 2H_2S(g) \rightleftharpoons CS_2(g) + 4H_2(g)$
  b. $2NO(g) \rightleftharpoons N_2(g) + O_2(g)$
  c. $2SO_3(g) + CO_2(g) \rightleftharpoons CS_2(g) + 4O_2(g)$
  d. $CH_4(g) + H_2O(g) \rightleftharpoons 3H_2(g) + CO(g)$

13.14 Write the equilibrium expression for each of the following reactions:
  a. $2HBr(g) \rightleftharpoons H_2(g) + Br_2(g)$
  b. $2BrNO(g) \rightleftharpoons Br_2(g) + 2NO(g)$
  c. $CH_4(g) + Cl_2(g) \rightleftharpoons CH_3Cl(g) + HCl(g)$
  d. $Br_2(g) + Cl_2(g) \rightleftharpoons 2BrCl(g)$

13.15 Write the equilibrium expression for the reaction in the diagram and calculate the numerical value of K_c. In the diagram, X atoms are orange and Y atoms are blue.

$$X_2(g) + Y_2(g) \rightleftharpoons 2XY(g)$$

Equilibrium
mixture

13.16 Write the equilibrium expression for the reaction in the diagram and calculate the numerical value of K_c. In the diagram, A atoms are red and B atoms are green.

$$2AB(g) \rightleftharpoons A_2(g) + B_2(g)$$

Equilibrium
mixture

13.17 What is the numerical value of K_c for the following reaction if the equilibrium mixture contains 0.030 M N_2O_4 and 0.21 M NO_2?

$$N_2O_4(g) \rightleftharpoons 2NO_2(g)$$

13.18 What is the numerical value of K_c for the following reaction if the equilibrium mixture contains 0.30 M CO_2, 0.033 M H_2, 0.20 M CO, and 0.30 M H_2O?

$$CO_2(g) + H_2(g) \rightleftharpoons CO(g) + H_2O(g)$$

13.19 What is the numerical value of K_c for the following reaction if the equilibrium mixture contains 0.51 M CO, 0.30 M H_2, 1.8 M CH_4, and 2.0 M H_2O?

$$CO(g) + 3H_2(g) \rightleftharpoons CH_4(g) + H_2O(g)$$

13.20 What is the numerical value of K_c for the following reaction if the equilibrium mixture contains 0.44 M N_2, 0.40 M H_2, and 2.2 M NH_3?

$$N_2(g) + 3H_2(g) \rightleftharpoons 2NH_3(g)$$

13.21 Identify each of the following as a homogeneous or heterogeneous equilibrium:
  a. $2O_3(g) \rightleftharpoons 3O_2(g)$
  b. $2NaHCO_3(s) \rightleftharpoons Na_2CO_3(s) + CO_2(g) + H_2O(g)$
  c. $C_6H_6(g) + 3H_2(g) \rightleftharpoons C_6H_{12}(g)$
  d. $4HCl(g) + Si(s) \rightleftharpoons SiCl_4(l) + 2H_2(g)$

13.22 Identify each of the following as a homogeneous or heterogeneous equilibrium:
  a. $CO(g) + H_2(g) \rightleftharpoons C(s) + H_2O(g)$
  b. $NH_4Cl(s) \rightleftharpoons NH_3(g) + HCl(g)$
  c. $CS_2(g) + 4H_2(g) \rightleftharpoons CH_4(g) + 2H_2S(g)$
  d. $Ti(s) + 2Cl_2(g) \rightleftharpoons TiCl_4(g)$

13.23 Write the equilibrium expression for each of the reactions in problem 13.21.

13.24 Write the equilibrium expression for each of the reactions in problem 13.22.

13.25 What is the numerical value of K_c for the following reaction if the equilibrium mixture at 750 °C contains 0.20 M CO and 0.052 M CO_2?

$$FeO(s) + CO(g) \rightleftharpoons Fe(s) + CO_2(g)$$

13.26 What is the numerical value of K_c for the following reaction if the equilibrium mixture at 800 °C contains 0.030 M CO_2?

$$CaCO_3(s) \rightleftharpoons CaO(s) + CO_2(g)$$

13.4 Using Equilibrium Constants

LEARNING GOAL Use an equilibrium constant to predict the extent of reaction and to calculate equilibrium concentrations.

The values of K_c can be large or small. The size of the equilibrium constant depends on whether equilibrium is reached with more products than reactants, or more reactants than products. However, the size of an equilibrium constant does not affect how fast equilibrium is reached.

Equilibrium with a Large K_c

When a reaction has a large equilibrium constant, it means that the forward reaction produced a large amount of products when equilibrium was reached. Then the equilibrium mixture contains mostly products, which makes the concentrations of the products in the numerator higher than the concentrations of the reactants in the denominator. Thus at equilibrium, this reaction has a large K_c. Consider the reaction of SO_2 and O_2, which has a large K_c. At equilibrium, the reaction mixture contains mostly product and few reactants (see **FIGURE 13.8**).

$$2SO_2(g) + O_2(g) \rightleftharpoons 2SO_3(g)$$

$$K_c = \frac{[SO_3]^2}{[SO_2]^2[O_2]} \quad \frac{\text{Mostly product}}{\text{Few reactants}} = 3.4 \times 10^2$$

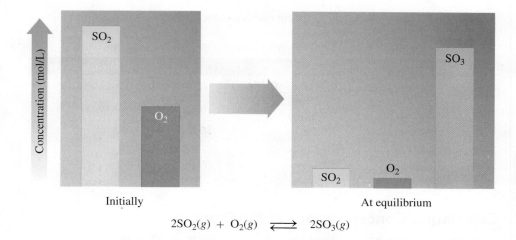

FIGURE 13.8 ▶ In the reaction of $SO_2(g)$ and $O_2(g)$, the equilibrium mixture contains mostly product $SO_3(g)$, which results in a large K_c.

Q Why does an equilibrium mixture containing mostly product have a large K_c?

$$2SO_2(g) + O_2(g) \rightleftharpoons 2SO_3(g)$$

Equilibrium with a Small K_c

When a reaction has a small equilibrium constant, the equilibrium mixture contains a high concentration of reactants and a low concentration of products. Then the equilibrium expression has a small number in the numerator and a large number in the denominator. Thus at equilibrium, this reaction has a small K_c. Consider the reaction for the formation of $NO(g)$ from $N_2(g)$ and $O_2(g)$, which has a small K_c (see **FIGURE 13.9**).

$$N_2(g) + O_2(g) \rightleftharpoons 2NO(g)$$

$$K_c = \frac{[NO]^2}{[N_2][O_2]} \quad \frac{\text{Few products}}{\text{Mostly reactants}} = 2 \times 10^{-9}$$

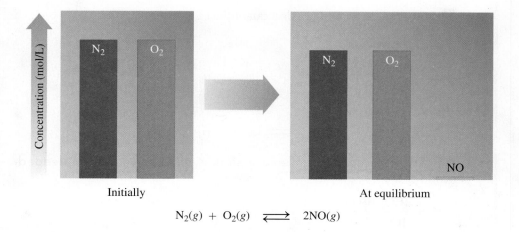

FIGURE 13.9 ▶ The equilibrium mixture contains a very small amount of the product NO and a large amount of the reactants N_2 and O_2, which results in a small K_c.

Q Does a reaction with a $K_c = 3.2 \times 10^{-5}$ contain mostly reactants or products at equilibrium?

$$N_2(g) + O_2(g) \rightleftharpoons 2NO(g)$$

A few reactions have equilibrium constants close to 1, which means they have about equal concentrations of reactants and products. Moderate amounts of reactants have been converted to products upon reaching equilibrium (see **FIGURE 13.10**).

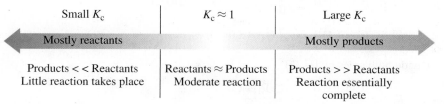

FIGURE 13.10 ▶ At equilibrium, a reaction with a large K_c contains mostly products, whereas a reaction with a small K_c contains mostly reactants.

Q Does a reaction with a $K_c = 1.2$ contain mostly reactants, mostly products, or about equal amounts of both reactants and products at equilibrium?

TABLE 13.3 lists some equilibrium constants and the extent of their reaction.

TABLE 13.3 Examples of Reactions with Large and Small K_c Values

Reactants	Products	K_c	Equilibrium Mixture Contains
$2CO(g) + O_2(g) \rightleftharpoons$	$2CO_2(g)$	2×10^{11}	Mostly product
$2H_2(g) + S_2(g) \rightleftharpoons$	$2H_2S(g)$	1.1×10^7	Mostly product
$N_2(g) + 3H_2(g) \rightleftharpoons$	$2NH_3(g)$	1.6×10^2	Mostly product
$PCl_5(g) \rightleftharpoons$	$PCl_3(g) + Cl_2(g)$	1.2×10^{-2}	Mostly reactant
$N_2(g) + O_2(g) \rightleftharpoons$	$2NO(g)$	2×10^{-9}	Mostly reactants

ENGAGE

Why does the equilibrium mixture for a reaction with a $K_c = 2 \times 10^{-9}$ contain mostly reactants and only a few products?

CORE CHEMISTRY SKILL
Calculating Equilibrium Concentrations

Calculating Concentrations at Equilibrium

When we know the numerical value of the equilibrium constant and all the equilibrium concentrations except one, we can calculate the unknown concentration as shown in Sample Problem 13.6.

SAMPLE PROBLEM 13.6 Calculating Concentration Using an Equilibrium Constant

For the reaction of carbon dioxide and hydrogen, the equilibrium concentrations are 0.25 M CO_2, 0.80 M H_2, and 0.50 M H_2O. What is the equilibrium concentration of $CO(g)$?

$$CO_2(g) + H_2(g) \rightleftharpoons CO(g) + H_2O(g) \qquad K_c = 0.11$$

TRY IT FIRST

SOLUTION

STEP 1 State the given and needed quantities.

	Given	Need	Connect
ANALYZE THE PROBLEM	0.25 M CO_2, 0.80 M H_2, 0.50 M H_2O	[CO]	K_c expression
	Equation		
	$CO_2(g) + H_2(g) \rightleftharpoons CO(g) + H_2O(g)$ $K_c = 0.11$		

STEP 2 Write the equilibrium expression, K_c, and solve for the needed concentration.

$$K_c = \frac{[CO][H_2O]}{[CO_2][H_2]}$$

We rearrange K_c to solve for the unknown [CO] as follows:

Multiply both sides by $[CO_2][H_2]$.

$$K_c \times [CO_2][H_2] = \frac{[CO][H_2O]}{\cancel{[CO_2][H_2]}} \times \cancel{[CO_2][H_2]}$$

$$K_c \times [CO_2][H_2] = [CO][H_2O]$$

Divide both sides by $[H_2O]$.

$$K_c \times \frac{[CO_2][H_2]}{[H_2O]} = \frac{[CO]\cancel{[H_2O]}}{\cancel{[H_2O]}}$$

$$[CO] = K_c \times \frac{[CO_2][H_2]}{[H_2O]}$$

Guide to Using the Equilibrium Constant

STEP 1
State the given and needed quantities.

STEP 2
Write the equilibrium expression, K_c, and solve for the needed concentration.

STEP 3
Substitute the equilibrium (molar) concentrations and calculate the needed concentration.

STEP **3** Substitute the equilibrium (molar) concentrations and calculate the needed concentration.

$$[CO] = K_c \times \frac{[CO_2][H_2]}{[H_2O]} = 0.11 \times \frac{[0.25][0.80]}{[0.50]} = 0.044 \text{ M}$$

STUDY CHECK 13.6

When ethene (C_2H_4) reacts with water vapor, ethanol (C_2H_6O) is produced. If an equilibrium mixture contains 0.020 M C_2H_4 and 0.015 M H_2O, what is the equilibrium concentration of C_2H_6O? At 327 °C, the K_c is 9.0×10^3.

$$C_2H_4(g) + H_2O(g) \rightleftharpoons C_2H_6O(g)$$

ANSWER

$[C_2H_6O] = 2.7 \text{ M}$

QUESTIONS AND PROBLEMS

13.4 Using Equilibrium Constants

LEARNING GOAL Use an equilibrium constant to predict the extent of reaction and to calculate equilibrium concentrations.

13.27 If the K_c for this reaction is 4, which of the following diagrams represents the molecules in an equilibrium mixture? In the diagrams, X atoms are orange and Y atoms are blue.

$$X_2(g) + Y_2(g) \rightleftharpoons 2XY(g)$$

      A   B   C

13.28 If the K_c for this reaction is 2, which of the following diagrams represents the molecules in an equilibrium mixture? In the diagrams, A atoms are orange and B atoms are green.

$$2AB(g) \rightleftharpoons A_2(g) + B_2(g)$$

      A   B   C

13.29 Indicate whether each of the following equilibrium mixtures contains mostly products or mostly reactants:
a. $Cl_2(g) + NO(g) \rightleftharpoons 2NOCl(g)$ $K_c = 3.7 \times 10^8$ pro
b. $2H_2(g) + S_2(g) \rightleftharpoons 2H_2S(g)$ $K_c = 1.1 \times 10^7$ pro
c. $3O_2(g) \rightleftharpoons 2O_3(g)$ $K_c = 1.7 \times 10^{-56}$ rea

13.30 Indicate whether each of the following equilibrium mixtures contains mostly products or mostly reactants:
a. $CO(g) + Cl_2(g) \rightleftharpoons COCl_2(g)$ $K_c = 5.0 \times 10^{-9}$
b. $2HF(g) \rightleftharpoons H_2(g) + F_2(g)$ $K_c = 1.0 \times 10^{-95}$
c. $2NO(g) + O_2(g) \rightleftharpoons 2NO_2(g)$ $K_c = 6.0 \times 10^{13}$

13.31 The equilibrium constant, K_c, for this reaction is 54.

$$H_2(g) + I_2(g) \rightleftharpoons 2HI(g)$$

If the equilibrium mixture contains 0.015 M I_2 and 0.030 M HI, what is the molar concentration of H_2?

13.32 The equilibrium constant, K_c, for the following reaction is 4.6×10^{-3}. If the equilibrium mixture contains 0.050 M NO_2, what is the molar concentration of N_2O_4?

$$N_2O_4(g) \rightleftharpoons 2NO_2(g)$$

13.33 The K_c for the following reaction at 100 °C is 2.0. If the equilibrium mixture contains 2.0 M NO and 1.0 M Br_2, what is the molar concentration of NOBr?

$$2NOBr(g) \rightleftharpoons 2NO(g) + Br_2(g)$$

13.34 The K_c for the following reaction at 225 °C is 1.7×10^2. If the equilibrium mixture contains 0.18 M H_2 and 0.020 M N_2, what is the molar concentration of NH_3?

$$3H_2(g) + N_2(g) \rightleftharpoons 2NH_3(g)$$

13.5 Changing Equilibrium Conditions: Le Châtelier's Principle

LEARNING GOAL Use Le Châtelier's principle to describe the changes made in equilibrium concentrations when reaction conditions change.

We have seen that when a reaction reaches equilibrium, the rates of the forward and reverse reactions are equal and the concentrations remain constant. Now we will look at what happens to a system at equilibrium when changes occur in reaction conditions, such as changes in concentration, volume, and temperature.

Le Châtelier's Principle

When we alter any of the conditions of a system at equilibrium, the rates of the forward and reverse reactions will no longer be equal. We say that a *stress* is placed on the equilibrium. Then the system responds by changing the rate of the forward or reverse reaction in the direction that relieves that stress to reestablish equilibrium. We can use **Le Châtelier's principle**, which states that when a system at equilibrium is disturbed, the system will shift in the direction that will reduce that stress.

> ### Le Châtelier's Principle
> When a stress (change in conditions) is placed on a reaction at equilibrium, the equilibrium will shift in the direction that relieves the stress.

Suppose we have two water tanks connected by a pipe. When the water levels in the tanks are equal, water flows in the forward direction from Tank A to Tank B at the same rate as it flows in the reverse direction from Tank B to Tank A. Suppose we add more water to Tank A. With a higher level of water in Tank A, more water flows in the forward direction from Tank A to Tank B than in the reverse direction from Tank B to Tank A, which is shown with a longer arrow. Eventually, equilibrium is reached as the levels in both tanks become equal, but higher than before. Then the rate of water flows equally between Tank A and Tank B.

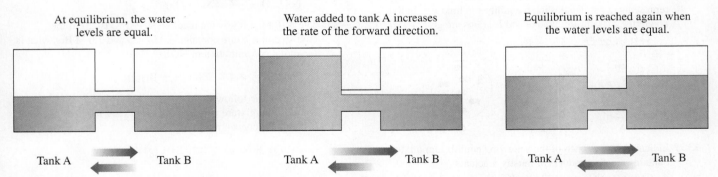

At equilibrium, the water levels are equal.

Water added to tank A increases the rate of the forward direction.

Equilibrium is reached again when the water levels are equal.

The stress of adding water to tank A increases the rate of the forward direction to reestablish equal water levels and equilibrium.

CORE CHEMISTRY SKILL
Using Le Châtelier's Principle

Effect of Concentration Changes on Equilibrium

We will now use the reaction of H_2 and I_2 to illustrate how a change in concentration disturbs the equilibrium and how the system responds to that stress.

$$H_2(g) + I_2(g) \rightleftharpoons 2HI(g)$$

Suppose that more of the reactant H_2 is added to the equilibrium mixture, which increases the concentration of H_2. Because a K_c cannot change for a reaction at a given

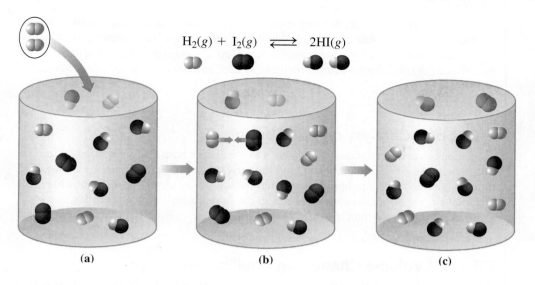

$$H_2(g) + I_2(g) \rightleftharpoons 2HI(g)$$

(a)   **(b)**   **(c)**

FIGURE 13.11 ▶ **(a)** The addition of H_2 places stress on the equilibrium system of $H_2(g) + I_2(g) \rightleftharpoons 2HI(g)$. **(b)** To relieve the stress, the forward reaction converts some reactants H_2 and I_2 to product HI. **(c)** A new equilibrium is established when the rates of the forward reaction and the reverse reaction become equal.

◉ If more product HI is added, will the equilibrium shift in the direction of the product or reactants? Why?

temperature, adding more H_2 places a stress on the system (see **FIGURE 13.11**). Then the system relieves this stress by increasing the rate of the forward reaction, which is indicated by the direction of the large arrow. Thus, more product is formed until the system is again at equilibrium. According to Le Châtelier's principle, adding more reactant causes the system to *shift* in the direction of the product until equilibrium is reestablished.

Add H_2

$$H_2(g) + I_2(g) \rightleftharpoons 2HI(g)$$

Suppose now that some H_2 is removed from the reaction mixture at equilibrium, which lowers the concentration of H_2 and slows the rate of the forward reaction. Using Le Châtelier's principle, we know that when some of the reactants are removed, the system will *shift* in the direction of the reactants until equilibrium is reestablished.

Remove H_2

$$H_2(g) + I_2(g) \rightleftharpoons 2HI(g)$$

The concentrations of the products of an equilibrium mixture can also increase or decrease. For example, if more HI is added, there is an increase in the rate of the reaction in the reverse direction, which converts some of the product to reactants. The concentration of the products decreases and the concentration of the reactants increases until equilibrium is reestablished. Using Le Châtelier's principle, we see that the addition of a product causes the system to *shift* in the direction of the reactants.

Add HI

$$H_2(g) + I_2(g) \rightleftharpoons 2HI(g)$$

In another example, some HI is removed from an equilibrium mixture, which decreases the concentration of the product. Then there is a *shift* in the direction of the product to reestablish equilibrium.

Remove HI

$$H_2(g) + I_2(g) \rightleftharpoons 2HI(g)$$

In summary, Le Châtelier's principle indicates that a stress caused by adding a substance at equilibrium is relieved when the equilibrium system shifts the reaction

TABLE 13.4 Effect of Concentration Changes on Equilibrium $H_2(g) + I_2(g) \rightleftharpoons 2HI(g)$	
Stress	Shift in the Direction of
Increasing $[H_2]$	Product
Decreasing $[H_2]$	Reactants
Increasing $[I_2]$	Product
Decreasing $[I_2]$	Reactants
Increasing $[HI]$	Reactants
Decreasing $[HI]$	Product

away from that substance. Adding more reactant causes an increase in the forward reaction to products. Adding more products causes an increase in the reverse reaction to reactants. When some of a substance is removed, the equilibrium system shifts in the direction of that substance. These features of Le Châtelier's principle are summarized in **TABLE 13.4**.

Effect of a Catalyst on Equilibrium

Sometimes a catalyst is added to a reaction to speed up a reaction by lowering the activation energy. As a result, the rates of both the forward and reverse reactions increase. The time required to reach equilibrium is shorter, but the same ratios of products and reactants are attained. Therefore, a catalyst speeds up the forward and reverse reactions, but it has no effect on the equilibrium mixture.

Effect of Volume Change on Equilibrium

If there is a change in the volume of a gas mixture at equilibrium, there will also be a change in the concentrations of those gases. Decreasing the volume will increase the concentration of gases, whereas increasing the volume will decrease their concentration. Then the system responds to reestablish equilibrium.

Let's look at the effect of decreasing the volume of the equilibrium mixture of the following reaction:

$$2CO(g) + O_2(g) \rightleftharpoons 2CO_2(g)$$

If we decrease the volume, all the concentrations increase. According to Le Châtelier's principle, the increase in concentration is relieved when the system shifts in the direction of the smaller number of moles.

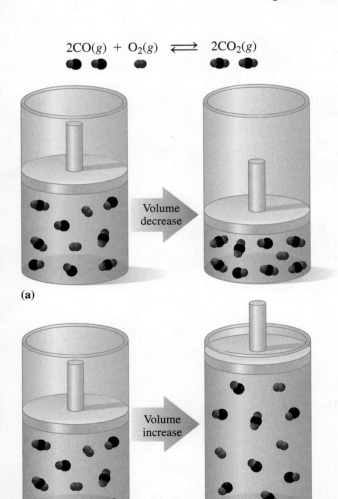

Decrease V

$2CO(g) + O_2(g) \rightleftharpoons 2CO_2(g)$
3 mol of gas   2 mol of gas

On the other hand, when the volume of the equilibrium gas mixture increases, the concentrations of all the gases decrease. Then the system shifts in the direction of the greater number of moles to reestablish equilibrium (see **FIGURE 13.12**).

Increase V

$2CO(g) + O_2(g) \rightleftharpoons 2CO_2(g)$
3 mol of gas   2 mol of gas

When a reaction has the same number of moles of reactants as products, a volume change does not affect the equilibrium mixture because the concentrations of the reactants and products change in the same way.

$H_2(g) + I_2(g) \rightleftharpoons 2HI(g)$
2 mol of gas   2 mol of gas

FIGURE 13.12 ▶ **(a)** A decrease in the volume of the container causes the system to shift in the direction of fewer moles of gas. **(b)** An increase in the volume of the container causes the system to shift in the direction of more moles of gas.

Q If you want to increase the product, should you increase or decrease the volume of the container?

CHEMISTRY LINK TO **HEALTH**
Oxygen–Hemoglobin Equilibrium and Hypoxia

The transport of oxygen involves an equilibrium between hemoglobin (Hb), oxygen, and oxyhemoglobin (HbO$_2$).

$$Hb(aq) + O_2(g) \rightleftharpoons HbO_2(aq)$$

When the O$_2$ level is high in the alveoli of the lung, the reaction shifts in the direction of the product HbO$_2$. In the tissues where O$_2$ concentration is low, the reverse reaction releases the oxygen from the hemoglobin. The equilibrium expression is written

$$K_c = \frac{[HbO_2]}{[Hb][O_2]}$$

At normal atmospheric pressure, oxygen diffuses into the blood because the partial pressure of oxygen in the alveoli is higher than that in the blood. At an altitude above 8000 ft, a decrease in the atmospheric pressure results in a significant reduction in the partial pressure of oxygen, which means that less oxygen is available for the blood and body tissues. The fall in atmospheric pressure at higher altitudes decreases the partial pressure of inhaled oxygen, and there is less driving pressure for gas exchange in the lungs. At an altitude of 18 000 ft, a person will obtain 29% less oxygen. When oxygen levels are lowered, a person may experience *hypoxia*, characterized by increased respiratory rate, headache, decreased mental acuteness, fatigue, decreased physical coordination, nausea, vomiting, and cyanosis. A similar problem occurs in persons with a history of lung disease that impairs gas diffusion in the alveoli or in persons with a reduced number of red blood cells, such as smokers.

According to Le Châtelier's principle, we see that a decrease in oxygen will shift the equilibrium in the direction of the reactants. Such a shift depletes the concentration of HbO$_2$ and causes the hypoxia.

Remove O$_2$
$$Hb(aq) + O_2(g) \rightleftharpoons HbO_2(aq)$$

Immediate treatment of altitude sickness includes hydration, rest, and if necessary, descending to a lower altitude. The adaptation to lowered oxygen levels requires about 10 days. During this time the bone marrow increases red blood cell production, providing more red blood cells and more hemoglobin. A person living at a high altitude can have 50% more red blood cells than someone at sea level. This increase in hemoglobin causes a shift in the equilibrium back in the direction of HbO$_2$ product. Eventually, the higher concentration of HbO$_2$ will provide more oxygen to the tissues and the symptoms of hypoxia will lessen.

Add O$_2$
$$Hb(aq) + O_2(g) \rightleftharpoons HbO_2(aq)$$

For some who climb high mountains, it is important to stop and acclimatize for several days at increasing altitudes. At very high altitudes, it may be necessary to use an oxygen tank.

Hypoxia may occur at high altitudes where the oxygen concentration is lower.

Effect of a Change in Temperature on Equilibrium

We can think of heat as a reactant or a product in a reaction. For example, in the equation for an endothermic reaction, heat is written on the reactant side. When the temperature of an endothermic reaction increases, the system responds by shifting in the direction of the products to remove heat.

Increase T
$$N_2(g) + O_2(g) + heat \rightleftharpoons 2NO(g)$$

If the temperature is decreased for an endothermic reaction, there is a decrease in heat. Then the system shifts in the direction of the reactants to add heat.

Decrease T
$$N_2(g) + O_2(g) + heat \rightleftharpoons 2NO(g)$$

In the equation for an exothermic reaction, heat is written on the product side. When the temperature of an exothermic reaction increases, the system responds by shifting in the direction of the reactants to remove heat.

Increase T
$$2SO_2(g) + O_2(g) \rightleftharpoons 2SO_3(g) + heat$$

If the temperature is decreased for an exothermic reaction, there is a decrease in heat. Then the system shifts in the direction of the products to add heat.

Decrease T

$$2SO_2(g) + O_2(g) \rightleftharpoons 2SO_3(g) + \text{heat}$$

TABLE 13.5 summarizes the ways we can use Le Châtelier's principle to determine the shift in equilibrium that relieves a stress caused by the change in a condition.

TABLE 13.5 Effects of Condition Changes on Equilibrium

Condition	Change (Stress)	Shift in the Direction of
Concentration	Adding a reactant	Products (forward reaction)
	Removing a reactant	Reactants (reverse reaction)
	Adding a product	Reactants (reverse reaction)
	Removing a product	Products (forward reaction)
Volume (container)	Decreasing the volume	Fewer moles of gas
	Increasing the volume	More moles of gas
Temperature	**Endothermic reaction**	
	Increasing the temperature	Products (forward reaction to remove heat)
	Decreasing the temperature	Reactants (reverse reaction to add heat)
	Exothermic reaction	
	Increasing the temperature	Reactants (reverse reaction to remove heat)
	Decreasing the temperature	Products (forward reaction to add heat)
Catalyst	Increasing the rates equally	No effect

ENGAGE

Why does adding $H_2O(g)$ to this reaction cause an equilibrium shift in the direction of the reactants?

SAMPLE PROBLEM 13.7 Using Le Châtelier's Principle

Methanol, CH_4O, is finding use as a fuel additive. Describe the effect of each of the following changes on the equilibrium mixture for the combustion of methanol:

$$2CH_4O(g) + 3O_2(g) \rightleftharpoons 2CO_2(g) + 4H_2O(g) + 1450 \text{ kJ}$$

a. adding more CO_2
b. adding more O_2
c. increasing the volume of the container
d. increasing the temperature
e. adding a catalyst

TRY IT FIRST

SOLUTION

a. When the concentration of the product CO_2 increases, the equilibrium shifts in the direction of the reactants.
b. When the concentration of the reactant O_2 increases, the equilibrium shifts in the direction of the products.
c. When the volume increases, the equilibrium shifts in the direction of the greater number of moles of gas, which is the products.
d. When the temperature is increased for an exothermic reaction, the equilibrium shifts in the direction of the reactants to remove heat.
e. When a catalyst is added, there is no change in the equilibrium mixture.

STUDY CHECK 13.7

Describe the effect of each of the following changes on the equilibrium mixture for the following reaction:

$$2HF(g) + Cl_2(g) + 357 \text{ kJ} \rightleftharpoons 2HCl(g) + F_2(g)$$

a. adding more Cl_2
b. decreasing the volume of the container
c. decreasing the temperature

ANSWER

a. When the concentration of the reactant Cl_2 increases, the equilibrium shifts in the direction of the products.
b. There is no change in equilibrium mixture because the moles of reactants are equal to the moles of products.
c. When the temperature for an endothermic reaction decreases, the equilibrium shifts to add heat, which is in the direction of the reactants.

CHEMISTRY LINK TO **HEALTH**
Homeostasis: Regulation of Body Temperature

In a physiological system of equilibrium called *homeostasis*, changes in our environment are balanced by changes in our bodies. It is crucial to our survival that we balance heat gain with heat loss. If we do not lose enough heat, our body temperature rises. At high temperatures, the body can no longer regulate our metabolic reactions. If we lose too much heat, body temperature drops. At low temperatures, essential functions proceed too slowly.

The skin plays an important role in the maintenance of body temperature. When the outside temperature rises, receptors in the skin send signals to the brain. The temperature-regulating part of the brain stimulates the sweat glands to produce perspiration. As perspiration evaporates from the skin, heat is removed and the body temperature is decreased.

In cold temperatures, epinephrine is released, causing an increase in metabolic rate, which increases the production of heat. Receptors on the skin signal the brain to constrict the blood vessels. Less blood flows through the skin, and heat is conserved. The production of perspiration stops, thereby lessening the heat lost by evaporation.

Blood vessels dilate
• sweat production increases
• sweat evaporates
• skin cools

Blood vessels constrict and epinephrine is released
• metabolic activity increases
• muscular activity increases
• shivering occurs
• sweat production stops

QUESTIONS AND PROBLEMS

13.5 Changing Equilibrium Conditions: Le Châtelier's Principle

LEARNING GOAL Use Le Châtelier's principle to describe the changes made in equilibrium concentrations when reaction conditions change.

13.35 In the lower atmosphere, oxygen is converted to ozone (O_3) by the energy provided from lightning.

$$3O_2(g) + \text{heat} \rightleftharpoons 2O_3(g)$$

For each of the following changes at equilibrium, indicate whether the equilibrium shifts in the direction of product, reactants, or does not change:
a. adding more $O_2(g)$
b. adding more $O_3(g)$
c. increasing the temperature

d. increasing the volume of the container
e. adding a catalyst

13.36 Ammonia is produced by reacting nitrogen gas and hydrogen gas.

$$N_2(g) + 3H_2(g) \rightleftharpoons 2NH_3(g) + 92 \text{ kJ}$$

For each of the following changes at equilibrium, indicate whether the equilibrium shifts in the direction of product, reactants, or does not change:
a. removing some $N_2(g)$
b. decreasing the temperature
c. adding more $NH_3(g)$
d. adding more $H_2(g)$
e. increasing the volume of the container

13.37 Hydrogen chloride can be made by reacting hydrogen gas and chlorine gas.

$$H_2(g) + Cl_2(g) + heat \rightleftharpoons 2HCl(g)$$

For each of the following changes at equilibrium, indicate whether the equilibrium shifts in the direction of product, reactants, or does not change:

a. adding more $H_2(g)$
b. increasing the temperature
c. removing some $HCl(g)$
d. adding a catalyst
e. removing some $Cl_2(g)$

13.38 When heated, carbon monoxide reacts with water to produce carbon dioxide and hydrogen.

$$CO(g) + H_2O(g) \rightleftharpoons CO_2(g) + H_2(g) + heat$$

For each of the following changes at equilibrium, indicate whether the equilibrium shifts in the direction of products, reactants, or does not change:

a. decreasing the temperature
b. adding more $H_2(g)$

c. removing $CO_2(g)$ as its forms
d. adding more $H_2O(g)$
e. decreasing the volume of the container

Applications

Use the following equation for the equilibrium of hemoglobin in the blood to answer problems 13.39 and 13.40:

$$Hb(aq) + O_2(g) \rightleftharpoons HbO_2(aq)$$

13.39 Athletes who train at high altitudes initially experience hypoxia, which causes their body to produce more hemoglobin. When an athlete first arrives at high altitude,

a. What is the stress on the hemoglobin equilibrium?
b. In what direction does the hemoglobin equilibrium shift?

13.40 A person who has been a smoker and has a low oxygen blood saturation uses an oxygen tank for supplemental oxygen. When oxygen is first supplied,

a. What is the stress on the hemoglobin equilibrium?
b. In what direction does the hemoglobin equilibrium shift?

13.6 Equilibrium in Saturated Solutions

LEARNING GOAL Write the solubility product expression for a slightly soluble ionic compound and calculate the K_{sp}; use K_{sp} to determine the solubility.

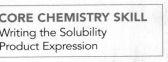

CORE CHEMISTRY SKILL
Writing the Solubility Product Expression

Until now, we have looked primarily at equilibrium systems that involve gases. However, there are also equilibrium systems that involve aqueous saturated solutions between solid solutes of slightly soluble ionic compounds and their ions. Everyday examples of solubility equilibrium in solution are found in the slightly soluble ionic compounds that are found in bone and kidney stones. Bone is composed of calcium phosphate, $Ca_3(PO_4)_2$, which produces ions during bone loss. Kidney stones are composed of compounds such as calcium oxalate, CaC_2O_4.

Solubility Product Expression

In a saturated solution, a solid slightly soluble ionic compound is in equilibrium with its ions. As long as the temperature remains constant, the concentration of the ions in the saturated solution is constant. Let us look at the solubility equilibrium equation for CaC_2O_4, which is written with the solid solute on the left and the ions in solution on the right.

Calcium oxalate produces small amounts of Ca^{2+} and $C_2O_4^{2-}$ in aqueous solution.

$$CaC_2O_4(s) \rightleftharpoons Ca^{2+}(aq) + C_2O_4^{2-}(aq)$$

The *solubility* of a substance is the quantity that dissolves to form a saturated solution. We represent the solubility in a saturated aqueous solution of solid CaC_2O_4 by the **solubility product expression (K_{sp})**, which is the product of the ion concentrations. As in other heterogeneous equilibria, the concentration of the solid CaC_2O_4 is constant and is not included in the solubility product expression.

$$K_{sp} = [Ca^{2+}][C_2O_4^{2-}]$$

In another example, we look at the equilibrium of solid calcium phosphate and its ions Ca^{2+} and PO_4^{3-}.

$$Ca_3(PO_4)_2(s) \rightleftharpoons 3Ca^{2+}(aq) + 2PO_4^{3-}(aq)$$

As with other equilibrium expressions, the molar concentration of each product ion is raised to a power that is equal to its coefficient in the balanced equilibrium equation. For this equilibrium equation, the solubility product expression consists of $[Ca^{2+}]$ raised to the power of 3 and $[PO_4^{3-}]$ raised to the power of 2.

$$K_{sp} = [Ca^{2+}]^3[PO_4^{3-}]^2$$

ENGAGE

Why is the solubility product expression for Cu_2CO_3 written $K_{sp} = [Cu^+]^2[CO_3^{2-}]$?

Solubility Product Constant

The numerical value of the solubility product expression is the **solubility product constant, K_{sp}**. In this text, solubility will be expressed as the molar solubility, which is the moles of solute that dissolve in 1 liter of saturated solution. The calculation of a solubility product constant is shown in Sample Problem 13.8.

CORE CHEMISTRY SKILL
Calculating a Solubility
Product Constant

SAMPLE PROBLEM 13.8 Calculating the Solubility Product Constant

We can make a saturated solution of $BaCO_3$ by adding solid $BaCO_3$ to water and stirring until equilibrium is reached. What is the numerical value of K_{sp} for $BaCO_3$ if the equilibrium mixture contains 5.1×10^{-5} M Ba^{2+} and 5.1×10^{-5} M CO_3^{2-}?

▶ **TRY IT FIRST**

SOLUTION

STEP 1 State the given and needed quantities.

ANALYZE THE PROBLEM	Given	Need	Connect
	$[Ba^{2+}] = 5.1 \times 10^{-5}$ M, $[CO_3^{2-}] = 5.1 \times 10^{-5}$ M	numerical value of K_{sp}	solubility product expression

STEP 2 Write the equilibrium equation for the dissociation of the slightly soluble ionic compound.

$$BaCO_3(s) \rightleftharpoons Ba^{2+}(aq) + CO_3^{2-}(aq)$$

STEP 3 Write the solubility product expression, K_{sp}.

$$K_{sp} = [Ba^{2+}][CO_3^{2-}]$$

STEP 4 Substitute the molar concentration of each ion into the K_{sp} expression and calculate.

$$K_{sp} = [5.1 \times 10^{-5}][5.1 \times 10^{-5}] = 2.6 \times 10^{-9}$$

STUDY CHECK 13.8

A saturated solution of AgBr contains 7.3×10^{-7} M Ag^+ and 7.3×10^{-7} M Br^-. What is the numerical value of K_{sp} for AgBr?

ANSWER

$K_{sp} = 5.3 \times 10^{-13}$

Terraces in Pamukkale, Turkey, are composed of calcium carbonate, which is slightly soluble.

Guide to Calculating K_{sp}

STEP 1
State the given and needed quantities.

STEP 2
Write the equilibrium equation for the dissociation of the slightly soluble ionic compound.

STEP 3
Write the solubility product expression, K_{sp}.

STEP 4
Substitute the molar concentration of each ion into the K_{sp} expression and calculate.

TABLE 13.6 gives values of K_{sp} for a selected group of slightly soluble ionic compounds at 25 °C.

SAMPLE PROBLEM 13.9 Calculating the Solubility Product Constant Using Coefficients

A saturated solution of strontium fluoride, SrF_2, contains 8.7×10^{-4} M Sr^{2+} and 1.7×10^{-3} M F^-. What is the numerical value of K_{sp} for SrF_2?

▶ **TRY IT FIRST**

SOLUTION

STEP 1 State the given and needed quantities.

ANALYZE THE PROBLEM	Given	Need	Connect
	$[Sr^{2+}] = 8.7 \times 10^{-4}$ M, $[F^-] = 1.7 \times 10^{-3}$ M	numerical value of K_{sp}	solubility product expression

TABLE 13.6 Examples of Solubility Product Constants (K_{sp})

Formula	K_{sp}
AgCl	1.8×10^{-10}
Ag_2SO_4	1.2×10^{-5}
$BaSO_4$	1.1×10^{-10}
$CaCO_3$	5.0×10^{-9}
CaF_2	3.2×10^{-11}
$Ca(OH)_2$	6.5×10^{-6}
$CaSO_4$	2.4×10^{-5}
$PbCl_2$	1.5×10^{-6}
$PbCO_3$	7.4×10^{-14}

STEP 2 Write the equilibrium equation for the dissociation of the slightly soluble ionic compound.

$$SrF_2(s) \rightleftharpoons Sr^{2+}(aq) + 2F^-(aq)$$

STEP 3 Write the solubility product expression, K_{sp}.

$$K_{sp} = [Sr^{2+}][F^-]^2$$

STEP 4 Substitute the molar concentration of each ion into the K_{sp} expression and calculate.

$$K_{sp} = [8.7 \times 10^{-4}][1.7 \times 10^{-3}]^2 = 2.5 \times 10^{-9}$$

STUDY CHECK 13.9

What is the numerical value of K_{sp} for silver oxalate, $Ag_2C_2O_4$, if a saturated solution contains 2.2×10^{-4} M Ag^+ and 1.1×10^{-4} M $C_2O_4^{2-}$?

ANSWER

$$K_{sp} = 5.3 \times 10^{-12}$$

Molar Solubility, S

The molar solubility, S, of a slightly soluble ionic compound is the number of moles of solute that dissolves in 1 liter of solution. For example, the molar solubility of cadmium sulfide, CdS, is found experimentally to be 1×10^{-12} M.

$$CdS(s) \rightleftharpoons Cd^{2+}(aq) + S^{2-}(aq)$$

Because CdS dissociates into Cd^{2+} and S^{2-} ions, they each have a concentration equal to the solubility S.

$$S = [Cd^{2+}] = [S^{2-}] = 1 \times 10^{-12} \text{ M}$$

If we know the K_{sp} of a slightly soluble ionic compound, we can determine its molar solubility, as shown in Sample Problem 13.10.

Cadmium sulfide is slightly soluble.

CORE CHEMISTRY SKILL

Calculating the Molar Solubility

Guide to Calculating Molar Solubility from K_{sp}

STEP 1
State the given and needed quantities.

STEP 2
Write the equilibrium equation for the dissociation of the slightly soluble ionic compound.

STEP 3
Write the solubility product expression, K_{sp}, using S.

STEP 4
Calculate the molar solubility, S.

SAMPLE PROBLEM 13.10 Calculating the Molar Solubility from K_{sp}

Calculate the molar solubility, S, of $PbSO_4$ if it has a $K_{sp} = 1.6 \times 10^{-8}$.

TRY IT FIRST

SOLUTION

STEP 1 State the given and needed quantities.

ANALYZE THE PROBLEM	Given	Need	Connect
	$K_{sp} = 1.6 \times 10^{-8}$	molar solubility (S) of $PbSO_4(s)$	solubility product expression

STEP 2 Write the equilibrium equation for the dissociation of the slightly soluble ionic compound.

$$PbSO_4(s) \rightleftharpoons Pb^{2+}(aq) + SO_4^{2-}(aq)$$

STEP 3 Write the solubility product expression, K_{sp}, using S.

$$K_{sp} = [Pb^{2+}][SO_4^{2-}] = S \times S = S^2 = 1.6 \times 10^{-8}$$

STEP **4** Calculate the molar solubility, S.

$$S^2 = 1.6 \times 10^{-8}$$
$$S = \sqrt{1.6 \times 10^{-8}} = 1.3 \times 10^{-4}\,M$$

Thus, 1.3×10^{-4} mol of $PbSO_4$ will dissolve in 1 L of solution.

STUDY CHECK 13.10

Calculate the molar solubility, S, of MnS if it has a $K_{sp} = 2.5 \times 10^{-10}$.

ANSWER

$S = 1.6 \times 10^{-5}\,M$

QUESTIONS AND PROBLEMS

13.6 Equilibrium in Saturated Solutions

LEARNING GOAL Write the solubility product expression for a slightly soluble ionic compound and calculate K_{sp}; use K_{sp} to determine the solubility.

13.41 For each of the following slightly soluble ionic compounds, write the equilibrium equation for dissociation and the solubility product expression:
  a. $MgCO_3$ **b.** CaF_2 **c.** Ag_3PO_4

13.42 For each of the following slightly soluble ionic compounds, write the equilibrium equation for dissociation and the solubility product expression:
  a. Ag_2S **b.** $Al(OH)_3$ **c.** BaF_2

13.43 A saturated solution of barium sulfate, $BaSO_4$, has $[Ba^{2+}] = 1 \times 10^{-5}\,M$ and $[SO_4^{2-}] = 1 \times 10^{-5}\,M$. What is the numerical value of K_{sp} for $BaSO_4$?

13.44 A saturated solution of copper(II) sulfide, CuS, has $[Cu^{2+}] = 1.1 \times 10^{-18}\,M$ and $[S^{2-}] = 1.1 \times 10^{-18}\,M$. What is the numerical value of K_{sp} for CuS?

13.45 A saturated solution of silver carbonate, Ag_2CO_3, has $[Ag^+] = 2.6 \times 10^{-4}\,M$ and $[CO_3^{2-}] = 1.3 \times 10^{-4}\,M$. What is the numerical value of K_{sp} for Ag_2CO_3?

13.46 A saturated solution of barium fluoride, BaF_2, has $[Ba^{2+}] = 3.6 \times 10^{-3}\,M$ and $[F^-] = 7.2 \times 10^{-3}\,M$. What is the numerical value of K_{sp} for BaF_2?

13.47 Calculate the molar solubility, S, of CuI if it has a K_{sp} of 1×10^{-12}.

13.48 Calculate the molar solubility, S, of SnS if it has a K_{sp} of 1×10^{-26}.

Follow Up

EQUILIBRIUM OF CO_2 IN THE OCEAN

One of Peter's research projects was to study the influence of a change in ocean water acidity on coral species, which have calcium carbonate skeletons. The acidity is affected by the concentration of carbon dioxide dissolved in the water. The carbon dioxide dissolves readily in water to form $H_2CO_3(aq)$, which breaks up into $H^+(aq)$ and $HCO_3^-(aq)$. The increase in H^+ makes the ocean water more acidic.

Applications

13.49 **a.** Write the equation for the dissociation of the slightly soluble ionic compound calcium carbonate.
  b. Write the solubility product expression, K_{sp}.

13.50 **a.** What is the numerical value of K_{sp} if the equilibrium mixture has $[Ca^{2+}]$ and $[CO_3^{2-}]$ equal to $7.1 \times 10^{-5}\,M$?
  b. If increasing the acidity of the ocean has the effect of decreasing the concentration of CO_3^{2-}, will calcium carbonate be more or less soluble if the acidity increases?

Coral reefs are affected by the changing acidity of the ocean.

REACTION RATES AND CHEMICAL EQUILIBRIUM

involves

Reaction Rates

are affected by

Concentrations of Reactants

Temperature

Catalyst

Reversible Reactions

when equal in rate give

Equilibrium Expression

is written as

$$K_c = \frac{[Products]}{[Reactants]}$$

Small K_c has Mostly Reactants

Large K_c has Mostly Products

Le Châtelier's Principle

indicates that equilibrium shifts for changes in

Concentration, Temperature, and Volume

in

Saturated Solutions

is called

Solubility Product Expression

is written as

$$K_{sp} = [cation][anion]$$

CHAPTER REVIEW

13.1 Rates of Reactions

LEARNING GOAL Describe how temperature, concentration, and catalysts affect the rate of a reaction.

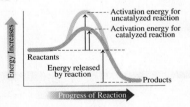

- The rate of a reaction is the speed at which the reactants are converted to products.
- Increasing the concentrations of reactants, increasing the temperature, or adding a catalyst can increase the rate of a reaction.

13.2 Chemical Equilibrium

LEARNING GOAL Use the concept of reversible reactions to explain chemical equilibrium.

- Chemical equilibrium occurs in a reversible reaction when the rate of the forward reaction becomes equal to the rate of the reverse reaction.
- At equilibrium, no further change occurs in the concentrations of the reactants and products as the forward and reverse reactions continue.

13.3 Equilibrium Constants

LEARNING GOAL Calculate the equilibrium constant for a reversible reaction given the concentrations of reactants and products at equilibrium.

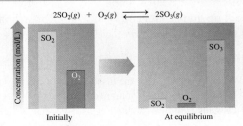

- An equilibrium constant, K_c, is the ratio of the concentrations of the products to the concentrations of the reactants, with each concentration raised to a power equal to its coefficient in the balanced chemical equation.
- For heterogeneous reactions, only the molar concentrations of gases are placed in the equilibrium expression.

13.4 Using Equilibrium Constants

LEARNING GOAL Use an equilibrium constant to predict the extent of reaction and to calculate equilibrium concentrations.

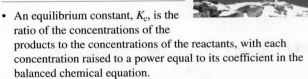

- A large value of K_c indicates that an equilibrium mixture contains mostly products and few reactants, whereas a small value of K_c indicates that the equilibrium mixture contains mostly reactants.
- Equilibrium constants can be used to calculate the concentration of a component in the equilibrium mixture.

13.5 Changing Equilibrium Conditions: Le Châtelier's Principle

LEARNING GOAL Use Le Châtelier's principle to describe the changes made in equilibrium concentrations when reaction conditions change.

- When reactants are removed or products are added to an equilibrium mixture, the system shifts in the direction of the reactants.
- When reactants are added or products are removed from an equilibrium mixture, the system shifts in the direction of the products.
- A decrease in the volume of a reaction container causes a shift in the direction of the smaller number of moles of gas.

- An increase in the volume of a reaction container causes a shift in the direction of the greater number of moles of gas.
- Increasing the temperature of an endothermic reaction or decreasing the temperature of an exothermic reaction will cause the system to shift in the direction of products.
- Decreasing the temperature of an endothermic reaction or increasing the temperature of an exothermic reaction will cause the system to shift in the direction of reactants.

13.6 Equilibrium in Saturated Solutions

LEARNING GOAL Write the solubility product expression for a slightly soluble ionic compound and calculate K_{sp}; use K_{sp} to determine the solubility.

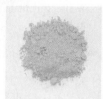

- In a saturated solution, a slightly soluble ionic compound is in equilibrium with its ions.
- In a saturated solution, the concentrations of the ions from the slightly soluble ionic compound are constant and can be used to calculate the solubility product constant, K_{sp}.
- If K_{sp} for a slightly soluble ionic compound is known, its solubility can be calculated.

KEY TERMS

activation energy The energy that must be provided by a collision to break apart the bonds of the reacting molecules.

catalyst A substance that increases the rate of reaction by lowering the activation energy.

chemical equilibrium The point at which the rate of forward and reverse reactions are equal so that no further change in concentrations of reactants and products takes place.

collision theory A model for a chemical reaction stating that molecules must collide with sufficient energy and proper orientation to form products.

equilibrium constant, K_c The numerical value obtained by substituting the equilibrium concentrations of the components into the equilibrium expression.

equilibrium expression The ratio of the concentrations of products to the concentrations of reactants, with each component raised to an exponent equal to the coefficient of that compound in the balanced chemical equation.

heterogeneous equilibrium An equilibrium system in which the components are in different states.

homogeneous equilibrium An equilibrium system in which all components are in the same state.

Le Châtelier's principle When a stress is placed on a system at equilibrium, the equilibrium shifts to relieve that stress.

rate of reaction The speed at which reactants form products.

reversible reaction A reaction in which a forward reaction occurs from reactants to products, and a reverse reaction occurs from products back to reactants.

solubility product constant, K_{sp} The product of the concentrations of the ions in a saturated solution of a slightly soluble ionic compound, with each concentration raised to a power equal to its coefficient in the balanced equilibrium equation.

solubility product expression The product of the ion concentrations, with each concentration raised to an exponent equal to the coefficient in the balanced chemical equation.

 ## CORE CHEMISTRY SKILLS

The chapter section containing each Core Chemistry Skill is shown in parentheses at the end of each heading.

Writing the Equilibrium Expression (13.3)

- An equilibrium expression for a reversible reaction is written by multiplying the concentrations of the products in the numerator and dividing by the product of the concentrations of the reactants in the denominator.
- Each concentration is raised to a power equal to its coefficient in the balanced chemical equation:

$$K_c = \frac{[\text{Products}]}{[\text{Reactants}]} = \frac{[\text{C}]^c [\text{D}]^d}{[\text{A}]^a [\text{B}]^b} \quad \text{Coefficients}$$

Example: Write the equilibrium expression for the following chemical reaction:

$$2NO_2(g) \rightleftharpoons N_2O_4(g)$$

Answer: $K_c = \dfrac{[N_2O_4]}{[NO_2]^2}$

Calculating an Equilibrium Constant (13.3)

- The equilibrium constant, K_c, is the numerical value obtained by substituting experimentally measured molar concentrations at equilibrium into the equilibrium expression.

Example: Calculate the numerical value of K_c for the following reaction when the equilibrium mixture contains 0.025 M NO_2 and 0.087 M N_2O_4:

$$2NO_2(g) \rightleftharpoons N_2O_4(g)$$

Answer: Write the equilibrium expression, substitute the molar concentrations, and calculate.

$$K_c = \frac{[N_2O_4]}{[NO_2]^2} = \frac{[0.087]}{[0.025]^2} = 140$$

Calculating Equilibrium Concentrations (13.4)

- To determine the concentration of a product or a reactant at equilibrium, we use the equilibrium expression to solve for the unknown concentration.

Example: Calculate the equilibrium concentration for CF_4 if $K_c = 2.0$, and the equilibrium mixture contains 0.10 M COF_2 and 0.050 M CO_2.

$$2COF_2(g) \rightleftharpoons CO_2(g) + CF_4(g)$$

Answer: Write the equilibrium expression.

$$K_c = \frac{[CO_2][CF_4]}{[COF_2]^2}$$

Rearrange the equation, substitute the molar concentrations, and calculate.

$$[CF_4] = K_c \times \frac{[COF_2]^2}{[CO_2]} = 2.0 \times \frac{[0.10]^2}{[0.050]} = 0.40 \text{ M}$$

Using Le Châtelier's Principle (13.5)

- Le Châtelier's principle states that when a system at equilibrium is disturbed by changes in concentration, volume, or temperature, the system will shift in the direction that will reduce that stress.

Example: Nitrogen and oxygen form dinitrogen pentoxide in an exothermic reaction.

$$2N_2(g) + 5O_2(g) \rightleftharpoons 2N_2O_5(g) + \text{heat}$$

For each of the following changes at equilibrium, indicate whether the equilibrium shifts in the direction of products, reactants, or does not change:

a. removing some $N_2(g)$
b. decreasing the temperature
c. increasing the volume of the container

Answer: a. Removing a reactant shifts the equilibrium in the direction of the reactants.

b. Decreasing the temperature shifts the equilibrium of an exothermic reaction in the direction of products, which produces heat.

c. Increasing the volume of the container shifts the equilibrium in the direction of the greater number of moles of gas, which is in the direction of the reactants.

Writing the Solubility Product Expression (13.6)

- In a saturated solution, a solid slightly soluble ionic compound is in equilibrium with its ions.

$$FeF_2(s) \rightleftharpoons Fe^{2+}(aq) + 2F^-(aq)$$

- We represent the solubility of solid FeF_2 in a saturated aqueous solution by the solubility product expression, K_{sp}, which is the product of the ion concentrations raised to the powers equal to the coefficients.

$$K_{sp} = [Fe^{2+}][F^-]^2$$

Example: Write the equilibrium equation for the dissociation of the slightly soluble ionic compound PbI_2, and its solubility product expression.

Answer: $PbI_2(s) \rightleftharpoons Pb^{2+}(aq) + 2I^-(aq)$ $K_{sp} = [Pb^{2+}][I^-]^2$

Calculating a Solubility Product Constant (13.6)

- The solubility product constant, K_{sp}, is calculated by substituting the molar concentrations of the ions into the solubility product expression.

Example: What is the numerical value of K_{sp} for a saturated solution of PbI_2 that contains 1.3×10^{-3} M Pb^{2+} and 2.6×10^{-3} M I^-?

Answer: The K_{sp} is calculated by substituting the molar concentrations into the solubility product expression.

$$K_{sp} = [Pb^{2+}][I^-]^2 = [1.3 \times 10^{-3}][2.6 \times 10^{-3}]^2$$
$$= 8.8 \times 10^{-9}$$

Calculating the Molar Solubility (13.6)

- The molar solubility, S, of a slightly soluble ionic compound is the number of moles of solute that dissolves in 1 liter of solution.
- If we know the K_{sp} of a slightly soluble ionic compound, we can calculate the molar solubility, S, of the slightly soluble ionic compound.

Example: What is the solubility product expression and the molar solubility, S, of NiS, which has a K_{sp} of 4.0×10^{-20}?

Answer: $K_{sp} = [Ni^{2+}][S^{2-}] = S \times S = S^2 = 4.0 \times 10^{-20}$

$$S = \sqrt{4.0 \times 10^{-20}} = 2.0 \times 10^{-10} \text{ M}$$

Therefore, 2.0×10^{-10} mol of NiS will dissolve in 1 L of saturated solution.

UNDERSTANDING THE CONCEPTS

The chapter sections to review are shown in parentheses at the end of each question.

13.51 Write the equilibrium expression for each of the following reactions: (13.3)

a. $CH_4(g) + 2O_2(g) \rightleftharpoons CO_2(g) + 2H_2O(g)$
b. $4NH_3(g) + 3O_2(g) \rightleftharpoons 2N_2(g) + 6H_2O(g)$
c. $C(s) + 2H_2(g) \rightleftharpoons CH_4(g)$

13.52 Write the equilibrium expression for each of the following reactions: (13.3)

a. $2C_2H_6(g) + 7O_2(g) \rightleftharpoons 4CO_2(g) + 6H_2O(g)$
b. $2KHCO_3(s) \rightleftharpoons K_2CO_3(s) + CO_2(g) + H_2O(g)$
c. $4NH_3(g) + 5O_2(g) \rightleftharpoons 4NO(g) + 6H_2O(g)$

13.53 Would the equilibrium constant, K_c, for the reaction in the diagrams have a large or small value? (13.4)

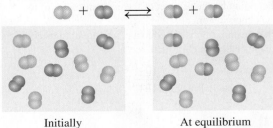

Initially   At equilibrium

13.54 Would the equilibrium constant, K_c, for the reaction in the diagrams have a large or small value? (13.4)

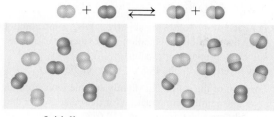

Initially   At equilibrium

13.55 Would T_2 be higher or lower than T_1 for the reaction shown in the diagrams? (13.5)

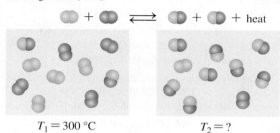

$T_1 = 300\ °C$ $T_2 = ?$

13.56 Would the reaction shown in the diagrams be exothermic or endothermic? (13.5)

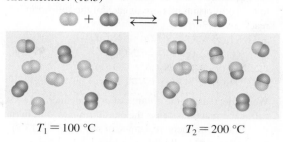

$T_1 = 100\ °C$ $T_2 = 200\ °C$

ADDITIONAL QUESTIONS AND PROBLEMS

13.57 For each of the following changes at equilibrium, indicate whether the equilibrium shifts in the direction of products, reactants, or does not change: (13.5)

$$C_2H_4(g) + Cl_2(g) \rightleftharpoons C_2H_4Cl_2(g) + heat$$

  a. increasing the temperature
  b. decreasing the volume of the container
  c. adding a catalyst
  d. adding more $Cl_2(g)$

13.58 For each of the following changes at equilibrium, indicate whether the equilibrium shifts in the direction of products, reactants, or does not change: (13.5)

$$N_2(g) + O_2(g) + heat \rightleftharpoons 2NO(g)$$

  a. increasing the temperature
  b. decreasing the volume of the container
  c. adding a catalyst
  d. adding more $N_2(g)$

13.59 For each of the following reactions, indicate if the equilibrium mixture contains mostly products, mostly reactants, or both reactants and products: (13.4)
  a. $H_2(g) + Cl_2(g) \rightleftharpoons 2HCl(g)$ $K_c = 1.3 \times 10^{34}$
  b. $2NOBr(g) \rightleftharpoons 2NO(g) + Br_2(g)$ $K_c = 2.0$
  c. $2H_2S(g) + CH_4(g) \rightleftharpoons CS_2(g) + 4H_2(g)$
                              $K_c = 5.3 \times 10^{-8}$
  d. $C(s) + H_2O(g) \rightleftharpoons CO(g) + H_2(g)$ $K_c = 6.3 \times 10^{-1}$

13.60 For each of the following reactions, indicate if the equilibrium mixture contains mostly products, mostly reactants, or both reactants and products: (13.4)
  a. $2H_2O(g) \rightleftharpoons 2H_2(g) + O_2(g)$ $K_c = 4 \times 10^{-48}$
  b. $N_2(g) + 3H_2(g) \rightleftharpoons 2NH_3(g)$ $K_c = 0.30$
  c. $2SO_2(g) + O_2(g) \rightleftharpoons 2SO_3(g)$ $K_c = 1.2 \times 10^9$
  d. $H_2(g) + S(s) \rightleftharpoons 2H_2S(g)$ $K_c = 7.8 \times 10^5$

13.61 Consider the reaction: (13.3)

$$2NH_3(g) \rightleftharpoons N_2(g) + 3H_2(g)$$

  a. Write the equilibrium expression.
  b. What is the numerical value of K_c for the reaction if the concentrations at equilibrium are 0.20 M NH_3, 3.0 M N_2, and 0.50 M H_2?

13.62 Consider the reaction: (13.3)

$$2SO_2(g) + O_2(g) \rightleftharpoons 2SO_3(g)$$

  a. Write the equilibrium expression.
  b. What is the numerical value of K_c for the reaction if the concentrations at equilibrium are 0.10 M SO_2, 0.12 M O_2, and 0.60 M SO_3?

13.63 The K_c for the following reaction is 5.0 at 100 °C. If an equilibrium mixture contains 0.50 M NO_2, what is the molar concentration of N_2O_4? (13.3, 13.4)

$$2NO_2(g) \rightleftharpoons N_2O_4(g)$$

13.64 The K_c for the following reaction is 15 at 220 °C. If an equilibrium mixture contains 0.40 M CO and 0.20 M H_2, what is the molar concentration of CH_4O? (13.3, 13.4)

$$CO(g) + 2H_2(g) \rightleftharpoons CH_4O(g)$$

13.65 According to Le Châtelier's principle, does the equilibrium shift in the direction of products or reactants when O_2 is added to the equilibrium mixture of each of the following reactions? (13.5)
  a. $3O_2(g) \rightleftharpoons 2O_3(g)$
  b. $2CO_2(g) \rightleftharpoons 2CO(g) + O_2(g)$
  c. $2SO_2(g) + O_2(g) \rightleftharpoons 2SO_3(g)$
  d. $2SO_2(g) + 2H_2O(g) \rightleftharpoons 2H_2S(g) + 3O_2(g)$

13.66 According to Le Châtelier's principle, does the equilibrium shift in the direction of products or reactants when N_2 is added to the equilibrium mixture of each of the following reactions? (13.5)

a. $2NH_3(g) \rightleftharpoons 3H_2(g) + N_2(g)$
b. $N_2(g) + O_2(g) \rightleftharpoons 2NO(g)$
c. $2NO_2(g) \rightleftharpoons N_2(g) + 2O_2(g)$
d. $4NH_3(g) + 3O_2(g) \rightleftharpoons 2N_2(g) + 6H_2O(g)$

13.67 Would decreasing the volume of the container for each of the following reactions cause the equilibrium to shift in the direction of products, reactants, or not change? (13.5)

a. $3O_2(g) \rightleftharpoons 2O_3(g)$
b. $2CO_2(g) \rightleftharpoons 2CO(g) + O_2(g)$
c. $P_4(g) + 5O_2(g) \rightleftharpoons P_4O_{10}(s)$
d. $2SO_2(g) + 2H_2O(g) \rightleftharpoons 2H_2S(g) + 3O_2(g)$

13.68 Would increasing the volume of the container for each of the following reactions cause the equilibrium to shift in the direction of products, reactants, or not change? (13.5)

a. $2NH_3(g) \rightleftharpoons 3H_2(g) + N_2(g)$
b. $N_2(g) + O_2(g) \rightleftharpoons 2NO(g)$
c. $N_2(g) + 2O_2(g) \rightleftharpoons 2NO_2(g)$
d. $4NH_3(g) + 3O_2(g) \rightleftharpoons 2N_2(g) + 6H_2O(g)$

13.69 The equilibrium constant, K_c, for the decomposition of $COCl_2$ to CO and Cl_2 is 0.68. If an equilibrium mixture contains 0.40 M CO and 0.74 M Cl_2, what is the molar concentration of $COCl_2$? (13.3, 13.4)

$$COCl_2(g) \rightleftharpoons CO(g) + Cl_2(g)$$

13.70 The equilibrium constant, K_c, for the reaction of carbon and water to form carbon monoxide and hydrogen is 0.20 at 1000 °C. If an equilibrium mixture contains solid carbon,

0.40 M H_2O, and 0.40 M CO, what is the molar concentration of H_2? (13.3, 13.4)

$$C(s) + H_2O(g) \rightleftharpoons CO(g) + H_2(g)$$

13.71 For each of the following slightly soluble ionic compounds, write the equilibrium equation for dissociation and the solubility product expression: (13.6)

a. $CuCO_3$ b. PbF_2 c. $Fe(OH)_3$

13.72 For each of the following slightly soluble ionic compounds, write the equilibrium equation for dissociation and the solubility product expression: (13.6)

a. CuS   b. Ag_2SO_4 c. $Zn(OH)_2$

13.73 A saturated solution of iron(II) sulfide, FeS, has $[Fe^{2+}] = 7.7 \times 10^{-10}$ M and $[S^{2-}] = 7.7 \times 10^{-10}$ M. What is the numerical value of K_{sp} for FeS? (13.6)

13.74 A saturated solution of copper(I) chloride, CuCl, has $[Cu^+] = 1.1 \times 10^{-3}$ M and $[Cl^-] = 1.1 \times 10^{-3}$ M. What is the numerical value of K_{sp} for CuCl? (13.6)

13.75 A saturated solution of manganese(II) hydroxide, $Mn(OH)_2$, has $[Mn^{2+}] = 3.7 \times 10^{-5}$ M and $[OH^-] = 7.4 \times 10^{-5}$ M. What is the numerical value of K_{sp} for $Mn(OH)_2$? (13.6)

13.76 A saturated solution of silver chromate, Ag_2CrO_4, has $[Ag^+] = 1.3 \times 10^{-4}$ M and $[CrO_4^{2-}] = 6.5 \times 10^{-5}$ M. What is the numerical value of K_{sp} for Ag_2CrO_4? (13.6)

13.77 What is the molar solubility, S, of CdS if it has a K_{sp} of 1.0×10^{-24}? (13.6)

13.78 What is the molar solubility, S, of $CuCO_3$ if it has K_{sp} of 1×10^{-26}? (13.6)

CHALLENGE QUESTIONS

The following groups of questions are related to the topics in this chapter. However, they do not all follow the chapter order, and they require you to combine concepts and skills from several sections. These questions will help you increase your critical thinking skills and prepare for your next exam.

13.79 The K_c at 100 °C is 2.0 for the decomposition reaction of NOBr. (13.3, 13.4, 13.5)

$$2NOBr(g) \rightleftharpoons 2NO(g) + Br_2(g)$$

In an experiment, 1.0 mol of NOBr, 1.0 mol of NO, and 1.0 mol of Br_2 were placed in a 1.0-L container.

a. Write the equilibrium expression for the reaction.
b. Is the system at equilibrium?
c. If not, will the rate of the forward or reverse reaction initially speed up?
d. At equilibrium, which concentration(s) will be greater than 1.0 mol/L, and which will be less than 1.0 mol/L?

13.80 Consider the following reaction: (13.3, 13.4, 13.5)

$$PCl_5(g) \rightleftharpoons PCl_3(g) + Cl_2(g)$$

a. Write the equilibrium expression for the reaction.
b. Initially, 0.60 mol of PCl_5 is placed in a 1.0-L flask. At equilibrium, there is 0.16 mol of PCl_3 in the flask. What are the equilibrium concentrations of PCl_5 and Cl_2?
c. What is the numerical value of the equilibrium constant, K_c, for the reaction?
d. If 0.20 mol of Cl_2 is added to the equilibrium mixture, will the concentration of PCl_5 increase or decrease?

13.81 Indicate how each of the following will affect the equilibrium concentration of CO in the following reaction: (13.3, 13.5)

$$C(s) + H_2O(g) + 31 \text{ kcal} \rightleftharpoons CO(g) + H_2(g)$$

a. adding more $H_2(g)$
b. increasing the temperature
c. increasing the volume of the container
d. decreasing the volume of the container
e. adding a catalyst

13.82 Indicate how each of the following will affect the equilibrium concentration of NH_3 in the following reaction: (13.3, 13.5)

$$4NH_3(g) + 5O_2(g) \rightleftharpoons 4NO(g) + 6H_2O(g) + 906 \text{ kJ}$$

a. adding more $O_2(g)$
b. increasing the temperature
c. increasing the volume of the container
d. adding more $NO(g)$
e. removing some $H_2O(g)$

13.83 Indicate if you would increase or decrease the volume of the container to *increase* the yield of the products in each of the following: (13.5)

a. $2C(s) + O_2(g) \rightleftharpoons 2CO(g)$
b. $2CH_4(g) \rightleftharpoons C_2H_2(g) + 3H_2(g)$
c. $2H_2(g) + O_2(g) \rightleftharpoons 2H_2O(g)$

13.84 Indicate if you would increase or decrease the volume of the container to *increase* the yield of the products in each of the following: (13.5)

    a. $Cl_2(g) + 2NO(g) \rightleftharpoons 2NOCl(g)$

    b. $N_2(g) + 2H_2(g) \rightleftharpoons N_2H_4(g)$

    c. $N_2O_4(g) \rightleftharpoons 2NO_2(g)$

13.85 The antacid milk of magnesia, which contains $Mg(OH)_2$, is used to neutralize excess stomach acid. If the solubility of $Mg(OH)_2$ in water is 9.7×10^{-3} g/L, what is the numerical value of K_{sp} for $Mg(OH)_2$? (13.6)

13.86 The slightly soluble ionic compound PbF_2 has a solubility in water of 0.74 g/L. What is the numerical value of K_{sp} for PbF_2? (13.6)

ANSWERS

Answers to Selected Questions and Problems

13.1 **a.** The rate of the reaction indicates how fast the products form or how fast the reactants are used up.

    b. At room temperature, the reactions involved in the growth of bread mold will proceed at a faster rate than at the lower temperature of the refrigerator.

13.3 The number of collisions will increase when the number of Br_2 molecules is increased.

13.5 **a.** increase   **b.** increase

    c. increase   **d.** decrease

13.7 A reversible reaction is one in which a forward reaction converts reactants to products, whereas a reverse reaction converts products to reactants.

13.9 **a.** not at equilibrium   **b.** at equilibrium

    c. at equilibrium

13.11 The reaction has reached equilibrium because the number of reactants and products does not change.

13.13 **a.** $K_c = \dfrac{[CS_2][H_2]^4}{[CH_4][H_2S]^2}$ **b.** $K_c = \dfrac{[N_2][O_2]}{[NO]^2}$

    c. $K_c = \dfrac{[CS_2][O_2]^4}{[SO_3]^2[CO_2]}$ **d.** $K_c = \dfrac{[H_2]^3[CO]}{[CH_4][H_2O]}$

13.15 $K_c = \dfrac{[XY]^2}{[X_2][Y_2]} = 36$

13.17 $K_c = 1.5$

13.19 $K_c = 260$

13.21 **a.** homogeneous equilibrium

    b. heterogeneous equilibrium

    c. homogeneous equilibrium

    d. heterogeneous equilibrium

13.23 **a.** $K_c = \dfrac{[O_2]^3}{[O_3]^2}$ **b.** $K_c = [CO_2][H_2O]$

    c. $K_c = \dfrac{[C_6H_{12}]}{[C_6H_6][H_2]^3}$ **d.** $K_c = \dfrac{[H_2]^2}{[HCl]^4}$

13.25 $K_c = 0.26$

13.27 Diagram **B** represents the equilibrium mixture.

13.29 **a.** mostly products   **b.** mostly products

    c. mostly reactants

13.31 $[H_2] = 1.1 \times 10^{-3}$ M

13.33 $[NOBr] = 1.4$ M

13.35 **a.** Equilibrium shifts in the direction of the product.

    b. Equilibrium shifts in the direction of the reactants.

    c. Equilibrium shifts in the direction of the product.

    d. Equilibrium shifts in the direction of the reactants.

    e. No shift in equilibrium occurs.

13.37 **a.** Equilibrium shifts in the direction of the product.

    b. Equilibrium shifts in the direction of the product.

    c. Equilibrium shifts in the direction of the product.

    d. No shift in equilibrium occurs.

    e. Equilibrium shifts in the direction of the reactants.

13.39 **a.** The oxygen concentration is lowered.

    b. Equilibrium shifts in the direction of the reactants.

13.41 **a.** $MgCO_3(s) \rightleftharpoons Mg^{2+}(aq) + CO_3^{2-}(aq)$;

$$K_{sp} = [Mg^{2+}][CO_3^{2-}]$$

    b. $CaF_2(s) \rightleftharpoons Ca^{2+}(aq) + 2F^-(aq)$; $K_{sp} = [Ca^{2+}][F^-]^2$

    c. $Ag_3PO_4(s) \rightleftharpoons 3Ag^+(aq) + PO_4^{3-}(aq)$;

$$K_{sp} = [Ag^+]^3[PO_4^{3-}]$$

13.43 $K_{sp} = 1 \times 10^{-10}$

13.45 $K_{sp} = 8.8 \times 10^{-12}$

13.47 $S = 1 \times 10^{-6}$ M

13.49 **a.** $CaCO_3(s) \rightleftharpoons Ca^{2+}(aq) + CO_3^{2-}(aq)$

    b. $K_{sp} = [Ca^{2+}][CO_3^{2-}]$

13.51 **a.** $K_c = \dfrac{[CO_2][H_2O]^2}{[CH_4][O_2]^2}$ **b.** $K_c = \dfrac{[N_2]^2[H_2O]^6}{[NH_3]^4[O_2]^3}$

    c. $K_c = \dfrac{[CH_4]}{[H_2]^2}$

13.53 The equilibrium constant for the reaction would have a small value.

13.55 T_2 is lower than T_1.

13.57 **a.** Equilibrium shifts in the direction of the reactants.

    b. Equilibrium shifts in the direction of the product.

    c. There is no change in equilibrium.

    d. Equilibrium shifts in the direction of the product.

13.59 **a.** mostly products

    b. both reactants and products

    c. mostly reactants

    d. both reactants and products

13.61 **a.** $K_c = \dfrac{[N_2][H_2]^3}{[NH_3]^2}$ **b.** $K_c = 9.4$

13.63 $[N_2O_4] = 1.3$ M

13.65 **a.** Equilibrium shifts in the direction of the product.

    b. Equilibrium shifts in the direction of the reactant.

    c. Equilibrium shifts in the direction of the product.

    d. Equilibrium shifts in the direction of the reactants.

13.67 **a.** Equilibrium shifts in the direction of the product.

    b. Equilibrium shifts in the direction of the reactant.

    c. Equilibrium shifts in the direction of the product.

    d. Equilibrium shifts in the direction of the reactants.

13.69 $[COCl_2] = 0.44$ M

13.71 a. $CuCO_3(s) \rightleftharpoons Cu^{2+}(aq) + CO_3{}^{2-}(aq)$;
$$K_{sp} = [Cu^{2+}][CO_3{}^{2-}]$$

b. $PbF_2(s) \rightleftharpoons Pb^{2+}(aq) + 2F^-(aq)$;
$$K_{sp} = [Pb^{2+}][F^-]^2$$

c. $Fe(OH)_3(s) \rightleftharpoons Fe^{3+}(aq) + 3OH^-(aq)$;
$$K_{sp} = [Fe^{3+}][OH^-]^3$$

13.73 $K_{sp} = 5.9 \times 10^{-19}$

13.75 $K_{sp} = 2.0 \times 10^{-13}$

13.77 $S = 1.0 \times 10^{-12}$ M

13.79 a. $K_c = \dfrac{[NO]^2[Br_2]}{[NOBr]^2}$

b. When the concentrations are placed in the expression, the result is 1.0, which is not equal to K_c. The system is not at equilibrium.

c. The rate of the forward reaction will increase.

d. The $[Br_2]$ and $[NO]$ will increase and $[NOBr]$ will decrease.

13.81 a. decrease   **b.** increase
c. increase   **d.** decrease
e. no change

13.83 a. increase   **b.** increase
c. decrease

13.85 $K_{sp} = 2.0 \times 10^{-11}$

Acids and Bases

14

A 30-YEAR-OLD MAN HAS been brought to the emergency room after an automobile accident. The emergency room nurses are tending to the patient, Larry, who is unresponsive. A blood sample is taken then sent to Brianna, a clinical laboratory technician, who begins the process of analyzing the pH, the partial pressures of O_2 and CO_2, and the concentrations of glucose and electrolytes.

Within minutes, Brianna determines that Larry's blood pH is 7.30 and the partial pressure of CO_2 gas is above the desired level. Blood pH is typically in the range of 7.35 to 7.45, and a value less than 7.35 indicates a state of acidosis. Respiratory acidosis occurs because an increase in the partial pressure of CO_2 gas in the bloodstream prevents the biochemical buffers in blood from making a change in the pH.

Brianna recognizes these signs and immediately contacts the emergency room to inform them that Larry's airway may be blocked. In the emergency room, they provide Larry with an IV containing bicarbonate to increase the blood pH and begin the process of unblocking his airway. Shortly afterward, Larry's airway is cleared, and his blood pH and partial pressure of CO_2 gas return to normal.

CAREER
Clinical Laboratory Technician

Clinical laboratory technicians, also known as medical laboratory technicians, perform a wide variety of tests on body fluids and cells that help in the diagnosis and treatment of patients. These tests range from determining blood concentrations of glucose and cholesterol to determining drug levels in the blood for transplant patients or a patient undergoing treatment. Clinical laboratory technicians also prepare specimens in the detection of cancerous tumors, and type blood samples for transfusions. Clinical laboratory technicians must also interpret and analyze the test results, which are then passed on to the physician.

⊗⊖⊕ KEY MATH SKILLS

- Solving Equations (1.4)
- Converting between Standard Numbers and Scientific Notation (1.5)

🔬 CORE CHEMISTRY SKILLS

- Writing Ionic Formulas (6.2)
- Balancing a Chemical Equation (8.2)
- Using Concentration as a Conversion Factor (12.4)
- Writing the Equilibrium Expression (13.3)
- Calculating Equilibrium Concentrations (13.4)
- Using Le Châtelier's Principle (13.5)

*These Key Math Skills and Core Chemistry Skills from previous chapters are listed here for your review as you proceed to the new material in this chapter.

14.1 Acids and Bases

LEARNING GOAL Describe and name acids and bases.

Citrus fruits are sour because of the presence of acids.

Acids and bases are important substances in health, industry, and the environment. One of the most common characteristics of acids is their sour taste. Lemons and grapefruits taste sour because they contain acids such as citric and ascorbic acid (vitamin C). Vinegar tastes sour because it contains acetic acid. We produce lactic acid in our muscles when we exercise. Acid from bacteria turns milk sour in the production of yogurt and cottage cheese. We have hydrochloric acid in our stomachs that helps us digest food. Sometimes we take antacids, which are bases such as sodium bicarbonate or milk of magnesia, to neutralize the effects of too much stomach acid.

The term *acid* comes from the Latin word *acidus*, which means "sour." We are familiar with the sour tastes of vinegar and lemons and other common acids in foods.

In 1887, the Swedish chemist Svante Arrhenius was the first to describe **acids** as substances that produce hydrogen ions (H^+) when they dissolve in water. Because acids produce ions in water, they are also electrolytes. For example, hydrogen chloride dissociates in water to give hydrogen ions, H^+, and chloride ions, Cl^-. The hydrogen ions give acids a sour taste, change the blue litmus indicator to red, and corrode some metals.

$$HCl(g) \xrightarrow{H_2O} H^+(aq) + Cl^-(aq)$$

Polar molecular compound · Dissociation · Hydrogen ion

Naming Acids

Acids dissolve in water to produce hydrogen ions, along with a negative ion that may be a simple nonmetal anion or a polyatomic ion. When an acid dissolves in water to produce a hydrogen ion and a simple nonmetal anion, the prefix *hydro* is used before the name of the nonmetal, and its *ide* ending is changed to *ic acid*. For example, hydrogen chloride (HCl) dissolves in water to form HCl(*aq*), which is named hydrochloric acid. An exception is hydrogen cyanide (HCN), which as an acid is named hydrocyanic acid.

When an acid contains oxygen, it dissolves in water to produce a hydrogen ion and an oxygen-containing polyatomic anion. The most common form of an oxygen-containing acid has a name that ends with *ic acid*. The name of its polyatomic anion ends in *ate*. If the acid contains a polyatomic ion with an *ite* ending, its name ends in *ous acid*.

The halogens in Group 7A (17) can form more than two oxygen-containing acids. For chlorine, the common form is chloric acid ($HClO_3$), which contains the chlorate polyatomic ion (ClO_3^-). For the acid that contains one more oxygen atom than the common form, the prefix *per* is used; $HClO_4$ is named *perchloric acid*. When the polyatomic ion in the acid has one oxygen atom less than the common form, the suffix *ous* is used. Thus, $HClO_2$ is named *chlorous acid*; it contains the chlorite ion (ClO_2^-). The prefix *hypo* is used for the acid that has two oxygen atoms less than the common form; $HClO$ is named *hypochlorous acid*. The names of some common acids and their anions are listed in **TABLE 14.1**.

TABLE 14.1 Names of Common Acids and Their Anions

Acid	Name of Acid	Anion	Name of Anion
HCl	**Hydro**chlor**ic acid**	Cl^-	Chlor**ide**
HBr	**Hydro**brom**ic acid**	Br^-	Brom**ide**
HI	**Hydro**iod**ic acid**	I^-	Iod**ide**
HCN	**Hydro**cyan**ic acid**	CN^-	Cyan**ide**
HNO_3	Nitr**ic acid**	NO_3^-	Nitr**ate**
HNO_2	Nitr**ous acid**	NO_2^-	Nitr**ite**
H_2SO_4	Sulfur**ic acid**	SO_4^{2-}	Sulf**ate**
H_2SO_3	Sulfur**ous acid**	SO_3^{2-}	Sulf**ite**
H_2CO_3	Carbon**ic acid**	CO_3^{2-}	Carbon**ate**
$HC_2H_3O_2$	Acet**ic acid**	$C_2H_3O_2^-$	Acet**ate**
H_3PO_4	Phosphor**ic acid**	PO_4^{3-}	Phosph**ate**
H_3PO_3	Phosphor**ous acid**	PO_3^{3-}	Phosph**ite**
$HClO_3$	Chlor**ic acid**	ClO_3^-	Chlor**ate**
$HClO_2$	Chlor**ous acid**	ClO_2^-	Chlor**ite**

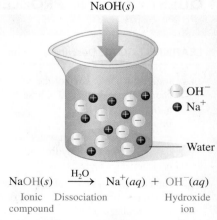

Sulfuric acid dissolves in water to produce one or two H^+ and an anion.

Bases

You may be familiar with some household bases such as antacids, drain openers, and oven cleaners. According to the Arrhenius theory, **bases** are ionic compounds that dissociate into cations and hydroxide ions (OH^-) when they dissolve in water. They are another example of strong electrolytes. For example, sodium hydroxide is an Arrhenius base that dissociates completely in water to give sodium ions (Na^+) and hydroxide ions (OH^-).

Most Arrhenius bases are formed from Groups 1A (1) and 2A (2) metals, such as NaOH, KOH, LiOH, and $Ca(OH)_2$. The hydroxide ions (OH^-) give Arrhenius bases common characteristics, such as a bitter taste and a slippery feel. A base turns litmus indicator blue and phenolphthalein indicator pink. **TABLE 14.2** compares some characteristics of acids and bases.

TABLE 14.2 Some Characteristics of Acids and Bases

Characteristic	Acids	Bases
Arrhenius	Produce H^+	Produce OH^-
Electrolytes	Yes	Yes
Taste	Sour	Bitter, chalky
Feel	May sting	Soapy, slippery
Litmus	Red	Blue
Phenolphthalein	Colorless	Pink
Neutralization	Neutralize bases	Neutralize acids

NaOH(*s*)

$$NaOH(s) \xrightarrow{H_2O} Na^+(aq) + OH^-(aq)$$

Ionic compound   Dissociation   Hydroxide ion

⊖ OH^-
⊕ Na^+

Water

An Arrhenius base produces cations and OH^- anions in an aqueous solution.

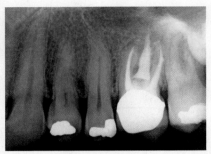

Calcium hydroxide, $Ca(OH)_2$, is used in the food industry to produce beverages, and in dentistry as a filler for root canals.

A soft drink contains H_3PO_4 and H_2CO_3.

Naming Bases

Typical Arrhenius bases are named as *hydroxides*.

Base	Name
LiOH	Lithium **hydroxide**
NaOH	Sodium **hydroxide**
KOH	Potassium **hydroxide**
$Ca(OH)_2$	Calcium **hydroxide**
$Al(OH)_3$	Aluminum **hydroxide**

SAMPLE PROBLEM 14.1 Names and Formulas of Acids and Bases

a. Identify each of the following as an acid or a base and give its name:
    1. H_3PO_4, ingredient in soft drinks
    2. NaOH, ingredient in oven cleaner
b. Write the formula for each of the following:
    1. magnesium hydroxide, ingredient in antacids
    2. hydrobromic acid, used industrially to prepare bromide compounds

TRY IT FIRST

SOLUTION

a. 1. acid, phosphoric acid   **2.** base, sodium hydroxide
b. 1. $Mg(OH)_2$ **2.** HBr

STUDY CHECK 14.1

a. Identify as an acid or a base and give the name for H_2CO_3.
b. Write the formula for iron(III) hydroxide.

ANSWER

a. acid, carbonic acid   **b.** $Fe(OH)_3$

QUESTIONS AND PROBLEMS

14.1 Acids and Bases

LEARNING GOAL Describe and name acids and bases.

14.1 Indicate whether each of the following statements is characteristic of an acid, a base, or both:
    a. has a sour taste
    b. neutralizes bases
    c. produces H^+ ions in water
    d. is named barium hydroxide
    e. is an electrolyte

14.2 Indicate whether each of the following statements is characteristic of an acid, a base, or both:
    a. neutralizes acids
    b. produces OH^- ions in water
    c. has a slippery feel
    d. conducts an electrical current in solution
    e. turns litmus red

14.3 Name each of the following acids or bases:
    a. HCl   **b.** $Ca(OH)_2$ **c.** $HClO_4$
    d. HNO_3 **e.** H_2SO_3 **f.** $HBrO_2$

14.4 Name each of the following acids or bases:
    a. $Al(OH)_3$ **b.** HBr   **c.** H_2SO_4
    d. KOH   **e.** HNO_2 **f.** $HClO_2$

14.5 Write formulas for each of the following acids and bases:
    a. rubidium hydroxide   **b.** hydrofluoric acid
    c. phosphoric acid   **d.** lithium hydroxide
    e. ammonium hydroxide   **f.** periodic acid

14.6 Write formulas for each of the following acids and bases:
    a. barium hydroxide   **b.** hydroiodic acid
    c. nitric acid   **d.** strontium hydroxide
    e. acetic acid   **f.** hypochlorous acid

14.2 Brønsted–Lowry Acids and Bases

LEARNING GOAL Identify conjugate acid–base pairs for Brønsted–Lowry acids and bases.

In 1923, J. N. Brønsted in Denmark and T. M. Lowry in Great Britain expanded the definition of acids and bases to include bases that do not contain OH^- ions. A **Brønsted–Lowry acid** can donate a hydrogen ion, H^+, and a **Brønsted–Lowry base** can accept a hydrogen ion.

A Brønsted–Lowry acid is a substance that donates H^+.

A Brønsted–Lowry base is a substance that accepts H^+.

A free hydrogen ion does not actually exist in water. Its attraction to polar water molecules is so strong that the H^+ bonds to a water molecule and forms a **hydronium ion, H_3O^+.**

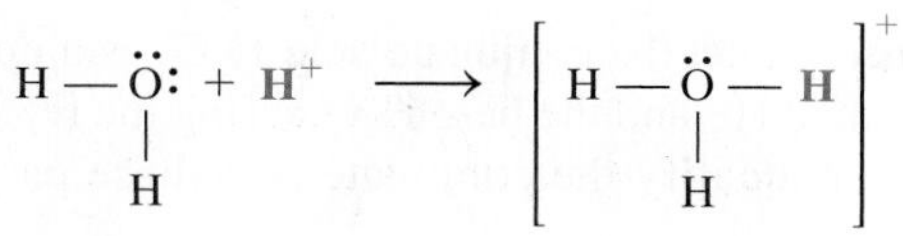

Water   Hydrogen ion   Hydronium ion

We can write the formation of a hydrochloric acid solution as a transfer of H^+ from hydrogen chloride to water. By accepting an H^+ in the reaction, water is acting as a base according to the Brønsted–Lowry concept.

$$HCl + H_2O \longrightarrow H_3O^+ + Cl^-$$

Hydrogen chloride   Water   Hydronium ion   Chloride ion

Acid   Base   Acidic solution
(H^+ donor) (H^+ acceptor)

In another reaction, ammonia (NH_3) acts as a base by accepting H^+ when it reacts with water. Because the nitrogen atom of NH_3 has a stronger attraction for H^+ than oxygen, water acts as an acid by donating H^+.

$$NH_3 + H_2O \rightleftharpoons NH_4^+ + OH^-$$

Ammonia   Water   Ammonium ion   Hydroxide ion

Base   Acid   Basic solution
(H^+ acceptor) (H^+ donor)

SAMPLE PROBLEM 14.2 Acids and Bases

In each of the following equations, identify the reactant that is a Brønsted–Lowry acid and the reactant that is a Brønsted–Lowry base:

a. $HBr(aq) + H_2O(l) \longrightarrow H_3O^+(aq) + Br^-(aq)$
b. $CN^-(aq) + H_2O(l) \rightleftharpoons HCN(aq) + OH^-(aq)$

TRY IT FIRST

SOLUTION

a. HBr, Brønsted–Lowry acid; H_2O, Brønsted–Lowry base
b. H_2O, Brønsted–Lowry acid; CN^-, Brønsted–Lowry base

STUDY CHECK 14.2

When HNO_3 reacts with water, water acts as a Brønsted–Lowry base. Write the equation for the reaction.

ANSWER

$HNO_3(aq) + H_2O(l) \longrightarrow H_3O^+(aq) + NO_3^-(aq)$

Conjugate Acid–Base Pairs

CORE CHEMISTRY SKILL
Identifying Conjugate
Acid–Base Pairs

According to the Brønsted–Lowry theory, a **conjugate acid–base pair** consists of molecules or ions related by the loss of one H^+ by an acid, and the gain of one H^+ by a base. Every acid–base reaction contains two conjugate acid–base pairs because an H^+ is transferred in both the forward and reverse directions. When an acid such as HF loses one H^+, the conjugate base F^- is formed. When the base H_2O gains an H^+, its conjugate acid, H_3O^+, is formed.

Because the overall reaction of HF is reversible, the conjugate acid H_3O^+ can donate H^+ to the conjugate base F^- and re-form the acid HF and the base H_2O. Using the relationship of loss and gain of one H^+, we can now identify the conjugate acid–base pairs as HF/F^- along with H_3O^+/H_2O.

Conjugate acid–base pair

Conjugate acid–base pair

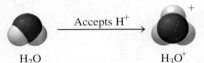

HF, an acid, loses one H^+ to form its conjugate base F^-. Water acts as a base by gaining one H^+ to form its conjugate acid H_3O^+.

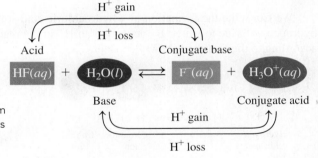

ENGAGE

Why is $HBrO_2$ the conjugate acid of BrO_2^-?

In another reaction, ammonia (NH_3) accepts H^+ from H_2O to form the conjugate acid NH_4^+ and conjugate base OH^-. Each of these conjugate acid–base pairs, NH_4^+/NH_3 and H_2O/OH^-, is related by the loss and gain of one H^+.

Ammonia, NH_3, acts as a base when it gains one H^+ to form its conjugate acid NH_4^+. Water acts as an acid by losing one H^+ to form its conjugate base OH^-.

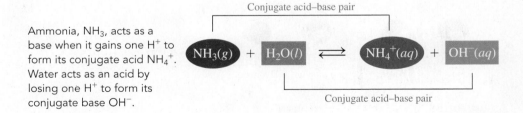

In these two examples, we see that water can act as an acid when it donates H^+ or as a base when it accepts H^+. Substances that can act as both acids and bases are **amphoteric** or *amphiprotic*. For water, the most common amphoteric substance, the acidic or basic behavior depends on the other reactant. Water donates H^+ when it reacts with a stronger base, and it accepts H^+ when it reacts with a stronger acid. Another example of an amphoteric substance is bicarbonate (HCO_3^-). With a base, HCO_3^- acts as an acid and donates one H^+ to give CO_3^{2-}. However, when HCO_3^- reacts with an acid, it acts as a base and accepts one H^+ to form H_2CO_3.

ENGAGE

Why can H_2O be both the conjugate base of H_3O^+ and the conjugate acid of OH^-?

Amphoteric substances act as both acids and bases.

H_3O^+
H_2CO_3 ◀ Acts as a base H_2O
HCO_3^- Acts as an acid ▶ OH^-
CO_3^{2-}

SAMPLE PROBLEM 14.3 Identifying Conjugate Acid–Base Pairs

Identify the conjugate acid–base pairs in the following reaction:

$$HBr(aq) + NH_3(aq) \longrightarrow Br^-(aq) + NH_4^+(aq)$$

TRY IT FIRST

SOLUTION

ANALYZE THE PROBLEM	Given		Need	Connect
	HBr	Br$^-$	conjugate acid–base pairs	lose/gain one H$^+$
	NH$_3$	NH$_4^+$		

STEP 1 **Identify the reactant that loses H$^+$ as the acid.** In the reaction, HBr donates H$^+$ to form the product Br$^-$. Thus HBr is the acid and Br$^-$ is its conjugate base.

STEP 2 **Identify the reactant that gains H$^+$ as the base.** In the reaction, NH$_3$ gains H$^+$ to form the product NH$_4^+$. Thus, NH$_3$ is the base and NH$_4^+$ is its conjugate acid.

STEP 3 **Write the conjugate acid–base pairs.**

HBr/Br$^-$ and NH$_4^+$/NH$_3$

STUDY CHECK 14.3

Identify the conjugate acid–base pairs in the following reaction:

$$HCN(aq) + SO_4^{2-}(aq) \rightleftharpoons CN^-(aq) + HSO_4^-(aq)$$

ANSWER

The conjugate acid–base pairs are HCN/CN$^-$ and HSO$_4^-$/SO$_4^{2-}$.

> ### Guide to Writing Conjugate Acid–Base Pairs
>
> **STEP 1**
> Identify the reactant that loses H$^+$ as the acid.
>
> **STEP 2**
> Identify the reactant that gains H$^+$ as the base.
>
> **STEP 3**
> Write the conjugate acid–base pairs.

QUESTIONS AND PROBLEMS

14.2 Brønsted–Lowry Acids and Bases

LEARNING GOAL Identify conjugate acid–base pairs for Brønsted–Lowry acids and bases.

14.7 Identify the reactant that is a Brønsted–Lowry acid and the reactant that is a Brønsted–Lowry base in each of the following:
   a. $HI(aq) + H_2O(l) \longrightarrow I^-(aq) + H_3O^+(aq)$
   b. $F^-(aq) + H_2O(l) \rightleftharpoons HF(aq) + OH^-(aq)$
   c. $H_2S(aq) + CH_3{-}CH_2{-}NH_2(aq) \rightleftharpoons$
     $\qquad HS^-(aq) + CH_3{-}CH_2{-}NH_3^+(aq)$

14.8 Identify the reactant that is a Brønsted–Lowry acid and the reactant that is a Brønsted–Lowry base in each of the following:
   a. $CO_3^{2-}(aq) + H_2O(l) \rightleftharpoons HCO_3^-(aq) + OH^-(aq)$
   b. $H_2SO_4(aq) + H_2O(l) \longrightarrow HSO_4^-(aq) + H_3O^+(aq)$
   c. $C_2H_3O_2^-(aq) + H_3O^+(aq) \rightleftharpoons HC_2H_3O_2(aq) + H_2O(l)$

14.9 Write the formula for the conjugate base for each of the following acids:
   a. HF   **b.** H_2O **c.** $H_2PO_3^-$
   d. HSO_4^- **e.** $HClO_2$

14.10 Write the formula for the conjugate base for each of the following acids:
   a. HCO_3^- **b.** $CH_3{-}NH_3^+$ **c.** HPO_4^{2-}
   d. HNO_2 **e.** HBrO

14.11 Write the formula for the conjugate acid for each of the following bases:
   a. CO_3^{2-} **b.** H_2O **c.** $H_2PO_4^-$
   d. Br^- **e.** ClO_4^-

14.12 Write the formula for the conjugate acid for each of the following bases:
   a. SO_4^{2-} **b.** CN^- **c.** NH_3
   d. ClO_2^- **e.** HS^-

14.13 Identify the Brønsted–Lowry acid–base pairs in each of the following equations:
   a. $H_2CO_3(aq) + H_2O(l) \rightleftharpoons HCO_3^-(aq) + H_3O^+(aq)$
   b. $NH_4^+(aq) + H_2O(l) \rightleftharpoons NH_3(aq) + H_3O^+(aq)$
   c. $HCN(aq) + NO_2^-(aq) \rightleftharpoons CN^-(aq) + HNO_2(aq)$
   d. $CHO_2^-(aq) + HF(aq) \rightleftharpoons HCHO_2(aq) + F^-(aq)$

14.14 Identify the Brønsted–Lowry acid–base pairs in each of the following equations:
   a. $H_3PO_4(aq) + H_2O(l) \rightleftharpoons H_2PO_4^-(aq) + H_3O^+(aq)$
   b. $CO_3^{2-}(aq) + H_2O(l) \rightleftharpoons HCO_3^-(aq) + OH^-(aq)$
   c. $H_3PO_4(aq) + NH_3(aq) \rightleftharpoons H_2PO_4^-(aq) + NH_4^+(aq)$
   d. $HNO_2(aq) + CH_3{-}CH_2{-}NH_2(aq) \rightleftharpoons$
     $\qquad CH_3{-}CH_2{-}NH_3^+(aq) + NO_2^-(aq)$

14.15 When ammonium chloride dissolves in water, the ammonium ion NH$_4^+$ donates an H$^+$ to water. Write a balanced equation for the reaction of the ammonium ion with water.

14.16 When sodium carbonate dissolves in water, the carbonate ion CO$_3^{2-}$ acts as a base. Write a balanced equation for the reaction of the carbonate ion with water.

14.3 Strengths of Acids and Bases

LEARNING GOAL Write equations for the dissociation of strong and weak acids; identify the direction of reaction.

In the process called **dissociation**, an acid or a base separates into ions in water. The *strength* of an acid is determined by the moles of H_3O^+ that are produced for each mole of acid that dissolves. The *strength* of a base is determined by the moles of OH^- that are produced for each mole of base that dissolves. Strong acids and strong bases dissociate completely in water, whereas weak acids and weak bases dissociate only slightly, leaving most of the initial acid or base undissociated.

Strong and Weak Acids

Strong acids are examples of strong electrolytes because they donate H^+ so easily that their dissociation in water is essentially complete. For example, when HCl, a strong acid, dissociates in water, H^+ is transferred to H_2O; the resulting solution contains essentially only the ions H_3O^+ and Cl^-. We consider the reaction of HCl in H_2O as going 100% to products. Thus, one mole of a strong acid dissociates in water to yield one mole of H_3O^+ and one mole of its conjugate base. We write the equation for a strong acid such as HCl with a single arrow.

$$HCl(g) + H_2O(l) \longrightarrow H_3O^+(aq) + Cl^-(aq)$$

There are only six common strong acids, which are stronger acids than H_3O^+. All other acids are weak. **TABLE 14.3** lists the relative strengths of acids and bases. **Weak**

Weak acids are found in foods and household products.

TABLE 14.3 Relative Strengths of Acids and Bases

Acid		Conjugate Base	
Strong Acids			
Hydroiodic acid	HI	I^-	Iodide ion
Hydrobromic acid	HBr	Br^-	Bromide ion
Perchloric acid	$HClO_4$	ClO_4^-	Perchlorate ion
Hydrochloric acid	HCl	Cl^-	Chloride ion
Sulfuric acid	H_2SO_4	HSO_4^-	Hydrogen sulfate ion
Nitric acid	HNO_3	NO_3^-	Nitrate ion
Hydronium ion	H_3O^+	H_2O	Water
Weak Acids			
Hydrogen sulfate ion	HSO_4^-	SO_4^{2-}	Sulfate ion
Phosphoric acid	H_3PO_4	$H_2PO_4^-$	Dihydrogen phosphate ion
Nitrous acid	HNO_2	NO_2^-	Nitrite ion
Hydrofluoric acid	HF	F^-	Fluoride ion
Acetic acid	$HC_2H_3O_2$	$C_2H_3O_2^-$	Acetate ion
Carbonic acid	H_2CO_3	HCO_3^-	Bicarbonate ion
Hydrosulfuric acid	H_2S	HS^-	Hydrogen sulfide ion
Dihydrogen phosphate ion	$H_2PO_4^-$	HPO_4^{2-}	Hydrogen phosphate ion
Ammonium ion	NH_4^+	NH_3	Ammonia
Hydrocyanic acid	HCN	CN^-	Cyanide ion
Bicarbonate ion	HCO_3^-	CO_3^{2-}	Carbonate ion
Methylammonium ion	$CH_3-NH_3^+$	CH_3-NH_2	Methylamine
Hydrogen phosphate ion	HPO_4^{2-}	PO_4^{3-}	Phosphate ion
Water	H_2O	OH^-	Hydroxide ion

Acid Strength Increases (left margin, arrow pointing up)

Base Strength Increases (right margin, arrow pointing down)

acids are weak electrolytes because they dissociate slightly in water, forming only a small amount of H_3O^+ ions. A weak acid has a strong conjugate base, which is why the reverse reaction is more prevalent. Even at high concentrations, weak acids produce low concentrations of H_3O^+ ions (see **FIGURE 14.1**).

FIGURE 14.1 ▶ A strong acid such as HCl is completely dissociated, whereas a weak acid such as $HC_2H_3O_2$ contains mostly molecules and a few ions.

Q What is the difference between a strong acid and a weak acid?

Many of the products you use at home contain weak acids. Citric acid is a weak acid found in fruits and fruit juices such as lemons, oranges, and grapefruit. The vinegar used in salad dressings is typically a 5% (m/v) acetic acid ($HC_2H_3O_2$) solution. In water, a few $HC_2H_3O_2$ molecules donate H^+ to H_2O to form H_3O^+ ions and acetate ions ($C_2H_3O_2^-$). The reverse reaction also takes place, which converts the H_3O^+ ions and acetate ions ($C_2H_3O_2^-$) back to reactants. The formation of hydronium ions from vinegar is the reason we notice the sour taste of vinegar. We write the equation for a weak acid in an aqueous solution with a double arrow to indicate that the forward and reverse reactions are at equilibrium.

$$HC_2H_3O_2(aq) + H_2O(l) \rightleftharpoons C_2H_3O_2^-(aq) + H_3O^+(aq)$$
Acetic acid   Acetate ion

ENGAGE

Why is NO_2^- a stronger base than NO_3^-?

Diprotic Acids

Some weak acids, such as carbonic acid, are *diprotic acids* that have two H^+, which dissociate one at a time. For example, carbonated soft drinks are prepared by dissolving CO_2 in water to form carbonic acid (H_2CO_3). A weak acid such as H_2CO_3 reaches equilibrium between the mostly undissociated H_2CO_3 molecules and the ions H_3O^+ and HCO_3^-.

$$H_2CO_3(aq) + H_2O(l) \rightleftharpoons H_3O^+(aq) + HCO_3^-(aq)$$
Carbonic acid   Bicarbonate ion
                                          (hydrogen carbonate)

Because HCO_3^- is also a weak acid, a second dissociation can take place to produce another hydronium ion and the carbonate ion (CO_3^{2-}).

$$HCO_3^-(aq) + H_2O(l) \rightleftharpoons H_3O^+(aq) + CO_3^{2-}(aq)$$
Bicarbonate ion   Carbonate ion
(hydrogen carbonate)

H_2CO_3 HCO_3^- CO_3^{2-}

Carbonic acid, a weak acid, loses one H^+ to form hydrogen carbonate ion, which loses a second H^+ to form carbonate ion.

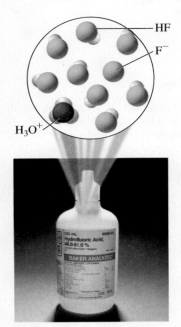

HF

F⁻

H_3O^+

Hydrofluoric acid is the only halogen acid that is a weak acid.

Sulfuric acid (H_2SO_4) is also a diprotic acid. However, its first dissociation is complete (100%), which means H_2SO_4 is a strong acid. The product, hydrogen sulfate (HSO_4^-) can dissociate again but only slightly, which means that the hydrogen sulfate ion is a weak acid.

$$H_2SO_4(aq) + H_2O(l) \longrightarrow H_3O^+(aq) + HSO_4^-(aq)$$

Sulfuric acid   Bisulfate ion
(hydrogen sulfate)

$$HSO_4^-(aq) + H_2O(l) \rightleftharpoons H_3O^+(aq) + SO_4^{2-}(aq)$$

Bisulfate ion   Sulfate ion
(hydrogen sulfate)

In summary, a strong acid such as HI in water dissociates completely to form an aqueous solution of the ions H_3O^+ and I^-. A weak acid such as HF dissociates only slightly in water to form an aqueous solution that consists mostly of HF molecules and a few H_3O^+ and F^- ions (see **FIGURE 14.2**).

Strong acid: $HI(aq) + H_2O(l) \longrightarrow H_3O^+(aq) + I^-(aq)$ Completely dissociated

Weak acid: $HF(aq) + H_2O(l) \rightleftharpoons H_3O^+(aq) + F^-(aq)$ Slightly dissociated

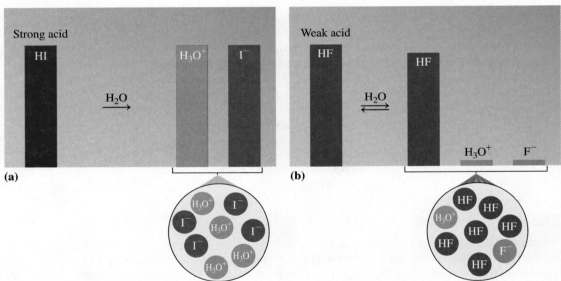

FIGURE 14.2 ▶ After dissociation in water, **(a)** the strong acid HI has high concentrations of H_3O^+ and I^-, and **(b)** the weak acid HF has a high concentration of HF and low concentrations of H_3O^+ and F^-.

Ⓠ How do the heights of H_3O^+ and F^- compare to the height of the weak acid HF in the bar diagram for HF?

Strong and Weak Bases

As strong electrolytes, **strong bases** dissociate completely in water. Because these strong bases are ionic compounds, they dissociate in water to give an aqueous solution of metal ions and hydroxide ions. The Group 1A (1) hydroxides are very soluble in water, which can give high concentrations of OH^- ions. A few strong bases are less soluble in water, but what does dissolve dissociates completely as ions. For example, when KOH forms a KOH solution, it contains only the ions K^+ and OH^-.

$$KOH(s) \xrightarrow{H_2O} K^+(aq) + OH^-(aq)$$

Strong Bases

Lithium hydroxide (LiOH)

Sodium hydroxide (NaOH)

Potassium hydroxide (KOH)

Rubidium hydroxide (RbOH)

Cesium hydroxide (CsOH)

Calcium hydroxide ($Ca(OH)_2$)*

Strontium hydroxide ($Sr(OH)_2$)*

Barium hydroxide ($Ba(OH)_2$)*

*Low solubility

Weak bases are weak electrolytes that are poor acceptors of hydrogen ions and produce very few ions in solution. A typical weak base, ammonia (NH_3) is found in window cleaners. In an aqueous solution, only a few ammonia molecules accept hydrogen ions to form NH_4^+ and OH^-.

$$NH_3(g) + H_2O(l) \rightleftharpoons NH_4^+(aq) + OH^-(aq)$$
Ammonia   Ammonium hydroxide

Bases in household products are used to remove grease and to open drains.

Bases Used in Household Products
Weak Bases
Window cleaner, ammonia, NH_3
Bleach, NaOCl
Laundry detergent, Na_2CO_3, Na_3PO_4
Toothpaste and baking soda, $NaHCO_3$
Baking powder, scouring powder, Na_2CO_3
Lime for lawns and agriculture, $CaCO_3$
Laxatives, antacids, $Mg(OH)_2$, $Al(OH)_3$

Strong Bases
Drain cleaner, oven cleaner, NaOH

Direction of Reaction

There is a relationship between the components in each conjugate acid–base pair. Strong acids have weak conjugate bases that do not readily accept H^+. As the strength of the acid decreases, the strength of its conjugate base increases.

In any acid–base reaction, there are two acids and two bases. However, one acid is stronger than the other acid, and one base is stronger than the other base. By comparing their relative strengths, we can determine the direction of the reaction. For example, the strong acid H_2SO_4 readily gives up H^+ to water. The hydronium ion H_3O^+ produced is a weaker acid than H_2SO_4, and the conjugate base HSO_4^- is a weaker base than water.

$$H_2SO_4(aq) + H_2O(l) \longrightarrow H_3O^+(aq) + HSO_4^-(aq)$$
Stronger   Stronger   Weaker   Weaker   Mostly products
acid   base   acid   base

Let's look at another reaction in which water donates one H^+ to carbonate (CO_3^{2-}) to form HCO_3^- and OH^-. From Table 14.3, we see that HCO_3^- is a stronger acid than H_2O. We also see that OH^- is a stronger base than CO_3^{2-}. To reach equilibrium, the stronger acid and stronger base react in the direction of the weaker acid and weaker base.

$$CO_3^{2-}(aq) + H_2O(l) \rightleftharpoons HCO_3^-(aq) + OH^-(aq)$$
Weaker   Weaker   Stronger   Stronger   Mostly reactants
base   acid   acid   base

SAMPLE PROBLEM 14.4 Direction of Reaction

Does the equilibrium mixture of the following reaction contain mostly reactants or products?

$$HF(aq) + H_2O(l) \rightleftharpoons H_3O^+(aq) + F^-(aq)$$

TRY IT FIRST

SOLUTION

From Table 14.3, we see that HF is a weaker acid than H_3O^+ and that H_2O is a weaker base than F^-. Thus, the equilibrium mixture contains mostly reactants.

$$HF(aq) + H_2O(l) \rightleftharpoons H_3O^+(aq) + F^-(aq)$$
Weaker   Weaker   Stronger   Stronger   Mostly reactants
acid   base   acid   base

STUDY CHECK 14.4

Does the equilibrium mixture for the reaction of nitric acid and water contain mostly reactants or products?

ANSWER

$$HNO_3(aq) + H_2O(l) \longrightarrow H_3O^+(aq) + NO_3^-(aq) \quad \text{Mostly products}$$

The equilibrium mixture contains mostly products because HNO_3 is a stronger acid than H_3O^+, and H_2O is a stronger base than NO_3^-.

QUESTIONS AND PROBLEMS

14.3 Strengths of Acids and Bases

LEARNING GOAL Write equations for the dissociation of strong and weak acids; identify the direction of reaction.

14.17 What is meant by the phrase "A strong acid has a weak conjugate base"?

14.18 What is meant by the phrase "A weak acid has a strong conjugate base"?

14.19 Identify the stronger acid in each of the following pairs:
   a. HBr or HNO_2 **b.** H_3PO_4 or HSO_4^-
   c. HCN or H_2CO_3

14.20 Identify the stronger acid in each of the following pairs:
   a. NH_4^+ or H_3O^+ **b.** H_2SO_4 or HCl
   c. H_2O or H_2CO_3

14.21 Identify the weaker acid in each of the following pairs:
   a. HCl or HSO_4^- **b.** HNO_2 or HF
   c. HCO_3^- or NH_4^+

14.22 Identify the weaker acid in each of the following pairs:
   a. HNO_3 or HCO_3^- **b.** HSO_4^- or H_2O
   c. $H_2SO_4^-$ or H_2CO_3

14.23 Predict whether each of the following reactions contains mostly reactants or products at equilibrium:
   a. $H_2CO_3(aq) + H_2O(l) \rightleftarrows HCO_3^-(aq) + H_3O^+(aq)$
   b. $NH_4^+(aq) + H_2O(l) \rightleftarrows NH_3(aq) + H_3O^+(aq)$
   c. $HNO_2(aq) + NH_3(aq) \rightleftarrows NO_2^-(aq) + NH_4^+(aq)$

14.24 Predict whether each of the following reactions contains mostly reactants or products at equilibrium:
   a. $H_3PO_4(aq) + H_2O(l) \rightleftarrows H_3O^+(aq) + H_2PO_4^-(aq)$
   b. $CO_3^{2-}(aq) + H_2O(l) \rightleftarrows OH^-(aq) + HCO_3^-(aq)$
   c. $HS^-(aq) + F^-(aq) \rightleftarrows HF(aq) + S^{2-}(aq)$

14.25 Write an equation for the acid–base reaction between ammonium ion and sulfate ion. Why does the equilibrium mixture contain mostly reactants?

14.26 Write an equation for the acid–base reaction between nitrous acid and hydroxide ion. Why does the equilibrium mixture contain mostly products?

14.4 Dissociation Constants for Acids and Bases

LEARNING GOAL Write the dissociation expression for a weak acid or weak base.

As we have seen, acids have different strengths depending on how much they dissociate in water. Because the dissociation of strong acids in water is essentially complete, the reaction is not considered to be an equilibrium situation. However, because weak acids in water dissociate only slightly, the ion products reach equilibrium with the undissociated weak acid molecules. For example, formic acid ($HCHO_2$) the acid found in bee and ant stings, is a weak acid. Formic acid is a weak acid that dissociates in water to form hydronium ion (H_3O^+) and formate ion (CHO_2^-).

$$HCHO_2(aq) + H_2O(l) \rightleftarrows H_3O^+(aq) + CHO_2^-(aq)$$

HCHO₂

CHO₂⁻

Formic acid, a weak acid, loses one H^+ to form formate ion.

Writing Dissociation Constants

An **acid dissociation expression, K_a,** can be written for weak acids that gives the ratio of the concentrations of products to the weak acid reactants. As with other dissociation expressions, the molar concentration of the products is divided by the molar concentration of the reactants. Because water is a pure liquid with a constant concentration, it is omitted. The numerical value of the acid dissociation expression is the acid dissociation constant. For example, the acid dissociation expression for the equilibrium equation of formic acid shown above is written

$$K_a = \frac{[H_3O^+][CHO_2^-]}{[HCHO_2]}$$

The numerical value of the K_a for formic acid at 25 °C is determined by experiment to be 1.8×10^{-4}. Thus, for the weak acid $HCHO_2$, the K_a is written

$$K_a = \frac{[H_3O^+][CHO_2^-]}{[HCHO_2]} = 1.8 \times 10^{-4} \quad \text{Acid dissociation constant}$$

The K_a for formic acid is small, which confirms that the equilibrium mixture of formic acid in water contains mostly reactants and only small amounts of the products. (Recall that the brackets in the K_a represent the molar concentrations of the reactants and products). Weak acids have small K_a values. However, strong acids, which are essentially 100% dissociated, have very large K_a values, but these values are not usually given. **TABLE 14.4** gives K_a and K_b values for selected weak acids and bases.

ENGAGE

Why is an acid with a $K_a = 1.8 \times 10^{-5}$ a stronger acid than an acid with a $K_a = 6.2 \times 10^{-8}$?

TABLE 14.4 K_a and K_b Values for Selected Weak Acids and Bases

Acids		K_a
Phosphoric acid	H_3PO_4	7.5×10^{-3}
Nitrous acid	HNO_2	4.5×10^{-4}
Hydrofluoric acid	HF	3.5×10^{-4}
Formic acid	$HCHO_2$	1.8×10^{-4}
Acetic acid	$HC_2H_3O_2$	1.8×10^{-5}
Carbonic acid	H_2CO_3	4.3×10^{-7}
Hydrosulfuric acid	H_2S	9.1×10^{-8}
Dihydrogen phosphate	$H_2PO_4^-$	6.2×10^{-8}
Hydrocyanic acid	HCN	4.9×10^{-10}
Hydrogen carbonate	HCO_3^-	5.6×10^{-11}
Hydrogen phosphate	HPO_4^{2-}	2.2×10^{-13}
Bases		K_b
Methylamine	CH_3-NH_2	4.4×10^{-4}
Carbonate	CO_3^{2-}	2.2×10^{-4}
Ammonia	NH_3	1.8×10^{-5}

Let us now consider the dissociation of the weak base methylamine:

$$CH_3-NH_2(aq) + H_2O(l) \rightleftharpoons CH_3-NH_3^+(aq) + OH^-(aq)$$

As we did with the acid dissociation expression, the concentration of water is omitted from the **base dissociation expression, K_b.** The base dissociation constant for methylamine is written

$$K_b = \frac{[CH_3-NH_3^+][OH^-]}{[CH_3-NH_2]} = 4.4 \times 10^{-4}$$

TABLE 14.5 summarizes the characteristics of acids and bases in terms of strength and equilibrium position.

TABLE 14.5 Characteristics of Acids and Bases

Characteristic	Strong Acids	Weak Acids
Equilibrium Position	Toward products	Toward reactants
K_a	Large	Small
$[H_3O^+]$ and $[A^-]$	100% of $[HA]$ dissociates	Small percent of $[HA]$ dissociates
Conjugate Base	Weak	Strong
Characteristic	**Strong Bases**	**Weak Bases**
Equilibrium Position	Toward products	Toward reactants
K_b	Large	Small
$[BH^+]$ and $[OH^-]$	100% of $[B]$ reacts	Small percent of $[B]$ reacts
Conjugate Acid	Weak	Strong

SAMPLE PROBLEM 14.5 Writing an Acid Dissociation Expression

Write the acid dissociation expression for the weak acid nitrous acid.

TRY IT FIRST

SOLUTION

STEP 1 Write the balanced chemical equation. The equation for the dissociation of nitrous acid is written

$$HNO_2(aq) + H_2O(l) \rightleftharpoons H_3O^+(aq) + NO_2^-(aq)$$

STEP 2 Write the concentrations of the products as the numerator and the reactants as the denominator. The acid dissociation expression is written as the concentration of the products divided by the concentration of the undissociated weak acid.

$$K_a = \frac{[H_3O^+][NO_2^-]}{[HNO_2]}$$

STUDY CHECK 14.5

Write the acid dissociation expression for hydrogen phosphate (HPO_4^{2-}).

ANSWER

$$K_a = \frac{[H_3O^+][PO_4^{3-}]}{[HPO_4^{2-}]}$$

Guide to Writing the Acid Dissociation Expression

STEP 1
Write the balanced chemical equation.

STEP 2
Write the concentrations of the products as the numerator and the reactants as the denominator.

QUESTIONS AND PROBLEMS

14.4 Dissociation Constants for Acids and Bases

LEARNING GOAL Write the dissociation expression for a weak acid or weak base.

14.27 Answer *true* or *false* for each of the following: A strong acid
  a. is completely dissociated in aqueous solution
  b. has a small value of K_a
  c. has a strong conjugate base
  d. has a weak conjugate base
  e. is slightly dissociated in aqueous solution

14.28 Answer *true* or *false* for each of the following: A weak acid
  a. is completely dissociated in aqueous solution
  b. has a small value of K_a
  c. has a strong conjugate base
  d. has a weak conjugate base
  e. is slightly dissociated in aqueous solution

14.29 Consider the following acids and their dissociation constants:

$$H_2SO_3(aq) + H_2O(l) \rightleftharpoons H_3O^+(aq) + HSO_3^-(aq)$$
$$K_a = 1.2 \times 10^{-2}$$

$$HS^-(aq) + H_2O(l) \rightleftharpoons H_3O^+(aq) + S^{2-}(aq)$$
$$K_a = 1.3 \times 10^{-19}$$

a. Which is the stronger acid, H_2SO_3 or HS^-?
b. What is the conjugate base of H_2SO_3?
c. Which acid has the weaker conjugate base?
d. Which acid has the stronger conjugate base?
e. Which acid produces more ions?

14.30 Consider the following acids and their dissociation constants:

$$HPO_4^{2-}(aq) + H_2O(l) \rightleftharpoons H_3O^+(aq) + PO_4^{3-}(aq)$$
$$K_a = 2.2 \times 10^{-13}$$

$$HCHO_2(aq) + H_2O(l) \rightleftharpoons H_3O^+(aq) + CHO_2^-(aq)$$
$$K_a = 1.8 \times 10^{-4}$$

a. Which is the weaker acid, HPO_4^{2-} or $HCHO_2$?
b. What is the conjugate base of HPO_4^{2-}?
c. Which acid has the weaker conjugate base?
d. Which acid has the stronger conjugate base?
e. Which acid produces more ions?

14.31 Phosphoric acid dissociates to form hydronium ion and dihydrogen phosphate. Phosphoric acid has a K_a of 7.5×10^{-3}. Write the equation for the reaction and the acid dissociation expression for phosphoric acid.

14.32 Aniline, $C_6H_5-NH_2$, a weak base with a K_b of 4.0×10^{-10}, reacts with water to form $C_6H_5-NH_3^+$ and hydroxide ion. Write the equation for the reaction and the base dissociation expression for aniline.

14.5 Dissociation of Water

LEARNING GOAL Use the water dissociation expression to calculate the $[H_3O^+]$ and $[OH^-]$ in an aqueous solution.

In many acid–base reactions, water is *amphoteric*, which means that it can act either as an acid or as a base. In pure water, there is a forward reaction between two water molecules that transfers H^+ from one water molecule to the other. One molecule acts as an acid by losing H^+, and the water molecule that gains H^+ acts as a base. Every time H^+ is transferred between two water molecules, the products are one H_3O^+ and one OH^-, which react in the reverse direction to re-form two water molecules. Thus, equilibrium is reached between the conjugate acid–base pairs of water.

Conjugate acid–base pair

$$:\ddot{O}-H + :\ddot{O}-H \rightleftharpoons H-\ddot{O}-H^+ + {}^-:\ddot{O}-H$$

| | | | |
| H | H | H | |

Conjugate acid–base pair

| **Acid** | **Base** | **Acid** | **Base** |
| (H^+ donor) | (H^+ acceptor) | (H^+ donor) | (H^+ acceptor) |

Writing the Water Dissociation Expression, K_w

Using the equation for water at equilibrium, we can write its equilibrium expression that shows the concentrations of the products divided by the concentrations of the reactants. Recall that square brackets around the symbols indicate their concentrations in moles per liter (M).

$$H_2O(l) + H_2O(l) \rightleftharpoons H_3O^+(aq) + OH^-(aq)$$

$$K = \frac{[H_3O^+][OH^-]}{[H_2O][H_2O]}$$

By omitting the constant concentration of pure water, we can write the **water dissociation expression, K_w.**

$$K_w = [H_3O^+][OH^-]$$

Experiments have determined, that in pure water, the concentration of H_3O^+ and OH^- at 25 °C are each 1.0×10^{-7} M.

Pure water $\quad [H_3O^+] = [OH^-] = 1.0 \times 10^{-7}$ M

ENGAGE

Why is the $[H_3O^+]$ equal to the $[OH^-]$ in pure water?

When we place the $[H_3O^+]$ and $[OH^-]$ into the water dissociation expression, we obtain the numerical value of K_w, which is 1.0×10^{-14} at 25 °C. As before, the concentration units are omitted in the K_w value.

$$K_w = [H_3O^+][OH^-]$$
$$= [1.0 \times 10^{-7}][1.0 \times 10^{-7}] = 1.0 \times 10^{-14}$$

Neutral, Acidic, and Basic Solutions

The K_w value (1.0×10^{-14}) applies to any aqueous solution at 25 °C because all aqueous solutions contain both H_3O^+ and OH^- (see **FIGURE 14.3**). When the $[H_3O^+]$ and $[OH^-]$ in a solution are equal, the solution is **neutral**. However, most solutions are not neutral; they have different concentrations of $[H_3O^+]$ and $[OH^-]$. If acid is added to water, there is an increase in $[H_3O^+]$ and a decrease in $[OH^-]$, which makes an acidic solution. If base is added, $[OH^-]$ increases and $[H_3O^+]$ decreases, which gives a basic solution. However, for any aqueous solution, whether it is neutral, acidic, or basic, the product $[H_3O^+][OH^-]$ is equal to K_w (1.0×10^{-14}) at 25 °C (see **TABLE 14.6**).

ENGAGE

Why is the product of the concentrations of H_3O^+ and OH^- in a solution equal to 1.0×10^{-14}?

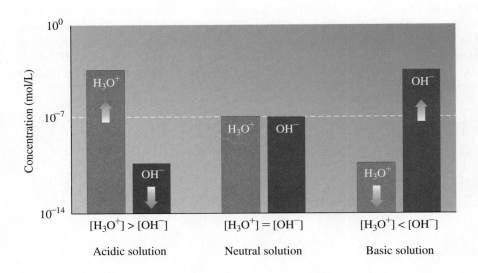

FIGURE 14.3 ▶ In a neutral solution, $[H_3O^+]$ and $[OH^-]$ are equal. In acidic solutions, the $[H_3O^+]$ is greater than the $[OH^-]$. In basic solutions, the $[OH^-]$ is greater than the $[H_3O^+]$.

Q Is a solution that has a $[H_3O^+]$ of 1.0×10^{-3} M acidic, basic, or neutral?

TABLE 14.6 Examples of $[H_3O^+]$ and $[OH^-]$ in Neutral, Acidic, and Basic Solutions

Type of Solution	$[H_3O^+]$	$[OH^-]$	K_w
Neutral	1.0×10^{-7} M	1.0×10^{-7} M	1.0×10^{-14}
Acidic	1.0×10^{-2} M	1.0×10^{-12} M	1.0×10^{-14}
Acidic	2.5×10^{-5} M	4.0×10^{-10} M	1.0×10^{-14}
Basic	1.0×10^{-8} M	1.0×10^{-6} M	1.0×10^{-14}
Basic	5.0×10^{-11} M	2.0×10^{-4} M	1.0×10^{-14}

Using the K_w to Calculate $[H_3O^+]$ and $[OH^-]$ in a Solution

If we know the $[H_3O^+]$ of a solution, we can use the K_w to calculate the $[OH^-]$. If we know the $[OH^-]$ of a solution, we can calculate $[H_3O^+]$ from their relationship in the K_w, as shown in Sample Problem 14.6.

$$K_w = [H_3O^+][OH^-]$$

$$[OH^-] = \frac{K_w}{[H_3O^+]} \qquad [H_3O^+] = \frac{K_w}{[OH^-]}$$

CORE CHEMISTRY SKILL
Calculating $[H_3O^+]$ and $[OH^-]$ in Solutions

ENGAGE

If you know the $[H_3O^+]$ of a solution, how do you use the K_w to calculate the $[OH^-]$?

SAMPLE PROBLEM 14.6 Calculating the $[H_3O^+]$ of a Solution

A vinegar solution has a $[OH^-] = 5.0 \times 10^{-12}$ M at 25 °C. What is the $[H_3O^+]$ of the vinegar solution? Is the solution acidic, basic, or neutral?

TRY IT FIRST

SOLUTION

STEP 1 State the given and needed quantities.

ANALYZE THE PROBLEM	Given	Need	Connect
	$[OH^-] = 5.0 \times 10^{-12}$ M	$[H_3O^+]$	$K_w = [H_3O^+][OH^-]$

STEP 2 Write the K_w for water and solve for the unknown $[H_3O^+]$.

$$K_w = [H_3O^+][OH^-] = 1.0 \times 10^{-14}$$

Solve for $[H_3O^+]$ by dividing both sides by $[OH^-]$.

$$\frac{K_w}{[OH^-]} = \frac{[H_3O^+][OH^-]}{[OH^-]} = \frac{1.0 \times 10^{-14}}{[OH^-]}$$

$$[H_3O^+] = \frac{1.0 \times 10^{-14}}{[OH^-]}$$

STEP 3 Substitute the known $[OH^-]$ into the equation and calculate.

$$[H_3O^+] = \frac{1.0 \times 10^{-14}}{[5.0 \times 10^{-12}]} = 2.0 \times 10^{-3} \text{ M}$$

Because the $[H_3O^+]$ of 2.0×10^{-3} M is larger than the $[OH^-]$ of 5.0×10^{-12} M, the solution is acidic.

Guide to Calculating $[H_3O^+]$ and $[OH^-]$ in Aqueous Solutions

STEP 1
State the given and needed quantities.

STEP 2
Write the K_w for water and solve for the unknown $[H_3O^+]$ or $[OH^-]$.

STEP 3
Substitute the known $[H_3O^+]$ or $[OH^-]$ into the equation and calculate.

STUDY CHECK 14.6

What is the $[H_3O^+]$ of an ammonia cleaning solution with $[OH^-] = 4.0 \times 10^{-4}$ M? Is the solution acidic, basic, or neutral?

ANSWER

$[H_3O^+] = 2.5 \times 10^{-11}$ M, basic

QUESTIONS AND PROBLEMS

14.5 Dissociation of Water

LEARNING GOAL Use the water dissociation expression to calculate the $[H_3O^+]$ and $[OH^-]$ in an aqueous solution.

14.33 Why are the concentrations of H_3O^+ and OH^- equal in pure water?

14.34 What is the meaning and value of K_w at 25 °C?

14.35 In an acidic solution, how does the concentration of H_3O^+ compare to the concentration of OH^-?

14.36 If a base is added to pure water, why does the $[H_3O^+]$ decrease?

14.37 Indicate whether each of the following solutions is acidic, basic, or neutral:
  a. $[H_3O^+] = 2.0 \times 10^{-5}$ M
  b. $[H_3O^+] = 1.4 \times 10^{-9}$ M
  c. $[OH^-] = 8.0 \times 10^{-3}$ M
  d. $[OH^-] = 3.5 \times 10^{-10}$ M

14.38 Indicate whether each of the following solutions is acidic, basic, or neutral:
  a. $[H_3O^+] = 6.0 \times 10^{-12}$ M
  b. $[H_3O^+] = 1.4 \times 10^{-4}$ M
  c. $[OH^-] = 5.0 \times 10^{-12}$ M
  d. $[OH^-] = 4.5 \times 10^{-2}$ M

14.39 Calculate the $[H_3O^+]$ of each aqueous solution with the following $[OH^-]$:
  a. coffee, 1.0×10^{-9} M
  b. soap, 1.0×10^{-6} M
  c. cleanser, 2.0×10^{-5} M
  d. lemon juice, 4.0×10^{-13} M

14.40 Calculate the $[H_3O^+]$ of each aqueous solution with the following $[OH^-]$:
  a. NaOH, 1.0×10^{-2} M
  b. milk of magnesia, 1.0×10^{-5} M
  c. aspirin, 1.8×10^{-11} M
  d. seawater, 2.5×10^{-6} M

Applications

14.41 Calculate the $[OH^-]$ of each aqueous solution with the following $[H_3O^+]$:
  a. stomach acid, 4.0×10^{-2} M
  b. urine, 5.0×10^{-6} M
  c. orange juice, 2.0×10^{-4} M
  d. bile, 7.9×10^{-9} M

14.42 Calculate the $[OH^-]$ of each aqueous solution with the following $[H_3O^+]$:
  a. baking soda, 1.0×10^{-8} M
  b. blood, 4.2×10^{-8} M
  c. milk, 5.0×10^{-7} M
  d. pancreatic juice, 4.0×10^{-9} M

14.6 The pH Scale

LEARNING GOAL Calculate pH from $[H_3O^+]$; given the pH, calculate the $[H_3O^+]$ and $[OH^-]$ of a solution.

In the environment, the acidity, or pH, of rain, water, and soil can have significant effects. When rain becomes too acidic, it can dissolve marble statues and accelerate the corrosion of metals. In lakes and ponds, the acidity of water can affect the ability of plants and fish to survive. The acidity of soil around plants affects their growth. If the soil pH is too acidic or too basic, the roots of the plant cannot take up some nutrients. Most plants thrive in soil with a nearly neutral pH, although certain plants, such as orchids, camellias, and blueberries, require a more acidic soil.

Personnel working in food processing, medicine, agriculture, spa and pool maintenance, soap manufacturing, and wine making measure the $[H_3O^+]$ and $[OH^-]$ of solutions. Although we have expressed H_3O^+ and OH^- as molar concentrations, it is more convenient to describe the acidity of solutions using the *pH scale*. On this scale, a number between 0 and 14 represents the H_3O^+ concentration for common solutions. A neutral solution has a pH of 7.0 at 25 °C. An acidic solution has a pH less than 7.0; a basic solution has a pH greater than 7.0 (see **FIGURE 14.4**).

If soil is too acidic, nutrients are not absorbed by crops. Then lime ($CaCO_3$), which acts as a base, may be added to increase the soil pH.

Acidic solution	pH < 7.0	$[H_3O^+] > 1.0 \times 10^{-7}$ M
Neutral solution	pH = 7.0	$[H_3O^+] = 1.0 \times 10^{-7}$ M
Basic solution	pH > 7.0	$[H_3O^+] < 1.0 \times 10^{-7}$ M

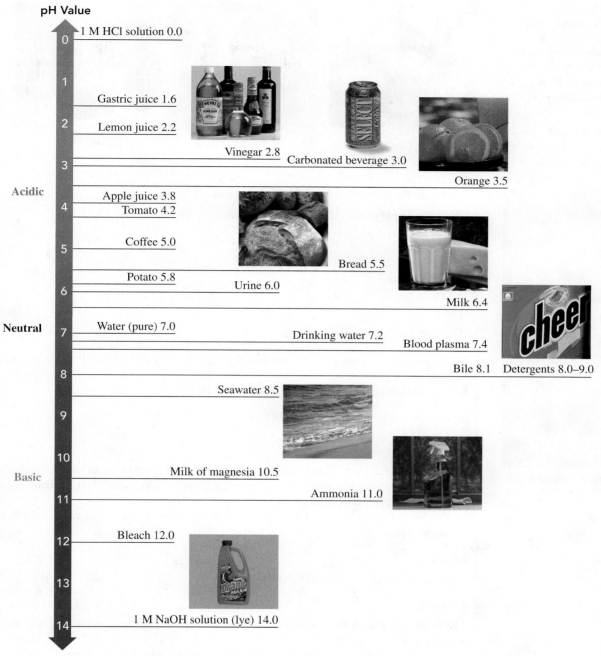

pH Value

- 0 — 1 M HCl solution 0.0
- 1
- Gastric juice 1.6
- 2 — Lemon juice 2.2
- Vinegar 2.8
- 3 — Carbonated beverage 3.0
- Orange 3.5
- Apple juice 3.8
- 4 — Tomato 4.2
- Coffee 5.0
- 5
- Bread 5.5
- Potato 5.8
- 6 — Urine 6.0
- Milk 6.4
- Water (pure) 7.0
- 7 — Drinking water 7.2
- Blood plasma 7.4
- Bile 8.1   Detergents 8.0–9.0
- 8
- Seawater 8.5
- 9
- 10
- Milk of magnesia 10.5
- 11 — Ammonia 11.0
- 12 — Bleach 12.0
- 13
- 14 — 1 M NaOH solution (lye) 14.0

Acidic

Neutral

Basic

FIGURE 14.4 ▶ On the pH scale, values below 7.0 are acidic, a value of 7.0 is neutral, and values above 7.0 are basic.

Q Is apple juice an acidic, a basic, or a neutral solution?

When we relate acidity and pH, we are using an inverse relationship, which is when one component increases while the other component decreases. When an acid is added to pure water, the $[H_3O^+]$ (acidity) of the solution increases but its pH decreases. When a base is added to pure water, it becomes more basic, which means its acidity decreases and the pH increases.

In the laboratory, a pH meter is commonly used to determine the pH of a solution. There are also various indicators and pH papers that turn specific colors when placed in solutions of different pH values. The pH is found by comparing the color on the test paper or the color of the solution to a color chart (see **FIGURE 14.5**).

FIGURE 14.5 ▶ The pH of a solution can be determined using (a) a pH meter, (b) pH paper, and (c) indicators that turn different colors corresponding to different pH values.

Q If a pH meter reads 4.00, is the solution acidic, basic, or neutral?

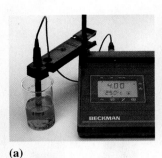
(a)

(b)

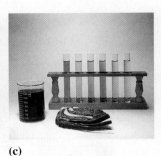
(c)

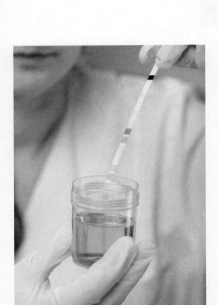

A dipstick is used to measure the pH of a urine sample.

SAMPLE PROBLEM 14.7 pH of Solutions

Consider the pH of the following body fluids:

Body Fluid	pH
Stomach acid	1.4
Pancreatic juice	8.4
Sweat	4.8
Urine	5.3
Cerebrospinal fluid	7.3

a. Place the pH values of the body fluids on the list in order of most acidic to most basic.
b. Which body fluid has the highest $[H_3O^+]$?

TRY IT FIRST

SOLUTION

a. The most acidic body fluid is the one with the lowest pH, and the most basic is the body fluid with the highest pH: stomach acid (1.4), sweat (4.8), urine (5.3), cerebrospinal fluid (7.3), pancreatic juice (8.4).
b. The body fluid with the highest $[H_3O^+]$ would have the lowest pH value, which is stomach acid.

STUDY CHECK 14.7

Which body fluid has the highest $[OH^-]$?

ANSWER

The body fluid with the highest $[OH^-]$ would have the highest pH value, which is pancreatic juice.

KEY MATH SKILL
Calculating pH from $[H_3O^+]$

Calculating the pH of Solutions

The pH scale is a logarithmic scale that corresponds to the $[H_3O^+]$ of aqueous solutions. Mathematically, **pH** is the negative logarithm (base 10) of the $[H_3O^+]$.

$$pH = -\log[H_3O^+]$$

Essentially, the negative powers of 10 in the molar concentrations are converted to positive numbers. For example, a lemon juice solution with $[H_3O^+] = 1.0 \times 10^{-2}$ M has a pH of 2.00. This can be calculated using the pH equation:

$$pH = -\log[1.0 \times 10^{-2}]$$
$$pH = -(-2.00)$$
$$= 2.00$$

The number of *decimal places* in the pH value is the same as the number of significant figures in the $[H_3O^+]$. The number to the left of the decimal point in the pH value is the power of 10.

$$[H_3O^+] = \mathbf{1.0} \times 10^{-2} \qquad pH = \mathbf{2.00}$$

Two SFs   Two SFs

Because pH is a log scale, a change of one pH unit corresponds to a tenfold change in $[H_3O^+]$. It is important to note that the pH decreases as the $[H_3O^+]$ increases. For example, a solution with a pH of 2.00 has a $[H_3O^+]$ that is ten times greater than a solution with a pH of 3.00 and 100 times greater than a solution with a pH of 4.00. The pH of a solution is calculated from the $[H_3O^+]$ by using the *log* key and changing the sign as shown in Sample Problem 14.8.

ENGAGE

Explain why 6.00 but not 6.0 is the correct pH for $[H_3O^+] = 1.0 \times 10^{-6}$ M.

SAMPLE PROBLEM 14.8 Calculating pH from $[H_3O^+]$

Aspirin, which is acetylsalicylic acid, was the first nonsteroidal anti-inflammatory drug (NSAID) used to alleviate pain and fever. If a solution of aspirin has a $[H_3O^+] = 1.7 \times 10^{-3}$ M, what is the pH of the solution?

Acidic H that dissociates in aqueous solution

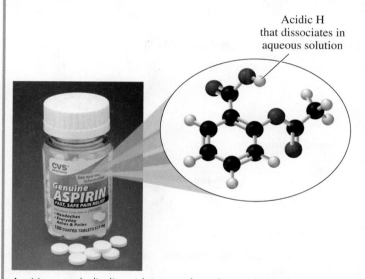

Aspirin, acetylsalicylic acid, is a weak acid.

TRY IT FIRST

SOLUTION

STEP 1 State the given and needed quantities.

ANALYZE THE PROBLEM	Given	Need	Connect
	$[H_3O^+] = 1.7 \times 10^{-3}$ M	pH	pH equation

STEP 2 Enter the $[H_3O^+]$ into the pH equation and calculate.

$$pH = -\log[H_3O^+] = -\log[1.7 \times 10^{-3}]$$

Calculator Procedure   **Calculator Display**

1.7 [EE or EXP] [+/−] 3 [log] [+/−] [=]   or   [+/−] [log] 1.7 [EE or EXP] [+/−] 3 [=]   `2.769551079`

Be sure to check the instructions for your calculator. Different calculators can have different methods for pH calculation.

Guide to Calculating pH of an Aqueous Solution

STEP 1
State the given and needed quantities.

STEP 2
Enter the $[H_3O^+]$ into the pH equation and calculate.

STEP 3
Adjust the number of SFs on the *right* of the decimal point.

STEP **3** Adjust the number of SFs on the *right* of the decimal point. In a pH value, the number to the *left* of the decimal point is an *exact* number derived from the power of 10. Thus, the two SFs in the coefficient determine that there are two SFs after the decimal point in the pH value.

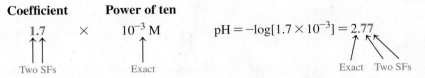

Coefficient   **Power of ten**

1.7   × 10^{-3} M $pH = -\log[1.7 \times 10^{-3}] = 2.77$

Two SFs   Exact   Exact   Two SFs

STUDY CHECK 14.8

What is the pH of bleach with $[H_3O^+] = 4.2 \times 10^{-12}$ M?

ANSWER

pH = 11.38

When we need to calculate the pH from $[OH^-]$, we use the K_w to calculate $[H_3O^+]$, place it in the pH equation, and calculate the pH of the solution as shown in Sample Problem 14.9.

SAMPLE PROBLEM 14.9 Calculating pH from $[OH^-]$

What is the pH of an ammonia solution with $[OH^-] = 3.7 \times 10^{-3}$ M?

> **TRY IT FIRST**

SOLUTION

STEP **1** State the given and needed quantities.

	Given	Need	Connect
ANALYZE THE PROBLEM	$[OH^-] = 3.7 \times 10^{-3}$ M	pH	$K_w = [H_3O^+][OH^-]$, pH equation

STEP **2** Enter the $[H_3O^+]$ into the pH equation and calculate. Because $[OH^-]$ is given for the ammonia solution, we have to calculate $[H_3O^+]$. Using the water dissociation expression, K_w, we divide both sides by $[OH^-]$ to obtain $[H_3O^+]$.

$$K_w = [H_3O^+][OH^-] = 1.0 \times 10^{-14}$$

$$\frac{K_w}{[OH^-]} = \frac{[H_3O^+][OH^-]}{[OH^-]}$$

$$[H_3O^+] = \frac{1.0 \times 10^{-14}}{[3.7 \times 10^{-3}]} = 2.7 \times 10^{-12} \text{ M}$$

Now, we enter the $[H_3O^+]$ into the pH equation.

$$pH = -\log[H_3O^+] = -\log[2.7 \times 10^{-12}]$$

Calculator Procedure   **Calculator Display**

2.7 [EE or EXP] [+/−] 12 [log] [+/−] [=]

or

[+/−] [log] 2.7 [EE or EXP] [+/−] 12 [=]

STEP **3** Adjust the number of SFs on the *right* of the decimal point.

2.7×10^{-12} M   pH = 11.57
Two SFs   Two SFs to the *right* of the decimal point

STUDY CHECK 14.9

Calculate the pH of a sample of bile that has $[OH^-] = 1.3 \times 10^{-6}$ M.

ANSWER

pH = 8.11

pOH

The **pOH** scale is similar to the pH scale except that pOH is associated with the $[OH^-]$ of an aqueous solution.

$$pOH = -\log[OH^-]$$

Solutions with high $[OH^-]$ have low pOH values; solutions with low $[OH^-]$ have high pOH values. In any aqueous solution, the sum of the pH and pOH is equal to 14.00, which is the negative logarithm of the K_w.

$$pH + pOH = 14.00$$

For example, if the pH of a solution is 3.50, the pOH can be calculated as follows:

$$pH + pOH = 14.00$$
$$pOH = 14.00 - pH = 14.00 - 3.50 = 10.50$$

A comparison of $[H_3O^+]$, $[OH^-]$, and their corresponding pH and pOH values is given in **TABLE 14.7**.

TABLE 14.7 A Comparison of pH and pOH Values at 25 °C, $[H_3O^+]$, and $[OH^-]$

pH	$[H_3O^+]$	$[OH^-]$	pOH
0	10^0	10^{-14}	14
1	10^{-1}	10^{-13}	13
2	10^{-2}	10^{-12}	12
3	10^{-3}	10^{-11}	11
4	10^{-4}	10^{-10}	10
5	10^{-5}	10^{-9}	9
6	10^{-6}	10^{-8}	8
7	10^{-7}	10^{-7}	7
8	10^{-8}	10^{-6}	6
9	10^{-9}	10^{-5}	5
10	10^{-10}	10^{-4}	4
11	10^{-11}	10^{-3}	3
12	10^{-12}	10^{-2}	2
13	10^{-13}	10^{-1}	1
14	10^{-14}	10^0	0

Acidic

Neutral

Basic

Acids produce the sour taste of the fruits we eat.

Calculating [H_3O^+] from pH

If we are given the pH of the solution and asked to determine the [H_3O^+], we need to reverse the calculation of pH.

$$[H_3O^+] = 10^{-pH}$$

For example, if the pH of a solution is 3.0, we can substitute it into this equation. The number of significant figures in [H_3O^+] is equal to the number of decimal places in the pH value.

$$[H_3O^+] = 10^{-pH} = 10^{-3.0} = 1 \times 10^{-3} \, M$$

For pH values that are not whole numbers, the calculation requires the use of the *10^x* key, which is usually a *2nd function* key. On some calculators, this operation is done using the inverse log equation as shown in Sample Problem 14.10.

SAMPLE PROBLEM 14.10 Calculating [H_3O^+] from pH

Calculate [H_3O^+] for a urine sample, which has a pH of 7.5.

TRY IT FIRST

SOLUTION

STEP 1 State the given and needed quantities.

ANALYZE THE PROBLEM	Given	Need	Connect
	pH = 7.5	[H_3O^+]	[H_3O^+] = 10^{-pH}

STEP 2 Enter the pH value into the inverse log equation and calculate.

$$[H_3O^+] = 10^{-pH} = 10^{-7.5}$$

Calculator Procedure   **Calculator Display**

[2nd] [log] [+/−] 7.5 [=]   or   7.5 [+/−] [2nd] [log] [=]   3.16227766 E −08

Be sure to check the instructions for your calculator. Different calculators can have different methods for this calculation.

STEP 3 **Adjust the SFs for the coefficient.** Because the pH value 7.5 has one digit to the *right* of the decimal point, the coefficient for [H_3O^+] is written with one SF.

$$[H_3O^+] = 3 \times 10^{-8} \, M$$
One SF

STUDY CHECK

What are the [H_3O^+] and [OH^-] of Diet Coke that has a pH of 3.17?

ANSWER

[H_3O^+] = 6.8×10^{-4} M, [OH^-] = 1.5×10^{-11} M

Guide to Calculating [H_3O^+] from pH

STEP 1
State the given and needed quantities.

STEP 2
Enter the pH value into the inverse log equation and calculate.

STEP 3
Adjust the SFs for the coefficient.

The pH of Diet Coke is 3.17.

CHEMISTRY LINK TO **HEALTH**
Stomach Acid, HCl

Gastric acid, which contains HCl, is produced by parietal cells that line the stomach. When the stomach expands with the intake of food, the gastric glands begin to secrete a strongly acidic solution of HCl. In a single day, a person may secrete 2000 mL of gastric juice, which contains hydrochloric acid, mucins, and the enzymes pepsin and lipase.

The HCl in the gastric juice activates a digestive enzyme from the chief cells called *pepsinogen* to form *pepsin*, which breaks down proteins in food entering the stomach. The secretion of HCl continues until the stomach has a pH of about 2, which is the optimum for activating the digestive enzymes without ulcerating the stomach lining. In addition, the low pH destroys bacteria that reach the stomach. Normally, large quantities of viscous mucus are secreted within the stomach to protect its lining from acid and enzyme damage. Gastric acid may also form under conditions of stress when the nervous system activates the production of HCl. As the contents of the stomach move into the small intestine, cells produce bicarbonate that neutralizes the gastric acid until the pH is about 5.

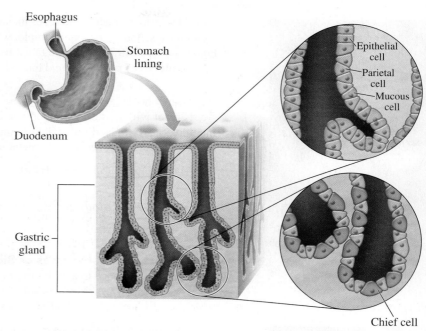

Parietal cells in the lining of the stomach secrete gastric acid HCl.

QUESTIONS AND PROBLEMS

14.6 The pH Scale

LEARNING GOAL Calculate pH from $[H_3O^+]$; given the pH, calculate the $[H_3O^+]$ and $[OH^-]$ of a solution.

14.43 Why does a neutral solution have a pH of 7.0?

14.44 If you know the $[OH^-]$, how can you determine the pH of a solution?

14.45 State whether each of the following solutions is acidic, basic, or neutral:
  a. blood plasma, pH 7.38
  b. vinegar, pH 2.8
  c. drain cleaner, pOH 2.8
  d. coffee, pH 5.52
  e. tomatoes, pH 4.2
  f. chocolate cake, pH 7.6

14.46 State whether each of the following solutions is acidic, basic, or neutral:
  a. soda, pH 3.22
  b. shampoo, pOH 8.3
  c. laundry detergent, pOH 4.56
  d. rain, pH 5.8
  e. honey, pH 3.9
  f. cheese, pH 4.9

14.47 A solution with a pH of 3 is 10 times more acidic than a solution with pH 4. Explain.

14.48 A solution with a pH of 10 is 100 times more basic than a solution with pH 8. Explain.

14.49 Calculate the pH of each solution given the following:
  a. $[H_3O^+] = 1 \times 10^{-4}$ M
  b. $[H_3O^+] = 3 \times 10^{-9}$ M
  c. $[OH^-] = 1 \times 10^{-5}$ M
  d. $[OH^-] = 2.5 \times 10^{-11}$ M
  e. $[H_3O^+] = 6.7 \times 10^{-8}$ M
  f. $[OH^-] = 8.2 \times 10^{-4}$ M

14.50 Calculate the pOH of each solution given the following:
  a. $[H_3O^+] = 1 \times 10^{-8}$ M
  b. $[H_3O^+] = 5 \times 10^{-6}$ M
  c. $[OH^-] = 1 \times 10^{-2}$ M
  d. $[OH^-] = 8.0 \times 10^{-3}$ M
  e. $[H_3O^+] = 4.7 \times 10^{-2}$ M
  f. $[OH^-] = 3.9 \times 10^{-6}$ M

Applications

14.51 Complete the following table:

$[H_3O^+]$	$[OH^-]$	pH	pOH	Acidic, Basic, or Neutral?
	1.0×10^{-6} M			
		3.49		
2.8×10^{-5} M				
			2.00	

14.52 Complete the following table:

$[H_3O^+]$	$[OH^-]$	pH	pOH	Acidic, Basic, or Neutral?
		10.00		
				Neutral
			5.66	
6.4×10^{-12} M				

14.53 A patient with severe metabolic acidosis has a blood plasma pH of 6.92. What is the $[H_3O^+]$ of the blood plasma?

14.54 A patient with respiratory alkalosis has a blood plasma pH of 7.58. What is the $[H_3O^+]$ of the blood plasma?

14.7 Reactions of Acids and Bases

LEARNING GOAL Write balanced equations for reactions of acids with metals, carbonates or bicarbonates, and bases.

Typical reactions of acids and bases include the reactions of acids with metals, carbonates or bicarbonates, and bases. For example, when you drop an antacid tablet in water, the bicarbonate ion and citric acid in the tablet react to produce carbon dioxide bubbles, water, and salt. A **salt** is an ionic compound that does not have H^+ as the cation or OH^- as the anion.

Acids and Metals

Acids react with certain metals to produce hydrogen gas (H_2) and a salt. Active metals include potassium, sodium, calcium, magnesium, aluminum, zinc, iron, and tin. In these single replacement reactions, the metal ion replaces the hydrogen in the acid.

$$\underset{\text{Metal}}{Mg(s)} + \underset{\text{Acid}}{2HCl(aq)} \longrightarrow \underset{\text{Hydrogen}}{H_2(g)} + \underset{\text{Salt}}{MgCl_2(aq)}$$

$$\underset{\text{Metal}}{Zn(s)} + \underset{\text{Acid}}{2HNO_3(aq)} \longrightarrow \underset{\text{Hydrogen}}{H_2(g)} + \underset{\text{Salt}}{Zn(NO_3)_2(aq)}$$

Magnesium reacts rapidly with acid and forms H_2 gas and a salt of magnesium.

Acids React with Carbonates or Bicarbonates

When an acid is added to a carbonate or bicarbonate, the products are carbon dioxide gas, water, and a salt. The acid reacts with CO_3^{2-} or HCO_3^- to produce carbonic acid (H_2CO_3), which breaks down rapidly to CO_2 and H_2O.

$$\underset{\text{Acid}}{2HCl(aq)} + \underset{\text{Carbonate}}{Na_2CO_3(aq)} \longrightarrow \underset{\substack{\text{Carbon}\\\text{dioxide}}}{CO_2(g)} + \underset{\text{Water}}{H_2O(l)} + \underset{\text{Salt}}{2NaCl(aq)}$$

$$\underset{\text{Acid}}{HBr(aq)} + \underset{\text{Bicarbonate}}{NaHCO_3(aq)} \longrightarrow \underset{\substack{\text{Carbon}\\\text{dioxide}}}{CO_2(g)} + \underset{\text{Water}}{H_2O(l)} + \underset{\text{Salt}}{NaBr(aq)}$$

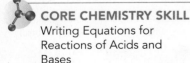

When sodium bicarbonate (baking soda) reacts with an acid (vinegar), the products are carbon dioxide gas, water, and a salt.

CORE CHEMISTRY SKILL
Writing Equations for Reactions of Acids and Bases

Acids and Hydroxides: Neutralization

Neutralization is a reaction between a strong or weak acid with a strong base to produce water and a salt. The H^+ of the acid and the OH^- of the base combine to form water. The salt is the combination of the cation from the base and the anion from the acid. We can write the following equation for the neutralization reaction between HCl and NaOH:

$$\underset{\text{Acid}}{HCl(aq)} + \underset{\text{Base}}{NaOH(aq)} \longrightarrow \underset{\text{Water}}{H_2O(l)} + \underset{\text{Salt}}{NaCl(aq)}$$

If we write the strong acid HCl and the strong base NaOH as ions, we see that H^+ combines with OH^- to form water, leaving the ions Na^+ and Cl^- in solution.

$$H^+(aq) + Cl^-(aq) + Na^+(aq) + OH^-(aq) \longrightarrow H_2O(l) + Na^+(aq) + Cl^-(aq)$$

When we omit the ions that do not change during the reaction (*spectator ions*), we obtain the *net ionic equation*.

$$H^+(aq) + \cancel{Cl^-(aq)} + \cancel{Na^+(aq)} + OH^-(aq) \longrightarrow H_2O(l) + \cancel{Na^+(aq)} + \cancel{Cl^-(aq)}$$

The net ionic equation for the neutralization of H^+ and OH^- to form H_2O is

$$H^+(aq) + OH^-(aq) \longrightarrow H_2O(l) \quad \text{Net ionic equation}$$

Balancing Neutralization Equations

In a neutralization reaction, one H^+ always reacts with one OH^-. Therefore, a neutralization equation may need coefficients to balance the H^+ from the acid with the OH^- from the base as shown in Sample Problem 14.11.

SAMPLE PROBLEM 14.11 Balancing Equations for Acids

Write the balanced equation for the neutralization of HCl(*aq*) and Ba(OH)$_2$(*s*).

TRY IT FIRST

SOLUTION

STEP 1 Write the reactants and products.

$$HCl(aq) + Ba(OH)_2(s) \longrightarrow H_2O(l) + salt$$

STEP 2 Balance the H$^+$ in the acid with the OH$^-$ in the base. Placing a coefficient of 2 in front of the HCl provides 2H$^+$ for the 2OH$^-$ from Ba(OH)$_2$.

$$2HCl(aq) + Ba(OH)_2(s) \longrightarrow H_2O(l) + salt$$

STEP 3 Balance the H$_2$O with the H$^+$ and the OH$^-$. Use a coefficient of 2 in front of H$_2$O to balance 2H$^+$ and 2OH$^-$.

$$2HCl(aq) + Ba(OH)_2(s) \longrightarrow 2H_2O(l) + salt$$

STEP 4 Write the formula of the salt from the remaining ions. Use the ions Ba^{2+} and 2Cl$^-$ and write the formula for the salt as BaCl$_2$.

$$2HCl(aq) + Ba(OH)_2(s) \longrightarrow 2H_2O(l) + BaCl_2(aq)$$

STUDY CHECK 14.11

Write the balanced equation for the neutralization of H$_2$SO$_4$(*aq*) and LiOH(*aq*).

ANSWER

$$H_2SO_4(aq) + 2LiOH(aq) \longrightarrow 2H_2O(l) + Li_2SO_4(aq)$$

> **Guide to Balancing an Equation for Neutralization**
> **STEP 1** Write the reactants and products.
> **STEP 2** Balance the H$^+$ in the acid with the OH$^-$ in the base.
> **STEP 3** Balance the H$_2$O with the H$^+$ and the OH$^-$.
> **STEP 4** Write the formula of the salt from the remaining ions.

CHEMISTRY LINK TO **HEALTH**
Antacids

Antacids are substances used to neutralize excess stomach acid (HCl). Some antacids are mixtures of aluminum hydroxide and magnesium hydroxide. These hydroxides are not very soluble in water, so the levels of available OH$^-$ are not damaging to the intestinal tract. However, aluminum hydroxide has the side effects of producing constipation and binding phosphate in the intestinal tract, which may cause weakness and loss of appetite. Magnesium hydroxide has a laxative effect. These side effects are less likely when a combination of the antacids is used.

$$Al(OH)_3(s) + 3HCl(aq) \longrightarrow 3H_2O(l) + AlCl_3(aq)$$
$$Mg(OH)_2(s) + 2HCl(aq) \longrightarrow 2H_2O(l) + MgCl_2(aq)$$

Antacids neutralize excess stomach acid.

Some antacids use calcium carbonate to neutralize excess stomach acid. About 10% of the calcium is absorbed into the bloodstream, where it elevates the level of serum calcium. Calcium carbonate is not recommended for patients who have peptic ulcers or a tendency to form kidney stones, which typically consist of an insoluble calcium salt.

$$CaCO_3(s) + 2HCl(aq) \longrightarrow CO_2(g) + H_2O(l) + CaCl_2(aq)$$

Still other antacids contain sodium bicarbonate. This type of antacid neutralizes excess gastric acid, increases blood pH, but also elevates sodium levels in the body fluids. It also is not recommended in the treatment of peptic ulcers.

$$NaHCO_3(s) + HCl(aq) \longrightarrow CO_2(g) + H_2O(l) + NaCl(aq)$$

The neutralizing substances in some antacid preparations are given in **TABLE 14.8**.

TABLE 14.8 Basic Compounds in Some Antacids

Antacid	Base(s)
Amphojel	Al(OH)$_3$
Milk of magnesia	Mg(OH)$_2$
Mylanta, Maalox, Di-Gel, Gelusil, Riopan	Mg(OH)$_2$, Al(OH)$_3$
Bisodol, Rolaids	CaCO$_3$, Mg(OH)$_2$
Titralac, Tums, Pepto-Bismol	CaCO$_3$
Alka-Seltzer	NaHCO$_3$, KHCO$_3$

QUESTIONS AND PROBLEMS

14.7 Reactions of Acids and Bases

LEARNING GOAL Write balanced equations for reactions of acids with metals, carbonates or bicarbonates, and bases.

14.55 Complete and balance the equation for each of the following reactions:
a. $ZnCO_3(s) + HBr(aq) \longrightarrow$
b. $Zn(s) + HCl(aq) \longrightarrow$
c. $HCl(aq) + NaHCO_3(s) \longrightarrow$
d. $H_2SO_4(aq) + Mg(OH)_2(s) \longrightarrow$

14.56 Complete and balance the equation for each of the following reactions:
a. $KHCO_3(s) + HBr(aq) \longrightarrow$
b. $Ca(s) + H_2SO_4(aq) \longrightarrow$
c. $H_2SO_4(aq) + Ca(OH)_2(s) \longrightarrow$
d. $Na_2CO_3(s) + H_2SO_4(aq) \longrightarrow$

14.57 Balance each of the following neutralization reactions:
a. $HCl(aq) + Mg(OH)_2(s) \longrightarrow H_2O(l) + MgCl_2(aq)$
b. $H_3PO_4(aq) + LiOH(aq) \longrightarrow H_2O(l) + Li_3PO_4(aq)$

14.58 Balance each of the following neutralization reactions:
a. $HNO_3(aq) + Ba(OH)_2(s) \longrightarrow H_2O(l) + Ba(NO_3)_2(aq)$
b. $H_2SO_4(aq) + Al(OH)_3(s) \longrightarrow H_2O(l) + Al_2(SO_4)_3(aq)$

14.59 Write a balanced equation for the neutralization of each of the following:
a. $H_2SO_4(aq)$ and $NaOH(aq)$
b. $HCl(aq)$ and $Fe(OH)_3(s)$
c. $H_2CO_3(aq)$ and $Mg(OH)_2(s)$

14.60 Write a balanced equation for the neutralization of each of the following:
a. $H_3PO_4(aq)$ and $NaOH(aq)$
b. $HI(aq)$ and $LiOH(aq)$
c. $HNO_3(aq)$ and $Ca(OH)_2(s)$

14.8 Acid–Base Titration

LEARNING GOAL Calculate the molarity or volume of an acid or base solution from titration information.

Suppose we need to find the molarity of a solution of HCl, which has an unknown concentration. We can do this by a laboratory procedure called **titration** in which we neutralize an acid sample with a known amount of base. In a titration, we place a measured volume of the acid in a flask and add a few drops of an **indicator**, such as phenolphthalein. An indicator is a compound that dramatically changes color when pH of the solution changes. In an acidic solution, phenolphthalein is colorless. Then we fill a buret with a NaOH solution of known molarity and carefully add NaOH solution to neutralize the acid in the flask (see **FIGURE 14.6**). We know that neutralization has taken place when the phenolphthalein in the solution changes from colorless to pink. This is called the neutralization **endpoint**. From the measured volume of the NaOH solution and its molarity, we calculate the number of moles of NaOH, the moles of acid, and use the measured volume of acid to calculate its concentration.

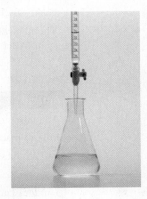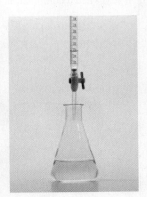

FIGURE 14.6 ▶ The titration of an acid. A known volume of an acid is placed in a flask with an indicator and titrated with a measured volume of a base solution, such as NaOH, to the neutralization endpoint.

Q What data is needed to determine the molarity of the acid in the flask?

SAMPLE PROBLEM 14.12 Titration of an Acid

If 16.3 mL of a 0.185 M Sr(OH)$_2$ solution is used to titrate the HCl in 0.0250 L of gastric juice, what is the molarity of the HCl solution?

$$Sr(OH)_2(aq) + 2HCl(aq) \longrightarrow 2H_2O(l) + SrCl_2(aq)$$

TRY IT FIRST

SOLUTION

STEP 1 State the given and needed quantities and concentrations.

ANALYZE THE PROBLEM	Given	Need	Connect
	0.0250 L of HCl solution, 16.3 mL of 0.185 M Sr(OH)$_2$ solution	molarity of the HCl solution	molarity, mole–mole factor
	Neutralization Equation $Sr(OH)_2(aq) + 2HCl(aq) \longrightarrow 2H_2O(l) + SrCl_2(aq)$		

STEP 2 Write a plan to calculate the molarity.

mL of Sr(OH)$_2$ solution → **Metric factor** → L of Sr(OH)$_2$ solution → **Molarity** → moles of Sr(OH)$_2$ → **Mole–mole factor** → moles of HCl → **Divide by liters** → molarity of HCl solution

STEP 3 State equalities and conversion factors, including concentrations.

1 L of Sr(OH)$_2$ solution	=	1000 mL of Sr(OH)$_2$ solution
$\dfrac{1000 \text{ mL Sr(OH)}_2 \text{ solution}}{1 \text{ L Sr(OH)}_2 \text{ solution}}$	and	$\dfrac{1 \text{ L Sr(OH)}_2 \text{ solution}}{1000 \text{ mL Sr(OH)}_2 \text{ solution}}$

1 L of Sr(OH)$_2$ solution	=	0.185 mol of Sr(OH)$_2$
$\dfrac{0.185 \text{ mol Sr(OH)}_2}{1 \text{ L Sr(OH)}_2 \text{ solution}}$	and	$\dfrac{1 \text{ L Sr(OH)}_2 \text{ solution}}{0.185 \text{ mol Sr(OH)}_2}$

2 mol of HCl	=	1 mol of Sr(OH)$_2$
$\dfrac{2 \text{ mol HCl}}{1 \text{ mol Sr(OH)}_2}$	and	$\dfrac{1 \text{ mol Sr(OH)}_2}{2 \text{ mol HCl}}$

STEP 4 Set up the problem to calculate the needed quantity.

$$16.3 \text{ mL Sr(OH)}_2 \text{ solution} \times \frac{1 \text{ L Sr(OH)}_2 \text{ solution}}{1000 \text{ mL Sr(OH)}_2 \text{ solution}} \times \frac{0.185 \text{ mol Sr(OH)}_2}{1 \text{ L Sr(OH)}_2 \text{ solution}} \times \frac{2 \text{ mol HCl}}{1 \text{ mol Sr(OH)}_2}$$

$$= 0.006\ 03 \text{ mol of HCl}$$

$$\text{molarity of HCl solution} = \frac{0.006\ 03 \text{ mol HCl}}{0.0250 \text{ L HCl solution}} = 0.241 \text{ M HCl solution}$$

STUDY CHECK 14.12

What is the molarity of an HCl solution if 28.6 mL of a 0.175 M NaOH solution is needed to titrate a 25.0-mL sample of the HCl solution?

ANSWER

0.200 M HCl solution

CORE CHEMISTRY SKILL
Calculating Molarity or Volume of an Acid or Base in a Titration

Guide to Calculations for an Acid–Base Titration

STEP 1
State the given and needed quantities and concentrations.

STEP 2
Write a plan to calculate the molarity or volume.

STEP 3
State equalities and conversion factors, including concentrations.

STEP 4
Set up the problem to calculate the needed quantity.

Interactive Video

Acid–Base Titration

QUESTIONS AND PROBLEMS

14.8 Acid–Base Titration

LEARNING GOAL Calculate the molarity or volume of an acid or base solution from titration information.

14.61 If you need to determine the molarity of a formic acid solution, $HCHO_2$, how would you proceed?

$$HCHO_2(aq) + H_2O(l) \rightleftharpoons H_3O^+(aq) + CHO_2^-(aq)$$

14.62 If you need to determine the molarity of an acetic acid solution, $HC_2H_3O_2$, how would you proceed?

$$HC_2H_3O_2(aq) + H_2O(l) \rightleftharpoons H_3O^+(aq) + C_2H_3O_2^-(aq)$$

14.63 What is the molarity of a solution of HCl if 5.00 mL of the HCl solution is titrated with 28.6 mL of a 0.145 M NaOH solution?

$$HCl(aq) + NaOH(aq) \longrightarrow H_2O(l) + NaCl(aq)$$

14.64 What is the molarity of an acetic acid solution if 25.0 mL of the $HC_2H_3O_2$ solution is titrated with 29.7 mL of a 0.205 M KOH solution?

$$HC_2H_3O_2(aq) + KOH(aq) \longrightarrow H_2O(l) + KC_2H_3O_2(aq)$$

14.65 If 38.2 mL of a 0.163 M KOH solution is required to neutralize completely 25.0 mL of a solution of H_2SO_4, what is the molarity of the H_2SO_4 solution?

$$H_2SO_4(aq) + 2KOH(aq) \longrightarrow 2H_2O(l) + K_2SO_4(aq)$$

14.66 A solution of 0.162 M NaOH is used to titrate 25.0 mL of a solution of H_2SO_4. If 32.8 mL of the NaOH solution is required to reach the endpoint, what is the molarity of the H_2SO_4 solution?

$$H_2SO_4(aq) + 2NaOH(aq) \longrightarrow 2H_2O(l) + Na_2SO_4(aq)$$

14.67 A solution of 0.204 M NaOH is used to titrate 50.0 mL of a 0.0224 M H_3PO_4 solution. What volume, in milliliters, of the NaOH solution is required?

$$H_3PO_4(aq) + 3NaOH(aq) \longrightarrow 3H_2O(l) + Na_3PO_4(aq)$$

14.68 A solution of 0.312 M KOH is used to titrate 15.0 mL of a 0.186 M H_3PO_4 solution. What volume, in milliliters, of the KOH solution is required?

$$H_3PO_4(aq) + 3KOH(aq) \longrightarrow 3H_2O(l) + K_3PO_4(aq)$$

14.9 Buffers

LEARNING GOAL Describe the role of buffers in maintaining the pH of a solution; calculate the pH of a buffer.

The lungs and the kidneys are the primary organs that regulate the pH of body fluids, including blood and urine. Major changes in the pH of the body fluids can severely affect biological activities within the cells. *Buffers* are present to prevent large fluctuations in pH.

The pH of water and most solutions changes drastically when a small amount of acid or base is added. However, when an acid or a base is added to a buffer solution, there is little change in pH. A **buffer solution** maintains the pH of a solution by neutralizing small amounts of added acid or base. In the human body, whole blood contains plasma, white blood cells and platelets, and red blood cells. Blood plasma contains buffers that maintain a consistent pH of about 7.4. If the pH of the blood plasma goes slightly above or below 7.4, changes in our oxygen levels and our metabolic processes can be drastic enough to cause death. Even though we obtain acids and bases from foods and cellular reactions, the buffers in the body absorb those compounds so effectively that the pH of our blood plasma remains essentially unchanged (see **FIGURE 14.7**).

In a buffer, an acid must be present to react with any OH^- that is added, and a base must be available to react with any added H_3O^+. However, that acid and base must not neutralize each other. Therefore, a combination of an acid–base conjugate pair is used in buffers. Most buffer solutions consist of nearly equal concentrations of a weak acid and a salt containing its conjugate base. Buffers may also contain a weak base and the salt of the weak base, which contains its conjugate acid.

For example, a typical buffer can be made from the weak acid acetic acid ($HC_2H_3O_2$) and its salt, sodium acetate ($NaC_2H_3O_2$). As a weak acid, acetic acid dissociates slightly in water to form H_3O^+ and a very small amount of $C_2H_3O_2^-$. The addition of its salt, sodium acetate, provides a much larger concentration of acetate ion ($C_2H_3O_2^-$), which is necessary for its buffering capability.

$$\underset{\text{Large amount}}{HC_2H_3O_2(aq)} + H_2O(l) \rightleftharpoons H_3O^+(aq) + \underset{\text{Large amount}}{C_2H_3O_2^-(aq)}$$

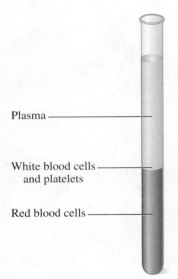

Plasma ——

White blood cells —— and platelets

Red blood cells ——

Whole blood consists of plasma, white blood cells and platelets, and red blood cells.

ENGAGE

Why does a buffer require the presence of a weak acid or weak base and the salt of that weak acid or weak base?

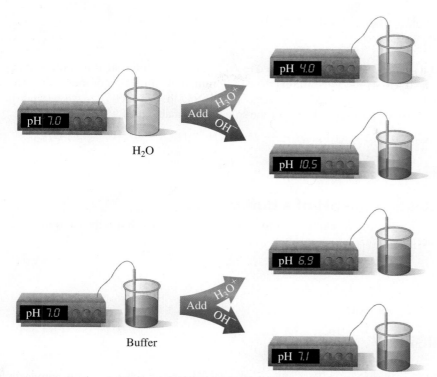

FIGURE 14.7 ▶ Adding an acid or a base to water changes the pH drastically, but a buffer resists pH change when small amounts of acid or base are added.

Q Why does the pH change several pH units when acid is added to water, but not when acid is added to a buffer?

We can now describe how this buffer solution maintains the $[H_3O^+]$. When a small amount of acid is added, the additional H_3O^+ combines with the acetate ion, $C_2H_3O_2^-$, causing the equilibrium to shift in the direction of the reactants, acetic acid and water. There will be a slight decrease in the $[C_2H_3O_2^-]$ and a slight increase in the $[HC_2H_3O_2]$, but both the $[H_3O^+]$ and pH are maintained (see **FIGURE 14.8**).

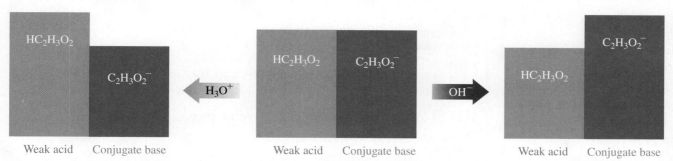

FIGURE 14.8 ▶ The buffer described here consists of about equal concentrations of acetic acid ($HC_2H_3O_2$) and its conjugate base acetate ion ($C_2H_3O_2^-$). Adding H_3O^+ to the buffer neutralizes some $C_2H_3O_2^-$, whereas adding OH^- neutralizes some $HC_2H_3O_2$. The pH of the solution is maintained as long as the added amount of acid or base is small compared to the concentrations of the buffer components.

Q How does this acetic acid–acetate ion buffer maintain pH?

$$HC_2H_3O_2(aq) + H_2O(l) \longleftarrow H_3O^+(aq) + C_2H_3O_2^-(aq)$$

Equilibrium shifts in the
direction of the reactants

If a small amount of base is added to this same buffer solution, it is neutralized by the acetic acid, $HC_2H_3O_2$, which shifts the equilibrium in the direction of the products acetate ion and water. The $[HC_2H_3O_2]$ decreases slightly and the $[C_2H_3O_2^-]$ increases slightly, but again the $[H_3O^+]$ and thus the pH of the solution are maintained.

$$HC_2H_3O_2(aq) + OH^-(aq) \longrightarrow H_2O(l) + C_2H_3O_2^-(aq)$$

Equilibrium shifts in the
direction of the products

CORE CHEMISTRY SKILL
Calculating the pH of a
Buffer

Calculating the pH of a Buffer

By rearranging the K_a expression to give $[H_3O^+]$, we can obtain the ratio of the acetic acid/acetate buffer.

$$K_a = \frac{[H_3O^+][C_2H_3O_2^-]}{[HC_2H_3O_2]}$$

Solving for $[H_3O^+]$ gives:

$$[H_3O^+] = K_a \times \frac{[HC_2H_3O_2]}{[C_2H_3O_2^-]} \quad \begin{array}{l} \longleftarrow \text{ Weak acid} \\ \longleftarrow \text{ Conjugate base} \end{array}$$

In this rearrangement of K_a, the weak acid is in the numerator and the conjugate base in the denominator. We can now calculate the $[H_3O^+]$ and pH for an acetic acid buffer as shown in Sample Problem 14.13.

SAMPLE PROBLEM 14.13 Calculating the pH of a Buffer

The K_a for acetic acid, $HC_2H_3O_2$, is 1.8×10^{-5}. What is the pH of a buffer prepared with $1.0\,M\,HC_2H_3O_2$ and $1.0\,M\,C_2H_3O_2^-$?

$$HC_2H_3O_2(aq) + H_2O(l) \rightleftharpoons H_3O^+(aq) + C_2H_3O_2^-(aq)$$

TRY IT FIRST

SOLUTION

STEP 1 State the given and needed quantities.

Guide to Calculating pH
of a Buffer

STEP 1
State the given and needed
quantities.

STEP 2
Write the K_a expression and
rearrange for $[H_3O^+]$.

STEP 3
Substitute [HA] and [A⁻] into
the K_a expression.

STEP 4
Use $[H_3O^+]$ to calculate pH.

	Given	Need	Connect
ANALYZE THE PROBLEM	$1.0\,M\,HC_2H_3O_2$, $1.0\,M\,C_2H_3O_2^-$	pH	K_a expression
	Equation		
	$HC_2H_3O_2(aq) + H_2O(l) \rightleftharpoons H_3O^+(aq) + C_2H_3O_2^-(aq)$		

STEP 2 Write the K_a expression and rearrange for $[H_3O^+]$.

$$K_a = \frac{[H_3O^+][C_2H_3O_2^-]}{[HC_2H_3O_2]}$$

$$[H_3O^+] = K_a \times \frac{[HC_2H_3O_2]}{[C_2H_3O_2^-]}$$

STEP 3 Substitute [HA] and [A⁻] into the K_a expression.

$$[H_3O^+] = 1.8 \times 10^{-5} \times \frac{[1.0]}{[1.0]}$$

$$[H_3O^+] = 1.8 \times 10^{-5}\,M$$

STEP 4 Use [H₃O⁺] to calculate pH. Placing the [H₃O⁺] into the pH equation gives the pH of the buffer.

$$\text{pH} = -\log[1.8 \times 10^{-5}] = 4.74$$

STUDY CHECK 14.13

One of the conjugate acid–base pairs that buffers the blood plasma is $H_2PO_4^-/HPO_4^{2-}$, which has a K_a of 6.2×10^{-8}. What is the pH of a buffer that is prepared from 0.10 M $H_2PO_4^-$ and 0.50 M HPO_4^{2-}?

ANSWER

pH = 7.91

Because K_a is a constant at a given temperature, the [H₃O⁺] is determined by the $[HC_2H_3O_2]/[C_2H_3O_2^-]$ ratio. As long as the addition of small amounts of either acid or base changes the ratio of $[HC_2H_3O_2]/[C_2H_3O_2^-]$ only slightly, the changes in [H₃O⁺] will be small and the pH will be maintained. If a large amount of acid or base is added, the *buffering capacity* of the system may be exceeded. Buffers can be prepared from conjugate acid–base pairs such as $H_2PO_4^-/HPO_4^{2-}$, HPO_4^{2-}/PO_4^{3-}, HCO_3^-/CO_3^{2-}, or NH_4^+/NH_3. The pH of the buffer solution will depend on the conjugate acid–base pair chosen.

Using a common phosphate buffer for biological specimens, we can look at the effect of using different ratios of $[H_2PO_4^-]/[HPO_4^{2-}]$ on the [H₃O⁺] and pH. The K_a of $H_2PO_4^-$ is 6.2×10^{-8}. The equation and the [H₃O⁺] are written as follows:

$$H_2PO_4^-(aq) + H_2O(l) \rightleftharpoons H_3O^+(aq) + HPO_4^{2-}(aq)$$

$$[H_3O^+] = K_a \times \frac{[H_2PO_4^-]}{[HPO_4^{2-}]}$$

K_a	$\dfrac{[H_2PO_4^-]}{[HPO_4^{2-}]}$	Ratio	[H₃O⁺]	pH
6.2×10^{-8}	$\dfrac{1.0\,\text{M}}{0.10\,\text{M}}$	$\dfrac{10}{1}$	6.2×10^{-7}	6.21
6.2×10^{-8}	$\dfrac{1.0\,\text{M}}{1.0\,\text{M}}$	$\dfrac{1}{1}$	6.2×10^{-8}	7.21
6.2×10^{-8}	$\dfrac{0.10\,\text{M}}{1.0\,\text{M}}$	$\dfrac{1}{10}$	6.2×10^{-9}	8.21

To prepare a phosphate buffer with a pH close to the pH of a biological sample, 7.4, we would choose concentrations that are about equal, such as 1.0 M $H_2PO_4^-$ and 1.0 M HPO_4^{2-}.

ENGAGE

How would a solution composed of HPO_4^{2-} and PO_4^{3-} act as a buffer?

CHEMISTRY LINK TO HEALTH
Buffers in the Blood Plasma

The arterial blood plasma has a normal pH of 7.35 to 7.45. If changes in H₃O⁺ lower the pH below 6.8 or raise it above 8.0, cells cannot function properly and death may result. In our cells, CO_2 is continually produced as an end product of cellular metabolism. Some CO_2 is carried to the lungs for elimination, and the rest dissolves in body fluids such as plasma and saliva, forming carbonic acid, H_2CO_3. As a weak acid, carbonic acid dissociates to give bicarbonate, HCO_3^-, and H_3O^+. More of the anion HCO_3^- is supplied by the kidneys to give an important buffer system in the body fluid— the H_2CO_3/HCO_3^- buffer.

$$CO_2(g) + H_2O(l) \rightleftharpoons H_2CO_3(aq) \rightleftharpoons$$
$$H_3O^+(aq) + HCO_3^-(aq)$$

Excess H_3O^+ entering the body fluids reacts with the HCO_3^-, and excess OH^- reacts with the carbonic acid.

$$H_2CO_3(aq) + H_2O(l) \longleftarrow H_3O^+(aq) + HCO_3(aq)$$

Equilibrium shifts in the
direction of the reactants

$$H_2CO_3(aq) + OH^-(aq) \longrightarrow H_2O(l) + HCO_3^-(aq)$$

Equilibrium shifts in the
direction of the products

For carbonic acid, we can write the equilibrium expression as

$$K_a = \frac{[H_3O^+][HCO_3^-]}{[H_2CO_3]}$$

To maintain the normal blood plasma pH (7.35 to 7.45), the ratio of $[H_2CO_3]/[HCO_3^-]$ needs to be about 1 to 10, which is obtained by the concentrations in the blood plasma of 0.0024 M H_2CO_3 and 0.024 M HCO_3^-.

$$[H_3O^+] = K_a \times \frac{[H_2CO_3]}{[HCO_3^-]}$$

$$[H_3O^+] = 4.3 \times 10^{-7} \times \frac{[0.0024]}{[0.024]}$$

$$= 4.3 \times 10^{-7} \times 0.10 = 4.3 \times 10^{-8} \text{ M}$$

$$pH = -\log[4.3 \times 10^{-8}] = 7.37$$

In the body, the concentration of carbonic acid is closely associated with the partial pressure of CO_2, P_{CO_2}. **TABLE 14.9** lists the normal values for arterial blood. If the CO_2 level rises, increasing $[H_2CO_3]$, the equilibrium shifts to produce more H_3O^+, which lowers the pH. This condition is called *acidosis*. Difficulty with ventilation or gas diffusion can lead to respiratory acidosis, which can happen in emphysema or when an accident or depressive drugs affect the medulla of the brain.

A lowering of the CO_2 level leads to a high blood pH, a condition called *alkalosis*. Excitement, trauma, or a high temperature may cause a person to hyperventilate, which expels large amounts of CO_2. As the partial pressure of CO_2 in the blood falls below normal, the equilibrium shifts from H_2CO_3 to CO_2 and H_2O. This shift decreases the $[H_3O^+]$ and raises the pH. The kidneys also regulate H_3O^+ and HCO_3^-, but they do so more slowly than the adjustment made by the lungs during ventilation.

TABLE 14.10 lists some of the conditions that lead to changes in the blood pH and some possible treatments.

TABLE 14.9 Normal Values for Blood Buffer in Arterial Blood

P_{CO_2}	40 mmHg
H_2CO_3	2.4 mmol/L of plasma
HCO_3^-	24 mmol/L of plasma
pH	7.35 to 7.45

TABLE 14.10 Acidosis and Alkalosis: Symptoms, Causes, and Treatments

Respiratory Acidosis: $CO_2\uparrow$ pH$\downarrow$	
Symptoms:	Failure to ventilate, suppression of breathing, disorientation, weakness, coma
Causes:	Lung disease blocking gas diffusion (e.g., emphysema, pneumonia, bronchitis, asthma); depression of respiratory center by drugs, cardiopulmonary arrest, stroke, poliomyelitis, or nervous system disorders
Treatment:	Correction of disorder, infusion of bicarbonate
Metabolic Acidosis: $H^+\uparrow$ pH$\downarrow$	
Symptoms:	Increased ventilation, fatigue, confusion
Causes:	Renal disease, including hepatitis and cirrhosis; increased acid production in diabetes mellitus, hyperthyroidism, alcoholism, and starvation; loss of alkali in diarrhea; acid retention in renal failure
Treatment:	Sodium bicarbonate given orally, dialysis for renal failure, insulin treatment for diabetic ketosis
Respiratory Alkalosis: $CO_2\downarrow$ pH$\uparrow$	
Symptoms:	Increased rate and depth of breathing, numbness, light-headedness, tetany
Causes:	Hyperventilation because of anxiety, hysteria, fever, exercise; reaction to drugs such as salicylate, quinine, and antihistamines; conditions causing hypoxia (e.g., pneumonia, pulmonary edema, heart disease)
Treatment:	Elimination of anxiety-producing state, rebreathing into a paper bag
Metabolic Alkalosis: $H^+\downarrow$ pH$\uparrow$	
Symptoms:	Depressed breathing, apathy, confusion
Causes:	Vomiting, diseases of the adrenal glands, ingestion of excess alkali
Treatment:	Infusion of saline solution, treatment of underlying diseases

QUESTIONS AND PROBLEMS

14.9 Buffers

LEARNING GOAL Describe the role of buffers in maintaining the pH of a solution; calculate the pH of a buffer.

14.69 Which of the following represents a buffer system? Explain.
  a. NaOH and NaCl   **b.** H_2CO_3 and $NaHCO_3$
  c. HF and KF   **d.** KCl and NaCl

14.70 Which of the following represents a buffer system? Explain.
  a. $HClO_2$ **b.** $NaNO_3$
  c. $HC_2H_3O_2$ and $NaC_2H_3O_2$ **d.** HCl and NaOH

14.71 Consider the buffer system of hydrofluoric acid, HF, and its salt, NaF.

$$HF(aq) + H_2O(l) \rightleftharpoons H_3O^+(aq) + F^-(aq)$$

  a. The purpose of this buffer system is to:
    1. maintain [HF]   **2.** maintain $[F^-]$
    3. maintain pH
  b. The salt of the weak acid is needed to:
    1. provide the conjugate base
    2. neutralize added H_3O^+
    3. provide the conjugate acid
  c. If OH^- is added, it is neutralized by:
    1. the salt   **2.** H_2O **3.** H_3O^+
  d. When H_3O^+ is added, the equilibrium shifts in the direction of the:
    1. reactants   **2.** products
    3. does not change

14.72 Consider the buffer system of nitrous acid, HNO_2, and its salt, $NaNO_2$.

$$HNO_2(aq) + H_2O(l) \rightleftharpoons H_3O^+(aq) + NO_2^-(aq)$$

  a. The purpose of this buffer system is to:
    1. maintain $[HNO_2]$ **2.** maintain $[NO_2^-]$
    3. maintain pH

  b. The weak acid is needed to:
    1. provide the conjugate base
    2. neutralize added OH^-
    3. provide the conjugate acid
  c. If H_3O^+ is added, it is neutralized by:
    1. the salt   **2.** H_2O
    3. OH^-
  d. When OH^- is added, the equilibrium shifts in the direction of the:
    1. reactants   **2.** products
    3. does not change

14.73 Nitrous acid has a K_a of 4.5×10^{-4}. What is the pH of a buffer solution containing 0.10 M HNO_2 and 0.10 M NO_2^-?

14.74 Acetic acid has a K_a of 1.8×10^{-5}. What is the pH of a buffer solution containing 0.15 M $HC_2H_3O_2$ and 0.15 M $C_2H_3O_2^-$?

14.75 Using Table 14.4 for K_a values, compare the pH of a HF buffer that contains 0.10 M HF and 0.10 M NaF with another HF buffer that contains 0.060 M HF and 0.120 M NaF.

14.76 Using Table 14.4 for K_a values, compare the pH of a H_2CO_3 buffer that contains 0.10 M H_2CO_3 and 0.10 M $NaHCO_3$ with another H_2CO_3 buffer that contains 0.15 M H_2CO_3 and 0.050 M $NaHCO_3$.

Applications

14.77 Why would the pH of your blood plasma increase if you breathe fast?

14.78 Why would the pH of your blood plasma decrease if you hold your breath?

14.79 Someone with kidney failure excretes urine with large amounts of HCO_3^-. How would this loss of HCO_3^- affect the pH of the blood plasma?

14.80 Someone with severe diabetes obtains energy by the breakdown of fats, which produce large amounts of acidic substances. How would this affect the pH of the blood plasma?

Follow Up

ACID REFLUX DISEASE

Larry has not been feeling well lately. He tells his doctor that he has discomfort and a burning feeling in his chest, and a sour taste in his throat and mouth. At times, Larry says he feels bloated after a big meal, has a dry cough, is hoarse, and sometimes has a sore throat. He has tried antacids, but they do not bring any relief.

The doctor tells Larry that he thinks he has acid reflux. At the top of the stomach there is a valve, the lower esophageal sphincter, that normally closes after food passes through it. However, if the valve does not close completely, acid produced in the stomach to digest food can move up into the esophagus, a condition called *acid reflux*. The acid, which is hydrochloric acid, HCl, is produced in the stomach to kill bacteria, microorganisms, and to activate the enzymes we need to break down food.

If acid reflux occurs, the strong acid HCl comes in contact with the lining of the esophagus, where it causes irritation and produces a burning feeling in the chest. Sometimes the pain in the chest is called *heartburn*. If the HCl reflux goes high enough to reach the throat, a sour taste may be noticed in the mouth. If Larry's symptoms occur three or more times a week, he may have a chronic condition known as *acid reflux disease* or *gastroesophageal reflux disease* (GERD).

Larry's doctor orders an *esophageal pH test* in which the amount of acid entering the esophagus from the stomach is measured over 24 h. A probe that measures the pH is inserted into the lower esophagus above the esophageal sphincter. The pH measurements indicate a reflux episode each time the pH drops to 4 or less.

In the 24-h period, Larry has several reflux episodes and his doctor determines that he has chronic GERD. He and Larry discuss treatment for GERD, which includes eating smaller meals, not lying down for 3 h after eating, making dietary changes, and losing weight. Antacids may be used to neutralize the acid coming up from the stomach. Other medications known as *proton pump inhibitors* (PPIs), such as Prilosec and Nexium, may be used to suppress the production of HCl in the stomach (gastric parietal cells), which raises the pH in the stomach to between 4 and 5, and gives the esophagus time to heal. Nexium may be given in oral doses

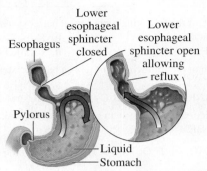

In acid reflux disease, the lower esophageal sphincter opens, allowing acidic fluid from the stomach to enter the esophagus.

of 40 mg once a day for 4 weeks. In severe GERD cases, an artificial valve may be created at the top of the stomach to strengthen the lower esophageal sphincter.

Applications

14.81 At rest, the $[H_3O^+]$ of the stomach fluid is 2.0×10^{-4} M. What is the pH of the stomach fluid?

14.82 When food enters the stomach, HCl is released and the $[H_3O^+]$ of the stomach fluid rises to 4×10^{-2} M. What is the pH of the stomach fluid while eating?

14.83 In Larry's esophageal pH test, a pH value of 3.60 was recorded in the esophagus. What is the $[H_3O^+]$ in his esophagus?

14.84 After Larry had taken Nexium for 4 weeks, the pH in his stomach was raised to 4.52. What is the $[H_3O^+]$ in his stomach?

14.85 Write the balanced chemical equation for the neutralization reaction of stomach acid HCl with $CaCO_3$, an ingredient in some antacids.

14.86 Write the balanced chemical equation for the neutralization reaction of stomach acid HCl with $Al(OH)_3$, an ingredient in some antacids.

14.87 How many grams of $CaCO_3$ are required to neutralize 100. mL of stomach acid HCl, which is equivalent to 0.0400 M HCl?

14.88 How many grams of $Al(OH)_3$ are required to neutralize 150. mL of stomach acid HCl with a pH of 1.5?

CONCEPT MAP

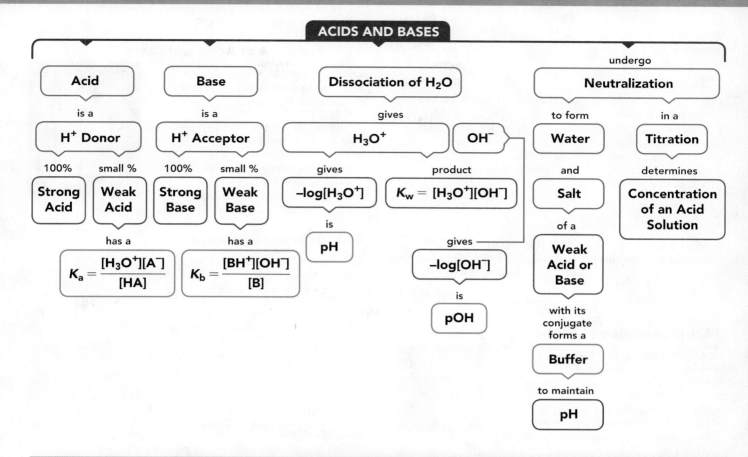

ACIDS AND BASES

- **Acid** — is a — **H⁺ Donor**
  - 100% → **Strong Acid**
  - small % → **Weak Acid** — has a — $K_a = \dfrac{[H_3O^+][A^-]}{[HA]}$

- **Base** — is a — **H⁺ Acceptor**
  - 100% → **Strong Base**
  - small % → **Weak Base** — has a — $K_b = \dfrac{[BH^+][OH^-]}{[B]}$

- **Dissociation of H₂O** — gives — **H₃O⁺** and **OH⁻**
  - gives → $-\log[H_3O^+]$ — is — **pH**
  - product → $K_w = [H_3O^+][OH^-]$
  - gives → $-\log[OH^-]$ — is — **pOH**

- undergo **Neutralization**
  - to form → **Water** and **Salt** — of a — **Weak Acid or Base** — with its conjugate forms a — **Buffer** — to maintain — **pH**
  - in a → **Titration** — determines — **Concentration of an Acid Solution**

CHAPTER REVIEW

14.1 Acids and Bases

LEARNING GOAL Describe and name acids and bases.

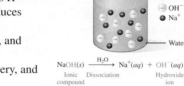

NaOH(s)

OH⁻
Na⁺

Water

$NaOH(s) \xrightarrow{H_2O} Na^+(aq) + OH^-(aq)$
Ionic compound · Dissociation · Hydroxide ion

- An Arrhenius acid produces H⁺ and an Arrhenius base produces OH⁻ in aqueous solutions.
- Acids taste sour, may sting, and neutralize bases.
- Bases taste bitter, feel slippery, and neutralize acids.
- Acids containing a simple anion use a *hydro* prefix, whereas acids with oxygen-containing polyatomic anions are named as *ic* or *ous* acids.

14.2 Brønsted–Lowry Acids and Bases

LEARNING GOAL Identify conjugate acid–base pairs for Brønsted–Lowry acids and bases.

$HCl + H_2O \longrightarrow H_3O^+ + Cl^-$
Hydrogen chloride · Water · Hydronium ion · Chloride ion

Acid (H⁺ donor) · Base (H⁺ acceptor) · Acidic solution

- According to the Brønsted–Lowry theory, acids are H⁺ donors and bases are H⁺ acceptors.
- A conjugate acid–base pair is related by the loss or gain of one H⁺.
- For example, when the acid HF donates H⁺, the F⁻ is its conjugate base. The other acid–base pair would be H_3O^+/H_2O.

$$HF(aq) + H_2O(l) \rightleftharpoons H_3O^+(aq) + F^-(aq)$$

14.3 Strengths of Acids and Bases

LEARNING GOAL Write equations for the dissociation of strong and weak acids; identify the direction of reaction.

- Strong acids dissociate completely in water, and the H⁺ is accepted by H_2O acting as a base.
- A weak acid dissociates slightly in water, producing only a small percentage of H_3O^+.
- Strong bases are hydroxides of Groups 1A (1) and 2A (2) that dissociate completely in water.
- An important weak base is ammonia, NH_3.

14.4 Dissociation Constants for Acids and Bases

LEARNING GOAL
Write the dissociation expression for a weak acid or weak base.

HCHO₂   CHO₂⁻

- In water, weak acids and weak bases produce only a few ions when equilibrium is reached.
- Weak acids have small K_a values whereas strong acids, which are essentially 100% dissociated, have very large K_a values.
- The reaction for a weak acid can be written as $HA + H_2O \rightleftharpoons H_3O^+ + A^-$. The acid dissociation expression is written as

$$K_a = \frac{[H_3O^+][A^-]}{[HA]}.$$

- For a weak base, $B + H_2O \rightleftharpoons BH^+ + OH^-$, the base dissociation expression is written as

$$K_b = \frac{[BH^+][OH^-]}{[B]}.$$

14.5 Dissociation of Water

LEARNING GOAL Use the water dissociation expression to calculate the $[H_3O^+]$ and $[OH^-]$ in an aqueous solution.

H_3O^+ OH^-

$[H_3O^+] = [OH^-]$
Neutral solution

- In pure water, a few water molecules transfer H^+ to other water molecules, producing small, but equal, amounts of $[H_3O^+]$ and $[OH^-]$.
- In pure water, the molar concentrations of H_3O^+ and OH^- are each 1.0×10^{-7} mol/L.
- The *water dissociation expression*, K_w, $[H_3O^+][OH^-] = 1.0 \times 10^{-14}$ at 25 °C.
- In acidic solutions, the $[H_3O^+]$ is greater than the $[OH^-]$.
- In basic solutions, the $[OH^-]$ is greater than the $[H_3O^+]$.

14.6 The pH Scale

LEARNING GOAL Calculate pH from $[H_3O^+]$; given the pH, calculate the $[H_3O^+]$ and $[OH^-]$ of a solution.

- The pH scale is a range of numbers typically from 0 to 14, which represents the $[H_3O^+]$ of the solution.
- A neutral solution has a pH of 7.0. In acidic solutions, the pH is below 7.0; in basic solutions, the pH is above 7.0.
- Mathematically, pH is the negative logarithm of the hydronium ion concentration,

$$pH = -\log[H_3O^+].$$

- The pOH is the negative log of the hydroxide ion concentration, $pOH = -\log[OH^-]$.
- The sum of the pH + pOH is 14.00.

14.7 Reactions of Acids and Bases

LEARNING GOAL Write balanced equations for reactions of acids with metals, carbonates, or bicarbonates, and bases.

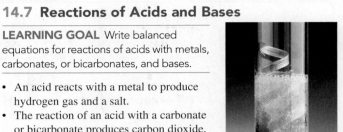

- An acid reacts with a metal to produce hydrogen gas and a salt.
- The reaction of an acid with a carbonate or bicarbonate produces carbon dioxide, water, and a salt.
- In neutralization, an acid reacts with a base to produce water and a salt.

14.8 Acid–Base Titration

LEARNING GOAL Calculate the molarity or volume of an acid or base solution from titration information.

- In a titration, an acid sample is neutralized with a known amount of a base.
- From the volume and molarity of the base, the concentration of the acid is calculated.

14.9 Buffers

LEARNING GOAL Describe the role of buffers in maintaining the pH of a solution; calculate the pH of a buffer.

- A buffer solution resists changes in pH when small amounts of an acid or a base are added.
- A buffer contains either a weak acid and its salt or a weak base and its salt.
- In a buffer, the weak acid reacts with added OH^-, and the anion of the salt reacts with added H_3O^+.
- Most buffer solutions consist of nearly equal concentrations of a weak acid and a salt containing its conjugate base.
- The pH of a buffer is calculated by solving the K_a expression for $[H_3O^+]$.

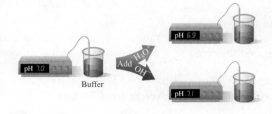

KEY TERMS

acid A substance that dissolves in water and produces hydrogen ions (H^+), according to the Arrhenius theory. All acids are hydrogen ion donors, according to the Brønsted–Lowry theory.

acid dissociation expression, K_a The product of the ions from the dissociation of a weak acid divided by the concentration of the weak acid.

amphoteric Substances that can act as either an acid or a base in water.

base A substance that dissolves in water and produces hydroxide ions (OH^-) according to the Arrhenius theory. All bases are hydrogen ion acceptors, according to the Brønsted–Lowry theory.

base dissociation expression, K_b The product of the ions from the dissociation of a weak base divided by the concentration of the weak base.

Brønsted–Lowry acids and bases An acid is a hydrogen ion donor; a base is a hydrogen ion acceptor.

buffer solution A solution of a weak acid and its conjugate base or a weak base and its conjugate acid that maintains the pH by neutralizing added acid or base.

conjugate acid–base pair An acid and a base that differ by one H^+. When an acid donates a hydrogen ion, the product is its conjugate base, which is capable of accepting a hydrogen ion in the reverse reaction.

dissociation The separation of an acid or a base into ions in water.

endpoint The point at which an indicator changes color. For the indicator phenolphthalein, the color change occurs when the number of moles of OH^- is equal to the number of moles of H_3O^+ in the sample.

hydronium ion, H_3O^+ The ion formed by the attraction of a hydrogen ion, H^+, to a water molecule.

indicator A substance added to a titration sample that changes color when the pH of the solution changes.

neutral The term that describes a solution with equal concentrations of $[H_3O^+]$ and $[OH^-]$.

neutralization A reaction between an acid and a base to form water and a salt.

pH A measure of the $[H_3O^+]$ in a solution; $pH = -\log[H_3O^+]$.

pOH A measure of the $[OH^-]$ in a solution; $pOH = -\log[OH^-]$.

salt An ionic compound that contains a metal ion or NH_4^+ and a nonmetal or polyatomic ion other than OH^-.

strong acid An acid that completely dissociates in water.

strong base A base that completely dissociates in water.

titration The addition of base to an acid sample to determine the concentration of the acid.

water dissociation expression, K_w The product of $[H_3O^+]$ and $[OH^-]$ in solution; $K_w = [H_3O^+][OH^-]$.

weak acid An acid that is a poor donor of H^+ and dissociates only slightly in water.

weak base A base that is a poor acceptor of H^+ and produces only a small number of ions in water.

 ## KEY MATH SKILLS

The chapter section containing each Key Math Skill is shown in parentheses at the end of each heading.

Calculating pH from $[H_3O^+]$ (14.6)

- The pH of a solution is calculated from the negative log of the $[H_3O^+]$.

$$pH = -\log[H_3O^+]$$

Example: What is the pH of a solution that has $[H_3O^+] = 2.4 \times 10^{-11}$ M?

Answer: We substitute the given $[H_3O^+]$ into the pH equation and calculate the pH.

$$pH = -\log[H_3O^+]$$
$$= -\log[2.4 \times 10^{-11} \text{ M}]$$
$$= 10.62 \quad \text{Two decimal places equal the two SFs in the } [H_3O^+] \text{ coefficient.}$$

Calculating $[H_3O^+]$ from pH (14.6)

- The calculation of $[H_3O^+]$ from the pH is done by reversing the pH calculation using the negative pH.

$$[H_3O^+] = 10^{-pH}$$

Example: What is the $[H_3O^+]$ of a solution with a pH of 4.80?

Answer: $[H_3O^+] = 10^{-pH}$
$$= 10^{-4.80}$$
$$= 1.6 \times 10^{-5} \text{ M}$$
   Two SFs in the $[H_3O^+]$ equal the two decimal places in the pH.

 ## CORE CHEMISTRY SKILLS

The chapter section containing each Core Chemistry Skill is shown in parentheses at the end of each heading.

Identifying Conjugate Acid–Base Pairs (14.2)

- According to the Brønsted–Lowry theory, a conjugate acid–base pair consists of molecules or ions related by the loss of one H^+ by an acid, and the gain of one H^+ by a base.
- Every acid–base reaction contains two conjugate acid–base pairs because an H^+ is transferred in both the forward and reverse directions.
- When an acid such as HF loses one H^+, the conjugate base F^- is formed. When H_2O acts as a base, it gains one H^+, which forms its conjugate acid, H_3O^+.

Example: Identify the conjugate acid–base pairs in the following reaction:

$$H_2SO_4(aq) + H_2O(l) \longrightarrow HSO_4^-(aq) + H_3O^+(aq)$$

Answer: $H_2SO_4(aq) + H_2O(l) \longrightarrow HSO_4^-(aq) + H_3O^+(aq)$
   Acid   Base   Conjugate base   Conjugate acid

Conjugate acid–base pairs: H_2SO_4/HSO_4^- and H_3O^+/H_2O

Calculating $[H_3O^+]$ and $[OH^-]$ in Solutions (14.5)

- For all aqueous solutions, the product of $[H_3O^+]$ and $[OH^-]$ is equal to the water dissociation expression, K_w.

$$K_w = [H_3O^+][OH^-]$$

- Because pure water contains equal numbers of OH^- ions and H_3O^+ ions each with molar concentrations of 1.0×10^{-7} M, the numerical value of K_w is 1.0×10^{-14} at 25 °C.

$$K_w = [H_3O^+][OH^-] = [1.0 \times 10^{-7}][1.0 \times 10^{-7}]$$
$$= 1.0 \times 10^{-14}$$

- If we know the $[H_3O^+]$ of a solution, we can use the K_w expression to calculate the $[OH^-]$. If we know the $[OH^-]$ of a solution, we can calculate the $[H_3O^+]$ using the K_w expression.

$$[OH^-] = \frac{K_w}{[H_3O^+]} \quad [H_3O^+] = \frac{K_w}{[OH^-]}$$

Example: What is the $[OH^-]$ in a solution that has $[H_3O^+] = 2.4 \times 10^{-11}$ M? Is the solution acidic or basic?

Answer: We solve the K_w expression for $[OH^-]$ and substitute in the known values of K_w and $[H_3O^+]$.

$$[OH^-] = \frac{K_w}{[H_3O^+]} = \frac{1.0 \times 10^{-14}}{[2.4 \times 10^{-11}]} = 4.2 \times 10^{-4}\,M$$

Because the $[OH^-]$ is greater than the $[H_3O^+]$, this is a basic solution.

Writing Equations for Reactions of Acids and Bases (14.7)

- Acids react with certain metals to produce hydrogen gas (H_2) and a salt.

$$Mg(s) + 2HCl(aq) \longrightarrow H_2(g) + MgCl_2(aq)$$
  Metal   Acid   Hydrogen   Salt

- When an acid is added to a carbonate or bicarbonate, the products are carbon dioxide gas, water, and a salt.

$$2HCl(aq) + Na_2CO_3(aq) \longrightarrow CO_2(g) + H_2O(l) + 2NaCl(aq)$$
  Acid   Carbonate   Carbon   Water   Salt
                             dioxide

- Neutralization is a reaction between a strong or weak acid and a strong base to produce water and a salt.

$$HCl(aq) + NaOH(aq) \longrightarrow H_2O(l) + NaCl(aq)$$
  Acid   Base   Water   Salt

Example: Write the balanced chemical equation for the reaction of $ZnCO_3(s)$ and hydrobromic acid $HBr(aq)$.

Answer:

$$ZnCO_3(s) + 2HBr(aq) \longrightarrow CO_2(g) + H_2O(l) + ZnBr_2(aq)$$

Calculating Molarity or Volume of an Acid or Base in a Titration (14.8)

- In a titration, a measured volume of acid is neutralized by a NaOH solution of known molarity.

- From the measured volume of the NaOH solution required for titration and its molarity, the number of moles of NaOH, the moles of acid, and the concentration of the acid are calculated.

Example: A 15.0-mL sample of a H_2SO_4 solution is titrated with 24.0 mL of a 0.245 M NaOH solution. What is the molarity of the H_2SO_4 solution?

$$H_2SO_4(aq) + 2NaOH(aq) \longrightarrow 2H_2O(l) + Na_2SO_4(aq)$$

Answer:

$$24.0\ \cancel{mL\ NaOH\ solution} \times \frac{1\ \cancel{L\ NaOH\ solution}}{1000\ \cancel{mL\ NaOH\ solution}}$$

$$\times \frac{0.245\ \cancel{mol\ NaOH}}{1\ \cancel{L\ NaOH\ solution}} \times \frac{1\ mol\ H_2SO_4}{2\ \cancel{mol\ NaOH}} = 0.002\ 94\ mol\ of\ H_2SO_4$$

$$Molarity\ (M) = \frac{0.002\ 94\ mol\ H_2SO_4}{0.0150\ L\ H_2SO_4\ solution} = 0.196\ M\ H_2SO_4\ solution$$

Calculating the pH of a Buffer (14.9)

- A buffer solution maintains pH by neutralizing small amounts of added acid or base.
- Most buffer solutions consist of nearly equal concentrations of a weak acid and a salt containing its conjugate base such as acetic acid, $HC_2H_3O_2$, and its salt $NaC_2H_3O_2$.
- The $[H_3O^+]$ is calculated by solving the K_a expression for $[H_3O^+]$, then substituting the values of $[H_3O^+]$, $[HA]$, and K_a into the equation.

$$K_a = \frac{[H_3O^+][C_2H_3O_2^-]}{[HC_2H_3O_2]}$$

Solving for $[H_3O^+]$ gives:

$$[H_3O^+] = K_a \times \frac{[HC_2H_3O_2]}{[C_2H_3O_2^-]} \quad \longleftarrow\ \text{Weak acid}$$
$$\qquad\qquad\qquad\qquad\qquad \longleftarrow\ \text{Conjugate base}$$

- The pH of the buffer is calculated from the $[H_3O^+]$.

$$pH = -\log[H_3O^+]$$

Example: What is the pH of a buffer prepared with 0.40 M $HC_2H_3O_2$ and 0.20 M $C_2H_3O_2^-$ if the K_a of acetic acid is 1.8×10^{-5}?

Answer: $[H_3O^+] = K_a \times \dfrac{[HC_2H_3O_2]}{[C_2H_3O_2^-]} = 1.8 \times 10^{-5} \times \dfrac{[0.40]}{[0.20]}$

$$= 3.6 \times 10^{-5}\,M$$

$$pH = -\log[3.6 \times 10^{-5}] = 4.44$$

UNDERSTANDING THE CONCEPTS

The chapter sections to review are shown in parentheses at the end of each question.

14.89 Determine if each of the following diagrams represents a strong acid or a weak acid. The acid has the formula HX. (14.3)

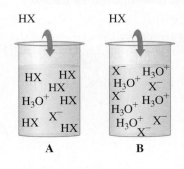

14.90 Adding a few drops of a strong acid to water will lower the pH appreciably. However, adding the same number of drops to a buffer does not appreciably alter the pH. Why? (14.9)

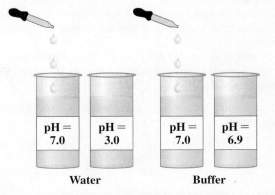

14.91 Identify each of the following as an acid or a base: (14.1)
    a. H_2SO_4 **b.** RbOH
    c. $Ca(OH)_2$ **d.** HI

14.92 Identify each of the following as an acid or a base: (14.1)
    a. $Sr(OH)_2$ **b.** H_2SO_3
    c. $HC_2H_3O_2$ **d.** CsOH

14.93 Complete the following table: (14.2)

Acid	Conjugate Base
H_2O	
	CN^-
HNO_2	
	$H_2PO_4^-$

14.94 Complete the following table: (14.2)

Base	Conjugate Acid
	HS^-
	H_3O^+
NH_3	
HCO_3^-	

Applications

14.95 Sometimes, during stress or trauma, a person can start to hyperventilate. Then the person might breathe into a paper bag to avoid fainting. (14.9)
    a. What changes occur in the blood pH during hyperventilation?

b. How does breathing into a paper bag help return blood pH to normal?

Breathing into a paper bag can help a person who is hyperventilating.

14.96 In the blood plasma, pH is maintained by the carbonic acid–bicarbonate buffer system. (14.9)
    a. How is pH maintained when acid is added to the buffer system?
    b. How is pH maintained when base is added to the buffer system?

14.97 State whether each of the following solutions is acidic, basic, or neutral: (14.6)
    a. sweat, pH 5.2   **b.** tears, pH 7.5
    c. bile, pH 8.1   **d.** stomach acid, pH 2.5

14.98 State whether each of the following solutions is acidic, basic, or neutral: (14.6)
    a. saliva, pH 6.8   **b.** urine, pH 5.9
    c. pancreatic juice, pH 8.0   **d.** blood, pH 7.45

ADDITIONAL QUESTIONS AND PROBLEMS

14.99 Identify each of the following as an acid, base, or salt, and give its name: (14.1)
    a. $HBrO_2$ **b.** CsOH
    c. $Mg(NO_3)_2$ **d.** $HClO_4$

14.100 Identify each of the following as an acid, base, or salt, and give its name: (14.1)
    a. HNO_2 **b.** $MgBr_2$
    c. NH_3 **d.** Li_2SO_3

14.101 Complete the following table: (14.2)

Acid	Conjugate Base
HI	
	Cl^-
NH_4^+	
	HS^-

14.102 Complete the following table: (14.2)

Base	Conjugate Acid
F^-	
	$HC_2H_3O_2$
	HSO_3^-
ClO^-	

14.103 Using Table 14.3, identify the stronger acid in each of the following pairs: (14.3)
    a. HF or HCN   **b.** H_3O^+ or H_2S
    c. HNO_2 or $HC_2H_3O_2$ **d.** H_2O or HCO_3^-

14.104 Using Table 14.3, identify the stronger base in each of the following pairs: (14.3)
    a. H_2O or Cl^- **b.** OH^- or NH_3
    c. SO_4^{2-} or NO_2^- **d.** CO_3^{2-} or H_2O

14.105 Determine the pH and pOH for each of the following solutions: (14.6)
    a. $[H_3O^+] = 2.0 \times 10^{-8}$ M
    b. $[H_3O^+] = 5.0 \times 10^{-2}$ M
    c. $[OH^-] = 3.5 \times 10^{-4}$ M
    d. $[OH^-] = 0.0054$ M

14.106 Determine the pH and pOH for each of the following solutions: (14.6)
    a. $[OH^-] = 1.0 \times 10^{-7}$ M
    b. $[H_3O^+] = 4.2 \times 10^{-3}$ M
    c. $[H_3O^+] = 0.0001$ M
    d. $[OH^-] = 8.5 \times 10^{-9}$ M

14.107 Are the solutions in problem 14.105 acidic, basic, or neutral? (14.6)

14.108 Are the solutions in problem 14.106 acidic, basic, or neutral? (14.6)

14.109 Calculate the $[H_3O^+]$ and $[OH^-]$ for a solution with each of the following pH values: (14.6)
    a. 3.00   **b.** 6.2
    c. 8.85   **d.** 11.00

14.110 Calculate the $[H_3O^+]$ and $[OH^-]$ for a solution with each of the following pH values: (14.6)
    a. 10.00   **b.** 5.0
    c. 6.54   **d.** 1.82

14.111 Solution A has a pH of 4.5, and solution B has a pH of 6.7. (14.6)
    a. Which solution is more acidic?
    b. What is the $[H_3O^+]$ in each?
    c. What is the $[OH^-]$ in each?

14.112 Solution X has a pH of 9.5, and solution Y has a pH of 7.5. (14.6)
    a. Which solution is more acidic?
    b. What is the $[H_3O^+]$ in each?
    c. What is the $[OH^-]$ in each?

14.113 What is the pH and pOH of a solution prepared by dissolving 2.5 g of HCl in water to make 425 mL of solution? (14.6)

14.114 What is the pH and pOH of a solution prepared by dissolving 1.0 g of $Ca(OH)_2$ in water to make 875 mL of solution? (14.6)

CHALLENGE QUESTIONS

The following groups of questions are related to the topics in this chapter. However, they do not all follow the chapter order, and they require you to combine concepts and skills from several sections. These questions will help you increase your critical thinking skills and prepare for your next exam.

14.115 For each of the following: (14.2, 14.3)
    1. H_2S **2.** H_3PO_4
    a. Write the formula for the conjugate base.
    b. Write the K_a expression.
    c. Which is the weaker acid?

14.116 For each of the following: (14.2, 14.3)
    1. HCO_3^- **2.** $HC_2H_3O_2$
    a. Write the formula for the conjugate base.
    b. Write the K_a expression.
    c. Which is the stronger acid?

14.117 Using Table 14.3, identify the conjugate acid–base pairs in each of the following equations and whether the equilibrium mixture contains mostly products or mostly reactants: (14.2, 14.3)
    a. $NH_3(aq) + HNO_3(aq) \rightleftharpoons NH_4^+(aq) + NO_3^-(aq)$
    b. $H_2O(l) + HBr(aq) \rightleftharpoons H_3O^+(aq) + Br^-(aq)$

14.118 Using Table 14.3, identify the conjugate acid–base pairs in each of the following equations and whether the equilibrium mixture contains mostly products or mostly reactants: (14.2, 14.3)
    a. $HNO_2(aq) + HS^-(aq) \rightleftharpoons H_2S(g) + NO_3^-(aq)$
    b. $Cl^-(aq) + H_2O(l) \rightleftharpoons OH^-(aq) + HCl(aq)$

14.119 Complete and balance each of the following: (14.7)
    a. $ZnCO_3(s) + H_2SO_4(aq) \longrightarrow$
    b. $Al(s) + HNO_3(aq) \longrightarrow$

14.120 Complete and balance each of the following: (14.7)
    a. $H_3PO_4(aq) + Ca(OH)_2(s) \longrightarrow$
    b. $KHCO_3(s) + HNO_3(aq) \longrightarrow$

14.121 Determine each of the following for a 0.050 M KOH solution: (14.6, 14.7, 14.8)
    a. $[H_3O^+]$
    b. pH
    c. pOH
    d. the balanced equation for the reaction with H_2SO_4
    e. milliliters of KOH solution required to neutralize 40.0 mL of a 0.035 M H_2SO_4 solution

14.122 Determine each of the following for a 0.10 M HBr solution: (14.6, 14.7, 14.8)
    a. $[H_3O^+]$
    b. pH
    c. pOH
    d. the balanced equation for the reaction with LiOH
    e. milliliters of HBr solution required to neutralize 36.0 mL of a 0.25 M LiOH solution

14.123 One of the most acidic lakes in the United States is Little Echo Pond in the Adirondacks in New York. Recently, this lake had a pH of 4.2, well below the recommended pH of 6.5. (14.6, 14.8)
    a. What are the $[H_3O^+]$ and $[OH^-]$ of Little Echo Pond?
    b. What are the $[H_3O^+]$ and $[OH^-]$ of a lake that has a pH of 6.5?
    c. One way to raise the pH of an acidic lake (and restore aquatic life) is to add limestone ($CaCO_3$). How many grams of $CaCO_3$ are needed to neutralize 1.0 kL of the acidic water from the lake if the acid is sulfuric acid?

$$H_2SO_4(aq) + CaCO_3(s) \longrightarrow CO_2(g) + H_2O(l) + CaSO_4(aq)$$

A helicopter drops calcium carbonate on an acidic lake to increase its pH.

Applications

14.124 The daily output of stomach acid (gastric juice) is 1000 mL to 2000 mL. Prior to a meal, stomach acid (HCl) typically has a pH of 1.42. (14.6, 14.7, 14.8)
    a. What is the $[H_3O^+]$ of stomach acid?
    b. One chewable tablet of the antacid Maalox contains 600. mg of $CaCO_3$. Write the neutralization equation, and calculate the milliliters of stomach acid neutralized by two tablets of Maalox.
    c. The antacid milk of magnesia contains 400. mg of $Mg(OH)_2$ per teaspoon. Write the neutralization equation, and calculate the number of milliliters of stomach acid that are neutralized by 1 tablespoon of milk of magnesia.

14.125 Calculate the volume, in milliliters, of a 0.150 M NaOH solution that will completely neutralize each of the following: (14.8)
    a. 25.0 mL of a 0.288 M HCl solution
    b. 10.0 mL of a 0.560 M H_2SO_4 solution

14.126 Calculate the volume, in milliliters, of a 0.215 M NaOH solution that will completely neutralize each of the following: (14.8)

  a. 3.80 mL of a 1.25 M HNO_3 solution
  b. 8.50 mL of a 0.825 M H_3PO_4 solution

14.127 A solution of 0.205 M NaOH is used to 20.0 mL of a H_2SO_4 solution. If 45.6 mL of the NaOH solution is required to reach the endpoint, what is the molarity of the H_2SO_4 solution? (14.8)

$$H_2SO_4(aq) + 2NaOH(aq) \longrightarrow 2H_2O(l) + Na_2SO_4(aq)$$

14.128 A 10.0-mL sample of vinegar, which is an aqueous solution of acetic acid, $HC_2H_3O_2$, requires 16.5 mL of a 0.500 M NaOH solution to reach the endpoint in a titration. What is the molarity of the acetic acid solution? (14.8)

$$HC_2H_3O_2(aq) + NaOH(aq) \longrightarrow H_2O(l) + NaC_2H_3O_2(aq)$$

14.129 A buffer solution is made by dissolving H_3PO_4 and NaH_2PO_4 in water. (14.9)

  a. Write an equation that shows how this buffer neutralizes added acid.
  b. Write an equation that shows how this buffer neutralizes added base.
  c. Calculate the pH of this buffer if it contains 0.50 M H_3PO_4 and 0.20 M $H_2PO_4^-$. The K_a for H_3PO_4 is 7.5×10^{-3}.

14.130 A buffer solution is made by dissolving $HC_2H_3O_2$ and $NaC_2H_3O_2$ in water. (14.9)

  a. Write an equation that shows how this buffer neutralizes added acid.
  b. Write an equation that shows how this buffer neutralizes added base.
  c. Calculate the pH of this buffer if it contains 0.20 M $HC_2H_3O_2$ and 0.40 M $C_2H_3O_2^-$. The K_a for $HC_2H_3O_2$ is 1.8×10^{-5}.

ANSWERS

Answers to Selected Questions and Problems

14.1 **a.** acid **b.** acid **c.** acid **d.** base **e.** both

14.3 **a.** hydrochloric acid **b.** calcium hydroxide **c.** perchloric acid **d.** nitric acid **e.** sulfurous acid **f.** bromous acid

14.5 **a.** RbOH **b.** HF **c.** H_3PO_4 **d.** LiOH **e.** NH_4OH **f.** HIO_4

14.7 **a.** HI is the acid (hydrogen ion donor), and H_2O is the base (hydrogen ion acceptor).
  b. H_2O is the acid (hydrogen ion donor), and F^- is the base (hydrogen ion acceptor).
  c. H_2S is the acid (hydrogen ion donor), and CH_3—CH_2—NH_2 is the base (hydrogen ion acceptor).

14.9 **a.** F^- **b.** OH^- **c.** HPO_3^{2-} **d.** SO_4^{2-} **e.** ClO_2^-

14.11 **a.** HCO_3^- **b.** H_3O^+ **c.** H_3PO_4 **d.** HBr **e.** $HClO_4$

14.13 **a.** The conjugate acid–base pairs are H_2CO_3/HCO_3^- and H_3O^+/H_2O.
  b. The conjugate acid–base pairs are NH_4^+/NH_3 and H_3O^+/H_2O.
  c. The conjugate acid–base pairs are HCN/CN^- and HNO_2/NO_2^-.
  d. The conjugate acid–base pairs are HF/F^- and $HCHO_2/CHO_2^-$.

14.15 $NH_4^+(aq) + H_2O(l) \rightleftharpoons NH_3(aq) + H_3O^+(aq)$

14.17 A strong acid is a good hydrogen ion donor, whereas its conjugate base is a poor hydrogen ion acceptor.

14.19 **a.** HBr **b.** HSO_4^- **c.** H_2CO_3

14.21 **a.** HSO_4^- **b.** HF **c.** HCO_3^-

14.23 **a.** reactants **b.** reactants **c.** products

14.25 $NH_4^+(aq) + SO_4^{2-}(aq) \rightleftharpoons NH_3(aq) + HSO_4^-(aq)$

The equilibrium mixture contains mostly reactants because NH_4^+ is a weaker acid than HSO_4^-, and SO_4^{2-} is a weaker base than NH_3.

14.27 **a.** true **b.** false **c.** false **d.** true **e.** false

14.29 **a.** H_2SO_3 **b.** HSO_3^- **c.** H_2SO_3 **d.** HS^- **d.** H_2SO_3

14.31 $H_3PO_4(aq) + H_2O(l) \rightleftharpoons H_3O^+(aq) + H_2PO_4^-(aq)$

$$K_a = \frac{[H_3O^+][H_2PO_4^-]}{[H_3PO_4]}$$

14.33 In pure water, $[H_3O^+] = [OH^-]$ because one of each is produced every time a hydrogen ion is transferred from one water molecule to another.

14.35 In an acidic solution, the $[H_3O^+]$ is greater than the $[OH^-]$.

14.37 **a.** acidic **b.** basic **c.** basic **d.** acidic

14.39 **a.** 1.0×10^{-5} M **b.** 1.0×10^{-8} M **c.** 5.0×10^{-10} M **d.** 2.5×10^{-2} M

14.41 **a.** 2.5×10^{-13} M **b.** 2.0×10^{-9} M **c.** 5.0×10^{-11} M **d.** 1.3×10^{-6} M

14.43 In a neutral solution, the $[H_3O^+]$ is 1.0×10^{-7} M and the pH is 7.00, which is the negative value of the power of 10.

14.45 **a.** basic **b.** acidic **c.** basic **d.** acidic **e.** acidic **f.** basic

14.47 An increase or decrease of one pH unit changes the $[H_3O^+]$ by a factor of 10. Thus a pH of 3 is 10 times more acidic than a pH of 4.

14.49 **a.** 4.0 **b.** 8.5 **c.** 9.0 **d.** 3.40 **e.** 7.17 **f.** 10.92

14.51

[H_3O^+]	[OH^-]	pH	pOH	Acidic, Basic, or Neutral?
1.0×10^{-8} M	1.0×10^{-6} M	8.00	6.00	Basic
3.2×10^{-4} M	3.1×10^{-11} M	3.49	10.51	Acidic
2.8×10^{-5} M	3.6×10^{-10} M	4.55	9.45	Acidic
1.0×10^{-12} M	1.0×10^{-2} M	12.00	2.00	Basic

14.53 1.2×10^{-7} M

14.55 a. $ZnCO_3(s) + 2HBr(aq) \longrightarrow$
$$CO_2(g) + H_2O(l) + ZnBr_2(aq)$$
b. $Zn(s) + 2HCl(aq) \longrightarrow H_2(g) + ZnCl_2(aq)$
c. $HCl(aq) + NaHCO_3(s) \longrightarrow$
$$CO_2(g) + H_2O(l) + NaCl(aq)$$
d. $H_2SO_4(aq) + Mg(OH)_2(s) \longrightarrow 2H_2O(l) + MgSO_4(aq)$

14.57 a. $2HCl(aq) + Mg(OH)_2(s) \longrightarrow 2H_2O(l) + MgCl_2(aq)$
b. $H_3PO_4(aq) + 3LiOH(aq) \longrightarrow 3H_2O(l) + Li_3PO_4(aq)$

14.59 a. $H_2SO_4(aq) + 2NaOH(aq) \longrightarrow 2H_2O(l) + Na_2SO_4(aq)$
b. $3HCl(aq) + Fe(OH)_3(s) \longrightarrow 3H_2O(l) + FeCl_3(aq)$
c. $H_2CO_3(aq) + Mg(OH)_2(s) \longrightarrow 2H_2O(l) + MgCO_3(s)$

14.61 To a known volume of a formic acid solution, add a few drops of indicator. Place a solution of NaOH of known molarity in a buret. Add base to acid until one drop changes the color of the solution. Use the volume and molarity of NaOH and the volume of a formic acid solution to calculate the concentration of the formic acid in the sample.

14.63 0.829 M HCl solution

14.65 0.124 M H_2SO_4 solution

14.67 16.5 mL

14.69 **b** and **c** are buffer systems. **b** contains the weak acid H_2CO_3 and its salt $NaHCO_3$. **c** contains HF, a weak acid, and its salt KF.

14.71 a. 3   **b.** 1 and 2
c. 3   **d.** 1

14.73 pH = 3.35

14.75 The pH of the 0.10 M HF/0.10 M NaF buffer is 3.46.
The pH of the 0.060 M HF/0.120 M NaF buffer is 3.76.

14.77 If you breathe fast, CO_2 is expelled and the equilibrium shifts to lower H_3O^+, which raises the pH.

14.79 If large amounts of HCO_3^- are lost, equilibrium shifts to higher H_3O^+, which lowers the pH.

14.81 pH = 3.70

14.83 2.5×10^{-4} M

14.85 $CaCO_3(s) + 2HCl(aq) \longrightarrow CO_2(g) + H_2O(l) + CaCl_2(aq)$

14.87 0.200 g of $CaCO_3$

14.89 a. This diagram represents a weak acid; only a few HX molecules separate into H_3O^+ and X^- ions.
b. This diagram represents a strong acid; all the HX molecules separate into H_3O^+ and X^- ions.

14.91 a. acid   **b.** base
c. base   **d.** acid

14.93

Acid	Conjugate Base
H_2O	OH^-
HCN	CN^-
HNO_2	NO_2^-
H_3PO_4	$H_2PO_4^-$

14.95 a. During hyperventilation, a person will lose CO_2 and the blood pH will rise.
b. Breathing into a paper bag will increase the CO_2 concentration and lower the blood pH.

14.97 a. acidic   **b.** basic
c. basic   **d.** acidic

14.99 a. acid, bromous acid
b. base, cesium hydroxide
c. salt, magnesium nitrate
d. acid, perchloric acid

14.101

Acid	Conjugate Base
HI	I^-
HCl	Cl^-
NH_4^+	NH_3
H_2S	HS^-

14.103 a. HF   **b.** H_3O^+
c. HNO_2 **d.** HCO_3^-

14.105 a. pH = 7.70; pOH = 6.30
b. pH = 1.30; pOH = 12.70
c. pH = 10.54; pOH = 3.46
d. pH = 11.73; pOH = 2.27

14.107 a. basic   **b.** acidic
c. basic   **d.** basic

14.109 a. [H_3O^+] = 1.0×10^{-3} M; [OH^-] = 1.0×10^{-11} M
b. [H_3O^+] = 6×10^{-7} M; [OH^-] = 2×10^{-8} M
c. [H_3O^+] = 1.4×10^{-9} M; [OH^-] = 7.1×10^{-6} M
d. [H_3O^+] = 1.0×10^{-11} M; [OH^-] = 1.0×10^{-3} M

14.111 a. Solution A
b. Solution A [H_3O^+] = 3×10^{-5} M;
Solution B [H_3O^+] = 2×10^{-7} M
c. Solution A [OH^-] = 3×10^{-10} M;
Solution B [OH^-] = 5×10^{-8} M

14.113 pH = 0.80; pOH = 13.20

14.115 a. 1. HS^-
2. $H_2PO_4^-$
b. 1. $\dfrac{[H_3O^+][HS^-]}{[H_2S]}$
2. $\dfrac{[H_3O^+][H_2PO_4^-]}{[H_3PO_4]}$
c. H_2S

14.117 a. HNO_3/NO_3^- and NH_4^+/NH_3; equilibrium mixture contains mostly products
b. HBr/Br^- and H_3O^+/H_2O; equilibrium mixture contains mostly products

14.119 a. $ZnCO_3(s) + H_2SO_4(aq) \longrightarrow$
$$CO_2(g) + H_2O(l) + ZnSO_4(aq)$$
b. $2Al(s) + 6HNO_3(aq) \longrightarrow 3H_2(g) + 2Al(NO_3)_3(aq)$

14.121 a. $[H_3O^+] = 2.0 \times 10^{-13}$ M
b. pH = 12.70
c. pOH = 1.30
d. $2KOH(aq) + H_2SO_4(aq) \longrightarrow 2H_2O(l) + K_2SO_4(aq)$
e. 56 mL of the KOH solution

14.123 a. $[H_3O^+] = 6 \times 10^{-5}$ M; $[OH^-] = 2 \times 10^{-10}$ M
b. $[H_3O^+] = 3 \times 10^{-7}$ M; $[OH^-] = 3 \times 10^{-8}$ M
c. 3 g of $CaCO_3$

14.125 a. 48.0 mL of NaOH solution
b. 74.7 mL of NaOH solution

14.127 0.234 M H_2SO_4 solution

14.129 a. acid:
$$H_2PO_4^-(aq) + H_3O^+(aq) \longrightarrow H_3PO_4(aq) + H_2O(l)$$
b. base:
$$H_3PO_4(aq) + OH^-(aq) \longrightarrow H_2PO_4^-(aq) + H_2O(l)$$
c. pH = 1.72

CI.21 Methane is a major component of purified natural gas used for heating and cooking. When 1.0 mol of methane gas burns with oxygen to produce carbon dioxide and water vapor, 883 kJ of heat is produced. At STP, methane gas has a density of 0.715 g/L. For transport, the natural gas is cooled to $-163\,°C$ to form liquefied natural gas (LNG) with a density of 0.45 g/mL. A tank on a ship can hold 7.0 million gallons of LNG. (2.7, 7.2, 7.3, 8.2, 8.3, 9.5, 10.1, 11.6)

An LNG carrier transports liquefied natural gas.

a. Draw the Lewis structure for methane, which has the formula CH_4.

b. What is the mass, in kilograms, of LNG (assume that LNG is all methane) transported in one tank on a ship?

c. What is the volume, in liters, of LNG (methane) from one tank when the LNG (methane) from one tank is converted to methane gas at STP?

d. Write the balanced chemical equation for the combustion of methane and oxygen in a gas burner, including the heat of reaction.

Methane is the fuel burned in a gas cooktop.

e. How many kilograms of oxygen are needed to react with all of the methane in one tank of LNG?

f. How much heat, in kilojoules, is released after burning all of the methane from one tank of LNG?

CI.22 Automobile exhaust is a major cause of air pollution. One pollutant is nitrogen oxide, which forms from nitrogen and oxygen gases in the air at the high temperatures in an automobile engine. Once emitted into the air, nitrogen oxide reacts with oxygen to produce nitrogen dioxide, a reddish brown gas with a sharp, pungent odor that makes up

Two gases found in automobile exhaust are carbon dioxide and nitrogen oxide.

smog. One component of gasoline is octane, C_8H_{18}, which has a density of 0.803 g/mL. In one year, a typical automobile uses 550 gal of gasoline and produces 41 lb of nitrogen oxide. (2.7, 7.2, 7.3, 8.2, 8.3, 9.2, 11.6)

a. Write balanced chemical equations for the production of nitrogen oxide and nitrogen dioxide.

b. If all the nitrogen oxide emitted by one automobile is converted to nitrogen dioxide in the atmosphere, how many kilograms of nitrogen dioxide are produced in one year by a single automobile?

c. Write a balanced chemical equation for the combustion of octane.

d. How many moles of C_8H_{18} are present in 15.2 gal of octane?

e. How many liters of CO_2 at STP are produced in one year from the gasoline used by the typical automobile?

CI.23 A mixture of 25.0 g of CS_2 gas and 30.0 g of O_2 gas is placed in 10.0-L container and heated to 125 °C. The products of the reaction are carbon dioxide gas and sulfur dioxide gas. (7.2, 7.3, 8.2, 9.2, 9.3, 11.1, 11.7)

a. Write a balanced chemical equation for the reaction.

b. How many grams of CO_2 are produced?

c. What is the partial pressure, in millimeters of mercury, of the remaining reactant?

d. What is the final pressure, in millimeters of mercury, in the container?

CI.24 In wine-making, glucose ($C_6H_{12}O_6$) from grapes undergoes fermentation in the absence of oxygen to produce ethanol and carbon dioxide. A bottle of vintage port wine has a volume of 750 mL and contains 135 mL of ethanol (C_2H_6O). Ethanol has a density of 0.789 g/mL. In 1.5 lb of grapes, there are 26 g of glucose. (2.7, 7.2, 7.3, 8.2, 9.2, 12.4)

Port is a type of fortified wine that is produced in Portugal.

When the glucose in grapes is fermented, ethanol is produced.

a. Calculate the volume percent (v/v) of ethanol in the port wine.

b. What is the molarity (M) of ethanol in the port wine?

c. Write the balanced chemical equation for the fermentation reaction of sugar in grapes.

d. How many grams of sugar from grapes are required to produce one bottle of port wine?

e. How many bottles of port wine can be produced from 1.0 ton of grapes (1 ton = 2000 lb)?

CI.25 Consider the following reaction at equilibrium:

$$2H_2(g) + S_2(g) \rightleftharpoons 2H_2S(g) + heat$$

In a 10.0-L container, an equilibrium mixture contains 2.02 g of H_2, 10.3 g of S_2, and 68.2 g of H_2S. (7.2, 7.3, 13.2, 13.3, 13.4, 13.5)

a. What is the numerical value of K_c for this equilibrium mixture?

b. If more H_2 is added to the equilibrium mixture, how will the equilibrium shift?

c. How will the equilibrium shift if the mixture is placed in a 5.00-L container with no change in temperature?

d. If a 5.00-L container has an equilibrium mixture of 0.300 mol of H_2 and 2.50 mol of H_2S, what is the $[S_2]$ if temperature remains constant?

CI.26 A saturated solution of silver hydroxide has a pH of 10.15. (7.2, 7.3, 13.2, 13.6)

a. Write the solubility product expression for silver hydroxide.

b. Calculate the numerical value of K_{sp} for silver hydroxide.

c. How many grams of silver hydroxide will dissolve in 2.0 L of water?

CI.27 A metal M with a mass of 0.420 g completely reacts with 34.8 mL of a 0.520 M HCl solution to form H_2 gas and aqueous MCl_3. (7.2, 7.3, 8.2, 9.2, 11.7)

a. Write a balanced chemical equation for the reaction of the metal M(s) and HCl(aq).

b. What volume, in milliliters, of H_2 at 720. mmHg and 24 °C is produced?

c. How many moles of metal M reacted?

d. Using your results from part **c**, determine the molar mass and name of metal M.

e. Write the balanced chemical equation for the reaction.

When a metal reacts with a strong acid, bubbles of hydrogen gas form.

CI.28 In a teaspoon (5.0 mL) of a liquid antacid, there are 400. mg of $Mg(OH)_2$ and 400. mg of $Al(OH)_3$. A 0.080 M HCl solution, which is similar to stomach acid, is used to neutralize 5.0 mL of the liquid antacid. (12.4, 14.6, 14.7, 14.8)

a. Write the equation for the neutralization of HCl and $Mg(OH)_2$.

b. Write the equation for the neutralization of HCl and $Al(OH)_3$.

c. What is the pH of the HCl solution?

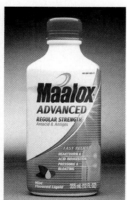

An antacid neutralizes stomach acid and raises the pH.

d. How many milliliters of the HCl solution are needed to neutralize the $Mg(OH)_2$?

e. How many milliliters of the HCl solution are needed to neutralize the $Al(OH)_3$?

CI.29 A KOH solution is prepared by dissolving 8.57 g of KOH in enough water to make 850. mL of KOH solution. (12.4, 14.6, 14.7, 14.8)

a. What is the molarity of the KOH solution?

b. What is the $[H_3O^+]$ and pH of the KOH solution?

c. Write the balanced chemical equation for the neutralization of KOH by H_2SO_4.

d. How many milliliters of a 0.250 M H_2SO_4 solution is required to neutralize 10.0 mL of the KOH solution?

CI.30 A solution of HCl is prepared by diluting 15.0 mL of a 12.0 M HCl solution with enough water to make 750. mL of HCl solution. (12.4, 14.6, 14.7, 14.8)

a. What is the molarity of the HCl solution?

b. What is the $[H_3O^+]$ and pH of the HCl solution?

c. Write the balanced chemical equation for the reaction of HCl and $MgCO_3$.

d. How many milliliters of the diluted HCl solution is required to completely react with 350. mg of $MgCO_3$?

CI.31 A volume of 200.0 mL of a carbonic acid buffer for blood plasma is prepared that contains 0.403 g of $NaHCO_3$ and 0.0149 g of H_2CO_3. At body temperature (37 °C), the K_a of carbonic acid is 7.9×10^{-7}. (7.2, 7.3, 8.4, 14.9)

$$H_2CO_3(aq) + H_2O(l) \rightleftharpoons HCO_3^-(aq) + H_3O^+(aq)$$

a. What is the $[H_2CO_3]$?

b. What is the $[HCO_3^-]$?

c. What is the $[H_3O^+]$?

d. What is the pH of the buffer?

e. Write a balanced chemical equation that shows how this buffer neutralizes added acid.

f. Write a balanced chemical equation that shows how this buffer neutralizes added base.

CI.32 In the kidneys, the ammonia buffer system buffers high H_3O^+. Ammonia, which is produced in renal tubules from amino acids, combines with H^+ to be excreted as NH_4Cl. At body temperature (37 °C), the $K_a = 5.6 \times 10^{-10}$. A buffer solution with a volume of 125 mL contains 3.34 g of NH_4Cl and 0.0151 g of NH_3. (7.2, 7.3, 8.4, 14.9)

$$NH_4^+(aq) + H_2O(l) \rightleftharpoons NH_3(aq) + H_3O^+(aq)$$

a. What is the $[NH_4^+]$?

b. What is the $[NH_3]$?

c. What is the $[H_3O^+]$?

d. What is the pH of the buffer?

e. Write a balanced chemical equation that shows how this buffer neutralizes added acid.

f. Write a balanced chemical equation that shows how this buffer neutralizes added base.

ANSWERS

CI.21 a.

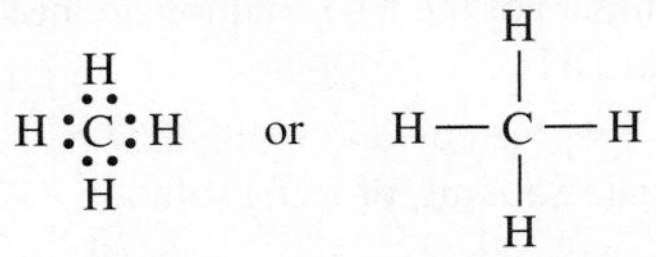

H:C:H   or   H—C—H (with H above and below)

  b. 1.2×10^7 kg of LNG (methane)
  c. 1.7×10^{10} L of LNG (methane)
  d. $CH_4(g) + 2O_2(g) \xrightarrow{\Delta} CO_2(g) + 2H_2O(g) + 883$ kJ
  e. 4.8×10^7 kg of O_2
  f. 6.6×10^{11} kJ

CI.23 a. $CS_2(g) + 3O_2(g) \xrightarrow{\Delta} CO_2(g) + 2SO_2(g)$
  b. 13.8 g of CO_2
  c. 37 mmHg
  d. 2370 mmHg

CI.25 a. $K_c = 248$
  b. If H_2 is added, the equilibrium will shift in the direction of the products.
  c. If the volume decreases, the equilibrium will shift in the direction of the products.
  d. $[S_2] = 0.280$ M

CI.27 a. $2M(s) + 6HCl(aq) \longrightarrow 3H_2(g) + 2MCl_3(aq)$
  b. 233 mL of H_2
  c. 6.03×10^{-3} mol of M
  d. 69.7 g/mol; gallium
  e. $2Ga(s) + 6HCl(aq) \longrightarrow 3H_2(g) + 2GaCl_3(aq)$

CI.29 a. 0.180 M
  b. $[H_3O^+] = 5.56 \times 10^{-14}$ M; pH = 13.255
  c. $2KOH(aq) + H_2SO_4(aq) \longrightarrow 2H_2O(l) + K_2SO_4(aq)$
  d. 3.60 mL

CI.31 a. 0.001 20 M
  b. 0.0240 M
  c. 4.0×10^{-8} M
  d. 7.40
  e. $HCO_3^-(aq) + H_3O^+(aq) \longrightarrow H_2CO_3(aq) + H_2O(l)$
  f. $H_2CO_3(aq) + OH^-(aq) \longrightarrow HCO_3^-(aq) + H_2O(l)$

Oxidation and Reduction

15

KIMBERLY'S TEETH HAVE become badly stained by food, coffee, and drinks that form a layer on top of the enamel of her teeth. Kimberly's dentist, Jane, has removed some of this layer by scraping her teeth, then brushing Kimberly's teeth with an abrasive toothpaste. After that, Jane begins the process of whitening Kimberly's teeth. She explains to Kimberly that she uses a gel containing 35% hydrogen peroxide (H_2O_2) that penetrates into the enamel of the tooth, where it undergoes a chemical reaction that whitens the teeth. The chemical reaction is referred to as an oxidation–reduction reaction where one chemical (hydrogen peroxide) is reduced and the other chemical (the coffee stain) is oxidized. During the oxidation, the coffee stains on the teeth become lighter or colorless, and therefore, the teeth are whiter.

CAREER

Dentist

Dentists are involved in assessing and maintaining the health of the teeth, gums, mouth, and bones of the jaw. They also examine X-rays to determine if there is disease or decay of the teeth and gums. Dentists remove decay and fill cavities, or they may extract teeth if necessary. Local anesthetics are often used for dental procedures. Dentists also educate patients on proper care of teeth and gums. Many dentists operate their own businesses with a staff that includes dental hygienists and a dental technician. A dentist may specialize in certain areas of dentistry. An orthodontist fits a patient's teeth with wires to straighten teeth or correct a bite. A periodontist treats problems with soft tissues of the gums. Pediatric dentists treat the teeth of children.

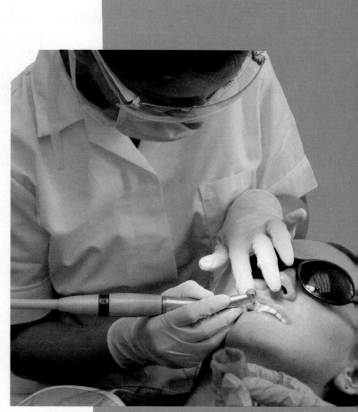

KEY MATH SKILLS

- Using Positive and Negative Numbers in Calculations (1.4)
- Solving Equations (1.4)

CORE CHEMISTRY SKILLS

- Writing Positive and Negative Ions (6.1)
- Writing Ionic Formulas (6.2)
- Balancing a Chemical Equation (8.2)
- Identifying Oxidized and Reduced Substances (8.4)

*These Key Math Skills and Core Chemistry Skills from previous chapters are listed here for your review as you proceed to the new material in this chapter.

LOOKING AHEAD

15.1 Oxidation and Reduction

LEARNING GOAL Identify a reaction as an oxidation or a reduction. Assign and use oxidation numbers to identify elements that are oxidized or reduced.

Oxidation and *reduction* are types of chemical reactions that are prevalent in our everyday lives. When we use natural gas to heat our home, turn on a computer, start a car, or eat food, we are using energy from the processes of oxidation and reduction.

In all of these processes, electrons are transferred during oxidation and reduction. In an **oxidation** reaction, a reactant loses one or more electrons. We say the substance is *oxidized*. In a **reduction** reaction, a reactant gains one or more electrons. We say the substance is *reduced*. When a substance undergoes oxidation, the electrons it loses must be gained by another substance, which is reduced. In an **oxidation–reduction reaction**, processes of oxidation and reduction must occur together. Sometimes we shorten this to *redox reactions*.

Writing Oxidation and Reduction Reactions

Glasses with photochromic lenses darken when exposed to sunlight.

Perhaps you have seen glasses with photochromic lenses that darken when exposed to UV light from the Sun. Once the wearer returns indoors, the lenses become clear again. The change from clear to dark and back to clear is the result of oxidation and reduction reactions. We can use the compounds in the photochromic lenses to illustrate some oxidation and reduction reactions.

To make the lenses photosensitive, silver chloride (AgCl) and copper(I) chloride (CuCl) are embedded in the lens material. Away from the Sun, visible light goes through the clear lens. However, when exposed to the Sun, there is an oxidation–reduction reaction when the UV light is absorbed by AgCl, and Cl atoms and Ag atoms are produced. It is the silver atoms that are black and darken the lenses. This redox reaction is written:

$$AgCl \longrightarrow Ag + Cl \quad \text{or}$$
$$Ag^+ + Cl^- \longrightarrow Ag + Cl \qquad \text{Oxidation–reduction reaction}$$

The chloride (Cl^-) ions lose electrons; they are oxidized. The silver (Ag^+) ions gain electrons; they are reduced.

Half-Reactions

To identify the oxidation and the reduction reactions, we can write each of the reactants with its product as a *half-reaction*. Then we observe that the Cl^- ion is oxidized because it loses an electron and the Ag^+ ion is reduced because it gains an electron. The half-reaction is called a reduction because the addition of a electron (negative charge) reduces the positive charge on Ag^+.

ENGAGE

How can you identify a half-reaction that is a reduction?

$$Ag^+ + e^- \longrightarrow Ag \qquad \text{Reduction \quad Gain of electron \quad Lenses darken}$$
$$Cl^- \longrightarrow Cl + e^- \qquad \text{Oxidation \quad Loss of electron}$$

To prevent the lenses from clearing immediately, some Cu^+ ions are added to the lenses. These Cu^+ ions convert the Cl atoms produced from the darkening process to Cl^- ions.

$$Cu^+ + Cl \longrightarrow Cu^{2+} + Cl^-$$
$$Cu^+ \longrightarrow Cu^{2+} + e^- \qquad \text{Oxidation}$$
$$Cl + e^- \longrightarrow Cl^- \qquad \text{Reduction}$$

When the wearer of the lenses moves out of the sunlight, the reactions take place in the opposite direction. The Cu^{2+} ions combine with the silver atoms to re-form Cu^+ and Ag^+, which causes the lens to become clear again.

$$Cu^{2+} + Ag \longrightarrow Cu^+ + Ag^+$$
$$Cu^{2+} + e^- \longrightarrow Cu^+ \qquad \text{Reduction}$$
$$Ag \longrightarrow Ag^+ + e^- \qquad \text{Oxidation \quad Lenses clear}$$

SAMPLE PROBLEM 15.1 Identifying Oxidation and Reduction Reactions

For each of the following, identify the reaction as an oxidation or a reduction:

a. $Cu^{2+} + 2 e^- \longrightarrow Cu$ **b.** $Fe \longrightarrow Fe^{3+} + 3 e^-$
c. $Cr^{2+} \longrightarrow Cr^{3+} + e^-$ **d.** $Li^+ + e^- \longrightarrow Li$

TRY IT FIRST

SOLUTION

a. Since Cu^{2+} gained electrons, this is a reduction.
b. Since Fe lost electrons, this is an oxidation.
c. Since Cr^{2+} lost an electron, this is an oxidation.
d. Since Li^+ gained an electron, this is a reduction.

STUDY CHECK 15.1

Is each of the following reactions an oxidation or a reduction?

a. $Na \longrightarrow Na^+ + e^-$
b. $Zn^{2+} + 2 e^- \longrightarrow Zn$

ANSWER

a. oxidation   **b.** reduction

Oxidation Numbers

In more complex oxidation–reduction reactions, the identification of the substances oxidized and reduced is not always obvious. To help identify the atoms or ions that are oxidized or reduced, we assign values called **oxidation numbers** (sometimes called *oxidation states*) to the elements of the reactants and products. It is important to recognize that oxidation numbers do not always represent actual charges, but they help us identify loss or gain of electrons.

Rules for Assigning Oxidation Numbers

The rules for assigning oxidation numbers to the atoms or ions in the reactants and products are given in **TABLE 15.1**.

CORE CHEMISTRY SKILL
Assigning Oxidation Numbers

ENGAGE

Why is the oxidation number 0 assigned to Mg and to Ba, whereas the oxidation number +2 is assigned to Mg^{2+} and to Ba^{2+}?

TABLE 15.1 Rules for Assigning Oxidation Numbers

1. The sum of the oxidation numbers in a molecule is zero (0), or for a polyatomic ion is equal to its charge.

2. The oxidation number of an element (monatomic or diatomic) is zero (0).

3. The oxidation number of a monatomic ion is equal to its charge.

4. In compounds, the oxidation number of Group 1A (1) metals is +1, and that of Group 2A (2) metals is +2.

5. In compounds, the oxidation number of fluorine is −1. Other nonmetals in Group 7A (17) are −1 except when combined with oxygen or fluorine.

6. In compounds, the oxidation number of oxygen is −2 except in OF_2, where O is +1; in H_2O_2 and other peroxides, O is −1.

7. In compounds with nonmetals, the oxidation number of hydrogen is +1; in compounds with metals, the oxidation number of hydrogen is −1.

We can now look at how these rules are used to assign oxidation numbers. For each formula, the oxidation numbers are written *below* the symbols of the elements (see **TABLE 15.2**).

TABLE 15.2 Examples of Using Rules to Assign Oxidation Numbers

Formula	Oxidation Numbers	Explanation
Br_2	Br_2 0	Each Br atom in diatomic bromine has an oxidation number of 0 (Rule 2).
Ba^{2+}	Ba^{2+} +2	The oxidation number of a monatomic ion is equal to its charge (Rule 3).
CO_2	CO_2 +4−2	In compounds, O has an oxidation number of −2 (Rule 6). Because CO_2 is neutral, the oxidation number of C is calculated as +4 (Rule 1). $C + 2O = 0$ $C + 2(-2) = 0$ $C = +4$
Al_2O_3	Al_2O_3 +3−2	In compounds, the oxidation number of O is −2 (Rule 6). For Al_2O_3 (neutral), the oxidation number of Al is calculated as +3 (Rule 1). $2Al + 3O = 0$ $2Al + 3(-2) = 0$ $2Al = +6$ $Al = +3$
$HClO_3$	$HClO_3$ +1+5−2	In compounds, the oxidation number of H is +1 (Rule 7), and O is −2 (Rule 6). For $HClO_3$ (neutral), the oxidation number of Cl is calculated as +5 (Rule 1). $H + Cl + 3O = 0$ $(+1) + Cl + 3(-2) = 0$ $Cl - 5 = 0$ $Cl = +5$
SO_4^{2-}	SO_4^{2-} +6−2	The oxidation number of O is −2 (Rule 6). For SO_4^{2-} (−2 charge), the oxidation number of S is calculated as +6 (Rule 1). $S + 4O = -2$ $S + 4(-2) = -2$ $S = +6$
CH_2O	CH_2O 0+1−2	In compounds, the oxidation number of H is +1 (Rule 7), and O is −2 (Rule 6). For CH_2O (neutral), the oxidation number of C is calculated as 0 (Rule 1). $C + 2H + O = 0$ $C + 2(+1) + (-2) = 0$ $C = 0$

ENGAGE

Why is the S in the formula H_2SO_3 assigned an oxidation number of +4, whereas the S in the formula H_2SO_4 is assigned an oxidation number of +6?

SAMPLE PROBLEM 15.2 Assigning Oxidation Numbers

Assign oxidation numbers to the elements in each of the following:

a. NCl_3 **b.** CO_3^{2-}

TRY IT FIRST

SOLUTION

a. NCl_3: The oxidation number of Cl is -1 (Rule 5). For NCl_3 (neutral), the sum of the oxidation numbers of N and 3Cl must be equal to zero (Rule 1). Thus, the oxidation number of N is calculated as $+3$.

$$N + 3Cl = 0$$
$$N + 3(-1) = 0$$
$$N = +3$$

The oxidation numbers are written as

$$\underset{+3\,-1}{NCl_3}$$

b. CO_3^{2-}: The oxidation number of O is -2 (Rule 6). For CO_3^{2-}, the sum of the oxidation numbers is equal to -2 (Rule 1). The oxidation number of C is calculated as $+4$.

$$C + 3O = -2$$
$$C + 3(-2) = -2$$
$$C = +4$$

The oxidation numbers are written as

$$\underset{+4\,-2}{CO_3^{2-}}$$

STUDY CHECK 15.2

Assign oxidation numbers to the elements in each of the following:

a. H_3PO_4 **b.** MnO_4^-

ANSWER

a. Because H has an oxidation number of $+1$, and O is -2, P will have an oxidation number of $+5$ to maintain a neutral charge.

$$\underset{+1\,+5\,-2}{H_3PO_4}$$

b. Because O has an oxidation number of -2, Mn must be $+7$ to give an overall charge of -1.

$$\underset{+7\,-2}{MnO_4^-}$$

Using Oxidation Numbers to Identify Oxidation–Reduction

Oxidation numbers can be used to identify the elements that are oxidized and the elements that are reduced in a reaction. In oxidation, the loss of electrons increases the oxidation number so that it is higher (more positive) in the product than in the reactant. In reduction, the gain of electrons decreases the oxidation number so that it is lower (more negative) in the product than in the reactant.

CORE CHEMISTRY SKILL
Using Oxidation Numbers

Reduction: oxidation number decreases

$$-7\;-6\;-5\;-4\;-3\;-2\;-1\;\;0\;\;+1\;+2\;+3\;+4\;+5\;+6\;+7$$

Oxidation: oxidation number increases

An oxidation reaction occurs when the charge becomes more positive. A reduction reaction occurs when the charge becomes more negative.

ENGAGE

Why is the reaction of the hydrogen atoms in H_2 to hydrogen ions ($2H^+$) called an oxidation reaction?

Guide to Using Oxidation Numbers

STEP 1
Assign oxidation numbers to each element.

STEP 2
Identify the increase in oxidation number as oxidation; the decrease as reduction.

SAMPLE PROBLEM 15.3 Using Oxidation Numbers to Determine Oxidation and Reduction

Identify the element that is oxidized and the element that is reduced in the following equation:

$$CO_2(g) + H_2(g) \longrightarrow CO(g) + H_2O(g)$$

TRY IT FIRST

SOLUTION

STEP 1 Assign oxidation numbers to each element. In H_2, the oxidation number of H is 0. In H_2O, the oxidation number of H is +1. In CO_2, CO, and H_2O, the oxidation number of O is −2. Using the oxidation number −2 for O, the oxidation number of C would be +4 in CO_2, and +2 in CO.

$$CO_2(g) + H_2(g) \longrightarrow CO(g) + H_2O(g)$$
$\;\;\;+4\,-2\qquad\;\;0\qquad\qquad\;\;+2\,-2\qquad+1\,-2\;\;$ Oxidation numbers

STEP 2 Identify the increase in oxidation number as oxidation; the decrease as reduction. H is oxidized because its oxidation number increases from 0 to +1. C is reduced because its oxidation number decreases from +4 to +2.

Oxidation number of
H increases; oxidation

$$CO_2(g) + H_2(g) \longrightarrow CO(g) + H_2O(g)$$
$\;\;\;+4\,-2\qquad\;\;0\qquad\qquad\;\;+2\,-2\qquad+1\,-2\;\;$ Oxidation numbers

Oxidation number of
C decreases; reduction

STUDY CHECK 15.3

In the following equation, which reactant is oxidized and which is reduced?

$$2Al(s) + 3Sn^{2+}(aq) \longrightarrow 2Al^{3+}(aq) + 3Sn(s)$$

ANSWER

The oxidation number of Al increases from 0 to +3; Al is oxidized.
The oxidation number of Sn^{2+} decreases from +2 to 0; Sn^{2+} is reduced.

CORE CHEMISTRY SKILL
Identifying Oxidizing and Reducing Agents

ENGAGE

In the following oxidation–reduction reaction, why is Zn the reducing agent, and Co^{2+} the oxidizing agent?

$Zn(s) + CoBr_2(aq) \longrightarrow$
$\qquad\qquad ZnBr_2(aq) + Co(s)$

Oxidizing and Reducing Agents

We have seen that an oxidation reaction must always be accompanied by a reduction reaction. The substance that loses electrons is oxidized, and the substance that gains electrons is reduced. For example, Zn is oxidized to Zn^{2+} by losing 2 electrons and Cl_2 reduced to $2Cl^-$ by gaining 2 electrons.

$$Zn(s) + Cl_2(g) \longrightarrow ZnCl_2(s)$$

In an oxidation–reduction reaction, the substance that is oxidized is called the **reducing agent** because it provides electrons for reduction. The substance that is reduced is called the **oxidizing agent** because it accepts electrons from oxidation. Because Zn is oxidized in this reaction, it is the *reducing agent*. In the same reaction Cl_2 is reduced, which makes it the *oxidizing agent*.

$Zn \longrightarrow Zn^{2+} + 2\,e^-$ Zn is oxidized; Zn is the *reducing agent*.
$Cl_2 + 2\,e^- \longrightarrow 2Cl^-$ Cl (in Cl_2) is reduced; Cl_2 is the *oxidizing agent*.

The terms we have used to describe oxidation and reduction are listed in the margin on the left.

Oxidation–Reduction Terminology

Loss of electrons	Gain of electrons
Oxidized	Reduced
Reducing agent	Oxidizing agent
Oxidation number increases	Oxidation number decreases

SAMPLE PROBLEM 15.4   Identifying Oxidizing and Reducing Agents

Identify the substance that is oxidized, the substance that is reduced, the oxidizing agent, and the reducing agent for the reaction of lead(II) oxide and carbon monoxide.

$$PbO(s) + CO(g) \longrightarrow Pb(s) + CO_2(g)$$

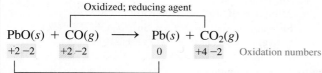

TRY IT FIRST

Lead oxides, once used in paint, are now banned due to the toxicity of lead.

SOLUTION

STEP 1 Assign oxidation numbers to the components in the equation.

Using the oxidation number -2 for O, the oxidation number of Pb in PbO is $+2$, the oxidation number of C in CO is $+2$., and the C in CO_2 would have an oxidation number of $+4$. The element Pb has an oxidation number of 0.

ANALYZE THE PROBLEM	Given	Need	Connect
	equation	oxidizing agent, reducing agent	oxidation numbers

Oxidized; reducing agent

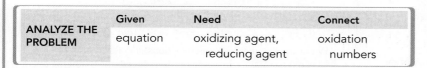

$$\underset{\substack{+2\ -2 \qquad +2\ -2 \qquad\quad 0 \qquad\ +4\ -2}}{PbO(s) + CO(g) \longrightarrow Pb(s) + CO_2(g)} \quad \text{Oxidation numbers}$$

Reduced; oxidizing agent

STEP 2 Identify the substance that is oxidized and the substance that is reduced. The C in CO is oxidized because the oxidation number of C increases from $+2$ to $+4$. The Pb in PbO is reduced because the oxidation number of Pb decreases from $+2$ to 0.

STEP 3 Identify the oxidized substance as the reducing agent and the reduced substance as the oxidizing agent. The compound CO is oxidized; it is the reducing agent. The compound PbO is reduced; it is the oxidizing agent.

Guide to Identifying Oxidizing and Reducing Agents

STEP 1
Assign oxidation numbers to the components in the equation.

STEP 2
Identify the substance that is oxidized and the substance that is reduced.

STEP 3
Identify the oxidized substance as the reducing agent and the reduced substance as the oxidizing agent.

STUDY CHECK 15.4

Identify the oxidizing agent and the reducing agent in the following reaction:

$$2Al(s) + 3CuO(s) \longrightarrow Al_2O_3(s) + 3Cu(s)$$

ANSWER

CuO is the oxidizing agent and Al is the reducing agent.

QUESTIONS AND PROBLEMS

15.1 Oxidation and Reduction

LEARNING GOAL Identify a reaction as an oxidation or a reduction. Assign and use oxidation numbers to identify elements that are oxidized or reduced.

15.1 Identify each of the following as an oxidation or a reduction reaction:
   a. $Al^{3+}(aq) + 3 e^- \longrightarrow Al(s)$
   b. $Ca(s) \longrightarrow Ca^{2+}(aq) + 2 e^-$
   c. $Fe^{3+}(aq) + e^- \longrightarrow Fe^{2+}(aq)$

15.2 Identify each of the following as an oxidation or a reduction reaction:
   a. $Ni^{2+}(aq) \longrightarrow Ni^{3+}(aq) + e^-$
   b. $K^+(aq) + e^- \longrightarrow K(s)$
   c. $Br_2(l) + 2 e^- \longrightarrow 2Br^-(aq)$

15.3 In each of the following reactions, identify the reactant that is oxidized and the reactant that is reduced:
   a. $4Al(s) + 3O_2(g) \longrightarrow 2Al_2O_3(s)$
   b. $Zn(s) + 2H^+(aq) \longrightarrow Zn^{2+}(aq) + H_2(g)$
   c. $F_2(g) + 2Br^-(aq) \longrightarrow 2F^-(aq) + Br_2(l)$

15.4 In each of the following reactions, identify the reactant that is oxidized and the reactant that is reduced:
a. $2Ag^+(aq) + Zn(s) \longrightarrow 2Ag(s) + Zn^{2+}(aq)$
b. $Ca(s) + S(s) \longrightarrow CaS(s)$
c. $Sn^{2+}(aq) + 2Cr^{3+}(aq) \longrightarrow Sn^{4+}(aq) + 2Cr^{2+}(aq)$

15.5 Assign oxidation numbers to each of the following:
a. Cu   b. F_2 c. Fe^{2+} d. Cl^-

15.6 Assign oxidation numbers to each of the following:
a. Al   b. Al^{3+} c. F^- d. N_2

15.7 Assign oxidation numbers to all the elements in each of the following:
a. KCl   b. MnO_2 c. NO   d. Mn_2O_3

15.8 Assign oxidation numbers to all the elements in each of the following:
a. H_2S b. NO_2 c. CCl_4 d. PCl_3

15.9 Assign oxidation numbers to all the elements in each of the following:
a. Li_3PO_4 b. SO_3^{2-} c. Cr_2S_3 d. NO_3^-

15.10 Assign oxidation numbers to all the elements in each of the following:
a. $C_2H_3O_2^-$ b. $AlCl_3$ c. NH_4^+ d. $HBrO_4$

15.11 Assign oxidation numbers to all the elements in each of the following:
a. HSO_4^- b. H_3PO_3 c. $Cr_2O_7^{2-}$ d. Na_2CO_3

15.12 Assign oxidation numbers to all the elements in each of the following:
a. N_2O b. LiOH   c. SbO_2^- d. IO_4^-

15.13 What is the oxidation number of the specified element in each of the following?
a. N in HNO_3 b. C in C_3H_6
c. P in K_3PO_4 d. Cr in CrO_4^{2-}

15.14 What is the oxidation number of the specified element in each of the following?
a. C in $BaCO_3$ b. Fe in $FeBr_3$
c. Cl in ClF_4^- d. S in $S_2O_3^{2-}$

15.15 Indicate whether each of the following describes the oxidizing agent or the reducing agent in an oxidation–reduction reaction:
a. the substance that is oxidized
b. the substance that gains electrons

15.16 Indicate whether each of the following describes the oxidizing agent or the reducing agent in an oxidation–reduction reaction:
a. the substance that is reduced
b. the substance that loses electrons

15.17 For each of the following reactions, identify the substance that is oxidized, the substance that is reduced, the oxidizing agent, and the reducing agent:
a. $2Li(s) + Cl_2(g) \longrightarrow 2LiCl(s)$
b. $Cl_2(g) + 2NaBr(aq) \longrightarrow 2NaCl(aq) + Br_2(l)$
c. $2Pb(s) + O_2(g) \longrightarrow 2PbO(s)$
d. $Al(s) + 3Ag^+(aq) \longrightarrow Al^{3+}(aq) + 3Ag(s)$

15.18 For each of the following reactions, identify the substance that is oxidized, the substance that is reduced, the oxidizing agent, and the reducing agent:
a. $2Li(s) + F_2(g) \longrightarrow 2LiF(s)$
b. $Cl_2(g) + 2KI(aq) \longrightarrow 2KCl(aq) + I_2(s)$
c. $Zn(s) + Ni^{2+}(aq) \longrightarrow Zn^{2+}(aq) + Ni(s)$
d. $Fe(s) + CuSO_4(aq) \longrightarrow FeSO_4(aq) + Cu(s)$

15.19 For each of the following reactions, identify the substance that is oxidized, the substance that is reduced, the oxidizing agent, and the reducing agent:
a. $2NiS(s) + 3O_2(g) \longrightarrow 2NiO(s) + 2SO_2(g)$
b. $Sn^{2+}(aq) + 2Fe^{3+}(aq) \longrightarrow Sn^{4+}(aq) + 2Fe^{2+}(aq)$
c. $CH_4(g) + 2O_2(g) \longrightarrow CO_2(g) + 2H_2O(g)$
d. $2Cr_2O_3(s) + 3Si(s) \longrightarrow 4Cr(s) + 3SiO_2(s)$

15.20 For each of the following reactions, identify the substance that is oxidized, the substance that is reduced, the oxidizing agent, and the reducing agent:
a. $2HgO(s) \longrightarrow 2Hg(l) + O_2(g)$
b. $Zn(s) + 2HCl(aq) \longrightarrow ZnCl_2(aq) + H_2(g)$
c. $2Na(s) + 2H_2O(l) \longrightarrow 2Na^+(aq) + 2OH^-(aq) + H_2(g)$
d. $14H^+(aq) + 6Fe^{2+}(aq) + Cr_2O_7^{2-}(aq) \longrightarrow$
$$6Fe^{3+}(aq) + 2Cr^{3+}_{,}(aq) + 7H_2O(l)$$

15.2 Balancing Oxidation–Reduction Equations Using Half-Reactions

LEARNING GOAL Balance oxidation–reduction equations using the half-reaction method.

CORE CHEMISTRY SKILL
Using Half-Reactions to Balance Redox Equations

In the **half-reaction method** for balancing equations, an oxidation–reduction reaction is written as two *half-reactions*. The half-reaction method uses ionic charges and electrons to balance each half-reaction. Oxidation numbers are not used. The loss of electrons by one reactant is used to identify the oxidized substance and the gain of electrons by another reactant is used to identify the reduced substance. Once the loss and gain of electrons are equalized for the half-reactions, they are combined to obtain the overall balanced equation. The half-reaction method is typically used to balance equations that are written as ionic equations. Let us consider the reaction between aluminum metal and a solution of Cu^{2+} as shown in Sample Problem 15.5.

SAMPLE PROBLEM 15.5 Using Half-Reactions to Balance Equations

Use half-reactions to balance the following equation:

$$Al(s) + Cu^{2+}(aq) \longrightarrow Al^{3+}(aq) + Cu(s)$$

SOLUTION

	Given	Need	Connect
ANALYZE THE PROBLEM	Equation: $Al(s) + Cu^{2+}(aq) \longrightarrow$ $Al^{3+}(aq) + Cu(s)$	balanced equation	half-reactions, electron balance

STEP 1 Write two half-reactions for the equation.

$$Al(s) \longrightarrow Al^{3+}(aq)$$
$$Cu^{2+}(aq) \longrightarrow Cu(s)$$

STEP 2 For each half-reaction, balance the elements other than H and O. In these half-reactions, the Al and Cu are already balanced.

$$Al(s) \longrightarrow Al^{3+}(aq)$$
$$Cu^{2+}(aq) \longrightarrow Cu(s)$$

STEP 3 Balance each half-reaction for charge by adding electrons. For the aluminum half-reaction, we need to add three electrons on the product side to balance the charge. With a loss of electrons, this is an oxidation.

$$Al(s) \longrightarrow \underbrace{Al^{3+}(aq) + 3\,e^-}_{} \quad \text{Oxidation}$$

0 charge = 0 charge

For the Cu^{2+} half-reaction, we need to add two electrons on the reactant side to balance the charge. With a gain of electrons, this is a reduction.

$$\underbrace{Cu^{2+}(aq) + 2\,e^-}_{} \longrightarrow Cu(s) \quad \text{Reduction}$$

0 charge = 0 charge

STEP 4 Multiply each half-reaction by factors that equalize the loss and gain of electrons. To obtain the same number of electrons in each half-reaction, we need to multiply the oxidation half-reaction by 2 and the reduction half-reaction by 3.

$$2 \times [Al(s) \longrightarrow Al^{3+}(aq) + 3\,e^-]$$
$$2Al(s) \longrightarrow 2Al^{3+}(aq) + 6\,e^- \quad 6\,e^- \text{ lost}$$

$$3 \times [Cu^{2+}(aq) + 2\,e^- \longrightarrow Cu(s)]$$
$$3Cu^{2+}(aq) + 6\,e^- \longrightarrow 3Cu(s) \quad 6\,e^- \text{ gained}$$

STEP 5 Add half-reactions and cancel electrons, and any identical ions or molecules. Check the balance of atoms and charge.

$$2Al(s) \longrightarrow 2Al^{3+}(aq) + 6\,e^-$$
$$3Cu^{2+}(aq) + 6\,e^- \longrightarrow 3Cu(s)$$
$$\overline{2Al(s) + 3Cu^{2+}(aq) + \cancel{6\,e^-} \longrightarrow 2Al^{3+}(aq) + \cancel{6\,e^-} + 3Cu(s)}$$

Final balanced equation:

$$2Al(s) + 3Cu^{2+}(aq) \longrightarrow 2Al^{3+}(aq) + 3Cu(s)$$

Guide to Balancing Redox Equations Using Half-Reactions

STEP 1
Write two half-reactions for the equation.

STEP 2
For each half-reaction, balance the elements other than H and O. If the reaction takes place in acid or base, balance O by adding H_2O, and H by adding H^+.

STEP 3
Balance each half-reaction for charge by adding electrons.

STEP 4
Multiply each half-reaction by factors that equalize the loss and gain of electrons.

STEP 5
Add half-reactions and cancel electrons, and any identical ions or molecules. In base, add OH^- to neutralize H^+. Check the balance of atoms and charge.

ENGAGE

How do we know that the final oxidation–reduction equation is balanced?

Check the balance of atoms and charge.

	Reactants		Products
	2Al	=	2Al
	3Cu	=	3Cu
Charge:	6 +	=	6 +

STUDY CHECK 15.5

Use the half-reaction method to balance the following equation:

$$Zn(s) + Fe^{3+}(aq) \longrightarrow Zn^{2+}(aq) + Fe^{2+}(aq)$$

ANSWER

$$Zn(s) + 2Fe^{3+}(aq) \longrightarrow Zn^{2+}(aq) + 2Fe^{2+}(aq)$$

Balancing Oxidation–Reduction Equations in Acidic Solution

When we use the half-reaction method for balancing equations for reactions in acidic solution, we include balancing O by adding H_2O and balancing H by adding H^+ as shown in Sample Problem 15.6.

SAMPLE PROBLEM 15.6 Using Half-Reactions to Balance Equations in Acidic Solution

Use half-reactions to balance the following equation for a reaction that takes place in acidic solution:

$$I^-(aq) + Cr_2O_7^{2-}(aq) \longrightarrow I_2(s) + Cr^{3+}(aq)$$

TRY IT FIRST

A dichromate solution (yellow) and an iodide solution (colorless) form a brown solution of Cr^{3+} and iodine (I_2).

SOLUTION

STEP 1 Write two half-reactions for the equation.

$$I^-(aq) \longrightarrow I_2(s)$$
$$Cr_2O_7^{2-}(aq) \longrightarrow Cr^{3+}(aq)$$

STEP 2 For each half-reaction, balance the elements other than H and O. If the reaction takes place in acid, balance O by adding H_2O, and H by adding H^+.

The two I atoms in I_2 are balanced with a coefficient of 2 for I^-.

$$2I^-(aq) \longrightarrow I_2(s)$$

The two Cr atoms are balanced with a coefficient of 2 for Cr^{3+}.

$$Cr_2O_7^{2-}(aq) \longrightarrow 2Cr^{3+}(aq)$$

Now balance O by adding H_2O to the product side.

$$Cr_2O_7^{2-}(aq) \longrightarrow 2Cr^{3+}(aq) + \mathbf{7H_2O}(l) \quad \text{H_2O balances O}$$

Balance H by adding H^+ to the reactant side.

$$\mathbf{14H^+}(aq) + Cr_2O_7^{2-}(aq) \longrightarrow 2Cr^{3+}(aq) + 7H_2O(l) \quad \text{H^+ balances H}$$

STEP 3 Balance each half-reaction for charge by adding electrons. A charge of -2 is balanced with two electrons on the product side.

$$\underset{-2}{2I^-(aq)} \longrightarrow \underset{-2}{I_2(s) + \mathbf{2}\,e^-} \quad \text{Oxidation}$$

A charge of $+6$ is obtained on the reactant side by adding six electrons.

$$\underset{+6}{\mathbf{6}\,e^- + 14H^+(aq) + Cr_2O_7^{2-}(aq)} \longrightarrow \underset{+6}{2Cr^{3+}(aq) + 7H_2O(l)} \quad \text{Reduction}$$

ENGAGE

Why are $7H_2O$ added to the product side of this half-reaction?

ENGAGE

Why are 6 e^- added to the reactant side of this half-reaction?

STEP 4 Multiply each half-reaction by factors that equalize the loss and gain of electrons. The half-reaction with I is multiplied by 3 to equal the gain of 6 e^- by the Cr half-reaction.

$$3 \times [2I^-(aq) \longrightarrow I_2(s) + 2e^-]$$
$$6I^-(aq) \longrightarrow 3I_2(s) + 6e^- \quad {\scriptstyle 6\,e^-\text{ lost}}$$
$$6e^- + 14H^+(aq) + Cr_2O_7^{2-}(aq) \longrightarrow 2Cr^{3+}(aq) + 7H_2O(l) \quad {\scriptstyle 6\,e^-\text{ gained}}$$

> **ENGAGE**
>
> Why is this half-reaction multiplied by a factor of 3?

STEP 5 Add half-reactions and cancel electrons, and any identical ions or molecules. Check the balance of atoms and charge.

$$6I^-(aq) \longrightarrow 3I_2(s) + 6e^-$$
$$6e^- + 14H^+(aq) + Cr_2O_7^{2-}(aq) \longrightarrow 2Cr^{3+}(aq) + 7H_2O(l)$$

$$\overline{\cancel{6e^-} + 14H^+(aq) + Cr_2O_7^{2-}(aq) + 6I^-(aq) \longrightarrow 2Cr^{3+}(aq) + 3I_2(s) + 7H_2O(l) + \cancel{6e^-}}$$

Final balanced equation:

$$14H^+(aq) + Cr_2O_7^{2-}(aq) + 6I^-(aq) \longrightarrow 2Cr^{3+}(aq) + 3I_2(s) + 7H_2O(l)$$

Check the balance of atoms and charge.

Reactants		Products
6I	=	6I
2Cr	=	2Cr
14H	=	14H
7O	=	7O
Charge: 6+	=	6+

STUDY CHECK 15.6

Use half-reactions to balance the following equation in acidic solution:

$$Fe^{2+}(aq) + IO_3^-(aq) \longrightarrow Fe^{3+}(aq) + I_2(s)$$

ANSWER

$$12H^+(aq) + 10Fe^{2+}(aq) + 2IO_3^-(aq) \longrightarrow 10Fe^{3+}(aq) + I_2(s) + 6H_2O(l)$$

Balancing Oxidation–Reduction Equations in Basic Solution

An oxidation–reduction reaction can also take place in basic solution. In that case, we use the same half-reaction method, but once we have the balanced equation, we will neutralize the H^+ with OH^- to form water. The H^+ is neutralized by adding OH^- to both sides of the equation to form H_2O as shown in Sample Problem 15.7.

SAMPLE PROBLEM 15.7 Using Half-Reactions to Balance Equations in Basic Solution

Use half-reactions to balance the following equation that takes place in basic solution:

$$Fe^{2+}(aq) + MnO_4^-(aq) \longrightarrow Fe^{3+}(aq) + MnO_2(s)$$

▶ **TRY IT FIRST**

SOLUTION

STEP 1 Write two half-reactions for the equation.

$$Fe^{2+}(aq) \longrightarrow Fe^{3+}(aq)$$
$$MnO_4^-(aq) \longrightarrow MnO_2(s)$$

STEP 2 For each half-reaction, balance the elements other than H and O. If the reaction takes place in base, balance O by adding H_2O, and H by adding H^+.

$$Fe^{2+}(aq) \longrightarrow Fe^{3+}(aq)$$

$$MnO_4^-(aq) \longrightarrow MnO_2(s) + \mathbf{2H_2O}(l) \quad \text{\small H_2O balances O}$$

$$\mathbf{4H^+}(aq) + MnO_4^-(aq) \longrightarrow MnO_2(s) + 2H_2O(l) \quad \text{\small H^+ balances H}$$

STEP 3 Balance each half-reaction for charge by adding electrons.

$$Fe^{2+}(aq) \longrightarrow \underbrace{Fe^{3+}(aq) + \mathbf{1\,e^-}}_{} \quad \text{\small Oxidation}$$
$$\underset{+2}{} \quad = \quad \underset{+2}{\phantom{Fe^{3+}}}$$

$$\underbrace{\mathbf{3\,e^-} + 4H^+(aq) + MnO_4^-(aq)}_{\text{0 charge}} \longrightarrow \underset{\text{0 charge}}{MnO_2(s) + 2H_2O(l)} \quad \text{\small Reduction}$$

STEP 4 Multiply each half-reaction by factors that equalize the loss and gain of electrons. The half-reaction with Fe is multiplied by 3 to equal the gain of 3 e^- by Mn.

$$\mathbf{3} \times [Fe^{2+}(aq) \longrightarrow Fe^{3+}(aq) + \mathbf{1\,e^-}]$$

$$3Fe^{2+}(aq) \longrightarrow 3Fe^{3+}(aq) + \mathbf{3\,e^-} \quad \text{\small 3 e^- lost}$$

$$\mathbf{3\,e^-} + 4H^+(aq) + MnO_4^-(aq) \longrightarrow MnO_2(s) + 2H_2O(l) \quad \text{\small 3 e^- gained}$$

STEP 5 Add half-reactions and cancel electrons, and any identical ions or molecules. In base, add OH^- to neutralize H^+. Check the balance of atoms and charge.

$$3Fe^{2+}(aq) \longrightarrow 3Fe^{3+}(aq) + \mathbf{3\,e^-}$$

$$\mathbf{3\,e^-} + 4H^+(aq) + MnO_4^-(aq) \longrightarrow MnO_2(s) + 2H_2O(l)$$

$$\overline{\mathbf{3\cancel{e^-}} + 4H^+(aq) + 3Fe^{2+}(aq) + MnO_4^-(aq) \longrightarrow 3Fe^{3+}(aq) + MnO_2(s) + 2H_2O(l) + \mathbf{3\cancel{e^-}}}$$

ENGAGE

Why is OH^- added to the oxidation–reduction equation when the reaction takes place in basic solution?

Final balanced equation:

$$4H^+(aq) + 3Fe^{2+}(aq) + MnO_4^-(aq) \longrightarrow 3Fe^{3+}(aq) + MnO_2(s) + 2H_2O(l)$$

To convert the equation to an oxidation–reduction reaction in basic solution, we neutralize H^+ with OH^- to form H_2O. For this equation, we add $4OH^-(aq)$ to both sides of the equation.

$$\mathbf{4OH^-}(aq) + 4H^+(aq) + 3Fe^{2+}(aq) + MnO_4^-(aq) \longrightarrow 3Fe^{3+}(aq) + MnO_2(s) + 2H_2O(l) + \mathbf{4OH^-}(aq)$$

Combining $4H^+$ and $4OH^-$ gives $4H_2O$ on the reactant side.

$$4H_2O(l) + 3Fe^{2+}(aq) + MnO_4^-(aq) \longrightarrow 3Fe^{3+}(aq) + MnO_2(s) + 2H_2O(l) + 4OH^-(aq)$$

Canceling $2H_2O$ on both the reactant and the product side gives the balanced equation in basic solution.

Final balanced equation in basic solution:

$$2H_2O(l) + 3Fe^{2+}(aq) + MnO_4^-(aq) \longrightarrow 3Fe^{3+}(aq) + MnO_2(s) + 4OH^-(aq)$$

Check the balance of atoms and charge.

Reactants		Products
3Fe	=	3Fe
1Mn	=	1Mn
4H	=	4H
6O	=	6O
Charge: 5+	=	**5+**

STUDY CHECK 15.7

Use half-reactions to balance the following equation in basic solution:

$$N_2O(g) + ClO^-(aq) \longrightarrow NO_2^-(aq) + Cl^-(aq)$$

ANSWER

$$N_2O(g) + 2ClO^-(aq) + 2OH^-(aq) \longrightarrow 2Cl^-(aq) + 2NO_2^-(aq) + H_2O(l)$$

QUESTIONS AND PROBLEMS

15.2 Balancing Oxidation–Reduction Equations Using Half-Reactions

LEARNING GOAL Balance oxidation–reduction equations using the half-reaction method.

15.21 Balance each of the following half-reactions in acidic solution:
   a. $Sn^{2+}(aq) \longrightarrow Sn^{4+}(aq)$
   b. $Mn^{2+}(aq) \longrightarrow MnO_4^-(aq)$
   c. $NO_2^-(aq) \longrightarrow NO_3^-(aq)$
   d. $ClO_3^-(aq) \longrightarrow ClO_2(aq)$

15.22 Balance each of the following half-reactions in acidic solution:
   a. $Cu(s) \longrightarrow Cu^{2+}(aq)$
   b. $SO_4^{2-}(aq) \longrightarrow SO_3^{2-}(aq)$
   c. $BrO_3^-(aq) \longrightarrow Br^-(aq)$
   d. $IO_3^-(aq) \longrightarrow I_2(s)$

15.23 Use the half-reaction method to balance each of the following in acidic solution:
   a. $Ag(s) + NO_3^-(aq) \longrightarrow Ag^+(aq) + NO_2(g)$
   b. $NO_3^-(aq) + S(s) \longrightarrow NO(g) + SO_2(g)$
   c. $S_2O_3^{2-}(aq) + Cu^{2+}(aq) \longrightarrow S_4O_6^{2-}(aq) + Cu(s)$

15.24 Use the half-reaction method to balance each of the following in acidic solution:
   a. $Mn(s) + NO_3^-(aq) \longrightarrow Mn^{2+}(aq) + NO_2(g)$
   b. $C_2O_4^{2-}(aq) + MnO_4^-(aq) \longrightarrow CO_2(g) + Mn^{2+}(aq)$
   c. $ClO_3^-(aq) + SO_3^{2-}(aq) \longrightarrow Cl^-(aq) + SO_4^{2-}(aq)$

15.25 Use the half-reaction method to balance each of the following in basic solution:
   a. $Fe(s) + CrO_4^{2-}(aq) \longrightarrow Fe_2O_3(s) + Cr_2O_3(s)$
   b. $CN^-(aq) + MnO_4^-(aq) \longrightarrow CNO^-(aq) + MnO_2(s)$

15.26 Use the half-reaction method to balance each of the following in basic solution:
   a. $Al(s) + ClO^-(aq) \longrightarrow AlO_2^-(aq) + Cl^-(aq)$
   b. $Sn^{2+}(aq) + IO_4^-(aq) \longrightarrow Sn^{4+}(aq) + I^-(aq)$

15.3 Electrical Energy from Oxidation–Reduction Reactions

CORE CHEMISTRY SKILL
Identifying Spontaneous Reactions

LEARNING GOAL Use the activity series to determine if an oxidation–reduction reaction is spontaneous. Write the half-reactions that occur in a voltaic cell and the cell notation.

When we placed a zinc metal strip in a solution of Cu^{2+}, reddish-brown Cu metal accumulated on the Zn strip according to the following spontaneous reaction.

$$Zn(s) + Cu^{2+}(aq) \longrightarrow Zn^{2+}(aq) + Cu(s) \quad \text{Spontaneous}$$

However, if we place a Cu metal strip in a Zn^{2+} solution, nothing will happen. The reaction does not run spontaneously in the reverse direction because Cu does not lose electrons as easily as Zn.

We can determine the direction of a spontaneous reaction from the *activity series*, which ranks the metals and H_2 in terms of how easily they lose electrons.

In the **activity series**, the metals that lose electrons most easily are placed at the top, and the metals that do not lose electrons easily are at the bottom. Thus the metals that are more easily oxidized are above the metals whose ions are more easily reduced (see **TABLE 15.3**). Active metals include K, Na, Ca, Mg, Al, Zn, Fe, and Sn. In single replacement reactions, the metal ion replaces the H in the acid. Metals listed below $H_2(g)$ will not react with H^+ from acids.

Copper strip

Zn^{2+} solution

Because Cu is below Zn on the activity series, no oxidation–reduction reaction occurs.

TABLE 15.3 Activity Series for Some Metals

	Metal		Ion
Most active oxidize easily	Li(s)	$\longrightarrow$	$Li^+(aq) + e^-$
	K(s)	$\longrightarrow$	$K^+(aq) + e^-$
	Ca(s)	$\longrightarrow$	$Ca^{2+}(aq) + 2\,e^-$
	Na(s)	$\longrightarrow$	$Na^+(aq) + e^-$
	Mg(s)	$\longrightarrow$	$Mg^{2+}(aq) + 2\,e^-$
	Al(s)	$\longrightarrow$	$Al^{3+}(aq) + 3\,e^-$
	Zn(s)	$\longrightarrow$	$Zn^{2+}(aq) + 2\,e^-$
	Cr(s)	$\longrightarrow$	$Cr^{3+}(aq) + 3\,e^-$
	Fe(s)	$\longrightarrow$	$Fe^{2+}(aq) + 2\,e^-$
	Ni(s)	$\longrightarrow$	$Ni^{2+}(aq) + 2\,e^-$
	Sn(s)	$\longrightarrow$	$Sn^{2+}(aq) + 2\,e^-$
	Pb(s)	$\longrightarrow$	$Pb^{2+}(aq) + 2\,e^-$
	$H_2(g)$	$\longrightarrow$	$2H^+(aq) + 2\,e^-$
	Cu(s)	$\longrightarrow$	$Cu^{2+}(aq) + 2\,e^-$
	Ag(s)	$\longrightarrow$	$Ag^+(aq) + e^-$
Least active oxidize with difficulty	Au(s)	$\longrightarrow$	$Au^{3+}(aq) + 3\,e^-$

According to the activity series, a metal will oxidize spontaneously when it is combined with the reverse of the half-reaction for any metal below it on the list. We use the activity series to predict the direction of the spontaneous reaction. Suppose we have two beakers. In one, we place a Mg strip in a solution containing Ni^{2+} ions. In the other, we place a Ni strip in a solution containing Mg^{2+} ions. Looking at the activity series we see that the half-reaction for the oxidation of Mg is listed above that for Ni, which means that Mg is the more active metal and loses electrons more easily than Ni. Using the activity series table, we write these two half-reactions as follows:

$$Mg(s) \longrightarrow Mg^{2+}(aq) + 2\,e^-$$
$$Ni(s) \longrightarrow Ni^{2+}(aq) + 2\,e^-$$

The reaction that will be spontaneous is the oxidation of Mg combined with the reverse (reduction) of Ni^{2+}.

$$Mg(s) \longrightarrow Mg^{2+}(aq) + 2\,e^-$$
$$Ni^{2+}(aq) + 2\,e^- \longrightarrow Ni(s)$$

Therefore, we combine the half-reactions, which gives the following overall reaction that occurs spontaneously:

$$Mg(s) + Ni^{2+}(aq) + 2\,e^- \longrightarrow Mg^{2+}(aq) + Ni(s) + 2\,e^-$$
$$Mg(s) + Ni^{2+}(aq) \longrightarrow Mg^{2+}(aq) + Ni(s) \qquad \text{Spontaneous}$$

However, a reaction between a Mg strip and a solution containing K^+ ions will not occur spontaneously. We determine this by looking at the two half-reactions needed.

$$Mg(s) \longrightarrow Mg^{2+}(aq) + 2\,e^-$$
$$K^+(aq) + e^- \longrightarrow K(s)$$

Because the half-reaction for the oxidation of K is above that for Mg, there will not be a spontaneous reaction between Mg and K^+.

$$Mg(s) + 2K^+(aq) \xcancel{\longrightarrow} Mg^{2+}(aq) + 2K(s) \qquad \text{No reaction takes place}$$

ENGAGE

Why would the following oxidation–reduction reaction be spontaneous?

$$Fe(s) + Pb^{2+}(aq) \longrightarrow Fe^{2+}(aq) + Pb(s)$$

SAMPLE PROBLEM 15.8 Predicting Spontaneous Reactions

Determine if the reaction for each of the following metals with an HCl (H^+) solution is spontaneous:

a. $Zn(s) + 2H^+(aq) \longrightarrow Zn^{2+}(aq) + H_2(g)$
b. $Cu(s) + 2H^+(aq) \longrightarrow Cu^{2+}(aq) + H_2(g)$

TRY IT FIRST

SOLUTION

a. Using the activity series (see Table 15.3), we see that Zn oxidizes more easily than H_2. Thus the oxidation half-reaction for Zn is combined with the reverse half-reaction for H_2.

$$Zn(s) \longrightarrow Zn^{2+}(aq) + 2\,e^-$$
$$2H^+(aq) + 2\,e^- \longrightarrow H_2(g)$$
$$Zn(s) + 2H^+(aq) \longrightarrow Zn^{2+}(aq) + H_2(g) \quad \text{Spontaneous}$$

b. Using the activity series (see Table 15.3), we see that H_2 oxidizes more easily than Cu. In this reaction, we would be combining an oxidation half-reaction with one that is above it in the activity series.

$$2H^+(aq) + 2\,e^- \longrightarrow H_2(g)$$
$$Cu(s) \longrightarrow Cu^{2+}(aq) + 2\,e^-$$
$$Cu(s) + 2H^+(aq) \xcancel{\longrightarrow} Cu^{2+}(aq) + H_2(g) \quad \text{Not spontaneous}$$

STUDY CHECK 15.8

Determine if the following reaction is spontaneous:

$$2Al(s) + 3Cu^{2+}(aq) \longrightarrow 2Al^{3+}(aq) + 3Cu(s)$$

ANSWER

Yes, this reaction is spontaneous.

Voltaic Cells

We can generate electrical energy from a spontaneous oxidation–reduction reaction by using an apparatus called a **voltaic cell**. The two half-reactions still take place, but in a cell the electrons must flow through an external circuit. For example, a piece of zinc metal in a Cu^{2+} solution becomes coated with a rusty-brown coating of Cu, while the blue color (Cu^{2+}) of the solution fades. The oxidation of the zinc metal provides electrons for the reduction of the Cu^{2+} ions. We can write the two half-reactions as

$$Zn(s) \longrightarrow Zn^{2+}(aq) + 2\,e^- \quad \text{Oxidation}$$
$$Cu^{2+}(aq) + 2\,e^- \longrightarrow Cu(s) \quad \text{Reduction}$$

The overall reaction is

$$Zn(s) + Cu^{2+}(aq) \longrightarrow Zn^{2+}(aq) + Cu(s)$$

As long as the Zn metal and Cu^{2+} ions are in the same container, the electrons are transferred directly from Zn to Cu^{2+}. However, when the components of the two half-reactions are placed in separate containers, called *half-cells*, the electrons flow from one half-cell to the other, producing an electrical current. In each half-cell, there is a strip of metal, called an *electrode*, in contact with the ionic solution. The electrode where oxidation takes place is called the **anode**; the **cathode** is the electrode where reduction takes place. In this example, the anode is a zinc metal strip placed in a Zn^{2+}($ZnSO_4$) solution. The cathode is a copper metal strip placed in a Cu^{2+}($CuSO_4$) solution. In this voltaic cell, the Zn anode and Cu cathode are connected by a wire that allows electrons to move from the oxidation half-cell to the reduction half-cell.

Anode is where
oxidation takes place
electrons are produced $\left.\right\}$ $Zn(s) \longrightarrow Zn^{2+}(aq) + 2\,e^-$

Cathode is where
reduction takes place
electrons are used up $\left.\right\}$ $Cu^{2+}(aq) + 2\,e^- \longrightarrow Cu(s)$

ENGAGE

What is the purpose of a salt bridge in a voltaic cell?

The circuit is completed by a *salt bridge* containing positive and negative ions that are placed in the half-cell solutions. The purpose of the salt bridge is to provide ions, such as Na^+ and SO_4^{2-} ions, to maintain an electrical balance in each half-cell solution. As oxidation occurs, there is an increase in Zn^{2+} ions, which is balanced by SO_4^{2-} anions from the salt bridge. At the cathode, there is a loss of Cu^{2+}, which is balanced by SO_4^{2-} moving into the salt bridge. The complete circuit involves the flow of electrons from the anode to the cathode and the flow of anions from the cathode solution to the anode solution (see **FIGURE 15.1**).

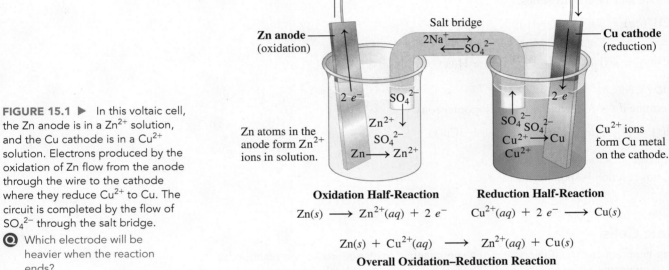

FIGURE 15.1 ▶ In this voltaic cell, the Zn anode is in a Zn^{2+} solution, and the Cu cathode is in a Cu^{2+} solution. Electrons produced by the oxidation of Zn flow from the anode through the wire to the cathode where they reduce Cu^{2+} to Cu. The circuit is completed by the flow of SO_4^{2-} through the salt bridge.

◉ Which electrode will be heavier when the reaction ends?

Oxidation Half-Reaction **Reduction Half-Reaction**
$Zn(s) \longrightarrow Zn^{2+}(aq) + 2\,e^-$ $Cu^{2+}(aq) + 2\,e^- \longrightarrow Cu(s)$

$$Zn(s) + Cu^{2+}(aq) \longrightarrow Zn^{2+}(aq) + Cu(s)$$
Overall Oxidation–Reduction Reaction

We can diagram the oxidation and reduction reactions that take place in the cell using a *shorthand notation* as follows:

$$Zn(s)\,|\,Zn^{2+}(aq)\,\|\,Cu^{2+}(aq)\,|\,Cu(s)$$

The components of the oxidation half-cell (anode) are written on the left side, and the components of the reduction half-cell (cathode) are written on the right. A single vertical line separates the solid Zn anode from the Zn^{2+} solution, and another vertical line separates the Cu^{2+} solution from the Cu cathode. A double vertical line separates the two half-cells.

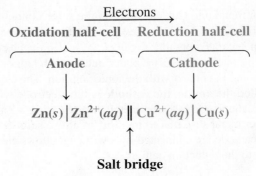

In some voltaic cells, there is no component in the half-reactions that can be used as an electrode. When this is the case, inert electrodes made of graphite or platinum are used for the transfer of electrons. If there are two ionic components in a cell, their symbols are separated by a comma. For example, suppose a voltaic cell consists of a platinum anode placed in a Sn^{2+} solution, and a silver cathode placed in a Ag^+ solution. The notation for the cell would be written as

$$Pt(s) \mid Sn^{2+}(aq), Sn^{4+}(aq) \parallel Ag^+(aq) \mid Ag(s)$$

The oxidation reaction at the anode is

$$Sn^{2+}(aq) \longrightarrow Sn^{4+}(aq) + 2\,e^-$$

The reduction reaction at the cathode is

$$Ag^+(aq) + e^- \longrightarrow Ag(s)$$

To balance the overall cell reaction, we multiply the cathode reduction by 2 and combine the two half-reactions.

$$2Ag^+(aq) + \mathbf{2\,e^-} \longrightarrow 2Ag(s) \qquad \text{Reduction}$$
$$\underline{Sn^{2+}(aq) \longrightarrow Sn^{4+}(aq) + \mathbf{2\,e^-}} \qquad \text{Oxidation}$$
$$Sn^{2+}(aq) + 2Ag^+(aq) \longrightarrow Sn^{4+}(aq) + 2Ag(s) \qquad \text{Oxidation–Reduction Reaction}$$

SAMPLE PROBLEM 15.9 Diagramming a Voltaic Cell

A voltaic cell consists of an iron (Fe) anode in a Fe^{2+} solution and a tin (Sn) cathode placed in a Sn^{2+} solution. Write the cell notation, the oxidation and reduction half-reactions, and the overall cell reaction.

TRY IT FIRST

SOLUTION

The notation for the cell would be written as

$$Fe(s) \mid Fe^{2+}(aq) \parallel Sn^{2+}(aq) \mid Sn(s)$$

The oxidation reaction at the anode is

$$Fe(s) \longrightarrow Fe^{2+}(aq) + 2\,e^-$$

The reduction reaction at the cathode is

$$Sn^{2+}(aq) + 2\,e^- \longrightarrow Sn(s)$$

To write the overall cell reaction, we combine the two half-reactions.

$$Fe(s) \longrightarrow Fe^{2+}(aq) + 2\,e^- \qquad \text{Oxidation}$$
$$\underline{Sn^{2+}(aq) + 2\,e^- \longrightarrow Sn(s)} \qquad \text{Reduction}$$
$$Fe(s) + Sn^{2+}(aq) \longrightarrow Fe^{2+}(aq) + Sn(s) \qquad \text{Oxidation–Reduction Reaction}$$

STUDY CHECK 15.9

Write the half-reactions and the overall cell reaction for the following notation of a voltaic cell:

$$Co(s) \mid Co^{2+}(aq) \parallel Cu^{2+}(aq) \mid Cu(s)$$

ANSWER

Anode reaction: $\qquad\qquad\qquad Co(s) \longrightarrow Co^{2+}(aq) + 2\,e^-$

Cathode reaction: $\qquad Cu^{2+}(aq) + 2\,e^- \longrightarrow Cu(s)$

Overall cell reaction: $\quad Co(s) + Cu^{2+}(aq) \longrightarrow Co^{2+}(aq) + Cu(s)$

Batteries come in many shapes and sizes.

Batteries

Batteries are needed to power your cell phone, watch, and calculator. Batteries also are needed to make cars start, and flashlights produce light. Within each of these batteries are voltaic cells that produce electrical energy. Let's look at some examples of commonly used batteries.

Lead Storage Battery

A lead storage battery is used to operate the electrical system in a car. We need a car battery to start the engine, turn on the lights, or operate the radio. If the battery runs down, the car won't start and the lights won't turn on. A car battery or a lead storage battery is a type of voltaic cell. In a typical 12-V battery, there are six voltaic cells linked together. Each of the cells consists of a lead (Pb) plate that acts as the anode and a lead(IV) oxide (PbO_2) plate that acts as the cathode. Both half-cells contain a sulfuric acid (H_2SO_4) solution. When the car battery is producing electrical energy (discharging), the following half-reactions take place:

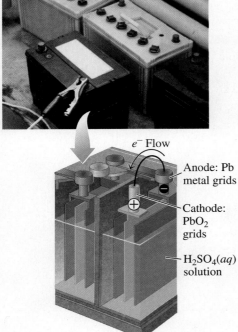

A 12-volt car battery is also known as a lead storage battery.

Anode (oxidation):

$$Pb(s) + SO_4^{2-}(aq) \longrightarrow PbSO_4(s) + 2\,e^-$$

Cathode (reduction):

$$2\,e^- + 4H^+(aq) + PbO_2(s) + SO_4^{2-}(aq) \longrightarrow PbSO_4(s) + 2H_2O(l)$$

Overall cell reaction:

$$4H^+(aq) + Pb(s) + PbO_2(s) + 2SO_4^{2-}(aq) \longrightarrow 2PbSO_4(s) + 2H_2O(l)$$

In both half-reactions, Pb^{2+} is produced, which combines with SO_4^{2-} to form an insoluble ionic compound $PbSO_4$. As a car battery is used, there is a buildup of $PbSO_4$ on the electrodes. At the same time, there is a decrease in the concentrations of the sulfuric acid components, H^+ and SO_4^{2-}. As a car runs, the battery is continuously recharged by an alternator, which is powered by the engine. The recharging reactions restore the Pb and PbO_2 electrodes as well as H_2SO_4. Without recharging, the car battery cannot continue to produce electrical energy.

Dry-Cell Batteries

Dry-cell batteries are used in calculators, watches, flashlights, and battery-operated toys. The term *dry cell* describes a battery that uses a paste rather than an aqueous solution. Dry cells can be acidic or alkaline. In an acidic dry cell, the anode is a zinc metal case that contains a paste of MnO_2, NH_4Cl, $ZnCl_2$, H_2O, and starch. Within this MnO_2 electrolyte mixture is a graphite cathode.

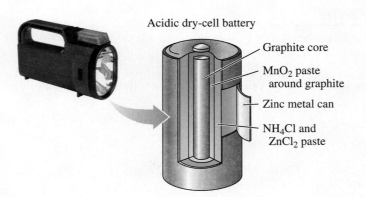

In an acidic dry-cell battery, the cathode is graphite, and the anode is a zinc case.

Anode (oxidation): $Zn(s) \longrightarrow Zn^{2+}(aq) + \mathbf{2\,e^-}$

Cathode (reduction): $\mathbf{2\,e^-} + 2MnO_2(s) + 2NH_4^+(aq) \longrightarrow Mn_2O_3(s) + 2NH_3(aq) + H_2O(l)$

Overall cell reaction: $Zn(s) + 2MnO_2(s) + 2NH_4^+(aq) \longrightarrow Zn^{2+}(aq) + Mn_2O_3(s) + 2NH_3(aq) + H_2O(l)$

An alkaline battery has similar components except that NaOH or KOH replaces the NH_4Cl electrolyte. Under basic conditions, the product of oxidation is zinc oxide (ZnO). Alkaline batteries tend to be more expensive, but they last longer and produce more power than acidic dry-cell batteries.

Anode (oxidation): $Zn(s) + 2OH^-(aq) \longrightarrow ZnO(s) + H_2O(l) + \mathbf{2\,e^-}$

Cathode (reduction): $\mathbf{2\,e^-} + 2MnO_2(s) + H_2O(l) \longrightarrow Mn_2O_3(s) + 2OH^-(aq)$

Overall cell reaction: $Zn(s) + 2MnO_2(s) \longrightarrow ZnO(s) + Mn_2O_3(s)$

Nickel–Cadmium (NiCad) Batteries

Nickel–cadmium (NiCad) batteries can be recharged. They use a cadmium anode and a cathode of solid nickel oxide NiO(OH)(s).

Anode (oxidation): $Cd(s) + 2OH^-(aq) \longrightarrow Cd(OH)_2(s) + \mathbf{2\,e^-}$

Cathode (reduction): $\mathbf{2\,e^-} + 2NiO(OH)(s) + 2H_2O(l) \longrightarrow$
$$2Ni(OH)_2(s) + 2OH^-(aq)$$

Overall cell reaction: $Cd(s) + 2NiO(OH)(s) + 2H_2O(l) \longrightarrow$
$$Cd(OH)_2(s) + 2Ni(OH)_2(s)$$

NiCad batteries are expensive, but they can be recharged many times. A charger provides an electrical current that converts the solid $Cd(OH)_2$ and $Ni(OH)_2$ products in the NiCad battery back to the reactants.

A NiCad battery in a cell phone can be recharged many times.

CHEMISTRY LINK TO THE **ENVIRONMENT**
Corrosion: Oxidation of Metals

Metals used in building materials, such as iron, eventually oxidize, which causes deterioration of the metal. This oxidation process, known as *corrosion*, produces rust on cars, bridges, ships, and underground pipes.

$$4Fe(s) + 3O_2(g) \longrightarrow 2Fe_2O_3(s)$$
Rust

The formation of rust requires both oxygen and water. The process of rusting requires an anode and cathode in different places on the surface of a piece of iron. In one area of the iron surface, called the *anode region*, the oxidation half-reaction takes place (see **FIGURE 15.2**).

Anode (oxidation): $Fe(s) \longrightarrow Fe^{2+}(aq) + 2\,e^-$
or
$$2Fe(s) \longrightarrow 2Fe^{2+}(aq) + 4\,e^-$$

The electrons move through the iron metal from the anode to an area called the *cathode region* where oxygen dissolved in water is reduced to water.

Cathode (reduction): $4\,e^- + 4H^+(aq) + O_2(g) \longrightarrow 2H_2O(l)$

By combining the half-reactions that occur in the anode and cathode regions, we can write the overall oxidation–reduction equation.

$$2Fe(s) + 4H^+(aq) + O_2(g) \longrightarrow 2Fe^{2+}(aq) + 2H_2O(l)$$

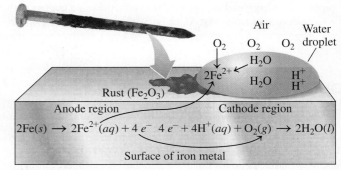

FIGURE 15.2 ▶ Rust forms when electrons from the oxidation of Fe flow from the anode region to the cathode region where oxygen is reduced. As Fe^{2+} ions come in contact with O_2 and H_2O, rust forms.

Why must both O_2 and H_2O be present for the corrosion of iron?

The formation of rust occurs as Fe^{2+} ions move out of the anode region and come in contact with dissolved oxygen (O_2). The Fe^{2+} oxidizes to Fe^{3+}, which reacts with oxygen to form rust.

$$4H_2O(l) + 4Fe^{2+}(aq) + O_2(g) \longrightarrow 2Fe_2O_3(s) + 8H^+(aq)$$
Rust

We can write the formation of rust starting with solid Fe reacting with O_2 as follows. There is no H^+ in the overall equation because H^+ is used and produced in equal quantities.

Corrosion of iron

$$4Fe(s) + 3O_2(g) \longrightarrow 2Fe_2O_3(s)$$
Rust

Other metals such as aluminum, copper, and silver also undergo corrosion, but at a slower rate than iron. The oxidation of Al on the surface of an aluminum object produces Al^{3+}, which reacts with oxygen in the air to form a protective coating of Al_2O_3. This Al_2O_3 coating prevents further oxidation of the aluminum underneath it.

$$Al(s) \longrightarrow Al^{3+}(aq) + 3\,e^-$$

When copper is used on a roof, dome, or a steeple, it oxidizes to Cu^{2+}, which is converted to a green patina of $Cu_2(OH)_2CO_3$.

$$Cu(s) \longrightarrow Cu^{2+}(aq) + 2\,e^-$$

When we use silver dishes and utensils, the Ag^+ ion from oxidation reacts with sulfides in food to form Ag_2S, which we call "tarnish."

$$Ag(s) \longrightarrow Ag^+(aq) + e^-$$

Prevention of Corrosion

Billions of dollars are spent each year to prevent corrosion and repair building materials made of iron. One way to prevent corrosion is to paint the bridges, cars, and ships with paints containing materials that seal the iron surface from H_2O and O_2. But it is necessary to repaint often; a scratch in the paint exposes the iron, which then begins to rust.

A more effective way to prevent corrosion is to place the iron in contact with a metal that substitutes for the anode region of iron. Metals such as Zn, Mg, or Al lose electrons more easily than iron. When one of these metals is in contact with iron, the metal acts as the anode instead of iron. For example, in a process called *galvanization*, an iron object is coated with zinc. The zinc becomes the anode because zinc is more easily oxidized than Fe. As long as Fe does not act as an anode, rust cannot form.

In a method called *cathodic protection*, structures such as iron pipes and underground storage containers are placed in contact with a piece of metal such as Mg, Al, or Zn, which is called the *sacrificial*

Oxidation (corrosion) is prevented by painting a metal surface.

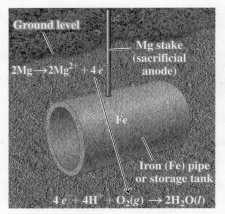

Rusting of an iron pipe is prevented by attaching a piece of magnesium, which oxidizes more easily than iron.

anode. Again, because these metals lose electrons more easily than Fe, they become the anode, thereby preventing the rusting of the iron. A magnesium plate that is welded or bolted to a ship's hull loses electrons more easily than iron or steel and protects the hull from rusting. Occasionally, a new magnesium plate is added to replace the magnesium as it is used up. Magnesium stakes placed in the ground are connected to underground pipelines and storage containers to prevent corrosion damage.

QUESTIONS AND PROBLEMS

15.3 Electrical Energy from Oxidation–Reduction Reactions

LEARNING GOAL Use the activity series to determine if an oxidation–reduction reaction is spontaneous. Write the half-reactions that occur in a voltaic cell and the cell notation.

15.27 Use the activity series in Table 15.3 to predict whether each of the following reactions will occur spontaneously:
   a. $2Au(s) + 6H^+(aq) \longrightarrow 2Au^{3+}(aq) + 3H_2(g)$
   b. $Ni^{2+}(aq) + Fe(s) \longrightarrow Ni(s) + Fe^{2+}(aq)$
   c. $2Ag(s) + Cu^{2+}(aq) \longrightarrow 2Ag^+(aq) + Cu(s)$

15.28 Use the activity series in Table 15.3 to predict whether each of the following reactions will occur spontaneously:
   a. $2Ag(s) + 2H^+(aq) \longrightarrow 2Ag^+(aq) + H_2(g)$
   b. $Mg(s) + Cu^{2+}(aq) \longrightarrow Mg^{2+}(aq) + Cu(s)$
   c. $2Al(s) + 3Cu^{2+}(aq) \longrightarrow 2Al^{3+}(aq) + 3Cu(s)$

15.29 Write the half-reactions and the overall cell reaction for each of the following voltaic cells:
   a. $Pb(s)\,|\,Pb^{2+}(aq)\,\|\,Cu^{2+}(aq)\,|\,Cu(s)$
   b. $Cr(s)\,|\,Cr^{2+}(aq)\,\|\,Ag^+(aq)\,|\,Ag(s)$

15.30 Write the half-reactions and the overall cell reaction for each of the following voltaic cells:
   a. $Al(s)\,|\,Al^{3+}(aq)\,\|\,Cd^{2+}(aq)\,|\,Cd(s)$
   b. $Sn(s)\,|\,Sn^{2+}(aq)\,\|\,Fe^{3+}(aq),\,Fe^{2+}(aq)\,|\,C$ (electrode)

15.31 Describe the voltaic cell and half-cell components and write the shorthand notation for the following oxidation–reduction reactions:
   a. $Cd(s) + Sn^{2+}(aq) \longrightarrow Cd^{2+}(aq) + Sn(s)$
   b. $Zn(s) + Cl_2(g) \longrightarrow Zn^{2+}(aq) + 2Cl^-(aq)$ C (electrode)

15.32 Describe the voltaic cell and half-cell components and write the shorthand notation for the following oxidation–reduction reactions:
   a. $Mn(s) + Sn^{2+}(aq) \longrightarrow Mn^{2+}(aq) + Sn(s)$
   b. $Ni(s) + 2Ag^{+}(aq) \longrightarrow Ni^{2+}(aq) + 2Ag(s)$

15.33 The following half-reaction takes place in a nickel–cadmium battery used in a cordless drill:

$$Cd(s) + 2OH^{-}(aq) \longrightarrow Cd(OH)_2(s) + 2\,e^{-}$$

   a. Is the half-reaction an oxidation or a reduction?
   b. What substance is oxidized or reduced?
   c. At which electrode would this half-reaction occur?

15.34 The following half-reaction takes place in a mercury battery used in hearing aids:

$$HgO(s) + H_2O(l) + 2\,e^{-} \longrightarrow Hg(l) + 2OH^{-}(aq)$$

   a. Is the half-reaction an oxidation or a reduction?
   b. What substance is oxidized or reduced?
   c. At which electrode would this half-reaction occur?

15.35 The following half-reaction takes place in a mercury battery used in pacemakers and watches:

$$Zn(s) + 2OH^{-}(aq) \longrightarrow ZnO(s) + H_2O(l) + 2\,e^{-}$$

   a. Is the half-reaction an oxidation or a reduction?
   b. What substance is oxidized or reduced?
   c. At which electrode would this half-reaction occur?

15.36 The following half-reaction takes place in a lead storage battery used in automobiles:

$$Pb(s) + SO_4^{2-}(aq) \longrightarrow PbSO_4(s) + 2\,e^{-}$$

   a. Is the half-reaction an oxidation or a reduction?
   b. What substance is oxidized or reduced?
   c. At which electrode would this half-reaction occur?

CHEMISTRY LINK TO THE ENVIRONMENT
Fuel Cells: Clean Energy for the Future

   Fuel cells are of interest to scientists because they provide an alternative source of electrical energy that is more efficient, does not use up oil reserves, and generates products that do not pollute the atmosphere. Fuel cells are considered to be a clean way to produce energy.

   Like other electrochemical cells, a fuel cell consists of an anode and a cathode connected by a wire. But unlike other cells, the reactants must continuously enter the fuel cell to produce energy; electrical current is generated only as long as the fuels are supplied. One type of hydrogen–oxygen fuel cell has been used in automobile prototypes. In this hydrogen cell, gas enters the fuel cell and comes in contact with a platinum catalyst embedded in a plastic membrane. The catalyst assists in the oxidation of hydrogen atoms to hydrogen ions and electrons (see **FIGURE 15.3**).

   The electrons produce an electric current as they travel through the wire from the anode to the cathode. The hydrogen ions flow through the plastic membrane to the cathode. At the cathode, oxygen molecules are reduced to oxide ions that combine with the hydrogen ions to form water. The overall hydrogen–oxygen fuel cell reaction can be written as

$$2H_2(g) + O_2(g) \longrightarrow 2H_2O(l)$$

   Fuel cells have already been used for power on the space shuttle and are now used to produce energy for cars and buses. An advantage of fuel cell cars is that the hydrogen fuel produces water and heat, which gives zero pollution.

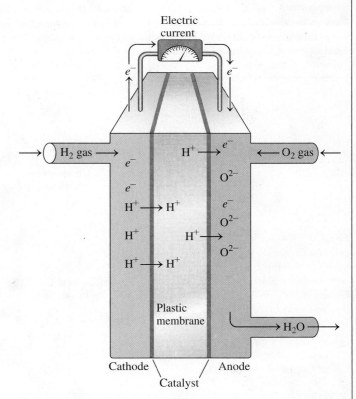

Oxidation
$$2H_2(g) \longrightarrow 4\,e^{-} + 4H^{+}(aq)$$

Reduction
$$4\,e^{-} + 4H^{+}(aq) + O_2(g) \longrightarrow 2H_2O(l)$$

FIGURE 15.3 ▶ With a supply of hydrogen and oxygen, a fuel cell can generate electricity continuously.

◉ In most electrochemical cells, the electrodes are eventually used up. Is this true for a fuel cell? Why or why not?

In homes, fuel cells may one day replace the batteries currently used to provide electrical power for cell phones, DVD players, and laptop computers. Fuel cell design is still in the prototype phase, although there is much interest in their development. We already know they can work, but modifications must still be made before they become reasonably priced and part of our everyday lives.

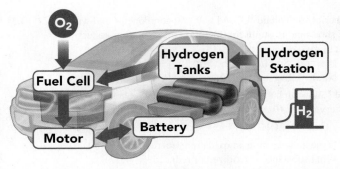

Fuel cells in cars use hydrogen as the fuel and produce water and heat.

15.4 Oxidation–Reduction Reactions That Require Electrical Energy

LEARNING GOAL Describe the half-cell reactions and the overall reactions that occur in electrolysis.

When we look at the activity series in Table 15.3, we see that the oxidation of Cu is below Zn. This means that the following oxidation–reduction reaction is not spontaneous:

$$Cu(s) + Zn^{2+}(aq) \longrightarrow Cu^{2+}(aq) + Zn(s) \quad \text{Not spontaneous}$$

Less active .... More active

To make this reaction take place, we need an **electrolytic cell** that uses an electrical current to drive a nonspontaneous oxidation–reduction reaction. This process is called **electrolysis** (see **FIGURE 15.4**).

ENGAGE

Why does the reaction in an electrolytic cell require an electrical current?

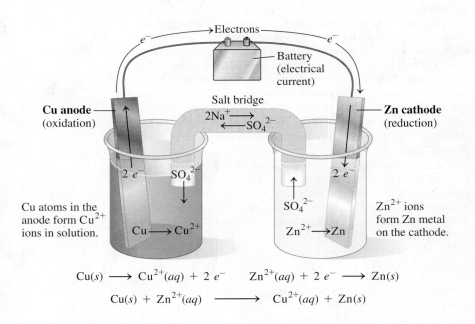

FIGURE 15.4 ▶ In this electrolytic cell, the Cu anode is in a Cu^{2+} solution, and the Zn cathode is in a Zn^{2+} solution. Electrons provided by a battery reduce Zn^{2+} to Zn and drive the oxidation of Cu to Cu^{2+} at the Cu anode.

Q Why is an electrical current needed to make the reaction of Cu(s) and $Zn^{2+}(aq)$ happen?

Cu atoms in the anode form Cu^{2+} ions in solution.

$Cu \longrightarrow Cu^{2+}$

Zn^{2+} ions form Zn metal on the cathode.

$Zn^{2+} \longrightarrow Zn$

$$Cu(s) \longrightarrow Cu^{2+}(aq) + 2\ e^- \qquad Zn^{2+}(aq) + 2\ e^- \longrightarrow Zn(s)$$

$$Cu(s) + Zn^{2+}(aq) \longrightarrow Cu^{2+}(aq) + Zn(s)$$

Electrolysis of Sodium Chloride

When molten sodium chloride is electrolyzed, the products are sodium metal and chlorine gas. In this electrolytic cell, electrodes are placed in the mixture of Na^+ and Cl^- and connected to a battery. The products are separated to prevent them from reacting spontaneously with each other. As electrons flow to the cathode, Na^+ is reduced to sodium metal. At the same time, electrons leave the anode as Cl^- is oxidized to Cl_2. The half-reactions and the overall reactions are

Anode (oxidation): $\qquad\qquad 2Cl^-(l) \longrightarrow Cl_2(g) + 2\,e^-$
Cathode (reduction): $\quad 2Na^+(l) + 2\,e^- \longrightarrow 2Na(l)$
$\overline{\qquad\qquad\qquad\quad 2Na^+(l) + 2Cl^-(l) \longrightarrow 2Na(l) + Cl_2(g) \quad}$ Not spontaneous

Electrical energy

Electroplating

In industry, the process of electroplating uses electrolysis to coat an object with a thin layer of a metal such as silver, platinum, or gold. Car bumpers are electroplated with chromium to prevent rusting. Silver-plated utensils, bowls, and platters are made by electroplating objects with a layer of silver.

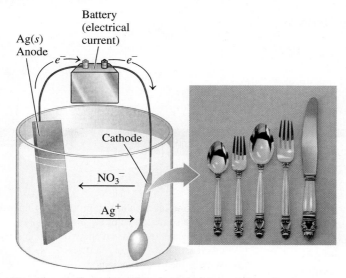

Utensils are silver-plated by electrolysis.

SAMPLE PROBLEM 15.10 Electrolysis

Electrolysis is used to chrome plate an iron hubcap by placing the hubcap in a Cr^{3+} solution.

a. What half-reaction takes place to plate the hubcap with metallic chromium?
b. Is the iron hubcap the anode or the cathode?

TRY IT FIRST

SOLUTION

a. The Cr^{3+} ions in solution would gain electrons (reduction).

$\quad Cr^{3+}(aq) + 3\,e^- \longrightarrow Cr(s)$

b. The iron hubcap is the cathode where reduction takes place.

STUDY CHECK 15.10

Why is energy needed to chrome plate the iron in Sample Problem 15.10?

ANSWER

Since Fe is below Cr in the activity series, the plating of Cr^{3+} onto Fe is not spontaneous. Energy is needed to make the reaction proceed.

Using electrolysis, a thin layer of chromium is plated onto a hubcap.

QUESTIONS AND PROBLEMS

15.4 Oxidation–Reduction Reactions That Require Electrical Energy

LEARNING GOAL Describe the half-cell reactions and the overall reactions that occur in electrolysis.

15.37 What we call "tin cans" are really iron cans coated with a thin layer of tin. The anode is a bar of tin and the cathode is the iron can. An electrical current is used to oxidize the Sn to Sn^{2+} in solution, which is reduced to produce a thin coating of Sn on the can.
  a. What half-reaction takes place to tin plate an iron can?
  b. Why is the iron can the cathode?
  c. Why is the tin bar the anode?

An iron can is coated with tin to prevent rusting.

15.38 Electrolysis is used to gold plate jewelry made of stainless steel.
  a. What half-reaction takes place when a Au^{3+} solution is used to gold plate a stainless steel ring?
  b. Is the ring the anode or the cathode?
  c. Why is energy needed to gold plate the ring?

15.39 When the tin coating on an iron can is scratched, rust will form. Use the activity series in Table 15.3 to explain why this happens.

15.40 When the zinc coating on an iron can is scratched, rust does not form. Use the activity series in Table 15.3 to explain why this happens.

Follow Up

WHITENING KIMBERLY'S TEETH

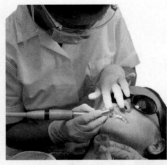

The reaction of hydrogen peroxide, $H_2O_2(aq)$, placed on Kimberly's teeth produces water, $H_2O(l)$, and oxygen gas. After Kimberly had her teeth cleaned and whitened, Jane recommended that Kimberly use a toothpaste containing tin(II) fluoride. Tooth enamel, which is composed of hydroxyapatite, $Ca_5(PO_4)_3OH$, is strengthened when it reacts with fluoride ions to form fluoroapatite, $Ca_5(PO_4)_3F$.

15.41 The simplified reaction for the production of aqueous H_2O_2 involves the combination of H_2 gas and O_2 gas.
  a. What reactant is oxidized and what reactant is reduced in this reaction?
  b. What is the oxidizing agent?
  c. What is the reducing agent?
  d. Write a balanced chemical equation for the reaction.

15.42 The reaction of solid Sn and F_2 gas forms solid SnF_2.
  a. What reactant is oxidized and what reactant is reduced in this reaction?
  b. What is the oxidizing agent?
  c. What is the reducing agent?
  d. Write a balanced chemical equation for the reaction.

CONCEPT MAP

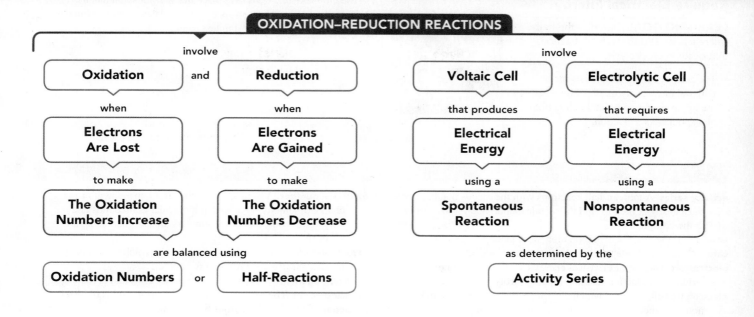

OXIDATION–REDUCTION REACTIONS

involve

Oxidation and **Reduction**

when **Electrons Are Lost**

when **Electrons Are Gained**

to make **The Oxidation Numbers Increase**

to make **The Oxidation Numbers Decrease**

are balanced using

Oxidation Numbers or **Half-Reactions**

involve

Voltaic Cell   **Electrolytic Cell**

that produces **Electrical Energy**

that requires **Electrical Energy**

using a **Spontaneous Reaction**

using a **Nonspontaneous Reaction**

as determined by the

Activity Series

CHAPTER REVIEW

15.1 Oxidation and Reduction

LEARNING GOAL Identify a reaction as an oxidation or a reduction. Assign and use oxidation numbers to identify elements that are oxidized or reduced.

- In an oxidation–reduction reaction, electrons are transferred from one reactant to another.
- The reactant that loses electrons is oxidized, and the reactant that gains electrons is reduced.
- Oxidation numbers assigned to elements keep track of the changes in the loss and gain of electrons.
- Oxidation is an increase in oxidation number; reduction is a decrease in oxidation number.
- The reducing agent is the substance that provides electrons for reduction.
- The oxidizing agent is the substance that accepts the electrons from oxidation.
- In molecular compounds and polyatomic ions, oxidation numbers are assigned using a set of rules.
- The oxidation number of an element is zero, and the oxidation number of a monatomic ion is the same as the ionic charge of the ion.
- The sum of the oxidation numbers for a compound is equal to zero and for a polyatomic ion is equal to the overall charge.
- Balancing oxidation–reduction equations using oxidation numbers involves the following:
  1. assigning oxidation numbers
  2. determining the loss and gain of electrons
  3. equalizing the loss and gain of electrons
  4. balancing the remaining substances by inspection

15.2 Balancing Oxidation–Reduction Equations Using Half-Reactions

LEARNING GOAL Balance oxidation–reduction equations using the half-reaction method.

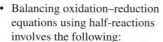

- Balancing oxidation–reduction equations using half-reactions involves the following:
  1. separating the equation into half-reactions
  2. balancing elements other than H and O
  3. if the reaction takes place in acid or base, balancing O with H_2O and H with H^+
  4. balancing charge with electrons
  5. equalizing the loss and gain of electrons
  6. combining half-reactions, canceling electrons, and combining H_2O and H^+
  7. using OH^- to neutralize H^+ to H_2O for an oxidation–reduction reaction in basic solution

15.3 Electrical Energy from Oxidation–Reduction Reactions

LEARNING GOAL Use the activity series to determine if an oxidation–reduction reaction is spontaneous. Write the half-reactions that occur in a voltaic cell and the cell notation.

- In a voltaic cell, the components of the two half-reactions of a spontaneous oxidation–reduction reaction are placed in separate containers called half-cells.
- With a wire connecting the half-cells, an electrical current is generated as electrons move from the anode where oxidation takes place to the cathode where reduction takes place.

15.4 Oxidation–Reduction Reactions That Require Electrical Energy

LEARNING GOAL Describe the half-cell reactions and the overall reactions that occur in electrolysis.

- The activity series, which lists metals with the most easily oxidized metal at the top, is used to predict the direction of a spontaneous reaction.

- In an electrolytic cell, electrical energy from an external source is used to make reactions take place that are not spontaneous.
- A method called electrolysis is used to plate chrome on hubcaps, zinc on iron, or gold on stainless steel jewelry.

 ### CORE CHEMISTRY SKILLS

The chapter section containing each Core Chemistry Skill is shown in parentheses at the end of each heading.

Assigning Oxidation Numbers (15.1)

- Oxidation numbers are assigned to atoms or ions in a reaction to determine the substance that is oxidized and the substance that is reduced.
- The rules for assigning oxidation numbers are as follows:

  1. for a molecule the sum is 0; for a polyatomic ion the sum is its charge
  2. for an element is 0
  3. for a monatomic ion is its charge
  4. for Group 1A (1) metals is +1, Group 2A (2) metals is +2
  5. for fluorine is −1; for other Group 7A (17) metals is −1, except when combined with O or F
  6. for oxygen is −2, except in OF_2 or peroxides
  7. for H in nonmetal compounds is +1; for H with a metal is −1

Example: Assign oxidation numbers to each of the elements in the following:

    a. Sn^{4+} **b.** Mn   **c.** MnO_2 **d.** $MgCO_3$

Answer: **a.** Sn^{4+} **b.** Mn   **c.** MnO_2 **d.** $MgCO_3$
        +4   0   +4−2   +2+4−2

Using Oxidation Numbers (15.1)

- An atom is oxidized when its oxidation number increases from reactant to product.
- An atom is reduced when its oxidation number decreases from reactant to product.

Example: Assign oxidation numbers to each element and identify which is oxidized and which is reduced.

$$Fe_2O_3(s) + 2Al(s) \longrightarrow Al_2O_3(s) + 2Fe(l)$$

Answer: $Fe_2O_3(s) + 2Al(s) \longrightarrow Al_2O_3(s) + 2Fe(l)$
       +3 −2   0   +3 −2   0

   $Fe(+3) \longrightarrow Fe(0)$ Fe is reduced.
   $Al(0) \longrightarrow Al(+3)$ Al is oxidized.

Identifying Oxidizing and Reducing Agents (15.1)

- In an oxidation–reduction reaction, the substance that is reduced is the oxidizing agent.
- In an oxidation–reduction reaction, the substance that is oxidized is the reducing agent.

Example: Identify the oxidizing agent and the reducing agent in the following:

$$SnO_2\,(s) + 2H_2(g) \longrightarrow Sn(s) + 2H_2O(l)$$

Answer: Oxidation numbers are assigned to identify the substance that is oxidized as the reducing agent and the substance that is reduced as the oxidizing agent.

$$SnO_2(s) + 2H_2(g) \longrightarrow Sn(s) + 2H_2O(l)$$
  +4 −2   0   0   +1 −2

$Sn(+4) \longrightarrow Sn(0)$
Sn is reduced; SnO_2 is the oxidizing agent.

$H(0) \longrightarrow H(+1)$
H is oxidized; H_2 is the reducing agent.

Using Half-Reactions to Balance Redox Equations (15.2)

- The half-reaction method for balancing equations involves
  1. separating the oxidation–reduction reaction into two *half-reactions*
  2. balancing the elements in each except H and O; then if the reaction takes place in acid or base balancing O with H_2O and H with H^+
  3. balancing charge by adding electrons
  4. multiplying each half-reaction by factors that equalize the loss and gain of electrons
  5. combining half-reactions and cancelling electrons, identical ions or molecules, and using OH^- in base to neutralize H^+

Example: Use the half-reaction method to balance the following oxidation–reduction equation in acidic solution:

$$NO_3^-(aq) + Sn^{2+}(aq) \longrightarrow NO(g) + Sn^{4+}(aq)$$

Answer:

1. $NO_3^-(aq) \longrightarrow NO(g)$
   $Sn^{2+}(aq) \longrightarrow Sn^{4+}(aq)$

2. $4H^+(aq) + NO_3^-(aq) \longrightarrow NO(g) + 2H_2O(l)$
   $Sn^{2+}(aq) \longrightarrow Sn^{4+}(aq)$

3. $3\,e^- + 4H^+(aq) + NO_3^-(aq) \longrightarrow NO(g) + 2H_2O(l)$
   $Sn^{2+}(aq) \longrightarrow Sn^{4+}(aq) + \mathbf{2\,e^-}$

4. $2 \times [3\,e^- + 4H^+(aq) + NO_3^-(aq) \longrightarrow NO(g) + 2H_2O(l)]$
   $\mathbf{6\,e^-} + 8H^+(aq) + 2NO_3^-(aq) \longrightarrow 2NO(g) + 4H_2O(l)$

   $\mathbf{3} \times [Sn^{2+}(aq) \longrightarrow Sn^{4+}(aq) + \mathbf{2\,e^-}]$
   $3Sn^{2+}(aq) \longrightarrow 3Sn^{4+}(aq) + \mathbf{6\,e^-}$

5. $8H^+(aq) + 2NO_3^-(aq) + 3Sn^{2+}(aq) \longrightarrow$
   $\qquad\qquad 2NO(g) + 3Sn^{4+}(aq) + 4H_2O(l)$

Identifying Spontaneous Reactions (15.3)

- The activity series in Table 15.3 places the metals that are easily oxidized at the top, and the metals that do not oxidize easily at the bottom.
- A metal will oxidize spontaneously when it is combined with the reverse of the half-reaction of any metal below it on the activity series in Table 15.3.

Example: Write a balanced redox equation for the spontaneous reaction for the following half-reactions on the activity series:

$$Ca(s) \longrightarrow Ca^{2+}(aq) + 2\,e^-$$
$$Ni(s) \longrightarrow Ni^{2+}(aq) + 2\,e^-$$

Answer: The spontaneous reaction is a combination of the oxidation half-reaction higher on the activity series with the reverse of the half-reaction below it.

$Ca(s) \longrightarrow Ca^{2+}(aq) + 2\,e^-$ Oxidation
$Ni^{2+}(aq) + 2\,e^- \longrightarrow Ni(s)$ Reduction
$Ca(s) + Ni^{2+}(aq) \longrightarrow Ca^{2+}(aq) + Ni(s)$
$\qquad\qquad\qquad\qquad$ Spontaneous reaction

UNDERSTANDING THE CONCEPTS

The chapter sections to review are shown in parentheses at the end of each question.

15.43 Classify each of the following as oxidation or reduction: (15.1)
  a. Electrons are lost.
  b. Reaction of an oxidizing agent.
  c. $O_2(g) \longrightarrow OH^-(aq)$
  d. $Br_2(l) \longrightarrow 2Br^-(aq)$
  e. $Sn^{2+}(aq) \longrightarrow Sn^{4+}(aq)$

15.44 Classify each of the following as oxidation or reduction: (15.1)
  a. Electrons are gained.
  b. Reaction of a reducing agent.
  c. $Ni(s) \longrightarrow Ni^{2+}(aq)$
  d. $MnO_4^-(aq) \longrightarrow MnO_2(s)$
  e. $Sn^{4+}(aq) \longrightarrow Sn^{2+}(aq)$

15.45 Assign oxidation numbers to the elements in each of the following: (15.1)
  a. VO_2 b. Ag_2CrO_4
  c. $S_2O_8^{2-}$ d. $FeSO_4$

15.46 Assign oxidation numbers to the elements in each of the following: (15.1)
  a. $NbCl_3$ b. NbO
  c. NbO_2 d. Nb_2O_5

15.47 Which of the following are oxidation–reduction reactions? (15.1)
  a. $Ca(s) + 2H_2O(l) \longrightarrow Ca(OH)_2(aq) + H_2(g)$
  b. $CaCO_3(s) \longrightarrow CaO(s) + CO_2(g)$
  c. $4Al(s) + 3O_2(g) \longrightarrow 2Al_2O_3(s)$

15.48 Which of the following are oxidation–reduction reactions? (15.1)
  a. $BaCl_2(aq) + Na_2SO_4(aq) \longrightarrow BaSO_4(s) + 2NaCl(aq)$
  b. $Cl_2(g) + 2NaBr(aq) \longrightarrow Br_2(l) + 2NaCl(aq)$
  c. $2KClO_3(s) \longrightarrow 2KCl(s) + 3O_2(g)$

15.49 For this reaction, identify each of the following: (15.1)

$$2Cr_2O_3(s) + 3Si(s) \longrightarrow 4Cr(s) + 3SiO_2(s)$$

  a. the substance reduced
  b. the substance oxidized
  c. the oxidizing agent
  d. the reducing agent

Chromium(III) oxide and silicon undergo an oxidation–reduction reaction.

15.50 For this reaction, identify each of the following: (15.1)

$$C_2H_4(g) + 3O_2(g) \xrightarrow{\Delta} 2CO_2(g) + 2H_2O(g)$$

a. the substance reduced
b. the substance oxidized
c. the oxidizing agent
d. the reducing agent

15.51 Balance each of the following half-reactions in acidic solution: (15.2)

a. $Zn(s) \longrightarrow Zn^{2+}(aq)$
b. $SnO_2^{2-}(aq) \longrightarrow SnO_3^{2-}(aq)$
c. $SO_3^{2-}(aq) \longrightarrow SO_4^{2-}(aq)$
d. $NO_3^-(aq) \longrightarrow NO(g)$

15.52 Balance each of the following half-reactions in acidic solution: (15.2)

a. $I_2(s) \longrightarrow I^-(aq)$
b. $MnO_4^-(aq) \longrightarrow Mn^{2+}(aq)$
c. $Br_2(l) \longrightarrow BrO_3^-(aq)$
d. $ClO_3^-(aq) \longrightarrow ClO_4^-(aq)$

15.53 Consider the following voltaic cell: (15.3)

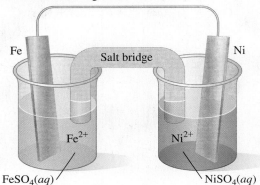

a. What is the oxidation half-reaction?
b. What is the reduction half-reaction?
c. What metal is the anode?
d. What metal is the cathode?
e. What is the direction of electron flow?
f. What is the overall reaction that takes place?
g. Write the shorthand cell notation.

15.54 Consider the following voltaic cell: (15.3)

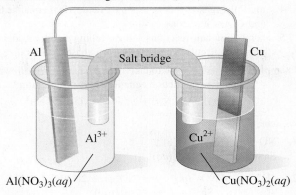

a. What is the oxidation half-reaction?
b. What is the reduction half-reaction?
c. What metal is the anode?
d. What metal is the cathode?
e. What is the direction of electron flow?
f. What is the overall reaction that takes place?
g. Write the shorthand cell notation.

ADDITIONAL QUESTIONS AND PROBLEMS

15.55 Which of the following are oxidation–reduction reactions? (15.1)

a. $AgNO_3(aq) + NaCl(aq) \longrightarrow AgCl(s) + NaNO_3(aq)$
b. $6Li(s) + N_2(g) \longrightarrow 2Li_3N(s)$
c. $Ni(s) + Pb(NO_3)_2(aq) \longrightarrow Ni(NO_3)_2(aq) + Pb(s)$
d. $2K(s) + 2H_2O(l) \longrightarrow 2KOH(aq) + H_2(g)$

15.56 Which of the following are oxidation–reduction reactions? (15.1)

a. $Ca(s) + F_2(g) \longrightarrow CaF_2(s)$
b. $Fe(s) + 2HCl(aq) \longrightarrow FeCl_2(aq) + H_2(g)$
c. $2NaCl(aq) + Pb(NO_3)_2(aq) \longrightarrow$
$$PbCl_2(s) + 2NaNO_3(aq)$$
d. $2CuCl(aq) \longrightarrow Cu(s) + CuCl_2(aq)$

15.57 In the mitochondria of human cells, energy is provided by the oxidation and reduction of the iron ions in the cytochromes. Identify each of the following reactions as an oxidation or reduction: (15.1)

a. $Fe^{3+} + e^- \longrightarrow Fe^{2+}$ b. $Fe^{2+} \longrightarrow Fe^{3+} + e^-$

15.58 Chlorine (Cl_2) is used as a germicide to kill microbes in swimming pools. If the product is Cl^-, was the elemental chlorine oxidized or reduced? (15.1)

15.59 Assign oxidation numbers to all the elements in each of the following: (15.1)

a. Co_2O_3 b. $KMnO_4$
c. $SbCl_5$ d. ClO_3^-
e. PO_4^{3-}

15.60 Assign oxidation numbers to all the elements in each of the following: (15.1)

a. PO_3^{2-} b. NH_4^+
c. $Fe(OH)_2$ d. HNO_3
e. H_2CO

15.61 Assign oxidation numbers to all the elements in each of the following reactions, and identify the reactant that is oxidized, the reactant that is reduced, the oxidizing agent, and the reducing agent: (15.1)

a. $2FeCl_2(aq) + Cl_2(g) \longrightarrow 2FeCl_3(aq)$
b. $2H_2S(g) + 3O_2(g) \longrightarrow 2H_2O(l) + 2SO_2(g)$
c. $P_2O_5(s) + 5C(s) \longrightarrow 2P(s) + 5CO(g)$

15.62 Assign oxidation numbers to all the elements in each of the following reactions, and identify the reactant that is oxidized, the reactant that is reduced, the oxidizing agent, and the reducing agent: (15.1)

a. $4Al(s) + 3O_2(g) \longrightarrow 2Al_2O_3(s)$
b. $I_2O_5(s) + 5CO(g) \longrightarrow I_2(g) + 5CO_2(g)$
c. $2Cr_2O_3(s) + 3C(s) \longrightarrow 4Cr(s) + 3CO_2(g)$

15.63 Write the balanced half-reactions and a balanced redox equation for each of the following reactions in acidic solution: (15.2)

a. $Zn(s) + NO_3^-(aq) \longrightarrow Zn^{2+}(aq) + NO_2(g)$
b. $MnO_4^-(aq) + SO_3^{2-}(aq) \longrightarrow Mn^{2+}(aq) + SO_4^{2-}(aq)$
c. $ClO_3^-(aq) + I^-(aq) \longrightarrow Cl^-(aq) + I_2(s)$
d. $Cr_2O_7^{2-}(aq) + C_2O_4^{2-}(aq) \longrightarrow Cr^{3+}(aq) + CO_2(g)$

15.64 Write the balanced half-reactions and a balanced redox equation for each of the following reactions in acidic solution: (15.2)

a. $Sn^{2+}(aq) + IO_4^-(aq) \longrightarrow Sn^{4+}(aq) + I^-(aq)$

b. $S_2O_3^{2-}(aq) + I_2(s) \longrightarrow S_4O_6^{2-}(aq) + I^-(aq)$

c. $Mg(s) + VO_4^{3-}(aq) \longrightarrow Mg^{2+}(aq) + V^{2+}(aq)$

d. $Al(s) + Cr_2O_7^{2-}(aq) \longrightarrow Al^{3+}(aq) + Cr^{3+}(aq)$

15.65 Use the activity series in Table 15.3 to predict whether each of the following reactions will occur spontaneously: (15.3)

a. $2Cr(s) + 3Ni^{2+}(aq) \longrightarrow 2Cr^{3+}(aq) + 3Ni(s)$

b. $Cu(s) + Zn^{2+}(aq) \longrightarrow Cu^{2+}(aq) + Zn(s)$

c. $Zn(s) + Pb^{2+}(aq) \longrightarrow Zn^{2+}(aq) + Pb(s)$

15.66 Use the activity series in Table 15.3 to predict whether each of the following reactions will occur spontaneously: (15.3)

a. $Zn(s) + Mg^{2+}(aq) \longrightarrow Zn^{2+}(aq) + Mg(s)$

b. $3Na(s) + Al^{3+}(aq) \longrightarrow 3Na^+(aq) + Al(s)$

c. $Mg(s) + Ni^{2+}(aq) \longrightarrow Mg^{2+}(aq) + Ni(s)$

15.67 In a voltaic cell, one half-cell consists of nickel metal in a Ni^{2+} solution, and the other half-cell consists of magnesium metal in a Mg^{2+} solution. Identify each of the following: (15.3)

a. the anode

b. the cathode

c. the half-reaction at the anode

d. the half-reaction at the cathode

e. the overall reaction

f. the shorthand cell notation

15.68 In a voltaic cell, one half-cell consists of zinc metal in a Zn^{2+} solution, and the other half-cell consists of copper metal in a Cu^{2+} solution. Identify each of the following: (15.3)

a. the anode

b. the cathode

c. the half-reaction at the anode

d. the half-reaction at the cathode

e. the overall reaction

f. the shorthand cell notation

15.69 Use the activity series in Table 15.3 to determine which of the following ions will be reduced when an iron strip is placed in an aqueous solution of that ion: (15.3)

a. $Ca^{2+}(aq)$ **b.** $Ag^+(aq)$

c. $Ni^{2+}(aq)$ **d.** $Al^{3+}(aq)$

e. $Pb^{2+}(aq)$

15.70 Use the activity series in Table 15.3 to determine which of the following ions will be reduced when an aluminum strip is placed in an aqueous solution of that ion: (15.3)

a. $Fe^{2+}(aq)$ **b.** $Au^{3+}(aq)$

c. $Mg^{2+}(aq)$ **d.** $H^+(aq)$

e. $Pb^{2+}(aq)$

15.71 In a lead storage battery, the following unbalanced half-reaction takes place: (15.3)

$$Pb(s) + SO_4^{2-}(aq) \longrightarrow PbSO_4(s)$$

a. Balance the half-reaction.

b. Is $Pb(s)$ oxidized or reduced?

c. Indicate whether the half-reaction takes place at the anode or cathode.

15.72 In an acidic dry-cell battery, the following unbalanced half-reaction takes place in acidic solution: (15.3)

$$MnO_2(s) \longrightarrow Mn_2O_3(s)$$

a. Balance the half-reaction.

b. Is $MnO_2(s)$ oxidized or reduced?

c. Indicate whether the half-reaction takes place at the anode or cathode.

15.73 Steel bolts made for sailboats are coated with zinc. Add the necessary components (electrodes, wires, batteries) to this diagram of an electrolytic cell with a zinc nitrate solution to show how it could be used to zinc plate a steel bolt. (15.3, 15.4)

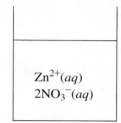

a. What is the anode?

b. What is the cathode?

c. What is the half-reaction that takes place at the anode?

d. What is the half-reaction that takes place at the cathode?

e. If steel is mostly iron, what is the purpose of the zinc coating?

15.74 Copper cooking pans are stainless steel pans plated with a layer of copper. Add the necessary components (electrodes, wires, batteries) to this diagram of an electrolytic cell with a copper(II) nitrate solution to show how it could be used to copper plate a stainless steel (iron) pan. (15.3, 15.4)

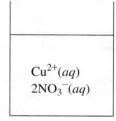

a. What is the anode?

b. What is the cathode?

c. What is the half-reaction that takes place at the anode?

d. What is the half-reaction that takes place at the cathode?

CHALLENGE QUESTIONS

The following groups of questions are related to the topics in this chapter. However, they do not all follow the chapter order, and they require you to combine concepts and skills from several sections. These questions will help you increase your critical thinking skills and prepare for your next exam.

15.75 Determine the oxidation number of Br in each of the following: (15.1)
   a. Br_2 **b.** $HBrO_2$
   c. BrO_3^- **d.** $NaBrO_4$

15.76 Determine the oxidation number of Cr in each of the following: (15.1)
   a. CrO **b.** $HCrO_4^-$
   c. CrO_3^- **d.** CrF_4

15.77 Use half-reactions to balance the following equation in acidic solution: (15.2)
$$Cr_2O_7^{2-}(aq) + NO_2^-(aq) \longrightarrow Cr^{3+}(aq) + NO_3^-(aq)$$

15.78 Use half-reactions to balance the following equation in acidic solution: (15.2)
$$Mn(s) + Cr^{3+}(aq) \longrightarrow Mn^{2+}(aq) + Cr(s)$$

15.79 The following unbalanced reaction takes place in acidic solution: (11.6, 15.2)
$$Ag(s) + NO_3^-(aq) \longrightarrow Ag^+(aq) + NO(g)$$
   a. Write the balanced equation.
   b. How many liters of $NO(g)$ are produced at STP when 15.0 g of silver reacts with excess nitric acid?

15.80 The following unbalanced reaction takes place in acidic solution: (12.6, 15.2)
$$MnO_4^-(aq) + Fe^{2+}(aq) \longrightarrow Mn^{2+}(aq) + Fe^{3+}(aq)$$
   a. Write the balanced equation.
   b. How many milliliters of a 0.150 M $KMnO_4$ solution are needed to react with 25.0 mL of a 0.400 M $FeSO_4$ solution?

15.81 A concentrated nitric acid solution is used to dissolve copper(II) sulfide. (12.6, 15.2)
$$CuS(s) + HNO_3(aq) \longrightarrow CuSO_4(aq) + NO(g) + H_2O(l)$$
   a. Write the balanced equation.
   b. How many milliliters of a 16.0 M HNO_3 solution are needed to dissolve 24.8 g of CuS?

15.82 The following unbalanced reaction takes place in acidic solution: (12.6, 15.2)
$$Cr_2O_7^{2-}(aq) + Fe^{2+}(aq) \longrightarrow Cr^{3+}(aq) + Fe^{3+}(aq)$$
   a. Write the balanced equation.
   b. How many milliliters of a 0.211 M $K_2Cr_2O_7$ solution are needed to react with 5.00 g of $FeSO_4$?

15.83 Balance the equation for the reaction that takes place in basic solution. (15.2)
$$Br^-(aq) + MnO_4^-(aq) \longrightarrow BrO_3^-(aq) + MnO_2(s)$$

15.84 Balance the equation for the reaction that takes place in basic solution. (15.2)
$$CN^-(aq) + IO_3^-(aq) \longrightarrow CNO^-(aq) + I^-(aq)$$

15.85 Using the activity series in Table 15.3, indicate whether each of the following reactions is spontaneous: (15.3)
   a. $Zn(s) + Ca^{2+}(aq) \longrightarrow Zn^{2+}(aq) + Ca(s)$
   b. $2Al(s) + 3Sn^{2+}(aq) \longrightarrow 2Al^{3+}(aq) + 3Sn(s)$
   c. $Mg(s) + 2H^+(aq) \longrightarrow Mg^{2+}(aq) + H_2(g)$

15.86 Using the activity series in Table 15.3, indicate whether each of the following reactions is spontaneous: (15.3)
   a. $Cu(s) + Ni^{2+}(aq) \longrightarrow Cu^{2+}(aq) + Ni(s)$
   b. $2Cr(s) + 3Fe^{2+}(aq) \longrightarrow 2Cr^{3+}(aq) + 3Fe(s)$
   c. $Fe(s) + Mg^{2+}(aq) \longrightarrow Fe^{2+}(aq) + Mg(s)$

15.87 Draw a diagram of a voltaic cell for $Ni(s)\,|\,Ni^{2+}(aq)\,\|\,Ag^+(aq)\,|\,Ag(s)$. (15.3)
   a. What is the anode?
   b. What is the cathode?
   c. What is the half-reaction that takes place at the anode?
   d. What is the half-reaction that takes place at the cathode?
   e. What is the overall reaction for the cell?

15.88 Draw a diagram of a voltaic cell for $Mg(s)\,|\,Mg^{2+}(aq)\,\|\,Al^{3+}(aq)\,|\,Al(s)$. (15.3)
   a. What is the anode?
   b. What is the cathode?
   c. What is the half-reaction that takes place at the anode?
   d. What is the half-reaction that takes place at the cathode?
   e. What is the overall reaction for the cell?

ANSWERS

Answers to Selected Questions and Problems

15.1 **a.** reduction   **b.** oxidation   **c.** reduction

15.3 **a.** Al is oxidized and O_2 is reduced.
   b. Zn is oxidized and H^+ is reduced.
   c. Br^- is oxidized and F_2 is reduced.

15.5 **a.** 0   **b.** 0   **c.** +2   **d.** −1

15.7 **a.** K is +1, Cl is −1.
   b. Mn is +4, O is −2.
   c. N is +2, O is −2.
   d. Mn is +3, O is −2.

15.9 **a.** Li is +1, P is +5, and O is −2.
   b. S is +4, O is −2.
   c. Cr is +3, S is −2.
   d. N is +5, O is −2.

15.11 **a.** H is +1, S is +6, and O is −2.
   b. H is +1, P is +3, and O is −2.
   c. Cr is +6, O is −2.
   d. Na is +1, C is +4, and O is −2.

15.13 **a.** +5   **b.** −2   **c.** +5   **d.** +6

15.15 **a.** The substance that is oxidized is the reducing agent.
   b. The substance that gains electrons is reduced and is the oxidizing agent.

15.17 **a.** Li is oxidized; Li is the reducing agent. Cl (in Cl_2) is reduced; Cl_2 is the oxidizing agent.
   b. Br^- (in NaBr) is oxidized; NaBr is the reducing agent. Cl (in Cl_2) is reduced; Cl_2 is the oxidizing agent.
   c. Pb is oxidized; Pb is the reducing agent. O (in O_2) is reduced; O_2 is the oxidizing agent.
   d. Al is oxidized; Al is the reducing agent. Ag^+ is reduced; Ag^+ is the oxidizing agent.

15.19 a. S^{2-} (in NiS) is oxidized; NiS is the reducing agent.
O (in O_2) is reduced; O_2 is the oxidizing agent.
  b. Sn^{2+} is oxidized; Sn^{2+} is the reducing agent. Fe^{3+} is reduced; Fe^{3+} is the oxidizing agent.
  c. C (in CH_4) is oxidized; CH_4 is the reducing agent. O (in O_2) is reduced; O_2 is the oxidizing agent.
  d. Si is oxidized; Si is the reducing agent. Cr^{3+} (in Cr_2O_3) is reduced; Cr_2O_3 is the oxidizing agent.

15.21 a. $Sn^{2+}(aq) \longrightarrow Sn^{4+}(aq) + 2\,e^-$
  b. $Mn^{2+}(aq) + 4H_2O(l) \longrightarrow MnO_4^-(aq) + 8H^+(aq) + 5\,e^-$
  c. $NO_2^-(aq) + H_2O(l) \longrightarrow NO_3^-(aq) + 2H^+(aq) + 2\,e^-$
  d. $e^- + 2H^+(aq) + ClO_3^-(aq) \longrightarrow ClO_2(aq) + H_2O(l)$

15.23 a. $2H^+(aq) + Ag(s) + NO_3^-(aq) \longrightarrow$
$Ag^+(aq) + NO_2(g) + H_2O(l)$
  b. $4H^+(aq) + 4NO_3^-(aq) + 3S(s) \longrightarrow$
$4NO(g) + 3SO_2(g) + 2H_2O(l)$
  c. $2S_2O_3^{2-}(aq) + Cu^{2+}(aq) \longrightarrow S_4O_6^{2-}(aq) + Cu(s)$

15.25 a. $2Fe(s) + 2CrO_4^{2-}(aq) + 2H_2O(l) \longrightarrow$
$Fe_2O_3(s) + Cr_2O_3(s) + 4OH^-(aq)$
  b. $3CN^-(aq) + 2MnO_4^-(aq) + H_2O(l) \longrightarrow$
$3CNO^-(aq) + 2MnO_2(s) + 2OH^-(aq)$

15.27 a. Since Au is below H_2 in the activity series, the reaction will not be spontaneous.
  b. Since Fe is above Ni in the activity series, the reaction will be spontaneous.
  c. Since Ag is below Cu in the activity series, the reaction will not be spontaneous.

15.29 a. Anode reaction: $Pb(s) \longrightarrow Pb^{2+}(aq) + 2\,e^-$
Cathode reaction: $Cu^{2+}(aq) + 2\,e^- \longrightarrow Cu(s)$
Overall cell reaction: $Pb(s) + Cu^{2+}(aq) \longrightarrow$
$Pb^{2+}(aq) + Cu(s)$
  b. Anode reaction: $Cr(s) \longrightarrow Cr^{2+}(aq) + 2\,e^-$
Cathode reaction: $Ag^+(aq) + e^- \longrightarrow Ag(s)$
Overall cell reaction: $Cr(s) + 2Ag^+(aq) \longrightarrow$
$Cr^{2+}(aq) + 2Ag(s)$

15.31 a. The anode is a Cd metal electrode in a Cd^{2+} solution. The anode reaction is
$Cd(s) \longrightarrow Cd^{2+}(aq) + 2\,e^-$
The cathode is a Sn metal electrode in a Sn^{2+} solution. The cathode reaction is
$Sn^{2+}(aq) + 2\,e^- \longrightarrow Sn(s)$
The shorthand notation for this cell is
$Cd(s)\,|\,Cd^{2+}(aq)\,\|\,Sn^{2+}(aq)\,|\,Sn(s)$
  b. The anode is a Zn metal electrode in a Zn^{2+} solution. The anode reaction is
$Zn(s) \longrightarrow Zn^{2+}(aq) + 2\,e^-$
The cathode is a C (graphite) electrode, where Cl_2 gas is reduced to Cl^-. The cathode reaction is
$Cl_2(g) + 2\,e^- \longrightarrow 2Cl^-(aq)$
The shorthand notation for this cell is
$Zn(s)\,|\,Zn^{2+}(aq)\,\|\,Cl_2(g),\,Cl^-(aq)\,|\,C$

15.33 a. The half-reaction is an oxidation.
  b. Cd metal is oxidized.
  c. Oxidation takes place at the anode.

15.35 a. The half-reaction is an oxidation.
  b. Zn metal is oxidized.
  c. Oxidation takes place at the anode.

15.37 a. $Sn^{2+}(aq) + 2\,e^- \longrightarrow Sn(s)$
  b. The reduction of Sn^{2+} to Sn occurs at the cathode, which is the iron can.
  c. The oxidation of Sn to Sn^{2+} occurs at the anode, which is the tin bar.

15.39 Since Fe is above Sn in the activity series, if the Fe is exposed to air and water, Fe will be oxidized and rust will form. To protect iron, Sn would have to be more active than Fe and it is not.

15.41 a. H_2 is oxidized and O_2 is reduced.
  b. O_2 is the oxidizing agent.
  c. H_2 is the reducing agent.
  d. $H_2(g) + O_2(g) \longrightarrow H_2O_2(aq)$

15.43 a. oxidation   **b.** reduction
  c. reduction   **d.** reduction
  e. oxidation

15.45 a. $V = +4, O = -2$
  b. $Ag = +1, Cr = +6, O = -2$
  c. $S = +7, O = -2$
  d. $Fe = +2, S = +6, O = -2$

15.47 Reactions **a** and **c** involve loss and gain of electrons; **a** and **c** are oxidation–reduction reactions.

15.49 a. Cr in Cr_2O_3 is reduced.
  b. Si is oxidized.
  c. Cr_2O_3 is the oxidizing agent.
  d. Si is the reducing agent.

15.51 a. $Zn(s) \longrightarrow Zn^{2+}(aq) + 2\,e^-$
  b. $SnO_2^{2-}(aq) + H_2O(l) \longrightarrow SnO_3^{2-}(aq) + 2H^+(aq) + 2\,e^-$
  c. $SO_3^{2-}(aq) + H_2O(l) \longrightarrow SO_4^{2-}(aq) + 2H^+(aq) + 2\,e^-$
  d. $3\,e^- + 4H^+(aq) + NO_3^-(aq) \longrightarrow NO(g) + 2H_2O(l)$

15.53 a. $Fe(s) \longrightarrow Fe^{2+}(aq) + 2\,e^-$
  b. $Ni^{2+}(aq) + 2\,e^- \longrightarrow Ni(s)$
  c. Fe is the anode.
  d. Ni is the cathode.
  e. The electrons flow from Fe to Ni.
  f. $Fe(s) + Ni^{2+}(aq) \longrightarrow Fe^{2+}(aq) + Ni(s)$
  g. $Fe(s)\,|\,Fe^{2+}(aq)\,\|\,Ni^{2+}(aq)\,|\,Ni(s)$

15.55 Reactions **b**, **c**, and **d** all involve loss and gain of electrons; **b**, **c**, and **d** are oxidation–reduction reactions.

15.57 a. Fe^{3+} is gaining electrons; this is a reduction.
  b. Fe^{2+} is losing electrons; this is an oxidation.

15.59 a. $Co = +3, O = -2$
  b. $K = +1, Mn = +7, O = -2$
  c. $Sb = +5, Cl = -1$
  d. $Cl = +5, O = -2$
  e. $P = +5, O = -2$

15.61 a. $2FeCl_2(aq) + Cl_2(g) \longrightarrow 2FeCl_3(aq)$
$\;\;\;\;+2-1 \qquad\;\; 0 \qquad\quad +3-1$
Fe in $FeCl_2$ is oxidized; $FeCl_2$ is the reducing agent.
Cl in Cl_2 is reduced; Cl_2 is the oxidizing agent.
  b. $2H_2S(g) + 3O_2(g) \longrightarrow 2H_2O(l) + 2SO_2(g)$
$\;\;\;\;+1-2 \qquad\; 0 \qquad\quad +1-2 \qquad +4-2$
S in H_2S is oxidized; H_2S is the reducing agent.
O in O_2 is reduced; O_2 is the oxidizing agent.
  c. $P_2O_5(s) + 5C(s) \longrightarrow 2P(s) + 5CO(g)$
$\;\;\;\;+5-2 \qquad\;\; 0 \qquad\quad 0 \qquad\; +2-2$
C is oxidized; C is the reducing agent.
P in P_2O_5 is reduced; P_2O_5 is the oxidizing agent.

15.63 a. $Zn(s) \longrightarrow Zn^{2+}(aq) + 2\,e^-$;
$$e^- + 2H^+(aq) + NO_3^-(aq) \longrightarrow NO_2(g) + H_2O(l)$$
Overall:
$$4H^+(aq) + Zn(s) + 2NO_3^-(aq) \longrightarrow$$
$$Zn^{2+}(aq) + 2NO_2(g) + 2H_2O(l)$$

b. $5\,e^- + 8H^+(aq) + MnO_4^-(aq) \longrightarrow$
$$Mn^{2+}(aq) + 4H_2O(l);$$
$$SO_3^{2-}(aq) + H_2O(l) \longrightarrow SO_4^{2-}(aq) + 2H^+(aq) + 2\,e^-$$
Overall:
$$6H^+(aq) + 2MnO_4^-(aq) + 5SO_3^{2-}(aq) \longrightarrow$$
$$2Mn^{2+}(aq) + 5SO_4^{2-}(aq) + 3H_2O(l)$$

c. $2I^-(aq) \longrightarrow I_2(s) + 2\,e^-$;
$$6\,e^- + 6H^+(aq) + ClO_3^-(aq) \longrightarrow Cl^-(aq) + 3H_2O(l)$$
Overall:
$$6H^+(aq) + ClO_3^-(aq) + 6I^-(aq) \longrightarrow$$
$$Cl^-(aq) + 3I_2(s) + 3H_2O(l)$$

d. $C_2O_4^{2-}(aq) \longrightarrow 2CO_2(g) + 2\,e^-$;
$$6\,e^- + 14H^+(aq) + Cr_2O_7^{2-}(aq) \longrightarrow$$
$$2Cr^{3+}(aq) + 7H_2O(l)$$
Overall:
$$14H^+(aq) + Cr_2O_7^{2-}(aq) + 3C_2O_4^{2-}(aq) \longrightarrow$$
$$2Cr^{3+}(aq) + 6CO_2(g) + 7H_2O(l)$$

15.65 a. Since Cr is above Ni in the activity series, the reaction will be spontaneous.
b. Since Cu is below Zn in the activity series, the reaction will not be spontaneous.
c. Since Zn is above Pb in the activity series, the reaction will be spontaneous.

15.67 a. The anode is Mg.
b. The cathode is Ni.
c. The half-reaction at the anode is
$$Mg(s) \longrightarrow Mg^{2+}(aq) + 2\,e^-$$
d. The half-reaction at the cathode is
$$Ni^{2+}(aq) + 2\,e^- \longrightarrow Ni(s)$$
e. The overall reaction is
$$Mg(s) + Ni^{2+}(aq) \longrightarrow Mg^{2+}(aq) + Ni(s)$$
f. The shorthand cell notation is
$$Mg(s)\,|\,Mg^{2+}(aq)\,\|\,Ni^{2+}(aq)\,|\,Ni(s)$$

15.69 a. $Ca^{2+}(aq)$ will not be reduced by an iron strip.
b. $Ag^+(aq)$ will be reduced by an iron strip.
c. $Ni^{2+}(aq)$ will be reduced by an iron strip.
d. $Al^{3+}(aq)$ will not be reduced by an iron strip.
e. $Pb^{2+}(aq)$ will be reduced by an iron strip.

15.71 a. $Pb(s) + SO_4^{2-}(aq) \longrightarrow PbSO_4(s) + 2\,e^-$
b. $Pb(s)$ is oxidized.
c. The half-reaction takes place at the anode.

15.73

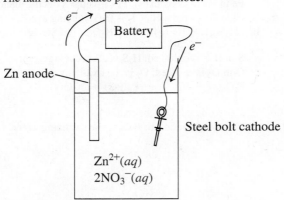

a. The anode is a bar of zinc.
b. The cathode is the steel bolt.
c. The half-reaction at the anode is
$$Zn(s) \longrightarrow Zn^{2+}(aq) + 2\,e^-$$
d. The half-reaction at the cathode is
$$Zn^{2+}(aq) + 2\,e^- \longrightarrow Zn(s)$$
e. The purpose of the zinc coating is to prevent rusting of the bolt by H_2O and O_2.

15.75 a. 0   **b.** +3   **c.** +5   **d.** +7

15.77 a. $8H^+(aq) + Cr_2O_7^{2-}(aq) + 3NO_2^-(aq) \longrightarrow$
$$2Cr^{3+}(aq) + 3NO_3^-(aq) + 4H_2O(l)$$

15.79 a. $4H^+(aq) + 3Ag(s) + NO_3^-(aq) \longrightarrow$
$$3Ag^+(aq) + NO(g) + 2H_2O(l)$$
b. 1.04 L of $NO(g)$ are produced at STP.

15.81 a. $3CuS(s) + 8HNO_3(aq) \longrightarrow$
$$3CuSO_4(aq) + 8NO(g) + 4H_2O(l)$$
b. 43.2 mL of HNO_3 solution

15.83 $H_2O(l) + Br^-(aq) + 2MnO_4^-(aq) \longrightarrow$
$$BrO_3^-(aq) + 2MnO_2(s) + 2OH^-(aq)$$

15.85 a. Since Zn is below Ca in the activity series, the reaction will not be spontaneous.
b. Since Al is above Sn in the activity series, the reaction will be spontaneous.
c. Since Mg is above H_2 in the activity series, the reaction will be spontaneous.

15.87

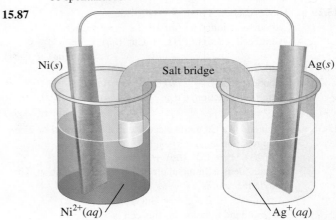

a. $Ni(s)$ is the anode.
b. $Ag(s)$ is the cathode.
c. The half-reaction at the anode is
$$Ni(s) \longrightarrow Ni^{2+}(aq) + 2\,e^-$$
d. The half-reaction at the cathode is
$$Ag^+(aq) + e^- \longrightarrow Ag(s)$$
e. The overall cell reaction is
$$Ni(s) + 2Ag^+(aq) \longrightarrow Ni^{2+}(aq) + 2Ag(s)$$

Nuclear Chemistry

SIMONE'S DOCTOR IS concerned about her elevated cholesterol, which could lead to coronary heart disease and a heart attack. He sends her to a nuclear medicine center to undergo a cardiac stress test. Dr. Paul, the radiologist, explains to Simone that a stress test measures the blood flow to her heart muscle at rest and then during stress. The test is performed in a similar way as a routine exercise stress test, but images are produced that show areas of low blood flow through the heart and areas of damaged heart muscle. It involves taking two sets of images of her heart, one when the heart is at rest, and one when she is exercising on a treadmill.

Dr. Paul tells Simone that he will inject thallium-201 into her bloodstream. He explains that Tl-201 is a radioactive isotope that has a half-life of 3.0 days. Simone is curious about the term "half-life."

Dr. Paul explains that a half-life is the amount of time it takes for one-half of a radioactive sample to break down. He assures her that after four half-lives, the radiation emitted will be almost zero. Dr. Paul tells Simone that Tl-201 decays to Hg-201 and emits energy similar to X-rays. When the Tl-201 reaches any areas within her heart with restricted blood supply, smaller amounts of the radioisotope will accumulate.

CAREER

Radiologist

A radiologist works in a hospital or imaging center where nuclear medicine is used to diagnose and treat a variety of medical conditions. In a diagnostic test, the radiologist uses a scanner, which converts radiation into images. The images are evaluated to determine any abnormalities in the body. A radiologist operates the instrumentation and computers associated with nuclear medicine such as computed tomography (CT), magnetic resonance imaging (MRI), and positron emission tomography (PET). A radiologist must know how to handle radioisotopes safely, use the necessary type of shielding, and give radioactive isotopes to patients. In addition, they must physically and mentally prepare patients for imaging. A patient may be given radioactive tracers such as technetium-99m, iodine-131, gallium-67, and thallium-201 that emit gamma radiation, which is detected and used to develop an image of the kidneys or thyroid or to follow the blood flow in the heart muscle.

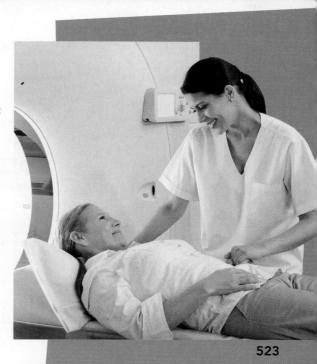

*These Key Math Skills and Core Chemistry Skills from previous chapters are listed here for your review as you proceed to the new material in this chapter.

LOOKING AHEAD

16.1 Natural Radioactivity

LEARNING GOAL Describe alpha, beta, positron, and gamma radiation.

Most naturally occurring isotopes of elements up to atomic number 19 have stable nuclei. Elements with atomic numbers 20 and higher usually have one or more isotopes that have unstable nuclei in which the nuclear forces cannot offset the repulsions between the protons. An unstable nucleus is *radioactive*, which means that it spontaneously emits small particles of energy called **radiation** to become more stable. Radiation may take the form of alpha (α) and beta (β) particles, positrons (β^+), or pure energy such as gamma (γ) rays. An isotope of an element that emits radiation is called a **radioisotope**. For most types of radiation, there is a change in the number of protons in the nucleus, which means that an atom is converted into an atom of a different element. This kind of nuclear change was not evident to Dalton when he made his predictions about atoms. Elements with atomic numbers of 93 and higher are produced artificially in nuclear laboratories and consist only of radioactive isotopes.

The atomic symbols for the different isotopes are written with the mass number in the upper left corner and the atomic number in the lower left corner. The mass number is the sum of the number of protons and neutrons in the nucleus, and the atomic number is equal to the number of protons. For example, a radioactive isotope of iodine used in the diagnosis and treatment of thyroid conditions has a mass number of 131 and an atomic number of 53.

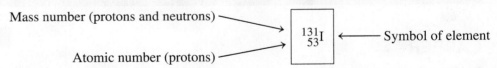

Mass number (protons and neutrons) → $^{131}_{53}\text{I}$ ← Symbol of element

Atomic number (protons) →

Radioactive isotopes are identified by writing the mass number after the element's name or symbol. Thus, in this example, the isotope is called iodine-131 or I-131. **TABLE 16.1** compares some stable, nonradioactive isotopes with some radioactive isotopes.

TABLE 16.1 Stable and Radioactive Isotopes of Some Elements

Magnesium	Iodine	Uranium
Stable Isotopes		
$^{24}_{12}\text{Mg}$	$^{127}_{53}\text{I}$	None
Magnesium-24	Iodine-127	
Radioactive Isotopes		
$^{23}_{12}\text{Mg}$	$^{125}_{53}\text{I}$	$^{235}_{92}\text{U}$
Magnesium-23	Iodine-125	Uranium-235
$^{27}_{12}\text{Mg}$	$^{131}_{53}\text{I}$	$^{238}_{92}\text{U}$
Magnesium-27	Iodine-131	Uranium-238

Types of Radiation

By emitting radiation, an unstable nucleus forms a more stable, lower energy nucleus. One type of radiation consists of *alpha particles*. An **alpha particle** is identical to a helium (He) nucleus, which has two protons and two neutrons. An alpha particle has a mass number of 4, an atomic number of 2, and a charge of 2+. The symbol for an alpha particle is the Greek letter alpha (α) or the symbol of a helium nucleus except that the 2+ charge is omitted.

Another type of radiation occurs when a radioisotope emits a *beta particle*. A **beta particle**, which is a high-energy electron, has a charge of 1−, and because its mass is so much less than the mass of a proton, it has a mass number of 0. It is represented by the Greek letter beta (β) or by the symbol for the electron including the mass number and the charge ($_{-1}^{0}e$). A beta particle is formed when a neutron in an unstable nucleus changes into a proton.

A **positron**, similar to a beta particle, has a positive (1+) charge with a mass number of 0. It is represented by the Greek letter beta with a 1+ charge, β^+, or by the symbol of an electron, which includes the mass number and the charge ($_{+1}^{0}e$). A positron is produced by an unstable nucleus when a proton is transformed into a neutron and a positron.

A positron is an example of *antimatter*, a term physicists use to describe a particle that is the opposite of another particle, in this case, an electron. When an electron and a positron collide, their minute masses are completely converted to energy in the form of *gamma rays*.

$$_{-1}^{0}e + {}_{+1}^{0}e \longrightarrow 2\,_{0}^{0}\gamma$$

Gamma rays are high-energy radiation, released when an unstable nucleus undergoes a rearrangement of its particles to give a more stable, lower energy nucleus. Gamma rays are often emitted along with other types of radiation. A gamma ray is written as the Greek letter gamma ($_{0}^{0}\gamma$). Because gamma rays are energy only, zeros are used to show that a gamma ray has no mass or charge.

TABLE 16.2 summarizes the types of radiation we will use in nuclear equations.

$_{2}^{4}\text{He}$
Alpha (α) particle

$_{-1}^{0}e$
Beta (β) particle

$$_{0}^{1}n \longrightarrow {}_{1}^{1}\text{H} + {}_{-1}^{0}e \text{ (or } \beta)$$

| Neutron in the nucleus | New proton remains in the nucleus | New electron emitted as a beta particle |

$$_{1}^{1}\text{H} \longrightarrow {}_{0}^{1}n + {}_{+1}^{0}e \text{ (or } \beta^+)$$

| Proton in the nucleus | New neutron remains in the nucleus | Positron emitted |

$_{0}^{0}\gamma$
Gamma (γ) ray

TABLE 16.2 Some Forms of Radiation

Type of Radiation	Symbol		Change in Nucleus	Mass Number	Charge
Alpha Particle	α	$_{2}^{4}\text{He}$	Two protons and two neutrons are emitted as an alpha particle.	4	2+
Beta Particle	β	$_{-1}^{0}e$	A neutron changes to a proton and an electron is emitted.	0	1−
Positron	β^+	$_{+1}^{0}e$	A proton changes to a neutron and a positron is emitted.	0	1+
Gamma Ray	γ	$_{0}^{0}\gamma$	Energy is lost to stabilize the nucleus.	0	0
Proton	p	$_{1}^{1}\text{H}$	A proton is emitted.	1	1+
Neutron	n	$_{0}^{1}n$	A neutron is emitted.	1	0

ENGAGE

What is the charge and the mass number of an alpha particle emitted by a radioactive atom?

SAMPLE PROBLEM 16.1 Radiation Particles

Identify and write the symbol for each of the following types of radiation:

a. contains two protons and two neutrons
b. has a mass number of 0 and a 1− charge

TRY IT FIRST

SOLUTION

a. An alpha (α) particle, $_{2}^{4}\text{He}$, has two protons and two neutrons.
b. A beta (β) particle, $_{-1}^{0}e$, is like an electron with a mass number of 0 and a 1− charge.

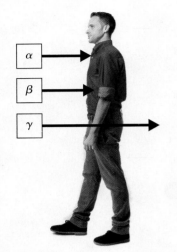

Different types of radiation penetrate the body to different depths.

Biological Effects of Radiation

When radiation strikes molecules in its path, electrons may be knocked away, forming unstable ions. If this *ionizing radiation* passes through the human body, it may interact with water molecules, removing electrons and producing H_2O^+, which can cause undesirable chemical reactions.

The cells most sensitive to radiation are the ones undergoing rapid division—those of the bone marrow, skin, reproductive organs, and intestinal lining, as well as all cells of growing children. Damaged cells may lose their ability to produce necessary materials. For example, if radiation damages cells of the bone marrow, red blood cells may no longer be produced. If sperm cells, ova, or the cells of a fetus are damaged, birth defects may result. In contrast, cells of the nerves, muscles, liver, and adult bones are much less sensitive to radiation because they undergo little or no cellular division.

Cancer cells are another example of rapidly dividing cells. Because cancer cells are highly sensitive to radiation, large doses of radiation are used to destroy them. The normal tissue that surrounds cancer cells divides at a slower rate and suffers less damage from radiation. However, radiation may cause malignant tumors, leukemia, anemia, and genetic mutations.

Radiation Protection

Nuclear medicine technologists, chemists, doctors, and nurses who work with radioactive isotopes must use proper radiation protection. Proper *shielding* is necessary to prevent exposure. Alpha particles, which have the largest mass and charge of the radiation particles, travel only a few centimeters in the air before they collide with air molecules, acquire electrons, and become helium atoms. A piece of paper, clothing, and our skin are protection against alpha particles. Lab coats and gloves will also provide sufficient shielding. However, if alpha emitters are ingested or inhaled, the alpha particles they give off can cause serious internal damage.

Beta particles have a very small mass and move much faster and farther than alpha particles, traveling as much as several meters through air. They can pass through paper and penetrate as far as 4 to 5 mm into body tissue. External exposure to beta particles can burn the surface of the skin, but they do not travel far enough to reach the internal organs. Heavy clothing such as lab coats and gloves are needed to protect the skin from beta particles.

Gamma rays travel great distances through the air and pass through many materials, including body tissues. Because gamma rays penetrate so deeply, exposure to gamma rays can be extremely hazardous. Only very dense shielding, such as lead or concrete, will stop them. Syringes used for injections of radioactive materials use shielding made of lead or heavy-weight materials such as tungsten and plastic composites.

When working with radioactive materials, medical personnel wear protective clothing and gloves and stand behind a shield (see **FIGURE 16.1**). Long tongs may be used to pick up vials of radioactive material, keeping them away from the hands and body. **TABLE 16.3** summarizes the shielding materials required for the various types of radiation.

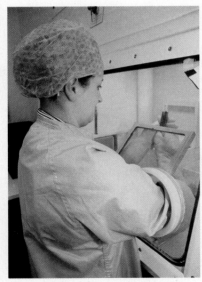

FIGURE 16.1 ▶ In a nuclear pharmacy, a person working with radioisotopes wears protective clothing and gloves and uses a lead glass shield on a syringe.

Q What types of radiation does a lead shield block?

TABLE 16.3 Properties of Radiation and Shielding Required

Property	Alpha (α) Particle	Beta (β) Particle	Gamma (γ) Ray
Travel Distance in Air	2 to 4 cm	200 to 300 cm	500 m
Tissue Depth	0.05 mm	4 to 5 mm	50 cm or more
Shielding	Paper, clothing	Heavy clothing, lab coats, gloves	Lead, thick concrete
Typical Source	Radium-226	Carbon-14	Technetium-99m

If you work in an environment where radioactive materials are present, such as a nuclear medicine facility, try to keep the time you spend in a radioactive area to a minimum. Remaining in a radioactive area twice as long exposes you to twice as much radiation.

Keep your distance! The greater the distance from the radioactive source, the lower the intensity of radiation received. By doubling your distance from the radiation source, the intensity of radiation drops to $(\frac{1}{2})^2$, or one-fourth of its previous value.

QUESTIONS AND PROBLEMS

16.1 Natural Radioactivity

LEARNING GOAL Describe alpha, beta, positron, and gamma radiation.

16.1 Identify the type of particle or radiation for each of the following:
a. ^4_2He
b. $^0_{+1}e$
c. $^0_0\gamma$

16.2 Identify the type of particle or radiation for each of the following:
a. $^0_{-1}e$
b. ^1_1H
c. ^1_0n

16.3 Naturally occurring potassium consists of three isotopes: potassium-39, potassium-40, and potassium-41.
a. Write the atomic symbol for each isotope.
b. In what ways are the isotopes similar, and in what ways do they differ?

16.4 Naturally occurring iodine is iodine-127. Medically, radioactive isotopes of iodine-125 and iodine-131 are used.
a. Write the atomic symbol for each isotope.
b. In what ways are the isotopes similar, and in what ways do they differ?

16.5 Identify each of the following:
a. $^0_{-1}\text{X}$
b. ^4_2X
c. ^1_0X
d. $^{38}_{18}\text{X}$
e. $^{14}_6\text{X}$

16.6 Identify each of the following:
a. ^1_1X
b. $^{81}_{35}\text{X}$
c. ^0_0X
d. $^{59}_{26}\text{X}$
e. $^0_{+1}\text{X}$

Applications

16.7 Write the symbol for each of the following isotopes used in nuclear medicine:
a. copper-64
b. selenium-75
c. sodium-24
d. nitrogen-15

16.8 Write the symbol for each of the following isotopes used in nuclear medicine:
a. indium-111
b. palladium-103
c. barium-131
d. rubidium-82

16.9 Supply the missing information in the following table:

Medical Use	Atomic Symbol	Mass Number	Number of Protons	Number of Neutrons
Heart imaging	$^{201}_{81}\text{Tl}$			
Radiation therapy		60	27	
Abdominal scan			31	36
Hyperthyroidism	$^{131}_{53}\text{I}$			
Leukemia treatment			32	17

16.10 Supply the missing information in the following table:

Medical Use	Atomic Symbol	Mass Number	Number of Protons	Number of Neutrons
Cancer treatment	$^{131}_{55}\text{Cs}$			
Brain scan			43	56
Blood flow		141	58	
Bone scan		85		47
Lung function	$^{133}_{54}\text{Xe}$			

16.11 Match the type of radiation (**1** to **3**) with each of the following statements:
1. alpha particle
2. beta particle
3. gamma radiation

a. does not penetrate skin
b. shielding protection includes lead or thick concrete
c. can be very harmful if ingested

16.12 Match the type of radiation (**1** to **3**) with each of the following statements:
1. alpha particle
2. beta particle
3. gamma radiation

a. penetrates farthest into skin and body tissues
b. shielding protection includes lab coats and gloves
c. travels only a short distance in air

16.2 Nuclear Reactions

LEARNING GOAL Write a balanced nuclear equation showing mass numbers and atomic numbers for radioactive decay.

In a process called **radioactive decay**, a nucleus spontaneously breaks down by emitting radiation. This process is shown by writing a *nuclear equation* with the atomic symbols of the original radioactive nucleus on the left, an arrow, and the new nucleus and the type of radiation emitted on the right.

CORE CHEMISTRY SKILL
Writing Nuclear Equations

$$\text{Radioactive nucleus} \longrightarrow \text{new nucleus} + \text{radiation} \ (\alpha, \beta, \beta^+, \gamma)$$

In a nuclear equation, the sum of the mass numbers and the sum of the atomic numbers on one side of the arrow must equal the sum of the mass numbers and the sum of the atomic numbers on the other side.

Alpha Decay

An unstable nucleus may emit an alpha particle, which consists of two protons and two neutrons. Thus, the mass number of the radioactive nucleus decreases by 4, and its atomic number decreases by 2. For example, when uranium-238 emits an alpha particle, the new nucleus that forms has a mass number of 234. Compared to uranium with 92 protons, the new nucleus has 90 protons, which is thorium.

ENGAGE

What happens to the U-238 nucleus when an alpha particle is emitted?

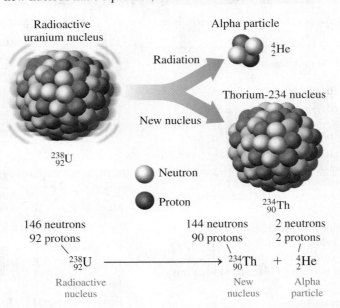

146 neutrons
92 protons

$^{238}_{92}\text{U}$
Radioactive
nucleus

144 neutrons
90 protons

$^{234}_{90}\text{Th}$
New
nucleus

2 neutrons
2 protons

$^{4}_{2}\text{He}$
Alpha
particle

$$^{238}_{92}\text{U} \longrightarrow {}^{234}_{90}\text{Th} + {}^{4}_{2}\text{He}$$

In the nuclear equation for alpha decay, the mass number of the new nucleus decreases by 4 and its atomic number decreases by 2.

We can look at writing a balanced nuclear equation for americium-241, which undergoes alpha decay as shown in Sample Problem 16.2.

SAMPLE PROBLEM 16.2 Writing an Equation for Alpha Decay

Smoke detectors that are used in homes and apartments contain americium-241, which undergoes alpha decay. When alpha particles collide with air molecules, charged particles are produced that generate an electrical current. If smoke particles enter the detector, they interfere with the formation of charged particles in the air, and the electrical current is interrupted. This causes the alarm to sound and warns the occupants of the danger of fire. Write the balanced nuclear equation for the decay of americium-241.

TRY IT FIRST

SOLUTION

A smoke detector sounds an alarm when smoke enters its ionization chamber.

	Given	Need	Connect
ANALYZE THE PROBLEM	Am-241, alpha decay	balanced nuclear equation	mass number, atomic number of new nucleus

STEP 1 Write the incomplete nuclear equation.

$$^{241}_{95}\text{Am} \longrightarrow ? + ^{4}_{2}\text{He}$$

STEP 2 Determine the missing mass number. In the equation, the mass number of the americium, 241, is equal to the sum of the mass numbers of the new nucleus and the alpha particle.

$$241 \quad = ? + 4$$
$$241 - 4 = ?$$
$$241 - 4 = 237 \text{ (mass number of new nucleus)}$$

STEP 3 Determine the missing atomic number. The atomic number of americium, 95, must equal the sum of the atomic numbers of the new nucleus and the alpha particle.

$$95 \quad = ? + 2$$
$$95 - 2 = ?$$
$$95 - 2 = 93 \text{ (atomic number of new nucleus)}$$

STEP 4 Determine the symbol of the new nucleus. On the periodic table, the element that has atomic number 93 is neptunium, Np. The nucleus of this isotope of Np is written as $^{237}_{93}\text{Np}$.

STEP 5 Complete the nuclear equation.

$$^{241}_{95}\text{Am} \longrightarrow ^{237}_{93}\text{Np} + ^{4}_{2}\text{He}$$

STUDY CHECK 16.2

Write the balanced nuclear equation for the alpha decay of Po-214.

ANSWER

$$^{214}_{84}\text{Po} \longrightarrow ^{210}_{82}\text{Pb} + ^{4}_{2}\text{He}$$

Guide to Completing a Nuclear Equation

STEP 1
Write the incomplete nuclear equation.

STEP 2
Determine the missing mass number.

STEP 3
Determine the missing atomic number.

STEP 4
Determine the symbol of the new nucleus.

STEP 5
Complete the nuclear equation.

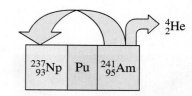

CHEMISTRY LINK TO HEALTH
Radon in Our Homes

The presence of radon gas has become a much publicized environmental and health issue because of the radiation danger it poses. Radioactive isotopes such as radium-226 are naturally present in many types of rocks and soils. Radium-226 emits an alpha particle and is converted into radon gas, which diffuses out of the rocks and soil.

$$^{226}_{88}\text{Ra} \longrightarrow ^{222}_{86}\text{Rn} + ^{4}_{2}\text{He}$$

Outdoors, radon gas poses little danger because it disperses in the air. However, if the radioactive source is under a house or building, the radon gas can enter the house through cracks in the foundation or other openings. Those who live or work there may inhale the radon. Inside the lungs, radon-222 emits alpha particles to form polonium-218, which is known to cause lung cancer.

$$^{222}_{86}\text{Rn} \longrightarrow ^{218}_{84}\text{Po} + ^{4}_{2}\text{He}$$

The U.S. Environmental Protection Agency (EPA) estimates that radon causes about 20 000 lung cancer deaths in one year. The EPA recommends that the maximum level of radon not exceed 4 picocuries (pCi) per liter of air in a home. One picocurie (pCi) is equal to 10^{-12} curies (Ci); curies are described in section 16.3. The EPA estimates that more than 6 million homes have radon levels that exceed this maximum.

A radon gas detector is used to determine radon levels in buildings.

Beta Decay

The formation of a beta particle is the result of the breakdown of a neutron into a proton and an electron (beta particle). Because the proton remains in the nucleus, the number of protons increases by one, whereas the number of neutrons decreases by one. Thus, in a nuclear equation for beta decay, the mass number of the radioactive nucleus and the mass number of the new nucleus are the same. However, the atomic number of the new nucleus increases by one, which makes it a nucleus of a different element (*transmutation*). For example, the beta decay of a carbon-14 nucleus produces a nitrogen-14 nucleus.

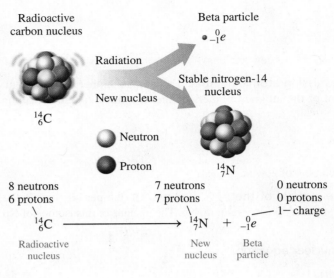

Radioactive carbon nucleus

Beta particle
$_{-1}^{0}e$

Radiation

New nucleus

Stable nitrogen-14 nucleus

$_{6}^{14}C$

○ Neutron

● Proton

$_{7}^{14}N$

8 neutrons	7 neutrons	0 neutrons
6 protons	7 protons	0 protons
		1− charge

$$_{6}^{14}C \longrightarrow \;_{7}^{14}N \;+\; _{-1}^{0}e$$

Radioactive nucleus   New nucleus   Beta particle

In the nuclear equation for beta decay, the mass number of the new nucleus remains the same and its atomic number increases by 1.

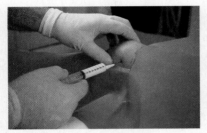

A radioisotope is injected into the joint to relieve the pain caused by arthritis.

SAMPLE PROBLEM 16.3 Writing a Nuclear Equation for Beta Decay

The radioactive isotope yttrium-90, a beta emitter, is used in cancer treatment and as a colloidal injection into large joints to relieve the pain caused by arthritis. Write the balanced nuclear equation for the beta decay of yttrium-90.

TRY IT FIRST

SOLUTION

ANALYZE THE PROBLEM	Given	Need	Connect
	Y-90, beta decay	balanced nuclear equation	mass number, atomic number of new nucleus

STEP 1 Write the incomplete nuclear equation.

$$_{39}^{90}Y \longrightarrow \; ? \;+\; _{-1}^{0}e$$

STEP 2 Determine the missing mass number. In the equation, the mass number of the yttrium, 90, is equal to the sum of the mass numbers of the new nucleus and the beta particle.

90 = ? + 0
90 − 0 = ?
90 − 0 = 90 (mass number of new nucleus)

STEP 3 Determine the missing atomic number. The atomic number of yttrium, 39, must equal the sum of the atomic numbers of the new nucleus and the beta particle.

39 = ? − 1
39 + 1 = ?
39 + 1 = 40 (atomic number of new nucleus)

STEP 4 Determine the symbol of the new nucleus. On the periodic table, the element that has atomic number 40 is zirconium, Zr. The nucleus of this isotope of Zr is written as $_{40}^{90}Zr$.

STEP 5 Complete the nuclear equation.

$$^{90}_{39}Y \longrightarrow {}^{90}_{40}Zr + {}^{0}_{-1}e$$

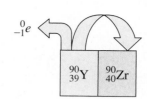

STUDY CHECK 16.3

Write the balanced nuclear equation for the beta decay of chromium-51.

ANSWER

$$^{51}_{24}Cr \longrightarrow {}^{51}_{25}Mn + {}^{0}_{-1}e$$

Positron Emission

In positron emission, a proton in an unstable nucleus is converted to a neutron and a positron. The neutron remains in the nucleus, but the positron is emitted from the nucleus. In a nuclear equation for positron emission, the mass number of the radioactive nucleus and the mass number of the new nucleus are the same. However, the atomic number of the new nucleus decreases by one, indicating a change of one element into another. For example, an aluminum-24 nucleus undergoes positron emission to produce a magnesium-24 nucleus. The atomic number of magnesium (12) and the charge of the positron (1+) give the atomic number of aluminum (13).

$$^{24}_{13}Al \longrightarrow {}^{24}_{12}Mg + {}^{0}_{+1}e$$

SAMPLE PROBLEM 16.4 Writing a Nuclear Equation for Positron Emission

Write the balanced nuclear equation for manganese-49, which decays by emitting a positron.

TRY IT FIRST

SOLUTION

ANALYZE THE PROBLEM	Given	Need	Connect
	Mn-49, positron emission	balanced nuclear equation	mass number, atomic number of new nucleus

STEP 1 Write the incomplete nuclear equation.

$$^{49}_{25}Mn \longrightarrow ? + {}^{0}_{+1}e$$

STEP 2 Determine the missing mass number. In the equation, the mass number of the manganese, 49, is equal to the sum of the mass numbers of the new nucleus and the positron.

$49 \quad = ? + 0$
$49 - 0 = ?$
$49 - 0 = 49$ (mass number of new nucleus)

STEP 3 Determine the missing atomic number. The atomic number of manganese, 25, must equal the sum of the atomic numbers of the new nucleus and the positron.

$25 \quad = ? + 1$
$25 - 1 = ?$
$25 - 1 = 24$ (atomic number of new nucleus)

STEP 4 Determine the symbol of the new nucleus. On the periodic table, the element that has atomic number 24 is chromium (Cr). The nucleus of this isotope of Cr is written as $^{49}_{24}Cr$.

STEP 5 Complete the nuclear equation.

$$^{49}_{25}Mn \longrightarrow {}^{49}_{24}Cr + {}^{0}_{+1}e$$

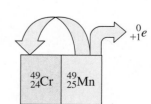

STUDY CHECK 16.4

Write the balanced nuclear equation for xenon-118, which undergoes positron emission.

ANSWER

$$^{118}_{54}\text{Xe} \longrightarrow ^{118}_{53}\text{I} + ^{0}_{+1}e$$

Radiation Source	Radiation	New Nucleus
Alpha emitter	$^{4}_{2}\text{He}$ +	New element
		Mass number − 4 Atomic number − 2
Beta emitter	$^{0}_{-1}e$ +	New element
		Mass number same Atomic number + 1
Positron emitter	$^{0}_{+1}e$ +	New element
		Mass number same Atomic number − 1
Gamma emitter	$^{0}_{0}\gamma$ +	Stable nucleus of the same element
		Mass number same Atomic number same

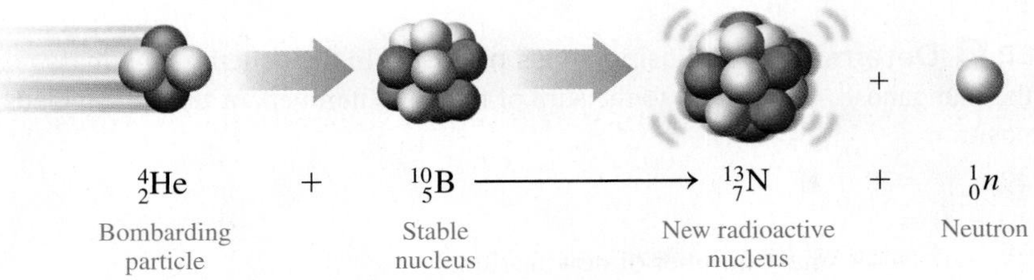

When nonradioactive B-10 is bombarded by an alpha particle, the products are radioactive N-13 and a neutron.

Gamma Emission

Pure gamma emitters are rare, although gamma radiation accompanies most alpha and beta radiation. In radiology, one of the most commonly used gamma emitters is technetium (Tc). The unstable isotope of technetium is written as the *metastable* (symbol m) isotope technetium-99m, Tc-99m, or $^{99m}_{43}\text{Tc}$. By emitting energy in the form of gamma rays, the nucleus becomes more stable.

$$^{99m}_{43}\text{Tc} \longrightarrow ^{99}_{43}\text{Tc} + ^{0}_{0}\gamma$$

FIGURE 16.2 summarizes the changes in the nucleus for alpha, beta, positron, and gamma radiation.

Producing Radioactive Isotopes

Today, many radioisotopes are produced in small amounts by bombarding stable, nonradioactive isotopes with high-speed particles such as alpha particles, protons, neutrons, and small nuclei. When one of these particles is absorbed, the stable nucleus is converted to a radioactive isotope and usually some type of radiation particle.

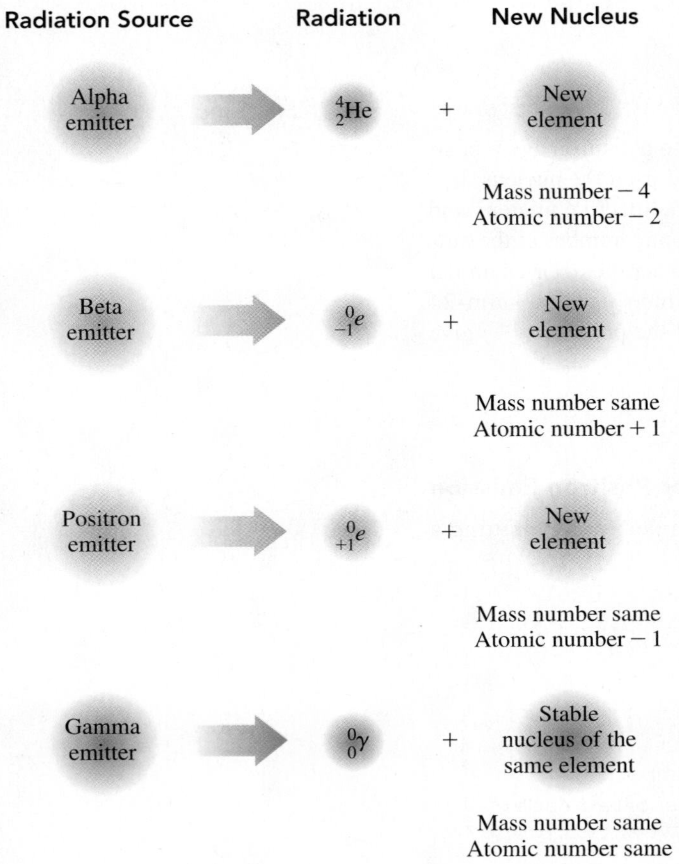

FIGURE 16.2 ▶ When the nuclei of alpha, beta, positron, and gamma emitters emit radiation, new, more stable nuclei are produced.

◉ What changes occur in the number of protons and neutrons when an alpha emitter gives off radiation?

$$^{4}_{2}\text{He} + ^{10}_{5}\text{B} \longrightarrow ^{13}_{7}\text{N} + ^{1}_{0}n$$

Bombarding particle | Stable nucleus | New radioactive nucleus | Neutron

All elements that have an atomic number greater than 92 have been produced by bombardment. Most have been produced in only small amounts and exist for only a short time, making it difficult to study their properties. For example, when californium-249 is bombarded with nitrogen-15, the radioactive element dubnium-260 and four neutrons are produced.

$$^{15}_{7}\text{N} + ^{249}_{98}\text{Cf} \longrightarrow ^{260}_{105}\text{Db} + 4^{1}_{0}n$$

Technetium-99m is a radioisotope used in nuclear medicine for several diagnostic procedures, including the detection of brain tumors and examinations of the liver and spleen. The source of technetium-99m is molybdenum-99, which is produced in a nuclear reactor by neutron bombardment of molybdenum-98.

$$^{1}_{0}n + ^{98}_{42}\text{Mo} \longrightarrow ^{99}_{42}\text{Mo}$$

Many radiology laboratories have small generators containing molybdenum-99, which decays to the technetium-99m radioisotope.

$$^{99}_{42}\text{Mo} \longrightarrow {}^{99\text{m}}_{43}\text{Tc} + {}^{0}_{-1}e$$

The technetium-99m radioisotope decays by emitting gamma rays. Gamma emission is desirable for diagnostic work because the gamma rays pass through the body to the detection equipment.

$$^{99\text{m}}_{43}\text{Tc} \longrightarrow {}^{99}_{43}\text{Tc} + {}^{0}_{0}\gamma$$

SAMPLE PROBLEM 16.5 Writing an Equation for an Isotope Produced by Bombardment

Write the balanced nuclear equation for the bombardment of nickel-58 by a proton, $^{1}_{1}\text{H}$, which produces a radioactive isotope and an alpha particle.

 TRY IT FIRST

SOLUTION

ANALYZE THE PROBLEM	Given	Need	Connect
	proton bombardment of Ni-58	balanced nuclear equation	mass number, atomic number of new nucleus

STEP 1 Write the incomplete nuclear equation.

$$^{1}_{1}\text{H} + {}^{58}_{28}\text{Ni} \longrightarrow ? + {}^{4}_{2}\text{He}$$

STEP 2 Determine the missing mass number. In the equation, the sum of the mass numbers of the proton, 1, and the nickel, 58, must equal the sum of the mass numbers of the new nucleus and the alpha particle.

1 + 58 = ? + 4
59 − 4 = ?
59 − 4 = 55 (mass number of new nucleus)

STEP 3 Determine the missing atomic number. The sum of the atomic numbers of the proton, 1, and nickel, 28, must equal the sum of the atomic numbers of the new nucleus and the alpha particle.

1 + 28 = ? + 2
29 − 2 = ?
29 − 2 = 27 (atomic number of new nucleus)

STEP 4 Determine the symbol of the new nucleus. On the periodic table, the element that has atomic number 27 is cobalt, Co. The nucleus of this isotope of Co is written as $^{55}_{27}\text{Co}$.

STEP 5 Complete the nuclear equation.

$$^{1}_{1}\text{H} + {}^{58}_{28}\text{Ni} \longrightarrow {}^{55}_{27}\text{Co} + {}^{4}_{2}\text{He}$$

STUDY CHECK 16.5

The first radioactive isotope was produced in 1934 by the bombardment of aluminum-27 by an alpha particle to produce a radioactive isotope and one neutron. Write the balanced nuclear equation for this bombardment.

ANSWER

$$^{4}_{2}\text{He} + {}^{27}_{13}\text{Al} \longrightarrow {}^{30}_{15}\text{P} + {}^{1}_{0}n$$

A generator is used to prepare technetium-99m.

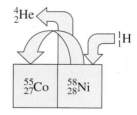

Interactive Video

Writing Equations for an Isotope Produced by Bombardment

QUESTIONS AND PROBLEMS

16.2 Nuclear Reactions

LEARNING GOAL Write a balanced nuclear equation showing mass numbers and atomic numbers for radioactive decay.

16.13 Write a balanced nuclear equation for the alpha decay of each of the following radioactive isotopes:
  a. $^{208}_{84}Po$
  b. $^{232}_{90}Th$
  c. $^{251}_{102}No$
  d. radon-220

16.14 Write a balanced nuclear equation for the alpha decay of each of the following radioactive isotopes:
  a. curium-243
  b. $^{252}_{99}Es$
  c. $^{251}_{98}Cf$
  d. $^{261}_{107}Bh$

16.15 Write a balanced nuclear equation for the beta decay of each of the following radioactive isotopes:
  a. $^{25}_{11}Na$
  b. $^{20}_{8}O$
  c. strontium-92
  d. iron-60

16.16 Write a balanced nuclear equation for the beta decay of each of the following radioactive isotopes:
  a. $^{44}_{19}K$
  b. iron-59
  c. potassium-42
  d. $^{141}_{56}Ba$

16.17 Write a balanced nuclear equation for the positron emission of each of the following radioactive isotopes:
  a. silicon-26
  b. cobalt-54
  c. $^{77}_{37}Rb$
  d. $^{93}_{45}Rh$

16.18 Write a balanced nuclear equation for the positron emission of each of the following radioactive isotopes:
  a. boron-8
  b. $^{15}_{8}O$
  c. $^{40}_{19}K$
  d. nitrogen-13

16.19 Complete each of the following nuclear equations and describe the type of radiation:
  a. $^{28}_{13}Al \longrightarrow ? + ^{0}_{-1}e$
  b. $^{180m}_{73}Ta \longrightarrow ^{180}_{73}Ta + ?$
  c. $^{66}_{29}Cu \longrightarrow ^{66}_{30}Zn + ?$
  d. $? \longrightarrow ^{234}_{90}Th + ^{4}_{2}He$
  e. $^{188}_{80}Hg \longrightarrow ? + ^{0}_{+1}e$

16.20 Complete each of the following nuclear equations and describe the type of radiation:
  a. $^{11}_{6}C \longrightarrow ^{11}_{5}B + ?$
  b. $^{35}_{16}S \longrightarrow ? + ^{0}_{-1}e$
  c. $? \longrightarrow ^{90}_{39}Y + ^{0}_{-1}e$
  d. $^{210}_{83}Bi \longrightarrow ? + ^{4}_{2}He$
  e. $? \longrightarrow ^{89}_{39}Y + ^{0}_{+1}e$

16.21 Complete each of the following bombardment reactions:
  a. $^{1}_{0}n + ^{9}_{4}Be \longrightarrow ?$
  b. $^{1}_{0}n + ^{131}_{52}Te \longrightarrow ? + ^{0}_{-1}e$
  c. $^{1}_{0}n + ? \longrightarrow ^{24}_{11}Na + ^{4}_{2}He$
  d. $^{4}_{2}He + ^{14}_{7}N \longrightarrow ? + ^{1}_{1}H$

16.22 Complete each of the following bombardment reactions:
  a. $? + ^{40}_{18}Ar \longrightarrow ^{43}_{19}K + ^{1}_{1}H$
  b. $^{1}_{0}n + ^{238}_{92}U \longrightarrow ?$
  c. $^{1}_{0}n + ? \longrightarrow ^{14}_{6}C + ^{1}_{1}H$
  d. $? + ^{64}_{28}Ni \longrightarrow ^{272}_{111}Rg + ^{1}_{0}n$

16.3 Radiation Measurement

LEARNING GOAL Describe the detection and measurement of radiation.

One of the most common instruments for detecting beta and gamma radiation is the Geiger counter. It consists of a metal tube filled with a gas such as argon. When radiation enters a window on the end of the tube, it forms charged particles in the gas, which produce an electrical current. Each burst of current is amplified to give a click and a reading on a meter.

$$Ar + radiation \longrightarrow Ar^{+} + e^{-}$$

Measuring Radiation

Radiation is measured in several different ways. When a radiology laboratory obtains a radioisotope, the *activity* of the sample is measured in terms of the number of nuclear disintegrations per second. The **curie (Ci)**, the original unit of activity, was defined as the

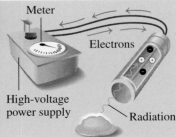

A radiation counter is used to determine radiation levels in workers at the Fukushima Daiichi nuclear power plant.

number of disintegrations that occur in 1 s for 1 g of radium, which is equal to 3.7×10^{10} disintegrations/s. The unit was named for the Polish scientist Marie Curie, who along with her husband, Pierre, discovered the radioactive elements radium and polonium. The SI unit of radiation activity is the **becquerel (Bq)**, which is 1 disintegration/s.

The **rad (radiation absorbed dose)** is a unit that measures the amount of radiation absorbed by a gram of material such as body tissue. The SI unit for absorbed dose is the **gray (Gy)**, which is defined as the joules of energy absorbed by 1 kg of body tissue. The gray is equal to 100 rad.

The **rem (radiation equivalent in humans)** is a unit that measures the biological effects of different kinds of radiation. Although alpha particles do not penetrate the skin, if they should enter the body by some other route, they can cause extensive damage within a short distance in tissue. High-energy radiation, such as beta particles, high-energy protons, and neutrons that travel into tissue, causes more damage. Gamma rays are damaging because they travel a long way through body tissue.

To determine the **equivalent dose** or rem dose, the absorbed dose (rad) is multiplied by a factor that adjusts for biological damage caused by a particular form of radiation. For beta and gamma radiation the factor is 1, so the biological damage in rems is the same as the absorbed radiation (rad). For high-energy protons and neutrons, the factor is about 10, and for alpha particles it is 20.

Biological damage (rem) = Absorbed dose (rad) × Factor

Often, the measurement for an equivalent dose will be in units of millirems (mrem). One rem is equal to 1000 mrem. The SI unit is the **sievert (Sv)**. One sievert is equal to 100 rem. **TABLE 16.4** summarizes the units used to measure radiation.

CHEMISTRY LINK TO HEALTH
Radiation and Food

Foodborne illnesses caused by pathogenic bacteria such as *Salmonella*, *Listeria*, and *Escherichia coli* have become a major health concern in the United States. *E. coli* has been responsible for outbreaks of illness from contaminated ground beef, fruit juices, lettuce, and alfalfa sprouts.

The U.S. Food and Drug Administration (FDA) has approved the use of 0.3 kGy to 1 kGy of radiation produced by cobalt-60 or cesium-137 for the treatment of foods. The irradiation technology is much like that used to sterilize medical supplies. Cobalt pellets are placed in stainless steel tubes, which are arranged in racks. When food moves through the series of racks, the gamma rays pass through the food and kill the bacteria.

It is important for consumers to understand that when food is irradiated, it never comes in contact with the radioactive source. The gamma rays pass through the food to kill bacteria, but that does not make the food radioactive. The radiation kills bacteria because it stops their ability to divide and grow. We cook or heat food thoroughly for the same purpose. Radiation, as well as heat, has little effect on the food itself because its cells are no longer dividing or growing. Thus irradiated food is not harmed, although a small amount of vitamin B_1 and C may be lost.

Currently, tomatoes, blueberries, strawberries, and mushrooms are being irradiated to allow them to be harvested when completely ripe and extend their shelf life (see **FIGURE 16.3**). The FDA has also approved the irradiation of pork, poultry, and beef to decrease potential infections and to extend shelf life. Currently, irradiated vegetable and meat products are available in retail markets in more than 40 countries. In the United States, irradiated foods such as tropical fruits, spinach, and ground meats are found in some stores. *Apollo 17* astronauts ate irradiated foods on the Moon, and some U.S. hospitals and nursing homes now use irradiated poultry to reduce the possibility of salmonella infections among residents. The extended shelf life of irradiated food also makes it useful for campers and military personnel. Soon, consumers concerned about food safety will have a choice of irradiated meats, fruits, and vegetables at the market.

(a)

(b)

FIGURE 16.3 ▶ **(a)** The FDA requires this symbol to appear on irradiated retail foods. **(b)** After two weeks, the irradiated strawberries on the right show no spoilage. Mold is growing on the nonirradiated ones on the left.

Q Why are irradiated foods used on spaceships and in nursing homes?

A dosimeter measures radiation exposure.

TABLE 16.4 Units of Radiation Measurement

Measurement	Common Unit	SI Unit	Relationship
Activity	curie (Ci) $1 \text{ Ci} = 3.7 \times 10^{10}$ disintegrations/s	becquerel (Bq) $1 \text{ Bq} = 1$ disintegration/s	$1 \text{ Ci} = 3.7 \times 10^{10} \text{ Bq}$
Absorbed Dose	rad	gray (Gy) $1 \text{ Gy} = 1 \text{ J/kg}$ of tissue	$1 \text{ Gy} = 100 \text{ rad}$
Biological Damage	rem	sievert (Sv)	$1 \text{ Sv} = 100 \text{ rem}$

People who work in radiology laboratories wear dosimeters attached to their clothing to determine any exposure to radiation such as X-rays, gamma rays, or beta particles. A dosimeter can be thermoluminescent (TLD), optically stimulated luminescence (OSL), or electronic personal (EPD). Dosimeters provide real time radiation levels measured by monitors in the work area.

SAMPLE PROBLEM 16.6 Radiation Measurement

One treatment for bone pain involves intravenous administration of the radioisotope phosphorus-32, which is incorporated into bone. A typical dose of 7 mCi can produce up to 450 rad in the bone. What is the difference between the units of mCi and rad?

TRY IT FIRST

SOLUTION

The millicuries (mCi) indicate the activity of the P-32 in terms of nuclei that break down in 1 s. The radiation absorbed dose (rad) is a measure of amount of radiation absorbed by the bone.

STUDY CHECK 16.6

For Sample Problem 16.6, what is the absorbed dose of radiation in grays (Gy)?

ANSWER

4.5 Gy

TABLE 16.5 Average Annual Radiation Received by a Person in the United States

Source	Dose (mrem)
Natural	
Ground	20
Air, water, food	30
Cosmic rays	40
Wood, concrete, brick	50
Medical	
Chest X-ray	20
Dental X-ray	20
Mammogram	40
Hip X-ray	60
Lumbar spine X-ray	70
Upper gastrointestinal tract X-ray	200
Other	
Nuclear power plants	0.1
Television	20
Air travel	10
Radon	200*

*Varies widely.

Exposure to Radiation

Every day, we are exposed to low levels of radiation from naturally occurring radioactive isotopes in the buildings where we live and work, in our food and water, and in the air we breathe. For example, potassium-40, a naturally occurring radioactive isotope, is present in any potassium-containing food. Other naturally occurring radioisotopes in air and food are carbon-14, radon-222, strontium-90, and iodine-131. The average person in the United States is exposed to about 360 mrem of radiation annually. Medical sources of radiation, including dental, hip, spine, and chest X-rays and mammograms, add to our radiation exposure. **TABLE 16.5** lists some common sources of radiation.

Another source of background radiation is cosmic radiation produced in space by the Sun. People who live at high altitudes or travel by airplane receive a greater amount of cosmic radiation because there are fewer molecules in the atmosphere to absorb the radiation. For example, a person living in Denver receives about twice the cosmic radiation as a person living in Los Angeles. A person living close to a nuclear power plant normally does not receive much additional radiation, perhaps 0.1 mrem in 1 yr. (One rem equals 1000 mrem.) However, in the accident at the Chernobyl nuclear power plant in 1986 in Ukraine, it is estimated that people in a nearby town received as much as 1 rem/h.

Radiation Sickness

The larger the dose of radiation received at one time, the greater the effect on the body. Exposure to radiation of less than 25 rem usually cannot be detected. Whole-body exposure of 100 rem produces a temporary decrease in the number of white blood cells. If the exposure to radiation is greater than 100 rem, a person may suffer the symptoms of radiation sickness: nausea, vomiting, fatigue, and a reduction in white-cell count. A whole-body dosage greater than 300 rem can decrease the white-cell count to zero. The person suffers diarrhea, hair loss, and infection. Exposure to radiation of 500 rem is expected to cause death in 50% of the people receiving that dose. This amount of radiation to the whole body is called the *lethal dose for one-half the population*, or the LD_{50}. The LD_{50} varies for different life forms, as **TABLE 16.6** shows. Whole-body radiation of 600 rem or greater would be fatal to all humans within a few weeks.

TABLE 16.6 Lethal Doses of Radiation for Some Life Forms

Life Form	LD_{50} (rem)
Insect	100 000
Bacterium	50 000
Rat	800
Human	500
Dog	300

QUESTIONS AND PROBLEMS

16.3 Radiation Measurement

LEARNING GOAL Describe the detection and measurement of radiation.

16.23 Match each property (**1** to **3**) with its unit of measurement.
1. activity
2. absorbed dose
3. biological damage

   a. rad   **b.** mrem
   c. mCi   **d.** Gy

16.24 Match each property (**1** to **3**) with its unit of measurement.
1. activity
2. absorbed dose
3. biological damage

   a. mrad   **b.** gray
   c. becquerel   **d.** Sv

Applications

16.25 Two technicians in a nuclear laboratory were accidentally exposed to radiation. If one was exposed to 8 mGy and the other to 5 rad, which technician received more radiation?

16.26 Two samples of a radioisotope were spilled in a nuclear laboratory. The activity of one sample was 8 kBq and the other 15 mCi. Which sample produced the higher amount of radiation?

16.27 a. The recommended dosage of iodine-131 is 4.20 μCi/kg of body mass. How many microcuries of iodine-131 are needed for a 70.0-kg person with hyperthyroidism?
   b. A person receives 50 rad of gamma radiation. What is that amount in grays?

16.28 a. The dosage of technetium-99m for a lung scan is 20. μCi/kg of body mass. How many millicuries of technetium-99m should be given to a 50.0-kg person (1 mCi = 1000 μCi)?
   b. Suppose a person absorbed 50 mrad of alpha radiation. What would be the equivalent dose in millirems?

16.4 Half-Life of a Radioisotope

LEARNING GOAL Given the half-life of a radioisotope, calculate the amount of radioisotope remaining after one or more half-lives.

The **half-life** of a radioisotope is the amount of time it takes for one-half of a sample to decay. For example, $^{131}_{53}I$ has a half-life of 8.0 days. As $^{131}_{53}I$ decays, it produces the non-radioactive isotope $^{131}_{54}Xe$ and a beta particle.

$$^{131}_{53}I \longrightarrow {}^{131}_{54}Xe + {}^{0}_{-1}e$$

Suppose we have a sample that initially contains 20. mg of $^{131}_{53}I$. In 8.0 days, one-half (10. mg) of all the I-131 nuclei in the sample will decay, which leaves 10. mg of I-131. After 16 days (two half-lives), 5.0 mg of the remaining I-131 decays, which leaves 5.0 mg of I-131. After 24 days (three half-lives), 2.5 mg of the remaining I-131 decays, which leaves 2.5 mg of I-131 nuclei still capable of producing radiation.

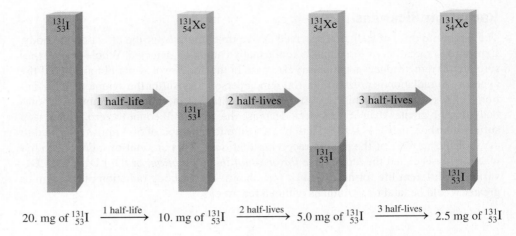

$$20. \text{ mg of } ^{131}_{53}\text{I} \xrightarrow{\text{1 half-life}} 10. \text{ mg of } ^{131}_{53}\text{I} \xrightarrow{\text{2 half-lives}} 5.0 \text{ mg of } ^{131}_{53}\text{I} \xrightarrow{\text{3 half-lives}} 2.5 \text{ mg of } ^{131}_{53}\text{I}$$

A **decay curve** is a diagram of the decay of a radioactive isotope. **FIGURE 16.4** shows such a curve for the $^{131}_{53}\text{I}$ we have discussed.

FIGURE 16.4 ▶ The decay curve for iodine-131 shows that one-half of the radioactive sample decays and one-half remains radioactive after each half-life of 8.0 days.

Q How many milligrams of the 20.-mg sample remain radioactive after two half-lives?

 CORE CHEMISTRY SKILL
Using Half-Lives

Guide to Using Half-Lives

STEP 1
State the given and needed quantities.

STEP 2
Write a plan to calculate the unknown quantity.

STEP 3
Write the half-life equality and conversion factors.

STEP 4
Set up the problem to calculate the needed quantity.

SAMPLE PROBLEM 16.7 Using Half-Lives of a Radioisotope

Phosphorus-32, a radioisotope used in the treatment of leukemia, has a half-life of 14.3 days. If a sample contains 8.0 mg of phosphorus-32, how many milligrams of phosphorus-32 remain after 42.9 days?

TRY IT FIRST

SOLUTION

STEP 1 State the given and needed quantities.

ANALYZE THE PROBLEM	Given	Need	Connect
	8.0 mg of P-32, 42.9 days elapsed, half-life = 14.3 days	milligrams of P-32 remaining	number of half-lives

STEP 2 Write a plan to calculate the unknown quantity.

days → Half-life → number of half-lives

milligrams of $^{32}_{15}\text{P}$ → Number of half-lives → milligrams of $^{32}_{15}\text{P}$ remaining

STEP 3 Write the half-life equality and conversion factors.

$$1 \text{ half-life} = 14.3 \text{ days}$$
$$\frac{14.3 \text{ days}}{1 \text{ half-life}} \quad \text{and} \quad \frac{1 \text{ half-life}}{14.3 \text{ days}}$$

STEP 4 Set up the problem to calculate the needed quantity. First, we determine the number of half-lives in the amount of time that has elapsed.

$$\text{number of half-lives} \quad = \quad 42.9 \; \cancel{\text{days}} \quad \times \quad \frac{1 \; \text{half-life}}{14.3 \; \cancel{\text{days}}} \quad = \quad 3.00 \; \text{half-lives}$$

Now we can determine how much of the sample decays in three half-lives and how many grams of the phosphorus remain.

$$8.0 \text{ mg of } {}^{32}_{15}\text{P} \xrightarrow{\text{1 half-life}} 4.0 \text{ mg of } {}^{32}_{15}\text{P} \xrightarrow{\text{2 half-lives}} 2.0 \text{ mg of } {}^{32}_{15}\text{P} \xrightarrow{\text{3 half-lives}} 1.0 \text{ mg of } {}^{32}_{15}\text{P}$$

STUDY CHECK 16.7

Iron-59 has a half-life of 44 days. If a nuclear laboratory receives a sample of 8.0 μg of iron-59, how many micrograms of iron-59 are still active after 176 days?

ANSWER

0.50 μg of iron-59

Interactive Video

Half-Lives

TABLE 16.7 Half-Lives of Some Radioisotopes

Element	Radioisotope	Half-Life	Type of Radiation
Naturally Occurring Radioisotopes			
Carbon-14	${}^{14}_{6}\text{C}$	5730 yr	Beta
Potassium-40	${}^{40}_{19}\text{K}$	1.3×10^9 yr	Beta, gamma
Radium-226	${}^{226}_{88}\text{Ra}$	1600 yr	Alpha
Strontium-90	${}^{90}_{38}\text{Sr}$	38.1 yr	Alpha
Uranium-238	${}^{238}_{92}\text{U}$	4.5×10^9 yr	Alpha
Some Medical Radioisotopes			
Carbon-11	${}^{11}_{6}\text{C}$	20 min	Positron
Chromium-51	${}^{51}_{24}\text{Cr}$	28 days	Gamma
Iodine-131	${}^{131}_{53}\text{I}$	8.0 days	Gamma
Oxygen-15	${}^{15}_{8}\text{O}$	2.0 min	Positron
Iron-59	${}^{59}_{26}\text{Fe}$	44 days	Beta, gamma
Radon-222	${}^{222}_{86}\text{Rn}$	3.8 days	Alpha
Technetium-99m	${}^{99m}_{43}\text{Tc}$	6.0 h	Beta, gamma

Naturally occurring isotopes of the elements usually have long half-lives, as shown in **TABLE 16.7**. They disintegrate slowly and produce radiation over a long period of time, even hundreds or millions of years. In contrast, the radioisotopes used in nuclear medicine have much shorter half-lives. They disintegrate rapidly and produce almost all their radiation in a short period of time. For example, technetium-99m emits half of its radiation in the first 6 h. This means that a small amount of the radioisotope given to a patient is essentially gone within two days. The decay products of technetium-99m are totally eliminated by the body.

CHEMISTRY LINK TO THE ENVIRONMENT
Dating Ancient Objects

Radiological dating is a technique used by geologists, archaeologists, and historians to determine the age of ancient objects. The age of an object derived from plants or animals (such as wood, fiber, natural pigments, bone, and cotton or woolen clothing) is determined by measuring the amount of carbon-14, a naturally occurring radioactive form of carbon. In 1960, Willard Libby received the Nobel Prize for the work he did developing carbon-14 dating techniques during the 1940s. Carbon-14 is produced in the upper atmosphere by the bombardment of $^{14}_{7}N$ by high-energy neutrons from cosmic rays.

$$^{1}_{0}n \quad + \quad ^{14}_{7}N \quad \longrightarrow \quad ^{14}_{6}C \quad + \quad ^{1}_{1}H$$

Neutron from cosmic rays   Nitrogen in atmosphere   Radioactive carbon-14   Proton

The carbon-14 reacts with oxygen to form radioactive carbon dioxide, $^{14}_{6}CO_2$. Living plants continuously absorb carbon dioxide, which incorporates carbon-14 into the plant material. The uptake of carbon-14 stops when the plant dies.

$$^{14}_{6}C \longrightarrow ^{14}_{7}N + ^{0}_{-1}e$$

As the carbon-14 decays, the amount of radioactive carbon-14 in the plant material steadily decreases. In a process called *carbon dating*, scientists use the half-life of carbon-14 (5730 yr) to calculate the length of time since the plant died. For example, a wooden beam found in an ancient dwelling might have one-half of the carbon-14 found in living plants today. Because one half-life of carbon-14 is 5730 yr, the dwelling was constructed about 5730 yr ago. Carbon-14 dating was used to determine that the Dead Sea Scrolls are about 2000 yr old.

The age of the Dead Sea Scrolls was determined using carbon-14.

A radiological dating method used for determining the age of much older items is based on the radioisotope uranium-238, which decays through a series of reactions to lead-206. The uranium-238 isotope has an incredibly long half-life, about 4×10^9 (4 billion) yr. Measurements of the amounts of uranium-238 and lead-206 enable geologists to determine the age of rock samples. The older rocks will have a higher percentage of lead-206 because more of the uranium-238 has decayed. The age of rocks brought back from the Moon by the *Apollo* missions, for example, was determined using uranium-238. They were found to be about 4×10^9 yr old, approximately the same age calculated for the Earth.

The age of a bone sample from a skeleton can be determined by carbon dating.

SAMPLE PROBLEM 16.8 Dating Using Half-Lives

Carbon material in the bones of humans and animals assimilates carbon until death. Using radiocarbon dating, the number of half-lives of carbon-14 from a bone sample determines the age of the bone. Suppose a sample is obtained from a prehistoric animal and used for radiocarbon dating. We can calculate the age of the bone or the years elapsed since the animal died by using the half-life of carbon-14, which is 5730 yr. A bone sample from the skeleton of a prehistoric animal has 25% of the activity of C-14 found in a living animal. How many years ago did the prehistoric animal die?

TRY IT FIRST

SOLUTION
STEP 1 State the given and needed quantities.

ANALYZE THE PROBLEM	Given	Need	Connect
	1 half-life of C-14 = 5730 yr, 25% of initial C-14 activity	years elapsed	number of half-lives

STEP 2 Write a plan to calculate the unknown quantity.

Activity: 100% $\xrightarrow{\text{1.0 half-life}}$ 50% $\xrightarrow{\text{2.0 half-lives}}$ 25%
(initial)

STEP 3 Write the half-life equality and conversion factors.

$$1 \text{ half-life} = 5730 \text{ yr}$$

$$\frac{5730 \text{ yr}}{1 \text{ half-life}} \quad \text{and} \quad \frac{1 \text{ half-life}}{5730 \text{ yr}}$$

STEP 4 Set up the problem to calculate the needed quantity.

$$\text{Years elapsed} = 2.0 \cancel{\text{ half-lives}} \times \frac{5730 \text{ yr}}{1 \cancel{\text{ half-life}}} = 11\,000 \text{ yr}$$

We would estimate that the animal died 11 000 yr ago.

STUDY CHECK 16.8

Suppose that a piece of wood found in a tomb had one-eighth of its original carbon-14 activity. About how many years ago was the wood part of a living tree?

ANSWER

17 000 yr

QUESTIONS AND PROBLEMS

16.4 Half-Life of a Radioisotope

LEARNING GOAL Given the half-life of a radioisotope, calculate the amount of radioisotope remaining after one or more half-lives.

16.29 For each of the following, indicate if the number of half-lives elapsed is:
1. one half-life
2. two half-lives
3. three half-lives

   a. a sample of Pd-103 with a half-life of 17 days after 34 days
   b. a sample of C-11 with a half-life of 20 min after 20 min
   c. a sample of At-211 with a half-life of 7 h after 21 h

16.30 For each of the following, indicate if the number of half-lives elapsed is:
1. one half-life
2. two half-lives
3. three half-lives

   a. a sample of Ce-141 with a half-life of 32.5 days after 32.5 days
   b. a sample of F-18 with a half-life of 110 min after 330 min
   c. a sample of Au-198 with a half-life of 2.7 days after 5.4 days

Applications

16.31 Technetium-99m is an ideal radioisotope for scanning organs because it has a half-life of 6.0 h and is a pure gamma emitter. Suppose that 80.0 mg were prepared in the technetium generator this morning. How many milligrams of technetium-99m would remain after each of the following intervals?
   a. one half-life   **b.** two half-lives
   c. 18 h   **d.** 24 h

16.32 A sample of sodium-24 with an activity of 12 mCi is used to study the rate of blood flow in the circulatory system. If sodium-24 has a half-life of 15 h, what is the activity after each of the following intervals?
   a. one half-life   **b.** 30 h
   c. three half-lives   **d.** 2.5 days

16.33 Strontium-85, used for bone scans, has a half-life of 65 days.
   a. How long will it take for the radiation level of strontium-85 to drop to one-fourth of its original level?
   b. How long will it take for the radiation level of strontium-85 to drop to one-eighth of its original level?

16.34 Fluorine-18, which has a half-life of 110 min, is used in PET scans.
   a. If 100. mg of fluorine-18 is shipped at 8:00 A.M., how many milligrams of the radioisotope are still active after 110 min?
   b. If 100. mg of fluorine-18 is shipped at 8:00 A.M., how many milligrams of the radioisotope are still active when the sample arrives at the radiology laboratory at 1:30 P.M.?

16.5 Medical Applications Using Radioactivity

LEARNING GOAL Describe the use of radioisotopes in medicine.

The first radioactive isotope was used to treat a person with leukemia at the University of California at Berkeley. In 1946, radioactive iodine was successfully used to diagnose thyroid function and to treat hyperthyroidism and thyroid cancer. Radioactive isotopes are now used to produce images of organs including the liver, spleen, thyroid gland, kidneys, brain, and heart.

To determine the condition of an organ in the body, a nuclear medicine technologist may use a radioisotope that concentrates in that organ. The cells in the body do not differentiate between a nonradioactive atom and a radioactive one, so these radioisotopes are easily incorporated. Then the radioactive atoms are detected because they emit radiation. Some radioisotopes used in nuclear medicine are listed in **TABLE 16.8**.

TABLE 16.8 Medical Applications of Radioisotopes

Isotope	Half-Life	Radiation	Medical Application
Au-198	2.7 days	Beta	Liver imaging; treatment of abdominal carcinoma
Ce-141	32.5 days	Beta	Gastrointestinal tract diagnosis; measuring blood flow to the heart
Cs-131	9.7 days	Gamma	Prostate brachytherapy
F-18	110 min	Positron	Positron emission tomography (PET)
Ga-67	78 h	Gamma	Abdominal imaging; tumor detection
Ga-68	68 min	Gamma	Detection of pancreatic cancer
I-123	13.2 h	Gamma	Treatment of thyroid, brain, and prostate cancer
I-131	8.0 days	Beta	Treatment of Graves' disease, goiter, hyperthyroidism, thyroid and prostate cancer
Ir-192	74 days	Gamma	Treatment of breast and prostate cancer
P-32	14.3 days	Beta	Treatment of leukemia, excess red blood cells, and pancreatic cancer
Pd-103	17 days	Gamma	Prostate brachytherapy
Sr-85	65 days	Gamma	Detection of bone lesions; brain scans
Tc-99m	6.0 h	Gamma	Imaging of skeleton and heart muscle, brain, liver, heart, lungs, bone, spleen, kidney, and thyroid; most widely used radioisotope in nuclear medicine
Y-90	2.7 days	Beta	Treatment of liver cancer

SAMPLE PROBLEM 16.9 Medical Applications of Radioactivity

In the treatment of abdominal carcinoma, a person is treated with "seeds" of gold-198, which is a beta emitter. Write the balanced nuclear equation for the beta decay of gold-198.

TRY IT FIRST

SOLUTION

ANALYZE THE PROBLEM	Given	Need	Connect
	gold-198, beta decay	balanced nuclear equation	mass number, atomic number of new nucleus

Titanium "seeds" filled with a radioactive isotope are implanted in the body to treat cancer.

STEP 1 Write the incomplete nuclear equation.

$$^{198}_{79}\text{Au} \longrightarrow ? + ^{0}_{-1}e$$

STEP 2 Determine the missing mass number. In beta decay, the mass number, 198, does not change.

$$^{198}_{79}\text{Au} \longrightarrow ^{198}? + ^{0}_{-1}e$$

STEP 3 Determine the missing atomic number. The atomic number of the new isotope increases by one.

$$^{198}_{79}\text{Au} \longrightarrow ^{198}_{80}? + ^{0}_{-1}e$$

STEP 4 Determine the symbol of the new nucleus. On the periodic table, mercury, Hg, has the atomic number of 80.

STEP 5 Complete the nuclear equation.

$$^{198}_{79}\text{Au} \longrightarrow ^{198}_{80}\text{Hg} + ^{0}_{-1}e$$

STUDY CHECK 16.9

In an experimental treatment, a person is given boron-10, which is taken up by malignant tumors. When bombarded with neutrons, boron-10 decays by emitting alpha particles that destroy the surrounding tumor cells. Write the balanced nuclear equation for the reaction for this experimental procedure.

ANSWER

$$^{1}_{0}n + ^{10}_{5}\text{B} \longrightarrow ^{7}_{3}\text{Li} + ^{4}_{2}\text{He}$$

Scans with Radioisotopes

After a person receives a radioisotope, the radiologist determines the level and location of radioactivity emitted by the radioisotope. An apparatus called a *scanner* is used to produce an image of the organ. The scanner moves slowly across the body above the region where the organ containing the radioisotope is located. The gamma rays emitted from the radioisotope in the organ can be used to expose a photographic plate, producing a *scan* of the organ. On a scan, an area of decreased or increased radiation can indicate conditions such as a disease of the organ, a tumor, a blood clot, or edema.

A common method of determining thyroid function is the use of *radioactive iodine uptake*. Taken orally, the radioisotope iodine-131 mixes with the iodine already present in the thyroid. Twenty-four hours later, the amount of iodine taken up by the thyroid is determined. A detection tube held up to the area of the thyroid gland detects the radiation coming from the iodine-131 that has located there (see **FIGURE 16.5**).

A person with a hyperactive thyroid will have a higher than normal level of radioactive iodine, whereas a person with a hypoactive thyroid will have lower values. If a person has hyperthyroidism, treatment is begun to lower the activity of the thyroid. One treatment involves giving a therapeutic dosage of radioactive iodine, which has a higher radiation level than the diagnostic dose. The radioactive iodine goes to the thyroid where its radiation destroys some of the thyroid cells. The thyroid produces less thyroid hormone, bringing the hyperthyroid condition under control.

(a)

(b)

FIGURE 16.5 ▶ **(a)** A scanner detects radiation from a radioisotope in an organ. **(b)** A scan shows radioactive iodine-131 in the thyroid.

Q What type of radiation would move through body tissues to create a scan?

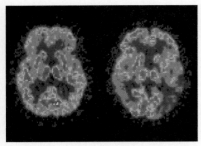

FIGURE 16.6 ▶ These PET scans of the brain show a normal brain on the left and a brain affected by Alzheimer's disease on the right.

⊙ When positrons collide with electrons, what type of radiation is produced that gives an image of an organ?

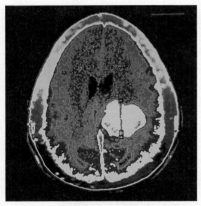

A CT scan shows a tumor (yellow) in the brain.

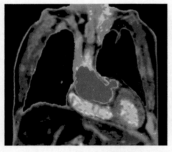

An MRI scan provides images of the heart and lungs.

Positron Emission Tomography

Positron emitters with short half-lives such as carbon-11, oxygen-15, nitrogen-13, and fluorine-18 are used in an imaging method called *positron emission tomography* (PET). A positron-emitting isotope such as fluorine-18 combined with substances in the body such as glucose is used to study brain function, metabolism, and blood flow.

$$^{18}_{9}F \longrightarrow {}^{18}_{8}O + {}^{0}_{+1}e$$

As positrons are emitted, they combine with electrons to produce gamma rays that are detected by computerized equipment to create a three-dimensional image of the organ (see **FIGURE 16.6**).

Computed Tomography

Another imaging method used to scan organs such as the brain, lungs, and heart is *computed tomography* (CT). A computer monitors the absorption of 30 000 X-ray beams directed at successive layers of the target organ. Based on the densities of the tissues and fluids in the organ, the differences in absorption of the X-rays provide a series of images of the organ. This technique is successful in the identification of hemorrhages, tumors, and atrophy.

Magnetic Resonance Imaging

Magnetic resonance imaging (MRI) is a powerful imaging technique that does not involve X-ray radiation. It is the least invasive imaging method available. MRI is based on the absorption of energy when the protons in hydrogen atoms are excited by a strong magnetic field. Hydrogen atoms make up 63% of all the atoms in the body. In the hydrogen nuclei, the protons act like tiny bar magnets. With no external field, the protons have random orientations. However, when placed within a strong magnetic field, the protons align with the field. A proton aligned with the field has a lower energy than one that is aligned against the field. As the MRI scan proceeds, radiofrequency pulses of energy are applied and the hydrogen nuclei resonate at a certain frequency. Then the radio waves are quickly turned off, and the protons slowly return to their natural alignment within the magnetic field and resonate at a different frequency. They release the energy absorbed from the radio wave pulses. The difference in energy between the two states is released, which produces the electromagnetic signal that the scanner detects. These signals are sent to a computer system, where a color image of the body is generated. Because hydrogen atoms in the body are in different chemical environments, different energies are absorbed. MRI is particularly useful in obtaining images of soft tissues, which contain large amounts of hydrogen atoms in the form of water.

CHEMISTRY LINK TO **HEALTH**
Brachytherapy

The process called *brachytherapy*, or seed implantation, is an internal form of radiation therapy. The prefix *brachy* is from the Greek word for short distance. With internal radiation, a high dose of radiation is delivered to a cancerous area, whereas normal tissue sustains minimal damage. Because higher doses are used, fewer treatments of shorter duration are needed. Conventional external treatment delivers a lower dose per treatment, but requires six to eight weeks of treatment.

Permanent Brachytherapy

One of the most common forms of cancer in males is prostate cancer. In addition to surgery and chemotherapy, one treatment option is to place 40 or more titanium capsules, or "seeds," in the malignant area. Each seed, which is the size of a grain of rice, contains radioactive iodine-125, palladium-103, or cesium-131, which decay by gamma emission. The radiation from the seeds destroys the cancer by interfering with the reproduction of cancer cells with minimal damage to adjacent normal tissues. Ninety percent (90%) of the radioisotopes decay within a few months because they have short half-lives.

Isotope	I-125	Pd-103	Cs-131
Radiation	Gamma	Gamma	Gamma
Half-Life	60 days	17 days	10 days
Time Required to Deliver 90% of Radiation	7 months	2 months	1 month

Almost no radiation passes out of the patient's body. The amount of radiation received by a family member is no greater than that received on a long plane flight. Because the radioisotopes decay to products that are not radioactive, the inert titanium capsules can be left in the body.

Temporary Brachytherapy

In another type of treatment for prostate cancer, long needles containing iridium-192 are placed in the tumor. However, the needles are removed after 5 to 10 min, depending on the activity of the iridium isotope. Compared to permanent brachytherapy, temporary brachytherapy can deliver a higher dose of radiation over a shorter time. The procedure may be repeated in a few days.

Brachytherapy is also used following breast cancer lumpectomy. An iridium-192 isotope is inserted into the catheter implanted in the space left by the removal of the tumor. The isotope is removed after 5 to 10 min, depending on the activity of the iridium source. Radiation is delivered primarily to the tissue surrounding the cavity that contained the tumor and where the cancer is most likely to recur. The procedure is repeated twice a day for five days to give an absorbed dose of 34 Gy (3400 rad). The catheter is removed, and no radioactive material remains in the body.

In conventional external beam therapy for breast cancer, a patient is given 2 Gy once a day for six to seven weeks, which gives a total absorbed dose of about 80 Gy or 8000 rad. The external beam therapy irradiates the entire breast, including the tumor cavity.

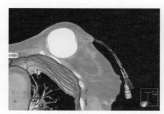 

A catheter placed temporarily in the breast for radiation from Ir-192.

QUESTIONS AND PROBLEMS

16.5 Medical Applications Using Radioactivity

LEARNING GOAL Describe the use of radioisotopes in medicine.

Applications

16.35 Bone and bony structures contain calcium and phosphorus.
  a. Why would the radioisotopes calcium-47 and phosphorus-32 be used in the diagnosis and treatment of bone diseases?
  b. During nuclear tests, scientists were concerned that strontium-85, a radioactive product, would be harmful to the growth of bone in children. Explain.

16.36 a. Technetium-99m emits only gamma radiation. Why would this type of radiation be used in diagnostic imaging rather than an isotope that also emits beta or alpha radiation?

  b. A person with *polycythemia vera* (excess production of red blood cells) receives radioactive phosphorus-32. Why would this treatment reduce the production of red blood cells in the bone marrow of the patient?

16.37 In a diagnostic test for leukemia, a person receives 4.0 mL of a solution containing selenium-75. If the activity of the selenium-75 is 45 μCi/mL, what dose, in microcuries, does the patient receive?

16.38 A vial contains radioactive iodine-131 with an activity of 2.0 mCi/mL. If a thyroid test requires 3.0 mCi in an "atomic cocktail," how many milliliters are used to prepare the iodine-131 solution?

16.6 Nuclear Fission and Fusion

LEARNING GOAL Describe the processes of nuclear fission and fusion.

During the 1930s, scientists bombarding uranium-235 with neutrons discovered that the U-235 nucleus splits into two smaller nuclei and produces a great amount of energy. This was the discovery of nuclear **fission**. The energy generated by splitting the atom was called *atomic energy*. A typical equation for nuclear fission is:

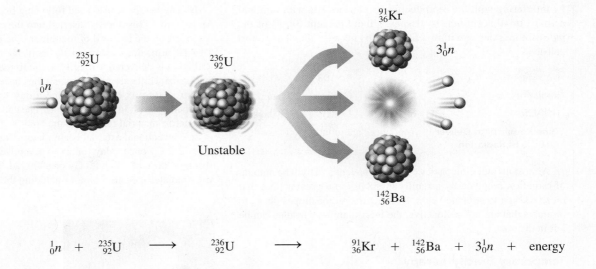

$$\,^1_0 n \;+\; \,^{235}_{92}U \;\longrightarrow\; \,^{236}_{92}U \;\longrightarrow\; \,^{91}_{36}Kr \;+\; \,^{142}_{56}Ba \;+\; 3\,^1_0 n \;+\; \text{energy}$$

If we could determine the mass of the products krypton, barium, and three neutrons, with great accuracy, we would find that their total mass is slightly less than the mass of the starting materials. The missing mass has been converted into an enormous amount of energy, consistent with the famous equation derived by Albert Einstein.

$$E = mc^2$$

where E is the energy released, m is the mass lost, and c is the speed of light, 3×10^8 m/s. Even though the mass loss is very small, when it is multiplied by the speed of light squared, the result is a large value for the energy released. The fission of 1 g of uranium-235 produces about as much energy as the burning of 3 tons of coal.

Chain Reaction

ENGAGE

Why is a critical mass of uranium-235 necessary to sustain a nuclear chain reaction?

Fission begins when a neutron collides with the nucleus of a uranium atom. The resulting nucleus is unstable and splits into smaller nuclei. This fission process also releases several neutrons and large amounts of gamma radiation and energy. The neutrons emitted have high energies and bombard other uranium-235 nuclei. In a **chain reaction**, there is a rapid increase in the number of high-energy neutrons available to react with more uranium. To sustain a nuclear chain reaction, sufficient quantities of uranium-235 must be brought together to provide a *critical mass* in which almost all the neutrons immediately collide with more uranium-235 nuclei. So much heat and energy build up that an atomic explosion can occur (see **FIGURE 16.7**).

Nuclear Fusion

In **fusion**, two small nuclei combine to form a larger nucleus. Mass is lost, and a tremendous amount of energy is released, even more than the energy released from nuclear fission. However, a fusion reaction requires a temperature of 100 000 000 °C to overcome the repulsion of the hydrogen nuclei and cause them to undergo fusion. Fusion reactions occur continuously in the Sun and other stars, providing us with heat and light. The huge amounts of energy produced by our sun come from the fusion of 6×10^{11} kg of hydrogen every second. In a fusion reaction, isotopes of hydrogen combine to form helium and large amounts of energy.

Scientists expect less radioactive waste with shorter half-lives from fusion reactors. However, fusion is still in the experimental stage because the extremely high temperatures needed have been difficult to reach and even more difficult to maintain. Research groups around the world are attempting to develop the technology needed to make the harnessing of the fusion reaction for energy a reality in our lifetime.

FIGURE 16.7 ▶ In a nuclear chain reaction, the fission of each uranium-235 atom produces three neutrons that cause the nuclear fission of more and more uranium-235 atoms.

Q Why is the fission of uranium-235 called a chain reaction?

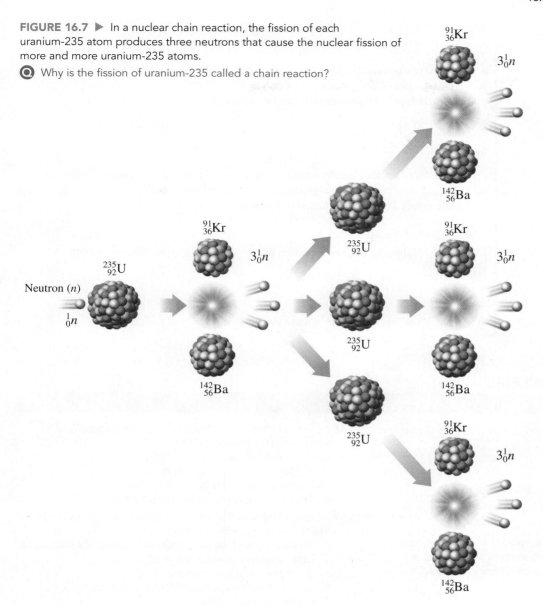

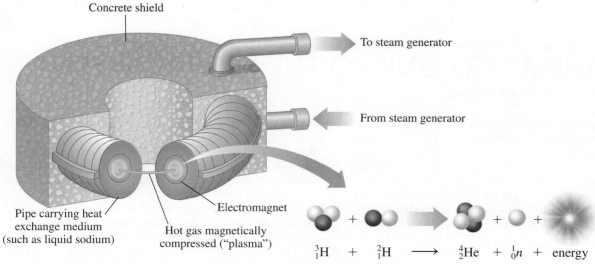

In a fusion reactor, high temperatures are needed to combine hydrogen atoms.

SAMPLE PROBLEM 16.10 Identifying Fission and Fusion

Classify the following as pertaining to fission, fusion, or both:

a. A large nucleus breaks apart to produce smaller nuclei.
b. Large amounts of energy are released.
c. Extremely high temperatures are needed for reaction.

TRY IT FIRST

SOLUTION

a. When a large nucleus breaks apart to produce smaller nuclei, the process is fission.
b. Large amounts of energy are generated in both the fusion and fission processes.
c. An extremely high temperature is required for fusion.

STUDY CHECK 16.10

Classify the following nuclear equation as pertaining to fission, fusion, or both:

$$^3_1H + {}^2_1H \longrightarrow {}^4_2He + {}^1_0n + energy$$

ANSWER

When small nuclei combine to release energy, the process is fusion.

QUESTIONS AND PROBLEMS

16.6 Nuclear Fission and Fusion

LEARNING GOAL Describe the processes of nuclear fission and fusion.

16.39 What is nuclear fission?

16.40 How does a chain reaction occur in nuclear fission?

16.41 Complete the following fission reaction:

$$^1_0n + {}^{235}_{92}U \longrightarrow {}^{131}_{50}Sn + ? + 2{}^1_0n + energy$$

16.42 In another fission reaction, uranium-235 bombarded with a neutron produces strontium-94, another small nucleus, and three neutrons. Write the balanced nuclear equation for the fission reaction.

16.43 Indicate whether each of the following is characteristic of the fission or fusion process, or both:
  a. Neutrons bombard a nucleus.
  b. The nuclear process occurs in the Sun.
  c. A large nucleus splits into smaller nuclei.
  d. Small nuclei combine to form larger nuclei.

16.44 Indicate whether each of the following is characteristic of the fission or fusion process, or both:
  a. Very high temperatures are required to initiate the reaction.
  b. Less radioactive waste is produced.
  c. Hydrogen nuclei are the reactants.
  d. Large amounts of energy are released when the nuclear reaction occurs.

CHEMISTRY LINK TO THE **ENVIRONMENT**
Nuclear Power Plants

In a nuclear power plant, the quantity of uranium-235 is held below a critical mass, so it cannot sustain a chain reaction. The fission reactions are slowed by placing control rods, which absorb some of the fast-moving neutrons, among the uranium samples. In this way, less fission occurs, and there is a slower, controlled production of energy. The heat from the controlled fission is used to produce steam. The steam drives a generator, which produces electricity. Approximately 20% of the electrical energy produced in the United States is generated in nuclear power plants.

Although nuclear power plants help meet some of our energy needs, there are some problems associated with nuclear power. One of the most serious problems is the production of radioactive byproducts that have very long half-lives, such as plutonium-239 with a half-life of 24 000 yr. It is essential that these waste products be stored safely in a place where they do not contaminate the environment.

Nuclear power plants supply about 20% of electricity in the United States.

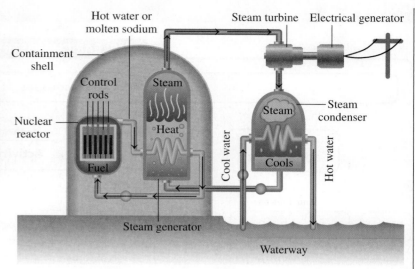

Heat from nuclear fission is used to generate electricity.

Follow Up

CARDIAC IMAGING USING A RADIOISOTOPE

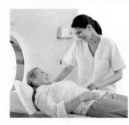

A thallium stress test can show narrowing of an artery during stress.

As part of her nuclear stress test, Simone starts to walk on the treadmill. When she reaches the maximum level, Dr. Paul injects a radioactive dye with Tl-201 with an activity of 74 MBq. The radiation emitted from areas of heart is detected by a scanner and produces images of her heart muscle. A thallium stress test can determine how effectively coronary arteries provide blood to the heart. If Simone has any damage to her coronary arteries, reduced blood flow during stress would show a narrowing of an artery or a blockage. After Simone rests for 3 h, Dr. Paul injects more dye with Tl-201 and she is placed under the scanner again. A second set of images of her heart muscle at rest are taken. When

Simone's doctor reviews her scans, he assures her that she had normal blood flow to her heart muscle, both at rest and under stress.

Applications

16.45 What is the activity of the radioactive dye injection for Simone
    a. in curies?
    b. in millicuries?

16.46 If the half-life of Tl-201 is 3.0 days, what is its activity, in megabecquerels
    a. after 3.0 days?
    b. after 6.0 days?

16.47 How many days will it take until the activity of the Tl-201 in Simone's body is one-eighth of the initial activity?

16.48 Radiation from Tl-201 to Simone's kidneys can be 24 mGy. What is this amount of radiation in rads?

CONCEPT MAP

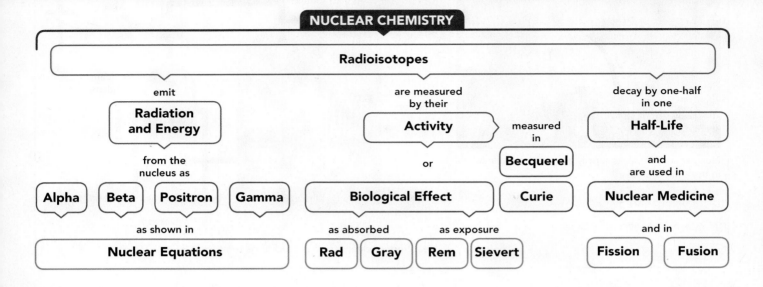

NUCLEAR CHEMISTRY

Radioisotopes

emit → **Radiation and Energy**

are measured by their → **Activity** — measured in → **Becquerel**

decay by one-half in one → **Half-Life**

from the nucleus as

Alpha | **Beta** | **Positron** | **Gamma**

or → **Curie**

and are used in → **Nuclear Medicine**

as shown in

Nuclear Equations

Biological Effect

as absorbed

as exposure

Rad | **Gray** | **Rem** | **Sievert**

and in → **Fission** | **Fusion**

CHAPTER REVIEW

16.1 Natural Radioactivity

LEARNING GOAL Describe alpha, beta, positron, and gamma radiation.

^4_2He
Alpha (α) particle

- Radioactive isotopes have unstable nuclei that break down (decay), spontaneously emitting alpha (α), beta (β), positron (β^+), and gamma (γ) radiation.
- Because radiation can damage the cells in the body, proper protection must be used: shielding, limiting the time of exposure, and distance.

16.2 Nuclear Reactions

LEARNING GOAL Write a balanced nuclear equation showing mass numbers and atomic numbers for radioactive decay.

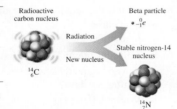

Radioactive carbon nucleus
Radiation
New nucleus
$^{14}_6\text{C}$

Beta particle
$^0_{-1}e$
Stable nitrogen-14 nucleus
$^{14}_7\text{N}$

- A balanced nuclear equation is used to represent the changes that take place in the nuclei of the reactants and products.
- The new isotopes and the type of radiation emitted can be determined from the symbols that show the mass numbers and atomic numbers of the isotopes in the nuclear equation.
- A radioisotope is produced artificially when a nonradioactive isotope is bombarded by a small particle.

16.3 Radiation Measurement

LEARNING GOAL Describe the detection and measurement of radiation.

- In a Geiger counter, radiation produces charged particles in the gas contained in a tube, which generates an electrical current.
- The curie (Ci) and the becquerel (Bq) measure the activity, which is the number of nuclear transformations per second.

- The amount of radiation absorbed by a substance is measured in rads or the gray (Gy).
- The rem and the sievert (Sv) are units used to determine the biological damage from the different types of radiation.

16.4 Half-Life of a Radioisotope

LEARNING GOAL Given the half-life of a radioisotope, calculate the amount of radioisotope remaining after one or more half-lives.

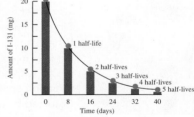

- Every radioisotope has its own rate of emitting radiation.
- The time it takes for one-half of a radioactive sample to decay is called its half-life.
- For many medical radioisotopes, such as Tc-99m and I-131, half-lives are short.
- For other isotopes, usually naturally occurring ones such as C-14, Ra-226, and U-238, half-lives are extremely long.

16.5 Medical Applications Using Radioactivity

LEARNING GOAL Describe the use of radioisotopes in medicine.

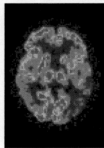

- In nuclear medicine, radioisotopes that go to specific sites in the body are given to the patient.
- By detecting the radiation they emit, an evaluation can be made about the location and extent of an injury, disease, tumor, or the level of function of a particular organ.
- Higher levels of radiation are used to treat or destroy tumors.

16.6 Nuclear Fission and Fusion

LEARNING GOAL

Describe the processes of nuclear fission and fusion.

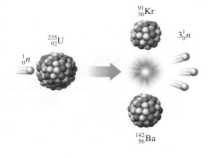

- In fission, the bombardment of a large nucleus breaks it apart into smaller nuclei, releasing one or more types of radiation and a great amount of energy.

- In fusion, small nuclei combine to form larger nuclei while great amounts of energy are released.

KEY TERMS

alpha particle A nuclear particle identical to a helium nucleus, symbol α or $_2^4He$.

becquerel (Bq) A unit of activity of a radioactive sample equal to one disintegration per second.

beta particle A particle identical to an electron, symbol $_{-1}^0e$ or β, that forms in the nucleus when a neutron changes to a proton and an electron.

chain reaction A fission reaction that will continue once it has been initiated by a high-energy neutron bombarding a heavy nucleus such as uranium-235.

curie (Ci) A unit of activity of a radioactive sample equal to 3.7×10^{10} disintegrations/s.

decay curve A diagram of the decay of a radioactive element.

equivalent dose The measure of biological damage from an absorbed dose that has been adjusted for the type of radiation.

fission A process in which large nuclei are split into smaller pieces, releasing large amounts of energy.

fusion A reaction in which large amounts of energy are released when small nuclei combine to form larger nuclei.

gamma ray High-energy radiation, symbol $_0^0\gamma$, emitted by an unstable nucleus.

gray (Gy) A unit of absorbed dose equal to 100 rad.

half-life The length of time it takes for one-half of a radioactive sample to decay.

positron A particle of radiation with no mass and a positive charge, symbol β^+ or $_{+1}^0e$, produced when a proton is transformed into a neutron and a positron.

rad (radiation absorbed dose) A measure of an amount of radiation absorbed by the body.

radiation Energy or particles released by radioactive atoms.

radioactive decay The process by which an unstable nucleus breaks down with the release of high-energy radiation.

radioisotope A radioactive atom of an element.

rem (radiation equivalent in humans) A measure of the biological damage caused by the various kinds of radiation (rad $\times$ radiation biological factor).

sievert (Sv) A unit of biological damage (equivalent dose) equal to 100 rem.

 ## CORE CHEMISTRY SKILLS

The chapter section containing each Core Chemistry Skill is shown in parentheses at the end of each heading.

Writing Nuclear Equations (16.2)

- A nuclear equation is written with the atomic symbols of the original radioactive nucleus on the left, an arrow, and the new nucleus and the type of radiation emitted on the right.
- The sum of the mass numbers and the sum of the atomic numbers on one side of the arrow must equal the sum of the mass numbers and the sum of the atomic numbers on the other side.
- When an alpha particle is emitted, the mass number of the new nucleus decreases by 4, and its atomic number decreases by 2.
- When a beta particle is emitted, there is no change in the mass number of the new nucleus, but its atomic number increases by one.
- When a positron is emitted, there is no change in the mass number of the new nucleus, but its atomic number decreases by one.
- In gamma emission, there is no change in the mass number or the atomic number of the new nucleus.

Example: **a.** Write a balanced nuclear equation for the alpha decay of Po-210.

b. Write a balanced nuclear equation for the beta decay of Co-60.

Answer: **a.** When an alpha particle is emitted, we calculate the decrease of 4 in the mass number (210) of the polonium, and a decrease of 2 in its atomic number.

$$_{84}^{210}Po \longrightarrow _{82}^{206}? + _2^4He$$

Because lead has atomic number 82, the new nucleus must be an isotope of lead.

$$_{84}^{210}Po \longrightarrow _{82}^{206}Pb + _2^4He$$

b. When a beta particle is emitted, there is no change in the mass number (60) of the cobalt, but there is an increase of 1 in its atomic number.

$$_{27}^{60}Co \longrightarrow _{28}^{60}? + _{-1}^0e$$

Because nickel has atomic number 28, the new nucleus must be an isotope of nickel.

$$_{27}^{60}Co \longrightarrow _{28}^{60}Ni + _{-1}^0e$$

Using Half-Lives (16.4)

- The half-life of a radioisotope is the amount of time it takes for one-half of a sample to decay.
- The remaining amount of a radioisotope is calculated by dividing its quantity or activity by one-half for each half-life that has elapsed.

Example: Co-60 has a half-life of 5.3 yr. If the initial sample of Co-60 has an activity of 1200 Ci, what is its activity after 15.9 yr?

Answer:

Years	Number of Half-Lives	Co-60 (Activity)
0	0	1200 Ci
5.3	1	600 Ci
10.6	2	300 Ci
15.9	3	150 Ci

In 15.9 yr, three half-lives have passed. Thus, the activity was reduced from 1200 Ci to 150 Ci.

$$1200 \text{ Ci} \xrightarrow{\text{1 half-life}} 600 \text{ Ci} \xrightarrow{\text{2 half-lives}} 300 \text{ Ci} \xrightarrow{\text{3 half-lives}} 150 \text{ Ci}$$

UNDERSTANDING THE CONCEPTS

The chapter sections to review are shown in parentheses at the end of each question.

In problems 16.49 to 16.52, a nucleus is shown with protons and neutrons.

 proton   neutron

16.49 Draw the new nucleus when this isotope emits a positron to complete the following figure: (16.2)

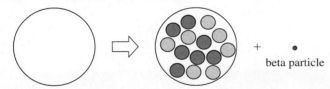

+ • positron

16.50 Draw the nucleus that emits a beta particle to complete the following figure: (16.2)

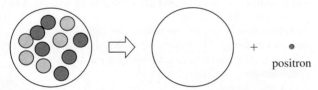

+ • beta particle

16.51 Draw the nucleus of the isotope that is bombarded in the following figure: (16.2)

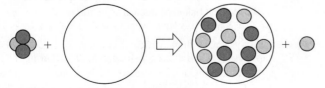

16.52 Complete the following bombardment reaction by drawing the nucleus of the new isotope that is produced in the following figure: (16.2)

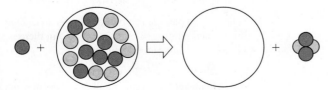

16.53 Carbon dating of small bits of charcoal used in cave paintings has determined that some of the paintings are from 10 000 to 30 000 yr old. Carbon-14 has a half-life of 5730 yr. In a 1 μg-sample of carbon from a live tree, the activity of carbon-14 is 6.4 μCi. If researchers determine that 1 μg of charcoal from a prehistoric cave painting in France has an activity of 0.80 μCi, what is the age of the painting? (16.4)

The technique of carbon dating is used to determine the age of ancient cave paintings.

16.54 Use the following decay curve for iodine-131 to answer questions **a** to **c**: (16.4)
  a. Complete the values for the mass of radioactive iodine-131 on the vertical axis.
  b. Complete the number of days on the horizontal axis.
  c. What is the half-life, in days, of iodine-131?

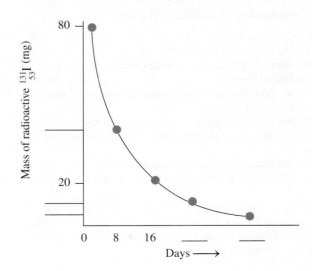

ADDITIONAL QUESTIONS AND PROBLEMS

16.55 Determine the number of protons and number of neutrons in the nucleus of each of the following: (16.1)
 a. sodium-25   **b.** nickel-61
 c. rubidium-84   **d.** silver-110

16.56 Determine the number of protons and number of neutrons in the nucleus of each of the following: (16.1)
 a. boron-10   **b.** zinc-72
 c. iron-59   **d.** gold-198

16.57 Identify each of the following as alpha decay, beta decay, positron emission, or gamma emission: (16.1, 16.2)
 a. $^{27m}_{13}Al \longrightarrow ^{27}_{13}Al + ^{0}_{0}\gamma$
 b. $^{8}_{5}B \longrightarrow ^{8}_{4}Be + ^{0}_{+1}e$
 c. $^{220}_{86}Rn \longrightarrow ^{216}_{84}Po + ^{4}_{2}He$

16.58 Identify each of the following as alpha decay, beta decay, positron emission, or gamma emission: (16.1, 16.2)
 a. $^{127}_{55}Cs \longrightarrow ^{127}_{54}Xe + ^{0}_{+1}e$
 b. $^{90}_{38}Sr \longrightarrow ^{90}_{39}Y + ^{0}_{-1}e$
 c. $^{218}_{85}At \longrightarrow ^{214}_{83}Bi + ^{4}_{2}He$

16.59 Write the balanced nuclear equation for each of the following: (16.1, 16.2)
 a. Th-225 (α decay)
 b. Bi-210 (α decay)
 c. cesium-137 (β decay)
 d. tin-126 (β decay)
 e. F-18 (β^+ emission)

16.60 Write the balanced nuclear equation for each of the following: (16.1, 16.2)
 a. potassium-40 (β decay)
 b. sulfur-35 (β decay)
 c. platinum-190 (α decay)
 d. Ra-210 (α decay)
 e. In-113m (γ emission)

16.61 Complete each of the following nuclear equations: (16.2)
 a. $^{4}_{2}He + ^{14}_{7}N \longrightarrow ? + ^{1}_{1}H$
 b. $^{4}_{2}He + ^{27}_{13}Al \longrightarrow ^{30}_{14}Si + ?$
 c. $^{1}_{0}n + ^{235}_{92}U \longrightarrow ^{90}_{38}Sr + 3^{1}_{0}n + ?$
 d. $^{23m}_{12}Mg \longrightarrow ? + ^{0}_{0}\gamma$

16.62 Complete each of the following nuclear equations: (16.2)
 a. $? + ^{59}_{27}Co \longrightarrow ^{56}_{25}Mn + ^{4}_{2}He$
 b. $? \longrightarrow ^{14}_{7}N + ^{0}_{-1}e$
 c. $^{0}_{-1}e + ^{76}_{36}Kr \longrightarrow ?$
 d. $^{4}_{2}He + ^{241}_{95}Am \longrightarrow ? + 2^{1}_{0}n$

16.63 Write the balanced nuclear equation for each of the following: (16.2)
 a. When two oxygen-16 atoms collide, one of the products is an alpha particle.
 b. When californium-249 is bombarded by oxygen-18, a new element, seaborgium-263, and four neutrons are produced.
 c. Radon-222 undergoes alpha decay.
 d. An atom of strontium-80 emits a positron.

16.64 Write the balanced nuclear equation for each of the following: (16.2)
 a. Polonium-210 decays to give lead-206.
 b. Bismuth-211 emits an alpha particle.
 c. A radioisotope emits a positron to form titanium-48.
 d. An atom of germanium-69 emits a positron.

16.65 A 120-mg sample of technetium-99m is used for a diagnostic test. If technetium-99m has a half-life of 6.0 h, how many milligrams of the technetium-99m sample remains active 24 h after the test? (16.4)

16.66 The half-life of oxygen-15 is 124 s. If a sample of oxygen-15 has an activity of 4000 Bq, how many minutes will elapse before it has an activity of 500 Bq? (16.4)

16.67 What is the difference between fission and fusion? (16.6)

16.68 **a.** What are the products in the fission of uranium-235 that make possible a nuclear chain reaction? (16.6)
 b. What is the purpose of placing control rods among uranium samples in a nuclear reactor?

16.69 Where does fusion occur naturally? (16.6)

16.70 Why are scientists continuing to try to build a fusion reactor even though the very high temperatures it requires have been difficult to reach and maintain? (16.6)

Applications

16.71 The activity of K-40 in a 70.-kg human body is estimated to be 120 nCi. What is this activity in becquerels? (16.3)

16.72 The activity of C-14 in a 70.-kg human body is estimated to be 3.7 kBq. What is this activity in microcuries? (16.3)

16.73 If the amount of radioactive phosphorus-32, used to treat leukemia, in a sample decreases from 1.2 mg to 0.30 mg in 28.6 days, what is the half-life of phosphorus-32? (16.4)

16.74 If the amount of radioactive iodine-123, used to treat thyroid cancer, in a sample decreases from 0.4 mg to 0.1 mg in 26.4 h, what is the half-life of iodine-123? (16.4)

16.75 Calcium-47, used to evaluate bone metabolism, has a half-life of 4.5 days. (16.2, 16.4)
 a. Write the balanced nuclear equation for the beta decay of calcium-47.
 b. How many milligrams of a 16-mg sample of calcium-47 remain after 18 days?
 c. How many days have passed if 4.8 mg of calcium-47 decayed to 1.2 mg of calcium-47?

16.76 Cesium-137, used in cancer treatment, has a half-life of 30 yr. (16.2, 16.4)
 a. Write the balanced nuclear equation for the beta decay of cesium-137.
 b. How many milligrams of a 16-mg sample of cesium-137 remain after 90 yr?
 c. How many years are required for 28 mg of cesium-137 to decay to 3.5 mg of cesium-137?

CHALLENGE QUESTIONS

The following groups of questions are related to the topics in this chapter. However, they do not all follow the chapter order, and they require you to combine concepts and skills from several sections. These questions will help you increase your critical thinking skills and prepare for your next exam.

16.77 Write the balanced nuclear equation for each of the following radioactive emissions: (16.2)
   a. an alpha particle from Hg-180
   b. a beta particle from Au-198
   c. a positron from Rb-82

16.78 Write the balanced nuclear equation for each of the following radioactive emissions: (16.2)
   a. an alpha particle from Gd-148
   b. a beta particle from Sr-90
   c. a positron from Al-25

16.79 All the elements beyond uranium, the transuranium elements, have been prepared by bombardment and are not naturally occurring elements. The first transuranium element neptunium, Np, was prepared by bombarding U-238 with neutrons to form a neptunium atom and a beta particle. Complete the following equation: (16.2)

$$^1_0n + {}^{238}_{92}U \longrightarrow ? + ?$$

16.80 One of the most recent transuranium elements ununoctium-294 (Uuo-294), atomic number 118, was prepared by bombarding californium-249 with another isotope. Complete the following equation for the preparation of this new element: (16.2)

$$? + {}^{249}_{98}Cf \longrightarrow {}^{294}_{118}Uuo + 3^1_0n$$

16.81 A 64-μCi sample of Tl-201 decays to 4.0 μCi in 12 days. What is the half-life, in days, of Tl-201? (16.3, 16.4)

16.82 A wooden object from the site of an ancient temple has a carbon-14 activity of 10 counts/min compared with a reference piece of wood cut today that has an activity of 40 counts/min. If the half-life for carbon-14 is 5730 yr, what is the age of the ancient wood object? (16.3, 16.4)

16.83 The half-life for the radioactive decay of calcium-47 is 4.5 days. If a sample has an activity of 1.0 μCi after 27 days, what was the initial activity, in microcuries, of the sample? (16.3, 16.4)

16.84 The half-life for the radioactive decay of Ce-141 is 32.5 days. If a sample has an activity of 4.0 μCi after 130 days have elapsed, what was the initial activity, in microcuries, of the sample? (16.3, 16.4)

16.85 Element 114 was recently named flerovium, symbol Fl. The reaction for its synthesis involves bombarding Pu-244 with Ca-48. Write the balanced nuclear equation for the synthesis of flerovium. (16.2)

16.86 Element 116 was recently named livermorium, symbol Lv. The reaction for its synthesis involves bombarding Cm-248 with Ca-48. Write the balanced nuclear equation for the synthesis of livermorium. (16.2)

Applications

16.87 A nuclear technician was accidentally exposed to potassium-42 while doing brain scans for possible tumors. The error was not discovered until 36 h later when the activity of the potassium-42 sample was 2.0 μCi. If potassium-42 has a half-life of 12 h, what was the activity of the sample at the time the technician was exposed? (16.3, 16.4)

16.88 The radioisotope sodium-24 is used to determine the levels of electrolytes in the body. A 16-μg sample of sodium-24 decays to 2.0 μg in 45 h. What is the half-life, in hours, of sodium-24? (16.4)

ANSWERS

Answers to Selected Questions and Problems

16.1 a. alpha particle
   b. positron
   c. gamma radiation

16.3 a. ${}^{39}_{19}K$, ${}^{40}_{19}K$, ${}^{41}_{19}K$
   b. They all have 19 protons and 19 electrons, but they differ in the number of neutrons.

16.5 a. β or ${}^{0}_{-1}e$ **b.** α or ${}^{4}_{2}He$
   c. n or ${}^{1}_{0}n$ **d.** ${}^{38}_{18}Ar$
   e. ${}^{14}_{6}C$

16.7 a. ${}^{64}_{29}Cu$ **b.** ${}^{75}_{34}Se$
   c. ${}^{24}_{11}Na$ **d.** ${}^{15}_{7}N$

16.9

Medical Use	Atomic Symbol	Mass Number	Number of Protons	Number of Neutrons
Heart imaging	${}^{201}_{81}Tl$	201	81	120
Radiation therapy	${}^{60}_{27}Co$	60	27	33
Abdominal scan	${}^{67}_{31}Ga$	67	31	36
Hyperthyroidism	${}^{131}_{53}I$	131	53	78
Leukemia treatment	${}^{32}_{15}P$	32	15	17

16.11 a. 1, alpha particle   **b.** 3, gamma radiation
   c. 1, alpha particle

16.13 a. $^{208}_{84}Po \longrightarrow ^{204}_{82}Pb + ^{4}_{2}He$

b. $^{232}_{90}Th \longrightarrow ^{228}_{88}Ra + ^{4}_{2}He$

c. $^{251}_{102}No \longrightarrow ^{247}_{100}Fm + ^{4}_{2}He$

d. $^{220}_{86}Rn \longrightarrow ^{216}_{84}Po + ^{4}_{2}He$

16.15 a. $^{25}_{11}Na \longrightarrow ^{25}_{12}Mg + ^{0}_{-1}e$

b. $^{20}_{8}O \longrightarrow ^{20}_{9}F + ^{0}_{-1}e$

c. $^{92}_{38}Sr \longrightarrow ^{92}_{39}Y + ^{0}_{-1}e$

d. $^{60}_{26}Fe \longrightarrow ^{60}_{27}Co + ^{0}_{-1}e$

16.17 a. $^{26}_{14}Si \longrightarrow ^{26}_{13}Al + ^{0}_{+1}e$

b. $^{54}_{27}Co \longrightarrow ^{54}_{26}Fe + ^{0}_{+1}e$

c. $^{77}_{37}Rb \longrightarrow ^{77}_{36}Kr + ^{0}_{+1}e$

d. $^{93}_{45}Rh \longrightarrow ^{93}_{44}Ru + ^{0}_{+1}e$

16.19 a. $^{28}_{14}Si$, beta decay

b. $^{0}_{0}\gamma$, gamma emission

c. $^{0}_{-1}e$, beta decay

d. $^{238}_{92}U$, alpha decay

e. $^{188}_{79}Au$, positron emission

16.21 a. $^{10}_{4}Be$ **b.** $^{132}_{53}I$

c. $^{27}_{13}Al$ **d.** $^{17}_{8}O$

16.23 a. 2, absorbed dose   **b.** 3, biological damage
c. 1, activity   **d.** 2, absorbed dose

16.25 The technician exposed to 5 rad received the higher amount of radiation.

16.27 a. 294 μCi   **b.** 0.5 Gy

16.29 a. two half-lives   **b.** one half-life
c. three half-lives

16.31 a. 40.0 mg   **b.** 20.0 mg
c. 10.0 mg   **d.** 5.00 mg

16.33 a. 130 days   **b.** 195 days

16.35 a. Because the elements Ca and P are part of the bone, the radioactive isotopes of Ca and P will become part of the bony structures of the body, where their radiation can be used to diagnose or treat bone diseases.

b. Strontium (Sr) acts much like calcium (Ca) because both are Group 2A (2) elements. The body will accumulate radioactive strontium in bones in the same way that it incorporates calcium. Radioactive strontium is harmful to children because the radiation it produces causes more damage in cells that are dividing rapidly.

16.37 180 μCi

16.39 Nuclear fission is the splitting of a large atom into smaller fragments with the release of large amounts of energy.

16.41 $^{103}_{42}Mo$

16.43 a. fission   **b.** fusion
c. fission   **d.** fusion

16.45 a. 2.0×10^{-3} Ci   **b.** 2.0 mCi

16.47 9.0 days

16.49

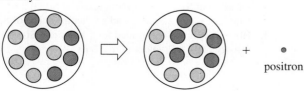

positron

16.51

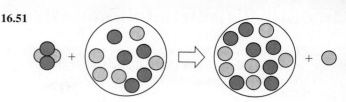

16.53 17 200 yr old

16.55 a. 11 protons and 14 neutrons
b. 28 protons and 33 neutrons
c. 37 protons and 47 neutrons
d. 47 protons and 63 neutrons

16.57 a. gamma emission
b. positron emission
c. alpha decay

16.59 a. $^{225}_{90}Th \longrightarrow ^{221}_{88}Ra + ^{4}_{2}He$

b. $^{210}_{83}Bi \longrightarrow ^{206}_{81}Tl + ^{4}_{2}He$

c. $^{137}_{55}Cs \longrightarrow ^{137}_{56}Ba + ^{0}_{-1}e$

d. $^{126}_{50}Sn \longrightarrow ^{126}_{51}Sb + ^{0}_{-1}e$

e. $^{18}_{9}F \longrightarrow ^{18}_{8}O + ^{0}_{+1}e$

16.61 a. $^{17}_{8}O$ **b.** $^{1}_{1}H$

c. $^{143}_{54}Xe$ **d.** $^{23}_{12}Mg$

16.63 a. $^{16}_{8}O + ^{16}_{8}O \longrightarrow ^{28}_{14}Si + ^{4}_{2}He$

b. $^{18}_{8}O + ^{249}_{98}Cf \longrightarrow ^{263}_{106}Sg + 4^{1}_{0}n$

c. $^{222}_{86}Rn \longrightarrow ^{218}_{84}Po + ^{4}_{2}He$

d. $^{80}_{38}Sr \longrightarrow ^{80}_{37}Rb + ^{0}_{+1}e$

16.65 7.5 mg of Tc-99m

16.67 In the fission process, an atom splits into smaller nuclei. In fusion, small nuclei combine (fuse) to form a larger nucleus.

16.69 Fusion occurs naturally in the Sun and other stars.

16.71 4.4×10^3 Bq

16.73 14.3 days

16.75 a. $^{47}_{20}Ca \longrightarrow ^{47}_{21}Sc + ^{0}_{-1}e$
b. 1.0 mg of Ca-47
c. 9.0 days

16.77 a. $^{180}_{80}Hg \longrightarrow ^{176}_{78}Pt + ^{4}_{2}He$

b. $^{198}_{79}Au \longrightarrow ^{198}_{80}Hg + ^{0}_{-1}e$

c. $^{82}_{37}Rb \longrightarrow ^{82}_{36}Kr + ^{0}_{+1}e$

16.79 $^{1}_{0}n + ^{238}_{92}U \longrightarrow ^{239}_{93}Np + ^{0}_{-1}e$

16.81 3.0 days

16.83 64 μCi

16.85 $^{48}_{20}Ca + ^{244}_{94}Pu \longrightarrow ^{292}_{114}Fl$

16.87 16 μCi

CI.33 Consider the reaction of sodium oxalate ($Na_2C_2O_4$) and potassium permanganate ($KMnO_4$) in acidic solution. The unbalanced equation is the following: (9.2, 9.3, 12.4, 12.5, 15.2)

$$MnO_4^-(aq) + C_2O_4^{2-}(aq) \longrightarrow Mn^{2+}(aq) + CO_2(g)$$

In an oxidation–reduction titration, the $KMnO_4$ from the buret reacts with $Na_2C_2O_4$.

a. What is the balanced oxidation half-reaction?
b. What is the balanced reduction half-reaction?
c. What is the balanced ionic equation for the reaction?
d. If 24.6 mL of a $KMnO_4$ solution is needed to titrate a solution containing 0.758 g of sodium oxalate ($Na_2C_2O_4$), what is the molarity of the $KMnO_4$ solution?

CI.34 A strip of magnesium metal dissolves rapidly in 6.00 mL of a 0.150 M HCl solution, producing hydrogen gas and magnesium chloride. (9.2, 9.3, 12.4, 12.6, 15.1, 15.2)

$$Mg(s) + HCl(aq) \longrightarrow H_2(g) + MgCl_2(aq) \quad \text{Unbalanced}$$

Magnesium metal reacts vigorously with hydrochloric acid.

a. Assign oxidation numbers to all of the elements in the reactants and products.
b. What is the balanced chemical equation for the reaction?
c. What is the oxidizing agent?
d. What is the reducing agent?
e. What is the pH of the 0.150 M HCl solution?
f. How many grams of magnesium can dissolve in the HCl solution?

CI.35 A piece of magnesium with a mass of 0.121 g is added to 50.0 mL of a 1.00 M HCl solution at a temperature of 22 °C. When the magnesium dissolves, the solution reaches a temperature of 33 °C. For the equation, see your answer to problem CI.34b. (9.2, 9.3, 9.4, 9.6, 12.4, 12.6)
a. What is the limiting reactant?
b. What volume, in milliliters, of hydrogen gas would be produced at 33 °C when the pressure is 750. mmHg?

c. How many joules were released by the reaction of the magnesium? Assume the density of the HCl solution is 1.00 g/mL and the specific heat of the HCl solution is the same as that of water.
d. What is the heat of reaction for magnesium in joules/gram? in kilojoules/mol?

CI.36 The iceman known as Ötzi was discovered in a high mountain pass on the Austrian–Italian border. Samples of his hair and bones had carbon-14 activity that was 50% of that present in new hair or bone. Carbon-14 undergoes beta decay and has a half-life of 5730 yr. (16.2, 16.4)

The mummified remains of Ötzi were discovered in 1991.

a. How long ago did Ötzi live?
b. Write a balanced nuclear equation for the decay of carbon-14.

CI.37 Some of the isotopes of silicon are listed in the following table: (4.4, 5.4, 10.1, 10.3, 16.2, 16.4)

Isotope	% Natural Abundance	Atomic Mass	Half-Life (radioactive)	Radiation
$^{27}_{14}Si$		26.987	4.9 s	Positron
$^{28}_{14}Si$	92.230	27.977	Stable	None
$^{29}_{14}Si$	4.683	28.976	Stable	None
$^{30}_{14}Si$	3.087	29.974	Stable	None
$^{31}_{14}Si$		30.975	2.6 h	Beta

a. In the following table, indicate the number of protons, neutrons, and electrons for each isotope listed:

Isotope	Number of Protons	Number of Neutrons	Number of Electrons
$^{27}_{14}Si$			
$^{28}_{14}Si$			
$^{29}_{14}Si$			
$^{30}_{14}Si$			
$^{31}_{14}Si$			

b. What are the electron configuration and the abbreviated electron configuration of silicon?
c. Calculate the atomic mass for silicon using the isotopes that have a natural abundance.
d. Write the balanced nuclear equations for the positron emission of Si-27 and the beta decay of Si-31.
e. Draw the Lewis structure and predict the shape of $SiCl_4$.
f. How many hours are needed for a sample of Si-31 with an activity of 16 μCi to decay to 2.0 μCi?

CI.38 K$^+$ is an electrolyte required by the human body and found in many foods as well as salt substitutes. One of the isotopes of potassium is potassium-40, which has a natural abundance of 0.012% and a half-life of 1.30×10^9 yr. The isotope potassium-40 decays to calcium-40 or to argon-40. A typical activity for potassium-40 is 7.0 μCi per gram. (16.2, 16.3, 16.4)

Potassium chloride is used as a salt substitute.

   a. Write a balanced nuclear equation for each type of decay.
   b. Identify the particle emitted for each type of decay.
   c. How many K$^+$ ions are in 3.5 oz of KCl?
   d. What is the activity of 25 g of KCl in becquerels?

CI.39 Uranium-238 decays in a series of nuclear changes until stable lead-206 is produced. Complete the following nuclear equations that are part of the uranium-238 decay series: (16.2, 16.3, 16.4)
   a. $^{238}_{92}\text{U} \longrightarrow \ ^{234}_{90}\text{Th} + ?$
   b. $^{234}_{90}\text{Th} \longrightarrow ? + \ ^{0}_{-1}e$
   c. $? \longrightarrow \ ^{222}_{86}\text{Rn} + \ ^{4}_{2}\text{He}$

CI.40 Of much concern to environmentalists is radon-222, which is a radioactive noble gas that can seep from the ground into basements of homes and buildings. Radon-222 is a product of the decay of radium-226 that occurs naturally in rocks and soil in much of the United States. Radon-222, which has a half-life of 3.8 days, decays by emitting an alpha particle. Radon-222, which is a gas, can be inhaled into the lungs where it is strongly associated with lung cancer. Radon levels in a home can be measured with a home radon-detection kit. Environmental agencies have set the maximum level of radon-222 in a home at 4 picocuries per liter (pCi/L) of air. (11.7, 16.2, 16.3, 16.4)

A home detection kit is used to measure the level of radon-222.

   a. Write the balanced nuclear equation for the decay of Ra-226.
   b. Write the balanced nuclear equation for the decay of Rn-222.
   c. If a room contains 24 000 atoms of radon-222, how many atoms of radon-222 remain after 15.2 days?
   d. Suppose a room has a volume of 72 000 L (7.2×10^4 L). If the radon level is the maximum allowed (4 pCi/L), how many alpha particles are emitted from Rn-222 in one day? ($1 \text{ Ci} = 3.7 \times 10^{10}$ disintegrations per second)

ANSWERS

CI.33 **a.** $\text{C}_2\text{O}_4{}^{2-}(aq) \longrightarrow 2\text{CO}_2(g) + 2\ e^-$
   b. $5\ e^- + 8\text{H}^+(aq) + \text{MnO}_4{}^-(aq) \longrightarrow$
$$\text{Mn}^{2+}(aq) + 4\text{H}_2\text{O}(l)$$
   c. $16\text{H}^+(aq) + 2\text{MnO}_4{}^-(aq) + 5\text{C}_2\text{O}_4{}^{2-}(aq) \longrightarrow$
$$10\text{CO}_2(g) + 2\text{Mn}^{2+}(aq) + 8\text{H}_2\text{O}(l)$$
   d. 0.0920 M KMnO$_4$ solution

CI.35 **a.** Mg is the limiting reactant.
   b. 127 mL of H$_2(g)$
   c. 2.3×10^3 J
   d. 1.90×10^4 J/g; 462 kJ/mol

CI.37 **a.**

Isotope	Number of Protons	Number of Neutrons	Number of Electrons
$^{27}_{14}\text{Si}$	14	13	14
$^{28}_{14}\text{Si}$	14	14	14
$^{29}_{14}\text{Si}$	14	15	14
$^{30}_{14}\text{Si}$	14	16	14
$^{31}_{14}\text{Si}$	14	17	14

   b. $1s^2 2s^2 2p^6 3s^2 3p^2$; [Ne]$3s^2 3p^2$
   c. 28.09 amu
   d. $^{27}_{14}\text{Si} \longrightarrow \ ^{27}_{13}\text{Al} + \ ^{0}_{+1}e$
$^{31}_{14}\text{Si} \longrightarrow \ ^{31}_{15}\text{P} + \ ^{0}_{-1}e$
   e.

$$:\!\ddot{\text{C}}\text{l}\!:$$
$$:\!\ddot{\text{C}}\text{l}\!-\!\text{Si}\!-\!\ddot{\text{C}}\text{l}\!: \qquad \text{Tetrahedral}$$
$$:\!\ddot{\text{C}}\text{l}\!:$$

   f. 7.8 h

CI.39 **a.** $^{238}_{92}\text{U} \longrightarrow \ ^{234}_{90}\text{Th} + \ ^{4}_{2}\text{He}$
   b. $^{234}_{90}\text{Th} \longrightarrow \ ^{234}_{91}\text{Pa} + \ ^{0}_{-1}e$
   c. $^{226}_{88}\text{Ra} \longrightarrow \ ^{222}_{86}\text{Rn} + \ ^{4}_{2}\text{He}$

17

Organic Chemistry

AT 4:35 A.M., A RESCUE crew responded to a call about a house fire. At the scene, Jack, a firefighter/emergency medical technician (EMT) found Diane, a 62-year old woman, lying in the front yard of her house. In his assessment, Jack reported that Diane had second- and third-degree burns over 40% of her body as well as a broken leg. He placed an oxygen re-breather mask on Diane to provide a high concentration of oxygen.

Another firefighter/EMT, Nancy, began dressing the burns with sterile water and cling film, a first aid material made of polyvinyl chloride, which does not stick to the skin and is protective. Jack and his crew transported Diane to the burn center for further treatment.

At the scene of the fire, arson investigators used trained dogs to find traces of accelerants and fuel. Gasoline, which is often found at arson scenes, is a mixture of organic molecules called alkanes. Alkanes or hydrocarbons are chains of carbon and hydrogen atoms. The alkanes present in gasoline consist of a mixture of 5 to 8 carbon atoms in a chain. Alkanes are extremely combustible; they react with oxygen to form carbon dioxide, water, and large amounts of heat. Because alkanes undergo combustion reactions, they can be used to start arson fires.

CAREER

Firefighter/Emergency Medical Technician

Firefighters/emergency medical technicians are first responders to fires, accidents, and other emergency situations. They are required to have an emergency medical technician certification in order to be able to treat seriously injured people. By combining the skills of a firefighter and an emergency medical technician, they increase the survival rates of the injured. The physical demands of firefighters are extremely high as they fight, extinguish, and prevent fires while wearing heavy protective clothing. They also train for and participate in firefighting drills, and maintain fire equipment so that it is always working and ready. Firefighters must also be knowledgeable about fire codes, arson, and the handling and disposal of hazardous materials. Since firefighters also provide emergency care for sick and injured people, they need to be aware of emergency medical and rescue procedures, as well as the proper methods for controlling the spread of infectious disease.

CORE CHEMISTRY SKILLS

- Balancing a Chemical Equation (8.2)
- Drawing Lewis Structures (10.1)
- Predicting Shape (10.3)

*These Core Chemistry Skills from previous chapters are listed here for your review as you proceed to the new material in this chapter.

17.1 Alkanes

LEARNING GOAL Write the IUPAC names and draw the condensed or line-angle structural formulas for alkanes.

At the beginning of the nineteenth century, scientists classified chemical compounds as inorganic or organic. An inorganic compound was a substance that was composed of minerals, and an organic compound was a substance that came from an organism, thus the use of the word *organic*. It was thought that some type of "vital force," which could only be found in living cells, was required to synthesize an organic compound. This perception was shown to be incorrect in 1828, when the German chemist Friedrich Wöhler synthesized urea, a product of protein metabolism, by heating an inorganic compound, ammonium cyanate. Organic compounds contain carbon and hydrogen and sometimes oxygen and other nonmetallic elements.

$$NH_4CNO \xrightarrow{\text{Heat}} H_2N-\overset{\displaystyle O}{\overset{\displaystyle \|}{C}}-NH_2$$

Ammonium cyanate (inorganic)

Urea (organic)

We use many organic compounds every day in products such as fuels, clothing, drugs, and cosmetics. The foods in our diets are composed of organic compounds such as carbohydrates, fats, and proteins that undergo digestion and metabolism to give us energy and small organic compounds that are used to build and repair the cells of our bodies.

The formulas of organic compounds are written with carbon first, followed by hydrogen, and then any other elements. Organic compounds typically have low melting and boiling points, are not soluble in water, and are less dense than water. For example, vegetable oil, which is a mixture of organic compounds, does not dissolve in water but floats on top. Many organic compounds undergo combustion and burn vigorously in air. By contrast, many inorganic compounds have high melting and boiling points. Inorganic compounds that are ionic are usually soluble in water, and most do not burn in air. **TABLE 17.1** contrasts some of the properties associated with organic and inorganic compounds, such as propane, C_3H_8, and sodium chloride, NaCl (see **FIGURE 17.1**).

Foods in the diet provide energy and materials for the cells of the body.

Vegetable oil, a mixture of organic compounds, is not soluble in water.

TABLE 17.1 Some Properties of Organic and Inorganic Compounds

Property	Organic	Example: C_3H_8	Inorganic	Example: NaCl
Elements Present	C and H, sometimes O, S, N, P, or Cl (F, Br, I)	C and H	Most metals and nonmetals	Na and Cl
Particles	Molecules	C_3H_8	Mostly ions	Na^+ and Cl^-
Bonding	Mostly covalent	Covalent	Many are ionic, some covalent	Ionic
Polarity of Bonds	Nonpolar, unless a strongly electronegative atom is present	Nonpolar	Most are ionic or polar covalent, a few are nonpolar covalent	Ionic
Melting Point	Usually low	$-188\,°C$	Usually high	$801\,°C$
Boiling Point	Usually low	$-42\,°C$	Usually high	$1413\,°C$
Flammability	High	Burns in air	Low	Does not burn
Solubility in Water	Not soluble unless a polar group is present	No	Most are soluble unless nonpolar	Yes

FIGURE 17.1 ▶ Propane, C_3H_8, is an organic compound, whereas sodium chloride, NaCl, is an inorganic compound.

Q Why is propane used as a fuel?

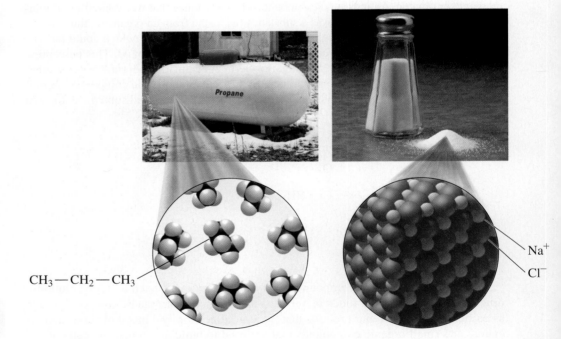

$CH_3 - CH_2 - CH_3$

Na$^+$
Cl$^-$

Representations of Carbon Compounds

Hydrocarbons are organic compounds that consist of only carbon and hydrogen. In organic molecules, every carbon atom has four bonds. In the simplest hydrocarbon, methane (CH_4), the carbon atom forms an octet by sharing its four valence electrons with four hydrogen atoms.

Methane

The most accurate representation of methane is the three-dimensional *space-filling model* (**a**) in which spheres show the relative size and shape of all the atoms. Another type of three-dimensional representation is the *ball-and-stick model* (**b**), where the atoms are shown as balls and the bonds between them are shown as sticks. In the ball-and-stick model of methane, CH_4, the covalent bonds from the carbon atom to each hydrogen atom are directed to the corners of a tetrahedron with bond angles of 109°. In the *wedge–dash model* (**c**), the three-dimensional shape is represented by symbols of the atoms with lines for bonds in the plane of the page, wedges for bonds that project out from the page, and dashes for bonds that are behind the page.

However, the three-dimensional models are awkward to draw and view for more complex molecules. Therefore, it is more practical to use their corresponding two-dimensional formulas. The **expanded structural formula (d)** shows all of the atoms and the bonds connected to each atom. A **condensed structural formula (e)** shows the carbon atoms each grouped with the attached number of hydrogen atoms.

ENGAGE

Why does methane have a tetrahedral shape?

 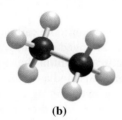 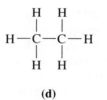

| (a) | (b) | (c) | (d) | (e) |

Three-dimensional and two-dimensional representations of methane: **(a)** space-filling model, **(b)** ball-and-stick model, **(c)** wedge–dash model, **(d)** expanded structural formula, and **(e)** condensed structural formula.

The hydrocarbon ethane with two carbon atoms and six hydrogen atoms can be represented by a similar set of three- and two-dimensional models and formulas in which each carbon atom is bonded to another carbon and three hydrogen atoms. As in methane, each carbon atom in ethane retains a tetrahedral shape. A hydrocarbon is referred to as a *saturated hydrocarbon* when all the bonds in the molecule are single bonds.

| (a) | (b) | (c) | (d) | (e) |

Three-dimensional and two-dimensional representations of ethane: **(a)** space-filling model, **(b)** ball-and-stick model, **(c)** wedge–dash model, **(d)** expanded structural formula, and **(e)** condensed structural formula.

Naming Alkanes

More than 90% of the compounds in the world are organic compounds. The large number of carbon compounds is possible because the covalent bond between carbon atoms is very strong, allowing carbon atoms to form long, stable chains.

The **alkanes** are a type of hydrocarbon in which the carbon atoms are connected only by single bonds. One of the most common uses of alkanes is as fuels. Methane, used in gas heaters and gas cooktops, is an alkane with one carbon atom. The alkanes ethane, propane, and butane contain two, three, and four carbon atoms, respectively, connected in a row or a *continuous chain*. As we can see, the names for alkanes end in *ane*. Such names are part of the **IUPAC (International Union of Pure and Applied Chemistry) system** used by chemists to name organic compounds. Alkanes with five or more carbon atoms in a chain are named using Greek prefixes: *pent* (5), *hex* (6), *hept* (7), *oct* (8), *non* (9), and *dec* (10) (see **TABLE 17.2**).

TABLE 17.2 IUPAC Names, Molecular Formulas, Condensed and Line-Angle Structural Formulas of the First Ten Alkanes

Carbon Atoms	IUPAC Name	Molecular Formula	Condensed Structural Formula	Line-Angle Structural Formula
1	Methane	CH_4	CH_4	—
2	Ethane	C_2H_6	$CH_3{-}CH_3$	—
3	Propane	C_3H_8	$CH_3{-}CH_2{-}CH_3$	⌃
4	Butane	C_4H_{10}	$CH_3{-}CH_2{-}CH_2{-}CH_3$	⌇
5	Pentane	C_5H_{12}	$CH_3{-}CH_2{-}CH_2{-}CH_2{-}CH_3$	⌇
6	Hexane	C_6H_{14}	$CH_3{-}CH_2{-}CH_2{-}CH_2{-}CH_2{-}CH_3$	⌇
7	Heptane	C_7H_{16}	$CH_3{-}CH_2{-}CH_2{-}CH_2{-}CH_2{-}CH_2{-}CH_3$	⌇
8	Octane	C_8H_{18}	$CH_3{-}CH_2{-}CH_2{-}CH_2{-}CH_2{-}CH_2{-}CH_2{-}CH_3$	⌇
9	Nonane	C_9H_{20}	$CH_3{-}CH_2{-}CH_2{-}CH_2{-}CH_2{-}CH_2{-}CH_2{-}CH_2{-}CH_3$	⌇
10	Decane	$C_{10}H_{22}$	$CH_3{-}CH_2{-}CH_2{-}CH_2{-}CH_2{-}CH_2{-}CH_2{-}CH_2{-}CH_2{-}CH_3$	⌇

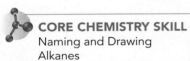

CORE CHEMISTRY SKILL
Naming and Drawing Alkanes

Condensed and Line-Angle Structural Formulas

In a condensed structural formula, each carbon atom and its attached hydrogen atoms are written as a group. A subscript indicates the number of hydrogen atoms bonded to each carbon atom.

$$
\begin{array}{cccc}
& H & & H \\
& | & & | \\
H-C- & = CH_3- & \qquad -C- & = -CH_2- \\
& | & & | \\
& H & & H \\
\text{Expanded} & \text{Condensed} & \text{Expanded} & \text{Condensed}
\end{array}
$$

ENGAGE

How does a line-angle structural formula represent an organic compound of carbon and hydrogen with single bonds?

When an organic molecule consists of a chain of three or more carbon atoms, the carbon atoms do not lie in a straight line. Rather, they are arranged in a zigzag pattern.

A simplified formula called the **line-angle structural formula** shows a zigzag line in which carbon atoms are represented as the ends of each line and as corners. For example, in the line-angle structural formula of pentane, each line in the zigzag drawing represents a single bond. The carbon atoms on the ends are bonded to three hydrogen atoms. However, the carbon atoms in the middle of the carbon chain are each bonded to two carbons and two hydrogen atoms as shown in Sample Problem 17.1.

SAMPLE PROBLEM 17.1 **Drawing Expanded, Condensed, and Line-Angle Structural Formulas for an Alkane**

Draw the expanded, condensed, and line-angle structural formulas for pentane.

 **TRY IT FIRST**

SOLUTION

	Given	Need	Connect
ANALYZE THE PROBLEM	pentane	structural formulas	carbon chain, H atoms, zigzag line

STEP 1 **Draw the carbon chain.** A molecule of pentane has five carbon atoms in a continuous chain.

$$C-C-C-C-C$$

STEP 2 Draw the expanded structural formula by adding the hydrogen atoms using single bonds to each of the carbon atoms.

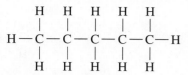

STEP 3 Draw the condensed structural formula by combining the H atoms with each C atom.

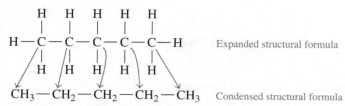

Expanded structural formula

Condensed structural formula

STEP 4 Draw the line-angle structural formula as a zigzag line in which the ends and corners represent C atoms.

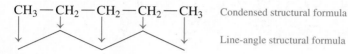

$CH_3-CH_2-CH_2-CH_2-CH_3$ Condensed structural formula

Line-angle structural formula

STUDY CHECK 17.1

Draw the condensed structural formula and write the name for the following line-angle structural formula:

ANSWER

$CH_3-CH_2-CH_2-CH_2-CH_2-CH_2-CH_3$ heptane

Structural Isomers

When an alkane has four or more carbon atoms, the atoms can be arranged so that a side group called a *branch* or **substituent** is attached to a carbon chain. For example, **FIGURE 17.2** shows two different ball-and-stick models for two compounds that have the molecular formula C_4H_{10}. One model is shown as a chain of four carbon atoms. In the other model, a carbon atom is attached as a branch or substituent to a carbon in a chain of three atoms. An alkane with at least one branch is called a *branched alkane*. When the two compounds have the same molecular formula but different arrangements of atoms, they are **structural isomers**.

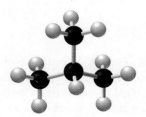

$CH_3-CH_2-CH_2-CH_3$

$CH_3-\overset{\displaystyle CH_3}{\underset{|}{CH}}-CH_3$

Guide to Drawing Structural Formulas for Alkanes

STEP 1
Draw the carbon chain.

STEP 2
Draw the expanded structural formula by adding the hydrogen atoms using single bonds to each of the carbon atoms.

STEP 3
Draw the condensed structural formula by combining the H atoms with each C atom.

STEP 4
Draw the line-angle structural formula as a zigzag line in which the ends and corners represent C atoms.

ENGAGE

How can two or more structural isomers have the same molecular formula?

FIGURE 17.2 ▶ The structural isomers of C_4H_{10} have the same number and type of atoms but are bonded in a different order.

Q What makes these molecules structural isomers?

In another example, we can draw the condensed and line-angle structural formulas for three different structural isomers with the molecular formula C_5H_{12} as follows:

Structural Isomers of C_5H_{12}		
Condensed	$CH_3-CH_2-CH_2-CH_2-CH_3$ 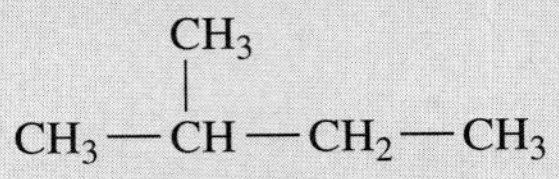 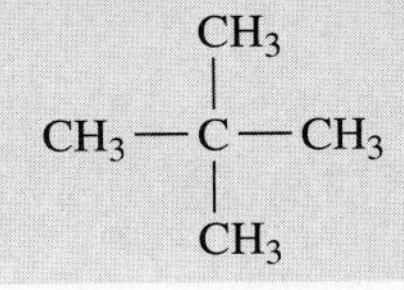	$CH_3-\overset{\overset{\displaystyle CH_3}{\vert}}{CH}-CH_2-CH_3$ $CH_3-\overset{\overset{\displaystyle CH_3}{\vert}}{\underset{\underset{\displaystyle CH_3}{\vert}}{C}}-CH_3$
Line-angle  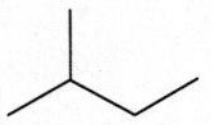 		

SAMPLE PROBLEM 17.2 Structural Isomers

Identify each pair of formulas as structural isomers or the same molecule.

a. 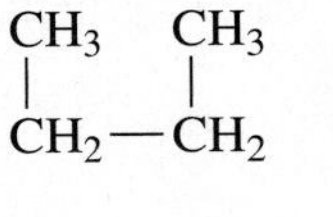 $\overset{\overset{\displaystyle CH_3}{\vert}}{CH_2}-\overset{\overset{\displaystyle CH_3}{\vert}}{CH_2}$ and $CH_2-CH_2-CH_3$ with CH_3 below first carbon

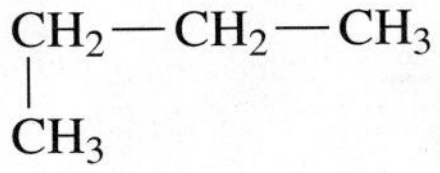

b. 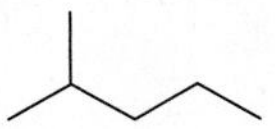   and

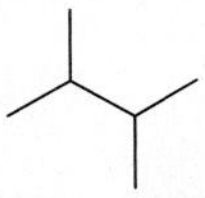

TRY IT FIRST

SOLUTION

a. When we add up the number of C atoms and H atoms, they give the same molecular formula C_4H_{10}. Both of the structures are continuous four-carbon chains even though one or more of the $-CH_3$ ends are drawn above or below the horizontal part of the chain. Thus, both condensed structural formulas represent the same molecule and are not structural isomers.

b. When we add up the number of C atoms and H atoms, they give the same molecular formula C_6H_{14}. The line-angle structural formula on the left has a five-carbon chain with a $-CH_3$ substituent on the second carbon of the chain. The line-angle structural formula on the right has a four-carbon chain with two $-CH_3$ substituents. Thus, there is a different order of bonding of atoms, which represents structural isomers.

STUDY CHECK 17.2

Why does the following formula represent a different structural isomer of the molecules in Sample Problem 17.2, part **b**?

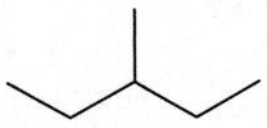

ANSWER

This formula represents a different structural isomer of C_6H_{14} because the $-CH_3$ substituent is on the third carbon of the chain.

Substituents in Alkanes

In the IUPAC names for alkanes, a carbon branch is named as an **alkyl group**, which is an alkane that is missing one hydrogen atom. The alkyl group is named by replacing the *ane* ending of the corresponding alkane name with *yl*. Alkyl groups cannot exist on their own: They must be attached to a carbon chain. When a halogen atom is attached to a carbon chain, it is named as a *halo* group: *fluoro, chloro, bromo,* or *iodo.* Some of the common groups attached to carbon chains are illustrated in **TABLE 17.3**.

TABLE 17.3 Formulas and Names of Some Common Substituents

Formula	CH_3-		CH_3-CH_2-		
Name	methyl		ethyl		

Formula	$CH_3-CH_2-CH_2-$		$CH_3-\overset{\mid}{C}H-CH_3$		
Name	propyl		isopropyl		

Formula	$CH_3-CH_2-CH_2-CH_2-$	$CH_3-\overset{\overset{\displaystyle CH_3}{\mid}}{C}H-CH_2-$	$CH_3-\overset{\mid}{C}H-CH_2-CH_3$	$CH_3-\overset{\overset{\displaystyle CH_3}{\mid}}{\underset{\underset{\displaystyle CH_3}{\mid}}{C}}-CH_3$
Name	butyl	isobutyl	*sec*-butyl	*tert*-butyl

Formula	$F-$	$Cl-$	$Br-$	$I-$
Name	fluoro	chloro	bromo	iodo

Naming Alkanes with Substituents

In the IUPAC system of naming, a carbon chain is numbered to give the location of the substituents. Let's take a look at how we use the IUPAC system to name the alkane shown in Sample Problem 17.3.

Interactive Video

Naming Alkanes

SAMPLE PROBLEM 17.3 Writing IUPAC Names for Alkanes with Substituents

Write the IUPAC name for the following alkane:

$$CH_3-\overset{\overset{\displaystyle CH_3}{\mid}}{C}H-CH_2-CH_2-\overset{\overset{\displaystyle Br}{\mid}}{\underset{\underset{\displaystyle CH_3}{\mid}}{C}}-CH_3$$

TRY IT FIRST

SOLUTION

		Given	Need	Connect
ANALYZE THE PROBLEM		six-carbon chain, two methyl groups, one bromo group	IUPAC name	position of substituents on the carbon chain

STEP 1 Write the alkane name for the longest chain of carbon atoms.

$$CH_3-\overset{\overset{\displaystyle CH_3}{\mid}}{C}H-CH_2-\overset{\overset{\displaystyle Br}{\mid}}{\underset{\underset{\displaystyle CH_3}{\mid}}{C}}-CH_2-CH_3 \qquad \text{hexane}$$

STEP 2 Number the carbon atoms from the end nearer a substituent.

$$\underset{1}{CH_3}-\underset{2}{\overset{\overset{\displaystyle CH_3}{\mid}}{C}H}-\underset{3}{CH_2}-\underset{4}{\overset{\overset{\displaystyle Br}{\mid}}{\underset{\underset{\displaystyle CH_3}{\mid}}{C}}}-\underset{5}{CH_2}-\underset{6}{CH_3} \qquad \text{hexane}$$

Guide to Naming Alkanes with Substituents

STEP 1
Write the alkane name for the longest chain of carbon atoms.

STEP 2
Number the carbon atoms from the end nearer a substituent.

STEP 3
Give the location and name for each substituent (alphabetical order) as a prefix to the name of the main chain.

STEP 3 Give the location and name for each substituent (alphabetical order) as a prefix to the name of the main chain. The substituents are listed in alphabetical order (bromo first, then methyl). A hyphen is placed between the number and the substituent name. When there are two or more of the same substituent, a prefix (*di, tri, tetra*) is used and commas separate the numbers. However, prefixes are not used to determine the alphabetical order of the substituents.

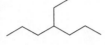

4-bromo-2,4-dimethylhexane

STUDY CHECK 17.3

Write the IUPAC name for the following compound:

ANSWER

4-ethylheptane

Drawing Structural Formulas for Alkanes

ENGAGE

What part of the IUPAC name gives **(a)** the number of carbon atoms in the carbon chain and **(b)** the substituents on the carbon chain?

The IUPAC name gives all the information needed to draw the condensed structural formula for an alkane. Suppose you are asked to draw the condensed structural formula for 2,3-dimethylbutane. The alkane name gives the number of carbon atoms in the longest chain. The names in the beginning indicate the substituents and where they are attached. We can break down the name in the following way:

2,3-Dimethylbutane				
2,3-	**Di**	**methyl**	**but**	**ane**
Substituents on carbons 2 and 3	two identical groups	—CH_3 alkyl groups	four C atoms in the main chain	single (C—C) bonds

SAMPLE PROBLEM 17.4 Drawing Condensed Structures from IUPAC Names

Draw the condensed and line-angle structural formulas for 2,3-dimethylbutane.

TRY IT FIRST

SOLUTION

	Given	Need	Connect
ANALYZE THE PROBLEM	2,3-dimethylbutane	condensed and line-angle structural formulas	four-carbon chain, two methyl groups

STEP 1 Draw the main chain of carbon atoms. For butane, we draw a chain of four carbon atoms or a zigzag line.

C—C—C—C

STEP 2 Number the chain and place the substituents on the carbons indicated by the numbers. The first part of the name indicates two methyl groups —CH₃: one on carbon 2 and one on carbon 3.

Methyl Methyl

CH₃ CH₃
  | |
C—C—C—C
1   2   3   4

STEP 3 For the condensed structural formula, add the correct number of hydrogen atoms to give four bonds to each C atom.

CH₃   CH₃
  | |
CH₃—CH—CH—CH₃

STUDY CHECK 17.4

Draw the condensed and line-angle structural formulas for 2-bromo-3-ethyl-4-methylpentane.

ANSWER

Br   CH₃
 | |
CH₃—CH—CH—CH—CH₃
         |
        CH₂
         |
        CH₃

Br

Guide to Drawing Structural Formulas for Alkanes with Substituents

STEP 1
Draw the main chain of carbon atoms.

STEP 2
Number the chain and place the substituents on the carbons indicated by the numbers.

STEP 3
For the condensed structural formula, add the correct number of hydrogen atoms to give four bonds to each C atom.

Properties and Uses of Alkanes

The first four alkanes—methane, ethane, propane, and butane—are gases at room temperature and are widely used as fuels.

Alkanes having five to eight carbon atoms (pentane, hexane, heptane, and octane) are liquids at room temperature. They are highly volatile, which makes them useful in fuels such as gasoline.

Liquid alkanes with 9 to 17 carbon atoms have higher boiling points and are found in kerosene, diesel, and jet fuels. Motor oil is a mixture of high-molecular-weight liquid hydrocarbons and is used to lubricate the internal components of engines. Mineral oil is a mixture of liquid hydrocarbons and is used as a laxative and a lubricant. Alkanes with 18 or more carbon atoms are waxy solids at room temperature. Known as paraffins, they are used in waxy coatings added to fruits and vegetables to retain moisture, inhibit mold, and enhance appearance. Petrolatum jelly, or Vaseline, is a semisolid mixture of hydrocarbons with more than 25 carbon atoms used in ointments and cosmetics and as a lubricant.

The solid alkanes that make up waxy coatings on fruits and vegetables help retain moisture, inhibit mold, and enhance appearance.

Reactions of Alkanes: Combustion

The carbon–carbon single bonds in alkanes are difficult to break, which makes them the least reactive family of organic compounds. However, alkanes burn readily (combustion) in oxygen to produce carbon dioxide, water, and energy.

$$\text{Alkane}(g) + O_2(g) \xrightarrow{\Delta} CO_2(g) + H_2O(g) + \text{energy}$$

Propane is the gas used in portable heaters and gas barbecues (see **FIGURE 17.3**). The equation for the combustion of propane (C_3H_8) is written:

$$C_3H_8(g) + 5O_2(g) \xrightarrow{\Delta} 3CO_2(g) + 4H_2O(g) + energy$$

Propane

Solubility and Density

Alkanes are nonpolar, which makes them insoluble in water. However, they are soluble in nonpolar solvents. Alkanes have densities from 0.62 g/mL to about 0.79 g/mL, which is less than the density of water (1.0 g/mL).

If there is an oil spill in the ocean, the alkanes in the oil, which do not mix with water, form a thin layer on the surface that spreads over a large area (see **FIGURE 17.4**). In April 2010, an explosion on an oil-drilling rig in the Gulf of Mexico caused the largest oil spill in U.S. history. At its maximum, an estimated 10 million liters of oil was leaked every day. If the crude oil reaches land, there can be considerable damage to beaches, shellfish, birds, and wildlife habitats. When animals such as birds are covered with oil, they must be cleaned quickly because ingestion of the hydrocarbons when they try to clean themselves is fatal.

FIGURE 17.3 ▶ The propane fuel in the tank undergoes combustion, which provides energy.

Ⓠ What is the balanced equation for the combustion of propane?

FIGURE 17.4 ▶ In oil spills, large quantities of oil spread out to form a thin layer on top of the ocean surface.

Ⓠ What physical properties cause oil to remain on the surface of water?

QUESTIONS AND PROBLEMS

17.1 Alkanes

LEARNING GOAL Write the IUPAC names and draw the condensed or line-angle structural formulas for alkanes.

17.1 Write the IUPAC name for each of the following alkanes:

a.
$$\begin{array}{c} CH_3 \\ | \\ CH_2-CH_2-CH_2 \\ | \\ CH_2-CH_3 \end{array}$$

b.

c.

17.2 Write the IUPAC name for each of the following alkanes:

a. $CH_3-CH_2-CH_2-CH_3$

b.

c.

17.3 Indicate whether each of the following pairs of formulas represents structural isomers or the same molecule:

a.
$$\begin{array}{cc} CH_3 & CH_3 \\ | & | \\ CH_3-CH-CH_3 \quad and \quad CH-CH_3 \\ & | \\ & CH_3 \end{array}$$

b.
$$\begin{array}{c} CH_3 \quad CH_3 \\ | \qquad | \\ CH_2-CH-CH_2-CH_3 \quad and \end{array}$$

$$\begin{array}{c} CH_3 \quad CH_3 \\ | \qquad | \\ CH_3-CH-CH-CH_3 \end{array}$$

c.  and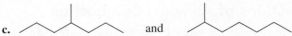

17.4 Indicate whether each of the following pairs of formulas represents structural isomers or the same molecule:

a.
$$\begin{array}{cc} CH_3 & CH_3 \\ | & | \\ CH_3-C-CH_3 \quad and \quad CH-CH_2-CH_3 \\ | & | \\ CH_3 & CH_3 \end{array}$$

b. $CH_3-CH-CH-CH_2$ and
with CH_3, CH_3, CH_3 substituents

$CH_3-CH-CH_2-CH-CH_3$
with CH_3, CH_3 substituents

c. (line-angle structure) and (line-angle structure)

17.5 Write the IUPAC name for each of the following:

a. $CH_3-CH-CH_2-CH_3$
with F substituent

b. CH_3-C-CH_3
with CH_3 (top) and CH_3 (bottom) substituents

c. (line-angle structure with Cl substituent)

17.6 Write the IUPAC name for each of the following:

a. $CH_3-CH-CH_2-CH_2-CH_2-Br$
with CH_3 substituent

b. $CH_3-CH-CH-CH_2-CH_2-CH_3$
with CH_3, CH_3 substituents

c. (line-angle structure with Br and Cl substituents)

17.7 Draw the condensed structural formula for **a** and **b** and the line-angle structural formula for **c** and **d**.
 a. 2-methylbutane b. 3,3-dichloropentane
 c. 2-chloro-2-methylhexane d. 2,3-dimethylheptane

17.8 Draw the condensed structural formula for **a** and **b** and the line-angle structural formula for **c** and **d**.
 a. 3-iodopentane
 b. 3-ethyl-2-methylpentane
 c. 1,5-dibromo-3-methylheptane
 d. 1,1,5-trichloropentane

Applications

17.9 Write the balanced chemical equation for the complete combustion of each of the following:
 a. methane
 b. 3-methylpentane

17.10 Write the balanced chemical equation for the complete combustion of each of the following:
 a. pentane
 b. methylpropane

17.2 Alkenes, Alkynes, and Polymers

LEARNING GOAL Write the IUPAC names and draw the condensed or line-angle structural formulas for alkenes and alkynes.

We organize organic compounds into *classes* or *families* by their **functional groups**, which are groups of specific atoms. Compounds that contain the same functional group have similar physical and chemical properties. Identifying functional groups allows us to classify organic compounds according to their structure, to name compounds within each family, predict their chemical reactions, and draw the structures for their products.

Alkenes and alkynes are classes of unsaturated hydrocarbons that contain *double* and *triple* bonds, respectively. The functional group of an **alkene** contains at least one double bond between carbons. The double bond forms when two adjacent carbon atoms share two pairs of valence electrons. The simplest alkene is ethene, C_2H_4, which is often called by its common name, ethylene. In ethene, each carbon atom is attached to two H atoms and the other carbon atom in the double bond. The resulting molecule has a flat geometry because the carbon and hydrogen atoms all lie in the same plane. The functional group of an **alkyne** contains a triple bond, which occurs when two carbon atoms share three pairs of valence electrons.

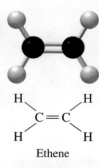

$$\begin{array}{cc} H & H \\ \diagdown & \diagup \\ C{=}C \\ \diagup & \diagdown \\ H & H \end{array}$$
Ethene

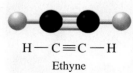

$$H-C{\equiv}C-H$$
Ethyne

A list of the common functional groups in organic compounds is shown in **TABLE 17.4**.

ENGAGE

How does the functional group in an alcohol differ from the functional group in a ketone?

TABLE 17.4 Classes of Organic Compounds

Class	Functional Group	Example/Name	Occurrence
Alkene	$\text{C}=\text{C}$	$H_2C=CH_2$ Ethene (ethylene)	Ripening of fruit, used to make polyethylene
Alkyne	$-C\equiv C-$	$H-C\equiv C-H$ Ethyne (acetylene)	Welding fuel
Alcohol	$-OH$	$CH_3-CH_2-\mathbf{OH}$ Ethanol (ethyl alcohol)	Solvent
Ether	$-O-$	$CH_3-CH_2-\mathbf{O}-CH_2-CH_3$ Diethyl ether	Solvent
Aldehyde	$-\overset{\overset{\text{O}}{\|\|}}{C}-H$	$CH_3-\overset{\overset{\text{O}}{\|\|}}{C}-\mathbf{H}$ Ethanal (acetaldehyde)	Preparation of acetic acid
Ketone	$-\overset{\overset{\text{O}}{\|\|}}{C}-$	$CH_3-\overset{\overset{\text{O}}{\|\|}}{C}-CH_3$ Propanone (acetone)	Solvent, paint and fingernail polish remover
Carboxylic acid	$-\overset{\overset{\text{O}}{\|\|}}{C}-OH$	$CH_3-\overset{\overset{\text{O}}{\|\|}}{C}-\mathbf{OH}$ Ethanoic acid (acetic acid)	Component of vinegar
Ester	$-\overset{\overset{\text{O}}{\|\|}}{C}-O-$	$CH_3-\overset{\overset{\text{O}}{\|\|}}{C}-O-CH_3$ Methyl ethanoate (methyl acetate)	Rum flavoring
Amine	$-\overset{\|}{N}-$	CH_3-NH_2 Methylamine	Odor of fish
Amide	$-\overset{\overset{\text{O}}{\|\|}}{C}-\overset{\|}{N}-$	$CH_3-\overset{\overset{\text{O}}{\|\|}}{C}-NH_2$ Ethanamide (acetamide)	Odor of mice

Fruit is ripened with ethene, a plant hormone.

A mixture of acetylene and oxygen undergoes combustion during the welding of metals.

Naming Alkenes and Alkynes

The IUPAC names for alkenes and alkynes are similar to those of alkanes. The IUPAC name of the simplest alkyne is ethyne, although it is often called by the common name, acetylene (see **TABLE 17.5**). When naming alkenes and alkynes, the longest carbon chain must contain the double or triple bond.

TABLE 17.5 Comparison of Alkanes, Alkenes, and Alkynes

Alkane	Alkene	Alkyne
CH_3-CH_3 Ethane	$H_2C=CH_2$ Ethene (ethylene)	$HC\equiv CH$ Ethyne (acetylene)
$CH_3-CH_2-CH_3$ Propane	$CH_3-CH=CH_2$ Propene (propylene)	$CH_3-C\equiv CH$ Propyne
$\wedge$ Propane	$\wedge$ Propene	$=\!=$ Propyne

An example of naming an alkene is shown in Sample Problem 17.5.

SAMPLE PROBLEM 17.5 Naming Alkenes and Alkynes

Write the IUPAC name for the following:

$$CH_3$$
$$|$$
$$CH_3-CH-CH=CH-CH_3$$

TRY IT FIRST

SOLUTION

ANALYZE THE PROBLEM	Given	Need	Connect
	five-carbon chain, double bond, methyl group	IUPAC name	replace the *ane* in the alkane name with *ene*

STEP 1 Name the longest carbon chain that contains the double bond. There are five carbon atoms in the longest carbon chain containing the double bond. Replacing the corresponding alkane ending with *ene* gives pentene.

$$CH_3$$
$$|$$
$$CH_3-CH-CH=CH-CH_3 \qquad \text{pentene}$$

STEP 2 Number the carbon chain from the end nearer the double bond. The number of the first carbon in the double bond is used to give the location of the double bond. Alkenes or alkynes with two or three carbons do not need numbers.

$$CH_3$$
$$|$$
$$\underset{5}{CH_3}-\underset{4}{CH}-\underset{3}{CH}=\underset{2}{CH}-\underset{1}{CH_3} \qquad \text{2-pentene}$$

STEP 3 Give the location and name of each substituent (alphabetical order) as a prefix to the alkene name. The methyl group is located on carbon 4.

$$CH_3$$
$$|$$
$$\underset{5}{CH_3}-\underset{4}{CH}-\underset{3}{CH}=\underset{2}{CH}-\underset{1}{CH_3} \qquad \text{4-methyl-2-pentene}$$

STUDY CHECK 17.5

Write the IUPAC name for the following:

$$CH_3-CH_2-C\equiv C-CH_2-CH_3$$

ANSWER

3-hexyne

> **Guide to Naming Alkenes and Alkynes**
>
> **STEP 1**
> Name the longest carbon chain that contains the double or triple bond.
>
> **STEP 2**
> Number the carbon chain from the end nearer the double or triple bond.
>
> **STEP 3**
> Give the location and name for each substituent (alphabetical order) as a prefix to the alkene or alkyne name.

Hydrogenation: Addition Reaction

In a reaction called *hydrogenation*, H atoms add to each of the carbon atoms in a double bond of an alkene. During hydrogenation, the double bonds are converted to single bonds in alkanes. A metal catalyst such as finely divided platinum (Pt), nickel (Ni), or palladium (Pd) is used to speed up the reaction.

$$CH_3-CH=CH-CH_3 + H_2 \xrightarrow{Pt} CH_3-CH_2-CH_2-CH_3$$
$$\text{2-Butene} \qquad\qquad\qquad\qquad \text{Butane}$$

CORE CHEMISTRY SKILL
Writing Equations for Hydrogenation and Polymerization

SAMPLE PROBLEM 17.6 Writing an Equation for Hydrogenation

Draw the condensed structural formula for the product of the following hydrogenation reaction:

$$CH_3-CH=CH_2 + H_2 \xrightarrow{\text{Pt}}$$

TRY IT FIRST

SOLUTION

ANALYZE THE PROBLEM	Given	Need	Connect
	alkene + H_2	structural formula of product	add 2H to double bond

$$CH_3-CH_2-CH_3$$

STUDY CHECK 17.6

Draw the condensed structural formula for the product of the hydrogenation of 2-methyl-1-butene, using a platinum catalyst.

ANSWER

$$\begin{array}{c} \quad\quad CH_3 \\ \quad\quad | \\ CH_3-CH-CH_2-CH_3 \end{array}$$

CHEMISTRY LINK TO **HEALTH**
Hydrogenation of Unsaturated Fats

Vegetable oils such as corn oil or safflower oil are unsaturated fats composed of fatty acids that contain double bonds. The process of hydrogenation is used commercially to convert the double bonds in the unsaturated fats in vegetable oils to saturated fats such as margarine, which are more solid. Adjusting the amount of added hydrogen produces partially hydrogenated fats such as soft margarine, solid margarine in sticks, and shortenings, which are used in cooking. For example, oleic acid is a typical unsaturated fatty acid in olive oil and has a double bond at carbon 9. When oleic acid is hydrogenated, it is converted to stearic acid, a saturated fatty acid.

The unsaturated fats in vegetable oils are converted to saturated fats to make a more solid product.

Double bond

Oleic acid (found in olive oil and other unsaturated fats)

Single bond

Stearic acid (found in saturated fats)

Polymers

A **polymer** is a large molecule that consists of small repeating units called **monomers**. In the past hundred years, the plastics industry has made synthetic polymers that are in many of the materials we use every day, such as carpeting, plastic wrap, nonstick pans, plastic cups, and rain gear (see **FIGURE 17.5**). In medicine, synthetic polymers are used to replace

Polyethylene

Polyvinyl chloride

Polypropylene

Polytetrafluoroethylene
(Teflon)

Polydichloroethylene
(Saran)

Polystyrene

FIGURE 17.5 ▶ Synthetic polymers provide a wide variety of items that we use every day.

Q What are some alkenes used to make the polymers in these plastic items?

diseased or damaged body parts such as hip joints, teeth, heart valves, and blood vessels. There are about 300 billion kg of plastics produced every year, which is about 40 kg for every person on Earth.

Many of the synthetic polymers are made by reactions of small alkene monomers. Polymerization reactions require high temperature, a catalyst, and high pressure (over 1000 atm). A polymer grows longer as each monomer is added at the end of the chain and contains as many as 1000 monomers. Polyethylene, a polymer made from ethylene monomers, is used in plastic bottles, film, and plastic dinnerware. More polyethylene is produced worldwide than any other polymer. Low-density polyethylene (LDPE) is flexible, breakable, less dense, and more branched than high-density polyethylene (HDPE). High-density polyethylene is stronger, more dense, and melts at a higher temperature than LDPE.

ENGAGE

What monomer is used to make polyethylene?

Monomer unit repeats

$$
\underset{\text{Ethene (ethylene) monomers}}{
\begin{array}{c}
H \\ | \\ C \\ | \\ H
\end{array}
{=}
\begin{array}{c}
H \\ | \\ C \\ | \\ H
\end{array}
+
\begin{array}{c}
H \\ | \\ C \\ | \\ H
\end{array}
{=}
\begin{array}{c}
H \\ | \\ C \\ | \\ H
\end{array}
+
\begin{array}{c}
H \\ | \\ C \\ | \\ H
\end{array}
{=}
\begin{array}{c}
H \\ | \\ C \\ | \\ H
\end{array}
}
\longrightarrow
\underset{\text{Polyethylene section}}{
\cdots {-}C{-}C{-}\!\!\left[{-}C{-}C{-}\right]_{\text{many}}\!\!C{-}C{-}\cdots
}
$$

TABLE 17.6 lists several alkene monomers that are used to produce common synthetic polymers. The alkane-like nature of these plastic synthetic polymers makes them unreactive. Thus, they do not decompose easily (they are not biodegradable). As a result, they have become significant contributors to pollution, on land and in the oceans. Efforts are being made to make them more degradable.

You can identify the type of polymer used to manufacture a plastic item by looking for the recycling symbol (arrows in a triangle) found on the label or on the bottom of the plastic container. For example, the number 5 or the letters PP inside the triangle is the code for a polypropylene plastic. There are now many cities that maintain recycling programs that reduce the amount of plastic materials that are transported to landfills.

Today, products such as lumber, tables and benches, trash receptacles, and pipes used for irrigation systems are made from recycled plastics.

Tables, benches, and trash receptacles can be manufactured from recycled plastics.

TABLE 17.6 Some Alkenes and Their Polymers

Recycling Code	Monomer	Polymer Section	Common Uses
2 HDPE **4** LDPE	$H_2C=CH_2$ Ethene (ethylene)	Polyethylene (PE)	Plastic bottles, film, insulation materials, shopping bags
3 V	$H_2C=CH$ —Cl Chloroethene (vinyl chloride)	Polyvinyl chloride (PVC)	Plastic pipes and tubing, garden hoses, garbage bags, shower curtains, credit cards, medical containers
5 PP	$H_2C=CH$ —CH₃ Propene (propylene)	Polypropylene (PP)	Ski and hiking clothing, carpets, artificial joints, plastic bottles, food containers, medical face masks, tubing
	$F—C=C—F$ Tetrafluoroethene	Polytetrafluoroethylene	Nonstick coatings, Teflon
	$H_2C=C—Cl$ 1,1-Dichloroethene	Polydichloroethylene (Saran)	Plastic film and wrap, Gore-Tex rainwear, medical implants
6 PS	$H_2C=CH$ Phenylethene (styrene)	Polystyrene (PS)	Plastic coffee cups and cartons, insulation

SAMPLE PROBLEM 17.7 Polymers

A firefighter/EMT arrives at a home where a premature baby has been delivered. To prevent hypothermia during transport to the neonatal facility, she wraps the baby in cling wrap, which is polydichloroethylene. Draw the monomer unit, which is 1,1-dichloroethene, and draw a portion of the polymer formed from three monomer units.

TRY IT FIRST ▶

SOLUTION

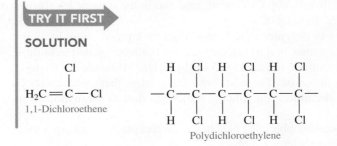

1,1-Dichloroethene

Polydichloroethylene

STUDY CHECK 17.7

Draw the condensed structural formula for the monomer used in the manufacturing of PVC.

ANSWER

$$\overset{\displaystyle Cl}{\underset{\displaystyle |}{H_2C = CH}}$$

QUESTIONS AND PROBLEMS

17.2 Alkenes, Alkynes, and Polymers

LEARNING GOAL Write the IUPAC names and draw the condensed or line-angle structural formulas for alkenes and alkynes.

17.11 Identify each of the following as an alkane, alkene, or alkyne:
  a. $CH_3 - CH = CH_2$
  b. $CH_3 - CH_2 - C \equiv CH$
  c. ⌇⌇⌇⌇⌇

17.12 Identify each of the following as an alkane, alkene, or alkyne:
  a. (line-angle structure)

  b. $CH_3 - \overset{\displaystyle CH_3}{\underset{\displaystyle \underset{\displaystyle CH_3}{|}}{C = C}} - CH_3$

  c. $CH_3 - CH_2 - \overset{\displaystyle CH_3}{\underset{\displaystyle |}{CH}} - CH_3$

17.13 Write the IUPAC name for each of the following:
  a. $H_2C = CH_2$
  b. $CH_3 - \overset{\displaystyle CH_3}{\underset{\displaystyle |}{C}} = CH_2$
  c. $CH_3 - CH_2 - C \equiv C - CH_3$

17.14 Write the IUPAC name for each of the following:
  a. $H_2C = CH - CH_2 - CH_3$
  b. $CH_3 - C \equiv C - CH_2 - \overset{\displaystyle CH_3}{\underset{\displaystyle |}{CH}} - CH_3$
  c. $CH_3 - CH_2 - CH = CH - CH_3$

17.15 Draw the condensed structural formula for each of the following compounds:
  a. propene   **b.** 1-hexyne
  c. 2-methyl-1-butene

17.16 Draw the condensed structural formula for each of the following compounds:
  a. 1-pentene   **b.** 3-methyl-1-butyne
  c. 3,4-dimethyl-1-pentene

17.17 Draw the condensed structural formula for the product in each of the following:
  a. $CH_3 - CH_2 - CH_2 - CH = CH_2 + H_2 \xrightarrow{Pt}$
  b. $CH_3 - CH = CH - CH_3 + H_2 \xrightarrow{Ni}$
  c. (line-angle structure) $+ H_2 \xrightarrow{Pt}$

17.18 Draw the condensed structural formula for the product in each of the following:
  a. $CH_3 - CH_2 - CH = CH_2 + H_2 \xrightarrow{Pt}$
  b. $CH_3 - \overset{\displaystyle CH_3}{\underset{\displaystyle |}{C}} = CH - CH_2 - CH_3 + H_2 \xrightarrow{Pt}$
  c. (line-angle structure) $+ H_2 \xrightarrow{Pt}$

17.19 What is a polymer?

17.20 What is a monomer?

17.21 Write an equation that represents the formation of a portion of polypropylene from three of its monomers.

17.22 Write an equation that represents the formation of a portion of polystyrene from three of its monomers.

Applications

17.23 The plastic polyvinylidene difluoride, PVDF, is made from monomers of 1,1-difluoroethene. Draw the expanded structural formula for a portion of the polymer formed from three monomers of 1,1-difluoroethene.

17.24 The polymer polyacrylonitrile, PAN, used in the fabric material Orlon is made from monomers of acrylonitrile. Draw the expanded structural formula for a portion of the polymer formed from three monomers of acrylonitrile.

$$\overset{\displaystyle CN}{\underset{\displaystyle |}{H_2C = CH}}$$

Acrylonitrile

17.3 Aromatic Compounds

LEARNING GOAL Describe the bonding in benzene; name aromatic compounds, and draw their line-angle structural formulas.

In 1825, Michael Faraday isolated a hydrocarbon called *benzene*, which had the molecular formula C_6H_6. A molecule of **benzene** consists of a ring of six carbon atoms with one hydrogen atom attached to each carbon. Because many compounds containing benzene had fragrant odors, the family of benzene compounds became known as **aromatic compounds**.

Benzene

In benzene, each carbon atom uses three valence electrons to bond to the hydrogen atom and two adjacent carbons. That leaves one valence electron, which scientists first thought was shared in a double bond with an adjacent carbon. In 1865, August Kekulé proposed that the carbon atoms in benzene were arranged in a flat ring with alternating single and double bonds between the carbon atoms. There are two possible structural representations of benzene due to resonance in which the double bonds can form between two different carbon atoms.

Today, we know that the six electrons are shared equally among the six carbon atoms and that all the carbon–carbon bonds in benzene are identical due to resonance. This unique feature of benzene makes it especially stable. Benzene is most often represented as a line-angle structural formula, which shows a hexagon with a circle in the center. Some of the ways to represent benzene are shown as follows:

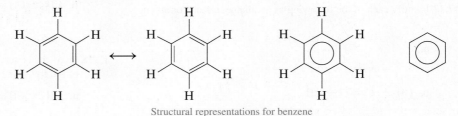

Structural representations for benzene

Some common examples of aromatic compounds that we use for flavoring are anisole from anise, estragole from tarragon, and thymol from thyme.

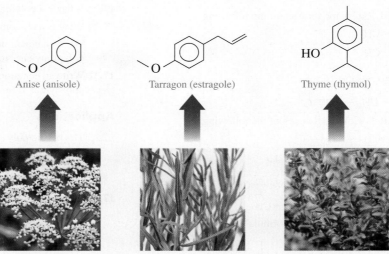

Anise (anisole) Tarragon (estragole) Thyme (thymol)

The aroma and flavor of the herbs anise, tarragon, and thyme are because of aromatic compounds.

Naming Aromatic Compounds

Many compounds containing benzene have been important in chemistry for many years and still use their common names. Names such as toluene, aniline, and phenol are allowed by IUPAC rules. When benzene has only one substituent, the ring is not numbered.

CH₃   NH₂   OH

Toluene
(methylbenzene)

Aniline
(aminobenzene)

Phenol
(hydroxybenzene)

ENGAGE

How is the structure of toluene similar to and different from that of phenol?

When there are two substituents, the benzene ring is numbered to give the lowest numbers to the substituents. When a common name can be used such as toluene, aniline, or phenol, the carbon atom attached to the methyl, amine, or hydroxyl group, is numbered as carbon 1.

1,2-Dichlorobenzene   1,3-Dichlorobenzene   1,4-Dichlorobenzene

CHEMISTRY LINK TO THE **ENVIRONMENT**
Some Common Aromatic Compounds

Aromatic compounds are common in nature and in medicine. Toluene is used as a reactant to make drugs, dyes, and explosives such as TNT (trinitrotoluene). The benzene ring is found in some amino acids (the building blocks of proteins), pain relievers such as aspirin, acetaminophen, and ibuprofen; and flavorings such as vanillin.

$$CH_3-CH-CH_2- \bigcirc -CH-C-OH$$

Ibuprofen

TNT (2,4,6-trinitrotoluene)

Aspirin

Acetaminophen

Vanillin

SAMPLE PROBLEM 17.8 Naming Aromatic Compounds

Write the IUPAC name for the following compound:

TRY IT FIRST

SOLUTION

STEP 1 Write the name for the aromatic compound. This aromatic compound contains a benzene ring with an amine group, which is aniline.

Guide to Naming Aromatic Compounds

STEP 1
Write the name for the aromatic compound.

STEP 2
If there are two substituents, number the aromatic ring.

STEP 3
Name a substituent as a prefix.

STEP 2 If there are two substituents, number the aromatic ring. The benzene ring is numbered to give the lower number when counting the carbon with the amine group as 1. The bromine atom is on carbon 2.

STEP 3 Name a substituent as a prefix.

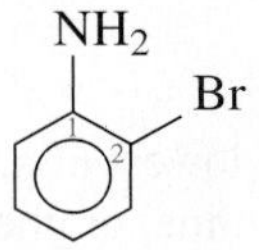

2-bromoaniline

STUDY CHECK 17.8
Write the IUPAC name for the following compound:

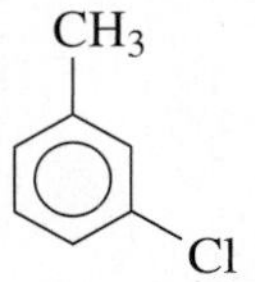

ANSWER
3-chlorotoluene

QUESTIONS AND PROBLEMS

17.3 Aromatic Compounds

LEARNING GOAL Describe the bonding in benzene; name aromatic compounds, and draw their line-angle structural formulas.

17.25 Write the IUPAC name for each of the following:

a.

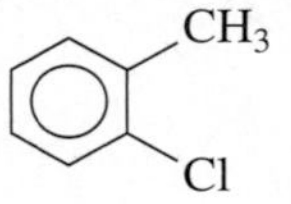

b.

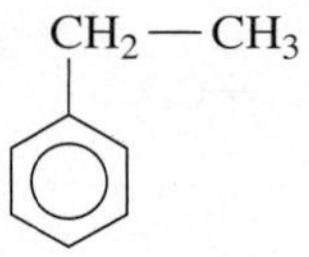

c.

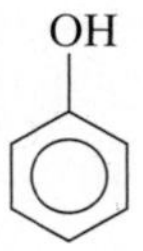

17.26 Write the IUPAC name for each of the following:

a.

b.

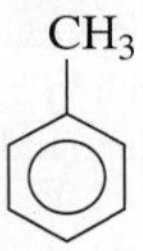

c.

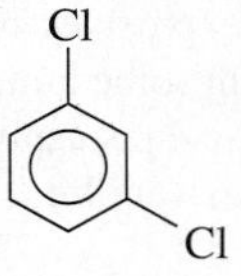

17.27 Draw the line-angle structural formula for each of the following compounds:
a. aniline
b. 1-chloro-4-fluorobenzene
c. 4-ethyltoluene

17.28 Draw the line-angle structural formula for each of the following compounds:
a. 1,3-dibromo-5-chlorobenzene
b. 2,4-dichlorotoluene
c. propylbenzene

17.4 Alcohols and Ethers

LEARNING GOAL Write the IUPAC and common names for alcohols and ethers; draw the condensed or line-angle structural formulas when given their names.

In the functional group of an **alcohol**, a hydroxyl group (—OH) replaces a hydrogen atom in a hydrocarbon. In a phenol, the hydroxyl group replaces a hydrogen atom attached to a benzene ring. Molecules of alcohols and phenols have a bent shape around the oxygen atom. In the functional group of an **ether**, an oxygen atom is attached by single bonds to two carbon atoms (see **FIGURE 17.6**).

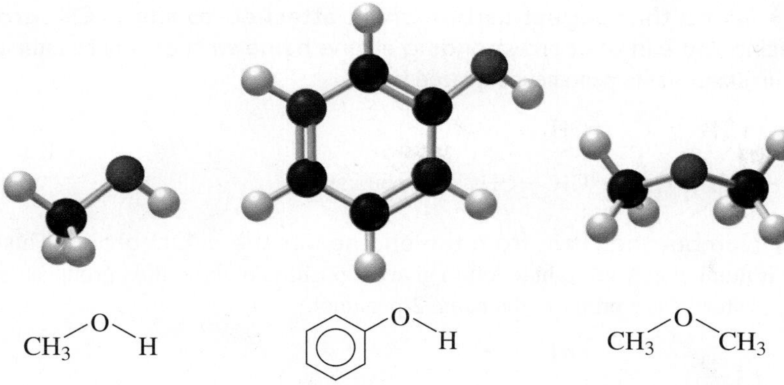

FIGURE 17.6 ▶ An alcohol or a phenol has a hydrogen atom replaced by a hydroxyl group (— OH); an ether contains an oxygen atom (— O —) bonded to two carbon groups.

Q How are the structures of alcohols and phenols similar?

CH₃—O—H   Methanol

⬡—O—H   Phenol

CH₃—O—CH₃   Dimethyl ether

Naming Alcohols

In the IUPAC system, an alcohol is named by replacing the final *e* in the corresponding alkane name with *ol*. The common name of a simple alcohol uses the name of the alkyl group followed by *alcohol*.

$CH_3—H$ $CH_3—OH$ $CH_3—CH_2—H$ $CH_3—CH_2—OH$
Methane   Methanol   Ethane   Ethanol
           (methyl alcohol)   (ethyl alcohol)

Alcohols with one or two carbon atoms do not require a number for the hydroxyl group. When an alcohol consists of a chain with three or more carbon atoms, the chain is numbered to give the position of the —OH group and any substituents on the chain.

$CH_3—CH_2—CH_2—OH$
3   2   1
1-Propanol
(propyl alcohol)

$CH_3—CH—CH_3$ with OH
1   2   3
2-Propanol
(isopropyl alcohol)

We can also draw the line-angle structural formulas for alcohols as shown for 2-propanol and 2-butanol.

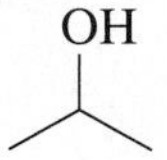

OH   2-Propanol

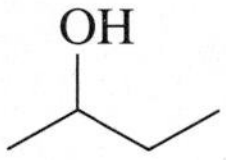

OH   2-Butanol

SAMPLE PROBLEM 17.9 Naming Alcohols

Write the IUPAC name for the following:

$CH_3—CH—CH_2—CH—CH_3$ with CH₃ and OH

TRY IT FIRST

SOLUTION

	Given	Need	Connect
ANALYZE THE PROBLEM	alcohol, five-carbon chain, methyl substituent	IUPAC name	position of —CH₃ and —OH groups, replace *e* in alkane name with *ol*

Guide to Naming Alcohols

STEP 1
Name the longest carbon chain attached to the —OH group by replacing the *e* in the corresponding alkane name with *ol*. Name an aromatic alcohol as a *phenol*.

STEP 2
Number the chain from the end nearer the —OH group.

STEP 3
Give the location and name for each substituent relative to the —OH group.

STEP 1 Name the longest carbon chain attached to the —OH group by replacing the *e* in the corresponding alkane name with *ol*. To name the alcohol, the *e* in alkane name pentane is replaced by *ol*.

$$CH_3—\underset{\underset{CH_3}{|}}{CH}—CH_2—\underset{\underset{OH}{|}}{CH}—CH_3 \qquad \text{pentanol}$$

STEP 2 Number the chain from the end nearer the —OH group. This carbon chain is numbered from right to left to give the position of the —OH group as carbon 2, which is shown as a prefix in the name 2-pentanol.

$$\underset{5}{CH_3}—\underset{4}{\underset{\underset{CH_3}{|}}{CH}}—\underset{3}{CH_2}—\underset{2}{\underset{\underset{OH}{|}}{CH}}—\underset{1}{CH_3} \qquad \text{2-pentanol}$$

STEP 3 Give the location and name of each substituent relative to the —OH group.

$$\underset{5}{CH_3}—\underset{4}{\underset{\underset{CH_3}{|}}{CH}}—\underset{3}{CH_2}—\underset{2}{\underset{\underset{OH}{|}}{CH}}—\underset{1}{CH_3} \qquad \text{4-methyl-2-pentanol}$$

STUDY CHECK 17.9

Write the IUPAC name for the following:

Cl
⋀⋁OH

ANSWER

3-chloro-1-butanol

Naming Ethers

An *ether* consists of an oxygen atom that is attached by single bonds to two carbon groups that are alkyl or aromatic groups. In the common name of an ether, the names of the alkyl or aromatic groups attached to the oxygen atom are written in alphabetical order, followed by the word *ether*.

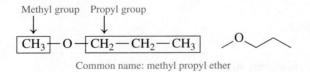

Methyl group   Propyl group

$$\boxed{CH_3}—O—\boxed{CH_2—CH_2—CH_3}$$

Common name: methyl propyl ether

CHEMISTRY LINK TO **HEALTH**
Some Important Alcohols, Phenols, and Ethers

Methanol (*methyl alcohol*), the simplest alcohol, is found in solvents and paint removers. If ingested, methanol is oxidized to formaldehyde, which can cause headaches, blindness, and death. Methanol is used to make plastics, medicines, and fuels. In car racing, it is used as a fuel because it is less flammable and has a higher octane rating than does gasoline.

Ethanol (*ethyl alcohol*) is used in hand sanitizers. In an alcohol-containing sanitizer, the amount of ethanol is typically 60% (v/v) but can be as high as 85% (v/v). This amount of ethanol can make hand sanitizers a fire hazard in the home because ethanol is highly

flammable. When using an ethanol-containing sanitizer, it is important to rub hands until they are completely dry. It is also recommended that sanitizers containing ethanol be placed in storage areas that are away from heat sources in the home.

1,2,3-Propanetriol (*glycerol* or *glycerin*), a trihydroxy alcohol, is a viscous liquid obtained from oils and fats during the production of soaps. The presence of several polar —OH groups makes it strongly attracted to water, a feature that makes glycerin useful as a skin softener in products such as skin lotions, cosmetics, shaving creams, and liquid soaps.

$$HO—CH_2—\underset{\underset{OH}{|}}{CH}—CH_2—OH$$

1,2,3-Propanetriol
(glycerol)

1,2-Ethanediol (*ethylene glycol*) is used as an antifreeze in heating and cooling systems. It is also a solvent for paints, inks, and plastics, and it is used in the production of synthetic fibers such as Dacron. If ingested, it is extremely toxic. In the body, it is oxidized to oxalic acid, which forms insoluble compounds in the kidneys that cause renal damage, convulsions, and death. Because its sweet taste is attractive to pets and children, ethylene glycol solutions must be carefully stored.

Antifreeze, containing ethylene glycol, raises the boiling point and decreases the freezing point of water in a radiator.

$$HO—CH_2—CH_2—OH \xrightarrow{[O]} HO—\underset{\underset{}{\overset{\overset{O}{\|}}{C}}}—\underset{\underset{}{\overset{\overset{O}{\|}}{C}}}—OH$$

1,2-Ethanediol
(ethylene glycol)

Oxalic acid

Phenols are found in several of the essential oils of plants, which produce the odor or flavor of the plant. Eugenol is found in cloves, vanillin in vanilla bean, isoeugenol in nutmeg, and thymol in thyme and mint. Thymol has a pleasant, minty taste and is used in mouthwashes and by dentists to disinfect a cavity before adding a filling compound.

Bisphenol A (BPA) is used to make polycarbonate, a clear plastic that is used to manufacture beverage bottles, including baby bottles. Washing polycarbonate bottles with certain detergents or at high temperatures disrupts the polymer, causing small amounts of BPA to leach from the bottles. Because BPA is an estrogen mimic, there are concerns about the harmful effects from low levels of BPA. In 2008, Canada banned the use of polycarbonate baby bottles. Plastic bottles and containers made of polycarbonate have the recycling symbol "7."

Bisphenol A (BPA)

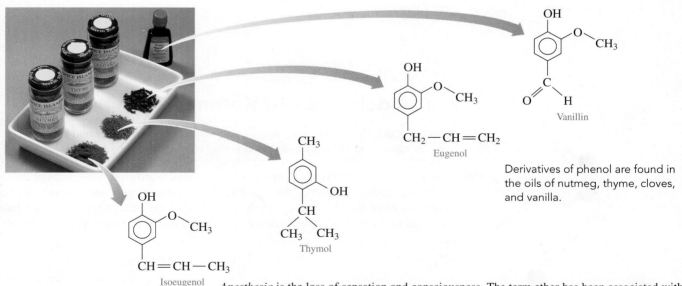

Eugenol

Vanillin

Derivatives of phenol are found in the oils of nutmeg, thyme, cloves, and vanilla.

Thymol

Isoeugenol

Anesthesia is the loss of sensation and consciousness. The term ether has been associated with anesthesia because diethyl ether was the most widely used anesthetic for more than a hundred years. Although it is easy to administer, ether is very volatile and highly flammable. A small spark in the operating room could cause an explosion. Since the 1950s, anesthetics such as Forane (isoflurane), Ethrane (enflurane), Suprane (desflurane), and Sevoflurane have been developed that are not as flammable. These anesthetics retain the ether group, but the addition of halogen atoms reduces the volatility and flammability of the ethers.

Isoflurane (Forane) is an inhaled anesthetic.

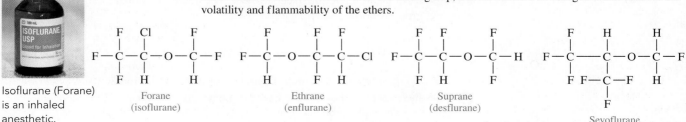

Forane
(isoflurane)

Ethrane
(enflurane)

Suprane
(desflurane)

Sevoflurane

QUESTIONS AND PROBLEMS

17.4 Alcohols and Ethers

LEARNING GOAL Write the IUPAC and common names for alcohols and ethers; draw the condensed or line-angle structural formulas when given their names.

17.29 Write the IUPAC name for each of the following:
a. $CH_3—CH_2—OH$

b.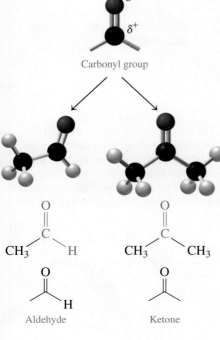

c.

17.30 Write the IUPAC name for each of the following:

a.

b. $CH_3—CH_2—\overset{\overset{\displaystyle CH_3}{|}}{CH}—\overset{\overset{\displaystyle CH_3}{|}}{CH}—CH_2—OH$

c.

17.31 Draw the condensed structural formula for each of the following:
a. 1-propanol
b. methyl alcohol
c. 3-pentanol
d. 2-methyl-2-butanol

17.32 Draw the condensed structural formula for each of the following:
a. ethyl alcohol
b. 3-methyl-1-butanol
c. 2,4-dichloro-3-hexanol
d. propyl alcohol

17.33 Write the common name for each of the following:
a. $CH_3—O—CH_2—CH_3$
b. $CH_3—CH_2—CH_2—O—CH_2—CH_2—CH_3$

c.

17.34 Write the common name for each of the following:
a. $CH_3—CH_2—O—CH_2—CH_2—CH_3$

b.

c. $CH_3—O—CH_3$

17.5 Aldehydes and Ketones

LEARNING GOAL Write the IUPAC and common names for aldehydes and ketones; draw the condensed or line-angle structural formulas when given their names.

The *carbonyl group* ($C=O$) has a carbon–oxygen double bond with two groups of atoms attached to the carbon atom at angles of 120°. Because the oxygen atom in the carbonyl group is much more electronegative than the carbon atom, the carbonyl group has a dipole with a partial negative charge (δ^-) on the oxygen and a partial positive charge (δ^+) on the carbon.

Carbonyl group

In the functional group of an **aldehyde**, the carbonyl group is bonded to at least one hydrogen atom. That carbon may also be bonded to another hydrogen, a carbon atom in an alkyl group, or an aromatic ring (see **FIGURE 17.7**). In the functional group of a **ketone**, the carbonyl group is bonded to two carbon atoms.

Aldehyde   Ketone

FIGURE 17.7 ▶ The carbonyl group is found in aldehydes and ketones.

◉ If aldehydes and ketones both contain a carbonyl group, how can you differentiate between compounds from each family?

Naming Aldehydes

In the IUPAC system, an aldehyde is named by replacing the *e* in the corresponding alkane name with *al*. No number is needed for the aldehyde group because it always appears at the end of the chain. The aldehydes with carbon chains of one to four carbon atoms are often referred to by their common names, which end in *aldehyde*. The roots (*form, acet, propion,* and *butyr*) of these common names are derived from Latin or Greek words (see **FIGURE 17.8**).

CORE CHEMISTRY SKILL
Naming Aldehydes and Ketones

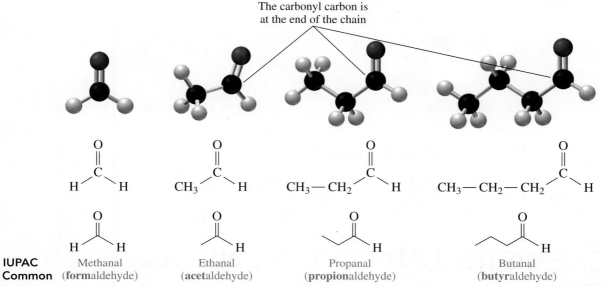

FIGURE 17.8 ▶ In the structures of aldehydes, the carbonyl group is always the end carbon.
◎ Why is the carbon in the carbonyl group in aldehydes always at the end of the chain?

The aldehyde of benzene is named benzaldehyde.

Benzaldehyde

SAMPLE PROBLEM 17.10 Naming Aldehydes

Write the IUPAC name for the following:

$$CH_3-CH_2-\overset{\overset{\displaystyle CH_3}{|}}{CH}-CH_2-\overset{\overset{\displaystyle O}{\|}}{C}-H$$

TRY IT FIRST

SOLUTION

	Given	Need	Connect
ANALYZE THE PROBLEM	five-carbon chain, methyl substituent	IUPAC name	position of —CH_3, replace *e* in alkane name with *al*

Guide to Naming Aldehydes

STEP 1
Name the longest carbon chain by replacing the *e* in the alkane name with *al*.

STEP 2
Name and number any substituents by counting the carbonyl group as carbon 1.

STEP 1 Name the longest carbon chain by replacing the *e* in the alkane name with *al*. The longest carbon chain containing the carbonyl group has five carbon atoms.

$$CH_3-CH_2-\underset{\underset{CH_3}{|}}{CH}-CH_2-\underset{\overset{O}{\|}}{C}-H \qquad \text{pentanal}$$

STEP 2 Name and number any substituents by counting the carbonyl group as carbon 1. Counting from the right, the methyl substituent is on carbon 3.

$$\underset{5}{CH_3}-\underset{4}{CH_2}-\underset{3}{\underset{\underset{CH_3}{|}}{CH}}-\underset{2}{CH_2}-\underset{1}{\underset{\overset{O}{\|}}{C}}-H \qquad \text{3-methylpentanal}$$

STUDY CHECK 17.10

What are the IUPAC and common names of the aldehyde with three carbon atoms?

ANSWER

propanal (propionaldehyde)

CHEMISTRY LINK TO THE **ENVIRONMENT**
Vanilla

Vanilla has been used as a flavoring for over a thousand years. After drinking a beverage made from powdered vanilla and cocoa beans with Emperor Montezuma in Mexico, the Spanish conquistador Hernán Cortés took vanilla back to Europe where it became popular for flavoring and for scenting perfumes and tobacco. Thomas Jefferson introduced vanilla to the United States during the late 1700s. Today much of the vanilla we use in the world is grown in Indonesia, Madagascar, Mexico, Papua New Guinea, and China.

The vanilla plant is a member of the orchid family and thrives under tropical conditions. There are many species of *Vanilla*, but *Vanilla planifolia* (or *V. fragrans*) is considered to produce the best flavor. The vanilla plant grows like a vine, which can attain a length of 100 feet. Its flowers are hand-pollinated to produce a green fruit that is picked in eight or nine months. The fruit is sun-dried so that it becomes a long, dark brown pod, which is called a "vanilla bean" because it looks like a string bean. The flavor and fragrance of the vanilla bean comes from the tiny black seeds found inside the dried bean.

The seeds and pod are used to flavor desserts such as custards and ice cream. The extract of vanilla is made by chopping up vanilla beans and mixing them with a 35% ethanol–water mixture. The liquid, which contains the aldehyde *vanillin*, is drained from the bean residue and used for flavoring.

Vanillin
(vanilla)

The vanilla bean is the dried fruit of the vanilla plant.

Vanilla flavoring is prepared by soaking vanilla beans in ethanol and water.

Naming Ketones

In the IUPAC system, the name of a ketone is obtained by replacing the *e* in the corresponding alkane name with *one*. However, the common names for unbranched ketones are still in use. In the common names, the alkyl groups bonded on either side of the carbonyl group are listed alphabetically, followed by *ketone*. Acetone, which is another name for propanone, has been retained by the IUPAC system. The naming of a ketone is shown in Sample Problem 17.11.

SAMPLE PROBLEM 17.11 **Naming Ketones**

Write the IUPAC name for the following ketone:

TRY IT FIRST

SOLUTION

	Given	Need	Connect
ANALYZE THE PROBLEM	five-carbon chain, methyl substituent	IUPAC name	position of —CH_3 and carbonyl, replace *e* in alkane name with *one*

STEP 1 Name the longest carbon chain by replacing the *e* in the alkane name with *one*.

   pentanone

STEP 2 Number the carbon chain from the end nearer the carbonyl group and indicate its location.

   2-pentanone

5 4 3 2 1

STEP 3 Name and number any substituents on the carbon chain. Counting from the right, the methyl substituent is on carbon 4.

   4-methyl-2-pentanone

5 4 3 2 1

STUDY CHECK 17.11

What is the common name of 3-hexanone?

ANSWER

ethyl propyl ketone

Guide to Naming Ketones

STEP 1
Name the longest carbon chain by replacing the *e* in the alkane name with *one*.

STEP 2
Number the carbon chain from the end nearer the carbonyl group and indicate its location.

STEP 3
Name and number any substituents on the carbon chain.

QUESTIONS AND PROBLEMS

17.5 Aldehydes and Ketones

LEARNING GOAL Write the IUPAC and common names for aldehydes and ketones; draw the condensed or line-angle structural formulas when given their names.

17.35 Write the common name for each of the following:

a. $CH_3 - \overset{\displaystyle O}{\overset{\displaystyle \|}{C}} - H$

b.

c. $H - \overset{\displaystyle O}{\overset{\displaystyle \|}{C}} - H$

17.36 Write the common name for each of the following:

a. $CH_3 - CH_2 - \overset{\displaystyle O}{\overset{\displaystyle \|}{C}} - CH_3$

b.

c. $CH_3 - CH_2 - \overset{\displaystyle O}{\overset{\displaystyle \|}{C}} - H$

17.37 Write the IUPAC name for each of the following:

a. $CH_3 - CH_2 - \overset{\displaystyle O}{\overset{\displaystyle \|}{C}} - H$

b.

c.

17.38 Write the IUPAC name for each of the following:

a. $CH_3 - CH_2 - CH_2 - \overset{\displaystyle O}{\overset{\displaystyle \|}{C}} - H$

b.

c.

17.39 Draw the condensed structural formula for each of the following:
a. acetaldehyde
b. 2-pentanone
c. butyl methyl ketone

17.40 Draw the condensed structural formula for each of the following:
a. propionaldehyde
b. 3,4-dimethylhexanal
c. 4-bromobutanone

17.6 Carboxylic Acids and Esters

LEARNING GOAL Write the IUPAC and common names for carboxylic acids and esters; draw the condensed or line-angle structural formulas when given their names.

Carbonyl group

$-\overset{\displaystyle O}{\overset{\displaystyle \|}{C}} - \boxed{OH}$ *Hydroxyl group*

Carboxyl group

In the functional group of a **carboxylic acid**, a carbonyl group is attached to a hydroxyl group (—OH), which forms a *carboxyl group*. Some ways to represent the carboxyl group in propanoic acid follow.

$CH_3 - CH_2 - \overset{\displaystyle O}{\overset{\displaystyle \|}{C}} - OH$ $\overset{\displaystyle O}{\diagup\diagdown\diagup} OH$ $CH_3 - CH_2 - COOH$

Propanoic acid
(propionic acid)

IUPAC Names of Carboxylic Acids

The IUPAC names of carboxylic acids replace the *e* in the corresponding alkane name with *oic acid*. If there are substituents, the carbon chain is numbered beginning with the carboxyl carbon.

$H - \overset{\displaystyle O}{\overset{\displaystyle \|}{C}} - OH$ $CH_3 - \overset{\displaystyle CH_3}{\overset{\displaystyle |}{CH}} - \overset{\displaystyle O}{\overset{\displaystyle \|}{C}} - OH$ $CH_3 - \overset{\displaystyle OH}{\overset{\displaystyle |}{CH}} - CH_2 - \overset{\displaystyle O}{\overset{\displaystyle \|}{C}} - OH$

Methanoic acid   2-Methylpropanoic acid   3-Hydroxybutanoic acid

The sour taste of vinegar is due to ethanoic acid (acetic acid).

As with the aldehydes, carboxylic acids with one to four carbon atoms have common names, which are derived from their natural sources. The common names use the prefixes of *form*, *acet*, *propion*, and *butyr*; these common names are derived from Latin or Greek words (see TABLE 17.7).

> **CORE CHEMISTRY SKILL**
> Naming Carboxylic Acids

TABLE 17.7 IUPAC and Common Names of Selected Carboxylic Acids

Condensed Structural Formula	Line-Angle Formula	IUPAC Name	Common Name	Ball-and-Stick Model
$H-\overset{\overset{O}{\|\|}}{C}-OH$		Methanoic acid	Formic acid	
$CH_3-\overset{\overset{O}{\|\|}}{C}-OH$		Ethanoic acid	Acetic acid	
$CH_3-CH_2-\overset{\overset{O}{\|\|}}{C}-OH$		Propanoic acid	Propionic acid	
$CH_3-CH_2-CH_2-\overset{\overset{O}{\|\|}}{C}-OH$		Butanoic acid	Butyric acid	

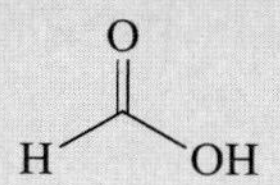

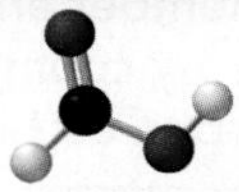

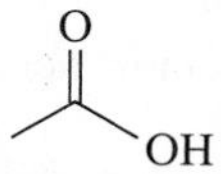

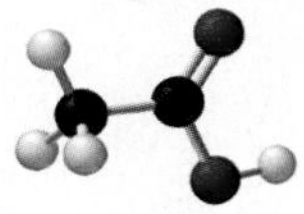

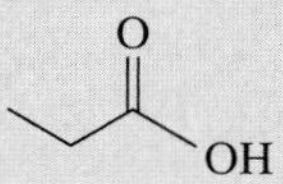

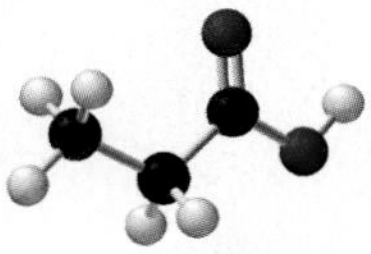

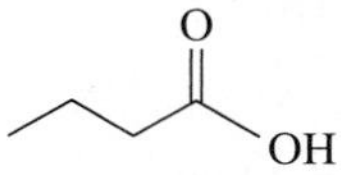

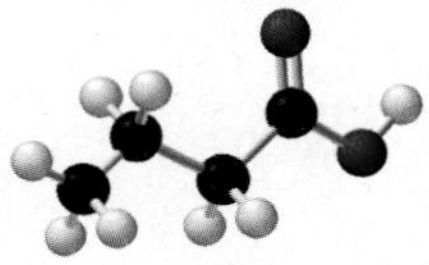

The simplest aromatic carboxylic acid is benzoic acid. The carbon of the carboxyl group is bonded to carbon 1 in the ring, and the ring is numbered to give the lowest numbers for any substituent.

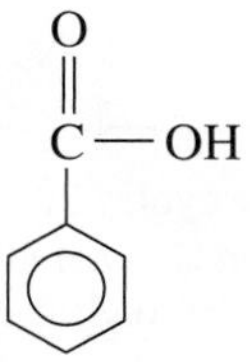

Benzoic acid 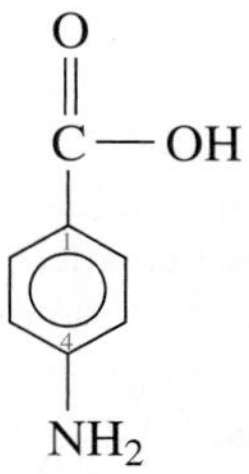   4-Aminobenzoic acid 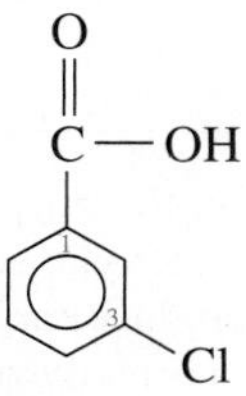   3-Chlorobenzoic acid

SAMPLE PROBLEM 17.12 Naming Carboxylic Acids

Write the IUPAC name for the following:

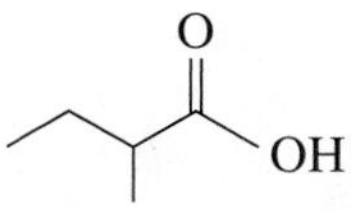

TRY IT FIRST

SOLUTION

ANALYZE THE PROBLEM	Given	Need	Connect
	four-carbon chain, methyl substituent	IUPAC name	position of $-CH_3$, replace *e* in alkane name with *oic acid*

Guide to Naming Carboxylic Acids

STEP 1
Identify the longest carbon chain and replace the *e* in the alkane name with *oic acid.*

STEP 2
Name and number any substituents by counting the carboxyl carbon as 1.

STEP 1 Identify the longest carbon chain and replace the *e* in the alkane name with *oic acid.* The IUPAC name of a carboxylic acid with four carbons is butanoic acid.

butanoic acid

STEP 2 Name and number any substituents by counting the carboxyl carbon as 1. Counting from the right, the methyl substituent is on carbon 2. The IUPAC name is 2-methylbutanoic acid.

2-methylbutanoic acid

4   3   2   1

STUDY CHECK 17.12

Write the IUPAC name for the following:

$$CH_3 - CH_2 - CH_2 - CH_2 - \overset{\overset{\displaystyle O}{\|}}{C} - OH$$

ANSWER
pentanoic acid

CHEMISTRY LINK TO **HEALTH**
Carboxylic Acids in Metabolism

Several carboxylic acids are part of the metabolic processes within our cells. For example, during glycolysis, a molecule of glucose is broken down into two molecules of pyruvic acid, or actually, its carboxylate ion, pyruvate. During strenuous exercise when oxygen levels are low (anaerobic), pyruvic acid is reduced to give lactic acid or the lactate ion.

$$CH_3 - \overset{\overset{\displaystyle O}{\|}}{C} - \overset{\overset{\displaystyle O}{\|}}{C} - OH + 2H \xrightarrow{\text{Reduction}} CH_3 - \overset{\overset{\displaystyle OH}{|}}{CH} - \overset{\overset{\displaystyle O}{\|}}{C} - OH$$

Pyruvic acid   Lactic acid

During exercise, pyruvic acid is converted to lactic acid in the muscles.

In the *citric acid cycle*, also called the Krebs cycle, di- and tricarboxylic acids are oxidized and decarboxylated (loss of CO_2) to produce energy for the cells of the body. These carboxylic acids are normally referred to by their common names. At the start of the citric acid cycle, citric acid with six carbons is converted to five-carbon α-ketoglutaric acid. Citric acid is also the acid that gives the sour taste to citrus fruits such as lemons and grapefruits.

$$\begin{array}{c} COOH \\ | \\ CH_2 \\ | \\ HO-C-COOH \\ | \\ CH_2 \\ | \\ COOH \end{array} \xrightarrow{[O]} \begin{array}{c} COOH \\ | \\ CH_2 \\ | \\ CH_2 \\ | \\ C=O \\ | \\ COOH \end{array} + CO_2$$

Citric acid   α-Ketoglutaric acid

The citric acid cycle continues as α-ketoglutaric acid loses CO_2 to give a four-carbon succinic acid. Then a series of reactions converts succinic acid to oxaloacetic acid. We see that some of the functional groups we have studied along with reactions such as hydration and oxidation are part of the metabolic processes that take place in our cells.

COOH   COOH   COOH   COOH

$$CH_2 \quad \xrightarrow{[O]} \quad C-H \quad \xrightarrow{H_2O} \quad HO-C-H \quad \xrightarrow{[O]} \quad C=O$$

$$CH_2 \qquad\qquad \| \qquad\qquad\qquad\qquad\quad | \qquad\qquad\qquad\quad |$$

$$CH_2 \qquad\qquad H-C \qquad\qquad\qquad CH_2 \qquad\qquad CH_2$$

$$COOH \qquad\quad COOH \qquad\qquad COOH \qquad\qquad COOH$$

Succinic acid   Fumaric acid   Malic acid   Oxaloacetic acid

At the pH of the aqueous environment in the cells, the carboxylic acids are ionized, which means it is actually the carboxylate ions that take part in the reactions of the citric acid cycle. For example, in water, succinic acid is in equilibrium with its carboxylate ion, succinate.

$$\begin{array}{ccc} COOH & & COO^- \\ | & & | \\ CH_2 & & CH_2 \\ | & +\ 2H_2O \rightleftharpoons & | & +\ 2H_3O^+ \\ CH_2 & & CH_2 \\ | & & | \\ COOH & & COO^- \end{array}$$

Succinic acid   Succinate ion

Citric acid gives the sour taste to citrus fruits.

Esters

In the functional group of an **ester**, the —H of the carboxylic acid is replaced by an alkyl group. Fats and oils in our diets contain esters of glycerol and fatty acids, which are long-chain carboxylic acids. The pleasant aromas and flavors of many fruits including bananas, oranges, and strawberries are due to esters.

Carboxylic Acid   **Ester**

$$\begin{array}{cc} \overset{\displaystyle O}{\overset{\|}{CH_3-C-OH}} & \overset{\displaystyle O}{\overset{\|}{CH_3-C-O-CH_3}} \end{array}$$

Ethanoic acid   Methyl ethanoate
(acetic acid)   (methyl acetate)

Esterification

In a reaction called *esterification*, an ester is produced when a carboxylic acid and an alcohol react in the presence of an acid catalyst (usually H_2SO_4) and heat. An excess of the alcohol reactant is used to shift the equilibrium in the direction of the formation of the ester product. In this esterification reaction, the —OH removed from the carboxylic acid and the —H removed from the alcohol combine to form water.

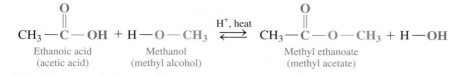

$$\overset{\displaystyle O}{\overset{\|}{CH_3-C-OH}} + H-O-CH_3 \; \overset{H^+,\ heat}{\rightleftharpoons} \; \overset{\displaystyle O}{\overset{\|}{CH_3-C-O-CH_3}} + H-OH$$

Ethanoic acid   Methanol   Methyl ethanoate
(acetic acid)   (methyl alcohol)   (methyl acetate)

CORE CHEMISTRY SKILL
Forming Esters

ENGAGE

Write the chemical equation for the esterification of 1-octanol and ethanoic acid to form the ester octyl ethanoate.

Esters such as methyl butanoate give the odor and flavor to many fruits.

SAMPLE PROBLEM 17.13 Writing Esterification Equations

An ester that has the smell of pineapples can be synthesized from butanoic acid and methanol. Write the balanced chemical equation for the formation of this ester.

TRY IT FIRST

SOLUTION

ANALYZE THE PROBLEM	Given	Need	Connect
	butanoic acid, methanol	esterification equation	ester and H_2O

$$CH_3-CH_2-CH_2-\overset{\overset{\displaystyle O}{\|}}{C}-OH \ + \ H-O-CH_3 \ \underset{}{\overset{H^+,\ heat}{\rightleftarrows}} \ CH_3-CH_2-CH_2-\overset{\overset{\displaystyle O}{\|}}{C}-O-CH_3 \ + \ \mathbf{H_2O}$$

Butanoic acid   Methanol   Methyl butanoate
(butyric acid)   (methyl alcohol)   (methyl butyrate)

STUDY CHECK 17.13

What are the IUPAC names of the carboxylic acid and alcohol that are needed to form the following ester, which has the odor of apples? (*Hint*: Separate the O and C=O of the ester group and add H— and —OH to give the original alcohol and carboxylic acid.)

$$CH_3-CH_2-\overset{\overset{\displaystyle O}{\|}}{C}-O-CH_2-CH_2-CH_2-CH_2-CH_3$$

ANSWER

propanoic acid and 1-pentanol

Naming Esters

The name of an ester consists of two words that are derived from the names of the alcohol and the acid in that ester. The first word indicates the *alkyl* part from the alcohol. The second word is the *carboxylate* part from the carboxylic acid. The IUPAC names of esters use the IUPAC names for the carbon chain of the acid, while the common names of esters use the common names of the acids. Let's take a look at the following ester, which has a fruity odor. We start by separating the ester bond into two parts, which gives us the alkyl of the alcohol and the carboxylate of the acid. Then we name the ester as an alkyl carboxylate (see **FIGURE 17.9**).

Methyl ethanoate
(methyl acetate)

FIGURE 17.9 ▶ The ester methyl ethanoate (methyl acetate) is made from methyl alcohol and ethanoic acid (acetic acid).

Ⓠ What change is made in the name of the carboxylic acid used to make the ester?

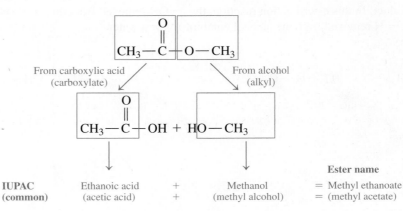

Esters in Plants

Many of the fragrances of perfumes and flowers and the flavors of fruits are due to esters. Small esters are volatile, so we can smell them, and soluble in water, so we can taste them. Several esters and their flavors and odors are listed in TABLE 17.8.

TABLE 17.8 Some Esters in Fruits and Flavorings

Condensed Structural Formula and Name	Flavor/Odor
$CH_3-\overset{\overset{\displaystyle O}{\|\|}}{C}-O-CH_2-CH_2-CH_3$ Propyl ethanoate (propyl acetate)	Pears
$CH_3-\overset{\overset{\displaystyle O}{\|\|}}{C}-O-CH_2-CH_2-CH_2-CH_2-CH_3$ Pentyl ethanoate (pentyl acetate)	Bananas
$CH_3-\overset{\overset{\displaystyle O}{\|\|}}{C}-O-CH_2-CH_2-CH_2-CH_2-CH_2-CH_2-CH_2-CH_3$ Oranges Octyl ethanoate (octyl acetate)	
$CH_3-CH_2-CH_2-\overset{\overset{\displaystyle O}{\|\|}}{C}-O-CH_2-CH_3$ Ethyl butanoate (ethyl butyrate)	Pineapples
$CH_3-CH_2-CH_2-\overset{\overset{\displaystyle O}{\|\|}}{C}-O-CH_2-CH_2-CH_2-CH_2-CH_3$ Pentyl butanoate (pentyl butyrate)	Apricots

SAMPLE PROBLEM 17.14 Naming Esters

Write the IUPAC and common names for the following:

$$CH_3-CH_2-\overset{\overset{\displaystyle O}{\|\|}}{C}-O-CH_2-CH_2-CH_3$$

TRY IT FIRST

SOLUTION

ANALYZE THE PROBLEM	Given	Need	Connect
	three-carbon carboxylate and propyl	IUPAC name	name alkyl, replace *ic acid* with *ate*

STEP 1 Write the name for the carbon chain from the alcohol as an *alkyl* group. The alcohol that is used for the ester is propanol, which is named as the alkyl group propyl.

$$CH_3-CH_2-\overset{\overset{\displaystyle O}{\|\|}}{C}-O-CH_2-CH_2-CH_3 \qquad \text{propyl}$$

> **Guide to Naming Esters**
>
> **STEP 1**
> Write the name for the carbon chain from the alcohol as an *alkyl* group.
>
> **STEP 2**
> Change the *ic acid* of the acid name to *ate*.

STEP **2** Change the *ic acid* of the acid name to *ate*. The carboxylic acid with three carbon atoms is propanoic acid. Replacing the *ic acid* with *ate* gives the IUPAC name propyl propanoate. The common name of propionic acid gives the common name for the ester of propyl propionate.

$$CH_3-CH_2-\overset{\overset{\displaystyle O}{\|}}{C}-O-CH_2-CH_2-CH_3$$

propyl propanoate

(propyl propionate)

STUDY CHECK 17.14

Write the IUPAC name for the following ester found in grapes:

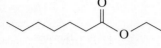

ANSWER

ethyl heptanoate

The odor of grapes is due to an ester.

QUESTIONS AND PROBLEMS

17.6 Carboxylic Acids and Esters

LEARNING GOAL Write the IUPAC and common names for carboxylic acids and esters; draw the condensed or line-angle structural formulas when given their names.

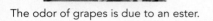

17.41 Write the IUPAC and common name (if any) for each of the following carboxylic acids:

a. $CH_3-\overset{\overset{\displaystyle O}{\|}}{C}-OH$

b. $CH_3-CH_2-\overset{\overset{\displaystyle O}{\|}}{C}-OH$

c. (line-angle structure with methyl branch and COOH)

d. (benzoic acid with Br substituent, C—OH group)

17.42 Write the IUPAC and common name (if any) for each of the following carboxylic acids:

a. $H-\overset{\overset{\displaystyle O}{\|}}{C}-OH$

b. (line-angle structure with Br substituent, OH)

c. (benzoic acid with Cl substituent, C—OH group)

d. $CH_3-CH_2-\overset{\overset{\displaystyle CH_3}{\underset{\displaystyle |}{}}}{C}H-\overset{\overset{\displaystyle O}{\|}}{C}-OH$

17.43 Draw the condensed structural formula for each of the following carboxylic acids:
a. butyric acid   b. benzoic acid
c. 2-chloroethanoic acid   d. 3-hydroxypropanoic acid

17.44 Draw the condensed structural formula for each of the following carboxylic acids:
a. 2-methylhexanoic acid   b. 3-ethylbenzoic acid
c. 2-hydroxyacetic acid   d. 2,4-dibromobutanoic acid

17.45 Draw the condensed structural formula for the ester formed when each of the following reacts with ethyl alcohol:
a. acetic acid   b. butyric acid

17.46 Draw the condensed structural formula for the ester formed when each of the following reacts with methyl alcohol:
a. formic acid   b. propionic acid

17.47 Draw the structural formula for the ester formed when the following carboxylic acids and alcohols react:

a. (line-angle acid C—OH) + HO—(propyl) $\overset{H^+, \text{ heat}}{\rightleftharpoons}$

b. $CH_3-CH_2-CH_2-CH_2-\overset{\overset{\displaystyle O}{\|}}{C}-OH$ +

$HO-\overset{\overset{\displaystyle CH_3}{\underset{\displaystyle |}{}}}{C}H-CH_3$ $\overset{H^+, \text{ heat}}{\rightleftharpoons}$

17.48 Draw the structural formula for the ester formed when the following carboxylic acids and alcohols react:

a. OH + HO $\xrightarrow{H^+, \text{heat}}$

b.
+ HO—CH$_2$—CH$_2$—CH$_2$—CH$_3$ $\xrightleftharpoons{H^+, \text{heat}}$

17.49 Write the IUPAC and common names, if any, for each of the following:

a. H—C(=O)—O—CH$_3$

b.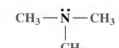

c. CH$_3$—CH$_2$—C(=O)—O—CH$_2$—CH$_3$

17.50 Write the IUPAC and common names, if any, for each of the following:

a.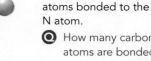

b. CH$_3$—CH$_2$—CH$_2$—CH$_2$—CH$_2$—C(=O)—O—CH$_3$

c. CH$_3$—CH$_2$—CH$_2$—C(=O)—O—CH$_3$

17.51 Draw the condensed structural formula for each of the following:
   a. methyl acetate   b. butyl formate
   c. ethyl pentanoate   d. propyl propanoate

17.52 Draw the condensed structural formula for each of the following:
   a. hexyl acetate   b. ethyl formate
   c. ethyl hexanoate   d. methyl benzoate

17.7 Amines and Amides

LEARNING GOAL Write the common names for amines and the IUPAC and common names for amides; draw the condensed or line-angle structural formulas when given their names.

In the functional group of a **amine**, a nitrogen atom is attached to one or more carbon atoms. In methylamine, a methyl group replaces one hydrogen atom in ammonia. The bonding of two methyl groups gives dimethylamine. In trimethylamine, methyl groups replace all three hydrogen atoms attached to the nitrogen atom (see **FIGURE 17.10**).

Naming Amines

Several systems are used for naming amines. For simple amines, the common names are often used. In the common name, the alkyl groups bonded to the nitrogen atom are listed in alphabetical order. The prefixes *di* and *tri* are used to indicate two and three identical substituents.

H—N̈—H   CH$_3$—N̈—H   CH$_3$—N̈—CH$_3$ CH$_3$—N̈—CH$_3$
   | | | |
   H   H   H   CH$_3$

Ammonia   Methylamine   Dimethylamine   Trimethylamine

FIGURE 17.10 ▶ Amines have one or more carbon atoms bonded to the N atom.

Ⓠ How many carbon atoms are bonded to the nitrogen atom in dimethylamine?

Aromatic Amines

The aromatic amines use the name *aniline*, which is approved by IUPAC.

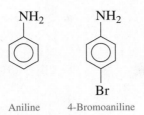

Aniline   4-Bromoaniline

Aniline is used to make many dyes, which give color to wool, cotton, and silk fibers, as well as blue jeans. It is also used to make the polymer polyurethane and in the synthesis of the pain reliever acetaminophen.

Indigo

Indigo used in blue dyes can be obtained from tropical plants such as *Indigofera tinctoria*.

SAMPLE PROBLEM 17.15 Naming Amines

Write the common name for each of the following amines:

a. $CH_3-CH_2-NH_2$ **b.**

TRY IT FIRST

SOLUTION

a. This amine has one ethyl group attached to the nitrogen atom; its name is ethylamine.
b. The common name for an amine with three methyl groups attached to the nitrogen atom is trimethylamine.

STUDY CHECK 17.15

What is the name of the following amine?

NH₂

ANSWER
aniline

CHEMISTRY LINK TO THE **ENVIRONMENT**
Alkaloids: Amines in Plants

Alkaloids are physiologically active nitrogen-containing compounds produced by plants. The term *alkaloid* refers to the "alkali-like" or basic characteristics of amines. Certain alkaloids are used in anesthetics, in antidepressants, and as stimulants, and many are habit forming.

As a stimulant, nicotine increases the level of adrenaline in the blood, which increases the heart rate and blood pressure. Nicotine is addictive because it activates pleasure centers in the brain. Coniine, which is obtained from hemlock, is extremely toxic.

Nicotine

Coniine

Caffeine is a central nervous system stimulant. Present in coffee, tea, soft drinks, energy drinks, chocolate, and cocoa, caffeine increases alertness, but it may cause nervousness and insomnia. Caffeine is also used in certain pain relievers to counteract the drowsiness caused by an antihistamine.

Caffeine

Caffeine is a stimulant found in coffee, tea, energy drinks, and chocolate.

Several alkaloids are used in medicine. Quinine, obtained from the bark of the cinchona tree, has been used in the treatment of malaria since the 1600s. Atropine from nightshade (belladonna) is used in low concentrations to accelerate slow heart rates and as an anesthetic for eye examinations.

Quinine

Atropine

For many centuries morphine and codeine, alkaloids found in the oriental poppy plant, have been used as effective painkillers. Codeine, which is structurally similar to morphine, is used in some prescription painkillers and cough syrups. Heroin, obtained by a chemical modification of morphine, is strongly addicting and is not used medically. The structure of the prescription drug OxyContin (oxycodone) used to relieve severe pain is similar to heroin. Today, there are an increasing number of deaths from OxyContin abuse because its physiological effects are also similar to those of heroin.

Morphine

Codeine

Heroin

OxyContin

The green, unripe poppy seed capsule contains a milky sap (opium) that is the source of the alkaloids morphine and codeine.

Amides

In the functional group of an **amide**, the hydroxyl group of a carboxylic acid is replaced by a nitrogen atom (see **FIGURE 17.11**).

Carboxylic Acid   **Amide**

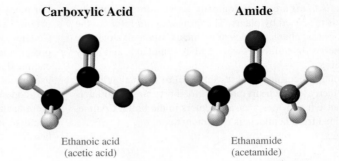

Ethanoic acid   Ethanamide
(acetic acid)   (acetamide)

FIGURE 17.11 ▶ Amides are derivatives of carboxylic acids in which a nitrogen atom replaces the hydroxyl group (—OH).

◎ What is the amide of pentanoic acid?

Amidation

CORE CHEMISTRY SKILL
Forming Amides

An amide is produced in a reaction called *amidation*, in which a carboxylic acid reacts with ammonia or an amine. A molecule of water is eliminated, and the fragments of the carboxylic acid and amine molecules join to form the amide, much like the formation of esters.

$$CH_3-CH_2-\overset{\overset{\displaystyle O}{\|}}{C}-OH + H-\overset{\overset{\displaystyle H}{|}}{N}-H \xrightarrow{\text{Heat}} CH_3-CH_2-\overset{\overset{\displaystyle O}{\|}}{C}-\overset{\overset{\displaystyle H}{|}}{N}-H + H_2O$$

Propanoic acid   Ammonia   Propanamide
(propionic acid)   (propionamide)

$$CH_3-CH_2-\overset{\overset{\displaystyle O}{\|}}{C}-OH + H-\overset{\overset{\displaystyle H}{|}}{N}-CH_3 \xrightarrow{\text{Heat}} CH_3-CH_2-\overset{\overset{\displaystyle O}{\|}}{C}-\overset{\overset{\displaystyle H}{|}}{N}-CH_3 + H_2O$$

Propanoic acid   Methylamine   *N*-Methylpropanamide
(propionic acid)   (*N*-methylpropionamide)

SAMPLE PROBLEM 17.16 Formation of Amides

Draw the condensed structural formula for the amide product in the following reaction:

$$CH_3-\overset{\overset{\displaystyle O}{\|}}{C}-OH + H_2N-CH_2-CH_3 \xrightarrow{\text{Heat}}$$

▶ **TRY IT FIRST**

SOLUTION

The condensed structural formula for the amide product can be drawn by attaching the carbonyl group from the acid to the nitrogen atom of the amine. The —OH group is removed from the acid and —H from the amine to form water.

$$CH_3-\overset{\overset{\displaystyle O}{\|}}{C}-\overset{\overset{\displaystyle H}{|}}{N}-CH_2-CH_3$$

STUDY CHECK 17.16

Draw the condensed structural formulas for the carboxylic acid and amine needed to prepare the following amide:

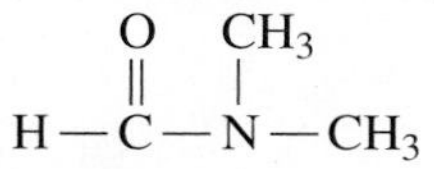

$$\underset{H}{\overset{\displaystyle O}{}} \; \overset{\displaystyle CH_3}{\underset{}{}}$$

H—C—N—CH₃

ANSWER

H—C—OH   and   H—N—CH₃

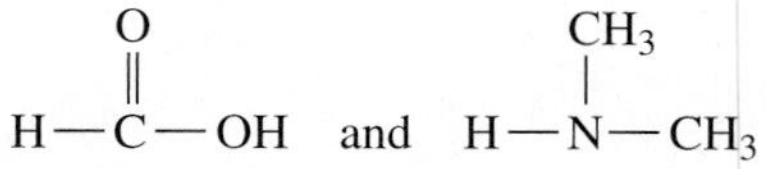

Naming Amides

In the IUPAC and common names for amides, the *oic acid* or *ic acid* from the carboxylic acid name is replaced with *amide*. We can diagram the name of an amide in the following way:

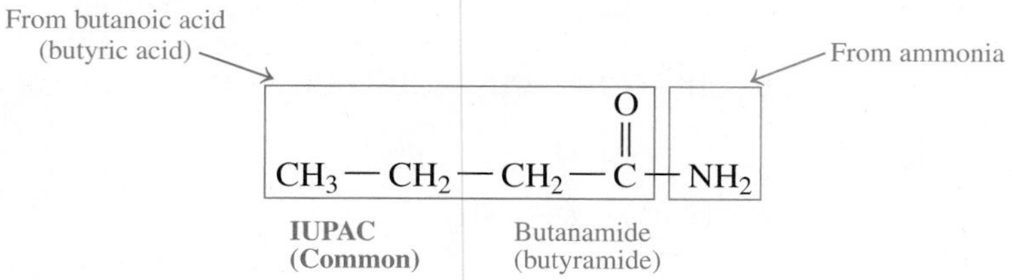

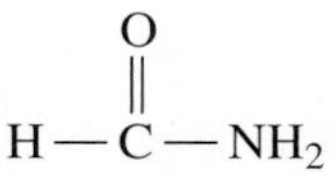

From butanoic acid
(butyric acid)   From ammonia

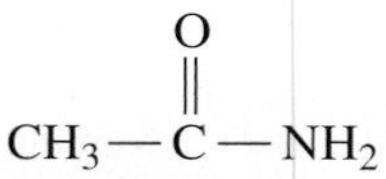

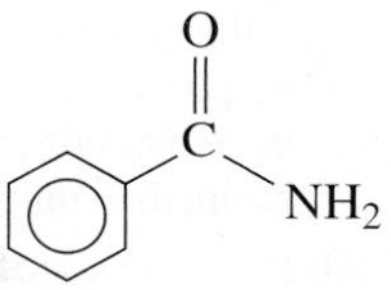

CH₃—CH₂—CH₂—C—NH₂

IUPAC   Butanamide
(Common)   (butyramide)

H—C—NH₂   CH₃—C—NH₂   C—NH₂

Methanamide   Ethanamide   Benzamide
(formamide)   (acetamide)

SAMPLE PROBLEM 17.17 Naming Amides

Write the IUPAC and common names for

$$CH_3—CH_2—\overset{\displaystyle O}{\overset{\|}{C}}—NH_2$$

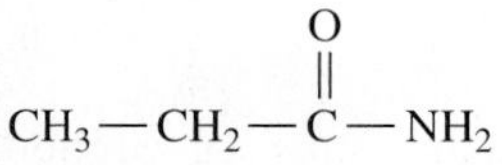

TRY IT FIRST

SOLUTION

The IUPAC name of the carboxylic acid is propanoic acid; the common name is propionic acid. Replacing the *oic acid* or *ic acid* ending with *amide* gives the IUPAC name of propanamide and common name of propionamide.

STUDY CHECK 17.17

Write the IUPAC name for

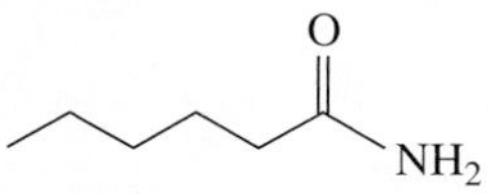

ANSWER

hexanamide

QUESTIONS AND PROBLEMS

17.7 Amines and Amides

LEARNING GOAL Write the common names for amines and the IUPAC and common names for amides; draw the condensed or line-angle structural formulas when given their names.

17.53 Write the common name for each of the following:

   a. CH_3—NH_2

   b. 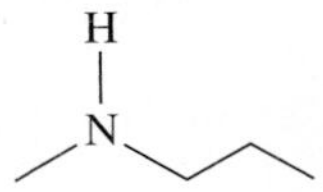 (line-angle structure with N—H)

   c. CH_3—CH_2—$\overset{\overset{\displaystyle CH_3}{|}}{N}$—$CH_2$—$CH_3$

17.54 Write the common name for each of the following:

   a. CH_3—CH_2—CH_2—NH_2

   b. CH_3—$\overset{\overset{\displaystyle H}{|}}{N}$—$CH_2$—$CH_3$

   c. 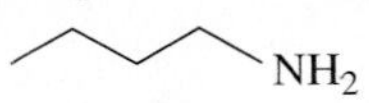 (line-angle structure with NH_2)

17.55 Draw the condensed structural formula for the amide formed in each of the following reactions:

   a. CH_3—$\overset{\overset{\displaystyle O}{||}}{C}$—$OH$ + NH_3 $\xrightarrow{\text{Heat}}$

   b. CH_3—$\overset{\overset{\displaystyle O}{||}}{C}$—$OH$ + H_2N—CH_2—CH_3 $\xrightarrow{\text{Heat}}$

   c. 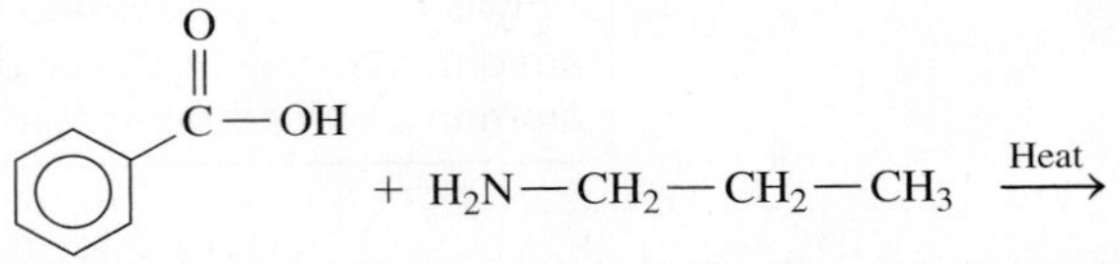 (benzene ring)$\overset{\overset{\displaystyle O}{||}}{C}$—$OH$ + H_2N—CH_2—CH_2—CH_3 $\xrightarrow{\text{Heat}}$

17.56 Draw the condensed structural formula for the amide formed in each of the following reactions:

   a. CH_3—CH_2—CH_2—CH_2—$\overset{\overset{\displaystyle O}{||}}{C}$—$OH$ + NH_3 $\xrightarrow{\text{Heat}}$

   b. CH_3—$\overset{\overset{\displaystyle CH_3}{|}}{CH}$—$\overset{\overset{\displaystyle O}{||}}{C}$—$OH$ + H_2N—CH_2—CH_3 $\xrightarrow{\text{Heat}}$

   c. 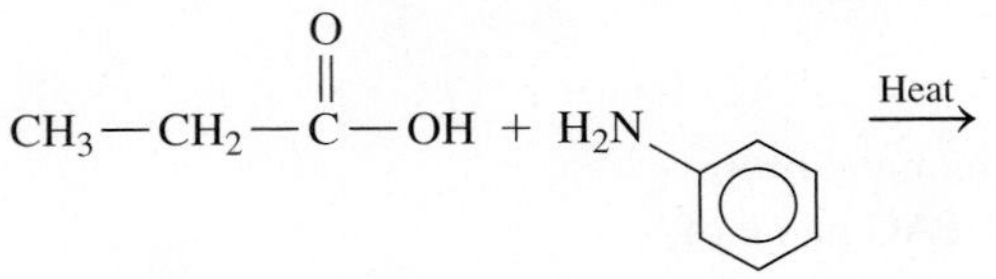 CH_3—CH_2—$\overset{\overset{\displaystyle O}{||}}{C}$—$OH$ + H_2N(benzene ring) $\xrightarrow{\text{Heat}}$

17.57 Write the IUPAC and common name (if any) for each of the following amides:

   a. CH_3—$\overset{\overset{\displaystyle O}{||}}{C}$—$NH_2$

   b. CH_3—CH_2—$\overset{\overset{\displaystyle Cl}{|}}{CH}$—$\overset{\overset{\displaystyle O}{||}}{C}$—$NH_2$

   c. H—$\overset{\overset{\displaystyle O}{||}}{C}$—$NH_2$

17.58 Write the IUPAC and common name (if any) for each of the following amides:

   a. CH_3—$\overset{\overset{\displaystyle Br}{|}}{CH}$—$\overset{\overset{\displaystyle O}{||}}{C}$—$NH_2$

   b. CH_3—CH_2—CH_2—CH_2—CH_2—$\overset{\overset{\displaystyle O}{||}}{C}$—$NH_2$

   c. 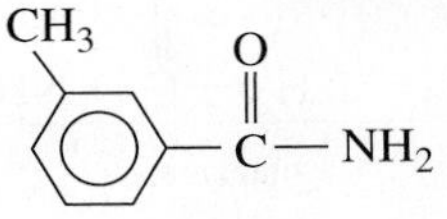 (benzene ring with CH_3)$\overset{\overset{\displaystyle O}{||}}{C}$—$NH_2$

17.59 Draw the condensed structural formula for each of the following amides:

   a. propionamide

   b. 2-methylpentanamide

   c. methanamide

17.60 Draw the condensed structural formula for each of the following amides:

   a. heptanamide

   b. benzamide

   c. 3-methylbutanamide

Follow Up

DIANE'S TREATMENT IN THE BURN UNIT

When Diane arrived at the hospital, she was transferred to the ICU burn unit. She had second-degree burns that caused damage to the underlying layers of skin and third-degree burns that caused damage to all the layers of the skin. Because body fluids are lost when deep burns occur, a lactated Ringer's solution was administered. The most common complications of burns are related to infection. To prevent infection, her skin was covered with a topical antibiotic. The next day Diane was placed in a tank to remove dressings, lotions, and damaged tissue. Dressings and ointments were changed every eight hours. Over a period of 3 months, grafts of Diane's unburned skin were used to cover burned areas. She remained in the burn unit for 3 months and then was discharged. However, Diane returned to the hospital for more skin grafts and plastic surgery.

The arson investigators determined that gasoline was the primary accelerant used to start the fire at Diane's house. Because there was a lot of paper and dry wood in the area, the fire spread quickly. Some of the hydrocarbons found in gasoline include hexane, heptane, octane, nonane, decane, and toluene.

Applications

17.61 The medications for Diane were contained in blister packs made of polychlorotrifluoroethylene (PCTFE). Draw the expanded structural formula for a portion of PCTFE polymer from three monomers of chlorotrifluoroethylene shown below:

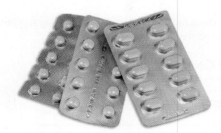

Chlorotrifluoroethylene

Tablets are contained in blister packs made from polychlorotrifluoroethylene (PCTFE).

17.62 New polymers have been synthesized to replace PVC in IV bags and cling film. One of these is EVA, ethylene polyvinylacetate. Draw the expanded structural formula for a portion of EVA using an alternating sequence that has two of each monomer shown below:

Ethylene

Vinyl acetate

17.63 Write the balanced chemical equation for the complete combustion of each of the following hydrocarbons found in gasoline used to start a fire:
a. nonane
b. toluene

17.64 Write the balanced chemical equation for the complete combustion of each of the following hydrocarbons found in gasoline used to start a fire:
a. hexane
b. octane

CONCEPT MAP

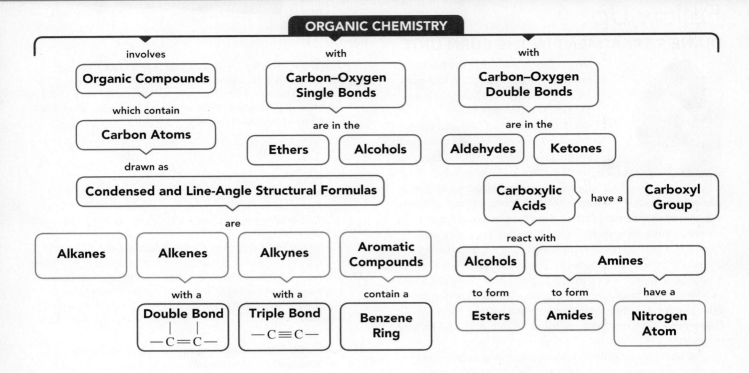

ORGANIC CHEMISTRY

involves → **Organic Compounds**
which contain → **Carbon Atoms**
drawn as → **Condensed and Line-Angle Structural Formulas**
are → **Alkanes** **Alkenes** **Alkynes** **Aromatic Compounds**
with a → **Double Bond** —C═C—
with a → **Triple Bond** —C≡C—
contain a → **Benzene Ring**

with → **Carbon–Oxygen Single Bonds**
are in the → **Ethers** **Alcohols**

with → **Carbon–Oxygen Double Bonds**
are in the → **Aldehydes** **Ketones**
Carboxylic Acids — have a → **Carboxyl Group**
react with → **Alcohols** **Amines**
to form → **Esters** to form → **Amides** have a → **Nitrogen Atom**

CHAPTER REVIEW

17.1 Alkanes

LEARNING GOAL Write the IUPAC names and draw the condensed or line-angle structural formulas for alkanes.

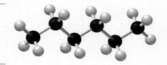

- Alkanes are hydrocarbons that have only C—C single bonds.
- In the expanded structural formula, a separate line is drawn for every bonded atom.
- A condensed structural formula depicts groups composed of each carbon atom and its attached hydrogen atoms.
- A line-angle structural formula represents the carbon skeleton as ends and corners of a zigzag line.
- Substituents such as alkyl groups and halogen atoms (named as fluoro, chloro, bromo, or iodo) can replace hydrogen atoms on the main chain.
- The IUPAC system is used to name organic compounds by indicating the number of carbon atoms and the position of any substituents.

17.2 Alkenes, Alkynes, and Polymers

LEARNING GOAL Write the IUPAC names and draw the condensed or line-angle structural formulas for alkenes and alkynes.

- Alkenes are unsaturated hydrocarbons that contain carbon–carbon double bonds (C═C).
- Alkynes contain a carbon–carbon triple bond (C≡C).
- The IUPAC names of alkenes end with *ene*; alkyne names end with *yne*.
- The main chain is numbered from the end nearer the double or triple bond.
- Alkenes react with hydrogen using a metal catalyst to form alkanes.
- Polymers are long-chain molecules that consist of many repeating units of smaller carbon molecules called monomers.

17.3 Aromatic Compounds

LEARNING GOAL Describe the bonding in benzene; name aromatic compounds, and draw their line-angle structural formulas.

- Most aromatic compounds contain benzene, C_6H_6, a cyclic structure containing six carbon atoms and six hydrogen atoms.
- The structure of benzene is represented as a hexagon with a circle in the center.
- The IUPAC system uses the names of benzene, toluene, aniline, and phenol.
- In the IUPAC name, two or more substituents are numbered and listed in alphabetical order.

17.4 Alcohols and Ethers

LEARNING GOAL Write the IUPAC and common names for alcohols and ethers; draw the condensed or line-angle structural formulas when given their names.

CH_3—O—H

Methanol

- The functional group of an alcohol is the hydroxyl group (—OH) bonded to a carbon chain.
- In a phenol, the hydroxyl group is bonded to an aromatic ring.
- In the IUPAC system, the names of alcohols have *ol* endings, and the location of the —OH group is given by numbering the carbon chain.
- In an ether, an oxygen atom (—O—) is connected by single bonds to two alkyl or aromatic groups.
- In the common names of ethers, the alkyl groups are listed alphabetically followed by the name *ether*.

17.5 Aldehydes and Ketones

LEARNING GOAL Write the IUPAC and common names for aldehydes and ketones; draw the condensed or line-angle structural formulas when given their names.

δ^-

δ^+

Carbonyl group

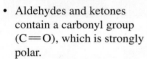
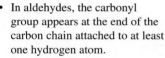

O‖C
CH_3 H
Aldehyde

O‖C
CH_3 CH_3
Ketone

- Aldehydes and ketones contain a carbonyl group (C=O), which is strongly polar.
- In aldehydes, the carbonyl group appears at the end of the carbon chain attached to at least one hydrogen atom.
- In ketones, the carbonyl group occurs between two carbon atoms.
- In the IUPAC system, the *e* in the alkane name is replaced with *al* for aldehydes and *one* for ketones.
- For ketones with more than four carbon atoms in the main chain, the carbonyl group is numbered to show its location.
- Many of the simple aldehydes and ketones use common names.

17.6 Carboxylic Acids and Esters

LEARNING GOAL Write the IUPAC and common names for carboxylic acids and esters; draw the condensed or line-angle structural formulas when given their names.

- A carboxylic acid contains the carboxyl functional group, which is a hydroxyl group connected to a carbonyl group.
- The IUPAC name of a carboxylic acid is obtained by replacing the *e* in the alkane name with *oic acid*.
- The common names of carboxylic acids with one to four carbon atoms are: formic acid, acetic acid, propionic acid, and butyric acid.
- In an ester, a carbon atom replaces the H of the hydroxyl group of a carboxylic acid.
- When heated in the presence of a strong acid, a carboxylic acid reacts with an alcohol to produce an ester. A molecule of water is removed: —OH from the carboxylic acid and —H from the alcohol molecule.
- The names of esters consist of two words, the alkyl group from the alcohol and the name of the carboxylate obtained by replacing the *ic acid* ending with *ate*.

17.7 Amines and Amides

LEARNING GOAL Write the common names for amines and the IUPAC and common names for amides; draw the condensed or line-angle structural formulas when given their names.

Dimethylamine

- A nitrogen atom attached to one or more carbon atoms forms an amine.
- In the common names of simple amines, the alkyl groups are listed alphabetically followed by the suffix *amine*.
- Amides are derivatives of carboxylic acids in which the hydroxyl group is replaced by a nitrogen atom. Amides are named by replacing the *ic acid* or *oic acid* ending with *amide*.
- Amides are formed by the reaction of a carboxylic acid with ammonia or an amine.

SUMMARY OF NAMING

Condensed Structural Formula	Family	IUPAC Name	Common Name
CH_3—CH_2—CH_3	Alkane	Propane	
CH_3—CH—CH_3 with CH_3 above	Alkane with a substituent	Methylpropane	
CH_3—CH=CH_2	Alkene	Propene	Propylene
CH_3—C≡CH	Alkyne	Propyne	
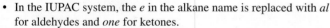	Aromatic	Benzene	

Condensed Structural Formula	Family	IUPAC Name	Common Name
$CH_3—OH$	Alcohol	Methanol	Methyl alcohol
$CH_3—O—CH_3$	Ether		Dimethyl ether
$\underset{}{H—\overset{\displaystyle O}{\overset{\|}{C}}—H}$	Aldehyde	Methanal	Formaldehyde
$CH_3—\overset{\displaystyle O}{\overset{\|}{C}}—CH_3$	Ketone	Propanone	Acetone; dimethyl ketone
$CH_3—\overset{\displaystyle O}{\overset{\|}{C}}—OH$	Carboxylic acid	Ethanoic acid	Acetic acid
$CH_3—\overset{\displaystyle O}{\overset{\|}{C}}—O—CH_3$	Ester	Methyl ethanoate	Methyl acetate
$CH_3—CH_2—NH_2$	Amine		Ethylamine
$CH_3—\overset{\displaystyle O}{\overset{\|}{C}}—NH_2$	Amide	Ethanamide	Acetamide

SUMMARY OF REACTIONS

The chapter sections to review are shown after the name of the reaction.

Combustion (17.1)

$$CH_3—CH_2—CH_3(g) + 5O_2(g) \xrightarrow{\Delta} 3CO_2(g) + 4H_2O(g) + \text{energy}$$
Propane

Hydrogenation (17.2)

$$H_2C{=}CH—CH_3 + H_2 \xrightarrow{Pt} CH_3—CH_2—CH_3$$
Propene — Propane

Esterification (17.2)

$$CH_3—\overset{\displaystyle O}{\overset{\|}{C}}—OH + HO—CH_3 \underset{}{\overset{H^+,\ heat}{\rightleftharpoons}} CH_3—\overset{\displaystyle O}{\overset{\|}{C}}—O—CH_3 + H_2O$$

Ethanoic acid   Methanol   Methyl ethanoate
(acetic acid)   (methyl alcohol)   (methyl acetate)

Amidation (17.7)

$$CH_3—CH_2—\overset{\displaystyle O}{\overset{\|}{C}}—OH + H—\overset{\displaystyle H}{\overset{\|}{N}}—H \xrightarrow{Heat} CH_3—CH_2—\overset{\displaystyle O}{\overset{\|}{C}}—\overset{\displaystyle H}{\overset{\|}{N}}—H + H_2O$$

Propanoic acid   Ammonia   Propanamide
(propionic acid)   (propionamide)

KEY TERMS

alcohol An organic compound that contains a hydroxyl group ($—OH$) attached to a carbon chain.

aldehyde An organic compound that contains a carbonyl group ($C{=}O$) bonded to at least one hydrogen atom.

alkane A type of hydrocarbon in which the carbon atoms are connected only by single bonds.

alkene A type of hydrocarbon that contains a carbon–carbon double bond ($C{=}C$).

alkyl group An alkane minus one hydrogen atom. Alkyl groups are named like the corresponding alkanes except a *yl* ending replaces *ane*.

alkyne A type of hydrocarbon that contains a carbon–carbon triple bond (C≡C).

amide An organic compound in which the hydroxyl group of a carboxylic acid is replaced by a nitrogen atom.

amine An organic compound that contains a nitrogen atom bonded to one or more carbon atoms.

aromatic compound A compound that contains the ring structure of benzene.

benzene A ring of six carbon atoms each of which is attached to one hydrogen atom, C_6H_6.

carboxylic acid An organic compound that contains the carboxyl group (—COOH).

condensed structural formula A formula that shows the carbon atoms grouped with the attached number of hydrogen atoms.

ester An organic compound in which the —H of a carboxyl group is replaced by a carbon atom.

ether An organic compound in which an oxygen atom is bonded to two carbon atoms.

expanded structural formula A formula that shows all of the atoms and the bonds connected to each atom.

functional group A group of atoms that determines the physical and chemical properties and naming of a class of organic compounds.

hydrocarbon A type of organic compound that contains only carbon and hydrogen.

IUPAC (International Union of Pure and Applied Chemistry) **system** A naming system used for organic compounds.

ketone An organic compound in which a carbonyl group (C=O) is bonded to two carbon atoms.

line-angle structural formula A formula which shows a zigzag line in which carbon atoms are represented as the ends of each line and as corners.

monomer The small organic molecule that is repeated many times in a polymer.

polymer A large molecule that is composed of many small, repeating structural units that are identical.

structural isomers Two compounds which have the same molecular formula but different arrangements of atoms.

substituent A group of atoms such as an alkyl group or a halogen bonded to the main chain of carbon atoms.

CORE CHEMISTRY SKILLS

The chapter section containing each Core Chemistry Skill is shown in parentheses at the end of each heading.

Naming and Drawing Alkanes (17.1)

- The alkanes ethane, propane, and butane contain two, three, and four carbon atoms, respectively, connected in a row or a *continuous* chain.
- Alkanes with five or more carbon atoms in a chain are named using the prefixes *pent* (5), *hex* (6), *hept* (7), *oct* (8), *non* (9), and *dec* (10).
- In the condensed structural formula, the carbon and hydrogen atoms on the ends are written as —CH_3 and the carbon and hydrogen atoms in the middle are written as —CH_2—.

Example: **a.** What is the name of CH_3—CH_2—CH_2—CH_3?
b. Draw the condensed structural formula for pentane.

Answer: **a.** An alkane with a four-carbon chain is named with the prefix *but* followed by *ane*, which is butane.
b. Pentane is an alkane with a five-carbon chain. The carbon atoms on the ends are attached to three H atoms each, and the carbon atoms in the middle are attached to two H each. CH_3—CH_2—CH_2—CH_2—CH_3

Writing Equations for Hydrogenation and Polymerization (17.2)

- The addition of small molecules to the double bond is a characteristic reaction of alkenes.
- Hydrogenation adds hydrogen atoms to the double bond of an alkene, using a metal catalyst, to form an alkane.
- In polymerization, many small molecules join together to form a polymer.

Example: **a.** Draw the condensed structural formula for 2-methyl-2-butene.
b. Draw the condensed structural formula for the product of the hydrogenation of 2-methyl-2-butene.

Answer: **a.**
$$CH_3-\underset{\underset{CH_3}{|}}{C}=CH-CH_3$$

b. When H_2 adds to a double bond, the product is an alkane with the same number of carbon atoms.
$$CH_3-\underset{\underset{CH_3}{|}}{CH}-CH_2-CH_3$$

Naming Aldehydes and Ketones (17.5)

- In the IUPAC system, an aldehyde is named by replacing the *e* in the alkane name with *al* and a ketone by replacing the *e* with one.
- The position of a substituent on an aldehyde is indicated by numbering the carbon chain from the carbonyl group and given in front of the name.
- For a ketone, the carbon chain is numbered from the end nearer the carbonyl group.

Example: Give the IUPAC name for the following:

Answer: 5-methyl-3-hexanone

Naming Carboxylic Acids (17.6)

- The IUPAC name of a carboxylic acid is obtained by replacing the *e* in the corresponding alkane name with *oic acid*.
- The common names of carboxylic acids with one to four carbon atoms are formic acid, acetic acid, propionic acid, and butyric acid.

Example: Write the IUPAC name for the following:

$$CH_3-\underset{\underset{Br}{|}}{CH}-CH_2-\overset{\overset{O}{||}}{C}-OH$$

Answer: 3-bromobutanoic acid

Forming Esters (17.6)

- Esters are formed when carboxylic acids react with alcohols in the presence of a strong acid and heat.

Example: Draw the condensed structural formula for the ester product of the reaction of propanoic acid and methanol.

Answer:
$$CH_3-CH_2-\overset{\overset{\displaystyle O}{\|}}{C}-O-CH_3$$

Forming Amides (17.7)

- Amides are formed when carboxylic acids react with ammonia or primary or secondary amines in the presence of heat.

Example: Draw the amide produced by the reaction of propanoic acid and ethylamine.

Answer:
$$CH_3-CH_2-\overset{\overset{\displaystyle O}{\|}}{C}-\overset{\overset{\displaystyle H}{|}}{N}-CH_2-CH_3$$

UNDERSTANDING THE CONCEPTS

The chapter sections to review are shown in parentheses at the end of each question.

17.65 Draw a portion of the polymer Teflon, which is made from 1,1,2,2-tetrafluoroethene (use three monomers). (17.2)

Teflon is used as a nonstick coating on cooking pans.

17.66 A garden hose is made of polyvinyl chloride (PVC), a polymer of chloroethene (vinyl chloride). Draw a portion of the polymer (use three monomers) for PVC. (17.2)

A plastic garden hose is made of PVC.

17.67 Citronellal, found in oil of citronella, lemon, and lemon grass, is used in perfumes and as an insect repellent. (17.2, 17.5)

An insect-repelling candle contains citronellal, which is found in natural sources.

Citronellal has the following structure:

$$CH_3-\overset{\overset{\displaystyle CH_3}{|}}{C}=CH-CH_2-CH_2-\overset{\overset{\displaystyle CH_3}{|}}{CH}-CH_2-\overset{\overset{\displaystyle O}{\|}}{C}-H$$

a. Complete the IUPAC name for citronellal:
___, ___-di _____-___-octenal.
b. What does the *en* in octenal signify?
c. What does the *al* in octenal signify?

17.68 Draw the condensed structural formula for each of the following: (17.5)
a. 2-heptanone, an alarm pheromone of bees
b. 2,6-dimethyl-3-heptanone, a communication pheromone of bees

Bees emit chemicals called pheromones to communicate.

ADDITIONAL QUESTIONS AND PROBLEMS

17.69 Draw the condensed structural formula for each of the following: (17.1, 17.2)
  a. 3-ethylhexane
  b. 2-pentene
  c. 2-hexyne

17.70 Draw the condensed structural formula for each of the following: (17.1, 17.2)
  a. 1-chloro-2-butyne
  b. 2,3-dimethylpentane
  c. 3-hexene

17.71 Write the IUPAC name for each of the following: (17.1, 17.2)

  a. CH₃—CH₂—C(CH₃)(CH₃)—CH₃

  b. CH₃—CH₂—C≡CH
  c. CH₃—CH=CH—CH₂—CH₃

17.72 Write the IUPAC name for each of the following: (17.1, 17.2)

  a. H₂C=C(CH₃)—CH₂—CH₂—CH₃
  b. CH₃—CH₂—Cl
  c. CH₃—CH₂—CH(CH₃CH₂)—CH₂—CH(Br)—CH₃

17.73 Classify each of the following according to its functional group: (17.2)
  a. CH₃—CH₂—CH₂—C(=O)—CH₃
  b. CH₃—CH=CH₂
  c. CH₃—C(=O)—O—CH₃
  d. CH₃—CH₂—CH₂—NH₂

17.74 Classify each of the following according to its functional group: (17.2)
  a. [structure with O and NH₂]
  b. CH₃—CH₂—CH₂—C(=O)—OH
  c. CH₃—CH(CH₃)—CH₂—OH
  d. CH₃—CH(CH₃)—C(=O)—H

17.75 Select the class of organic compound that matches each of the following definitions (**a** to **d**): alkane, alkene, alkyne, alcohol, ether, aldehyde, ketone, carboxylic acid, ester, amine, or amide. (17.2)
  a. contains a hydroxyl group
  b. contains one or more carbon–carbon double bonds
  c. contains a carbonyl group bonded to a hydrogen
  d. contains only carbon–carbon single bonds

17.76 Select the class of organic compound that matches each of the following definitions (**a** to **d**): alkane, alkene, alkyne, alcohol, ether, aldehyde, ketone, carboxylic acid, ester, amine, or amide. (17.2)
  a. contains a carboxyl group in which the hydrogen atom is replaced by a carbon atom
  b. contains a carbonyl group bonded to a hydroxyl group
  c. contains a nitrogen atom bonded to one or more carbon atoms
  d. contains a carbonyl group bonded to two carbon atoms

17.77 Classify each of the following according to its functional group(s): (17.2)

  a.

Almonds

  b.

Cinnamon sticks

17.78 Classify each of the following according to its functional group(s): (17.2)
  a. BHA is an antioxidant used as a preservative in foods such as baked goods, butter, meats, and snack foods.

BHA

Baked goods contain BHA as a preservative.

  b. CH₃—C(=O)—C(=O)—CH₃

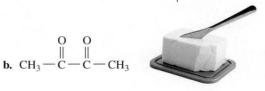

Butter

17.79 Name each of the following aromatic compounds: (17.3)

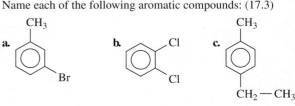

17.80 Name each of the following aromatic compounds: (17.3)

a. (benzene ring with NH₂ top, Br bottom)

b. (benzene ring with CH₃ and Cl)

c. (benzene ring with F, F)

17.81 Draw the structural formula for each of the following: (17.3)
  a. ethylbenzene
  b. 4-chlorotoluene
  c. 1, 4-dibromobenzene

17.82 Draw the structural formula for each of the following: (17.3)
  a. 3-fluorotoluene
  b. 2-methylaniline
  c. 1,4-dimethylbenzene

17.83 Write the IUPAC name for **a** and **b** and the common name for **c**. (17.4)

a. $CH_3 - CH_2 - \overset{\overset{\displaystyle CH_3}{|}}{CH} - OH$

b. (structure with OH)

c. $CH_3 - CH_2 - O - CH_2 - CH_2 - CH_2 - CH_3$

17.84 Write the IUPAC name for **a** and **b** and the common name for **c**. (17.4)

a. (structure with Cl and OH)

b. $CH_3 - \overset{\overset{\displaystyle CH_3}{|}}{CH} - CH_2 - CH_2 - OH$

c. (ether structure)

17.85 Draw the condensed structural formula for each of the following: (17.4)
  a. 4-chlorophenol
  b. 2-methyl-3-pentanol
  c. 2-methyl-2-propanol

17.86 Draw the condensed structural formula for each of the following: (17.4)
  a. 3-hexanol
  b. 2-pentanol
  c. ethyl propyl ether

17.87 Write the IUPAC name for each of the following: (17.5)

a. (benzene ring with C(=O)H and Cl)

b. $Cl - CH_2 - CH_2 - \overset{\overset{\displaystyle O}{\|}}{C} - H$

c. $CH_3 - \overset{\overset{\displaystyle Cl}{|}}{CH} - \overset{\overset{\displaystyle O}{\|}}{C} - CH_2 - CH_3$

17.88 Write the IUPAC name for each of the following: (17.5)

a. $CH_3 - CH_2 - \overset{\overset{\displaystyle O}{\|}}{C} - CH_3$

b. (benzene ring with C(=O)H and Cl)

c. $CH_3 - \overset{\overset{\displaystyle CH_3}{|}}{CH} - \overset{\overset{\displaystyle OH}{|}}{CH} - CH_2 - \overset{\overset{\displaystyle O}{\|}}{C} - H$

17.89 Draw the condensed structural formula for each of the following: (17.5)
  a. 2-bromobenzaldehyde
  b. 3-chloropropionaldehyde
  c. ethyl methyl ketone

17.90 Draw the condensed structural formula for each of the following: (17.5)
  a. butyraldehyde
  b. 2-bromobutanal
  c. 3,5-dimethylhexanal

17.91 Write the IUPAC name for each of the following: (17.6)

a. $CH_3 - \overset{\overset{\displaystyle CH_3}{|}}{CH} - CH_2 - \overset{\overset{\displaystyle O}{\|}}{C} - OH$

b. (benzene ring with C(=O)-O-CH₃)

c. $CH_3 - CH_2 - \overset{\overset{\displaystyle O}{\|}}{C} - O - CH_2 - CH_3$

17.92 Write the IUPAC name for each of the following: (17.6)

a. $CH_3 - \overset{\overset{\displaystyle CH_3}{|}}{CH} - CH_2 - CH_2 - \overset{\overset{\displaystyle O}{\|}}{C} - OH$

b. (benzene ring with C(=O)-OH and Cl)

c. (benzene ring with C(=O)-O-CH₂-CH₃)

17.93 Draw the condensed structural formula for each of the following: (17.7)
   a. ethylamine
   b. hexanamide
   c. triethylamine

17.94 Draw the condensed structural formula for each of the following: (17.7)
   a. formamide
   b. ethylpropylamine
   c. diethylmethylamine

17.95 Write the name for each of the following: (17.7)

   a. $CH_3-CH_2-CH_2-CH_2-N(CH_3)-$ (structure with N)

   b. $CH_3-CH_2-CH_2-CH_2-\overset{\overset{\textstyle O}{\|}}{C}-NH_2$

   c. $CH_3-CH_2-CH_2-CH_2-CH_2-NH_2$

17.96 Write the name for each of the following: (17.7)

   a. $Cl-CH_2-CH_2-CH_2-\overset{\overset{\textstyle O}{\|}}{C}-NH_2$

   b. $CH_3-CH_2-\overset{\overset{\textstyle O}{\|}}{C}-NH_2$

   c. $CH_3-CH_2-CH_2-\overset{\overset{\textstyle CH_2-CH_3}{|}}{N}-CH_2-CH_3$

CHALLENGE QUESTIONS

The following groups of questions are related to the topics in this chapter. However, they do not all follow the chapter order, and they require you to combine concepts and skills from several sections. These questions will help you increase your critical thinking skills and prepare for your next exam.

17.97 Draw the condensed structural formulas and write the IUPAC names for all eight alcohols that have the molecular formula $C_5H_{12}O$. (17.4)

17.98 There are four amine isomers with the molecular formula C_3H_9N. Draw their condensed structural formulas. (17.7)

17.99 The insect repellent DEET can be made from an amidation reaction of 3-methylbenzoic acid and diethylamine. Draw the condensed structural formula for DEET. (17.7)

Many insect repellents contain DEET, which is an amide.

17.100 One of the compounds that give blackberries their odor and flavor can be made by heating hexanoic acid and 1-propanol with an acid catalyst. Draw the condensed structural formula for this compound. (17.6)

17.101 Complete and balance each of the following reactions: (17.1, 17.2)
   a. $CH_3-CH_2-C\equiv CH + O_2 \xrightarrow{\Delta}$
   b. $CH_3-CH_2-CH=CH-CH_2-CH_3 + H_2 \xrightarrow{Pt}$

17.102 Complete and balance each of the following reactions: (17.1, 17.2)
   a. $H_2C=CH_2 + H_2 \xrightarrow{Pt}$
   b. $CH_3-CH_2-CH_2-CH_2-CH=CH_2 + O_2 \xrightarrow{\Delta}$

ANSWERS

Answers to Selected Questions and Problems

17.1 **a.** hexane   **b.** heptane   **c.** pentane

17.3 **a.** same molecule
   b. isomers of C_6H_{14}
   c. isomers of C_8H_{18}

17.5 **a.** 2-fluorobutane
   b. dimethylpropane
   c. 2-chloro-3-methylpentane

17.7 **a.** $CH_3-\overset{\overset{\textstyle CH_3}{|}}{CH}-CH_2-CH_3$

   b. $CH_3-CH_2-\overset{\overset{\textstyle Cl}{|}}{\underset{\underset{\textstyle Cl}{|}}{C}}-CH_2-CH_3$

c. [structure: branched chain with Cl]

d. [structure: branched chain]

17.9 a. $CH_4(g) + 2O_2(g) \xrightarrow{\Delta} CO_2(g) + 2H_2O(g) + energy$

b. $2C_6H_{14}(l) + 19O_2(g) \xrightarrow{\Delta} 12CO_2(g) + 14H_2O(g) + energy$

17.11 a. An alkene has a double bond.
b. An alkyne has a triple bond.
c. An alkene has a double bond.

17.13 a. ethene
b. 2-methylpropene
c. 2-pentyne

17.15 a. $CH_3-CH=CH_2$
b. $HC\equiv C-CH_2-CH_2-CH_2-CH_3$
c. $H_2C=C(CH_3)-CH_2-CH_3$

17.17 a. $CH_3-CH_2-CH_2-CH_2-CH_3$
b. $CH_3-CH_2-CH_2-CH_3$
c. [structure: branched chain]

17.19 A polymer is a large molecule composed of small units that are repeated many times.

17.21 $3\ H_2C=CH(CH_3) \longrightarrow$ [polymer structure chain]

17.23 [polymer structure chain with F and H]

17.25 a. 2-chlorotoluene
b. ethylbenzene
c. phenol

17.27 a. [benzene ring with NH$_2$] **b.** [benzene ring with Cl and F] **c.** [benzene ring with CH$_3$ and CH$_2$—CH$_3$]

17.29 a. ethanol   **b.** 2-butanol   **c.** 2-chlorophenol

17.31 a. $CH_3-CH_2-CH_2-OH$
b. CH_3-OH
c. $CH_3-CH_2-CH(OH)-CH_2-CH_3$
d. $CH_3-C(OH)(CH_3)-CH_2-CH_3$

17.33 a. ethyl methyl ether   **b.** dipropyl ether
c. butyl propyl ether

17.35 a. acetaldehyde   **b.** methyl propyl ketone
c. formaldehyde

17.37 a. propanal   **b.** 2-methyl-3-pentanone
c. benzaldehyde

17.39 a. $CH_3-\overset{\overset{\displaystyle O}{\|}}{C}-H$

b. $CH_3-\overset{\overset{\displaystyle O}{\|}}{C}-CH_2-CH_2-CH_3$

c. $CH_3-\overset{\overset{\displaystyle O}{\|}}{C}-CH_2-CH_2-CH_2-CH_3$

17.41 a. ethanoic acid (acetic acid)
b. propanoic acid (propionic acid)
c. 3-methylhexanoic acid
d. 3-bromobenzoic acid

17.43 a. $CH_3-CH_2-CH_2-\overset{\overset{\displaystyle O}{\|}}{C}-OH$

b. [benzene ring]$-\overset{\overset{\displaystyle O}{\|}}{C}-OH$

c. $Cl-CH_2-\overset{\overset{\displaystyle O}{\|}}{C}-OH$

d. $HO-CH_2-CH_2-\overset{\overset{\displaystyle O}{\|}}{C}-OH$

17.45 a. $CH_3-\overset{\overset{\displaystyle O}{\|}}{C}-O-CH_2-CH_3$

b. $CH_3-CH_2-CH_2-\overset{\overset{\displaystyle O}{\|}}{C}-O-CH_2-CH_3$

17.47 a. [ester structure]

b. $CH_3-CH_2-CH_2-CH_2-\overset{\overset{\displaystyle O}{\|}}{C}-O-\overset{\overset{\displaystyle CH_3}{|}}{CH}-CH_3$

17.49 a. methyl methanoate (methyl formate)
b. ethyl ethanoate (ethyl acetate)
c. ethyl propanoate (ethyl propionate)

17.51 a. $CH_3-\overset{\overset{\displaystyle O}{\|}}{C}-O-CH_3$

b. $H-\overset{\overset{\displaystyle O}{\|}}{C}-O-CH_2-CH_2-CH_2-CH_3$

c. $CH_3-CH_2-CH_2-CH_2-\overset{\overset{\displaystyle O}{\|}}{C}-O-CH_2-CH_3$

d. $CH_3-CH_2-\overset{\overset{\displaystyle O}{\|}}{C}-O-CH_2-CH_2-CH_3$

17.53 a. methylamine   **b.** methylpropylamine
    c. diethylmethylamine

17.55 a.

$$CH_3 - \overset{\overset{\textstyle O}{\|}}{C} - NH_2$$

b.

$$CH_3 - \overset{\overset{\textstyle O}{\|}}{C} - \overset{\overset{\textstyle H}{|}}{N} - CH_2 - CH_3$$

c.

 $- \overset{\overset{\textstyle O}{\|}}{C} - \overset{\overset{\textstyle H}{|}}{N} - CH_2 - CH_2 - CH_3$

17.57 a. ethanamide (acetamide)   **b.** 2-chlorobutanamide
    c. methanamide (formamide)

17.59 a.

$$CH_3 - CH_2 - \overset{\overset{\textstyle O}{\|}}{C} - NH_2$$

b.

$$CH_3 - CH_2 - CH_2 - \overset{\overset{\textstyle CH_3}{|}}{CH} - \overset{\overset{\textstyle O}{\|}}{C} - NH_2$$

c.

$$H - \overset{\overset{\textstyle O}{\|}}{C} - NH_2$$

17.61

$$-\overset{\overset{\textstyle Cl}{|}}{\underset{\underset{\textstyle F}{|}}{C}} - \overset{\overset{\textstyle F}{|}}{\underset{\underset{\textstyle F}{|}}{C}} - \overset{\overset{\textstyle Cl}{|}}{\underset{\underset{\textstyle F}{|}}{C}} - \overset{\overset{\textstyle F}{|}}{\underset{\underset{\textstyle F}{|}}{C}} - \overset{\overset{\textstyle Cl}{|}}{\underset{\underset{\textstyle F}{|}}{C}} - \overset{\overset{\textstyle F}{|}}{\underset{\underset{\textstyle F}{|}}{C}} -$$

17.63

a. $C_9H_{20}(l) + 14O_2(g) \xrightarrow{\Delta} 9CO_2(g) + 10H_2O(g) + energy$

b. $C_7H_8(l) + 9O_2(g) \xrightarrow{\Delta} 7CO_2(g) + 4H_2O(g) + energy$

17.65

$$-\overset{\overset{\textstyle F}{|}}{\underset{\underset{\textstyle F}{|}}{C}} - \overset{\overset{\textstyle F}{|}}{\underset{\underset{\textstyle F}{|}}{C}} - \overset{\overset{\textstyle F}{|}}{\underset{\underset{\textstyle F}{|}}{C}} - \overset{\overset{\textstyle F}{|}}{\underset{\underset{\textstyle F}{|}}{C}} - \overset{\overset{\textstyle F}{|}}{\underset{\underset{\textstyle F}{|}}{C}} - \overset{\overset{\textstyle F}{|}}{\underset{\underset{\textstyle F}{|}}{C}} -$$

17.67 a. 3,7-dimethyl-6-octenal
    b. The *en* signifies that a double bond is present.
    c. The *al* signifies that an aldehyde is present.

17.69 a.

$$CH_3 - CH_2 - \overset{\overset{\textstyle CH_2 - CH_3}{|}}{CH} - CH_2 - CH_2 - CH_3$$

b. $CH_3 - CH = CH - CH_2 - CH_3$

c. $CH_3 - C \equiv C - CH_2 - CH_2 - CH_3$

17.71 a. 2,2-dimethylbutane   **b.** 1-butyne
    c. 2-pentene

17.73 a. ketone   **b.** alkene
    c. ester   **d.** amine

17.75 a. alcohol   **b.** alkene
    c. aldehyde   **d.** alkane

17.77 a. aromatic, aldehyde
    b. aromatic, aldehyde, alkene

17.79 a. 3-bromotoluene
    b. 1,2-dichlorobenzene
    c. 4-ethyltoluene

17.81 a.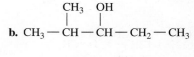

17.83 a. 2-butanol
    b. 3-methyl-2-pentanol
    c. butyl ethyl ether

17.85 a.

OH

Cl

b.

$$CH_3 - \overset{\overset{\textstyle CH_3}{|}}{CH} - \overset{\overset{\textstyle OH}{|}}{CH} - CH_2 - CH_3$$

c.

$$CH_3 - \overset{\overset{\textstyle OH}{|}}{\underset{\underset{\textstyle CH_3}{|}}{C}} - CH_3$$

17.87 a. 4-chlorobenzaldehyde
    b. 3-chloropropanal
    c. 2-chloro-3-pentanone

17.89 a.

$\overset{\overset{\textstyle O}{\|}}{C} - H$, Br

b. $Cl - CH_2 - CH_2 - \overset{\overset{\textstyle O}{\|}}{C} - H$

c. $CH_3 - CH_2 - \overset{\overset{\textstyle O}{\|}}{C} - CH_3$

17.91 a. 3-methylbutanoic acid
    b. methyl benzoate
    c. ethyl propanoate

17.93 a. $CH_3 - CH_2 - NH_2$

b. $CH_3 - CH_2 - CH_2 - CH_2 - CH_2 - \overset{\overset{\textstyle O}{\|}}{C} - NH_2$

c. $CH_3 - CH_2 - \overset{\overset{\textstyle CH_2 - CH_3}{|}}{N} - CH_2 - CH_3$

17.95 a. butyldimethylamine   **b.** pentanamide
    c. pentylamine

17.97

$CH_3-CH_2-CH_2-CH_2-CH_2-OH$ 1-Pentanol

$CH_3-\overset{\underset{|}{OH}}{CH}-CH_2-CH_2-CH_3$ 2-Pentanol

$CH_3-CH_2-\overset{\underset{|}{OH}}{CH}-CH_2-CH_3$ 3-Pentanol

$HO-CH_2-\overset{\underset{|}{CH_3}}{CH}-CH_2-CH_3$ 2-Methyl-1-butanol

$HO-CH_2-CH_2-\overset{\underset{|}{CH_3}}{CH}-CH_3$ 3-Methyl-1-butanol

$CH_3-\overset{\overset{CH_3}{|}}{\underset{\underset{OH}{|}}{C}}-CH_2-CH_3$ 2-Methyl-2-butanol

$CH_3-\overset{\underset{|}{OH}}{CH}-\overset{\overset{CH_3}{|}}{CH}-CH_3$ 3-Methyl-2-butanol

$CH_3-\overset{\overset{CH_3}{|}}{\underset{\underset{CH_3}{|}}{C}}-CH_2-OH$ 2,2-Dimethyl-1-propanol

17.99

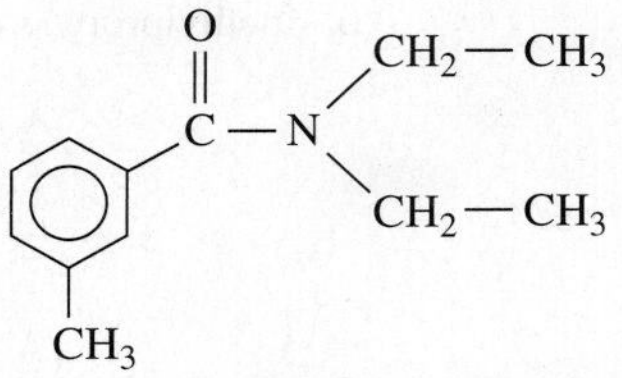

(structure: 3-methylbenzene ring bearing C(=O)–N(CH₂–CH₃)(CH₂–CH₃); a CH₃ substituent on the ring)

17.101 a. $2CH_3-CH_2-C\equiv CH + 11O_2 \xrightarrow{\Delta}$

$8CO_2 + 6H_2O + energy$

b. $CH_3-CH_2-CH=CH-CH_2-CH_3 + H_2 \xrightarrow{Pt}$

$CH_3-CH_2-CH_2-CH_2-CH_2-CH_3$

Biochemistry

REBECCA LEARNED OF her family's hypercholesterolemia after her mother died of a heart attack at 44. Her mother had a total cholesterol level of 600 mg/dL, which is three times the normal level. Rebecca learned that the lumps under her mother's skin and around the cornea of her eyes, called *xanthomas*, were caused by cholesterol that was stored throughout her body. In familial hypercholesterolemia (FH), genetic mutations prevent the removal of cholesterol from the bloodstream. As a result, low-density lipoprotein (LDL) accumulates in the blood and a person with FH can develop cardiovascular disease early in life. At the lipid clinic, Rebecca's blood tests showed that she had a total cholesterol level of 420 mg/dL and an LDL-cholesterol level of 275 mg/dL. Rebecca was diagnosed with FH.

She learned that from birth, cholesterol has been accumulating throughout her arteries.

After her diagnosis, Rebecca met with Susan, a clinical lipid specialist at the lipid clinic where she informed Rebecca about managing risk factors that can lead to heart attack and stroke. Susan told Rebecca about medications she would prescribe, and then followed the results.

CAREER

Clinical Lipid Specialist

Clinical lipid specialists work with patients who have lipid disorders such as high cholesterol, high triglycerides, coronary heart disease, obesity, and FH. At a lipid clinic, a clinical lipid specialist reviews a patient's lipid profile, which includes total cholesterol, high-density lipoprotein (HDL), and low-density lipoprotein (LDL). If a lipid disorder is identified, the lipid specialist assesses the patient's current diet and exercise program. The lipid specialist diagnoses and determines treatment including dietary changes such as reducing salt intake, increasing the amount of fiber in the diet, and lowering the amount of fat. He or she also discusses drug therapy using lipid-lowering medications that remove LDL-cholesterol to help patients achieve and maintain good health.

Allied health professionals such as nurses, nurse practitioners, pharmacists, and dietitians are certified by a clinical lipid specialist program for specialized care of patients with lipid disorders.

- Writing Equations for Hydrogenation and Polymerization (17.2)
- Naming Carboxylic Acids (17.6)

- Forming Esters (17.6)
- Forming Amides (17.7)

*These Core Chemistry Skills from previous chapters are listed here for your review as you proceed to the new material in this chapter.

LOOKING AHEAD

Carbohydrates contained in foods such as pasta and bread provide energy for the body.

Energy

Photosynthesis

$CO_2 + H_2O$

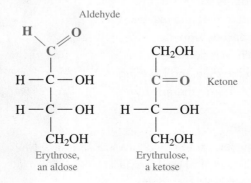

Energy

Carbohydrates + O_2

Respiration

FIGURE 18.1 ▶ During photosynthesis, energy from the Sun combines CO_2 and H_2O to form glucose, $C_6H_{12}O_6$, and O_2. During respiration in the body, carbohydrates are oxidized to CO_2 and H_2O, while energy is produced.

Q What are the reactants and products of respiration?

18.1 Carbohydrates

LEARNING GOAL Classify a carbohydrate as an aldose or a ketose; draw the open-chain and Haworth structures for glucose, galactose, and fructose.

Each day, you may enjoy the polysaccharides called starches in bread and pasta. The table sugar used to sweeten cereal, tea, or coffee is sucrose, a disaccharide that consists of two simple sugars, glucose, and fructose. **Carbohydrates** such as table sugar, lactose in milk, and cellulose are all made of carbon, hydrogen, and oxygen. Simple sugars, which have formulas of $C_n(H_2O)_n$, were once thought to be hydrates of carbon, thus the name *carbohydrate*. In a series of reactions called *photosynthesis*, energy from the Sun is used to combine the carbon atoms from carbon dioxide (CO_2) and the hydrogen and oxygen atoms of water into the carbohydrate glucose.

$$6CO_2 + 6H_2O + \text{energy} \underset{\text{Respiration}}{\overset{\text{Photosynthesis}}{\rightleftharpoons}} \underset{\text{Glucose}}{C_6H_{12}O_6} + 6O_2$$

In the body, glucose is oxidized in a series of metabolic reactions known as *respiration*, which releases chemical energy to do work in the cells. Carbon dioxide and water are produced and returned to the atmosphere. The combination of photosynthesis and respiration is the *carbon cycle*, in which energy from the Sun is stored in plants by photosynthesis and made available to us when the carbohydrates in our diets are metabolized (see **FIGURE 18.1**).

Monosaccharides

The simplest carbohydrates are the **monosaccharides**. A monosaccharide cannot be split into smaller carbohydrates. A monosaccharide has a chain of three to seven carbon atoms, one in a carbonyl group and all the others are attached to hydroxyl groups (—OH). In an **aldose**, the carbonyl group is on the first carbon (—CHO); a **ketose** contains the carbonyl group on the second carbon atom as a ketone (C=O).

Aldehyde

$$\begin{array}{c} \text{H} \diagdown \text{O} \\ \text{C} \\ | \\ \text{H—C—OH} \\ | \\ \text{H—C—OH} \\ | \\ \text{CH}_2\text{OH} \end{array}$$

Erythrose, an aldose

$$\begin{array}{c} \text{CH}_2\text{OH} \\ | \\ \text{C=O} \quad \text{Ketone} \\ | \\ \text{H—C—OH} \\ | \\ \text{CH}_2\text{OH} \end{array}$$

Erythrulose, a ketose

Monosaccharides are also classified by the number of carbon atoms. A monosaccharide with three carbon atoms is a *triose*, one with four carbon atoms is a *tetrose*; a *pentose* has five carbons, and a *hexose* contains six carbons. We can use both classification systems to indicate the aldehyde or ketone group and the number of carbon atoms. An aldopentose is a five-carbon monosaccharide that is an aldehyde; a ketohexose is a six-carbon monosaccharide that is a ketone.

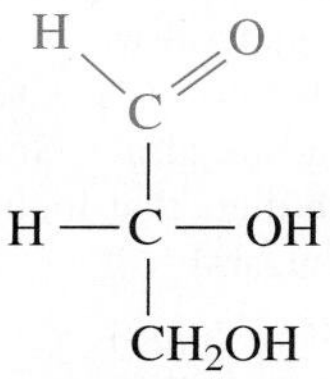

ENGAGE

Why is ribulose classified as a ketopentose and glucose as an aldohexose?

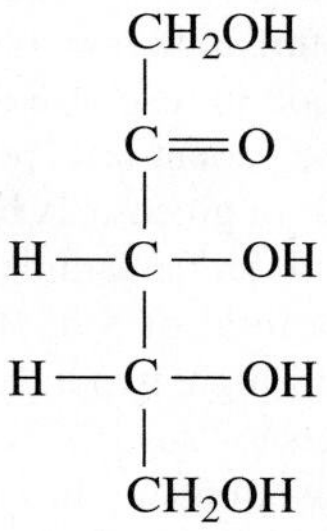

Glyceraldehyde
(aldotriose)

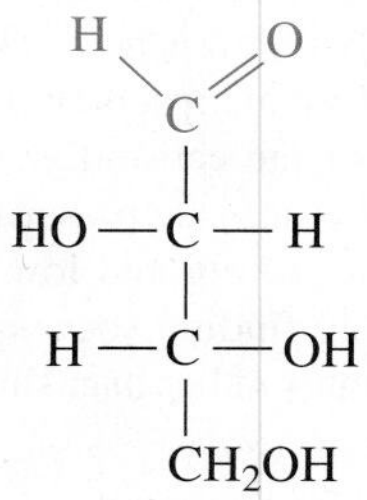

Threose
(aldotetrose)

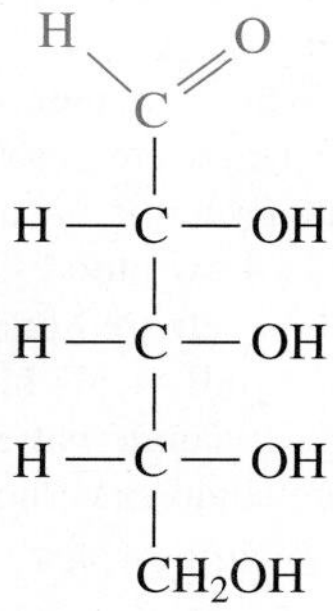

Ribose
(aldopentose)

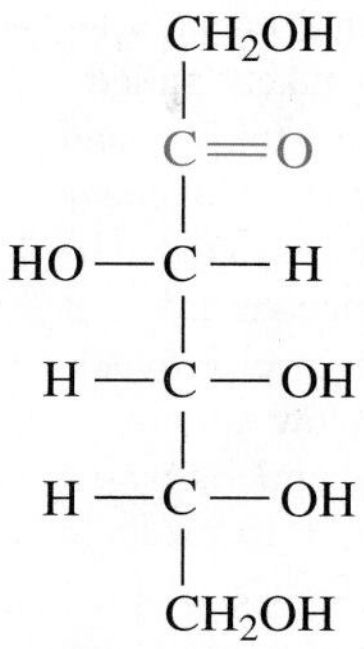

Fructose
(ketohexose)

Ribulose

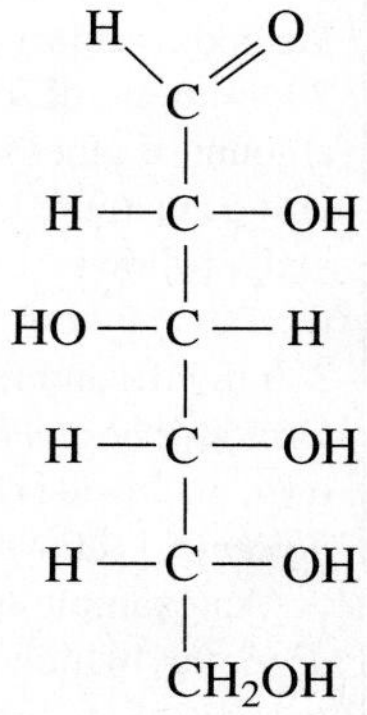

Glucose

Open-Chain Structures of Some Important Monosaccharides

Glucose, galactose, and fructose are the most important monosaccharides. They are all hexoses with the molecular formula $C_6H_{12}O_6$. In nature and the cells of the body, their most common form is the D isomer, which has the —OH group attached to carbon 5 on the right side of the chain.

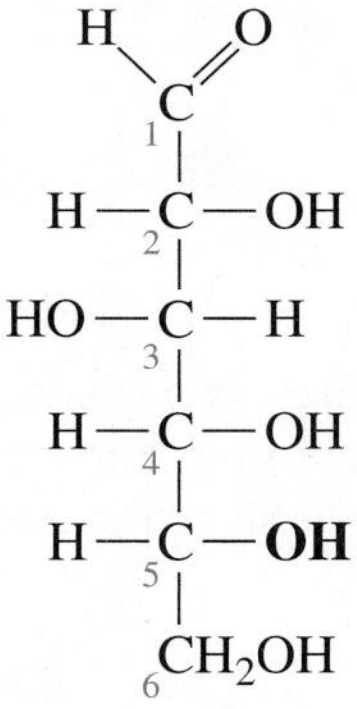

D-Glucose

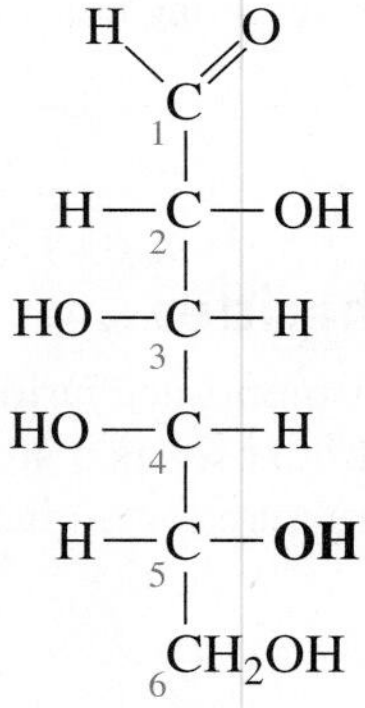

D-Galactose

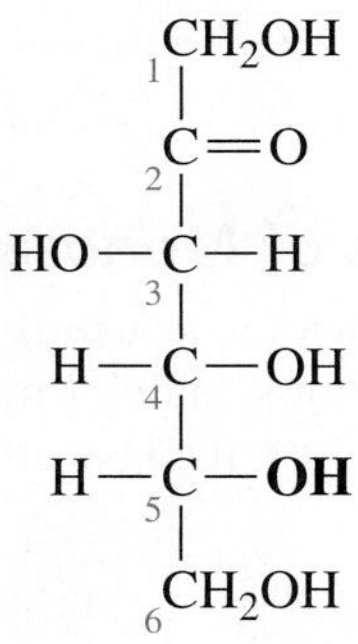

D-Fructose

The most common hexose, **D-glucose**, $C_6H_{12}O_6$, an aldohexose also known as dextrose and blood sugar, is found in fruits, vegetables, corn syrup, and honey. It is a building block of the disaccharides sucrose, lactose, and maltose, and polysaccharides such as starch, cellulose, and glycogen.

D-Galactose is an aldohexose that does not occur in the free form in nature. It is obtained from the disaccharide lactose, a sugar found in milk and milk products. D-Galactose is important in the cellular membranes of the brain and nervous system. The only difference in the structures of D-glucose and D-galactose is the arrangement of the —OH group on carbon 4.

In contrast to glucose and galactose, **D-fructose** is a ketohexose. The structure of D-fructose differs from D-glucose at carbons 1 and 2 by the location of the carbonyl group. D-Fructose is the sweetest of the carbohydrates, twice as sweet as sucrose (table sugar). This makes D-fructose popular with dieters because less D-fructose and, therefore, fewer calories are needed to provide a pleasant taste. D-Fructose is found in fruit juices and honey.

ENGAGE

What are the differences in the open-chain structural formulas of D-galactose and D-fructose?

CHEMISTRY LINK TO HEALTH
Hyperglycemia and Hypoglycemia

A doctor may order a glucose tolerance test to evaluate the body's ability to return to normal blood glucose concentrations of 70 to 90 mg/dL of blood in response to the ingestion of a specified amount of glucose. The patient fasts for 12 h and then drinks a solution containing 75 g/dL of glucose. A blood sample is taken immediately, followed by more blood samples each half-hour for 2 h, and then every hour for a total of 5 h. If the blood glucose exceeds 200 mg/dL and remains high, hyperglycemia may be indicated. The term *glyc* or *gluco* refers to "sugar." The prefix *hyper* means above or over, and *hypo* is below or under. Thus the blood sugar level in *hyperglycemia* is above normal and in *hypoglycemia* it is below normal.

An example of a disease that can cause hyperglycemia is *type 2 diabetes*, which occurs when the pancreas is unable to produce sufficient quantities of insulin. As a result, glucose levels in the body fluids can rise as high as 350 mg/dL. Symptoms of diabetes include thirst, excessive urination, and increased appetite. In older people, type 2 diabetes is sometimes a consequence of excessive weight gain.

When a person is hypoglycemic, the blood glucose level rises and then decreases rapidly to levels below 40 mg/dL. In some cases, hypoglycemia is caused by overproduction of insulin by the pancreas. Low blood glucose can cause dizziness, general weakness, and muscle tremors. A diet may be prescribed that consists of several small meals high in protein and low in carbohydrate. Some hypoglycemic patients are finding success with diets that include more complex carbohydrates rather than simple sugars.

A glucose solution is given to determine blood glucose levels.

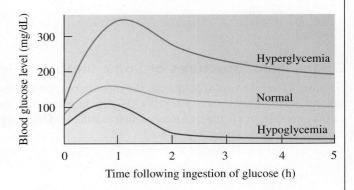

Haworth Structures of Monosaccharides

CORE CHEMISTRY SKILL
Drawing Haworth Structures

Up until now, we have drawn the structures for monosaccharides such as D-glucose as open chains. However, the most stable form of hexoses is a six-atom ring, known as a *Haworth structure*. We can draw the Haworth structure for D-glucose as shown in Sample Problem 18.1.

Guide to Drawing Haworth Structures

STEP 1
Turn the open-chain structure clockwise by 90°.

STEP 2
Fold the horizontal carbon chain into a hexagon, rotate the groups on carbon 5, and bond the O on carbon 5 to carbon 1.

STEP 3
Draw the new —OH group on carbon 1 below the ring to give the α form or above the ring to give the β form.

SAMPLE PROBLEM 18.1 Drawing the Haworth Structure for D-Glucose

D-Glucose has the following open-chain structure. Draw the Haworth structure for D-glucose.

D-Glucose

TRY IT FIRST

SOLUTION

STEP **1** Turn the open-chain structure clockwise by 90°. The —H and —OH groups on the right of the vertical carbon chain are now below the horizontal carbon chain. Those on the left of the open chain are now above the horizontal carbon chain.

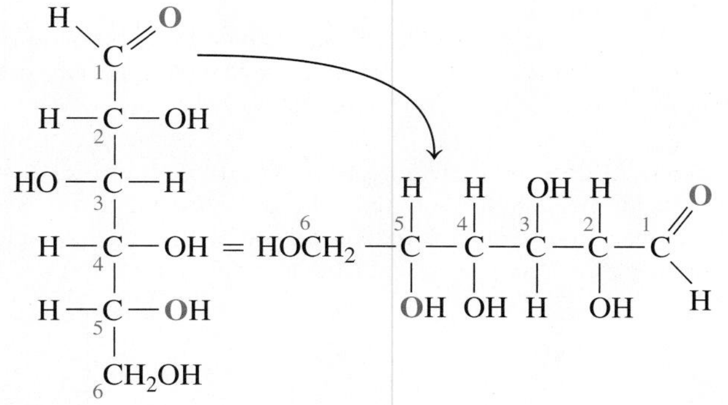

D-Glucose (open chain)

STEP **2** Fold the horizontal carbon chain into a hexagon, rotate the groups on carbon 5, and bond the O on carbon 5 to carbon 1. With carbons 2 and 3 as the base of a hexagon, move the remaining carbons upward. Rotate the groups on carbon 5 so that the —OH group is close to carbon 1. To complete the Haworth structure, draw a bond between the oxygen of the —OH group on carbon 5 to carbon 1. By convention, the carbon atoms in the ring are drawn as corners.

ENGAGE

How does a new hydroxyl group form on carbon 1 when the oxygen atom on carbon 5 reacts with carbon 1?

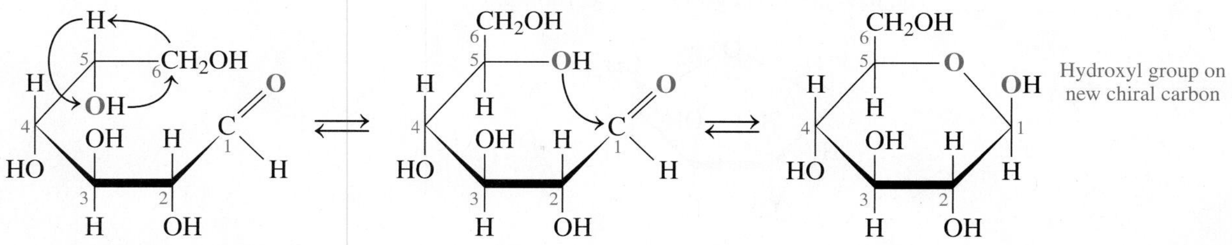

Rotation of groups on carbon 5   Carbon-5 oxygen bonds to carbon 1   Cyclic structure

STEP **3** Draw the new —OH group on carbon 1 below the ring to give the α form or above the ring to give the β form. Because the new —OH group can form above or below the plane of the Haworth structure, there are two forms of D-glucose, which differ only by the position of the —OH group at carbon 1.

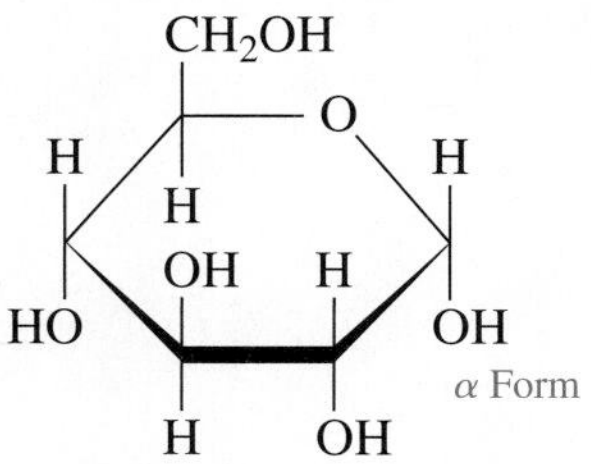

α-D-Glucose 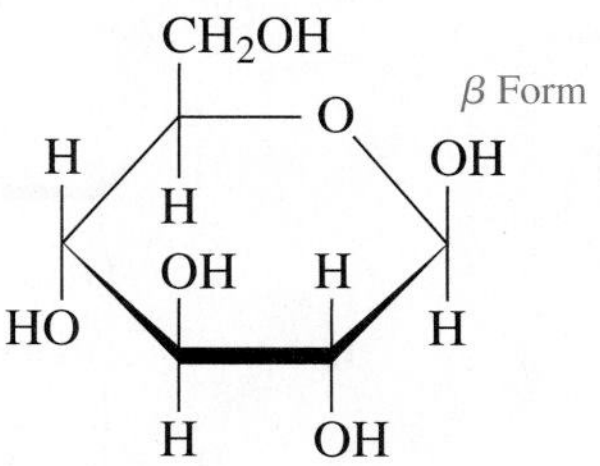   β-D-Glucose

STUDY CHECK 18.1

D-Galactose is an aldohexose like D-glucose, differing only in the arrangement of the —OH group on carbon 4. Draw the Haworth structure for α-D-galactose.

D-Galactose

ANSWER

In contrast to D-glucose and D-galactose, D-fructose is a ketohexose. It forms the five-atom ring when the hydroxyl group on carbon 5 reacts with the carbon of the ketone group. The new hydroxyl group is on carbon 2.

α-D-Fructose

QUESTIONS AND PROBLEMS

18.1 Carbohydrates

LEARNING GOAL Classify a carbohydrate as an aldose or a ketose; draw the open-chain and Haworth structures for glucose, galactose, and fructose.

18.1 What functional groups are found in all monosaccharides?

18.2 What is the difference between an aldose and a ketose?

18.3 What are the functional groups and number of carbons in a ketopentose?

18.4 What are the functional groups and number of carbons in an aldohexose?

18.5 What are the kind and number of atoms in the ring portion of the Haworth structure of D-glucose?

18.6 What are the kind and number of atoms in the ring portion of the Haworth structure of D-fructose?

18.7 Identify each of the following Haworth structures as the α or β form:

a.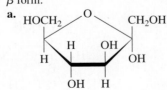

b.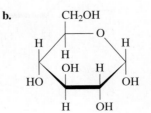

18.8 Identify each of the following Haworth structures as the α or β form:

a.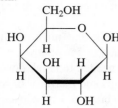

b.

Applications

18.9 Classify each of the following monosaccharides as an aldopentose, aldohexose, ketopentose, or ketohexose:

a. Psicose is present in low levels in foods.

b. Lyxose is a component of bacterial lipids.

Psicose

Lyxose

18.10 Classify each of the following monosaccharides as an aldopentose, aldohexose, ketopentose, or ketohexose:

a. A solution of xylose is given to test absorption by the intestines.

b. Tagatose, found in fruit, is similar in sweetness to sugar.

Xylose

Tagatose

18.11 An infant with galactosemia can utilize D-glucose in milk but not D-galactose. How does the open-chain structure of D-galactose differ from that of D-glucose?

18.12 D-Fructose is the sweetest monosaccharide. How does the open-chain structure of D-fructose differ from that of D-glucose?

18.2 Disaccharides and Polysaccharides

LEARNING GOAL Describe the monosaccharide units and linkages in disaccharides and polysaccharides.

A **disaccharide** is composed of two monosaccharides linked together. The most common disaccharides are maltose, lactose, and sucrose.

Maltose, or malt sugar, is a dissacharide obtained from starch and is found in germinating grains. Maltose is used in cereals, candies, and the brewing of beverages. Maltose has a *glycosidic bond* between two glucose molecules. To form maltose, the —OH group

α Form

α-D-Glucose

α-D-Glucose

+ **H₂O**

α Form

α(1→4)-Glycosidic bond

α-Maltose, a disaccharide

on carbon 1 in the first glucose forms a bond with the —OH group on carbon 4 in a second glucose molecule, which is designated as an $\alpha(1{\rightarrow}4)$ linkage. Because the second glucose molecule in maltose has a free —OH on carbon 1, there are α and β forms of maltose.

Lactose, milk sugar, is a disaccharide found in milk and milk products (see **FIGURE 18.2**). It makes up 6 to 8% of human milk and about 4 to 5% of cow's milk. Some people do not produce sufficient quantities of the enzyme needed to break down lactose, and the sugar remains undigested, causing abdominal cramps and diarrhea. In some commercial milk products, an enzyme called lactase is added to break down lactose. The bond in lactose is a $\beta(1{\rightarrow}4)$-glycosidic bond because the β form of galactose links to the hydroxyl group on carbon 4 of glucose.

ENGAGE

How does an $\alpha(1{\rightarrow}4)$-glycosidic bond differ from a $\beta(1{\rightarrow}4)$-glycosidic bond?

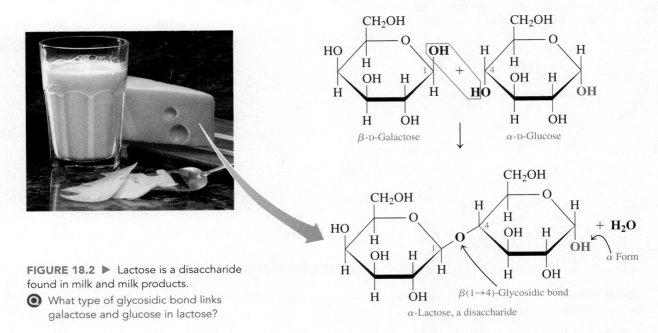

FIGURE 18.2 ▶ Lactose is a disaccharide found in milk and milk products.

◉ What type of glycosidic bond links galactose and glucose in lactose?

Sucrose, ordinary table sugar, is the most abundant disaccharide in the world. Most of the sucrose for table sugar comes from sugar cane (20% by mass) or sugar beets (15% by mass) (see **FIGURE 18.3**). Both the raw and refined forms of sugar are sucrose. Some estimates indicate that each person in the United States consumes an average of 68 kg (150 lb) of sucrose every year either by itself or in a variety of food products. Sucrose consists of an α-D-glucose and a β-D-fructose molecule joined by a bond called an $\alpha,\beta(1{\rightarrow}2)$-glycosidic bond.

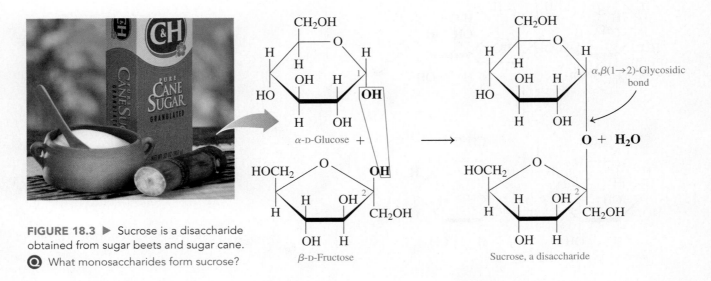

FIGURE 18.3 ▶ Sucrose is a disaccharide obtained from sugar beets and sugar cane.

◉ What monosaccharides form sucrose?

SAMPLE PROBLEM 18.2 Glycosidic Bonds in Disaccharides

Melibiose is a disaccharide that is 30 times sweeter than sucrose.

Melibiose

a. What are the monosaccharide units in melibiose?
b. What type of glycosidic bond links the monosaccharides?
c. Identify the structure as α- or β-melibiose.

> TRY IT FIRST

SOLUTION

a. First monosaccharide (left)	When the —OH group on carbon 4 is above the plane, it is D-galactose. When the —OH group on carbon 1 is below the plane, it is α-D-galactose.
Second monosaccharide (right)	When the —OH group on carbon 4 is below the plane, it is D-glucose.
b. Type of glycosidic bond	The —OH group at carbon 1 of α-D-galactose bonds with the —OH group on carbon 6 of glucose, which makes it an $\alpha(1\rightarrow6)$-glycosidic bond.
c. Name of disaccharide	The —OH group on carbon 1 of glucose is below the plane, which is α-melibiose.

STUDY CHECK 18.2

Cellobiose is a disaccharide composed of two D-glucose molecules connected by a $\beta(1\rightarrow4)$-glycosidic bond. Draw the Haworth structure for β-cellobiose.

ANSWER

CHEMISTRY LINK TO **HEALTH**
How Sweet is My Sweetener?

Although many of the monosaccharides and disaccharides taste sweet, they differ considerably in their degree of sweetness. Dietetic foods contain sweeteners that are noncarbohydrate or carbohydrates that are sweeter than sucrose. Some examples of sweeteners compared with sucrose are shown in **TABLE 18.1**.

Sucralose, which is known as Splenda, is made from sucrose by replacing some of the hydroxyl groups with chlorine atoms.

Sucralose (Splenda)

Aspartame, which is marketed as NutraSweet and Equal, is used in a large number of sugar-free products. It is a noncarbohydrate sweetener made of aspartate and a methyl ester of phenylalanine. It does have some caloric value, but it is so sweet that only a very small quantity is needed. However, phenylalanine, one of the breakdown products, poses a danger to anyone who cannot metabolize it properly, a condition called *phenylketonuria* (PKU).

From aspartate   From phenylalanine
Aspartame (NutraSweet)

Another artificial sweetener, Neotame, is a modification of the aspartame structure. The addition of a large alkyl group to the amine group prevents enzymes from breaking the amide bond between aspartate and phenylalanine. Thus, phenylalanine is not produced when Neotame is used as a sweetener. Very small amounts of Neotame are needed because it is about 10 000 times sweeter than sucrose.

Large alkyl group
to modify Aspartame   Neotame

Saccharin, which is marketed as Sweet'N Low, has been used as a noncarbohydrate artificial sweetener for the past 35 years.

Saccharin (Sweet'N Low)

Artificial sweeteners are used as sugar substitutes.

Stevia is a sugar substitute obtained from the leaves of a plant *Stevia rebaudiana*. It is composed of steviol glycosides that are about 150 times sweeter than sucrose. Stevia has been used to sweeten tea and medicines in South America for 1500 years.

Steviol

TABLE 18.1 Relative Sweetness of Sugars and Artificial Sweeteners	
	Sweetness Relative to Sucrose ($=100$)
Monosaccharides	
Galactose	30
Glucose	75
Fructose	175
Disaccharides	
Lactose	16
Maltose	33
Sucrose	100
Sugar Alcohols	
Sorbitol	60
Maltitol	80
Xylitol	100
Artificial Sweeteners (Noncarbohydrate)	
Aspartame	18 000
Saccharin	45 000
Sucralose	60 000
Neotame	1 000 000

Polysaccharides

A **polysaccharide** is a polymer of many monosaccharides joined together. Four biologically important polysaccharides—amylose, amylopectin, glycogen, and cellulose—are all polymers of D-glucose that differ only in the type of glycosidic bonds and the amount of branching in the molecule.

Starch, a storage form of glucose in plants, is found as insoluble granules in rice, wheat, potatoes, beans, and cereals. Starch is composed of two kinds of polysaccharides, *amylose* and *amylopectin*. **Amylose**, which makes up about 20% of starch, consists of α-D-glucose molecules connected by $\alpha(1\rightarrow4)$-glycosidic bonds in a continuous chain. A typical polymer of amylose may contain from 250 to 4000 α-D-glucose molecules. Sometimes called a straight-chain polymer, polymers of amylose are actually coiled in helical fashion.

Amylopectin, which makes up as much as 80% of plant starch, is a branched-chain polysaccharide. Like amylose, the glucose molecules are connected by $\alpha(1\rightarrow4)$-glycosidic bonds. However, at about every 25 glucose units, there is a branch of glucose molecules attached by an $\alpha(1\rightarrow6)$-glycosidic bond between carbon 1 of the branch and carbon 6 in the main chain (see **FIGURE 18.4**).

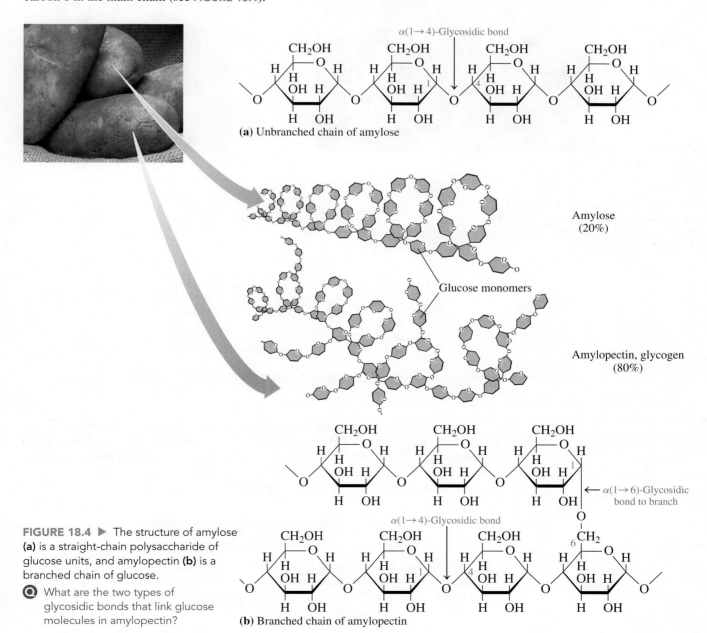

FIGURE 18.4 ▶ The structure of amylose **(a)** is a straight-chain polysaccharide of glucose units, and amylopectin **(b)** is a branched chain of glucose.

Q What are the two types of glycosidic bonds that link glucose molecules in amylopectin?

(a) Unbranched chain of amylose

Amylose (20%)

Glucose monomers

Amylopectin, glycogen (80%)

$\alpha(1\rightarrow4)$-Glycosidic bond

$\alpha(1\rightarrow6)$-Glycosidic bond to branch

$\alpha(1\rightarrow4)$-Glycosidic bond

(b) Branched chain of amylopectin

Starches hydrolyze easily in water and acid to give smaller saccharides called *dextrins*, which then hydrolyze to maltose and finally glucose. In our bodies, these complex carbohydrates are digested by the enzymes amylase (in saliva) and maltase. The glucose obtained provides about 50% of our nutritional calories.

$$\text{Amylose, amylopectin} \xrightarrow{\text{H}^+ \text{ or amylase}} \text{dextrins} \xrightarrow{\text{H}^+ \text{ or amylase}} \text{maltose} \xrightarrow{\text{H}^+ \text{ or maltase}} \text{many D-glucose units}$$

Glycogen, or animal starch, is a polymer of glucose that is stored in the liver and muscle of animals. It is used in our cells at a rate that maintains the blood level of glucose and provides energy between meals. The structure of glycogen is very similar to that of amylopectin except that glycogen is more highly branched.

Cellulose is the major structural material of wood and plants. Cotton is almost pure cellulose. In cellulose, glucose molecules form a long unbranched chain similar to that of amylose. However, the glucose units in cellulose are linked by $\beta(1{\rightarrow}4)$-glycosidic bonds (see **FIGURE 18.5**).

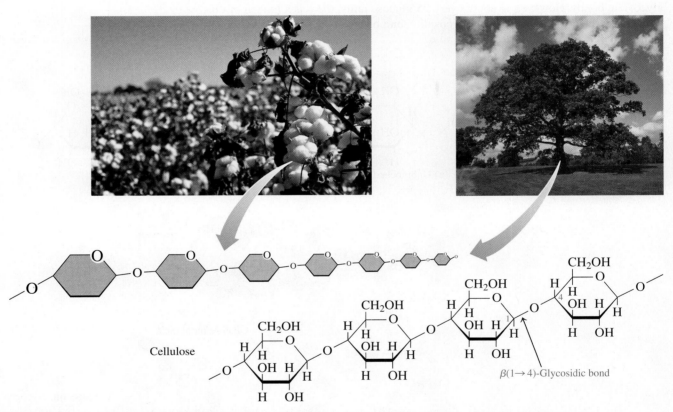

FIGURE 18.5 ▶ The polysaccharide cellulose is composed of glucose units linked by $\beta(1{\rightarrow}4)$-glycosidic bonds.

Q Why are humans unable to digest cellulose?

Humans have enzymes in saliva and pancreatic juices that break apart the $\alpha(1{\rightarrow}4)$-glycosidic bonds of starches but not the $\beta(1{\rightarrow}4)$-glycosidic bonds of cellulose. Thus, humans cannot digest cellulose. Animals such as horses, cows, and goats can obtain glucose from cellulose because their digestive systems contain bacteria that provide enzymes to break apart $\beta(1{\rightarrow}4)$-glycosidic bonds.

SAMPLE PROBLEM 18.3 Structures of Polysaccharides

Identify the polysaccharide described by each of the following:

a. a polysaccharide that is stored in the liver and muscle tissues
b. an unbranched polysaccharide containing $\beta(1{\rightarrow}4)$-glycosidic bonds
c. a starch containing $\alpha(1{\rightarrow}4)$- and $\alpha(1{\rightarrow}6)$-glycosidic bonds

TRY IT FIRST

SOLUTION

a. glycogen **b.** cellulose **c.** amylopectin, glycogen

STUDY CHECK 18.3

Cellulose and amylose are both unbranched glucose polymers. How do they differ?

ANSWER

Cellulose contains glucose units connected by $\beta(1\rightarrow4)$-glycosidic bonds, whereas the glucose units in amylose are connected by $\alpha(1\rightarrow4)$-glycosidic bonds.

QUESTIONS AND PROBLEMS

18.2 Disaccharides and Polysaccharides

LEARNING GOAL Describe the monosaccharide units and linkages in disaccharides and polysaccharides.

18.13 For each of the following, state the monosaccharide units, the type of glycosidic bond, and the name of the disaccharide, including the α or β form:

a.

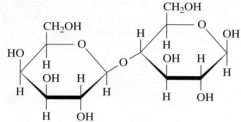

b.

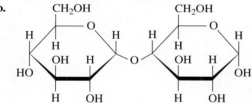

18.14 For each of the following, state the monosaccharide units, the type of glycosidic bond, and the name of the disaccharide, including the α or β form:

a.

b.

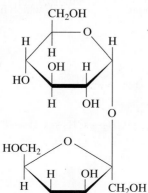

Applications

18.15 Isomaltose, obtained from the breakdown of starch, has the following Haworth structure:

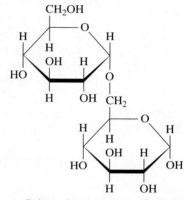

a. Is isomaltose a mono-, di-, or polysaccharide?
b. What are the monosaccharides in isomaltose?
c. What is the glycosidic link in isomaltose?
d. Is this the α or β form of isomaltose?

18.16 Sophorose, found in certain types of beans, has the following Haworth structure:

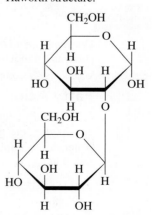

a. Is sophorose a mono-, di-, or polysaccharide?
b. What are the monosaccharides in sophorose?
c. What is the glycosidic link in sophorose?
d. Is this the α or β form of sophorose?

18.17 Identify the disaccharide that fits each of the following descriptions:
a. ordinary table sugar
b. found in milk and milk products
c. also called malt sugar
d. contains galactose and glucose

18.18 Identify the disaccharide that fits each of the following descriptions:
  a. used in brewing
  b. composed of two glucose units
  c. also called milk sugar
  d. contains glucose and fructose

18.19 Give the name of one or more polysaccharides that matches each of the following descriptions:
  a. not digestible by humans
  b. the storage form of carbohydrates in plants

  c. contains only $\alpha(1\rightarrow4)$-glycosidic bonds
  d. the most highly branched polysaccharide

18.20 Give the name of one or more polysaccharides that matches each of the following descriptions:
  a. the storage form of carbohydrates in animals
  b. contains only $\beta(1\rightarrow4)$-glycosidic bonds
  c. contains both $\alpha(1\rightarrow4)$- and $\alpha(1\rightarrow6)$-glycosidic bonds
  d. produces maltose during digestion

18.3 Lipids

LEARNING GOAL Draw the condensed or line-angle structural formula for a fatty acid, a triacylglycerol, and the products of hydrogenation or saponification. Identify the steroid nucleus.

Lipids are a family of biomolecules that have the common property of being soluble in organic solvents but not very soluble in water. The word *lipid* comes from the Greek word *lipos*, meaning "fat" or "lard." Within the lipid family, there are certain structures that distinguish the different types of lipids. Lipids that contain *fatty acids* include waxes and triacylglycerols, commonly known as fats and oils, which are esters of glycerol and fatty acids. Lipids that are *steroids* do not contain fatty acids but are characterized by the steroid nucleus of four fused carbon rings.

Lipids are naturally occurring compounds in cells and tissues, which are soluble in organic solvents and not in water.

Fatty Acids

CORE CHEMISTRY SKILL
Identifying Fatty Acids

A **fatty acid** contains a long unbranched chain of carbon atoms, usually 12 to 20, with a carboxylic acid group at one end. The long carbon chain makes fatty acids insoluble in water. An example is lauric acid, a 12-carbon acid found in coconut oil, which has a structure that can be drawn in several forms as follows:

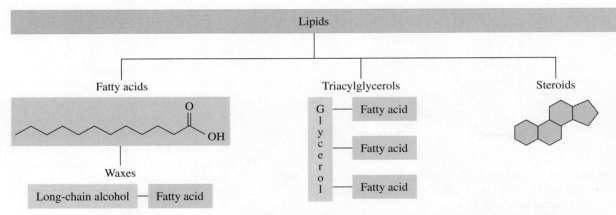

Condensed structural formulas

Line-angle structural formula

Saturated fatty acids, such as lauric acid, contain only single bonds between carbons. *Monounsaturated fatty acids* have one double bond in the carbon chain, and *polyunsaturated fatty acids* have two or more double bonds. **TABLE 18.2** lists some of the typical fatty acids in lipids.

ENGAGE

Why is stearic acid a saturated fatty acid and oleic acid a monounsaturated fatty acid?

TABLE 18.2 Structures and Melting Points of Common Fatty Acids

Name	Carbon Atoms: Double Bonds	Source	Melting Point (°C)	Structural Formulas
Saturated Fatty Acids				
Lauric acid	12:0	Coconut	44	$CH_3-(CH_2)_{10}-COOH$
Myristic acid	14:0	Nutmeg	55	$CH_3-(CH_2)_{12}-COOH$
Palmitic acid	16:0	Palm	63	$CH_3-(CH_2)_{14}-COOH$
Stearic acid	18:0	Animal fat	69	$CH_3-(CH_2)_{16}-COOH$
Monounsaturated Fatty Acids				
Palmitoleic acid	16:1	Butter	0	$CH_3-(CH_2)_5-CH=CH-(CH_2)_7-COOH$
Oleic acid	18:1	Olive, pecan, grapeseed	14	$CH_3-(CH_2)_7-CH=CH-(CH_2)_7-COOH$
Polyunsaturated Fatty Acids				
Linoleic acid	18:2	Soybean, sunflower	−5	$CH_3-(CH_2)_4-CH=CH-CH_2-CH=CH-(CH_2)_7-COOH$
Linolenic acid	18:3	Corn	−11	$CH_3-CH_2-CH=CH-CH_2-CH=CH-CH_2-CH=CH-(CH_2)_7-COOH$

Cis–Trans Isomers

Compounds with double bonds can be drawn as two structures known as *cis* and *trans* isomers. In 2-butene, which has a double bond, the cis isomer has the carbon groups on the same side of the double bond, but the trans isomer has the carbon groups on opposite sides of the double bond.

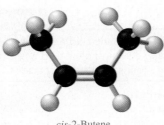

cis-2-Butene

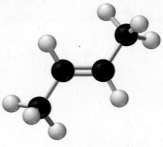

trans-2-Butene

$$CH_3 — CH = CH — CH_3$$
2-Butene

Same side

$$
\begin{array}{c}
CH_3 \quad\quad CH_3 \\
C = C \\
H \quad\quad\quad H
\end{array}
$$

cis-2-Butene
(mp −139 °C; bp 3.7 °C)

Opposite sides

$$
\begin{array}{c}
H \quad\quad CH_3 \\
C = C \\
CH_3 \quad\quad H
\end{array}
$$

trans-2-Butene
(mp −106 °C; bp 0.3 °C)

We can draw the unsaturated fatty acid oleic acid as cis and trans isomers using its line-angle structural formula. The cis structure is the more prevalent isomer found in naturally occurring unsaturated fatty acids. In the cis isomer, the carbon chain has a "kink" at the double bond site. In contrast, the trans isomer, called elaidic acid, is a straight chain without a kink at the double bond site. Saturated fatty acids fit close together in a regular pattern, which allows strong attractions to occur between the carbon chains. However, unsaturated fatty acids have irregular shape, which leads to fewer interactions between carbon chains. Thus, saturated fatty acids have high melting points because more energy is required to separate the fatty acid molecules.

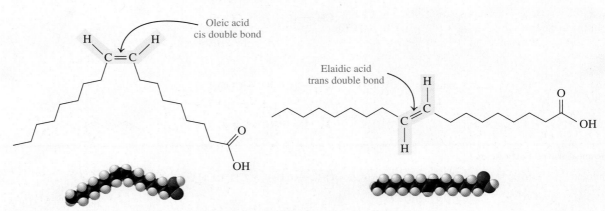

Almost all naturally occurring unsaturated fatty acids have one or more cis double bonds.

CHEMISTRY LINK TO **HEALTH**
Trans Fatty Acids and Hydrogenation

During the early 1900s, margarine became a popular replacement for highly saturated fats such as butter and lard. Margarine is produced by partially hydrogenating the unsaturated fats in vegetable oils such as safflower oil, corn oil, canola oil, cottonseed oil, and sunflower oil.

Hydrogenation and Trans Fats
In vegetable oils, the unsaturated fats usually contain cis double bonds. As hydrogenation occurs, double bonds are converted to single bonds. However, a small number of the cis double bonds are converted to trans double bonds. If the label on a product states that the oils have been "partially hydrogenated," that product will contain trans fatty acids.

The concern about trans fatty acids is that their altered structure may make them behave like saturated fatty acids in the body. Several studies reported that trans fatty acids raise the levels of LDL-cholesterol, low-density lipoproteins containing cholesterol that can accumulate in the arteries. Some studies also report that trans fatty acids lower HDL-cholesterol, high-density lipoproteins that carry cholesterol to the liver to be excreted.

Foods containing trans fatty acids include baked goods, stick margarine, soft margarine, cookies, crackers, and vegetable shortening. The American Heart Association recommends that margarine should have

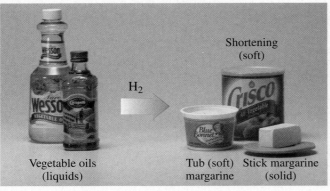

Many soft margarines, stick margarines, and solid shortenings are produced by the partial hydrogenation of vegetable oils.

no more than 2 g of saturated fat per tablespoon. It also recommends the use of soft margarine, which is lower in trans fatty acids because soft margarine is only slightly hydrogenated, and light margarine because it has less fat and therefore fewer trans fatty acids. Currently, the amount of trans fats is included on the Nutritional Facts label on food products. In the United States, a food label for a product that has less than 0.5 g of trans fat in one serving can read "0 g of trans fat."

The best advice may be to reduce total fat in the diet by using fats and oils sparingly, cooking with little or no fat, substituting olive oil or canola oil for other oils, and limiting the use of coconut oil and palm oil, which are high in saturated fatty acids.

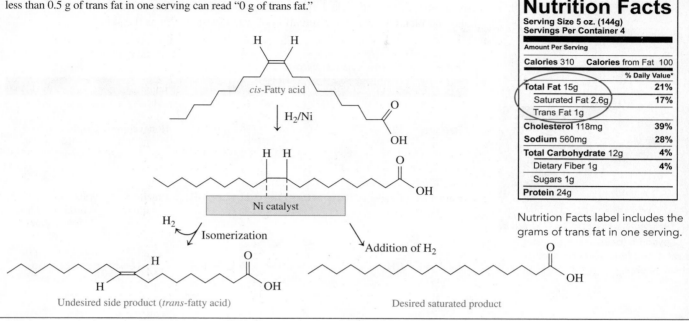

Nutrition Facts

Serving Size 5 oz. (144g)
Servings Per Container 4

Amount Per Serving

Calories 310   Calories from Fat 100

	% Daily Value*
Total Fat 15g	**21%**
Saturated Fat 2.6g	**17%**
Trans Fat 1g	
Cholesterol 118mg	**39%**
Sodium 560mg	**28%**
Total Carbohydrate 12g	**4%**
Dietary Fiber 1g	**4%**
Sugars 1g	
Protein 24g	

Nutrition Facts label includes the grams of trans fat in one serving.

SAMPLE PROBLEM 18.4 Structures and Properties of Fatty Acids

The line-angle structural formula for vaccenic acid, a fatty acid found in dairy products and human milk, is shown below.

a. Why is this substance an acid?
b. How many carbon atoms are in vaccenic acid?
c. Is the fatty acid saturated, monounsaturated, or polyunsaturated?
d. Would it be soluble in water?

 **TRY IT FIRST**

SOLUTION

a. Vaccenic acid contains a carboxylic acid group.
b. It contains 18 carbon atoms.
c. It is a monounsaturated fatty acid.
d. No; its long hydrocarbon chain makes it insoluble in water.

STUDY CHECK 18.4

Palmitoleic acid is a fatty acid with the following condensed structural formula:

$$CH_3-(CH_2)_5-CH=CH-(CH_2)_7-\overset{\overset{\displaystyle O}{\|}}{C}-OH$$

a. How many carbon atoms are in palmitoleic acid?
b. Is the fatty acid saturated, monounsaturated, or polyunsaturated?
c. Would it be soluble in water?

ANSWER

a. 16   **b.** monounsaturated   **c.** no

Waxes

Waxes are found in many plants and animals. A wax is an ester of a long-chain fatty acid and a long-chain alcohol. The formulas of some common waxes are given in **TABLE 18.3**. Beeswax obtained from honeycombs and carnauba wax from palm trees are used to give a protective coating to furniture, cars, and floors. Jojoba wax is used in making candles and cosmetics such as lipstick. Lanolin, a mixture of waxes obtained from wool, is used in hand and facial lotions to aid retention of water, which softens the skin.

Honeycomb (beeswax) is an ester of a saturated fatty acid and a long-chain alcohol.

TABLE 18.3 Some Typical Waxes

Type	Condensed Structural Formula	Source	Uses
Beeswax	$CH_3-(CH_2)_{14}-\overset{\overset{\textstyle O}{\|}}{C}-O-(CH_2)_{29}-CH_3$	Honeycomb	Candles, shoe polish, wax paper
Carnauba wax	$CH_3-(CH_2)_{24}-\overset{\overset{\textstyle O}{\|}}{C}-O-(CH_2)_{29}-CH_3$	Brazilian palm tree	Waxes for furniture, cars, floors, shoes
Jojoba wax	$CH_3-(CH_2)_{18}-\overset{\overset{\textstyle O}{\|}}{C}-O-(CH_2)_{19}-CH_3$	Jojoba bush	Candles, soaps, cosmetics

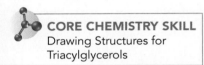

CORE CHEMISTRY SKILL

Drawing Structures for Triacylglycerols

Fats and Oils: Triacylglycerols

In the body, fatty acids are stored as fats and oils known as **triacylglycerols**. These substances, also called *triglycerides*, are triesters of glycerol (a trihydroxy alcohol) and fatty acids. In a triacylglycerol, three hydroxyl groups on glycerol form ester bonds with the carboxyl groups of three fatty acids. For example, glycerol and three molecules of stearic acid form glyceryl tristearate (tristearin).

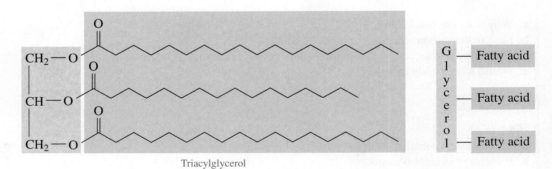

Triacylglycerol

$$CH_2-O-H + HO-\overset{\overset{\textstyle O}{\|}}{C}-(CH_2)_{16}-CH_3$$
$$CH-O-H + HO-\overset{\overset{\textstyle O}{\|}}{C}-(CH_2)_{16}-CH_3 \longrightarrow$$
$$CH_2-O-H + HO-\overset{\overset{\textstyle O}{\|}}{C}-(CH_2)_{16}-CH_3$$

Glycerol 3 Stearic acid molecules

Ester bond

$$CH_2-O-\overset{\overset{\textstyle O}{\|}}{C}-(CH_2)_{16}-CH_3$$
$$CH-O-\overset{\overset{\textstyle O}{\|}}{C}-(CH_2)_{16}-CH_3 + \boxed{3H_2O}$$
$$CH_2-O-\overset{\overset{\textstyle O}{\|}}{C}-(CH_2)_{16}-CH_3$$

Glyceryl tristearate (tristearin)

Triacylglycerols are the major form of energy storage for animals. Animals that hibernate eat large quantities of plants, seeds, and nuts that are high in calories. Prior to hibernation, these animals, such as polar bears, gain as much as 14 kg a week. As the external temperature drops, the animal goes into hibernation. The body temperature drops to nearly freezing, and there is a dramatic reduction in cellular activity, respiration, and heart rate. Animals that live in extremely cold climates will hibernate for 4 to 7 months. During this time, stored fat is the only source of energy.

Melting Points of Fats and Oils

A **fat** is a triacylglycerol that is solid at room temperature, such as fats in meat, whole milk, butter, and cheese. An **oil** is a triacylglycerol that is usually liquid at room temperature and is usually obtained from a plant source (see **FIGURE 18.6**).

The amounts of saturated, monounsaturated, and polyunsaturated fatty acids in some typical fats and oils are shown in **FIGURE 18.7**. Saturated fatty acids have higher melting points than unsaturated fatty acids because they pack together more tightly. Animal fats usually contain more saturated fatty acids than do vegetable oils. Therefore the melting points of animal fats are higher than those of vegetable oils.

Prior to hibernation, a polar bear eats food with a high caloric content.

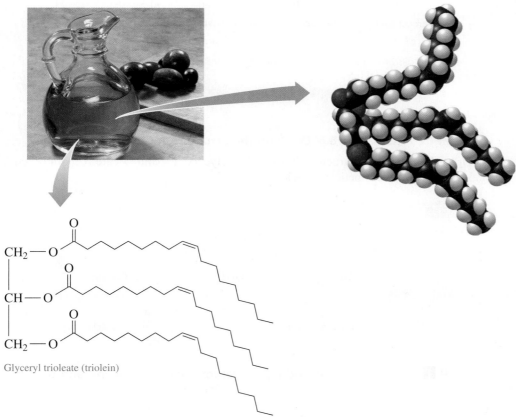

Glyceryl trioleate (triolein)

FIGURE 18.6 ▶ Vegetable oils such as olive oil, corn oil, and safflower oil contain unsaturated fats.

Q Why is olive oil a liquid at room temperature?

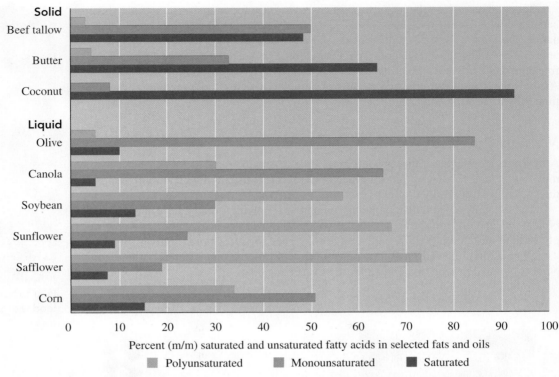

FIGURE 18.7 ▶ Vegetable oils are liquids at room temperature because they have a higher percentage of unsaturated fatty acids than do animal fats.

◉ Why is butter a solid at room temperature, whereas canola oil is a liquid?

Triacylglycerols are used to thicken creams and lotions.

Guide to Drawing Triacylglycerols

STEP 1
Draw the condensed structural formulas for glycerol and the fatty acids.

STEP 2
Form ester bonds between the hydroxyl groups on glycerol and the carboxyl groups on each fatty acid.

SAMPLE PROBLEM 18.5 Drawing the Structure for a Triacylglycerol

Draw the condensed structural formula for glyceryl tripalmitoleate (tripalmitolein), which is used in cosmetic creams and lotions.

TRY IT FIRST

SOLUTION

	Given	Need	Connect
ANALYZE THE PROBLEM	glyceryl tripalmitoleate (tripalmitolein)	condensed structural formula	ester of glycerol and three palmitoleic acid molecules

STEP 1 Draw the condensed structural formulas for glycerol and the fatty acids.

$$CH_2-OH \ + \ HO-\overset{\overset{\displaystyle O}{\|}}{C}-(CH_2)_7-CH=CH-(CH_2)_5-CH_3 \quad \text{Palmitoleic acid}$$

$$CH-OH \ + \ HO-\overset{\overset{\displaystyle O}{\|}}{C}-(CH_2)_7-CH=CH-(CH_2)_5-CH_3 \quad \text{Palmitoleic acid}$$

$$CH_2-OH \ + \ HO-\overset{\overset{\displaystyle O}{\|}}{C}-(CH_2)_7-CH=CH-(CH_2)_5-CH_3 \quad \text{Palmitoleic acid}$$

Glycerol

STEP 2 Form ester bonds between the hydroxyl groups on glycerol and the carboxyl groups on each fatty acid.

$$
\begin{array}{l}
\text{CH}_2-\text{O}-\overset{\overset{\displaystyle O}{\|}}{\text{C}}-(\text{CH}_2)_7-\text{CH}=\text{CH}-(\text{CH}_2)_5-\text{CH}_3 \\[4pt]
\text{CH}-\text{O}-\overset{\overset{\displaystyle O}{\|}}{\text{C}}-(\text{CH}_2)_7-\text{CH}=\text{CH}-(\text{CH}_2)_5-\text{CH}_3 \\[4pt]
\text{CH}_2-\text{O}-\overset{\overset{\displaystyle O}{\|}}{\text{C}}-(\text{CH}_2)_7-\text{CH}=\text{CH}-(\text{CH}_2)_5-\text{CH}_3
\end{array}
$$

Glyceryl tripalmitoleate (tripalmitolein)

STUDY CHECK 18.5

Draw the line-angle structural formula for the triacylglycerol containing three molecules of myristic acid (14:0).

ANSWER

Reactions of Triacylglycerols

The double bonds in unsaturated fatty acids react with hydrogen to produce saturated fatty acids. For example, when hydrogen is added to glyceryl trioleate (triolein) using a nickel catalyst, the product is the saturated fat glyceryl tristearate (tristearin).

> **CORE CHEMISTRY SKILL**
> Drawing the Products for the Hydrogenation and Saponification of a Triacylglycerol

Double bonds

$$
\begin{array}{l}
\text{CH}_2-\text{O}-\overset{\overset{\displaystyle O}{\|}}{\text{C}}-(\text{CH}_2)_7-\text{CH}=\text{CH}-(\text{CH}_2)_7-\text{CH}_3 \\[4pt]
\text{CH}-\text{O}-\overset{\overset{\displaystyle O}{\|}}{\text{C}}-(\text{CH}_2)_7-\text{CH}=\text{CH}-(\text{CH}_2)_7-\text{CH}_3 + 3\text{H}_2 \\[4pt]
\text{CH}_2-\text{O}-\overset{\overset{\displaystyle O}{\|}}{\text{C}}-(\text{CH}_2)_7-\text{CH}=\text{CH}-(\text{CH}_2)_7-\text{CH}_3
\end{array}
$$

Glyceryl trioleate (triolein)

$\xrightarrow{\text{Ni}}$

Single bonds

$$
\begin{array}{l}
\text{CH}_2-\text{O}-\overset{\overset{\displaystyle O}{\|}}{\text{C}}-(\text{CH}_2)_7-\text{CH}_2-\text{CH}_2-(\text{CH}_2)_7-\text{CH}_3 \\[4pt]
\text{CH}-\text{O}-\overset{\overset{\displaystyle O}{\|}}{\text{C}}-(\text{CH}_2)_7-\text{CH}_2-\text{CH}_2-(\text{CH}_2)_7-\text{CH}_3 \\[4pt]
\text{CH}_2-\text{O}-\overset{\overset{\displaystyle O}{\|}}{\text{C}}-(\text{CH}_2)_7-\text{CH}_2-\text{CH}_2-(\text{CH}_2)_7-\text{CH}_3
\end{array}
$$

Glyceryl tristearate (tristearin)

Saponification occurs when a fat is heated with a strong base such as sodium hydroxide to give glycerol and the sodium salts of the fatty acids, which is soap. When NaOH is used, a solid soap is produced that can be molded into a desired shape; KOH produces a

Fat or oil + strong base $\xrightarrow{\text{Heat}}$ glycerol + salts of fatty acids (soap)

$$
\begin{array}{l}
\text{CH}_2-\text{O}-\overset{\overset{\displaystyle O}{\|}}{\text{C}}-(\text{CH}_2)_{14}-\text{CH}_3 \\[4pt]
\text{CH}-\text{O}-\overset{\overset{\displaystyle O}{\|}}{\text{C}}-(\text{CH}_2)_{14}-\text{CH}_3 + \textbf{3NaOH} \\[4pt]
\text{CH}_2-\text{O}-\overset{\overset{\displaystyle O}{\|}}{\text{C}}-(\text{CH}_2)_{14}-\text{CH}_3
\end{array}
$$

Glyceryl tripalmitate (tripalmitin)

$\xrightarrow{\text{Heat}}$

$$
\begin{array}{l}
\text{CH}_2-\text{OH} \qquad \text{Na}^+\ {}^-\text{O}-\overset{\overset{\displaystyle O}{\|}}{\text{C}}-(\text{CH}_2)_{14}-\text{CH}_3 \\[4pt]
\text{CH}-\text{OH} + \text{Na}^+\ {}^-\text{O}-\overset{\overset{\displaystyle O}{\|}}{\text{C}}-(\text{CH}_2)_{14}-\text{CH}_3 \\[4pt]
\text{CH}_2-\text{OH} \qquad \text{Na}^+\ {}^-\text{O}-\overset{\overset{\displaystyle O}{\|}}{\text{C}}-(\text{CH}_2)_{14}-\text{CH}_3
\end{array}
$$

Glycerol   3 Sodium palmitate (soap)

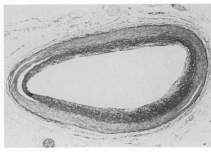

(a)

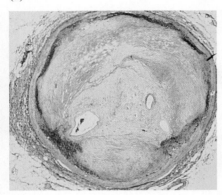

(b)

FIGURE 18.8 ▶ Excess cholesterol forms plaque that can block an artery, resulting in a heart attack. **(a)** A cross section of a normal, open artery shows no buildup of plaque. **(b)** A cross section of an artery that is almost completely clogged by atherosclerotic plaque.

🔍 What property of cholesterol would cause it to form deposits along the coronary arteries?

softer, liquid soap. Oils that are polyunsaturated produce softer soaps. Names like "coconut soap" or "avocado shampoo" tell you the sources of the oil used in the reaction.

Steroids: Cholesterol

Steroids are compounds containing the steroid nucleus, which consists of four carbon rings fused together. Although they are lipids, steroids do not contain fatty acids.

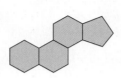

Steroid nucleus

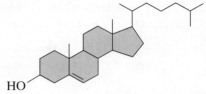

HO

Cholesterol

Attaching other atoms and groups of atoms to the steroid nucleus forms a wide variety of steroid compounds. **Cholesterol**, which is one of the most important and abundant steroids in the body, is a *sterol* because it contains an oxygen atom as a hydroxyl group (—OH). Like many steroids, cholesterol has methyl groups, a double bond, and a carbon side chain. In other steroids, the hydroxyl group is replaced by a carbonyl group (C=O). Cholesterol is obtained from eating meats, milk, and eggs, and it is also synthesized by the liver. There is no cholesterol in vegetable and plant products.

Cholesterol in the Body

If a diet is high in cholesterol, the liver produces less cholesterol. A typical daily American diet includes 400 to 500 mg of cholesterol, one of the highest in the world. The American Heart Association has recommended that we consume no more than 300 mg of cholesterol a day. Researchers suggest that saturated fats and cholesterol are associated with diseases such as diabetes; cancers of the breast, pancreas, and colon; and atherosclerosis. In atherosclerosis, deposits of a protein–lipid complex (plaque) accumulate in the coronary blood vessels, restricting the flow of blood to the tissue and causing necrosis of the tissue (see **FIGURE 18.8**). In the heart, plaque accumulation could result in a *myocardial infarction* (heart attack). Other factors that may also increase the risk of heart disease are family history, lack of exercise, smoking, obesity, diabetes, gender, and age.

QUESTIONS AND PROBLEMS

18.3 Lipids

LEARNING GOAL Draw the condensed or line-angle structural formula for a fatty acid, a triacylglycerol, and the products of hydrogenation or saponification. Identify the steroid nucleus.

18.21 Classify each of the following fatty acids as saturated, monounsaturated, or polyunsaturated (see Table 18.2):
a. lauric acid **b.** linolenic acid **c.** stearic acid

18.22 Classify each of the following fatty acids as saturated, monounsaturated, or polyunsaturated (see Table 18.2):
a. linoleic acid **b.** palmitoleic acid **c.** myristic acid

18.23 Caprylic acid is an 8-carbon saturated fatty acid that occurs in coconut oil (10%) and palm kernel oil (4%). Draw the condensed structural formula for glyceryl tricaprylate (tricaprylin).

18.24 Draw the condensed structural formula for glyceryl trilaurate (trilaurin).

Applications

18.25 Safflower oil is called a polyunsaturated oil, whereas olive oil is a monounsaturated oil. Explain.

18.26 Why does olive oil have a lower melting point than butter fat?

18.27 Write the balanced chemical equation for the hydrogenation of glyceryl tripalmitoleate, a fat containing glycerol and three palmitoleic acid molecules.

18.28 Write the balanced chemical equation for the hydrogenation of glyceryl trilinolenate, a fat containing glycerol and three linolenic acid molecules.

18.29 Write the balanced chemical equation for the NaOH saponification of glyceryl trimyristate, a fat containing glycerol and three myristic acid molecules.

18.30 Write the balanced chemical equation for the NaOH saponification of glyceryl trioleate, a fat containing glycerol and three oleic acid molecules.

18.31 Draw the structure for the steroid nucleus.

18.32 What are the functional groups on the cholesterol molecule?

18.4 Amino Acids and Proteins

LEARNING GOAL Describe protein functions and draw structures for amino acids and peptides.

Proteins perform different functions in the body. Some proteins form structural components such as cartilage, muscles, hair, and nails. Wool, silk, feathers, and horns in animals are made of proteins. Proteins that function as enzymes regulate biological reactions such as digestion and cellular metabolism. Other proteins, such as hemoglobin and myoglobin, transport oxygen in the blood and muscle. **TABLE 18.4** gives examples of proteins that are classified by their functions in biological systems.

TABLE 18.4 Classification of Some Proteins and Their Functions

Class of Protein	Function	Examples
Structural	Provide structural components	*Collagen* is in tendons and cartilage. *Keratin* is in hair, skin, wool, horns, and nails.
Contractile	Make muscles move	*Myosin* and *actin* contract muscle fibers.
Transport	Carry essential substances throughout the body	*Hemoglobin* transports oxygen. *Lipoproteins* transport lipids.
Storage	Store nutrients	*Casein* stores protein in milk. *Ferritin* stores iron in the spleen and liver.
Hormone	Regulate body metabolism and the nervous system	*Insulin* regulates blood glucose level. *Growth hormone* regulates body growth.
Enzyme	Catalyze biochemical reactions in the cells	*Sucrase* catalyzes the hydrolysis of sucrose. *Trypsin* catalyzes the hydrolysis of proteins.
Protection	Recognize and destroy foreign substances	*Immunoglobulins* stimulate immune responses.

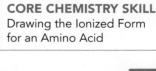

The horns of animals are made of proteins.

Amino Acids

Proteins are composed of molecular building blocks called *amino acids*. Every **amino acid**, which is ionized in biological environments, has a central alpha carbon atom (α carbon) bonded to an ammonium group ($-NH_3^+$), a carboxylate group ($-COO^-$), a hydrogen atom, and an R group. The differences in the 20 α-amino acids present in human proteins are due to the unique characteristics of the R groups.

Ionized Structure of Alanine, an α-Amino Acid

We classify amino acids according to their R groups, which determine their properties in aqueous solution. The *nonpolar amino acids* have hydrogen, alkyl, or aromatic R groups, which make them *hydrophobic* ("water fearing"). A *polar amino acid* interacts with water because it has a polar R group, which is *hydrophilic* ("water loving"). *Neutral polar amino acids* contain hydroxyl ($-OH$), thiol ($-SH$), or amide ($-CONH_2$) R groups. The R group of a polar *acidic amino acid* contains a carboxylate group ($-COO^-$). The R group of a polar *basic amino acid* contains an ammonium group ($-NH_3^+$). The ionized structures and common names of the 20 α-amino acids commonly found in proteins with their R groups highlighted in tan, and their three-letter and one-letter abbreviations, are listed in **TABLE 18.5**.

CORE CHEMISTRY SKILL
Drawing the Ionized Form for an Amino Acid

ENGAGE

Why is valine classified as a nonpolar amino acid whereas histidine is classified as a polar amino acid?

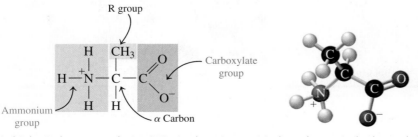

In biological systems, alanine is ionized to give positively and negatively charged groups.

TABLE 18.5 Structures, Names, and Abbreviations of 20 Common Amino Acids at Physiological pH (7.4)

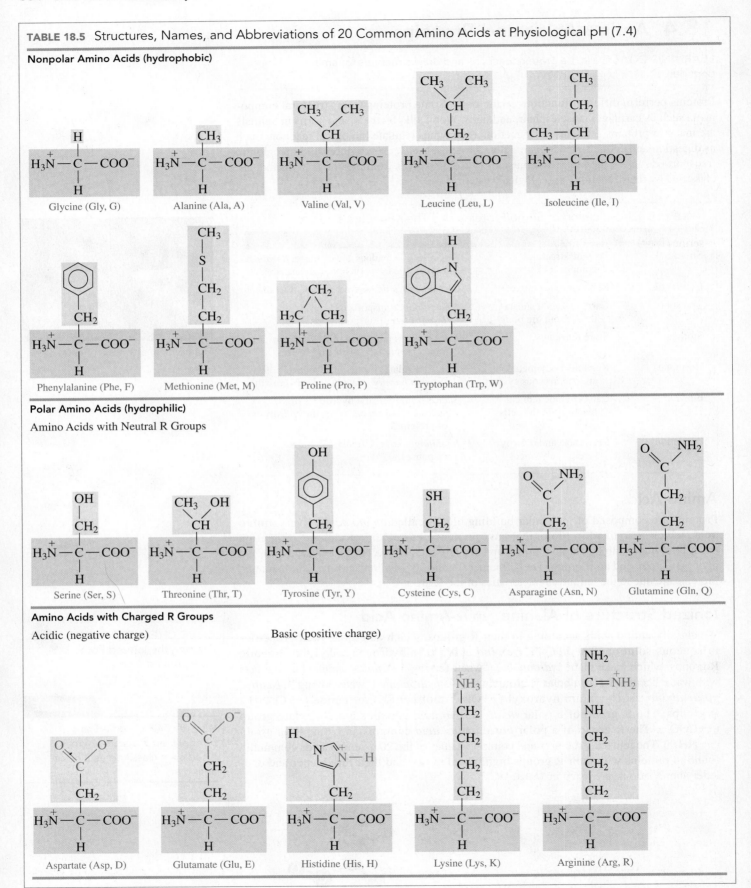

Nonpolar Amino Acids (hydrophobic)

Glycine (Gly, G)

Alanine (Ala, A)

Valine (Val, V)

Leucine (Leu, L)

Isoleucine (Ile, I)

Phenylalanine (Phe, F)

Methionine (Met, M)

Proline (Pro, P)

Tryptophan (Trp, W)

Polar Amino Acids (hydrophilic)

Amino Acids with Neutral R Groups

Serine (Ser, S)

Threonine (Thr, T)

Tyrosine (Tyr, Y)

Cysteine (Cys, C)

Asparagine (Asn, N)

Glutamine (Gln, Q)

Amino Acids with Charged R Groups

Acidic (negative charge)

Basic (positive charge)

Aspartate (Asp, D)

Glutamate (Glu, E)

Histidine (His, H)

Lysine (Lys, K)

Arginine (Arg, R)

A summary of the classification of amino acids follows:

Type of Amino Acid	Number	Type of R Groups	Interaction with Water
Nonpolar	9	Nonpolar	Hydrophobic
Polar, neutral	6	Contain O and S atoms, but no charge	Hydrophilic
Polar, acidic	2	Contain carboxylate groups, negative charge	Hydrophilic
Polar, basic	3	Contain ammonium groups, positive charge	Hydrophilic

SAMPLE PROBLEM 18.6 Structural Formulas of Amino Acids

Draw the condensed structural formula and give the abbreviations for serine.
serine $(R = -CH_2-OH)$

TRY IT FIRST

SOLUTION

The structure of a specific amino acid is drawn by attaching the side group (R) to the central carbon atom of the general structure of an amino acid.

The abbreviations for serine are Ser and S.

STUDY CHECK 18.6

Classify serine in Sample Problem 18.6 as polar or nonpolar and hydrophobic or hydrophilic.

ANSWER

polar, hydrophilic

CHEMISTRY LINK TO HEALTH
Essential Amino Acids

Of the 20 amino acids used to build the proteins in the body, 11 can be synthesized in the body. The other nine amino acids, listed in **TABLE 18.6**, are *essential amino acids* that cannot be synthesized and must be obtained from the proteins in the diet. In addition to the essential amino acids required by adults, infants and growing children also require arginine, cysteine, and tyrosine.

Complete proteins, which contain all of the essential amino acids, are found in animal products such as eggs, milk, meat, fish, and poultry. However, gelatin and plant proteins such as grains, beans, and nuts are *incomplete proteins* because they are deficient in one or more of the essential amino acids. Diets that rely on plant foods for protein must contain a variety of protein sources to obtain all the essential amino acids. Some examples of *complementary protein*

TABLE 18.6 Essential Amino Acids for Adults

Histidine (His, H)	Phenylalanine (Phe, F)
Isoleucine (Ile, I)	Threonine (Thr, T)
Leucine (Leu, L)	Tryptophan (Trp, W)
Lysine (Lys, K)	Valine (Val, V)
Methionine (Met, M)	

sources include rice and beans, or a peanut butter sandwich on whole grain bread. Rice and wheat contain the amino acid methionine that is deficient in beans and peanuts, whereas beans and peanuts are rich in lysine that is lacking in grains (see **TABLE 18.7**).

Complete proteins such as eggs, milk, meat, and fish contain all of the essential amino acids. Incomplete proteins from plants such as grains, beans, and nuts are deficient in one or more essential amino acids.

TABLE 18.7 Amino Acid Deficiencies in Selected Vegetables and Grains

Food Source	Amino Acid Deficiency
Wheat, rice, oats	Lysine
Corn	Lysine, tryptophan
Beans	Methionine, tryptophan
Peas, peanuts	Methionine
Almonds, walnuts	Lysine, tryptophan
Soy	Methionine

ENGAGE

In the dipeptide Val–Thr, how do you know that valine is the N-terminus and threonine the C-terminus?

Peptides

A *peptide bond* is an amide bond that forms when the $-COO^-$ group of one amino acid reacts with the $-NH_3^+$ group of the next amino acid. We can write the formation of the dipeptide glycylalanine (Gly–Ala, GA) between glycine and alanine. In this dipeptide, glycine, which is written on the left with a free $-NH_3^+$ group, is the *N-terminal amino acid* or *N-terminus*. Alanine, which is written on the right with a free $-COO^-$ group, is called the *C-terminal amino acid* or *C-terminus*. In the name of a peptide, each amino acid beginning from the N-terminus has the *ine*, *an*, or *ate* replaced by *yl*. The last amino acid at the C-terminus uses its full name.

A peptide bond between the ionized structures of glycine and alanine forms the dipeptide glycylalanine.

SAMPLE PROBLEM 18.7 Drawing a Peptide

Draw the structure and give the name for the tripeptide Gly–Ser–Met.

TRY IT FIRST

SOLUTION

ANALYZE THE PROBLEM	Given	Need	Connect
	Gly–Ser–Met	peptide structure, name	peptide bond, amino acid names

STEP 1 Draw the structures for each amino acid in the peptide, starting with the N-terminus.

Guide to Drawing a Peptide

STEP 1
Draw the structures for each amino acid in the peptide, starting with the N-terminus.

STEP 2
Remove the O atom from the carboxylate group and two H atoms from the ammonium group in the adjacent amino acid. Use peptide bonds to connect the amino acids.

STEP 2 Remove the O atom from the carboxylate group and two H atoms from the ammonium group in the adjacent amino acid. Use peptide bonds to connect the amino acids. Repeat this process until the C-terminus is reached.

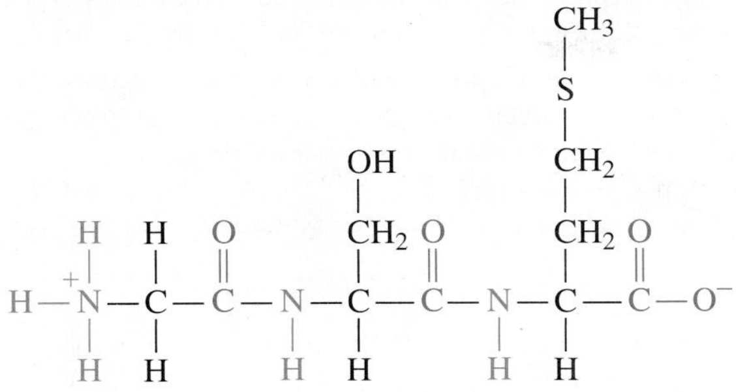

N-terminus   Gly–Ser–Met, GSM   C-terminus

The tripeptide is named by replacing the last syllable of each amino acid name with *yl*, starting with the N-terminus. The C-terminus retains its complete amino acid name.

N-terminus   glycine is named glycyl

serine is named seryl

C-terminus   methionine keeps its full name

The tripeptide is named glycylserylmethionine.

STUDY CHECK 18.7

Draw the structure and give the name for Phe–Thr, a section in glucagon, which is a peptide hormone that increases blood glucose levels.

ANSWER

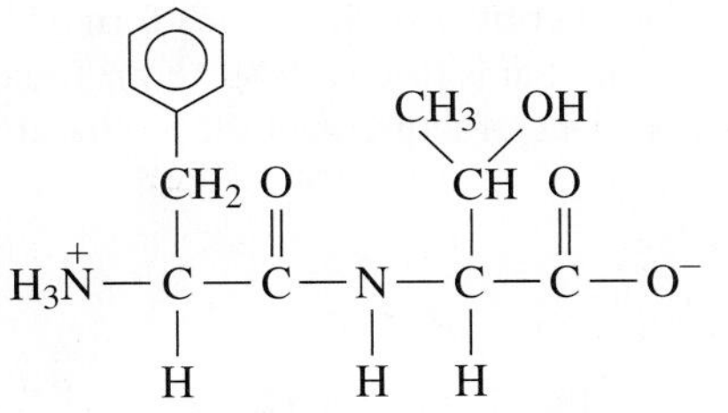

Phenylalanylthreonine

QUESTIONS AND PROBLEMS

18.4 Amino Acids and Proteins

LEARNING GOAL Describe protein functions and draw structures for amino acids and peptides.

18.33 Draw the ionized form for each of the following amino acids:
   a. glycine   **b.** threonine   **c.** phenylalanine

18.34 Draw the ionized form for each of the following amino acids:
   a. tyrosine   **b.** leucine   **c.** methionine

18.35 Classify each of the amino acids in problem 18.33 as nonpolar or polar. If polar, indicate if the R group is neutral, acidic, or basic. Indicate if each would be hydrophobic or hydrophilic.

18.36 Classify each of the amino acids in problem 18.34 as nonpolar or polar. If polar, indicate if the R group is neutral, acidic, or basic. Indicate if each would be hydrophobic or hydrophilic.

18.37 Give the name of the amino acid represented by each of the following abbreviations:
   a. Ala   **b.** V
   c. Lys   **d.** C

18.38 Give the name of the amino acid represented by each of the following abbreviations:
   a. Trp   **b.** M
   c. Pro   **d.** G

18.39 Draw the structure for each of the following peptides, and give the three-letter and one-letter abbreviations for their names:
   a. alanylcysteine   **b.** serylphenylalanine
   c. glycylalanylvaline

18.40 Draw the structure for each of the following peptides, and give the three-letter and one-letter abbreviations for their names:
   a. prolylaspartate   **b.** threonylleucine
   c. methionylglutaminyllysine

Applications

18.41 Peptides isolated from rapeseed that may lower blood pressure have the following sequence of amino acids. Draw the structure for each peptide and write the one-letter abbreviations.
   a. Arg–Ile–Tyr   **b.** Val–Trp–Ile–Ser

18.42 Peptides from sweet potato with antioxidant properties have the following sequence of amino acids. Draw the structure for each peptide and write the one-letter abbreviations.
   a. Asp–Cys–Gly–Tyr   **b.** Asn–Tyr–Asp–Glu–Tyr

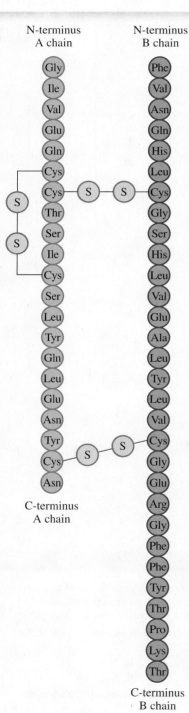

FIGURE 18.9 ▶ The sequence of amino acids in human insulin is the primary structure.

🔍 What kinds of bonds occur in the primary structure of a protein?

18.5 Protein Structure

LEARNING GOAL Identify the levels of structure of a protein.

A **protein** is a polypeptide of 50 or more amino acids that has biological activity. Each protein in our cells has a unique sequence of amino acids that determines its three-dimensional structure and biological function.

Primary Structure

The **primary structure** of a protein is the particular sequence of amino acids held together by peptide bonds. The first protein to have its primary structure determined was insulin, which was accomplished by Frederick Sanger in 1953. Since that time, scientists have determined the amino acid sequences of thousands of proteins. Insulin is a hormone that regulates the glucose level in the blood. In the primary structure of human insulin, there are two polypeptide chains. In chain A, there are 21 amino acids, and in chain B there are 30 amino acids. The polypeptide chains are held together by *disulfide bonds* formed by the thiol groups of the cysteine amino acids in each of the chains (see **FIGURE 18.9**). Today, human insulin, for the treatment of diabetes, with this exact structure is produced in large quantities through genetic engineering.

Secondary Structure

The **secondary structure** of a protein describes the structure that forms when amino acids form hydrogen bonds between the atoms in the backbone within a single polypeptide chain or between polypeptide chains. The most common types of secondary structures are the *alpha helix* and the *beta-pleated sheet*.

In an alpha helix (α helix), hydrogen bonds form between each N—H group and the oxygen of a C=O group in the next turn of the α helix (see **FIGURE 18.10**). Because there are many hydrogen bonds along the peptide, it has a helical or coiled shape.

In the secondary structure known as the beta-pleated sheet (β-pleated sheet), hydrogen bonds hold polypeptide chains together side by side. The hydrogen bonds holding the sheets tightly in place account for the strength and durability of proteins such as silk (see **FIGURE 18.11**).

Collagen

Collagen, the most abundant protein in the body, makes up as much as one-third of all the protein in vertebrates. It is found in connective tissue, blood vessels, skin, tendons, ligaments, the cornea of the eye, and cartilage. The strong structure of collagen is a result of three polypeptides woven together by hydrogen bonds to form a *triple helix*. When several triple helices wrap together, they form the fibrils that make up connective tissues and tendons. In a young person, collagen is elastic. As a person ages, additional bonds form between the fibrils, which make collagen less elastic. Cartilage and tendons become more brittle, and wrinkles are seen in the skin.

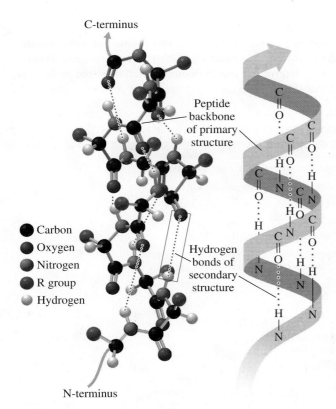

C-terminus

Peptide
backbone
of primary
structure

Hydrogen
bonds of
secondary
structure

- Carbon
- Oxygen
- Nitrogen
- R group
- Hydrogen

N-terminus

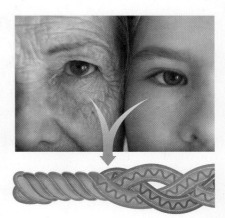

Triple helix   3 peptide chains

Collagen fibers are triple helices of
polypeptide chains held together by
hydrogen bonds.

FIGURE 18.10 ▶ The α (alpha) helix acquires a coiled
shape from hydrogen bonds between the N—H of the
peptide bond in one turn of the polypeptide and the C=O
of the peptide bond in the next turn.

Q What are partial charges of the H in N—H and the
O in C=O that allow for hydrogen bonding?

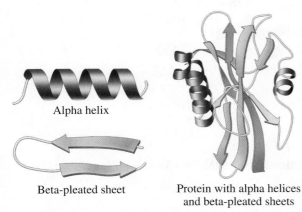

Alpha helix

Beta-pleated sheet

Protein with alpha helices
and beta-pleated sheets

The ribbon model of a protein shows regions of alpha
helices and beta-pleated sheets.

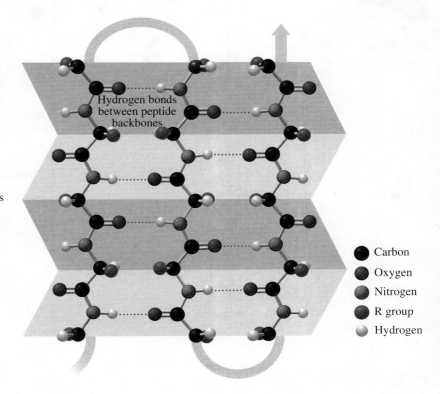

Hydrogen bonds
between peptide
backbones

- Carbon
- Oxygen
- Nitrogen
- R group
- Hydrogen

FIGURE 18.11 ▶ In a β (beta)-pleated sheet
secondary structure, hydrogen bonds form between
side-by-side sections of the peptide chains.

Q How do the hydrogen bonds differ between
a β-pleated sheet and an alpha helix?

Tertiary Structure

The **tertiary structure** of a protein involves attractions and repulsions between the R groups of the amino acids in the polypeptide chain. It is the unique three-dimensional shape of the *tertiary structure* that determines the biological function of the molecule. **TABLE 18.8** lists the interactions that stabilize the tertiary structures of proteins.

TABLE 18.8 Some Interactions That Stabilize Tertiary Structures

Interaction	Nature of Bonding
Hydrophobic	Interactions between nonpolar groups
Hydrophilic	Attractions between polar groups and water
Salt bridges (ionic bonds)	Ionic interactions between acidic and basic amino acids
Hydrogen bonds	Attractions between H and O or N
Disulfide bonds	Strong covalent links between sulfur atoms of two cysteine amino acids

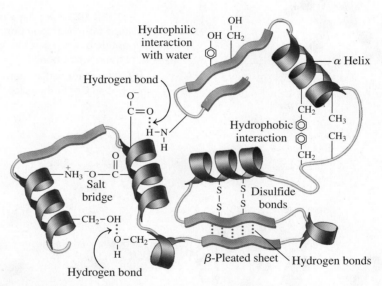

Interactions between amino acid R groups fold a polypeptide into a specific three-dimensional shape called its tertiary structure.

ENGAGE

Why is the interaction between arginine and aspartate in a tertiary structure called a salt bridge?

CORE CHEMISTRY SKILL
Identifying the Primary, Secondary, Tertiary, and Quaternary Structures of Proteins

SAMPLE PROBLEM 18.8 Interactions That Stabilize Tertiary Structures

What interaction would you expect between the following amino acids?

ANALYZE THE PROBLEM	Given	Need	Connect
	pairs of amino acids	interaction between amino acids	specific R group

a. cysteine and cysteine
b. glutamate and lysine

▶ TRY IT FIRST

SOLUTION

a. Two cysteines, each containing —SH in their R group, will form a disulfide bond.
b. An ionic bond (salt bridge) can form by the interaction of the —COO⁻ in the R group of glutamate and the —NH₃⁺ in the R group of lysine.

STUDY CHECK 18.8

What interaction would you expect between valine and leucine in the tertiary structure of a protein?

ANSWER

Both valine and leucine have nonpolar R groups and would have a hydrophobic interaction.

Myoglobin, a single polypeptide chain of 153 amino acids, stores oxygen in skeletal muscle. It forms a compact tertiary structure that contains a pocket of amino acids and a heme group that binds and stores one oxygen (O_2) molecule (see **FIGURE 18.12**).

Quaternary Structure

When a biologically active protein consists of two or more polypeptide subunits, the structural level is referred to as a **quaternary structure**. Hemoglobin, a protein that transports oxygen in blood, consists of four polypeptide chains or subunits. The subunits are held together in the quaternary structure by the same kinds of interactions that stabilize the tertiary structures. In the hemoglobin molecule, the quaternary structure of hemoglobin can bind and transport four molecules of oxygen. **TABLE 18.9** and **FIGURE 18.13** summarize the structural levels of proteins.

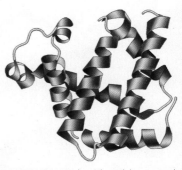

FIGURE 18.12 ▶ The ribbon model represents the tertiary structure of the polypeptide chain of myoglobin that binds oxygen.

◉ Would hydrophilic amino acids be found on the outside or inside of the myoglobin structure?

TABLE 18.9 Summary of Structural Levels in Proteins

Structural Level	Characteristics
Primary	Peptide bonds join amino acids in a specific sequence in a polypeptide.
Secondary	Hydrogen bonds along or between peptide chains form α helix, or a β-pleated sheet.
Tertiary	A protein folds into a compact, three-dimensional shape stabilized by interactions between R groups of amino acids.
Quaternary	Two or more protein subunits combine to form a biologically active protein stabilized by the same interactions as in the tertiary structure.

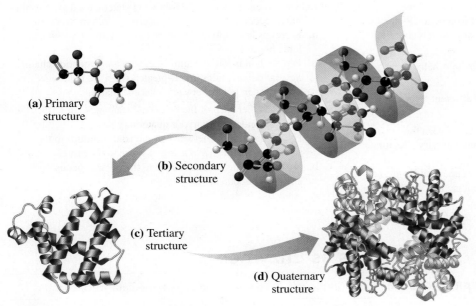

(a) Primary structure

(b) Secondary structure

(c) Tertiary structure

(d) Quaternary structure

FIGURE 18.13 ▶ Proteins consist of **(a)** primary, **(b)** secondary, **(c)** tertiary, and often **(d)** quaternary structural levels.

◉ What is the difference between a tertiary structure and a quaternary structure?

SAMPLE PROBLEM 18.9 Identifying Protein Structure

Indicate whether the following interactions are responsible for the primary, secondary, tertiary, or quaternary structures of proteins:

a. disulfide bonds between portions of a protein chain
b. peptide bonds that form a chain of amino acids
c. hydrophobic interactions

> **TRY IT FIRST**

SOLUTION

ANALYZE THE PROBLEM	Given	Need	Connect
	interactions between amino acids	structural level	identify characteristics

a. Disulfide bonds help to stabilize the tertiary and quaternary levels of protein structure.
b. Peptide bonds form between amino acids in the primary structure of a protein.
c. Hydrophobic interactions between two nonpolar R groups occur in the tertiary and quaternary structures of levels of protein structure.

STUDY CHECK 18.9

What structural level is represented by the grouping of two subunits in insulin?

ANSWER

quaternary structural level

Interactive Video

Different Levels of Protein Structure

QUESTIONS AND PROBLEMS

18.5 Protein Structure

LEARNING GOAL Identify the levels of structure of a protein.

18.43 Two peptides each contain one molecule of valine and two molecules of serine. Write the three-letter and one-letter abbreviations for their possible primary structures.

18.44 Two peptides each contain one molecule of alanine, one molecule of glycine, and one molecule of isoleucine. Write the three-letter and one-letter abbreviations for their possible primary structures.

18.45 What is the difference in bonding between an α helix and a β-pleated sheet?

18.46 How is the structure of a β-pleated sheet different from that of a triple helix?

18.47 What type of interaction would you expect between the R groups of the following amino acids in the tertiary structural level of a protein?
a. cysteine and cysteine   **b.** glutamate and arginine
c. serine and aspartate   **d.** leucine and leucine

18.48 What type of interaction would you expect between the R groups of the following amino acids in the tertiary structural level of a protein?
a. phenylalanine and isoleucine   **b.** aspartate and histidine
c. asparagine and tyrosine   **d.** alanine and proline

18.49 Indicate whether each of the following statements describes the primary, secondary, tertiary, or quaternary protein structure:
a. R groups in amino acids interact to form disulfide bonds or ionic bonds.
b. Peptide bonds join the amino acids in a polypeptide chain.
c. Several polypeptides in a β-pleated sheet are held together by hydrogen bonds between adjacent chains.

18.50 Indicate whether each of the following statements describes the primary, secondary, tertiary, or quaternary protein structure:
a. Hydrogen bonding between amino acid R groups in the same polypeptide gives a coiled shape to the protein.
b. Hydrophilic amino acids move to the polar aqueous environment outside the protein.
c. An active protein contains four subunits.

18.6 Proteins as Enzymes

LEARNING GOAL Describe the role of an enzyme in an enzyme-catalyzed reaction.

An **enzyme** has a unique three-dimensional shape that recognizes and binds a small group of reacting molecules called *substrates*. In a catalyzed reaction, an enzyme must first bind to a substrate in a way that favors catalysis.

A typical enzyme is much larger than its substrate. However, within its tertiary struc-ture is a region called the **active site** that binds a substrate or substrates and catalyzes the reaction. This active site is often a small pocket within the larger tertiary structure that closely fits the substrate (see **FIGURE 18.14**).

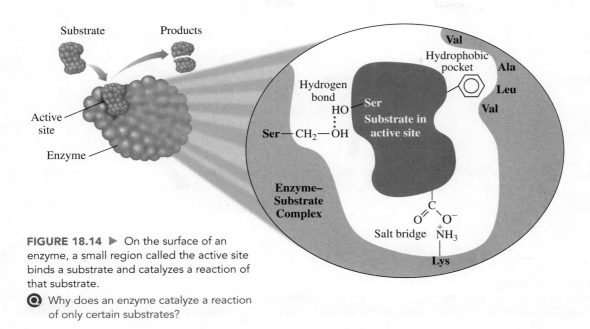

FIGURE 18.14 ▶ On the surface of an enzyme, a small region called the active site binds a substrate and catalyzes a reaction of that substrate.

◉ Why does an enzyme catalyze a reaction of only certain substrates?

Enzyme-Catalyzed Reaction

The combination of an enzyme (E) and a substrate (S) within the active site forms an *enzyme–substrate* (ES) *complex* that provides an alternative pathway for the reaction with lower activation energy. Within the active site, the amino acid R groups catalyze the reac-tion to give an *enzyme–product* (EP) *complex*. Then the products are released, and the enzyme is available to bind to another substrate molecule.

$$\underset{\text{Enzyme and Substrate}}{\textbf{E + S}} \quad \rightleftharpoons \quad \underset{\text{Enzyme–Substrate Complex}}{\textbf{ES complex}} \quad \longrightarrow \quad \underset{\text{Enzyme–Product Complex}}{\textbf{EP complex}} \quad \longrightarrow \quad \underset{\text{Enzyme and Product}}{\textbf{E + P}}$$

Models of Enzyme Action

An early theory of enzyme action, called the *lock-and-key model*, described the active site as having a rigid, nonflexible shape. According to the lock-and-key model, the shape of the active site was analogous to a lock, and its substrate was the key that specifically fit that lock. However, this model was a static one that did not allow for the flexibility of the tertiary shape of an enzyme and the way we now know that the active site can adjust to the shape of a substrate.

In the dynamic model of enzyme action, called the **induced-fit model**, the flexibility of the active site allows it to adapt to the shape of the substrate. At the same time, the shape of the substrate may be modified to better fit the geometry of the active site. As a result, the fit of both the active site and the substrate provides the best alignment for the catalysis of the reaction of the substrate. In the induced-fit model, substrate and enzyme work together to acquire a geometrical arrangement that lowers the activation energy.

In the hydrolysis of the disaccharide lactose by the enzyme lactase, a molecule of lac-tose binds to the active site of lactase. As the lactose binds to the enzyme, both the active site and the substrate lactose change shape. In this ES complex, the glycosidic bond of lac-tose is in a position that is favorable for *hydrolysis*, which is the splitting by water of a large molecule into smaller parts. The R groups on the amino acids in the active site then cata-lyze the hydrolysis of lactose which produces the monosaccharides glucose and galactose. Because the structures of the products are no longer attracted to the active site, they are released, which allows lactase to react with another lactose molecule (see **FIGURE 18.15**).

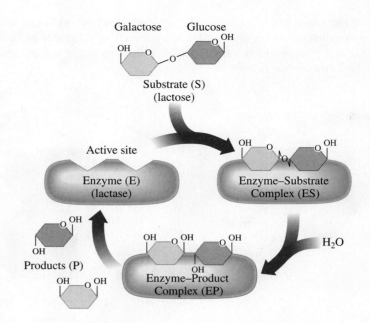

Galactose   Glucose

Substrate (S)
(lactose)

FIGURE 18.15 ▶ In the induced-fit model, a flexible active site and substrate both adjust to provide the best fit for the reaction. Lactose binds to the active site to align the glycosidic bond for hydrolysis. The monosaccharide products are released, and the enzyme binds to another lactose.

Q Why does the enzyme-catalyzed hydrolysis of lactose go faster than the hydrolysis of lactose in the chemistry laboratory?

Active site

Enzyme (E)
(lactase)

Enzyme–Substrate
Complex (ES)

H_2O

Products (P)

Enzyme–Product
Complex (EP)

QUESTIONS AND PROBLEMS

18.6 Proteins as Enzymes

LEARNING GOAL Describe the role of an enzyme in an enzyme-catalyzed reaction.

18.51 Match the terms, (1) enzyme, (2) enzyme–substrate complex, and (3) substrate, with each of the following:
   a. has a tertiary structure that recognizes the substrate
   b. has a structure that fits the active site of an enzyme
   c. the combination of an enzyme with the substrate

18.52 Match the terms, (1) active site, (2) lock-and-key model, and (3) induced-fit model, with each of the following:
   a. the portion of an enzyme where catalytic activity occurs
   b. an active site that adapts to the shape of a substrate
   c. an active site that has a rigid shape

18.53 **a.** Write an equation that represents an enzyme-catalyzed reaction.
   b. How is the active site different from the whole enzyme structure?

18.54 **a.** Why does an enzyme speed up the reaction of a substrate?
   b. After the products have formed, what happens to the enzyme?

18.7 Nucleic Acids

LEARNING GOAL Describe the structure of the nucleic acids in DNA and RNA.

There are two closely related types of nucleic acids: deoxyribonucleic acid (**DNA**) and ribonucleic acid (**RNA**). Both are polymers of repeating monomer units known as *nucleotides*. A DNA molecule may contain several million nucleotides; smaller RNA molecules may contain up to several thousand. Each nucleotide has three components: a base that contains nitrogen, a five-carbon sugar, and a phosphate group (see **FIGURE 18.16**).

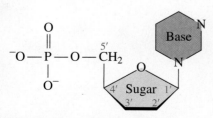

FIGURE 18.16 ▶ The general structure of a nucleotide includes a nitrogen-containing base, a sugar, and a phosphate group.

Q In a nucleotide, what types of groups are bonded to a five-carbon sugar?

Bases

The nitrogen-containing *bases* in nucleic acids are derivatives of *purine* or *pyrimidine*. In DNA, the purine bases with double rings are adenine (A) and guanine (G), and the pyrimidine bases with single rings are cytosine (C) and thymine (T). RNA contains the same bases, except thymine (5-methyluracil) is replaced by uracil (U) (see **FIGURE 18.17**).

Purines

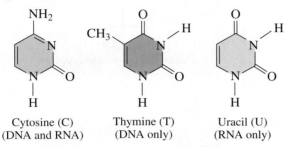

Adenine (A)
(DNA and RNA)

Guanine (G)
(DNA and RNA)

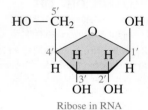

Purine   Pyrimidine

Pyrimidines

Cytosine (C)
(DNA and RNA)

Thymine (T)
(DNA only)

Uracil (U)
(RNA only)

FIGURE 18.17 ▶ DNA contains the bases A, G, C, and T; RNA contains A, G, C, and U.

🅠 Which bases are found in DNA?

Ribose and Deoxyribose Sugars

In RNA, the five-carbon sugar is *ribose*, which gives the letter R in the abbreviation RNA. The atoms in the pentose sugars are numbered with primes (1′, 2′, 3′, 4′, and 5′) to differentiate them from the atoms in the bases. In DNA, the five-carbon sugar is *deoxyribose*, which is similar to ribose except that there is no hydroxyl group (—OH) on C2′. The *deoxy* prefix means "without oxygen" and provides the D in DNA (see **FIGURE 18.18**).

Nucleosides and Nucleotides

A **nucleoside** is produced when a pyrimidine or purine forms a glycosidic bond to C1′ of a sugar, either ribose or deoxyribose. For example, adenine, a purine, and ribose form a nucleoside called adenosine.

Pentose Sugars in RNA and DNA

Ribose in RNA

No oxygen is bonded to this carbon

Deoxyribose in DNA

FIGURE 18.18 ▶ The five-carbon pentose sugar found in RNA is ribose and deoxyribose in DNA.

🅠 What is the difference between ribose and deoxyribose?

Sugar   +   Base   ⟶   Nucleoside   +   H₂O

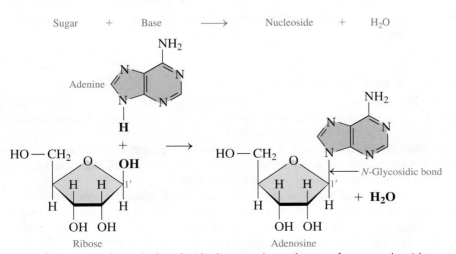

Ribose

Adenosine

A base forms an *N*-glycosidic bond with ribose or deoxyribose to form a nucleoside.

Nucleotides are produced when the C5′—OH group of ribose or deoxyribose in a nucleoside forms a phosphate ester. All the nucleotides in RNA and DNA are shown in **FIGURE 18.19**.

Adenosine monophosphate (AMP)
Deoxyadenosine monophosphate (dAMP)

Guanosine monophosphate (GMP)
Deoxyguanosine monophosphate (dGMP)

Cytidine monophosphate (CMP)
Deoxycytidine monophosphate (dCMP)

Uridine monophosphate (UMP)

Deoxythymidine monophosphate (dTMP)

FIGURE 18.19 ▶ The nucleotides of RNA (shown in black) are similar to those of DNA (shown in magenta), except in DNA the sugar is deoxyribose and deoxythymidine replaces uridine.

Q What are two differences in the nucleotides of RNA and DNA?

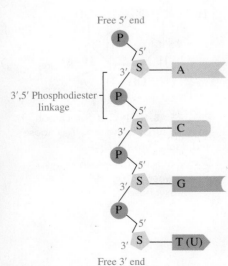

In the primary structure of nucleic acids, each sugar in a sugar–phosphate backbone is attached to a base.

The name of a nucleoside that contains a purine ends with *osine*, whereas a nucleoside that contains a pyrimidine ends with *idine*. The names of nucleosides of DNA add *deoxy* to the beginning of their names. The corresponding nucleotides in RNA and DNA are named by adding *monophosphate* to the end of the nucleoside name. Although the letters A, G, C, U, and T represent the bases, they are often used in the abbreviations of the respective nucleotides.

Structure of Nucleic Acids

The **nucleic acids** are polymers of many nucleotides in which the 3′ hydroxyl group of the sugar in one nucleotide bonds to the phosphate group on the 5′ carbon atom in the sugar of the next nucleotide. This connection between the sugars in adjacent nucleotides is referred to as a **phosphodiester linkage**. As more nucleotides are added, a backbone forms that consists of alternating sugar and phosphate groups. The bases, which are attached to each sugar, extend out from the sugar–phosphate backbone.

In any nucleic acid, the sugar at one end has a free 5′ phosphate group, and the sugar at the other end has a free 3′ hydroxyl group. A nucleic acid sequence is read from the free 5′ phosphate end to the free 3′ hydroxyl end using only the letters of the bases. For example, the nucleotide sequence in the section of RNA shown in **FIGURE 18.20** is —A C G U—.

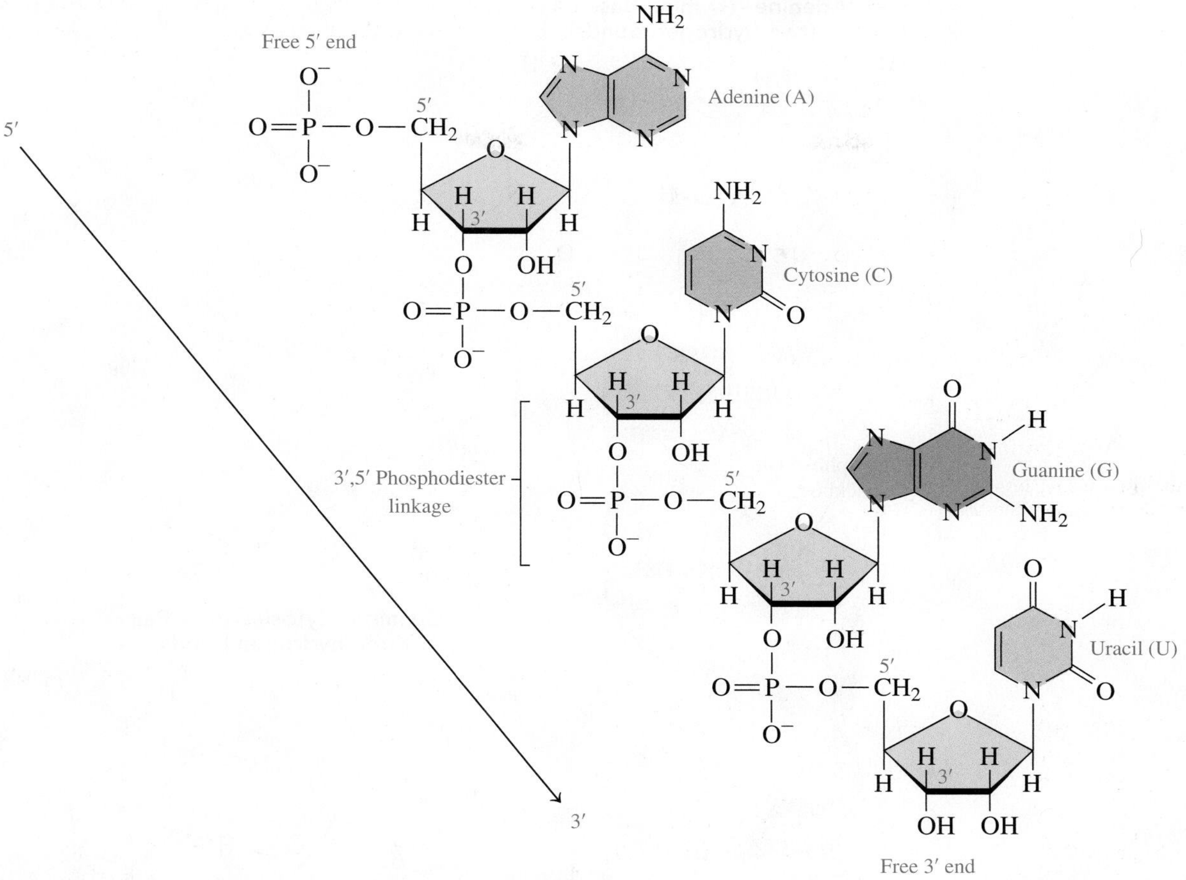

FIGURE 18.20 ▶ In the primary structure of an RNA, the nucleotides are linked by 3′,5′ phosphodiester linkages.
Ⓠ What is the abbreviation for the sequence of nucleotides in this RNA section?

DNA Double Helix: A Secondary Structure

In 1953, James Watson and Francis Crick proposed that DNA was a **double helix** that consists of two polynucleotide strands winding about each other like a spiral staircase. The sugar–phosphate backbones are analogous to the outside railings, with the bases arranged like steps along the inside.

Complementary Base Pairs

Each of the bases along one polynucleotide strand forms hydrogen bonds to a specific base on the opposite DNA strand. Adenine only bonds to thymine, and guanine only bonds to cytosine (see **FIGURE 18.21**). The pairs AT and GC are called **complementary base pairs**. The specific pairing of the bases occurs because adenine and thymine form two hydrogen bonds, while cytosine and guanine form three hydrogen bonds.

CORE CHEMISTRY SKILL
Writing the Complementary
DNA Strand

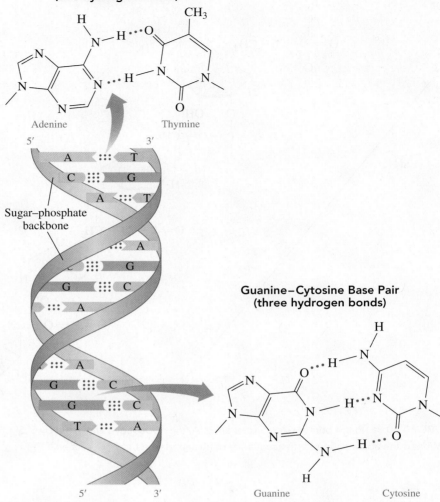

FIGURE 18.21 ▶ In the model shown, the sugar–phosphate backbone is represented by a ribbon with hydrogen bonds between complementary base pairs.

Q Why are GC base pairs more stable than AT base pairs?

SAMPLE PROBLEM 18.10 Complementary Base Pairs

Write the complementary base sequence for the following segment of a strand of DNA:

A C G A T C T

TRY IT FIRST

SOLUTION

In the complementary strand of DNA, the complementary base pairs are AT and CG.

Original segment of DNA:	A C G A T C T
	: : : : : : :
Complementary segment:	T G C T A G A

STUDY CHECK 18.10

What sequence of bases is complementary to a DNA sequence of G G T T A A C C ?

ANSWER

C C A A T T G G

DNA Replication

In DNA **replication**, the strands in the parent DNA separate, which allows the synthesis of complementary strands of DNA. The replication process begins when an enzyme catalyzes the unwinding of a portion of the double helix by breaking the hydrogen bonds between the complementary bases. These single strands or *parent DNA* now act as templates for the synthesis of new complementary strands of DNA (see **FIGURE 18.22**). Within the nucleus, nucleotides for each base form hydrogen bonds with their complementary bases.

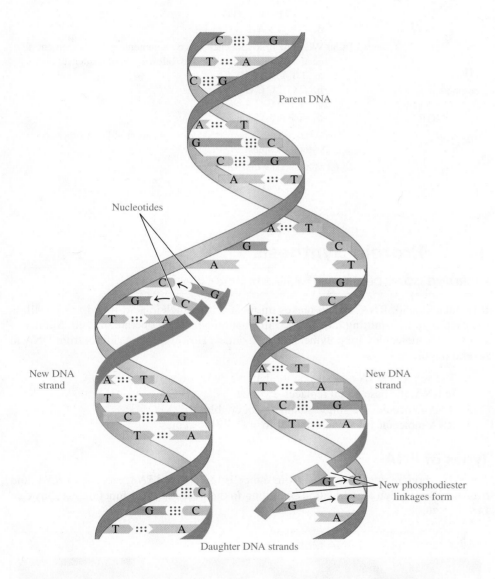

FIGURE 18.22 ▶ In DNA replication, the separate strands of the parent DNA are the templates for the synthesis of complementary strands, which produces two exact copies of DNA.

◉ How many strands of the parent DNA are in each of the new double-stranded copies of DNA?

Eventually, the entire double helix of the parent DNA is copied. In each new DNA molecule, one strand of the double helix is from the original DNA and one is a newly synthesized strand. This process produces two new DNAs called *daughter DNA* that are identical to each other and exact copies of the original parent DNA. In DNA replication, complementary base pairing ensures the correct placement of bases in the daughter DNA strands.

QUESTIONS AND PROBLEMS

18.7 Nucleic Acids

LEARNING GOAL Describe the structure of the nucleic acids in DNA and RNA.

18.55 Identify each of the following bases as a component of RNA, DNA, or both:
  a. thymine
  b.

18.56 Identify each of the following bases as a component of RNA, DNA, or both:
  a. guanine
  b.

18.57 How are the two strands of nucleic acid in DNA held together?

18.58 What is meant by complementary base pairing?

18.59 Write the base sequence in a complementary DNA segment if each original segment has the following base sequence:
  a. A A A A A A
  b. G G G G G G
  c. A G T C C A G G T
  d. C T G T A T A C G T T

18.60 Write the base sequence in a complementary DNA segment if each original segment has the following base sequence:
  a. T T T T T T
  b. C C C C C C C C C
  c. A T G G C A
  d. A T A T G C G C T A

18.61 What process ensures that the replication of DNA produces identical copies?

18.62 What is daughter DNA?

18.8 Protein Synthesis

LEARNING GOAL Describe the synthesis of protein from mRNA.

Ribonucleic acid, RNA, which makes up most of the nucleic acid found in the cell, is involved with transmitting the genetic information needed to operate the cell. Similar to DNA, RNA molecules are polymers of nucleotides. However, RNA differs from DNA in several important ways:

1. The sugar in RNA is ribose rather than the deoxyribose found in DNA.
2. In RNA, the base uracil replaces thymine.
3. RNA molecules are single stranded, not double stranded.
4. RNA molecules are much smaller than DNA molecules.

Types of RNA

There are three major types of RNA in the cells: *messenger RNA*, *ribosomal RNA*, and *transfer RNA*, which are classified according to their location and function, as shown in **TABLE 18.10**.

TABLE 18.10 Types of RNA Molecules in Humans

Type	Abbreviation	Function in the Cell	Percentage of Total RNA
Ribosomal RNA	rRNA	major component of the ribosomes; site of protein synthesis	80
Messenger RNA	mRNA	carries information for protein synthesis from the DNA to the ribosomes	5
Transfer RNA	tRNA	brings specific amino acids to the site of protein synthesis	15

In replication, the genetic information in DNA is reproduced by making identical copies of DNA. In **transcription**, the information contained in DNA is transferred to mRNA molecules. In **translation**, the genetic information now present in the mRNA is used to build the sequence of amino acids of the desired protein (see **FIGURE 18.23**).

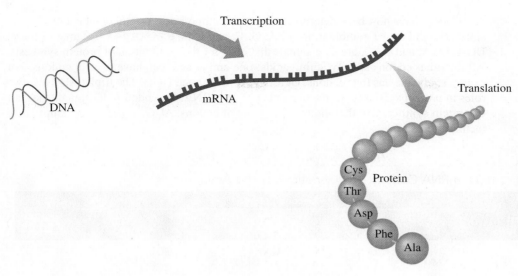

Transcription

DNA

mRNA

Translation

Cys
Thr   Protein
Asp
Phe
Ala

FIGURE 18.23 ▶ The genetic information in DNA is replicated in cell division and used to produce messenger RNAs that code for the amino acids needed for protein synthesis.

Q What is the difference between transcription and translation?

Transcription begins when the section of DNA to be copied unwinds. One strand of DNA acts as a template as bonds are formed to each complementary base: C is paired with G, T pairs with A, and A pairs with U (not T).

CORE CHEMISTRY SKILL
Writing the mRNA Segment for a DNA Template

SAMPLE PROBLEM 18.11 RNA Synthesis

The sequence of bases in a segment of the DNA template for mRNA is C G A T C A. What corresponding mRNA is produced?

TRY IT FIRST

SOLUTION

To form the corresponding section of mRNA, each base in the DNA template is paired with its complementary base: G with C, C with G, T with A, and A with U.

DNA template: C G A T C A

Transcription   ↓ ↓ ↓ ↓ ↓ ↓

Complementary base sequence in mRNA: G C U A G U

STUDY CHECK 18.11

What is the DNA template that codes for the mRNA segment with the nucleotide sequence G G G U U U A A A?

ANSWER

C C C A A A T T T

The Genetic Code

The **genetic code** consists of a series of three nucleotides (triplet) in mRNA, called a **codon**. Each codon specifies an amino acid and its sequence in a protein. Early work on protein synthesis showed that repeating triplets of uracil, UUU, produced a polypeptide that contained only phenylalanine. Therefore, a sequence of UUU UUU UUU codes for three phenylalanines.

Interactive Video

Protein Synthesis

Codons in mRNA: UUU UUU UUU

Translation   ↓   ↓   ↓

Amino acid sequence: Phe — Phe — Phe

Codons have now been determined for all 20 amino acids. A total of 64 codons is possible from the triplet combinations of A, G, C, and U (see TABLE 18.11). Three of these, UGA, UAA, and UAG, are stop signals that code for the termination of protein synthesis. All the other codons specify amino acids; one amino acid can have several codons. For example, glycine has four codons: GGU, GGC, GGA, and GGG. The triplet AUG has two roles in protein synthesis. At the beginning of an mRNA, the codon AUG signals the start of protein synthesis. In the middle of a series of codons, AUG codes for the amino acid methionine.

TABLE 18.11 mRNA Codons: The Genetic Code for Amino Acids

First Letter	Second Letter				Third Letter
	U	**C**	**A**	**G**	
U	UUU ⎱ Phe (F) UUC ⎰ UUA ⎱ Leu (L) UUG ⎰	UCU ⎱ UCC ⎱ Ser (S) UCA ⎰ UCG ⎰	UAU ⎱ Tyr (Y) UAC ⎰ UAA STOP[b] UAG STOP[b]	UGU ⎱ Cys (C) UGC ⎰ UGA STOP[b] UGG Trp (W)	U C A G
C	CUU ⎱ CUC ⎱ Leu (L) CUA ⎰ CUG ⎰	CCU ⎱ CCC ⎱ Pro (P) CCA ⎰ CCG ⎰	CAU ⎱ His (H) CAC ⎰ CAA ⎱ Gln (Q) CAG ⎰	CGU ⎱ CGC ⎱ Arg (R) CGA ⎰ CGG ⎰	U C A G
A	AUU ⎱ AUC ⎱ Ile (I) AUA ⎰ AUG START[a]/ Met (M)	ACU ⎱ ACC ⎱ Thr (T) ACA ⎰ ACG ⎰	AAU ⎱ Asn (N) AAC ⎰ AAA ⎱ Lys (K) AAG ⎰	AGU ⎱ Ser (S) AGC ⎰ AGA ⎱ Arg (R) AGG ⎰	U C A G
G	GUU ⎱ GUC ⎱ Val (V) GUA ⎰ GUG ⎰	GCU ⎱ GCC ⎱ Ala (A) GCA ⎰ GCG ⎰	GAU ⎱ Asp (D) GAC ⎰ GAA ⎱ Glu (E) GAG ⎰	GGU ⎱ GGC ⎱ Gly (G) GGA ⎰ GGG ⎰	U C A G

START[a] codon signals the initiation of a peptide chain.

STOP[b] codons signal the end of a peptide chain.

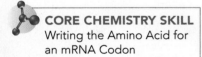

CORE CHEMISTRY SKILL

Writing the Amino Acid for an mRNA Codon

Protein Synthesis: Translation

Once the mRNA is synthesized, it migrates out of the nucleus into the cytoplasm to the ribosomes. At the ribosomes, the *translation* process converts the codons on mRNA into amino acids to make a protein.

Protein synthesis begins when the mRNA combines with a ribosome. There, tRNA molecules, which carry amino acids, align with mRNA, and a peptide bond forms between the amino acids. After the first tRNA detaches from the ribosome, the ribosome shifts to the next codon on the mRNA. Each time the ribosome shifts and the next tRNA aligns with the mRNA, a peptide bond joins the new amino acid to the growing polypeptide chain. After all the amino acids for a particular protein have been linked together by peptide bonds, the ribosome encounters a stop codon. Because there are no tRNAs to complement the termination codon, protein synthesis ends and the completed polypeptide chain is released from the ribosome. Then interactions between the amino acids in the chain form the protein into the three-dimensional structure that makes the polypeptide into a biologically active protein (see FIGURE 18.24).

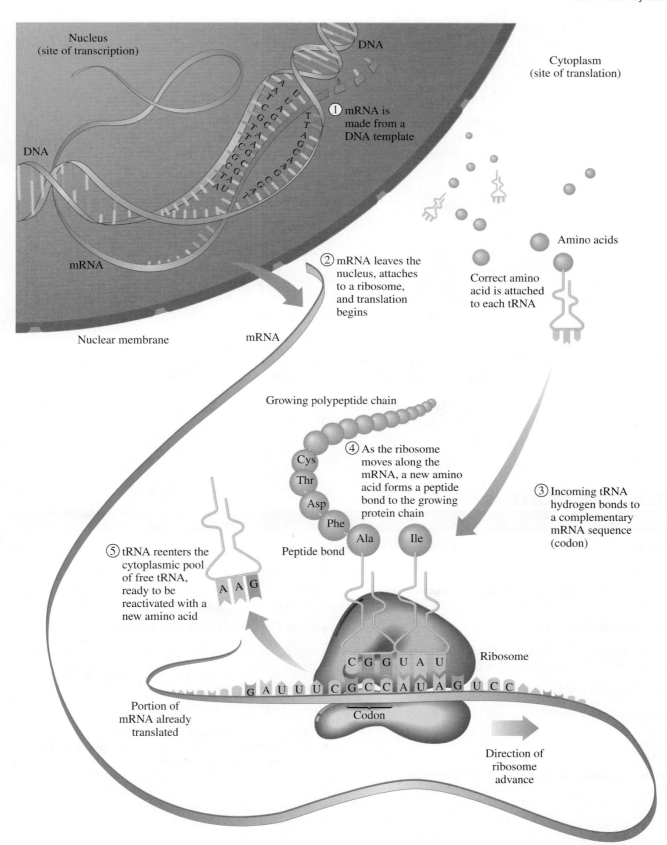

FIGURE 18.24 ▶ In the translation process, the mRNA synthesized by transcription attaches to a ribosome, and tRNAs pick up their amino acids, bind to the appropriate codon, and place them in a growing peptide chain.

Q How is the correct amino acid placed in the peptide chain?

SAMPLE PROBLEM 18.12 Protein Synthesis: Translation

Use three-letter and one-letter abbreviations to write the amino acid sequence for the peptide from the mRNA sequence of UCA AAA GCC CUU.

SOLUTION

Each of the codons specifies a particular amino acid. Using Table 18.11, we write the codons and the amino acids in the peptide.

ANALYZE THE PROBLEM	Given	Need	Connect
	mRNA UCA AAA GCC CUU	amino acid sequence	genetic code

mRNA codons:

Amino acid sequence: Ser—Lys—Ala—Leu, SKAL

STUDY CHECK 18.12

Use three-letter and one-letter abbreviations to write the amino acid sequence for the peptide from the mRNA sequence of GGG AGC AGU GAG GUU.

ANSWER

Gly–Ser–Ser–Glu–Val, GSSEV

QUESTIONS AND PROBLEMS

18.8 Protein Synthesis

LEARNING GOAL Describe the synthesis of protein from mRNA.

18.63 Write the corresponding section of mRNA produced from the following section of DNA template:

    C C G A A G G T T C A C

18.64 Write the corresponding section of mRNA produced from the following section of DNA template:

    T A C G G A A G C T A

18.65 What amino acid is coded for by each of the following RNA codons?
  a. CUG   **b.** UCC
  c. GGU   **d.** AGG

18.66 What amino acid is coded for by each of the following RNA codons?
  a. AAG   **b.** GUC
  c. CGG   **d.** GCA

Applications

18.67 The following is a segment of the DNA template that codes for human insulin:

    TTT GTG AAC CAA CAC CTG

  a. Write the corresponding mRNA segment.
  b. Write the three-letter and one-letter abbreviations for the corresponding peptide segment.

18.68 The following is a segment of the DNA template that codes for human insulin:

    TGC GGC TCA CAC CTG GTG

  a. Write the corresponding mRNA segment.
  b. Write the three-letter and one-letter abbreviations for the corresponding peptide segment.

Follow Up

REBECCA'S PROGRAM TO LOWER CHOLESTEROL IN FAMILIAL HYPERCHOLESTEROLEMIA (FH)

At the lipid clinic, Susan told Rebecca how to maintain a diet with less beef and chicken, more fish with omega-3 oils, low-fat dairy products, no egg yolks, and no coconut or palm oils. Rebecca maintained her new diet containing lower quantities of fats and increased quantities of fiber for the next year. She also increased her exercise using a treadmill and cycling and lost 35 lb. When Rebecca returned to the lipid clinic to check her progress with Susan, a new set of blood tests indicated that her total cholesterol had dropped from 420 to 390 mg/dL and her LDL had dropped from 275 to 255 mg/dL.

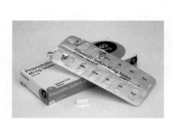

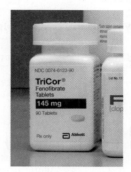

Pravastatin (Pravachol)

Because Rebecca's changes in diet and exercise did not sufficiently lower her cholesterol level, Susan prescribed a medication to help lower blood cholesterol levels. The most common medications used for treating high LDL-cholesterol are the statins lovastatin (Mevacor), pravastatin (Pravachol), simvastatin (Zocor), atorvastatin (Lipitor), and rosuvastatin (Crestor). For some FH patients, statin therapy may be combined with fibrates such as gemfibrozil or fenofibrate that reduce the synthesis of enzymes that break down fats in the blood.

Susan prescribed pravastatin (Pravachol), 80. mg once a day, which was effective. Later, Susan added fenofibrate (TriCor), one 145-mg tablet, each day to the Pravachol. Rebecca understands that her medications and diet and exercise plan is a lifelong process and its impact is to help her live a healthier, longer life.

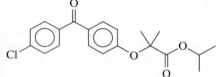

Fenofibrate (TriCor)

Applications

18.69 If Rebecca's prescription for Pravachol was 80. mg daily, how many grams of Pravachol did Rebecca consume in one week?

18.70 Five months later, Rebecca added the fibrate TriCor, one 145-mg tablet daily. How many grams of TriCor did Rebecca consume in one week?

CONCEPT MAP

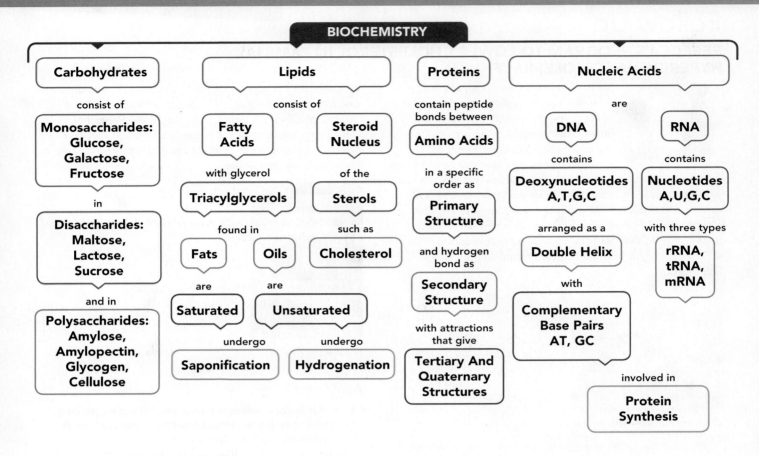

BIOCHEMISTRY

Carbohydrates

consist of

Monosaccharides:
Glucose,
Galactose,
Fructose

in

Disaccharides:
Maltose,
Lactose,
Sucrose

and in

Polysaccharides:
Amylose,
Amylopectin,
Glycogen,
Cellulose

Lipids

consist of

Fatty Acids — **Steroid Nucleus**

with glycerol

Triacylglycerols

found in

Fats — **Oils**

are — are

Saturated — **Unsaturated**

undergo — undergo

Saponification — **Hydrogenation**

of the

Sterols

such as

Cholesterol

Proteins

contain peptide bonds between

Amino Acids

in a specific order as

Primary Structure

and hydrogen bond as

Secondary Structure

with attractions that give

Tertiary And Quaternary Structures

Nucleic Acids

are

DNA — **RNA**

contains — contains

Deoxynucleotides A,T,G,C — **Nucleotides A,U,G,C**

arranged as a — with three types

Double Helix — **rRNA, tRNA, mRNA**

with

Complementary Base Pairs AT, GC

involved in

Protein Synthesis

CHAPTER REVIEW

18.1 Carbohydrates

LEARNING GOAL Classify a carbohydrate as an aldose or a ketose; draw the open-chain and Haworth structures for glucose, galactose, and fructose.

- Carbohydrates are composed of carbon, hydrogen, and oxygen.
- Monosaccharides are polyhydroxy aldehydes (aldoses) or ketones (ketoses).
- Monosaccharides are also classified by their number of carbon atoms: triose, tetrose, pentose, or hexose. Important monosaccharides are glucose, galactose, and fructose.
- The predominant form of monosaccharides is the cyclic form of five or six atoms.
- The Haworth structure forms when an —OH group (usually the one on carbon 5 in hexoses) reacts with the carbonyl group of the same molecule.

18.2 Disaccharides and Polysaccharides

LEARNING GOAL Describe the monosaccharide units and linkages in disaccharides and polysaccharides.

- Disaccharides are two monosaccharide units joined together by a glycosidic bond.
- In the most common disaccharides maltose, lactose, and sucrose, there is at least one glucose unit.
- Polysaccharides are polymers of monosaccharide units. Amylose is an unbranched polymer of glucose, and amylopectin is a branched polymer of glucose. Glycogen, the storage form of glucose in animals, is similar to amylopectin with more branching. Cellulose is also a polymer of glucose, but in cellulose the glycosidic bonds are β bonds rather than α bonds.

18.3 Lipids

LEARNING GOAL Draw the condensed or line-angle structural formula for a fatty acid, a triacylglycerol, and the products of hydrogenation or saponification. Identify the steroid nucleus.

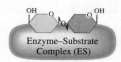

- Lipids are nonpolar compounds that are not soluble in water.
- Classes of lipids include waxes, fats, oils, and steroids.
- Fatty acids are long-chain carboxylic acids that may be saturated or unsaturated.
- Triacylglycerols are esters of glycerol with three fatty acids.
- Fats contain more saturated fatty acids and have higher melting points than most vegetable oils.
- The hydrogenation of unsaturated fatty acids converts double bonds to single bonds.
- In saponification, a fat heated with a strong base produces glycerol and the salts of the fatty acids (soap).
- Steroids are lipids containing the steroid nucleus, which is a fused structure of four rings.

18.4 Amino Acids and Proteins

LEARNING GOAL Describe protein functions and draw structures for amino acids and peptides.

- Some proteins are enzymes or hormones, whereas others are important in structure, transport, protection, storage, and muscle contraction.
- A group of 20 amino acids provides the molecular building blocks of proteins.
- Attached to the central (alpha) carbon of each amino acid are an ammonium group, a carboxylate group, and a unique R group.
- Peptides form when an amide bond links the carboxylate group of one amino acid and the ammonium group of a second amino acid.
- Long chains of amino acids that are biologically active are called proteins.

18.5 Protein Structure

LEARNING GOAL Identify the levels of structure of a protein.

- The primary structure of a protein is its sequence of amino acids joined by peptide bonds.
- In the secondary structure, hydrogen bonds between atoms in the peptide bonds produce a characteristic shape such as an α helix or, a β-pleated sheet.
- The most abundant protein in the body is collagen, which is composed of fibrils of triple helices that are hydrogen bonded.
- A tertiary structure is stabilized by interactions between R groups of amino acids in one region of the polypeptide chain with R groups in different regions of the protein.
- In a quaternary structure, two or more subunits combine for biological activity.

18.6 Proteins as Enzymes

LEARNING GOAL Describe the role of an enzyme in an enzyme-catalyzed reaction.

Enzyme–Substrate Complex (ES)

- Enzymes are proteins that act as biological catalysts by accelerating the rate of cellular reactions.
- Within the tertiary structure of an enzyme, a small pocket called the active site binds the substrate.
- In the lock-and-key model, an early theory of enzyme action, a substrate precisely fits the shape of the active site.
- In the induced-fit model, both the active site and the substrate undergo changes in their shapes to give the best fit for efficient catalysis. In the enzyme–substrate complex, catalysis takes place when the amino acid R groups in the active site of an enzyme react with a substrate.
- When the products of catalysis are released, the enzyme can bind another substrate molecule.

18.7 Nucleic Acids

LEARNING GOAL Describe the structure of the nucleic acids in DNA and RNA.

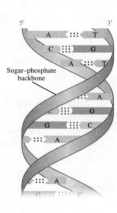
Sugar–phosphate backbone

- Nucleic acids, such as deoxyribonucleic acid (DNA) and ribonucleic acid (RNA), are polymers of nucleotides.
- A nucleotide is composed of three parts: a base, a pentose sugar, and a phosphate group.
- In DNA, the sugar is deoxyribose and the base can be adenine, thymine, guanine, or cytosine. In RNA, the sugar is ribose and uracil replaces thymine.
- Each nucleic acid has its own unique sequence of bases.
- A DNA molecule consists of two strands of nucleotides that are wound around each other like a spiral staircase.
- The two strands are held together by hydrogen bonds between complementary base pairs AT and GC.

18.8 Protein Synthesis

LEARNING GOAL Describe the synthesis of protein from mRNA.

- The bases in the mRNA are complementary to the DNA, except A in DNA is paired with U in RNA.
- The genetic code consists of a series of codons, which are sequences of three bases that specify the amino acids in a protein.
- Proteins are synthesized at the ribosomes.
- During translation, tRNAs bring the appropriate amino acids to the mRNA at the ribosome and peptide bonds form.
- When the polypeptide is released, it takes on its secondary and tertiary structures and becomes a functional protein in the cell.

KEY TERMS

active site A pocket in a part of the tertiary enzyme structure that binds substrate and catalyzes a reaction.

aldose A monosaccharide that contains an aldehyde group.

amino acid The building block of proteins, consisting of a hydrogen atom, an ammonium group, a carboxylate group, and a unique R group attached to the alpha carbon.

amylopectin A branched-chain polymer of starch composed of glucose units joined by $\alpha(1 \rightarrow 4)$- and $\alpha(1 \rightarrow 6)$-glycosidic bonds.

amylose An unbranched polymer of starch composed of glucose units joined by $\alpha(1 \rightarrow 4)$-glycosidic bonds.

carbohydrate A simple or complex sugar composed of carbon, hydrogen, and oxygen.

cellulose An unbranched polysaccharide composed of glucose units linked by $\beta(1 \rightarrow 4)$-glycosidic bonds that cannot be hydrolyzed by the human digestive system.

cholesterol The most prevalent of the steroid compounds found in cellular membranes.

codon A sequence of three bases in mRNA that specifies a certain amino acid to be placed in a protein. A few codons signal the start or stop of protein synthesis.

complementary base pairs In DNA, adenine is always paired with thymine (A and T or T and A), and guanine is always paired with cytosine (G and C or C and G). In forming RNA, adenine is paired with uracil (A and U).

disaccharide A carbohydrate composed of two monosaccharides joined by a glycosidic bond.

DNA Deoxyribonucleic acid; the genetic material of all cells containing nucleotides with deoxyribose, phosphate, and the four bases: adenine, thymine, guanine, and cytosine.

double helix The helical shape of the double chain of DNA that is like a spiral staircase with a sugar–phosphate backbone on the outside and base pairs like stair steps on the inside.

enzyme A protein that catalyzes a biological reaction.

fat A triacylglycerol that is solid at room temperature and usually comes from animal sources.

fatty acid A long-chain carboxylic acid found in many lipids.

fructose A ketohexose which is combined with glucose in sucrose.

galactose An aldohexose that occurs combined with glucose in lactose.

genetic code The sequence of codons in mRNA that specifies the amino acid order for the synthesis of protein.

glucose An aldohexose found in fruits, vegetables, corn syrup, and honey that is also known as blood sugar and dextrose. The most prevalent monosaccharide in the diet. Most polysaccharides are polymers of glucose.

glycogen A polysaccharide formed in the liver and muscles for the storage of glucose as an energy reserve. It is composed of glucose in a highly branched polymer joined by $\alpha(1 \rightarrow 4)$- and $\alpha(1 \rightarrow 6)$-glycosidic bonds.

induced-fit model A model of enzyme action in which the shape of a substrate and the active site of the enzyme adjust to give an optimal fit.

ketose A monosaccharide that contains a ketone group.

lactose A disaccharide consisting of glucose and galactose found in milk and milk products.

lipids A family of compounds that is nonpolar in nature and not soluble in water; includes fats, oils, waxes, and steroids.

maltose A disaccharide consisting of two glucose units; it is obtained from the hydrolysis of starch.

monosaccharide A polyhydroxy compound that contains an aldehyde or a ketone group.

nucleic acid A large molecule composed of nucleotides; found as a double helix in DNA and as the single strands of RNA.

nucleoside The combination of a pentose sugar and a base.

nucleotide Building block of a nucleic acid consisting of a base, a pentose sugar (ribose or deoxyribose), and a phosphate group.

oil A triacylglycerol that is usually a liquid at room temperature and is obtained from a plant source.

phosphodiester linkage The phosphate link that joins the hydroxyl group in one nucleotide to the phosphate group on the next nucleotide.

polysaccharide A polymer of many monosaccharide units, usually glucose. Polysaccharides differ in the types of glycosidic bonds and the amount of branching in the polymer.

primary structure The specific sequence of the amino acids in a protein.

protein A term used for biologically active polypeptides that have many amino acids linked together by peptide bonds.

quaternary structure A protein structure in which two or more protein subunits form an active protein.

replication The process of duplicating DNA by pairing the bases on each parent strand with their complementary bases.

RNA Ribonucleic acid; a type of nucleic acid that is a single strand of nucleotides containing ribose, phosphate, and the four bases: adenine, cytosine, guanine, and uracil.

saponification The reaction of a fat with a strong base to form glycerol and salts of fatty acids (soaps).

secondary structure The formation of an α helix or β-pleated sheet by hydrogen bonds.

steroids Types of lipid composed of a multicyclic ring system.

sucrose A disaccharide composed of glucose and fructose; commonly called table sugar or "sugar."

tertiary structure The folding of the secondary structure of a protein into a compact structure that is stabilized by the interactions of R groups such as salt bridges and disulfide bonds.

transcription The transfer of genetic information from DNA by the formation of mRNA.

translation The interpretation of the codons in mRNA as amino acids in a peptide.

triacylglycerols A family of lipids composed of three fatty acids bonded through ester bonds to glycerol, a trihydroxy alcohol.

CORE CHEMISTRY SKILLS

The chapter section containing each Core Chemistry Skill is shown in parentheses at the end of each heading.

Drawing Haworth Structures (18.1)

- The Haworth structure shows the ring structure of a monosaccharide.
- Groups on the right side of the open-chain structure of the monosaccharide are below the plane of the ring, those on the left are above the plane.
- The new —OH group is the α form if it is below the plane of the ring, or the β form if it is above the plane.

Example: Draw the Haworth structure for β-D-idose.

Answer: In the β-form, the new —OH group is above the plane of the ring.

Identifying Fatty Acids (18.3)

- Fatty acids are unbranched carboxylic acids that typically contain an even number (12 to 20) of carbon atoms.
- Fatty acids may be saturated, monounsaturated with one double bond, or polyunsaturated with two or more double bonds.

Example: State the number of carbon atoms, saturated or unsaturated, and name of the following:

Answer: 16 carbon atoms, saturated, palmitic acid

Drawing Structures for Triacylglycerols (18.3)

- Triacylglycerols are esters of glycerol with three long-chain fatty acids.

Example: Draw the line-angle structural formula for the triacylglycerol containing three molecules of palmitic acid (16:0) and give the name.

Answer:

Glyceryl tripalmitate (tripalmitin)

Drawing the Products for the Hydrogenation and Saponification of a Triacylglycerol (18.3)

- The hydrogenation of unsaturated fatty acids of a triacylglycerol in the presence of a Pt or Ni catalyst converts double bonds to single bonds.
- In saponification, a triacylglycerol heated with a strong base produces glycerol and the salts of the fatty acids (soap).

Example: Identify each of the following as hydrogenation or saponification and state the products:

- **a.** the reaction of palm oil with NaOH
- **b.** the reaction of corn oil and H_2 using a nickel catalyst

Answer: **a.** The reaction of palm oil with NaOH is saponification, and the products are glycerol and the potassium salts of the fatty acids, which is soap.

 b. In hydrogenation, H_2 adds to double bonds in corn oil, which produces a more saturated, and thus more solid, fat.

Drawing the Ionized Form for an Amino Acid (18.4)

- The central α carbon of each amino acid is bonded to an ammonium group ($-NH_3^+$), a carboxylate group ($-COO^-$), a hydrogen atom, and a unique R group.
- The R group gives an amino acid the property of being nonpolar, polar, acidic, or basic.

Example: Draw the condensed structural formula for cysteine.

Answer:

Identifying the Primary, Secondary, Tertiary, and Quaternary Structures of Proteins (18.5)

- The primary structure of a protein is the sequence of amino acids joined by peptide bonds.
- In the secondary structures of proteins, hydrogen bonds between atoms in the peptide bonds produce an alpha helix or a beta-pleated sheet.
- The tertiary structure of a protein is stabilized by R groups that form hydrogen bonds, disulfide bonds, and salt bridges, and hydrophobic groups that move to the center and hydrophilic R groups that move to the surface.
- In a quaternary structure, two or more subunits are combined for biological activity, held by the same interactions found in tertiary structures.

Example: Identify the following as characteristic of the primary, secondary, tertiary, or quaternary structure of a protein:

a. The R groups of two amino acids interact to form a salt bridge.
b. Eight amino acids form peptide bonds.
c. A polypeptide forms an alpha helix.
d. Two amino acids with hydrophobic R groups move toward the inside of the folded protein.
e. A protein with biological activity contains four polypeptide subunits.

Answer:

a. tertiary, quaternary
b. primary
c. secondary
d. tertiary
e. quaternary

Writing the Complementary DNA Strand (18.7)

- During DNA replication, new DNA strands are made along each of the original DNA strands.
- The new strand of DNA is made by forming hydrogen bonds with the bases in the template strand: A with T and T with A; C with G and G with C.

Example: Write the complementary base sequence for the following DNA segment:
   A T T C G G T A C

Answer: T A A G C C A T G

Writing the mRNA Segment for a DNA Template (18.8)

- Transcription is the process that produces mRNA from one strand of DNA.
- The bases in the mRNA are complementary to the DNA except that A in DNA is paired with U in RNA.

Example: What is the mRNA produced for the DNA segment
   C G C A T G T C A ?

Answer: G C G U A C A G U

Writing the Amino Acid for an mRNA Codon (18.8)

- The genetic code consists of a sequence of three bases (codons) that specifies the order for the amino acids in a protein.
- The codon AUG signals the start of transcription and codons UAG, UGA, and UAA signal it to stop.

Example: Use three-letter and one-letter abbreviations to write the amino acid sequence for the peptide from the mRNA sequence CCG UAU GGG .

Answer: Pro–Tyr–Gly, PYG

UNDERSTANDING THE CONCEPTS

The chapter sections to review are shown in parentheses at the end of each question.

18.71 Melezitose is a saccharide with the following structure: (18.1, 18.2)

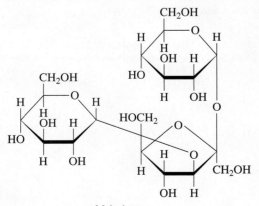

Melezitose

a. Is melezitose a mono-, di-, tri-, or polysaccharide?
b. What ketohexose and aldohexose are present in melezitose?

18.72 What are the disaccharides and polysaccharides present in each of the following? (18.1, 18.2)

(a)   (b)

(c)   (d)

18.73 One of the triacylglycerols in palm oil is glyceryl tripalmitate. Draw the condensed structural formula for glyceryl tripalmitate. (18.3)

Palm fruit from palm trees are a source of palm oil.

18.74 One of the compounds in jojoba wax used to make candles consists of an 18-carbon saturated fatty acid and a 22-carbon saturated alcohol. Draw the condensed structural formula for jojoba wax. (18.3)

Candles contain jojoba wax.

18.75 Identify each of the following as a saturated, monounsaturated, or polyunsaturated fatty acid: (18.3)
a. $CH_3-(CH_2)_4-(CH=CH-CH_2)_2-(CH_2)_6-COOH$
b. linolenic acid

Salmon is a good source of unsaturated fatty acids.

18.76 Identify each of the following as a saturated, monounsaturated, or polyunsaturated fatty acid: (18.3)
a. $CH_3-(CH_2)_{14}-COOH$
b. $CH_3-(CH_2)_7-CH=CH-(CH_2)_7-COOH$

18.77 Seeds and vegetables are often deficient in one or more essential amino acids. The table below shows which essential amino acids are present in each food. (18.4)

Source	Lysine	Tryptophan	Methionine
Oatmeal	No	Yes	Yes
Rice	No	Yes	Yes
Garbanzo beans	Yes	No	Yes
Lima beans	Yes	No	No
Cornmeal	No	No	Yes

Use the table to decide if each food combination provides the essential amino acids lysine, tryptophan, and methionine.
a. rice and garbanzo beans
b. lima beans and cornmeal
c. a salad of garbanzo beans and lima beans

18.78 Use the table in problem 18.77 to decide if each food combination provides the essential amino acids lysine, tryptophan, and methionine. (18.4)

Oatmeal is deficient in the essential amino acid lysine.

a. rice and lima beans
b. rice and oatmeal
c. oatmeal and lima beans

18.79 For each of the following pairs of R groups, identify the amino acids and the type of interaction that forms between them: (18.4, 18.5)

a. $-CH_2-\overset{\overset{\displaystyle O}{\|}}{C}-NH_2$ and $HO-CH_2-$

b. $-CH_2-\overset{\overset{\displaystyle O}{\|}}{C}-O^-$ and $H_3\overset{+}{N}-(CH_2)_4-$

18.80 For each of the following pairs of R groups, identify the amino acids and the type of interaction that forms between them: (18.4, 18.5)
a. $-CH_2-SH$ and $HS-CH_2-$

b. $-CH_2-\overset{\overset{\displaystyle CH_3}{|}}{CH}-CH_3$ and CH_3-

The proteins in hair contain many cysteine R groups.

18.81 Answer the following questions for the given section of DNA: (18.7, 18.8)
a. Complete the bases in the parent and new strands.

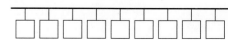

Parent strand: | | A | T | | | G | T | |

New strand: | C | | | G | G | | | C | C |

b. Using the new strand as a template, write the mRNA sequence.

| | | | | | | | |

c. Write the 3-letter abbreviations for the amino acids that would go into the peptide from the mRNA you wrote in part **b**.

18.82 Answer the following questions for the given section of DNA: (18.7, 18.8)

    a. Complete the bases in the parent and new strands.

Parent strand:

A		G	T			C	T

New strand:

	C			G	C	G	

b. Using the new strand as a template, write the mRNA sequence.

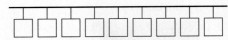

c. Write the 3-letter abbreviations for the amino acids that would go into the peptide from the mRNA you wrote in part **b.**

ADDITIONAL QUESTIONS AND PROBLEMS

18.83 What are the structural differences in D-glucose and D-galactose? (18.1)

18.84 What are the structural differences in D-glucose and D-fructose? (18.1)

18.85 Draw the Haworth structure for α-D- and β-D-gulose, whose open-chain structure is shown below. (18.1)

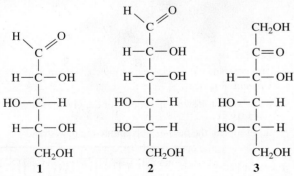

D-Gulose

18.86 From the compounds shown, select those that match the following: (18.1)

    a. an aldohexose   **b.** an aldopentose   **c.** a ketohexose

18.87 Gentiobiose, a carbohydrate found in saffron, contains two glucose molecules linked by a β(1→6)-glycosidic bond. Draw the Haworth structure for α-gentiobiose. (18.1, 18.2)

18.88 β-Cellobiose is a disaccharide obtained from the hydrolysis of cellulose. It contains two glucose molecules linked by a β(1→4)-glycosidic bond. Draw the Haworth structure for β-cellobiose. (18.1, 18.2)

18.89 Draw the condensed structural formula for Ser–Lys–Asp at physiological pH. (18.4)

18.90 Draw the condensed structural formula for Val–Ala–Leu at physiological pH. (18.4)

18.91 Identify the base and sugar in each of the following nucleosides: (18.7)

    a. deoxythymidine   **b.** adenosine

    c. cytidine   **d.** deoxyguanosine

18.92 Identify the base and sugar in each of the following nucleotides: (18.7)

    a. CMP   **b.** dAMP

    c. dGMP   **d.** UMP

18.93 Write the complementary base sequence for each of the following parent DNA segments: (18.7)

    a. G A C T T A G G C

    b. T G C A A A C T A G C T

    c. A T C G A T C G A T C G

18.94 Write the complementary base sequence for each of the following parent DNA segments: (18.7)

    a. T T A C G G A C C G C

    b. A T A G C C C T T A C T G G

    c. G G C C T A C C T T A A C G

18.95 Match the following statements with rRNA, mRNA, or tRNA: (18.8)

    a. carries genetic information from the nucleus to the ribosomes

    b. found in the ribosome

18.96 Match the following statements with rRNA, mRNA, or tRNA: (18.8)

    a. acts as a template for protein synthesis

    b. brings amino acids to the ribosomes for protein synthesis

CHALLENGE QUESTIONS

The following groups of questions are related to the topics in this chapter. However, they do not all follow the chapter order, and they require you to combine concepts and skills from several sections. These questions will help you increase your critical thinking skills and prepare for your next exam.

18.97 Raffinose is a trisaccharide found in green vegetables such as cabbage, asparagus, and broccoli. It is composed of three different monosaccharides. Identify the monosaccharides in raffinose. (18.1, 18.2)

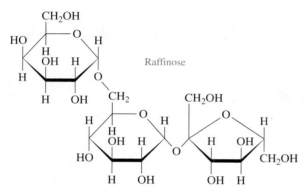

Raffinose

18.98 The disaccharide trehalose found in mushrooms is composed of two α-D-glucose molecules joined by an $\alpha(1\rightarrow1)$-glycosidic bond. Draw the Haworth structure for trehalose. (18.1, 18.2)

18.99 Sunflower oil can be used to make margarine. A triacylglycerol in sunflower oil contains two linoleic acids and one oleic acid. (18.3)

Sunflower oil is obtained from the seeds of the sunflower.

a. Draw the condensed structural formulas for two isomers for the triacylglycerol in sunflower oil.
b. Using one of the isomers, write the reactants and products for the reaction that would be used when sunflower oil is used to make solid margarine.

18.100 What type of interaction would you expect between the following amino acids in a tertiary structure? (18.4, 18.5)
a. threonine and asparagine
b. valine and alanine
c. arginine and aspartate

18.101 What are some differences between each of the following pairs? (18.4, 18.5)
a. secondary and tertiary protein structures
b. essential and nonessential amino acids
c. polar and nonpolar amino acids

18.102 What are some differences between each of the following pairs? (18.4, 18.5)
a. an ionic bond (salt bridge) and a disulfide bond
b. α helix and β-pleated sheet
c. tertiary and quaternary structures of proteins

Applications

18.103 Endorphins are polypeptides that reduce pain. Use three-letter abbreviations to write the amino acid sequence for the endorphin leucine enkephalin (leu-enkephalin), which has the following mRNA: (18.4, 18.5, 18.8)

AUG UAC GGU GGA UUU CUA UAA

18.104 Endorphins are polypeptides that reduce pain. Use three-letter abbreviations to write the amino acid sequence for the endorphin methionine enkephalin (met-enkephalin), which has the following mRNA: (18.4, 18.5, 18.8)

AUG UAC GGU GGA UUU AUG UAA

18.105 Aspartame, which is used in artificial sweeteners, contains the following dipeptide: (18.4, 18.5)

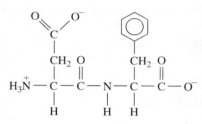

Aspartame is an artificial sweetener.

a. What are the amino acids in the dipeptide?
b. What is the name of the dipeptide in aspartame?
c. Give the three-letter and one-letter abbreviations for the dipeptide in aspartame.

18.106 The tripeptide shown has a strong attraction for Cu^{2+} and is present in human blood and saliva. (18.4, 18.5)

a. What are the amino acids in the tripeptide?
b. What is the name of this tripeptide?
c. Give the three-letter and one-letter abbreviations for the tripeptide.

18.107 The following sequence is a portion of a DNA template: (18.4, 18.5, 18.8)

GCT TTT CAA AAA
a. Write the corresponding mRNA segment.
b. Write the three-letter and one-letter abbreviations for the corresponding peptide segment.

18.108 The following sequence is a portion of a DNA template: (18.4, 18.5, 18.8)

TGT GGG GTT ATT
a. Write the corresponding mRNA segment.
b. Write the three-letter and one-letter abbreviations for the corresponding peptide segment.

ANSWERS

Answers to Selected Questions and Problems

18.1 Hydroxyl groups are found in all monosaccharides along with a carbonyl on the first or second carbon.

18.3 A ketopentose contains hydroxyl and ketone functional groups and has five carbon atoms.

18.5 In the ring portion of the Haworth structure of glucose, there are five carbon atoms and an oxygen.

18.7 a. α form   **b.** α form

18.9 a. ketohexose   **b.** aldopentose

18.11 In galactose, the —OH group on carbon 4 extends to the left. In glucose, this —OH group goes to the right.

18.13 a. one molecule of glucose and one molecule of galactose; $\beta(1\rightarrow4)$-glycosidic bond; β-lactose
b. two molecules of glucose; $\alpha(1\rightarrow4)$-glycosidic bond; α-maltose

18.15 a. disaccharide
b. two molecules of glucose
c. $\alpha(1\rightarrow6)$-glycosidic bond
d. α

18.17 a. sucrose   **b.** lactose
c. maltose   **d.** lactose

18.19 a. cellulose   **b.** amylose, amylopectin
c. amylose   **d.** glycogen

18.21 a. saturated   **b.** polyunsaturated   **c.** saturated

18.23

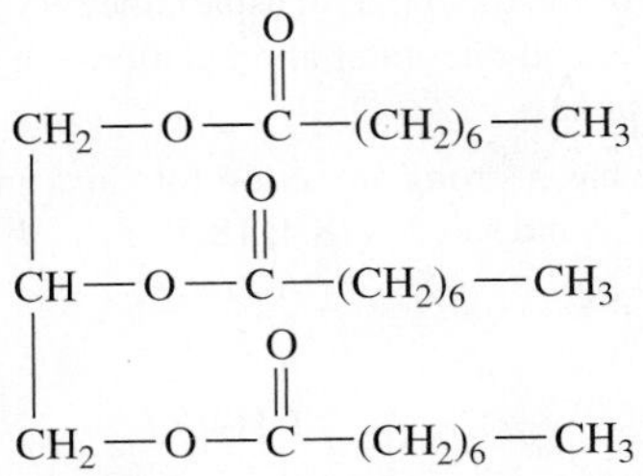

18.25 Safflower oil contains fatty acids with two or three double bonds; olive oil contains a large amount of oleic acid, which has only one (monounsaturated) double bond.

18.27

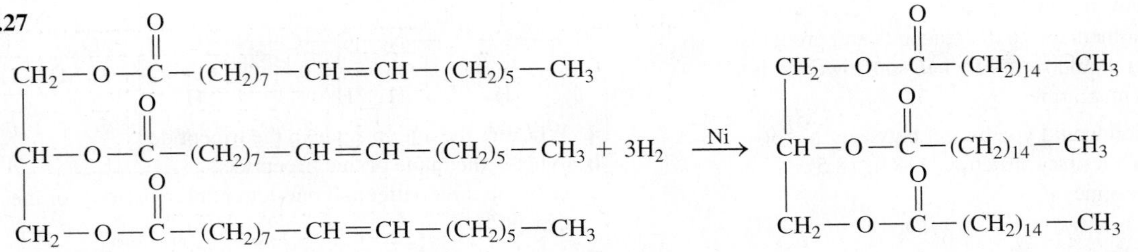

18.29

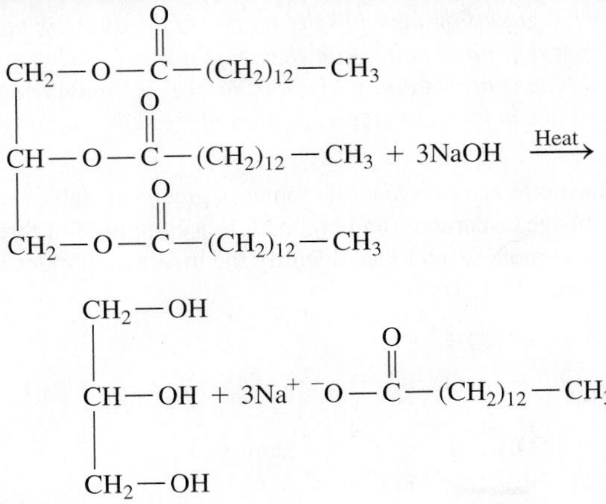

18.31

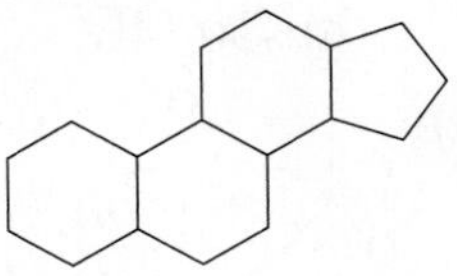

18.33 a.

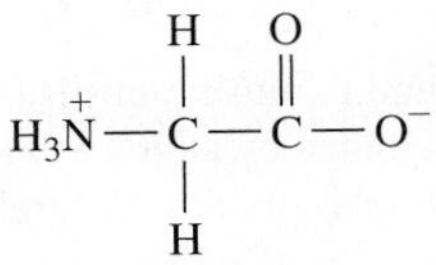

b.

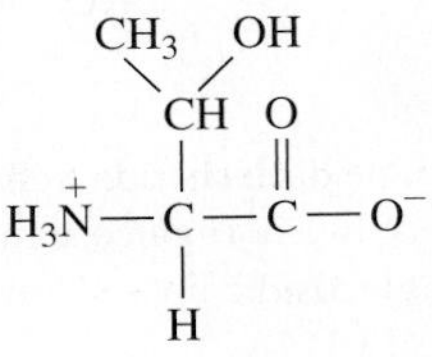

c.

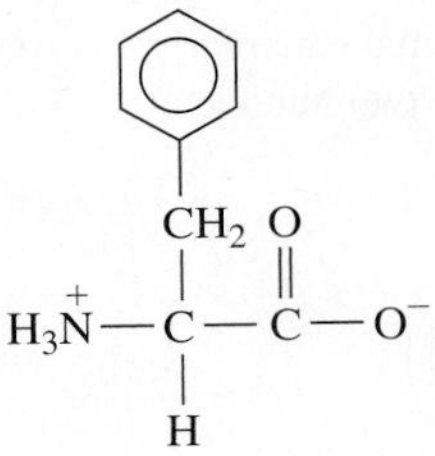

18.35 a. nonpolar, hydrophobic
b. polar, neutral, hydrophilic
c. nonpolar, hydrophobic

18.37 a. alanine   **b.** valine
c. lysine   **d.** cysteine

18.39 a.

$$
\underset{\text{Ala–Cys, AC}}{
\overset{+}{H_3N}
-\underset{\underset{H}{|}}{\overset{\overset{CH_3}{|}}{C}}
-\overset{\overset{O}{\|}}{C}
-\underset{\underset{H}{|}}{N}
-\underset{\underset{H}{|}}{\overset{\overset{\overset{SH}{|}}{CH_2}}{C}}
-\overset{\overset{O}{\|}}{C}
-O^-
}
$$

Ala–Cys, AC

b.

$$
\overset{+}{H_3N}
-\underset{\underset{H}{|}}{\overset{\overset{\overset{OH}{|}}{\overset{CH_2}{|}}}{C}}
-\overset{\overset{O}{\|}}{C}
-\underset{\underset{H}{|}}{N}
-\underset{\underset{H}{|}}{\overset{\overset{CH_2}{|}}{C}}
-\overset{\overset{O}{\|}}{C}
-O^-
$$

Ser–Phe, SF

c.

$$
\overset{+}{H_3N}
-C
-C
-N
-C
-C
-N
-C
-C
-O^-
$$

Gly–Ala–Val, GAV

18.41 a.

RIY

b.

VWIS

18.43 Val–Ser–Ser, VSS; Ser–Val–Ser, SVS; or Ser–Ser–Val, SSV

18.45 In the α helix, hydrogen bonds form between the oxygen atom in the C=O group and hydrogen in the N—H group in the next turn of the chain. In the β-pleated sheet, side-by-side hydrogen bonds occur between parallel peptides or across sections of a long polypeptide chain.

18.47 a. disulfide bond   **b.** salt bridge
      c. hydrogen bond   **d.** hydrophobic interaction

18.49 a. tertiary and quaternary   **b.** primary
      c. secondary

18.51 a. (1) enzyme   **b.** (3) substrate
      c. (2) enzyme–substrate complex

18.53 a. $E + S \rightleftharpoons ES \longrightarrow EP \longrightarrow E + P$
      b. The active site is a region or pocket within the tertiary structure of an enzyme that accepts the substrate, aligns the substrate for reaction, and catalyzes the reaction.

18.55 a. DNA   **b.** both DNA and RNA

18.57 The two DNA strands are held together by hydrogen bonds between the complementary bases in each strand.

18.59 a. T T T T T T
      b. C C C C C C
      c. T C A G G T C C A
      d. G A C A T A T G C A A

18.61 The DNA strands separate to allow each of the bases to pair with its complementary base, which produces two exact copies of the original DNA.

18.63 G G C U U C C A A G U G

18.65 a. leucine (Leu)   **b.** serine (Ser)
      c. glycine (Gly)   **d.** arginine (Arg)

18.67 a. AAA CAC UUG GUU GUG GAC
      b. Lys–His–Leu–Val–Val–Asp, KHLVVD

18.69 0.56 g of Pravachol

18.71 a. Melezitose is a trisaccharide.
      b. Melezitose contains two glucose molecules and a fructose molecule.

18.73

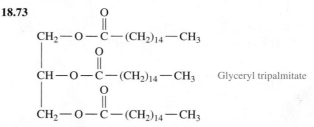

Glyceryl tripalmitate

18.75 a. polyunsaturated

18.77 a. yes   **b.** no   **c.** no

18.79 a. asparagine and serine; hydrogen bond
      b. aspartate and lysine; salt bridge

18.81 a.

Parent strand: G A T C C G T G G
New strand: C T A G G C A C C

b.

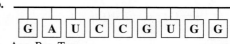

c. Asp–Pro–Trp

18.83 They differ only at carbon 4 where the —OH group in glucose is on the right side and in galactose it is on the left side.

18.85

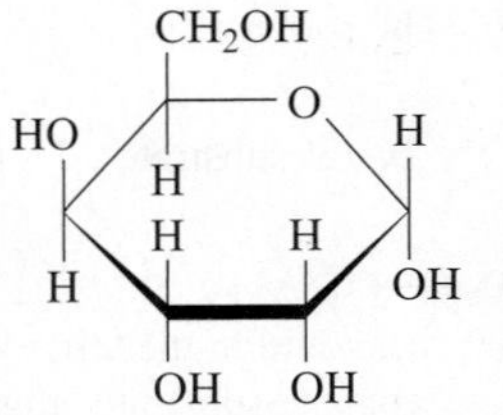

α-D-Gulose 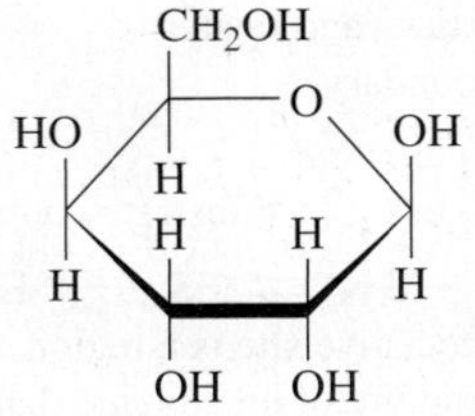   β-D-Gulose

18.87

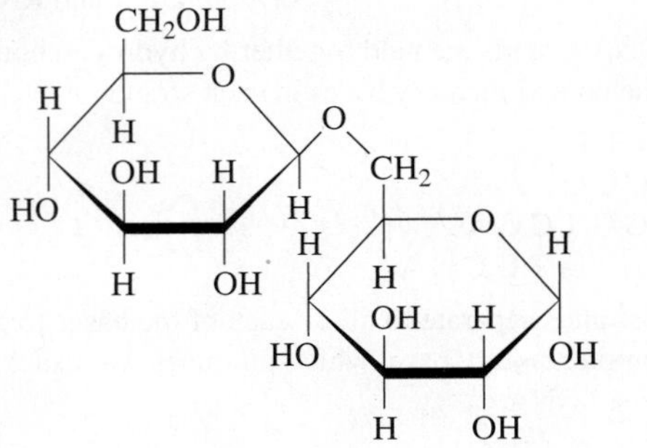

18.89

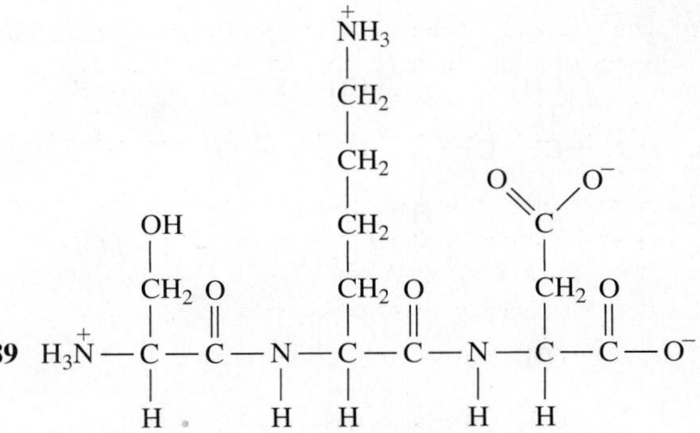

18.91 a. thymine and deoxyribose   **b.** adenine and ribose
      c. cytosine and ribose   **d.** guanine and deoxyribose

18.93 a. A C G T T T G A T C G A
      b. A C G T T T G A T G A
      c. T A G C T A G C T A G C

18.95 a. mRNA   **b.** rRNA

18.97 galactose, glucose, and fructose

18.99 a.

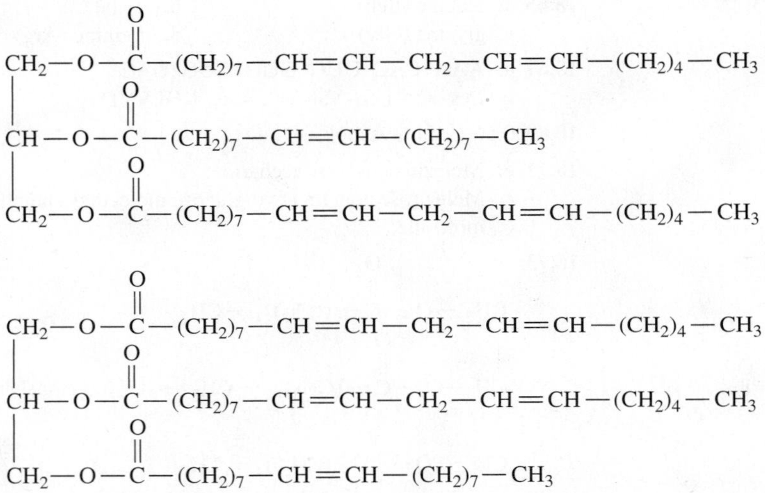

b.

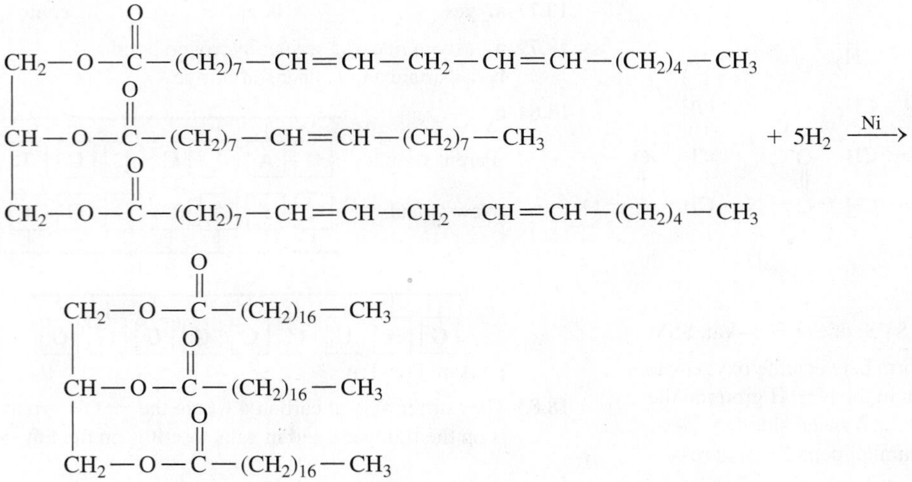

$+ \ 5H_2 \xrightarrow{\text{Ni}}$

18.101 a. The secondary structure of a protein depends on hydrogen bonds to form a helix or a pleated sheet. The tertiary structure is determined by the interaction of R groups, which determine the three-dimensional structure of the protein.

b. Nonessential amino acids are synthesized by the body, but essential amino acids must be supplied by the diet.

c. Polar amino acids have hydrophilic side groups; nonpolar amino acids have hydrophobic side groups.

18.103 START−Tyr−Gly−Gly−Phe−Leu−STOP

18.105 a. aspartate and phenylalanine
b. aspartylphenylalanine
c. Asp–Phe, DF

18.107 a. CGA AAA GUU UUU
b. Arg–Lys–Val–Phe, RKVF

CI.41 A compound called butylated hydroxytoluene, or BHT, with a molecular formula $C_{15}H_{24}O$, is added to preserve foods such as cereal. As an antioxidant, BHT reacts with oxygen in the cereal container, which protects the food from spoilage. (2.5, 2.6, 7.1, 7.2, 15.1, 17.4)

BHT is an antioxidant added to preserve foods such as cereal.

    a. What functional group is present in BHT?
    b. Why is BHT referred to as an "antioxidant"?
    c. The U.S. Food and Drug Administration (FDA) allows a maximum of 50. ppm of BHT added to cereal. How many milligrams of BHT could be added to a box of cereal that contains 15 oz of dry cereal?

CI.42 Olive oil contains a high percentage of glyceryl trioleate (triolein). (7.2, 11.7, 12.6, 18.3)

One of the triacyl-glycerols in olive oil is glyceryl trioleate (triolein).

    a. Draw the condensed structural formula for glyceryl trioleate (triolein).
    b. How many liters of H_2 gas at STP are needed to completely react with the double bonds in 100. g of triolein?
    c. How many milliliters of a 6.00 M NaOH solution are needed to completely saponify 100. g of triolein?

CI.43 A sink drain can become clogged with solid fat such as glyceryl tristearate (tristearin). (8.2, 12.6, 18.3)

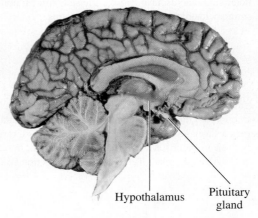

A sink drain can become clogged with saturated fats.

    a. How would adding lye (NaOH) to the sink drain remove the blockage?
    b. Write a balanced equation for the reaction that occurs.
    c. How many milliliters of a 0.500 M NaOH solution are needed to completely saponify 10.0 g of tristearin?

CI.44 In response to signals from the nervous system, the hypothalamus secretes a polypeptide hormone known as gonadotropin-releasing factor (GnRF), which stimulates the pituitary gland to release other hormones into the bloodstream.

Hypothalamus   Pituitary gland

Gonadotropin-releasing factor (GnRF) is secreted by the hypothalamus.

Two of the hormones are known as gonadotropins, which are the luteinizing hormone (LH) in males, and the follicle-stimulating hormone (FSH) in females. GnRF is a decapeptide with the following primary structure: (18.4, 18.5)
Glu–His–Trp–Ser–Tyr–Gly–Leu–Arg–Pro–Gly. (18.4, 18.5)

    a. What amino acid is the N-terminus in GnRF?
    b. What amino acid is the C-terminus in GnRF?
    c. Which amino acids in GnRF are nonpolar?
    d. Draw the condensed structural formulas at physiological pH for the acidic and basic amino acids in GnRF.
    e. Draw the condensed structural formulas for the primary structure of the first three amino acids starting from the N-terminus of GnRF at physiological pH.

CI.45 The plastic known as PETE (**p**oly**e**thylene**te**rephthalate) is a polymer of terephthalic acid and ethylene glycol. PETE is used to make plastic soft drink bottles and containers for salad dressings, shampoos, and dishwashing liquids. Today, PETE is the most widely recycled of all the plastics; in a single year, 1.5×10^9 lb of PETE are recycled. After PETE is separated from other plastics, it can be used in polyester fabric, fill for sleeping bags, door mats, and tennis ball containers. The density of PETE is 1.38 g/mL. (2.5, 2.6, 17.6)

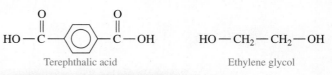

Terephthalic acid   Ethylene glycol

Plastic bottles made of PETE are ready to be recycled.

a. Draw the condensed structural formula for the ester formed from one molecule of terephthalic acid and one molecule of ethylene glycol.
b. Draw the condensed structural formula for the product formed when a second molecule of ethylene glycol reacts with the ester you drew for the answer in part **a**.
c. How many kilograms of PETE are recycled in one year?
d. What volume, in liters, of PETE is recycled in one year?
e. Suppose a landfill with an area of a football field and a depth of 5.0 m holds 2.7×10^7 L of recycled PETE. If all of the PETE that is recycled in a year were placed in landfills, how many would it fill?

CI.46 Thalassemia is an inherited genetic mutation that limits the production of hemoglobin. If less hemoglobin is produced, there is a shortage of red blood cells (anemia). As a result, the body does not have sufficient amounts of oxygen. In one form of thalassemia, thymine (T) is deleted from section 91 (bold) in the following segment of normal DNA: (18.4, 18.5, 18.8)

89	90	91	92	93	94
AGT	CAG	**CTG**	CAC	TGT	GAC A....

a. Write the complementary (template) strand for this normal DNA segment.
b. Write the mRNA sequence using the (template) strand in part **a**.
c. What amino acids are placed in the beta chain using the portion of mRNA in part **b**?
d. What is the order of nucleotides after T is deleted?
e. Write the template strand for the mutated DNA segment.
f. Write the mRNA sequence from the mutated DNA segment using the template strand in part **e**.
g. What amino acids are placed in the beta chain by the mutated DNA segment?
h. How might the properties of this segment of the beta chain be different from the properties of the normal protein?
i. How might the level of structure in hemoglobin be affected if beta chains are not produced?

ANSWERS

CI.41 a. The —OH group in BHT is bonded to a carbon atom in an aromatic ring, which means BHT has a phenol functional group.
b. BHT is referred to as an "antioxidant" because it reacts with oxygen in the food container, rather than the food, thus preventing or retarding spoilage of the food.
c. 21 mg of BHT

CI.43 a. Adding NaOH will saponify the glyceryl tristearate (fat), breaking it up into fatty acid salts and glycerol, which are soluble and will wash down the drain.

b.

$$CH_2-O-\overset{\overset{O}{\|}}{C}-(CH_2)_{16}-CH_3$$
$$CH-O-\overset{\overset{O}{\|}}{C}-(CH_2)_{16}-CH_3 + 3NaOH \longrightarrow$$
$$CH_2-O-\overset{\overset{O}{\|}}{C}-(CH_2)_{16}-CH_3$$

$$CH_2-OH$$
$$CH-OH + 3Na^+ \ ^-O-\overset{\overset{O}{\|}}{C}-(CH_2)_{16}-CH_3$$
$$CH_2-OH$$

c. 67.3 mL of a 0.500 M NaOH solution

CI.45 a.

$$HO-\overset{\overset{O}{\|}}{C}-\bigcirc-\overset{\overset{O}{\|}}{C}-O-CH_2-CH_2-OH$$

b.

$$HO-CH_2-CH_2-O-\overset{\overset{O}{\|}}{C}-\bigcirc-\overset{\overset{O}{\|}}{C}-O-CH_2-CH_2-OH$$

c. 6.8×10^8 kg of PETE
d. 4.9×10^8 L of PETE
e. 30. landfills

Credits

p. 588   JR/JDP/Reuters
p. 589   Pearson Education, Inc.
p. 590   Pearson Education, Inc.
p. 592   Lynn Watson/Shutterstock
p. 594   Imagestate Media Partners Limited -
Impact Photos/Alamy
p. 594   Koenig/Blickwinkel/Alamy
p. 595   Pearson Education, Inc.
p. 595   De Meester Johan/Arterra Picture
Library/Alamy
p. 599   Corepics/Fotolia
p. 599   Sumroeng/Fotolia
p. 600   Pearson Education, Inc.
p. 601   Pearson Education, Inc.
p. 604   Siede Preis/Photodisc/Getty Images
p. 604   Pearson Education, Inc.
p. 604   Frank Greenaway/Dorling Kinders-
ley, Ltd.
p. 604   Parpalea Catalin/Shutterstock
p. 605   Pavlo Kucherov/Fotolia
p. 605   C Squared Studios/Photodisc/Getty
Images
p. 605   Multiart/Shutterstock
p. 605   riderfoot/Fotolia
p. 607   Pearson Education, Inc.

CHAPTER 18

p. 612   Pearson Education, Inc.
p. 612   AZP Worldwide/Shutterstock
p. 614   Pearson Education
p. 618   Pearson Education, Inc.
p. 618   Pearson Education, Inc.
p. 618   Pearson Education, Inc.
p. 621   Pearson Education, Inc.
p. 622   Danny E Hooks/Shutterstock
p. 622   Getty Images
p. 626   Pearson Education, Inc.
p. 628   Tischenko Irina/Shutterstock
p. 629   Dawn Wilson/Getty Images
p. 629   Pearson Education, Inc.
p. 630   Kuleczka/Shutterstock
p. 632   National Heart, Lung and Blood
Institute
p. 632   National Heart, Lung, and Blood
Institute
p. 633   Fuse/Getty Images
p. 636   Pearson Education, Inc.
p. 639   Laurence Mouton/PhotoAlto sas/
Alamy
p. 655   Malcom Park/RGB Ventures /
SuperStock/Alamy

p. 655   © Robert Clare/Alamy
p. 655   Matt Rourke/AP Images
p. 656   Pearson Education, Inc.
p. 657   Dawn Wilson/Getty Images
p. 657   Fuse/Getty Images
p. 660   Pearson Education, Inc.
p. 660   Danny E Hooks/Shutterstock
p. 660   Pearson Education, Inc.
p. 660   Pearson Education, Inc.
p. 661   Ekkachai/Shutterstock
p. 661   Pearson Education, Inc.
p. 661   margo555/Fotolia
p. 661   3355m/Fotolia
p. 661   Pearson Education, Inc.
p. 661   Pearson Education, Inc.
p. 663   Pearson Education, Inc.
p. 668   Pearson Education, Inc.
p. 668   Lauree Feldman/Getty Images
p. 668   Martin M. Rotker/Science Source
p. 668   David Zaitz/Alamy

Glossary/Index

A

Abbreviated electron configuration, 144

Absolute zero, 78, 78t

Acetaldehyde, 212t

Acetic acid, 212t, 305
  buffers, 472, 473

Acetone, 585

Acetylene (ethyne), 570

Acetylsalicylic acid (aspirin), 52t, 463

Acid A substance that dissolves in water and produces hydrogen ions (H^+), according to the Arrhenius theory. All acids are hydrogen ion donors, according to the Brønsted–Lowry theory, 443–487
  bases and (neutralization), 468
  Brønsted–Lowry, 447–449
  characteristics of, 445t, 456t
  conjugate acid–base pair, 448–449
  diprotic, 451–452
  naming, 444–445, 445t
  reactions of, 468–470
  stomach, 467
  strength of, 450–454
  strong, 450–451, 450t, 451, 456t
  titration, 470–472
  weak, 450–451, 450t,

Acid–base titration, 470–472

Acid dissociation constant, 455

Acid dissociation expression, K_a The product of the ions from the dissociation of a weak acid divided by the concentration of the weak acid, 455–457, 455t

Acidic amino acid, 634

Acidic solution, 458
  balancing oxidation–reduction equations in, 500–501
  examples of, 458t
  pH of, 461
  stomach acid, 467

Acidosis, 476, 476t

Acid rain, 287, 387

Activation energy The energy that must be provided by a collision to break apart the bonds of the reacting molecules, 411–412

Active learning, 8, 8t

Active site A pocket in a part of the tertiary enzyme structure that binds substrate and catalyzes a reaction, 643

Activity series A table of half-reactions with the metals that oxidize most easily at the top, and the metals that do not oxidize easily at the bottom, 503, 503t

Actual yield The actual amount of product produced by a reaction, 263

Addition (mathematical)
  negative numbers, 11
  positive numbers, 11
  significant figures and, 37–39

Adenine (A), 644, 645

Adenosine monophosphate (AMP), 645, 646

Air
  as gas mixture, 349–350, 361
  typical composition, 349t

Alanine, 633, 636t, 637

Alchemists, 5

Alcohol An organic compound that contains a hydroxyl group ($—OH$) attached to a carbon chain, 570t, 578–582
  ball-and-stick model, 579
  functional group, 570t
  hydroxyl groups in, 578
  important, 580–581
  naming, 579–580

Aldehyde An organic compound that contains a carbonyl group ($C=O$) bonded to at least one hydrogen atom, 570t, 582–586
  naming, 583–584

Aldohexose, 614

Aldose A monosaccharide that contains an aldehyde group, 613

Alkali metal An element in Group 1A (1), except hydrogen, 108

Alkaline battery, 509

Alkaline earth metal An element in Group 2A (2), 108

Alkaloid, 595

Alkalosis, 476, 476t

Alkane A type of hydrocarbon in which the carbon atoms are connected only by single bonds, 561
  ball-and-stick model, 563
  branch, substituents in, 559–569, 565t
  combustion, 567–568
  comparing name, 570t
  condensed structural formulas for, 562–563
  drawing structural formulas for, 566–567
  naming, 561–562, 562t, 565–566
  properties and use of, 567
  structural formulas, 562, 562t, 567
  substituents used in naming, 565–566

Alkene A type of hydrocarbon that contains a carbon–carbon double bond ($C=C$), 569
  double bond, 569–572, 570t
  hydrogenation of, 571–572
  identifying, 570–571
  naming, 570–571, 570t, 576–578
  polymers, 572–575, 574t
  structure, 569, 570

Alkyl group An alkane minus one hydrogen atom. Alkyl groups are named like the corresponding alkanes except a *yl* ending replaces *ane*, 564

Alkyne A type of hydrocarbon that contains a carbon–carbon triple bond ($C≡C$), 569–575
  identifying, 569–570
  naming, 570–571, 570t
  structure, 569

Alpha decay, 528–529

α (alpha) helix A secondary level of protein structure, in which hydrogen bonds connect the $N—H$ of one peptide bond with the $C=O$ of a peptide bond farther down the chain to form a coiled or corkscrew structure, 638, 639

Alpha particle A nuclear particle identical to a helium nucleus, symbol α or ^4_2He, 525, 525t
  protecting from, 526, 527t

Altitude
  atmospheric pressure and, 328t
  boiling point and, 336, 337
  oxygen–hemoglobin equilibrium and, 429

Amethyst, 73

Amidation, 596–597, 602

Amide An organic compound in which the hydroxyl group of a carboxylic acid is replaced by a nitrogen atom, 570t, 596–597

Amine An organic compound that contains a nitrogen atom bonded to one or more carbon atoms, 570t
  aromatic, 594
  definition of, 593
  naming of, 593

Amino acid The building block of proteins, consisting of a hydrogen atom, an ammonium group, a carboxylate group, and a unique R group attached to the alpha carbon, 633–638
  acidic, 634t
  abbreviations, 634t
  basic, 634t
  codons for, 651–652
  definition of, 633
  essential, 635
  genetic code and, 651–652
  names, 634t
  N- and C-terminal, 636
  polar and nonpolar, 634t
  in proteins, 634t
  protein synthesis (translation), 652, 653
  structure of, 633, 634t
  types of, 635

Ammonia, 593

Ammonium ion, 180t

AMP (adenosine monophosphate), 645, 646

Amphoteric Substances that act as either an acid or a base in water, 448

Amylase, 622

Amylopectin A branched-chain polymer of starch composed of glucose units joined by $\alpha(1\rightarrow4)$- and $\alpha(1\rightarrow6)$-glycosidic bonds, 621

Amylose An unbranched polymer of starch composed of glucose units joined by $\alpha(1\rightarrow4)$-glycosidic bonds, 508, 621

Anesthesia, 581

Aniline, 576, 594

Anion A negatively charged ion such as Cl^-, O^{2-}, or SO_4^{2-}, 167

Anode The electrode where oxidation takes place, 505–506, 508–509

Antacids, 2, 469, 469t

Antihistamine, 595

Antimatter, 525

Aqueous solution, 364, 365t

Area, 35

Arginine, 635t

Aromatic amines, 594

Aromatic compound A compound that contains the ring structure of benzene, 570t, 576–578

Nutrition, 89–92
caloric value of foods, 89–91, 90*t*
Nutrition Facts label, 90

O

Observation Information determined by noting and recording a natural phenomenon, 4
Octet A set of eight valence electrons, 166
Octet rule Elements in Groups 1A to 7A (1, 2, 13 to 17) react with other elements by forming ionic or covalent bonds to produce a stable electron configuration, usually eight electrons in the outer shell, 166
exceptions, 285
Octyl ethanoate, 591*t*
Oil A triacylglycerol that is usually a liquid at room temperature and is obtained from a plant source, 628
hydrogenation, 626–627
melting point, 629–631
polyunsaturated, 629
at room temperature, 630
vegetable, 626
See also Crude oil
Oil spills, 568
Oleic acid, 572, 625*t*, 626
Omega-3 fatty acids in fish oils, 655
Opium, 595
Orbital The region around the nucleus where electrons of certain energy are most likely to be found: *s* orbitals are spherical; *p* orbitals have two lobes, 139
capacity, 141
shape of, 139–141
Orbital diagram A diagram that shows the distribution of electrons in the orbitals of the energy levels, 142–144
Period 1, 143–144
Period 2, 144–145
Period 3, 145–146
Organic compounds, 559, 560
bonding, 560*t*
carbon compounds, 560–561
classification of, 569, 570*t*
functional groups, 569–575
naming, 561, 562*t*, 563, 565–566
properties of, 560*t*
Osmosis The flow of a solvent, usually water, through a semipermeable membrane into a solution of higher solute concentration, 395–396
properties of solutions, 395–399
Osmotic pressure The pressure that prevents the flow of water into the more concentrated solution, 395–396
Osteoporosis, 57
Overweight, 69
Oxaloacetic acid, 588–589
Oxidation The loss of electrons by a substance. Oxidation may involve the addition of oxygen or the loss of hydrogen, 239–242, 492
Oxidation number A number equal to zero in an element or the charge of a monatomic ion; in molecular compounds and polyatomic ions, oxidation numbers are assigned using a set of rules, 493, 494*t*
assigning of, 494*t*
oxidation–reduction identified using, 495–496

Oxidation–reduction reaction A reaction in which electrons are transferred from one reactant to another, 239–242
balancing of, 498–503
in biological systems, 240–241
characteristics of, 241, 241*t*
definition of, 239
electrical energy and, 503–510
half-reactions, 492–493, 498–500
oxidation numbers used to identify, 495–496
writing, 492
zinc and copper(II) sulfate, 240, 241
Oxidizing agent The reactant that gains electrons and is reduced, 496
Oxycodone (OxyContin), 595
Oxygen-15, 539
Oxygen–hemoglobin equilibrium, 429
Ozone, 286

P

Pain relievers, 595
Palladium-103, 542*t*, 544
Palmitic acid, 625*t*
Palmitoleic acid, 625*t*
Paracelsus, 5, 52
Paraffins, 567
Partial pressure The pressure exerted by a single gas in a gas mixture, 348–352
Pascal (Pa), 327, 328*t*
Pauli exclusion principle, 141
p **block** The elements in Groups 3A (13) to 8A (18) in which electrons fill the *p* orbitals, 147
Pentose sugars, 645
Pentyl butanoate, 591*t*
Pentyl ethanoate, 591*t*
Peptide, 636–637
Peptide bond, 636–637
Peptide chain, 638
Percentage, 12, 46–47
Percent concentration, 374–376, 380*t*
Percent yield The ratio of the actual yield for a reaction to the theoretical yield possible for the reaction, 263–265
Period A horizontal row of elements in the periodic table, 106
Periodic properties, 151–156
Periodic table An arrangement of elements by increasing atomic number such that elements having similar chemical behavior are grouped in vertical columns, 106–111, 107
electron configurations and, 147–151
periods and groups in, 106–107
trends in, 151–156
Per (prefix), 445
Pesticides, 52*t*, 102, 249
Peta (P), 40*t*
pH A measure of the $[H_3O^+]$ in a solution; $pH = -log[H_3O^+]$, 460–467
calculating, 462–465, 474–477
indicators, 470
scale, 461
Phenol, 576, 580–581
plant sources of, 580
structure of, 576–577
types of, 581
Phenolphthalein, 470
Phenylalanine, 620, 634*t*
Phenylketonuria (PKU), 620

pH meter, 462
Phosphodiester linkage The phosphate link that joins the hydroxyl group in one nucleotide to the phosphate group on the next nucleotide, 646
Phosphorus-32, 542*t*
Photon A packet of energy that has both particle and wave characteristics and travels at the speed of light, 135
Photosynthesis, 612
Physical change A change in which the physical properties of a substance change but its identity stays the same, 74–75, 75*t*
Physical properties The properties that can be observed or measured without affecting the identity of a substance, 74, 74*t*, 75*t*
Pico (p), 40*t*
PKU (phenylketonuria), 620
Place values, 10–16
Plaque, 632
Plasma, 363
Plastics, 573–574
recycling, 573
pOH A measure of the $[OH^-]$ in a solution; $pOH = -log[OH^-]$, 465
Polar amino acid An amino acid that is soluble in water because its R group is polar: hydroxyl (—OH), sulfur group (—SH), carbonyl (C=O), ammonium (—NH$_3^+$), or carboxylate (—COO$^-$), 633, 634*t*
Polar covalent bond A covalent bond in which the electrons are shared unequally between atoms, 294
Polarity A measure of the unequal sharing of electrons, indicated by the difference in electronegativities, 294–296
of molecules, 297–298
Polar molecule A molecule containing bond dipoles that do not cancel, 297–298
Polar solute, 364
Polar solvent, 362
Polyatomic ion A group of covalently bonded nonmetal atoms that has an overall electrical charge, 179–183
compounds containing, 180, 182*t*
molecule shapes and, 288–293
name of, 179–180, 180*t*, 182
products containing, 179
writing formula, 180–181
Polydichloroethylene (Saran), 573, 574*t*
Polyethylene, 573, 573, 574*t*
Polymer A large molecule that is composed of many small, repeating structural units that are identical, 569, 572–575
alkene, 569, 572–575, 574*t*
definition of, 572
plastics, recycle, 573
recycling, 573
synthetic, 572–573
Polypropylene, 573, 574*t*
Polysaccharide A polymer of many monosaccharide units, usually glucose. Polysaccharides differ in the types of glycosidic bonds and the amount of branching in the polymer, 617–624
cellulose, 622–623
definition of, 621–623
digesting, 622

Metric and SI Units and Some Useful Conversion Factors

Length — SI Unit Meter (m)	Volume — SI Unit Cubic Meter (m^3)	Mass — SI Unit Kilogram (kg)
1 meter (m) = 100 centimeters (cm)	1 liter (L) = 1000 milliliters (mL)	1 kilogram (kg) = 1000 grams (g) 1 mg = 1000 mcg
1 m = 1000 millimeters (mm)	1 mL = 1 cm^3 1 dl = 100 mL	1 g = 1000 milligrams (mg) 1 lb = 16 oz
1 cm = 10 mm	1 L = 1.057 quart (qt) 1 mL = 1cc	1 kg = 2.205 lb
1 kilometer (km) = 0.6214 mile (mi)	1 qt = 946.4 mL	1 lb = 453.6 g
1 inch (in.) = 2.54 cm (exact)	1 pint (pt) = 473.2 mL	1 mol = 6.022×10^{23} particles
	1 gal = 3.785 L	**Water**
	1 tsp = 5 mL	density = 1.00 g/mL (at 4 °C)
	1 tbsp (T) = 15 mL	

Temperature — SI Unit Kelvin (K)	Pressure — SI Unit Pascal (Pa)	Energy — SI Unit Joule (J)
$T_F = 1.8(T_C) + 32$	1 atm = 760 mmHg	1 calorie (cal) = 4.184 J (exact)
$T_C = \dfrac{T_F - 32}{1.8}$	1 atm = 101.325 kPa	1 kcal = 1000 cal
$T_K = T_C + 273$	1 atm = 760 Torr	
	1 mol of gas = 22.4 L (STP)	**Water**
	$R = 0.0821$ L·atm/mol·K	Heat of fusion = 334 J/g
	$R = 62.4$ mmHg·atm/mol·K	Heat of vaporization = 2260 J/g
		Specific heat (SH) = 4.184 J/g °C; 1.00 cal/g °C

Metric and SI Prefixes

Prefix	Symbol	Power of Ten
Prefixes That Increase the Size of the Unit		
peta	P	10^{15}
tera	T	10^{12}
giga	G	10^{9}
mega	M	10^{6}
kilo	k	10^{3}
Prefixes That Decrease the Size of the Unit		
deci	d	10^{-1}
centi	c	10^{-2}
milli	m	10^{-3}
micro	μ (mc)	10^{-6}
nano	n	10^{-9}
pico	p	10^{-12}
femto	f	10^{-15}

Formulas and Molar Masses of Some Typical Compounds

Name	Formula	Molar Mass (g/mol)	Name	Formula	Molar Mass (g/mol)
Ammonia	NH_3	17.03	Hydrogen chloride	HCl	36.46
Ammonium chloride	NH_4Cl	53.49	Iron(III) oxide	Fe_2O_3	159.70
Ammonium sulfate	$(NH_4)_2SO_4$	132.15	Magnesium oxide	MgO	40.31
Bromine	Br_2	159.80	Methane	CH_4	16.04
Butane	C_4H_{10}	58.12	Nitrogen	N_2	28.02
Calcium carbonate	$CaCO_3$	100.09	Oxygen	O_2	32.00
Calcium chloride	$CaCl_2$	110.98	Potassium carbonate	K_2CO_3	138.21
Calcium hydroxide	$Ca(OH)_2$	74.10	Potassium nitrate	KNO_3	101.11
Calcium oxide	CaO	56.08	Propane	C_3H_8	44.09
Carbon dioxide	CO_2	44.01	Sodium chloride	NaCl	58.44
Chlorine	Cl_2	70.90	Sodium hydroxide	NaOH	40.00
Copper(II) sulfide	CuS	95.62	Sulfur trioxide	SO_3	80.07
Hydrogen	H_2	2.016	Water	H_2O	18.02